1993

A Study of Enzymes

Volume II
Mechanism of Enzyme Action

Editor
Stephen A. Kuby
Professor of Biochemistry
Research Professor of Medicine
Laboratory for Study of Hereditary and Metabolic Disorders
Departments of Biochemistry and Medicine
University of Utah
Salt Lake City, Utah

CRC Press
Boca Raton Ann Arbor Boston

Library of Congress Cataloging-in-Publication Data

Mechanism of enzyme action.

(A Study of enzymes; v. 2)
Includes bibliographical references and index.
1. Enzymes. I. Kuby, Stephen Allen, 1925—
II. Series.
QP601.S677 vol. 2 574.19′25 s 90-2277
ISBN 0-8493-6988-6

Direct all inquiries to CRC Press, Inc., 2000 Corporate Blvd., Boca Raton, Florida 33431.

Library of Congress Number 90-2277
International Standard Book Number 0-8493-6988-6
Printed in the United States

Lorsqu'il n'est pas en notre pouvoir de discerner les plus vraies opinions, nous devons suivre les plus probables. (When it is not in our power to determine what is true, we ought to act according to what is most probable.)

René Descartes (1596—1650)

Dedicated to all students of enzymology, both past and present, whose fascination with enzyme mechanisms made this treatise possible.

PREFACE

Although there have been several excellent theses on Enzyme Mechanisms, the time seems unusually ripe for a definitive and selected series of reviews on the subject.

In the past several years, there has been a remarkable resurgence of activity and expansion of knowledge in the area of enzyme mechanisms, both in theory and in data acquisition. Many enzyme proteins, whose primary sequence seemed almost beyond the realm of existing technology have been now had their covalent structures deduced from their complementary DNA's. This then permitted a host of chemical derivatization and modification studies designed to shed light on structure-function relationships. With the recent advent of new techniques, especially in the X-ray deduced macromolecular crystal structures and in the NMR deduced solution structures, and coupled with some specific chemically synthesized and in some cases site-directed mutagenesis experiments (where critical, or noncritical, residues may be substituted in the enzyme molecule, or in peptide fragments of the enzyme molecule), an explosion in enzyme mechanistic studies seems almost imminent.

Structure-function relationships in the future will undoubtedly be assisted by computer modeling, where computer techniques making use of the X-ray crystal and NMR solution structure data, will permit an extension of ideas beyond those of conventional methods.

Together with the application of new or updated spectroscopic approaches, e.g., in nuclear magnetic resonance, in fluorescence, in infra-red Fourier analyses, one will be able to study in greater detail the solution structure of the enzyme-substrate complex(s). A hint of these tremendous advances to be made may be found, e.g., a number of the studies reported here (e.g., on superoxide dismutases, Chapter 21 and on adenylate kinase, Chapter 17).

Molecular dynamics of enzyme proteins are now under vigorous investigation. Interactions of the enzyme molecule with its substrate(s) or modifiers or with specific inhibitors may induce changes in shape and in characteristics. Conformational (or fluctuational) changes on a micro scale lead to ideas concerning the dynamics of such systems (cf. Chapter 1).

Metals in enzyme catalysis have long been the serious study a number of investigators, and ideas concerning complexation in biological systems are only now beginning to be codified (see Chapter 2).

Sound Bioorganic and physical chemical approaches to the study of enzymatic mechanisms are still the preferred method of approach in certain problems of enzyme mechanisms (as illustrated e.g., in Chapters 3 and 4, and in Chapters 19 and 20).

The time is ripe for a detailed review of recent studies on allosteric enzymes (e.g., Chapter 5 on glycogen phosphorylase) and on multienzyme complexes (Chapters 4 and 9).

A remarkable headway has been made in the kinetic studies of certain dehydrogenases, whose X-ray structures are now well known (e.g., dihydrofolate reductase, Chapter 8). Very elegant modification and derivatization experiments, together with the application of a fluorometric approach, is shedding much light on the mechanism of glutamate dehydrogenase (e.g., Chapter 7).

Medical advances are to be anticipated as a result of in vivo and in vitro mechanistic studies (e.g., Chapters 6, 22, 23, and 24).

Since the early days of Michaelis' pioneering experiments, studies of flavo-protein catalysis have been at the forefront of investigations into the mechanisms of redox reaction (Chapters 10, 11, and 12); and ideas concerning semi-quinones and free radicals originated in these studies. Free-radical oxygen (e.g., superoxide anion) has also absorbed the attention of those interested in the dismutation H_2O_2 (e.g., Chapter 21), and the oxygen utilizing reactions, e.g., the monooxygenases and dioxygenases (Chapter 13), will continue to receive the attention they deserve. Now, however, new approaches have permitted deeper insight into the role that the very short-lived intermediates (e.g., with oxygen) play in their mechanisms.

The mechanism of action of the ATP-utilizing enzymes and carriers will continue to occupy much attention especially because of their intrinsic biological importance (Chapters 14, 15, 17, and 18). Their modes of action and those of the phosphotransferases (Chapters 17 and 18) and the nucleotidyl transferases (Chapter 16) are fascinating and absorbing topics for the mechanistic bent of mind.

It, therefore, was the original goal to follow Volume I, which deals with a sound description on enzyme kinetics and ligand binding with Volume II, to deal in detail, with carefully selected topics in enzyme mechanisms, as written by those investigators with a special interest in the particular enzyme mechanism(s) they discuss here.

We hope that we have not been unsuccessful in both of these goals.

THE EDITOR

Stephen A. Kuby, Ph.D., is a Professor of Biochemistry and Research Professor of Medicine at the University of Utah School of Medicine, Salt Lake City, Utah. He is also the head of the Biochemical Division of the Laboratory for the Study of Hereditary and Metabolic Disorders, University of Utah Research Park, Salt Lake City, Utah 84108. He graduated *Summa cum Laude* from New York University in 1948 with an A.B. in Chemistry. He obtained his M.S. in Biochemistry in 1951, and his Ph.D. in Biochemistry in 1953 at the University of Wisconsin under Professor H. A. Lardy. As a recipient of the U.S. Public Health Service Fellowship, he spent his post-doctorate period with Professor B. Chance at the University of Pennsylvania (Johnson Foundation for Medical Physics) and with Professor H. Theorell at the Karolinska Institute (Medical Nobel Institute, Stockholm, Sweden). Following a period at the Enzyme Institute (University of Wisconsin) as an Assistant Professor, in 1963 he joined the faculty at the University of Utah, and has held his present positions since 1969. His research interests have dealt with many aspects of enzyme and protein chemistry and certain aspects of medicinal chemistry and inherited disorders (including the muscular dystrophies). Mechanistic enzymology is his current interest, and "state of the art" approaches are being applied to study of enzyme action, e.g., of the kinases.

He is a member of the American Society of Biochemistry and Molecular Biology, the American Chemical Society Division of Biochemistry (Chemistry), Sigma Xi, Phi Beta Kappa, American Association for the Advancement of Science, and the New York Academy of Sciences. He has been a member of the Subcommittee on Enzymes, Committee on Biological Chemistry, Division of Chemistry and Chemical Technology, National Academy of Sciences and National Research Council; and also a member of the Physiological Chemistry Study Section, Division of Research Grants, National Institutes of Health, Publich Health Service, Department of Health, Education and Welfare. His long list of publications reflect these research interests.

CONTRIBUTORS

Gilbert Arthur
Assistant Professor
Department of Biochemistry and Molecular Biology
University of Manitoba
Winnipeg, Manitoba, Canada

Patrick C. Choy
Professor
Department of Biochemistry and Molecular Biology
University of Manitoba
Winnipeg, Manitoba, Canada

Roberta F. Colman
Professor
Department of Chemistry and Biochemistry
University of Delaware
Newark, Delaware

Cindy L. Fisher
Postdoctoral Fellow
Department of Molecular Biology
Research Institute of Scripps Clinic
La Jolla, California

David C. Fry
Department of Biological Chemistry
Johns Hopkins School of Medicine
Baltimore, Maryland
Research Scientist
Department of Physical Chemistry
Hoffmann LaRoche
Nutley, New Jersey

Elizabeth D. Getzoff
Assistant Member
Department of Molecular Biology
Research Institute of Scripps Clinic
La Jolla, California

Robert W. Gracy
Professor and Chairman
Department of Biochemistry
University of North Texas/Texas College of Osteopathic Medicine
Fort Worth, Texas

Clark J. Gubler
Professor Emeritus of Biochemistry
Department of Chemistry
Brigham Young University
Provo, Utah

Robert Hallewell
Director, Protein Engineering
Chiron Research Laboratories
Chiron Corporation
Emeryville, California

Minoru Hamada
Professor and Chairman
Department of Hygiene
Miyazaki Medical College
Miyazaki-gun, Miyazaki, Japan

D. J. Hupe
Senior Research Fellow
Merck Institute for Therapeutic Research
Rahway, New Jersey

Kazutomo Imahori
President
Mitsubishi-Kasei Institute of Life Sciences
Tokyo, Japan

Yuzuru Ishimura
Professor and Chairman
Department of Biochemistry
School of Medicine
Keio University
Tokyo, Japan

Mark S. Johnson
Research Associate
Department of Biochemistry
Loma Linda University
Loma Linda, California

Nobuhiko Katunuma
Professor and Director
Institute for Enzyme Research
School of Medicine
The University of Tokushima
Tokushima, Japan

Seiichi Kawashima
Chief
Department of Biochemistry
Tokyo Metropolitan Institute of Gerontology
Tokyo, Japan

Martin Klingenberg
Professor
Institute for Physical Biochemistry
University of Munich
Munich, West Germany

Stephen A. Kuby
Professor of Biochemistry
Research Professor of Medicine
Laboratory for Study of Hereditary and Metabolic Disorders and the Departments of Biochemistry and Medicine
University of Utah
Salt Lake City, Utah

Rufus Lumry
Professor
Department of Chemistry
University of Minnesota
Minneapolis, Minnesota

Neil B. Madsen
Professor
Department of Biochemistry
University of Alberta
Edmonton, Alberta, Canada

Vincent Massey
Professor
Department of Biological Chemistry
University of Michigan
Ann Arbor, Michigan

Albert S. Mildvan
Professor
Department of Biological Chemistry and Chemistry
Johns Hopkins School of Medicine
Baltimore, Maryland

Mikund J. Modak
Professor
Department of Biochemistry
UMDNJ/New Jersey Medical School
Newark, New Jersey

John F. Morrison
Professor
Department of Biochemistry
John Curtin School of Medical Research
Australian National University
Canberra A. C. T., Australia

Yoichi Nakamura
Instructor
Central Laboratory of Clinical Investigation
Miyazaki Medical College
Miyazaki-gun, Miyazaki, Japan

Hiromichi Okuda
Professor
Department of Medical Biochemistry
School of Medicine
Ehime University
Onsen-gun, Ehime, Japan

Virendra Nath Pandey
Scientific Officer D
Department of Biochemistry
Bhabha Atomic Research Center
Bombay, India

Victoria A. Roberts
Research Associate
Department of Molecular Biology
Research Institute of Scripps Clinic
La Jolla, California

Lawrence M. Schopfer
Assistant Research Scientist
Department of Biological Chemistry
University of Michigan
Ann Arbor, Michigan

Engin H. Serpersu
Assistant Professor
Department of Biochemistry
University of Tennessee
Knoxville, Tennessee

Michihiro Sumida
Associate Professor
Department of Medicine
Ehime University
Onsen-gun, Ehime, Japan

Koichi Suzuki
Chief
Department of Molecular Biology
The Tokyo Metropolitan Institute of
Medical Sciences
Tokyo, Japan

John A. Tainer
Assistant Member
Department of Molecular Biology
Research Institute of Scripps Clinic
La Jolla, California

Hitoshi Takenaka
Instructor
Department of Hygiene
Miyazaki Medical College
Miyazaki-gun, Miyazaki, Japan

Kosaku Uyeda
Professor
Department of Biochemistry
University of Texas Health Science Center
Dallas, Texas

R. J. P. Williams
Professor
Inorganic Chemistry Laboratory
University of Oxford
Oxford, United Kingdom

Kunio Yagi
Director
Institute of Applied Biochemistry
Mitake, Gifu, Japan

Shozo Yamamoto
Professor
Department of Biochemistry
School of Medicine
Tokushima University
Tokushima, Japan

K. Ümit Yüksel
Director, Biopolymer Analysis Laboratory
Department of Biochemistry
University of North Texas
Texas College of Osteopathic Medicine
Fort Worth, Texas

TABLE OF CONTENTS

Volume I

Section I: Enzyme Catalysis and Steady-State Kinetics

Chapter 1

Historical Introduction, Theories of Enzyme Catalysis, and Some Elementary Considerations of Enzyme Kinetics

Chapter 2

A Description of Steady-State Kinetics and Quasi- or Rapid Eqilibrium Kinetics by a Development of the Rate Expressions for Several Selected Mechanisms and Their Characteristics

Chapter 3
Inhibition and Product Inhibition

Chapter 4
Effects of pH and Temperature

TABLE OF CONTENTS

Volume II

Section I
Some Theoretical Aspects of Enzyme Mechanisms

One of the principal objects of theoretical research in any department of knowledge is to find the point of view from which the subject appears in its greatest simplicity. J. Willard Gibbs in a letter to the *American Academy of Arts and Sciences* (1981) (See Collected Works, Vol. I, 1928) on the occasion of his receiving the Rumford Medal for his treatise *On the Equilibrium of Heterogeneous Substances*.

Chapter 1

MECHANICAL FORCE, HYDRATION, AND CONFORMATIONAL FLUCTUATIONS IN ENZYMIC CATALYSIS

Rufus Lumry

TABLE OF CONTENTS

ABSTRACT

Ligand binding and its regulation in hemoglobin are achieved by redistributions of conformational free energy. The conformation is not only an active participant in mechanism, but also provides most of the machinery. When the first ligand is bound, tension is developed but quickly relaxed by rearrangements of the soft protein conformation. When the last three ligands are removed, tension is replaced by compression. The conformation interconverts electronic potential energy, mechanical force, and conformational free energy in the several steps of ligation. Such mechanisms are called rack mechanisms. A nearly identical rack mechanism was postulated for chymotrypsin and many other enzymes at a time when conformations were thought to be very stiff, secondary-bond-rearrangement conformers were not known, and the X-ray-diffraction structure was not available. The existence and functional role of domains, first proposed in this mechanism, have since been well established and there is abundant evidence for active participation of conformation in catalytic function. However, there continue to be improvements in our understanding of protein physical properties, conformational fluctuations, and interaction with water which require constant updating of the original mechanism. In this chapter the mechanism is elaborated in more detail as required by this new information in order to provide a hypothesis of mechanism as complete as is currently possible. The improved and corrected hypothesis is suitable for testing and the testing can make the applications of site-specific mutagenesis to the study of enzyme function efficient, as they are not now.

As postulated, two domains in chymotrypsin move together in a spontaneous fluctuation, compressing the protein functional groups (the "charge-relay system") and the substrate into a short-lived metastable state closely resembling the transition states for primary-bond rearrangement. This is a delicately constructed process which raises the electronic potential energy of the chemically reacting assembly of groups, positions these groups relative to each other, and distorts substrates. This process, originally envisioned for a relatively hard protein, is unlikely in proteins as soft as most current evidence suggests, but the difficulties appear now to be removed by the finding of activated transitions between conformers and recent evidence that proteins have two kinds of substructure. One, called a "knot", is strong, dense, and highly cooperative. The other, called a "matrix," is soft and has relatively high motility. The knots determine the folding, the kinetic stability, and the fluctuation characteristics of the matrices. A knot and its associated matrix comprise a functional domain. Enzymes have two domains per function so they resemble a basket with knot handles. Substrates are held in the basket which conforms to the substrate on initial binding and continues to readjust as the process continues. Each knot carries at least one of the functional groups and as the basket contracts these groups are brought together into the pretransition-state metastable state in which the functional groups are activated and the substrate distorted toward the transition-state configuration. The mechanism is completed by addition of "transition-state stabilization" and by improved protein-substrate interaction at the catalytic site, entropy management, and probably some of the other devices proposed to participate in function. The domain-closure process reflects contractions and expansions of the protein along a functional conformation coordinate, which includes all participating subprocesses coupled directly or indirectly so as to maintain the phase relations necessary to establish the desired overall pattern of thermodynamic changes. The mechanism is essentially fluctuational, but the fluctuations appear to be those having high probability in the protein system at equilibrium. This is expected for proteins with an equilibrium function, but somewhat surprising in enzymes. Enzyme function and apparently much of biology manifests first-order linear-response behavior.

An omnipresent principle in protein evolution is the improved survival value associated with more efficient utilization of food without loss of speed. The principle, called free-energy complementarity, produces flattening of the free-energy profiles of reactions which increases overall rates and allows the processes to progressively approach closer and closer to thermodynamic efficiency. Joined with the domain-closure process, the principle rationalizes the evolutionary development and cooperativity of complex enzyme systems.

A very useful tool in analyzing linkage systems is the apparently ubiquitous, nearly linear relationships between enthalpy and entropy changes measured from the advancement of one subprocess when another subprocess to which it is coupled directly or indirectly is advanced using the appropriate independent variable. To each such relationship there corresponds a linear-free-energy relationship and the latter requires that the coupling between the subprocesses be independent of advancement. The enthalpy-entropy plots, called compensation plots and characterized by their slopes called the compensation temperatures, T_c, provide a number of new insights. For example, the T_c values frequently found in physiologically important processes are near 300K, which implies that at physiological temperatures the protein is poised near the midpoint of the functional conformation coordinate where fluctuations along this coordinate are maximal. T_c values near operating temperatures also imply that domain-closure and other compression-expansion processes of the protein occur at nearly constant free energy. If so, there is no special requirement for dynamical activation by momentum waves arising in the solvent as is currently suggested by several authors.

Compensation theory is applied in a primitive way to isolate the contributions to the thermodynamic changes along the reaction profiles for chymotryptic hydrolysis from the α-carbon sidechains of a family of ester substrates. The sidechain feature of the mechanism is shown to be a supplementary feature added to the fundamental domain-closure mechanism. Some results of the analysis are discussed in terms of protein activity coefficients and the A and B hydration shells. These interactions with water are considered in some detail in relation to current ideas about molecular processes of water and the A and B hydration shells as described by Rupley and Careri and their respective co-workers. The hydration shells may be at least as important in free-energy management as the protein conformation but one is forced to conclude that there is as yet too little information about any of the devices woven into the catalytic mechanism to support serious attempts to rationalize enzymic mechanisms quantitatively.

The rack mechanism is updated and, though neither complete nor fully established, it is the best — at least the most complete — hypothesis of mechanism available. Thus far it provides the only consistent explanation for data which have been obtained in site-specific-mutagenesis studies of enzymes.

I. INTRODUCTION

Although the molecular mechanisms which support enzymic catalysis remain elusive, there is general optimism that they will be revealed quickly and unambiguously by systematic exchange of amino acid residues. It is probably true that, in principle, site-specific mutagenesis has this power, but optimism should be tempered by the failure of skillful exploitation of the many hemoglobin mutants over many years to establish the mechanism of heme-heme interaction even at the descriptive level. The natural mutants and normal species variants have been a source of qualitative information, but the determination of mechanism is primarily a quantitative undertaking and mutants have major importance because they expand the range of quantitative investigation. With new techniques any exchange can be made at any residue position, but that is not entirely a blessing since, for practical reasons, only a small fraction of the astronomical number of potential mutants can be made and characterized in the detail required. To make the use of mutants efficiently, it is obviously necessary to have hypotheses to test which are comprehensive in the sense that they integrate in an internally consistent way all the known factors suspected to participate in mechanism. Of course, there is no reason to believe that all important factors have been discovered, and the continuing high rate at which new factors of established or potential importance are discovered in studies of the hemoglobins suggests that for enzymes, which have been much less intensively studied than the hemoglobins, efficient exploitation of site-specific mutagenesis may not yet be possible. Then the best hypotheses which can be constructed will fail and we can go no farther until a new idea or observation leads to the discovery of a new factor and thence to a new hypothesis.

Not all of this discussion is obvious. A good fraction of the recent converts to site-specific mutagenesis are going to be discouraged to discover both the depth and detail necessary in useful hypotheses and the high level of experimental precision the testing requires. Again, the history of hemoglobin research provides a realistic example of the tortuous pathways to be expected in applications to enzymes.

It is the purpose of this chapter to illustrate what we mean by a ''best hypothesis'' using the ''rack mechanism'' for chymotryptic catalysis. Rack mechanism is the general name for those protein mechanisms which depend on interconversions of mechanical and electronic work.[1] In the rack mechanism for hemoglobin, the electronic changes on ligand binding favor a new geometry for the heme complex ion,[2] but relaxation of the complex ion is harnessed to conformation so that the free energy of the latter is raised as that of the complex ion is lowered. In this way, as compared with the unharnessed reactions of small molecules, little free energy is lost to heat and so remains available to facilitate ligand binding to other subunits. The conservation of free energy is perhaps the most important feature of rack mechanisms, as was originally suggested by the thermodynamics of protein conformations.[1a] Specifically, the enthalpy and entropy are equally important in determining conformation so that these quantities change in a parallel fashion as conformations change. The net free energy change is thus small, even though the change in conformation, the associated conformational relaxation causes redistribution of the enthalpy and entropy in amounts and places in the total protein found in evolution to effect a useful process.

A second important feature is that free energy stored in conformations is distributed over many small changes in conformational geometry (the ''Lilliput principle''[3]). The sharp spectral isosbestic points in oxygenation show that there are two and only two electronic states of the heme complex ion despite large variations in tertiary conformations. This is a consequence of the weakness of the secondary interactions responsible for protein folding. Many weak bonds can cooperate to develop transient strain and longer-term stress in a small set of primary bonds, e.g., those which undergo change during reaction. The term ''rack'' comes from the picture of strong-bond systems distorted as a result of their attachment to the protein. In fact, we now know that the stress is so broadly delocalized in both conformation and heme complex ion in hemoglobin that no primary bond is much distorted. A total free-energy difference from true equilibrium in conformation and primary-bond system greater than about 10 kcal is unusual. Larger differences favor unfolding and must be accommodated by primary bonds. In transition-state formation, conformations may provide larger amounts of free energy than this but only on a transient basis.

In rack mechanisms, conformations play an ''active'' role in contrast to the essentially passive rearrangement of functional groups postulated in Koshland's ''induced fit''.[4] The ''transition-state'' postulate of Wolfenden[5] and Leinhard[6] does not explicitly include conformation changes as an essential feature of mechanism, but certainly could. Most other ''physical-organic'' theories such as those of Jencks[7] do not ignore conformation changes, but usually treat them as local events in an otherwise quite rigid protein. A major objective of most hypotheses of enzymic catalysis is to explain the reduction in activation free energy for primary-bond-rearrangement steps. Rack mechanisms are advanced on the assumption that any or all of these hypotheses may be correct, but all are incomplete. Further reduction of activation free energies is postulated to be a consequence of the survival value of free-energy conservation, as is discussed in several other sections. An additional objective is to explain in molecular terms how protein conformations contribute free energy to active-site groups to reduce the thermal requirements for activation. Most frequently, this translates into the provision of devices which allow temporary increase in potential energy at active sites by utilizing free-energy sources unique to protein systems. Of these, the currently most popular are conformational fluctuations which focus momentum fluctuations arising in the heat bath at the reaction site. This class of "dynamic rack mechanisms"[1a] is most completely described by Gavish[8a] (Reference 8 and papers by Gavish, Welch, Careri, Somogyi, and Volkenstein in "The Fluctuating Enzyme"[15]). The other class depends on free-energy redistributions which do not utilize internal-energy and pressure fluctuations in the heat bath in any way unique to proteins. The first class is discussed only briefly in this chapter. The second class is the major topic.

For the sake of simplicity in exposition, an ordinary desk stapler can be used as a primitive model for the rack mechanisms for enzymes. The substrate and protein are positioned by the initial closing of the jaws of the stapler. The formation of the activated complex, usually for a primary-bond rearrangement step, is then the click step in which the staple is driven into the packet of pages. The critical part of the model is the source of free energy for the click step. In small-molecule reactions, the activation energy is provided by a very brief conversion of thermal energy into potential energy. If the interactions responsible for stable folding were sufficiently strong, the whole protein or large fraction of it would undergo thermal equilibration as a single unit. Energy fluctuations would be large and those equal to enzyme activation energies very frequent.[146] Although secondary interactions might produce such gestalt behavior in the best built proteins, large parts of most proteins and certainly of enzymes are soft. In those regions, each small collection of atoms achieves temperature equilibrium independently and in picosecond times, that is, as though each were a small molecule in a liquid environment.[28] As a result, the large internal-energy part of the thermal energy is no more effective as a source of activation energy than in a homogeneous fluid.

The pressure-volume fluctuations which form a small part of "enthalpy heat" are not so simply swept away. Although cohesive forces are weak and incomplete in large parts of proteins, repulsion potentials produce pressure fluctuations because they enforce strong correlations among the motions of atoms. If one atom in a protein moves, some of its neighbors must move. Most disturbances of this kind occur in small collections of atoms so that the return to mechanical equilibrium is rapid. However, some involve many atoms and thus maintain coherence for longer times. These "communal" modes obviously can facilitate transition-state formation by mechanical compression or the tension developed as the conformation relaxes from transient compression. Perhaps, in small regions of the protein in which cohesive energy is high, strain is generated by transient expansion. Such very short-lived conversions of P-V "heat" to free energy may be sufficient to drive the stapler jaws in our model through the click step. Such "thermal" or dynamical enzymic rack mechanisms are actively championed,[8] but their testing is difficult because it turns on quantitative distinctions and also because there are many complexities for which as yet no models exist (see Section VIII.D). It is not possible with the limited information available about the mechanical properties of protein conformations and the coupling between water and protein to exclude such mechanisms. However, some experimental information and alternative explanations of viscosity data generally advanced in support of such mechanisms make them unattractive as will be discussed at various points in this chapter.

The rack mechanism which we favor, and to which we shall specifically apply the name rack mechanism, includes the feature of mechanical activation, but the force is generated in conformational fluctuations which occur at nearly constant protein free energy. The justification was first provided by experimental and theoretical evidence for the conservation of free energy in physiologically important protein processes which depend on protein conformational changes. Free-energy conservation seems to be a general consequence of evolutionary improvements in the thermodynamic efficiency of food utilization, and those improvements also improve catalytic rates since the higher-lying free-energy barriers along the reaction coordinates of enzymes tend to drop and free energies of the lower-lying metastable states tend to increase, as shown in Figure 1, and discussed later.

''Transition-state stabilization'' is a much better known mechanism for lowering the free-energy barriers to primary-bond rearrangement.[5] This is the chemical stabilization which results from protein conformations tailored to interact more favorably with the substrate in its transition-state configuration than in its configuration in the most stable enzyme-substrate complex. This type of stabilization appears to be well established, and is generally supposed to be the major kind of transition-state stabilization. To a first approximation it is independent of the rack mechanism and, like the other ''physical-organic'' hypotheses of mechanism, need not be discussed in detail here, but it will become clear that the data supporting this hypothesis are at least as well explained by rack mechanisms.

Both of these mechanisms conform to Pauling's famous dictum about enzymic catalysis that the catalyst reduces the activation free energy, that is, the chemical transition-state stabilization reduces the standard free-energy of formation of the transition state.[18] Rack mechanisms have this effect, but it is generally secondary to the transient increase in free energy at the catalytic site in the mechanically produced metastable ground state from which the transition state is formed. The total free energy of the protein conformation system, which also includes all significant water-protein interactions, is constantly redistributed in a large set of equilibrium fluctuations, described by the enthalpy probability-density distribution function. The mechanism is constructed from this noise by enhancing fluctuations which produce short-lived metastable states very similar in conformation, configuration, and free energy to the higher-lying transition states, usually those for primary bond rearrangement. The efficiency of these fluctuations is high, not only because they occur at nearly constant free energy, but also because the protein is constructed to favor them and to be unfavorable to nonproductive fluctuations. The lifetime of the metastable state cannot be long because the conformational relaxation processes which destroy the state are rapid. In principle, the lifetime need be no longer than that of the activated complex itself, but then the reaction would be very inefficient. This lifetime problem, once a matter of considerable concern, has been alleviated and perhaps removed entirely by the discovery that protein conformers are often separated by free-energy barriers[9,41a] (see Figure 2). The times for transitions between such conformers are determined by the heights of the barriers. Judging from the barriers found in myoglobin[9a,9c] some of these can be remarkably high. However, some regions of protein conformations are quite soft, which means that the mechanical behavior is dominated by the faster rather than the slower relaxation processes. The distinction between hard and soft in this connection is primarily that between slow and fast relaxation. Relaxation studies of group motion suggest that the relevant times are 1 to 10 ns,[187] but enzymes appear to become harder as they contract, so even longer times may obtain near the transition-state configuration. Group motions are found to be slower when inhibitor or substrates are present.

Because the fluctuations are those of a system of equilibrium, the activation mechanism can be called fluctuational but the technical name is first-order linear-response mechanism[65] (*vide infra*). First-order linear-response behavior is closely related to ''irreversible thermodynamics'' but in a more general way also to fluctuations occurring in reacting systems at any steady-state condition.

II. MOLECULAR DESCRIPTION OF THE RACK MECHANISM FOR CHYMOTRYPSIN

Like ester and amide hydrolysis in solution, the exact details of mechanism in enzymic

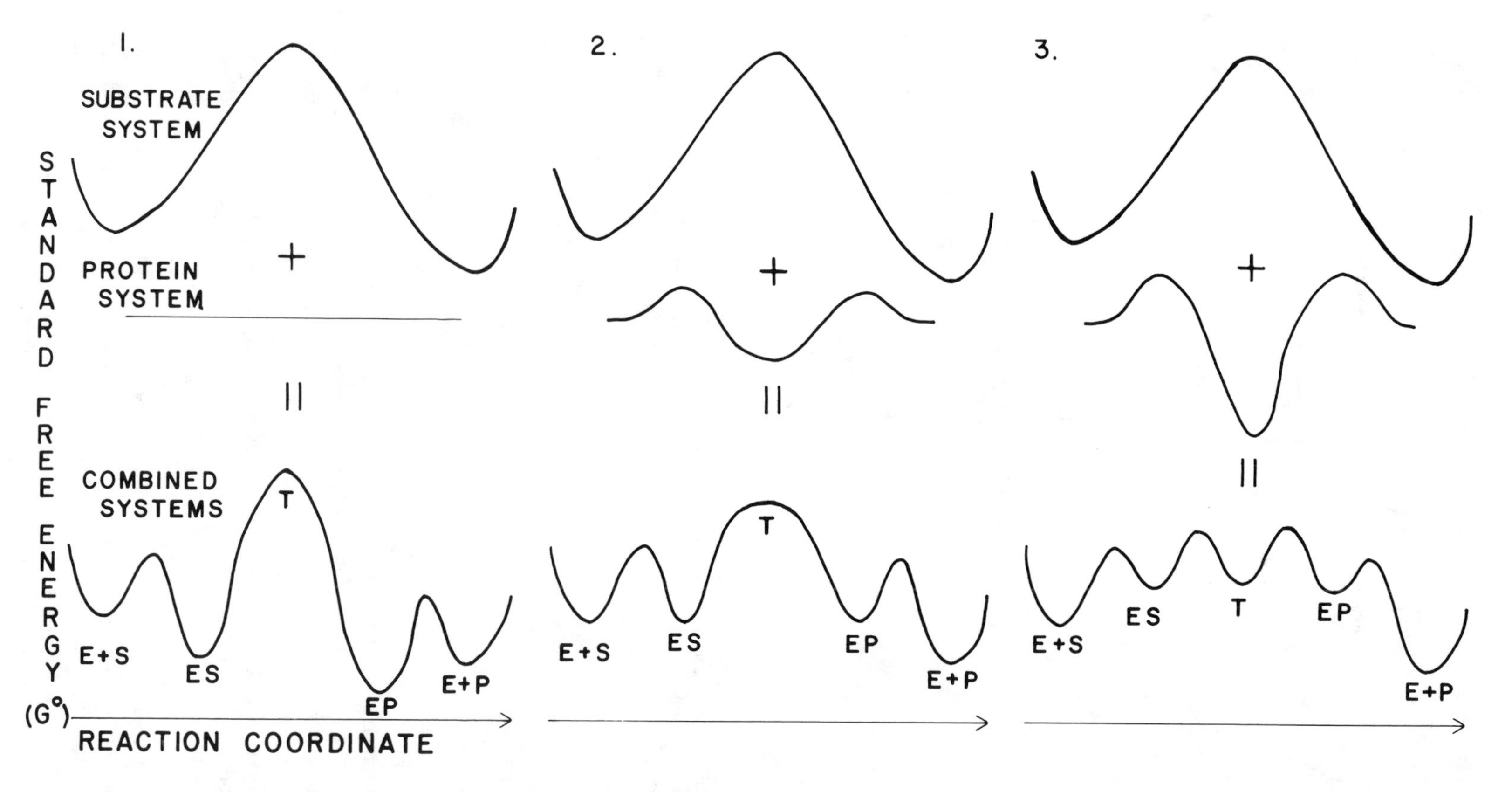

FIGURE 1. Development of complementarity particularized for chymotryptic catalysis of ester hydrolysis but abbreviated to include only the first primary-bond-rearrangment transition state. (1) Applies to the hydrolysis of the ester in homogeneous solution. (2) Applies to an intermediate step in evolution. (3) Describes a more recent state. T is assumed to be a tetrahedral-carbon intermediate and its conversion from an activated complex to a metastable intermediate is due at least in part to the development of "transition-state stabilization" in the complex of substrate and protein groups participating directly in the bond-rearrangement process. Additional stabilization of T is attributed in the text to the development of more extensive association of the two protein domains in part due to increased interaction between α-carbon sidechain of substrate and protein. The acylenzyme intermediate is added and the profiles are further elaborated in Figure 14.

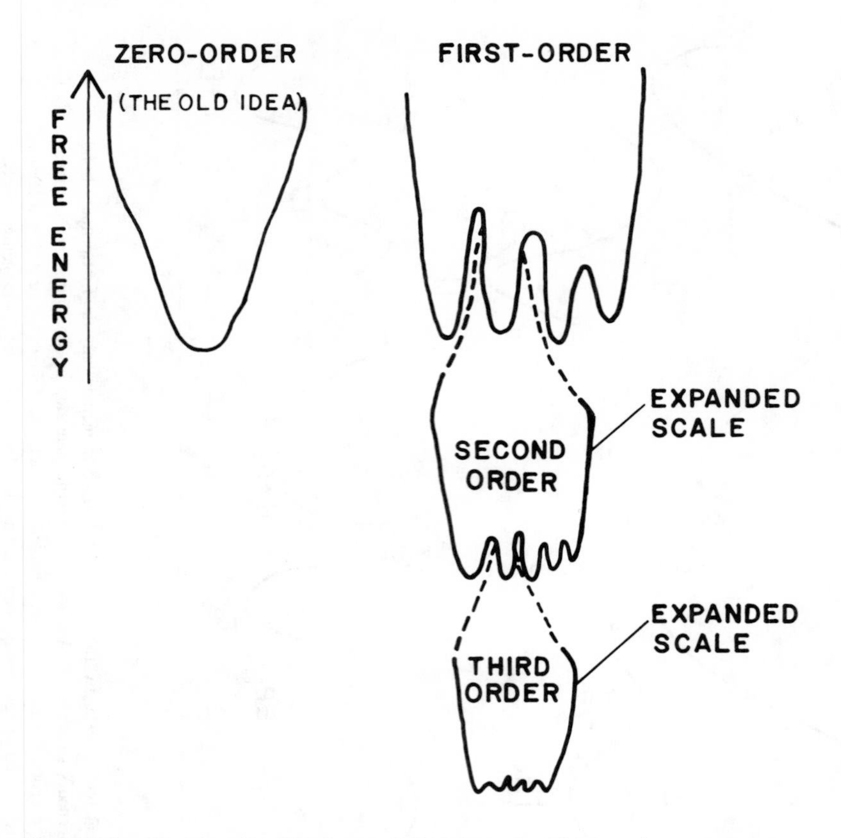

FIGURE 2. The hierarchy of conformers in folded proteins. These are due to rearrangements of secondary interactions and, because these rearrangements redistribute the free volume, the conformers were originally called "mobile-defect" conformers.[41a] In the first-order diagram the barriers between substates are high, so the lifetime in the substates is relatively long.[9] In the second-order diagram, a single substate is further analyzed into its substates and these are separated by smaller barriers. In the third-order diagram, further detail is shown but the barrier heights are of the order of kT so that transitions between third-order substates are not limited by the barriers. Transitions between conformers of the first and second order are called "activated" transitions for obvious reasons. The only quantitative information about barrier heights and their distribution available are those for small heme-proteins obtained by Frauenfelder, et al.[9c]

hydrolysis can vary not only with type of substrate, but also from substrate to substrate in a single family. In ester hydrolysis, the major variation is the exact position of the transition state and the importance of the tetrahedral-carbon configuration[22] but the steady-state limiting step can change from acylenzyme formation to acylenzyme destruction as well. These variations and other fine details of the electronic mechanism continue to be the focus of much research on chymotryptic hydrolysis, demonstrating the complexity involved but perhaps also the inadequacy of our understanding of proteins. A reasonably well-accepted description of the events taking place between protein functional groups and ester[173] is shown in Figure 3. No attempt is made to distinguish the transition states, but a tetrahedral intermediate appears, and for convenience we shall assume that it is a mestastable state.

The only high-resolution, three-dimensional difference X-ray-diffraction study of chymotrypsin relevant to the conformational changes in catalysis is that of the phenylethane boronic acid derivative obtained using the crystals of dimeric α-CT[101] (*vide infra*). In one molecule of each dimer, a tetrahedral intermediate is found. In the other, the reaction stops in what is probably the Michaelis-Menten complex. Considerable cooperativity between

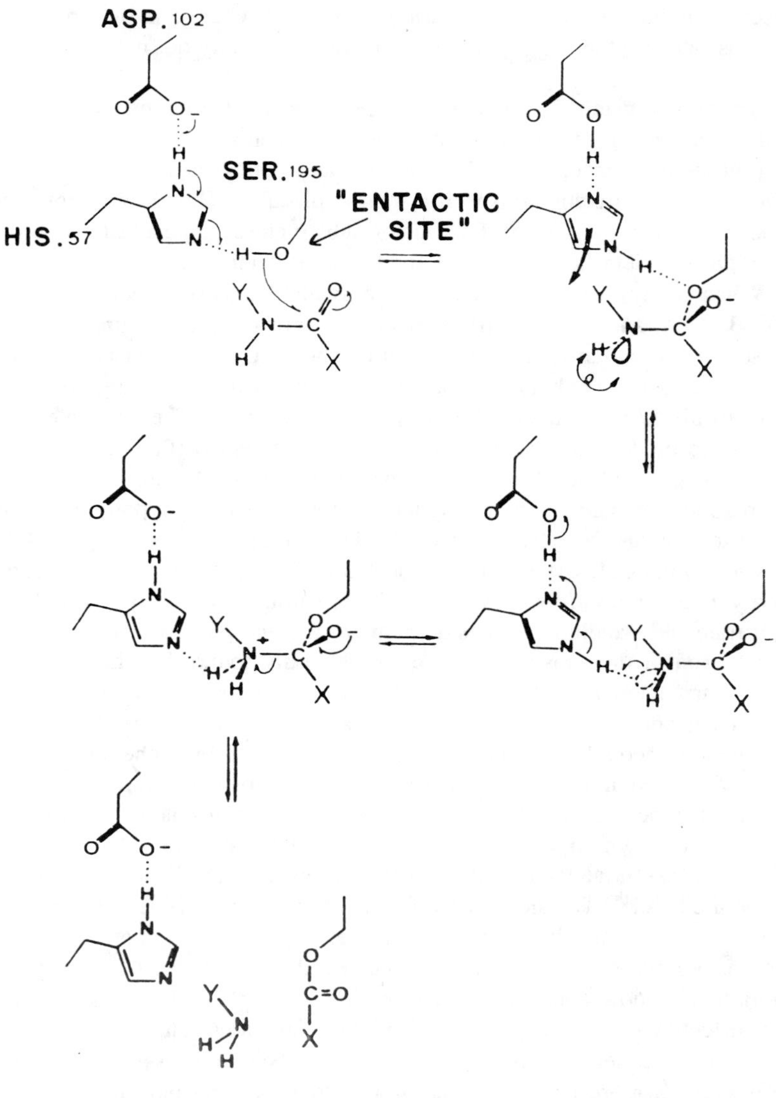

FIGURE 3. Chemical events in chymotryptic hydrolysis of ester substrates. Many detailed descriptions of these events have been advanced. This one, due to Bizzozero and Dutler,[173] appears to be the most consistent with available data and the principles of organic reactions which apply (Deslongchamps, Woodward and Hoffman, etc.), but most of the details are irrelevant for purposes of this chapter. The functional groups forming the "charge-relay system" are named. The "entactic" site shown is the hydroxyl group of Ser.195, but the imidazole group of His.57 also forms chemical derivatives in entactic processes (see text). The figure is a slight modification of Scheme 2 from page 56 of an article published in *Bioorganic Chemistry*, 10, 46, 1981 and is reproduced with the permission of the authors and the copyright owners, Academic Press, Inc.

monomers of the dimer was found in the parent structure, so this difference in virtual substrate binding was not unexpected. Earlier studies of inhibitor binding and of unnatural acylenzyme species, e.g., the diisopropylphosphoryl chymotrypsins, in two-dimensional, low-resolution studies, suggested a variety of small conformational rearrangements in regions well removed from the active site.[10] These findings were unreliable so that conformational events in chymotryptic catalysis have had to be inferred from other types of information, which can also produce unreliable conclusions. A variety of kinds of change in the protein have been detected, but associating them with structural changes is usually not much more than guesswork.

That chymotryptic catalysis depends on changes in relative geometry of major parts of the molecule became apparent quite early as a consequence of studies on denaturation,[12] optical-rotation, fluorescence, circular-dichroism, and optical absorption[13] as they change on interaction with specific and nonspecific small molecules. This information was to some extent supported by kinetics data.[2a] Two major substructures appeared to be required, and these were originally identified with the B and C chains determined by amino acid sequence analysis.[2b] When X-ray diffraction data became available, these substructures were found to be the A (B chain) and B (C chain) cylinders,[14] confirming the original guess. We shall call the substructures domains, and this definition should not be confused with that used in sorting introns and exons. Each domain carries one of the two functional groups then known: the imidazole of his.57 on A and the hydroxyl of ser.195 on B. The available data strongly suggested that the binding of substrate favors the movement of the domains toward each other in such a way as to "activate" these functional groups. Domain closure is a complex conformational and free-energy rearrangement[35,158] which in our hypotheses leads at maximum closure to a metastable state very similar to the transition state (Figure 4).

As shown in Figure 4, the α-carbon sidechain of the substrate moves deeper into its binding cavity as domain closure advances. The resulting increase in favorable interaction between substrate and protein facilitates domain closure and probably produces increasingly favorable distortion of the substrate. It is quite generally found that the rate constants for primary-bond rearrangement steps in enzymic catalysis demonstrate a quantitative sensitivity to the size, shape, and chemical nature of substrate sidechains, even when the latter have virtually no inductive coupling to the substrate groups undergoing chemical change. Thus, specificity is manifested in both the formation of enzyme-substrate complexes and in catalytic efficiency. This has been a valuable guide to the search for mechanism and in the original rack mechanism for chymotrypsin it was thought that the role of sidechain in "kinetic" specificity was an essential feature of the chymotrypsin mechanism.[123c] As we shall see, that cannot be the case.[137] The sidechain effect is rather an added feature exploited initially in evolution to enhance specificity in primitive enzymes like chymotrypsin — primitive because they have only a degratory role and thus need not conserve free energy released in product formation. It now appears that this originally simple add-on has been extensively elaborated to effect much more sophisticated enzymic functions. These matters are discussed in some detail in subsequent sections. At this point, we begin discussion of the topics which now make possible a more detailed description of the rack mechanism for chymotrypsin, and which also show that the same mechanism is very likely applicable to at least one family of nucleases and to the lysozymes.

III. THERMODYNAMIC FOUNDATIONS OF THE CHYMOTRYPSIN RACK MECHANISM

Haldane proposed the first strictly mechanical mechanism from analogy with catalysis on solid surfaces.[16] Substrates are bound at surface sites which are so spatially arranged that the substrate becomes distorted. Any distortion of primary bonds partially uncouples bonding electron pairs and so enhances reactivity. This is essentially an energy argument, but suggests immediately the corresponding free-energy argument applicable to proteins, that is, proteins contain a mechanism by which free energy can be redistributed in such a way that free energy released in the formation of enzyme-substrate complexes is stored until it can be used to labilize the transition state, or it is released by improved enzyme-substitute interactions as the system moves toward a transition state. Then, it follows, in a rough way, that

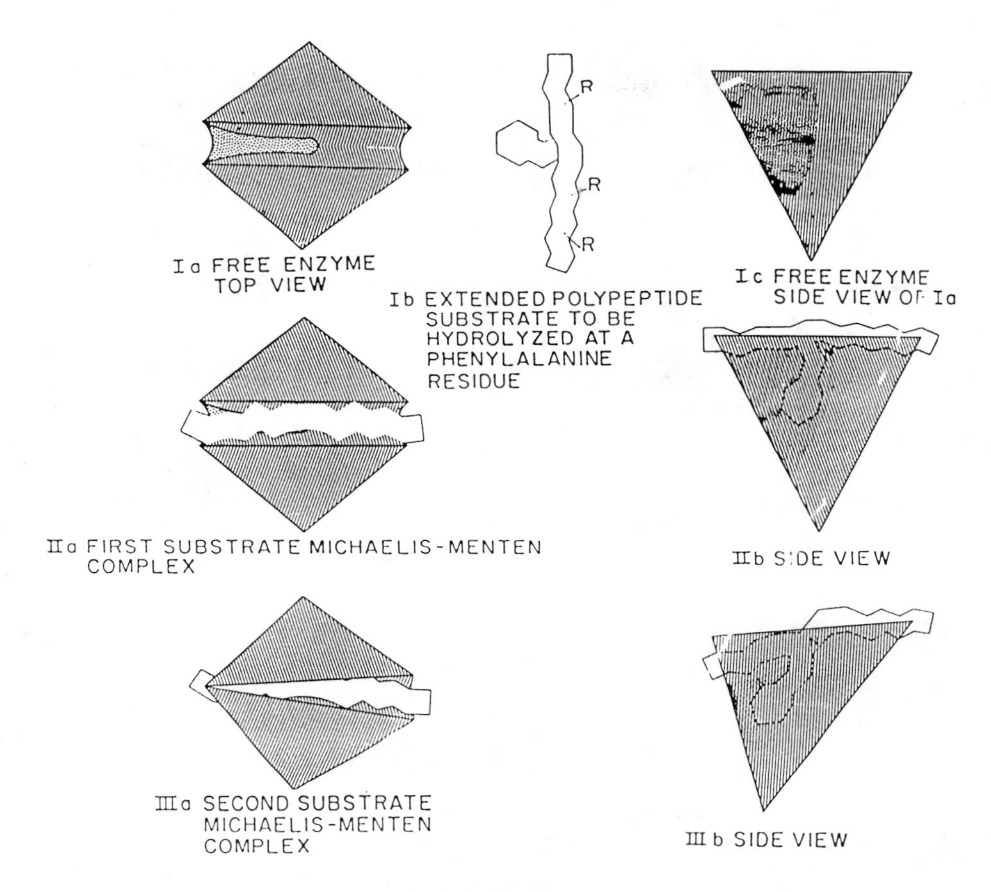

FIGURE 4. Original pictorial description of the rack mechanism for chymotryptic catalysis.[91a] The essential features are shown: domain closure (grossly exaggerated), activation of functional groups by their compression into substrate, and progressive improvement in the interaction of substrate α-carbon sidechain with protein. This figure, from *Structure and Stability of Biological Macromolecules,* S. Timasheff and G. Fasman, Eds. Marcel Dekker, New York, is reprinted with permission of the editors and the publisher (copyright 1969, Marcel Dekker Inc.).

''the better the binding, the better the catalysis'', in which better catalysis means increased rate constant. Obviously, this well-known statement contains a compromise, since too much free-energy conservation will increase the Michaelis constant and reduce the overall rate at any fixed substrate concentration. Albery and Knowles[17] have discussed the efficiency problems which are involved. Our concern is with the variety of ways that the conservation of free energy may be exploited, particularly in rack mechanisms.

To that end, consider the family of *N*-acetyl-L-amino acid methyl or ethyl ester substrates for chymotrypsin, the only family of substrates for any enzyme for which nearly complete thermodynamic profiles exist and these only at one pH value. Enzyme-substrate affinity and catalytic rate constant increase monotonically with size and shape of the α-carbon sidechain of these substrates, but only up to a limit set somehow by the binding pocket.[75] Haldane's mechanism provides the principle by which this behavior is to be explained, but the choice of a rigid solid as model for the protein precludes much useful deduction about real proteins, since the latter are now known to behave more like liquids than solids. Of course, free energy can be stored in primary bonds, but this is where the resemblance ends. Chemical storage is important in proteins as a means for preventing denaturation when the amounts

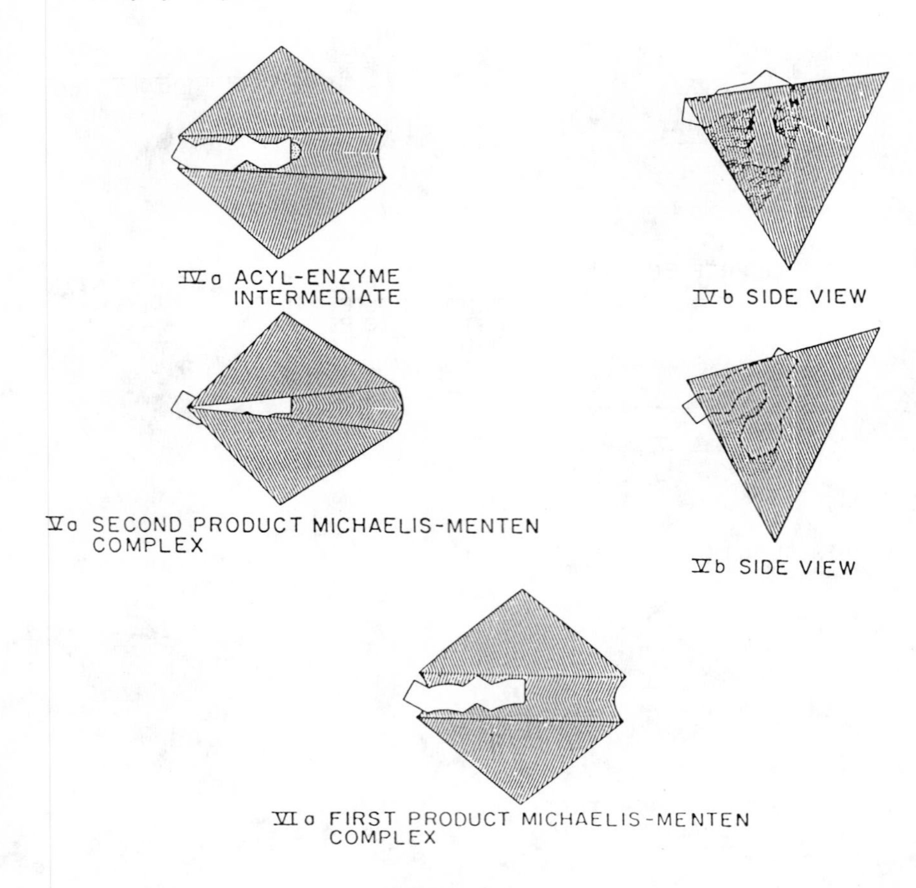

FIGURE 4 (continued).

of free energy which must be stored are significant relative to folded stability. It is also important as an alternative storage device in evolutionary progress toward free-energy conservation. In this chapter there is no need to discuss this kind of storage as a special topic. Instead, we must concentrate on the conformational devices for storage and transfer. In particular, consider three ways in which such devices provide versatility not possible with Haldane's model. (1) The free energy released in substrate-protein interactions can be stored by conformational rearrangements so as to be available to facilitate any of the subsequent steps of the catalytic reaction. Long-term storage of conformational free energy involves many small conformational rearrangements so that there are no potential-energy ''hot spots'' nor, for that matter, entropic ''hot spots.'' However, transition-state formation requires fast conformational rearrangements which do produce hot spots in the assembly of reacting groups of substrate and protein. Such rearrangements can be rare if they are not into states of the protein lying in the normal range of fluctuations at equilibrium. Alternatively, they can be frequent if the transition states have free energies not too much higher than those which occur in the normal equilibrium fluctuation spectrum. The distinction is not a sharp one, but becomes so when the characteristic times are examined. (2) Large fractions of the free energy released as a result of the chemical rearrangements which occur in passing through transition states can be conserved. Some trade-off is required since the substrate must bind to the enzyme and catalytic rate constants have to be kept high, which may require some dissipation — although in principle it need not. In any event, the underlying requirement that free energy must be conserved as completely as possible without loss in temporal efficiency is well established in evolution, and it is shown in this chapter that this requirement plays a major role in the evolution of enzymic mechanisms. (3) Some or all of the substrate specificity manifested as changes in true single-step rate constants need not be an intrinsic or essential feature of the catalytic mechanism, but rather an elaboration of that mechanism or a new process added on in evolution.

Figure 1 contains a hypothetical free-energy description of the evolution of an enzymic process. It is based on experimental information for chymotrypsin and the few other proteins for which the free-energy profiles are known, but it is more generally rooted in the obvious consequences of the pressure in evolution towards increased efficiency in the use of free energy. These consequences are the proliferation of metastable intermediates and thus of elementary steps, the flattening of free-energy profiles, and the enlargement of cooperative units.[8b,91] We may suppose for convenience that the substrate configuration at the barrier top in the uncatalyzed reaction (substrate system in Figure 1.1) is a tetrahedral-carbon intermediate in basic ester hydrolysis. The latter, labeled T, is not usually the activated complex in the homogeneous reactions, but there is no loss of generality if we assume it to be. The barrier to its formation remains high in the combined systems profile of Figure 1.2, but splits with further progress, making T a metastable intermediate sandwiched between two new barriers, as shown in Figure 1.3. Each time a new metastable intermediate is developed the nearby peaks are drawn down and the nearby metastable states are drawn up. There is an increase in overall reaction velocity whenever the net result is to lower the highest barrier. In principle, the proliferation of mestastable states can go on indefinitely.

The net overall change in free energy of the protein component is, of course, zero, but along the reaction profile the protein component gains and loses free energy in amounts which tend to complement the losses and gains in the substrate system. This is the source of the term ''free-energy complementarity''. Its importance is obvious. If complementarity is close, little free energy is lost until the products leave the protein. The chemical free energy released by the substrate system is trapped as conformational free energy, and such trapping becomes more efficient the greater the number of elementary steps.[91]

It seems probable that the final combined-systems profile could be produced by non-chemical conformational adjustments alone, but it is now generally thought that the major cause for the splitting of the T barrier is chemical stabilization of the transition-state configuration of the substrate, the transition-state-stabilization hypothesis. However, the free-energy profiles are not of much help in distinguishing between that hypothesis and the rack hypothesis, and the data offered in support of the former appear to be as well and perhaps more generally explained by the latter. Virtually by definition the binding of transition-state analogs as they are described favors and is favored by some degree of domain closure. Thus, transition-state stabilization would be inevitable in rack mechanisms even without the additional stabilization described by the transition-state hypothesis.

The free-energy profiles in Figure 1 can be extended to include the acylenzyme intermediate in chymotryptic catalysis by reflecting the profile back from T whatever the description of T, but this is only a formal exercise unless the partitioning of total free energy of EA between chemical bond and protein conformation has been determined. An enzymic reaction can be said to be described when the quantitative free-energy profile for the protein has been established. It can be said to be explained when the free energy at any point on that profile can be factored into contributions from each of the participating molecular processes. However, the latter is much more complicated than it sounds, since each protein species is actually a distribution over conformers with different intrinsic catalytic parameters. To understand many complex time-dependent biological reactions, it is necessary to know the extent of averaging over conformers which occurs at each step. This aspect of fluctuational behavior is too complicated to be considered in this chapter so we must often settle for somewhat simplistic descriptions.

IV. KNOTS AND MATRICES

Using Provencher's extraordinarily important numerical method for Laplace transforms,[28] Gregory has succeeded in generating probability-density distribution functions for the rate constants measured in proton-exchange experiments.[29] Attempts[31] to generate these functions using closed-form expressions were not successful. Gregory's method supplements the earlier finding by Woodward and Rosenberg[32] that the rank order of proton exchange from proteins is preserved under a wide range of changes in pH, temperature, and solvent-composition. These two features of the proton-exchange method make it a tool of enormous power to be exploited in gross isotope measurement in solution, NMR, and other methods for measuring single-site exchange and neutron-diffraction measurements of exchange in crystals. In studies of gross isotope exchange of enzymes in solution these distributions at temperatures below 55° consist of three overlapping cusps (Figure 5). Peak III contains the large rate constants for sites, generally at protein-water interfaces where the local conformational fluctuations are largest. Peak II contains rate constants for slowly-exchanging sites, and peak I is due to a small family of very slowly exchanging sites. Claims that II sites exchange via partial unfolding and I sites always exchange via complete reversible thermal denaturation have become untenable.[33] For the serine proteases, because their derivatives are unstable and their autolysis rates high, useful exchange data are in short supply. Exchange data from crystalline monoisopropylphosphoryltrypsin measured with neutron diffraction[34] (Figure 6) have, however, proved invaluable (*vide infra*).

For HEW lysozyme and RNase A, exchange from I and II sites occurs without partial or full unfolding below about 55°. The exchange catalysts diffuse through the protein fabric. Although exchangeable sites can be identified by their rank order and so studied individually for temperature, pH, and solvent dependence, they cannot be assigned to protein residues by the gross solution isotopic method, that is, by measurements of isotope appearance in solutions of the labeled proteins. Fortunately, the I sites can be located as a group using neutron diffraction to measure deuterium exchange by the crystals. For RNase A and HEW lysozyme,[174] the numbers of I sites obtained by the two methods are identical within small experimental errors. The sites for RNase A and HEW lysozyme are shown in Figure 7. Using methods which measure exchange rates for single sites in solution, primarily NMR methods, the specific residue sites can be assigned their position in the rank order as has already been accomplished with BPTI.[36]

The activation enthalpies and entropies for exchange of II families, determined using the solution isotope method, were found to be small, consistent with the rate-limiting step being diffusion of the catalyst through the protein.[37] The I sites demonstrated much larger values which suggest that the limiting step is local distortion required for attack by the catalyst.[37] The exchange behavior is qualitatively different for the two families, demonstrating that the substructures have some qualitative differences. For example, the existence of rank-order conservation in proton exchange of the I and II site families requires that the activation enthalpies and entropies for these sites demonstrate compensation behavior[296] (see Section V). With HEW lysozyme two compensation plots are found, both linear within errors. The characteristic parameter, the compensation temperature T_c, was found to be about 470K for II sites and 350 ± 10K for III sites.[29] The aggregate of such information on exchange of the I sites forced Gregory and Lumry[29,38] to conclude that these very slowly exchanging sites are parts of distinct protein substructures which are highly cooperative and act like strong springs. These must derive their strength from a synergism of electrostatic factors focused on a core of hydrogen bonds probably inductively coupled through the α-carbon atoms. The buried parts of these cores are surrounded by alkyl and aryl sidechain groups which mutually interact through dispersion forces but which also reduce the local dielectric constant to increase the strength of the hydrogen bonds and all other interactions in the substructures.

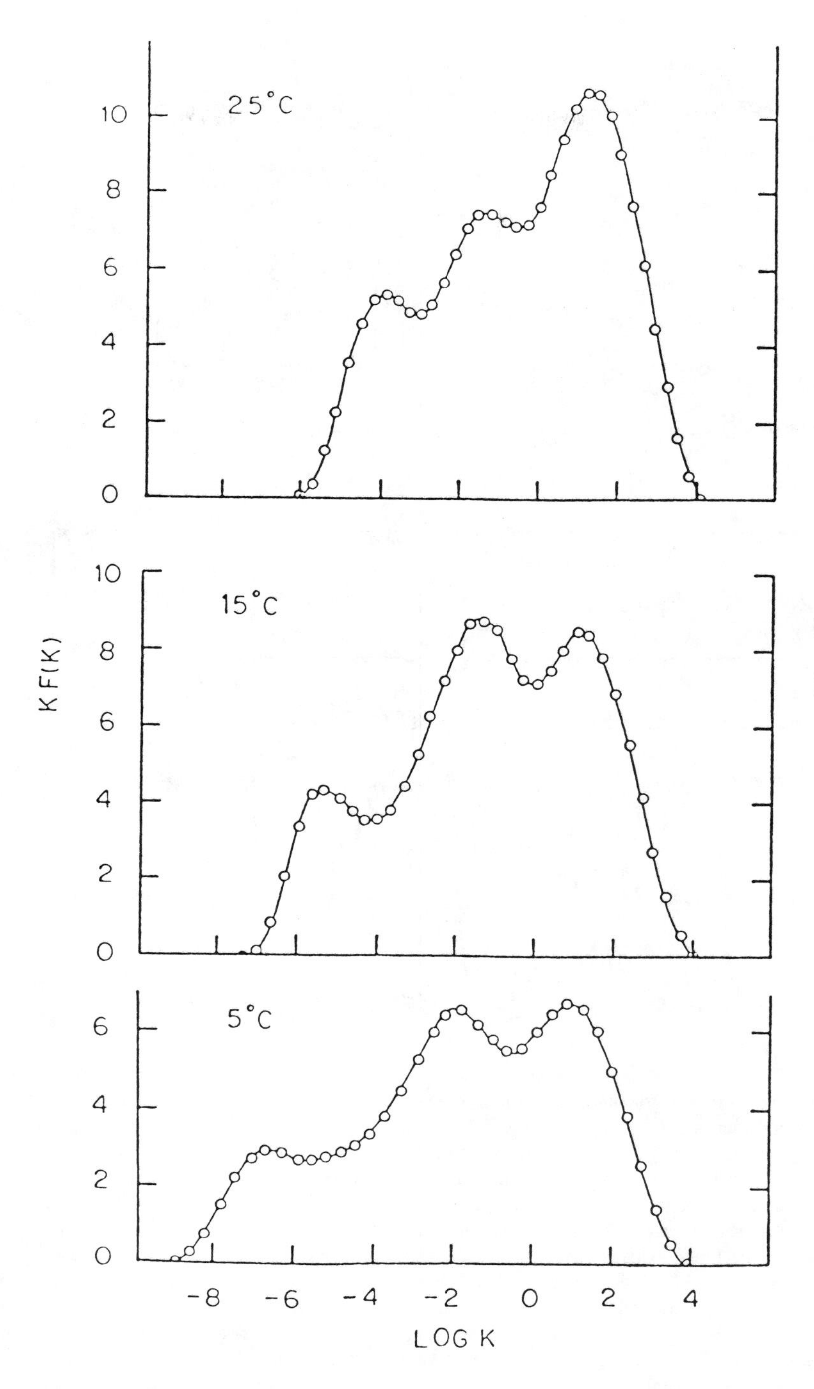

FIGURE 5. Exchange-rate probability-density distribution functions for HEW lysozyme determined by numerical Laplace inversion of the data of Gregory, et al.[29a,29b] Three peaks are observed at each temperature. The first peak on the left contains the very slowly exchanging sites (peak I family) attributed to knots; the middle peak contains the slowly exchanging sites (peak II family) most of which are matrix sites. The rapidly exchanging sites (peak III family) lie in regions of highest motility, most being at or very close to protein-water interface. This figure is from *The Fluctuating Enzyme*, G. R. Welch, Ed., Wiley-Interscience, New York, 1986, and is reprinted with the permission of the publisher and the author (copyright 1986, Wiley-Interscience).

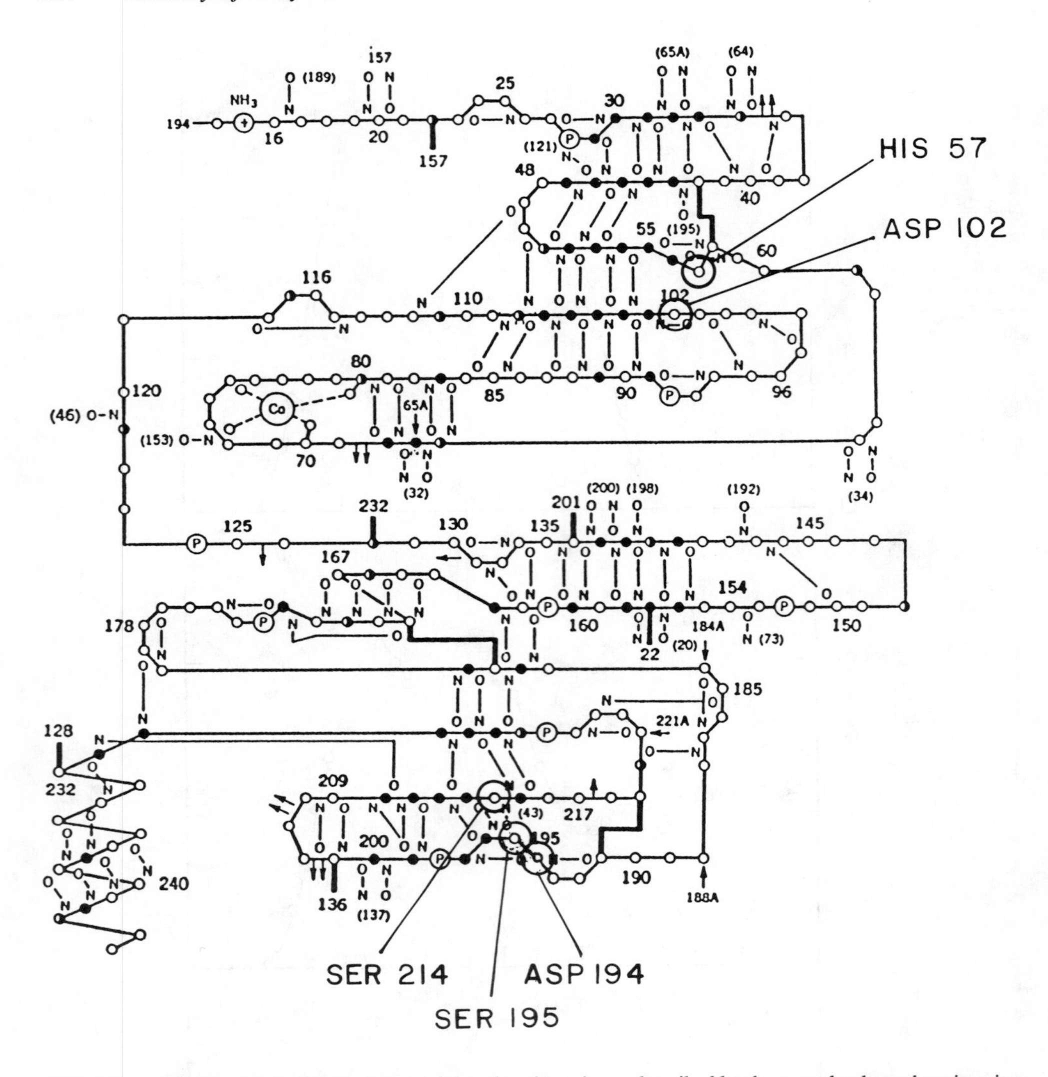

FIGURE 6. The two knots of monoisopropylphosphoryltrypsin, as described by the very slowly exchanging sites. Kossiakoff, using neutron diffraction, found that protons had not exchanged from the residues indicated by black dots in the figure after about 1 year at pH 7 and 20°C. There was a significant time demarcation between these sites and all others, suggesting that all the sites which did not exchange are knot sites.[34] The black dots (excluding those in the pigtail) then identify two knots which can be clearly recognized in the three-dimensional structure. It can be seen that the hydrogen-bonded arrays conform to the requirements proposed for β-sheet knots but such well-ordered knots may be rare. Compared with other proteins for which there is information about knot sites, the knots are large. The functional residues of the "charge-relay system" (his.57, ser.195, and asp.102) are first residues on the polypeptide as it emerges from knots. These residues and ser.214 (lysine in chymotrypsin), which may also be part of the charge-relay system, are palendromically related as can be seen in the figure. Diffraction experiments show lowest B factors in all polypeptide segments with black dots in the figure.[101a,182] This figure, modified to identify functional residues, is reprinted by permission of the publisher from *Nature,* 296, 713 (copyright 1982, Macmillan Journals Limited).

The electrostatic factors produce the I type structures only when the steric restrictions on cooperative contraction are minor. The relatively high degree of contraction produces high-density structures with small-amplitude, high-frequency fluctuations.

Core surfaces exposed to solvent appear to be so constructed by positioning of charged groups that they enhance the stability of the protein-water interface and perhaps directly the nearby intrapeptide hydrogen bonds, and thus gain strength and thermodynamic stability for the core.[47]

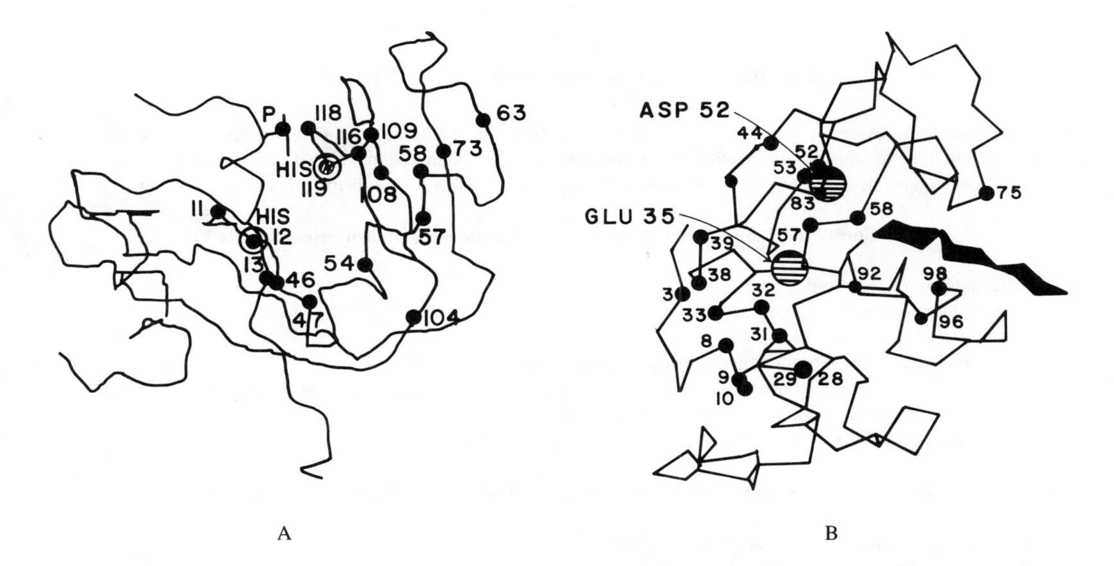

FIGURE 7. Knot residues in ribonuclease A and HEW lysozyme detected by proton-exchange methods. The black dots identify very slowly exchanging sites (peak III family) in RNase A (A) and in lysozyme (B). The functional groups so far identified in catalysis studies are circled in A and cross-hatched in B. Each knot carries one of these residues either on a very slowly exchanging (knot) site or on a first residue emerging from a knot. Proton-exchange provides only a partial description of knots. More detailed descriptions may show that all functional groups belong to knot residues. The inhibitor (N-acetylglucosamine)$_3$ is shown in B to identify the interdomain binding region. It is bound in the matrix ''basket'' formed by the two domains.

These substructures are called ''knots''. They are, of course, not knots in the mathematical sense but rather in the sense that when they are broken, which is a cooperative process, they ''untie'' the protein conformation. At present, the knots can be described reliably only in terms of the peak I sites. Almost certainly some peak II sites lie in knots but exchange too rapidly to be identified by the exchange method. The participating side-chains can be identified only provisionally using X-ray-diffraction data. The high densities and small fluctuations of these compact structures should reveal knots in the diffraction data as regions with these properties. The few knots so far identified by proton exchange also have the highest atom number density[84] indicating a high proportion of aliphatic and aromatic groups in the knots. High density appears to be a useful way to get some idea of what knots may look like in the many proteins for which adequate proton-exchange data are not available.

The peak II sites identify softer substructures called ''matrices'' for lack of a better term. Although exceptions are perhaps to be expected, each knot with its tethered matrix, comprises a semiautonomous structural unit which has been defined earlier in this chapter as a domain. The knot determines the conformation of its domain and many of the properties. Table 1 is a provisional list of the properties determined by the two kinds of substructure. In the few proteins for which knot descriptions are well advanced, it is found that knots are formed from the most highly conserved exons.[39] This suggests that the discovery of a new knot is a rare event in evolution and that protein evolution consists of a few such discoveries followed by their tailoring and combination by matrix mutations so as to support a variety of useful functions, usually of increasing complexity. One general feature of great importance achieved by the discovery of knots is ''kinetic stability'', i.e., the reduction in rate of unfolding necessary to prevent destruction of proteins with small net thermodynamic stability. Low thermodynamic stability is necessary because relatively high matrix motility is essential for function. The transition states for protein denaturation now appear to be formed by expanding knots just to the point at which they lose cooperativity. This is supported for

TABLE 1
The Roles of Knots and Matrices in Proteins

There is some evidence for involvement of both matrices and knots in most of the items listed below but generally one type of substructure is more important than the other. A provisional classification of some properties of enzymic proteins according to the relative importance of the two kinds of substructures is given below.

Highly Conserved Exons Code for Knots and Knots are Primarily Responsible For

Genetic stability of function[a]
Average conformation
Kinetic stability[b]
Low compressibility (their true compressibility is small)
Rigid platforms for electromechanical events in bond rearrangement catalysis; the functional groups for catalysis are either in knots or on residues just emerging from knots
High-frequency, low-amplitude fluctuations

Weakly Conserved Exons Code for Matrices and Matrices are Primarily Responsible For

Low-frequency, medium to large amplitude fluctuations
Predenaturation behavior (including thermal-expansion coefficients)
Free-volume rearrangements both activated and nonactivated
Secondary substrate specificity (e.g., α-carbon sidechains of protease substrates)
Coordinated motions of the knots in catalysis (in response to substrate binding and changes in substrates)
Low compressibility (artifact of permeability of proteins to water which makes matrices mechanically transparent)
Specificity in protein-protein association (e.g., immunological specificity, subunit association in hemoglobin and associated free-energy redistribution) (cooperative with segments containing peak III exchange sites)
Differences among individuals of a single species resulting from change in single residues

[a] Hemoglobin and myoglobin appear to have either no knots or only small ''pins'' rather than knots.[179,30,66] If so, genetic stability ought to lie in the preservation of fold, which is, of course, entirely consistent with the great variability of residues found in all but a very few positions in those proteins having the ''myoglobin fold''. On closer examination the fluctuational characteristics of these proteins are likely to be found to be quite different from those of enzymes and protease inhibitors.

[b] Knots are also very likely the major factor in thermodynamic stability (see text.) Residue changes in knots which do not preserve linear compensation behavior are apparently lethal mutations. This is unlikely to be true for matrices, which appear to easily accommodate a variety of substitutions.

HEW lysozyme by the similarity of the melting temperature and the T_c value for proton exchange from knot sites. The old expansion (''subtle change'') explanation[185] is thus detailed to focus on knots. This modification is strongly supported for this protein by clever studies of Segawa and Sughiara[197b] of the effect of inhibitor binding on the unfolding and folding rate constants. Their data show that the binding site does not exist until knot-contraction occurs. Studies of the folding of cytochrome c suggest that the final electronic state of the heme complex ion is established by knot contraction.[87]

Matrix expansion and contraction is necessary for functions such as catalysis, but little volume increase occurs in knot expansion into the unfolding transition state. The ''minimum stability'' requirement for catalysis[41a] is primarily required for matrices. It may be that matrices contribute little to the thermodynamic stability,[42] although they make major contributions to the standard enthalpy and entropy changes in unfolding. It begins to appear probable that knots are primarily responsible (see Section IV). This would be in part consistent with a very important consideration of the meaning of enthalpy and entropy changes with special reference to denaturation given by Benzinger.[77] This is a complex matter discussed in detail elsewhere.[78] However, there is confusion about the description of thermally denatured states of proteins. The commonly held idea that in pure water solvent these states resemble random-coil homopolypeptides is probably wrong. Proteins with even a few disulfide bonds do not much expand on denaturation in water and new studies, particularly those using high-precision gel chromatography and newer light scattering methods, suggest that expansion in water is generally small.[43] The picture which is developing for denaturation in pure water is that of expansion to about twice the volume of the folded form with some persistent structure, which in some cases is found in the native form and in others consists of new organizations with generally high motility. The effective external surface is larger than that of the folded form, but the major difference is the many water molecules which have moved into the expanded netlike structure and the increase in dielectric constant. The physical parameter which reflects these changes most usefully is probably the dielectric constant, which changes from a variety of small but variable local values to a large value uniform, on average, throughout all but the regions of persistent structure. The large, nearly temperature-independent, positive heat-capacity change has been attributed to increased exposure of hydrophobic sidechains, but we suspect it will be found to be due as much or more to onset of high motility in the "net" and thus a direct consequence of the large increase in dielectric constant. A third source is undoubtedly change in A and B hydration sheets (see Section VIII.D).

It is becoming increasingly well established that the domain rather than the total protein is the smallest building block. The enzymes so far examined have two domains, but studies of such complex enzymes as aspartyltranscarbamylase show that the domains forming a catalytically active site do not have to be on the same protein subunit[44] (*vide infra*).

The stability of multidomain proteins appears to depend on the domains acting more independently than dependently. Removal of the S peptide from Rnase A removes one of the knots (Figure 7), but the remaining protein behaves in unfolding experiments and in tests of structural integrity as though it is about half denatured.[45] The second knot lies near the C-terminus, and Taniuchi et al.[50] have shown that, on sequential removal of residues from the C-terminus of Rnase A, a dramatic change in protein properties occurs when residue 120, a phenylalanyl residue (which, judging from the X-ray picture, is probably part of that knot), is removed. Again, the product behaves as though it is partially denatured but still retains a typical cooperative unit. Residue 119 contains one of the two functional imidazole groups. The S and C peptides form unusually stable helices,[46] perhaps because of the distribution of the relatively large number of charged groups,[47,188] as already discussed. If the knot-matrix description is correct, the residues in these peptides have been selected to establish unusual helix stability, so that detailed study of residue variation, already in progress,[48,188] should provide valuable information about some of the ways the special strength and stability of knots is achieved. The several observations suggest that to a first approximation the two domains in the intact protein behave independently in thermal denaturation.[49] Such independence has been established for a number of proteins by Privalov et al.[168] That this cannot be the case for RNase A is established by the enthalpy change which shows that the cooperative unit in unfolding for the intact protein must contain all or major parts of both domains. However, this does not require that the coupling between domains be strong, only that their intrinsic stabilities be quite similar. It has been demonstrated for a number of enzymes that the domains can be isolated often in stable, folded form.[57,50d,177]

Proton-exchange studies demonstrate several subtle but important indications of structural cooperatively. A particularly interesting example is the finding that removal of the RNase A C-terminal sequence 119-124 greatly accelerates proton-exchange in all of the remainder of the protein.[50a] This group has the lowest motility found in the protein by analysis of X-ray-diffraction temperature factors.[51] Other types of experiment show similar influence of residues carrying functional groups on structure (see his57 in chymotrypsin, Section VIII.C).

Another example of what might be called "knot tricks" is provided by HEW lysozyme and α-lactalbumin. These proteins have about the same size and much sequence homology but quite different function.[52] The latter protein is not itself catalytic, but combines with galactosyl transferase to form an active enzyme but does so only when a relatively weak calcium-binding site is empty. That site is thought to lie in the bottom of the cleft corresponding to the intermatrix substrate-binding site of HEW lysozyme.[53] Thermal denaturation behavior, as reported by Ikeguchi and co-workers,[52d,52e] suggests a single knot in the absence of calcium ion and two knots with behavior very similar to that of HEW lysozyme[115e] when the ion is bound. The picture is somewhat clouded by the different interpretations given by different groups,[50f] but it nevertheless appears clear that an enzyme — lysozyme — has been converted into an ion-binding protein and the affinity for the ion adjusted by coupling knot formation to the ion-binding process.

In view of the uncertainty about the extent of unfolding in thermal and denaturant-produced denaturation, the term ''unfolding transition'' is ambiguous. The frequently used term ''melting'' is probably more accurately descriptive, but it too implies knowledge we do not yet have. We therefore have chosen in the remainder of the chapter to describe the weak first-order phase changes of proteins using the implication-free term ''denaturation.''

Although the proton-exchange data necessary to establish knot substructures exist for only a few proteins, there is considerable evidence of other types to suggest that knots are common, perhaps ubiquitous, in enzymes. A particularly well-studied example is thermolisin, from which a large knot or polyknot apparently free of matrix has been excised without causing its denaturation.[57] Its thermal-denaturation behavior demonstrates a cooperative unit which is about half of that of the intact protein.

The existence of semirigid knots to support functional groups and fix the details of domain closure, and of flexible matrices to accommodate substrates and make domain closure possible is considerable support for enzymic rack mechanisms. Further support from ''structure'' is provided by the finding that the protein groups established as being direct participants in the chemical events are distributed so that each of the two knots carries at least one functional group.[58] This is shown for trypsin in Figure 6 and for RNase A and HEW lysozyme in Figure 7. Any domain closure or contraction of matrices will force the two domains together in these proteins and in any other proteins constructed in the same way. Because the functional groups are opposed, any change in knot-knot orientation must alter the electronic situation in the chemically reacting assembly. This construction directly connects the electronic events to conformation — as it does in all rack mechanisms. It provides a conformational reaction coordinate subject to a variety of fine tuning through pH, solvent composition, etc., and it provides the essential element of mutability in large through residue changes in knots, and in small through residue exchange in matrices and surface residues.

Since there is ample evidence for HEW lysozyme, RNase A and several serine proteins, for contractions and expansions, hardening and softening, and so on during catalytic function,[9e,35,72,98,176] rack mechanisms are indicated for these enzymes. Put more accurately, the rack machines are suggested, but proof that they facilitate catalysis by mechanical force still rests on inference and the elimination of alternatives, rather than on unequivocal experimental fact. The only nonmechanical alternative which we can suggest is that domain closure generates a series of electric-field changes at the reactive site which stimulate rate-limiting proton migration, and do so without appreciably raising the potential energy of the reacting assembly or any development of mechanical strain. This is Warshel's dynamic version of the catalytic mechanism of Stearn and Eyring.[55] If it is important, it is very likely a useful sophistication added in later evolution.

It is clear from the construction of these enzymes that domain closure can occur spontaneously whether it is driven by momentum waves arising in the thermal reservoirs or by isoergonic fluctuations of conformation. In either case, the useful fluctuations are consequences of delicate evolutionary adjustment, and thus quantitatively sensitive to almost all nondestructive residue changes since both geometry and fluctuation times are important. Enthalpy-entropy compensation (Section V) provides an essential element of stability, which makes large quantitative adjustments in functional parameters possible at relatively low overall thermodynamic stability. If compensation temperatures for the processes in which fluctuations occur are near operating temperatures, e.g., body temperature in thermostatted organisms, perturbations of the fluctuation spectrum have small free-energy consequences, even though the enthalpy and entropy changes may be large. This appears to be the reason why changes in charge, almost all of which must be expected to modify the fluctuation spectrum, produce negligible effects, or effects which can be rationalized in terms of simple electrostatic models. Classic examples are the disappearance of the Bohr effect in ferric hemoglobins when investigated near 24°[56,62] and the acid Bohr effect of ferrous hemoglobins.[67a]

The structural picture which has emerged shows that domain-closure fluctuations can occur without assistance (see Section VIII.B). Specifically in the case of the N-acetyl amino acid methyl esters with chymotrypsin, they can occur without fundamental dependence on the α-carbon sidechain. As is shown below, this follows from the existence of a linear compensation pattern when sidechain is varied but it is more obviously implicit in the finding that all but three orders of magnitude of the total enhancement in the catalytic rate in this family is achieved with N-acetylglycine methyl ester. The additional enhancement observed with the other substrates in this family must be a consequence of an elaboration of the fundamental mechanism or an added process discovered subsequently in evolution.[63] Obviously, in this case the latter is correct. Sidechain specificity is discussed in Section VIII.B.

It is possible that knots have special electronic properties which have been woven into the catalytic mechanism. The original model for knots was the unusual cooperative behavior of hydrogen-bonded water clusters in liquid water. This behavior (Section VIII.D) has been attributed to cooperative electron rearrangements involving at least four water molecules,[191] and the clusters have recently been shown to be unusually effective as traps for protons and electrons.[59] If the same kind of cooperativity is important in the strong hydrogen-bond systems of protein knots, it may give knots special acid-base or other electronic properties. At this time there is not sufficient relevant information to make its discussion worthwhile.

V. ENTHALPY-ENTROPY COMPENSATION PHENOMENA — A NEW TOOL FOR ANALYSIS OF LINKAGE SYSTEMS

Processes studied in closely related chemical families are very often found to yield enthalpy and entropy changes which are linearly related within experimental error. Earlier called isokinetic (isoequilibrium) phenomena in the classic monograph by Leffler and Grunwald,[189] they are now called enthalpy-entropy compensation phenomena to better emphasize the mutual compensation of enthalpy and entropy changes which leaves the free-energy changes small.[62] The linear-free-energy phenomena (LFE) are better known, but Leffler showed that compensation behavior was an implicit consequence of LFE behavior[61,175] (see Equation 11). Compensation behavior is a powerful tool for the analysis of kinetic and equilibrium data obtained in systems of coupled subprocesses (linkage systems) when coupling is weak.[64] Although strict linearity is not possible except in the trivial limit of zero coupling, plots of enthalpy changes against the corresponding entropy changes obtained with chemical alteration of reactants in congener series or systematic change of experimental conditions are often linear within experimental precision, even when the precision level is only a few percent. Compensation phenomena is the rule rather than the exception in biology, indicating either ubiquitous weak coupling in biological linkage systems or good compliance with mean-field models[64] (*vide infra*). A very important implication of these phenomenological observations is that many biological machines can be described quantitatively using first-order linear-response theory, which is an elaborate way of saying that machine operation is supported by changes in the machine which occur with significant probability when the

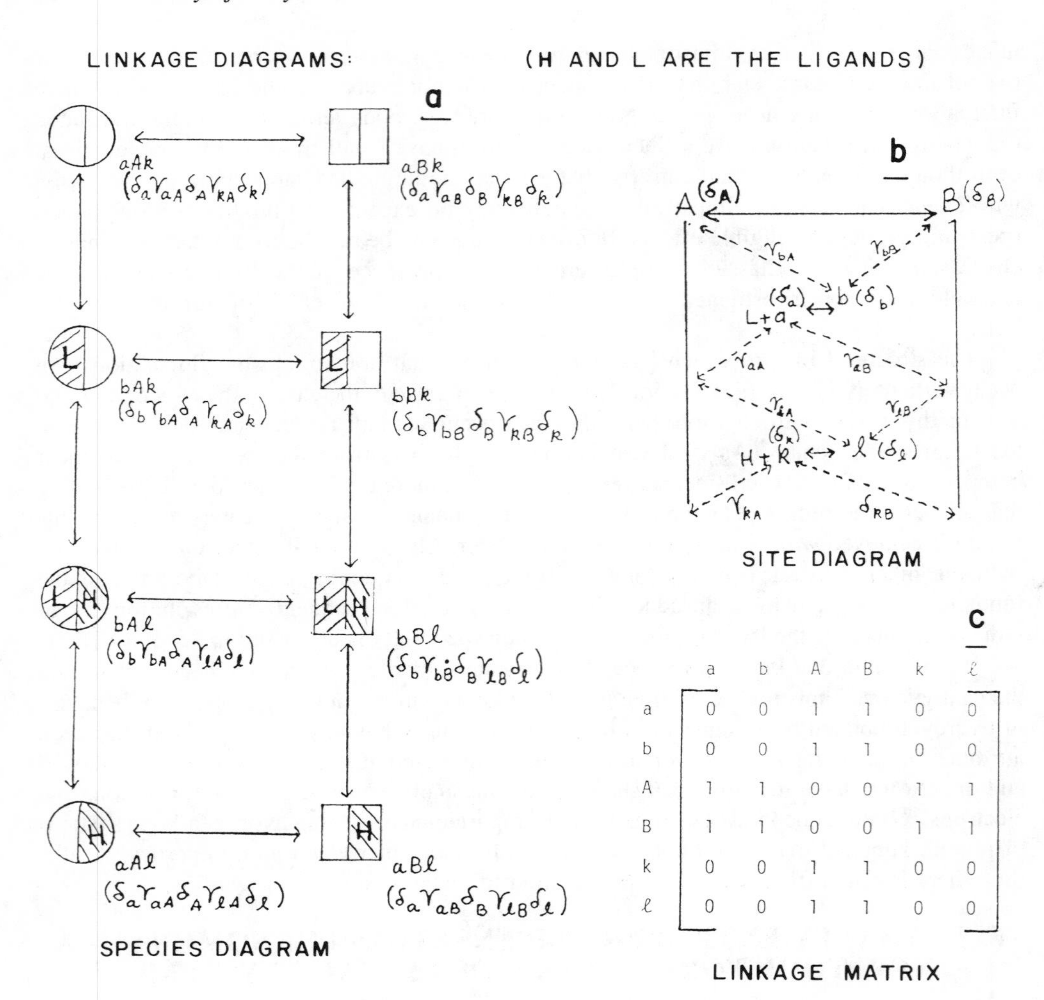

FIGURE 8.

machine is at rest,[65] that is, the changes in the machine which support function are members of the distribution of fluctuations at equilibrium. The machines under consideration here are enzymes for which an old review[62a] is still the best introduction to their compensation phenomenology. The major sources of compensation and some of the theory are discussed in a recent monograph[11f] which supplements but does not replace the monograph by Leffler and Grunwald.[189]

The simple theory for compensation phenomena arising in a typical linkage system is given below (from Reference 76). The conventional species diagram and the site ''linkage diagram'' to which it is converted are shown in Figure 8. The approximation which yields linear compensation behavior is that the coupling coefficients, γ_{ij}, are independent of the degree of advancement of the subprocesses.

$$P_{aAk} = \bar{n}_{aAk} = \Pi_a \Pi_A \Pi_k$$

$$\underset{\text{species}}{\Sigma\Sigma\Sigma}\ P_{ijm} = 1$$

$$\left.\begin{aligned} \Pi_a + \Pi_b &= 1 \\ \Pi_A + \Pi_B &= 1 \\ \Pi_k + \Pi_l &= 1 \end{aligned}\right\} \text{sites}$$

$$G_{aAk} = -\kappa T \ln\left(\frac{\Delta aAk^{N_{aAK}}}{N_{aAk}!}\right)$$

$$= -n_{aAk} RT \ln\left(\frac{\Delta_{aAk} e}{n_{aAk} N}\right)$$

$$\text{Let } \Delta_{aAk} \rightarrow \delta_a \gamma_{aA} \delta_A \gamma_{kA} \delta_k \quad (1)$$

$$G = \sum_i \sum_j \sum_m G_{ijm} \rightarrow \sum_i \sum_j \sum_m \left[-\Pi_i \Pi_j \Pi_m RT \ln\left(\frac{\delta_i \gamma_{ij} \delta_j \gamma_{mj} \delta_m}{\Pi_i \Pi_j \Pi_k}\right)\right] + RT \ln \frac{N}{e}$$
$$-\bar{n}_L RT \ln\left(\frac{\Delta_L e}{\bar{n}_L N}\right) - \bar{n}_H RT \ln\left(\frac{\Delta_H e}{\bar{n}_H N}\right) \quad (2)$$

From Equation 2 the chemical-potential changes for the subprocesses can be determined using site partition functions δ_i, and the parameter γ_{ab}, etc., which couple the subprocesses, thus using equations such as (3).

$$\mu_a = \left(\frac{\partial G}{\partial \Pi_a}\right)_{\Pi_b, \Pi_A, \Pi_B, \Pi_k, \Pi_l} \quad (3)$$

$$(\mu_b - \mu_a - \mu_L) = \left[-RT \ln\left(\frac{\delta_b}{\delta_a \Delta_L}\right) + RT \ln\left(\frac{\Pi_b}{\Pi_a \bar{n}_L}\right) + RT \ln N \right.$$
$$\left. -\Pi_B RT \ln\left(\frac{\gamma_{bB}}{\gamma_{aB}}\right) - \Pi_A RT \ln\left(\frac{\gamma_{bA}}{\gamma_{bA}}\right)\right] = 0 \quad (4)$$

Δ_L is the single particle partition function for ligand, L, Δ_H for ligand H. $\bar{n}_L$ is the concentration of L per mole of protein, so also for $\bar{n}_j$H. *N* is Avogadros number.

$$(\mu_B - \mu_A) = \left[-RT \ln\left(\frac{\delta_B}{\delta_A}\right) + RT \ln\left(\frac{\Pi_B}{\Pi_A}\right) - \Pi_a RT \ln\left(\frac{\gamma_{aB}}{\gamma_{aA}}\right)\right.$$
$$\left. -\Pi_b RT \ln\left(\frac{\gamma_{bB}}{\gamma_{bA}}\right) - \Pi_k \ln\left(\frac{\gamma_{kB}}{\gamma_{kA}}\right) - \Pi_l \ln\left(\frac{\gamma_{lB}}{\gamma_{lA}}\right)\right] = 0 \quad (5)$$

$$(\mu_l - \mu_k - \mu_H) = \left[-RT\ln\left(\frac{\delta_l}{\delta_R\Delta_L}\right) + RT\ln\left(\frac{\Pi_l}{\Pi_R\bar{n}_H}\right) + RT\ln N \right.$$
$$\left. -\Pi_B RT\ln\left(\frac{\gamma_{lb}}{\gamma_{kB}}\right) - \Pi_A RT\ln\left(\frac{\gamma_{lA}}{\gamma_{kA}}\right)\right] = 0 \qquad (6)$$

$$\overline{\Delta G}^\circ_{ba,app}(\Pi_B) = -RT\ln\left(\frac{\Pi_b}{\Pi_a\bar{n}_L}\right) = \left[-RT\ln\left(\frac{\delta_b}{\delta_a\Delta_L N}\right) - RT\ln\left(\frac{\gamma_{ba}}{\gamma_{aA}}\right)\right] - \Pi_B RT\ln\left(\frac{\gamma_{bB}\gamma_{aA}}{\gamma_{aB}\gamma_{bA}}\right)$$
$$= \Delta\mu^\circ_{ba}(0) + \Pi_B\Delta\Delta\mu^c\begin{pmatrix} b & a \\ B & A \end{pmatrix} \qquad (7)$$

$$\overline{\Delta G}^\circ_{BA,app}(\Pi_b,\Pi_l) = \left[-RT\ln\left(\frac{\Pi_B}{\Pi_A}\right) - RT\ln\left(\frac{\gamma_{aB}}{\gamma_{aA}}\right) - RT\ln\left(\frac{\gamma_{kB}}{\gamma_{kA}}\right)\right]$$
$$-\Pi_b RT\ln\left(\frac{\gamma_{bB}\gamma_{aA}}{\gamma_{bA}\gamma_{aB}}\right) - \Pi_l RT\ln\left(\frac{\gamma_{lB}\gamma_{kA}}{\gamma_{lA}\gamma_{kB}}\right)$$
$$= \Delta\mu^\circ_{BA}(0,0) + \Pi_b\Delta\Delta\mu + \Pi_b\Delta\Delta\mu^c\begin{pmatrix} b & a \\ B & A \end{pmatrix} + \Pi_l\Delta\Delta\mu^c\begin{pmatrix} l & k \\ B & A \end{pmatrix} \qquad (8)$$

$$\overline{\Delta G}^\circ_{lk}(\Pi_B) = -RT\ln\left(\frac{\Pi_l}{\Pi_R\bar{n}_H}\right) = -RT\ln\left(\frac{\delta_l}{\delta_k\Delta_L N}\right) - RT\ln\left(\frac{\gamma_{lA}}{\gamma_{kA}}\right)$$
$$-\Pi_B RT\ln\left(\frac{\gamma_{lB}\gamma_{kA}}{\gamma_{kB}\gamma_{lA}}\right)$$
$$= \Delta\mu^\circ_{lk}(0) + \Pi_B\Delta\Delta\mu^c\begin{pmatrix} l & k \\ B & A \end{pmatrix} \qquad (9)$$

Equations 7, 8, and 9 demonstrate the linear-free energy relationships between apparent subprocesses, standard free energy changes, and the site probabilities. These are a consequence of the assumption that the site probabilities are independent of the degrees of advancement of the subprocesses. The assumption is validated by the linearity of the experimental compensation plots (see text).

Each of the apparent standard free-energy changes is a linear function of one or more Π values. Hence, in general, the corresponding standard enthalpy and entropy changes must also be linear functions of the same Π values. Thus, if α, β, σ, ξ and ζ depend only on T,

$$\overline{\Delta G}^\circ_{app}(\Pi) = \alpha + \beta\Pi = \overline{\Delta H}^\circ_{app}(\Pi) - T\overline{\Delta S}^\circ_{app}(\Pi) = \sigma + \rho\Pi - T\xi - T\zeta\Pi$$
$$= (\sigma - T\xi) + (\rho - T\zeta)\Pi \qquad (10)$$

$$\Pi = \frac{\overline{\Delta H}^\circ_{app}(\Pi) - \sigma}{\rho} = \frac{\overline{\Delta S}^\circ_{app}(\Pi) - \xi}{\zeta}$$
$$\overline{\Delta H}^\circ_{app}(\Pi) = \left[\frac{\rho}{\zeta}\xi + \sigma\right] + \frac{\rho}{\zeta}\overline{\Delta S}^\circ_{app}(\Pi)$$
$$\overline{\Delta H}^\circ_{app}(\Pi) = C + T_c\,\overline{\Delta S}^\circ_{app}(\Pi) \qquad (11)$$

Equation 11 is the conventional enthalpy-entropy compensation relationship with slope T_c, the compensation temperature.

Alternatively,

$$\overline{\Delta G}^{\circ}_{app}(\Pi) = C + (T_c - T)\,\overline{\Delta S}^{\circ}_{app}(\Pi) \tag{12a}$$

$$\overline{\Delta G}^{\circ}_{app}(\Pi) = C + \left(1 - \frac{T}{T_c}\right)\overline{\Delta H}^{\circ}_{app}(\Pi) \tag{12b}$$

If in the above example $\bar{n}_H$ is varied, $\overline{\Delta G}_{ba,app}$ (Π_B) will be linear in Π_B although it will not be linear in $\bar{n}_H$. The progress of the conformation change, A → B, can often be followed to give Π_B values in which case if $\bar{n}_L$ is varied at several fixed values of $\bar{n}_H$, $\Delta\mu_{ba}$ (0), $\Delta\Delta\mu^c_v\begin{pmatrix} b & a \\ B & A \end{pmatrix}$, RT ln $\left(\frac{\delta_b}{\delta_a\Delta_L}\right)$, $\Delta\mu^{o}_{lk}(0)$, $\Delta\Delta\mu^c\begin{pmatrix} l & k \\ B & A \end{pmatrix}$ and RT ln $\left(\frac{\delta_l}{\delta_k\Delta_H}\right)$ can be determined.

The most extensive collection of linear compensation plots for a single protein was found[62] in the data for ligand-binding by ferric hemoglobins and myoglobins amassed by Beetlestone and co-workers,[67a] and in recent years by Anusiem and co-workers.[68] Their studies of species variants produced the oldest example we have found in which change in amino acid residue is the independent variable.[67b] Recent applications of site-specific mutagenesis have demonstrated that compensation is a frequent consequence of residue variation even when residue substitutions are made in a random fashion. Beetlestone et al. studied pH effects on azide-ion binding using more than 30 species variants of hemoglobin and, with a smaller sample, studied ionic-strength dependence, some effects of solvent additives, alkaline ionization, and a number of different ligands. Data reported by other investigators considerably extends this collection of compensation patterns.[69] The only parameter for characterizing compensation plots available with existing data is the slope of the ΔH° vs. ΔS° plot, known as the ''compensation temperature'' (T_c, Equation 11). The T_c values for all ferrihemoglobin patterns lie with 20° of 295K and usually much closer.[2c,62] Consideration of the theory suggests that these plots reflect a common subprocess to which the measured subprocess — usually ligand binding — and the driving subprocesses — those varied by change in residue composition, pH, or other effector concentration and solvent composition — are coupled.[73] This is consistent with the similarity of the compensation patterns derived from studies of myoglobin,[62,67a,69c] which reasonably well establish that the central subprocess is a characteristic of the confirmation of the single subunits of hemoglobin. Roughton's data for oxygen binding to ferrous hemoglobins in which oxygen pressure was the driving variable demonstrates compensation temperatures in the same range[73b] and, more recently, compensation patterns have been demonstrated for a number of ferrous hemoglobins, although the T_c values tend to be closer to physiological temperatures.[71] The driving variable in the latter studies was also ligand concentration and in some studies several ligands were used. The Beetlestone group[67a] showed that the acid Bohr effect in ferrous HbA is probably a manifestation of the pH dependence of the central subprocess, by which we mean the subprocess responsible for compensation.

The central subprocess has been called the functional conformation process and is described in terms of motion along the (functional) conformational coordinate.[2c,62,73] Much of hemoglobin research is devoted to attempts to describe that coordinate in molecular terms. Only recently has its fluctuational nature become generally appreciated. Many other examples of compensation phenomena for protein processes and for processes of small molecules in aqueous solution demonstrate T_c values in this same range.[62] This led to the proposal that the functional conformation process is a common feature of proteins attributable to their interaction with water and specifically to a cluster process unique to bulk water and responsible for most of the abnormal behavior of water.[62] The proposal proved to be somewhat naive, but we show in Section VIII.D that water-protein interactions probably play a major role in the complicated set of events occurring along the conformational coordinates of proteins.

The binding of competitive inhibitors to chymotrypsin demonstrates compensation with T_c values near 294K.[74a] pH was varied with single inhibitors and also a large series of inhibitors was studied at constant pH.[74b] Unfortunately, inhibitor binding to chymotrypsin has since been found to be complicated by a large number of processes, usually unsuspected and uncontrolled, some of which do not have direct connection to specific function, while others may. In steady-state catalytic studies at pH 8, these complications are either surpressed or adjusted to their normal role in function. Reliable quantitative descriptions of compensation behavior in chymotryptic hydrolysis of *N*-acetyl-L-amino acid ethyl esters have been reported by Dorovska-Taran et al.[75a] The T_c values were determined for the formation of the metastable intermediates, enzyme-substrate complex (ES), and acylenzyme (EA), and are discussed in Section VIII.B. Where the data for other enzymes are sufficient to allow testing for compensation patterns, the patterns are usually found.[183] Exceptions have proved to be rare.

Reversible denaturation of the proteins studied for dependence on solvent composition often yields compensation behavior with T_c values characteristic of the class but exceptions are common and the entire subject remains confused (see Section VIII.D). Thermal denaturation as a function of residue change demonstrates compensation with T_c values lying between 285 and 360K. Compensation behavior is not spurious and is due to linkage since the T_c values are significantly different from the mean experimental temperature. The more rigorous form of the "temperature test" is that T_c be independent of experimental temperature within the small errors due to nonzero heat capacity changes.[78] *A priori* there are good reasons why T_c values for biological processes might appear near 300K: (1) "second-law compensation",[78] when detectable, must demonstrate T_c values equal within errors to the mean experimental temperature. (2) When compensation behavior is due to the anomolous low-temperature subprocess of normal liquid water, T_c values are near 295K.[62] (3) Regulatory subprocesses such as the Bohr effect in hemoglobin, in order to allow fine regulation at physiological temperature must be poised between their extreme states, and since at $T_{expt} = T_c$, the advancement is near 50%, it has been necessary in evolution to find ways to fix the T_c values for regulatory processes close to physiological temperature. The last reason also applies to (central) functional conformation subprocesses. Furthermore, T_c values for the latter near-physiological temperatures are consistent with the possibility that many biological processes are of the first-order linear-response type since fluctuations in subprocesses involving two extreme states are largest when the temperature equals their T_c values.[78] However, that is not the case for any subprocesses which consist of a continuum of states. It also may not be the case in the formation of transitions states for members of congener families of substrates, since the varying thermal contributions to such processes can produce spurious T_c values and thus obscure the true compensation pattern of the linkage system.

A fifth reason T_c values may be found near 298K is the statistical instability associated with determining ΔH and ΔS from the same van't Hoff or Arrhenius plot. Linear-regression fitting tends to force the slope toward the value of the experimental mean harmonic temperature. Spurious compensation behavior produced in this way is undoubtedly common so that many examples of compensation in which T_c is found to be close to the mean harmonic temperature remain suspect. Verification by the temperature test, the finding that T_c is significantly different from the mean experimental temperature, use of two independent experimental methods to obtain ΔG and ΔS or ΔH, and application of the statistical tests of Exner[190a,190b] or Krug et al.[190b] have somewhat diminished concern about spurious compensation behavior particularly since many reliable values have been found to be near 298K. Nevertheless, it remains unwise to rely heavily on untested T_c values near 298K obtained by analysis of single van't Hoff or Arrhenius plots.

It is important to note that linear compensation behavior will persist so long as the coupling free energies (see Equation 3) have a very weak dependence on the degree of advancement of the subprocess. This is certainly the case for sufficiently weak coupling, and that is why we use that term. However, protein conformations are the basis for most coupling of subprocesses in proteins and the rearrangements of conformation even with large advancements in the subprocesses are likely to be often small. In such cases, the coupling free energies can be constant within small experimental errors even if they are quite large.[64] It is difficult to explain the virtually ubiquitous appearance of compensation behavior in biology on any other basis. The existence of compensation behavior can greatly simplify experiment and analysis because the thermodynamic quantities are linear in concentrations or site probabilities. Much of biology is thus reduced to systems of linear equations, and this makes it relatively easy to establish the correct linkage diagram and the details of coupling among subprocesses, i.e., direct interaction of two subprocesses vs. indirect coupling through one or more subprocesses to which the two of interest are directly coupled, and thus provides a method for dissecting the thermodynamics into contributions from the individual subprocesses.[79] The linkage theory for many systems is thus remarkably simple. ''Linkage maps''[80] become no more than matrices of the coefficients of the site probabilities. The homotropic coupling constants are on the diagonal, and the off-diagonal terms are the heterotropic coupling constants.[64] This state of affairs makes it possible to define orthogonal subprocesses using the conventional secular-equation method of normal vibrations in molecular structure and orthogonal chemical concentrations in relaxation theory.[79] In time, these ''normal coordinates'' will become important since they describe in the simplest way the responses of even very complex linkage systems to perturbations and thus provide a basis for relating the effects of perturbations to molecular changes, e.g., the molecular consequences of change in concentrations of single effectors. The application of chemical relaxation theory to include temporal behavior also becomes tractable. The fine details of linkage, that is, the indirect and direct coupling, can be handled mathematically using tensors of low rank. These simplicities suggest that much nonlinear behavior in living systems is due to feedback rather than to any intrinsic properties of single proteins.

Compensation temperatures may provide a diagnostic test for the subprocesses or substructures responsible for the compensation pattern. Thus, all the compensation temperatures so far found for compensation behavior reasonably attributable to knots lie between 335 and 365K.[183] Patterns which might reflect participation of bulk water yield values from 285 to 300K. Strong denaturants usually produce values around 380K (see Section VIII.D). Central functional subprocesses also have values near 300K. This diagnostic use of T_c must be examined in more detail before it can become reliable but its consistency so far is excellent as demonstrated in several cases discussed later. However, T_c values can depend on the standard state and with some standard states it is necessary to make corrections for cratic entropies unless ΔG_i is plotted vs. ΔS_i. There are several statistical procedures for determining the best value of T_c, and the most reliable procedure has not always been used. There is thus considerable uncertainty (perhaps as much as 20% in worst cases) in some of the values reported in this chapter and in our previous publications, and relatively few sets of data precise enough to set a narrow limit on T_c accuracy are as yet available.

An important example of potential diagnostic use of much interest to the emerging field of site-specific mutagenesis is the compensation plot shown in Figure 9. The points were calculated from the data of Matsumura et al.[54] for the denaturation rate constant of kanamycin nucleotidyltransferase in which residue changes were made at three positions. Substitution only at residue 252 (see figure caption) increased the rate constant slightly. Substitution at all three produced a negligible change in the constant. The double substitutions greatly decreased the constant. As shown in the figure, the activation enthalpy increased by as much as 70 kcal/mol relative to the wild type, a 64% increase. The thermodynamic stability

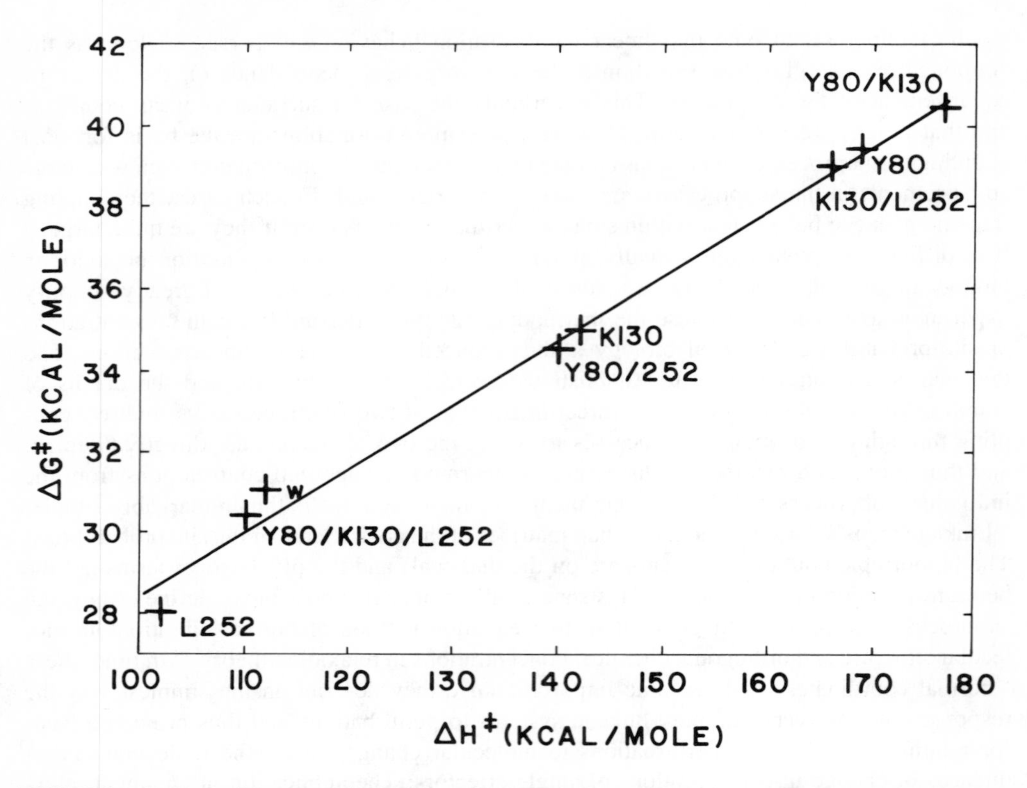

FIGURE 9. Compensation pattern obtained from the denaturation-rate data of some man-made mutants of kanamycin nucleotidyltransferase.[54] The activation free energies at 298 K were calculated from the original Arrhenius plots supplied by Dr. M. Matsumura. T_c is 354 K, which is 13° higher than the highest experimental temperature used in any of the experiments. The point labels mean the following: W is the wild type; Y80, tyr for asp at position 80; K130, lys for thr at 130; and L252, leu for pro at 252. The protein is a monomer containing only 253 residues.

increases as the rate constant decreases. The high kinetic stability and large activation enthalpy and entropy changes as well as their variation with single-residue changes appear remarkable for a protein with only 250 residues. They would be less remarkable if the residue-exchanges enhancing kinetic stability were in knots. If so, the T_c value might be expected to lie between 335 and 365K. It is 354K.

As just suggested, knot strengths can vary greatly. Knots in proteins which normally operate at high temperatures if they do not obtain high thermodynamic stability by the special solvent composition in which they normally exist (see Section VIII.D) must have unusually strong or large knots. Bovine pancreatic trypsin inhibitor has a very small knot but high kinetic and high thermodynamic stability. The matrix regions of this protein are thought to be unusually soft, but the large density of disulfide bonds in these regions may nevertheless considerably enhance stability. The knots in myoglobin as estimated by neutron-diffraction measurements of very slowly exchanging protons are small, no more than 15 sites in 2 groups and exchange rates are 2 orders of magnitude more rapid than for enzyme knot sites.[179] Metmyoglobin and carbonyl myoglobin did not demonstrate identical sets of slowest-exchanging sites. This result either casts doubt on the existence of knots in myoglobin or introduces a new factor of potential importance. However, kinetics and thermodynamics of ligand-binding do not change on removal of N- and C-terminal polypeptide up to the knot regions.[198] Further residue removal destroys function. Domain integrity is apparently sufficiently well established to support function[30,60] (*vide infra*). In general, knot size and strength vary from function to function, which is not at all surprising.

VI. PROTEIN ACTIVITY COEFFICIENTS AND THEIR IMPLICATIONS

Water is rarely a good solvent for proteins, although a few highly charged proteins, e.g., myoglobin, are soluble in water in all proportions. The relative proportion of polar and nonpolar residues in protein surfaces has been shown by many investigators to be important in determining the solubility and response to solvent additives. However, a satisfactory rationalization of protein-water interactions in terms of the solubility behavior of small molecules has not yet been established and is unlikely to be established until liquid water is better understood. An equally important problem is the absence of a reliable description of the state and behavior of water molecules in the B-hydration shells (see Section VIII.D). In A shells, most water molecules are temporarily held by uncharged polar groups of proteins and thus are restricted from making most of the usual strong water-water interactions found in bulk water.[117a,122] Not only is cluster formation inhibited in A shells and perhaps in B shells but also the distances between water molecules in A shells may often limit the stability and the number of water-water hydrogen bonds. Regardless of the protein, many water molecules fall into this interface class, so that water is probably an even poorer solvent for proteins than is generally thought. Many proteins may be soluble in water only because their lattice energies are small.

A considerable amount of thermodynamic information about protein-solvent interactions exists; the earlier work by Bull and Breeze, the more recent work by Timasheff, Gekko, and the Schriers and their respective co-workers on solutions[81] and that of Lüscher, Noguchi, Likhtenshtein, Rupley, and Careri and their respective co-workers on water adsorption by proteins are particularly noteworthy,[117a, 120,121] but still only a fraction of the total. We need consider only a few of the findings in this section.

The first, due primarily to Timasheff and co-workers, is that the most effective precipitating agents, e.g., the polyols and sulfate salts, function not by dehydration, as would be reflected by large decreases in water activity, but rather by raising the activity coefficients of dissolved proteins. There is an excellent correlation between precipitating effect and stabilization against denaturation in solution as might be expected since the activity coefficients of denatured proteins are raised more than those of the corresponding folded forms. However, as shown later, a stabilizing agent for a given protein may be a denaturant in another range of temperature.

Such familiar destabilizing solvent additives as urea, guanadinum ion, and $CaCl_2$ decrease the activity coefficients of folded and denatured proteins, but do so more for the latter than the former. They also increase solubility of native forms. Very generally the logarithms of the protein activity coefficients are linear functions of the molar concentration of solvent additive, so that virtually all studies in mixed aqueous solution must demonstrate compensation behavior. This is sometimes surprising. Urea is soluble in some familiar proteins but guanidinium ion is not.[136] Furthermore, urea and guandinium salts are supposed to interact directly with protein surfaces at high additive concentrations, but not at low. The changes in protein free energy with additive concentration even at high concentrations are generally only a few kilocalories at 25°, but the enthalpy and entropy changes are large, often very large, so that the free-energy changes they produce at 37° can be quite large. Although the solutions are far from physiological, this behavior establishes that protein activity coefficients are far from unity so the use of concentrations *in vivo* as *in vitro* can yield very misleading conclusions.

The second is that the ratio of nonpolar to polar residues in protein-water interfaces is quantitatively important, but polar residues appear to be as sensitive to solvent composition as nonpolar residues. Judging only from the behavior of small-solute probes of solution characteristics, the major effect of solvent additives is an indirect consequence of their effect on water,[62] a point of view which becomes increasingly attractive as water and protein solutions are studied in increasing detail. Additives are effective at all concentration levels, and there is often an excellent correlation[145] of their effect with the "structure-forming" and "structure-breaking" effectiveness of the additives extending to several other classes, such as the Hofmeister series of ions[143] (but see References 114,1176, and 163 on β-lactoglobin and other "atypical" proteins). A direct relationship between this behavior and the hydrophobic hydration of nonpolar groups in the protein surface has not yet been established, in part because the hydrophobic hydration and hydrophobic interaction of small molecules is not yet understood[82,153g] and in part because A shells appear to be unique and as yet have not been modeled (see Section VIII.D).

For any dissolved protein, a change in solvent composition, temperature, or pressure must produce changes in the surface area, A, volume, V, and internal concentration of water and any other component of the solvent which can dissolve in the protein. This is a consequence of the change in protein activity coefficient and the Gibbs-Duhem relationship,[83,122a] Equation 13. Of particular practical importance are the magnitudes of the changes in volume and associated conformational distortion occurring in *in vitro* experiments since, if these are significant, deductions from the experiment which do not take these changes into account will be wrong or at least misleading.

The third finding of special interest has to do with the anomolous isothermal compressibilities of proteins. The experimental values are very small, sometimes only a few times larger than those for block tin or solid NaCl, although the thermal expansion coefficients are not obviously unusual.[11c,85] These low values cannot be taken as evidence that volume changes under pressure are small since they reflect changes in protein-solvent interfaces and conformational rearrangement, as well as protein volume changes. Furthermore, changes in protein volumes of 1 to 6% accompanying function have been reported,[87,176d,199] and if the apparent compressibilities were true values, the free energies would rise by some hundreds of kilocalories in compressions of this order. Kundrot and Richards determined the X-ray-diffraction structure of HEW lysozyme at 1000 atm.[20] The knots, as described in Figure 6, were essentially unchanged,

$$[P]\left\{\left(\frac{\partial \mu_p}{\partial \overline{V}_p}\right)_{a_w,\overline{A}_p} d\overline{V}_p + \left(\frac{\partial \mu_p}{\partial \overline{A}_p}\right)_{a_w,\overline{V}_p} d\overline{A}_p + \left(\frac{\partial \mu_p}{\partial a_w}\right)_{\overline{V}_p,\overline{A}_p} da_w\right\} + [w]RTd\ln\gamma_w = 0 \quad (13)$$

but considerable deformation occurred, particularly in a section of the matrix at the bottom of the substrate cleft. These results suggest that the low apparent compressibilities are partially artifactual in that, although the knots have true compressibilities of the order of 10^{-6} atm^{-1}, the matrices are mechanically transparent and so contribute only small relaxation terms to the total compressibility. The relaxation term has been measured for one enzyme[192] and its size is consistent with this explanation. However, there are several relaxation processes which make interpretation of compressibility data difficult: migration of water from protein interior to bulk solvent, polypeptide rearrangement, relaxation events in interfacial water, and pressure-dependent ionization. There has accumulated considerable evidence that water moves into proteins as pressure is increased.[21] Compressibility data can be a major source of information about protein fluctuations in relation to protein function,[116] but much more systematic work is required.

The fourth finding illustrates the potential for confusion introduced by the dependence of protein volumes on solvent composition. The PV work done on HEW lysozyme in the elegant pressure study by Kundrot and Richards[20] was estimated to be about 1 kcal at 1000 atm. Judging from the dependence of protein activity coefficients on solvent composition, considerably larger amounts of PV work can be done on proteins by solvent changes which raise protein activity coefficients. Thus, 40 vol % glycerol solutions may effect protein compression equivalent to as much as 10 kcal, and may produce displacements larger than those found by Kundrot and Richards. Since the mother liquors in which protein crystals are bathed during the collection of X-ray-diffraction data are usually those most effective in compressing proteins, the X-ray-diffraction structures may be significantly different in some detail from structures which obtain in pure water.[176d,178]

Some studies of proton-exchange in the crystal and in solution show very similar values of structural parameters thus indicating little difference in the structural integrity in solution and in crystal.[35a,88] However, most studies suggest geometric difference and all studies appear to show lower motility in the crystals. Solution-crystal comparisons are particularly hazardous for proteins which undergo volume changes of one or a few percent during function. These may have relatively high effective compressibilities but in any event can be expected to be significantly perturbed by changes in solvent, pH, etc. experienced in conventional studies of function. Hemoglobin presents another kind of example because the protein cap over the distal side of the heme group is soft.[9a] Through Equation 13 changes in experimental variables will alter the shape and size of the cavity in which the ligand sites. The changes are unlikely to be large, but the binding process appears to be very sensitive to cavity geometry.[9a] Although pH was neglected in writing Equation 13, it is another variable which can be used to perturb protein volume, shape, and fluctuation spectrum. It is particularly important in the hemoglobins and myoglobins because of the high surface charge densities and the broad fluctuation spectra.[124]

Viscosity dependencies of ligand binding in myoglobin and rate parameters for some enzymes are the principal support for the hypothesis of mechanical activation by momentum fluctuations arising in the solvent.[8] Since viscosity is varied by using glycerol-water mixtures or similar mixtures giving a wide range of viscosities, the support is equivocal. It is likely that the effects are intrinsically thermodynamic since equilibrium compression effected by the solvents will narrow the fluctuation spectrum at equilibrium and thus increase frequencies and decrease magnitudes of functional fluctuations. The study of the dependence of proton exchange of lysozyme in glycerol-water mixtures is very informative.[33] Exchange from II sites (matrix sites) of HEW lysozyme is sensitive to glycerol mole fraction because catalyst diffusion through the protein is slowed. The overall activation energies for exchange are increased which is good evidence that the reduction in gated diffusion rates is due to reduced protein volume rather than to viscous damping of fluctuations in protein and solvent. Proton exchange from I sites (knot sites) is slightly slowed. T_c is the same within error for both families and the activation energies are little changed. The functional dependence on viscosity of exchange from I sites is inconsistent with a kinetic damping mechanism.

Experiments such as the measurement of depolarization of tryptophan fluorescence detect directly the influence of solvent viscosity on matrix fluctuations in protein conformations. Such experiments show that this influence is relatively small[89] and this result is supported by other kinds of data (see Section VIII.C). These results are consistent with observed changes in proton-exchange rates of matrices which can be attributed to direct viscosity effects or to changes in protein free volume.[33b]

VII. FREE-ENERGY COMPLEMENTARITY

As discussed in the first section, Haldane's enzyme model explains in principle how free energy is conserved so that the free energy released in one step can be used to facilitate formation of the subsequent transition state. Proteins have many more ways to effect conservation, and it can be argued that this is a primary source of their utility, since free-energy conservation appears to be a major theme in evolution. For example, this characteristic is responsible for efficient catalysis, as is demonstrated by Figures 1 and 12 for chymotryptic catalysis, and very likely the basis for the evolution of more sophisticated enzymes and enzyme systems (*vide infra*). Free-energy reaction profiles for other enzymes offer considerable support for this "evolutionary" force, since rate enhancement is usually found to be achieved by coupling those events releasing free energy to those which require free energy in order to occur. Protein conformations provide devices for long-term free-energy storage which one expects to be limited by protein stability, but the work of Bolen and co-workers, discussed in Section VIII.B, suggests that very large amounts can be so stored — 23 kcal in their studies of the formation of an acylenzyme from a sultone substrate having a very large ring-strain energy.[90]

The utility of free-energy conservation has been discussed in considerable detail elsewhere under the title "free-energy complementarity",[8b,91] and the case is made that in addition to the reduction in dependence on thermal energy in transition-state formation there is an omnipresent evolutionary force arising from the competitive advantage of approaching equilibrium efficiency in the utilization of free energy without loss in speed.[91] Regardless of evolutionary changes which enhance the ability of an organism to get and keep food as well as adapt itself to use new foods, there is always an underlying advantage in improving the efficiency in the use of food (Figure 1). Further developments to improve specificity and reject undesirable reactions are described below. These developments appear to be the basis for the successful evolution of multienzyme systems,[86] which improve physiological sophistication but is also often accompanied by improvements in free-energy complementarity. These arise from enlarging cooperative units by introducing coupling among more and more originally independent enzymes as well as from further flattening of the reaction profiles of single enzymes. Within each enzyme and within each enzyme system, whether the constituent enzymes are coupled to each other or not, complementarity forces a flattening of the free-energy profile. This results in an ever-increasing number of metastable intermediates and a reduction in the amplitude of the oscillations in the total free energy of each enzyme-substrate system which improves its thermodynamic efficiency.

Conformational devices for storage and manipulation of free energy are technically mechanical, since they involve change of atom coordinates in response to a force, whether that force be due to chemical, mechanical, or electrical potentials. Storage as "spring" energy was an important feature of the original rack mechanism,[1a,1b] but it is not simply consistent with the parallel-change behavior of enthalpy and entropy characteristic of protein conformations.[5] Nor is it consistent with the softness of protein conformations which is a consequence of soft matrices. As already discussed, short-lived mechanical storage is possible and necessary for enzymic catalysis if the rack hypothesis is correct. Long-term conformational storage must depend primarily on transitions, usually activated transitions, between "mobile-defect" conformers (see Section I and Figure 2).

The degree of development of complementarity in a total enzymic system can be assessed by the degree of flattening of the total free-energy profile. When well advanced, the free-energy maxima are all of about the same height and there is then no single slowest step. By this assessment, complementarity is highly developed in some multienzyme systems, suggesting that improvement in complementarity has been a major factor in natural selection for these systems for some time. Multienzyme particles are similar to hemoglobin in their manipulation of free energy so that complementarity in such particle systems is not surprising. It is, however, initially surprising that there is some evidence that intracellular enzyme systems constructed to achieve some single goal such as synthesis, but not associated in particles, also have highly developed complementarity. On further consideration it becomes obvious that it does no good to improve the complementarity of one enzyme in the system without improving it in the others. At one time in evolution the greatest improvement was to be achieved by improving one enzyme, at another time by improving another enzyme, and so on. Such evolutionary achievements in systems of enzymes constructed to produce a single product has been called "supercomplementarity".[86,91a] The development of super-

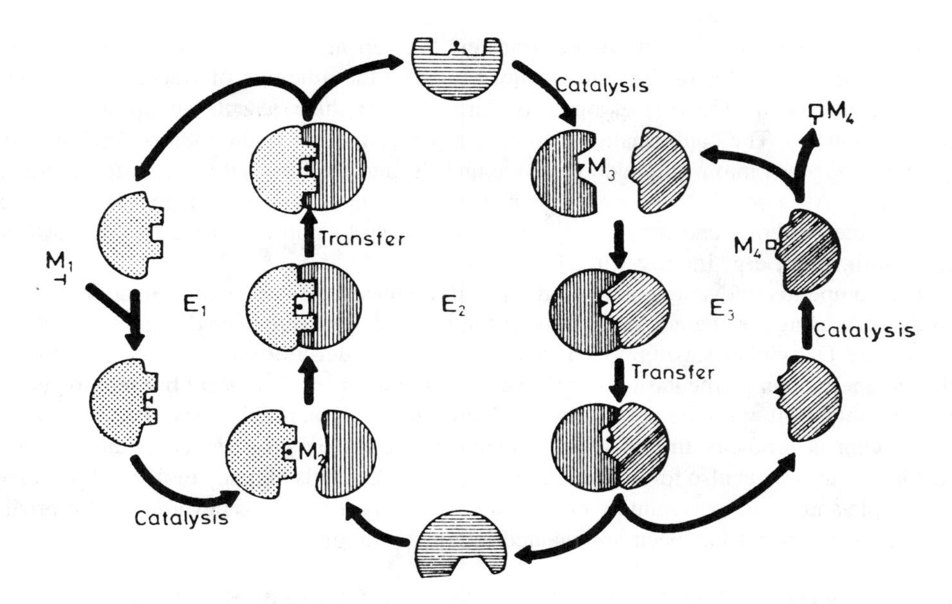

FIGURE 10. Friedrich's "metabolite compartmentation".[93] Three intracellular enzymes are shown coupled into a cooperative unit for the transformations of the metabolites M. In this example, E_2 undergoes a conformation change during the transfer of M_2 from E_1 and another conformation change when M_3 is transferred to E_3. The enzymes are not tightly associated but diffusion is not rate limiting. Calculations by Srivastava and Bernhard[94c] suggest that the hydration of the enzymes is limited to the A and B shells. According to Friedrich, the shape changes are required for recognition, but, since these changes are coordinated by metabolite transfer and driven by free energy released in metabolite transformations, conformational free energy is also redistributed between proteins in each transfer (see text). The figure is reprinted from P. Friedrich, *Supramolecular Enzyme Organization*, Pergamon Press, Oxford, by permission of the author and publisher (copyright 1984, Pergamon Press).

complementarity would be severely restricted if free-energy transfer from enzyme to enzyme were limited by exchange of substrates and products, that is, were restricted in amount by the invariant free energies of formation of the small chemical species. Friedrich[93] seems to have been the first to propose that, in systems in which the proteins are not permanently associated, the product of one enzyme which is the substrate for the next in the series is directly transferred from the first protein to the second, rather than passing into and out of free solution. As shown in Figure 10 from Friedrich, the enzyme conformations are supposed to change shape during catalysis so that an enzyme carrying its product recognizes the free form of the next enzyme, and only that form. Recognition triggers the transfer of the product and the simultaneous changes in the conformations of the two enzymes. Although relevant data have been difficult to accumulate, there is sufficient supporting evidence to make this topic a current major focus of multienzyme studies.[94] The "nonassociated" multienzyme systems are intracellular, e.g., the glycolysis enzymes, and one advantage is thought to be the diminished dependence on water, there being only about two monolayers of water per intracellular enzyme on average.[100] The overall free-energy profile for the glycolytic system is thought to be very flat.[92] If so, its evolution has been much influenced by supercomplementarity. So far as data are available, the enzymes in this system have quantitatively different behavior *in vivo* and *in vitro*. *In vivo* the maximum free-energy peaks are probably all of same height within experimental error, so that no one enzyme provides the rate-limiting step. This is exactly the behavior expected from advanced development of supercomplementarity[91a] and can only arise if, in addition to the transfer of reactant/product, there is a transfer of free energy from the conformation of one protein to that of the other in the recognition and reactant/product transfer. We repeat that, were direct conformation to conformation transfer of free energy not possible, the amounts of free-energy transfer would be limited by the free energies of formation of the reactants and products, i.e., chemical transfer. The conformational free-energy transfer allows the total collection of the glycolytic enzymes into a single cooperative unit despite the fact that there is no long-lived association. As a result these multienzymes systems have the advantages of associated multienzyme systems, and may have found other ways to improve the efficiency of free-energy utilization, e.g, improved feedback.[92]

The competitive advantage in evolution from increased efficiency in utilization of food, without increasing the characteristics times of utilization, was appreciated even in the earliest days of the Darwinian revolution. Although the proper definition of efficiency (what we call free-energy complementarity) may have been given earlier, it was not brought to general attention until 1946 with the publication of Schrodinger's remarkable book, *What is Life?*.[95] The concept is probably the most fundamental basis not only for understanding survival value in evolution, but also for understanding the specific details of many biological processes both simple and complex. Granting that 1946 is relatively recent, it is nevertheless surprising how little the concept has been appreciated let alone exploited.

VIII. UPDATING THE RACK MECHANISM FOR CHYMOTRYPTIC CATALYSIS

A. DOMAIN CONSTRUCTION AND FUNCTION

As the knots can be defined by proton-exchange data, the functional groups are usually on the first residue of the polypeptide chain where it emerges from a knot (Figures 5, 6, and 7). Thus, for trypsin and chymotrypsin of the three functional groups forming the "charge-relay system", ser. 195 is on one knot, and asp. 102 and his. 57 on the other (Figure 4). Residue 214, which is the palendromic partner for 102, may be another member of the charge-relay system and, judging from the study of trypsin by Kossiakoff,[34] it is on a II site, on a knot which is otherwise composed of I sites. More complete descriptions of knots may show that the functional-group residues are actually in knots, as indeed a few already seem to be.

Regions of high density, high number density, and small fluctuation amplitudes, low B factors as detected in X-ray-diffraction studies, so far appear to provide reliable though rough descriptions of knots to be further supported by high degrees of residue homology when comparing enzymes in families. Thus, hen and goose egg-white lysozymes and T-4 phage lysozyme are indicated to have very similar knots although their matrices demonstrate large differences.[79] Judging from the trypsin inhibitors, proteinase inhibitors have single strong knots. The BPTI knot is constructed about a β-sheet core. Another trypsin inhibitor is built from α-helix core.[97] One wonders if this is an example of convergent evolution or sequential modification of knots. Enzymes or zymogens with more than two knots may have two catalytic functions or, perhaps, represent a greater degree of sophistication in catalysis. Zymogens like plasminogen appear to be grab bags of knots with different but not necessarily directly related functions. Aspartate transcarbamylase,[44] aspartate amino transferase,[176c] and glutathione reductase[98c] have the two knots supporting catalytic function on separate protein subunits (vide infra).

The pictorial description of the rack mechanism for chymotrypsin shown in Figure 1 includes the motion of two domains toward each other[117f,184] to pinch the susceptible substrate bond between the imidazole group of his 57 and the hydroxyl group of ser 195, a process in which the change-relay system is activated. This stapler-like motion is made more probable by substrate binding, and at least partially driven by increasing interaction of substrate α-carbon sidechain with its binding pocket. The description can be extended to include the participation of the oxyanion binding site, the ile. 16 to asp. 194 ion pair, and most of the other possible factors in mechanism which have been suggested. Such detail will have to be considered at a later date. The domain-closure and sidechain specificity for chymotrypsin has been discussed in some detail in previous papers, sometimes, somewhat incorrectly.[11,137] Here we correct the errors and consider new detail in the model arising from the topics discussed in the preceeding sections.

The discovery of knots is particularly fortuitious since the original rack proposal was postulated on the basis of a more or less isotropic protein having considerable rigidity, an assumption not entirely consistent with the high motility proposed in the same papers and one since proved untenable for matrices. Knots greatly reduce the number of degrees of freedom so that the entropy loss in transition-state formation is minimized.[167] To a first approximation there is a single mode, the domain-closure mode, since most of the supporting subprocesses such as activation of the charge-relay system are coupled to this conformational mode. The solid knots are very effective in the activation subprocess but the matrices probably also help establish rigidity since their integrity is improved by substrate binding and by reduction in protein volume as the system moves toward the transition state. In effect this is a built-in device to lower entropy without much cost since bonding improves as contraction advances. Rack mechanisms require transient rigidity in matrices during transition-state passage to preserve mechanical stress. Matrices might be able to sustain stress but they appear to be too soft. Knots somewhat remedy this weakness of matrices. On the other hand, matrix flexibility allows domain closure and, through intimate adjustments to substrate sidechains as domain-closure advances, enhances both equilibrium and dynamic specificity. Rates of proton-exchange from II sites are particularly sensitive probes of hardening in matrices, and exchange from I sites can supply essential information about knot-matrix interaction. Many other probes also reflect changes in matrix integrity, usually in terms of altered fluctuational behavior. One general conclusion from such indirect studies is that the domain properties they reflect are closely tied to functional state.[3,11,12] Unfortunately, the only direct source of this kind of information for transition states is activation volume, which is an ambiguous consequence of changes in water, substrates, and protein.

X-ray-diffraction methods applied systematically to substrate and inhibitor binding at highest resolution supplemented by studies of mutants will provide important average geometry and B-factor information about domain integrity and domain closure. Diffraction data have not had noteworthy success in rationalizing probe data which reflect widespread structural reorganization, e.g., proton-exchange data. However, major readjustments on substrate and inhibitor binding have been detected and gross "hinge bending" is now frequently detected.[98,158] Changes in interdomain geometry on activation of trypsinogen have been studied (e.g., See Reference 99). Numerous small-density changes throughout most of the chymotrypsin dimer on binding inhibitors were suggested by X-ray-diffraction data obtained at relatively low precision[10] but have not been reinvestigated using three-dimensional difference comparisons at the higher resolution now possible. There is only one state-of-the-art study of chymotrypsin. Blevins and Tulinsky compared the free protein with the phenylethaneboronic-acid derivative, in which one molecule of each dimer forms what is thought to be a tetrahedral-carbon intermediate.[101] At a precision index value of 1.8 A no differences could be detected, but moderately extensive conformational differences less than about 2 A could not be detected at this precision so the study neither confirms nor denies the occurrence of some domain closure.[102] The comparison may nevertheless provide an estimate of the geometric extent of domain closure and is consistent with the small positional changes necessary to activate the functional groups, <0.5 A. The situation is somewhat less equivocal for HEW lysozyme in which the hinge bending mode of domain closure has been studied.[158] An interesting example is the capturing of a sodium dodecyl sulfate molecule in the knots-matrix "basket".[98h] In this connection the computer simulations of the hinge-bending mode are interesting insofar as they give some idea of the complexity of domain closure.[103] The construction of domain-closure coordinates from mobile-defect conformers is discussed in Section D.

The remarkable thermodynamics changes required by the chymotrypsin rack mechanism are achieved by several subprocesses, all of which are coordinated through the domain-closure subprocess (Section D). Fine tuning in evolution, usually small changes in fluctuation spectra as much as in average interatomic distances, is most often accomplished by residue changes in matrices and a very large fraction of the total of matrix and even peak III (fast proton exchange) residues can participate in these fine adjustments.

Knots have little conformational flexibility, but considerable functional flexibility in the sense that the electronic tricks they support can be applied to different chemical processes, an example for chymotrypsin being the amide and ester hydrolyses, rarer carbon-carbon bond cleavage,[104] and nonspecific processes discussed in Section VII. In those cases in which it has been possible to relate exons to knots, knots have been found to be coded by the most conserved exons[39] (Table 1). The variation of residue composition among knots in the lysozyme and serine-protease families is considerable, but their geometries change very little, as demonstrated by the similarity of the conformations and B factors of chymotrypsin, subtilisn, *Steptomyces griseus* protease A and α-lytic protease in the regions containing the functional groups and only in these regions.[105] There are significant quantitative differences in catalytic efficiency and specificity among the enzymes, but the similarity of their knots is a measure of the importance of conservation of knot-forming residues whether in sequences or widely separated. The similarities appear to be more consistent with divergence from a common ancestral knot than with the generally accepted ''convergent evolution'' of chymotrypsin and subtilisin. Has a given kind of knot been discovered in evolution, once, twice, or many times?

It is unlikely that the charge-relay system, ubiquitous in the serine proteases, can support the observed catalytic efficiency without some sort of activation.[137] In the rack mechanism proposed for these proteins, activation is a major consequence of compression, which alters the geometries and the electronic characteristics of the charge-relay groups and the substrate simultaneously.[137] It is difficult to imagine any activated complex of a small-molecule reaction having such coordinated complexity. In chymotrypsin, activation greatly increases the probability of proton migration, and it is reasonable to expect that with the best subtrates proton tunneling is efficient. Whether or not nature has exploited tunneling, it is probable that proton migration is not rate limiting with good substrates. Poorer substrates are probably poor because they effect less closure of domains and thus less effective force fields. With these, proton transfer should be rate limiting. Not only will different substrates, good to poor, for example, have different transition states, but, the better the substrate, the larger the volume of the transitions state in phase space. The length and angle differences are not large but the consequences can be, as has been best demonstrated in recent years by Zundel and Scheiner and their respective co-workers.[107] The hydrogen bond is a very versatile chemical device easily perturbed to give efficient proton transfer in one enzyme, hydride-ion transfer in another, and hydrogen-atom transfer in a third. Efficiently directed compression appears to be ideal means for fine tuning hydrogen bonds, but can be much more versatile. Thus, for example, the electronic properties of many co-enzymes, including metal ions simply complexed by protein groups, are susceptible to transient distortion in transition states, just as when mechanical distortion forces transitions from one stable state to another in hemoglobin and the cytochromes c.

Finally, we note once again that the domain pairs need not be parts of the same protein, so that this discussion of chymotrypsin is applicable to enzymes in which the functional groups of the catalytic site are distributed over knots from two proteins rather than one.

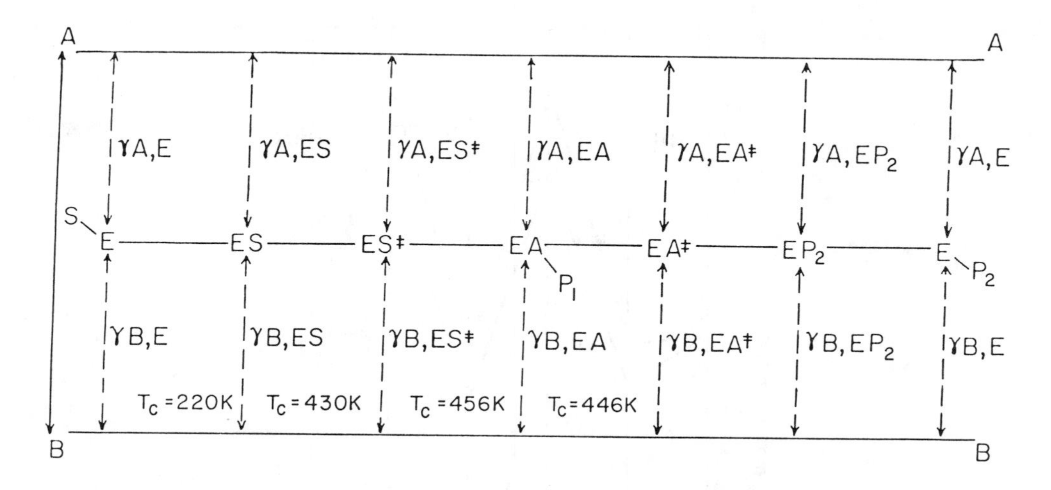

FIGURE 11. Linkage diagram appropriate to the compensation patterns observed in steady-state chymotryptic hydrolysis of N-acetyl-L-amino acid methyl esters by Dorovska-Taran, et al.[75a] The subprocesses participating in catalysis can be separated into two groups, A and B, whenever a compensation pattern is observed. In this application, the driving variable which produced compensation behavior was the α-carbon substrate sidechain. Group A then becomes the total of subprocesses coordinated in the catalytic process through the interaction of the sidechain with water or protein. This is the (sidechain) dependent group D. Group B includes all other subprocesses. It is the (sidechain) independent group I. See caption for Figure 12 for more detail.

B. SPECIFICITY AND EFFICIENCY

The relationship between specificity and efficiency, which is manifested in the rate constant zero order in substrate concentration, k_{cat}, has been a valuable guide in the search for mechanism but, as already mentioned, it is nevertheless probable that the fundamental catalytic mechanism of chymotrypsin and similar enzymes does not require participation of the α-carbon sidechain of substrates, a conclusion which makes Figures 12 and 13 valuable. Rather, the side chain increases or, when of inappropriate size, shape, or chemical characteristics, diminishes catalytic rates through modulation of the fundamental domain-closure mechanism.[12d] The modulation is large with good substrates for chymotrypsin but small for good substrates of elastase, a less efficient primitive proteinase with catalytic rates no greater than those obtained in chymotryptic hydrolysis of the methyl esters of *N*-acetylglycine and *N*-acetyl-L-alanine. For chymotrypsin, even with the best substrates, at least five orders of magnitude in catalytic efficiency must be attributed to the sidechain-independent subprocesses, the I group in Figures 11, 12, and 13. Thus, in the rack mechanism, the fundamental process is spontaneous domain closure unassisted by interactions of the protein with α-carbon sidechain of substrate.

Although domain-closure fluctuations in the absence of a specific substrate may be neither large nor frequent, they provide one probable explanation for the high nonspecific chemical reactivity of the serine-195, imidazole-57, and methanine-192 groups of the serine proteases. Other enzymes also demonstrate this physiologically unrelated high reactivity, the ''entactic-site'' reactivity of Williams and Vallee.[108] Our explanation is that potentially reactive molecules, even when they do not resemble natural substrates, wander in to the jaws of enzymes and are destroyed by random occurrence of the electronic trick of the enzyme. An example from the serine proteinase literature is the study of the reaction of iodoacetamide with the imidazole group of the his-57 imidazole group of trypsin.[109] This is an especially good example since activation by an analog of substrate sidechain was found to be quantitatively the same for the iodoacetamide reaction and hydrolysis of *N*-acetylglycine ethyl ester. The activation behavior implies normal participation of domain closure in the absence of specific binding of reactant. Entactic sites may be one explanation for some

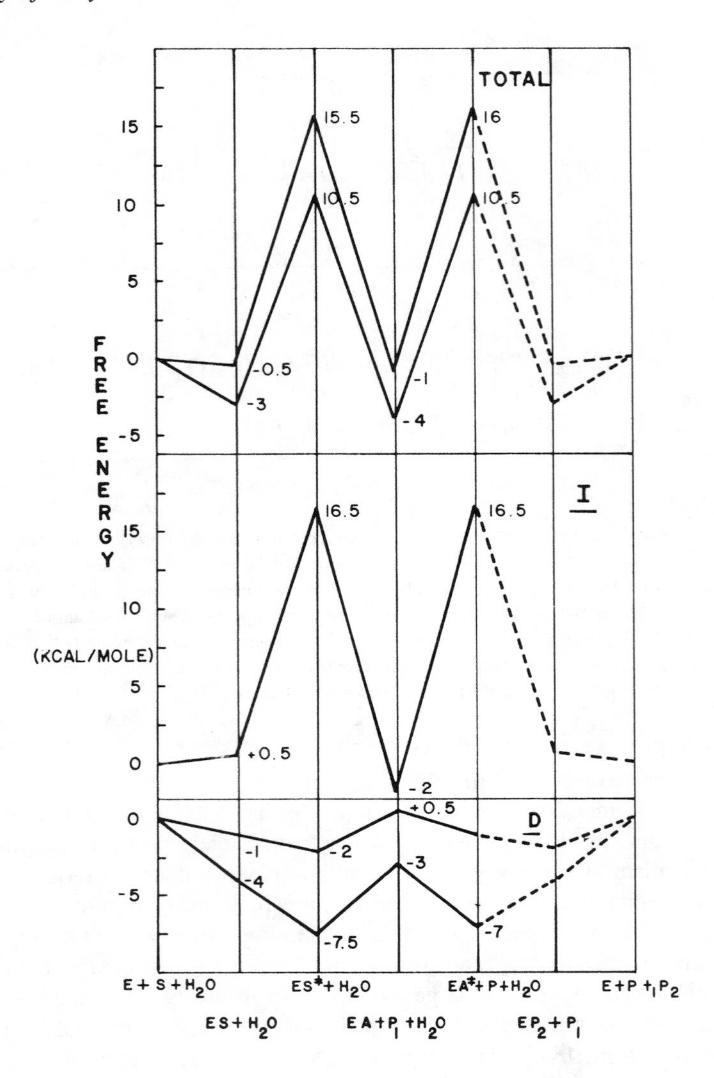

FIGURE 12. Free-energy profiles for α-chymotryptic hydrolysis of the methyl esters of N-acetyl-L-α-amino butyric acid and N-acetyl-L-phenylalanine at 25°C, pH 8, and 0.1 *M* KCl. Values for the total free-energy changes in each step are from Dorovska-Taran, et al.[75a] The T_c values were 200 K for formation of ES and EA from E + S and 510 and 490 K for the formation of the activated complexes ES‡ and EA‡, respectively. These were determined by Dorovska-Taran, et al., using Krug's method. The total of subprocesses involved in these steps are divided into two groups: group I, which are independent of substrate sidechain, and group D, which are dependent on substrate sidechain (see caption Figure 11). The free-energy changes forming the D profile are the contributions to the measured catalytic behavior due only to the sidechain calculated on the assumption that for a given sidechain all are governed by a single advancement parameter, ΠD′, as is consistent with the linear compensation pattern observed. The relevant equation is

$$\Delta G_{BA,app} = \Delta\mu_{BA}(0,0,\ldots) + \sum_{i=1}^{m} \Pi_{I_i'} \Delta\Delta\mu^c \begin{bmatrix} B & A \\ I_i' & I_i \end{bmatrix} + \Pi_{D'} \cdot \sum_{i=1}^{m} \Delta\Delta\mu^c \begin{bmatrix} B & A \\ D_i' & D_i \end{bmatrix}$$

$$= \Delta\mu_{BA}(\Pi_{I_1'},\Pi_{I_2'},\ldots,\Pi_{D'} = 0) + \Pi_{D'} \sum_{i=1}^{m} \Delta\Delta\mu^c \begin{bmatrix} B & A \\ D_i' & D_i \end{bmatrix}$$

Data for the reference substrate, N-acetyglycine methyl ester, are from Martinek and co-workers.[110] The standard state for the substrates in water is 1 *M*, but their standard free-energies, enthalpies, and entropies of formation in water are undoubtedly different. The values have not been reported so the authors have assumed all substrates to have zero enthalpies and free energies of formation in water at pH 8 and 25°C. This must be taken into account in comparing the profiles for the two substrates (see text).

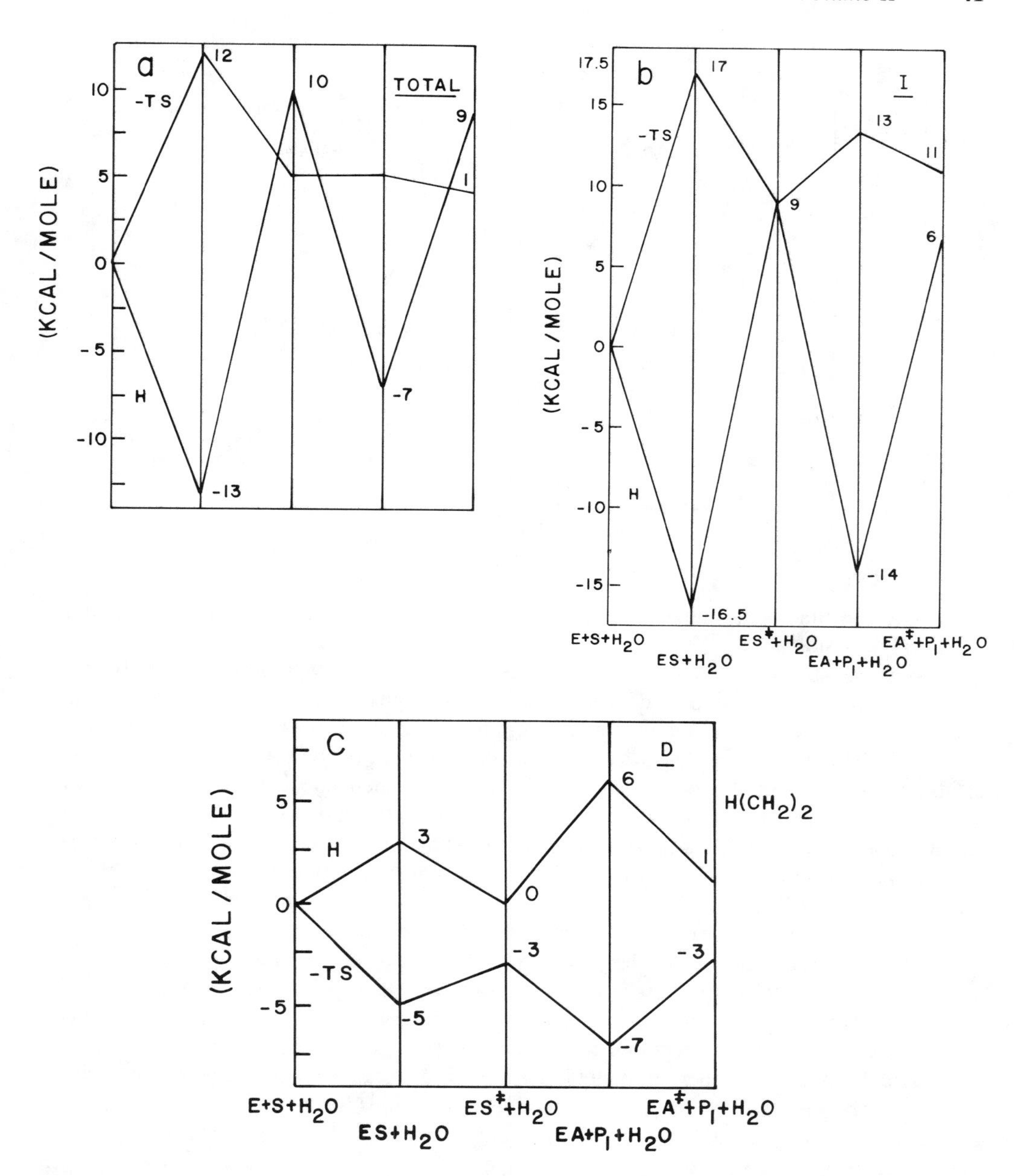

FIGURE 13. Enthalpy and (entropy × temperature) profiles for the α-chymotryptic catalysis of the methyl esters of N-acetyl-L-α-aminobutyric acid and N-acetyl-L-phenylalanine at 25° (see Figure 12 for free-energy profiles and terminology). Figures a and b are for the N-acetyl-L-α-aminobutyric acid substrate; d and e are for the N-acetyl-L-phenylalanine substrate. Figure c is for the I subprocesses and can be taken as those reflected in hydrolysis of N-acetylglycine methyl ester (see text). The enthalpy and entropy data were calculated for the parent from the free-energy data of Martinek and coworkers using compensation temperatures from Dorovska-Taran, et al.[75a] By comparing Figures c and e, it can be seen that the qualitative features of Figure b are not dependent on the use of the N-acetylglycine methyl ester data.

nonphysiological free-radical production, and the phenomenon itself provides justification for attempts to modify enzymes for nonphysiological commercial use. It is already established that nature utilizes knots of the same kind for quite different chemical functions by combining them in different ways. Also, as illustrated by the comparison of HEW lysozyme and α-lactalbumin (see Section IV), the same pair of domains can support catalysis in one protein

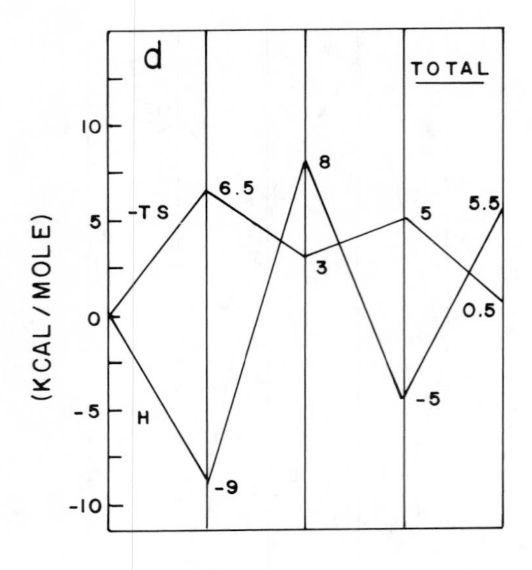

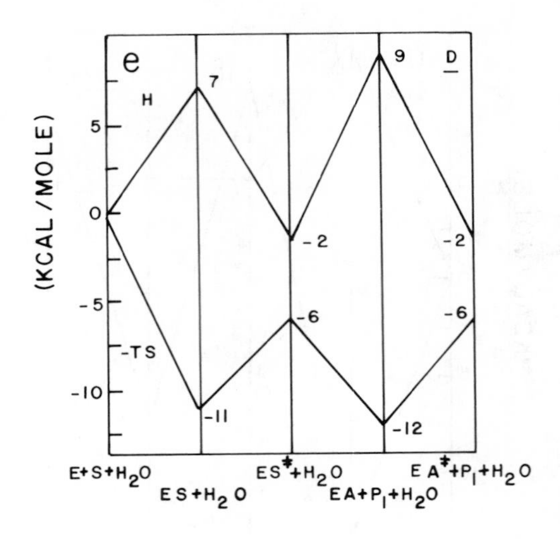

FIGURE 13d and e.

and metal-ion binding in another. The potential versatility of any given kind of knot is apparently considerable.

As already discussed, in a steady-state kinetics study of the chymotryptic hydrolysis of a series of *N*-acetyl-L-amino acid methyl esters with aliphatic or aromatic α-carbon side-chains, it was found that all the parameters demonstrated statistically reliable compensation patterns.[75a] The driving variable was the α-carbon sidechain of the substrates. This finding establishes that the total process can be split into two independent parts: the side-chain-independent subprocessess (I) common to all the substrates and the side-chain-dependent subprocesses (D). This separation is consistent with the conclusion that the specificity enhancement achieved by D is a supplement to the intrinsic catalytic mechanism,[63,73a,185] in which case the glycine substrate should be a suitable model for I free of D. In principle, the I and D separation can be made without data for the parent substrate, but in this case there is some uncertainty about the values of the index of advancement for the different sidechains so it is necessary to assume that hydrolysis of the glycine substrate depends only on I. Experimental values of the several enthalpy and entropy changes for the glycine substrate do not appear to have been reported for the desired pH and temperature, but these can be calculated from the free-energy changes[110] using compensation analysis. The appropriate linkage diagram is Figure 11, and the thermodynamic profiles for the substrate with smallest side chain — the α-aminobutyric-acid substrate — and for the largest sidechain — the phenylalanine substrate — are presented in Figures 12 and 13. The standard states for the substrates were 1 M, and the free-energies, enthalpies, and entropies of formation in this standard state were normalized to 0. This throws contributions from differences in solvation of the sidechains of the substrates into D and is required because the values for the enthalpies and entropies of formation in solution have not been found in the literature. It has two undesirable consequences. The first is that the vertical positions of the profiles are wrong. The second is that the thermodynamic quantities for ES formation are somewhat deceptive. Were corrections made for the differences in formation thermodynamics, the correct vertical positions of the profiles would allow comparisons of the absolute values of thermodynamic quantities between substrates. The differences in solubility of the substrates, which appears only in the step of ES formation, would be eliminated, which is desirable because these differences are irrelevant to the function of the enzyme.

The most important deductions from Figures 12 and 13 appear to be the following:

1. The enthalpy and entropy contributions from the I subprocesses to the formation of ES are large and negative. They nearly compensate each other at 25°, so the stability of ES is due primarily to the D contributions and varies with the solubility of the substrate. Since we know the sign and roughly the values of the thermodynamic differences in solubility, it is possible to estimate the corrections. The result is that the standard free energies, energies and entropies of ES formation for the substrates up to that with C-5 sidechain are about the same and also about the same for those for the glycine substrate, that is, the values shown for I in Figures 12 and 13. The large values of $\Delta H°$ and $\Delta S°$ are consistent with significant contraction of the protein in ES formation, but do not establish this to be the case. Large sidechains are not efficiently accommodated and as a result impair binding and catalysis, a frequently observed pattern.
2. The free-energy changes in ES formation, uncorrected for standard free energy of formation of the substrate in water, are consistent with expectation from transfer of models for the side chains from octanol or other amphipathic solvent to water.[111] When the values for the entire series studied are plotted against the negative of the transfer free energies, the plot is linear with slope of 1.05.[75a] The free energies for the same series using the methionine-192-sulfoxide form of α-chymotrypsin give a linear plot with slope of 0.54.[75b] The contributions from D contain only fractions of the total free energy of desolvation of the sidechains. The remainders are either stored in the protein or dissipated and can be calculated using the compensation temperature. As discussed below, additional free energy becomes available from further improvement in enzyme-substrate interaction as domain closure advances. The comparison of the protein and its methionine-sulfoxide form[75b] shows that the utilization of the free energy released in ES formation depends on subtle differences in the protein and consequently forces the conclusion that use of the transfer free energies in estimating the "hydrophobic-hydration" contributions to ES formation is dangerous. In this case, since the activation free energies for ES‡ are significantly lower with the methionine-sulfoxide derivative, about 2 kcal with the phenylalanine substrate, this form of the enzyme is better able to utilize the free energy derived from desolvation of substrate sidechains. The role of sidechain in ES formation is clearly not a passive one. Similar behavior is found in the plots of ΔG°_{EA} against $^{-}\Delta G^{\circ}_{transfer}$ for the formation of the acylenzyme, EA, from E and S: but with α-CT the slope is 1.01, with the methionine-sulfoxide derivative it is 0.72.
3. The striking observation for ES‡ formation is the large positive contribution to the entropy of activation from the I. In the total, $\Delta S^{\ddagger}_{ES}$, this is partially masked by the D contribution so that $\Delta S^{\ddagger}_{ES}$ does not appear to be unusual. In both I and D processes, the enthalpy and entropy changes in each step of the reaction tend to cancel each other. Compensation is nearly exact in D and probably close in I. In I it does not appear to be close for the formation of the transition states, but this is because there are large and somewhat different thermal contributions to the activation enthalpies. The positive contribution to the entropy of activation of ES‡ and the smaller positive contribution for EA‡ are consistent with complementarity, but superficially not consistent with the generation of force by domain closure (see Section VIII.D).
4. D and I show well-developed free-energy complementarity in the sense that, in each step along the reaction coordinate, their free-energy changes are in the opposite direction. Notice how this flattens the free-energy profile for the total system (Figure 12) as expected (see Figure 1). This behavior is also consistent with domain closure, and as such demonstrates the utility of domain closure in establishing complementarity by exploiting the natural balance of enthalpy and entropy in protein conformations.
5. The free-energy complementarity is mimicked in the H and S profiles. That is an

aspect of complementarity in proteins which has not been previously apparent. It is nevertheless to be expected since D facilitates each step in the mechanism and it can hardly do so except by making negative contributions to enthalpy change when I makes positive contributions and vice versa. One can then say that the behavior of the entropy contributions is required by the existence of compensation behavior in each step, but this puts the cart before the horse in the sense that compensation behavior is a consequence of the fundamental drive toward complementarity.

The molecular nature of EA is not suggested by the analysis, but the profiles do suggest that the domain-closure process relaxes only part way from ES‡. If the restrictions preventing relaxation are centered in the acyl bond, that bond would be expected to have abnormal characteristics, and that is what has been found. The simplest example is the acyl derivative formed with the hydrocinnamate group.[112] The spectral changes on acylation can be separated from those due to changes in optical absorption of indole and phenol groups, and are found to be very similar to the changes taking place in the formation of hydrocinnamoylimidazole. On denaturation by heat or urea,[13h] the protein difference spectrum reverts to that of the methylhydrocinnamate ester in a reversible process. Since there is considerable evidence that this process is not group transfer from his57 to ser195 or relaxation from a tetrahedral-carbon species, the odd spectrum of EA reflects an odd electronic situation, due either to mechanical perturbation or perturbation by nearby charges and dipoles.

A more complicated but potentially more important set of results already mentioned have been reported by Bohlen and co-workers[90] for the acylation of chymotrypsin by a sultone which has 23 kcal of ring-strain energy. The specific hydrolysis reaction between this sultone and the protein has an equilibrium constant near unity, and the off reaction is considerably faster than the hydrolysis reaction. Since there are possibilities for preserving this large amount of energy in some primary-bond rearrangement, it is not possible to conclude that this much energy, sufficient to denature the protein, can be stored in the protein conformation. Nevertheless, the data may set a new upper limit to the amount of energy which can be stored as free energy in protein conformations using some lock-in device to slow the denaturation rate.

Just as the total free energies of formation of ES and EA and their D contributions are linearly related to the free energies of transfer of sidechain models from water to octanol, there is linearity between the total activation free energies as well as their contributions from D and the transfer free energies values. The plots have slopes near 2 for both α-chymotrypsin[75a] and its monomethioninesulfoxide derivative.[75b] These results are typical of the LFE relationships reported by Hansch and co-workers for many congener families of substrates with many different enzymes.[113] This behavior and a large slope would be expected as a result of dispersion interaction between polypeptide and substrate sidechain since the differences in their electronic polarizability are small. It can also reflect the displacement of water from the binding site[166] but the thermodynamic consequences of such displacements are not yet known. However, according to our hypothesis, it also reflects a more general tightening of conformation and the latter, through the domain-closure mechanism, is the major basis for the variation of rate with sidechain size and type. The currently favored alternative is additional "transition-state stabilization" by the generation of strain in the assembly of reacting groups.

The seeming ubiquity of LFE relationships in enzymology suggests that both linear compensation behavior and the weak-coupling approximation are universal in enzymology. Unfortunately, without compensation temperatures for the LFE relationship in enzymes, it is impossible to determine whether or not there are still more remarkable underlying simplicities hidden in the LFE data.

It is important to note that our separation of the D subprocesses from the total as an evolutionary adaptation of the fundamental mechanism to achieve more sophisticated specificity may be somewhat oversimplified. Any rate process which depends on protein conformation will generally have some or all of its transition states determined by minimum free energy rather than minimum potential energy. It must then be expected that the position of a given transition state along the reaction coordinate[12d] and sometimes the reaction path in phase space will vary from substrate to substrate even when their differences are only in sidechains electronically isolated from the reacting groups. Reactant and solvent changes can produce the same results in reactions in homogeneous solution, but these have become more obvious and are likely to be of lesser magnitude since entropy effects, except in water, rarely equal enthalpy effects in importance. A chymotrypsin example is the proposal based on kinetic-isotope-effect measurements, that hydrolysis of acetylchymotrypsin has a simpler, perhaps more primitive, mechanism than hydrolysis of larger acyl-chymotrypsins resembling those formed from polypeptide substrates.[23] If the proposal is correct, the separation of I and D groups of subprocesses we have made is still correct, but the interpretations of the thermodynamic consequences of this separation are oversimplified. Thus, it is possible that without significant interaction of sidechain with matrices, as is the case with the glycine substrate, the charge-relay system is not fully activated or there is insufficient domain closure to take advantage of this system in one of two proton-transfer steps. The latter is more consistent with the isotope effects (see Figure 3). In either case, the D subprocesses have implications for mechanism which have not been taken into account in this chapter. However, the points for the glycine substrate fall on the compensation lines generated by data for the larger substrates, so the mechanistic effects are continuously variable and thus probably a manifestation of changes in the degree of domain closure.

C. FLUCTUATIONAL AND DYNAMICAL ASPECTS OF CATALYSIS

Small distortions of knots are required to explain their proton exchange behavior,[37] but these are expensive and thus provide additional evidence that a high degree of hardness, i.e., semirigid character, is an evolutionary requirement in knots for kinetic stability and catalytic function. Domain fluctuations and adaptation to substrate require flexible and adjustable regions of conformation. The softness of these matrices explains the requirement for minimum (thermodynamic, folded) stability[41a] in enzymes and the frequent but apparently not inevitable evidence for predenaturation behavior[115] found in the free proteins. In free enzymes, the intramatrix secondary interactions are incomplete and easily rearranged within the rather severe limits ultimately set by the knots. Matrix properties are by no means uniform, as shown, for example, by the predenaturation behavior of RNase A,[115d] but tend to have relatively low densities, relatively high motilities, and should have relatively high local dielectric constants. There is considerable evidence not only that contraction and distortion occur when substrates and inhibitors are bound,[9e,72,176] but also that the contraction can be quite large. Even small shrinkage will reduce motility and dielectric constant dramatically since these properties are dominated by the cooperativity enforced by packing. In enzymes, domain closure and contraction are two faces of the same coin.

The thermal-expansion coefficients of enzymic proteins at least free of substrates and inhibitors are similar to those of unstructured polymers.[11d,85] They do not appear to be unusual and probably can be attributed to matrices although there is still much to be learned about the A- and B-hydration shells (*vide infra*) which can complicate interpretation of the thermal-expansion behavior (see Section VIII.D). The apparent isothermal compressibilities, as already noted, are remarkably small and their explanation requires either a special behavior of A- and B-shell water[85b,116,122] or that the matrices are very nearly mechanically transparent, an explanation consistent with the uptake of water under pressure,[21] the solubility of urea, trichloroethane and other small polar and nonpolar molecules in proteins, and the mobility of proton-exchange catalysts (*vide infra*). The small compressibilities are not consistent with matrix properties which are unusual, certainly not much like those of "harmonic-oscillator" solid states we propose for knots. The heat-capacity behavior of matrices should be especially informative and the few heat-capacity data obtained at high resolution tend to bear this out.[118] In particular Suurkuusk found a special increment of heat capacity when dry or slightly damp proteins are fully hydrated.[118a] Rupley and co-workers[118c] found abnormal specific heat capacities for the first hydration shell, consisting of about 0.25 g water per gram of protein in their study of HEW lysozyme. They also found that catalytic activity appeared with the development of this shell.[121a] The hydration studies of Ripley and Careri and their co-workers[117a,121] complement these of Likhtenshtein et al.[122e] using α-chymotrypsin but the latter workers found the onset of catalytic activity to occur at about 0.5 g water per grams polyprotein. Since Hnojewyj and Reyerson[119] in proton-exchange experiments found unexchangeable cores of knot size in dry or very nearly dry proteins, one reasonably suspects that the knots have intrinsic stability even in totally dry proteins, very likely more so than in wet proteins since Rupley and co-workers and others[117] have found greater stability in thermal denaturation the lower the hydration. Then the ''turning-on'' of the protein by A-shell water[115c] can be attributed to the relaxation of matrices into functional forms by conformational adjustment to the A-shell as well as to the abnormal properties of A-shell water. Rupley et al. (Section VIII.D) attribute the very high apparent heat capacity found in A-shell formation to the A-shell water itself but the A-shell and the polypeptide are thermodynamically inseparable in relaxation of the functional enzyme, so that the heat capacity cannot be subdivided and it is realistic to identify it with the ''communal heat capacity,'' which was introduced at an earlier point.

The communal heat capacity is a critical consideration in the construction and rationalization of enzymic mechanisms based on fluctuations since it measures the width of the fluctuation spectrum as represented by the enthalpy or entropy probability-density distribution functions.[11e,146] The value must be large enough to be consistent with the size of the enthalpy and entropy changes in function attributable to protein conformation. Using Suurkuusk's data,[118a] we estimate the total communal heat capacity of chymotrypsinogen measured in complete solution, which includes any B-shell abnormality, to be about 10% of the total. If this is due to simple changes in soft modes, it is too small to be of much use. If it is due to the labilizing of transitions among conformers, the order of magnitude becomes more promising. In the extreme case of a rather major change between two conformers poised near the midpoint of the transition, the standard deviation of the enthalpy would be about 15 kcal/mol for chymotrypsinogen. This is an unlikely maximum, but the order of magnitude is consistent with the numbers involved in applying fluctuation hypotheses and appears to demonstrate that, if any of these hypotheses is correct, it must depend on fluctuations among mobile-defect conformers. It then follows that the functional conformation processes discussed in more detail below depend on two, or more likely, a few mobile defect conformers.

Water-soluble proteins have been found to be stable in solvents ranging from pure hydrocarbons to pure glycerol without loss of structure, and with some retention of function if traces of water are present.[162] An interesting solvent example is ethylammonium nitrate, a fused salt,[139] but a small amount of water seems necessary with it, as with other solvents.

Lihktenshtein and co-workers studied the mobility of spin labels of various lengths attached to the proteins at various points.[147] These studies revealed characteristics of the hydration shells and showed the A- and B-shells to be sensitive to conformation processes which they attributed to domain closure.[122e,124d] They detected the A-shell in which spin-label mobility was relatively slow, relaxation time of about 10^{-7}, the B-shell, with glycerol-like high viscosity, and the C or bulk-water shell beginning about 12 A out from the protein ''surface''. These three shells[122a] are now generally recognized. The B-shell is now roughly described (*vide infra*) and it has both normal-water and abnormal properties. Most of the charged groups lie in this shell and within small errors they have the same ionization thermodynamics found in bulk water.[193] Thermodynamically the B-shell is a natural consequence of the fact that the activity of water must be the same in the A- and C-shells. Otherwise this shell is still an enigma (see Section VIII.D). In the hydration of HEW lysozyme it was not found necessary for catalytic activity nor for development of the proton-exchange kinetics found in complete solution.[121] There is evidence from hydration studies, as mentioned above, that little hydration is necessary for normal function. Furthermore, it is believed that intracellular water is in such small amounts that intracellular enzymes have no more than A- and B-shells.[100] Of course, intramembrane proteins have very little water and certainly no need for a bulk aqueous phase, and, of course, there is little need for an A-shell if a protein has a negligible number of polar groups at its interface with its environment. In such cases we must deal with a nonaqueous interface, which may be as complicated as aqueous A-shells.

For better coverage of protein hydration the reader is referred to reviews[144,145] and summary papers[117a,120c,121a,140] which include the important newer information. The major new development is the characterization of the A-shell. In agreement with earlier proposals (e.g., Reference 122a), the A-shell is found to have two quite-different parts, perhaps three. For HEW lysozyme, the first 0.07 g of water per gram of dry-weight protein goes to the ionizable groups.[117a] It does not "bring the protein to life", but the next 0.18 g/g does. This water is held by the noncharged polar groups at the surface of the protein. Additional water solvates the nonpolar groups in the interface but is not critical to function. Whether one considers it to be a third part of the A-shell water or part of the B-shell appears to be a matter of taste but this water is not essential for catalytic function (however, see Reference 122e on chymotrypsin). Judging from the proton-exchange patterns the first 0.25 g per gram restores important normal-solution properties. One may thus conclude that the additional water in contact with hydrophobic protein surface is functionally better considered part of the B-shell. Several groups have studied the "normalization" of a dry protein[117e,122c,122f,124d,124e] and Lüscher et al. demonstrated compensation behavior in the formation of the A-shell of chymotrypsin[120a] (*vide infra*).

For convenience in the subsequent discussion we call the water held by ionizing groups A″ and that tethered to nonionized polar groups A′. The total comprise all or most of Kuntz' "nonfreezing water".[145] A′ water has a high specific heat and other properties indicating extreme differences from bulk water.[118c] On the average A′ water molecules are too far apart to make the full complement of water-water hydrogen bonds and, as a consequence, the surface free energy is on average high and delicately determined by the exact distribution of unionized polar groups in the water-protein interface.[122] Small changes in shape and size of the interface can be expected to be, in effect, amplified by A′ water rearrangements. In detail the groups to which they are tethered must allow some patches of nearly complete bonding and some local surface regions with very poor water-water interactions. Rupley and co-workers[117a] have shown that formation of the A-shell from bulk water has thermodynamic consequences of the same order of magnitude as the total overall changes in G, H, S, and C_p in denaturation so adjustments resulting from changes in surface size and shape may also be very significant. This feature has been incorporated in rack mechanisms and more recently by Rupley, Gratton and Careri[121a] in a similar domain-closure catalytic mechanism which, however, emphasizes the fluctuational significance of A′ water more than its thermodynamic consequences.[140]

The importance of hydration shells or at least their close connection to catalytic function has been demonstrated by Lüscher and co-workers in comparative studies of hydration of chymotrypsin and an acyl derivative tosylchymotrypsin.[120b] Both the amounts of water in the shells and the thermodynamics associated with formation of the shells are almost remarkably different. It is also worthwhile noting that, although the linearity of most enzymic Arrhenius plots tends to generate complacency about such unwanted complications as solvent dependencies, nearly all enzymes show dependencies on solvent composition of rate-limiting steps as well as binding steps, often in both cases demonstrating linear compensation behavior. Nor for that matter are the Arrhenius and van't Hoff plots always found to be linear and there have been few rate studies carried out with sufficient precision to establish the absence of important activation- or binding-step heat capacities. Such studies as those of Rupley and Careri and their co-workers raise questions that will be very hard to answer but one can scarcely expect to make the necessary utilization of thermodynamic information to describe mechanism without quantitative accommodation to all the hydration shells.

Protein hydration offers possibilities for explaining some long-standing puzzles, e.g., the solubility differences between albumins and globulins. Similarly a few well-studied proteins like β-lactoglobulin and the serum albumins fall into one or more classes which are quite different from the enzymes, the proteinase inhibitors, or the hemoglobins.[114] Arakawa and Timasheff[163d] have shown a qualitatively different dependence of the activity coefficient of β-lactoglobulin on solvent compositions from that demonstrated by other familiar enzymes although this may prove to be an artifact of the experimental temperature range (*vide infra*).

In view of the highly significant compensation pattern found by Lüscher and co-workers in A′ hydration-shell formation,[120a] the results of Rupley et al., and those of Fujita and Noda,[117c,117d] the responsible process must be the normalization of the dry protein to its functional state (see also Reference 124d). It is thus interesting that the T_c is 430K which lies in the phenomenological range associated with matrices.

As already discussed, spin labels provide a sensitive probe of hydration shells and the way they are influenced by events in the protein proper. In fully hydrated hemoglobins and myoglobins the labels reflect a compensation process in the hydration water in response to ligand binding with T_c of about 300K.[147] The amount of nonfreezing water is reported to vary with the state of ligation. Enzymes have not yet been investigated in equivalent detail but spin labels also demonstrate compensation behavior in response to functional changes.[147]

Water-protein interactions were added to the original rack mechanisms when it was found that the compensation temperature observed in hemoglobin and myoglobin processes was 300K since this is in the range found with small-molecule processes in water.[62] However, the epr studies of Likhtenshtein and co-workers[147] demonstrated that the spin labels reflect a process occurring in the protein as it influences the A and B hydration so that it is most unlikely that the compensation behavior can be attributed to subprocesses of bulk water and certainly not in any direct way. In these cases the local hydration is coupled to processes occurring within the protein proper but does not directly determine the compensation temperature. There may be no distinct subprocesses of water involved but the device effecting coupling remains important and unknown. Perturbations of bulk water and the A and B shell were subsequently attributed to change in protein volume and shape[2a,62] and we have found no realistic alternatives since then.

The evidence that chymotrypsin has a phenomenologically simple functional conformation coordinate is equivocal. On pH variation all steady-state rate parameters show compensation behavior,[126c] but there is considerable uncertainty about the T_c values found in the study.[127] The compensation behavior reflects primarily changes in ionization of protein functional groups, but these are not exposed to bulk water so the results may reflect the functional conformation coordinate. At constant pH of 8 with substrate sidechain as driving variable, T_c values of 200 and 460K were found by Dorovska et al. (see captions for Figures 12 and13). Many studies of the rates of hydrolysis of acylchymotrypsins demonstrate T_c values in the range 430 to 480K. In other enzymes, free of substrate, T_c values are usually near 300K[125] and, to the extent these values are reliable, they are consistent with the expectation that progress along the functional conformation coordinate is in general nearly isoergonic at physiological temperature (*vide infra*).

As already mentioned, domain closure is a complicated set of conformational events which only to a first approximation resembles the motions of two rigid protein subunits. There is rather a progression of steps in which the domains readjust to each other and grow together usually about parts of the substrate. To make catalysis efficient the persistence time at maximum closure for any substrate must be relatively long, 1ns would appear to be ample. However, this is very long relative to the time for simple displacements so one or a few activated substate transitions must be involved to prevent too rapid relaxation. Although very fast local motions have been detected, recent evidence, mostly from fluorescence studies, demonstrates a ''breathing'' relaxation time of 1ns.[187] Much longer times are well known for major transconformation processes. Times longer than 1ns would have efficiency advantages but may be difficult to achieve since the geometric changes are small and there are more and more alternative pathways for relaxation as lifetimes increase.

Although knots are likely always to be distinguishable as separate structural elements, matrices may not be. The sharing of matrix regions by the knots may often be so extensive that domains cannot be recognized and the term ''domain closure'' become inappropriate. The process is then better described as volume contraction since appropriate contraction will force knots together.

Domain closure is limited at the open end of its coordinate by the increase in enthalpy due to increased interatomic distances and increased dielectric constant. At the closed end it is limited perhaps by the increase in potential energy but the true situation is probably more complicated than this. There is no *a priori* reason why closure cannot be limited by the entropy decrease but it is difficult to envision mechanisms of activation which depend on such losses. On the other hand the domain-closure process can activate the chemically reacting assembly by the transient tension which develops as a compressive fluctuation relaxes. In chymotrypsin the T state (Figure 1) is probably forced first into the ''forward'' and then into the ''backward'' transitions states in this way. One can argue on the basis of microscopic reversibility and efficient complementarity that maximum closure in the domain-closure processes must occur at T. The mechanism of chymotryptic catalysis is nicely symmetrical because of the similarity of the major reactant and products and the acylenzyme intermediate but most enzymes do not have this characteristic so some bond-rearrangement transition states must be facilitated by compression and others by tension (see Section VIII.D). Compression has been emphasized to simplify the discussion.

A high-precision study of noncovalently bound competitive-inhibitor binding as a function of pH using the shielding of the imidazole group of his-57 by inhibitor demonstrated exact linearity with T_c values in the range 275 to 295K, depending on inhibitor.[26,74,128] The results were inconsistent with those obtained by some other workers, yet consistent with those obtained by still other workers. Inhibitor binding is now known to be very complicated for the chymotrypsins and is not yet understood. Some of the many sources of complication are the ''active-to-inactive'' transition at high and low pH, strong interactions with buffer components, ion binding, polymerization, the instability of α-chymotrypsin relative to γ-chymotrypsin above pH 5.5, and conformation processes sensitive to ionizations near neutral pH. The last is the least understood and at present the most important. The most completely studied of the complicating processes is the active-to-inactive transition.[129] The transition is a very slow refolding process with remarkably small overall G, H, and S changes but a large heat-capacity change. The functional ion-pair formed by the ile-16 and asp-194 ionic groups is progressively protonated and then disrupted by refolding. At low ionic strength there is weak electrostatic coupling of the process to his-57, but the transition never makes significant contributions to inhibitor-binding thermodynamics between pH 8 and 4.5. It is therefore necessary to assume the existence of one or more important processes occurring near neutrality which complicate the noncovalent binding of inhibitors.

The covalently bound inhibitors, natural acyl enzymes like the aliphatic acyl derivatives, and nonnatural acyl derivatives of which the diisopropylphosphoryl derivative (DIPCT) is

TABLE 2
First-Order Denaturation Rate Constants for Several Chymotrypsins and their Derivatives in 8 *M* Urea at pH 7.3[a] and 30°C

Protein	k × 10³ s⁻¹	Structure
α-CT[α]	32	
γ-CT	63	
δ-CT	22	
Succinylated-α-CT[b]	50	
Phenylmethane Sulfonyl-α-CT	40	E–S–C–⟨phenyl⟩
Tosyl-α-CT	38	E–S–⟨phenyl⟩–C
Acetyl-α-CT	27	
Diphenyl Carbamyl-α-CT	7.2	E–C–N(⟨phenyl⟩)₂
Diphenyl Carbamyl-γ-CT	14	
Diphenyl Carbamyl-δ-CT	4.8	
ChymotrypsinogenA[d][c]	3.7	
Succinylated CGN A	3.7	
Methionine 192 Sulfoxide-CGNA	3.3	
Cinnamoyl-α-CT	3.0	E–C–C=C–⟨phenyl⟩
Diisopropyl Phosphoryl-α-CT	1.7	E–P(O<)(O<)
Trimethyl Acetyl-α-CT	1.7	E–C–C(C)(C)(C)

[a] Phosphate buffer
[b] Completely succinylated except for ile.16.
[c] Completely succinylated.
[d] From Reference 134.

the historical prototype, are experimentally simpler to study. These have been extensively studied through their effect on optical rotation, fluorescence, denaturation, etc.[12,13,131] Large effects reflecting size, shape, and chemical differences were observed as has been already discussed. The free protein in such studies showed large pH-dependent changes in some observables in the neutral-pH region. Enthalpy, entropy, and heat-capacity changes in thermal denaturation were found to be sensitive to the specific nonnatural acyl species[148] but the interpretation is complicated by the presence of the denatured species. The rate constants for denaturation give less complicated information about native protein forms and systematic studies of the acyl derivatives have been carried out. Relevant results are the following: Hopkins and co-workers determined the (first-order) denaturation rate constants for a number of acyl derivatives by following the time course of fluorescence-intensity change at pH 7.3 after rapid dilution of small samples of aqueous protein solutions with 8 *M* urea.[132,134] Their results are given in Table 2, and it can be seen that the rate constant is reduced and the degree of reduction roughly distributes the derivatives among two or three classes. The differences in sidechains of representative examples of the classes may suggest the basis for the differences among the classes, but the stabilization effect, measured by the rate constant, is small even with DIPCT. Milville and Hopkins applied the method to some acyl derivatives in which the acyl groups were straight-chain hydrocarbons.[130] They found a rate-constant ratio of 40 for free protein to the acyl derivative with a carbon-6 (C-6) chain, and a ratio of unity with the C-2 derivative. They also found inverted bell-shaped pH dependencies of the rate constant with minima between pH 5.5 and 6.5. Pohl using acid denaturation found even smaller effects between CT and DIPCT than were found in Hopkins' urea denaturation study.[133] In his work the differences in activation enthalpy and entropy were zero within error, suggesting no significant difference in position of DIPCT and CT along a functional conformation coordinate. In the urea-denaturation experiments of Hopkins and Spikes,[134] large differences were found: about 20 kcal/mol in the activation enthalpy and 65 kcal/mol K in the activation entropy. Comparison of the results from acid and urea denaturation suggest the possibility that all the differences in rate constants in the urea experiments are due to differences in permeability of the matrices to urea. Pohl found that the activation enthalpy for CT was little altered when the denaturant was guanidinium.HCl, and this is consistent with results obtained with HEW lysozyme using guanidinium.HCl.[117i,197] As is now well known, α-chymotrypsin in the crystal dissolves urea even at low solution concentrations of urea but does not dissolve guanidinum.HCl.[136a] In sum, the relative insensitivity of the denaturation rate constant itself to differences in acylating group does not indicate much difference between free protein and the acyl derivatives, but the urea dependence suggests major alterations in matrix volume and integrity which are much the same in ground and transition states. These suggest that knot properties are only weakly altered by changes in their associated matrices. Pohl found that α-, γ, and δ-chymotrypsins have very similar unfolding rate constants.[149] He measured the activation enthalpies and entropies for the unfolding rate constants of α-CT and chymotrypsinogens A and B. The results were identical for the three proteins,[133a] again indicating identical fully formed knots.

Although compensation behavior with T_c values near 300 K possibly spurious has been reported in the chymotrypsin literature, a functional conformation coordinate is unlikely to be established until the pH-dependent behavior is better understood.

Belleau and co-workers[151a] determined the effect on the rate of methane sulfonylation of acetylcholinesterase as it is influenced by binding analogs of substrate sidechain. They used two series of alkyl quaternary ammonium salts having the general formula $RCH_2N^+(CH_3)_3$, R being either a straight-chain alkyl group or a cyclic alkyl group. The equilibrium binding constants for 35 tetralkyammonium salts were also determined.[151b] Several of the more important conclusions follow. First, the standard equilibrium binding data demonstrated compensation behavior with a T_c about 290K; the errors are considerable. The binding enthalpy and entropy changes were sensitive to size and shape of the alkyl group shape being as important as size demonstrating several clearly differentiable families. Second, the methane-sulfonylation rates were increased by binding of the salt "inhibitors" by factors as large as 60 and then fell as salt size increased. The logarithm of the rate-enhancement factor was closely correlated with the equilibrium binding data indicating that the subprocess responsible for enhancement must show similar compensation behavior.

This study uses Inagami's "separated substrate side chain" method[24,25] in perhaps its most powerful application to demonstrate the indirect influence of detached substrate sidechains on the entactic reactivity of his.57. The effects are not large but compensation behavior with a T_c value near 290K is indicated if not firmly established. The authors attribute the effects to varying amounts of water displaced from the pocket in a domain-closure mechanism.[122c,122d] Water displacement is probably important but the dependence of the rate on salt size and shape requires a mechanism to harness some of the free-energy released on salt binding. The standard enthalpy and entropy changes reflect that mechanism and only indirectly the water displacement. Finally the common T_c value of 290K suggests the existence of a single functional conformation process and in so doing provides support for similar apparent functional simplicity in other serine esterases and proteinases. However, as with chymotrypsin,[73b] inhibitor binding, here the binding of the quaternary ammonium salts, may be complex involving greater motion along this conformational coordinate than occurs in the catalysis of the normal substrates. Competitive inhibitors in rack mechanisms are often transition-state analogs.

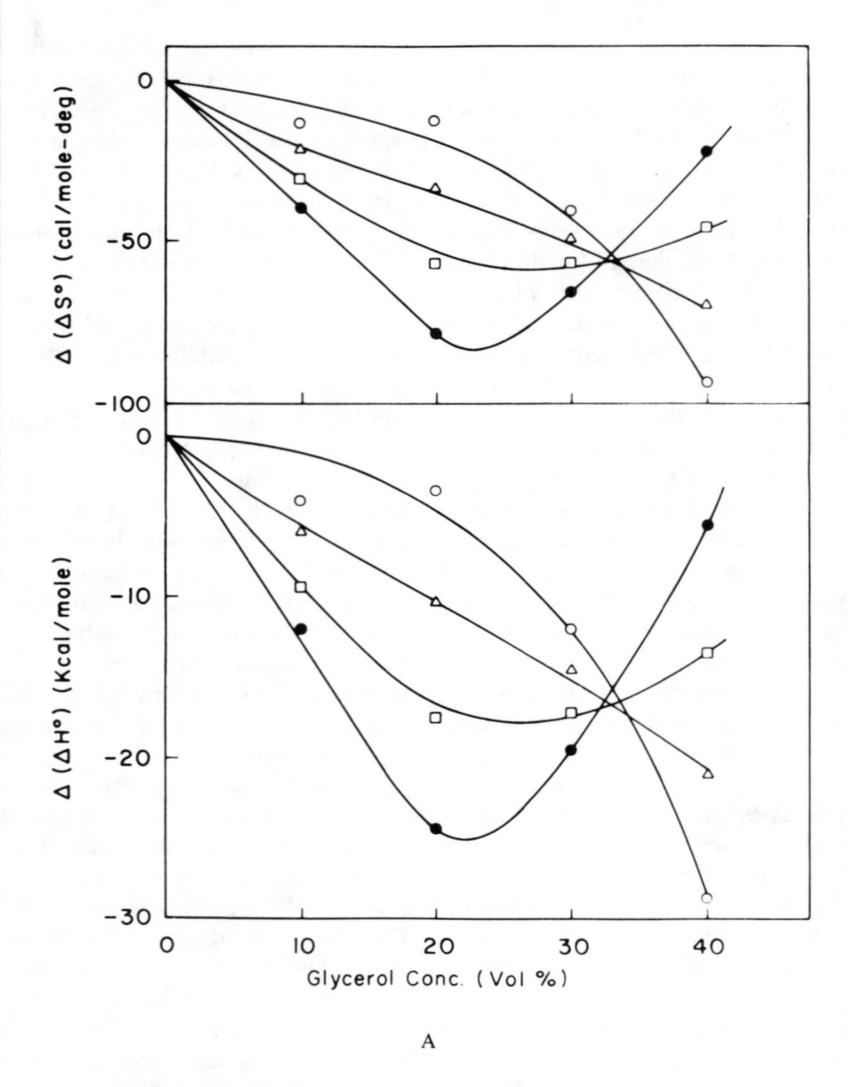

A

FIGURE 14. The free-energy profile for chymotryptic hydrolysis of small ester substrates. This qualitative figure extends Figure 1 to include the metastable form of EA, $(EA)_2$; a conformational adjustment step for ES ($(ES)_1 \rightarrow (ES)_2$); and four small peaks to represent the pre-equilibrium domain-closure processes which appear in the experimental transition-state-formation data and in which the free energy of the protein system is temporarily re-distributed. These metastable states are $(ES)_3$, $(EA)_1$, $(EA)_3$, and $(EP_2H)_3$. T and T′ can be assumed to be true tetrahedral intermediates; with very good substrates it is likely that each pair of free-energy barriers, e.g., ES‡ and T‡, are fused into a single barrier.

D. THERMODYNAMICS OF DOMAIN CLOSURE

The free-energy reaction profile for chymotryptic hydrolysis of simple ester substrates taking into account conformational events at least to a first approximation is shown in Figure 14. The descriptions of those species appearing along the reaction coordinate which are not specifically discussed in the text are reasonably obvious as are the reasons for including them. Thus, the first enzyme-substrate complex is the preliminary association complex and its formation includes most of the desolvation of substrate and protein. The second complex is included to take into account the relaxation of the protein to accommodate substrate. T and T′ (see Figure 1). for convenience without loss of generality are explicitly tetrahedral

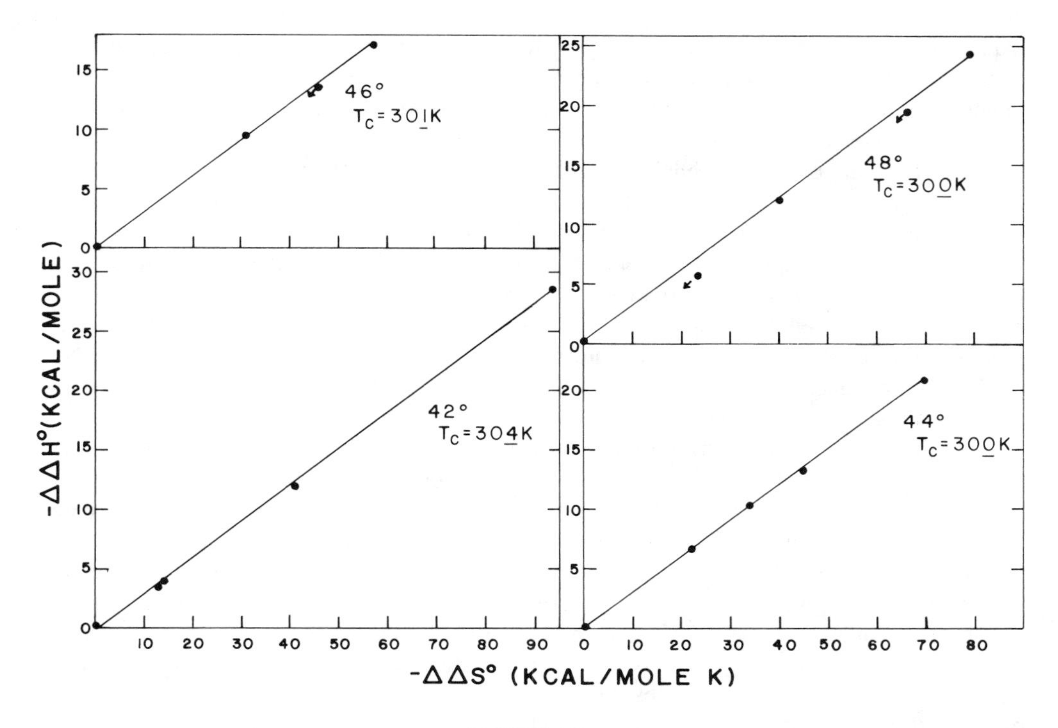

FIGURE 14B.

intermediates. For substrates in which this is not the case the close symmetry of the profile about T and T′ disappears but such details appear to be irrelevant to the purposes of this chapter. What this chapter should do is rationalize the thermodynamic changes in the domain-closure process so as to explain how domain closure can raise the potential energy by mechanical compression. Obviously this is now impossible. There are many subprocesses which participate in domain closure and at this time the quantitative consequences of none of them can be estimated with confidence. On the other hand, a good many useful qualitative deductions have become possible since a good fraction of the protein literature is now susceptible to more detailed interpretation and integration using the knots-matrices description of protein structure and the generalized concept of domain closure. The few examples presented in this chapter barely scratch the surface.

For chymotrypsin most of the interesting thermodynamic information is concealed in the positive enthalpy and entropy contributions from the I subprocesses to the formation of ES‡ from ES. These numbers are the net consequences of the several contributing factors and their resolution into individual contributions is extremely difficult. It will require detailed studies of series of substrates, competitive inhibitors and other transition-state analogs complemented by residue changes, studies of binding and catalysis as a function of hydration, and systematic studies of observables reflecting conformational state. This is a big order and the undertaking can be made tractable only if it is guided by adequately sophisticated hypotheses. Whether or not computers of sufficient size to accommodate realistic models with their more sophisticated potential-energy functions will be developed in time to yield more reliable and thus more useful theoretical results remains to be seen. Computer treatments of liquid water have not been particularly successful and proteins are even more complicated. Larger samples, more parameters in potential-energy functions, and very likely calculations which start at the heat-capacity level rather than the energy level are required for calculations on water to have the necessary power. Protein fluctuations can be examined by molecular-dynamics calculations but the results to date suggest that the potential-energy functions are inadequate and that the distance and angle information derived by applications of the X-ray diffraction method is also inadequate. The highest resolution now available is at best marginally useful, but there is the more serious problem that the structural information consists of averages over conformers. Low-temperature and high-speed diffraction studies should improve the situation. At present, the most useful information would be clear descriptions of knots and matrices. Diffraction studies and calculations in their present state of development can be of considerable value in obtaining these descriptions.

Any serious attempt to explain enzymic catalysis must explain its relationship to complementarity. The contraction-expansion of conformation provides the connection in rack mechanisms but it is not of much quantitative use until fully described by inclusion of water-protein interactions. There is, in fact, no hard evidence that the contributions to the thermodynamic changes in globular protein function from conformational changes per se are any larger than those from interactions of conformation with water.

The direct protein-water interactions are the following: displacement of water on substrate binding and the A- and B-shells and their changes resulting from rearrangements of protein surface and charge redistribution. The indirect interactions are those in bulk solvent mediated by the direct interactions. The latter are complicated and confusing but of demonstrated importance in protein function. There is some possibility that their effects are in part due to alterations in coupling of protein with solvent which alter the efficiency of momentum transfer from bulk solvent to protein but it more likely that the effects are more thermodynamic in nature even though they are likely to alter the fluctuation spectra of proteins and the penetration rates of small molecules into proteins. They focus attention on B-shells and on protein activity coefficients but do not yet provide explanations for the coupling involved. It is necessary to consider representative effects and we are forced to do so in a cavalier fashion since water behavior is not yet understood at the molecular level. Thus, the description of water is essentially our own.[152,191] It is probably as accurate as any other and, in contrast to the alternatives, it has become increasingly well supported as new publications on water appear.

The use of co-solvents to alter water and solutes as probes in the study of water properties, always profitable, has become a major experimental tool. One important discovery by Roux et al.[194] is that uncharged monofunctional amphiphiles except methanol and ethanol form micelle-like entities with critical micelle concentration (cmc) values often so low as to be below the usual experimental range. The cmc values decrease as the size of the hydrophobic parts of these solutes increase so that for many larger amphiphiles there is little information about the properties of homogeneous solutions and extrapolations to infinite dilution may be more often incorrect than correct. Monofunctional amphiphiles, except quaternary ammonium ions and similar charged species, are thus poor co-solvents with which to explore the clustering in water but hydrazine, hydrogen peroxide, urea, formamide, and guanidinum ion are "simple" cluster breakers (*vide infra*) and can be identified as a particularly simple class of "structure breakers". Ethylene glycol is a member of more complicated and more general multifunctional amphiphile family. At very low concentrations it stimulates the formation of clathrate-like water cages about its oily parts.[152] The number of water molecules in the smallest cages, that is, those about the smallest amphiphiles (and methane, ethane, and argon), are thought to contain about 15 to 25 water molecules. Amphiphiles with larger hydrophobic parts and the larger alkanes are caged by increasingly larger number of water molecules.[153e,153f] Ethylene-glycol molecules at mole fractions greater than about 0.04 complete with each other for water molecules so that caging becomes incomplete and eventually they become cluster breakers.[152a] This behavior is consistent with cages of about 18 water molecules. At even higher concentrations EG destroys the connectivity characteristic of water behaving like hydrazine. That these effects are very subtle is shown by the fact that in the low-concentration regime 1,3 dioxane behaves like hydrazine, and 1,4 dioxane like ethylene glycol.[145,195]

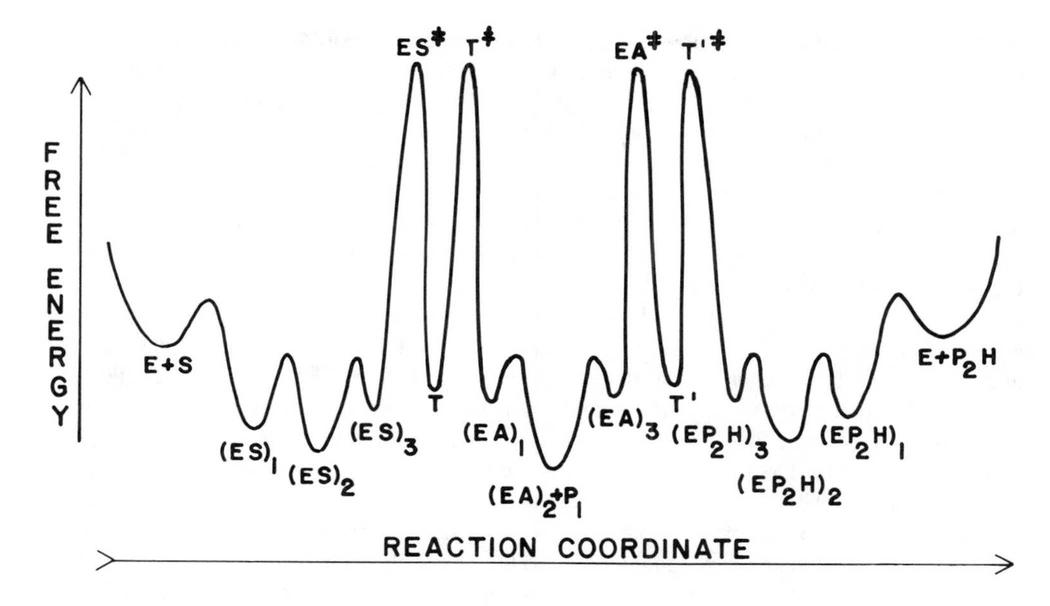

FIGURE 15. Equilibrium denaturation of chymotrypsinogen A in solutions of increasing glycerol content.[81] (a) Plots of the standard enthalpy and entropy changes per mole of protein against glycerol concentration: O, 42°; Δ, 44°; □, 46°; ●, 48°. This figure is taken from H. Gekko and S. Timasheff, *Biochemistry,* 20, 4681, and is reproduced by permission of the publisher (copyright 1981, American Chemical Society). (b) The compensation plots of the same data. The increments, $\Delta\Delta H°$ and $\Delta\Delta S°$, are plotted. "Turn around" behavior (see test) was observed at 46° and 48° as indicated by the arrows on the points from highest glycerol concentrations. The compensation temperatures are identical within errors. Despite the apparent variety of effects of gycerol there is a single compensation temperature.

Systematic study of the solubility thermodynamics of probe molecules such as argon as a function of co-solvents from the two families invariably demonstrates enthalpy-entropy compensation behavior with temperature-independent T_c values near 295K. This phenomenological behavior is, of course, paralleled by the corresponding set of linear free-energy relationships. When micelle formation and protein denaturation have been used to probe these mixtures, LFE behavior is also generally observed although with charged micelles the charge effects must be first eliminated. Where accurate enthalpy and entropy data exist, the expected compensation behavior is observed but the T_c values for protein systems are not always near 295K. A single compensation line is, however, usually found except at very high co-solvent mole fractions. That this is surprising becomes clear in examining one particularly good example, the effect of glycerol on denaturation of chymotrypsinogen A studied by Gekko and Timasheff.[81a] Figure 15a shows the standard enthalpy and entropy changes for this protein as a function of glycerol concentration. The complicated forms of these curves suggest that glycerol has several effects on the protein. Nevertheless, as shown in Figure 15b, there is a single compensation pattern with a temperature-independent T_c value of 302K. Additional interesting features can also be seen. (1) There appears to be a single composition at which enthalpy and entropy increments are independent of temperature. (2) Below 302K glycerol is a denaturant rather than a stabilizer. As already discussed, glycerol has been found to stabilize proteins by raising the activity coefficients of native forms less than those of denatured forms. Whenever polyols raise protein activity coefficients they are particularly useful probes for protein hydration shells because they are excluded from these shells. This exclusion follows from the Chatalier-principle but has been established experimentally by Lehmann and Zaccai,[156] whose small-angle neutron-diffraction data show that glycerol is excluded from 3 l of A and B water per mole of protein in a 1-*M* solution of RNase A 20% glycerol by volume. The alternative interpretation, that there is less complete exclusion from a larger volume, is unlikely. The activity of water is not much altered by glycerol at 25°, a 9% reduction in 35% glycerol. The effect of glycerol in the chymotrypsinogen denaturation experiments cannot be assigned to the denatured form since the activity coefficients of both native and denatured forms rise about equally rapidly with increasing glycerol. As we shall see below, glycerol may lower protein activity coefficients in some temperature ranges. Although this has not yet been established,[180] one cannot be certain that the denaturing effect of glycerol below 302K is due to a larger increase in the activity of the native form relative to that of the denatured form.

Although the ethylene-glycol data for chymotrypsinogen A and RNase A denaturation demonstrate the same high T_c, 370K,[161] Gekko and Morikawa[81b] comparing this co-solvent and larger polyols in denaturation of these proteins as well as HEW lysozyme found qualitatively different patterns almost certainly consequence of the different surface polarity ratios of the proteins.[81e,82h] These findings and others from similar experiments[186] indicate that folded stability depends as much on interactions of water with protein polar groups as on water-hydrophobe interactions. That there is some underlying simplicity in the dependence on water is demonstrated by examples discussed in the next few paragraphs which illustrate the fact that only two narrow ranges of T_c values have thus far been found using polyols and alkanols as co-solvents. This result taken together with the relatively recently established characteristics of protein hydration shells (*vide infra*) suggest not only that these shells interact in different ways with the cooperative subsystems of water but also that such interactions are one major missing factor in the attempts which have been made to rationalize the thermodynamics of protein stability.

Very similar compensation behavior was found[62] in the data obtained in studies of ethanol effects in thermal denaturation of RNase A.[157] There was a single temperature-independent compensation temperature of 285K, and, as in Figure 15b, true compensation behavior is indicated by the temperature independence of T_c. Ethanol is a stabilizer below 285K and a denaturant above for ethanol concentrations lower than mole fraction 0.09. Higher concentrations were not studied. This is superficially consistent with the limited experimental information from studies of small solute processes. Ethanol appears to be in the same co-solvent class as glycerol and ethylene glycol. The compensation line doubles back on itself at higher concentrations and higher temperatures without much change in slope. This behavior is observed in the glycerol-chymotrypsinogen study (Figure 15b) but the signs of the increments in $\Delta H°$ and $\Delta S°$ are all positive in ethanol-water mixture and all negative in the polyol solutions. Furthermore, ethanol at temperatures above 290K has thus far been found to lower protein activity coefficients so that it is preferentially concentrated at the protein surface. It may also have some solubility in some proteins at higher concentrations, as suggested by the solubility of methanol in chymotrypsin and trichloroacetate and trichloroethanol in a number of proteins.

The ratio of nonpolar to polar groups in the protein-water interface is smaller for HEW lysozyme than for chymotrypsinogen and even smaller for RNase A. The effects of polyols is different on the denaturation of the three proteins. Compensation behavior cannot be found in the data for RNase A in glycerol solutions,[81a] although it was found in the ethylene-glycol studies[81h] and in ethanol-water in which T_c is 285K as noted alone. Gekko found a T_c value for HEW lysozyme of about 370K with methanol, the same value found with all polyols except ethylene glycol.[81h] He studied the transfer of the alkyl and aryl amino acids from water to aqueous solutions of the linear polyols: glycerol, sorbitol, and xylitol.[155] Using the transfer data for glycine to correct for the ionized groups, he estimated the thermodynamic transfer quantities for the sidechains, a customary procedure but not an entirely accurate one. The T_c values were 300 to 320K and these values would be adjusted down toward 295K on correcting for the small cratic entropy differences in the reference and polyol solutions. These polyols thus enhanced solubility at temperatures greater than 300K and lowered solubility below.

Individually and collectively these observations on the polyol-protein interactions are very confusing. They do, however, suggest that the effects are mediated by changes in the B hydration shell. This is not inconsistent with the data from protein-hydration studies indicating that the B-shell, although not essential for the development of normal conformational and functional properties, is nevertheless important in making fine adjustments in both [117a,121a] (*vide infra*). It is apparent that the polarity ratio of the surfaces of both folded and denatured forms of proteins is as important as temperature in the polyol effects. In many of the studies attempts were made to explain the effects in terms of the polarity ratio alone, but this approach is oversimplified at least insofar as it ignores temperature.

Urea, Gu.HCl, and $CaCl_2$ are more complicated additives than the small polyols insofar as they are preferentially concentrated near protein surfaces and their effects through water cannot be readily distinguished from true complex formation with protein groups. It will be remembered that these denaturants lower activities of native and denatured forms and do so more for denatured form than native forms. Almog et al.[81g,81i] measured the thermodynamic changes in transfer of HEW lysozyme from dilute buffers to aqueous solutions of urea and Gu.HCl so as to obtain information on both native and denatured protein forms. Again the native form was influenced quantitatively only slightly less than the denatured form. Both forms yield the same T_c value within error but the compensation pattern disappears at higher co-solvent concentrations so the T_c from the denatured protein is not reliable. For these additives the T_c values for the native form clustered about 370K implying that, if experiments were possible at 100° and higher, the additives would be found to be stabilizers.

The several studies just discussed are sufficient to illustrate the complexity of the interactions of water with proteins. We now return to the consideration of simpler solutes.

With respect to solvation of hydrophobic molecules and atoms, water appears to have two characteristic temperatures or narrow temperature ranges centered near 295 and 375K. The one commonly observed, near 295K, is, we suggest, the crossover point at which the free-energy of solvation of hydrophobic molecules is independent of the cluster phenomena, that is, the free energies of the lower-temperature form and intermediate-temperature form of water are the same. At 295K water appears to be a mixture of clusters and connectivity water, as described below. Since different kinds of solutes will interact in somewhat different ways with the clusters and with connectivity water (*vide supra*), this temperature will vary although little variation has so far been observed. There is good evidence that there is a characteristic T_c for solvation changes in ionization processes near 290K[106] and some that proton solvation in water is dependent on preformed water tetramers, a dependence also indicated for electron solvation.[59c] Consider, for example, the ionization processes of aliphatic acids and amines. As Kauzmann first noticed,[160] the standard enthalpy and standard entropy changes are not consistent with simple release of the proton from the acids nor with the transfer of the proton from an amine to water. The standard free-energy changes are consistent.[106] To rationalize these observations on the basis of current information we should have to conclude that despite the fact that ions are structure breakers to greater or lesser degree depending on the particular ion, e.g., the Hofmeister series,[143] the hydrated ions are cluster formers. This rationalization is important because it suggests that protein-charged groups and perhaps even well-hydrated uncharged polar groups of proteins alter water and are themselves altered by the same kinds of cluster effects responsible for hydrophobic hydration.

Molecular descriptions for several forms or species of pure water and the events in "hydrophobic hydration" and the "hydrophobic effect" have been given by many authors[145] (*vide infra*). At present for cold and supercooled water some sort of cooperative clustering idea is considered to be particularly attractive.[135,159e] For higher temperatures a "connectivity'' model currently is most consistent with the data[159d,159f] and at still higher temperatures water has the properties, including the solvent properties, of a strongly associated, hydrogen bonding liquid with only pairwise cooperativity. Hydrazine seems to be a good model for high-temperature water.[154] For example, the thermodynamic quantities for solvation of some surfactants in water at 160° are virtually identical with those for solvation of the same molecules in hydrazine at 25°.[154]

Let us call the three kinds I (cold), II (connectivity), and III (''simple''). Proportions of I to II and II to III change very slowly with temperature or pressure so all three co-exist but there is little III when I is a major constituent and vice versa. The abnormalities of the II water appear to be due to a high degree of connectivity. The percentage of all possible hydrogen bonds made in II water is very high. Just how high will have to be determined by an accurate percolation model and this will be difficult because the characteristics of the average hydrogen bond change with temperature. With increasing temperature the average hydrogen bond grows increasingly long, weak, and flexible.[170]

With our description of water the characteristic temperature near 290K is explained in terms of a model for I water as a collection of very short-lived (0.6 ps) clusters of varying size in an environment of II water all formed from a single building block for which several models have been proposed: a stiff tetrahedral pentamer,[159d,191] a flat pentagon,[159a] a stiff nearly tetrahedral tetramer,[59a] a five-coordinated central water molecule,[159b] and a structure consisting of bifurcating hydrogen bonds.[159c] None of these has been established and they all may in time prove no replacement for the framework-interstial model of water first proposed by Somoilev[181a] and until recently the most attractive of the structural models for water.[181b] In order to have a chance to explain the abnormal properties of water the clusters must have some degree of cooperatively, actually quite a high degree. The source of this cooperativity has become a matter of much interest in recent years despite the fact that none of the small-cluster models has received general acceptance. There is a greater agreement about the large clathrate-like cages surrounding the hydrophobic parts of solutes since many new data tend to support this early proposal by Glew. The smallest of such cages is perhaps sufficient in size to accommodate argon, methane, ethane, methanol, ethanol, ethylene glycol, 1,4 dioxane, dioxygen, carbon monoxide, and probably the nitrogen molecule. The hydroxyl groups and ether oxygens are thought to replace water molecules in the cage structure. A good guess is that the structure is the pentagonal dodecahedron, the smallest cavity structure found in the ice clathrates. It can be noted that the flat and tetrahedral pentamers and the tetrahedral tetramer are all possible structural subunits of pentagonal dodecahedra. All or most of the ice clathrate structures consist of such units, but which of these units, if any, is the smallest water building block from which the larger clusters which dominate water near 0° and below are formed? At these low temperatures a variety of large, often very large, clusters are accommodated in much reduced amounts of connectivity or nonclustered water molecules. These two types of water, I and II, appear to be about equally responsible for the abnormalities of water. Co-solvents like hydrazine break up the clusters at lower concentrations and the more-adjustable and less-cooperative connectivity system at higher concentrations. The cluster abnormalities have entirely disappeared in 0.23-*M* fraction hydrazine or hydrogen peroxide[135] and further additions of hydrazine mimic the effect of increasing temperature.[154]

The clathrate-like cages which form about oily parts of solutes appear to be superimposed on the cluster and connectivity systems of pure water but they may also occur in pure water either as empty structures or with water molecules as guests. These large clusters are not ice fragments and may be assemblies of a different building block than that of ice, e.g., distorted tetrahedral tetramers. These structures would explain the ability of water to supercool, a property not well explained by most proposed building blocks. Nevertheless, there is little evidence that the clusters even with hydrophobic guests have large stability and this is of considerable interest because there is other evidence that they are formed in an all-or-none way.[153] Since the abnormal features of hydrophobic hydration persist up to temperatures somewhat above 100°, it is probable that clathrate-like structures can form even when the average population of building-block clusters is nearly zero. If so, they would appear to be more stable than is generally supposed.

The potential implications of water clusters for proteins have been given little attention but the clusters provide one of the few obvious ways to explain co-solvent effects on protein activity coefficients. For example, B-hydration shells may be as much a consequence of inhibited cluster formation as of poor matching of A-shells to bulk water. However, if so, the thermodynamics of ionization of protein acid and base groups in B-shells should be abnormal, yet within small experimental error they are identical with values in bulk water as already noted.[193] Explanations given for processes dependent on protein solubilities in terms of small-molecule solvation have become increasingly unsatisfactory. The small volume changes on denaturation are usually cited as evidence against the classical interpretation but equally important is the finding, originally suggested by Shinoda and Fujita,[82] that hydrophobic hydration is favored rather than unfavored by the Frank and Evans "icebergs", now tentatively identifiable as clathrate-like cages. The negative enthalpy and entropy changes in hydrophobic hydration nearly compensate each other and the poor solubility is a consequence of a weak reaction field just as in all other cases of poor solubility.[82d,191]

The characteristic temperature of water near 290K is apparently not only the temperature at which hydrophobic molecules are equally soluble in I and II water but also that at which acid and base ionization is independent of composition in highly aqueous solutions. It is also the midpoint temperature when two-state models are used to interpret a wide variety of data from pure water, at least it is within the errors. If, as suspected, there is none of the larger clathrate clusters in pure water, this does not comfortably square with its implication that the small clusters and the clathrate clusters have similar compensation behavior but the aggregate of errors is such that in time quantitative differences may be detected and found to be small.

It has been shown earlier in the section that near 290K the denaturation of several proteins become independent of the concentration of stabilizing or destabilizing cosolvents, i.e., the compensation temperatures in solubility and denaturation generated by change in co-solvent concentration are near 290K, but with some proteins in urea or guanidinium or ethylene-glycol mixtures with water the compensation temperature and thus the temperature at which denaturation is independent of solvent composition is about 370K. Up to 370K the Henry's law coefficients for argon and methane solubility rise with temperature.[171] Above 370K, they decrease. According to our model for water the solubility behavior can be attributed to change in dominance of II and III water. Recall that III resembles hydrazine with predominate interactions being pairwise and thus insufficient to support cooperativity. This suggests a second two-state behavior of water with a characteristic temperature near 370K which would explain the loss of solubility dependence on solvent composition near 370K.[172] It is thus a potential explanation for Privalov's finding that, if the standard heat capacity change in denaturation is independent of temperature,[148] the specific enthalpy of denaturation of a number of proteins is the same at about 110°.[114]

One final matter is internal pressure which may be important if enzymes depend significantly on momentum fluctuations arising in the solvent. Thermodynamically the internal pressure has little influence on the free energy of solvation of proteins and thus little influence on protein activity coefficients. Changes in internal pressure will produce changes in the sizes of the enthalpy and entropy of formation of the protein in the solvent but they must nearly compensate each other. On the other hand, momentum fluctuations are zero at the temperature of maximum density. It is highly desirable that studies of enzymic reactions be carried out around the TMDs of H_20 and D_20 to see if fluctuational effects are after all important. Brief and unfortunately equivocal studies were carried out by Abugo,[196] who found sharp breaks in the temperature plots of the Michaelis constant and k_{cat} at the respective TMDs. The rate constants were lower below the TMDs. We look forward to more extensive studies of this type.

The hydration study of lysozyme by Rupley, Careri, and their co-workers [117a,121] is one of those rare undertakings which changes the entire perspective in a major field. The multivariable experiment they devised complements site-specific mutagenesis and adds an important element of stability to such studies in that its basis is fundamental chemistry, the basis on which protein function must be explained if it is to be understood. The original ''Rupley experiment'' can be refined in several ways. In particular proteins are found most difficult to denature when dry or having only the A-shell (see Reference 117); samples can be dispersed in nonpolar liquids. The work of Klibanov et al. using hydrocarbons[162] is a major step in that direction. At the next level of complexity liquids simulating inhibited water can be used. Unfortunately, hydrazine, the best of these, is somewhat difficult to work with but measurements can be made rapidly before disulfide reduction occurs. Data obtained in urea and guanidinium-ion solutions may contain the information we need but are probably otherwise complicated. Studies of the effects of more convenient solvent additives, e.g., 1,3 dioxane, probably can provide a good fraction of the missing information but they must be systematic and include all the quantities measured by Rupley: catalytic rate, substrate and inhibitor-binding thermodynamics, proton-exchange, and the (apparent partial molar) heat capacity, free energy, and enthalpy and entropy of the protein. Such experiments are expensive, tedious, and currently unpopular but they can greatly multiply the value of each dollar spent on SSM. Obviously, large economies of money and time can be achieved by well-integrated collaboration. The precedent has been established by ''big physics'' and the goals are more obviously beneficial.

We have neglected a number of proposals about the dependence of protein function on water. These fall into two general classes. The first includes other effects generated at protein-water interfaces and these are very similar to those discussed in the previous paragraphs but supported by different kinds of data, e.g., the pioneering work of Low and Somero,[164] who measured water effects in catalysis and found a large effect of pressure-relating hydration and function. The second class concerns water displaced from substrate-binding regions and from interdomain regions. Representative examples are the proposals of Martinek,[110] Käiväräinen[141] and Loftfield[165] and their respective co-workers. Käiväräinen has published a monograph[141] on his proposal in which he also discusses a large amount of important work from Eastern bloc countries on domain motion, protein dynamics and hydration. This work has attracted little attention in the West. It describes in some detail the possibilities for enzymic mechanisms based on fluctuation phenomena and in so doing provides considerable support for the rack mechanism as a common basis for enzymic catalysis.

It is obvious that some water molecules must be displaced when substrates or inhibitors are bound. Some of these molecules are detectable in X-ray diffraction pictures but primarily because water molecules not held in a fixed geometry cannot be detected by diffraction methods. As a result, information about total water in binding sites and in other parts of proteins is in short supply.[19] The thermodynamic consequences of all such displacements are undoubtedly significant and must eventually be factored into quantitative descriptions of enzyme function. However, it is doubtful such water displacements are fundamental in catalysis. Geometric changes in domain closure are small relative to the volume of a water molecule but total volume changes are large relative to the volume of a water molecule and many water molecules may be displaced in the contractions of proteins. However, protein matrices are liquid-like so some large fractions of the volume changes must be due to free-volume changes.

Water displacement from internal regions is another of the many devices we must suspect to have been woven into protein mechanisms. There are many of these devices including propinquity effects, local electric fields, and other physical-organic adjustments which can facilitate catalysis. The excellent review by Berezin and co-workers[166] covers most of these currently popular as they may be involved in chymotryptic catalysis. In addition it contains a detailed discussion of the role of substrate sidechains in catalysis illustrated with free-energy reaction profiles for other enzymes demonstrating the very considerable advancement in complementarity which has thus far been achieved in evolution.

There is little reason to suspect that most of the devices so far proposed, at least those not obviously foolish, have been ignored in evolution and there must be more yet to be discovered. So far as we now know, Nature has no philosophy in mutation experiments other than that arising from survival pressure so it is difficult to establish which, if any, of the well-known devices is most fundamental, i.e., the major discovery making controlled, efficient catalysis and the necessary evolutionary mutability possible. There may be no single device responsible, rather a hodge-podge with some combination supporting function in one enzyme and other combinations in other enzymes. If one adds up the catalytic advantages estimated by the most enthusiastic proponents of each of the popular devices, the total is much larger than is observed. This is a meaningless exercise but it suggests that more rapid rates are possible but less important than regulation and thermodynamic efficiency. The high cooperativity of multienzyme systems obviously did not arise simply to enhance rates. It depends primarily on the internal and external dynamics of once-independent domains. The pairwise interaction of domains found in simple single-protein enzymes is retained but, as discussed earlier, the catalytic domain pairs need not be partners in single protein subunits. The catalytic pairs have been linked through free-energy redistribution and feedback control into single cooperative systems because this improves survival value. The underlying theme is free-energy complementarity and it is the ability of enzymes and other proteins to support and improve complementarity which makes them successful vehicles for physiology. Their primitive ancestors may have achieved the marginal catalytic and specificity effectiveness necessary at that time using devices incapable of supporting complementarity, and we include in this class those devices which do not depend on conformation for free-energy redistribution. If so, the ancestors must have been rapidly displaced once the value of complementarity was discovered. This argument leads us to conclude that there is indeed a fundamental mechanism of enzymic function and that it is domain closure or more basically contraction and expansion.[1c,2a,88a,88c,122c,184a,184b,184c]

IX. THE FLUCTUATING PROTEIN

This chapter concludes with Figure 16 which is a simplified description of the linkage system of a single protein with a catalytic, binding, or strictly regulatory function. The diagram is simplified sufficiently to be tractable and general. Coupling among the subprocesses can be direct, as indicated by the edges of the parallelopiped, or, more often, indirect through protein conformation, as shown by the lines extending from the vertices into the protein. The entire system is in rapid dynamic equilibrium. The distribution of fluctuations is wide and complicated by normal modes which selectively link groups of subprocesses with similar characteristic frequencies as discussed by Careri.[167] Beat frequencies between modes may have special importance for function but it is unlikely that we shall know much about such fine details for some years to come. Some modes involve most of the system and thus enforce an underlying coherence on all the subprocesses which participate in function. These are expansion-contraction and deformation modes and from them has developed a single composite mode having the necessary phase relations among the subprocesses and the necessary compensation temperature to support the physiological function of

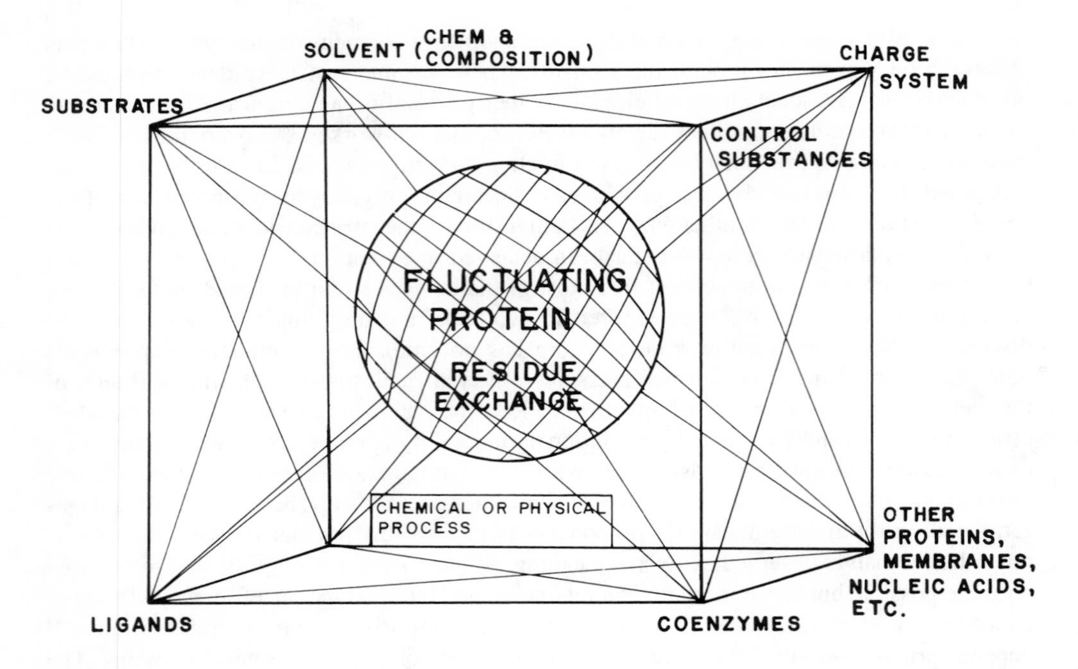

FIGURE 16. The fluctuating protein. The nonconformational subprocesses are indicated by the vertices of the rectangular box. The lines connecting the vertices represent direct coupling, but some lines have been omitted to avoid confusion. The couplings of these subprocesses to the conformational subprocesses are represented by lines running from the vertices into the protein. Quaternary couplings among protein or domain substructures of the total protein are not shown. The average and dynamical properties of the protein can be modified by residue change and, because the operation of the protein machine is dynamical regardless of function, changes in most residues will produce quantitative changes in the operating characteristics. As noted in the text, hemoglobin and myoglobin are first-order linear-response machines and many enzymes may also fall in this class.

the protein. The compensation temperature is usually near the operating temperature of the system as is required in enzymes to prevent dephasing and to allow frequent large-amplitude displacements in the gross functional mode. In binding proteins like hemoglobin it is required to preserve regulatory sensitivity and oxygen affinity by maintaining the poised positions of the subprocesses. Were the compensation temperature far from the operating temperature the system would be sluggish because of the complex pattern of thermal-energy supplementation which would be required in function. Nevertheless, the requirement is not a strict one. In enzymes the effective heat-capacity changes in functional events are thought to be small but better data may present some surprises. In the hemoglobins the compensation temperatures of the liganded and unliganded forms are very similar. It is to be noted, however, that as experimental temperatures deviated from T_c the observed free-energy behavior becomes increasingly complex suggesting deviation from most efficient operation. Compensation behavior of enzymes is more difficult to understand primarily because of the high T_c values associated with transition-state formation.

The hemoglobin machine is a first-order linear-response machine or very nearly so. Machine operation consists of fluctuations which occur with high probability in the protein at "rest". The mechanism arises from the noise described by the fluctuation spectrum but the noise is not necessarily chaotically correlated since the tailoring of the spectrum in evolution may limit major stochastic events to those which effect the physiological process.

Most proteins with only equilibrium physiological functions are likely to be found to be members of the "linear-response" class. Those subprocesses or collections of coupled subprocesses in enzymes with T_c values near operating temperatures are similar manifestations of linear-response behavior. Those with high T_c values may or may not be. It has been noted that T_c values above 430K appear to be associated with matrices. This identification can also be made for transition-state formation in enzymic rack mechanisms but only provisionally since there are complications from "transition-state stabilization" and from normal thermal contributions to transition-state formation which may make the T_c values found in transition-state formation misleadingly high.

It has not been possible to consider the fluctuations of protein conformations and associated water in detail in this chapter. In particular there is no space to consider the possible role of such fluctuations in the association of proteins with other proteins, other kinds of macromolecules, and surfaces such as those of cells and plasma membranes. This matter has been discussed in a preliminary way elsewhere.[150] The fundamental idea is that the fluctuation spectra associated with the partition functions for proteins must be closely matched for proteins to associate strongly. Although charge-charge interactions, which have a long-range potential, and water displacement and rearrangement at contacting surface must be important, the potential functions for hydrogen bonds and dispersion interactions are short range. Unless the fluctuation spectra are well matched, the fraction of total time in which the latter interactions persist is small. If matching is poor, then proteins can associate only if the free-energy required to promote them into new states in which matching over the contact surfaces is adequate must be smaller than the free energy of association in the promoted states. Thus, in hemoglobin free of ligands the matching is good and the association strong. Addition of the first oxygen molecule upsets this balance and requires promotion which results in a large decrease in association strength. We have proposed that the free-energy redistribution responsible for linkage among the ligand sites occurs in the promotion process. Increasingly it appears to us improbable that any other explanation for protein associations can explain the high efficiency found in immune processes, the association of subunits of deoxyhemoglobin, etc.

It is unlikely that all the ideas advanced in this chapter will be found to be correct. They are the best explanations we have found for a large number of enzyme data which are poorly if at all explicable on the basis of the most currently fashionable hypotheses of mechanism. A few of the topics are well-established, though as yet little used, tools of considerable power in the study of proteins. These we have tried to develop toward the stage necessary for everyday use. Compensation behavior in particular can greatly simplify the analysis of data from linkage systems and together with feedback regulation it provides a simple but probably adequate foundation for theoretical considerations of nonlinear behavior arising from the coupling of the subprocesses of single enzymes and of enzyme systems.

The major goal in preparing this chapter was to present and defend the basic conformational hypothesis of enzyme mechanism as completely as is now possible for use in designing experiments, particularly site-specific-mutagenesis experiments. This undertaking ran downhill becoming little more than speculation when faced with the obviously important but little investigated participation of B hydration shells and bulk water, features of mechanism not likely to be efficiently explained by SSM studies alone. The hypothesis provides potential explanations for effects of matrix and knot alterations, but only qualitative explanations will be possible until the interactions of polypeptide conformations with their environments are much better understood.

ACKNOWLEDGMENTS

I am indebted to Drs. R. Gregory, R. Biltonen, G. Welch, T. Hopkins, D. F. Evans, C. Jolicoeur, H. Frank, and J. Rifkind for their suggestions and advice. Most of the newer ideas presented in this chapter arose as a result of collaborations with Drs. H. Frank, R. Gregory, C. Jolicoeur, E. Battistel, and D. F. Evans. They were elaborated into final forms while the author was Visiting Professor at the University of Granada, in 1985 under support provided by the Ministry of Education of the Spanish Government. I am deeply indebted to my hosts, Drs. M. Cortijo and P. Matero. Drs. T. Hopkins, J. Sturtevant, K. Gekko, and M. Matsumura provided unpublished data for which I am grateful as I am to Ms. J. Johnson and J. Alexander for their forbearance in typing the manuscript. The study of rack mechanisms has been supported in this laboratory by the Office of Naval Research, the National Science Foundation, the National Institutes of Health, the American Cancer Society, and the Graduate School of the University of Minnesota at various times over a period of 34 years. The preparation of this chapter was underwritten by the Graduate School of the University of Minnesota. This is publication number 223 from the Laboratory for Biophysical Chemistry (received February 1988).

ADDENDA (June 1989)

The last 2 years has produced a large number of references which are useful in evaluating the proposals in this paper but few which require significant revisions. We cannot revise the test to include many of the new references but some items and some references require comment. Reference 200 is primarily concerned with the fluctuational bases for protein-protein association, domain closure, etc. but also contains a summary of major topics in this chapter some of which have been updated.

None of the new references we have found resolve some of the major disagreements between the interpretations given in the text and the currently accepted interpretations. Thus, although there is new evidence favoring domain closure and relatively soft matrices, observations favoring our description of the events in enzymic catalysis, the relative importance of that mechanism, and the one based on transition-state stabilization remains obscure. On the other hand support for knot-matrix construction in single-knot and double-knot proteins is accumulating particularly rapidly in site-directed-mutagenesis studies and will soon be sufficient for a critical evaluation of this construction principle. A particularly instructive example are the results obtained by Lim and Sauer with a repressor protein.[201] In a single study the authors were able to determine which of the residues in the ''hydrophobic cluster'' actually were in the knot and by substitutions for several of these residues to demonstrate in considerable detail the delicacy of residue choice necessary to maintain the high packing density necessary to preserve a given kind of knot. That such an extensive collection of information about a knot can be produced in a single study suggests that the mutagenesis methodology when applied efficiently may be the quickest as well as the most reliable tool for describing knots.

The relative contributions to knots from their hydrogen-bonded systems and from the dispersion interactions also remains uncertain. Gregory and Lumry[29b] originally suggested that the high density of knots primarily through inductive electron-density displacements induces a special kind of electronic cooperativity in their hydrogen-bonded system and thus added additional strength. Electrostatic cooperativity is a major factor in knots with high helix content when a large fraction of knot surface is exposed to solvent.[47] Such knots might be called ''polar knots'' to contrast them with the ''dispersion-interaction'' knots in which all or most of the sidechains are aliphatic or aromatic, but, of course, the helices are at least partially embedded in the remainder of the domain knots and thus also depend to some extent on dispersion interactions. One knot of ribonuclease A is of the polar type and the helix-forming residues when isolated in a single peptide, the C peptide, retain a weak helix-forming ability at least at lower temperatures.[208] The latter observation suggests that the synergism of charged groups, helix dipolar groups, and water dominates the dispersion interactions of the helix in the intact protein. On the other hand in some knots there are long or broken hydrogen bonds suggesting that the hydrogen-bonded system has less than maximum strength because it is dominated by the interactions among sidechains. The electronic-cooperativity proposal has not been tested and the incomplete hydrogen bonding suggests that it is not important. However, there is no reason why one type of stabilizing device need always dominate the other.

It is becoming increasingly difficult to avoid the conclusion that in knot-matrix proteins the knots are of predominate importance in determining thermodynamic stability in folding. This has interesting rate implications since it implies that matrices are not usually in their most stable states and may sometimes make unfavorable contributions to folding. It also suggests that all the states of denaturation past knot disruption can have about the same free energy with the result that the detailed description of the ''denatured'' state is very sensitive to perturbations by solvent variables, pH, and temperature.

Because enzymic function involved depends on interdomain motility, the coupling between domains cannot be so large that the two-state model for denaturation is more than a first approximation to the true model. Brandts and co-workers have estimated the deviations to be expected.[207] One implication of their findings is that experimental conditions can usually be found which allow denaturation of single domains in two-domain proteins even when the domains have nearly equal thermodynamic stability under normal conditions. Several examples of this behavior have already been reported. Only in rare cases will both domains have equal stability so that thermodynamic quantities calculated using the two-state approximation uncorrected for coupling are a more ambiguous basis for comparing mutants, solvent conditions, etc. than has been supposed.

Two knots of HEW lysozyme are coded by a single exon. This suggests that there is only one knot in this protein. However, two domains are formed one using one part of the total set of residues and one the other. Domain closure is well established for this protein so the coding is misleading. There are two knots. They probably arose independently in evolution and their separate exons have become accidentally fused.

The work of Klibanov and co-workers on the behavior of damp and dry proteins in organic solvents was given considerably less attention in the text than it deserves. This work confirms the findings by Rupley and Careri and their respective co-workers on the importance of the A hydration shell in establishing catalytic activity and normal motility in enzymes and greatly expands our knowledge of the role of solvent in catalysis. Of particular importance is the inescapable conclusion primarily enforced by the work of Zaks and Klibanov[204] that a considerable fraction of the A′ hydration shell is dissolved in the protein rather than situated at the protein-water interface. The dissolved water apparently is an essential plastcizer. In view of such behavior as proton exchange from folded states which depends on the diffusion through the conformation of catalysts containing water or on water itself it is not surprising that proteins dissolve water. Several other types of evidence for the solution of water in proteins have been discussed in the text but none of these has previously been interpreted to indicate that the amounts of dissolved water can be quite large, i.e., a large fraction of the apparent A′ shell. The amount must depend on the relative polarity of the interior of the protein and the average integrity of folding. The plasticizing water may be entirely responsible for the development of normal solution dynamics and function but it seems unlikely and especially unlikely for proteins with a partially exposed knot.

Cytochrome c has a high helix content in a small protein. The oxidized form demonstrates in its magnetic susceptibility both a continuous change as it becomes ''normalized'' to its value in the functional protein by water and a cooperative process occurring at 0.18 g of water per gram of protein.[2b] In the reduced form the cooperative process is absent. For both there was a large amount of hysteresis as would be expected if the normalization depends on dissolved water acting as a plasticizer. It is not surprising that proteins dried by lyophilization and by slow dehydration often demonstrate quite different behavior.

In the cytochrome c studies as in other early hydration studies and in Klibanov's recent studies of damp protein aggregates in organic solvents there is rarely evidence for an essential role of B hydration shells in the normalization process. These results may be deceptive since under conditions of complete hydration B-shells appear to have considerable quantitative importance. Furthermore Rupley et al.[33a] found the B-shell of HEW lysozyme to have properties distinctly different from those of bulk water. If this water is perturbed by the protein, the protein must be perturbed by the water. Although some of these observations may be due to the aggregated state of the proteins which obtains as low hydration levels, it is not unlikely that some of the effects of co-solvents on protein activity coefficients and functions will be traced ultimately to the B-shells. Similarly the effects of viscosity on ligand-protein recombination rates in myoglobin and hemoglobin and on enzymic rate parameters may be mediated by B-shells. A major fraction of the dependence of the effects of changes in solvent viscosity produced by additions of such co-solvents as glycerol must be attributed to thermodynamic changes produced by these additions as discussed in the text. The high catalytic activity of enzyme aggregates hydrated only to the extent that they have A-shells or significant A-shell development eliminates any essential role for viscosity and thus eliminates activation by momentum waves in these processes as essential in function. Nevertheless some viscosity-dependent effects in fully dissolved proteins appear to be established and those for enzymic processes may reflect supplementary activation by PV fluctuations in the manner proposed by Gavish and others.[8] We suspect, however, that the correct explanation is the perturbation of the conformational fluctuation spectrum by solvent fluctuations mediated through the B-shell.

As noted in the footnotes for Table 1 and discussed in Reference 150 there is evidence that myoglobins lack major knots and as a result must depend on some construction principle other than the knot-matrix principle. The alternative principle was proposed to be the strong, long-range electrostatic interactions among the helices. This is not a new idea and has been explored using somewhat different electrostatic models without as yet any unambiguous conclusions. The principle becomes very attractive if the knot-matrix construction principle has to be excluded. The high variability of hemoglobin preparations made attempts in our laboratory to detect knots in hemoglobin using proton exchange fruitless, so for the present the only evidence against knots in this protein is the absence of significant knots in myoglobin. Our rough considerations of the structural differences among hemoglobin and of myoglobin mutants suggest that the most important feature of folding preserved in the evolution of the myoglobin fold is the distance and angle relationships among the helices rather than any major knots.

The functional mechanism of hemoglobin remains a mystery perhaps even more of a mystery than has been appreciated in the past since Doyle, Cera, and Gill[203] have established that three of the isosbestic points appearing in the visible spectrum during the spectrophotometric determination of oxygen-binding isotherms are sharp within the instrumental errors of the highest-performance commercial spectrophotometer. The authors have rationalized the slightly less sharp fourth isosbestic point terms of intrinsic spectral differences between α and β subunits. This is an old story, one basis for the "two-state" controversy which still rages, but it becomes one of central importance now that the sharpness has been established. A full discussion of the implications of the sharp isosbestic points cannot be given here but its major peculiarity rests on the fact that many different conformational states and substates have been established for myoglobin and hemoglobin with probabilities of occurrence dependent on degree of oxygenation. Furthermore, easily detected differences in the positions of the absorption peaks responsible for the isosbestic points are found in comparing hemoglobins from different species, Barcroft's "span".[205] In recent considerations of the basis of linear enthalpy-entropy compensation behavior and its associate linear-free-energy behavior Gregory and Lumry[202] have concluded that these phenomena reflect a very simple potential-function behavior, that given by the mean-field approximation. The linearity of the compensation plots establishes it a very good approximation. These authors propose that the best explanation for the adequacy of the approximation is that the potentials involved in the conformational processes which participate in function are averaged by fluctuations. It is possible that the sharp isosbestic points are somehow a consequence of such fluctuational averaging but it may also be true that there is some other characteristic of the conformations of these proteins which imposes restrictions which average the interactions between heme group and conformation so that the spectra reflect only two electronic states. Even though the stress at the heme group in equilibrium states is small, it is very surprising that the porphyrin, which is primarily responsible for the large bands in the visible spectrum, does not reflect in its spectra these small stresses and their changes. The absence of major spectral changes resulting from the even larger changes in heme geometry known to occur in more major changes of state is remarkable.

The fluctuational basis for the redistribution of free energy in linkage systems mentioned in the text is discussed in some detail in Reference 200.

As shown in the text, the compensation plots from the study by Timasheff and Gekko on the effect of glycerol on the thermal denaturation of chymotrypsinogen A (Figure 15) imply that effect of glycerol on protein activity coefficients should be the same for native and denatured forms at about 302K. This suggests that at lower temperatures additions of glycerol should raise the activity coefficient of the native form more rapidly than that of the denatured form and thus destabilize the native form. Although we have not found any specific study of this matter for chymotrypsinogen since the text was written, there is an abundance of evidence to suggest that glycerol is a stabilizer at low temperatures. This implies either a new effect of glycerol at lower temperatures or, more reasonably, that glycerol at such temperatures decreases the activities of both native and denatured forms. This is consistent with the effects of many co-solvents on small-molecule reactions in aqueous media and appears to be due to change in relative proportions of two macrostates of water. Benson and Krause have now provided very strong evidence that these states exist and that in pure water the transition temperature at which the standard chemical potentials of the water in the two states are equal is 287K.[206] This behavior of water must be taken into account in interpreting the effects of other co-solvents on denaturation processes. For full resolution of this complication it is necessary to have data for the activity coefficient of water in solutions of varying co-solvent composition at temperatures both higher and lower than 287K. To our considerable surprise no accurate data of this kind appears to have been reported for glycerol and ethylene glycol.

REFERENCES

In the following an attempt has often been made to group a few articles which complement or supplement each other under a single group number even though no more than one reference in the group may be directly required at a position in the test.

1a. **Lumry, R. and Eyring, H.,** Conformation changes of proteins, *J. Phys. Chem.,* 58, 110, 1954.

1b. **Eyring, H., Lumry, R., and Spikes, J. D.,** Kinetic and thermodynamic aspects of enzyme-catalyzed reactions, in *Mechanisms of Enzyme Action,* McElroy, W. and Glass, G., Eds., Johns Hopkins University Press, Baltimore, 1954, 123.

1c. **Lumry, R.,** The determination of electronic properties by protein structure in heme-proteins, *Biophysics,* 1, 3, 1961.

1d. **Lumry, R.,** Substrate control of protein conformation, *Abhandl. Deut. Akad. Wiss. Berlin. Kl. Chem. Geol. Biol.,* 6, 125, 1964.

1e. **Lumry, R.,** Some aspects of the thermodynamics and mechanism of enzymic catalysis, *Enzymes,* 1, 157, 1959.

2a. **Lumry, R. and Biltonen, R.,** Thermodynamic and kinetic aspects of protein conformations in relation to physiological function, in *Structure and Stability of Biological Macromolecules,* Timasheff, S. and Fasman, G., Eds., Marcel Dekker, New York, 1969, chap. 2.

2b. **Lumry, R., Solbakken, A., Sullivan, J., and Reyerson, L.,** Studies of rack mechanisms in heme proteins. I. The magnetic susceptibility of cytochrome c in relation to hydration, *J. Am. Chem. Soc.,* 84, 142, 1962.

2c. **Lumry, R.,** Fundamental problems in the physical chemistry of protein behavior, in *Electron and Coupled Energy Transfer in Biological Systems,* King, T. and Klingenberg, M., Eds., Marcel Dekker, New York, 1971, chap. 1.

3. **Lumry, R.,** *Electron and Coupled Energy Transfer in Biological Systems,* King, T. and Lingenberg, M., Eds., Marcel Dekker, New York, 1971, 78.

4. **Koshland, Jr., D.,** Application of a theory of enzyme specificity to protein synthesis, *Proc. Natl. Acad. Soc. U.S.A.,* 44, 1958, 98; *Cold Spring Harbor Symp. Quant. Biol.* 28, 1963, 473.

5a. **Wolfenden, R.,** Analog approaches to the structure of the transition state in enzyme reactions, *Accts. Chem. Res.,* 5, 10, 1972.

5b. **Wolfenden, R. and Frick, L.,** Mechanisms of enzyme action and inhibition: transition-state analogues for acid-base catalysis, *J. Protein Chem.,* 5, 147, 1986.

6a. **Lienhard, G.,** Enzymatic catalysis and transition-state theory, *Science,* 180, 149, 1973.

6b. **Leinhard, G., Secemski, I., Koehler, K., and Lindquist, R.,** Enzymatic catalysis and the transition-state theory of reaction rates: transition-state analogues, *Cold Spring Harbor Symp. Quant. Biol.,* 36, 45, 1972.

7a. **Jencks, W.,** *Adv. Enzymol.,* 51, 75, 1980.

7b. **Jencks, W.,** On the attribution and additivity of binding energies, *Proc. Natl. Acad. Sci.* U.S.A., 78, 1981, 4046.

8a. **Gavish, B.,** Molecular dynamics and the transient strain model of enzyme catalysis, in *The Fluctuating Enzyme,* Welch, G., Eds., Wiley-Interscience, New York, 1986, 263.

8b. **Somogyi, B., Welch, G. R., and Damjanovich, S.,** The dynamic basis of energy transduction in enzymes, *Biochem. Biophys, Acta.,* 768, 81, 1984.

8c. **Doster, W.,** Viscosity scaling and protein dynamics, *Biophys. Chem.,* 17, 97, 1983.

8d. **Beece, D., Eisenstein, L., Frauenfelder, H., Good, D., Marden, M., Reinisch, L., Reynold, A., Sorenson, L., and Yue, K.,** Solvent viscosity and protein dynamics, *Biochemistry,* 19, 5147, 1980.

8e. **Gavish, B. and Weber, M. M.,** Viscosity-dependent structural fluctuations in enzyme catalysis, *Biochemistry,* 18, 1269, 1979.

9a. **Ansari, A., Berendzen, J., Braunstein, D., Cowen, B. R., and Frauenfelder, H.,** Rebinding and relaxation in the myoglobin pocket, *Biophys. Chem.,* 26, 337, 1987.

9b. **Frauenfelder, H., Petsko, G., and Tsernoglu, D.,** Temperature-dependent x-ray diffraction as a probe of protein structural dynamics, *Nature,* 280, 558, 1979.

9c. **Frauenfelder, H.,** Ligand binding and protein dynamics, in *Structure and Motion: Membranes, Nucleic Acids and Proteins,* Lementi, E., Corongiu, G., Sarma, M., and Sarma, R., Eds., Adenine Press, 1985, 204.

9d. **Kanehisa, M. and Ikegami, A.,** Structural changes and fluctuations of proteins. II. Analysis of the denaturation of globular proteins, *Biophys. Chem.,* 6, 131, 1977.

9e. **Smith, J., Hendrickson, W., Honzatko, R., and Sheriff, S.,** Structural heterogeneity in protein crystals, *Biochemistry,* 25, 5018, 1986.

9f. **Knapp, E., Fischer, S., and Parak, F.,** The influence of protein dynamics on mössbauer spectra, *J. Chem. Phys.,* 78, 4701, 1983.

10. **Tulinksy, A.,** in *Biomolecular Structure, Conformation, Function and Evolution,* Vol 1, Srinivasan, R., Ed., Pergamon Press, Elmsford, NY, 1980, 183.

11a. **Lumry, R.,** The determination of electronic properties by protein structure in heme-proteins, *Biophysics,* 1, 3, 1961.

11b. **Lumry, R. and Biltonen, R.,** Thermodynamic and kinetic aspects of protein conformations in relation to physiological functions, in *Structure and Stability of Biological Macromolecules,* Timasheff, S. and Fasman, G., Eds., Marcel Dekker, New York, 1969, chap. 2.

11c. **Lumry, R.,** Conformational mechanisms for free-energy transduction in protein systems, in *Ann. NY Acad. Sci.,* 227, 46, 1974.

11d. **Lumry, R.,** Some recent developments in the search for mechanisms of enzymic catalysis, in *Enzymology in the Practice of Laboratory Medicine,* Freier, E. and Blume, P., Eds., Academic Press, New York, 1974, 3.

11e. **Lumry, R.,** Interpretation of calorimeter data from cooperative systems, in *Bioenergetics and Thermodynamics: Model Systems,* Braibanti, A., Ed., D. Reidel, Dordrecht, Netherlands, 1980, 405.

11f. **Lumry, R. and Gregory, R.,** Free-energy management in protein reactions: concepts, complications, and compensation, in *The Fluctuating Enzyme,* Welch and G. Rickey, Ed., John Wiley & Sons, New York, 1986, 116.

12a. **Biltonen, R. and Lumry, R.,** Variability of state A for some members of the chymotrypsinogen family of proteins, *J. Am. Chem. Soc.,* 91, 4251, 4256, 1969; 92, 1970.

12b. **Biltonen, R., Lumry, R., Madison, V., and Parker, H.,** The conversion of chymotrypsinogen A to α-chymotrypsin, *Proc. Natl. Acad. Sci. U.S.A.,* 54, 1412, 1965.

12c. **Brandts, J. F. and Lumry, R.,** The reversible thermal denaturation of chymotrypsinogen, *J. Phys. Chem.,* 67, 1484, 1963.

12d. **Parker, H. and Kim, Y. D.,** *Biophysics,* 1, 3, 1961.

13a. **Parker, H. and Lumry, R.,** Substrate control of conformation characteristics in chymotrypsin, *J. Am. Chem. Soc.,* 85, 483, 1962.

13b. **Moon, A., Mercouroff, J., and Hess, G.,** Characterization of a reversibly formed enzyme complex in the reaction of chymotrypsin with diisopropyl fluorophosphate, *J. Biol. Chem.,* 240, 717, 1965.

13c. **Parker, H.,** Effects of Substrates and Inhibitors on the Optical Rotatory Dispersion and Ultra-Violet Absorption of α-Chymotrypsin in State A, Dissertation, University of Minnesota, Minneapolis, 1967.

13d. **Moon, A., Mercouroff, J., and Hess, G.,** Characterization of a reversibly formed enzyme complex in the reaction of chymotrypsin with diisopropyl fluorophosphate, *J. Biol. Chem.,* 240, 717, 1965.

13e. **Cane, W.,** Studies of Homologous Acyl-Chymotrypsins, Dissertation, University of Minnesota, Minneapolis, 1966.

13f. **Havsteen, B. and Hess, G.,** Chymotrypsin-substrate complexes. Hydrogen ion equilibria in chymotrypsin and diisopropylphosphoryl-chymotrypsin, *Biochem. Biophys. Res. Commun.,* 14, 313, 1964.

13g. **Moon, A., Mercouroff, J., and Hess, G.,** *J. Biol. Chem.,* 240, 717, 1965.

13h. **Havsteen, H. and Hess, G. P.,** Evidence for conformation changes in α-chymotrypsin-catalyzed reactions. VI. Changes in optical rotatory dispersion, *J. Am. Chem. Soc.,* 85, 791, 1963.

14a. **Matthews, B., Sigler, P., Henderson, R., and Blow, D.,** Three-dimensional structure of tosyl-α-chymotrypsin, *Nature,* 214, 652, 1967.

15. **Welch, G. R., Ed.,** *The Fluctuating Enzyme,* Wiley-Interscience, New York, 1986.

16. **Haldan, J.,** *Enzymes,* 1930, MIT Press, Cambridge, 1965.

17. **Albery, W. and Knowles, J.,** Free-energy profile for the reaction catalyzed by triosephosphate, *Biochemistry,* 15, 5627, 5631, 1965.

18. **Pauling, L.,** Molecular architecture and biological reactions, *Chem. Eng. News,* 24, 1375, 1946.

19. **Rashin, A., Iofin, M., and Honig, B.,** Internal cavities and buried waters in globular proteins, *Biochemistry,* 25, 3619, 1986.

20. **Kundrot, C. E. and Richards, F.,** Crystal structure of hen egg-white lysozyme at a hydrostatic pressure of 1000 atmospheres, *J. Mol. Biol.,* 193, 157, 1987.

21a. **Chryssomallis, C., Drickamer, G., and Weber, G.,** The measurement of fluorescence polarization at high pressures, *J. Appl Phys.,* 49, 3084, 1978.

21b. **Chryssomallis, G., Rogergson, P., Drickamer, H., and Weber, G.,** Effect of hydrostatic pressure upon lysozyme and chymotrypsinogen detected by fluorescence polarization, *Biochemistry,* 20, 3955, 1981.

21c. **Nystrom, B. and Roots, J.,** Proteins in mixed solvents. Study of pressure-induced denaturation of bovine serum albumin by photon correlation spectroscopy, *Chem. Phys. Lett.,* 91, 236, 1982.

22. **Mishra, A. and Klapper, M.,** Kinetic isotope effects associated with α-chymotrypsin deacylation: evidence for reaction mechanism plasticity, *Biochemistry,* 25, 7328, 1986.

23. **Stein, R. L., Elrod, J. P., and Schowen, R. L.,** Correlative variations in enzyme-derived and substrate-derived structures of catalytic transition states. Implications for the catalytic strategy of acyl-transfer enzymes, *J. Am. Chem. Soc.,* 105, 2446, 1983.

24. **Koshland, D. and Neet, K.,** The catalytic and regulatory properties of enzymes, *Annu. Rev. Biochem.,* 37, 359, 1968.

25a. **Inagami, T. and Murachi, T.,** *J. Biol. Chem.,* 239, 1395, 1964.

25b. **Bender, M. and Kezdy, F.,** Mechanism of action of proteolytic enzymes, *Annu. Rev. Biochem.,* 34, 49, 1965.

25c. **Seydoux, F., Coutouly, G., and Yon, J.,** Amines as modifiers of the tryptic hydrolysis of neutral substrates, *Biochemistry,* 10, 2284, 1971.

26. **Lumry, R. and Biltonen, R.,** *Structure and Stability of Biological Macromolecules,* Timasheff, S. and Fasman, G., Eds., Marcel Dekker, New York, 1969, 156.

27a. **Genberg, L., Heisel, F., McLendon, G., and Miller, R.,** Vibrational energy relaxation processes in heme proteins: model systems of vibrational energy dispersion in disordered systems, *J. Phys. Chem.,* 91, 5521, 1987.

27b. **Petrich, J., Martin, J., Houde, D., Poyart, C., and Orszag, A.,** Time-resolved raman spectroscopy with subpicosecond resolution: vibrational cooling and delocalization of strain energy in photodissociated (carbonmonoxy)hemoglobin. *Biochemistry,* 26, 7914, 1987.

28. **Provencher, S.,** A constrained regularization method for inverting data represented by linear algebraic or integral equations, *Compt. Phys. Commun.,* 27, 213, 1982.

29a. **Gregory, R.,** A comparison of analytically and numerically derived hydrogen-exchange rate distribution functions, *Biopolymers,* 22, 896, 1983.

29b. **Gregory, R. and Lumry, R.,** Hydrogen-exchange evidence for distinct structural classes in globular proteins, *Biopolymers,* 24, 301, 1985.

29c. **Lumry, R. and Gregory, R.,** A "knots" and "matrices" model for globular proteins, *Biophys, J.,* 45, 259a, 1984.

30. **Makarov, A., Mgeladze, G., Monaselidze, D., and Esipova, N.,** Manifestations of intraglobular dynamics: microcalorimetry of leghemoglobin crystals, *J. Polymer Sci.: Polymer Symp.,* 69, 101, 1981.

31a. **Knox, D., Lumry, R., and Rosenberg, A.,** The use of the hydrogen-exchange method in the characterization of conformational fluctuations, *Biophys. J.,* 188, 1979.

31b. **Knox, D. and Rosenberg, A.,** *Biopolymers,* 19, 1049, 1980.

32. **Woodward, C. and Rosenberg, A.,** Studies of hydrogen exchange in proteins, VI. The correlation of ribonuclease exchange kinetics with the temperature-induced transition, *J. Biol. Chem.,* 246, 4105, 1971.

33a. **Shinkel, J., Downer, N., and Rupley, J.,** Hydrogen exchange of lysozyme powders. Hydration dependence of internal motions, *Biochemistry,* 24, 352, 1985.

33b. **Gregory, R.,** The influence of glycerol on hydrogen isotope exchange in lysozyme, *Biopolymers,* submitted.

33c. **Hilton, B., Trudeau, K., and Woodward, C.,** Hydrogen exchange rates in pancreatic tryptin inhibitor are not correlated to thermal stability in urea, *Biochemistry,* 20, 4697, 1981.

33d. **Woodward, C. and Hilton, B.,** Hydrogen isotope exchange kinetics of single protons in bovine pancreatic trypsin inhibitor, *Biophys. J.,* 32, 561, 1980.

33e. **Eftink, M. and Ghiron, C.,** Fluoresence quenching of the buried tryptophan residue of cod parvalbumin, *Biophys. Chem.,* 32, 173, 1985.

33f. **Delepierre, M., Dobson, C., Karplus, M., Poulson, F., States, D., and Wedlin, R.,** Electrostatic effects and hydrogen-exchange behavior in proteins. The pH dependence of exchange rates in lysozyme, *J. Mol. Biol.,* 197, 111, 1987.

34a. **Kossiakoff, A.,** Protein dynamics investigated by the neutron diffraction hydrogen-exchange technique, *Nature,* 296, 713, 1982.

34b. **Kossiakoff, A.,** Neutron protein crystallography: advances in methods and applications, *Annu. Rev. Biophys. Bioeng.,* 12, 159, 1983.

34c. **Kossiakoff, A.,** Use of neutron diffraction - H/D exchange technique to determine the conformational dynamics of trypsin, in *Neutrons in Biology,* Schoernborn, B. P., Ed., Plenum Press, New York, 1984, 281.

35a. **Huang, T., Bachovchin, W., Griffin, R., and Dobson, C.,** High-resolution nitrogen-15 nuclear magnetic resonance studies of α-lytic protease in solid state, *Biochemistry,* 23, 5933, 1984.

35b. **Huber, T. and Bennett, W. S., Jr.,** Functional significance of flexibility in proteins, *Biopolymers,* 22, 261, 1983.

36. **Richarz, R., Sehr, P., Wagner, G., and Wüthrich, K.,** Kinetics of the exchange of individual amide protons in the basic pancreatic trypsin inhibitor, *J. Mol. Biol.,* 130, 19, 1979.

37. **Welch, G. R., Ed.,** *The Fluctuating Enzyme,* Wiley-Interscience, New York, 1986, 33.

38. **Welch, G. R., Ed.,** *The Fluctuating Enzyme,* Wiley-Interscience, New York, 1986, 36.

39a. **Welch, G. R., Ed.,** *The Fluctuating Enzyme,* Wiley-Interscience, New York, 1986, 15.

39b. **Rogers, J.,** Exon shuffling and intron insertion in serine protease genes, *Nature,* 315, 458, 1985.

40a. **Lumry, R. and Biltonen, R.,** *Structure and Stability of Biological Macromolecules,* Timasheff, S. and Fasman, G., Eds., Marcel Dekker, New York, 1969, 159.

40b. **Lumry, R.,** *Electron and Coupled Energy Transfer in Biological Systems,* King, T. and Klingenberg, M., Eds., Marcel Dekker, New York, 1971, chap. 1.

41a. **Lumry, R. and Rosenberg, A.,** The water basis for mobile defects in proteins and the role of these defects in function, *Colloqu. Int. C.N.R.S.,* 246, 53, 1975.

41b. **Richards, F.,** Packing defects, cavities, volume fluctuations and access to the interior of proteins, *Carlberg Res. Commun.,* 44, 47, 1979.

41c. **Lumry, R.,** Storage and migration of free energy via bonding defects in protein conformation. Importance in function of normally variable and abnormal hemoglobins, *Proc. 1st Natl. Symp. on Sickle-Cell Disease,* DHEW Publ. NIH75, 1974, 165.

41d. **Tilton, R., Jr., Singh, U., Kuntz, I., Jr., and Kollman, P.,** Protein-ligand dynamics, *J. Mol. Biol.,* 199, 195, 1988.

42. **Welch, G. R., Ed.,** *The Fluctuating Enzyme,* Wiley-Interscience, New York, 1986, 45.

43a. **Corbett, R. and Roche, R.,** Use of high-speed size-exclusion chromatography for the study of protein folding and stability, *Biochemistry,* 23, 1888, 1984.

43b. **Wang, C., Cook, K., and Pecora, R.,** Dynamic light scattering studies of ribonuclease A, *Biophys. Chem.,* 11, 439, 1980.

43c. **Amir, D. and Haas, E.,** BPTI is compact when the disulfide bonds are reduced, *Biophys. J.,* 51, 556a, 1987.

43d. **Sullivan, J.,** personal communication, 1987.

43e. **Nicoli, D. and Benedek, G.,** Study of thermal denaturation of lysozyme and other globular proteins by light-scattering spectroscopy, *Biopolymers,* 15, 2421, 1976.
44a. **Robey, E. and Schachman, H.,** Regeneration of active enzyme by formation of hybrids from inactive derivatives: Implications for active sites shared between polypeptide chains of aspartate transcarbamylase, *Proc. Natl. Acad. Sci. U.S.A.,* 82, 361, 1985.
44b. **Wente, S. and Schachman, H.,** Shared active sties in oligomeric enzymes: model studies with defective mutants of aspartate transcarbamoylase produced by site-directed mutagenesis, *Proc. Natl. Acad. Sci. U.S.A.,* 84, 31, 1987.
44c. **Krause, K., Volz, K., and Lipscomb, W.,** 2.5 Å Structure of aspartate carbamoyltransferase complexed with the bisubstrate analog *N*-(phosphonacetyl)-L-aspartate, *J. Mol. Biol.,* 193, 527, 1987.
44d. **Blomberg, S. and Allewell, N.,** Thermodynamic and kinetic linkages between ligand binding, folding and assembly of the catalytic subunit of *E. coli* aspartate transcarbamoylase, *Biophys. J.,* 53, 451a, 1988.
45. **Hearn, R., Richards, F., Sturtevant, J., and Watt, G.,** Thermodynamics of the binding of S-peptide to S-protein to form ribonuclease A. *Biochemistry,* 10, 806, 1971.
46a. **Brown, J. E. and Klee, W. A.,** Helix-coil transition of the isolated amino terminus of ribonuclease, *Biochemistry,* 10, 470, 1971.
47a. **Sundaralingam, M., Sekharudu, Y., Yathindra, N., and Ravichandran, V.,** *Protein: Structure, Function, Genetics,* 2, 64, 1987.
47b. **Sundaralingam, M., Sekharudu, Y., Yathindra, N., and Ravichandran, V.,** *Intern. J. Quant. Chem.: Quant. Biol. Symp.,* 14, 289, 1987.
47c. **Creighton, T. E.,** Stability of alpha-helices, *Nature,* 386, 547, 1987.
47d. **Shoemaker, K. R., Kim, P. S., York, E. J., Stewart, J. M., and Baldwin, R. L.,** Tests of the helix dipole model for stabilization of α-helices, *Nature,* 326, 563, 1987.
48. **Rico, M., Santoro, J., Bermejo, F. J., Herranz, J., Mieto, J. L., Gallego, E., and Jimenez, M. A.,** Thermodynamic parameters for the helix-coil thermal transition of ribonuclease-S peptide and derivatives from H-nmr data, *Biopolymers,* 25, 1031, 1986.
49a. **Brazhnikov. E. V., Chirgadze, Y. N., Dolgikh, D. A., and Ptitsyn, O. B.,** Noncooperative temperature melting of a globular protein without specific tertiary structure: acid form of bovine carbonic anhydrase b, *Biopolymers,* 24, 1899, 1985.
49b. **Ohgushi, M. and Wada, A.,** Molten-globule state: a compact form of globular proteins with mobile side-chains, *FEBS Lett.,* 164, 21, 1983.
50a. **Roy, S., Dibello, C., and Taniuchi, H.,** Extremely fast hydrogen exchange of ribonuclease-(1-118) as compared with native ribonuclease A and its implication for the conformational energy state, *Int. J. Peptide Protein Res.,* 27, 165, 1986.
50b. **Lin, M. C.,** The structural roles of amino acid residues near the carboxyl terminus of bovine pancreatic ribonuclease a, *J. Biol. Chem.,* 245, 6726, 1970.
50c. **Gutte, B., Lin, M., Caldi, D., and Merrifield, R.,** Reactivation of des(119-120, or 121-124) ribonuclease A by mixture with synthetic COOH-terminal peptides of varying length, *J. Biol. Chem.,* 247, 4763, 1972.
50d. **Andria, G. and Taniuchi, H.,** The complementing fragment-dependent renaturation by enzyme-catalyzed disulfide interchange of RNase-(1-118) containing non-native disulfide bonds, *J. Biol. Chem.,* 253, 2262, 1978.
50e. **Kuwajima, K., Hiraoka, Y., Ikeguchi, M., and Sugai, S.,** Comparison of the transient folding intermediates in lysozyme and α-lactalbumin, *Biochemistry,* 24, 874, 1985.
50f. **Pfeil, W. and Adowski, M.,** A scanning calorimetric study of bovine and human apo-alpha-lactalbumin, *Studia Biophysica,* 109, 163, 1983.
51. **Wlodawer, A., Bott, R., and Sjolin, L.,** Proton-exchange rates in ribonuclease A, *Proc. Natl. Acad. Sci. U.S.A.,* 79, 1418, 1982.
52a. **Permyakov, E. A., Yarmolenko, V. V., Kalinichenko, L. P., Morozova, L. A., and Burstein, E. A.,** Calcium binding to α-lactalbumin: structural rearrangement and association constant by means of intrinsic protein fluorescence changes, *Biochem. Biophys. Res. Commun.,* 100, 191, 1981.
52b. **Segawa, T. and Sugai, S.,** Interactions of covalent metal ions with bovine, human, and goat α-lactalbumin, *J. Biochem.,* 93, 1321, 1983.
52c. **Dolgikh, D. A., Gilmanshin, R. I., Branzhnikov, E. V., Bychkova, V. E., Semisotnov, G. V., Venyaminov, S. Y., and Ptitsyn, O. B.,** α-Lactalbumin: compact state with fluctuating tertiary structure?, *Fed. Eur. Biochem. Soc.,* 136, 311, 1981.
52d. **Ikeguchi, M., Kuwajima, K., and Sugai, S.,** Ca^{2+} induces alteration in the unfolding behavior of α-lactalbumin, *J. Biochem.,* 99, 1191, 1986.
52e. **Ikeguchi, M., Kuwajima, K., Mitani, M., and Sugai, S.,** Evidence for identity between the equilibrium unfolding intermediate and a transient intermediate: a comparative study of the folding reactions of α-lactalbumin and lysozyme, *Biochemistry,* 25, 6965, 1986.
53. **Murakami, K., Andree, P., and Berliner, L.,** Metal-ion binding to α-lactalbumin species, *Biochemistry,* 21, 5488, 1982.

54a. **Matsumura, M., Yasumura, S., and Aiba, S.,** Cumulative effect of intragenic amino-acid replacements on the thermostability of a protein, *Nature,* 323, 356, 1986.
54b. **Matsumura, M.,** personal communication, 1987.
55a. **Warshel, A. and Russell, S.,** Theoretical correlation of structure and nergetics in the catalytic reactions of trypsin, *J. Am. Chem. Soc.,* 108, 659, 1986.
55b. **Eyring, H. and Stearn, A.,** Application of the theory of absolute reaction rates to proteins, *Chem. Rev.,* 24, 253, 1939.
55c. **Stearn, A .,** The application of quantum mechanics to certain cases of homogeneous catalysis. II. Certain aspects of enzyme action, *J. Gen Physiol.,* 18, 301, 1935; Kinetics of biological reactions with special reference to enzymic processes, *Adv. Enzymology,* 9, 25, 1949.
56. **Welch, G. R., Ed.,** *The Fluctuating Enzyme,* Wiley-Interscience, New York, 1986, 33.
57a. **Vita, C., Dalzoppa, D., and Fontana, A.,** Independent folding of the carboxyl-terminal fragment 228-316 of thermolysin, *Biochemistry,* 23, 5512, 1984.
57b. **Vita, C., Dalzoppo, D., and Fontana, A.,** Folding of thermolysin fragments: identification of the minimum size of a carboxyl-terminal fragment that can fold into a stable native-like structure, *J. Mol. Biol.,* 182, 331, 1985.
57c. **Vita, C., Fontana, A., and Chaiken, I. M.,** Folding of thermolysin fragments: correlation between conformational stability and antigenicity of carboxyl-terminal fragments, *Biochemistry,* 151, 191, 1985.
57d. **Dalzoppo, D., Vita, C., and Fontana, A.,** Domain characteristics of the carboxyl terminal fragment 206-316 of thermolysin: pH and ionic strength dependence of conformation.
57e. **Corbett, R. J. T. and Roche, R. S.,** The unfolding mechanism of thermolysin, *Biopolymers,* 22, 101, 1983.
58a. **Gregory, R. and Lumry, R.,** Association of slowly exchanging protons with catalytic functional groups, *Biophys. J.,* 49, 441a, 1986.
58b. **Welch, G. R., Ed.,** *The Fluctuating Enzyme,* Wiley-Interscience, New York, 1986, 45.
59a. **Robinson, W., Senanayake, P. and Freeman, G.,** Water clusters, *J. Phys. Chem.,* 91, 2123, 1987.
59b. **Shizuka, H., Ohiwara, T., Narita, A., Sumitani, M., and Yoshihara, K.,** NaCl effect on the excited-state proton-dissociation reaction of naphthols: water structure in the presence of NaCl, *J. Phys. Chem.,* 90, 6708, 1986.
59c. **Hameka, H., Robinson, G. W., and Mareden, C.,** Structure of the hydrated electron, *J. Phys. Chem.,* 91, 3150, 1987.
59d. **Robinson, G. W.,** in preparation.
59e. **Lee, J., Giffin, R., and Robinson, G. W.,** 2-Naphthol: a simple example of proton transfer effected by water structure, *J. Chem. Phys.,* 82, 4920, 1985.
59f. **Lee, J. and Robinson, G. W.,** Time-resolved studies of "salt effects" on weak acid dissociation, *J. Am. Chem. Soc.,* in press.
60. **Ragon, R., Colonna, G., Bismuto, E., and Irace, G.,** Unfolding pathway of myoglobin: effect of denaturants on solvent accessibility of tyrosyl residues, *Biochemistry,* 26, 2130, 1987.
61. **Leffler, J.,** Isokinetic relationships form linear-free-energy relationships, *J. Org. Chem.,* 20, 1202, 1955.
62. **Lumry, R. and Rajender, S.,** Enthalpy-entropy compensation phenomena in water solutions of proteins and small molecules a ubiquitous property of water, *Biopolymers,* 9, 1125, 1970.
63. **Lumry, R. and Rajender, S.,** *Biopolymers,* 9, 1206, 1970.
64a. **Gregory, R. and Lumry, R.,** Simplifications in analysis of linkage systems arising from their enthalpy-entropy compensation behavior, *Biophys. J.,* 49, 440a, 1986.
64b. **Lumry, R. and Gregory, R.,** to be published.
65a. **Kubo, R.,** Some aspects of the statistical mechanical theory of irreversible processes, lectures in *Theoretical Physics,* Vol. 1, Wiley-Interscience, New York, 1961, 120.
65b. **Richter, P. and Ross, J.,** The efficiency of engines operating around a steady state at finite frequencies, *J. Chem. Phys.,* 69, 5521, 1978.
65c. **Zwanzig, R.,** Time-correlation functions and transport coefficients in statistical mechanics, *Annu. Rev. Phys. Chem.,* 1667, 1965.
65d. **Dogonadze, R., Kuznetsov, A., and Ulstrup, J.,** Conformational dynamics in biological electron and atom-transfer reactions, *J. Theor. Biol.,* 69, 239, 1977.
65e. **Kosioff, R. and Ratner, M.,** Beyond linear response: line shapes for coupled spins or oscillators via direct calculation of dissipated power, *J. Chem. Phys.,* 80, 2352, 1984.
66. **Welch, G. R., Ed.,** *The Fluctuating Enzyme,* Wiley-Interscience, New York, 1986, 58.
67a. **Beetlestone, J., Adeosun, O., Goddard, J., Kushimo, J., Ogunlesi, M., Ogunmola, G., Okongo, K., and Seamonds, B.,** Reactivity differences between hemoglobins. XIX, *J. Chem. Soc.,* 1215, 1976.
67b. **Lumry, R. and Rajender, S.,** *Biopolymers,* 9, 1170, 1970.
67c. **Welch, G. R., Ed.,** *The Fluctuating Enzyme,* Wiley-Interscience, New York, 1986, 104.
68a. **Anusiem, A., Beetlestone, J., Kushimo, J., and Oshodi, A.,** Spectral evidence for two forms of acid ferrihemoglobin, *Arch. Biochem. Biophys.,* 175, 138, 1976.

68b. **Anusiem, A. C. I. and Oshodi, A. A.,** Hemoglobin, role of water in protein reactions: binding of azide ion to ferrihemoglobin in water and ethylene glycol mixtures, *Arch. Biochem. Biophys.,* 189, 392, 1978.

68c. **Anusiem, A. and Lumry, R.,** Calorimetric determination of azide-ion binding to ferrihemoglobin A in water and 5% butanol, *J. Am. Chem. Soc.,* 95, 904, 1973.

69a. **Blanck, J. and Scheler, W.,** Ligand exchange reactions of methemoglobins: mechanism and parameters of activation, *Acta Biol Med. Germ.,* 20, 721, 1968.

69b. **Kotani, M. and Morimoto, H.,** *Magnetic Resonance in Biological Systems,* Ehrenberg, A., Malmstrom, B., and Vanngard, T., Eds., Pergamon Press, Elmsford, NY, 1967, 135.

69c. **Gurd, F., Hoffman, B., Wang, M.-Y., and Shire, S.,** Oxygen binding to myoglobins and their cobalt analogs, *J. Am. Chem. Soc.,* 101, 7394, 1979.

69d. **Ascenzi, P., Brunori, M., and Giacometti, G.,** Thermodynamics of the reaction of ferric myoglobin from *Aplysia limacina* with azid and fluoride, *J. Mol. Biol.,* 182, 607, 1985.

70. **Welch, G. R., Ed.,** *The Fluctuating Enzyme,* Wiley-Interscience, New York, 1986, 59.

71a. **Imai, K.,** *Alosteric Effects in Haemoglobin,* Cambridge University Press, Cambridge, 1982, chap. 6.

71b. **Imai, K. and Yonetoni, T.,** Thermodynamical studies of oxygen equilibrium of hemoglobin. Nonuniform heats and entropy changes for the individual oxygenation steps and enthalpy-entropy compensation, *J. Biol. Chem.,* 250, 7093, 1975.

71c. **Ikeda-Saito, M., Yonetoni, T., Chiancone, E., Ascoli, F., Verzili, D., and Antonini, E.,** Thermodynamic properties of oxygen equilibria of dimeric and tetrameric hemoglobins form *Scapharca inaequivalvis, J. Mol. Biol.,* 170, 1009, 1983.

71d. **Stephanos, J. and Addision, A.,** Thermochromism of monomeric heme proteins, *J. Biol. Chem.,* in press.

72. **Maliwal, B. and Lakowicz, J.,** Effect of ligand binding and conformational changes in proteins on oxygen quenching and fluorescence depolarization of tryptophan residues, *Biophys. Chem.,* 19, 337, 1984.

72b. **Citri, N.,** Conformation changes of proteins, *Adv. Enzy.,* 37, 397, 1973.

73a. **Lumry, R. and Rajender, S.,** *Biopolymers,* 9, 1206, 1970.

73b. **Lumry, R.,** Protein conformations, "rack" mechanisms and water, *Adv. Chem. Phys.,* 21, 567, 1971.

74a. **Yapel, A.,** A Kinetics Study of the Imidazole Groups of Chymotrypsin, Dissertation, University of Minnesota, Minneapolis, 1967.

74b. **Shiao, D. and Sturtevant, J.,** Calorimetric investigations of the binding of inhibitors to α-chymotrypsin. I, *Biochemistry,* 8, 4910, 1969; **Shiao, D.,** *Biochemistry,* 9, 1083, 1970.

75a. **Dorovska-Taran, V., Momtcheva, R., Gulubova, N., and Martinek, K.,** The specificity in the elementary steps of α-chymotrypsin catalysis. A temperature study with a series of *n*-acetyl-L-amino acid methyl esters, *Biochim. Biophys. Acta,* 702, 37, 1982.

75b. **James, D.,** Specificity Determination in Chymotryptic Catalysis-Thermodynamic Basis, Dissertation, University of Minnesota, Minneapolis, 1981.

76. **Welch, G. R., Ed.,** *The Fluctuating Enzyme,* Wiley-Interscience, New York, 1986, 77.

77a. **Benzinger, T. H.,** in *Adolescent Nutrition and Growth,* Heald, F., Ed., Appleton-Century-Crofts, New York, 1969; *Nature,* 229, 100, 1971.

77b. **Benzinger, T.,** Thermodynamics, chemical reactions and molecular biology, *Nature,* 229, 100, 1971.

77c. **Benzinger, T. H. and Hammer, C.,** Unwinding the double helix; complete equation for chemical equilibrium, *Curr. Top. Cell. Regul.* 18, 475, 1981.

77d. **Welch, G. R., Ed.,** *The Fluctuating Enzyme,* Wiley-Interscience, New York, 1986, 58.

77e. **Slater, J.** *Introduction to Chemical Physics,* McGraw-Hill, New York, 1939, 178.

78. **Welch, G. R., Ed.,** *The Fluctuating Enzyme,* Wiley-Interscience, New York, 1986, 149.

79. **Lumry, R. and Gregory, R.,** to be published.

80a. **Keyes, M.,** Structure and Function of Sperm Whale Myoglobin, Dissertation, University of Minnesota, Minneapolis, 1968.

80b. **Lumry, R.,** *Electron and Coupled Energy Transfer in Biological Systems,* King, T. and Klingenberg, M., Eds., Marcel Dekker, New York, 1971, 7.

80c. **Lumry, R.,** *Electron and Coupled Energy Transfer,* King, T. and Klingenberg, M., Eds., Marcel Dekker, New York, 1971, 173.

80d. **Wyman, J.,** Linkage graphs: a study in the thermodynamics of macromolecules, *Q. Rev. Biophys.,* 17, 453, 1984.

81a. **Gekko, K. and Timasheff, S. N.,** Thermodynamic and kinetic examination of protein stabilization by glycerol, *Biochemistry,* 20, 4677, 1981.

81b. **Gekko, K. and Morikawa, T.,** Thermodynamics of polyol-induced thermal stabilization of chymotrypsinogen, *J. Biochem.,* 90, 51, 1981.

81c. **Arakawa, T. and Timasheff, S. N.,** Theory of protein solubility, *Meth. Enzymol.,* 114, 49, 1985.

81d. **Lee, J. C. and Timasheff, S. N.,** Partial specific volumes and interactions with solvent components of proteins in guanidine hydrochloride, *Biochemistry,* 13, 257, 1974.

81e. **Gekko, K.,** Calorimetric study on thermal denaturation of lysozyme in polyol-water mixtures, *J. Biochem.,* 91, 1197, 1982.

81f. **Lee, J. C. and Timasheff, S. N.,** The stabilization of proteins by sucrose, *J. Biol. Chem.*, 256, 7193, 1981.
81g. **Almog, R., Schrier, M. Y., and Schrier, E.,** Thermodynamic quantities of interaction and unfolding in the transfer of ribonuclease A from dilute buffer to aqueous cosolute solutions, *J. Phys. Chem.*, 82, 1703, 1978.
81h. **Gekko, K.,** Mechanism of protein stabilization by polyols. Thermodynamics of transfer of amino acids and proteins from water to aqueous polyol solutions, *Studies Phys. Theor. Chem.*, 27, 339, 1982.
81i. **Schrier, M. Y. and Schrier, E. E.,** Transfer free energies and average static accessibilities for ribonuclease a in guanidinium hydrochloride and urea solutions, *Biochemistry,* 15, 2607, 1976.
81j. **Schrier, M. Y., Ying, A. H. C., Ross, M. E., and Schrier, E. E.,** Free energy changes and structural consequences for the transfer of urea from water and ribonuclease A from buffer to aqueous salt solutions, *J. Phys. Chem.*, 81, 674, 1977.
82a. **Shinoda, K. and Fujihara, M.,** Analysis of the solubility of hydrocarbons in water, *Bull. Chem. Soc. Jpn.*, 41, 2612, 1968.
82b. **Shinoda, K.,** Iceberg formation and solubility, *J. Phys. Chem.*, 81, 1300, 1977.
82c. **Barbe, M. and Patterson, D.,** Enthalpy-entropy compensation and order in alkane and aqueous systems, *J. Phys. Chem.*, 80, 2435, 1976.
82d. **Lumry, R. and Frank, H.,** Enthalpy-entropy compensation patterns resolved, *Proc. 6th Int. Biophys. Cong.*, Vol. 2, 1978, 554.
82e. **Welch, G. R., Ed.,** *The Fluctuating Enzyme,* Wiley-Interscience, New York, 1986, 156.
83. **Lumry, R. and Rajender, S.,** *Biopolymers,* 9, 1187, 1970.
84. **Ponder, J. W. and Richards, F. M.,** Tertiary templates for proteins. Use of packing criteria in the enumeration of allowed sequences for different structural classes, *J. Mol. Biol.*, 193, 775, 1987.
85. **Welch, G. R., Ed.,** *The Fluctuating Enzyme,* Wiley-Interscience, New York, 1986, 49.
86. **Welch, G. R., Somogyi, B., and Damjanovich, S.,** The role of protein fluctuations in enzyme action: a review, *Prog. Biophys. Mol. Biol.*, 39, 109. 1982.
87. **Parr, G. and Taniuchi, H.,** A thermodynamic study of ordered complexes of cytochrome c fragments, *J. Biol. Chem.*, 257, 10103, 1982.
88a. **Green, D.,** A framework of principles for the unification of bioenergetics, *Ann. N.Y. Acad. Sci.*, 227, 6, 1974.
88b. **Ji, S.,** Entropy and negentropy in enzymic catalysis, *Ann. N.Y. Acad. Sci.*, 227, 419, 1974.
89a. **Eftink, M. R. and Hagaman, K. A.,** Viscosity dependence of the solute quenching of the tryptophanyl fluorescence of proteins, *Biophys. Chem.*, 25, 277, 1986.
89b. **Borson, J., Novak, E., and Makinen, M.,** Solvent structure and enzyme study, *J. Biomol. Struc. Dyn.*, 3, 197, 1985.
90. **Bolen, D. W., Kumura, T., and Nitta, Y.,** Energetics of α-chymotrypsin-mediated hydrolysis of a strained cyclic ester, *Biochemistry,* 26, 146, 1987.
91a. **Lumry, R. and Biltonen, R.,** *Structure and Stability of Biological Macromolecules,* Timasheff, S. and Fasman, G., Eds., Marcel Dekker, New York, 1969, 167 and 187.
91b. **Welch, G. R., Ed.,** *The Fluctuating Enzyme,* Wiley-Interscience, New York, 1986, 3.
92. **Richter, P. H. and Ross, J.,** Concentration oscillations in efficiency: glycolysis, *Science,* 211, 715, 1981.
93a. **Friedrich, P.,** *Supramolecular Enzyme Organization,* Pergamon Press, Oxford, 1984. chap. 7.
93b. **Srivastava, D. and Bernhard, S.,** Metabolic transfer via enzyme-enzyme complexes, *Science,* 234, 1081, 1986.
94a. **Srere, P.,** Complexes of sequential metabolic enzymes, *Annu. Rev. Biochem.*, 56, 89, 1987.
94b. **Welch, G. R.,** *Dynamics of Biochemical Systems,* Ricard, J. and Cornish-Bowden, A., Eds., Plenum Press, Elmford, NY, 1984, 84.
94c. **Srivastava, D. and Bernhard, S.,** Enzyme-enzyme interactions and the regulation of metabolic reaction pathways, *Curr. Top. Cell. Regul.*, 28, 1, 1985.
94d. **Keleti, T. and Ovadi, J.,** Control of metabolism by dynamic macromolecular interactions, *Curr. Top. Cell. Regul.*, 29, 1, 1987.
95. **Schrödinger, E.,** *What is life?,* Cambridge University Press, Cambridge, 1946.
96. **Tilton, R., Jr., Kuntz, I., Jr., and Petsko, G.,** Cavities in proteins: Structure of a metmyoglobom-xenon complex solved to 1.9 Å, *Biochemistry,* 23, 2849, 1984.
97. **Wüthrich, K., Strop, P., Ebina, S., and Williamson, M. P.,** A globular protein with slower amide proton exchange from an α helix than from antiparallel β sheets, *Biochem. Biophys. Res. Commun.*, 122, 1174, 1984.
98a. **Steitz, T., Harrison, R., Weber, I., and Leahy, M.,** Ligand-induced conformational changes in proteins, *Ciba Found. Symp.*, 93, 25, 1983.
98b. **Anderson, C., Zucker, F., and Steitz, T.,** Space-filling models of kinase clefts and conformation changes, *Science,* 204, 375, 1979.

98c. **Janin, J. and Wodak, S.,** Structural domains in proteins and their role in dynamics of protein function, *Prog. Biophys. Mol. Biol.,* 42, 21, 1983.

98d. **Remington, S., Wiegand, G., and Huber, R.,** Crystallographic refinement and atomic models of two different forms of citrate synthase at 2.7 and 1.7 Å resolution, *J. Mol. Biol.,* 158, 111, 1982.

98e. **Bennett, W. and Steitz, T.,** Structure of a complex between yeast hexokinase A and glucose. II. Detailed comparisons of conformation and active site configuration with the native hexokinase B monomer and dimer, *J. Mol. Biol.,* 140, 211, 1980.

98f. **Chothia, C. and Lesk, A.,** Helix movements in proteins, *TIBS,* 110, March, 1985.

98g. **Alber, T., Gilbert, W., Ponzi, D., and Petsko, G.,** The role of mobility in the substrate binding and catalytic mechanism of enzymes, *Ciba Found. Symp.,* 93, 4, 1982.

98h. **Yonath, A., Padjarny, A., Honig, B., Sielecki, A., and Traub, W.,** Crystallographic studies of protein denaturation and renaturation. II. Sodium dodecyl sulfate induced structural changes in triclinic lysozyme, *Biochemistry,* 16, 1418, 1977.

98i. **Steitz, T., Harrison, R., Weber, I., and Leahy, M.,** Ligand-induced conformational changes in proteins, *Ciba Found. Symp.,* 93, 25, 1983.

99. **Bode, W.,** The transition of bovine trypsinogen to a trypsinlike state upon strong-ligand binding, *J. Mol. Biol.,* 127, 367, 1974.

100. **Srivastava, D. and Bernhard, S.,** Biophysical chemistry of metabolic reaction sequences in concentrated enzyme solution and in the cell, *Annu. Rev. Biophys. Chem.,* 16, 175, 1987.

101a. **Blevins, R. and Tulinsky, A.,** The refinement and the structure of the dimer of α-chymotrypsin at 1.67 angstroms, *J. Biol. Chem.,* 260, 4264, 1985; **Tulinsky, A. and Blevins, R.,** *Acta Cryst.,* A40 (Suppl. C-17), 1984.

101b. **Tulinsky, A. and Blevins, R.,** Structure of a tetrahedral transition state complex of α-chymotrypsin dimer at 1.8Å resolution, *Acta Cryst.,* 262, 3827, 1987.

101c. **Blevins, R. and Tulinsky, A.,** Comparison of the independent solvent structures of dimeric α-chymotrypsins with themselves and with α-chymotrypsin, *Acta Cryst.,* 260, 8865, 1985.

102. **Tulinsky, A.,** personal communication, 1986, 1987.

103a. **Bruccoleri, R., Karplus, M., and McCammon, A.,** The hinge-bending mode of a lysozyme-inhibitor complex, *Biopolymers,* 25, 1767, 1986.

103b. **Brünger, A., Brooks, B., and Karplus, M.,** Active-site dynamics of ribonuclease A, *Proc. Natl. Acad. Sci. U.S.A.,* 82, 8458, 1985.

104. **Doherty, D. and Shapia, R.,** A kinetic study of the hydrolysis of C-C bonds by α-chymotrypsin, *Third Congress on Biochemistry,* Brussels, 1955, 37.

105. **Fujinaga, M., Delbaere, L. T. J., Brayer, G. D., and James, M. N.,** Chymotrypsin, Refined structure of a-lytic protease at 1.7 Å resolution analysis of hydrogen bonding and solvent structure, *J. Mol. Biol.,* 183, 479, 1985.

106a. **Ives, D. and Marsden, P.,** The ionization function of di-isopropylcyanoacetic acid in relation to hydration equilibria and the compensation line, *J. Chem. Soc.,* 649, 1965.

106b. **Hepler, L.,** Thermodynamic analysis of the Hammett equation, the temperature dependence of the isoequilibrium (isokinetic) relationship, *Can. J. Biochem.,* 49, 2803, 1971.

106c. **Lumry, R. and Rajender, S.,** *Biopolymers,* 9, 1134, 1970.

107a. **Zundel, G.,** Proton polarizability of hydrogen bonds: infrared methods, relevance to electrochemical and biological systems, *Methods Enzymol.,* 127, 31, 1986.

107b. **Hillenbrand, E. and Scheiner, S.,** Analysis of the principles governing proton-transfer reactions. Comparison of the imine and amine groups, *J. Am. Chem. Soc.,* 107, 7690, 1985.

107c. **Szczesniak, M. and Scheiner, S.,** Effects of external ions on the dynamics of proton transfer across a hydrogen bond, *J. Phys. Chem.,* 89, 1835, 1985.

107d. **Steiner, S., Redfern, P., and Hillenbrand, E.,** Factors influencing proton positions in biomolecules, *Int. J. Quantum Chem.,* 29, 817, 1986.

108a. **Vallee, B. and Williams, R. J. P.,** Metalloenzymes. Entactic nature of their active sites, *Proc. Natl. Acad. Sci. U.S.A.,* 59, 498, 1968.

108b. **Williams, R. J. P.,** Entactic state, *Cold Spring Harbor Symp. Quant. Biol.,* 36, 53, 1971.

109. **Inagami, T. and Hatano, H.,** Effect of alkylguanidines on the inactivation of trypsin by alkyation and phosphorylation, *J. Biol. Chem.,* 244, 1176, 1969.

110. **Martinek, K., Klyosov, A. A., Kazanskaya, N. F., and Berezin, I. V.,** The free-energy-reaction coordinate profile for α-chymotryptic hydrolysis of a series of *n*-acetyl-α-L-amino acid methyl esters, *Int. J. Chem. Kinetics,* 6, 801, 1974.

111a. **Hansch, C. and Coats, E.,** Alpha-chymotrypsin: case study of substituent constants and regression analysis in enzymic structure-activity relations, *J. Pharm. Sci.,* 59, 731, 1970.

111b. **Berezin, I., Kazanskya, N., and Klyosov, A.,** Determination of the individual rate constants of α-chymotrypsin-catalyzed hydrolysis with the added nucleophilic agent, 1,4-butanediol, *FEBS Lett.,* 15, 121, 1971.

112. **Parker, H.**, unpublished material from this laboratory, 1967.
113a. **Hansch, C.**, Structure-activity relationship in the chymotrypsin hydrolysis of p-nitrophenyl esters, *J. Org. Chem.*, 37, 92, 1972.
113b. **Helmer, F., Kiehs, K., and Hansch, C.**, The linear free-energy relationship between partition coefficients and the binding and conformational perturbation of macromolecules by small organic compounds, *Biochemistry*, 7, 2858, 1968.
113c. **Hansch, C., Deutsch, E., and Smith, R.**, The use of substituent constants and regression analysis in the study of enzymic mechanisms, *J. Am. Chem. Soc.*, 87, 2738, 1965.
113d. **Hansch, C.**, Quantitative structure-activity relationship of chymotrypsin-ligand interactions, *J. Med. Chem.*, 20, 1420, 1977.
114. **Privalov, P.**, Stability of proteins, *Adv. Protein Chem.*, 33, 167, 1979.
115a. **Wuthrich, K., Roder, H., and Wagner, G.**, Internal mobility and unfolding of globular peins, *Protein Folding*, 549, 1980.
115b. **Doster, W., Simon, B., Schmidt, G., and Mayr, W.**, Compressibility of lysozyme in solution from time-resolved Brillouin difference spectroscopy, *Biopolymers*, 24, 1543, 1985.
115c. **Lecomte, J. and Llinas, M.**, H NMR spectral patterns of rapidly flipping tyrosyl rings: a study of crambin in organic solvents, *J. Am. Chem. Soc.*, 106, 2741, 1984.
115d. **Klee, W. A.**, Intermediate stages in the thermally-induced transconformation reactions of bovine pancreatic ribonuclease A, *Biochemistry*, 6, 3736, 1967.
115e. **Bull, H., and Breese, K.**, Temperature dependence of partial molar volumes of proteins, *Biopolymers*, 12, 2351, 1973.
115f. **Miller, J. and Bolen, W. D.**, A guanidine hydrochloride induced change in ribonuclease without gross unfolding, *Biochem. Biophys. Res. Commun.*, 81, 610, 1978.
116. **Gekko, K. and Hasegawa, Y.**, Compressibility-structure relationship of globular proteins, *Biochemistry*, 25, 6563, 1986.
117a. **Rupley, J., Yang, P.-H., and Tollin, G.**, Thermodynamic and related studies of water interacting with proteins, *Water in Polymers*, (Advances in Chemistry Series, Vol. 127), 111, 1980.
117b. **Rüegg, M., Moor, U., and Blanc, B.**, Hydration and thermal denaturation of β-lactoglobulin, *Biochim. Biophys. Acta*, 400, 334, 1975.
117c. **Fujita, Y. and Noda, Y.**, Effect of hydration on the thermal denaturation of lysozyme measured by differential scanning calorimetry, *Bull. Chem. Soc. Jpn.*, 51, 1567, 1978.
117d. **Fujita, Y. and Noda, Y.**, Effect of hydration on the thermal stability of protein as measured by differential scanning calorimetry: chymotrypsinogen A, *Int. J. Peptide Protein Res.*, 18, 12, 1981.
117e. **Poole, P. and Finney, J.**, Sequential hydration of dry proteins: a direct difference IR investigation of sequence homologs lysozyme and α-lactalbumin, *Biopolymers*, 23, 1647, 1984.
117f. **Likhtenshtein, G.**, Water and dynamics of proteins and membranes, *Studia Biophysica*, 111, 89, 1986.
117g. **Scalon, W. J. and Eisenberg, D.**, Solvent bound in sperm whale metmyoglobin type A crystals at 6.1 and 23.5 C, *J. Phys. Chem.*, 85, 3251, 1981.
117h. **Berlin, E., Kliman, P. G., Segawa, S. I., and Kume, K.**, Comparison between the unfolding rate and structural fluctuations in native lysozyme. Effects of denaturants, ligand binding, and intrachain cross-linking on hydrogen exchange and unfolding kinetics, *Biopolymers*, 25, 1981, 1986.
118a. **Suurkuusk, J.**, Specific heat measurements on lysozyme, chymotrypsinogen, and ovalbumin in aqueous solution and in solid state, *Acta Chem. Scand. B*, 28, 409, 1974.
118b. **Bull, H. and Breese, K.**, Protein hydration. II. Specific heat of egg albumin, *Arch. Biochem. Biophys.*, 128, 497, 1968.
118c. **Yang, P.-H. and Rupley, J. A.**, Heat capacity of the lysozyme-water system, *Biochemistry*, 12, 2654, 1979.
118d. **Berlin, E., Kliman, P., and Pallansch, M.**, Effect of sorbed water on the heat capacity of crystalline proteins, *Thermochim. Acta*, 4, 11, 1972.
119. **Hnojewyj, W. S. and Reyerson, L. H.**, Further studies on the absorption of H_2O and D_2O vapors by lysozyme and the deuterium-hydrogen exchange effect, *J. Phys. Chem.*, 65, 1694, 1961.
120a. **Lüscher, M., Rüegg, M., and Schindler, P.**, Thermodynamic studies of the interaction of α-chymotrypsin with water. II. Statistical analyses of the enthalpy-entropy compensation effect, *Biochim. Biophys. Acta*, 536, 27, 1978.
120b. **Lüscher, M., Schindler, P., Rüegg, M., and Rottenberg, M.**, Effect of inhibitor complex formation on the hydration properties of α-chymotrypsin. Changes induced in protein hydration by tosylation of the native enzymes, *Biopolymers*, 18, 1775, 1979.
120c. **Lüscher-Matli, M. and Rëgg, M.**, Thermodynamic functions of biopolymer hydration. I. Investigation of a specific primary hydration process, *Biopolymers*, 21, 419, 1982.
120d. **Lüscher-Matli, M. and Ruegg, M.**, Thermodynamic functions of biopolymer hydration. II. Enthalpy-entropy compensation in hydrophilic hydration processes, *Biopolymers*, 21, 419, 1982.
121a. **Rupley, J. A., Gratton, E., and Careri, G.**, Water and globular proteins, *TIBS*, January, 18, 1983.

121b. **Careri, G., Giansanti, A., and Gratton, E.,** Lysozyme film hydration events: an IR and gravimetric study, *Biopolymers,* 18, 1187, 1979.
122a. **Lumry, R.,** Participation of water in protein reactions, *Ann. N. Y. Acad. Sci.,* 227, 471, 1974.
122b. **Lumry, R.,** *Electron and Coupled Energy Transfer in Biological Systems,* King, T. and Klingenberg, M., Eds., Marcel Dekker, New York, 1971, 62.
122c. **Belleau, B.,** Patterns of ligand-induced conformational changes on a receptor surface, in *Physicochemical Aspects of Drug Action,* Ariens, E., Ed., Pergamon Press, Elmsford, NY, 1966.
122d. **Belleau, B.,** Water as the determinant of thermodynamic transitions in the interaction of aliphatic chains with acetylcholinesterase and the cholinergic receptors, *Ann. N. Y. Acad. Sci.,* 144, 705, 1967.
122e. **Likhtenshtein, G.,** The water-protein interactions and dynamic structure of proteins, *Colloq. Int., C.N.R.S.,* No. 246, 45, 1976.
122f. **Likhtenshtein, G.,** Water and dynamics of proteins and membranes, *Biofizika,* 11, 23, 1968.
123a. **Lumry, R.,** *Electron and Coupled Energy Transfer in Biological Systems,* King, T. and Klingenberg, M., Eds., Marcel Dekker, New York, 1971, 84.
123b. **Lumry, R. and Biltonen, R.,** *Structure and Stability of Biological Macromolecules,* Timasheff, S. and Fasman, G., Eds., Marcel Dekker, New York, 1969, 184.
123c. **Lumry, R.,** Structure-function relationships in proteins and their possible bearing on the photosynthetic process, in *Photosynthesis Mechanisms in Green Plants,* Publ. 1145, National Academy of Sciences and National Research Council, Washington, DC, 1963, 625.
124a. **G. Petsko and D. Ringe,** Fluctuations in protein structure from x-ray diffraction, *Annu. Rev. Biophys. Bioeng.,* 13, 31, 1984.
124b. **Ringe, D., Kurijan, J., Petsko, G., Karplus, M., Frauenfelder, H., Tilton, R., and Kuntz, I.,** Temperature dependence of protein structure and mobility, *Trans. Am. Crystallogr. Assoc.,* 20, 109, 1985.
124d. **Belonogova, O. V., Folov, E. N., Krasnopol'skaya, S. A., Atanasov, B. P., Gins, V. K., Mukhin, E. N., Levina, A. A., Andreeva, A. P., Likhtenshtein, G. I., and Gol'danskii, V. I.,** Effect of degree of hydration on mobility of Mössbauer atoms in active centers of metalloenzymes and carriers, *Dokl. Akad. Nauk,* 241, 219, 1978.
124e. **Brown, W. E. III, Sutcliffe, J. W., and Pulsinelli, P. D.,** Multiple internal reflectance infrared spectra of variably hydrated hemoglobin and myoglobin films: effects of globin hydration on ligand binding, conformer dynamics and reactivity at the heme, *Biochemistry,* 22, 2914, 1983.
125. unpublished results from this laboratory.
126a. **Lumry, R. and Rajender, S.,** *Biopolymers,* 9, 1165, 1970.
126b. **Cohen, S., Vaidya, V., and Schultz, R.,** Active site of α-chymotrypsin, *Proc. Natl. Acad. Sci. U.S.A.,* 66, 249, 1970.
126c. **Rajender, S., Lumry, R., and Han, M.,** Steady-state kinetics of the chymotryptic hydrolysis of *N*-acetyl-L-tryptophan ethyl ester at pH 8.0, *J. Am. Chem. Soc.,* 92, 1378, 1970.
127. **Barksdale, A.,** unpublished results from this laboratory, 1973.
128a. **Lumry, R. and Biltonen, R.,** *Structure and Stability of Biological Macromolecules,* Timasheff, S. and Fasman, G., Eds., Marcel Dekker, New York, 1969, 156.
129a. **Stoesz, J. and Lumry, R.,** Refolding transition of α-chymotrypsin: pH and salt dependence, *Biochemistry,* 17, 3, 693, 1979.
129b. **Stoesz, J., Lumry, R., and Shiao, D.,** The effects of chemical modification on the refolding transition of α-chymotrypsin; pH and salt dependence, *J. Biophys. Chem.,* 10, 105, 1979.
129c. **Welch, G. R., Ed.,** *The Fluctuating Enzyme, Wiley-Interscience, New York,* 1986, 118.
129d. **Fersht, A. and Requina, Y.,** Equlibria and rate constants for the interconversion of two conformers of α-chymotrypsin, *J. Mol. Biol.,* 60, 279, 1971.
130. **Milville, R. and Hopkins, T.,** Denaturation rates of acylderivatives of α-chymotrypsin, to be published.
131a. **Biltonen, R., Lumry, R., Madison, V., and Parker, H.,** The optical rotatory dispersion of α-chymotrypsin, *Proc. Natl. Acad. Sci. U.S.A.,* 54, 1018, 1965.
132a. **Hopkins, T. and Shiao, D.,** Variation in the urea denaturation rate of some members of the chymotrypsin family, *Biophys. J.,* Abstr. ME2, 1968.
132b. unpublished work from this laboratory.
133a. **Pohl, F.,** Kinetics of the Reversible Conformation Changes of Globular Proteins, Dissertation, Universities of Göttingen and Constance, 1969.
133b. **Pohl, F.,** Temperature dependence of the kinetics of folding of chymotrypsinogen A, *FEBS Lett.,* 65, 293, 1976.
133c. **Pohl, F.,** On the kinetics of structural transition I of some pancreatic proteins, *FEBS Lett.,* 3, 60, 1969.
133d. **Pohl, F.,** Simple temperature-jump method in the second and hour range and the reversible denaturation of chymotrypsin, *Eur. J. Biochem.,* 4, 373, 1968. Pohl, F., *Eur. J. Biochem.,* 7, 146, 1968.
134. **Hopkins, T. R. and Spikes, J. D.,** Denaturation of proteins in 8 M urea as monitored by tryptophan fluorescence: α-chymotrypsin, chymotrypsinogen and some derivatives, *Biochem. Biophys. Res. Commun.,* 28, 480, 1967.

135. **Oguni, M. and Angell, C.,** Heat capacities of $H_2O + H_2O_2$ and $H_2O + N_2H_4$, binary solutions: isolation of a singular component of C_p of supercooled water, *J. Chem. Phys.*, 73, 1948, 1980.

136a. **Hibbard, L. and Tulinsky, A.,** Expression of functionally of α-chymotrypsin. Effects of guanidine hydrochloride and urea on the onset of denaturation, *Biochemistry,* 17, 5460, 1978.

136b. **Strambini, G. B. and Gonelli, M.,** Proteins in mixed solvents, effects of urea and guanidine hydrochloride on the activity and dynamical structure of equine liver alcohol dehydrogenase, *Biochemistry,* 25, 2471, 1986.

137a. **Lumry, R., Rajender, S., and Han, M. H.,** Studies of the chymotrypsinogen family of proteins. XV. pH and temperature dependence of the α-chymotryptic hydrolysis of *N*-acetyl-L-tryptophan ethyl ester, *J. Phys. Chem.*, 75, 1375, 1971.

137b. **Lumry, R. and Rajender, S.,** Enthalpy-entropy compensation phenomena of α-chymotrypsin and the temperature of minimum sensitivity. XVI, *J. Phys. Chem.*, 75, 1387, 1971.

138a. **Battistel, E., Lumry, R., and Jolicoeur, C.,** Heat capacity changes accompanying oxygenation of human hemoglobin, *Biophys. J.*, 41, 9a, 1983.

139. **Magnuson, D., Bodley, J., and Evans, D. F.,** Activity and stability of phosphatase in alkaline solutions of water and the fused salt ethylammonium nitrate, *J. Soln. Chem.*, 13, 583, 1984.

140. **Careri, G.,** Molecular hydration and its possible role in enzymes, in *Biophysics of Water,* Franks, F. and Mathias, S. F., Eds., Wiley-Interscience, New York, 1982, 58.

141a. **Käiväräinen, A.,** in *Solvent-Dependent Flexibility of Proteins and Principles of their Function,* D. Rediel, Dordrecht, Netherlands, 1983, chap. 7.

142. **Harvey, S. C. and Hoekstra, P.,** Dielectric relaxation spectra of water adsorbed on lysozyme, *J. Phys. Chem.*, 76, 2989, 1972.

143a. **Von Hippel, P. and Schleich, T.,** The effects of neutral salts on the structure and conformational stability of macromolecules in solution, in *Structure and Stability of Biological Macromolecules,* Timasheff, S. and Fasman, G., Eds., Marcel Dekker, New York, 1969, chap. 6.

143b. **Collins, K. and Washabaugh, W.,** The Hofmeister effect and the behavior of water at interfaces, *Q. Rev. Biophys.*, 18, 323, 1985.

144. **Kuntz, I. and Kauzmann, W.,** Hydration of proteins and polypeptides, *Adv. Protein Chem.*, 28, 239, 1974.

145. **Franks, F. and Eagland, D.,** The role of solvent interactions in protein conformation, *Crit. Rev. Biochem.*, 165, 1975.

146a. **Cooper, A.,** Protein fluctuations and the thermodynamic uncertainty principle, *Prog. Biophys. Mol. Biol.*, 44, 181, 1984.

146b. **Welch, G. R., Ed.,** *The Fluctuating Enzyme,* Wiley-Interscience, New York, 1986, 133.

147a. **Likhtenshtein, G.,** in *Spin Labeling Methods in Molecular Biology,* John Wiley & Sons, New York, 1976, chap. 6 and 7.

147b. **Belonogova, O. V., Frolov, E. N., Illustrov, N. V., and Likhtenshtein, G.,** Effect of temperature and degree of hydration on the mobility of spin labels in surface layers of proteins, *Mol. Biol.*, 13, 576, 1979.

147c. **Ivanov, L., Kinizkaja, L., and Kokhanov, Yu.,** Lysozyme macromolecule studied by the spin-label technique, *Mol. Biol.*, 8, 48, 1974.

147d. **Likhtenshtein, G., Grebenshchikov, Yu., and Avilova, T.,** An investigation of the microrelief and conformational mobility of proteins by the esr method., *Mol. Biol.*, 6, 67, 1972.

148a. **Lumry, R.,** *Electron and Coupled Energy Transfer in Biological Systems,* King, T. and Klingenberg, M., Eds., Marcel Dekker, New York, 1971, 36, 43.

148b. **Shiao, D., Lumry, R., and Fahey, J.,** Studies of the chymotrypsinogen family of proteins. XI. Heat capacity changes accompanying reversible thermal unfolding of proteins, *J. Am. Chem. Soc.*, 93, 2024, 1971.

149. **Pohl, F.,** personal communication, 1969.

150a. **Lumry, R.,** Dynamic basis of macromolecule association and tissue recognition, *Biophys. J.*, 21, 113a, 1978.

150b. **Welch, G. R.,** *The Fluctuating Enzyme,* Wiley-Interscience, New York, 1986, 109.

150c. **Lumry, R.,** Boundary-layer control of protein conformations and the dynamical basis of protein-protein interaction, in *Dynamic Aspects of Bioelectrolytes and Biomembranes,* Oosawa, F. and Imai, N., Eds., Kodanska, Tokyo, 1982.

150d. **Lumry, R.,** Dynamics of small molecule-protein interactions, in *Bioenergetics and Thermodynamics: Model systems,* D. Reidel, Dordrecht, Netherlands, 1979, 435.

151a. **Belleau, B. and DiTullio, V.,** Kinetic effects of alkyl quaternary ammonium salts on the methanesulfonylation of the acetylcholinesterase catalytic center. Significance of substitutent volumes and binding enthalpies, *J. Am. Chem. Soc.*, 92, 6320, 1970.

151b. **Belleau, B. and Lavoie, J.,** A biophysical basis of ligand-induced activation of excitable membranes and associated enzymes. A thermodynamic study using acetylcholinesterase as a model receptor, *Can. J. Biochem.*, 46, 1397, 1968.

152a. **Huot, J.-Y., Battistel, E., Jolicoeur, C., Lumry, R., Villeneuve, G., Lavallee, J.-F. and Anusiem, A.,** Analogies and differences between water and ethylene glycol: a comprehensive investigation of water-ethylene glycol mixtures at 5°, 25° and 45°, *J. Solution Chem.*, in press.

152b. **Welch, G. R., Ed.,** *The Fluctuating Enzyme,* Wiley-Interscience, New York 1986, 166.

153a. **Nishikawa, S., Mashima, M., and Yasunaga, T.,** Ultrasonic absorption mechanism in an aqueous solution of n-propyl alcohol, *Bull Chem. Soc. Jpn.*, 48, 61, 1975.

153b. **Atkinson, G., Rajagopalan, S., and Atkinson, B.,** Ultrasonic absorption in aqueous binary mixtures. II. p-Dioxanewater at 11 and 25°, *J. Chem. Phys.*, 75, 3511, 1980.

153c. **Green, J., Sceats, M., and Lacey, A.,** Hydrophobic effects in the water network structure of aqueous solutions of a semiclathrate molecules, *J. Chem. Phys.*, 87, 3603, 1987.

153d. **Matteoli, E. and Lepori, L.,** solute-solute interactions in water. II. An analysis through the Kirkwood-Buff integrals of 14 organic solutes, *J. Chem. Phys.*, 80, 2856, 1984.

153e. **Hvidt, A., Moss, R., and Nielsen, G.,** volume properties of aqueous solutions of tert.-butyl alcohol at temperatures between 5 and 25°, *Acta Chem. Scand.*, B32, 274, 1978.

153f. **Bruun, S. and Hvidt, A.,** Volume properties of binary mixtures of water with 2-propanol, *Bunsen Berichte,* 81, 930, 1977.

153g. **Nakanishi, K., Ikari, K., Okazaki, S., and Touhara, H.,** *J. Chem. Phys.*, 80, 1656, 1984.

154a. **Evans, D. F. and Wightman, P.,** Micelle formation above 100°C, *J. Coll. Interface Sci.*, 86, 515, 1982.

154b. **Ramadan, M., Evans, D. F., Lumry, R., and Philson, S.,** Micelle formation in hydrazine-water mixtures, *J. Phys. Chem.*, 89, 3405, 1985.

154c. **Ramadan, M., Evans, D. F., and Lumry, R.,** Why micelles form in water and hydrazine: a reexamination of the origins of hydrophobicity, *J. Phys. Chem.*, 87, 5020, 1983.

154e. **Ramadan, M., Evans, D. F., and Lumry, R.,** Why micelles form in water and hydrazine: a reexamination of the origins of hydrophobicity, *J. Phys. Chem.*, 87, 5020, 1983.

155. **Gekko, K.,** Mechanism of polyol-induced protein stabilization: solubility of amino acids and diglycine in aqueous polyol solutions, *J. Biochem.*, 90, 1633, 1981.

156. **Lehmann, M. S. and Zaccai, G.,** Neutron small-angle scattering studies of ribonuclease in mixed aqueous solutions and determination of the preferentially bound water, *Biochemistry,* 23, 1939, 1984.

157. **Brandts, J. F. and Hunt, L.,** Thermodynamics of protein denaturation. III. Denaturation of ribonuclease in water and in aqueous urea and aqueous ethanol mixtures, *J. Am. Chem. Soc.*, 89, 4826, 1967.

158a. **Blake, C., Mair, G., North, A., Phillips, D., and Sarma, V.,** Domain motions in HEW lysozyme. *Proc. R. Soc. London (Ser.B),* 167, 365, 378, 1967.

158b. **Schultz, G. and Schirmer, R.,** *Principles of Protein Structure,* Springer-Verlag, Berlin, 1979, 95.

158c. See also references for other enzymes given in 103a.

159a. **Speedy, R.,** Pentagon-pentagon correlations in water, *J. Phys. Chem.*, 89, 171, 1985.

159b. **Grunwald, E.,** Model for the structure of the liquid water network, *J. Am. Chem. Soc.*, 108, 5719, 1986.

159c. **Giguere, P.,** Bifurcated hydrogen bonds in water, *J. Raman Spec.*, 15, 354, 1984.

159d. **Stanley, H., Teixeira, J., and Geiger, A.,** Interpretation of the unusual behavior of water and water-d2 at low temperature: are concepts of percolation relevant to the "puzzle of liquid water"?, *Physica,* 106A, 260, 1981.

159e. **Ohtomo, N., Tokiwano, K., and Arakawa, K.,** The structure of liquid water by neutron scattering, *Bull. Chem. Soc. Jpn.*, 54, 1802, 1981.

160. **Kauzmann, W.,** Some factors in the interpretation of protein denaturation, *Adv. Protein Chem.*, 14, 1, 1959.

161. **Fujita, Y. and Noda, Y.,** The effect of ethylene glycol on the thermal denaturation of ribonuclease A and chymotrypsinogen A as measured by differential scanning calorimetry, *Bull. Chem. Soc. Jpn.*, 57, 2177, 1984.

162a. **Klibanov, A.,** in *The Biological Basis of New Developments in Biotechnology,* Laskin, A. and Rogers, P., Eds., Plenum Press, Elmsford, NY, 1983, 497.

162b. **Zaks, P. and Klibanov, A.,** Enzymic catalysis in organic media at 100°, *Science,* 224, 1249, 1984.

162c. **Weetall, H. H. and Vann, W. P.,** Studies on immobilized trypsin in high concentrations of organic solvents, *Biotech. Bioeng.*, 18, 105, 1976.

162d. **Waks, M.,** Proteins and peptides in water-restricted environments, *Proteins: Struc. Func. Gen.*, 1, 4, 1986.

163. **Arakawa, T. and Timasheff, S.,** Abnormal solubility behavior of β-lactoglobulin: salting-in by glycine and sodium chloride, *Biochemistry,* 26, 5147, 1987.

164a. **Low, P. S. and Somero, G. N.,** Protein hydration changes during catalysis: a new mechanism of enzymic rate-enhancement and ion activation/inhibition of catalysis, *Proc. Natl. Acad. Sci. U.S.A.*, 72, 3304, 1975.

164b. **Low, P. S. and Somero, G. N.,** Activation volumes in enzymic catalysis: their sources and modification by low-molecular-weight solutes, *Proc. Natl. Acad. Sci. U.S.A.*, 72, 3014, 1975.

165. **Loftfield, R. B., Eigner, E. A., Pastuszyn, A., Erik Lovgren, T. N., and Jakubowski, H.,** Conformational changes during enzyme catalysis: role of water in the transition state, *Proc. Natl. Acad. Sci. U.S.A.*, 77, 3374, 1980.

166. **Berezin, I., Klyosov, A., and Martinek, K.,** General principles of enzymatic catalysis, *Soviet Sci. Rev. Section B, Chem. Rev.* 1, 205, 1970.
167. **Careri, G.,** in *Quantum Statistical Mechanics in the Natural Sciences,* Kornyshev, B. and Widmayer, S., Eds., Plenum Press, New York, 1974, 15.
168. **Privalov, P.,** Stability of proteins. Proteins which do not present a single cooperative system, *Adv. Protein Chem.,* 35, 47, 1982.
169. **Welch, G. R., Ed.,** *The Fluctuating Enzyme,* Wiley-Interscience, New York, 1986, 156.
170. **Frank, H.,** Personal communication, 1978.
171. **Crovetto, R., Fernandez-Prini, R., and Japas, M.,** Solubilities of inert gases and methane in H_2O and D_2O in the temperature range of 300 to 600K, *J. Chem. Phys.,* 76, 1077, 1982.
172. **Lumry, R.,** to be published.
173. **Bizzozero, S. A. and Dutler, H.,** Stereochemcial aspects of peptide hydrolysis catalyzed by serine proteases of the chymotrypsin type, *Bioorg. Chem.,* 10, 46, 1981.
174. **Mason, S., Bentley, G., and McIntryre, G.,** Deuterium exchange in lysozyme at 1.4 Å resolution, in *Neutrons in Biology,* Schoenborn, B., Ed., Plenum Press, New York, 1983, 323.
175. **Grunwald, E.,** Thermodynamic properties, propensity laws and solvent models in solutions in self-associating solvents: application to aqueous alcohol solutions, *J. Am. Chem. Soc.,* 106, 5414, 1984.

176a. **Fukuda, M. and Kunugi, S.,** Pressure dependence of thermolysin catalysis, *Eur. J. Biochem.,* 142, 565, 1984.

176b. **Teichberg, V. I. and Shinitzky, M.,** Fluorescence polarization studies of lysozyme and lysozyme-saccharide complexes, *J. Mol. Biol.,* 74, 519, 1973.

176c. **Pfister, K., Sandmeier, E., Berchtold, W., and Christen, P.,** Conformation changes in asparate aminotransferase, *J. Biol. Chem.,* 260, 11414, 1985.

176d. **Pavlov, M. Y.,** Determination of mutual arrangement of domains in bidomain proteins by diffuse x-ray scattering, *Doklady,* 281, 458, 1985.

177. **Light, A., Duda, C. T., Odorzynski, T. W., and Moore, W. G. I.,** Refolding of serine proteinases, *Biochemistry,* 31, 19, 1986.
178. **Frauenfelder, H., Hartmenn, H., Karplus, M., Kuntz, I., Jr., Kuriyan, J., Parak, F., Petsko, G., Ringe, D., Tilton, R., Jr., Connolly, M., and Max, N.,** Thermal expansion of a protein, *Biochemistry,* 26, 254, 1986.

179a. **Phillips, A.,** Hydrogen bonding and exchange in oxymyoglobin, in *Neutrons in Biology,* Schoenborn, B., Ed., Plenum Press, New York, 1983, 305.

179b. **Raghavan, A. and Schoenborn, B. S.,** The structure of bond water and refinement of acid metmyoglobin, in *Neutrons in Biology,* Schoenborn, B., Ed., Plenum Press, New York, 1983, 247.

180a. **Dipaola, G. and Belleau, B.,** Apparent metal volumes and heat capacities of several electrolytes in aqueous glycerol solutions, *Colloq. Internatl. CNRS,* 246, 255, 1976.

180b. **DiPaola, G. and Belleau, B.,** Thermodynamic evidence for a selective solvation of glycerol and hexitols by aqueous β-lactoglobulin, *Can. J. Chem.,* 56, 848, 1978.

181a. **Somoilov, O.,** Structure of Aqueous Electrolyte Solutions and the Hydration of Ions, Consultants Bureau, New York, (translated by D. Ives), 1965.

181b. **Frank, I. I.,** in *Water, A Comprehensive Treatise,* Vol. 1, Franks, F., Ed., 1972, 515.

182. **Tsukuda, H. and Blow, D.,** Structure of α-chymotrypsin refined at 1.68 Å resolution, *J. Mol. Biol.,* 184, 703, 1985.
183. **Lumry, R.,** unpublished observations from this laboratory.

184a. **Sukhorukov, B. and Likhtenshtein, G.,** The kinetics and the mechanism of denaturation of biopolymers, *Biofizika,* 10, 935, 1965.

184b. **Ptitsyn, O.,** Interdomain mobility in proteins and its probably functional role, *FEBS Lett.,* 93, 14, 1978.

184c. **Likhtenshtein, G.,** Principles of the entropy and energy properties of enzymic processes, *Biofizica,* 11, 24, 1966.

184c. **Lumry, R. and Biltonen, R.,** *Structure and Stability of Biological Macromolecules,* Timasheff, S. and Fasman, G., Eds., Marcel Dekker, New York, 1969, 83.

184d. **Lumry, R. and Biltonen, R.,** Interpretation of kinetic data for cooperative transitions of proteins, Abstrs. 150th Meet. Am. Chem. Soc., Sept. 1965, C197.

186a. **Gerlsma, S. and Stuur, E.,** The effect of polyhydric and monohydric alcohols on the heat-induces reversible denaturation of lysozyme and ribonuclease, *Int. J. Peptide Protein Res.,* 4, 377, 1972.

186b. **Fujita, Y., Iwaa, Y., and Noda, Y.,** The effect of polyhydric alcohols on the thermal denaturation of lysozyme as measured by differential scanning calorimetry, *Bull. Chem. Soc. Jpn.,* 55, 1896, 1982.

187a. **Demchenko, A. P.,** Fluorescence molecular relaxation studies of protein dynamics. The probe binding site of melittin is rigid on the nanosecond time scale, *FEBS Lett.,* 2305, 182, 99, 1985.

187b. **Cohen, S., Vaidya, V., and Schultz, R.,** *Proc. Natl. Acad. Sci. U.S.A.,* 66, 249, 1970.

187c. **Tanaka, F., Kaneda, N., Mataga, N., Tamai, N., Yamazaki, I., and Hayashi, K.,** Analysis of nonexponential fluorescence decays functions of a single tryptophan residue in erabutoxin b, *J. Phys. Chem.,* 91, 6344, 1987.

187d. **Chen, L.X.-Q, Longworth, J. W., and Fleming G. R.,** Picosecond time-resolved fluorescence of ribonuclease T1. A pH and substrate analogue binding study, *Biophys. J.,* 51, 865, 1987.

187e. **MacKerell, A., Jr., Rigler, R., Nillson, L., Hahn, U., and Saenger, W.,** Protein dynamics: a time-resolved fluorescence, energetic and molecular dynamics study of ribonuclease T_1, *Biophys. Chem.,* 26, 247, 1987.

188a. **Kuwajima, K. and Baldwin, R. L.,** Exchange behavior of the H-bonded amide protons in the 3 to 13 helix of ribonuclease S, *J. Mol. Biol.,* 169, 299, 1983.

188b. **Shoemaker, K. R., Kim, P. S., York, E. J., Stewart, J. M., and Baldwin, R. L.,** Tests of the helix dipole model for stabilization of α-helices.

188c. **Loftus, D., Gbenle, G. O., Kim, P. S., and Baldwin, R. L.,** Effects of denaturants on amide protein exchange rates: a test for structure in protein fragments and folding intermediates, *Biochemistry,* 25, 1428, 1986.

189. **Leffler, J. and Grunwald, E.,** *Rates and Equilibria of Organic Reactions,* John Wiley & Sons, New York, 1963, chap. 6.

190a. **Exner, O.,** Statistic of the enthalpy-entropy relationship, I, *Coll. Czech. Chem. Commun.* 37, 1425, 1972;

190b. **Exner, O.,** Statistic of the enthalpy-entropy relationship. II, *Coll. Czech. Chem. Commun.,* 38, 781, 1973.

190b. **Exner, O.,** The enthalpy-entropy relationship, *Prog. Phys. Org. Chem.,* 10, 411, 1973.

190c. **Krug, R., Hunter, W., and Greiger-Block, R.,** Enthalpy-entropy compensation: an example of the misuse of least-squares and correlation analysis, *Adv. Chem. Ser. (Am. Chem. Soc.),* 52, 192, 1977.

191. **Lumry, R., Battistel, E., and Jolicoeur, C.,** Geometric relaxation in water, *Faraday Symp. Chem. Soc.,* 17, 93, 1982 and discussion sections of this volume.

192. **Kanda, H., Ookubo, N., Nakajima, H., Suzuki, Y., Minato, M., Ihara, T., and Wada, Y.,** Ultrasonic absorption in aqueous solutions of lysozyme, *Biopolymers,* 15, 785, 1976.

193. **Shiao, D. and Sturtevant, J.,** Heats of binding protons to globular proteins, *Biopolymers,* 15, 1201, 1976.

194. **Roux, G., Roberts, D., Perron, G., and Desnoyers, J.,** Microheterogeneity in aqueous-organic solutions: heat capacities, volumes and expansibilities of some alcohols, aminoalcohols and tertiary amines in water, *J. Sol'n Chem.,* 9, 629, 1980.

195. **Remerie, K.,** Solvent Structure and Solvation Effects in Aqueous Solutions of 1,3-1,4 Dioxane, Dissertation, University of Groningen, Netherlands, 1984.

196. **Abugo, O.,** Studies of the Effect of Solvent on the Thermodynamics and Kinetic Reactivity of Genetic Variants of Human Erythrocyte Glucose-6-Phosphate Dehydrogenase with C6P, Dissertation, University of Ibadan, 1984.

197a. **Segawa, S. and Sugihara, M.,** Characterization of the transition state of lysozyme folding. I. Effect of protein-solvent interactions on the transition state, *Biopolymers,* 23, 2473, 1984.

197b. **Segawa, S. and Sugihara, M.,** Characterization of the transition state of lysozyme folding. II. Effects of the intrachain crosslinking and the inhibitor binding on the transition state, *Biopolymers,* 23, 2489, 1984.

198a. **Brunori, M.,** personal communication, 1985.

198b. **De Sanctis, G., Falconi, G., Giardina, B., Ascoli, F., and Brunori, M.,** Mini-myoglobin, *J. Mol. Biol.,* 200, 725, 1988.

199. **Howlett, G. and Schachman, H.,** Allosteric regulation of aspartate transcarbamolylasse. Changes in the sedimentation coefficient promoted by the bisubstrate analogues *N*-(phosphonacetyl)-L-aspartate, *Biochemistry,* 16, 507, 1977.

200. **Lumry, R.,** Dynamic factors in protein-protein association, *J. Mol. Liquids,* 42, 113, 1989.

201. **Lim, W. and Sauer, R.,** Alternative packing arrangements in the hydrophobic core of γ repressor, *Nature,* 339, 31, 1989.

202. **Gregory, R. and Lumry, R.,** Efficient thermodynamic treatment of protein linkage systems, *Biophys. J.,* 55, 345a, 1989.

203. **Doyle, M., Cera, E., and Gill, S.,** Effect of differences in optical properties of intermediate oxygenated species of hemoglobin A_0 on Adair-constant determination, *Biochemistry,* 27, 820, 1988.

204. **Zaks, A. and Klibanov, A.,** The effect of water on enzyme action in organic media, *J. Biol. Chem.,* 283, 22283, 1988.

205. **Barcroft, J.,** *The respiratory function of the blood, Part II,* Cambridge University Press, London, 1928.

206. **Benson, B. and Krause, D.,** The solubility and isotopic fractionation of gases in dilute aqueous solutions. II. Solubilities of the noble gases, *J. Solution Chem.,* 18, 803, 1989.

207. **Brandts, J., Hy, C.-Q., Lin, L., and Mas, M.,** A simple model for proteins with interacting domains. Applications to experimental DSC data, *Biochemistry,* 28, 8588, 1989.

208. **Kuwajima, K. and Baldwin, R.,** Exchange behavior of the H-bonded protons in the 3 to 13 helix of ribonuclease, *J. Mol. Biol.,* 169, 281, 1983; **Bierzynski, A., Kim, P. S., and Baldwin, R.,** A salt bridge stabilizes the helix formed by isolated C-peptide of RNase A., *Proc. Natl. Acad. Sci. U.S.A.,* 79, 2470, 1982.

Chapter 2

COMPLEXATION AND CATALYSIS IN BIOLOGY

R. J. P. Williams

TABLE OF CONTENTS

I. GENERAL INTRODUCTION

In this introductory chapter to biological catalysis involving metal ions, I have divided the material into two parts. Part A deals with the nature of the formed metal/protein complexes and their relation to catalysis. This is in essence a study of isolated molecules. Part B treats the problem of the formation of the molecules within biological systems and reveals that the true problem of biological speciation is very complicated. It is not a thermodynamic problem but one of the kinetics of locally controlled transport of both metal ion and protein and of the synthesis of the protein or other ligand.

II. PART A. INTRODUCTION TO CATALYSIS BY METAL PROTEINS

The catalytic act involves the introduction of a rate enhancement by a body which interacts with reactants in a cyclic manner. As well as stating that the catalyst is unchanged from before the beginning to after the end of the cycle we can state that it is changed during the cycle. Catalysis is not a static act but involves flow of electrons and atoms not only of the substrate but of the catalyst too. The catalyst flows in cyclic fashion while the reaction flows in a pathway direction, downhill usually in a thermodynamic sense. On a molecular scale then catalysis is about the dynamics of molecules, as well as atomic positions and attacking power, whether they be simple ions or large heterogeneous-phase surfaces or part of enzymes in biology. In this article I shall be concerned, therefore, with both the structures and dynamics of biological metal complexes in order to give a background to their catalytic activity. It follows immediately that an inhibitor of a catalyst can be a blocking agent which can be described by a binding description, say the key and lock hypotheses, but see below, or it can be a compound which alters the rates of cycling of the catalyst by restricting ease of motion, a kind of viscous drag. We return to these points later after we have outlined some basic features of catalysis especially by simple ions and complex ions.

A. THE NATURE OF CHEMICAL CHANGE

Catalysed chemical change such as the hydrolysis of an ester involves several steps. The first is the coming together of the reactants which is controlled by diffusion. The second is the binding of the two reactants which we usually look upon as involving minor distortion of the ground state of the reactants only. This assumption is convenient in the first step of docking but can be quite wrong. The free reactants often exist in an ensemble of states, one or more of which is selected by the catalyst. The next steps are taken to be a rearrangement of electrons and atoms to form a transition state and then an intermediate. This pair of steps may occur several times before bound products are obtained. The bound products are released in a step not unlike the binding step and the final process is diffusional separation of products. Intervention in any of these steps will either catalyze or inhibit the reaction. Any part played by a molecular catalyst involves exactly the same set of steps by the catalyst as those just outlined for the reactants except that the catalyst must complete a cycle of changes. We ask next how can a catalyst intervene in the steps of the reactants and it is immediately obvious that intervention is possible at the level of diffusion, binding of substrate and product, and electron and atom rearrangement during the formation of transition states and intermediates. Diffusion catalysts can only operate upon diffusion-limited reactions which are unusual and which we shall therefore leave until last but notice that three dimensional reactions can be reduced to two or one dimension. Catalysis by binding implies that the effect of the catalyst is just to allow reactants to rest close to one another giving an effective increase in concentration and/or a dominance of a good orientation of molecules to assist reaction. These effects could be looked upon as purely entropic and the role of attacking groups of the

TABLE 1
Some Small Molecule Substrates of Metalloenzymes

Small molecule substrate	Enzyme
CO_2	Carbonic anhydrase (Zn)
O_2	Many oxidases (Fe, Cu), e.g., (cyt. P-450)
H_2O_2	Catalase (Fe, Mn), peroxidases (Fe)
$O_2{\cdot}^-$ (superoxide)	Superoxide dismutases (Cu, Fe, Mn)
CH_4	Fe/S enzymes
N_2	Nitrogenase (Fe, Mo, V)
$CO(NH_2)_2$	Urease (Ni)
Radical rearrangements (Diols)	B_{12}(Co) enzymes
Electron transfer	Fe, Cu, centers
O_2 production from H_2O	Mn protein
H_2	Hydrogenases (Ni, Fe)

TABLE 2
Some Site-Specific Metal Substitutions

Enzyme	Substitutions
Carboxypeptidase (Zn)[a]	Mn, Fe, Co, Ni, Cu, Cd
Catalase (Fe)	Mn
Hemoglobin (Fe)	Co
Alkali phosphatase (Mg, Zn)	Mn, Fe, Co, Ni, Cu, Cd (both sites)
Phospholipase A.2 (Ca)[b]	Lanthanide series
Pyruvate kinase (K, Mg)	Tl(K), Mn(Mg)

[a] Replacements made on a 1:1 basis have very selective effects on activity.
[b] Replacements here are inactive but this is not true for all calcium enzymes.

catalyst is not necessarily involved. Finally the catalyst can act so as to perturb the relative positions of atoms or electrons in the complex between it and the reactants. This is the most conventional view of catalysis and involves attacking and leaving groups but note again that this step involves *movement* of atoms and electrons in the catalyst as well as in the reactants.

B. METALS IN PROTEINS

There is no new principle involved in catalysis by metalloproteins but they provide excellent subjects for study. In the first place metalloproteins handle virtually all the reactions of the primary *small* molecules of biology, Table 1. This is because they provide the only centers which can bind such molecules with any strength and they are really good attacking atoms. Organic chemicals (biochemicals) in contrast with metal ions provide very poor redox centers, poor local charge density, poor Lewis acid centers, and even not such very good bases in biology in comparison with say $M.OH^-$. That the molecular substrates of metalloenzymes, as opposed to organic enzymes, are small reduces the complexity of the catalysis. In the second place the metal ions in enzymes are exceptionally good probes of their own and the neighboring structure so that their intermediates are readily observed. This possibility arises from their electronic and nuclear structures as well as their redox properties. In the third place very skilful use has been made of site-specific *substitution* of metals (isomorphous replacement, no mutagenesis necessary) which can lead to telling *series* of observations where the *expected* effect of isomorphous substitution (contrast site specific mutagenesis) is known in advance. Examples are given in Table 2.

C. COMPLEXATION: PRIMARY CONSIDERATIONS

The catalysis of concern in this article is by metal ions bound by protein ligands to some

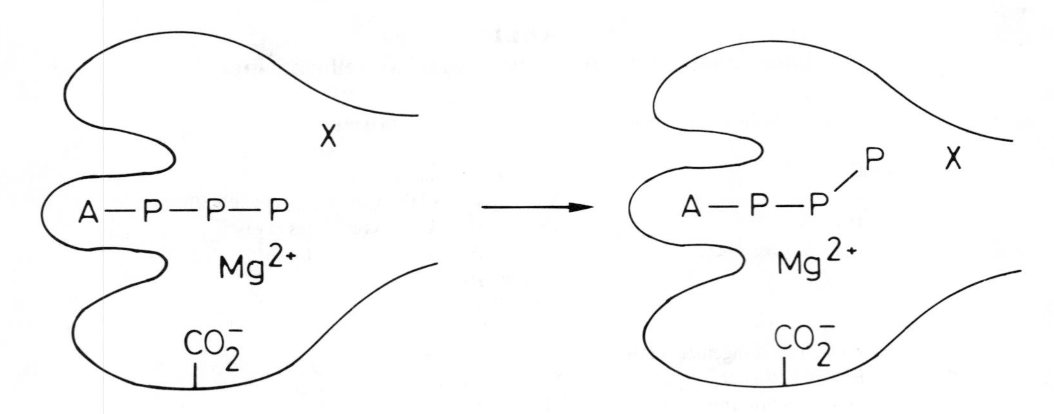

FIGURE 1. A schematic illustration of the way in which Mg^{2+} ions can aid phosphate transfer from ATP to X on an enzyme surface.

degree in a metalloenzyme. In this case the description and understanding of the structure and the energetics of the structure before entrance of the substrate requires a detailed analysis of protein fold energies and dynamics (see below). Complexation between a metal and a large cooperative body such as a protein is quite different from complexation of the kind.

$$[Cu(H_2O)_6]^{2+} + 4NH_3 \rightleftarrows [Cu(NH_3)_4(H_2O)_2]^{2+} + 4H_2O$$

Here the ammonia undergoes little more than a small change in electronic or molecular structure and the stereochemistry is dominated by the metal ion. In metalloenzymes this is not so. Depending on circumstance the metal or the protein can dominate the stereochemistry. Examples are given below.

In a second extreme type of catalysis the metal ion is not much linked to the enzyme and is more or less a rearrangeable part of the substrate. This is the case for the so-called enzyme-metal complexes. An example would be the hydrolysis of Mg^{2+}.ATP. Here the reaction is

$$Mg^{2+}.ATP + H_2O \rightarrow Mg^{2+}.ADP + P$$

In the most naive view the magnesium just moves from the $\beta\gamma$ phosphates of ATP to the α,β phosphates of ADP in the reaction so assisting the release of γ-P. Notice the description here is in terms of facilitated dynamics of magnesium ion along a path provided by the protein surface and the chain of phosphates of A.P.P.P. (Figure 1). Of course the protein can bias the relative energies of the two positions for Mg^{2+}. All intermediate cases between the extremes are anticipated.

D. THE ENTATIC STATE

The use of the word "entatic" by Vallee and Williams in 1968[1] was to bring to the attention of both chemists and biochemists the fact that metal ion sites in metalloenzymes must be expected to be distorted away from the observed ground state structures as seen in *simple* inorganic complexes. The reasoning went as follows. Catalysis requires a special series of reaction steps. Where a catalytic atom is involved in these steps, i.e., an inorganic or organic atom, it is highly likely that at some stage there are considerable energy barriers to the reaction to be overcome. The catalyst can then be improved if the ground state of the atom as found in the catalyst, enzyme, is already of a geometry which will reduce this barrier (Figure 2).

It is highly unlikely that this will be a commonly observed prefered stereochemistry of

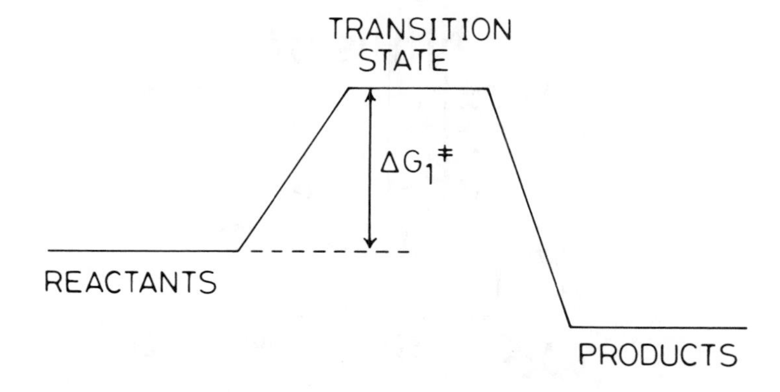

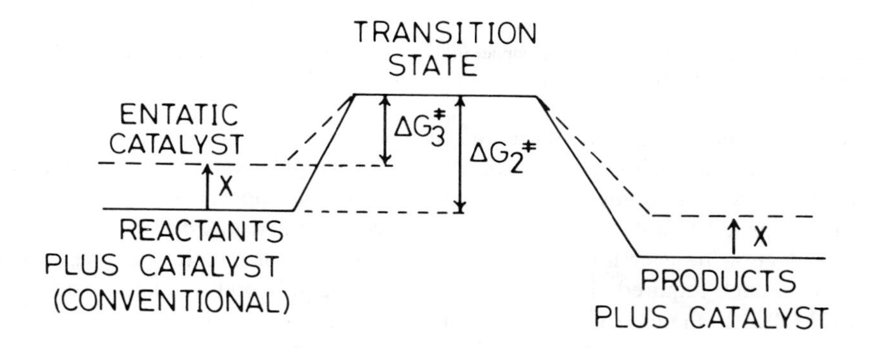

FIGURE 2. The introduction of strain energy at the catalytic site is illustrated by moving the energy of the catalyst by an energy, X, toward the transition state. The other points in the diagram are conventional, compare the top (usual diagram for catalysis) and the bottom figures.

the metal in monodentate or bidentate ligand complexes since in these complexes metal ions enter a relatively deep potential well dominated by their own stereochemical preference. An example is given by electron transfer in models and enzymes. The stereochemical cycle for electron transfer in copper bound to four ammonia or two ethylenediamine ligands is

$$\underset{\text{tetragonal}}{[Cu(II)(NH_3)_4(H_2O)_2]^{2+}} \rightleftarrows \underset{\text{tetrahedral}}{[Cu(I)(NH_3)_4]^{+}} + 2H_2O$$

This has two problem steps (1) the loss of $2H_2O$, i.e., change in coordination number, and (2) stereochemical changes. Such changes generate considerable kinetic barriers both at the level of entropy and heat energy in the reactions of small complex ions. In the copper blue proteins it has been pointed out for many years[2] that this situation is avoided by the constraints placed on the geometry of the copper *by the protein,* Figure 3. The copper in both oxidation state is in a four-coordinate site close to tetrahedral though one bond is very long.[3] The only change on reaction is a slight movement of the center of gravity of the copper relative to the protein ligand, i.e., bond length change. The result is a considerable rate enhancement of electron transfer catalysis by the copper ions relative to small complex ion electron transfer. We may say that there arose a design during evolution of a ground state site for copper related to its function. This is an entatic state and we expect it to appear frequently in

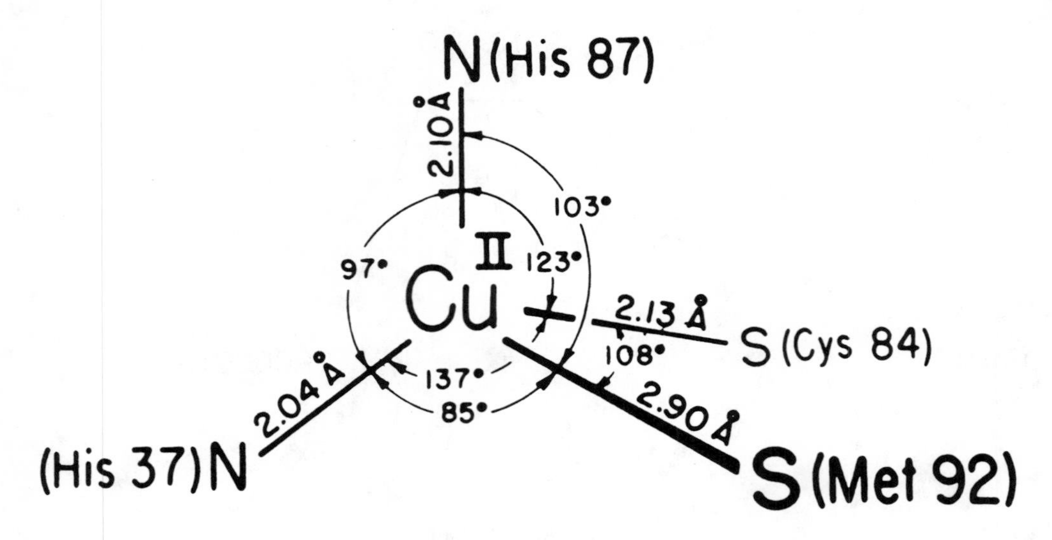

FIGURE 3. The coordination sphere around copper in the "blue" centers, for example, of azurin. The geometry is that expected for optimal value in electron transfer reactions.

enzymes. Notice that it has been described in relation to the rate process only. The thermodynamics at the site, redox potential, can also be manipulated by varying somewhat the geometry, the liganding atoms, or the protein surround or even the strain energy. If such use of strain was not found it would appear that evolution had failed.

Where electron transfer uses iron not copper this type of entatic state stereochemical control is not much required since with small donor ligands both oxidation states are octahedral (N,O) as is the case in low-spin cytochrome *c* and with large donor ligands both oxidation states are tetrahedral (S) as in the iron-sulfur proteins. Even here the protein has a role to play since the rate of electron transfer is not just related to the ease of valence state change per se; it is related to the redox potential of the donor and acceptor and internally to the complex, entatic or not, it will always require some relaxation of nuclear positions. Again the protein controls the distance of the redox center from its redox partners.

As far as the redox potential is concerned this is under the control of much more than the geometry and the donor atoms since the protein is the solvent, the second and third coordination spheres in an enzyme, and is specifically selected. The ordered "solvent" can readily introduce strains even in the coordination sphere so as to adjust the redox potential and since this is also related to long-range electrostatic interactions it can be used to actually trigger the reaction, allosteric control. The ordered "solvent" (protein) can also exert kinetic influence since the relaxation properties of this solvent, e.g., benzene ring flipping of phenylalanine, is controlled by the fold (Figure 4).[4]

[In passing the inorganic chemist will note that a metalloenzyme is very like a heterogeneous catalyst where the solid matrix is the protein which generates the special site and the special solvent conditions for catalysis.]

E. CONTROLLED DYNAMICS INTERNAL TO PROTEINS

As far as relaxation and triggering are concerned the important points can be illustrated by reference to *controlled* electron transfer. If the rate of the electron transfer step is to be controlled it can be managed by the manipulation of the allowed motions in the catalytic cycle, i.e., by a conformational adjustment which either assists or prevents reaction. Consider a reaction which proceeds as follows:

FIGURE 4. The fold of cytochrome *c* showing also the mobilities of protein side chains. The filled-in residues are immobile and hatched residues are some which are known to be mobile. The protein is the controlled solvent for the reaction and is neither a solid nor a liquid matrix but somewhere in between.

$$
\begin{array}{ccccc}
 & & M_X + S & \longrightarrow & M_YS \\
 & & & & \downarrow e \\
M_X + S^- & \longleftarrow & [M_YS']^- & \leftarrow & [M_YS]^-
\end{array}
$$

This is a compulsory order mechanism, *e*-transfer follows S binding (not the opposite), and reaction follows that. Relaxation is considerable in the first $M_X \rightarrow M_Y$ and in the final cycle-closing step. There is no electron transfer here without prior manipulation of the catalyst by the substrate, S. This is the path for electron transfer in cytochrome P-450 and it is effectively irreversibly linked to a series of large conformational changes. A simple electron transfer protein is not of this kind, e.g., copper blue sites and cytochrome *c* (but note CN^- addition reactions). In fact the iron in P-450 cycles through a series of spin-state as well as of oxidation state changes and the protein geometry must cycle too. Two other enzymes of this kind are diol dehydratases (B_{12}) and citrate synthetase.

Now we have used the above illustrations to show the following points about complexation of metal ions by proteins.

1. It can control the geometry around the metal of an observed (entatic) ground state.
2. It can control the physical and chemical properties of that ground state without necessary major effects on the coordination sphere, e.g., redox or acid/base properties or distances

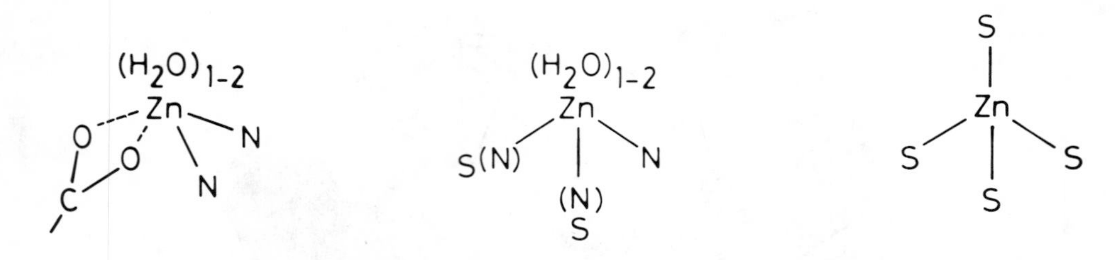

FIGURE 5. The coordination site of carboxypeptidase, carbonic anhydrase (3N), and alcohol dehydrogenase (N,2S) which are closely maintained no matter which metal ion is substituted. The third structure is that of cross-linking zinc in several enzymes and is not found at active sites.

to neighbors. These effects are in the second cordination sphere or in the solvent (ordered protein) sphere.

3. It can control the dynamics allowable around the site either disallowing major change as in simple electron transfer proteins or allowing major change before an activity can appear as in cytochrome P-450. This is the control over intermediates and transition states. Obviously we must ask if there are special properties of certain protein structures which can be linked in different ways to the coordination spheres of metal ions so as to produce very static or very dynamic proteins to aid very different functions (see below).

F. MOTIONS INVOLVED IN CATALYSIS: SMALL COMPLEXES AND PROTEINS

The danger faced by the inorganic chemist when he designs a small molecule catalyst using as the basis for design the physical or chemical properties of the metal site in the protein are that he will forget the role of the motions controlled by the protein fold. The design can include the correct stereochemistry of the ground state but in order to accommodate the steric strain around the metal (entatic state) it will be necessary to strap the ligands around the metal. Much flexibility in the coordination sphere is then lost. For electron transfer reactions this may not be a great loss but for substitution and complex insertion reactions it could be very damaging. More subtle are the mobilities in the second and other coordination spheres of the metal formed by protein amino acid side chains. It has to be remembered that an enzyme active site has a somewhat ''ordered'' cooperative solvent around it which is quite unlike a disordered liquid such as water or an organic solvent. An example which illustrates the peculiarity of the ''cooperative semiordered solvent'' is given by cytochrome *c* in Figure 4.

G. A SIMPLE TEST OF ENTASIS

The use of metal site-specific changes should reveal the fact that the protein creates the geometry or series of geometries for the metal ions compatible with the optimal catalytic condition. Excellent examples of the *enforced* ground state geometries of metal ions are found in carbonic anhydrase and carboxypeptidase where all metal ions give a similar ground state structure which is not the expected structure for most of the metal ions concerned (Figure 5). Now, however, the further factor of the somewhat dynamic nature of the active site cannot be created equally for all the metal ions since the different metal ions are in no way able with equal ease to change from the enforced preferred (by evolution) ground state geometry to some other geometry; yet this other geometry may well be required in an intermediate. The consequence is that metal ion activity not binding is highly selective. This is reflected in the well-known fact that Co(II), but not Ni(II) or Cu(II), is almost as good as Zn(II) in zinc enzymes (Figure 6). La(III) is a better replacement for Ca(II) than is Mg(II).

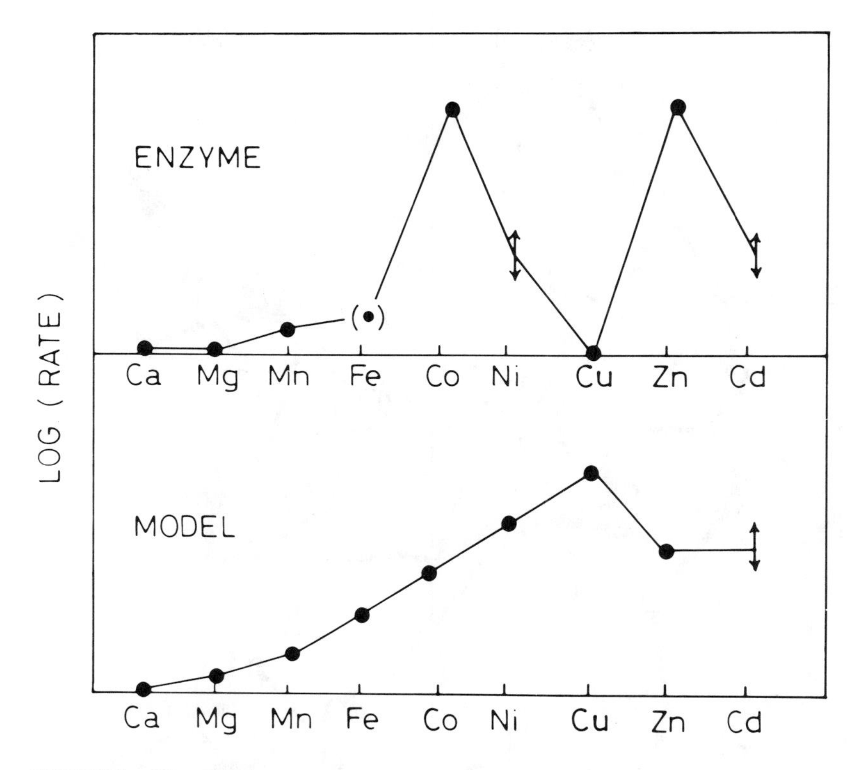

FIGURE 6. The catalytic power of the metal ions of the Irving-Williams series (below) as from stability constants and as in model reactions, i.e., Lewis acid strength, and an (above) observed order in an enzyme, carbonic anhydrase, illustrating the control exerted by the protein over the metal function.

Of course where oxidation state changes are needed in enzymes the principle follows most obviously since redox potential changes are metal specific.

H. THE PROPERTIES OF PROTEIN FOLDS

The fundamental problem we have before us is the link in metalloproteins between composition (sequence): structure (fold): dynamics: function.

The sequence controls everything so long as we include any such strongly bound agents as large organic molecules (coenzymes) and metal ions essential to the final function. Cytochrome *c* even folds around metal-free porphyrins but the fold is stabilized in metalloporphyrin complexes of the protein. Not all proteins fold on their own and this becomes important as we shall see.

Now there are two major classes of fold, α-helices and β-sheets, (Figure 7A and B). Proteins comprised overwhelming of one or the other class can be useful in letting us see why a sheet might have advantages over a helix and vice versa. In principle it might be thought that either could generate a similar site geometry and this is likely to be so but they will be of rather different stability and mobility. Table 3 shows that to our knowledge enzymes, and this includes metalloenzymes, are very rarely based on a folded set of helices with but one or two striking exceptions, Table 4. However, oxygen carriers (all of them), DNA-binding fingers, calcium trigger proteins, histones, some virus coat proteins, and many proteins in mechanical devices in muscle are derived from helices. Now a bundle of helices is known to undergo relatively easy helix/helix motion (no strong cross-links by H-bonds) while a sheet, especially a barrel sheet, does not deform easily (very strong, nonexchanging, H-bonds). The suggestion is strong and a clear logic supports it that while mechanical

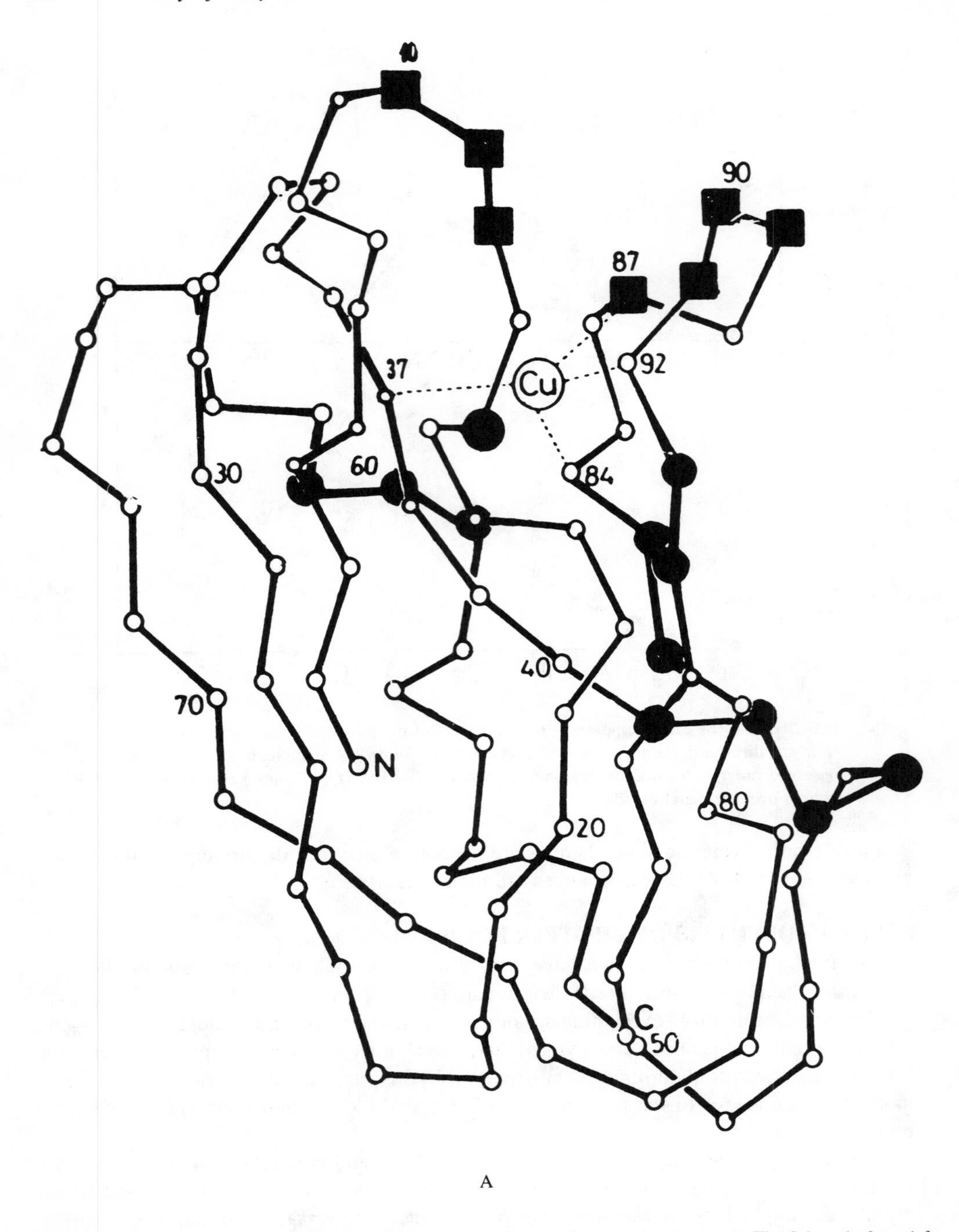

A

FIGURE 7. Examples of β-sheet and α-helical structures in metal-protein complexes. (A) The β-barrel of azurin[3] where the copper atom sits in an entatic site, Figure 3 and (B) the α-helical structure of calmodulin (Ca), a trigger protein (see Reference 5 and 6). Note the way the Cu cross-links distant parts of azurin but Ca in calmodulin does not. Note also in (B) that the triggering has been illustrated as a change in coordinating ligands which drives a set of helices into a relative slipping transformation. This is the same mechanism as in hemoglobin.

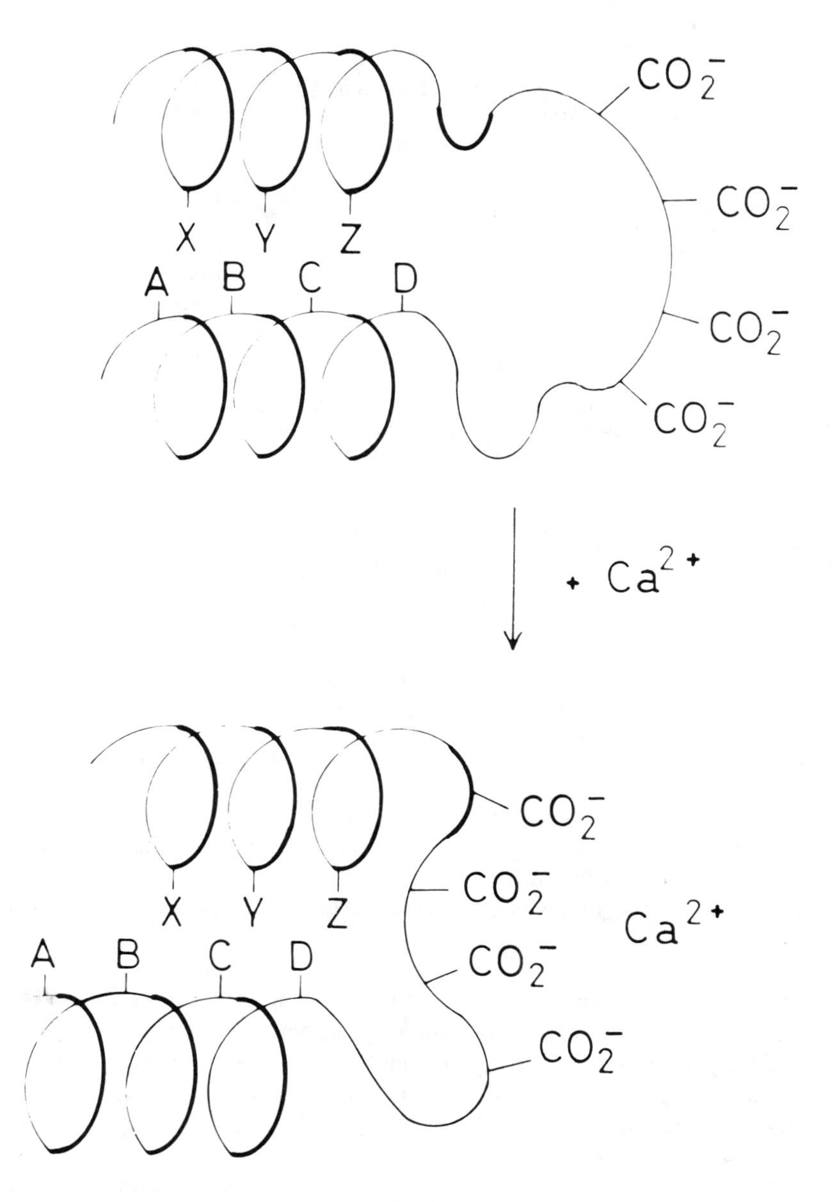

FIGURE 7B

allosteric devices must have the possibility of considerable motion (of helices) an enzyme must not. An allosteric enzyme should be composed of both helices and sheets (Table 4). In effect these points follow from the simple idea that enzymes must not be just catalytic they must be highly selective, not only to generate the catalytic site, but also the substrate binding zone. Selectivity of activity requires recognition to be as nearly specific as possible within the limitations of the dynamics of the reaction pathway. The whole active site should then require the stability of the β-sheet structure. Specificity and mobility are at opposite extremes, however, hence all proteins represent some compromise in order to generate activity. Let us look at specificity in terms of recognition.

I. RECOGNITION OF SUBSTRATE AT PROTEIN BINDING SITES

We can write a series of recognition devices of decreasing binding selection

TABLE 3
β-Sheet Enzymes

Enzyme	Metal	Function
Superoxide dismutase	Zn, Cu	O_2^- disproportionation
Plastocyanin	Cu	Electron transfer
Carbonic anhydrase	Zn	Hydration of CO_2
Many peptidases	Zn	Peptide hydrolysis
Alcohol dehydrogenase	Zn	Alcohol oxidation
Rubredoxin	Fe	Electron transfer
Ferredoxins Fe_2S_2, Fe_4S_4	Fe	Electron transfer

Note: Apart from the general point that enzymes are generally based on β-sheets Brandén has shown that the active site is frequently close to emergent loops, more than one, of the sheet and controlled by the sheet (personal communication).

TABLE 4
Largely Helical Enzymes

Protein (enzyme)	Special feature
Citrate synthetase	Compulsory order of substrate addition
Cytochrome P-450	Compulsory order of substrate addition
Cytochrome *c*	Immobilized by cross-linking
Cytochrome *b* and a_3	Compulsory connection with proton or electron pumps with a compulsory order

1. A dye fitting a mold (perfect). No dynamics.
2. A key fitting a lock (some adjustment of the moving parts of the lock, therefore less perfect). Dynamics during fitting.
3. A hand fitting a glove (adjustment of hand and glove, less perfect still). Here dynamics remain after binding.

Catalytic value is very low in (1) since it takes a long time for the dye to get into the perfect fit of the mold and no transformation in the bound combination is permitted (no dynamics). Catalytic strength may be quite good locally in (3) but a large number of hands can enter a glove to a good approximation. It is not so selective in binding but exclusion of large molecules can be possible. The reaction can be fast (good attacking groups) but only by increasing the number of fingers both of the hand and the glove is high selectivity achieved. This is probably the way protein surfaces come together with one another but it is not the best construction for a catalyst. The situation in (2) is much more close to ideal for catalysis. A series of very closely related states is permitted during certain transformations (of the lock). Selectivity is high. [The key, the substrate, should be a little mobile too for optimal value.]

It seems to follow almost logically against this background that β-sheets are better than helices for enzymes. They are cross-linked between chains of sequence by H-bonds. They are then of low but not zero mobility of side chains and will generate case (2) above. It follows that it is also a good idea to cross-link sheets (and helices) to give extra design possibilities for extracellular enzymes (–S–S– bridges or calcium-cross links) and to make intracellular enzymes (Zn-cross links). These structures are then not so open to hydrolysis. It is an equally good idea to make mechanical devices for transfer of material or energy from helical rods since these slip and rotate relative to one another through hydrophobic contacts in water and poor H-bonds in membranes. They should not be cross-linked. This is recognition case (3). The problem of metal complexation in biology is now to get the

TABLE 5
Calcium-binding Sites in some Proteins

	Sequence of ligands
Extracellular Enzymes	
Phospholipase A_2	Tyr-28, Glu-30, Gly-32, Asp-49, $2H_2O$
Staphyloccal nuclease	Asp-19, Asp-21, Asp-40, Thr-41, Glu-43, $1H_2O$
Trypsin (not at active site)	Glu-70, Asn-72, Glu-80, $2H_2O$
Thermolysin	Asp-138, Glu-177, Asp-185, Glu-187, Glu-190, $1H_2O$
	Asp-57, Asp-59, Glu-61, $3H_2O$
Intracellular Mechanical Devices	
Parvalbumin	Asp-51, Asp-53, Ser-55, Phe-57, Glu-59, Glu-62
	Asp-90, Asp-92, Asp-94, Lys-96, Glu-101, $1H_2O$
Calmodulin	Asp-20, Asp-22, Asn-24, Thr-26, Thr-28, Glu-31
	Asp-56, Asp-58, Asn-60, Thr-62, Asp-64, Glu-67
	Asp-93, Asp-95, Asn-97, Tyr-99, Ser-101, Glu-104
	Asn-129, Asp-131, Asp-133, Glu-135, Asn-137, Glu-140
Intestinal calcium binding protein	Ala-15, Glu-17, Asp-19, Gln-22, Ser-24, Glu-27
	Asp-54, Asn-56, Asp-58, Glu-60, Ser-62, Glu-65
Troponin C	Asp-27, Asp-29, Gly-31, Asp-33, Ser-35, Glu-38
	Asp-63, Asp-65, Ser-67, Thr-19, Asp-71, Glu-74
	Asp-103, Asn-105, Asp-107,Tyr-109, Asp-111, Glu-114
	Asp-139, Asn-141, Asp-143, Arg-145, Asp-147, glu-150

From Levine, B. A. and Dalgarno, D. C., *Biochim. Biophys. Acta,* 726, 187, 1983. With permission.

right metal with the right kind of protein for a particular function. For example calcium, will always be a poor catalytic group while zinc will be a good one.

It follows from this description that we should look on a metal ion in a protein in a new way. We should ask first does it bind to a local pocket along the length of a short stretch of sequence or does it cross-link widely separate parts (see Table 5[7] and Figure 8). For example, in hemoglobin and calmodulin the metal only binds to a local stretch of sequence while in phospholipase A_2 or rubredoxin it actually cross-links widely separated parts of the sequence. We must expect that the former style of interaction allows greater possibility for directed and possibly larger conformational transitions associated with metal-binding sites. The second allows the protein fold energy to play a large part in metal binding. An examination of protein structures along these lines helps to appreciate why metal enzymes are frequently large molecules even when they handle small substrates e.g., O_2 and CO_2.

Now when we consider small substrates we may more nearly revert to case (1) since we can exclude say all but diatomic molecules by restricting the mold. The binding to O_2 centers usually allows binding of O_2, CO, NO, CN^-, and so on but exclude all large molecules by physical exclusion not chemical recognition. In all events it is clear that we are required to examine the metal site of a protein and the mobility of the metal and the protein both locally at the site and at long range.

III. PART B. METAL COMPLEX CATALYSTS IN BIOLOGICAL SPACE AND TIME

A. THE SELECTIVITY OF METAL AND LIGANDS

To a coordination chemist one of the biggest mysteries about the sites of metal ions in enzymes is the selectivity of the combination. Table 6 gives some known liganding atoms

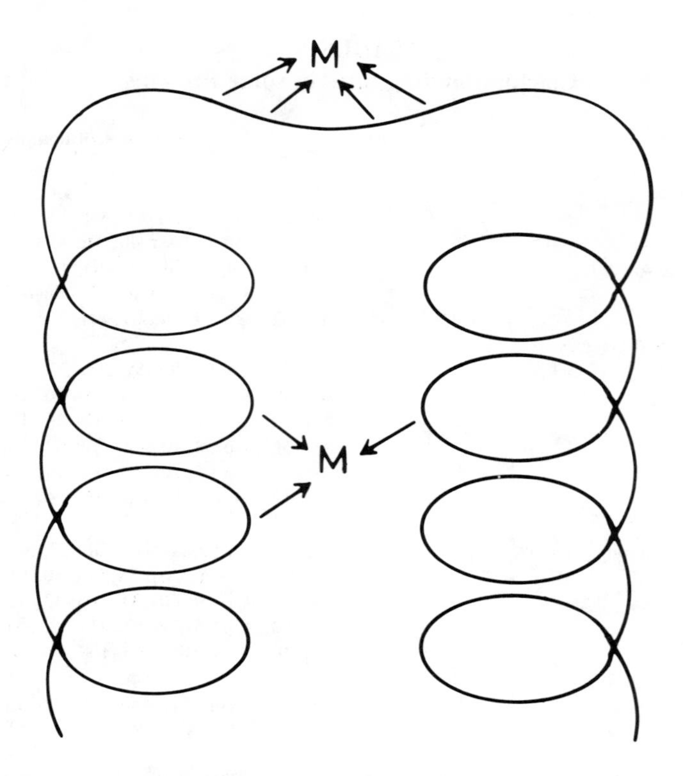

FIGURE 8. An illustration of the way in which a regulatory metal ion might bind along a sequence, see Figure 5b also, and contrast a cross-linking metal site as found in enzymes.

TABLE 6
Metal Ion Coordination in Proteins

Metal protein	Coordinating ligands
Chlorophyll proteins (Mg)	5 N donors
Haemoglobin (Fe)	5 N donors
Vitamin B_{12} enzymes (Co)	5 N donors + carbon
F-430 enzymes (Ni)	4 N donors + ?
Hydrolytic enzymes (Zn)	3 N donors or 2 N and 1 O donor
Alcohol dehydrogenase (Zn)	4 S donors
Rubredoxin (Fe)	4 S donors
Calcium enzymes	7—9 O donors
Magnesium enzymes	5—6 O donors + 1—2 N donors

of single metal ion sites. It is clear that although we can state generally that the strongest donors, N or S, bind to transition metal ions and not to Ca^{2+} or Mg^{2+} even this rule is broken in chlorophyll. Leaving this exceptional case aside we can write that at equilibrium with a given single site.

$$\text{Amount of metal bond} = K_{ML}^{eff}\,[M][L]$$

K^{eff} is an effective stability constant given the pH and salt of the surrounds, [M] is the free ion concentration, and [L] is the free ligand, protein, concentration. Given the observations of Table 6 and given all we know today about K_{ML}, which is a huge literature, we are forced

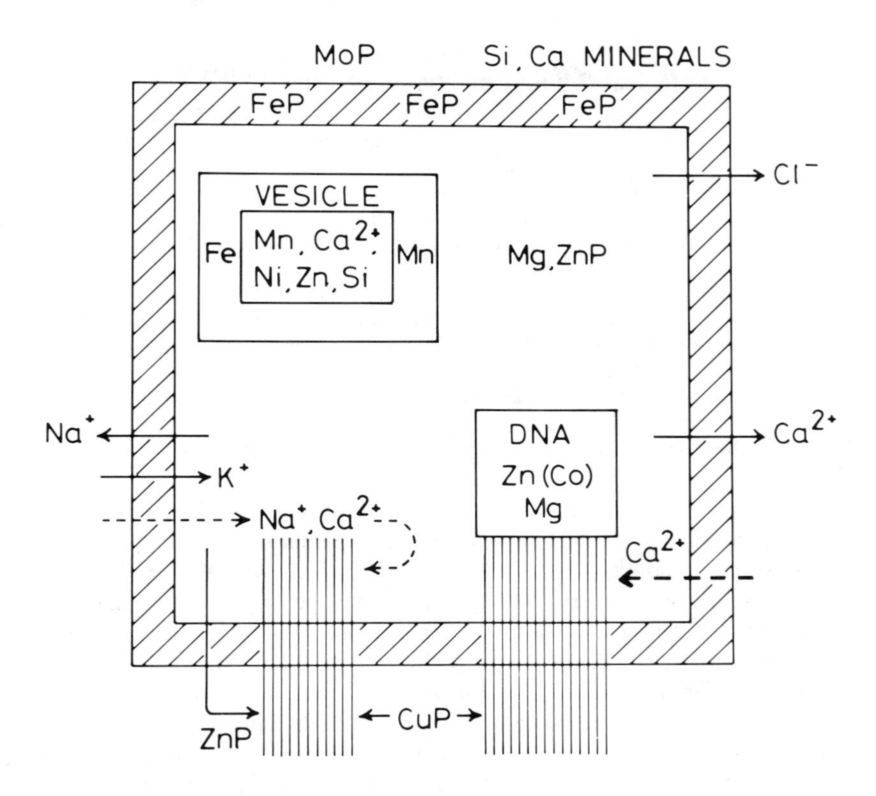

FIGURE 9. A schematic representation of the energized compartments for metals found generally in cellular systems. Each step of transfer and many of the modes of retention of the metal ions require complex formation. Dotted lines represent inward pulses used in signaling.

to say, I believe, that it is not possible to explain the selectivity observed in terms of stability constants alone. Copper(II) would bind preferentially everywhere following the Irving-Williams series.[8] In my opinion there is no way that Zn^{2+} rather than Cu^{2+} can be maintained in the observed sites of enzymes in cells unless $[Cu^{2+}]$ is very much less than $[Zn^{2+}]$ in these cellular regions. It must be the kinetic manipulation of [M] in different compartments due to various metal and ligand, transport, synthesizing, and pumping systems that yields the final observed distributions. In other words the distribution of complexed metals in biology is a ''deliberately'' manipulated kinetic system arising because [M] and [L] are set up in compartments by control of diffusion and synthesis. What are these compartments? A schematic example is given in Figure 9.

Inspection of a large number of biological systems leads to Table 7 which makes some clear general postulates. The copper of biology is extracellular or in a few special vesicles and is extremely rarely found in cells. There is an extremely low concentration of free copper in the cytoplasm. The zinc of biology is in another set of vesicles or is intracellular but very rarely freely extracellular except in pulses, i.e., release from vesicles. The iron, Fe(II), in eukaryotic cells is in organelles or is in outer membranes, but is extremely rarely in the cytoplasm or outside cells. Fe(III) is also low but it is frequently irreversibly trapped by chemical centers, e.g., porphyrin. The way in which this is managed is in a designated vesicle. This vesicle fractionation is managed by synthesis in the Golgi apparatus. Thus, the problem of selective association of metal and ligand is solved in biology by control over space (diffusion) as well as through the stability constants. The biochemist and chemist is fooled into thinking that he is dealing with an open competition simply because he uses

TABLE 7
Biological Compartmentation of Some Metal Ions

Metal	Compartment
Copper	Usually extracellular or in vesicles Very rare in cytoplasm of eukaryotic cells
Calcium	Usually extracellular or in vesicles Low in cytoplasm
Magnesium	Universally around 10^{-3} *M* except in special vesicles
Iron	Often in membraneous systems and organelles; low in cytoplasm except ferritin vesicles
Zinc	Universally distributed but high in certain vesicles
Nickel	Not in eukaryote cells except some vesicles
Manganese	Mainly confined to organelles and vesicles when in eukaryotic cells

Note: Every vesicle has selective uptake devices for one or two but not all metal ions.

analysis of broken up systems instead of local microanalysis. His application of simple acid/base equilibria is incorrect. It is movement across membranes and trapping in irreversible chelate traps which is critical to the particular ML which will appear. For example the mitochondria is a chemical factory for synthesis of protoporphyrin and uptake and insertion of iron while the chloroplast looks after Mg and chlorophyll. In supplying both organelles the proteins which will bind these cofactors are also controlled. There is virtually no spatial confusion and therefore no chemical confusion.

The clearest comparative example I can give of this kinetic control over complexation is as follows. (Kinetic control means that association is biased by the control over diffusion.) All the ring chelates for Ni, Co, Mg, and Fe (and an odd one for Cu) which form the coenzymes F-430, vitamin B_{12}, chlorophyll, and porphyrin, respectively (and uroporphyrin for Cu in some pigments), are formed from the same precursor, Uroprophyrin, Figure 10. The metals are all put into the rings as divalent cations. With such rings there is only one thermodynamic binding order for them all as for all other organic centers, i.e., the Irving-Williams order

$$\mathrm{Cu} > \mathrm{Ni} > \mathrm{Co} > \mathrm{Fe} > \mathrm{Mg}$$

This order even applies when Ni and Co are low spin. Iron is high spin here without nitrogen donors in the 5th and 6th positions. The conclusion has to be that the observed compounds are formed by kinetic traps, based on insertion steps which depend on the available free metal ion [M] in parts of space where the final syntheses of the ligands occur. The particular speciation, an M with a particular L, is due to the locality of the M and L and the insertion steps and not due to thermodynamics.

Now selectivity of combination in evolution is a reflection of selectivity of function as listed in Table 1. Different metal ions are forced to do different things. Following this rule with the ideas of the entatic state we conclude, for example, that not only does nickel go with the F-430 ring but that nickel and the F-430 ring have some functional strength not possessed by nickel with any other of the five possible rings nor with other metal ions and the F-430 ring. Magnesium chlorophyll properties are another evolved choice making chlorin an evolved choice for Mg^{2+} not Ni^{2+}, Fe^{2+}, or Co^{2+} (or Cu^{2+}). Again no special cofactor system has appeared yet for Mn^{2+} but there is one special to molybdenum. There is only one other world familiar to the chemist where this type of kinetic selection is refined — the organic reagents used in (qualitative) analytical chemistry. It is worth looking again at analytical procedures which use phase separations.

All those who were fortunate enough to suffer as students the agonies of classical qualitative analysis know some golden rules. An outstanding one is to apply group (analytical

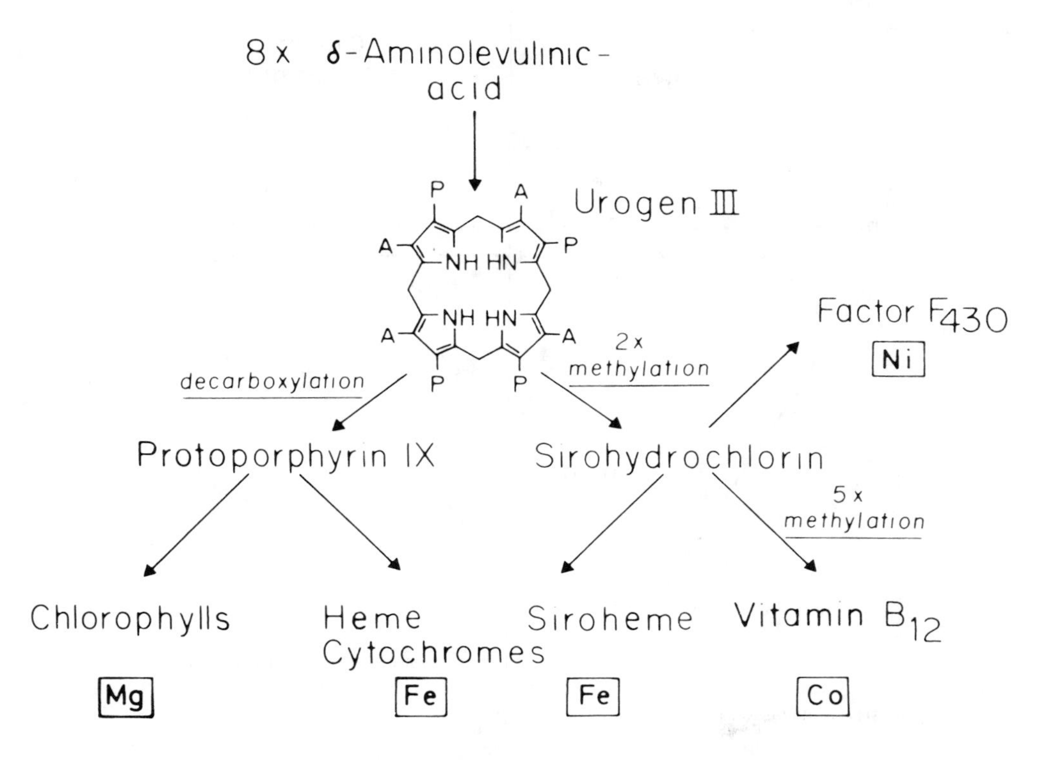

FIGURE 10. A diagram showing the relationship between the organic ligands for different metals Mg, Fe, Co, Ni, and Cu derived from a uroporphyrin. There is no known special cyclic ligand for Mn or Zn but there is a special noncyclic chelate for Mo.

not periodic table grouping) separation before the use of selective organic reagents. This limits the confusion which would arise through free competition for the organic reagents. Here the methods follow the practice in biology — separate, before making final combinations, by moving different metal ions into different vesicles (test tubes).

In analysis the separations are managed in fact by transfer following phase separations. Thus, early in qualitative analysis copper is removed from Ni, Co, Fe, and Mn by precipitation in acid media as a sulfide. Then the solution is transferred to a new environment (test tube) where iron is precipitated as a hydroxide. Meanwhile, the organic reagent test for copper is carried out in a separate test tube. Each metal is precipitated, separated, and then redissolved in isolation to combine with a selected reagent. This is the description of biological speciation as well as of analytical group-table analysis.

Separation however needs selectivity too and here arises the power of transfer steps. Thermodynamics gives the equilibrium

$$M + L_1 \rightleftarrows ML_1$$

Transfer adds to this

$$M + L_1 \rightleftarrows ML_1 \xrightarrow{k_{tr}} ML_2 + L_1$$

where k_{tr} is the rate of transfer. No matter what the strength of formation of ML its transfer can be made to depend not on this strength but on the stereochemistry of ML_1 since the to-be-transferred ML_1 is recognized not M or L_1. Now metal ions generate differently shaped complexes so that from a pool of say Zn and Cu we have

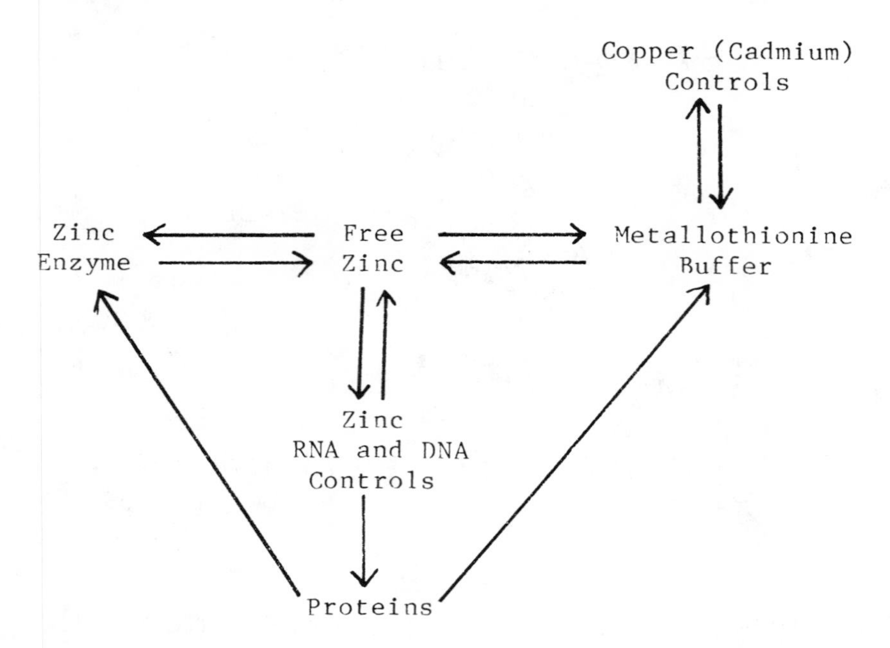

FIGURE 11. The cross-talk from free zinc to different control systems which regulate protein synthesis, metal-ion buffering, homeostasis, and so on. There is an intense feedback between synthesis and metal ion levels.

$$\text{Zn, Cu} + L_1 \begin{array}{l} \swarrow\!\nearrow\; ZnL_1 \nearrow \begin{array}{l} ZnL_1 \text{ in a space A} \rightarrow ZnL_2 \text{ catalyst} \\ \text{(tetrahedral)} \end{array} \\ \qquad + \\ \nwarrow\!\searrow\; CuL_1 \searrow \begin{array}{l} \text{(tetragonal)} \\ CuL_1 \text{ in space B} \rightarrow CuL_3 \text{ catalyst} \end{array} \end{array}$$

where L_1 is a ligand, e.g., metallothionine, which binds the two different metal ions. In this case copper goes outside the cell or to one set of vesicles, e.g., chromaffin granules, and zinc goes inside the cell or to another set of vesicles, e.g., the islets of Langerhan (Figure 11). In the vesicles the specific enzymes are then made from proteins L_2 and L_3 which are in those selected vesicles. The secret of success is that apometallothionine should be a random coil protein (which it is) or a loose set of helices and when folded it should take up a structure dependent on the metal ion to which it is bound, which is true, but that carboxypeptidase in pancreatic vesicles (Zn) and β-hydroxylase in chromaffin vesicles (Cu) should fold in β-sheets so as to create cavities for entatic Zn and Cu in different vesicles or other parts of space. In part this is the opposite principle to the use of the entatic state to give enzyme sites, since here the transfer protein metallothionine is dominated in its fold by the metal ion coordination geometry and not the reverse.

The same principles apply in the context of the competition between calcium and magnesium for the helical protein calmodulin. Binding is such that logK for calcium greatly exceeds logK for magnesium but in cells [Mg] > [Ca] by 10^4. Free calcium is pumped out of the cell. It follows that Mg^{2+} binds to calmodulins only to some extent but even then it does not act as a trigger since MgL (L = calmodulin) has a different fold (somewhat) from CaL. The relationship here is

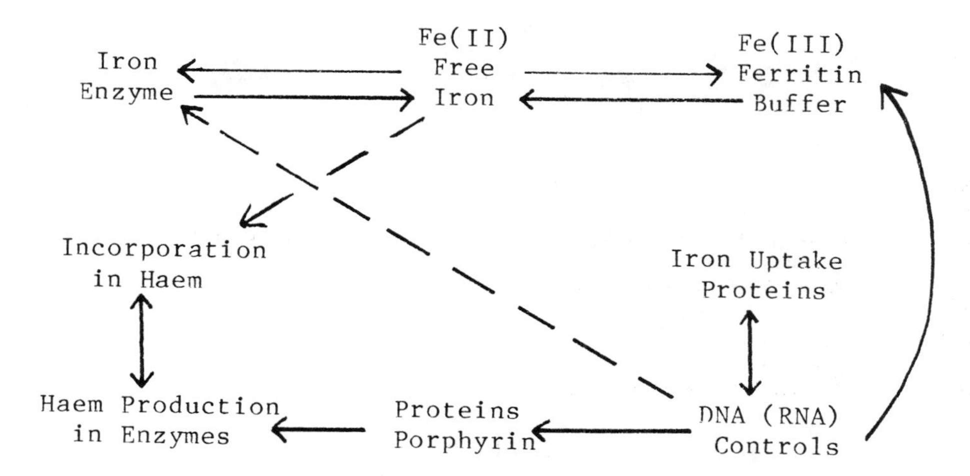

FIGURE 12. The cross-talk between the iron complexes and synthesis in order to maintain iron levels differentially in different compartments.

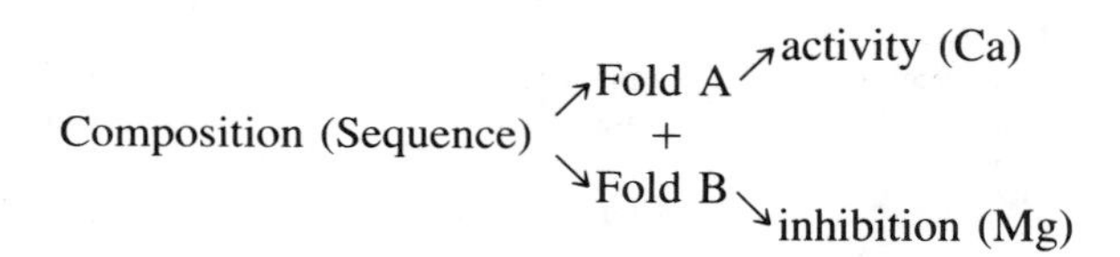

In both these cases the selection in thermodynamic terms is not so important as the carryover of the structural specificity into the recognitional and dynamical potential of the product, all of which are controlled by kinetic switching of concentration gradients, i.e., letting Ca^{2+} into a cell.

By way of contrast let us take proteins where selection in the initial state ML helps to create the functional property. A good example is Cu/Zn superoxide dismutase, a β-sheet protein which is in the cytoplasm of all eukaryotic cells.* (Note this very exceptional distribution of copper.) Here the binding constant for Cu in the Cu site is very high and relatively much greater than in the zinc site. The Zn site is stereochemically adverse for Cu. In fact the copper site in superoxide dismutase is remarkably strongly built. The whole protein is based on a tight β-sheet barrel, Figure 7a, compare the loose structures of calmodulin and metallothionine, and the sites for the two metals are engineered so that zinc binding and cross-linking generates a hole of great binding power for copper almost specifically. Zinc binds where it does because copper is so effectively removed. The protein generates the selectivity and the entatic conditioning of the copper is aided by the zinc.

B. REGULATION

It is hoped that at this stage biological systems of metal ion complexes are seen not as simple chemical systems but as regulated syntheses in compartments far from general equilibrium (Figure 11). Another illustration is given in Figure 12. Of course the system obeys chemical principles but it is locally energized. Moreover the state of the system, the concentration of species, is fed back to DNA/RNA particles which respond (Figures 11 and 12). This type of response has been discussed elsewhere. We do not know to how many metal ions it applies but probably it exceeds Cu, Zn, Fe, Mn, Mg, and Ca.

* See Chapter 21 by Tainer *et al.*, for an extensive discussion of the Cu/Zn superoxide dismutase — editorial note.

TABLE 8
Element Pulses

Element	Pulse (message)
Na^+	Nerve conduction
K^+	Nerve conduction
Ca^{2+}	Muscle, synapse, general signal for metabolic control via hormones, fertilization, death at high concentrations
Mg^{2+}	Release from chloroplast: plant metabolic control (dark → light)
Zn^{2+}	Fertilization, growth
Fe/Mn	None known
Cu^{2+}	None known
H^+	Digestion, plant metabolism control

TABLE 9
Systems for Homeostasis

Metal ion	Buffer	Pump	Additional pump
Na^+	None but high concentration	ATP-ase	Kidney
K^+	None but high concentration	Electroneutrality (Na^+ OUT)	Kidney
Ca^{2+}	Parvalbumin Calcium phosphate	ATP-ase	Bone Cells
Fe^{3+}/Fe^{2+}	Ferritin	Transferrin (Ferroxamin) Receptor	Mitochondria
Mn^{3+}/Mn^{2+}	(Ferritin?) Phosphate Ppt.	ATP-ase (Ca^{2+}) Transferrin	Mitochondria, golgi, chloroplast
Cu^{2+}	Metallothionine	Receptor (?) (Albumin)	Lysozome
Zn^{2+}	Metallothionine	Receptor (?) (Albumin)	Various vehicles (Golgi)
Mg^{2+}	ATP(?)	?	Chloroplast
H^+	Proteins, RP^{2-}, HCO_3^-	ATP-ase	Very general in membranes

C. PULSED SYSTEMS AND HOMEOSTASIS

That the inorganic ions or their complexes are in energized environments either due to concentration gradients or electric fields means that they can be pulsed from their resting environment by a breakdown in the constraints which maintain it. Such rapid pulsing forms the basis of much signaling in biology (Table 8), where the time scale can go from nerve pulse (K^+), muscle triggers (Ca^{2+}), energy capture (H^+) on a fast scale to the slower effects likened here to hormonal control of variations in Mn^{2+}, Fe^{2+}, and Zn^{2+}.

Against this background and over a longer time scale cell compartments in steady states have to keep metal ion levels fixed. This is called homeostasis and depends on complex feedback controls to ligand synthesis. Homeostasis[10] cannot be described in detail here but Table 9 gives some examples.

IV. CONCLUSION

This chapter is intended to outline the basic features of metal ion coordination chemistry in biology. The expected (best thermodynamic) combination of metal and ligand based on selected chemical donor atoms or stereochemistry is not often found. Some compromise to generate appropriate function not only at a site but also more largely in biological space and time have been entered into. The observed condition of a metal ion is then difficult to understand except in a local context. Magnesium is not quite the same "element" in chloroplasts as in mitochondria because of the manipulations by which it is handled. Thus, all essential elements have a special biochemistry due to the energized condition of living systems (see Figure 13).

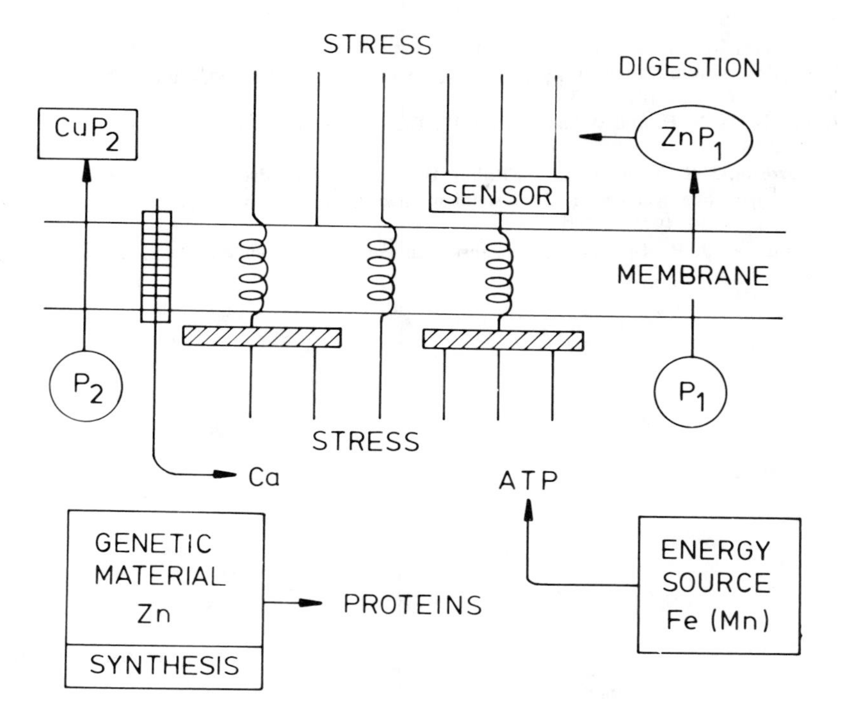

FIGURE 13. The coupling of energy (bottom right) due to metal enzyme energy capture devices into the functioning of metal ions where ATP is used to transport the ions and to generate stress in filaments. The direct uses are triggered by calcium complex formation. However, the stress fibers are regularly degraded or cross-linked by zinc and copper proteins, respectively, so as to allow expansion of cell systems. The zinc anc copper catalysts are moved out of cells to interact with the filaments under time-controlled and energized systems. The whole system responds to physical stress and to chemical signals at sensors. How far the feedback to DNA and protein synthesis goes is unknown yet (bottom left-hand).

REFERENCES

This is a general article. There are now two long series of books on metal ions in biological systems and these give references to specific points. I therefore give these two general reference works plus the contents of the chapters of the present volume and the following specific references.

1. **Spiro, T. G., Ed.,** *Metal Ions in Biology,* Vol. 1 to 10, Wiley-Interscience, New York.
2. **Sigel, H., Ed.,** *Metal Ions in Biological Systems,* Vol. 1 to 23, Marcel Dekker, New York.

1. **Vallee, B. L. and Williams, R. J. P.,** Metallo-enzymes: the entatic nature of their active sites, *Proc. Natl. Acad. Sci. U.S.A.,* 59, 498, 1968.
2. **Williams, R. J. P.,** Heavy metals in biological systems, *Endeavour,* 26, 96, 1967.
3. **Colman, P. M., Freeman, H. C., Guss, J. M., Murata, M., Norris, V. A., Ramshaw, J. A. M., and Vantatappa, M. P.,** X-ray crystal structure analysis of plastocyanin at 2.7Å resolution, *Nature,* 272, 319, 1978.
4. **Williams, R. J. P., Moore, G. R., Porteous, R., Robinson, M. N., Soffe, N., and Williams, R. J. P.,** Solution structure of cytochrome *c*, *J. Mol. Biol.,* 183, 409, 1985.
5. **Williams, R. J. P.,** The physics and chemistry of calcium-binding proteins, *Ciba Found. Symp.,* 122, 145, 1986.

6. **Kretsinger, R. H.,** Calcium-binding proteins, *Annu. Rev. Biochem.,* 45, 239, 1976.
7. **Levine, B. A. and Dalgarno, D. C.,** The dynamics and function of calcium-binding proteins, *Biochim. Biophys. Acta,* 726, 187, 1983.
8. **Irving, H. M. N. H. and Williams, R. J. P.,** The stability of transition metal complexes, *J. Chem. Soc.,* 3182, 1953.
9. **Richardson, J. S., Thomas, K. A., Rubin, D. H., and Richardson, D. C.,** Crystal structure of bovine Cu,Zn superoxide dismutase at 3Å resolution: chain tracing and metal ligands, *Proc. Natl. Acad. Sci. U.S.A.,* 72, 1349, 1975.
10. **Williams, R. J. P.,** Bio-inorganic chemistry: its conceptual evolution, *Coord. Chem. Rev.,* 100, 573, 1990.

Chapter 3

MECHANISTIC ROLE OF BIOTIN IN ENZYMATIC CARBOXYLATION REACTIONS

A. S. Mildvan, D. C. Fry, and E. H. Serpersu

TABLE OF CONTENTS

I. INTRODUCTION

The prosthetic group D-biotin serves as an intermediate carrier of CO_2 on several carboxylating enzymes.[1,2] An example is pyruvate carboxylase, a metallobiotin enzyme,[3] which catalyzes the following reactions:

$$\text{ATP} + \text{HCO}_3^- + \text{E-biotin} \overset{\text{Mg}^{2+}}{\rightleftharpoons} \text{ADP} + \text{P}_i + \text{E-biotin-CO}_2 \quad (1)$$

$$\text{E-biotin-CO}_2 + \text{pyruvate} \rightleftharpoons \text{oxaloacetate} + \text{E-biotin} \quad (2)$$

Equation 1 is a synthetase reaction, common to most biotin enzymes and Equation 2 is a transcarboxylation. The enzyme, transcarboxylase, is unique in that it catalyzes two consecutive reactions of type (2).[2] This review will consider recent experiments bearing on the mechanisms of these reactions.

II. MECHANISM OF CARBOXYBIOTIN SYNTHETASE

Three mechanisms have been proposed for Reaction 1 and are shown in Figure 1. Mechanism 3, which predicts double inversion or net retention of configuration at phosphorus, is ruled out by the detection of stereochemical inversion at phosphorus in the pyruvate carboxylase reaction.[4] Although mechanism 2 and nonconcerted variants of this mechanism[5] cannot be ruled out by the stereochemistry of the reaction, mechanism 1 is suggested by the formation of ATP from ADP and carbamoyl phosphate (an analog of carboxyphosphate) catalyzed by sheep liver pyruvate carboxylase and acetyl-CoA carboxylase, in the latter case, in the absence of biotin.[6,7] As pointed out by Hansen and Knowles,[4] mechanism 1 is also supported by Wimmer's failure to detect positional isotope exchange in ATP[8] and by Ashman and Keech's failure to detect ADP-ATP exchange with pyruvate carboxylase, in the absence of bicarbonate,[6] suggesting that bicarbonate rather than biotin is directly phosphorylated by ATP.

III. MODEL REACTIONS RELEVANT TO ENZYMATIC CARBOXYLATION OF BIOTIN

A long-standing mechanistic question has been how a very weak nucleophile like biotin, a derivative of urea, can attack and get carboxylated either by carboxyphosphate to complete step 1, or by a very weak electrophile or carboxylating agent like oxaloacetate, as in the reverse of equation 2, a transcarboxylation reaction. Retey and Lynen proposed that biotin is enolized, presumably by the enzyme, to form a more nucleophilic species, analogous to imidazole[9] (Equation 3), and Bruice and co-workers obtained evidence for the nucleophilicity of enolbiotin analogs in model reactions.[10,11] This reasonable assumption was widely incorporated into a number of proposed mechanisms for biotin enzymes.[12-14] No direct evidence for the enolization of biotin existed until an NMR study of the exchange of the NH protons of biotin itself and of several biotin derivatives with water protons[13-15] revealed that under mild conditions (25°C, pH 7.4) the enzymatically reactive N-1 proton exchanges with water protons at a rapid rate, comparable to the rates of carboxylation of enzyme-bound biotin. Hence, in principle, no additional catalysis by the enzyme over that provided by 10^{-7} *M* $[OH^-]$ at neutral pH is necessary to deprotonate the amide NH group in the enolization of biotin. The enzymatically reactive N-1 proton of biotin and of the biotin methyl ester exchanged with water protons six and three times faster respectively than did the N-3 proton, presumably due to steric hindrance from the side chain at N-3 (Figure 2),[13-15] and possibly

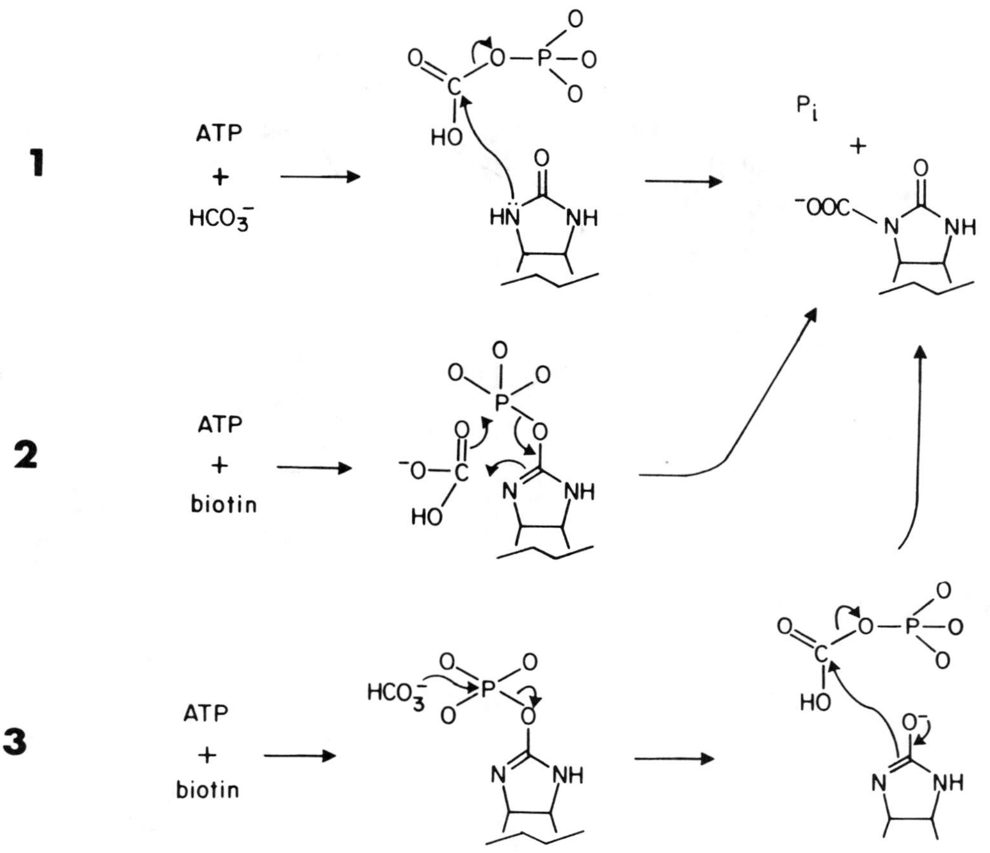

1 ATP + HCO_3^- → → P_i + ^-OOC

2 ATP + biotin →

3 ATP + biotin → →

FIGURE 1. Three alternative mechanisms proposed for the ATP-dependent carboxylation of biotin (From Reference 4 with permission). A nonconcerted variant of mechanism 2 has also been proposed. (From Berkessel, A. and Breslow, R., *Bioorg. Chem.*, 14, 249, 1986. With permission.)

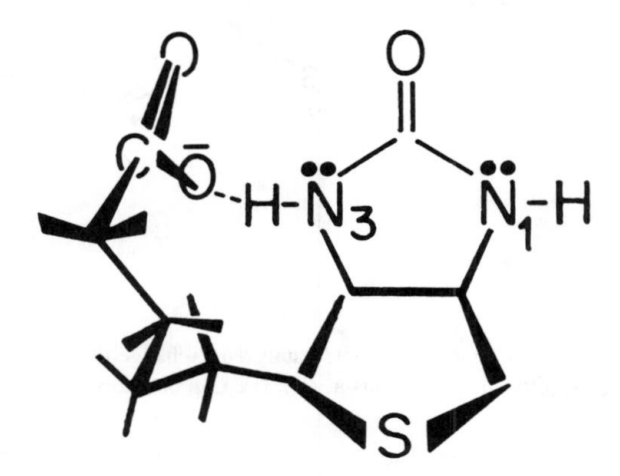

FIGURE 2. Conformation of D-biotin in solution as suggested by the chemical shifts of its NH protons.[13,14]

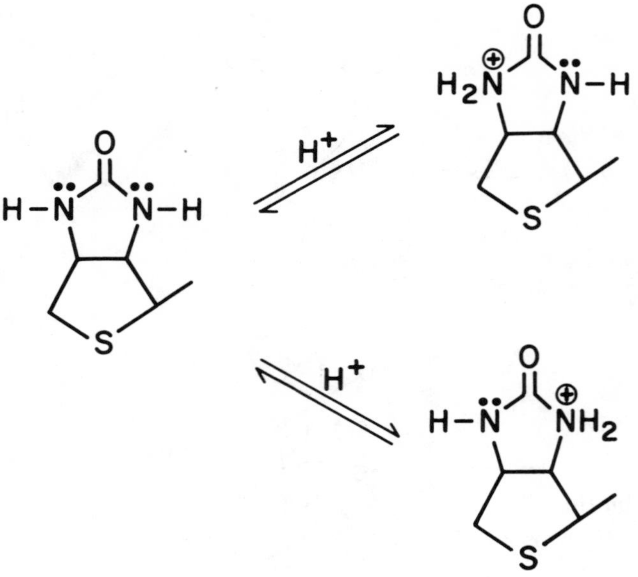

A

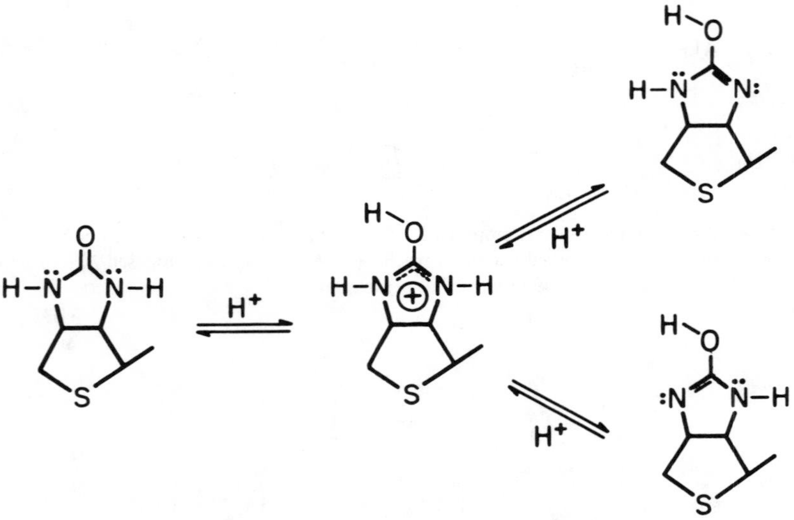

B

FIGURE 3. Alternative mechanisms of acid-catalyzed exchange of the amide NH protons of biotin. (A) The N-protonation mechanism. (B) The O-protonation mechanism.

related to the fact that N-1 is out of the plane of the imidazolone ring of biotin.[16] As in other amides and peptides, both acid and base catalysis of proton exchange in biotin were observed, first order in $[H^+]$ below pH 6 and first order in $[OH^-]$ above pH 6. Mechanisms for acid and base catalysis of proton exchange are shown in Figures 3 and 4, respectively.

Two alternative mechanisms for acid-catalyzed NH exchange are possible, initial N protonation (Figure 3A) or initial O protonation, followed by NH deprotonation (Figure 3B), the latter mechanism leading directly to the enol form of biotin. The N-protonation mechanism

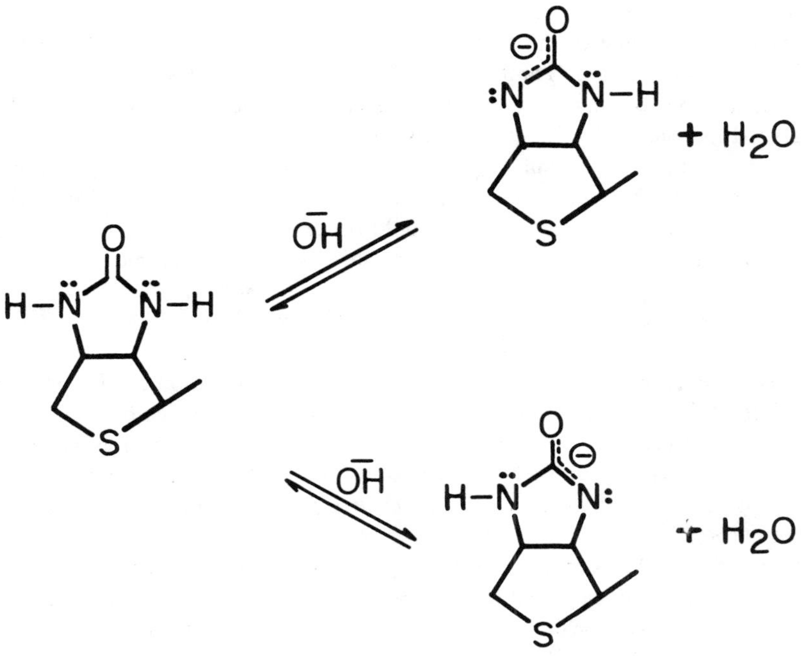

FIGURE 4. Mechanism of base-catalyzed exchange of the ar nide NH protons of biotin.[13,14]

is favored in amides with substituents of low electrophilicity, as in biotin, while O-protonation is favored with more electrophilic substituents.[17] Nevertheless, recent solvent deuterium and primary ^{13}C isotope effects on the decarboxylation of ^{13}C-carboxybiotin have been interpreted in terms of an O-protonation mechanism.[18]

Another mechanistic proposal relevant to the O-protonation mechanism made by one of us was that of transannular interaction of the sulfur of biotin with the carbonyl carbon to facilitate carbonyl polarization and thereby to increase the nucleophilicity of the ureido system.[12] Indirect evidence for this suggestion was provided by the significantly reduced activity of O-heterobiotin in enzyme-catalyzed carboxylations.[19] In the crystalline state, the distance from the sulfur to the carbonyl carbon of biotin exceeds van der Waals contact by only 0.13 Å.[14,16] However, early IR and ^{13}C NMR data failed to detect significant transannular interaction in D-biotin at neutral pH,[20] and more recent X-ray[5] and ^{13}C NMR studies[5,21] have failed to detect such interaction at very low pH or even when the electrophile BF_3 had added to the carbonyl oxygen. The detection of acid catalyzed proton exchange of the 1′-NH of biotin, and its methyl ester, second order in $[H^+]$ at 25°C, was interpreted as resulting from such transannular interaction.[14] However, this second-order dependence could not be confirmed at the lower temperature of 10°C where the slowed rates could be measured more accurately,[15,22] or by the independent NMR method of NH line broadening at 25°C,[23] although the low-temperature studies did yield a kinetic order in $[H^+]$ slightly greater than unity at N-1 of biotin (1.23 ± 0.07) and its methyl ester (1.30 ± 0.06), and an order indistinguishable from unity at N-3.[15,22] Hence, while a second-order mechanism may begin to contribute at extremely low pH, we must conclude that transannular interaction is a hypothesis for the role of sulfur in biotin which remains unsupported by existing physical or kinetic data.

Base-catalyzed NH exchange in biotin is dominant above pH 6 and is more rapid than in almost all other urea derivatives studied,[13,14,23] including O-heterobiotin.[14] The only exception is the D-sulfoxide of biotin in which base-catalyzed NH exchange is about fivefold faster than that in biotin, presumably due to stabilization of the anionic imidate or "enolate" intermediate by the partially positive sulfoxide sulfur.[15] Indeed, at pH 7.4 and 25°C, the rate constant for exchange of the 1′-NH of biotin (58 s^{-1}) is similar to k_{cat} values for the carboxylation of enzyme-bound biotin (15 to 74 s^{-1}).[13,14] Hence the biotin enolate, although low in concentration, is clearly kinetically competent to function in catalysis,[13,14,23] and the base-catalyzed mechanism is therefore likely to be relevant to the enzymatic mechanism of carboxylation of biotin. Base-catalyzed exchange consists of simple proton abstraction by hydroxyl ion to yield the biotin enolate anion, which is then reprotonated by water (Figure 4). Reprotonation of the enolate of biotin could occur on the nitrogen, regenerating biotin, or on the oxygen, yielding the enol tautomer, which is thermodynamically less stable.

IV. MECHANISM OF BIOTIN TRANSCARBOXYLATION

The model reactions of biotin establish its facile deprotonation at N-1 and enolization under mild conditions, at rates consistent with those of enzymatic carboxylations of biotin. Of course, on an enzyme, the tautomerization of biotin could be further assisted by a proton donor to the carbonyl oxygen or by a basic group other than OH^-, deprotonating the 1′-NH, although the latter is kinetically unnecessary and could interfere sterically with the subsequent carboxylation of biotin. While the conversion of the biotin enolate to the enol by O-protonation would decrease the nucleophilicity of the nitrogen,[23] it would properly position a proton donor to carry out the reverse of Reaction 2, which is the overall exchange of a proton of biotin[24] with the carboxyl group of the substrate, with retention of configuration at the leaving carbon atom.[9,25] A mechanism based on these considerations is shown in Figure 5. Mechanisms of this type have been criticized[26] on the grounds that the high kinetic lability of the enolic-OH proton (>5000 s^{-1}) would be incompatible with the known overall conservation of 10% of ^3H in the transcarboxylase reaction[24] which consists of two consecutive reactions of the form of Equation 2. This criticism is unconvincing, since the two subsites on transcarboxylase between which the enol of biotin would migrate are, at most, 7 Å apart,[27] requiring only allowed rotations about two single bonds of the biotin side chain.[16] Rotation rates about single bonds can greatly exceed 5000 s^{-1}, even on enzyme-bound species.[28]

Combined ^{13}C and deuterium kinetic isotope effect studies of Reaction 2 catalyzed by transcarboxylase,[29] and by pyruvate carboxylase[26] confirm the nonconcertedness of the proton transfer and carboxyl transfer steps in the overall process. The primary ^{13}C and solvent ^2H_2O kinetic isotope effects on Reaction 2 observed in the pyruvate carboxylase reaction are consistent with a scheme in which rate-limiting proton transfer steps both precede and follow the carboxyl transfer, as in the mechanism of Figure 5.[26] An alternative mechanism was proposed (Figure 6) also consistent with these observations, which makes use of a basic group to deprotonate the 1′-NH of biotin.[26] As pointed out above, this basic group must be properly positioned to avoid steric interference with carboxyl transfer. It must also have unusual, but not unprecedented properties. For example, a sulfhydryl group with a pK_A below 5 would satisfy the analysis of the isotope effects.[26,30] Moreover, ^3H conservation in the transcarboxylase reaction would have to survive an additional proton transfer to and from such a group at each biotin site.

The mechanisms of Figures 5 and 6 are also consistent with the following key observations on biotin and biotin enzymes. Hence neither can be excluded at present.

1. In model proton exchange reactions, steric factors decrease the assessibility of N-3 of biotin, rendering enzyme mechanisms involving the close approach of a basic group toward this position less favorable.[14]
2. Isotopic exchange studies of the transcarboxylase reaction, which detected ^3H conservation, suggest that biotin itself is the proton carrier.[24]

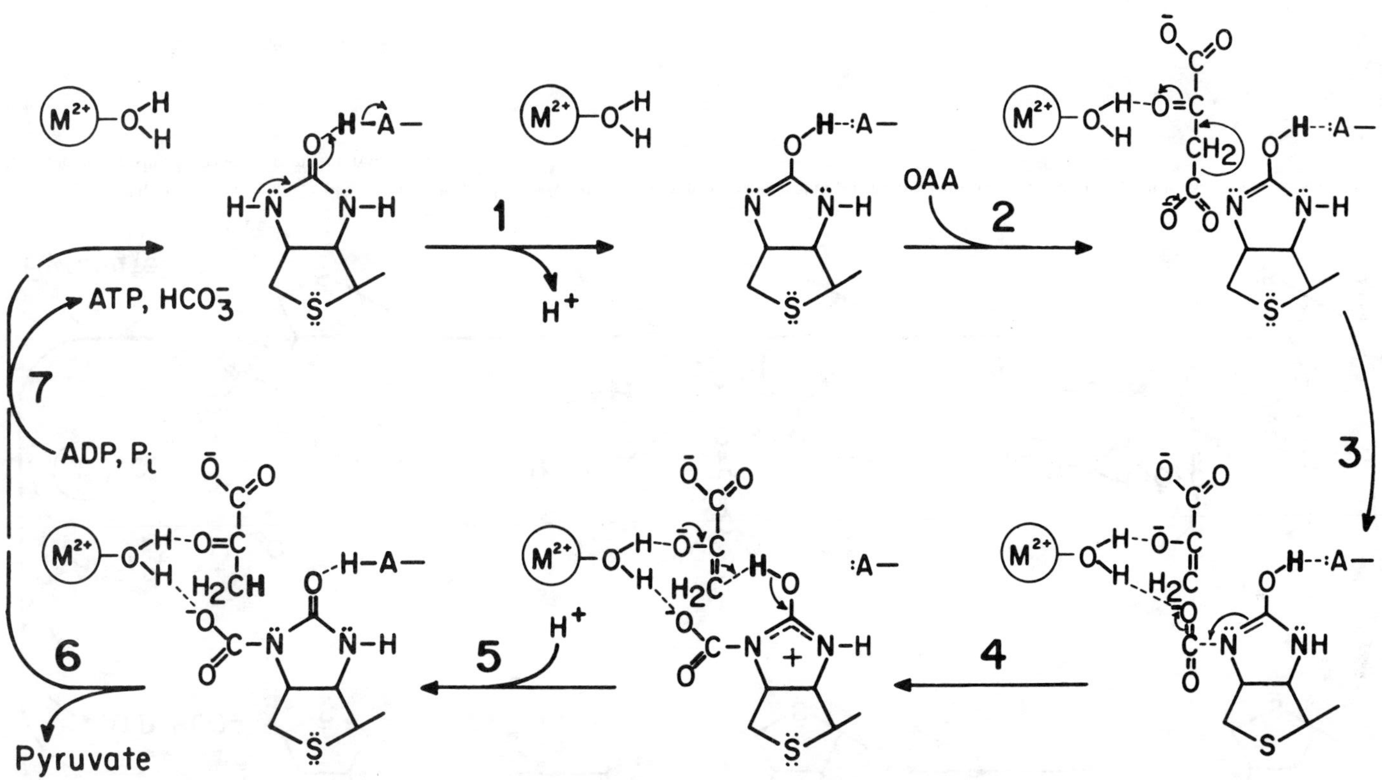

FIGURE 5. Mechanism of enzyme-catalyzed carboxylation of biotin by oxaloacetate using the enol of biotin as the proton carrier. The mechanism is based on model reactions and enzymatic studies discussed in the text. All of the indicated steps are revesible. Alternative to reactions 3 and 4, a tetrahedral intermediate may form.[37] Reaction 7 consists of several steps, the mechanisms of which are considered in Figure 1, and in the text.

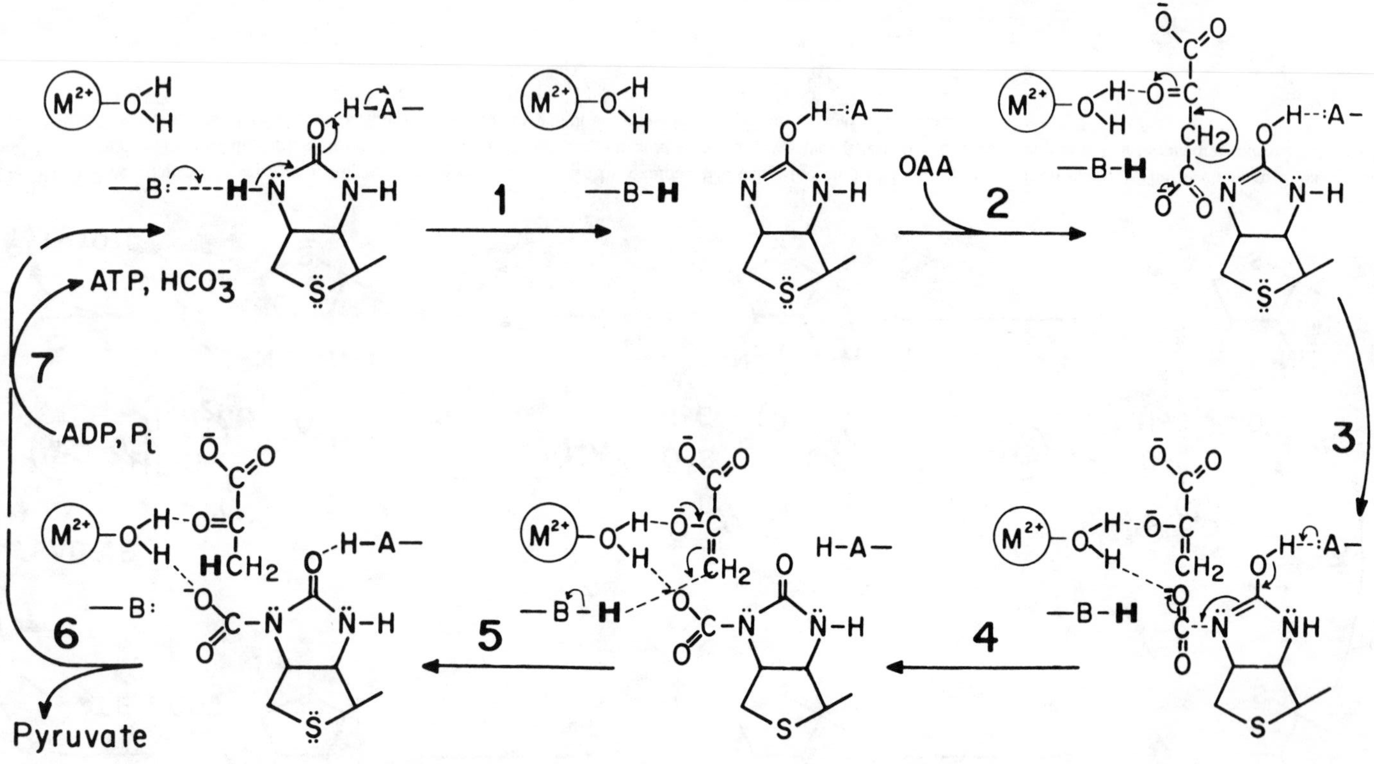

FIGURE 6. Alternative mechanism of enzyme catalyzed carboxylation of biotin by oxaloacetate, using the keto form of biotin as proton carrier and a basic group B at the active site. Expanded from Reference 26.

3. The biotin enzymes pyruvate carboxylase and transcarboxylase, which use oxaloacetate to carboxylate biotin, have been shown to be metalloenzymes, with the bound divalent cation very near the oxaloacetate (pyruvate) binding site, presumably promoting the activity of oxaloacetate as a carboxylating agent.[3,12,31-33]
4. The concertedness of carboxylation and proton transfer is ruled out by independent experiments of three types. Transcarboxylase and propionyl-CoA carboxylase catalyze the elimination of HF from β-fluoropropionyl CoA without concomitant carboxylation of this substrate.[34] The carboxybiotin forms of transcarboxylase and pyruvate carboxylase catalyze the stereospecific protonation of preformed enolpyruvate.[35] The primary ^{13}C kinetic isotope effect in the transcarboxylase reaction is decreased by the use of deuteropyruvate,[29] and similarly, the primary ^{13}C isotope effect in pyruvate carboxylase is decreased in $^{2}H_2O$,[26] indicating separate, rate-determining, carboxyl, and proton-transferring steps.
5. A nonconcerted reaction requires one or more intermediates. Two alternative intermediates have been suggested in the carboxyl transfer. One involves the carboxylation of biotin via enzyme-bound CO_2.[36] The other proposes a tetrahedral intermediate resulting from the addition of N-1 of biotin to the protonated C-4 carboxyl group of oxaloacetate, followed by the departure of the enolate of pyruvate.[37] Theoretical considerations indicate that both processes, and their respective microscopic reverses would be facilitated stereoelectronically, by distortion of the carboxylate plane out of the urea plane of carboxybiotin.[38] While we arbitrarily show the first alternative, without such carboxylate distortion, a choice cannot as yet be made between these intermediates. Evidence suggesting an enzyme-bound CO_2 intermediate in the related reaction catalyzed by P-enolpyruvate carboxylase, which is not a biotin enzyme, has been noted by R. H. Abeles,[39] based on studies of Fujita et al. with the partial substrate P-enol-α-ketobutyrate.[40] Although this partial substrate is not carboxylated by (^{18}O) HCO_3^-, the enzyme does catalyze the transfer of *more than one* atom of ^{18}O to phosphate. Such incorporation of more than one ^{18}O into phosphate requires the reversible cleavage of the enzyme-bound carboxyphosphate intermediate, either into CO_2 and phosphate, or by a reversible carboxylation of the enzyme.
6. Finally, both pyruvate carboxylase and propionyl CoA carboxylase fail to catalyze the detritiation of their respective substrates, pyruvate and propionyl CoA, unless ATP and bicarbonate are present.[12,41] The absence of such partial reactions, previously explained by a concerted mechanism, now requires a new explanation. Two possible explanations are substrate synergy, that is, the conformational requirement for all substrates to be present before the detritiation can take place, or the functioning of a buried proton carrier, that is, the shielding of biotin from the solvent. The failure to detect enzyme-catalyzed elimination of HF from fluoropropionyl CoA unless all components are present supports the former alternative.[34]

ACKNOWLEDGMENT

We are grateful to J. R. Knowles, C. L. Perrin, W. P., Jencks, and R. H. Abeles for helpful comments.

REFERENCES

1. **Moss, J. and Lane, M. D.,** The biotin dependent enzymes, *Adv. Enzymol.*, 35, 321, 1971.
2. **Wood, H. G. and Barden, R. E.,** Biotin enzymes, *Annu. Rev. Bichem.*, 46, 385, 1977.

3. **Scrutton, M. C., Utter, M. F., and Mildvan, A. S.,** Pyruvate carboxylase. VI. The presence of tightly bound manganese, *J. Biol. Chem.*, 241, 3480, 1966.
4. **Hansen, D. E. and Knowles, J. R.,** N-Carboxybiotin formation by pyruvate carboxylase: the stereochemical consequence at phosphorus, *J. Am. Chem. Soc.*, 107, 8304, 1985.
5. **Berkessel, A. and Breslow, R.,** On the structures of some adducts of biotin with electrophiles: does sulfur transannular interaction with the carbonyl group play a role in the chemistry or biochemistry of biotin?, *Bioorg. Chem.*, 14, 249, 1986.
6. **Ashman, L. K. and Keech, D. B.,** Sheep kidney pyruvate carboxylase, *J. Biol. Chem.*, 250, 14, 1975.
7. **Polakis, S. E., Guchhait, R. B., Zwergel, E. E., Lane, M. D., and Cooper, T. G.,** Acetyl coenzyme A carboxylase system of *Escherichia coli, J. Biol. Chem.*, 249, 6657, 1974.
8. **Wimmer, M.,** personal communication, 1985.
9. **Retey, J. and Lynen, F.,** Zur biokemischen funktion des biotins, *Biochem. Z.*, 342, 256, 1965.
10. **Hegarty, A. F., Bruice, T. C., and Benkovic, S. J.,** Biotin and nucleophilicity of 2-methoxy-2-imidazoline toward SP^2 carbonyl carbon, *J. Chem. Soc. Chem. Commun.*, 1173, 1969.
11. **Bruice, T. C.,** Some pertinent aspects of mechanism as determined with small molecules, *Annu. Rev. Biochem.*, 45, 331, 1976.
12. **Mildvan, A. S., Scrutton, M. C., and Utter, M. F.,** Pyruvate carboxylase. VII. A possible role for tightly bound manganese, *J. Biol. Chem.*, 241, 3488, 1966.
13. **Fry, D. C., Fox, T. L., Lane, M. D., and Mildvan, A. S.,** NMR studies of the exchange of the amide protons of d-biotin and its derivatives, *Ann. N.Y. Acad. Sci.*, 447, 140, 1985.
14. **Fry, D. C., Fox, T. L., Lane, M. D., and Mildvan, A. S.,** Exchange characteristics of the amide protons of D-biotin and derivatives. Implications for the mechanism of biotin enzymes and the role of sulfur in biotin, *J. Am. Chem. Soc.*, 107, 7659, 1985.
15. **Fox, T. L., Tipton, P. A., Cleland, W. W., and Mildvan, A. S.,** Exchange rates of the amide protons of d-biotin sulfoxide stereoisomers, *J. Am. Chem. Soc.*, 109, 2127, 1987.
16. **DeTitta, G. T., Parthasarathy, R., Blessing, R. H., and Stallings, W.,** Carboxybiotin translocation mechanisms suggested by diffraction studies of biotin and its vitamers, *Proc. Natl. Acad. Sci. U.S.A.*, 77, 333, 1980.
17. **Perrin, C. L. and Arrhenius, G. M. L.,** Mechanisms of acid-catalyzed proton exchange in N-methyl amides, *J. Am. Chem. Soc.*, 104, 6693, 1982.
18. **Tipton, P. A. and Cleland, W. W.,** Mechanisms of decarboxylation of carboxybiotin, *J. Am. Chem. Soc.*, 110, 5866, 1988.
19. **Lane, M. D., Young, D. L., and Lynen, F.,** The enzymatic synthesis of holotranscarboxylase from apotranscarboxylase and (+) biotin, *J. Biol. Chem.*, 239, 2858, 1964.
20. **Bowen, C. E., Rauscher, E., and Ingraham, L. L.,** The basicity of biotin, *Arch. Biochem. Biophys.*, 125, 865, 1968.
21. **Fry, D. C. and Mildvan, A. S.,** unpublished data, 1985.
22. **Serpersu, E. H., Mildvan, A. S., Fox, T., Fry, D. C., and Lane, M. D.,** A re-examination of the acid catalyzed exchange rates of the amide protons of D-biotin and its methyl ester, *Biochemistry,* 26, 4160, 1987.
23. **Perrin, C. L. and Dwyer, T. J.,** Proton exchange in biotin: a reinvestigation with implications for the mechanism of CO_2 transfer, *J. Am. Chem. Soc.*, 109, 5163, 1987.
24. **Rose, I. A., O'Connell, E. L., and Solomon, F.,** Intermolecular tritium transfer in the transcarboxylase reaction, *J. Biol. Chem.*, 251, 902, 1976.
25. **Rose, I. A.,** Stereochemistry of pyruvate kinase, pyruvate carboxylase, and malate enzyme reactions, *J. Biol. Chem.*, 245, 6052, 1970.
26. **Attwood, P. V., Tipton, P. A., and Cleland, W. W.,** Carbon-13 and deuterium isotope effects on oxaloacetate decarboxylation by pyruvate carboxylase, *Biochemistry,* 25, 8197, 1986.
27. **Fung, C. H., Gupta, R. K., and Mildvan, A. S.,** Magnetic resonance studies of the proximity and spatial arrangement of propionyl CoA and pyruvate on a biotin metalloenzyme, transcarboxylase, *Biochemistry,* 15, 85, 1976.
28. **Nowak, T. and Mildvan, A. S.,** NMR studies of selectively hindered internal motion of substrate analogs at the active site pyruvate kinase, *Biochemistry,* 11, 2813, 1972.
29. **O'Keefe, S. J. and Knowles, J. R.,** Enzymatic biotin-mediated carboxylation is not a concerted process, *J. Am. Chem. Soc.*, 108, 328, 1986.
30. **Attwood, P. V. and Cleland, W. W.,** Decarboxylation of oxaloacetate by pyruvate carboxylase, *Biochemistry,* 25, 8191, 1986.
31. **Northrop, D. and Wood, H. G.,** Transcarboxylase V. The presence of bound zinc and cobalt, *J. Biol. Chem.*, 244, 5801, 1969.
32. **Fung, C. H., Mildvan, A. S., Allerhand, A., Komoroski, R., and Scrutton, M. C.,** Interaction of pyruvate with pyruvate carboxylase and pyruvate kinase as studied by paramagnetic effects on ^{13}C relaxation rates, *Biochemistry,* 12, 620, 1973.

33. **Fung, C. H., Mildvan, A. S., and Leigh, J. S.,** Electron and nuclear magnetic resonance studies of the interaction of pyruvate with transcarboxylase, *Biochemistry,* 13, 1160, 1974.
34. **Stubbe, J. and Abeles, R. H.,** Are carboxylations involving biotin concerted or nonconcerted?, *J. Biol. Chem.,* 255, 236, 1980.
35. **Kuo, D. J. and Rose, I. A.,** Utilization of enolpyruvate by the carboxybiotin form of transcarboxylase: evidence for a non-concerted mechanism, *J. Am. Chem. Soc.,* 104, 3235, 1982.
36. **Sauers, C. K., Jencks, W. P., and Groh, S.,** The alcohol-bicarbonate-water system. Structure-reactivity studies of alkyl monocarbonates and on the rates of their decomposition in aqueous alkali, *J. Am. Chem. Soc.,* 97, 5546, 1975.
37. **Rose, I. A.,** The enzymology of enol pyruvate, in *NMR and Biochemistry,* Opella, S. J. and Lu, P., Eds., Marcel Dekker, New York, 1979, 323.
38. **Thatcher, G. R. J., Poirier, R., and Kluger, R.,** Enzymic carboxyl transfer from N-carboxybiotin. A molecular orbital evaluation of conformational effects in promoting reactivity, *J. Am. Chem. Soc.,* 108, 2699, 1986.
39. **Abeles, R. H.,** personal communication, 1987.
40. **Fujita, N., Katsura, I., Nishino, T., and Katsuki, H.,** Reaction mechanism of P-enolpyruvate carboxykinase. Bicarbonate dependent dephosphorylation of P-enol-α-ketobutyrate, *Biochemistry,* 23, 1774, 1984.
41. **Prescott, D. and Rabinowitz, J. L.,** The enzymatic carboxylation of propionyl coenzyme A, *J. Biol. Chem.,* 243, 1551, 1968.

Chapter 4

THIAMIN-DEPENDENT REACTIONS-MECHANISMS

Clark J. Gubler

TABLE OF CONTENTS

I. INTRODUCTION

Vitamin B_1, later named thiamin, was among the earliest of the vitamins to be isolated, characterized, and its nutritional significance discovered. It was shown as early as 1937[1] that the coenzyme form of thiamin was the diphosphate:

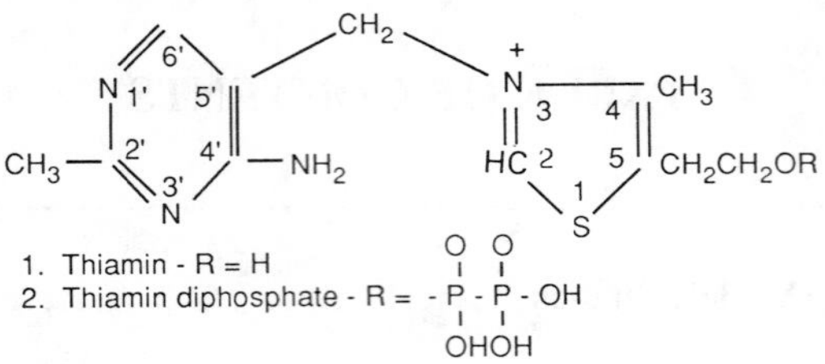

1. Thiamin - R = H
2. Thiamin diphosphate - R = - P(O)(OH) - P(O)(OH) - OH

This unusual heterocyclic compound is a rather bizarre combination of pyrimidine and a thiazole. Though we have long known its important physiological functions as the coenzyme in pyruvic decarboxylase, the three known α-keto acid dehydrogenases, and transketolase and phosphoketolase, it took many years to work out the mechanism of action of these enzymes. They all belong to the lyase group of enzymes and, as such, catalyze the cleavage of a –C–C–bond immediately adjacent to an α-keto group, eliminating CO_2 in the case of the α-keto acids and a 2-carbon moiety, glycoaldehyde, in the case of transketolase. The mechanisms of these thiamin diphosphate (ThDP)-dependent enzymes will be discussed from the standpoint of the results of model nonenzymatic reactions catalyzed by thiamin and thiamin diphosphate and studies of the individual enzyme-catalyzed reactions.

II. NONENZYMATIC (MODEL) STUDIES

Possibly the earliest clue to the mode of action of ThDP came when Ugai et al.[2] showed that *N*-ethylthiazolium bromide catalyzed the formation of benzoin from benzaldehyde. Mizuhara and co-workers[3] then demonstrated that pyruvate was decarboxylated at pH 8.4 by thiamin and that pyruvate and acetaldehyde formed acetoin in the presence of thiamin under the same conditions. It was shown early by Breslow's group[4-6] and by Krampitz et al.[7,8] that the reactions catalyzed by the ThDP-dependent enzymes could also be catalyzed, though at much slower rates by thiamin diphosphate or thiamin alone and even by appropriate thiazolium compounds. Model studies with these compounds have contributed much to our understanding of the mechanism of action of thiamin-dependent enzymes. Many early theories proposed various parts of the thiamin molecule as the site participating directly in splitting the –C–C– bond; but Breslow[4] set these at rest when he showed, in model systems, that the substrate (pyuvate) was bound at the C_2 of the thiazolium ring. This was then shown to be the case also for the ThDP-enzyme catalyzed reactions.[5] As shown by Breslow, the C_2 of the thiazole in thiamin has a dissociable hydrogen:

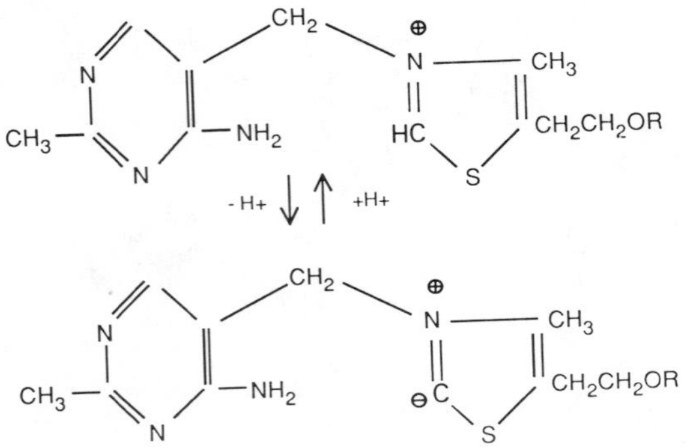

- H+ ↓↑ +H+

Thus, when thiamin or ThDP are placed in a D_2O-containing solution at pH 7, there is a rapid exchange of D for H on the C_2 of the thiazole. The negative charge of the ylid formed when a hydrogen dissociates from the active $-C_2$ can then react with the appropriate carbonyl compounds at C_2, where the bond adjacent to the carbonyl is activated and split. Imidazole and oxazole show some exchanges, but thiazole shows a greater exchange rate due to the properties of the S in the ring. When thiazole is substituted with an alkyl group and quaternized, as in thiamin, this exchange rate is enhanced manyfold, thus endowing ThDP with its unique properties as a lyase coenzyme. Electron withdrawal by the positive charge on the nitrogen makes the proton on C_2 much more labile. It follows then that any substitution on C_2, other than hydrogen, renders the analog inactive in both model and enzymatic systems.

When pyruvate reacts with thiamin the carbonyl carbon binds with the ylid at the negatively charged C_2, forming a covalent adduct in which the –C–C– bond is destabilized and CO_2 is split off. The resultant thiamin adduct is 2-α-hydroxyethyl thiamin (HET). The synthesis of HET by Krampitz et al.[10] and Miller et al.[11] and comparison with the product obtained on reaction of thiamin with pyruvate or acetaldehyde confirmed the Breslow mechanism. The subsequent isolation of this adduct from the reaction of pyruvate with pyruvate decarboxylase[9] indicated that nonenzymatic and enzymatic catalysis had a common mechanism. The sequence of reaction in this mechanism is shown in Figure 1 with pyruvate as the substrate. That this reaction sequence also holds for the known enzymatic decarboxylations of α-keto acids has been confirmed.[7,12] A discussion of the factors which contribute to the unique catalytic power of ThDP would be in order at this point. The ability of the C_2 carbon to ionize to form the ylid (carbanion) has already been discussed. The + charge on the N at position 3 aids in this process. The peculiar characteristics of the S-atom also play a role, since imidazolium and oxazolium combinations are inactive. The aromatic character of the thiazolium ring must also be preserved. Substitution of any bulky group such as a phenyl group on N_3 greatly decreases the catalytic activity. This is probably due to steric hindrance with the binding of the substrate at C_2. The role of the aminopyrimidine group is not as clear. However, Breslow and co-workers[3,4] were able to show that thiamin was much more effective than compounds in which the aminopyrimidine had been replaced by other substituents. Jordan et al.[13] and Schellenberger[14] have further indicated the importance of this part of the molecule. It was suggested that protonation of the $N_{1'}$ ring nitrogen might lead to the formation of a strongly bonded dimer between the $N_{3'}$-ring N and the 4′-amino group, which may have importance in the stabilization of the transition state in pyruvate decarboxylation. According to Schellenberger[14] the 4′-NH_2 group assists in the first step, i.e., addition of the substrate at the C_2^-, by protonating the 2 oxo group and also in the elimination of the aldehyde adduct product from the C_2 position by deprotonating the corresponding α-OH group of the HETPP. This elimination of the aldehyde adduct and regeneration of thiamin diphosphate is the rate-limiting step in the process. Hence the 4′-NH_2 of the pyrimidine ring serves as an intramolecular proton relais (Figure 2). Further evidence of the importance of this group is that all modifications or elimination of this group leads to total loss of catalytic activity.

In relation to the quaternizing group, the methylene bridge between the pyrimidine and the thiazole is necessary for maximum activity. If the methylene group is eliminated or contains a methyl substitution for one of the hydrogens, the activity is greatly reduced. Replacement of the methylene by an ethylene group, so that a 2-carbon bridge separates the two rings, results in loss of only 40% of the activity.[15-17] The 2′-methyl group on the pyrimidine is not an absolute requirement, since the 2′-ethyl and 2′-propyl analogs possess considerable activity. However, the 2′-butyl analog is not only catalytically inactive, but is indeed a potent antagonist. The 2′-H analog is also totally inactive. Thus, Yount and Metzler[18] showed that the pyrimidine ring was not essential in model reactions, but structures which had inductive effects similar to those of the pyrimidine moiety had the highest catalytic activity.

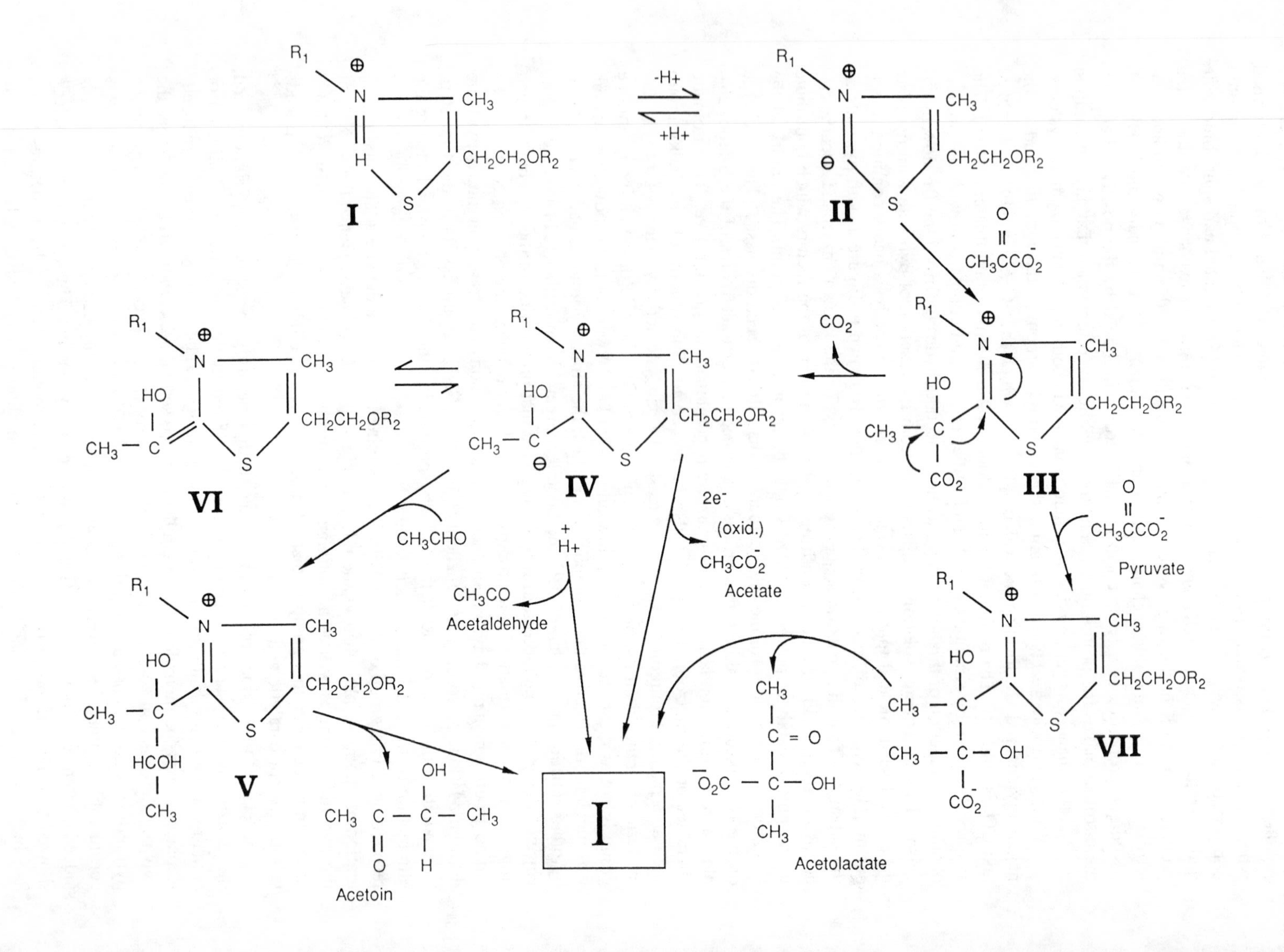
I
-H+
+H+
II
III
IV
VI
V
VII
CO2
2e-
(oxid.)
Pyruvate
Acetate
Acetaldehyde
Acetoin
Acetolactate

FIGURE 1. Reaction sequence in the decarboxylation of pyruvate with subsequent reactions. str 4I = thiamin, II = thiamin ylid, III = pyruvate adduct, IV = α-hydroxyethyl thiamin adduct-carbanion (active acetaldehyde or HET), V = Acetoin adduct, VI = equilibrium form of IV, and VII = acetolactate adduct.

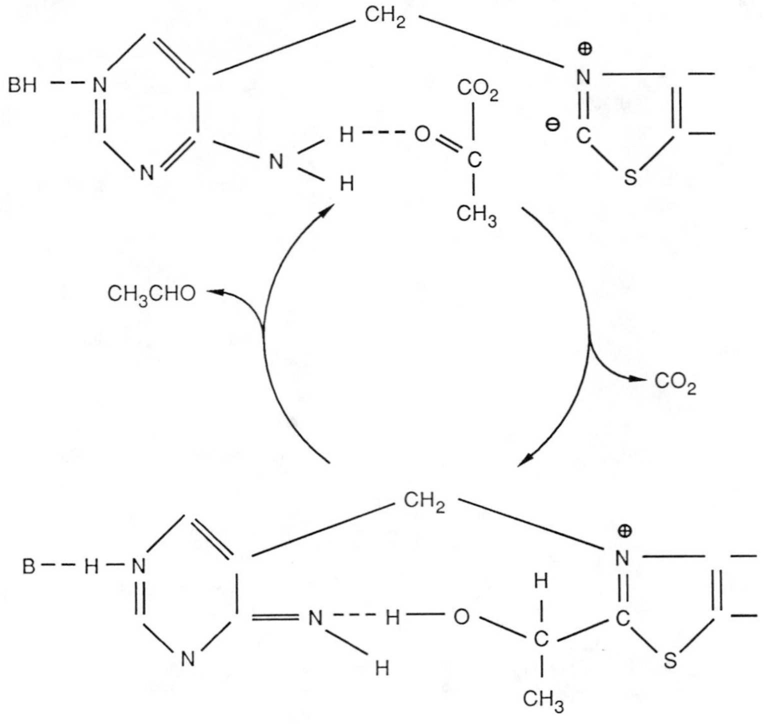

FIGURE 2. Proton relais function of the 4′-NH_2-pyrimidine system of thiamin. (Adapted from Reference 14.)

Krampitz et al.[19] made a detailed study of the acetaldehyde-thiamin and the acetaldehyde-ThDP reaction. They found the corresponding α-hydroxy ethyl derivative (HET and HETDP) and were able to isolate the acetoin adduct of both. With pyruvate in excess and thiamin or its diphosphate, the reaction forms α-acetolactate. Holzer and Beaucamp[12] were able to detect the pyruvate adduct.

The reaction involved in the decarboxylation of pyruvate and subsequent conversion to various products are presented in Figure 1. The first and most essential step is the conversion of thiazolium C_2 to the ylid (zwitterion) (II) which makes possible the attack on the carbonyl group –C–C– bond with the formation of the pyruvate adduct (III). This destabilizes the –C–CO_2^- bond so that CO_2 is eliminated (leaving group) with the ''active acetaldehyde'' left on C_2 as the carbanion. This is stabilized somewhat by its resonating forms:

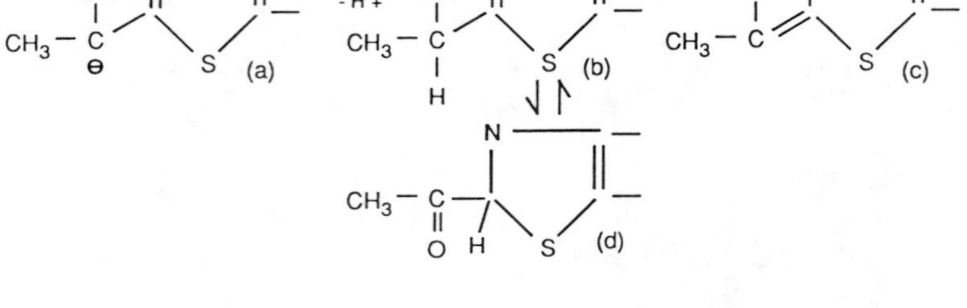

The (c) form as an enol could ketonize as in (d) and block the further transformations, but this is avoided since it would destroy the aromaticity of the thiazolium ring. The most likely resonance form for the subsequent reactions is (a), the carbanion. As indicated earlier and in Figure 1, Schellenberger attributes an interaction of the carbanion with the 4'-NH_2 group on the pyrimidine (base catalysis) with the destabilization of the HET and liberation of the acetaldehyde from the thiamin (see Reference 14 and Figure 1). The carbanion (a) can also react further with other aldehydes to form acyloins, with pyruvate to form acetolactate or with appropriate oxidants to be oxidized to acetate.

For further information see References 6, 7, 8, 14, 20, and 21.

III. PYRUVATE DECARBOXYLASE

Of all the ThDP-dependent enzymes, pyruvate decarboxylase has been the most thoroughly studied, particularly the yeast and wheat germ enzymes. These are cytoplasmic enzymes and hence are much more easily purified than the membrane-bound enzymes.

The sequence and mechanism of the enzyme-catalyzed decarboxylation of pyruvate appears to be the same as that shown in the model reactions[9] (Figure 1), the influence of the apoenzyme resulting in a manyfold increase in reaction rate (catalytic activity). The active intermediate after decarboxylation of pyruvate in the enzymatic reaction appears to be the α-carbanion of 2-α-hydroxyethyl ThDP rather than the protonated compound isolated. In order for the α-carbanion to have a sufficiently long lifetime for further conversion a strongly lipophilic environment would be required around the active site.[21-23] This had been suggested from several angles. Both enzymes have a very high content of lipophilic amino acids.[21,23,24] Fluorescence studies have indicated that the active site is associated with a lipophilic area of the protein, in a pocket.[25-27] Lehmann et al.,[28] have also shown that α-oxoacids with longer or branched chains, i.e., more lipophilic, have a lower Km for binding to pyruvate decarboxylase than pyruvate.

Holopyruvate decarboxylase from yeast is a tetramer of 230,000 to 250,000 Da, that from wheat germ about 275,000.[29,30] They are composed of two dimers with slightly different chain length. The Michaelis-Menten kinetics curves are sigmoidal in shape, thus indicating positive cooperativity between the active sites of the subunits. The measured Hill coefficient is between 1.7 and 1.8. When substrate is added to the resting holoenzyme a lag phase is observed before the actual enzymatic reaction starts. This may be due to allosteric activation sites which first bind substrate and then activate the substrate conversion sites.[23,31-34] It is now known that eight tryptophan residues located at the active site are essential for coenzyme binding and hence for activity and also contribute to the lipophilicity of the active site. Two –SH groups are also essential for substrate binding.[23]

IV. PYRUVATE DEHYDROGENASE COMPLEX

The decarboxylation of the α-keto acids, in contrast to decarboxylation of β-keto acids and α-amino acids, is accomplished by ThDP-requiring enzymes. Whereas the simple decarboxylation discussed in Sections II and III is carried out by relatively simple ThDP-dependent enzymes and is confined mostly to procaryotic type cells, the oxidative decarboxylation of α-keto acids is accomplished by a much more complex multienzyme system which requires CoASH, NAD^+, FAD, and oxidized lipoic acid as cofactors in addition to ThDP, and contains at least three enzymes in an organized complex. The sequence of reactions have been shown to proceed as follows:

1. $R\text{–}COCOOH + E_1\text{–}ThDP \rightarrow E_1\text{–}RCHOH\text{–}ThDP + CO_2$.

2. $E_1\text{–RCHOH–ThDP} + E_2\text{–Lip}\langle S{-}S \rangle \rightleftarrows E_1\text{–ThDP} + E_2\text{–Lip}\langle \begin{smallmatrix} SH \\ SC(=O)\text{–}R \end{smallmatrix}$

3. $E_2\text{–Lip}\langle \begin{smallmatrix} SH \\ SC(=O)\text{–}R \end{smallmatrix} + \text{CoASH} \rightleftarrows E_2\text{–Lip}\langle \begin{smallmatrix} SH \\ SH \end{smallmatrix} + \text{RCOSCoA}$

4. $E_2\text{Lip}\langle \begin{smallmatrix} SH \\ SH \end{smallmatrix} + E_3\text{–FAD} \rightleftarrows E_2\text{Lip}\langle S{-}S \rangle + E_3\text{–FADH}_2$

5. $E_3\text{–FADH}_2 + \text{NAD}^+ \rightleftarrows E_3\text{–FAD} + \text{NADH} + \text{H}^+$

Net: 6. $\text{RCOCOOH} + \text{CoASH} + \text{NAD}^+ \rightarrow \text{RCOSCoA} + \text{NADH} + \text{H}^+ + \text{CO}_2$

where E_1 = α-ketoacid dehydrogenase (EC 1·2·4·1)
E_2 = lipoic acyl transferase (EC 2·3·1·12)
E_3 = dihydrolipoyl dehydrogenase (EC 1·6·4·3).

Of the three dehydrogenase enzymes, the one, α-ketoglutarate dehydrogenase complex, is essentially specific for α-ketoglutarate and operates within the tricarboxylic acid (TCA) cycle. This will be treated in Chapter 8. Another is responsible for the oxidative decarboxylation of the three branched-chain α-keto acids derived from the deamination of the three essential branched chain amino acids, leucine, isoleucine, and valine. The third is pyruvate dehydrogenase, which is primarily responsible for the oxidative decarboxylation of pyruvate for entry into the TCA cycle. The reaction sequences and mechanisms are essentially the same for all three of these enzymes. This discussion will therefore confine itself chiefly to pyruvate dehydrogenase complex. (For more complete coverage consult References 36 to 39.)

The pyruvate dehydrogenase (PDH) complex is located in the mitochondrial inner membrane matrix space.[40,41] It catalyzes the oxidative decarboxylation of pyruvate to acetyl-CoA according to the net equation 6. The reaction proceeds quasiirreversibly to the right and occupies a key position in cellular metabolism by controlling the entrance of C_2 fragments arising from carbohydrates (glucose) and amino acids into the TCA cycle with subsequent oxidation or conversion to fatty acids. The interrelationships of these pathways are illustrated in Figure 3. Because of this key role in energy metabolism, the PDH-complex has been studied in detail in a number of laboratories (see References 36, 39, 42, and 43). The two most widely studied PDH-complexes have been isolated from *Escherichia coli* and beef heart mitochondria. The composition of these two complexes is shown in Table 1. PDH complex from *E. coli* has a molecular weight of 4.8×10^6 Da and contains 24 subunits of E_1 (PDH) of 100,000 Da each, 24 subunits of E_2 (dihydrolipoyl transacetylase) of 80,000 Da each, and 12 subunits of E_3 (dihydrolipoyl dehydrogenase) of 56,000 subunits each. E_2 serves structurally as the core of the complex with E_1 and E_3 bound to it with noncovalent bonds forming an octahedral structure which appears to allow a favorable geometrical placement of the subunits. Conformational changes in the complex during the reaction seem to be minimal. The more complex mammalian (beef heart) PDH complex has a molecular weight of 8.5×10^6 Da which contains 30 E_1 subunits of 154,000 Da each, 60 E_2 subunits of 52,000 Da each, and 12 E_3 subunits of 110,000 Da. E_1 is a tetrameric protein with two subunits of 41,000 Da and two of 36,000 Da, i.e., $\alpha_2\beta_2$ structure. It shows typical allosteric kinetics. E_2 is monomeric, while E_3 is dimeric with each subunit 55,000 Da.

The reaction sequence and mechanism for pyruvate oxidation by the PDH complex is shown in Figure 4. The first reaction, corresponding to Equation 1, above, occurs in two

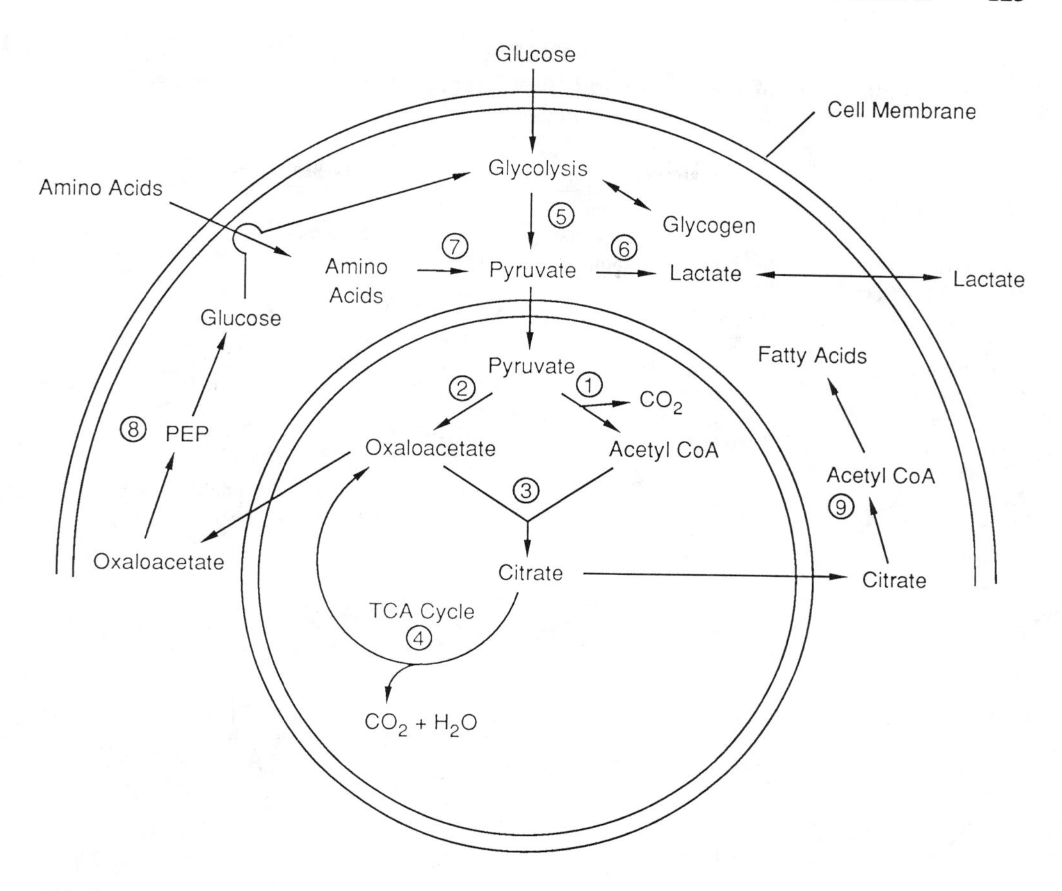

FIGURE 3. Major routes of pyruvate metabolism in mammalian cells: 1, pyruvate dehydrogenase; 2, pyruvate carboxylase; 3, citrate synthase; 4, citric acid cycle (TCA); 5, glycolysis, glycogenolysis; 6, lactate dehydrogenase; 7, transamination; 8, gluconeogenesis; 9, lipogenesis. (Adapted from Reference 37b.)

steps as with pyruvate decarboxylase as shown in Figure 1, first forming the pyruvate adduct (III) then decarboxylating to leave the aldehyde adduct (IV), hydroxyethyl thiamin diphosphate (HETPP). This then reacts with an oxidized lipoyl moiety as in Equation 2, above, by which the lipoyl is reduced and the aldehyde is oxidized to acetyl. The acetyl is transferred to the E_2-lipoyl in the process and E_1-ThDP is freed. In Reaction 3, the acetyl is transferred from Lipoyl$\langle^{SH}_{S}$ acetyl to CoASH to give acetyl-SCoA, which can then enter the TCA cycle. In Reaction 4 the reduced (dihydrolipoyl) E_2–L$\langle^{SH}_{SH}$ is reoxidized by E_3, which contains FAD and S-S groups in a complex transfer of electrons in which the E_2–L$\langle^{SH}_{SH}$ is oxidized to E_2–L$\langle^{S}_{S}$ again with the reduction of E_3–$\diagdown^{[FAD]}_{S\text{-}S}\diagup$ to E_3–$\diagdown^{[FADH_2]}_{S\text{-}S}\diagup$, which is then reoxidized by NAD^+, and the system is restored for another cycle. E_3 has been shown to have an essential –S–S– bridge at the active site, which is believed to be involved in the reoxidation of the L$\langle^{SH}_{SH}$ on E_2,[50] but the mechanism has not been worked out.

TABLE 1
Composition of the *E. coli* and Beef Heart Pyruvate Dehydrogenase Complexes[36,37,40-45]

	E. coli		Beef heart			
Enzyme	Mr	No. subunits/ complex	Mr	Subunits no.	Mr	Subunits/ complex
Native complex	4.8×10^6		8.5×10^6			
E_1 Pyruvate dehydrogenase	100,000	24	154,000	2 α 2 β	41,000 36,000	60
E_2Dihydrolipoyl *trans*-acetylase	80,000	24	3.1×10^6	60	52,000	60
E_3Dihydrolipoyl dehydrogenase	56,000	12	110,000	2	55,000	12
Kinase			~100,000	1	48,000 45,000	(5)
Phosphatase			~150,000	1	97,000	(5)

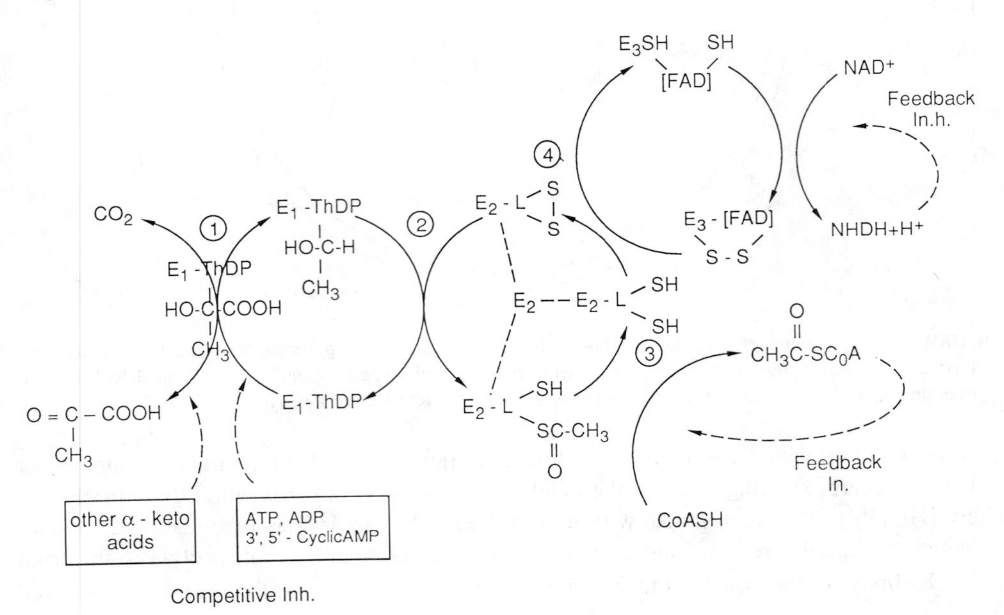

FIGURE 4. Classical reaction sequence for pyruvate oxidation by pyruvate dehydrogenase complex. E_1–ThDP = pyruvate dehydrogenase, E_2–L(S–S) = dihydrolipoyl transacetylase, and E_3–[FAD] = dihydrolipoyl dehydrogenase. S–S (adapted from References 37a and 51.)

Since each of the 24 subunits of the core dihydrolipoyl transacetylase (E_2) has been shown to contain 2 lipoyl moieties made up of lipoic acid bound to a lysyl residue, that makes 48 such chains per PDH-complex. The question then arises, what function do the extra 24 chains serve, when presumably only 24 can react with the 24 PDH chains. The classical approach[51] has been to assign a mobility to this long lysyllipoic acid chain which would enable it to rotate and pick up the aldehyde adduct from E_1, oxidize it to acetyl, transfer this to CoASH, then interact with the FAD on E_3, thus restoring the dihydrolipoyl to the disulfide again as indicated in Figure 4. However, this requires that the catalytic sites lie within 28 to 30 Å of each other. Measurement of the distances from E_1 and E_2 to E_3 by

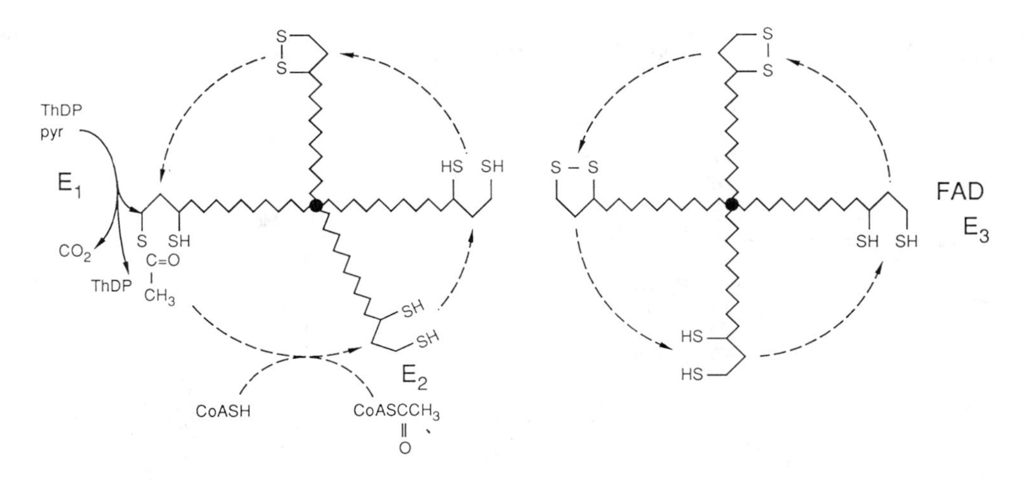

FIGURE 5. A schematic representation of a suggested mechanism in which two lipoyl residues are involved in the transfer of intermediates from E_1 to E_3 in the pyruvate dehydrogenase complex. (Adapted from Reference 49.)

fluorescence resonance energy transfer[52] has shown it to be 50Å or more. This, along with the existence of 48 lipoyl residues, suggests that the simple mechanism may not be adequate. Hammes[49] and others[36,48,54] have suggested an alternative mechanism as shown in Figure 5. This fits better with the number of lipoyl residues and the distances observed. In this suggested mechanism, the acetyl group and electrons would be transferred from one lipoyl to a second one before transfer to the FAD on E_3. Thus, the four 14Å α-lipoyl-*N*-lysyl chains would span the total 56Å distance required. Kinetic studies of the overall reaction (Equation 6) of pig heart PyDH complex suggest that it proceeds at the three sites (Figure 4) by a ping-pong mechanism.[34]

Regulation of the PDH complex activity depends on two mechanisms, i.e., product inhibition and covalent modification. Since the decarboxylation of Py catalyzed by E_1(PDH) is essentially irreversible, this is the logical site for the regulation. Thus, PDH is inhibited by acetyl CoA and NADH, the end products of Reactions 1 to 5 shown above, and this is competitive with respect to CoA and NAD^+, respectively. The reaction can thus be regulated by the intramitochondrial acetylCoA/CoA and NADH/NAD^+ ratios.[36,42,56] Kinetic studies with PDH complexes have given values for intramitochondrial acetyl-CoA of 5 to 10 μM and for NADH of about 50 μM. These are in the same range as the Km values for CoASH ($\sim 5\mu M$) and NAD^+ ($\sim 40\mu M$).[39]

Covalent modification involves the phosphorylation and dephosphorylation of specific seryl residues at three sites on the α-chain of the PDH tetramer by a specific kinase and phosphatase, respectively. Phosphorylation as a regulatory mechanism for mammalian PDH complex was first demonstrated by Reed and co-workers[57] and Wieland et al.[58] Phosphorylation leads to inactivation of PDH, and hence of the PDH complex and dephosphorylation leads to reactivation. Phosphorylation at what has been designated as site 1 is much more rapid than at sites 2 and 3, and phosphorylation at site 1 correlates well with the inactivation of PDH(E_1).[59] It has been claimed that further phosphorylation of sites 2 and 3 inhibits the subsequent reactivation by the phosphatase.[60]

The pyruvate dehydrogenase system is well designed for fine regulation and its regulation depends on the complex interaction of multiple factors. Both the kinase and the phosphatase require Mg^{++}, the apparent Km of the phosphatase for Mg^{++} is about 2 mM, while that of the kinase is on the order of 20 μM. Thus, the intramitochondrial concentrations of Mg^{++} play a regulatory role in the activation-inactivation of PDH. The actual substrate for the kinase is $MgATP^{2-}$. ADP inhibits the kinase competitively with ATP only if K^+ or NH_4^+

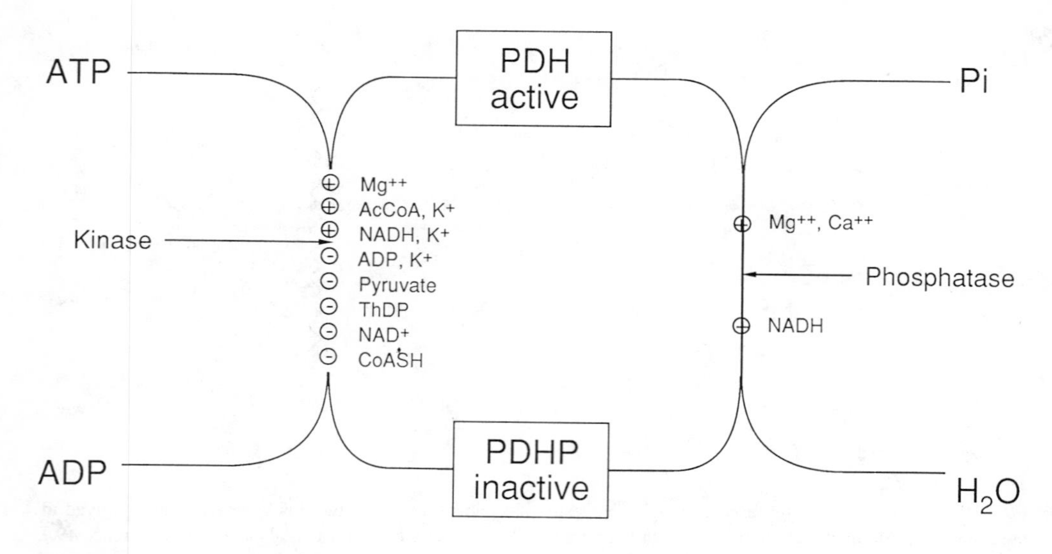

FIGURE 6. Schematic representation of the covalent modification of pyruvate dehydrogenase and its control by various substances. (Adapted from Reference 37b.)

are present. AMP, CTP, GTP, and UTP are ineffective. Though pyruvate is the substrate for PDH, it also inhibits the kinase, noncompetitively with respect to ATP. The apparent Ki values are about 80 and 900 μM for heart and kidney PDH, respectively. Cyclic AMP and GMP apparently play no role in this phosphorylation mechanism. Ca^{++} is required in addition to Mg^{++} for activity of the phosphatase. It appears that it plays a structural role in the binding of the phosphatase to the transacetylase (E_2). Phosphatase activity is inhibited by NADH and this is reversed by NAD^+. The kinase activity is inhibited by acetyl-CoA and NADH if K^+ or NH_4^+ are present. The coenzyme of PDH, ThDP, inhibits kinase activity. This would appear to be due to its binding at the active site of PDH, thus changing the conformation around the phosphorylation site #1 making the serine-OH less accessible to the kinase. (For more complete treatment of the phosphorylation - dephosphorylation regulatory mechanism see References 37, 39, and 47.) The mechanism and factors involved in regulation of PDH by phosphorylation-dephosphorylation are summarized in Figure 6. In addition to these factors there is ample evidence that hormonal and other factors also play a role under physiological conditions (*in vivo*). Activity of the PDH complex is decreased in diabetes and starvation as a result of insulin deficiency.[62,63] The mechanism presumably involves activation of PDH kinase with an increase in the inactive phosphorylated form (PDHP). Insulin has been shown to increase the percentage of the active form of PDH complex in adipose tissue, which appears to be mediated through activation of the PDH phosphatase, possibly by altering the concentration of free Ca^{++} in mitochondria.[63,64] This insulin effect is most pronounced in adipose tissue. It has also been shown[65] that exercise increased the proportion of PDH.

V. TRANSKETOLASE

A. HISTORICAL

Warburg and co-workers[66,67] were the first to recognize a second pathway for glucose metabolism in the oxidation of glucose-6-phosphate by glucose-6-phosphate dehydrogenase (Zwischenferment) using a new coenzyme, coenzyme II (now designated as $NADP^+$).

The various steps of this pathway, now known as the pentose phosphate or hexose monophosphate shunt pathway, were subsequently elucidated by the work of Horecker and co-workers, Racker and co-workers, and others. It was first thought to be an alternate pathway for the oxidation of glucose-6-phosphate to produce energy, but it is now considered to be a multifunctional metabolic pathway.

The extensive work of Racker and co-workers indicated that transketolase (TK) plays an integral role in this pathway and its role and mechanism of action have since been the subject of many studies by various groups around the world. It is now well recognized that in tissues, starting with ribose-5-phosphate (R-5-P) as substrate, TK is the rate-limiting enzyme and hence plays a vital role in the operation of the pathway. For this reason it has been the focus of a large amount of research on the isolation, purification, characteristics, and mechanism of action of this enzyme.

B. FUNCTIONS OF PATHWAY

The generally accepted classical pentose phosphate pathway is shown in Figure 7.[66,67] This pathway occurs in the cytosol along with the glycolytic pathway, with which it exchanges several intermediates. It is composed of three systems: (1) an oxidizing or dehydrogenase-decarboxylating system which oxidizes G-6-P with $NADP^+$ to produce NADPH and CO_2 + ribulose-5-P (Ru-5-P). The $\Delta G^{o\prime}$ of this conversion is -30.8 kJ/mol, which is enough to drive the NADPH/$NADP^+$ ratio to an equilibrium value of over 2000 at 0.05 atm CO_2 pressure. This emphasizes the importance and participation of this pathway in biosynthesis. (2) An isomerizing system produces an equilibrium mixture of Ru-5-P, ribose-5-phosphate (R-5-P), and xylulose-5-phosphate (Xu-5-P). (3) A sugar rearrangement system uses TK and transaldolase (TA) to produce C_3, C_4, C_5, C_6, and C_7 (and possibly C_8[69]) intermediates by transfer of C_2 and C_3 ketol fragments, respectively, to appropriate aldose acceptors. Since these enzymes function with a variety of donor ketoses and acceptor aldoses, this allows considerable versatility to serve the multifunctional needs of the pathway.

A summary of the reactions of the pentose phosphate pathway follows:

6-Glucose-6-phosphate + 6 H_2O + 12 NADP → 6-Ribose-5-phosphate + 6 CO_2 + 12 NADPH + 12 H^+
6-Ribose-5-phosphate → 4-Fructose-6-phosphate + 2-Glyceraldehyde-3-phosphate
2-Glyceraldehyde-3-phosphate → Glyceraldehyde-3-phosphate + Dihydroxyacetone phosphate
Glyceraldehyde-3-phosphate + Dihydroxyacetone phosphate → Fructose-1,6-diphosphate
H_2O + Fructose-1,6-diphosphate → Fructose-6-phosphate + Pi
5-Fructose-6-phosphate → 5-Glucose-6-phosphate

Glucose-6-phosphate + 12 NADP + 7 H_2O → 6 CO_2 + 12 NADPH + 12 H^+ + Pi
(STR 4)

If these reactions are added, the net reaction seems to be the oxidation of one molecule of G-6-P to CO_2 and H_2O with the production of 12 NADPH. However, this is a minor function of the pathway. It is obvious that pentoses, GA-3-P, G-6-P, Fr-6-P, etc. can be siphoned off at various steps without jeopardizing the operation of the pathway, the chief functions of which are:[70]

1. Supply of reducing power as NADPH + H^+ for anabolic processes (biosynthesis) (a) of fatty acids, (b) of cholesterol and steroids, and (c) for reductive carboxylation of pyruvate to malate for reversal of glycolysis and regeneration of C_4-dicarboxylic acids for the citric acid cycle.
2. Supply of R-5-P for production of nucleotides and nucleic acids.
3. Supply of glycolytic intermediates Fr-6-P, Fr-1, 6-DiP, and Ga-3-P.
4. Supply of Ru-1,5-di-P in photosynthetic production of carbohydrate.[71]

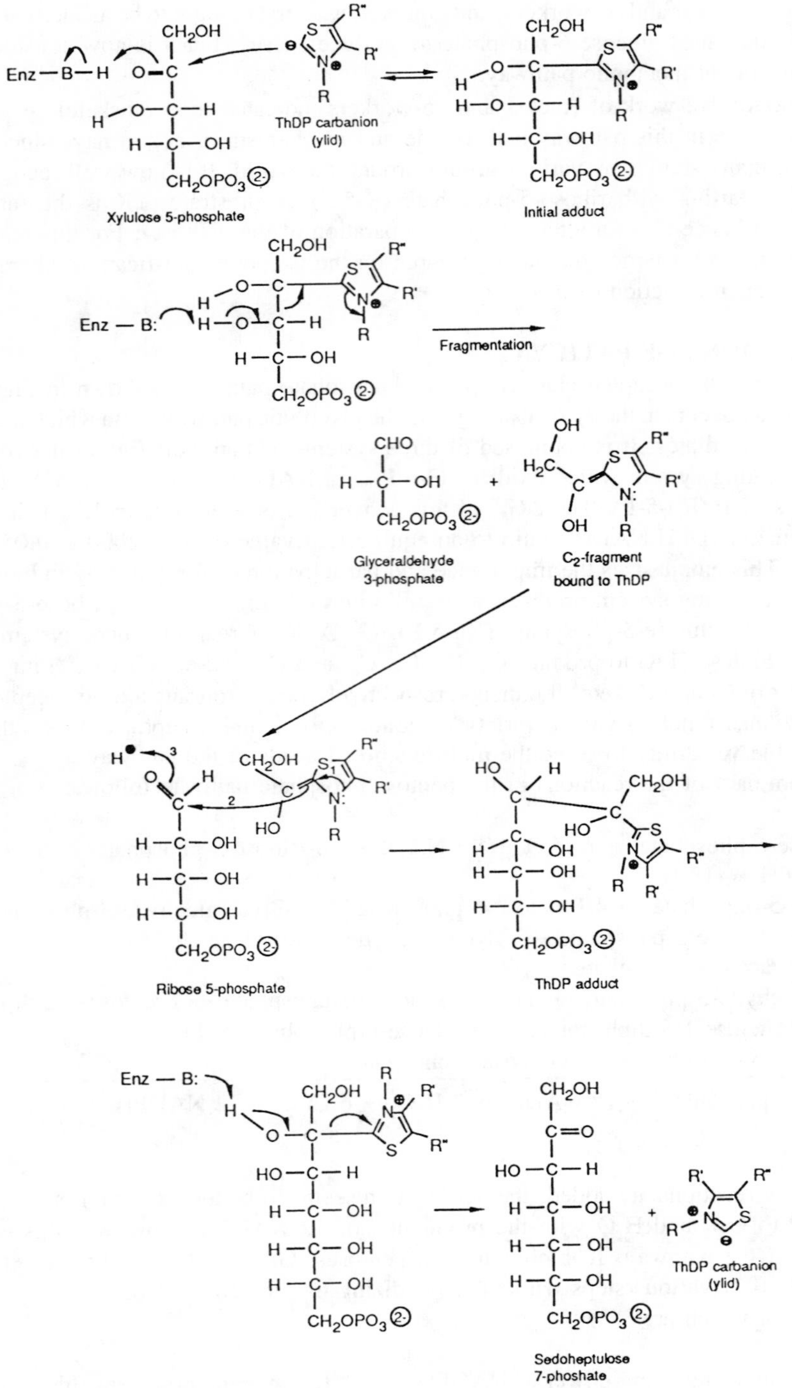

FIGURE 7. Mechanism of action of transketolase. The initial step in the reaction catalyzed by transketolase is formation of a C_2 fragment that is bound to thiamine diphosphate. Glyceraldehyde 3-phosphate is produced in this step. The second step in transketolase catalysis is nucleophilic attack of the C_2-adduct of ThDP upon the carbonyl carbon of ribose 5-phosphate. The product is sedoheptulose 7-phosphate.

5. Role of integrity of red cells.
6. Myelin production in the central nervous system and brain functions.[72,74]
7. Supply of C_3, C_4, C_5, C_6, C_7, and C_8 intermediates for use in a variety of functions.
8. Also plays a role in specialized fermentation schemes in several bacterial species.

C. MECHANISM OF TK ACTION

As with other ThDP-catalyzed reactions the first step is the binding of the donor substrate, in this case Xu-5-P, at the C_2 of the ThDP carbanion to give the initial adduct, in which the $-C_2-C_3-$ bond is activated leading to the splitting of the substrate into an aldose moiety (acetaldehyde-3-phosphate), which is the leaving product, and the remaining 2-carbon ketol moiety, which remains bound to the ThDP. This active C_2 fragment can then react with the carbonyl carbon of the acceptor aldose (in this case R-5-P) to form the 7-carbon adduct on ThDP, which is then released as sedoheptulose-7-phosphate, leaving the ThDP-carbanion free to initiate another reaction (see Figure 7). The initial adduct is the Xu-5-P-ThDP adduct. With the splitting, GA-3P leaves and the 2-carbon adduct is 2-hydroxy-ethylthiamindi-phosphate(dHEThDP) or "active glycoaldehyde". In order to remove this 2-carbon moiety, it must first bind to an appropriate aceptor aldose. Active ketol donors must have the –OH group on C_3 and C_4 trans to each other with the C_3–OH in the L-configuration and the C_4-OH in the D. Thus, D-Xu-5-P, D-Fr-6-P, and D-sedoheptulose-7-P are the most active donors. L-erythrulose, D-xylulose, hydroxypyruvate, and D-glycero-D-ido-octulose-8-P have also been found to act as donors under certain circumstances. As acceptors, D-GA-3-P, D-erythrose-4-P, and D-R-5-P are the most efficient, but D-G-6-P, D-allose-6-P, D-arabinose-5-P, glycoaldehyde, D-glyceraldehyde, and formaldehyde have also been shown to act in certain systems (for review see Reference 75).

The K_m for R-5-P with Xu-5-P as second fixed substrate is around 4×10^{-4} *M* for yeast TK,[75] and 3×10^{-4} *M* for erythrocyte TK,[77,78] but much lower (5.6×10^{-5} *M*) for rat liver TK.[79] For the ketol donors, Xu-5-P, with R-5-P as fixed substrate, was reported to have a K_m of 2.1×10^{-4} *M* and 7.7×10^{-5} *M* for yeast,[75,76] 2.5×10^{-5} *M* for rat liver,[79] and 2.0×10^{-4} *M* for erythrocyte TK.[77,78] The K_m for Fr-6-P is in the range of 1.8 to 1.9×10^{-4} *M* for both yeast[75] and erythrocyte TK.[78] A K_m of 4.9×10^{-3} *M* was reported for yeast TK[80] with GA-3-P as fixed substrate. The kinetics have apparently not been determined for the other substrates.

All purified TK preparations, except spinach TK, have been shown to be dimeric proteins. TK from brewers,[80] baker's,[81] and *C. utilis*[76] yeasts, as well as rat liver[79] are dimeric with identical subunits. On the other hand, pork liver TK is tetrameric with nonidentical subunits.[82] Hence, they all have two active centers with two coenzyme (ThDP) molecules tightly bound with noncovalent bonds.

Kremer et al.[83] and Kochetov[84] demonstrated that each subunit has one arginine at the active site which is essential for activity. It is not active in binding of the coenzyme, hence must play a role in the catalytic process. Kochetov[84] also demonstrated that the two active centers each contain one histidine residue which is involved in the binding of ThDP with the apoTK by interacting with the terminal phosphate residue of the coenzyme. Kochetov and co-workers[85,86] have also shown that each active center of TK contains a carboxyl group which is essential for catalytic activity. It is not involved in the binding of substrate or cofactor. Since the pKa of the C_2-proton on the thiazolium ring of ThDP is 12.7,[87] there must be some mechanism which facilitates the dissociation of this proton during enzymatic catalysis at physiological pH values. It was suggested that the presence of the carboxyl group increases the acidity of the proton on C_2 and thus contributes to ylid formation, which is the essential first step in catalysis for all of the ThDP-dependent enzymes, and which makes possible the formation of the ThDP-substrate adduct.

D. INHIBITORS

It is well known[78,88,89] that TK activity from various sources is inhibited by arsenate, phosphate, and sulfate anions at millimolar concentrations. Hence, when these are used in preparation steps they must be removed for subsequent activity assay. Klein and Brand[89] reported mixed competitive inhibition with Ki′ of 2.7×10^{-2}, 5.5×10^{-3}, and 4.9×10^{-3} *M* for arsenate, sulfate, and phosphate, respectively, for TK from *C. utilis*.

The K_m for ThDP with baker's yeast TK is 6.0×10^{-7} *M*.[88] Some other analogues of ThDP were shown to be competitive inhibitors with respect to ThDP.[88] Thus, with Th, ThP, pyrophosphate, and thiazoleDP, the inhibitory constants (Ki,M) were reported to be 3.4×10^{-2}, 2.0×10^{-3}, 2.8×10^{-4}, and 5.4×10^{-6} M, respectively. The Ki values were calculated by the formula:

$$Ki = [I]\frac{Km'}{Km} - 1$$

where [I] is the concentration of the inhibitor, Km = the Michaelis constant for ThDP in the absence of inhibitor, and Km′ = the effective constant in the presence of inhibitor. This indicates that the pyrophosphate residue on ThDP makes a significant contribution to the binding of the coenzyme with the apoenzyme, and hence to the catalytic function. Shreve et al[90] reported that thiamin thiazoloneDP showed competitive inhibition of TK with respect to ThDP. Since it behaves differently with TK than with PDH complex and PDC, they conclude that it cannot serve as a transition state analog for TK as it seems to for PDH complex and PDC. This is attributed to the relatively hydrophilic environment around the active center for TK in contrast to the hydrophobic for the other two.

D-Arabinose-5-P (AR-5-P) and D-Glucose-6-p (G-6-P) have been shown to be competitive inhibitors of *C. utilis* TK by Wood[76] with a Ki of 2.6×10^{-3} and 3.6×10^{-3} M for Ar-5-P with Ru-5-P and Xu-5-P, respectively, as substrates and 1.7×10^{-4} and 1.3×10^{-4} for G-6-P with Ru-5-P and Xu-5-P, respectively, as substrates. Ar-5-P is not able to serve as the acceptor, whereas G-6-P can serve as a weak acceptor with some TK preparations. Oxythiamin (OTh) and pyrithiamin were shown to be competitive inhibitors (Ki = 1.4×10^{-3} for Oth and 4.3×10^{-3} for PTh). This is in contrast to OThDP and PThDP where the Ki^1 values are 3×10^{-8} and 1.1×10^{-4} M, respectively.

REFERENCES

1. **Lohmann, K. and Schuster, P.,** Cocarboxylase, *Biochem. Z.*, 294, 188, 1937.
2. **Ugai, T., Dokawa, T., and Tsabokawa, S.,** *J. Pharm. Soc., Jpn,* 64, 3, 1943.
3. **Mizuhara, S. and Handler, P.,** Thiamine-catalyzed reactions, *J. Am. Chem. Soc.*, 76, 571, 1954.
4. **Breslow, R.,** On the mechanism of thiamine action. IV. Evidence from studies on model systems, *J. Am. Chem. Soc.*, 80, 3719, 1958.
5. **Breslow, R. and McNelis, E.,** Studies on model systems for enzyme action. Synthesis of reactive intermediates and evidence on the function of the pyrimidine ring, *J. Am. Chem. Soc.*, 81, 3080, 1959.
6. **Breslow, R.,** Mechanism of thiamine action: prediction from model experiments, *Ann. N.Y. Acad.Sci.*, 98, 445, 1962.
7. **Krampitz, L. O., Suzuki, I., and Greull, G.,** Mechanism of action of thiamine diphosphate in enzyme reactions, *Ann. N.Y. Acad. Sci.*, 98, 466, 1962.
8. **Krampitz, L. O.,** *Thiamin Diphosphate and its Catalytic Functions,* Marcel Dekker, New York, 1970.
9. **Holzer, H. and Beaucamp, K.,** Nachweis und Characterisierung von Zwischenprodukten der Decarboxylierung und Oxidation von Pyruvat: ''Aktiviertes'' Pyruvat und ''Aktivierter'' Acetaldehyd, *Angew. Chem.*, 71, 776, 1959.

10. **Krampitz, L. O., Gruell, G., Miller, C. S., Bicking, J. B., Skeggs, H., and Sprague, J. M.,** Active AcH-thiamine intermediate, *J. Am. Chem. Soc.*, 80, 5893, 1958.
11. **Miller, C. S., Sprague, J. M., and Krampitz, L. O.,** The reaction of thiamine with carbonyl compounds, *Ann. N.Y. Acad. Sci.*, 98, 401, 1982.
12. **Holzer, H. and Beaucamp, K.,** Nachweis und Charakterisierung von α-Lactyl-Thiaminpyrophosphat (''Aktives Pyruvat'') und α-Hydroxyethyl-Thiaminpyrophosphat (''Aktiver Acetaldehyde'') als Zwischenprodukte der Decarboxylierung von Pyruvat mit Pyruvatdecarboxylase aus Bierhefe, *Biochim. Biophys. Acta*, 46, 225, 1961.
13. **Jordan, F., Chen, G., Nishikawa, S., and Wu, B. S.,** Potential roles of the amino-pyrimidine ring in thiamin-catalyzed reactions, *Ann. N.Y. Acad. Sci.*, 378, 14, 1982.
14. **Schellenberger, A.,** The amino group and steric factors in thiamin catalysis, *Ann. N.Y. Acad. Sci.*, 378, 51, 1982.
15. **Rogers, E. G.,** Thiamine antagonists, *Ann. N.Y. Acad. Sci.*, 98, 412, 1962.
16. **Rogers, E. F.,** Thiamine antagonists, in *Methods in Enzymology*, Volume XVIII-A, McCormick, D. B. and Wright, L. D., Eds., Academic Press, New York, 1970, 245.
17. **Metzler, D. E.,** Thiamine coenzymes, in *The Enzymes*, 2nd ed., Vol. 2, Boyer, P. D., Lardy, H., and Myrback, K., Eds., Academic Press, New York, 1960, 295.
18. **Yount, R. G. and Metzler, D.,** Decarboxylation of pyruvate by thiamine analogues, *J. Biol. Chem.*, 234, 738, 1959.
19. **Krampitz, L. O., Suzuki, I., and Gruel, G.,** Role of thiamine diphosphate in catalysis, *Fed. Proc.*, 20, 971, 1961.
20. **Krampitz, L. O.,** Catalytic functions of thiamin diphosphate, *Annu. Rev. Biochem.*, 38, 692, 1969.
21. **Ullrich, J., Ostrovsky, Y. M., Ezyaguirre, J., and Holzer, H.,** Thiamin pyrophosphate-catalyzed decarboxylation of α keto acids. *Vitamins Hormones*, 28, 365, 1971.
22. **Schellenberger, A.,** Structure and mechanism of action of the active center of yeast pyruvate decarboxylase, *Angew. Chem. Int. Ed. English*, 6, 1024, 1967.
23. **Ullrich, J.,** Structure-function relationships in pyruvate decarboxylase of yeast and wheat germ, *Ann. N.Y. Acad. Sci.*, 378, 287, 1982.
24. **Ullrich, J., Wittorf, J. H., and Gubler, C. J.,** Amino acid composition of cytoplasmic yeast pyruvate decarboxylase, *FEBS Lett.*, 4, 275, 1969.
25. **Wittorf, J. H. and Gubler, C. J.,** Coenzyme binding in yeast pyruvate decarboxylase: a fluorescent enzyme-inhibitor complex, *Eur. J. Biochem.*, 14, 53, 1970.
26. **Ullrich, J. and Donner, I.,** Flurometric study of 2-P-toluidinonaphthalene-6-sulfonite binding to cytoplasmic yeast pyruvate decarboxylase, *Hoppe-Seylers Z. Physiol. Chem.*, 357, 1030, 1970.
27. **Ostrovskii, Yu. M., Ullrich, J., and Holzer, H.,** Heterogeneity of hydrophobic structures of yeast pyruvate decarboxylase. *Biokhimiya*, 36, 739, 1971.
28. **Lehmann, H., Fischer, G., Huebner, G., Kohnert, K. D., and Schellenberger, A.,** *Eur. J. Biochem.*, 32, 83, 1973.
29. **Gounaris, A. D., Turkenkoff, I., Civerchia, I. D., and Greenlie, J.,** Pyruvate decarboxylase. III. Specificity restrictions for thiamine pyrophosphate in the protein association step, subunit structure, *Biochim. Biophys. Acta*, 405, 492, 1975.
30. **Ullrich, J. and Freisler, H.,** Gehalt und Effekte von Proteinasen bei Pyruvat Decarboxylase aus Bierhefe, *Hoppe-Seylers Z. Physiol. Chem.*, 358, 318, 1977.
31. **Ullrich, J. and Donner, I.,** Kinetic evidence for two sites in cytoplasmic yeast pyruvate decarboxylase, *Hoppe-Seylers Z. Physiol. Chem.*, 351, 1026, 1970.
32. **Boiteux, A. and Hess, B.,** Allostevic properties of yeast pyruvate decarboxylase, *FEBS Lett.*, 9, 293, 1970.
33. **Hübner, G., Fischer, G., and Schellenberger, A.,** Theories of thiamine pyrophosphate action. XI. Effect of carbonyl compounds on the rate of the yeast pyruvate decarboxylase reaction, *Z. Chem.*, 10, 436, 1970.
34. **Hübner, G., Weidhase, R., and Schellenberger, A.,** The mechanism of substrate activation of pyruvate decarboxylase: a first approach, *Eur. J. Biochem.*, 92, 175, 1978.
35. **Hamada, M. and Takenaka,** This Book 4C α-Ketoglutarate Dehydrogenase.
36. **Koike, M. and Koike, K.,** Biochemical properties of mammalian 2-oxoacid dehydrogenase multienzyme complexes and clinical relevancy with chronic lactic acidosis, *Ann. N.Y. Acad. Sci.*, 378, 225, 1982.
37. **Reed, L. J.,** Regulation of mammalian pyruvate dehydrogenase complex by a phosphorylation-dephosphorylation cycle, (a) in *Thiamine*, Gubler, C. J., Fujiwara, M., and Draufns, P. M., Eds., John Wiley & Sons, New York, 1976; (b) *Curr. Topics Cell. Regul.*, 18, 95, 1981.
38. **Reed, L. J. and Pettit, F. H.,** Phosphorylation and dephosphorylation of pyruvate dehydrogenase, in *Protein Phosphorylation*, Cold Spring Harbor Conference on Cell Proliferation 8, 701, 1981.
39. **Wieland, O. H.,** The mammalian pyruvate dehydrogenase complex: structure and regulation, *Rev. Physiol. Biochem. Pharmacol.*, 96, 123, 1983.

40. **Addink, A. D. F., Boer, P., Wakabayashi, T., and Green, D. E.,** Enzyme localization on beef heart mitochondria. A biochemical and election microscopic study, *Eur. J. Biochem.*, 29, 47, 1972.
41. **Nosterescu, M. L., Siess, E. A., and Wieland, O. H.,** Ultrastructural localization of pyruvate dehydrogenase in rat heart muscle, *Histochemie,* 34, 355, 1973.
42. **Reed, L. J., Linn, T. C., Pettit, F. H., Oliver, R. M., Hucho, F., Pelly, J. F., Randall, D. D., and Roche, T. E.,** Pyruvate dehydrogenase complex: structure, function and regulation, in *Energy Metabolism and the Regulation of Metabolic Processes in Mitochondria,* Mehlman, M. A. and Hanson, R. E., Eds., Academic Press, New York, 1972, 253.
43. **Randle, P. J., Sugden, P. H., Kerbey, A. L., Radcliffe, P. M., and Hutson, N. J.,** Regulation of pyruvate oxidation and the conversion of glucose, *Biochem. Soc. Symp.*, 47, 1978.
44. **Akiyama, S. K. and Hammes, G. G.,** Elementary steps in the reaction mechanism of pyruvate dehydrogenase multienzyme complex from *Escherichia coli:* kinetics of flavin reduction, *Biochemistry,* 20, 1491, 1981.
45. **CaJacob, C. A., Gavino, G. R., and Frey, P. A.,** Pyruvate dehydrogenase complex of *Escherichia coli.* Thiamin pyrophosphate and NADH-dependent hydrolysis of acetyl Co A, *J. Biol. Chem.*, 260, 14610, 1985.
46. **Danson, M. J., Hale, G., and Perham, R. N.,** The role of lipoic acid residues in the pyruvate dehydrogenase multienzyme complex of *Escherichia coli, Biochem. J.*, 199, 505, 1981.
47. **Reed, L. J.,** Regulation of pyruvate dehydrogenase complex by phosphorylation and dephosphorylation, in *Thiamine,* Gubler, C. J., Fujiwara, M., and Dreyfus, P. M., Eds., John Wiley & Sons, New York, 1978, 19.
48. **Frey, P. A.,** Mechanism of coupled electron and group transfer in *Escherichia coli* pyruvate dehydrogenase, *Ann. N.Y. Acad. Sci.*, 378, 250, 1982.
49. **Hammes, G. G.,** Multienzyme complexes, in *Enzyme Catalysis and Regulation.*, (Mol. Biol. Intern. Series of Monographs and Textbooks), Horecker, B., Kaplan, N. O., Marmur, J., and Sheraga, H. A., Eds., Academic Press, New York, 1982, 209.
50. **Bosma, H. J., deGraaf-Hess, A. C., deKok, A., Veeger, C., Visser, A. J. W. G., and Voordouw, G.,** *Ann. N.Y. Acad. Sci.*, 378, 265, 1982.
51. **Koike, M., Hamada, M., Koike, K., Hiraoka, T., and Nakaula, Y.,** Purification and function of α-ketoacid dehydrogenases in mammalian multienzyme complexes, in *Thiamine,* Gubler, C. J., Fujiwara, M., and Dreyfus, P. M., Eds., Wiley-Interscience, New York, 1976, 5.
52. **Angelides, K. J. and Hammes, G. G.,** Fluorescence studies of the pyruvate dehydrogenase multienzyme complex from *Escherichia coli, Biochemistry,* 18, 1223, 1979.
53. **Shephert, G. B. and Hammes, G. G.,** Fluorescence energy transfer measurements in the pyruvate dehydrogenase multienzyme complex from *Escherichia coli* with chemically modified lipoic acid, *Biochemistry,* 16, 5234, 1977.
54. **Collins, J. H. and Reed, L. J.,** Acyl group and electron pair relay system: a network of interacting lipoyl moieties in the pyruvate and α-ketoglutarate dehydrogenase complexes from *Escherichia coli, Proc. Natl. Acad. Sci. U.S.A.*, 74, 4223, 1977.
55. **Tsai, C. S., Burgett, M. W., and Reed, L. J.,** α-Keto acid dehydrogenase complexes. XX. A kinetic study of the pyruvate dehydrogenase complex from bovine kidney, *J. Biol. Chem.*, 248, 8348, 1973.
56. **Randle, P. J.,** Phosphorylation-dephosphorylation cycles and the regulation of fuel selection in mammals, *Curr. Topics Cell. Regul.*, 18, 107, 1981.
57. **Linn, T. C., Pettit, F. H., and Reed, L. J.,** α-Ketoacid dehydrogenase complexes. X. Regulation of the activity of the pyruvate dehydrogenase complex from beef kidney mitochondria by phosphorylation and dephosphorylation, *Proc. Natl. Acad. Sci. U.S.A.*, 62, 234, 1969.
58. **Wieland, O. H. and Jagow-Westermann, B. V.,** ATP-dependent inactivation of heart muscle pyruvate dehydrogenase and reactivation by Mg^{++}, *FEBS. Lett.*, 3, 271, 1969.
59. **Yeaman, S. J., Hutcheson, E. T., Roche, T. E., Pettit, F. H., Brown, J. R., Reed, L. J., Watson, D. C., and Dixon, G. H.,** Sites of phosphorylation on pyruvate dehydrogenase from bovine kidney and heart, *Biochemistry,* 17, 2364, 1978.
60. **Sugden, P. H., Hutson, N. J., Kerbey, A. L., and Randle, P. J.,** Phosphorylation of additional sites on pyruvate dehydrogenase inhibits reactivation by pyruvate dehydrogenase phosphatase, *Biochem. J.*, 169, 433, 1978.
61. **Randle, P. J., Sale, G. J., Kerbey, A. L., and Kearns, A.,** Regulation of pyruvate dehydrogenase complex by phosphorylation and dephosphorylation, in *Protein Phosphorylation,* Cold Spring Harbor Conf. on Cell Proliferation, Roven, O. M. and Krebs, E. G., Eds., Book A, Vol. 8, 1981.
62. **Randle, P. J.,** α-Ketoacid dehydrogenase complexes and respiratory fuel utilization in diabetes, *Diabetalogia,* 28, 479, 1985.
63. **Denton, R. M. and Hughes, W. A.,** Pyruvate dehydrogenase and the hormonal regulation of fat synthesis in mammalian tissues, *Int. J. Biochem.*, 9, 545, 1978.

64. **Parker, J. C. and Jarett, L.,** Insulin mediator stimulates pyruvate dehydrogenase of intact liver mitochondria, *Diabetes,* 34, 92, 1985.
65. **Dohm, G. L., Patel, V. K., and Kasperek, G. J.,** Regulation of muscle pyruvate metabolism during exercise, *Biochem. Med. Metabolic Biol.,* 35, 260, 1986.
66. **Warburg, O., Christian, W., and Griese, A.,** Wasserstoffübertragendes Coferment, Siene Zusammensetzung und Wirkungsweise, *Biochem. Z.,* 282, 157, 1935.
67. **Warburg, O. and Christian, W.,** Gärungs-Coferment aus roten Blutzellen, *Biochem. Z.,* 285, 156, 1936.
68. **Williams, J. F.,** A critical examination of the evidence for the reactions of the pentose pathway in animal tissues, *TIBS,* 5, 315, 1980.
69. **Horecker, B. L., Paoletti, F., and Williams, J. F.,** Occurrence and significance of octulose phosphates in liver, in thiamin, twenty years of progress, *Ann. N.Y. Acad. Sci.,* 378, 25, 1982.
70. **Gubler, C. J.,** Physiological functions of transketolase, in *Thiamin Pyrophosphate Biochemistry,* Vol. 1, Showen, R. L. and Schellenberger, A., Eds., CRC Press, Boca Raton, FL, 1988, chap. 16.
71. **Calvin, M. and Basshan, J. A.,** *The Photosynthesis of Carbon Compounds,* Benjamin, New York, 1962.
72. **Dreyfus, P. M. and Hansen, G.,** The effect of thiamin deficiency on the pyruvate decarboxylase system of the central nervous sytem, *Biochem. Biophys. Acta,* 104, 78, 1965.
73. **Plaitakis, A., Hwang, E. C., Van Woert, M. H., Szilagyi, P. I. A., and Berl, S.,** Effect of thiamin deficiency on brain neurotransmitter systems, *Ann. N.Y. Acad. Sci.,* 378, 367, 1982.
74. **Gibson, G., Barclay, L., and Blass, J.,** The role of the cholinergic system in thiamin deficiency, *Ann. N.Y. Acad. Sci.,* 378, 382, 1982.
75. **Racker, E.,** Transketolase, in *The Enzymes,* Vol. 5, 2nd ed., Boyer, P. D., Lardy, H., and Myrback, K., Eds., Academic Press, New York, 1961, 397.
76. **Wood, T.,** The preparation of transketolase free from D-ribulose-5-phosphate-3-epimerase, *Biochim. Biophys. Acta,* 659, 233, 1981.
77. **Warnock, L. and Prudhomme, C.,** The isolation and preliminary characterization of apotransketolase from human erythrocytes, *Biochem. Biophys. Res. Commun.,* 106, 719, 1982.
78. **Himmo, S. D., Thomson, M., and Gubler, C. J.,** Purification of transketolase from human erythrocytes, *Prep. Biochem.,* 18, 261, 1988.
79. **Paoletti, F.,** Purification and properties of transketolase from fresh rat liver, *Arch. Biochem. Biophys.,* 222, 489, 1983.
80. **Saitow, S., Ozawa, T., and Tomita, Y.,** The purification and some properties of brewer's yeast apotransketolase, *FEBS Lett.,* 40, 114, 1974.
81. **Cavalieri, S., Neet, K. E., and Sable, H. Z.,** Enzymes of pentose biosynthesis. The quaternary structure and reacting form of transketolase from baker's yeast, *Arch. Biochem. Biophys.,* 171, 527, 1975.
82. **Philippov, P. P., Shestakova, I. K., Tikhomirova, N. K., and Kochetov, G. A.,** Characterization and properties of pig liver transketolase, *Biochem. Biophys. Acta,* 613, 359, 1980.
83. **Kremer, A. B., Egan, R. M., and Sable, H. Z.,** The active site of transketolase. Two arginine residues are essential for activity, *J. Biol. Chem.,* 255, 2405, 1980.
84. **Kochetov, G. A.,** Structure of the active center of transketolase, *Ann. N.Y. Acad. Sci.,* 378, 306, 1982.
85. **Meshkalkina, L. E., Kuimov, A. N., Kabakov, A. N., Tsorina, O. N., and Kochtov, G. A.,** *Biochem. Int.,* 9, 9, 1984.
86. **Kuimov, A. N., Meshkalkina, L. E., and Kochitov, G. A.,** An investigation of the carboxyl group function in the active center of transketolase, *Biochem. Int.,* 11, 913, 1985.
87. **Hopmann, R. F. W. and Brugnoni, G. P.,** PK of thiamine C(2) hydrogen. *Nature New Biol.,* 246, 147, 1973.
88. **Kochetov, G. A., Izotova, A. E., and Meshalkina, L. E.,** Inhibition of transketolase by analogues of the coenzyme, *Biochem. Biophys. Res. Commun.,* 43, 11980, 1971.
89. **Klein, H. and Brand, K.,** Reinigung und Egenschafften der Transketolase aus Candida utilis, *Hoppe-Seyler's Z. Physiol. Chem.,* 358, 1325, 1977.
90. **Shreve, D. S., Holloway, M. P., Haggerty, J. C., III, and Sable, H. Z.,** The catalytic mechanism of transketolase, *J. Biol. Chem.,* 258, 12405, 1983.
91. **Heinrich, P. G., Steffen, H., Janser, P., and Wiss, O.,** Studies on the reconstruction of apotransketolase with thiamine pyrophosphate and analogs of the coenzyme, *Eur. J. Biochem.,* 30, 533, 1972.

Section II
Allostery and Non-Michaelis-Menten Behavior

Until a problem has been logically defined it cannot be experimentally solved. G. N. Lewis and M. Randall (1923), *Thermodynamics and the Free Energy of Chemical Substances*, McGraw-Hill Book Co., Inc., New York, p. 324.

Chapter 5

GLYCOGEN PHOSPHORYLASE AND GLYCOGEN SYNTHETASE

Neil B. Madsen

TABLE OF CONTENTS

I. GLYCOGEN PHOSPHORYLASE

A. INTRODUCTION AND HISTORICAL BACKGROUND

This enzyme has been the subject of a number of recent reviews which treat certain aspects of this chapter in greater detail. For example, the allosteric phenomena are reviewed extensively,[1-3] as is the molecular structure,[4] the regulation by covalent modification[5] and the catalytic mechanism.[6,7] This chapter will attempt to integrate all of these areas, together with a reassessment of some aspects in the light of more recent knowledge.

We can now look back on more than 50 years of research on phosphorylase since Carl and Gerti Cori announced their discovery of the formation of glucose 1-phosphate, from inorganic phosphate added to minced muscle, at the 1935 International Congress of Physiology, with a full written account appearing a year later.[8] They noted that the addition of small amounts of AMP provided a great enhancement of the activity. By 1940 the enzyme had been purified, its kinetics explored, and AMP recognized as an essential dissociable coenzyme.[9] Inhibition by glucose was determined to be specific for that sugar and to be competitive with glucose 1-P. Indeed, data published in that paper can be plotted to show sigmoid kinetics for glucose 1-P in the presence of glucose. In 1941 a second form of phosphorylase was isolated which did not require AMP for activity and when the kinetics of this enzyme was explored, it was found that the reciprocal plot for glucose 1-P concentrations was concave upwards in the presence of glucose, but became linear when the substrate concentrations were squared.[10] It was clearly recognized that glucose caused the kinetics to change to second order. Thus, phosphorylase may represent the first example of an enzyme which has an allosteric activator and cooperative substrate kinetics, although the full significance of the findings had to await a theoretical framework into which they could be fitted.

Several discoveries during the 1950s laid the basis for our current knowledge of the molecular structure and regulation of phosphorylase. In 1955 the enzyme was found to contain two identical subunits in the case of phosphorylase *b* (the form which requires AMP for activity) while the AMP independent form (*a*) contains four.[11] This was the first discovery of the subunit structure of an enzyme and, when the number of binding sites for AMP was equated with the number of subunits,[12] an important clue to the nature of homotropic cooperativity of ligand binding was exposed to view. Unfortunately we did not understand it then, or do the relevant comparison with hemoglobin. We did find that AMP lowers the energy of activation for phosphorylase *a* and suggested that it binds to a site near the active site and makes the enzyme into a more efficient catalyst.[13]

In 1959 Fischer et al.[14] discovered that the enzyme which converts phosphorylase *a* to *b* is a protein phosphatase which removes phosphate from a single serine residue, later shown to be number 14. Thus, this simple chemical change is responsible for the profound differences in enzymatic properties of the two forms. Phosphorylase is not only one of the archtypal allosteric enzymes but it was the first shown to undergo reversible covalent modification which is under metabolic, hormonal, and neural control.

Certain mysterious chemical and spectroscopic observations on phosphorylase were explained by the discovery, in 1957, that each subunit contains a molecule of pyridoxal-5′-phosphate.[15] The coenzyme is covalently bound through a Schiff base to a lysine residue,[16] but, unlike the situation in other PLP enzymes, this linkage plays no role in catalysis since it may be reduced to an irreversible bond without destroying the activity. As will be discussed later, the phosphate group of PLP appears to perform a unique function in catalysis.

B. BIOLOGICAL REGULATION

Phosphorylase catalyzes the first step in the intracellular degradation of glycogen by adding inorganic phosphate across the α-1,4-glucosidic links of the glucose termini to form α-glucose 1-P. Although the equilibrium constant for this reaction is 0.28, phosphorylase catalyzes only a degradative reaction under physiological conditions because the concentration of Pi greatly exceeds that of glucose 1-P while phosphoglucomutase converts the latter continually to glucose 6-P (equilibrium constant = 20). In muscle the role of glycogen phosphorylase is to provide phosphorylated glucose units for glycolysis in response to the energy demands of contraction while in liver it is to provide free glucose (via phosphoglucomutase and glucose 6-phosphatase) for the maintenance of blood glucose levels during fasting. In other tissues or organisms, net glycogen degradation may occur only in emergency situations such as anoxia or starvation.

Glycogen *in vivo* exists as large molecular weight aggregates known as glycogen particles, which are probably single giant molecules.[17,18] When isolated under benign conditions, these particles contain all the enzymes necessary for the degradation and synthesis of glycogen, as well as the kinases and phosphatases which carry out the covalent modifications of these enzymes.[19,20] While glycogen phosphorylase and synthetase are both quite soluble, both are usually in particulate form *in vivo*, bound to glycogen by "glycogen storage sites" which are distinct from the active sites.[21-23] This binding mode orients the active sites of phosphorylase to the glycogen molecule while the "control face" is on the opposite side of the enzyme and permits Ser-14 to be phosphorylated or dephosphorylated while the enzyme is bound to the glycogen, and whether or not the phosphorylase is active. We may predict a similar structural motive for the synthetase.[23]

Phosphorylase manifests several levels of regulation, some of which have been superceded or altered in the course of time. The inhibition by glucose, while obviously deriving from the structural similarity to glucose 1-P, is not a simple competitive situation because, although the sugar moieties share most of the ligand interactions, these two molecules bind to different conformational states.[24,25] The inhibition by glucose may represent an ancient feedback control mechanism, but it no longer applies to the muscle system because there is virtually no free glucose in muscle cells. In liver, where the intracellular concentration of glucose is equivalent to that of the blood, feedback inhibition may have been replaced by a more sophisticated control of metabolic interconversion based on the conformational change caused by the glucose. The "T" state conformation stabilized by glucose is the preferred substrate for phosphorylase phosphatase and hence glucose activates the conversion of the more active α form to the less active *b* form. Hers and colleagues elaborated this hypothesis and term phosphorylase *a* the glucose receptor of the liver cell.[26]

The next level of regulation is via energy state and consists of the activation of phosphorylase *b* by AMP and its inhibition by ATP, ADP, and glucose 6-P. All of these effectors compete for the same site in the N-terminal domain, some 30 Å from the active site.[27-29] Comparison of the amino acid sequences from the rabbit muscle, *Escherichia coli,* and potato enzymes show a high homology in the core of the protein and around the active site but virtually none for the first 80 residues which contain the phosphorylatable serine and parts of the nucleotide binding site in the mammalian enzyme.[5] Thus, we may assume that this energy linked control of regulation evolved after the glucose control. In muscle cells, and most other cells, the nucleotide concentrations do not change enough to activate phosphosylase *b* during normal muscle contraction or other functions.[30] A more sophisticated level of regulation has evolved to allow phosphorylase to "escape" from allosteric regulation. Phosphorylation of Ser-14 results in only minimal inhibition by ATP, ADP, and glucose 6-P and therefore activates the enzyme at a constant level of activators and inhibitors. This permits an "on-off" control of glycogen phosphorolysis which is linked to muscle contraction by the neural release of calcium and, for example, results in an increased rate of up to 1000-fold in the leg muscles of a sprinter.[31] When phosphorylase kinase is absent for genetic reasons, there is a pronounced lag period in glycogen breakdown and the maximal rate is decreased.[32] Phosphorylase *b* activation is thus inefficient, slow, and may depend on increased levels of IMP, Pi, and other factors.[5,30]

The situation is less clear in liver, where the enzyme is poised at a partially phosphorylated state which is determined by a complex interplay between various hormones, especially glucagon and insulin, and metabolites such as glucose.[26] In addition, allosteric control of both forms of phosphorylase by a variety of effectors may play a more important role than in muscle, since the function here is the modulated response to relatively slow changes in blood sugar levels rather than to the quick and violent contraction of muscle.

C. MOLECULAR STRUCTURE

As discussed above, phosphorylase was the first enzyme shown to exist in oligomeric form. The early hydrodynamic data was interpreted to indicate that the subunits are identical but this was not proven until amino acid sequence data negated several contrary reports. While phosphorylase *a* from mammalian skeletal muscle exists in tetrameric form *in vitro* and the *b* form is a dimer, this difference is not true for phosphorylases from liver, lobster muscle, and rabbit heart (isozyme I).[5] Furthermore, both forms of the skeletal muscle enzyme probably function as a dimer *in vivo* since the latter has a higher specific activity and a greater affinity for glycogen. The existence of the tetrameric form would not appear to have too much significance for explaining structure-function relationships in phosphorylase and, indeed, has caused a great deal of confusion over the years.

The complete amino acid sequence of phosphorylase *a* was announced in 1977,[33] and was absolutely essential for the elucidation of the crystal structure. Refinement of the X-ray derived crystal structure together with elucidation of the cDNA sequence[34] confirmed the amino acid sequence with only a few minor changes. The most important of these was the insertion of one new residue, Ile-307A, to make 842 in all. These residues plus the coenzyme, the phosphate on Ser-14, and the acetyl group on Ser-1 make up a molecular weight of 97,511 for the subunit. Amino acid sequence data for glycogen phosphorylases from such widely separated organisms as *E. coli,* potato, and partial sequences from dogfish muscle and yeast are available and comparisons will be commented on when appropriate, or an earlier review may be consulted.[5]

High resolution (2 Å) structures of phosphorylase *a* and phosphorylase *b* are now being refined at San Francisco[4,24,35] and Oxford,[36-39] respectively, and the two laboratories are also engaged in comparing the two structures. The glucose of crystallization is found in the active site of the phosphorylase *a* molecules and the derived structure is therefore of the inactive T conformation. The phosphorylase *b* crystals are grown in the presence of IMP but the conformation is still predominantly of the T type and is isomorphous with the *a* form. The chief differences are found in the N-terminus near Ser-14 and the nucleotide binding site.

The phosphorylase subunit contains two major domains, each consisting of a core of twisted β-sheet surrounded by α-helices. The N-terminus, residues 1-489, forms a broad interface with the C-terminus and the PLP coenzyme lies sandwiched between, approximately parallel with the two faces. It is surrounded by predominantly nonpolar residues but its phosphate is thrust into the hydrophilic active site cavity and is coordinated by positive charges. The active site is a large cavity approximately 15 Å from the protein surface and formed by residues from both domains. The two subunits are associated closely through the N-termini, which associate as two doughnuts might, leaving a large cavity between the two subunits.[40] The cavity is lined with the sort of hydrophilic residues usually found on a protein surface and contains approximately 150 water molecules, which might form a hydrogen-bonding network. This type of association may allow two large subunits to unite closely without burying too many charges. The interface between the two subunits is the region which transmits homotropic allosteric interactions between ligand binding sites of the two subunits, as well as heterotropic interactions within each subunit.[22,39,41,42]

Although the two subunits in the dimer are related by a twofold axis of symmetry which, in the crystal, is coincident with a crystallographic twofold axis, the dimer is anything but symmetric in the orientation of its ligand binding sites and other surface features. As demonstrated by computer-drawn space-filling models,[22] the entrances to the two active sites are within a concave "catalytic face" which is oriented toward the glycogen substrate by the two glycogen storage sites found on the outer rim of this face. The latter sites bind the terminal glucose chains of glycogen some 22 times as tightly as do the active sites, so that the enzyme remains bound to the glycogen particle whether or not it is catalytically active.[43] Furthermore, several catalytic cycles involving several different terminal chains of glycogen may occur during each binding event.

On the opposite side of the dimer may be described a convex "control face" which contains the binding sites for AMP, Ser-14 on the flexible N-termini, and the binding sites (still undescribed in detail) for phosphorylase kinase and protein phosphatase. The phosphate moiety of AMP appears to anchor the nucleotide by binding to a relatively nonspecific anion binding site composed of two to three arginine residues.[29] The purine base may "stack" against Tyr-75 in the activated "R" conformation, as seen in the phosphorylase *b* crystals,[29] but not in the T conformation observed in the *a* crystals.[27] Thus, the movement of Tyr-75 to complex the purine base of AMP may account for the movement of Helix 49 to 75 and, since this helix connects the two AMP binding sites, for the homotropic cooperativity of AMP binding. The nucleoside portion of AMP interacts with residues from both subunits of the dimer and this probably accounts for the increased strength of interaction of the subunits upon AMP binding. ATP and glucose 6-P also bind to the AMP site with their phosphates binding to Arg-309 which is in a different position than when AMP binds to this residue.[28,39] Thus, while the binding site for glucose 6-P does not overlap that of AMP, their requirement for a common residue explains why the binding of one excludes the other, as observed by kinetic and physical studies.

The first 18 N-terminal residues are not detectable in the crystal structure of phosphorylase *b* and are assumed to be flexible and not bound to the main body of the subunit.[44] When Ser-14 becomes phosphorylated, all except the first four residues are clearly distinguished.[35] The N-terminus now lies across the subunit interface and makes several contacts with residues from both. Thus, the Ser-14 phosphate forms salt bridges with Arg-69 and with Arg-43′ from the symmetry related subunit. The former contact is just to the outside of helix 49 to 75, on the inside of which is the binding site for AMP, thus providing a link between the two control sites. Arg-10 and Lys-11 form hydrogen bonds with residues from the other subunit and there may be hydrophobic interactions as well. The greater strength of subunit interactions in phosphorylase *a* as compared to phosphorylase *b* is readily explained by this new structure, which is well illustrated in Figure 1 of Reference 5. As Sprang and Fletterick[40] have pointed out, the phosphorylation of the Ser-14 of phosphorylase *b* forms a convalently attached ligand which now binds to a specific ligand binding site as well as facilitating other binding interactions of the hitherto detached N-terminus.

D. ALLOSTERIC CONTROL

Since any investigation of allostery involves enzyme kinetics, an understanding of the kinetic mechanism is important. For glycogen phosphorylase from a variety of sources, this has been established as rapid equilibrium random bi bi.[45-47] Isotope exchange at equilibrium studies have confirmed the mechanism for some cases,[48,49] as well as demonstrating that the kinetic mechanism is not changed when ATP or glucose alters the kinetics from Michaelian to the sigmoid allosteric type.[50] In this mechanism the apparent Michaelis constants should be equivalent to dissociation constants, a useful feature which has been confirmed for some ligands under specified conditions.

Parmeggiani and Morgan discovered the inhibition of phosphorylase *b* by ATP and its relief by AMP in 1962.[51,52] They suggested that phosphorylase *b* activity in the heart was regulated by the concentrations of the activator, AMP, and the inhibitors ATP and glucose

6-P. In 1964, Helmreich and Cori showed that AMP lowers the Km for the substrate, Pi, and vice versa.[53] The same year, Madsen demonstrated the same effect for the other substrate, glucose 1-P, and also that ATP was competitive for AMP and presumably bound to the same site.[54] In addition, while ATP appeared to compete with glucose 1-P, it caused the velocity vs. substrate concentration curve to assume a sigmoid shape. The slope of the Hill plot was 1.0 for the uninhibited reaction but in the presence of ATP, the slope was 1.8. The data were interpreted in terms of the hypothesis on allosteric transitions (conformational changes) recently published by Monod et al.[55] Also in 1964, Ullman, Vagelos, and Monod showed that AMP improved the binding of bromthymol blue to phosphorylase *b*, assumed to be due to conformational changes induced by the activator, and that the binding of AMP shows homotropic cooperativity.[56]

Only a year later, in 1965, Monod et al. published their plausible model for allosteric transitions and thus provided all workers in the field with a theoretical framework within which to plan data for testing each system.[57] While their concerted model for allosteric transitions does not apply too well to phosphorylase *b*, it is convenient to define a static inactive T-state conformation and an active R-state conformation. The latter may not be a single structure if conformational changes occurs in the course of each catalytic cycle. The concerted model does apply more closely to phosphorylase *a*.

Substrate cooperativity in phosphorylase *b* does not manifest itself clearly until an inhibitor such as ATP is present, and then the Hill constant, n, increases with increasing inhibitor concentration until the maximal expected value of two is reached,[58] indicating an increasing strength of active site interaction. Cooperativity between activator sites is seen at low concentrations of AMP but not at high concentrations unless ATP or glucose 6-P is present. The data were shown to fit the concerted model for allosteric transitions with some significant discrepancies, but an allosteric constant L (ratio of T to R states) of approximately 3000 could be obtained.[58,59] Cooperativity for binding of substrates, AMP, and ATP was also demonstrated by their effects on the fluorescence of PLP,[1,60] while direct binding experiments by classical methods also confirmed the cooperativity of AMP binding.[61,62] As discussed in an earlier review,[1] measured parameters for most of the conformational changes caused by the binding of AMP are coincident with the binding as determined by direct methods. These indicators of conformational changes include catalytic activity and changes in PLP fluorescence, which have to be transmitted some 30 Å, and changes in the spectroscopic properties of nitroxide or nitrobenzo-oxa-diazole moieties bound to Cys-317, near the AMP binding site.[63] Since all these experiments indicate that the change of state function ($\bar{R}$) is equivalent to the ligand saturation function ($\bar{Y}$), it was argued that the model for concerted transition changes does not apply to phosphorylase *b*.[1] Buc and Buc[64] obtained evidence which conflicts with this conclusion when measuring different parameters, and the reasons for this discrepancy are still not clear.

It is likely that several, or even a continuum of, conformational states may exist between inactive and unliganded phosphorylase *b* and a fully liganded and active enzyme. Radda and colleagues first showed that glucose 1-P caused a major conformational change in phosphorylase *b* saturated with AMP.[2] Thus, AMP is not able to provide the optimal conformation for the binding of glucose 1-P and part of the binding energy of the latter must be used for a conformational change, accounting for the limiting Km being ten times that found with phosphorylse *a*.[47]

Another form of allosteric control, whose physiological significance is not yet understood, arises from the location of an inhibitor site on the solvent side of a peptide loop which forms part of the active site and which accommodates purine derivatives such as inosine, adenosine, and caffeine. AMP and IMP also bind here at very high concentrations, accounting for the inhibitory effects of these activators. The binding of compounds to this site is synergistic with glucose binding, both stabilizing the T state, because Asn-284 forms a hydrogen bond with the 2-hydroxyl of glucose while, on the solvent side of the loop, the purine base is intercalated between the phenyl rings of Phe-285 and Tyr-612.[24,65] This loop must move when glucose 1-P is bound in order that the phosphate of the substrate can occupy part of the position previously taken by the carboxy group of Asp-283, so that both glucose and caffeine, for example, are competitive inhibitors with respect to substrate but caffeine does not occupy the same site.[25] A recent study has derived the interaction constants for the various sites on phosphorylase and shown, for example, that the synergism between glucose and caffeine has an interaction constant of 0.2.[66] On the other hand, AMP activation is inhibited noncompetitively by caffeine and glucose, with interaction constants of approximately six. The substrate analogue, UDP-glucose, is a noncompetitive inhibitor for AMP but these two compounds have an interaction constant of 0.25 (improve each other's binding) because both stabilize the R conformation.

While phosphorylase *a* is still an allosteric protein, it has lost most of the properties of phosphorylase *b* in this regard. The allosteric constant, L, has changed from 3000 to 10 or less,[67] representing a change in the energy involved in the $T \rightarrow R$ transition of at least 3.5 kcal. This energy has been provided by the binding of the Ser-14 phosphate and associated residues, and is also reflected in a 200-fold tighter binding of AMP. Furthermore, the *a* form is not subject to significant inhibition by ATP and glucose 6-P, which seem only to inhibit the extra activity provided by the AMP. On the other hand, glucose and caffeine exhibit the same pattern of inhibitions and synergy noted for the *b* form.[5]

As discussed in detail earlier,[5] the binding of the phosphorylated N-terminus across the subunit interface leads to much stronger association and, as a predictable consequence, intermediate conformations between the T and R forms are less likely to occur. One might therefore suggest that symmetry must be conserved and phosphorylase *a* should follow the concerted transition model of Monod et al.[57] In fact, Griffiths, Price, and Radda[68] have shown that conformational changes, monitored by a spin-label attached to (presumably) the Cys-317 of each subunit, precede the binding of AMP ($\bar{R} > \bar{Y}$ throughout the titration with AMP). Thus, a simple chemical modification has increased the subunit interactions so strongly that the same enzyme has changed from the sequential model of Koshland et al.[69] to the concerted transition or symmetry model. This agrees with the observation by Koshland[70] that when intersubunit interactions are too strong to permit stable intermediate conformations, the general sequential model changes to the limiting case of the concerted transition model.

E. REGULATION BY COVALENT MODIFICATION

The author has recently published an extensive review on this subject,[5] so only a brief summary will be given here. In the case of skeletal muscle, it has been shown that less than 5% of the phosphorylase is present as the active phosphorylase *a* form. The phosphorylase *b* is likely to be inactive because most of it will be liganded by the allosteric inhibitors ATP, ADP, and glucose 6-P, according to the analysis of Busby and Radda.[2] This situation accounts for the very low level of net glycogenolysis in resting muscle, although this may underestimate the extent of phosphorolysis masked by turnover. When the enzyme is phosphorylated, a large part of it will be liganded with AMP while that part which is still liganded with the allosteric inhibitors will nevertheless exhibit significant partial activity since these inhibitors now have little effect.[5] Regulation of phosphorylase kinase in muscle would appear to be primarily via the calcium released in response to the neural turn-on of contraction, thus coordinating glycogenolysis with mechanical work, while hormonal (adrenalin) stimulation of cAMP release potentiates the calcium effect because the covalently activated kinase has a much lower requirement for calcium.[71] It is possible also that catecholamines cause an increase in the cytosolic level of calcium which is sufficient to activate the phosphorylase kinase but not to cause muscle contraction. The authoritative review by Pickett-Gies and Walsh should be consulted for details of this complicated situation.[71]

In the case of the liver, it would appear that most hormones except glucagon cause activation of phosphorylase kinase and hence phosphorylase by increasing the cytosolic calcium concentration without affecting the cAMP.[71] Glucagon, on the other hand, activates the system via increased cAMP levels and subsequent covalent phosphorylation of the kinase, the latter then needing less calcium for activity.[71]

The classical experiments carried out by Danforth, Helmreich, and Cori[72] on the effect of electrical stimulation on the interconversion of phosphorylases *a* and *b* in frog muscle suggested that protein phosphatase activity was constant under all conditions and therefore not subject to regulation. Since then a bewildering plethora of experimental detail has obscured this simple picture but we cannot yet arrive at an acceptable model. The extensive review by Ballou and Fischer[73] should be consulted for details and in this space I will try to summarize the current understanding. Protein phosphatase-1 is the predominant activity in muscle, and the only phosphatase found in the glycogen particle. It is found in solution in association with inhibitor-2, which must be phosphorylated by glycogen synthetase kinase-3 in order to activate the phosphatase. The regulation of this MgATP-dependent enzyme is not understood. Another protein, inhibitor-1, is phosphorylated by cAMP-dependent kinase and it then inhibits the cytosolic phosphatase and thus provides, in principle, a hormonally controlled inhibition to complement the hormonally activated kinase activities. The phosphatase found in the glycogen particle is associated with a G-protein which is needed for binding to the glycogen and which also protects it from regulation by inhibitors-1 or -2, so that this phosphatase-1 form is spontaneously active and may account for the earlier observations of Danforth, Helmreich, and Cori.[72] Substrate-directed regulation (activation by glucose or caffeine, inhibition by AMP) can be demonstrated in *in vitro* model systems but are probably not physiologically significant since, for example, little free glucose is found in the muscle cell.

As noted earlier, substrate level regulation of protein phosphatase-1 may play an important role in the regulation of glycogen metabolism in the liver. Hers and colleagues have shown that increased blood sugar levels result in the binding of glucose to phosphorylase, making it a much better substrate for the phosphatase and hence causing the conversion of *a* to *b* forms.[26] After the phosphorylase *a* level has decreased sufficiently to relieve inhibition of the phosphatase, the latter can dephosphorylate and hence activate glycogen synthetase. This mechanism accounts for the lag in glycogen formation after a large increase in glucose concentration. Caffeine is synergistic with glucose in accelerating the decrease in phosphorylase *a* and shortening the lag period before activation of synthetase when isolated hepatocytes are studied.[74] Other studies, using isolated enzymes *in vitro,* have provided additional support for the substrate directed control of protein phosphatases in the liver,[75,76] and the caffeine effects suggest that the nucleoside inhibitor site may have a physiological role to play but the significance is as yet unclear.[77] However, phosphorylase *a*, with several ligand binding sites subject to a variety of physiological effectors, can inhibit both phosphatase-1 and the various forms of phosphatase-2A under some circumstances, thus preventing their action on other phosphoproteins. Relief of this inhibition by the binding of glucose and/or other ligands allows the phosphatases to first dephosphorylate phosphorylase *a* and subsequently other phosphoproteins. Other forms of regulation of the phosphatases, discussed above for the muscle situation, may also apply in the liver but the details are still unclear.[73]

F. CATALYTIC MECHANISM

Retention of the α-configuration at the anomeric carbon of glucose during the transfer of glucose from glucose 1-P to the 4-hydroxy of the terminal glucose in the acceptor polysaccharide suggests a double displacement mechanism similar to those found in glycosidases.[7] We might expect donation of a proton from an acidic group to the oxygen of the glycosidic bond, with stabilization of the resultant glucosyl moiety, possibly by a carboxylate group, followed by attack on the anomeric carbon by the 4-hydroxy group of the acceptor glucose unit, possibly aided by general base-catalyzed deprotonation of the 4-hydroxy group. None of the functional groups involved have been identified but a role has been suggested for the phosphate moiety of PLP.

As reviewed previously,[7] several laboratories have performed the difficult syntheses of an impressive number of PLP analogues which they used to reconstitute the enzyme from apophosphorylase, with subsequent testing of properties and activity. By this means, all parts of the PLP molecule have been ruled out as having any direct role in catalysis except the phosphate moiety, and it must be capable of assuming a dianionic form and have a pK near neutrality. The crystallographic studies confirmed the chemical work by showing that nearly the entire PLP molecule is buried in a hydrophobic pocket between the N-terminal and C-terminal domains so that only part of the phosphate moiety is in the active site, oriented toward the phosphate of the substrate. The crystals of both forms of phosphorylase yield the structures of the inactive T conformation and the phosphate-phosphate distances (P atom-P atom) observed are slightly too great for direct interaction. Thus, using glucose 1,2-cyclic phosphate as a substrate analogue, this distance is 6.8 and 5.4 Å for the *a* and *b* forms, respectively. Solution studies, however, suggest strongly that direct contact does occur in the fully active state during the catalytic cycle.

The first such evidence came from Graves' laboratory where it was shown that the inactive enzyme resulting from reconstitution with pyridoxal was active in the presence of phosphate, phosphite, or fluorophosphate, suggesting that the latter compounds bound in the site normally occupied by the phosphate moiety of PLP.[78] Pyrophosphate is a competitive inhibitor of both phosphite and glucose 1-P, but only 1 mol of this inhibitor bound per subunit, suggesting that it simultaneously occupied the sites for the phosphate of both substrate and coenzyme. Inhibition by a series of diphosphonates separated by 1, 2, or 3 methylenes was most effective for the first, with a P-P separation of 3 Å, while the others, with P-P distances of 5 and 6 Å were progressively less effective.[25] Finally, it was shown that a compound which combines substrate and coenzyme through a pyrophosphate bond, pyridoxal pyrophosphate glucose, could be used to reconstitute the enzyme.[79,80] When oligosaccharide was added to this enzyme analogue, the glucose was transferred to the 4-hydroxy group of the terminal glucose in the correct configuration. The analogy with the substrate for glycogen synthetase, uridine diphosphate glucose, may be more than coincidental.

There is general agreement that the phosphates of the coenzyme and substrate interact during catalysis and discussion now centers on the precise role of the coenzyme phosphate, as outlined more extensively in our previous review.[7] Helmreich and colleagues argue that it acts as a Bronsted acid base, transferring a proton directly to the substrate phosphate in order to labilize the glycosidic bond.[81,82] They have shown that the coenzyme phosphate is a dianion in activated phosphorylase and is protonated in the inactive form.[83] Phosphorylase can use various compounds as substrates, including D-glucal, α-D-glucosyl fluoride, and heptenitol, and the authors have shown that stereo-specific protonations occur which may involve the reversible protonation of the coenzyme phosphate.[81,82,84,85] They have also provided evidence for the glucosyl-enzyme intermediate.[81,84]

Madsen, Withers, and Fukui have suggested a slightly different role for the coenzyme phosphate, namely that of a Lewis acid (electrophile).[25,79,80] They suggest that when the substrate is bound the coenzyme is tightly coordinated by positive charges and distorted to a trigonal bipyramidal configuration with the empty apical position carrying a positive charge which is attacked by a nucleophilic oxygen of the substrate. The formation of the resultant pseudo pyrophosphate bond is analogous to an abortive phosphate transfer and similar to the situation seen with uridine diphosphate glucose and glycogen synthetase (as well as with pyridoxal pyrophosphate glucose, discussed above) and would be expected to withdraw electrons from the substrate phosphate and thus labilize the glycosidic bond. Stabilization of the glucosyl carbonium ion intermediate would not differ in principle from that proposed for the alternative hypothesis while protonation and deprotonation would be catalyzed by protein functional groups, as suggested at the beginning of this section.

As discussed extensively before,[7] it is difficult to distinguish between the two proposed roles for the coenzyme phosphate, and both groups have interpreted various experimental results in alternative ways. It would not appear profitable to speculate interminably over minor details and hopefully a clever experiment will eventually clarify the situation. One other aspect of the mechanism concerns the steric problem deriving from the approach of a bulky substrate constituent, the terminal glucose of the saccharide, to the same α-side of the glucose oxocarbonium ion from which the bulky inorganic phosphate has just been removed but which is still bound to the enzyme. We proposed the ''sliding domain'' hypothesis to explain this phenomenon.[86] The N-terminal domain, binding the glucose of glucose 1-P, would move the substrate close to the PLP, the phosphate would be removed and sequestered, and the N-terminus, bearing the stabilized glucosyl moiety, would slide back to bring the latter close to the 4-hydroxy group of the acceptor glucan. This hypothesis, also, has yet to be tested in any meaningful way.

II. GLYCOGEN SYNTHETASE

A. INTRODUCTION

Leloir and Cardini demonstrated the synthesis of glycogen from UDP-glucose by a liver extract in 1957.[87] UDP-glucose had already been shown to be the glucosyl donor for the synthesis of other compounds while the previously assumed biosynthesis of glycogen from glucose 1-P by phosphorylase was becoming suspect because the stimulation of liver glycogenolysis by hormones also caused activation of phosphorylase.[88] Furthermore, Larner et al. later showed that the concentration of inorganic phosphate *in vivo* greatly exceeds that of glucose 1-P and therefore phosphorylase must catalyze only glycogen phosphorolysis under physiological conditions, as noted in above Section I.B.[89] Leloir et al. also discovered the activation of glycogen synthetase by glucose 6-P.[90,91] This characteristic led Larner and colleagues to the discovery of two forms of the enzyme, one dependent and one independent of the activator, because insulin caused the muscle enzyme to become independent of glucose 6-P for activity.[92,93] Intensive research in several laboratories led to the conclusion that cAMP-dependent protein kinase caused the phosphorylation and inactivation (dependence on glucose 6-P) of glycogen synthase as well as the phosphorylation of phosphorylase kinase, leading to the activation of glycogen phosphorylase. The glycogenolytic and glycogen synthetic cascade systems and their reciprocal relationships have become the classical example of cellular regulation. The glycogen synthetase system has been reviewed extensively and the earlier work, some of which has been revised later, can be accessed in three representative reviews.[94-96] Two articles have just appeared which deal primarily with the control of the enzyme in muscle[97] and liver.[98]

The subunit molecular weight of the enzyme from muscle is of the order of 86,000 and the isolated enzyme is a tetramer.[97] Amino acid sequences have been determined for the N-terminal 29 residues and the C-terminal 124 residues, confirming that the subunits are identical.[97] All of the phosphorylation sites are contained in these two sequences. The liver enzyme has a similarly sized subunit although some studies suggest a somewhat larger size, and proteolysis has caused isolation of smaller subunits.[93] A recent study by Rulfs et al.[99] used antibody to rat heart glycogen synthetase for Western blotting analyses of rat heart and liver extracts and obtained an M_r of 93,000 in both cases, confirmed with a similar M_r for a polypeptide formed by in vitro translation of RNA from either tissue. Since the species had a slightly lower electrophoretic mobility than phosphorylase, and an M_r of 92,500 was used for the latter rather than the correct 97,400, this work suggests strongly that the true M_r is probably 98,000 or at least 10,000 more than that usually obtained for purified preparations by most workers, including Rulfs et al.[99] The implications that the true native form of the enzyme may not have yet been isolated and characterized, or that there is some other explanation, have been discussed by these authors but do raise some anxieties about our current understanding of the enzyme.

The oligomeric form of the liver enzyme would appear to be a dimer, and several lines of evidence indicate that it is an isozyme of the muscle enzyme, although quite homologous in structure and most properties.[98] Glycogen synthetase binds more tightly to glycogen than does phosphorylase and nearly all of it in the cell is associated with the glycogen particles, so that most purification procedures isolate a glycogen pellet as a first step.[95] It does not yet seem to have been established if the tetrameric muscle enzyme binds to glycogen as a dimer, as is the case with phosphorylase, although there is evidence for similar ''glycogen storage'' sites separate from the active sites.[23]

B. KINETICS AND MECHANISM

In contrast to the formation of a glycosidic bond from glucose 1-P by phosphorylase, with an equilibrium constant of 3.5, the formation from UDP-glucose by the synthetase is essentially irreversible under physiological conditions. Kornfeld and Brown[100] demonstrated slight reversibility at pH 7.5, and their data can be used to calculate a change in free energy of −5 kcal/mol. Based on estimates of free energies of hydrolysis of the glycosidic bond in glycogen of −4 kcal and of UDP-glucose of −8 kcal, Gold[101] estimated −4 kcal per mol for the reaction at pH 7.4, in reasonable agreement with the experimental result. The equilibrium constant would therefore be approximately 830 at pH 7.4,or 400 at pH 7.0 and 25°.[101] The extra energy in the synthetase reaction is due to the ionization of UDP, which also accounts for the sensitivity to pH, and makes UDP-glucose an ideal donor for synthesis.

After some initial confusion, it is now generally accepted that the kinetic mechanism is rapid equilibrium random bi bi, the same as for phosphorylase. No exchange could be demonstrated between UDP and UDP-glucose in the absence of glycogen in the 1962 study by Kornfeld and Brown,[100] and this result has since been confirmed (for the liver enzyme) by McVerry and Kim[102] and by Salsas and Larner.[103] A ping-pong mechanism, with a stable glucosyl-enzyme intermediate formed with the release of UDP, is therefore ruled out but, as with phosphorylase, a glucosyl-enzyme intermediate may be part of the reaction mechanism, there being no release of products until glucosyl transfer is complete. For the enzymes from both liver and muscle, the kinetic patterns indicate the formation of ternary complexes of enzyme and the two substrates.[102,103] The product UDP inhibits competitively with UDPG and noncompetitively with glycogen, a result consistent with the random addition of substrates.[103,104] The definitive study on this question was carried out by Gold,[101] who proved the mechanism by the method of isotope-exchange at chemical equilibrium.

Salsas and Larner[103] used both glycogen and maltose as substrates and demonstrated that the Km for UDP-glucose was essentially the same with either acceptor saccharide, being 45 and 48 μM, respectively. The limiting Km for maltose was 230 mM but that for glycogen was much lower, being 1.5 μg/ml or 8×10^{-7} M in terms of glucosyl endgroups. The limiting Km for phosphorylase and glycogen is at least 20 times more, but if we consider the binding to glycogen in the absence of substrates the difference is even larger.[47] That for phosphorylase is 135 μM while that for the synthetase can be estimated at 2 μM.[103] As is the case for phosphorylase, the liver synthetase binds much less tightly to glycogen, with a dissociation constant of 30 μg/ml calculated by McVerry and Kim.[102]

The very tight binding of the muscle synthetase to glucogen was reflected in the measurement of a slow exchange of the enzyme between glycogen molecules with a rate constant of 0.3 min^{-1}.[101] Considering the multichain action pattern, Gold suggested that hundreds of individual chains in the glycogen molecule must be available to the enzyme during the average lifetime of the complex, and that the enzyme may migrate over the surface of the glycogen before dissociating.[101]

While glycogen synthetase can use saccharides as small as glucose and maltose as acceptors,[23,103] the Km's indicate a preference for large glycogen-like molecules which can presumably bind to the "storage" sites, and the short outer chains of various limit dextrins also increase the Km's. Thus, as was true for phosphorylase, the active site *per se* has a low affinity for oligosaccharides, with a Km of 98 μ*M* reported for maltoheptaose.[23] The enzyme can use other glucosyl donors, with ADP-glucose showing half the rate obtained with UDP-glucose.[95]

Very little is known about the catalytic mechanism of glycogen synthetase but in section I.F above we discussed the possible similarities with phosphorylase. Since phosphorylase can transfer glucose from an analogue of UDP-glucose, namely pyridoxal pyrophosphate glucose, one may suggest that the synthetase, containing no PLP, may employ its substrate in a fashion similar to that in which phosphorylase employs its coenzyme-phosphate combination. As was the case with phosphorylase, the synthetase is inhibited by 1,5-gluconolactone, suggesting that at some stage in its transfer from UDP-glucose to glycogen the glucosyl moiety may exist as a planar oxonium ion.[102] For reasons discussed for phosphorylase, the latter may be stabilized by a covalent link to the enzyme. The only definitive information available about the structure of the active site comes from the use of an affinity label by Tagaya, Nakano, and Fukui.[105] They synthesized uridine diphosphopyridoxal and showed that it bound to the muscle enzyme with a dissociation constant of 25 μ*M* and then formed a Schiff base with a lysine residue. UDP-glucose or UDP but not glucose 6-P provided protection and the effect of pH on inactivation suggested a pK of 8.85. The authors suggested a role for this lysine similar to that of Lys-573 in phosphorylase, but a ten-residue peptide containing the lysine bore no sequence homology to any in phosphorylase.

C. ALLOSTERIC CONTROL

Earlier studies on the enzyme suggested that, like phosphorylase, glycogen synthetase exists in a phosphorylated and a dephosphorylated form, with the tacit assumption that a single site was involved.[94] The phosphorylated (*b*-form) was dependent on glucose 6-P and was inhibited by ATP, ADP, phosphate, and UDP but glucose 6-P could antagonize these inhibitions. The dephosphorylated (*a*-form) had "escaped" these allosteric controls because it did not require glucose 6-P for activity and, while ATP was still an inhibitor, only very low concentrations of the activator were needed to reverse this.[106] Larner and associates[107,108] carried out a kinetic study on these effectors with nine muscle synthetase preparations which had phosphate contents from 0.27 to 3.5 per subunit. In the absence of ATP there was little effect of either phosphorylation state or glucose 6-P on the maximal velocity. However, the "Km" for UDP-glucose in the absence of glucose increased from 1 m*M* to more than 1 *M* as the phosphate content increased to 3.5 per subunit. Glucose 6-P greatly decreased the Km at all phosphorylation states, so that it ranged from 45 to 250 μ*M*. Similarly, the Km for glucose 6-P also increased with phosphorylation, from 3.3 to 4100 μ*M* over the range of 0.27 to 3.5 mol P per subunit. Phosphorylation also caused an increased sensitivity to inhibition by ATP and phosphate. Knowing now that there are multiple sites for phosphorylation by a variety of kinases, Cohen has suggested the need for further studies with synthetase preparations which are precisely defined molecular species with respect to the phosphate contents of their various sites.[97]

Synthetase from the liver is probably subject to even more complex controls than that from the muscle but phosphorylation again decreases activity.[94,98] This time, however, the Km for UDP-glucose is not altered appreciably but the maximal velocity is decreased markedly, nor does glucose 6-P restore the activity fully. It does, however, decrease the

Km for UDP-glucose while its own Km is increased up to 70-fold upon full (*in vivo*) phosphorylation. The effect of ATP on the *a* and *b* forms of the liver enzyme, and the antagonism by glucose 6-P, are similar in general with the picture discussed above for the muscle system.

D. REGULATION BY COVALENT MODIFICATION

This subject is so complex and extensive that it is treated in two major reviews, one each for the muscle and liver systems.[97,98] Because it is possible to isolate much more of the muscle synthetase, work on it is at a more advanced level and will be emphasized in this section. The difficulties of working with this multisite-phosphorylated enzyme have resulted in a large body of complex literature but we are fortunate that Cohen, one of the chief workers in the field, has recently integrated the material into a comprehensive review which should be consulted for details.[97] In this section, work on the muscle system will not be referenced to the original literature but only to Cohen's review.

A total of ten serine residues has been found to be phosphorylated *in vitro* by one or more of at least ten protein kinases. However, "only" 7 of the serines have been found to be phosphorylated *in vivo*, while Cohen has suggested that only six kinases are likely to have a physiological role in muscle, although some may be significant in other tissues. In Table 1 are summarized the phosphorylatable serines, grouped according to the "sites" as originally isolated in various proteolytically derived peptides. Thus, site-1 is subdivided into a and b, site 3 into a, b, and c. Specific residue numbers have now been assigned to these sites and it is possible to determine the specificity and rates for each kinase acting on the individual serines, as summarized in the table. Most effects on activity have been determined as a ratio of activity in the absence and presence of saturating glucose 6-P. The activity ratio often gives an inverse linear relationship to the phosphorylation of a particular site but site-1b has no discernible effect. Site-5 has no effect on activity but must be phosphorylated in order for glycogen synthetase kinase-3 to have significant activity. Since muscle protein phosphatases exhibit a very slow action at this site, most isolated preparations are at least partially phosphorylated at site-5. Removal of phosphate with potato acid phosphatase prevents the action of the kinase-3 without affecting the other kinases listed in Table 1. The significance of this synergism between kinases-5 and -3 is not yet known.

Linkage between sites is also manifested in the activity of the muscle phosphatases. Thus activities on the individual sites-2, -3, and -1a are comparable, yet when sites-1a, -1b, and -2 are all phosphorylated, protein phosphatase-1 acts on site-2 from 5 to 10-fold faster than on site-1a, and 100-fold faster than on site 1-b. Action on the latter two sites is accelerated in sequence as each preceding site is dephosphorylated. The order of both the phosphorylation and dephosphorylation of the three serines comprising site-3 is as yet unknown.

Several laboratories have examined the effects of adrenalin and insulin on the phosphate content of glycogen synthetase and the distribution among the various sites, as summarized by Cohen.[97] Data from Cohen's laboratory is shown in Table 1. The enzyme in the resting muscle of normally fed rabbits exhibits partial phosphorylation of all the sites which affect activity, and the activity ratio is approximately 0.2. Injection of adrenalin lowers the activity ratio to 0.04 with the addition of just over 2 mol P per subunit, distributed over all sites except site-5. The major change in the activity ratio may be attributed to the phosphorylation of sites-2 and 3. Since adrenalin has a β-adrenergic effect only, one might expect the phosphorylation of sites 2- and -1 to be mediated via cAMP-dependent kinase but site-2 may also be affected by phosphorylase kinase (activated under these conditions) or subject to less dephosphorylation because phosphatase is inhibited by the phosphorylated inhibitor-1 or by phosphorylase *a*. The latter two effects may explain the phosphorylation of site-3 since the kinase-3 specific for that site is not presently known to be subject to regulation.

TABLE 1

Summary of the Phosphorylation Sites in Glycogen Synthetase, Their Major Kinases, Phosphatase Action, Enzymatic Effects of their Phosphorylation and Physiological Effects on Phosphorylation

	Site 2	Site 3			Site 5	Site 1	
		a	b	c		a	b
Residue number[a]	**N7**	**C30**	**C34**	**C38**	**C46**	**C87**	**C100**
A. Kinase Action and Rate[b]							
cAMP-dependent kinase	10%					100%	5%
Phosphorylase kinase	Specific for N7						
Calmodulin dependent	100%						10%
Glyc. Syn. Kinase-3		Acts on all three serines					
Glyc. Syn. Kinase-4	Specific for N7						
Glyc. Syn. Kinase-5					Specific for C46		
B. Effect on Activity Ratio (−/+ G-6-P)	Greater than C87	Greater than 1a + 1b + 2 but additive to them (all 6 = maximal effect)			No effect on activity. Required for action of Glyc. Syn. Kinase-3	Yes	No effect on acitivity
C. Phosphatase Action							
Individual sites	Comparable rate	Comparable rate			Very slow	Campar.	
if 1 + 2 phosphor.	Fastest rate					Medium	Slowest
D. Physiological Effects on Phosphate/mol[c]							
Resting muscle	0.39	1.27			0.65	0.33	0.38
+ epinephrine (change)	+0.6	+1.2			0	+0.26	+0.27
+ insulin (change)	+0.06	−0.45			+0.11	−0.06	−0.06

[a] Numbering with respect to the N-terminal 29 residues and the C-terminal 124 residues.

[b] Approximate initial rate of action on each site expressed as a percentage of the maximal rate for each kinase.

[c] "Resting muscle" is from fed rabbits injected with propanolol, "+ epinephrine" refers to fed rabbits injected with epinephrine, "+ insulin" refers to 24-hour fasted rabbits also injected with propanolol but the control values were similar to "resting muscle".

The classical effect of insulin on the activity of muscle glycogen synthetase, discovered some 25 years ago by Larner,[92] is manifested by an increase in the activity ratio from approximately 0.2 to 0.35.[97] Data in Table 1 indicates that the major effect is caused by a decrease in the phosphorylation of site-3. The mechanism for this insulin effect is unknown but may involve activation of either phosphatase-1 or -2A.

Other unsolved problems with respect to the regulation of glycogen synthetase concern the changes in the activity ratio during muscle contraction and in response to glycogen concentration. In the former situation, the activity ratio begins to increase during a tetanus, continuing to rise for several minutes into the rest period until a gradual decline to the resting level ensues.[109] The mechanism for this effect or the phosphorylation sites involved are not yet known but may be related to the changes in glycogen content which occur during contraction. Danforth demonstrated that the activity ratio of glycogen synthetase rises with a decrease in the glycogen concentration in muscle, and falls to a minimal value as the glycogen level approaches its maximal value.[109] Insulin displaces the curve describing this relationship, so that an increased activity ratio is seen for each glycogen concentration. While this effect provides a satisfying explanation for the limitation of muscle glycogen concentration to a reasonably constant maximum, the mechanism has so far eluded experimental investigation.[97]

The glycogen synthetase found in liver is an isozyme of the muscle enzyme and shares some similarities but exhibits greater complexity with respect to regulation by phosphorylation, as summarized in the comprehensive review by Roach.[98] The major kinases found in muscle have their counterparts in liver but their relative importance may not be the same. cAMP-dependent protein kinase phosphorylates a site which is homologous to site-2 in the muscle enzyme and, more slowly, a site homologous to site-3a, inactivation being correlated better with the second phosphorylation. *In vivo* studies indicate that glucagon mediates its action through cAMP but some controversy exists over whether or not this is solely via the cAMP-dependent kinase. It is possible that cAMP exercises an indirect control over the protein phosphatases, either by phosphorylation of inhibitor-1 or by promoting an increased level of phosphorylase *a*, which is an inhibitor of the major protein phosphatases. A second mechanism by which liver glycogen synthetase becomes phosphorylated and hence inactivated involves α-adrenergic agonists, angiotensin II, and vasopressin, all of which may modulate intracellular distributions of calcium. The three major calcium-dependent kinases, phosphorylase kinase, calmodulin dependent kinase, and protein kinase C, may all be activated by this route. The latter kinase is also activated by diacylglycerols, which permits another form of regulation. Again, in contrast to direct activation of synthetase kinases, inhibition of protein phosphatases by the increased level of phosphorylase *a*, resulting from calcium stimulation of phosphorylase kinase, has been suggested to account for the effects of the hormones mentioned in this paragraph. Clarification of this situation will require considerably more experimentation.

Glucose and insulin, alone or together, are known to activate liver glycogen synthetase *in vivo* and to cause dephosphorylation. The insulin effect is small in the absence of glucose. The hypothesis elaborated by Hers, Stallmans, and their colleagues, in which phosphorylase *a* is termed the glucose receptor of the liver cell, has been discussed earlier in Section I.F. As reviewed by Roach,[98] this hypothesis is attractive but cannot be considered proven nor is it likely to be applicable under all conditions. The mechanism by which the effect of insulin is mediated is still quite obscure and merits the intensive investigations now underway in a number of laboratories.

III. CONCLUDING REMARKS

Glycogen phosphorylase and synthetase are the yin and yang of glycogen metabolism, their similarities and differences providing an enduring source of fascination. They both act on the same polysaccharide substrate, with only minor differences in preferences for structural details, and both appear to bind this substrate at "storage sites" separate from the catalytic sites, although the synthetase exhibits much tighter binding. They employ similar donor compounds for the glucose moiety which is to be transferred to the acceptor polysaccharides, so much so that UDP-glucose is a good competitive inhibitor for the glucose 1-P substrate in the reaction catalyzed by phosphorylase, although the reverse is not true. The synthetase does not utilize the pyridoxal phosphate coenzyme found in phosphorylase although the UMP moiety of UDP-glucose may play a similar role, as discussed above. The oligomeric structures of the two enzymes are also very similar. Some of these considerations, and others, led Leloir to speculate that one enzyme might be convertible into the other, although, as he said, no evidence was available to support this assumption.[110] Amino acid sequences totaling 163 residues are now available for the synthetase and comparisons with the phosphorylase sequences do not encourage one to speculate on their evolutionary relationships, but this could change with further information.

When we consider the characteristics of their regulation, we have seen an almost mirror image of form and content, starting with the activation of phosphorylase by phosphorylation and the opposite effect on the synthetase. Covalent modification of both enzymes does result in their "escape" from allosteric regulation by cellular metabolites, but these are usually different molecules. Thus, glucose 6-P plays much the same role for the synthetase as does AMP for phosphorylase, for which the sugar phosphate is an inhibitor. ATP, however, inhibits the *b* forms of both enzymes.

Phosphorylase is subject to covalent modification at a single serine, conveniently near the N-terminal, by a single kinase, while the synthetase exhibits multisite phosphorylation by a variety of kinases. In retrospect, it was fortunate that the first phosphorylatable enzyme to be studied was relatively simple in its chemistry so that definitive results could be obtained with the relatively primitive technology available 20 to 30 years ago. Glycogen synthetase represents a much more complex situation with respect to its chemistry and a much more sophisticated system of regulation with respect to its control by covalent modification. A three-dimensional structure would be of immmense help but may have to await availability of a chemically pure protein from a cDNA expression vector, since variable phosphorylation and possible proteolysis now hinder crystallization attempts. The full elucidation of the structure-function relationships of glycogen synthetase is a worthy goal for the excellent laboratories now studying this intriguing enzyme.

ACKNOWLEDGMENTS

Research conducted in the author's laboratory was supported by grant MT-1414 from the Medical Research Council of Canada. It is a pleasure to thank Miss Susan Smith for assistance in constructing the manuscript.

REFERENCES

1. **Madsen, N. B., Avramovic-Zikic, O., Lue, P. F., and Honikel, P. O.,** Studies on allosteric phenomena in glycogen phosphorylase *b. Mol. Cell Biochem.*, 11, 35, 1976.
2. **Busby, S. J. W. and Radda, G. K.,** Regulation of the glycogen phosphorylase system — from physical measurements to biological speculations, *Curr. Top. Cell. Regul.*, 10, 89, 1976.
3. **Dombrádi, V.,** Structural aspects of the catalytic and regulatory function of glycogen phosphorylase, *Int. J. Biochem.*, 13, 125, 1981.
4. **Fletterick, R. J. and Madsen, N. B.,** The structures and related functions of phosphorylase *a*, *Annu. Rev. Biochem.*, 49, 31, 1980.

5. **Madsen, N. B.,** Glycogen phosphorylase, in *The Enzymes,* Vol. 17, 3rd ed., Boyer, P. B. and Krebs, E. G., Eds., Academic Press, New York, 1986, 365.
6. **Helmreich, E. J. and Klein, H. W.** The role of pyridoxal phosphate in the catalysis of glycogen phosphorylases, *Angew. Chem.,* 19, 441, 1980.
7. **Madsen, N. B. and Withers, S. G.,** Glycogen phosphorylase, in *Coenzymes and Cofactors,* Vol. 1, Dolphin, D., Poulson, R., and Avramovic, O., Eds., John Wiley & Sons, New York, 1986, 355.
8. **Cori, C. F. and Cori, G. T.,** Mechanism of formation of hexose monophosphate in muscle and isolation of a new phosphate ester, *Proc. Soc. Exp. Biol. Med.,* 34, 702, 1936.
9. **Cori, G. T. and Cori, C. F.,** The kinetics of the enzymatic synthesis of glycogen from glucose-1-phosphate, *J. Biol. Chem.,* 135, 733, 1940.
10. **Cori, C. F., Cori, G. T., and Green, A. A.,** Crystalline muscle phosphorylase. III. Kinetics, *J. Biol. Chem.,* 151, 39, 1943.
11. **Madsen, N. B. and Cori, C. F.,** The inhibition of muscle phosphorylase by p-chloromercuribenzoate, *Biochim. Biophys. Acta,* 18, 156, 1955.
12. **Madsen, N. B. and Cori, C. F.,** The binding of adenylic acid by muscle phosphorylase, *J. Biol. Chem.,* 224, 899, 1957.
13. **Madsen, N. B. and Cori, C. F.,** The interaction of phosphorylase with protamine, *Biochim. Biophys. Acta,* 15, 516, 1954.
14. **Fischer, E. H., Graves, D. J., Crittenden, E. R. S., and Krebs, E. G.,** Structure of the site phosphorylated in the phosphorylase *b* to *a* reaction, *J. Biol. Chem.,* 234, 1698, 1959.
15. **Baranowski, T., Illingworth, B., Brown, O. H., and Cori, C. F.,** The isolation of pyridoxal-5-phosphate from crystalline muscle phosphorylase, *Biochim. Biophys. Acta,* 25, 16, 1957.
16. **Fischer, E. H., Kent, A. B., Snyder, E. R., and Krebs, E. G.,** The reaction of sodium borohydride with muscle phosphorylase, *J. Am. Chem. Soc.,* 80, 2906, 1958.
17. **Wanson, J. C. and Drochmans, P.,** Rabbit skeletal muscle glycogen, *J. Cell Biol.,* 38, 130, 1968.
18. **Porter, K. R. and Bruni, C.,** An electron microscope study of the early effects of 3′-Me-DAB on rat liver cells, *Cancer Res.,* 19, 997, 1959.
19. **Meyer, F., Heilmeyer, L. M. G., Haschke, R. H., and Fischer, E. H.,** Control of phosphorylase activity in a muscle glycogen particle. I. Isolation and characterization of the protein-glycogen complex, *J. Biol. Chem.,* 245, 6642, 1970.
20. **Caudwell, B., Antoniw, J. F., and Cohen, P.,** Calsequestrin, myosin, and the components of the protein-glycogen complex in rabbit skeletal muscle, *Eur. J. Biochem.,* 86, 511, 1978.
21. **Fletterick, R. J., Sygusch, J., Semple, H., and Madsen, N. B.,** Structure of glycogen phosphorylase *a* at 3.0 Å resolution and its ligand binding sites at 6 Å, *J. Biol. Chem.,* 251, 6142, 1976.
22. **Fletterick, R. J., Sprang, S., and Madsen, N. B.,** Analysis of the surface topography of glycogen phosphorylase *a*: implications for metabolic interconversion and regulatory machanism, *Can. J. Biochem.* 57, 789, 1979.
23. **Larner, J., Takeda, Y., and Hizukuri, S.,** The influence of chain size and molecular weight on the kinetic constants for the span glucose to polysaccharide for rabbit muscle glycogen synthase, *Mol. Cell. Biochem.,* 12, 131, 1976.
24. **Sprang, S. R., Goldsmith, E. J., Fletterick, R. J., Withers, S. G., and Madsen, N. B.,** Catalytic site of glycogen phosphorylase: structure of the T state and specificity for α-D-glucose, *Biochemistry,* 21, 5364, 1982.
25. **Withers, S. G., Madsen, N. B., Sprang, S. R., and Fletterick, R. J.,** Catalytic site of glycogen phosphorylase: structural changes during activation and mechanistic implications, *Biochemistry,* 21, 5372, 1982.
26. **Hers, H. G.,** The control of glycogen metabolism in the liver, *Annu. Rev. Biochem.,* 45, 167, 1976.
27. **Kasvinsky, P. J., Madsen, N. B., Sygush, J., and Fletterick, R. J.,** The regulation of phosphorylase *a* by nucleotide derivatives, *J. Biol. Chem.,* 253, 3343, 1978.
28. **Johnson, L. N., Stura, E. A., Wilson, K. S., Sansom, M. S. P., and Weber, I. T.,** Nucleotide binding to glycogen phosphorylase *b* in the crystal, *J. Mol. Biol.,* 134, 639, 1979.
29. **Stura, E. A., Zanotti, G., Babu, Y. S., Sansom, M. S. P., Stuart, D. I., Wilson, K. S., Johnson, L. N., and Van De Werve, G.,** Comparison of AMP and NADPH binding to glycogen phosphorylase *b*, *J. Mol. Biol.,* 170, 529, 1983.
30. **Rahim, Z. H. A., Perrett, D., and Griffiths, J. R.,** Skeletal muscle purine nucleotide levels in normal and phosphorylase kinase deficient mice, *FEBS Lett.,* 69, 203, 1976.
31. **Newsholme, E. A. and Leech, A. R.,** *Biochemistry for the Medical Sciences,* John Wiley & Sons, New York, 1983, 362.
32. **Danforth, W. H. and Lyon, J. B.,** Glycogenolysis during tetanic contraction of isolated mouse muscles in the presence and absence of phosphorylase *a, J. Biol. Chem.,* 239, 4047, 1964.
33. **Titani, K., Koide, A., Hermann, J., Ericson, L. H., Kumar, S., Wade, R., Walsh, K. A., Neurath, H., and Fischer, E. H.,** Complete amino acid sequence of rabbit skeletal muscle glycogen phosphorylase, *Proc. Natl. Acad. Sci. U.S.A.,* 74, 4762, 1977.

34. **Nakano, K., Hwana, P. K., and Fletterick, R. J.,** Complete cDNA sequence for rabbit muscle glycogen phosphorylase, *FEBS Lett.*, 204, 283, 1986.
35. **Sprang, S. R. and Fletterick, R. J.,** The structure of glycogen phosphorylase *a* at 2.5 Å resolution, *J. Mol. Biol.*, 131, 523, 1979.
36. **Jenkins, J. A., Johnson, L. N., Stuart, D. I., Stura, E. A., Wilson, K. S., and Zanotti, G.,** Phosphorylase: control and activity, *Phil. Trans. R. Soc. B.*, 293, 23, 1981.
37. **Lorek, A., Wilson, K. S., Stura, E. A., Jenkins, J. A., Zanotti, G., and Johnson, L. N.,** Allosteric inhibition of glycogen phosphorylase *b*: a crystallographic study, *J. Mol. Biol.*, 140, 565, 1980.
38. **Sansom, M. S. P., Stura, F. A., Babu, Y. S., McLaughlin, P., and Johnson, L. N.,** Pyridoxal phosphate in glycogen phosphorylase b: conformation, environment and role in catalysis, in *Chemical and Biological Aspects of Vitamin B6 Catalysis, Part A,* Evangelopolous, A. E., Ed., Alan R. Liss, New York, 1984, 127.
39. **Lorek, A., Wilson, K. S., Sansom, M. S. P., Stuart, D. I., Stura, E. A., Jenkins, J. A., Zanotti, G., Hajdr, J., and Johnson, L. N.,** Allosteric interactions of glycogen phosphorylase *b*, a crystallographic study, *Biochem. J.*, 218, 45, 1984.
40. **Sprang, S. and Fletterick, R. J.,** Subunit interactions and the allosteric response in phosphorylase, *Biophys. J.*, 32, 175, 1980.
41. **Madsen, N. B., Kasvinsky, P. J., and Fletterick, P. J.,** Allosteric transitions of phosphorylase *a* and the regulation of glycogen metabolism, *J. Biol. Chem.*, 253, 9097, 1978.
42. **Fletterick, R. J. and Madsen, N. B.,** X-rays reveal phosphorylase architecture, *TIBS,* 2, 145, 1977.
43. **Kasvinsky, P. J., Madsen, N. B., Fletterick, R. J., and Sygusch, J.,** X-ray crystallographic and kinetic studies of oligosaccharide binding to phosphorylase, *J. Biol. Chem.*, 253, 1290, 1978.
44. **Weber, I. T., Johnson, L. N., Wilson, K. S., Ycates, D. G. R., Wild, D. L., and Jenkins, J. A.,** Crystallographic studies on the activity of glycogen phosphorylase *b, Nature,* 274, 433, 1978.
45. **Maddaiah, V. T. and Madsen, N. B.,** Kinetics of purified liver phosphorylase, *J. Biol. Chem.*, 241, 3873, 1966.
46. **Engers, H. D., Bridger, W. A., and Madsen, N. B.,** The kinetic mechanism of phosphorylase *b, J. Biol. Chem.*, 244, 5936, 1969.
47. **Engers, H. D., Shechosky, S., and Madsen, N. B.,** Kinetic mechanism of phosphorylase *a*. I. Initial velocity studies, *Can. J. Biochem.*, 48, 746, 1970.
48. **Engers, H. D., Bridger, W. A., and Madsen, N. B.,** Kinetic mechanism of phosphorylase *a*. II. Isotope exchange studies at equilibrium, *Can. J. Biochem.*, 48, 755, 1969.
49. **Gold, A. M., Johnson, R. M., and Tseng, J. K.,** Kinetic mechanism of rabbit muscle glycogen phosphorylase *a*. *J. Biol. Chem.*, 245, 2564, 1970.
50. **Engers, H. D., Bridger, W. A., and Madsen, N. B.,** Isotope exchange at equilibrium as a test for homotropic cooperativity of allosteric enzymes, *Biochemistry,* 9, 3281, 1970.
51. **Parmeggiani, A. and Morgan, H. E.,** The effect of adenine nucleotides and inorganic phosphate on muscle phosphorylase activity, *Biochem. Biophys. Res. Commun.*, 9, 252, 1962.
52. **Morgan, H. E. and Parmeggiani, A.,** Regulation of glycogenolysis in muscle. III. Control of muscle glycogen phosphorylase activity, *J. Biol. Chem.*, 239, 2440, 1964.
53. **Helmreich, E. and Cori, C. F.,** The role of adenylic acid in the activation of phosphorylase, *Proc. Natl. Acad. Sci. U.S.A.*, 51, 131, 1964.
54. **Madsen, N. B.,** Allosteric properties of phosphorylase *b, Biochem. Biophys. Res. Commun.*, 15, 390, 1964.
55. **Monod, J., Changeaux, J. P., and Jacob, F.,** Allosteric proteins and cellular control systems, *J. Mol. Biol.*, 6, 306, 1963.
56. **Ullman, A., Vagelos, P. R., and Monod, J.,** The effect of 5′-adenylic acid upon the association between bromthymal blue and muscle phosphorylase *b, Biochem. Biophys. Res. Commun.*, 17, 86, 1964.
57. **Monod, J., Wyman, J., and Changeux, J.-P.,** On the nature of allosteric transitions: a plausible model, *J. Mol. Biol.*, 12, 88, 1965.
58. **Madsen, N. B. and Shechosky, S.,** Allostric properties of phosphorylase *b*. II. Comparison with a kinetic model, *J. Biol. Chem.*, 242, 3301, 1967.
59. **Kastenschmidt, L. L., Kastenschmidt, J., and Helmreich, E.,** Subunit interactions and their relationship to the allosteric properties of rabbit skeletal muscle phosphorylase *b*. *Biochemistry,* 7, 3590, 1968.
60. **Mott, D. M. and Bieber, A. L.,** Structural specificity of the adenosine 5′-phosphate site of glycogen phosphorylase *b*. *J. Biol. Chem.*, 245, 4058, 1970.
61. **Avramovic, O. and Madsen, N. B.,** Allosteric properties of phosphorylase *b*. III. Inactivation by cyanate and binding studies, *J. Biol. Chem.*, 243, 1656, 1968.
62. **Mateo, P. L., Baron, C., Lopez-Mayorga, O., Jimenez, J. S., and Cortijo, M.,** AMP and IMP binding to glycogen phosphorylase *b, J. Biol. Chem.*, 259, 9384, 1984.
63. **Campbell, I. D., Dwek, R. A., Price, N. C. and Radda, G. K.,** Studies on the interaction of ligands with phosphorylase *b* using a spin-label probe, *Eur. J. Biochem.*, 30, 339, 1972.

64. **Buc, M. H. and Buc, H.,** Allosteric interactions between AMP and orthophosphate sites on phosphorylase *b* from rabbit muscle, in *4th Fed. Eur. Biochem. Soc. Symposium on Regulation of Enzyme Acivity and Allosteric Interactions,* Academic Press, New York, 1968, 109.
65. **Sprang, S. R., Fletterick, R. J., Stern, M., Yang, D., Madsen, N. B., and Sturtevant, J. M.,** Analysis of an allosteric binding site: the nucleoside inhibitor site of phosphorylase *a, Biochemistry,* 21, 2036, 1984.
66. **Madsen, N. B., Shechosky, S., and Fletterick, R. J.,** Site-site interactions in glycogen phosphorylase *b, Biochemistry,* 22, 4460, 1983.
67. **Helmreich, E., Michaelides, M. C., and Cori, C. F.,** Effect of substrates and substrate analogs on the binding of 5′-adenylic acid to muscle phosphorylase *a, Biochemistry,* 6, 3695, 1967.
68. **Griffiths, J. R., Price, N. C., and Radda, G. K.,** Conformational changes in phosphorylase *a* studied by a spin label probe, *Biochim. Biophys. Acta,* 358, 275, 1974.
69. **Koshland, D. E., Nemethy, G., and Filmer, D.,** Comparison of experimental data and theoretical models in proteins containing subunits, *Biochemistry,* 5, 365, 1966.
70. **Koshland, D. E.,** Conformational aspects of enzyme regulation, *Curr. Top. Cell. Reg.,* 1, 1, 1969.
71. **Pickett-Gies, C. A. and Walsh, D.,** Phosphorylase kinase, in *The Enzymes,* Vol. 17, 3rd ed., Boyer, P. B. and Krebs, E. G., Eds., Academic Press, New York, 1986, 395.
72. **Danforth, W. H., Helmreich, E., and Cori, C. F.,** The effect of contraction and of epinephrine on the phosphorylase activity of frog sartoreus muscle, *Proc. Natl. Acad. Sci. U.S.A.,* 48, 1191, 1962.
73. **Ballou, L. M. and Fischer, E. H.,** Phosphoprotein phosphatases, in *The Enzymes,* Vol. 17, 3rd ed., Boyer, P. B. and Krebs, E. G., Eds., Academic Press, New York, 1986, 311.
74. **Kasvinsky, P. J., Fletterick, R. J., and Madsen, N. B.,** Regulation of the dephosphorylation of glycogen phosphorylase *a* and synthase *b* by glucose and caffeine in isolated hepatocytes, *Can. J. Biochem.,* 59, 387, 1981.
75. **Monanu, M. O. and Madsen, N. B.,** Rabbit liver phosphorylase *a* phosphatase: regulation by glucose and caffeine, *Can. J. Biochem. Cell Biol.,* 63, 115, 1985.
76. **Monanu, M. O. and Madsen, N. B.,** Distinction between substrate and enzyme-directed effects of modifiers of rabbit liver phosphorylase *a* phosphatases, *Biochem. Cell Biol.,* 69, 293, 1987.
77. **Madsen, N. B. Fletterick, R. J., and Kasvinsky, P. J.,** Regulation of protein phosphatase I via glycogen phosphorylase, in *Protein Phosphorylation,* Rosen, O. M. and Krebs, E. G., Eds., Cold Spring Harbor Laboratory, Cold Spring Harbor, NY, 1981, 483.
78. **Parrish, R. J., Uhing, R. J., and Graves, D. J.,** Effect of phosphate analogues on the activity of pyridoxal reconstituted glycogen phosphorylase, *Biochemistry,* 16, 4824, 1977.
79. **Withers, S. G., Madsen, N. B., Sykes, B. D., Takagi, M., Shimomura, S., and Fukui, T.,** Evidence for direct phosphate-phosphate interaction between pyridoxal phosphate and substrate in the glycogen phosphorylase catalytic mechanism, *J. Biol. Chem.,* 256, 10759, 1981.
80. **Takagi, M., Fukui, T., and Shimomura, S.,** Catalytic mechanism of glycogen phosphorylase: pyridoxal (5′) diphospho(1)-α-glucose as a transition-state analogue, *Proc. Natl. Acad. Sci. U.S.A.,* 79, 3716, 1982.
81. **Klein, H. W., Palm, D., and Helmreich, E. J. M.,** General acid-base catalysis of α-glucan phosphorylases, *Biochemistry,* 21, 6675, 1982.
82. **Klein, H. W., Im, M. J., and Helmreich, E. J. M.,** The role of pyridoxal 5′-phosphate and orthophosphate in general acid-base catalysis by α-glucan phosphorylases, in *Chemical and Biological Aspects of Vitamin B6 Catalysis, Part A,* Evangelopoulos, A. E., Ed., Alan R. Liss, New York, 1984, 117.
83. **Feldman, K. and Hull, W. E.,** ^{31}P nuclear magnetic resonance studies of glycogen phosphorylase from rabbit skeletal muscle: ionization state of pyridoxal 5′-phosphate, *Proc. Natl. Acad. Sci. U.S.A.,* 74, 856, 1977.
84. **Klein, H. W., Schlitz, E., and Helmreich, E. J. M.,** A catalytic role of the dianionic 5′-phosphate of pyridoxal 5′-phosphate in glycogen phosphorylases: formation of a covalent glucosyl intermediate, in *Protein Phosphorylation,* Rosen, O. M. and Krebs, E. G., Eds., Cold Spring Harbor Laboratory, Cold Spring Harbor, NY, 1981, 305.
85. **Palm, D., Blumenauer, G., Klein H. W., and Blanc-Meusser, M.,** α-glucan phosphorylases catalyse the glucosyl transfer from α-D-glucosyl fluoride to oligosaccharides, *Biochem. Biophys. Res. Commun.,* 111, 530, 1983.
86. **Madsen, N. B. and Withers, S. G.,** The catalytic mechanism of phosphorylase: novel role of the coenzyme phosphate, in *Chemical and Biological Aspects of Vitamin B6 Catalysis, Part A,* Evangelopoulos, A. E., Ed., Alan R. Liss, New York, 1984, 117.
87. **Leloir, L. F. and Cardini, C. E.,** Biosynthesis of glycogen from uridine diphosphate glucose, *J. Am. Chem. Soc.,* 79, 6340, 1957.
88. **Sutherland, E. W. and Cori, C. F.,** The effect of hyperglycemic-glycogenolytic factor and epinephrine on liver phosphorylase, *J. Biol. Chem.,* 188, 531, 1951.
89. **Larner, J., Pillar-Palasi, C., and Rechman, D. J.,** Insulin-stimulated glycogen formation in rat diaphragm, *Arch. Biochem. Biophys.,* 86, 56, 1960.

90. **Leloir, L. F., Olavarria, J. M., Goldemberg, S. H., and Carminatii, H.,** Biosynthetis of glycogen from uridine diphosphate glucose, *Arch. Biochem. Biophys.*, 81, 508, 1959.
91. **Leloir, L. F. and Goldemberg, S. H.,** Synthesis of glycogen from uridine diphosphate glucose in liver, *J. Biol. Chem.*, 235, 919, 1960.
92. **Villar-Palasi, C. and Larner, J.,** Insulin-mediated effect on the activity of UDPG-glycogen transglucosylase of muscle, *Biochim. Biophys. Acta*, 39, 171, 1960.
93. **Friedman, D. L. and Larner, J.,** Studies on UDPG-α-glucan transglucosylase, *Biochemistry*, 2, 669, 1963.
94. **Larner, J. and Villar-Palasi, C.,** Glycogen synthetase and its control, *Curr. Top. Cell. Regul.*, 3, 195, 1971.
95. **Stalmans, W. and Hers, H. G.,** Glycogen synthesis from UDPG, in *The Enzymes*, Vol. 2, 3rd ed., Boyer, P. B., Ed., Academic Press, New York, 1973, 310.
96. **Roach, P. J. and Larner, J.,** Regulation of glycogen synthase, *TIBS*, 1, 110, 1976.
97. **Cohen, P.,** Muscle glycogen synthase, in *The Enzymes*, Vol. 17, 3rd ed., Boyer, P. B. and Krebs, E. G., Eds., Academic Press, New York, 1986, 462.
98. **Roach, P. J.,** Liver glycogen synthase, in *The Enzymes*, Vol. 17, 3rd ed., Boyer, P. B. and Krebs, E. G., Eds., Academic Press, New York, 1986, 500.
99. **Rulfs, J., Wolleben, C. D., Miller, T. B., and Johnson, G. L.,** Immunologic identification of a glycogen synthase 93,000-dalton subunit from rat heart and liver, *J. Biol. Chem.*, 260, 1203, 1985.
100. **Kornfeld, R. and Brown, D. H.,** Preparation and properties of uridine diphosphate glucose-glycogen transferase, *J. Biol. Chem.*, 237, 1771, 1962.
101. **Gold, A. M.,** Kinetic mechanism of rabbit muscle glycogen synthase I, *Biochemistry*, 19, 3766, 1980.
102. **McVerry, P. H. and Kim, K.-H.,** Purification and kinetic mechanism of rat liver glycogen synthase, *Biochemistry*, 13, 3505, 1974.
103. **Salsas, E. and Larner, J.,** Kinetic studies on muscle glycogen synthetase, *J. Biol. Chem.*, 250, 3471, 1975.
104. **Plesner, L., Plesner, I. W., and Esmann, V.,** Kinetic mechanism of glycogen synthase D from human polymorphonuclear leukocytes, *J. Biol. Chem.*, 249, 1119, 1974.
105. **Tagaya, M., Nakano, K., and Fukui, T.,** A new affinity label for the active site of glycogen synthase, *J. Biol. Chem.*, 260, 6670, 1985.
106. **Piras, R., Rothman, L. B., and Cabib, E.,** Regulation of muscle glycogen synthetase by metabolites, *Biochemistry*, 7, 56, 1968.
107. **Roach, P. J., Takeda, Y., and Larner, J.,** Rabbit skeletal muscle glycogen synthase I, *J. Biol. Chem.*, 251, 1913, 1976.
108. **Roach, P. J. and Larner, J.,** Rabbit skeletal muscle glycogen synthase. II, *J. Biol. Chem.*, 251, 1920, 1976.
109. **Danforth, W. H.,** Glycogen synthetase activity in skeletal muscle, *J. Biol. Chem.*, 240, 588, 1965.
110. **Leloir, L. F.,** Role of uridine diphosphate glucose in the synthesis of glycogen, in *Control of Glycogen Metabolism*, Whelan, W. J. and Cameron, M. P., Eds., Little, Brown, Boston, 1964, 68.

Section III
Proteases — Structure and Mechanisms

In order to determine the precise nature of the action of an enzyme on a protein molecule, one needs to know the exact structure of the protein, and it would also be necessary to identify the split products produced through the action of the enzyme on the protein. Max Bergmann and Joseph S. Fruton, The Specificity of Proteinases, p. 66, in *Advances in Enzymology and Related Subjects*, (F. F. Nord and C. H. Werkman, editors), Vol. I., Interscience Publishers, Inc., New York (1941).

Chapter 6

CALCIUM-ACTIVATED NEUTRAL PROTEASE

Kazutomo Imahori, Koichi Suzuki, and Seiichi Kawashima

TABLE OF CONTENTS

I. INTRODUCTION

Calcium-dependent neutral protease was first identified in rabbit skeletal muscle as kinase activating factor (KAF),[1] which means that it has the activity to convert phosphorylase b kinase from the inactive form to the active one. Later on, several reports were published suggesting that the enzyme has physiological roles in myofibrillar protein turnover and in the processing of receptor proteins.[2-4] However, no one had succeeded in complete purification of this enzyme, which precluded further characterization. In 1978 we purified the enzyme to homogeneity[5] and characterized it as follows. The enzyme is a neutral thiol protease. It is a simple protein comprising large (80,000 mol wt) and small (30,000 mol wt) subunits. The role of calcium ion is to change the conformation of the molecule so as to expose an essential thiol group. Recently, however, it was revealed that limited autolysis, brought on by this induced conformational change, is crucial for the activation of the enzyme.[6] After characterization of this enzyme we named it calcium-activated neutral protease (CANP). Later, Murachi named it calpain.[7] However, as described above, the activation of this enzyme requires its autolysis. This strongly suggests that the enzyme so far called CANP or calpain is not the enzyme itself but its zymogen or proenzyme. Accordingly, it will be necessary to change the nomenclature of the enzyme in the future.

The enzyme so far studied was the one which requires millimolar order Ca^{2+} for its activation. However, Mellgren[8] noticed the existence of another type of CANP which requires micromolar Ca^{2+}. Accordingly we named the two forms mCANP and μCANP, respectively (Figure 1); these correspond to calapin II and I, respectively, according to the nomenclature of Murachi.

Although the existence of CANP is ubiquitous its physiological role has not been elucidated. However, it is clear that the activity of CANP is finely regulated by the concentration of Ca^{2+} on one hand and also by an endogenous inhibitor on the other. In the following sections, we will explain the molecular architecture of CANPs, the roles of Ca^{2+} in the activation of the enzyme, and finally the endogenous inhibitor of this enzyme.

II. MOLECULAR ARCHITECTURE

A. DOMAIN STRUCTURE

Both μ and mCANP are heterodimers consisting of 80K and 30K subunits. Since their 30K subunits are identical,[9] the differences in their properties are ascribed to structural differences in the 80K subunits. The amino acid sequence of CANP has been determined by analyzing the cDNA base sequence. The complete structures of chicken,[10] human μ[11] and m[12] 80K subunits, and human,[13] rabbit,[14] and porcine[15] 30K subunits have been determined. Partial structures of rabbit μ and m80K subunits have also been reported.[16]

The 80K subunits of μ and mCANPs are composed of about 700 residues and the μ80K subunit is usually larger than the m80K subunit. The 30K subunit consists of about 270 residues. The primary translation products of the human μ80K, m80K, and 30K subunits are composed of 714, 700, and 268 residues, respectively, and their calculated molecular weights are 81,889, 80,005, and 28,315, respectively. Analysis of the structure of the mature 80K subunit indicates that the N-terminus is blocked and the sequence always starts from residue 2 of the primary translation product.[10,17] N-terminal processing does not occur with the 30K subunit, though the amino group of Met-1 is blocked, probably by an acetyl group.[18] Clear evidence for other posttranslational modification is not available at present. An inspection of the sequence and a search for sequence homology reveal clear domain structures in the 30K and 80K subunits (Figure 2).[10,14] The domain structure of CANP is discussed below using the chicken 80K and human 30K subunits as examples. This fundamental domain structure is common to all CANPs from various sources.

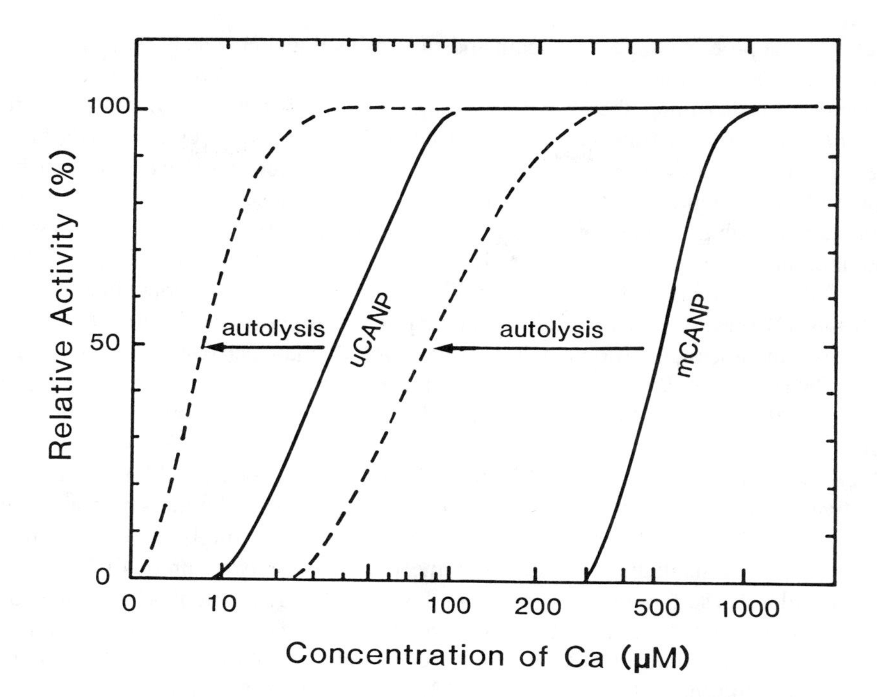

FIGURE 1. Ca^{2+}-requirements of μCANP, mCANP, and their autolyzed forms.

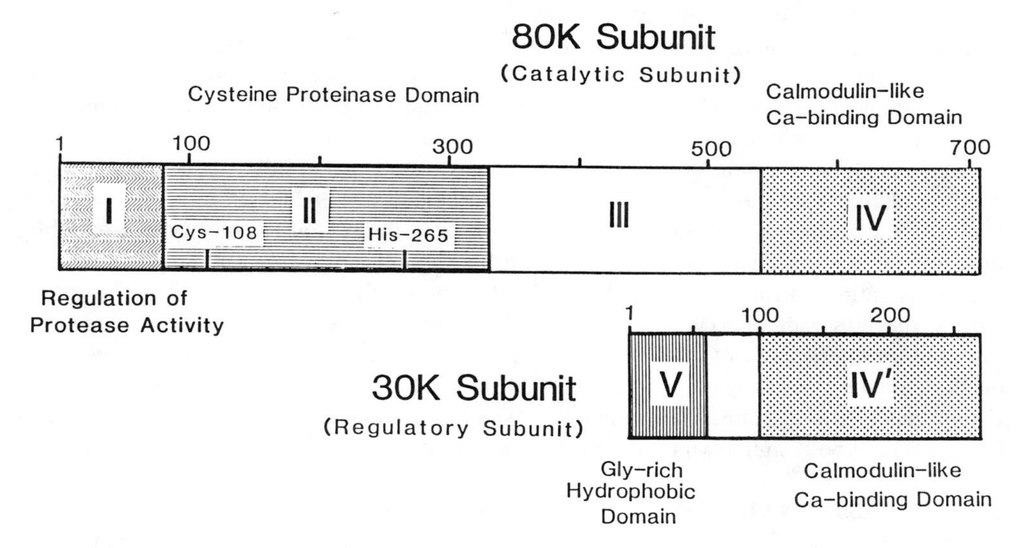

FIGURE 2. Schematic domain structure of chicken CANP.

Domain I (residues 1 to 79) — This domain has no structural homology to other proteins. However, during the autocatalytic activation, several residues are removed from the N-terminus[10] as described below. Upon removal of a short peptide from the N-terminus of the 80K subunit, the buried active site SH group is exposed to the surface and CANP becomes active.[18] This domain masks the active site and is important for the regulation of CANP activity.

Domain II (residues 80 to 330) — The sequence of about 40 residues from Pro-81, which contains the active site Cys-108, is highly homologous to the N-terminal sequence of other cysteine proteinases, e.g., papain and cathepsins B and H. The N-terminal sequence, especially that around the active site cysteine residue, is well conserved among cysteine proteinases.[19] Furthermore, the sequence around His-265 in CANP is homologous to that around the active site histidine residue in cysteine proteinases, a region that is also well conserved.[19] His-265 could be one of the active site amino acid residues, though it has not been proved experimentally. For this reason and others, this region is regarded as a protease domain corresponding to other cysteine proteinases. Judging from the sequence homology, this domain should be active by itself without Ca^{2+}. This potential protease activity may be repressed by interaction with other domains in the whole CANP molecule.

Domain III (residues 331 to 560) — Since proteins homologous to this domain are not known, the function of this domain is not clear yet. It may participate in the interaction with the 30K subunit and/or an endogenous inhibitor, or it may be involved in the regulation of activity by transducing the signal produced in the calmodulin domain to the protease domain.

Domain IV (residues 561 to 705) and domain IV′ (residues 99 to 268) — Four consecutive EF hand structures exist in domain IV of the 80K subunit. Their spacings and sequences are highly homologous to those in calmodulin. Interestingly, another calmodulin-like domain (IV′) exists in the 30K subunit, namely, CANP has two calmodulin-like domains in the molecule. These two domains are homologous, sharing about 50% sequence identity when their C-termini are aligned. These calmodulin-like domains when synthesized in *Escherichia coli* by recombinant DNA techniques can bind two or four Ca^{2+} ions per mole,[20] indicating that 4 to 6 mol of Ca^{2+} bind to CANP. This agrees with a previous report on the binding of Ca^{2+} to CANP[18] and suggests that the CA^{2+} binding sites exist only in domains IV and IV′. The affinities of Ca^{2+} for these domains parallel the concentration of Ca^{2+} required for activity (Ka).[20] Thus, the calcium sensitivity of CANP is essentially determined by the affinity of Ca^{2+} for the calmodulin-like domain. A conformational change induced upon binding of Ca^{2+} to IV and IV′ activates the repressed protease activity of domain II.

Domain V (residues 1 to 98) — This domain has a peculiar sequence rich in glycine. Clusters of glycine residues are found, the longest one is 20 residues and a block of 11 residues also exists. The content of glycine in the region of residues 10 to 56 is more than 80%. As the amino acid residues other than glycine in the region are mostly hydrophobic, this domain is referred to as the glycine-rich hydrophobic domain. The sequence of this domain is entirely different from that of the 80K subunit, though the succeeding C-terminal half of the 30K subunit (IV′) shows marked sequence homology to IV. This domain is essential for interaction of CANP with the cell membranes or phospholipids but has no effect on the enzyme activity itself.[17] There are proline-rich and glutamic acid-rich regions near C-terminus of this domain. The former may be a hinge connecting domains V and IV′, and the latter may be a signal which terminates the interaction of V with the cell membrane.

B. COMPARISON OF STRUCTURES OF VARIOUS CANPs

A comparison of the amino acid sequences of some typical CANP is summarized in Table 1. The sequences of the same isozyme from different sources show more than 90% identity, whereas different isozymes from the same source show much lower sequence homology (50 to 60%). A comparison of the sequence of each domain between μ and mCANPs indicates that the sequence homology in domain II (about 75%) is significantly higher than those of other domains (50 to 60%). Presumably, μ and mCANPs have essentially the same protease domain and differ only in their calcium sensitivity, a property that is ascribed to other domains, especially domains I and IV.

The known sequences of the 30K subunits are almost identical. It is not clear whether the 30K subunits are conserved among CANPs from divergent sources. The sequences known at present are all from mammalian sources.

TABLE 1
Sequence Homology of CANP

Combination		Homology (%)
80K Subunit		
Human μ	Rabbit μ	95
Human m	Rabbit m	94
Human m	Human μ	61
domain I		51
domain II		73
domain III		62
domain IV		51
Rabbit m	Rabbit μ	51
Chicken	Human μ	69
Chicken	Human m	66
30K Subunit		
Human	Rabbit	97
Rabbit	Porcine	97
Porcine	Human	96
Calmodulin in 30K and 80K		
Rabbit 30K	Rabbit μ80K	51
Rabbit 30K	Rabbit m80K	50

C. GENE STRUCTURE AND MOLECULAR EVOLUTION

The gene structures of the chicken 80K[21] and human 30K[22] subunits have been determined. Both are about 10 kb long and consist of 22 and 11 exons, respectively. Intron-exon junctions are always found at or very close to the domain junctions, supporting the domain structure predicted on the basis of the sequence homology. Of particular interest are the gene structures of domain IV and IV′. Each of the four EF hand structures in these domains is encoded by one exon. It has been proposed on the basis of the sequence homology that calmodulin arose from a minigene coding for one EF hand structure by a two-step gene duplication. This theory is also applicable to CANP. Furthermore, since the positions of intron-exon junctions in domains IV and IV′ are exactly the same at the amino acid sequence level as in the base sequence level, the two calmodulin-like domains in the 30 and 80K domains have evolved apparently from the same ancestor. In contrast, the gene structure of the protease domain does not show any clear correlation with the functional unit. Two active site amino acid residues, Cys-108 and His-265, are divided by four introns. Whether this situation is unique to CANP or common to other cysteine proteinases remains to be answered.

Summarizing, the 80K subunit is a chimeric protein that probably arose by fusion of at least four genes coding for a cysteine proteinase, calmodulin, and two of the domains. The same calmodulin gene fused with a gene encoding the glycine-rich domain to give the 30K subunit.

III. ROLE OF CALCIUM IONS

A. CONFORMATIONAL CHANGE

A structural change in CANP conformation was the first event observed to be caused by Ca^{2+}.[23] CANP was first inactivated by modification of the SH groups with monoiodoacetic acid in order to avoid autolytic degradation and its conformations in the presence and absence of Ca^{2+} were compared. Differences were found in the CD and UV spectra in the 250- to 340-nm region, indicating a change in environment around some Trp and/or Tyr residues. The pH and Ca^{2+} dependencies of these changes were found to coincide with those of native CANP activity. On the other hand, little change was observed in CD spectrum in the 210- to 250-nm region, an area that reflects the secondary structure of proteins. Thus, Ca^{2+}

induces local conformational changes in the CANP molecule without any gross structural change in the polypeptide backbone.

Another approach to the study of the Ca^{2+}-induced conformational changes in CANP was the titration of the sulfhydryl groups with DTNB. On addition of Ca^{2+}, some buried SH groups became exposed,[24,25] one of which was specifically modified by monoiodoacetic acid with an accompanying loss of enzyme activity.[24] This SH group was identified later to be the active site of the thiol protease domain.[26] The same conformational change was induced by Mn^{2+} which does not activate mCANP.[18] Inclusion of a low concentration of Ca^{2+}, which alone is not enough to activate mCANP, could activate mCANP in the presence of Mn^{2+}. These observations suggest a two-step activation mechanism; a conformational change from a "closed form" to an "open form" which is induced by high concentrations of Ca^{2+} or Mn^{2+}, and a subsequent Ca^{2+}-specific activation of the exposed active site by low concentrations of Ca^{2+}.

B. AUTOLYTIC ACTIVATION

On incubation of CANP in the presence of Ca^{2+}, CANP autolyzes in a very limited manner and the autolyzed CANP acquires an increased Ca^{2+} sensitivity (Figure 1).[27,28] By comparing the autolyzed products with the native protein, it has been shown that 17, 27, and 19 amino acid residues are removed from the N-terminus of chicken CANP and rabbit μCANP and mCANP, respectively. In all cases the small subunit is degraded simultaneously to 18- to 20-kDa polypeptide;[29,30] however, the increase in Ca^{2+} sensitivity is attributed to the degradation of the large subunit.[17]

The autolysis occurs in the presence of substrate.[28] but can not be a side reaction of protease activity since the pH and Ca^{2+}-dependencies for the autolytic and proteolytic processes are different, and the rate of autolysis is accelerated rather than inhibited by the presence of substrate.[6] Furthermore, the initiation of substrate degradation by native CANP is delayed when compared to that by autolyzed CANP and the whole protease activity can be accounted for by autolyzed CANP.[31] Thus, native CANP is an inactive proenzyme that must be autolyzed to be active as a protease towards substrates.

C. INTERACTION WITH MEMBRANES

On subcellular fractionation, most of the CANP activity is recovered in the cytosolic fraction.[32,33] Therefore, it seems rather contradictory that CANP is expected to be involved in the degradation of membrane proteins,[34-36] and cytoskeletal proteins.[37-39] However, it was found recently that CANP interacts with cytoplasmic membranes in a Ca^{2+}-dependent manner.[40,41] Moreover, the Ca^{2+} requirement for autolytic activation is much reduced in the presence of some phospholipids, probably components of cytoplasmic membranes.[42,43] For interaction with phospholipids, the N-terminal hydrophobic portion of the small subunit is essential.[44]

CANP may bind to membranes by recognizing membrane proteins that act as substrates and/or serve as loci for CANP activation. Since membranes are an important subcellular site for calcium mobilization, CANP bound to a membrane would be exposed to locally high concentrations of Ca^{2+} sufficient to induce autolytic activation. Once CANP is activated, it is active at the lower physiological Ca^{2+} concentrations of the cytoplasm. On autolytic activation, the hydrophobic N-terminal portion of the small subunit is removed, with loss of affinity for membranes, and the activated CANP, released to the cytoplasm, may degrade cytoplasmic substrates. Alternatively, the activated CANP remains at the cytoplasmic membrane and acts on proteins such as protein kinase C, which is bound to the membrane in a Ca^{2+}-dependent manner,[45] and denatured hemoglobin, which is adsorbed to membranes.[46] From any respect, the CANP-membrane interaction should be taken into consideration when the physiological function of CANP *in vivo* is examined.

IV. ENDOGENOUS CANP INHIBITOR

In most cells, an endogenous CANP inhibitor co-exists with CANP and is believed to be involved, along with Ca^{2+}, in the modulation of CANP activity. The inhibitor has been purified from various animal tissues. Inhibitor isolated from different sources have common features such as high stability to heat and acid, low content of aromatic amino acids, and the ability to inhibit CANP specifically. However, there are arguments about its molecular weight. Many different values are reported ranging from 25 to 110 kDa. A systematic examination has demonstrated, however, that intact CANP inhibitor is an oligomeric protein composed of subunits with a molecular weight of 110 kDa and that smaller inhibitors are fragments derived from the intact protein by proteolysis during purification.[47]

One intact subunit can inhibit several CANP molecules while the smallest fragment inhibits only one molecule.[48-50] These results suggest that the inhibitor is made up of several functional domains. This proposition was recently verified by the determination of the cDNA sequence.[51] In the primary structure there are 4 repetitive domains comprising 140 amino acid residues; each domain possibly represents one functional unit.

Since the amount of CANP inhibitor usually is greater than that of CANP[7], a simple calculation implies that it is impossible for CANP to function in cells. Therefore, there must be a more complex mechanism *in vivo* for the inhibition of CANP by the inhibitor; differences in cellular localization or differences in the Ca^{2+} dependency of the CANP-inhibitor interaction vs. CANP activation may be involved in this mechanism.

In addition to CANP inhibitor, a heat-stable, low molecular weight CANP activator has been identified in cytoplasmic and microsomal fractions of brain tissues.[52,53] This activator does not change the Ca^{2+} sensitivity of CANP but enhances its catalytic activity. The significance of CANP activator in cell function, especially of brain, is not known and awaits further investigation.

REFERENCES

1. **Huston, H. R. and Krebs, E. G.,** Activation of skeletal muscle phosphorylase kinase by Ca^{2+}. II. Identification of the kinase activating factor as a proteolytic enzyme, *Biochemistry,* 7, 2116, 1968.
2. **Busch, W. A., Stromer, M. H., Goll, D. E., and Suzuki, A.,** Ca^{2+}-specific removal of Z lines from rabbit skeletal muscle, *J. Cell Biol.,* 52, 367, 1972.
3. **Reddy, M. K., Etlinger, J. D., Rabinowitz, M., Fishman, D. A., and Zak, R.,** Removal of Z-lines and α-actinin from isolated myofibrils by a calcium-activated neutral protease, *J. Biol. Chem.,* 250, 4278, 1975.
4. **Dayton, W. R., Goll, D. E., Zeece, M. G., Robson, R. M., and Reville, W. J.,** A Ca^{2+}-activated protease possibly involved in myofibrillar protein turnover. Purification from porcine muscle, *Biochemistry,* 15, 2150, 1976.
5. **Ishiura, S., Murofushi, H., Suzuki, K., and Imahori, K.,** Studies of a calcium-activated neutral protease from chicken skeletal muscle. I. Purification and characterization, *J. Biochem.,* 84, 225, 1978.
6. **Kawashima, S., Inomata, M., and Imahori, K.,** Autolytic and proteolytic processes of calcium-activated neutral protease are independent from each other, *Biomed. Res.,* 7, 327, 1986.
7. **Murachi, T., Tanaka, K., Hatanaka, M., and Murakami, T.,** Intracellular Ca^{2+}-dependent protease (calpain) and its high-molecular-weight endogenous inhibitor (calpastatin), *Adv. Enz. Reg.,* 19, 407, 1981.
8. **Mellgren, R. L.,** Canine cardiac calcium-dependent proteases: resolution of two forms with different requirements for calcium, *FEBS Lett.,* 109, 129, 1980.
9. **Kawasaki, H., Imajoh, S., Kawashima, S., Hayashi, H., and Suzuki, K.,** The small subunit of calcium dependent proteases with different calcium sensitivities are identical, *J. Biochem.,* 99, 1525, 1986.
10. **Ohno, S., Emori, Y., Imajoh, S., Kawasaki, H., Kisaragi, M., and Suzuki, K.,** Evolutional origin of calcium dependent protease by fusion of genes for a thiol protease and a calcium binding protein?, *Nature,* 312, 566, 1984.

11. **Aoki, K., Imajoh, S., Ohno, S., Emori, Y., Koike, M., Kosaki, G., and Suzuki, K.,** Complete amino acid sequence of the large subunit of the low-Ca^{2+}-requiring form of human Ca^{2+}-activated neutral protease (μCANP) deduced from its cDNA sequence, *FEBS Lett.*, 205, 313, 1986.
12. **Imajoh, S. et al.,** in preparation.
13. **Ohno, S., Emori, Y., and Suzuki, K.,** Nucleotide sequence of a cDNA for the small subunit of human calcium-dependent protease, *Nucleic Acids Res.*, 14, 5559, 1986.
14. **Emori, Y., Kawasaki, H., Imajoh, S., Kawashima, S., and Suzuki, K.,** Isolation and sequence analysis of cDNA clones for the small subunit of rabbit calcium-dependent protease, *J. Biol. Chem.*, 261, 9472, 1986.
15. **Sakihama, T., Kakidani, H., Zenita, K., Yumoto, N., Kikuchi, T., Sasaki, T., Kannagi, R., Nakanishi, S., Ohmori, M., Takio, K., Titani, K., and Murachi, T.,** A putative Ca^{2+}-binding protein: structure of the light chain of porcine calpain elucidated by molecular cloning and protein sequence analysis, *Proc. Natl. Acad. Sci. U.S.A.*, 82, 6075, 1985.
16. **Emori, Y., Kawasaki, H., Sugihara, H., Imajoh, S., Kawashima, S., and Suzuki, K.,** Isolation and sequence analysis of cDNA clones for the large subunits of two isozymes of rabbit calcium-dependent protease, *J. Biol. Chem.*, 261, 9465, 1986.
17. **Imajoh, S., Kawasaki, H., and Suzuki, K.,** Limited autolysis of calcium-activated neutral protease (CANP): reduction of the Ca^{2+}-requirement is due to the NH_2-terminal processing of the large subunit, *J. Biochem.*, 100, 633, 1986.
18. **Suzuki, K. and Ishiura, S.,** Effect of metal ions on the structure and activity of calcium-activated neutral protease (CANP), *J. Biochem.*, 93, 1463, 1983.
19. **Takio, K., Towatari, T., Katunuma, N., Teller, D. C., and Titani, K.,** Homology of amino acid sequences of rat liver cathepsin B and H with that of papain, *Proc. Natl. Acad. Sci. U.S.A.*, 80, 3606, 1983.
20. **Minami, Y., Emori, Y., and Suzuki, K.,** EF-hand structure-domain of calcium activated neutral protease (CANP) can bind Ca^{2+} ions, *J. Biochem.*, 101, 889, 1987.
21. **Emori, Y., Ohno, S., Tobita, M., and Suzuki, K.,** Gene structure of calcium-dependent protease retains the ancestral organization of the calcium binding protein gene, *FEBS Lett.*, 194, 249, 1986.
22. **Miyake, S., Emori, Y., and Suzuki, K.,** Gene organization of the small subunit of human calcium-dependent neutral protease, *Nucleic Acids Res.*, 14, 8805, 1986.
23. **Tsuji, S., Ishiura, S., Takahashi-Nakamura, M., Katamoto, T., Suzuki, K., and Imahori, K.,** Studies on a Ca^{2+}-activated neutral protease of rabbit skeletal muscle. II. Characterization of sulfhydral groups and a role of Ca^{2+} ions in this enzyme, *J. Biochem.*, 90, 1405, 1981.
24. **Suzuki, K.,** Reaction of calcium-activated neutral protease (CANP) with an epoxysuccinyl derivative (E64c) and iodoacetic acid, *J. Biochem.* 93, 1305, 1983.
25. **Inomata, M., Nomoto, M., Hayashi, M., Nakamura, M., Imahori, K., and Kawashima, S.,** Comparison of low and high calcium requiring forms of the calcium-activated neutral protease (CANP) from rabbit skeletal muscle, *J. Biochem.*, 95, 1661, 1984.
26. **Suzuki, K., Hayashi, H., Hayashi, T., and Iwai, K.,** Amino acid sequence around the active site cysteine residue of calcium-activated neutral protease (CANP), *FEBS Lett.*, 152, 67, 1983.
27. **Suzuki, K., Tsuji, S., Kubota, S., Kimura, Y., and Imahori, K.,** Limited autolysis of Ca^{2+}-activated neutral protease (CANP) changes its sensitivity to Ca^{2+} ions, *J. Biochem.*, 90, 275, 1981.
28. **Inomata, M., Hayashi, M., Nakamura, M., Imahori, K., and Kawashima, S.,** Hydrolytic and autolytic behavior of two forms of calcium-activated neutral protease (CANP), *J. Biochem.*, 98, 407, 1985.
29. **Hathaway, D. R., Werth, D. K., and Haeberle, J. R.,** Limited autolysis reduces the Ca^{2+} requirement of a smooth muscle Ca^{2+}-activated protease, *J. Biol. Chem.*, 257, 9072, 1982.
30. **Coolican, S. A., Haiech, J., and Hathaway, D. R.,** The role of subunit autolysis in activation of smooth muscle Ca^{2+}-dependent proteases, *J. Biol. Chem.*, 261, 4170, 1986.
31. **Inomata, M., Imahori, K., and Kawashima, S.,** Autolytic activation of calcium-activated neutral protease, *Biochem. Biophys. Res. Commun.*, 138, 638, 1986.
32. **Guroff, G.,** A neutral, calcium-activated protease from the soluble fraction of rat brain, *J. Biol. Chem.*, 239, 149, 1964.
33. **Reville, W. J., Goll, D. E., Stromer, M. H., Robson, M. R., and Dayton, W. R.,** A Ca^{2+}-activated protease possibly involved in myofibrillar protein turnover. Subcellular localization of the protease in porcine skeletal muscle, *J. Cell Biol.*, 70, 1, 1976.
34. **Gates, R. E. and King, L., Jr.,** Proteolysis of the epidermal growth factor receptor by endogenous calcium-activated neutral protease from rat liver, *Biochem. Biophys. Res. Commun.*, 113, 255, 1983.
35. **Pant, H. C., Virmani, M., and Gallant, P. E.,** Calcium-induced proteolysis of spectrin and band 3 protein in rat erythrocyte membranes, *Biochem. Biophys. Res. Commun.*, 117, 372, 1983.
36. **Siman, R., Baudry, M., and Lynch, G.,** Brain fodrin: substrate for calpain I, an endogenous calcium-activated protease, *Proc. Natl. Acad. Sci. U.S.A.*, 81, 3572, 1984.

37. **Nelson, W. J. and Traub, P.,** Properties of a Ca^{2+}-activated protease specific for the intermediate-sized filament protein vimentin in Ehrlich-ascites-tumor cells, *Eur. J. Biochem.*, 116, 51, 1981.
38. **Zimmerman, U.-J. and Schlaepfer, W. W.,** Characterization of a brain calcium-activated protease that degrades neurofilament proteins, *Biochemistry,* 21, 3977, 1982.
39. **Klein, I., Lehotay, D., and Gondek, M.,** Characterization of a calcium-activated protease that hydrolyzes microtubule-associated protein, *Arch. Biochem. Biophys.*, 208, 520, 1981.
40. **Pontremoli, S., Melloni, E., Sparatore, B., Salamino, F., Michetti, M., Sacco, O., and Horecker, B. L.,** Binding of erythrocyte membrane is the physiological mechanism for activation of Ca^{2+}-dependent neutral proteinase, *Biochem. Biophys. Res. Commun.*, 128, 311, 1985.
41. **Gopalakrishna, R. and Barsky, S. H.,** Hydrophobic association of calpains with subcellular organells. Compartmentalization of calpains and the endogenous calpastatin in tissues, *J. Biol. Chem.*, 261, 13936, 1986.
42. **Coolican, S. A. and Hathaway, D. R.,** Effect of L-α-phosphotidy-inositol on a vascular smooth muscle Ca^{2+}-dependent protease. Reduction of the Ca^{2+} requirement for autolysis, *J. Biol. Chem.*, 259, 11627, 1984.
43. **Pontremoli, S., Melloni, E., Sparatore, B., Salamino, F., Michetti, M., Sacco, O., and Horecker, B. L.,** Role of phospholipids in the activation of the Ca^{2+}-dependent proteinase of human erythrocyes, *Biochem. Biophys. Res. Commun.*, 129, 389, 1985.
44. **Imajoh, S., Kawasaki, H., and Suzuki, K.,** The amino-terminal hydrophobic region is essential for its activation by phosphatidylinositol., *J. Biochem.*, 99, 1281, 1986.
45. **Melloni, E., Pontremoli, S., Michetti, M., Sacco, O., Sparatore, B., Salamino, F., and Horecker, B. L.,** Binding of protein kinase C to neutrophil membranes in the presence of Ca^{2+} and its activation by a Ca^{2+}-requiring proteinase, *Proc. Natl. Acad. Sci. U.S.A.*, 82, 6435, 1985.
46. **Pontremoli, S., Melloni, E., Sparatore, B., Michetti, M., and Horecker, B. L.,** A dual role for the Ca^{2+}-requiring proteinase in the degradation of hemoglobin by erythrocyte membrane proteinases, *Proc. Natl. Acad. Sci. U.S.A.*, 81, 6714, 1984.
47. **Nakamura, M., Inomata, M., Hayashi, M., Imahori, K., and Kawashima, S.,** Purification and characterization of 210,000-dalton inhibitor of calcium-activated neutral protease from rabbit skeletal muscle and its relation to 50,000-dalton inhibitor, *J. Biochem.*, 98, 757, 1985.
48. **DeMartino, G. N. and Croall, D. E.,** Purification and characterization of a protein inhibitor of calcium-dependent proteases from rat liver, *Arch. Biochem. Biophys.*, 232, 713, 1984.
49. **Imajoh, S., Kawasaki, H., Kisaragi, M., Mukai, M., Sugita, H., and Suzuki, K.,** A 107-kDa inhibitor for calcium-activated neutral protease (CANP), *Biomed. Res.*, 5, 481, 1984.
50. **Nakamura, M., Inomata, M., Hayashi, M., Imahori, K., and Kawashima, S.,** Purification and characterization of an inhibitor of calcium-activated neutral protease from rabbit skeletal muscle: purification of 50,000-dalton inhibitor, *J. Biochem.*, 96, 1399, 1984.
51. **Emori, Y., Kawasaki, H., Imajoh, S., Imahori, K., and Suzuki, K.,** Endogenous inhibitor for calcium-dependent cysteine protease contains four internal repeats that could be responsible for its multiple reactive sites, *Proc. Natl. Acad. Sci. U.S.A.*, in press.
52. **DeMartino, G. N. and Blumenthal, D. K.,** Identification and partial purification of a factor that stimulates calcium-dependent proteases, *Biochemistry,* 21, 4297, 1982.
53. **Takeyama, Y., Nakanishi, H., Uratsuji, Y., Kishimoto, A., and Nishizuka, Y.,** A calcium-protease activator associated with brain microsomal-insoluble elements, *FEBS Lett.*, 194, 110, 1986

Section IV
Dehydrogenases — Structure and Catalytic Mechanism

For the best and safest method of philosophizing seems to be, first diligently to investigate the properties of things and then to seek hypotheses to explain them. For hypotheses ought to be fitted merely to explain the properties of things and not attempt to predetermine them except in so far as they can be an aid to experiments. Isaac Newton, *Philosophiae Naturalis Principia Mathematica* (1687), translated by Andrew Motte (1729), reviewed by Cajari, Vol. 2, University of California Press, Berkeley, Calif., 1934, p. 673.

Chapter 7

GLUTAMATE DEHYDROGENASE (BOVINE LIVER)

Roberta F. Colman

TABLE OF CONTENTS

I. INTRODUCTION

Glutamate dehydrogenases, catalyzing the reversible reductive amination of α-ketoglutarate to form L-glutamate, has been isolated from a wide variety of sources from mammalian liver to fungi, bacteria and plants.[1] In general, the reaction catalyzed by all of these enzymes is similar, except that the vertebrate enzymes function with either NAD or NADP, whereas those from lower organisms are coenzyme specific.[2] In addition, the enzymes isolated from vertebrates, such as chicken, bovine and human liver, exhibit allosteric regulation by purine nucleotides which is absent in enzymes from lower organisms, and the vertebrate enzymes undergo a reversible molecular association-dissociation reaction which may influence their sensitivity to regulation. In recent years, the order of amino acids in several glutamate dehydrogenases has been determined either directly by protein sequencing or by deduction from the nucleotide sequence of glutamate dehydrogenase genes; studies of the crystal structure of the enzyme have been initiated; analyses of the subunit association-dissociation reaction under denaturing and nondenaturing conditions have been extended; evidence on the chemical mechanism of the glutamate dehydrogenase-catalyzed reaction has been obtained from isotope exchange experiments and model reactions; and measurements of the number of ligand binding sites and the structural specificity for binding to catalytic and regulatory sites have been clarified. In addition, particular amino acid residues have been implicated in the catalytic or regulatory functions of the enzyme by chemical modification studies using group specific reagents or affinity labels. This chapter will concentrate on the more recent developments on the structure and function of bovine liver glutamate dehydrogenase. Other aspects are discussed more extensively in earlier reviews.[1-6]

II. PHYSICOCHEMICAL STUDIES

A. UNIMER

The structure of bovine liver glutamate dehydrogenase has been extensively investigated and reviewed.[2,4,5] The enzyme is composed of identical peptide chains of molecular weight 56,000[7-10] and the amino acid sequence of this peptide chain has been published.[1,10,11] However, the individual polypeptide chain is only obtained under denaturing conditions and the molecular weight of the smallest species of glutamate dehydrogenase with enzymatic activity appears to be approximately 320,000.[8] This hexameric species has variously been termed ''monomer,'' ''hexamer'', and ''oligomer'' in the extant literature on glutamate dehydrogenase, but for purposes of clarity this chapter will refer to the six-polypeptide structure as ''unimer'' in accordance with the proposal of Sund et al.[4]

Electron microscopic and small-angle X-ray scattering studies have been conducted which suggest that the six subunits are arranged in two groups of three, in the form of two triangles layered on one another.[12,13] Each layer is viewed as composed of three prolate ellipsoids of rotation with the major axes of the ellipsoids pointing in the same direction. The length of an individual subunit along its major axis is estimated at about 66.5 Å with an overall length for the unimer of 133 Å. From this model, the diameter for the subunit along its minor axis is 43 Å.[14] The model of six identical polypeptide chains arranged in two groups of three is consistent with crosslinking studies in which adipimidate converts almost all the protein to crosslinked trimers, observed under denaturing conditions in sodium dodecyl sulfate.[15]

More recently, Tashiro et al. have examined in detail the time dependence of unfolding of glutamate dehydrogenase over a guanidine hydrochloride concentration range of 0.5 to 6.0 *M* using light scattering and circular dichroism.[16] With increasing concentrations of guanidine hydrochloride, two transition regions were observed: the first between 0.6 and 1.0 *M* guanidine hydrochloride and the second between 2.0 and 2.8 *M* denaturant. It was proposed that the enzyme exists as a hexamer below 0.6 *M*, as a trimer between 1.0 and 2.0 *M* and a monomer above 2.8 *M* guanidine hydrochloride; and that furthermore there exists a stable trimer with native structure as an intermediate in the pathway of dissociation.[16] The kinetic studies indicated that the dissociation of hexamer to trimer occurs ten times more rapidly than the dissociation of trimer to monomer and that subunit unfolding occurs more slowly than dissociation.[16] A follow-up paper reported the reassociation of trimers from 0 to 1.0 *M* guanidine hydrochloride.[17] These important studies provide the groundwork needed for the preparation of various types of hybrid glutamate dehydrogenases to evaluate interactions among subunits. In fact, it has already been found that enzyme which is dissociated in 1.5 *M* guanidine hydrochloride to give inactive trimers can then be reassociated to the hexamer form with full restoration of activity.[18]

B. REVERSIBLE POLYMERIZATION OF UNIMERS

It has long been known that the 320,000-Da unimer of glutamate dehydrogenase undergoes a reversible association-dissociation reaction which is influenced by protein concentration, pH, buffer, coenzymes, sterols, and regulatory purine nucleotides.[19-22] For example, at an enzyme concentration of 2 mg/ml, average molecular weights of 1.7×10^6 have been observed. It is thought that the unimer associates lengthwise, along the major axis of the subunits, to yield long rod-like polymers of indefinite length[2,4,5,14] and the reversible polymerization has been well fit by an indefinite linear polymerization model.[23,24] It is known now that the state of aggregation does not affect the specific activity of glutamate dehydrogenase as measured in the absence of allosteric regulators.[25-27] However, as will be discussed in Section IV, several of the regulatory compounds have been shown to dissociate aggregated forms of the enzyme and to bind preferentially to the unimer. In those cases, the state of aggregation influences the affinity of the enzyme for its regulator, as well as the shape of its binding curve.[24,27-29] These aspects of the enzyme have been well reviewed.[2,6]

C. TERTIARY STRUCTURE

The tertiary structure of the bovine liver glutamate dehydrogenase has not been solved and indeed crystals suitable for X-ray crystallographic analysis have yet to be obtained. Because it was considered that the difficulties in preparing suitable crystals might be due to the polymerization of the bovine liver enzyme, several attempts have been made to examine crystals from species whose glutamate dehydrogenases were not characterized by a reversible association-dissociation. The rat liver enzyme does not polymerize extensively, and preliminary data were presented for crystals of the native enzyme as well as enzyme chemically modified with pyridoxal phosphate.[30] However, the crystals were fragile and although large, they only diffracted to 6 Å. Crystals were prepared of a glutamate dehydrogenase from tuna liver,[31] which does not polymerize to species larger than a hexamer. Despite promising initial results, structural information has not been forthcoming. The most recent and successful attempt to obtain appropriate crystals has been made using a nonallosteric bacterial enzyme, the NAD^+-dependent glutamate dehydrogenase from *Clostridium symbiosum*.[32,33] This enzyme has a subunit molecular weight of 49,000 and normally exists as a hexamer of identical subunits, similar to the vertebrate glutamate dehydrogenases. The subunit size of the *C. symbiosum* enzyme is about 10% smaller than that of the bovine liver glutamate dehydrogenase, but the bacterial hexamer has been described (as based on the electron-density map[33]) in terms of a cylindrical structure with an overall length of 108 Å and radius of 44 Å, similar to the dimensions estimated for the bovine liver enzyme. It has been reported that the crystals of *C. symbiosum* glutamate dehydrogenase diffract to at least 2-Å resolution,[32] although the analysis of only the 6-Å resolution map has thus far appeared.[33] The interpretation of the electron density map is hindered by the unavailability of information on the amino acid sequence of the enzyme from this species;[32,34] yet, the general structural features of this molecule have been described.[33] Since all of the glutamate dehydrogenases catalyze the same reaction, it is anticipated that crystallographic analysis of the *C. symbiosum* enzyme will reveal the common structural components of the active site, although it cannot yield direct information on the structures of the regulatory sites.

Spectroscopic studies have been conducted on bovine liver glutamate dehydrogenase, including those on the optical rotatory dispersion, circular dichroism, ultraviolet absorption, difference spectra, and fluorescence spectra of the enzyme under various conditions; these have been summarized in earlier reviews.[2-5] Some have been interpreted to indicate that glutamate dehydrogenase, when in solution, has a high content of α-helix.[1] In most studies, however, rather than being analyzed as an index of the solution conformation of the enzyme, the spectral perturbations have been used as experimental tools to monitor the binding of specific ligands by the enzyme.[35]

D. AMINO ACID SEQUENCE

The primary structure of glutamate dehydrogenase from the liver of several vertebrates has been reported, based directly on the sequence of amino acid residues in the proteins. The 505 amino acid sequence of the human enzyme[11] is strikingly similar to the 501 residue polypeptide of the bovine[10,11,36] and the 504-residue polypeptide of the chicken enzyme.[37] Of the 501 amino acids which they have in common, the bovine and human enzymes differ only in 24, the bovine and chicken enzymes in 27, while the human and chicken enzymes differ in 41 residues. In addition, the rat liver enzyme[38] is remarkably homologous in amino acid sequence. Since these enzymes all exhibit similar catalytic activity as well as ability to be regulated by a variety of compounds including purine nucleotides, it is apparent that the nonidentical residues cannot be essential for either catalysis or regulation. The chicken enzyme has three and the human enzyme four extra amino acids beyond the alanine that has usually been found as the amino terminal of the bovine enzyme.[11] McCarthy et al. presented evidence for four extra residues at the amino terminal end of the bovine enzyme when that glutamate dehydrogenase was isolated by a more rapid procedure from brain and liver.[39] Presumably, limited proteolysis is responsible for the shorter bovine liver enzyme, although the two types of enzyme are not markedly different in catalytic or regulatory functions.

The amino acid sequences have now been published of several glutamate dehydrogenases from fungi or bacteria; these enzymes differ from those of the vertebrates in being specific for either NAD^+ or $NADP^+$ and in not being subject to allosteric regulation. The sequence of 453 amino acids has been determined for the NADP-specific enzyme of the fungus *Neurospora crassa* directly by fragmentation of the protein[40,41] or by deduction from the nucleotide sequence of the gene.[42] The nucleotide sequence of the *Saccharomyces cerevisiae* (yeast) gene for NADP-dependent glutamate dehydrogenase was ascertained,[43,44] from which the sequence of the 454-amino acid residue protein was deduced; and the order of nucleotides was determined in the *Escherichia coli* K12 gene encoding for the 447-amino acid subunit of the NADP-specific glutamate dehydrogenase.[45,46] Homology is stronger within types of glutamate dehydrogenases (NADP-specific or dual coenzyme specificity) than between types. Thus, the yeast enzyme is 63% identical to the *N. crassa* enzyme,[43] 51% homologous to the *E. coli* enzyme,[44] but only 24% homologous to the bovine liver glutamate dehydrogenase.[44] The strongest conservation is in the amino terminal half of the molecule. On the basis of secondary structure predictions and sequence comparisons with dehydrogenases of crystallographically known structure, Wootton has designated amino acid sequences for two coenzyme binding sites in both the bovine liver and *N. crassa* NADP-specific enzymes.[47] In the bovine enzyme, one domain was predicted in the approximate region of amino acid residues 9 to 126, and the second from amino acid residues 245 to 356. The region between these two domains was proposed as that for substrate binding and catalysis. In contrast,

Moye et al.[43] have only located one nucleotide binding domain on the basis of secondary structure predictions for the yeast enzyme, that analogous to the C-terminal designated coenzyme domain of the *N. crassa* enzyme (lysine-218 to threonine-353). In addition to these structural studies, sequence information has been reported[48] for the much larger, discrete, NAD-specific glutamate dehydrogenase of *N. crassa* which has more than 1000 amino acid residues, has a subunit molecular weight of 116,000 and exists as a tetramer.[49]

III. CATALYTIC REACTION

Glutamate dehydrogenase catalyzes the reversible oxidative deamination of L-glutamate to α-ketoglutarate and ammonium ion:

$$\begin{array}{c} COO^- \\ | \\ H_2N\text{-}C\text{-}H \\ | \\ CH_2 \\ | \\ CH_2 \\ | \\ COO^- \end{array} + NAD(P) \rightleftharpoons \begin{array}{c} COO^- \\ | \\ C{=}O \\ | \\ CH_2 \\ | \\ CH_2 \\ | \\ COO^- \end{array} + NAD(P)H + NH_4^+ + H^+$$

At 25° (μ = 0.3), the equilibrium constant[6]

$$K = \frac{[NAD(P)H][\alpha\text{-ketoglutarate}][NH_4^+][H^+]}{[NAD(P)][\text{L-glutamate}]}$$

is 7×10^{-14} M^2, indicating the great predominance of oxidized coenzyme and amino acid at equilibrium. The general features of the reaction have been thoroughly reviewed.[1-4,6] The bovine liver enzyme can utilize either NAD^+ or $NADP^+$ in the transfer of hydrogen to the 4-position of the nicotinamide ring with B stereospecificity. Transferred nuclear Overhauser enhancement studies used to examine the conformation of the coenzyme bound to glutamate dehydrogenase[50] revealed that both NAD^+ and $NADP^+$ bind with syn conformation about the nicotinamide-ribose bond, as may be general for dehydrogenases of B stereospecificity. The bovine enzyme is relatively nonspecific at the coenzyme site, using in the catalytic reaction the nucleotide fragments NMN and NMNPR (albeit at low rates) as well as nucleotide analogues such as 3-acetylpyridine adenine dinucleotide, nicotinamide-6-mercaptopurine dinucleotide and nicotinamide 1,N^6-ethenoadenine dinucleotide.[4] More recently, it has been shown that the thionicotinamide analogues of NAD^+ and $NADP^+$ are effective alternate coenzymes.[51,52]

The specificity for the substrate site has been extensively explored by the kinetic evaluation of alternate substrates and competitive inhibitors,[1-3] with a paper by Hornby et al.[53] representing the most recent extension of these approaches. In general, the dicarboxylic acid glutamate is the best substrate and the most effective competitive inhibitors have either two carboxylates 7.5 Å apart or a single carboxylate and an electronegative group, such as bromo or nitro.[1,2] The enzyme also catalyzes the oxidative deamination of L-α-amino monocarboxylates, although the optimum pH is 1 to 1.5 pH units higher;[1-3] this difference has been attributed to the requirement of the protonated form of an enzymic group of pK 7.8 for binding dicarboxylic acids, and of the unprotonated form of that group for binding monocarboxylates.[54] In contrast to the carboxylate substrate, the requirement for ammonium ion is highly specific.

The analysis of the kinetic mechanism of glutamate dehydrogenase was initially controversial,[1-3] but it is now thought that the enzyme can bind substrates in at least a partially random order, with certain pathways being preferred.[55,56] The mechanism has also been examined by the transient kinetics of formation of spectrally distinguishable enzyme-ligand complexes.[3,6] Fisher[6] has summarized four experimentally observable phases of the oxidative deamination of glutamate:

[I] Enzyme + NADP + glutamate ⇌ (enzyme − NADP − glutamate)

[II] (Enzyme − NADP − glutamate) ⇌ (enzyme − NADPH − α-iminoglutarate) $\overset{H_2O}{\rightleftharpoons}$

(enzyme − NADPH − α-ketoglutarate) + NH_3

[III] (Enzyme − NADPH − α-ketoglutarate) + glutamate ⇌

(enzyme − NADPH − glutamate) + α-ketoglutarate

[IV] (Enzyme − NADPH − glutamate) ⇌ enzyme + NADPH + glutamate

Since the first two phases are faster than the last two, replacement of the α-hydrogen of glutamate by deuterium does not appreciably decrease the steady state velocity, although there is an isotope effect of 1.5 to 1.7 on the "initial burst" of the reaction which reflects the hydride transfer step (II). In the two ternary complexes, Enzyme-NADP-glutamate and Enzyme-NADPH-α-ketoglutarate, the substrate and coenzyme are each bound to the enzyme much more tightly than in the absence of the other.[6,57] The steady state rate of the oxidative deamination of glutamate is limited by the rate of release of NADPH or α-ketoglutarate (phases II or IV). As will be discussed in Section IV, it is well known that ADP activates the steady state glutamate dehydrogenase reaction, whereas GTP inhibits this overall reaction. This observation is now understandable since ADP weakens the binding of reduced coenzyme (thereby increasing the slowest step of the oxidative deamination), while GTP tightens the binding of reduced coenzyme (thus slowing the rate-determining step of the reaction).

It has frequently been proposed that α-iminoglutarate is an enzyme-bound intermediate in the reversible conversion of glutamate to α-ketoglutarate. Evidence was provided by Hochreiter et al.[58] who showed that when $NaBH_4$ was added to an incubation mixture of glutamate dehydrogenase, [^{14}C]-α-ketoglutarate and ammonia (but no coenzyme), radioactive glutamate was generated. The isolated amino acid was predominantly in the L-configuration, rather than the DL-form characteristic of the nonenzymatic reaction, suggesting that reduction occurred while the α-iminoglutarate was bound to the enzyme. Furthermore, Fisher and Viswanathan[59] found that the rapid exchange of ^{18}O from the carbonyl of α-ketoglutarate requires the simultaneous presence of glutamate dehydrogenase, NADPH and ammonia and is not due to reversal of the overall reaction. These results were interpreted as supporting an enzyme-NADPH-α-iminoglutarate complex as an obligatory intermediate in the enzymatic reaction, while the enzyme-α-iminoglutarate binary complex was thought to lack kinetic competence.[59] Further support for the involvement of α-iminoglutarate as an intermediate comes from the demonstration that glutamate dehydrogenase catalyzes the reduction of Δ′-pyrroline-α-carboxylic acid by NADPH to the amino acid proline.[60] This reaction has been studied[60-64] as a simpler model of the enzymatic reduction of α-iminoglutarate, which is only part of the kinetically more complex reductive amination of α-ketoglutarate.

Smith et al.[1] have suggested that a lysine residue of glutamate dehydrogenase could participate directly in the catalytic reaction by condensing with α-ketoglutarate to form a Schiff base with release of water. Subsequently, attack of ammonia on the enzyme-bound Schiff base would yield the α-iminoglutarate intermediate bound to the enzyme. This intermediate could then be reduced by NADPH to generate glutamate. In the reverse direction, the first step would be abstraction from glutamate of a hydride ion (with transfer to coenzyme) to form the α-iminoglutarate intermediate; this step would be followed by reaction of the intermediate with an α-amino group of lysine to yield the proposed Schiff base. Finally, hydrolysis of the Schiff base could result in α-ketoglutarate. There is ample evidence implicating lysine-126 in the reaction catalyzed by glutamate dehydrogenase (see Section V.A), but no direct experimental support for the existence of an enzyme-bound Schiff base intermediate. Were such an intermediate to be significant, the addition of $NaBH_4$ to an incubation mixture of enzyme and [^{14}C]-α-ketoglutarate should have led to trapping of the radioactive substrate covalently bound to enzyme; rather, the radioactivity was recovered by Hochreiter et al.[58] as glutamate (predominantly L-, with no labeled enzyme being reported). Furthermore, reduction of a putative enzyme-bound Schiff base intermediate should have led to inactivation of the enzyme in the presence of α-ketoglutarate and $NaBH_4$; on the contrary, no such substrate-dependent inactivation was observed (Colman, R. F., unpublished observations). While a negative result may not be definitive, the isolation of radioactive L-glutamate upon the addition of $NaBH_4$ to enzyme and [^{14}C]-α-ketoglutarate suggests that the existent intermediate was accessible to $NaBH_4$. Therefore, it is unlikely that the reaction proceeds through a Schiff base formed by an enzymic lysine residue and α-ketoglutarate.

An alternate reaction mechanism has been proposed by Rife and Cleland[54] involving direct attack by ammonia on α-ketoglutarate to form a carbinolamine, which then yields α-iminoglutarate upon dehydration. Rife and Cleland have assigned to a reactive lysine residue the role of general acid in the transfer of a proton to the hydroxyl group of the carbinolamine prior to the expulsion of water to form iminoglutarate. Another lysine has been postulated to participate as the protonated species in the binding of the carboxylate of substrates;[54] and indeed there are studies in the literature implicating two reactive lysines, residues 126 and 27 (see Section V.A). Reduction of iminoglutarate by NADPH has been proposed to produce the product L-glutamate.

IV. REGULATION AND LIGAND BINDING SITES

The activity of bovine liver glutamate dehydrogenase is influenced by many substances[1-3] including purine nucleotides,[65,66] several estrogens,[67] zinc ion,[68] and palmitoyl coenzyme A.[69] The regulatory effects of the purine nucleotides have been studied in greatest detail. As measured by the steady state glutamate dehydrogenase reaction at pH 8, ADP activates, GTP inhibits and NADH (at relatively high concentrations) inhibits by binding to a site distinct from the catalytic site;[2,65] however, the type and magnitude of the allosteric effects depend on pH.[70] Although the adenosine and guanosine nucleotides appear to compete kinetically, chemical modification studies have demonstrated that the sites which they occupy are not identical. Cross and Fisher[71] described the binding of allosteric ligands to glutamate dehydrogenase in terms of occupation of a combination of subsites, thereby accounting for the apparent interactions among certain ligand sites and the mutual exclusion of other ligands. The thermodynamics of the binding of adenosine, cAMP, AMP, ADP, and ATP have been compared[72] and the ability of a wide variety of purine nucleotide analogues to activate or inhibit the enzyme have been evaluated.[73] In general, adenosine, AMP and ADP are increasingly effective in activating and binding to the enzyme.[72] The activating ADP site is much more specific for the natural purine structure than is the GTP inhiibitory site.[73] Thus, 1,N^6-etheno ATP and 1,N^6-etheno ADP function as allosteric inhibitors,[74] as do 2′,3′-*O*-(2,4,6-trinitrocyclohexadienylidene)-GTP (TNP-GTP) and 3′-*O*-anthraniloyl-GTP.[75] In contrast, 2′,3′-*O*-(2,4,6-trinitrocyclohexadienylidene)-ADP (TNP-ADP) functions as an activator.[76]

The reversible binding to liver glutamate dehydrogenase has been measured for ligands including the pyridine nucleotide coenzymes, GTP, TNP-GTP, and ADP. Although this glutamate dehydrogenase can utilize either NADH or NADPH as coenzyme, reactions in which NADH is coenzyme are more complex, since inhibition is observed at high concentrations of the reduced coenzyme, whereas activation is noted at elevated levels of NAD.[2] It has been postulated that the activating site for NAD is the same regulatory site occupied by other purine nucleotides.[2,65] There has been some controversy about the relationship between the inhibitory NADH site and the ''purine nucleotide'' site occupied by ADP and GTP. Although it was initially proposed that these sites were distinct,[2,65] more recent experiments implied that ADP, but not GTP, may compete for binding to the adenyl portion of the inhibitory site for NADH.[77-84] Two binding sites per enzyme subunit have been found for NADH; and GTP tightens the binding of NADH at both of its sites.[80] Indeed, a mutual interaction between the GTP and NADH sites has been proposed on the basis of the combined effects of these purine nucleotides in protecting glutamate dehydrogenase against specific chemical modification.[85] Although NADPH also can occupy two sites per peptide chain, the affinity of the enzyme for the second set of sites is about ten times weaker than for NADH, thus accounting for the different kinetic behavior of the two coenzymes.[80,81] In contrast, the NAD fluorescent derivative 1,N^6-etheno-NAD (ϵNAD) has been found to occupy only one set of sites, presumed to be the active site since ϵNAD functions as a coenzyme in the glutamate dehydrogenase reaction.[86]

Measurements of the reversible binding of the activator, [^{14}C]-ADP, to the enzyme in the absence of coenzyme have revealed the availability of two sites per enzyme subunit, albeit with a tenfold difference in affinity.[87] The addition of NADH weakens the binding at the tight site so the K_d values for the two ADP sites are indistinguishable; this result corresponds to the effect of ADP in weakening the binding of NADH by the enzyme.

Furthermore, studies of the equilibrium binding of radioactive GTP by glutamate dehydrogenase have revealed that NADH strengthens the binding of GTP.[27,28] In the presence of reduced coenzyme, two GTP sites per enzyme subunit are detected while only the GTP site per enzyme subunit is measurable in the absence of NADH.[88] Hiratsuka has reported that glutamate dehydrogenase also binds reversibly a chromophoric analogue of GTP, [2′,3′-O-(2,4,6-trinitrocyclohexadienylidene)-GTP] at two sites per enzyme subunit, only one of which is influenced by NADH.[75]

The reversible polymerization of glutamate dehydrogenase, which depends on the enzyme concentration in the range 0.1 to 4.0 mg/ml, has been described (Section II.B) and thoroughly reviewed.[2,4] Although the specific activity of the catalytic glutamate dehydrogenase reaction is independent of the state of aggregation, it is well known that allosteric inhibitors of the enzyme, such as GTP in the presence of reduced coenzyme, promote dissociation of the enzyme as measured at high protein concentrations.[2] Reciprocally, the sensitivity of the enzyme to several allosteric inhibitors is influenced by the enzyme concentration and therefore by its state of polymerization. Thus, the enzyme-GTP dissociation constant increases markedly and the binding curves exhibit greater cooperativity as the protein concentration is increased.[27] These results have been demonstrated to result from the preferential binding of GTP to the enzyme unimer as compared with the polymer.[28] More recent studies have indicated that it is only binding at the high affinity (not the low affinity) GTP site observed in the presence of NADH that is affected by the extent of enzyme aggregation.[88] Apparent cooperativity dependent on the reversible association-dissociation reaction of the enzyme has also been reported for the inhibitory zinc ions.[68] Since the concentration of glutamate dehydrogenase in bovine liver is high (>1 to 2 mg/g), the state of enzyme aggregation may be an important determinant of its responsiveness to allosteric regulation under physiological conditions.

V. AMINO ACID RESIDUES IMPLICATED IN CATALYTIC AND REGULATORY SITES

Since this aspect of the enzyme was last reviewed,[1,4] progress has been made in designating certain amino acid residues as participants in the catalytic or regulatory sites of glutamate dehydrogenase. Thus, residues important in the active site as well as the regulatory sites for NADH, GTP, ADP, and steroids have been identified. (The positional numbers cited in this article in all cases refer to those corresponding to the revised sequence of the bovine enzyme,[11] rather than necessarily to those given in the original paper.)

A. CATALYTIC SITE

Lysine-126 has been designated as important for catalysis on the basis of inactivation by covalent reaction with several compounds. Glutamate dehydrogenase is irreversibly inactivated by 4-(iodoacetamido) salicyclic acid (ISA) at pH 7.5 to 7.6.[89,90] The reaction is limited to about 1 mol per enzyme subunit, and the inclusion of the substrate α-ketoglutarate protects against the inactivation. Furthermore, although the modified enzyme exhibits no appreciable decrease in its ability to bind NAD, NADH, ADP, or GTP, it shows reduced binding of radioactive α-ketoglutarate.[90] A single modified peptide was isolated and the incorporation of ISA was attributed to modification of lysine-126.[91] The ISA-modified enzyme exhibits a fluorescence emission, due to the covalently bound acetamidosalicylate moiety, which is maximal at 400 nm[89,92] and which has allowed estimates to be made of the distances between the catalytic site and other enzymic sites by the technique of resonance energy transfer.[92]

Additional evidence for the involvement of lysine-126 in catalysis comes from studies of the inactivation of glutamate dehydrogenase by pyridoxal 5′-phosphate. Reduction of the pyridoxal 5′-phosphate-inactivated enzyme with $NaBH_4$ allowed the isolation of a hexadecapeptide containing the modified lysine-126.[93] This lysine is considered to have the abnormally low pK of 7.7 to 8.0,[1,94] which may account for the relatively specific modification of lysine-126 by several compounds normally reactive with lysines in general. Although there was subsequent controversy about whether modification of lysine-126 led to total or only partial inactivation, and whether α-ketoglutarate or NADH protects completely against inactivation,[95-97] the issue now appears to be resolved. Pyridoxal-5′-phosphate actually modifies two residues: lysine-126 and lysine-333.[98] Only lysine-126 is important for activity since NADH plus α-ketoglutarate preserve activity, but allow the modification of lysine-333 while protecting against reaction with lysine-126.[98] (In contrast, modification of lysine-333 results in depolymerization of the enzyme into catalytically active hexamers.[98]) The inactivation of enzyme is cooperative since the residual activity of the completely modified enzyme was reached after an average of three subunits per hexamer had been modified by pyridoxal 5′-phosphate at lysine-126. Similar cooperativity was also reported in the inactivation by reaction of lysine-126 with 4-amino-6-chloro-5-oxohexanoic acid: 50% inactivation is caused by modification of 17 to 20% of lysine-126.[99] These results suggest that interactions among subunits can influence the effects of chemical modification, a theme frequently encountered in covalent reactions of glutamate dehydrogenase.

Another lysine, residue 27, has been implicated in the catalytic function of the enzyme.[100] Glyoxal is a reversible inhibitor competitive with α-ketoglutarate. Upon trapping the Schiff base intermediate by reduction with NaB^3H_4, the inactivation becomes irreversible. Although small amounts of radioactivity were distributed in several peptides (~10% each), the predominant incorporation (44%) occurred in the peptide containing lysine-27. These results were interpreted to indicate that lysine-27 participates in the binding of the dicarboxylate substrate, α-ketoglutarate.[100]

Differential reaction of enzyme with iodoacetate at pH 7.0 led to inactivation concomitant with the incorporation of about 1 mol reagent per peptide chain.[101] In contrast to many dehydrogenases, glutamate dehydrogenase has not been shown to have essential, reactive cysteines. Use of radioactive iodoacetate led to recovery of 50% of the radioactivity in a peptide including modified methionine-169, with only 7% of the label being incorporated into cysteine-115 and lesser amounts in other peptides. Since striking protection against inactivation was provided by NADH plus α-ketoglutarate, the authors concluded that methionine-169 is also located in the active site.[101] This may be the same methionine residue earlier reported to be responsible for the inactivation produced by 4-iodoacetamidosalicyclic acid at pH 6.0.[102]

B. NADH REGULATORY SITE

Regulatory sites would not be expected to feature amino acid residues of unusual reactivity since they are not direct participants in the catalytic reaction. Therefore, it is understandable that successful labeling of regulatory sites in glutamate dehydrogenase has predominantly been accomplished by affinity labeling involving the use of compounds structurally related to the natural regulatory compounds, each of which has an additional reactive functional group capable of covalent reaction with amino acid side chains.[103] One such compound is 5′-*p*-fluorosulfonyl benzoyl adenosine (5′-FSBA) used to label specifically the NADH inhibitory site of glutamate dehydrogenase.[104,105] This compound may be considered as an analogue of NADH: in addition to the adenine and ribose moieties, it has a carbonyl group adjacent to the 5′-position which is structurally similar to the first phosphoryl group of natural purine nucleotides and, if the molecule is arranged in an extended conformation, the sulfonyl fluoride may be located in a position equivalent to the ribose proximal to the nicotinamide of NaDH. The sulfonylfluoride can act as an electrophile in covalent reactions with several classes of amino acids including tyrosine and lysine. 5′-FSBA reacts irreversibly with glutamate dehydrogenase with incorporation of up to 1 mol reagent per mole enzyme subunit.[104] Modified enzyme exhibits total loss of the inhibition by high concentrations of NADH, but retains full catalytic activity as measured in the absence of allosteric ligands. This modified enzyme is still inhibited more than 90% by GTP and is activated normally by ADP, indicating that the loss of NADH inhibition is the only major aberration in its kinetic characteristics. The rate constant for reaction is decreased markedly by high concentrations of NADH and completely by high levels of NADH plus low concentrations of GTP,[104,105] consistent with modification of the NADH regulatory site. The enzyme becomes unresponsive to NADH inhibition when an average of only 0.5 mol reagent per mol enzyme subunit (or 3 mol/mol hexamer) have been incorporated; and under these conditions equal amounts of modified tyrosine and lysine were found throughout the reaction.[105] It was concluded that both tyrosine and lysine are present in the NADH inhibitory site close to the sulfonyl fluoride of the bound reagent and that covalent modification of either residue on three of the six peptides of hexameric enzyme is sufficient to eliminate NADH inhibition. The two nucleosidyl peptides isolated from modified enzyme were identified as containing derivatized tyrosine-190 and lysine-420, respectively.[106] Thus, although these residues are separated by almost half the protein in the linear sequence, they appear to be close to each other in the folded structure.

An earlier study using the general lysine reagent trinitrobenzenesulfonate reported that rapid modification of lysine-423 on three of the six enzyme subunits caused complete loss of NADH inhibition, along with small changes in other kinetic parameters;[107,108] reaction with lysine-420 on the other three subunits occurred more slowly. These experiments are consistent with those involving 5′-FSBA[105,106] in implicating the region of residues 420 to 423 in the NADH regulatory site. A recent study using the 2′,3′-dialdehyde derivative of NADPH (oNADPH) as a reagent for glutamate dehydrogenase showed that about three sites per enzyme hexamer were specifically modified along with loss of NADH inhibition.[85] This

TABLE 1
Amino Acid Residues Identified with Particular Sites of Glutamate Dehydrogenase

Catalytic site	NADH inhibitory site	GTP inhibitory site	ADP activating site	Steroid inhibitory site
Lys-27				
			His-82	
				Cys-89
Lys-126				
Met-169				
	Tyr-190			
		Tyr-262		
	Cys-319			
		Tyr-407		
	Lys-420			Lys-420
	Lys-423			Lys-423

modified enzyme exhibited a decreased affinity for GTP, but was still inhibited 93% by saturating concentrations of GTP. Protection against the change in kinetic properties caused by oNADPH was provided by high concentrations of NADH in the presence of GTP, as in the case of 5′-FSBA,[105] suggesting that oNADPH might be labeling the same NADH regulatory sites. Since the dialdehyde should be proximal to the amino acid residues in the binding site of the *cis*-hydroxyls of the nicotinamide ribose, it is likely that characterization of the peptide specifically labeled by oNADPH (currently in progress) will lead to an independent identification of a lysine residue in this part of the NADH regulatory site.

Incubation of glutamate dehydrogenase with another type of reactive nucleotide analogue has allowed identification of a cysteinyl residue in a different region of the NADH inhibitory site.[109,110] The compound 6-(4-bromo-2,-3-dioxobutyl)-thioadenosine 5′-diphosphate (6-BDB-TADP) is closely related to the adenine nucleotide structure, is negatively charged at neutral pH and has a reactive bromodioxobutyl group capable of reaction with several nucleophilic side chains of amino acids. Because of the location of the reactive functional group adjacent to the 6-position, the compound might be expected to react with amino acid residues in the purine subsite of nucleotide binding sites. Glutamate dehydrogenase reacts covalently with up to one mol 6-BDB-TADP/peptide chain with specific elimination of NADH inhibition.[109] The rate constant for loss of NADH inhibition exhibits a nonlinear dependence on reagent concentration, suggesting a reversible binding of reagent prior to irreversible modification. Isolation of a 19-membered cysteinyl peptide from glu-311 to lys-329 indicated that cysteine-319 was the target of 6-BDB-TADP.[110] Thus, specific chemical modification studies, as summarized in Table 1, have implicated four amino acid residues from different parts of the linear sequence as participants in the NADH regulatory site. This type of data may provide important constraints for assessing in solution the three-dimensional structure of glutamate dehydrogenase.

C. GTP REGULATORY SITE

The enzyme has been selectively desensitized to GTP inhibition both by general chemical modification and by affinity labeling. Incubation with tetranitromethane led to a time-dependent formation of 3 mol of 3-nitrotyrosine per mole subunit after 1 h; however, one tyrosine was considered to react more rapidly than the others.[10,111,112] Reaction with the most rapidly reacting tyrosine correlated with an increased velocity measured in the presence of GTP, although the activity, as measured in the absence of GTP, decreased as a result of the slower modifications. Derivatization of the most reactive tyrosine, identified as tyrosine-

407,[111] was held responsible for the partial loss of GTP inhibition, but since there was neither extensive evaluation of the requirements for protection against the modification nor characterization of the affinity of the modified enzyme for GTP, it is difficult to assess whether one of the two GTP sites of the enzyme was attacked. It should also be pointed out that this enzyme was not tested for sensitivity to NADH inhibition and therefore one cannot exclude the possibility of a primary modification of the NADH regulatory site with an indirect effect on GTP affinity.

Glutamate dehydrogenase is inhibited allosterically both by GTP and by 1,N^6-ethenoadenosine 5′-triphosphate (εATP); thus, it is not surprising that the enzyme is covalently modified in a functionally similar manner by the GTP analogue, 5′-*p*-fluorosulfonylbenzoyl guanosine[113,88] (FSBG) and the fluorescent εATP analogue, 5′-*p*-flurosulfonylbenzoyl-1,N^6-ethenoadenosine[74] (FSBεA). The enzyme is not inactivated by reaction with FSBG[88] or FSBεA,[74] as measured in the absence of regulatory compounds, but exhibits a time-dependent increase in the activity assayed in the presence of 1 μ*M* GTP. As compared to native glutamate dehydrogenase, the modified enzyme exhibits a decreased affinity for and diminished maximum inhibition by saturating concentrations of GTP, a decreased maximum extent of activation with no change in affinity for ADP, and a normal ability to be inhibited by high NADH concentrations. In contrast to the 2 mol of GTP bound reversibly by native enzyme, only one mol of GTP is bound per peptide chain of the modified enzyme in the absence or presence of NADH. This result implies that one of the natural GTP sites is eliminated as a result of reaction with either FSBG or FSBεA.[74,88] A nonlinear dependence of the rate of reaction on the FSBεA concentration suggests a reversible binding prior to irreversible modification. The rate constant is not appreciably decreased by the substrate α-ketoglutarate, by high concentrations of NADH or by ADP. A decrease in the rate constant is caused by added GTP, and complete protection is provided by 100 μM NADH plus GTP concentrations as low as 10 to 25 μ*M*, indicating that the high affinity GTP site is attacked by these reagents. The change in sensitivity to GTP inhibition correlates with incorporation of FSBεA, with the complete functional change occurring at about 1 mol reagent per enzyme subunit,[74,76] which has been attributed quantitatively to the *O*-[(4-carboxyphenyl)sulfonyl]tyrosine detected after acid hydrolysis.[76] The tyrosyl peptide labeled by 5′-FSBεA has been isolated and the modified residue identified as tyrosine-262.[114] Studies of the conformation of the fluorescent 5′-FSBεA bound to enzyme indicate that the purine and the benzoyl groups are stacked.[115] It has therefore been proposed that the essential tyrosine-262 modified by FSBεA on glutamate dehydrogenase is involved in the binding of the guanine moiety of GTP.

The distance between two chromophores on a protein can be determined by fluorescence energy transfer provided that the emission spectrum of the donor chromophore overlaps the absorption spectrum of the acceptor chromorphore and that the two species are sufficiently close to each other.[116] For glutamate dehydrogenase, the -SBεA group, with its fluorescence emission maximum at 418 nm, can be used as the energy donor located at the GTP site.[76] The nucleotide analogue 2′,3′-*O*-(2,4,6-trinitrocyclohexadienylidene)-ADP (TNP-ADP) is a satisfactory probe of the natural ADP activating site of glutamate dehydrogenase since it activates the enzyme twofold and competes for the binding of radioactive ADP. TNP-ADP absorbs maximally at 405 nm and, as the energy acceptor, exhibits excellent spectral overlap with the -SBεA-modified enzyme. The fluorescence at 405 nm of the FSBεA-modified glutamate dehydrogenase is progressively quenched by increasing concentrations of TNP-ADP bound reversibly to the ADP activating site. The efficiency of energy transfer is 77%, which allows the calculation of 18 Å as an average distance between the -SBεA and TNP-ADP sites.[76] Clearly, the enzyme labeled at the high affinity GTP site can still bind an ADP analogue, but it is apparent that these GTP and ADP sites are relatively close together.[76]

Glutamate dehydrogenase modified by iodoacetamidosalicylic acid at lysine-126 is also

TABLE 2
Summary of Distances Between Energy Donor-Acceptor Pairs on Glutamate Dehydrogenase Measured by Resonance Energy Transfer[42,44]

Donor[a]	Location	Acceptor[b]	Location	Distance (Å)
-SBεA	GTP Site (Tyr-262)	TNP-ADP	ADP Site	18
ISA	Catalytic site (Lys-126)	TNP-ADP	ADP Site	33
ISA	Catalytic site (Lys-126)	-SBaεA	GTP Site	23

[a] -SBεA, 5′-*p*-sulfonylbenzoyl-1,N[6]-ethenoadenosine; ISA, 4-(iodoacetamido)salicylic acid.
[b] TNP-ADP, 2′,3′-*o*-(2,4,6-trinitrocyclohexadienylidene)-ADP; -SBaεA, 5′-*p*-sulfonylbenzoyl-1,N[6]-2-azaethenoadenosine.

fluorescent, with an emission peak at 400 nm; this fluorescence overlaps the absorption spectrum of TNP-ADP.[92] From the quenching of the acetamidosalicylate donor fluorescence upon addition of TNP-ADP, an average distance of 33 Å was estimated between the catalytic and ADP sites.[92] Furthermore, glutamate dehydrogenase reacts covalently with another fluorescent nucleotide analogue, 5′-*p*-fluorosulfonylbenzoyl-2-aza-ethenoadenosine (FSBaεA) to yield enzyme with about one mol reagent incorporated per peptide chain. As compared to native enzyme, this -SBaεA-enzyme exhibits decreased sensitivity to GTP inhibition, but retains its catalytic activity as well as its ability to be activated by ADP and inhibited by high concentrations of NADH.[92] It was concluded that FSBaεA, like FSBεA, labels a GTP site in glutamate dehydrogenase, but the absorption spectrum of the -SBaεA- modified enzyme occurs at longer wavelength and overlaps the fluorescence emission spectrum of the ISA-modified enzyme, allowing the estimation of 23 Å between the GTP and catalytic sites in the doubly labeled enzyme.[92] The distance relationships among the labeled sites determined by resonance energy transfer measurements are summarized in Table 2. The published experiments do not provide information on the question of whether these distances are between sites of a single subunit or whether, instead, they span subunits. It is notable, however, that all the distances are smaller than the diameter of both the major and minor axes of a subunit,[12,14] making it possible that these represent distances for intrasubunit sites. The distance of 33 Å between the ADP and catalytic sites and 23 Å between the GTP and catalytic sites suggests that the regulatory sites are closer to each other as compared with their distances from the catalytic site. These distance measurements are consistent with the allosteric model of regulation of glutamate dehydrogenase whereby the regulatory nucleotides are bound at distinct sites and indirectly influence the catalytic reaction.

D. ADP REGULATORY SITE

Less has been learned directly, thus far, by covalent labeling of the ADP site. Glutamate dehydrogenase reacts irreversibly with the adenine nucleotide analogue 2-(4-bromo-2,3-dioxobutylthio)adenosine 5′-monophosphate (2-BDB-TAMP) with incorporation of about 1 mol of reagent per mole of enzyme subunit.[87] Reaction with the 5′-diphosphate form of the compound proceeds similarly, but reflects greater affinity between the enzyme and reagent. The modified enzyme is not inactivated by this reaction as measured in the absence of allosteric effectors, and is still inhibited 92% by added GTP. However, this altered enzyme exhibits a marked decrease in the extent of activation by ADP, as well as a decrease in the number of ADP sites (measured by reversible binding of [^{14}C]ADP) from two to one per enzyme subunit.[87] In addition to the defect in its response to ADP, this modified enzyme is not inhibited by high concentrations of NADH. It is considered that 2-BDB-TAMP acts as an affinity label of an ADP site and indirectly influences the NADH inhibitory site because ADP (but not NADH) protects the enzyme against all the functional changes produced by

2-BDB-TAMP.[87] The high affinity ADP site is thought to be the target of 2-BDB-TAMP since protection is observed in the low ADP concentration range expected for $K_d = 0.67$ μM.[87] That the site of labeling of glutamate dehydrogenase by 6-(4-bromo-2,3-dioxobutyl)-thioadenosine 5′-diphosphate is different from the ADP site attacked by 2-BDB-TAMP is indicated by the lack of complete protection by ADP against 6-BDB-TADP.[109] Indeed, the difference between the modes of action of 2-BDB-TAMP and 6-BDB-TADP toward glutamate dehydrogenase reflect the great specificity for binding at the activating site, as illustrated by the observation that a compound in which the free 6-NH_2 is altered, such as 1,N^6-ethenoadenosine 5′-diphosphate, appears to bind to the GTP inhibitory rather than the ADP activating site.[73,74] Recently a peptide has been isolated from enzyme modified by 2-BDB-TAMP, and its amino acid sequence has been ascertained.[121] The modified residue has been identified as histidine-82. (In the course of this work it was also found that the positions of glutamine-84 and histidine-85 had been given as reversed in the revised sequence of bovine liver glutamate dehydrogenase.[11])

It may be relevant to the labeling of histidine-82 by 2-BDB-TAMP that ethoxyformylation of a single histidine per subunit by diethylpyrocarbonate was previously reported to mimic the activating effects of ADP.[117] An attempt to conduct photoaffinity labeling of the ADP site using 8-azido-ADP has been described.[118] However, the total amount of labeling was limited to about 6% of the polypeptide chains, so that the modified enzyme could not readily be characterized.

E. STEROID REGULATORY SITE

Several steroid hormones, including estradiol and diethylstilbesterol, have been found to reversibly inhibit glutamate dehydrogenase activity and were described as allosteric inhibitors which could be reversed by ADP, although their relationship to other regulatory sites and their physiological importance has remained controversial.[2] Incubation of glutamate dehydrogenase with 3-[2′-^{14}C]acetyl-2-nitroestrone leads to covalent incorporation of the acetyl group into the enzyme, probably by transacetylation.[119] Enzyme with 1.3 mol acetyl incorporated per enzyme subunit retained full glutamate dehydrogenase activity, but exhibited about a twofold weakening of its affinity for ADP and GTP as well as a partial desensitization to estradiol. About 80 to 90% of the radioactivity was located in a peptide containing both lysine-420 and lysine-423 as modified residues.[119] The effect of this acetylation appears to be similar to that produced by the reaction of glutamate dehydrogenase with trinitrobenzene sulfonate,[107,108] but the acetyl derivatives have not been characterized as well.

Iodoacetylstilbesterol was evaluated as an affinity label of the putative estrogen binding site.[120] The reagent causes inactivation, but the protection by estradiol or diethylstilbesterol was not remarkable. At an incorporation of ≤ 1 mol reagent per peptide chain, 60 to 70% of the radioactivity was recovered in a peptide containing modified cysteine-89.[120] Although the cysteine residues of glutamate dehydrogenase do not react with iodoacetic acid or iodoacetamide under these conditions,[101] it is unclear whether the diethylstilbesterol directs the reaction because it is a hydrophobic carrier of a reactive group or because there is a specific steroid binding site.

VI. CONCLUDING REMARKS

Compiled in Table 1 are the amino acid residues which have been associated by chemical modification experiments with particular sites of glutamate dehydrogenase. The three residues identified with the catalytic site are located in the amino terminal portion of the enzyme. This observation is consistent with the finding of greatest homology in the amino terminal half of the molecule when the known sequences of glutamate dehydrogenases from all sources are compared. A necessary (albeit not sufficient) criterion for a contribution of a given amino acid to catalysis is that the residue be conserved in the glutamate dehydrogenases of all species. A lysine equivalent to the bovine lysine-126 has been recognized in the enzymes from *E. coli*, yeast and *N. crassa*, as well as from the vertebrates human, rat and chicken.[1,43,46] A lysine equivalent to lysine-27 can be located in the vertebrate and *Neurospora* enzymes, and is replaced conservatively by an arginine in the *E. coli* enzyme. A role for these residues in the binding site of the negatively charged substrate seems reasonable. This criterion is less supportive of an essential role for methionine-169: although this methionine is present in all the vertebrate enzymes,[1] it is missing in the bacterial and fungal enzymes.

The amino acid residues associated with the regulatory sites are predominantly found in the middle or carboxyl terminal half of the molecule, the exceptions being histidine-82 and cysteine-89, which are close to each other. The identification of four residues in the NADH inhibitory site suggests that this regulatory domain is constituted by the folding of different regions of the linear sequence of the enzyme. The relative proximity between residues in the GTP and NADH inhibitory sites is interesting in view of the mutual interactions between these sites; and labeling of lysine-420 and -423 by a putative steroid affinity label raises the possibility that the steroid and NADH inhibitory sites are identical or overlapping.

The vertebrate glutamate dehydrogenases are all subject to allosteric regulation by the purine nucleotides, whereas the bacterial and fungal enzymes are not so regulated. Thus, all of the amino acid residues required for these allosteric functions must be conserved in the vertebrate enzymes, but may differ in the bacterial and fungal enzymes. The residues equivalent to histidine-82, cysteine-89, tyrosine-190, tyrosine-262, cysteine-319, tyrosine-407, lysine-420, and lysine-423 are all present in bovine, chicken and human glutamate dehydrogenases,[1,11] but they cannot be located in the enzymes from *Neurospora*[1] or *E. coli*.[46] These eight residues, implicated by chemical modification in the regulatory functions of the enzyme, thus fulfill the necessary (but not sufficient) criterion of conservation among species that exhibit allosteric regulation of glutamate dehydrogenase.

The sites listed in Table 1 are so designated for simplicity; however, they should not be viewed as rigid, discrete loci. There are numerous interactions among sites, the perturbation (reversible) or covalent chemical modification of one site frequently produces complex effects on other binding sites. Furthermore, interactions between subunits can be important, as in the cases where modification of only some of the subunits by 4-amino-6-chloro-5-oxohexanoic acid,[99] trinitrobenzenesulfonate,[107,108] 5′-*p*-fluorosulfonylbenzoyl adenosine,[105] or the 2′,3′-dialdehyde derivative of NADPH[85] cause the total change in the function of all subunits. Any role proposed for a particular amino acid residue will have to be evaluated in terms of the emerging information on the crystal structure of glutamate dehydrogenase and should also be reexamined by site directed mutagenesis coupled with detailed characterization of the catalytic, binding and regulatory properties of the altered enzymes.

REFERENCES

1. **Smith, E. L., Austen, B. M., Blumenthal, K. M., and Nyc, J. F.,** Glutamate Dehydrogenases, in *The Enzymes,* Vol. 9, 3rd ed., Boyer, P. D., Ed., Academic Press, New York, 1975, 293.
2. **Goldin, B. R. and Frieden, C.,** L-Glutamate dehydrogenases, *Curr. Top. Cell. Regul.,* 4, 77, 1972.
3. **Fisher, H. F.,** Glutamate dehydrogenase — ligand complexes and their relationship to the mechanism of the reaction, *Adv. Enzymol.,* 39, 369, 1973.
4. **Sund, H., Markau, K., and Koberstein, R.,** Glutamate dehydrogenase, in *Subunits in Biological Systems,* Vol. 7c, Timasheff, S. N. and Fasman, G. D., Eds., Marcel Dekker, New York, 1975, 225.
5. **Eisenberg, H., Josephs, R., and Reisler, E.,** Bovine liver glutamate dehydrogenase, *Adv. Protein Chem.,* 30, 101, 1976.
6. **Fisher, H. F.,** L-Glutamate dehydrogenase from bovine liver, *Methods Enzymol.,* 113, 16, 1985.

7. **Appella, E. and Tomkins, G. M.,** The subunits of bovine liver glutamate dehydrogenase: demonstration of a single peptide chain, *J. Mol. Biol.*, 18, 77, 1966.
8. **Cassman, M. and Schachman, H. K.,** Sedimentation equilibrium studies on glutamic dehydrogenase, *Biochemistry,* 10, 1015, 1971.
9. **Marler, E. and Tanford, C.,** The molecular weight of the polypeptide chains of L-glutamate dehydrogenase, *J. Biol. Chem.*, 239, 4217, 1964.
10. **Smith, E. L., Landon, M., Piszkiewicz, D., Brattin, W. J., Langley, T. J., and Melamed, M. D.,** Bovine liver glutamate dehydrogenase: tentative amino acid sequence; identification of a reactive lysine; nitration of a specific tyrosine and loss of allosteric inhibition by guanosine triphosphate, *Proc. Natl. Acad. Sci. U.S.A.*, 67, 724, 1970.
11. **Julliard, J. H. and Smith, E. L.,** Partial amino acid sequence of the glutamate dehydrogenase of human liver and a revision of the sequence of the bovine enzyme, *J. Biol. Chem.*, 254, 3427, 1979. (This is the most recent revision of the sequence and the one that is referred to in this chapter.)
12. **Josephs, R.,** Electron microscope studies on glutamic dehydrogenase: subunit structure of individual molecules and linear associates, *J. Mol. Biol.*, 55, 147, 1971.
13. **Fiskin, A. M., Bruggen, E. F. J., and Fisher, H. F.,** Structure and function of oligomeric dehydrogenases. Electron microscopic studies of beef liver glutamate dehydrogenase quaternary structure, *Biochemistry,* 10, 2396, 1971.
14. **Eisenberg, H.,** Glutamate dehydrogenase: anatomy of a regulatory enzyme, *Acc. Chem. Res.*, 4, 379, 1971.
15. **Hucho, F., Rasched, I., and Sund, H.,** Studies of glutamate dehydrogenase: analysis of functional areas and functional groups, *Eur. J. Biochem.*, 52, 221, 1975.
16. **Tashiro, R., Inoue, T., and Shimozawa, R.,** Subunit dissociation and unfolding of bovine liver glutamate dehydrogenase induced by guanidine hydrochloride, *Biochim. Biophys. Acta,* 706, 129, 1982.
17. **Inoue, T., Fukushima, K., Tastumoto, T., and Shimozawa, R.,** Light scattering study on subunit association-dissociation equilibria of bovine liver glutamate dehydrogenase, *Biochim. Biophys. Acta,* 786, 144, 1984.
18. **Bell, E. T. and Bell, J. E.,** Catalytic activity of bovine glutamate dehydrogenase requires a hexamer structure, *Biochem. J.*, 217, 327, 1984.
19. **Frieden, C.,** Glutamic dehydrogenase. I. The effect of coenzyme on the sedimentation velocity and kinetic behavior, *J. Biol. Chem.*, 234, 809, 1959.
20. **Frieden, C.,** Glutamic dehydrogenase. II. The effect of various nucleotides on the association-dissociation and kinetic properties, *J. Biol. Chem.*, 234, 815, 1959.
21. **Frieden, C.,** The effect of pH and other variables on the dissociation of beef liver glutamic dehydrogenase, *J. Biol. Chem.*, 237, 2396, 1962.
22. **Frieden, C.,** Different structural forms of reversibly dissociated glutamic dehydrogenase: relation between enzymatic activity and molecular weight, *Biochem. Biophys. Res. Commun.*, 10, 410, 1963.
23. **Cohen, R. J. and Benedek, G. B.,** The functional relationship between polymerization and catalytic activity of beef liver glutamate dehydrogenase I. Theory, *J. Mol. Biol.*, 108, 151, 1976.
24. **Cohen, R. J., Jedziniak, J. A., and Benedek, G. B.,** The functional relationship between polymerization and catalytic activity of beef liver glutamate dehydrogenase II. Experiment, *J. Mol. Biol.*, 108, 179, 1976.
25. **Fisher, H. F., Cross, D. G., and McGregor, L. L.,** Catalytic activity of subunits of glutamic dehydrogenase, *Nature,* 196, 895, 1962.
26. **Fisher, H. F., Cross, D. G., and McGregor, L. L.,** The independence of the substrate specificity of glutamate dehydrogenase on the state of aggregation, *Biochim. Biophys. Acta,* 99, 165, 1965.
27. **Frieden, C. and Colman, R. F.,** Glutamate dehydrogenase concentration as a determinant in the effect of purine nucleotides on enzymatic activity, *J. Biol. Chem.*, 242, 1705, 1967.
28. **Colman, R. F. and Frieden, C.,** Cooperative interaction between the GTP binding sites of glutamate dehydrogenase, *Biochem. Biophys. Res. Commun.*, 22, 100, 1966.
29. **Huang, C. Y. and Frieden, C.,** Rates of GDP-induced and GTP-induced depolymerization of glutamate dehydrogenase: a possible factor in metabolic regulation, in *Proc. Natl. Acad. Sci. U.S.A.*, 64, 338, 1969.
30. **Birktoft, J. J., Miake, F., Banaszak, L. J., and Frieden, C.,** Crystallographic studies of glutamate dehydrogenase. Preliminary crystal data, *J. Biol. Chem.*, 254, 4915, 1979.
31. **Birktoft, J. J., Miake, F., Frieden, C., and Banaszak, L. J.,** Crystallographic studies of glutamate dehydrogenase. II. Preliminary crystal data for the tuna liver enzyme, *J. Mol. Biol.*, 138, 145, 1980.
32. **Rice, D. W., Hornby, D. P., and Engel, P. C.,** Crystallization of an NAD^+-dependent glutamate dehydrogenase from *Clostridium symbiosum, J. Mol. Biol.*, 181, 147, 1985.
33. **Rice, D. W., Baker, P. J., Farrants, G. W., and Hornby, D. P.,** The crystal structure of glutamate dehydrogenase from *Clostridium symbiosum* at 0.6 nm resolution, *Biochem. J.*, 242, 789, 1987.
34. **Birktoft, J. J., Bradshaw, R. A. and Banaszak, L. J.,** Structure of porcine heart cytoplasmic malate dehydrogenase: Combining X-ray diffraction and chemical sequence data in structural studies, *Biochemistry,* 26, 2722, 1987.

35. **Delabar, J. M., Martin, S. R., and Bayley, P. M.,** The binding of NADH and NADPH to bovine liver glutamate dehydrogenase. Spectroscopic characterization, *Eur. J. Biochem.,* 127, 367, 1982.
36. **Moon, K. and Smith, E. L.,** Sequence of bovine liver glutamate dehydrogenase. VIII. Peptides produced by specific chemical cleavages; the complete sequence of the protein, *J. Biol. Chem.,* 248, 3082, 1973.
37. **Moon, K., Piszkiewicz, D., and Smith, E. L.,** Amino acid sequence of chicken liver glutamate dehydrogenase, *J. Biol. Chem.,* 248, 3093, 1973.
38. **Coffee, C. J. and Frieden, C.,** unpublished data, as reported in Reference 1.
39. **McCarthy, A. D., Walker, J. M., and Tipton, K. F.,** Purification of glutamate dehydrogenase from ox brain and liver, *Biochem. J.,* 191, 605, 1980.
40. **Holder, A. A., Wootton, J. C., Baron, A. J., Chambers, G. K., and Fincham, J. R. S.,** The amino acid sequence of *Neurospora* NADP-specific glutamate dehydrogenase, *Biochem. J.,* 149, 757, 1975.
41. **Blumenthal, K. M., Moon, K., and Smith, E. L.,** NADP-specific glutamate dehydrogenase of *Neurospora, J. Biol. Chem.,* 250, 3644, 1975.
42. **Kinnard, J. H. and Fincham, J. R. S.,** The complete nucleotide sequence of the *Neurospora crassa* am (NADP-specific glutamate dehydrogenase) gene, *Gene,* 26, 253, 1983.
43. **Moye, W. S., Amuro, N., Rao, J. K., and Zalkin, H.,** Nucleotide sequence of yeast GDH 1 encoding NADP-dependent glutamate dehydrogenase, *J. Biol. Chem.,* 260, 8502, 1985.
44. **Nagasu, T. and Hall, B. D.,** Nucleotide sequence of the GDH gene coding for the NADP-specific glutamate dehydrogenase of *Saccharomyces cerevisiae, Gene,* 37, 247, 1985.
45. **McPherson, M. J. and Wootton, J. C.,** Complete nucleotide sequence of the *Escherichia coli* gdhA gene, *Nucleic Acids Res.,* 11, 5257, 1983.
46. **Valle, F., Becerril, B., Chen, E., Seeburg, P., Heyneker, H., and Bolivar, F.,** Complete nucleotide sequence of the glutamate dehydrogenase gene from *Escherichia coli* K-12, *Gene,* 27, 193, 1984.
47. **Wootton, J. C.,** The coenzyme-binding domains of glutamate dehydrogenases, *Nature,* 252, 542, 1974.
48. **Austen, B. M., Haberland, M. E., and Smith, E. L.,** Secondary structure predictions for the NAD-specific glutamate dehydrogenase of *Neurospora crassa, J. Biol. Chem.,* 255, 8001, 1980.
49. **Uno, I., Matsumoto, K., Adachi, K., and Ishikawa, T.,** Regulation of NAD-dependent glutamate dehydrogenase by protein kinases in *Saccharomyces cerevisiae, J. Biol. Chem.,* 259, 1288, 1984.
50. **Levy, H. R., Ejchart, A., and Levy, G. C.,** Conformations of nicotinamide coenzymes bound to dehydrogenases determined by transferred nuclear Overhauser effects, *Biochemistry,* 22, 2792, 1983.
51. **Alex, S. and Bell, J. E.,** Dual nucleotide specificity of bovine glutamate dehydrogenase, *Biochem. J.,* 191, 299, 1980.
52. **Male, K. B. and Storey, K. B.,** Regulation of coenzyme utilization by bovine liver glutamate dehydrogenase: investigations using thionicotinamide analogues of NAD and NADP in a dual wavelength assay, *Int. J. Biochem.,* 14, 1083, 1982.
53. **Hornby, D. P., Engel, P. C., and Hatanaka, S.-I.,** Beef liver glutamate dehydrogenase: a study of the oxidation of various alternative amino acid substrates retaining the correct spacing of the two carboxylate groups, *Int. J. Biochem.,* 15, 495, 1983.
54. **Rife, J. E. and Cleland, W. W.,** Determination of the chemical mechanism of glutamate dehydrogenase from pH studies, *Biochemistry,* 19, 2328, 1980.
55. **Rife, J. E. and Cleland, W. W.,** Kinetic mechanism of glutamate dehydrogenase, *Biochemistry,* 19, 2321, 1980.
56. **Li Muti, C. and Bell, J. E.,** A steady-state random order mechanism for the oxidative deamination of norvaline by glutamate dehydrogenase, *Biochem. J.,* 211, 99, 1983.
57. **Bell, E. T., Li Muti, C., Renz, C. L., and Bell, J. E.,** Negative cooperativity in glutamate dehydrogenase, *Biochem. J.,* 225, 209, 1985.
58. **Hochreiter, M. C., Patek, D. R., and Schellenberg, K. A.,** Catalysis of α-iminoglutarate formation from α-ketoglutarate and ammonia by bovine glutamate dehydrogenase, *J. Biol. Chem.,* 247, 6271, 1972.
59. **Fisher, H. F. and Viswanathan, T. S.,** Carbonyl oxygen exchange evidence of imine formation in the glutamate dehydrogenase reaction and identification of the "occult role" of NADPH, *Proc. Natl. Acad. Sci. U.S.A.,* 81, 2747, 1984.
60. **Fisher, H. F., Srinivasan, R., and Rougvie, A. E.,** Glutamate dehydrogenase catalyzes the reduction of a Schiff base (Δ'-pyrroline-2-carboxylic acid) by NADPH, *J. Biol. Chem.,* 287, 13208, 1982.
61. **Srinivasan, R. and Fisher, H. F.,** Configurational, conformational and solvent effects on the reduction of a Schiff base by reduced pyridine nucleotide analogs, *Arch. Biochem. Biophys.,* 223, 453, 1983.
62. **Srinivasan, R. and Fisher, H. F.,** Comparison of the energetics of the uncatalyzed and glutamate dehydrogenase catalyzed α-imino acid-α-amino acid interconversion, *Bichemistry,* 24, 5356, 1985.
63. **Srinivasan, R. and Fisher, H. F.,** Reversible reduction of an α-imino acid to an α-amino acid catalyzed by glutamate dehydrogenase: effect of ionizable functional groups, *Biochemistry,* 24, 618, 1985.
64. **Srinivasan, R. and Fisher, H. F.,** Structural features facilitating the glutamate dehydrogenase catalyzed α-imino acid-α-amino acid interconversion, *Arch. Biochem. Biophys.,* 246, 743, 1986.

65. **Frieden, C.,** Glutamate dehydrogenase. V. The relation of enzyme structure to catalytic function, *J. Biol. Chem.,* 238, 3286, 1963.
66. **Frieden, C.,** Glutamate dehydrogenase. VI. Survey of purine nucleotide and other effects on the enzyme from various sources, *J. Biol. Chem.,* 240, 2028, 1965.
67. **Pons, M., Michel, F., Descomps, B., and Crastes de Paulet, A.,** Structural requirements for maximal inhibitory allosteric effect of estrogens and estrogen analogues on glutamate dehydrogenase, *Eur. J. Biochem.,* 84, 257, 1978.
68. **Colman, R. F. and Foster, D. F.,** The absence of zinc in bovine liver glutamate dehydrogenase, *J. Biol. Chem.,* 245, 6190, 1970.
69. **Fahien, L. A. and Kmiotek, E.,** Regulation of glutamate dehydrogenase by palmitoyl-coenzyme A, *Arch. Biochem. Biophys.,* 212, 247, 1981.
70. **Bailey, J., Bell, E. T., and Bell, J. E.,** Regulation of bovine glutamate dehydrogenase. The effects of pH and ADP, *J. Biol. Chem.,* 259, 5579, 1982.
71. **Cross, D. G. and Fisher, H. F.,** The mechanism of glutamate dehydrogenase reaction. III. The binding of ligands at multiple subsites and resulting kinetic effects, *J. Biol. Chem.,* 245, 2612, 1970.
72. **Subramanian, S., Stickel, D. C., and Fisher, H. F.,** Thermodynamics of complex formation between bovine liver glutamate dehydrogenase and analogs of ADP, *J. Biol. Chem.,* 250, 5885, 1975.
73. **Lascu, I., Bârzu, T., Ty, N. G., Ngoc, L. D., Bârzu, O., and Mantsch, H. H.,** Regulatory effects of purine nucleotide analogs with liver glutamate dehydrogenase, *Biochim. Biophys. Acta,* 482, 251, 1977.
74. **Jacobson, M. A. and Colman, R. F.,** Affinity labeling of a GTP site of glutamate dehydrogenase by a fluorescent nucleotide analogue, 5′-[*p*-(fluorosulfonyl)benzoyl]-1,N^6-ethenoadenosine, *Biochemistry,* 21, 2177, 1982.
75. **Hiratsuka, T.,** A chromophoric and fluorescent analog of GTP, 2′,3′-*O*-(2,4,6-trinitrocyclohexadienylidene)-GTP, as a spectroscopic probe for the GTP inhibitory site of liver glutamate dehydrogenase, *J. Biol. Chem.,* 260, 4784, 1985.
76. **Jacobson, M. A. and Colman, R. F.,** Resonance energy transfer between the ADP site of glutamate dehydrogenase and a GTP site containing a tyrosine labeled with 5′-[*p*-fluorosulfonyl)benzoyl]-1,N^6-ethenoadenosine, *Biochemistry,* 22, 4247, 1983.
77. **Pantaloni, D. and Dessen, P.,** Glutamate déshydrogénase. Fixations des coenzymes NAD et NADP et d'autres nucleotides derives de l'adénosine-5′-phosphate, *Eur. J. Biochem.,* 11, 510, 1969.
78. **Koberstein, R., Krause, J., and Sund, H.,** Studies of glutamate dehydrogenase. The interaction of glutamate dehydrogenase with α-NADH, *Eur. J. Biochem.,* 40, 543, 1973.
79. **Pantaloni, D. and Lecuyer, B.,** Glutamate dehydrogenase. Caracterisation et etude thermodynamique des differents complexes formés avec les coenzymes et substrats: role de effecteurs ADP et GTP, *Eur. J. Biochem.,* 40, 381, 1973.
80. **Koberstein, R. and Sund, H.,** Studies of glutamate dehydrogenase. The influence of ADP, GTP and L-glutamate on the binding of the reduced coenzyme to beef-liver glutamate dehydrogenase, *Eur. J. Biochem.,* 36, 545, 1973.
81. **Krause, J., Buhner, M., and Sund, A.,** Studies of glutamate dehydrogenase. The binding of NADH and NADPH to beef liver glutamate dehydrogenase, *Eur. J. Biochem.,* 41, 593, 1974.
82. **Kempfle, M. A., Muller, R. F., and Winkler, H. A.,** Relaxation kinetic studies of coenzyme binding to glutamate dehydrogenase from beef liver, *Biochem. Biophys. Res. Commun.,* 85, 593, 1978.
83. **Dieter, H., Koberstein, R., and Sund, H.,** Studies of glutamate dehydrogenase. The interaction of ADP, GTP and NADPH in complexes with glutamate dehydrogenase, *Eur. J. Biochem.,* 115, 217, 1981.
84. **Delabar, J. M., Martin, S. R., and Bayley, P. M.,** The binding of NADH and NADPH to bovine liver glutamate dehydrogenase. Spectroscopic characterization, *Eur. J. Biochem.,* 127, 367, 1982.
85. **Lark, R. H. and Colman, R. F.,** Reaction of the 2′,3′-dialdehyde derivative of NADPH at a nucleotide site of bovine liver glutamate dehydrogenase, *J. Biol. Chem.,* 261, 10659, 1986.
86. **Favilla, R. and Mazzini, A.,** The binding of 1,N^6-etheno-NAD to bovine liver glutamate dehydrogenase, *Biochim. Biophys. Acta,* 788, 48, 1984.
87. **Batra, S. P. and Colman, R. F.,** Affinity labeling of an allosteric ADP site of glutamate dehydrogenase by 2-(4-bromo-2,3-dioxobutylthio)adenosine 5′-monophosphate, *J. Biol. Chem.,* 261, 15565, 1986.
88. **Pal, P. K. and Colman, R. F.,** Affinity labeling of an allosteric GTP site of bovine liver glutamate dehydrogenase by 5′-*p*-fluorosulfonylbenzoyl-guanosine, *Biochemistry,* 18, 838, 1979.
89. **Malcolm, A. D. B. and Radda, G. K.,** The reaction of glutamate dehydrogenase with 4-iodoacetamido salicylic acid, *Eur. J. Biochem.,* 15, 555, 1970.
90. **Wallis, R. B. and Holbrook, J. J.,** The effect of modifying lysine-126 on the physical, catalytic and regulatory properties of bovine liver glutamate dehydrogenase, *Biochem. J.,* 133, 173, 1973.
91. **Holbrook, J. J., Roberts, P. A., and Wallis, R. B.,** The site at which 4-iodoacetamidosalicylate reacts with glutamate dehydrogenase, *Biochem. J.,* 133, 165, 1973.
92. **Jacobson, M. A. and Colman, R. F.,** Distance relationships between the catalytic site labeled with 4-(iodoacetamido)salicylic acid and regulatory sites of glutamate dehydrogenase, *Biochemistry,* 23, 3789, 1984.

93. **Piszkiewicz, D., Landon, M., and Smith, E. L.,** Bovine liver glutamate dehydrogenase. Sequence of a hexadecapeptide containing a lysyl residue reactive with pyridoxal 5′-phosphate, *J. Biol. Chem.,* 245, 2622, 1970.
94. **Piszkiewicz, D. and Smith, E. L.,** Bovine liver glutamate dehydrogenase equilibria and kinetics of imine formation by lysine-97 with pyridoxal 5′-phosphate, *Biochemistry,* 24, 4544, 1971. (In the original publication the lysine modified was identified as lysine-97; as a result the subsequent revision of the amino acid sequence of the bovine liver enzyme, the residue is correctly designated lysine-126.)
95. **Goldin, B. R. and Frieden, C.,** The effect of pyridoxal phosphate modification on the catalytic and regulatory properties of bovine liver glutamate dehydrogenase, *J. Biol. Chem.,* 247, 2139, 1972.
96. **Chen, S.-S. and Engel, P. C.,** The equilibrium position of the reaction of bovine liver glutamate dehydrogenase with pyridoxal 5′-phosphate. A demonstration that covalent modification with this reagent completely abolishes catalytic activity, *Biochem. J.,* 147, 351, 1975.
97. **Chen, S.-S. and Engel, P. C.,** Ox liver glutamate dehydrogenase. The role of lysine-126 reappraised in the light of studies of the inhibition and inactivation by pyridoxal 5′-phosphate, *Biochem. J.,* 149, 619, 1975.
98. **Talbot, J.-C., Gros, C., Cosson, M.-P., and Pantaloni, D.,** Physicochemical evidence for the existence of two pyridoxal 5′-phosphate binding sites on glutamate dehydrogenase and characterization of their functional role, *Biochim. Biophys. Acta,* 494, 19, 1977.
99. **Rasool, C. G., Nicolaidis, S., and Akhtar, M.,** The asymmetric distribution of enzymic activity between the six subunits of bovine liver glutamate dehydrogenases, use of D- and L-glutamyl α-chloromethyl ketones (4-amino-6-chloro-5-oxohexanoic acid, *Biochem. J.* 157, 675, 1976.
100. **Rasched, I., Jornvall, H., and Sund, H.,** Studies of glutamate dehydrogenase. Identification of an amino group involved in the substrate binding, *Eur. J. Biochem.,* 41, 603, 1974.
101. **David, M., Rashed, I., and Sund, H.,** Studies of glutamate dehydrogenase. Methionine 169: the preferentially carboxymethylated residue, *Eur. J. Biochem.,* 74, 379, 1977.
102. **Rosen, N. L., Bishop, L., Burnett, J. B., Bishop, M., and Colman, R. F.,** Methionyl residue critical for activity and regulation of bovine liver glutamate dehydrogenase, *J. Biol. Chem.,* 248, 7359, 1973.
103. **Colman, R. F.,** Affinity labeling of purine nucleotide sites in proteins, *Annu. Rev. Biochem.,* 52, 67, 1983.
104. **Pal, P. K., Wechter, W. J., and Colman, R. F.,** Affinity labeling of the inhibitory DPNH site of bovine liver glutamate dehydrogenase by 5′-fluorosulfonylbenzoyl adenosine, *J. Biol. Chem.,* 250, 8140, 1975.
105. **Saradambal, K. V., Bednar, R. A., and Colman, R. F.,** Lysine and tyrosine in the NADH inhibitory site of bovine liver glutamate dehydrogenase, *J. Biol. Chem.,* 256, 11866, 1981.
106. **Schmidt, J. A. and Colman, R. F.,** Identification of the lysine and tyrosine peptides labeled by 5′-*p*-fluorosulfonylbenzoyladenosine in the NADH inhibitory site of glutamate dehydrogenase, *J. Biol. Chem.,* 259, 14515, 1984.
107. **Coffee, C. J., Bradshaw, R. A., Goldin, B. R., and Frieden, C.,** Identification of the sites of modification of bovine liver glutamate dehydrogenase reacted with trinitrobenzenesulfonate, *Biochemistry,* 19, 3516, 1971.
108. **Goldin, B. R. and Frieden, C.,** Effect of trinitrophenylation of specific lysyl residues on the catalytic, regulatory and molecular properties of bovine liver glutamate dehydrogenase, *Biochemistry,* 19, 3527, 1971.
109. **Batra, S. P. and Colman, R. F.,** Affinity labeling of the DPNH inhibitory site of glutamate dehydrogenase by 6-[(4-bromo-2,3-dioxobutyl)thio]-6-deaminoadenosine 5′-diphosphate, *Biochemistry,* 23, 4940, 1984.
110. **Batra, S. P. and Colman, R. F.,** Isolation and identification of a cysteinyl peptide labeled by 6-[(4-bromo-2,3-dioxobutyl)thio]-6-deaminoadenosine 5′-diphosphate in the DPNH inhibitory site of glutamate dehydrogenase, *Biochemistry,* 25, 3508, 1986.
111. **Piszkiewicz, D., Landon, M., and Smith, E. L,** Bovine glutamate dehydrogenase. Loss of allosteric inhibition by GTP and nitration of tyrosine-412, *J. Biol. Chem.,* 246, 1324, 1971. (This residue is correctly designated Tyr-407 in accordance with the revised amino acid sequence.)
112. **Smith, E. L. and Piszkiewicz, D.,** Bovine glutamate dehydrogenase. The pH dependence of native and nitrated enzyme in the presence of allosteric modifiers, *J. Biol. Chem.,* 248, 3089, 1973.
113. **Pal, P. K., Reischer, R. J., Wechter, W. J., and Colman, R. F.,** A new affinity label for guanosine sites in proteins, *J. Biol. Chem.,* 253, 6644, 1978.
114. **Jacobson, M. A. and Colman, R. F.,** Isolation and identification of a tyrosyl peptide labeled by 5′-[(*p*-fluorosulfonyl)benzoyl]-1,N^6-ethenoadenosine at a GTP site of glutamate dehydrogenase, *Biochemistry,* 23, 6377, 1984.
115. **Jacobson, M. A. and Colman, R. F.,** Evaluation of the intramolecular stacking of the fluorosulfonylbenzoyl derivatives of 1,N^6-ethenoadenosine, adenosine and guanosine, *J. Biol. Chem.,* 259, 1454, 1984.
116. **Stryer, L.,** Fluorescence energy transfer as a spectroscopic ruler, *Annu. Rev. Biochem.,* 47, 819, 1978.
117. **George, A. and Bell, J. E.,** Effects of ADP on glutamate dehydrogenase: diethyl pyrocarbonate modification, *Biochemistry,* 19, 6057, 1980.

118. **Koberstein, R., Cobianchi, L., and Sund, H.,** Interaction of the photoaffinity label 8-azido-ADP with glutamate dehydrogenase, *FEBS Lett.*, 64, 176, 1976.
119. **Michel, F., Pons, M., Julliard, J., Descomps, B., and Crastes de Paulet, A.,** Enzymic hydrolysis of 3-acetyl-estrogens or analogues by glutamate dehydrogenase with specific acylation of the estrogen binding site, *Eur. J. Biochem.*, 84, 275, 1978.
120. **Michel, F., Pons, M., Descomps, B., and Crastes de Paulet, A.,** Affinity labeling of the estrogen binding site of glutamate dehydrogenase with iodoacetyldiethylstilbesterol., *Eur. J. Biochem.*, 84, 267, 1978.
121. **Batra, S. P., Lark, R. H., and Colman, R. F.,** Identification of itistidyl peptide labeled by 2-(4-bromo-2,3-dioxobutylthio) adenosine 5′-monophosphate in an ADP regulatory site of glutamate dehydrogenase, *Arch. Biochem. Biophys.*, 270, 277, 1989.

Chapter 8

DIHYDROFOLATE REDUCTASE

John F. Morrison

TABLE OF CONTENTS

I. INTRODUCTION

Dihydrofolate reductase (DHFR; EC 1.5.1.3) is one of over 250 pyridine nucleotide-dependent dehydrogenases. The enzyme from all sources catalyzes the reaction:

$$\text{7,8-dihydrofolate} + \text{NADPH} + \text{H}^+ \longrightarrow \text{5,6,7,8-tetrahydrofolate} + \text{NADP} \quad (1)$$

while enzymes from several sources also catalyze the reaction:

$$\text{folate} + \text{NADPH} + \text{H}^+ \rightarrow \text{7,8-dihydrofolate} + \text{NADP} \quad (2)$$

(The structures of folate and its reduced forms are given in Figure 1). Reaction 2 is relatively slow, especially at neutral pH[1] and with bacterial enzymes.[2,3] Considerable interest has been taken in the enzyme over the past 25 years. The initial interest related to the important metabolic role played by dihydrofolate reductase (DHFR) in the biosynthesis of nucleic acids. It became the target for the action of inhibitory analogues of folate, which included pteridines, pyrimidines, triazines, and quinazolines, and which proved to be useful clinically as anticancer drugs.[4,5] Recent interest relates strongly to the small size of the enzyme which has facilitated the determination of its three-dimensional structure at a level of resolution greater than that for any other dehydrogenase.[6,7] It is the availability of detailed structural information that has stimulated studies on the kinetic and chemical mechanisms of the reaction.

The aims of mechanistic enzymology are to explain in stereochemical terms the way in which an enzyme catalyzes a particular reaction in such an efficient manner. To achieve these aims it is necessary to obtain structural, thermodynamic and kinetic information through the application of the techniques of X-ray crystallography, chemistry and biochemistry. Data obtained by these techniques can be complemented considerably by the use of site-directed mutagenesis which allows the generation of specific mutant enzyme forms and of computer-controlled graphics systems which allow the display and manipulation of the structures of enzyme-ligand complexes. Studies on DHFR provide an excellent example of the way in

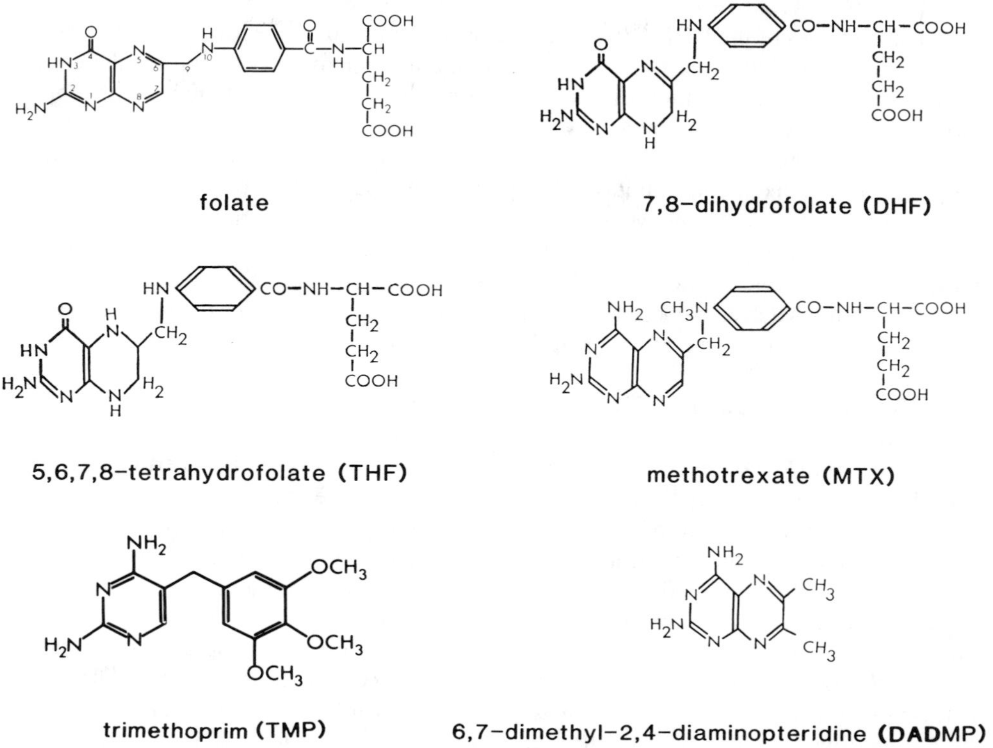

folate

7,8-dihydrofolate (DHF)

5,6,7,8-tetrahydrofolate (THF)

methotrexate (MTX)

trimethoprim (TMP)

6,7-dimethyl-2,4-diaminopteridine (DADMP)

FIGURE 1. Structures of derivatives and analogues of folate.

which the aforementioned techniques can come together to yield insights into the structure-function relationships for an enzyme. It is the enzyme from *Escherichia coli* that has been subjected to the most extensive investigations so far and thus it can be regarded as the prototype bacterial enzyme. Structural information is available for DHFR from *Lactobacillus casei*,[6,7] chicken liver,[8] and for the enzyme that is coded for by the R-67 plasmid of *E. coli*.[9] Structural studies are also in progress with human DHFR.[10] For this reason, and because the enzyme is available in relatively large quantities as a result of cloning of the human *fol* gene,[10] human DHFR might be regarded as the prototype vertebrate enzyme.

II. GENERAL PROPERTIES OF DHFR

In the following sections most emphasis will be placed on the results of studies with DHFR from *E. coli*. Such a restriction is made because of the limited amount of space available and because most mechanistic investigations have been performed with the enzyme from this source. However, reference will be made to enzymes from other sources for comparative purposes or whenever it is necessary to illustrate a particular point about catalytic behavior.

A. MOLECULAR COMPOSITION

Most bacterial enzymes are monomeric with molecular weights in the vicinity of 18,000.[1] Exceptions are the types I and II R-plasmid-encoded enzymes with double the molecular weight and either two or four subunits.[9] Vertebrate enzymes are also monomeric with molecular weights of about 22,000.[1]

B. KINETIC PARAMETERS WITH NATURAL SUBSTRATES

Kinetic studies of the DHFR reaction have been hampered by the low Michaelis constant values for both NADPH and DHF which are in the low micromolar region for the enzyme from *E. coli*.[1,11] Circumvention of this problem has been achieved by analysis of progress curve data.[11] More recently, there has been used in this laboratory for studies on chicken liver DHFR an alternative procedure that involves the measurement of initial velocities in the presence of a fixed concentration of a classical competitive inhibitor (I). The fitting of the initial velocity data to the Michaelis-Menten equation yields an apparent value for the Michaelis constant (appK_a) which is equal to K_a $(1 + I/K_i)$ where K_i is the dissociation constant for the inhibitor. The latter value can be determined by use of an alternative substrate that has a higher Michaelis constant and the true Michaelis constant for the natural substrate (K_a) calculated by using the relationship $K_a = \text{app}K_a/(1 + I/K_i)$.

C. SUBSTRATE SPECIFICITY

1. Pterin Substrates

DHFR from *E. coli* exhibits a narrow specificity in relation to the pterin substrate as judged by the facts that only DHF has a relatively low K_m value of about 1 μM and undergoes reaction at a reasonable rate of about 18 s^{-1} at neutral pH.[11] Folate will undergo reaction under the same conditions with a K_m value of 6 μM, but the maximum velocity of the reaction is lower by almost five orders of magnitude.[3] Since the rate of folate reduction is so slow, it can be used as an inhibitor of the reaction with DHF as substrate. As the K_i value for folate of 11 μM is similar to its K_m value, catalysis or an isomerization reaction associated with catalysis, must be rate-limiting in the conversion of folate to DHF at pH 7.4. At pH 5.2, there is a 14-fold increase in the rate of folate reduction and an increase in the K_m for folate to 28 μM.[3] DHFR from chicken liver is far better able to reduce folate with a pH-independent maximum velocity which is 10% of that with DHF.[12] Further, it is capable of reducing 7,8-dihydro-6-methylpterin which is not a substrate for the *E. coli* enzyme. No information is yet available about the binding of folate at the active site of any DHFR and little attention has been paid to the overall reduction of folate to THF. The question of the release of DHF from the enzyme as a function of the concentration of NADPH has not been addressed.

2. Pyridine Nucleotide Substrates

Apart from NADPH, the reduced forms of nicotinamide hypoxanthine dinucleotide phosphate (NHDP), nicotinamide-1,N^6-ethenoadenine dinucleotide phosphate (ϵ-NADP) and 3-acetylpyridine adenine dinucleotide phosphate (APADP) behave as substrates for *E. coli* DHFR[13] (Table 1). The identity of the pyridine nucleotide has little effect on the magnitude of the Michaelis constant for DHF. There are only small differences in the dissociation constants for the binding of the pyridine nucleotides to the free enzyme and the presence of DHF on the enzyme has no marked effect on the values for their Michaelis constants. This lack of synergism for substrate binding contrasts with the very marked influence that NADPH can have on the binding of folate analogues.[3] (The enhancement of the binding of one ligand to an enzyme by another ligand has been referred to as cooperativity.[14-16] The use of the latter term to describe such effects is confusing as it has a well-established meaning in the area of allosteric enzymes. Ligand binding to monomeric DHFR cannot involve subunit interactions. As synergism aptly describes the ability of one ligand to bring about the increased binding of a second ligand, it should be the term of choice).

In contrast to the small variation in the values of the kinetic constants associated with the pyridine nucleotides listed in Table 1, the maximum velocity is dependent on the identity of the nucleotide substrate. The turnover number with NADPH is some sevenfold greater than with its acetylpyridine derivative. The enzyme is unable to utilize reduced thionicotin-

TABLE 1
Kinetic Parameters for *E. Coli* DHFR with Reduced Pyridine Nucleotide Substrates

Pyridine nucleotide (NH)	Parameter: Maximum velocity (s^{-1})	Dissociation constant E-NH (μM)	Michaelis constant (NH, μM)	Dissociation constant (E-DHF) (μM)	Michaelis constant (DHF, μM)
NADPH	16.5	3.8 (0.5)[a]	2.5	0.8	0.5
NHDPH[b]	6.4	4.8 (2.8)	2.1	0.6	0.3
ε-NADPH	4.3	2.4 (1.4)	1.6	0.8	0.5
APADPH	2.4	1.6 (1.2)	3.0	0.6	1.1

Note: Values were obtained at pH 7.4, 30°C (13).

[a] Figures in brackets are dissociation constants determined by fluorescence titration.
[b] NHDP, nicotinamide hypoxanthine dinucleotide phosphate; ε-NADP, nicotinamide-1, N^6-ethenoadenine dinucleotide phosphate; APADP, 3-acetylpyridine adenine dinucleotide phosphate.

amide adenine dinucleotide phosphate (TNADPH) as a hydride donor. On the basis of data obtained with DHFR from *Lactobacillus casei,*[17,18] this could be due to the fact that the thionicotinamide moiety of TNADPH does not form any specific interactions with the enzyme, but rather extends out into solution.

D. STEREOCHEMISTRY OF HYDRIDE TRANSFER

DHFR is an A-side specific dehydrogenase as the hydride transfer occurs from the 4-pro-R hydrogen of NADPH. The transfer is made to the *si*-face of folate for the formation of both DHF and THF and the absolute stereochemistry at the C-6 and C-7 carbons is S.[19,20] The results imply that folate and DHF are bound to DHFR in a similar manner.

E. EQUILIBRIUM CONSTANT FOR DHFR REACTION

The equilibrium constant (K_{eq}) for the conversion of NADPH and DHF to NADP and THF is described by the relationship

$$K_{eq} = \frac{[NADP][THF]}{[NADPH][DHF][H^+]}$$

The equilibrium lies very far to the right as judged by the reported values of 5.6×10^{11} M^{-1}, $8.4 \times 10^{10}\, M^{-1}$, and $1.3 \times 10^{11}\, M^{-1}$ using DHFR from chicken liver,[21] *Streptococcus faecium,*[22] and *E. coli,*[23] respectively. These values are very much greater than that of 1.6×10^3 for the internal equilibrium between the ternary complexes enzyme-NADPH-DHF and enzyme-NADP-THF.[23] The magnitude of the overall equilibrium constant emphasizes the essential irreversibility of the DHFR reaction.

III. INHIBITION OF DHFR BY SUBSTRATE ANALOGUES

A. INHIBITION BY FOLATE ANALOGUES

1. Slow-Binding Inhibition

Because of the central role played by DHFR in the biosynthesis of DNA, it has been the target for both anticancer and antibacterial drugs.[4,5] Indeed, over the past 30 years a wide range of folate analogues has been synthesized and tested as inhibitors of the enzyme. The classical example of an anticancer drug is methotrexate (MTX; Figure 1) which has

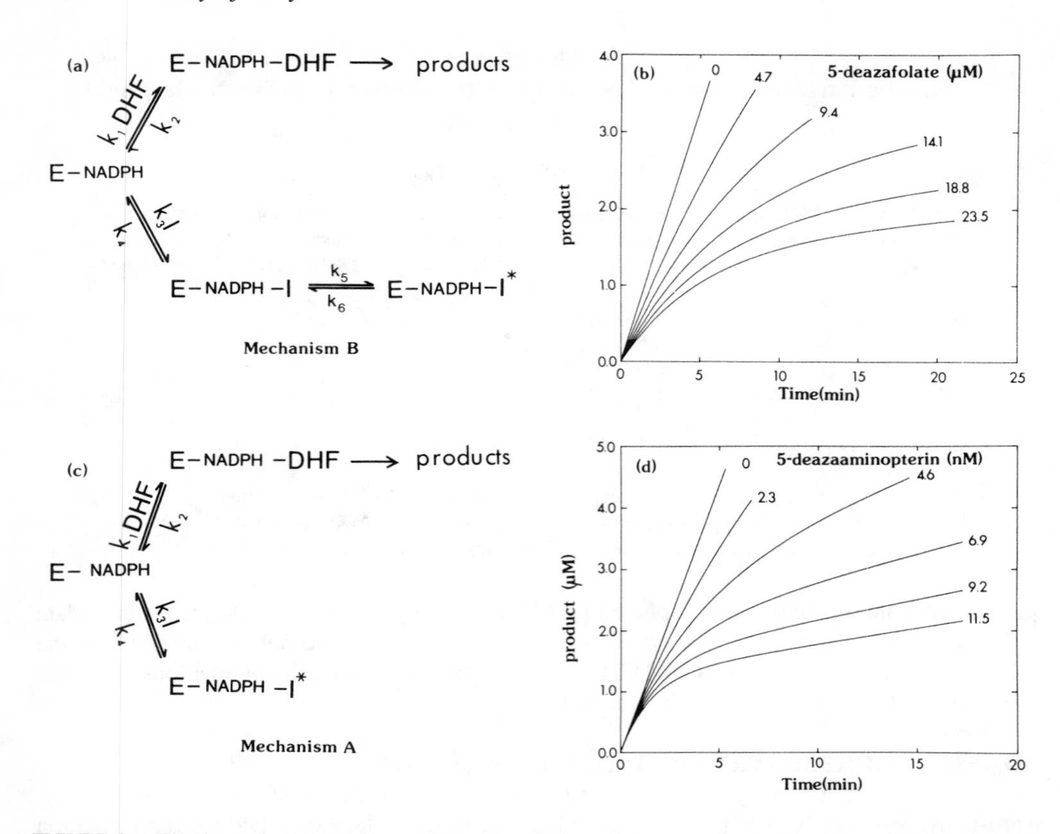

FIGURE 2. Kinetic mechanisms (a and c) and reaction kinetics (b and d) for the inhibition of DHFR by slow-binding inhibitors.

been in clinical use for nearly 25 years while trimethoprim (TMP, Figure 1) is used clinically as an antibacterial drug.

Although it has been recognized for some time that MTX and TMP are potent inhibitors of DHFR from several sources, detailed steady-state kinetic studies of the inhibition have been undertaken only over the past decade. Investigations with the enzyme from *E. coli*[3,24] have shown that the inhibition by MTX is time-dependent and involves the rapid formation of an enzyme-NADPH-MTX complex which undergoes a relatively slow, reversible isomerization reaction to form a more thermodynamically stable ternary enzyme complex. This sequence of reactions is illustrated in Figure 2a as Mechanism B[25,26] with the symbol, I representing MTX. The isomerization reaction is, in effect, a conformational change which has been observed to occur on the binding of ligands to DHFR from various sources.[1] It also provides the means of markedly enhancing the potency of an inhibitor and produces a complex from which the inhibitor dissociates at a rate which is independent of substrate concentration. The enhancement of the inhibition with MTX is such that the K_i value associated with the formation of the initial collision complex (E-NADPH-I, Figure 2a) is lowered by a factor of 270 (Table 2). Thus the concentration of MTX required to observe inhibition becomes comparable to that at which the enzyme is used for steady-state kinetic studies and the inhibition falls into the tight-binding category. As the inhibition by MTX is also time dependent on a steady-state time scale, it is referred to as being a slow, tight-binding inhibitor.[25] Recent stopped-flow studies with DHFR from *Streptococcus faecium*[27] have shown that the initial collision complex formed between enzyme-NADPH and MTX undergoes a further isomerization reaction which precedes, and is much faster than, that observed in steady-state kinetic investigations. It is not yet known if the binding of MTX to the binary complex of NADPH with *E. coli* DHFR involves two isomerizations.

TABLE 2
Values of Parameters Associated with the Slow-Binding Inhibition of *E. coli* DHFR by Folate Analogues at pH 7.4

Type of Inhibition and Mechanism	Compound	Parameter				
		K_i (n*M*)	k_f[a]	k_r (min^{-1})	K_i* (p*M*)	k_5/k_6
Slow, tight-binding B	Methotrexate	3.6	6.9	0.026	13	270
	5-Deazamethotrexate	2.9	12.0	0.015	3.6	800
	Aminopterin	8.2	2.9	0.006	16	480
	Trimethoprim	0.49	2.0	0.086	20	23
Slow, tight-binding A	5-Deazaaminopterin	0.0066	5.9	0.039		
Slow-binding B	5-Deazafolate	190	0.33	0.009	5000	37

Note: Values are for the interaction of the folate analogues with the enzyme-NADPH complex.[3,24]

[a] k_f and k_r refer, respectively, to k_3 and k_4 (mechanism A) or to k_5 and k_6 (mechanism B). The units of k_f are nm^{-1} min^{-1} (mechanism A) or min^{-1} (mechanism B).

Other folate analogues that exhibit the same slow, tight-binding behavior as MTX are 5-deazamethotrexate, aminopterin and trimethoprim (TMP; Table 2). However, there are significant differences in the contribution made by the isomerization reaction, as reflected in the value of k_5/k_6, to the overall strength of inhibitor binding. TMP is similar to MTX with respect to its behavior as an inhibitor of DHFR from *E. coli*. By contrast, TMP is a relatively poor, classical inhibitor of the enzyme from chicken liver.[3] The difference appears to be due to subtle geometrical differences between the binding sites of the two enzymes.[28,29] The inhibition of DHFR by 5-deazafolate is also described by Mechanism B (Figure 2a). But as the concentration required to cause inhibition is high relative to the enzyme concentration, 5-deazafolate is simply a slow-binding inhibitor.[25] For tight-binding inhibition, folate analogues must be 2,4-diamino derivatives of pyrimidines or pteridines.

For an inhibition reaction conforming to Mechanism B, progress curves obtained by starting the reaction with enzyme will show that

1. The initial velocity decreases to a slower steady-state velocity.
2. The rate of change increases with inhibitor concentration.
3. Both the initial and steady-state velocities vary as a function of inhibitor concentration.

These effects are seen clearly for the slow-binding inhibition of *E. coli* DHFR by 5-deazafolate (Figure 2b). However, under conditions where the steady-state concentration of the initial collision complex becomes negligibly small, the variation of the initial velocity with inhibitor concentration will not be observed. It would appear then that there is simply a direct, slow formation of the final E-NADPH-I* complex (Mechanism A, Figure 2c). This type of inhibition is given by the slow, tight-binding inhibitor, 5-deazaaminopterin (Figure 2d, Table 2).

For DHFR from *E. coli*, all folate analogues that act as tight-binding inhibitors exhibit slow-binding characteristics (Table 2). However, slow-binding does occur without tight-binding as observed with 5-deazafolate. At present, there are no indications of what type of reaction(s) or amino acid residues are responsible for the isomerizations that are induced in DHFR by several folate analogues. Perhaps the slow reaction involves the elimination of a specific water molecule as appears to be the case with slow-binding inhibitors of thermolysin.[30]

TABLE 3
Effect of NADPH on the Binding of Folate and Folate Analogues to *E. coli* DHFR at pH 7.4[3]

Compound (I)	Dissociation constant (n*M*) for E-I	E-NADPH-I[a]	Synergistic effect of NADPH
Trimethoprim (TMP)	20	0.020	1000
Phenyltriazine[b]	>30000	35	>900
2,4-Diamino-6,7-dimethyl pterin (DADMP)	1100	21	50
Methotrexate (MTX)	0.62	0.013	50
Folate	32000	11000	3
Pyrimethamine[b]	13	10	1

[a] With the slow-binding inhibitors, trimethoprim and methotrexate, the dissociation constant given is that for the overall inhibition reaction (see Table 2).

[b] Phenyltriazine; 4,6-diamino-1,2-dihydro-2,2-dimethyl–1-(4-phenyl-thiomethylphenyl)-1,3,5-triazine; pyrimethamine, 2,4-diamino-5-(p-chlorophenyl)-6-ethylpyrimidine.

2. Effects of NADPH on the Binding of Folate Analogues

NADPH can have a marked synergistic effect on the binding of folate analogues to DHFR[3] and the effect is observed with compounds that function either as slow-binding (MTX, TMP) or as classical (phenyltriazine, DADMP) inhibitors (Table 3). However, the binding of folate or pyrimethamine to the enzyme is little influenced by the presence of NADPH and so the degree of synergism can vary by as much as 1000-fold. It has been proposed that synergism arises with TMP because of the interaction of one of its methoxy groups with the nicotinamide ring of NADPH.[14,15] The substitution of Leu-28 by Arg-28 does not reduce the synergistic effect of NADPH, but does decrease the binding of TMP to DHFR.[31]

B. INHIBITION BY ANALOGUES OF NADP

The reaction catalyzed by DHFR is inhibited by fragments of NADP which include adenosine-5′-phosphate, adenosine-2′-phosphate, ADP-ribose, adenosine-2′, 5′-diphosphate, and ATP-ribose.[13] Each of the compounds gives rise to inhibitions that are linear competitive with respect to NADPH and linear noncompetitive in relation to DHF. Such results are expected for a kinetic mechanism that involves the random addition of substrates (see Section IV) and they indicate that each of the compounds combines with both the free enzyme and the enzyme-DHF complex. The presence of DHF on the enzyme has only a small effect on the binding of the fragments whose strength of binding increases in the order that is given above.

The oxidized forms of several pyridine nucleotides including NADP, ϵ-NADP, NAD, 3-acetylpyridine adenine dinucleotide phosphate, nicotinamide 1,N^6-etheneoadenine dinucleotide phosphate, and nicotinamide hypoxanthine dinucleotide phosphate act as inhibitors in a similar manner to fragments of NADP.[13] However, thionicotinamide adenine dinucleotide phosphate (TNADP) and its reduced form (TNADPH) exhibit different behavior in that they cause inhibition that is noncompetitive with respect to both NADPH and DHF.[13] This result has led to the conclusion that these nucleotides bind not only to the free enzyme and the enzyme-DHF complex, but also to an enzyme-THF complex. Such a conclusion is confirmed

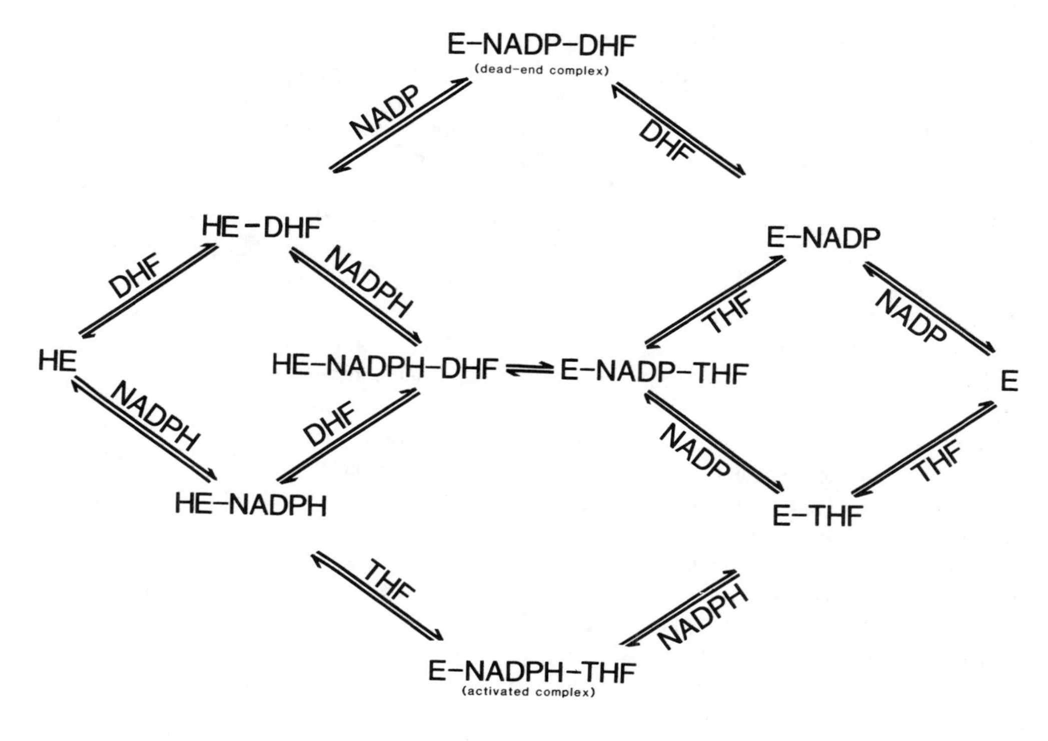

FIGURE 3. Kinetic mechanism of the reaction catalyzed by DHFR from *Escherichia coli*.

by the results of double inhibition experiments with TNADP and THF which show that a dead-end enzyme-TNADP-THF complex can form. This finding implies that, during reaction, there is present a kinetically significant steady-state concentration of an enzyme-THF complex. It is also in agreement with the conclusion that substrate activation by NADPH is due to the formation of an enzyme-NADPH-THF complex from which THF is released more rapidly than from the enzyme-THF complex.[23,32]

IV. KINETIC MECHANISM OF DHFR REACTION

A. STEADY STATE-KINETICS

1. Steady-State Velocity Studies

The kinetic mechanism of a reaction refers to the order in which substrates add to, and products dissociate from, an enzyme. Its determination must be regarded as the mandatory first step in the investigation of any enzyme-catalyzed reaction as the kinetic mechanism provides the framework for the interpretation of structural, thermodynamic and other kinetic data.

It is now well established that, at neutral pH, the reaction catalyzed by *E. coli* DHFR, with NADPH and DHF as substrates, conforms to a random mechanism with the formation of a dead-end, enzyme-NADP-DHF complex.[11] (see Figure 3) Thus, at the active site of the enzyme, there is one sub-site for the binding of the pterin substrate and another for the binding of the pyridine nucleotide. There is no evidence for the formation of the other possible dead-end complex *viz* enzyme-NADPH-THF.[11] This is not surprising in view of the finding that higher concentrations of NADPH can facilitate the release of the product, THF.[23,32] The kinetic mechanism for the reaction in the reverse direction could not be determined directly by application of the techniques used to study the forward reaction as the reaction is essentially irreversible.

The data obtained for the forward reaction from steady-state velocity, as well as product and dead-end inhibition, studies could be analyzed adequately on the basis that rapid equilibrium conditions apply to the random mechanism.[11] However, this can be only an approximate description of the mechanism. In a truly rapid equilibrium, random mechanism, catalysis is slow relative to all other steps. Therefore the reaction of the substrates with either the free enzyme or with the appropriate binary enzyme-substrate complex would be at thermodynamic equilibrium and kinetic data would yield true thermodynamic values for the dissociation constants of the binary enzyme-substrate complexes. Further, the steady-state concentrations of ternary and binary enzyme-product complexes, on the release side of the reaction sequence, would be negligible. These criteria are not met by the DHFR reaction. The kinetically determined value for the dissociation constant of the enzyme-NADPH complex is about sevenfold greater than the thermodynamic value (Table 1) and thus the reaction of NADPH with free enzyme is not in rapid equilibrium. Neither is the reaction of NADPH with the enzyme-DHF complex in rapid equilibrium as it is slowed down by increasing viscosity.[32] The viscosity data also indicate that NADPH behaves as a sticky substrate. This conclusion is confirmed by the results of deuterium isotope effect experiments which show that the value for $^{D}(V/K)_{NADPH}$ is only slightly greater than one.[33,34] The value for ^{D}V is also about one so that product release must be at least partly rate-limiting. Thus, it can be expected that a binary exzyme-product complex would be present under steady-state conditions. The behavior of reduced thio-NADPH (TNADPH) as a non-competitive inhibitor with respect to NADPH[13] and the activation of DHFR by higher concentrations of NADPH[32] led to the proposals that there is a steady-state concentration of the binary enzyme-THF complex and that there are alternative pathways for the release of THF from this complex (Figure 3). One pathway simply involves dissociation of THF from the enzyme-THF complex while the other involves the formation of a ternary enzyme-NADPH-THF complex from which THF is released at a faster rate than from the binary enzyme-THF complex. Similar conclusions have been reached by Benkovic and his colleagues[23] from direct measurements of the rates of dissociation of these complexes (see Section IV.B).

2. pH Dependence of Kinetic Parameters

The fundamental kinetic parameters for any enzyme-catalyzed reaction are V, the maximum velocity at saturating concentrations of the substrates, and $V/K_{substrate}$, the apparent first order rate constant for the interaction of substrate with the enzyme as the concentration of the substrate tends to zero.[35] The variation of these parameters as a function of pH can provide information about the number of ionizing amino acid residues at the active site of the enzyme that are involved in the reaction. Such data can also indicate the ionization state of the residue that is required for catalysis or substrate binding. The $V/K_{substrate}$ data do not necessarily yield the true pK_a value for the enzyme form with which the variable substrate reacts, as the pH profile will be displaced outwards by the stickiness of the substrate. A sticky substrate is one that reacts to give products faster than it dissociates from the enzyme.[35] However, the true pK_a value of the enzyme form can be determined from studies of the variation with pH of the K_i value for an inhibitory analogue of the varied substrate. This interaction is a thermodynamic one and thus there are no kinetic effects. The nature of the ionizing group, whether it is a cationic or neutral acid, might also be elucidated by determining the shift of the pK_a value when the reaction is performed in the presence of an organic solvent.[35]

The aforementioned techniques have been utilized for the elucidation of the chemical mechanism of the DHFR reaction.

a. Variation with pH of V and $(V/K)_{DHF}$

In the presence of a saturating concentration of NADPH, a plot of log $(V/K)_{DHF}$ against pH shows that only a single ionizing group is observed[36] (Figure 4a). Analysis of the data

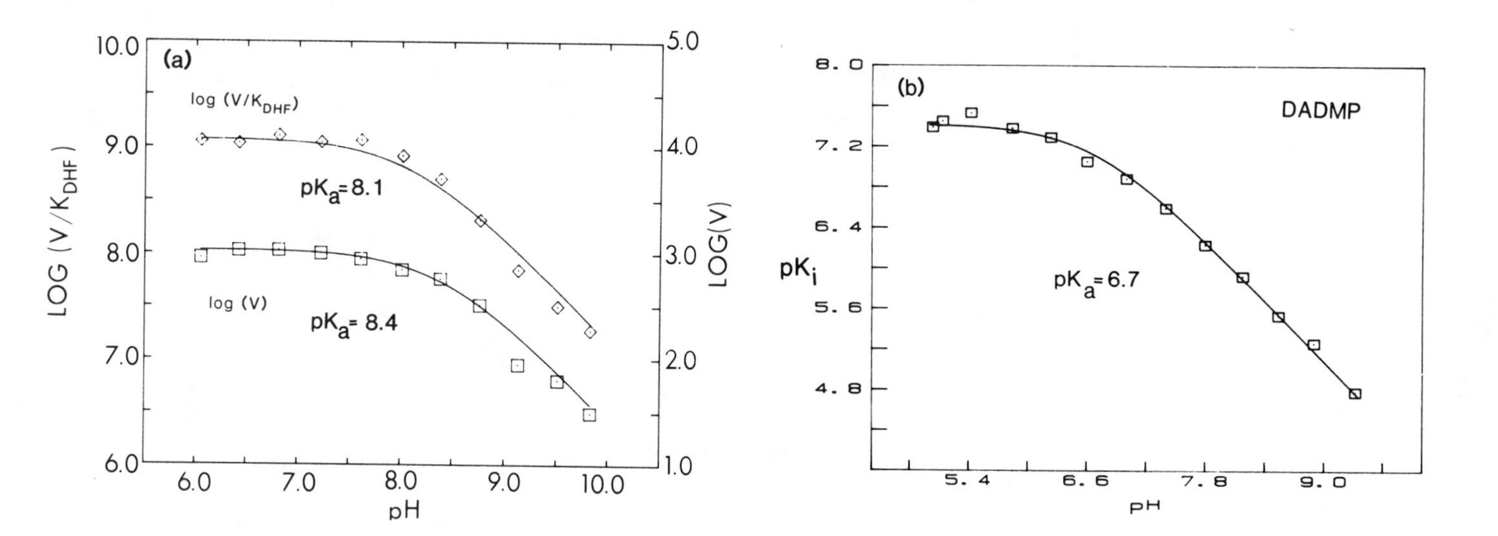

FIGURE 4. Variation with pH of (a), log V and log $(V/K)_{DHF}$ with NADPH at 100 μM; (c), log V and log $(V/K)_{NADPH}$ with DHF at 40 μM; (b) pK_i for binding of 2,4-diamino-6,7-dimethylpteridine (DAMP) to the enzyme-NADPH complex; (d) pK_i for binding of NADPH to free enzyme and of ATP-ribose to the enzyme-DHF complex.

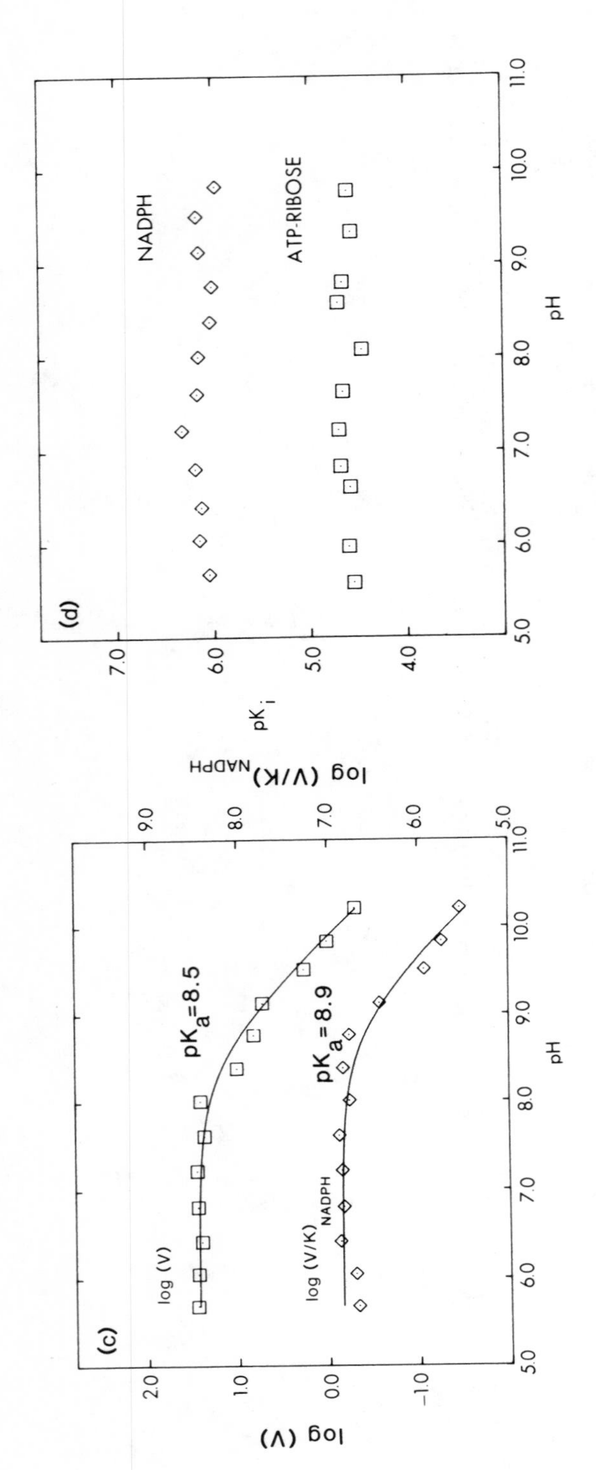

FIGURE 4. (continued)

TABLE 4
Values for pK_a and Kinetic Parameters Associated with the DHFR Reaction[32,36]

Variable substrate	Reaction	pH-independent parameter: V/K (μM^{-1} s^{-1})	pH-independent parameter: V (s^{-1})	Observed pK_a value: V/K profile	Observed pK_a value: V profile
DHF	E-NADPH + DHF	20	17.7	8.1	8.4
NADPH	E-DHF + NADPH	3.4	13.2	8.9	8.5
APADPH[a]	E-DHF + APADPH	0.5	1.6	7.4	8.0

[a] APADPH, reduced 3-acetylpyridine adenine dinucleotide phosphate.

yielded the results of Table 4. As DHF does not have an ionizing group in the region of pH 8, the ionizing group observed in the V/K_{DHF} profile must be on the enzyme and be of importance for either the binding of DHF or for catalysis. The V profile is similar to the V/K_{DHF} profile (Figure 4a) and shows that a group on the enzyme with an observed pK_a value of 8.4 must be protonated to achieve the maximum reaction velocity (Table 4).

The similarity of the V/K_{DHF} and V profiles suggests that the single ionizing group on the enzyme is involved with catalysis. This follows from the fact that, when the variable substrate combines with both the protonated and unprotonated enzyme forms, the ionizing group of an amino acid residue that plays a role in catalysis must be observed in both profiles. The pK_a value of 8.1 from the V/K profile (Table 4) does not necessarily represent a true pK_a value as the true value can be displaced outwards if the variable substrate, DHF is sticky.[35] Fortuitously, the ionizing group, which is observed in the V/K profile and whose function is associated with catalysis rather than substrate binding, is involved with the binding of 2,4-diaminopteridines (see Section V.B.1 and 2). Thus, this type of folate analogue can be used to determine the true pK_a value of the ionizing group of the enzyme-NADPH complex. With 2,4-diamino-6,7-dimethylpteridine (DADMP, Figure 1) as the analogue of DHF, the value was determined to be 6.7[33] (Figure 4b). This value, which is somewhat lower than that of 7.9 reported previously,[36] is some 1.4 pH units below the pK_a value of the V/K_{DHF} profile. However, it is comparable to that of about 6.3 as obtained from fluorescence quenching data for the binding of DADMP to the free enzyme[37] and that of 6.4 to 6.7 from NMR studies of substrate binding.[38]

b. Variation with pH of V and $(V/K)_{NADPH}$

The variation of V/K_{NADPH} and V as a function of pH, with DHF present at a saturating concentration, also shows the presence of only a single ionizing group at the active site of DHFR (Figure 4c). The pK_a value from the V profile is, as expected, similar to that obtained with DHF as the variable. But the pK_a from the V/K_{NADPH} profile is even higher at 8.9 (Table 4). Again the similarity of the two profiles suggests that the ionizing group is involved with, and must be protonated for, catalysis. Confirmation of this conclusion comes from the finding that the binding of NADPH to the free enzyme and of ATP-ribose to the enzyme-DHF complex is pH-independent[33] (Figure 4d). From these results it follows that the pK_a value for the enzyme-NADPH and enzyme-DHF complexes must be identical with that of the free enzyme. As the binding of substrate to form binary and ternary complexes is also pH independent,[23,37,38] the intrinsic pK_a values for all forms of the enzyme are identical (see Figure 7).

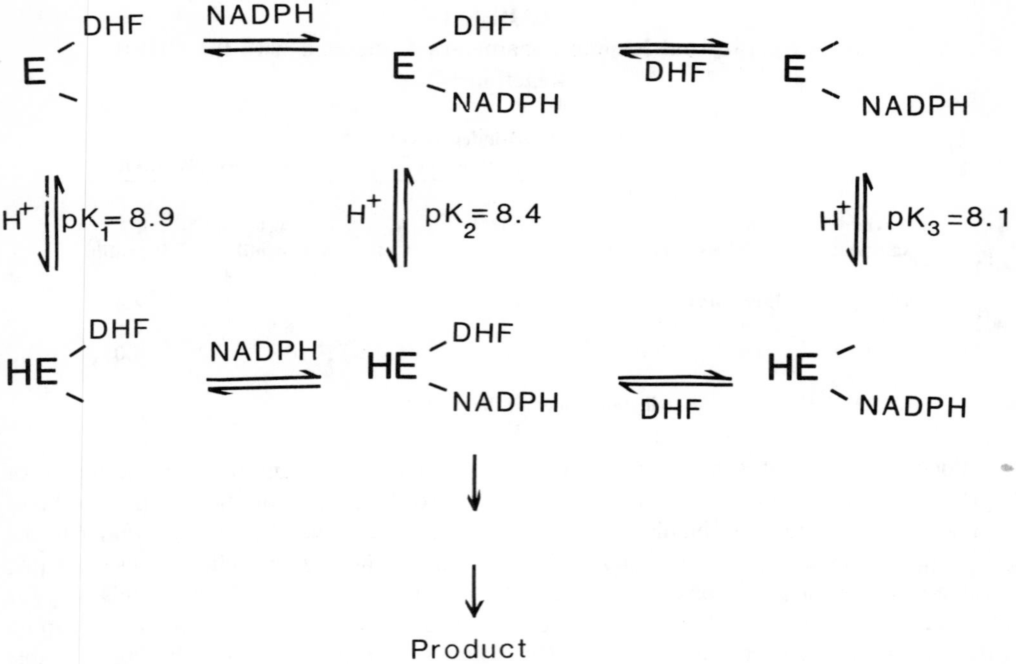

E–DHF ⇌(NADPH) E(DHF)(NADPH) ⇌(DHF) E–NADPH

H^+ ⇅ $pK_1 = 8.9$ H^+ ⇅ $pK_2 = 8.4$ H^+ ⇅ $pK_3 = 8.1$

HE–DHF ⇌(NADPH) HE(DHF)(NADPH) ⇌(DHF) HE–NADPH

↓

↓

Product

FIGURE 5. Pathways for the pH-dependent formation of a productive protonated enzyme-NADPH-DHF complex.

3. General Kinetic Reaction Scheme

On the basis of the pH data and the kinetic behavior of NADPH and DHF, a scheme can be formulated for the DHFR reaction in the presence of a saturating concentration of one or other of the substrates (Figure 5). The essential features are that (1) the single ionizing group at the active site of the binary enzyme-DHF (or enzyme-NADPH) complex can exist in unprotonated and protonated forms which are in rapid protonic equilibrium. (2) NADPH (or DHF) can react with both unprotonated and protonated forms of the binary enzyme complex, but it is only the reactions with the protonated species that give rise to a productive ternary complex. If the formation of a nonproductive complex did not occur, V would not decrease as a function of pH (see Figure 4a and c). (3) The protonated and unprotonated forms of the ternary complex are in rapid protonic equilibrium. If this were not so, the PH profiles would be expected to exhibit a hump or a hollow.[35]

The observed (or conditional) pK_a values for the enzyme-DHF complex of 8.9, for the enzyme-NADPH complex of 8.1 and for the enzyme-NADPH-DHF complex of 8.4 are elevated well above the intrinsic pK_a value of about 6.5 for these complexes. The kinetic effects of NADPH and DHF ensure that, at neutral pH, the various enzyme species exist in their productive, protonated forms. These findings suggest strongly that the function of the single ionizing group at the active site of DHFR is to provide a proton for the reduction of the 5,6-imine double bond of DHF.

In the presence of APADPH, which is a much poorer substrate than NADPH (Table 1), the observed pK_a values for the corresponding binary and ternary complexes are lower (Table 4). Further, APADPH reacts with the enzyme-DHF complex more slowly than does NADPH. The latter reaction, in turn, is slower than the interaction of DHF with the enzyme-NADPH complex which approaches the diffusion rate[32] (see Figure 6).

B. RAPID REACTION KINETICS

A more detailed understanding of the kinetic mechanism of an enzyme-catalyzed reaction

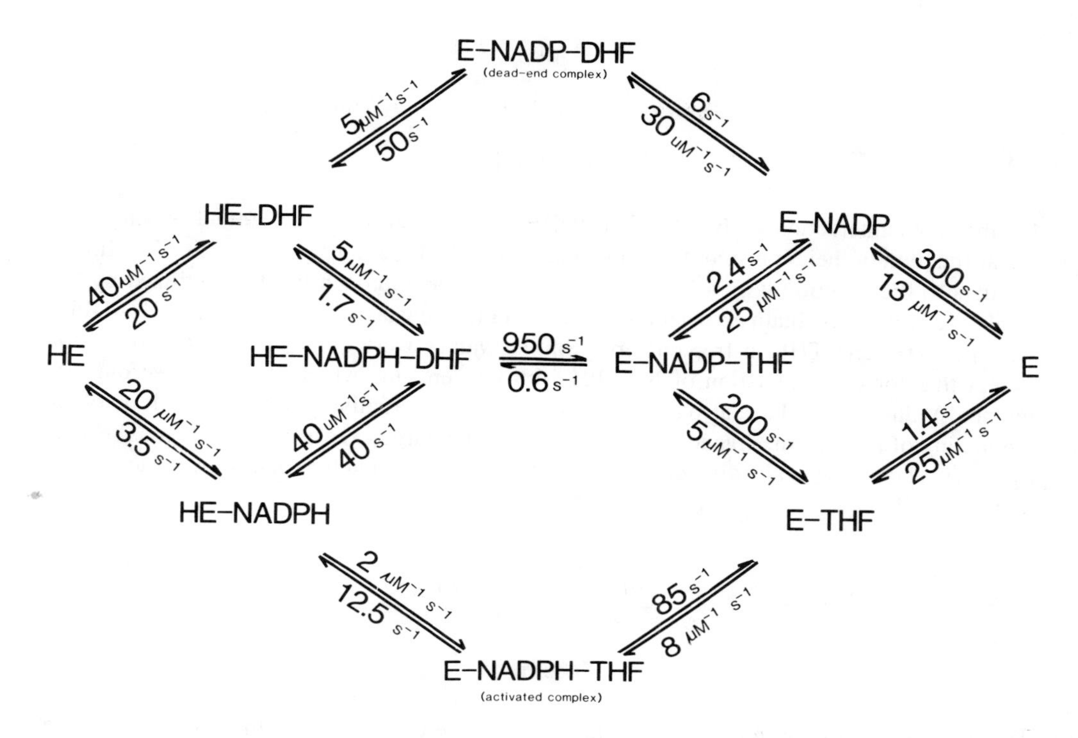

FIGURE 6. Rate constants associated with the various steps of the kinetic scheme for the DHFR reaction.

requires data for the variation with pH of the rate constants associated with each of the steps along the reaction pathway. With such information in hand, it becomes possible to reach conclusions about the flux along any reaction pathway and to identify rate-limiting steps at any pH value. Predictions can be made about the shift in the pK_a value of an ionizing group at the active site from a true to an apparent value because of the kinetic effects of a variable substrate. In addition, values of kinetic parameters and of kinetic isotope effects at any pH can be calculated.

1. Rate Constants Associated with the Formation of Binary and Ternary Enzyme Complexes

Determinations of the bimolecular and unimolecular rate constants for the interaction of NADPH, NADP, DHF and THF with the free form of DHFR and the appropriate binary complexes have been made by the use of stopped-flow fluorescence.[23,38] The technique requires relatively large amounts of DHFR as the enzyme and substrate are used at comparable concentrations. Studies of this type with DHFR have been facilitated recently by the availability of strains of *E. coli* which overproduce the enzyme[23,39] and by the fact that the binding of substrate causes a high degree of quenching of the intrinsic fluorescence of DHFR.

a. Formation of Binary Complexes

The formation of the enzyme-NADPH and enzyme-DHF complexes has been found to occur in two phases.[23,38] The first phase is rapid and dependent on the ligand concentration while the slower phase is independent of the ligand concentration. The data suggest that the enzyme exists in two forms (E_1 and E_2) which are in equilibrium, that rapid binding of the substrates occurs with one form only (E_1) and that there is then a slow conversion of E_2 to E_1 (Scheme I). The observed rate constants

$$E_2 \rightleftharpoons E_1 \underset{R_{off}}{\overset{k_{on}L}{\rightleftharpoons}} E_1L$$

Scheme I

for the rapid interaction of NADPH and DHF with the enzyme were found to vary as a linear function of their concentrations and thus values for k_{on} and k_{off} were obtained by fitting the data to the function $k_{obs} = k_{on} [L] + k_{off}$ The same approach has been utilized by the Benkovic group[23] to obtain association and dissociation constants for the binding of NADPH, NADP, DHF, and THF to free enzyme. Similar values were obtained by the two groups except that for the association of NADPH with free enzyme which showed a tenfold difference. Values for k_{off} have also been determined by an alternative procedure.[23,38] It involves the mixing of the enzyme-ligand complex (E-L_1) with a large excess of an analogous second ligand (L_2) so that when L_1 dissociates from E-L_1, it is immediately replaced by L_2 according to Scheme II.

$$E\text{–}L_1 \underset{k_1L_1}{\overset{k_2}{\rightleftharpoons}} E \underset{k_4}{\overset{k_3L_2}{\rightleftharpoons}} EL_2$$

Scheme II

The dissociation rate is monitored by the fluorescence change due to the difference in the fluorescence quenching by the two ligands. Under conditions where $k_3[L_2] \gg k_1[E]\ [L_1]\ k_2$, the observed rate is equal to k_2. The two methods yield comparable values for the dissociation constants. However, the dissociation constants calculated from the ratio of k_{off} to k_{on} are higher than the experimentally determined values with DHF and NADPH[38] or with DHF.[23] It appears that this is due to subsequent isomerizations of the binary enzyme-NADPH and enzyme-DHF complexes which are not detected by the technique of fluorescence quenching. The binding of NADP and THF to DHFR cannot involve isomerizations of the initial collision complexes as the kinetically and thermodynamically determined values of the dissociation constants are not significantly different. Values for the uni- and bimolecular rate constants are shown in Figure 6.

b. Formation of Ternary Complexes

Dissociation constants for the release of NADPH and DHF from the enzyme-NADPH-DHF complex cannot be measured directly because the ternary complex undergoes reaction. However, as thionicotinamide-adenine-dinucleotide phosphate (TNADPH) is a nonreacting analogue of NADPH,[13] it can be utilized with stopped flow fluorescence techniques to obtain both association and dissociation constants for DHF which might be expected to be similar to those which apply with NADPH.[23] The association rate constant of 25 $\mu M^{-1}\ s^{-1}$ observed for the reaction of DHF with the enzyme-TNADPH complex is similar to that of 40 $\mu M^{-1}\ s^{-1}$, as determined in single turnover experiments[23] (see Figure 6), while the dissociation rate constant is 40 s^{-1}. No direct determinations have been made of the rate of dissociation of NADPH from a ternary enzyme-NADPH-folate analogue complex. It was considered that a 2,4-diaminopteridine, such as MTX, would not be a suitable analogue of DHF.[23] This is because NADPH markedly enhances the binding of 2,4-diaminopteridines to DHFR (Table 3), but has little synergistic effect on the binding of pterins.[3] As the reaction rate with folate is so low, it may well be a good analogue of DHF for measurements of the release of NADPH from a ternary complex. Because the DHFR reaction is essentially irreversible, rate constants associated with the formation of an enzyme-NADPH-THF complex could be determined (Figure 6).

Rate constants for the formation of a dead-end enzyme-NADP-DHF complex[23] have been determined and are included in Figure 6. The ratio of the constants for the binding of DHF to the enzyme-NADP complex and of NADP to the enzyme-DHF complex yield dissociation constants of comparable magnitude to those determined kinetically.[11] These data indicate that the presence of one ligand on the enzyme causes a three- to fourfold enhancement of the binding of the other.

The formation of ternary enzyme-NADP(H)-THF complexes has the effect of enhancing the rate at which THF dissociates from the enzyme (Figure 6). While the dissociation rate is 1.4 s^{-1} for the enzyme-THF complex, it is 2.4 s^{-1} for the enzyme-NADP-THF complex and 12.5 s^{-1} for the enzyme-NADPH-THF complex. Such findings have implications with respect to the kinetic mechanism of the reaction. No isomerization reactions for the aforementioned ternary complexes were observed.[23]

C. PRE-STEADY-STATE KINETICS

1. Occurrence of More than One Form of DHFR

In stopped flow experiments it was observed that, when the reaction is started by the addition of DHFR, there is a time-dependent increase of enzymic activity.[40] However, when either NADPH or DHF is pre-incubated with the enzyme, the lag period is eliminated and there is a 2.3-fold increase in enzymic activity. This activation, which has been interpreted in terms of the relatively slow interconversion of two enzyme forms (Scheme I), draws attention to the need for preincubation of one substrate with the enzyme before undertaking pre-steady-state kinetics. Such preincubation is not necessary for most steady-state kinetic experiments as the $t_{1/2}$ value for the interconversion process is 9 s.[40]

2. Burst of Product Formation

On the rapid mixing of DHF with the enzyme-NADPH complex, or of NADPH with the enzyme-DHF complex, there is a pre-steady-state burst of product formation.[23] The burst is best monitored by fluorescence energy transfer which permits measurement of the interconversion of the enzyme-NADPH-DHF and the enzyme-NADP-THF complexes. The result indicates that the rate-determining step must occur after the catalytic reaction. Both the rate and amplitude of the burst decrease as the pH increases. Analysis of the data shows that the pH dependency is due to the ionization of a single group at the active site of the enzyme with a pK_a value of 6.5 and that the pH-independent rate of the interconversion is 950 s^{-1} (see Figure 8). The pK_a value for the enzyme-NADPH-DHF complex is similar to that for the free enzyme and the binary enzyme-substrate complexes as is to be expected when the binding of substrates to DHFR is pH independent.[23,37,38] The interconversion of the ternary complexes in the forward direction exhibits a kinetic deuterium isotope effect of three. The value is similar to that observed at high pH in steady-state kinetic studies (see Table 6), but low compared to those for several other dehydrogenases.[34]

D. COMPLETE KINETIC MECHANISM FOR REACTION

The kinetic mechanism of the DHFR reaction, in the pH-independent region where the enzyme exists in its fully protonated form, is illustrated quantitatively in Figure 6. The scheme allows for the random addition of substrates as well as the random release of products. However, the magnitudes of the rate constants make it clear that there would be a preferred pathway of product release. This would involve the rapid release of NADP from the ternary enzyme-NADP-THF complex and the replacement of NADP by NADPH to produce an activated complex from which THF dissociates at a faster rate than from either the enzyme-THF or the enzyme-NADP-THF complexes. The catalytic cycle could then commence again through the interaction of DHF with the resulting enzyme-NADPH complex. It has been considered that the release of THF is the rate-limiting step of the reaction and that it is only through the activation of its release by NADPH that the rate of this step would correspond

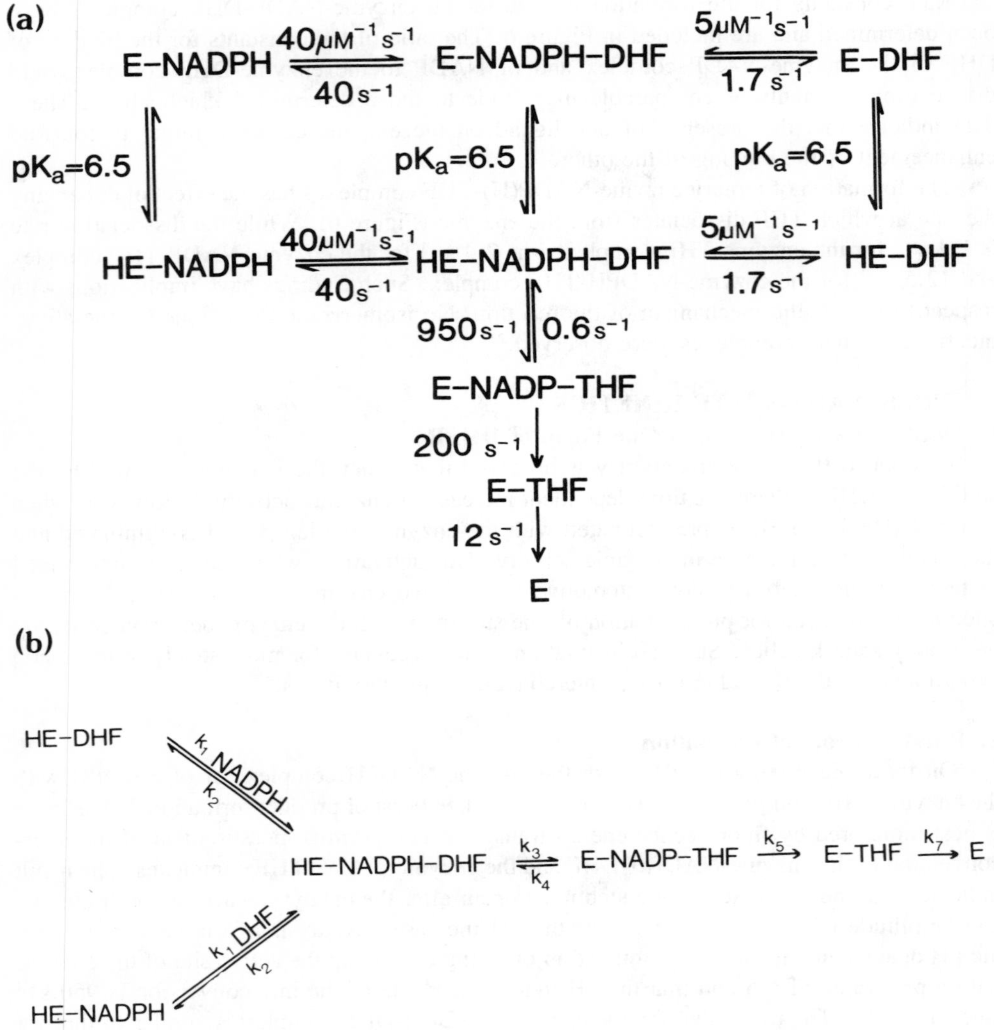

(a)

$$\text{E-NADPH} \underset{40\,s^{-1}}{\overset{40\,\mu M^{-1}s^{-1}}{\rightleftharpoons}} \text{E-NADPH-DHF} \underset{5\,\mu M^{-1}s^{-1}}{\overset{1.7\,s^{-1}}{\rightleftharpoons}} \text{E-DHF}$$

pK_a=6.5 (each vertical equilibrium)

$$\text{HE-NADPH} \underset{40\,s^{-1}}{\overset{40\,\mu M^{-1}s^{-1}}{\rightleftharpoons}} \text{HE-NADPH-DHF} \underset{5\,\mu M^{-1}s^{-1}}{\overset{1.7\,s^{-1}}{\rightleftharpoons}} \text{HE-DHF}$$

$$\text{HE-NADPH-DHF} \underset{0.6\,s^{-1}}{\overset{950\,s^{-1}}{\rightleftharpoons}} \text{E-NADP-THF} \xrightarrow{200\,s^{-1}} \text{E-THF} \xrightarrow{12\,s^{-1}} \text{E}$$

(b)

HE-DHF (k_1 NADPH, k_2); HE-NADPH (k_1 DHF, k_2)

$$\text{HE-NADPH-DHF} \underset{k_4}{\overset{k_3}{\rightleftharpoons}} \text{E-NADP-THF} \xrightarrow{k_5} \text{E-THF} \xrightarrow{k_7} \text{E}$$

FIGURE 7. (a), Rate constants and pK_a values of the reaction steps to be considered for the calculation of observed kinetic parameters listed in Table 6. (b), Rate constant numbers which were used to obtain the relationships given in Table 5.

to the maximum velocity of 12.5 s^{-1}.[23] While it is clear that NADPH causes substrate activation, there is no general agreement that such activation is an essential part of THF release. A higher maximum velocity of 17.7 s^{-1} has been reported under conditions where there is no downward curvature of a plot of 1/v against 1/NADPH.[32] This value agrees well with that of 17.6 s^{-1} which has been given recently.[41]

Simulation procedures are useful in showing that the experimental progress curves obtained under a variety of conditions can be described by theoretical progress curves constructed by use of experimentally determined rate constants and pK_a values. This procedure[42] has been utilized to show that the data of Figure 6 can predict well the progress curves obtained for the DHFR reaction.[23] A further test of the validity of any kinetic scheme is to use the rate constants and pK_a values to calculate the values of the kinetic parameters, under any chosen conditions, for comparison with those determined experimentally. To outline the way in which such calculations are performed, there is presented in Figure 7a a reduced

TABLE 5
Relationships for Calculation of Predicted Values for Kinetic Parameters

Parameter		Relationship
V/E_t		$\frac{k_3k_5k_7}{k_3(k_5 + k_7) + k_7(k_4 + k_5)(1 + K_2/H)}$
V/KE_t		$\frac{k_1k_3k_5}{k_3k_5 + k_2(k_4 + k_5)(1 + K_2/H)}$
K_m (low pH)[a]		$\frac{k_7(k_2k_4 + k_2k_5 + k_3k_5)}{k_1 [k_7 (k_4 + k_5) + k_3(k_5 + k_7)]}$
App pK_a from[a] V/K profile		$pK_1 + \log \quad 1 + \left[\frac{k_3/k_2}{1 + k_4/k_5}\right]$
App pK_a from V profile		$pK_1 + \log \quad 1 + \left[\frac{k_3(k_5 + k_7)}{k_7(k_4 + k_5)}\right]$
$D_{V/K)}$[b]	low pH	$\frac{{}^Dk + k_3/k_2 + {}^DK_{eq}\, k_4/k_5}{1 + k_3/k_2 + k_4/k_5}$
	high pH	$\frac{{}^Dk + {}^DK_{eq}\, k_4/k_5}{1 + k_4/k_5}$
D_V	low pH	$\frac{{}^Dk + k_3/k_5 + k_3/k_7 + {}^DK_{eq}\, k_4/k_5}{1 + k_3/k_5 + k_3/k_7 + k_4/k_5}$
	high pH	$\frac{{}^Dk + {}^DK_{eq}\, k_4/k_5}{1 + k_4/k_5}$

Note: The rate constants are those used in Figure 7.

[a] k_1 and k_2 are used for both the interaction of DHF with E-NADPH and for the interaction of NADPH with E-DHF

[b] D_k is the intrinsic deuterium isotope effect.

version of the complete kinetic scheme (Figure 6) which includes only those steps which must be taken into account. Further, Figure 7b shows the rate constants which are used for the relationships given in Table 5. The results of the calculations, which are listed in Table 6, illustrate the good agreement between the predicted and pH-independent (at low pH) values for V, V/K_{NADPH} and V/K_{DHF}. The data also predict well the pK_a values that are observed in the pH profiles (Figure 4). It might be noted that the relationships for the Michaelis constants which contain k_7 are not very sensitive to the magnitude of this rate constant and that the rate of interaction of DHF with the enzyme-NADPH complex approaches the diffusion rate.

The rate-limiting step of an enzyme-catalyzed reaction can change with pH whenever its chemistry involves an ionizing group at the active center of the enzyme. Such a change is illustrated graphically for the DHFR reaction (Figure 8). The upper curve represents the pH profile for the conversion of HE-NADPH-DHF to E-NADP-THF. It exhibits a pK_a value of 6.5 and a pH-independent rate of 950 s^{-1} (cf Figures 6 and 7a). The lower straight line represents the pH-independent release of THF at 12 s^{-1}. As the pH increases, the rate of the chemical step slows down because of the decrease in the proportion of total enzyme that

TABLE 6
Comparison of Predicted and Experimental Values for Kinetic Parameters

Parameter	Predicted value	Experimental value	Ref.
V (s^{-1})	11.8	12.5[a]	23
$(V/K_{NADPH}(\mu M^{-1}\ s^{-1})$	5.0	3.4	32
K_{NADPH} (μM)	2.3	2.5	11
$(V/K)_{DHF}$ ($\mu M^{-1}\ s^{-1}$)	38	20	36
K_{DHF} (μM)	0.3	0.5	11
Apparent pK_a from V profile	8.4	8.4	32, 36
Apparent pK_a from $(V/K)_{NADPH}$ profile	9.3	8.9	32
Apparent pK_a from $(V/K)_{DHF}$ profile	7.9	8.1	32
D_V			
Low pH	1.02	1.0—1.3	12, 23, 33
High pH	3.0	2.9	12, 23, 33
$D_{(V/K)NADPH}$			
Low pH	1.0	1.1	12, 33
High pH	3.0	3.0	12, 33
$D_{(V/K)DHF}$			
Low pH	1.08	1.8	33
High pH	3.0	3.0	33

Note: Values for the kinetic parameters were calculated by using the rate constants given in Figure 7 and the relationships listed in Table 5. Values for V, V/K, and the Michaelis constants are those that apply in the pH-independent region (Figure 4a and c). It was assumed that the equilibrium isotope effect, $D_{K_{eq}}$ is equal to one and that the intrinsic isotope effect, D_k is 3 (Table 5).

[a] Values of 17.7 s^{-1},[36] 17.0 s^{-1},[41] and 16.5 s^{-1},[11,13] have also been reported for V.

is present in protonated form (see Figure 7a). Theoretically the point can be reached where it becomes equal to and then less than the pH-independent release of product. The intersection of the two curves occurs at pH 8.4 which is the observed pK_a value from the V profile. Above this pH, it is the chemistry, rather than product release, that is rate limiting. Deuterium isotope effects are therefore observed at higher, but not lower, pH values (see Section E).

E. KINETIC DEUTERIUM ISOTOPE EFFECTS

1. Reduction of DHF: Order of Protonation and Hydride Transfer

Data from pH profiles do not allow any conclusions to be reached about the order of protonation and hydride transfer from NADPH for the reduction of the 5,6-imine double bond of DHF. Protonation could occur before, after or at the same time as hydride transfer. However, conclusions can be reached on the basis of the variation with pH of deuterium isotope effects.

Kinetic deuterium isotope effects have been determined by varying one substrate in the presence of a saturating concentration of the other.[33] The value for ^{D}V varies from 1.1 ± 0.1 at pH 7.0 to 2.9 ± 0.2 at pH 10.2. These results, like those of Figure 7, indicate that product release is largely rate limiting at neutral pH while catalysis becomes more rate limiting as the pH increases. Over the same pH range the value for $^{D}(V/K)_{NADPH}$ increases from 1.1 ± 0.1 to 3.0 ± 0.2 whereas that for $^{D}(V/K)_{DHF}$ increases from 1.8 ± 0.1 to 3.0 ± 0.2. The data with NADPH confirm the conclusion reached from viscosity experiments[32] that NADPH behaves as a sticky substrate at neutral pH. The deuterium isotope data also indicate that DHF exhibits stickiness which is less than that with NADPH and which is not observed in viscosity experiments.

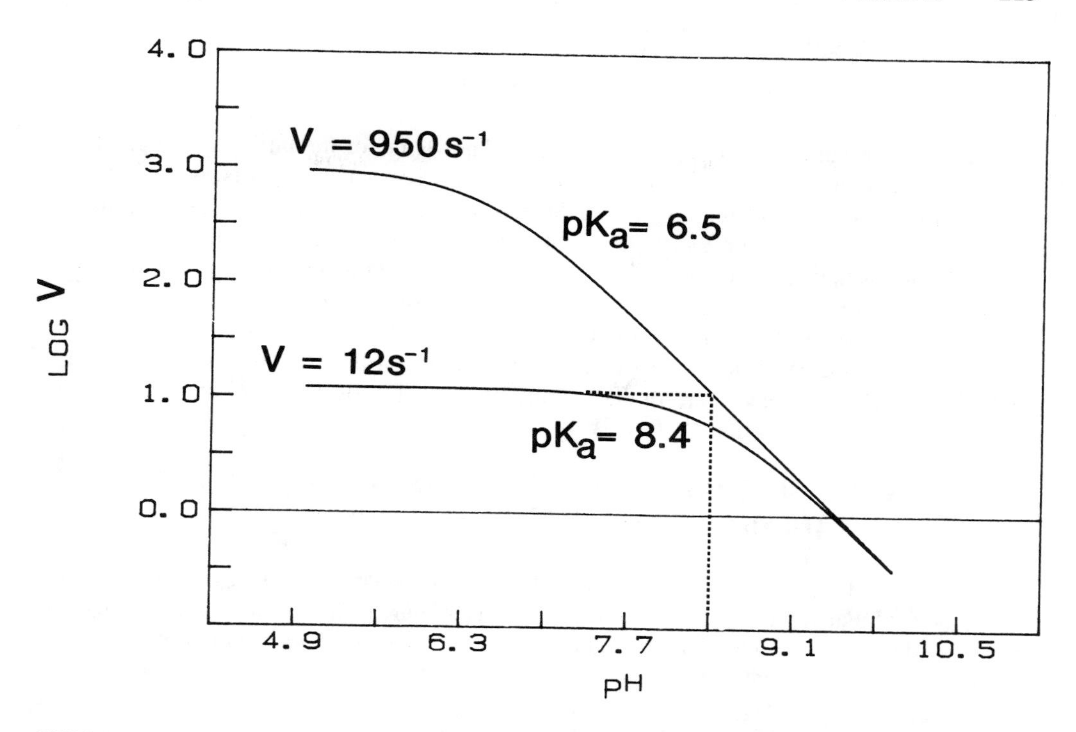

FIGURE 8. Graphic illustration of the change with increasing pH of the rate-limiting step of the DHFR reaction from pH-independent release of THF (12 s^{-1}) to pH-dependent chemical step (pH-independent rate, 950 s^{-1}; pK_a = 6.5). The lower curve represents the V profile. The maximum velocity of the reaction falls when the chemical step becomes rate limiting.

The increase with pH of the values for $^{D}(V/K)_{NADPH}$, $^{D}(V/K)_{DHF}$, and ^{D}V towards a limiting value of 3 is in accord with the three possible mechanisms for reduction of the double bond of DHF. Differentiation between the mechanisms would be facilitated by a knowledge of the intrinsic isotope effect (^{D}k) for the reaction.[34] However, such information is not available and is likely to be difficult to obtain because of the inherent instability of THF. If 3 were the value for ^{D}k, then it would have to be concluded that protonation and hydride transfer are concerted. However, this sequence cannot be mandatory as a mutant form of DHFR, which cannot catalyze the protonation step, is able to catalyze hydride transfer from NADPH to protonated DHF[43] (Section V.E). If ^{D}k is assumed to be greater than 3, calculations indicate that the value of the internal commitment for $^{D}(V/K)_{NADPH}$ is low irrespective of whether protonation precedes or follows hydride transfer.[33] The low value is not consistent with hydride transfer occurring first. Thus it must be concluded from the deuterium isotope effect data that the reduction of DHF proceeds by protonation of the substrate, followed by hydride transfer (or by an essentially concerted reaction). This conclusion is in accord with the demands of chemistry[44] and the original proposal for the reaction.[45] Theoretical calculations favor this mechanism although they also provide some support for an alternative mechanism that involves protonation of the 0-4 oxygen of DHF[46] (see Section VI).

2. Calculation of Intrinsic pK_a Values for Binary Complexes

Theoretically, it is possible to use deuterium isotope effects to calculate the intrinsic pK_a values of binary enzyme-substrate complexes which have been shifted outwards in pH profiles because of substrate stickiness. This is done by subtracting from the observed pK_a value of the V/K profile, the log value of the stickiness term. For a reaction which exhibits a pH-independent value for $^{D}(V/K)$ at low pH, the stickiness term is defined as

$$\text{stickiness term} = \frac{{}^{D}(V/K)_{\text{high pH}}{}^{-1}}{{}^{D}(V/K)_{\text{low pH}}{}^{-1}}$$

It will be noted that calculations do not require a knowledge of the intrinsic isotope effect. However, such calculations are not likely to be precise because of errors associated with the determination of values for $^{D}(V/K)$ and the need to have limiting values of this parameter. This is especially true when the variable substrate exhibits considerable stickiness. Under these conditions the value of $^{D}(V/K)_{\text{low pH}}$ approaches one and the magnitude of the stickiness term becomes very sensitive to the experimentally-determined $^{D}(V/K)_{\text{low pH}}$ value which can be expected to have a standard error of at least 5%. Calculations using the deuterium isotope effect data (Section IV.E.1) suggest that the intrinsic pK_a values for the enzyme-NADPH and enzyme-DHF complexes are in the vicinity of 7.5 rather than 6.5 as obtained by more direct measurements (Section IV.B.1.b).

V. STRUCTURE OF DHFR-LIGAND COMPLEXES AND CHEMICAL MECHANISM OF REACTION

The complete kinetic scheme for the DHFR reaction that is outlined in Figures 6 and 7a can be formulated without any knowledge of the structure of the active site of the enzyme or of the identity of the amino acid residue that participates in the protonation of DHF. But such information is essential for the presentation of a realistic chemical mechanism for the conversion of DHF to THF.

A. OVERALL STRUCTURE OF COMPLEXES

Three-dimensional structures have been determined, at high resolution, for complexes of two bacterial chromosomal enzymes with folate analogues either in the absence or presence of NADPH. The complexes are enzyme-MTX,[6,7] enzyme-TMP,[15,47] and enzyme-NADPH-TMP[15] with *E. coli* DHFR as well as enzyme-NADPH-MTX with *L. casei* DHFR.[6,7] Suitable crystals have not been obtained for structural determinations of the free bacterial enzymes or for their complexes with one of the two substrates. The overall structures of the enzymes from *E. coli* and *L. casei* are similar. Their basic backbones are folded into an eight-stranded β sheet. (Figure 9). Seven of the strands, which begin at the N-terminus, are parallel while the last, at the C-terminus, is antiparallel. Four α helices are inserted into the long loops of chain connecting the successive β strands.

B. STRUCTURE OF ACTIVE SITE

The folding of the *E. coli* DHFR in the binary enzyme-TMP complex differs little from that of the ternary enzyme-NADPH-TMP complex.[15] Further, there is considerable similarity between the structures of the ternary *L. casei* DHFR-NADPH-MTX complex and the binary enzyme-MTX complex.[6,7] Consequently it will be assumed that any conclusions reached about the binding of NADPH or folate analogues to the *L. casei* enzyme are also valid for the binding of the same ligands to the enzyme from *E. coli*. The X-ray data indicate that the active site of DHFR is a well-defined cleft which is about 15Å deep and which runs across one whole face of the enzyme (Figure 9). There are distinct pockets or subsites for the binding of the pyridine nucleotides and pterin substrates (or substrate analogues). Thus the structural data are in accord with the conclusion that NADPH and DHF add randomly to the enzyme (see Section IV.A).

1. Binding of MTX at Pterin Subsite

MTX is bound at the active site in extended conformation with the pteridine ring almost perpendicular to the aromatic ring of the p-aminobenzoyl moiety and pointing into a hy-

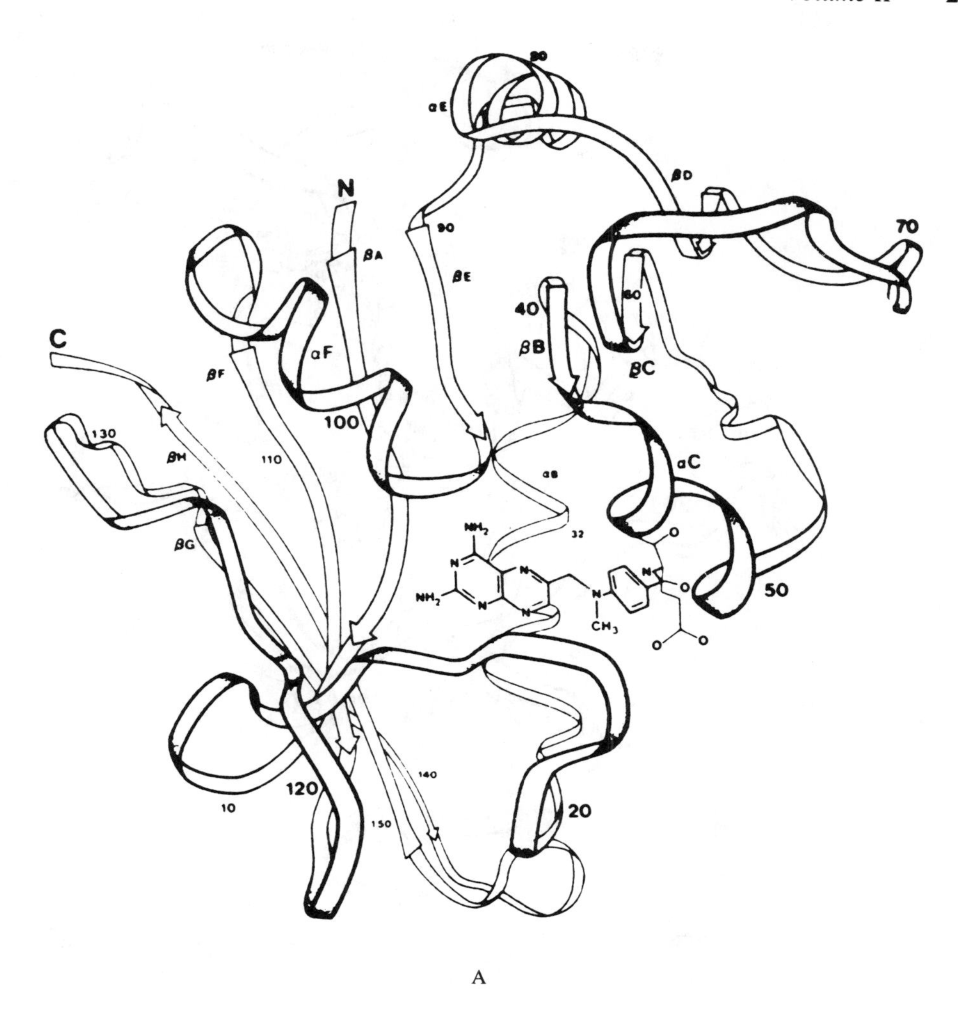

FIGURE 9. Backbone ribbon drawing of (A) the binary *E. coli* DHFR-methotrexate complex and (B) the ternary *L. casei* DHFR-NADPH-methotrexate complex. (From Bolin, J. T., Filman, D. J., Matthews, D. A., Hamlin, R. C., and Kraut, J., *J. Biol. Chem.*, 257, 13650, 1982. With permission.)

drophobic cavity. The residues involved in the binding of the pteridine moiety are highly conserved in both bacterial and vertebrate dihydrofolate reductases.[6] All are hydrophobic except Asp-27 which is the only ionizing residue at the sub-site (Figure 9a). Hence it must be the one whose ionization is observed in pH profiles and which is the source of the proton for the reduction of DHF.[36,43] The hydrophobic environment would account for the elevated pK_a value of about 6.5 for a carboxyl group. Asp-27 is close to the N-1 nitrogen of MTX and appears to interact through the formation of a $E\text{-}COO^-\text{-}H\text{-}MTX^+$ ion pair complex. There is considerable spectroscopic and NMR data to indicate that, at neutral pH, the N-1 nitrogen of MTX with a pK_a value of 5.7 in solution, is protonated on the enzyme.[31,48-50] The pK_a of bound MTX is raised to about 10. The pteridine moiety also makes hydrophobic or van der Waals interactions with several amino acid residues of which Ala-7 and Phe-31 are conserved. H-bonding occurs either directly or through water molecules to conserved Thr-113, Ile-5 and Ile-94. The pyrazine moiety is not directly hydrogen-bonded to the enzyme. The N-5 nitrogen makes no hydrogen bonds at all while that to N-8 is weak and indirect. The p-aminobenzoyl moiety, which occupies a hydrophobic pocket, interacts with Leu-28 and Phe-31 (conserved) and makes van der Waals contacts with Ile-50, Ile-94, and

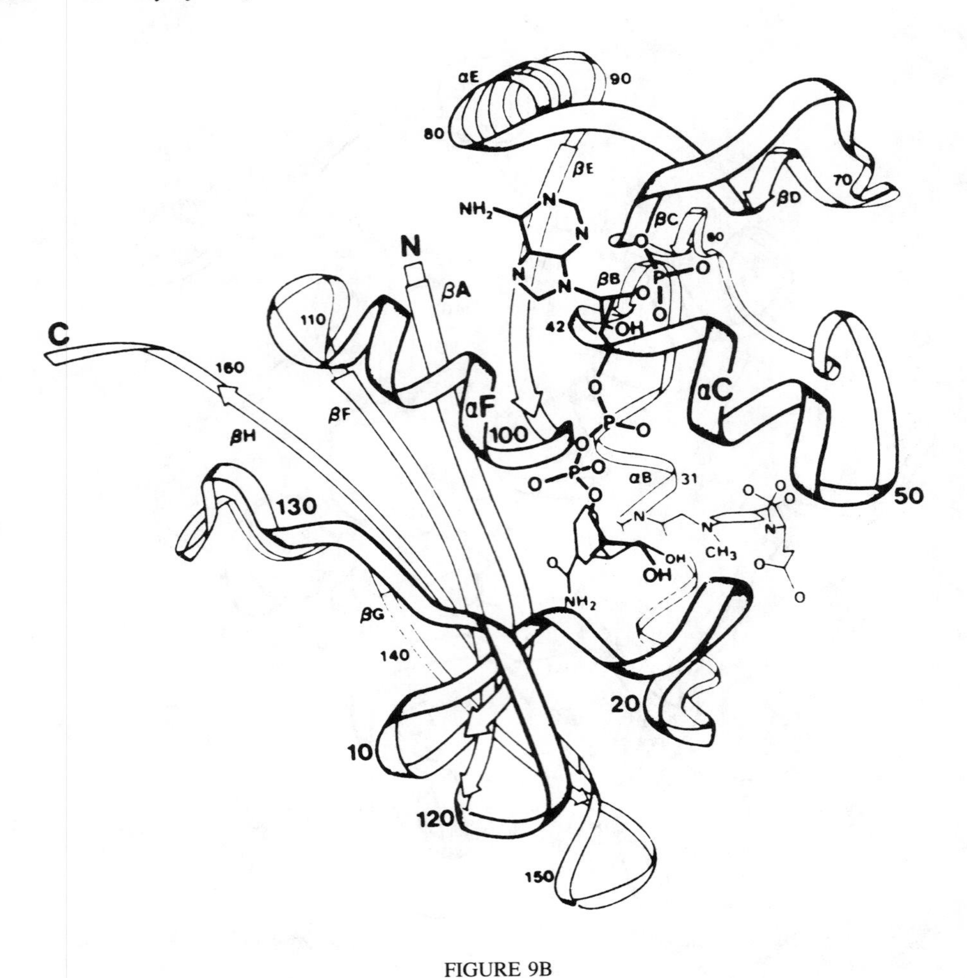

FIGURE 9B

Leu-54 (conserved). The α-carboxyl group of the glutamate moiety reacts with Arg-57 (conserved) while the binding of the γ-carboxyl group of the same moiety appears to be weak and nonspecific.

2. Binding of NADPH at the Pyridine Nucleotide Subsite

No three-dimensional structures are available for any complex of *E. coli* DHFR that contains NADPH. By contrast, the structure of the ternary enzyme-NADPH-MTX complex with DHFR from *Lactobacillus casei* has been determined a high resolution.[6,7] Thus, to complete the picture of the binding of ligands at the active site of DHFR, brief mention will be made of the binding of NADPH to the enzyme from this source.[7] NADPH is bound to DHFR in an extended conformation within a long cleft on the enzyme surface that stretches across five strands of β sheet (Figure 9b). The nicotinamide moiety lies within the hydrophobic cavity with a cluster, on the β-side of the ring, of hydrophobic side chains which excludes water molecules. The A-side of the ring is in contact with the pteridine at its binding site and this observation is consistent with the finding that hydride transfer for DHFR-catalyzed reactions is A-side specific. The adenine moiety binds directly to the enzyme through hydrophobic interactions with several residues of which only Leu-62 is conserved. The pyrophosphate moiety is bound through extensive hydrogen bonds to backbone amido groups as well as to the side chain of Arg-44 and Thr-46 (conserved). The 2′-phosphate group forms hydrogen-bonds and salt bridges with four side chains and this is almost certainly

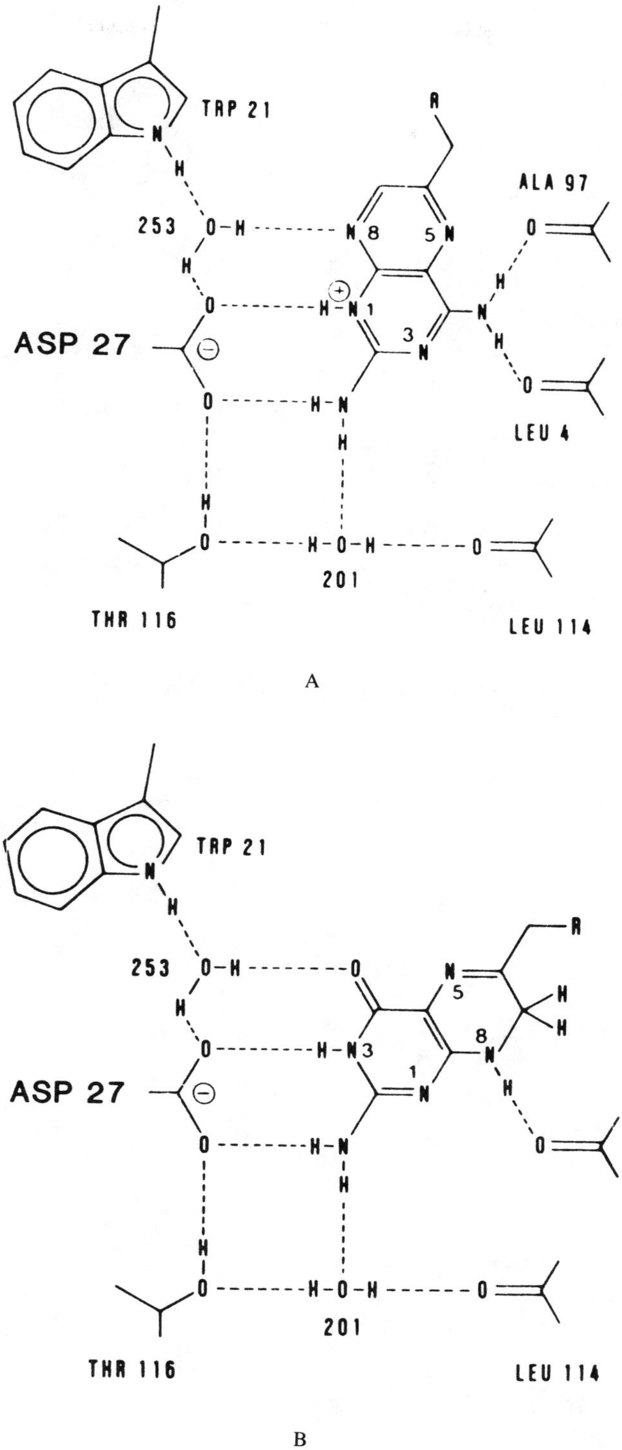

A

B

FIGURE 10. Representation of the binding at the pterin subsite of DHFR: of the pteridine moiety of methotrexate to (a), wild type enzyme (c), Asn-27 mutant enzyme (d), Ser-27 mutant enzyme and of the pterin moiety of dihydrofolate to (b), wild-type enzyme. The structure given in panel (b) was predicted by computer modeling.[6,43]

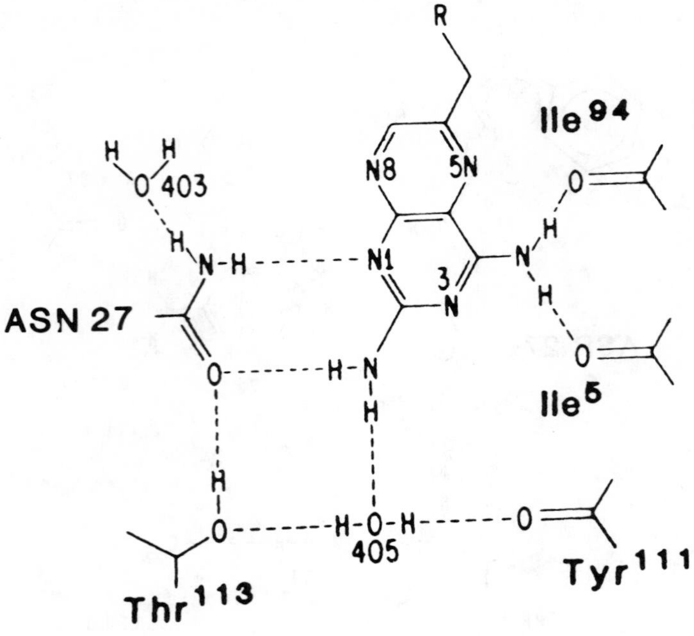

FIGURE 10C

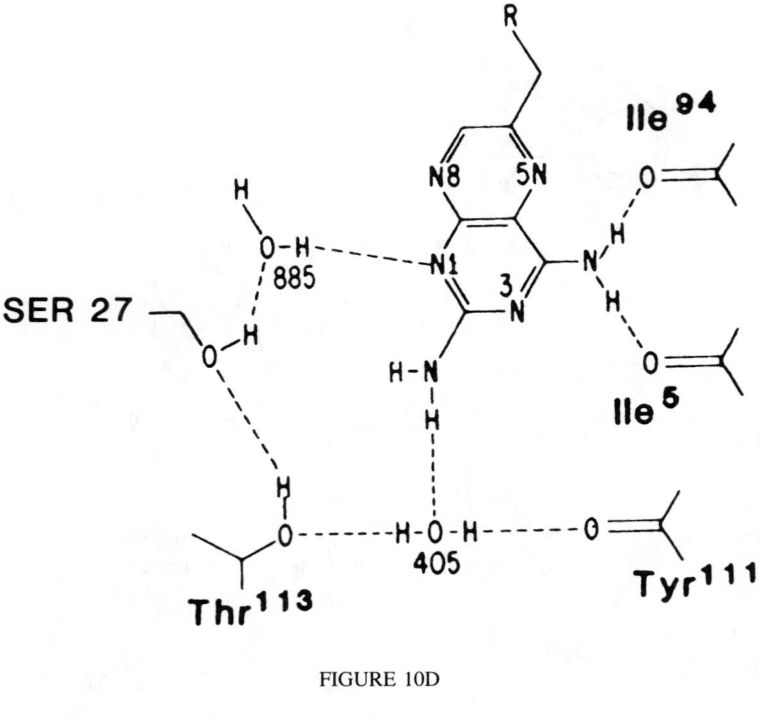

FIGURE 10D

the reason why NADPH is bound more strongly than NADH. The ribosyl moieties occupy solvent-exposed regions of the sub-site and their binding is by means of hydrophobic contacts and hydrogen-bonding of the 2′- and 3′-hydroxyl groups.

C. POSTULATED BINDING OF DHF AT PTERIN SUBSITE

From the three-dimensional structure of the enzyme-MTX complex, it becomes apparent that the pteridine ring of DHF cannot be bound at the active site with the same geometry as MTX.[6] If this were so, hydride transfer from NADPH to the C-6 carbon of DHF would predict an absolute configuration of R rather than S as observed experimentally (Figure 10a). Thus, the pterin ring of bound DHF, as well as bound folate (see Section II.D), must be turned upside down from the way in which the pteridine ring of MTX is bound. As yet there have been no reports on the binding of a pterin substrate to DHFR, but a model structure for the binding of the 4-keto tautomer of DHF to the enzyme has been proposed[6] (Figure 10b). The model predicts that the N-3 nitrogen of DHF, instead of the N-1 nitrogen of MTX, will be within hydrogen-bonding distance of the carboxyl oxygen atom of Asp-27. Further, it shows that there is a loss of one hydrogen bond in the enzyme-DHF complex relative to the enzyme-MTX complex which could contribute towards the weaker binding of DHF.

D. CHEMICAL MECHANISM OF REACTION

From the catalytic point of view, it is of particular interest that the proposed model places the N-5 nitrogen of DHF at such a distance from Asp-27 that protonation of the nitrogen atom by transfer of a proton from Asp-27 could not be direct. However, there has been postulated a mechanism which allows the indirect transfer of a proton[33] and which is illustrated in Figure 11. The mechanism envisions that the water molecule Wat-253, which is bound to MTX and Trp-21 in the enzyme-MTX complex (Figure 10a) is hydrogen-bonded to Trp-21, protonated Asp-27 and the carbonyl group of DHF in the ternary enzyme-NADPH-DHF complex. Then, as a result of electron and hydrogen transfers (panel a), the enolic form of DHF is produced with concomitant formation of a negative charge on Asp-27 and a positive charge on the N-3 nitrogen of DHF (panel b). Rotation of the enolic group allows the lone pair of electrons on the N-5 nitrogen of DHF to make a nucleophilic attack on the hydrogen atom of the group and this attack would be enhanced by the movement of electrons towards the positively charged N-3 nitrogen (panel c). The formation of a positive charge on the N-5 nitrogen would cause the C-6 carbon to develop carbonium ion character and thereby facilitate nueclophilic attack by the hydride ion from NADPH to form THF (panel d). The hydrogen-bonding network (Figures 10 and 11) undoubtedly plays an important role in the unfavorable transfer of a proton from a carboxyl group (Asp-27) with a pK_a value of 6.5 (Figure 7) to one with a pK_a in solution of 3.8 (DHF).[51] The proton transfer would be driven by the setting up of a stable hydrogen-bond network which can form with a negatively charged carboxyl group. Certainly the formation of such a structure would lower the pK_a value of the carboxyl group and could raise that of the N-5 nitrogen of DHF. It seems likely that protonated DHF would be detected only in the ternary enzyme complex and that the steady-state concentration of the complex containing protonated DHF would be low. Such a conclusion is in accord with reports that DHF is not protonated when bound to DHFR at neutral pH.[50] Theoretical aspects of the binding of DHF at the active site of DHFR and the mechanism of its conversion to THF have been considered recently.[52]

E. MUTANT FORMS OF DHFR

Site-directed mutagenesis provides an ideal method for the elucidation of the function of conserved amino acid residues at the active site of an enzyme. However, when an amino acid residue has been altered, it is important that the structural integrity of the mutant enzyme be confirmed. This is necessary to ensure that the alteration has not resulted in distortions

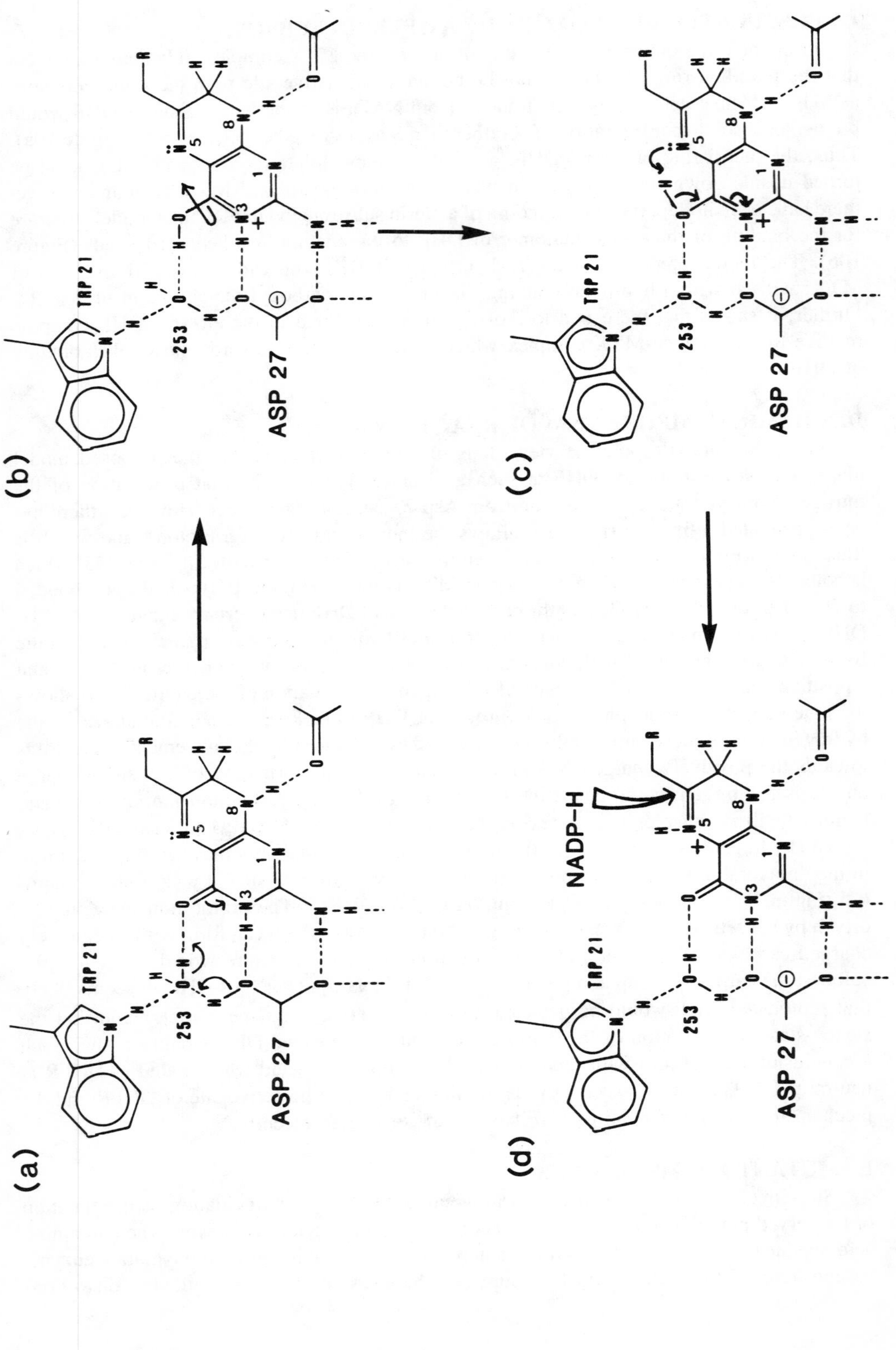

(a)

TRP 21

253

ASP 27

(b)

TRP 21

253

ASP 27

(c)

TRP 21

253

ASP 27

(d)

NADP-H

TRP 21

253

ASP 27

FIGURE 11. Proposed chemical mechanism for the protonation of, and transfer of a hydride ion to, dihydrofolate as catalyzed by DHFR. (From Morrison, J. F. and Stone, S. R., *Biochemistry*, 27, 5499, 1988. With permission.)

in other parts of the molecule which cause unexpected changes at the active site and which preclude the interpretation of kinetic data. Particular attention has been paid to this point for the studies on DHFR mutants.[43]

1. Functional Role of Asp-27

The results of kinetic investigations have pointed strongly to the function of Asp-27 as a proton donor. This conclusion can be checked by changing the amino acid residue to one which is incapable of acting as a proton donor and such a change has been made. Two mutant forms of *E. coli* DHFR have been obtained in which Asp-27 is replaced by Asn-27 or Ser-27.[43] The Asn-27 mutant was produced by oligonucleotide-directed mutagenesis of the cloned wild-type *E. coli* gene while the Ser-27 mutant was obtained as a spontaneous revertant of the Asn-27 mutant. The two mutant enzyme forms were crystallized as binary enzyme-MTX complexes and their structures determined by X-ray crystallography (Figure 10c and d). Their crystal structures were isomorphous with each other and with the wild-type enzyme. Electron-density difference maps of the DHFR (Asn-27)-MTX and DHFR (Ser-27)-MTX complexes relative to the DHFR (Asp-27)-MTX complex showed that, essentially, there were changes only in the positions of two water molecules close to the MTX binding site.

In accord with predictions, the activity at neutral pH of the Asn-27 mutant ($0.10\ s^{-1}$) and of the Ser-27 mutant ($0.44\ s^{-1}$) is low compared with that of the wild type enzyme under the same conditions ($30\ s^{-1}$).[43] The K_{DHF} values for the mutant enzymes are also elevated about 40- and 140-fold, respectively. The fact that some activity was observed suggested that the function of Asp-27 might be circumvented and this suggestion was confirmed by the finding that both the V and V/K_{DHF} values increase as the pH decreases. It appears that the protonated form of DHF, which arises at low pH ($pK_a = 3.8$),[51] can function as a substrate for the mutant form of DHFR which is still capable of catalyzing hydride transfer. Indeed, the activity of the mutant enzymes at low pH is, at least, comparable to that of the wild-type enzyme in the pH-independent region. The true V/K value of about $80\ \mu M^{-1}\ s^{-1}$ for the reaction of protonated DHF with the DHFR (Ser-27)-NADPH complex is similar to that of $40\ \mu M^{-1}\ s^{-1}$ for the reaction of DHF with the wild-type DHFR-NADPH complex (Table 6, Figure 6). However, it should be mentioned that both the V and V/K profiles with each mutant form of enzyme exhibit slopes which are greater than the expected value of -1 for the protonation of DHF.

The protonated form of DHF ($DHFH^+$) functions as a substrate for the mutant forms of DHFR and thus, it must be bound at the pterin sub-site of each enzyme prior to hydride transfer. This conclusion contrasts with the finding that there is no formation of an enzyme-$MTXH^+$ complex either through interaction of a mutant enzyme and $MTXH^+$ or through reaction of a proton with a neutral mutant enzyme-MTX complex.[53] As an enzyme-$MTXH^+$ complex does form when MTX binds to the wild-type DHFR with an ionizing Asp-27 residue (see Section E.2), the failure to observe similar complexes with the mutant forms of DHFR may be due to the lack of charge neutralization. By contrast, charge neutralization of $DHFH^+$ undergoing reaction at the active site of a mutant DHFR would occur because of hydride transfer from NADPH to yield neutral THF.

2. Role of Asp-27 in MTX Binding

Asp-27 is not only invoked with the catalytic function of DHFR, but also with the binding of MTX.[43] The dissociation constants at neutral pH have been reported as 0.07 n*M* for wild-type enzyme, 1.9 n*M* for the Asn-27 mutant enzyme and 210 n*M* for the Ser-27 mutant enzyme. The replacement of the ion pair between the N-1 nitrogen of MTX and Asp-27 by a hydrogen bond between the same nitrogen and Asn-27 reduces the binding energy by about 1.8 kcal/mol. This decrease is much less than had been expected. Indeed, it was generally considered that the ion pair formation was a considerable factor in the strong binding of MTX to DHFR. However, the result is in agreement with the report that there is only about a tenfold decrease in the binding of MTX to protonated DHFR as compared with the binding of protonated MTX to the ionized enzyme.[37] The further decrease in the binding of MTX to the Ser-27 mutant enzyme is due to structural changes as well as to the loss of the ionic interaction and the hydrogen bond to the 2-amino group (Figure 10).

3. Functions of Thr-113, His-45, and Leu-54

Thr-113 is a conserved amino acid residue whose function appears to be the maintenance of a hydrogen bond network (Figure 10) and a relatively low value for K_{DHF}. When this residue is replaced by Val-113, the maximum velocity of the reaction does not change, but the V/K_{DHF} value is reduced about 40-fold to 0.56 μM^{-1} s^{-1} while the K_{DHF} value is increased from 1 μM to 25 μM.[41] It has been argued that the increase in the Michaelis constant for DHF is likely to be due to a decrease in the binding of this substrate. The suggestion arises from the finding that the binding of MTX to this mutant enzyme form is reduced because of the increase in the off rate relative to that for the wild-type enzyme. Replacement by Gln-45 of the His-45 residue, which would be expected to interact with the nicotinamide 5′-phosphate moiety of NADPH, does result in the weaker binding of the pyridine nucleotide.[41] However, it does not alter the value for K_{DHF}. Replacement by glycine of the conserved Leu-54 residue, which interacts with the benzoylglutamate moiety of DHF, does cause an elevation of the K_{DHF} value.[54] The maximum velocity of the reaction is not altered. However, there is a change in the rate-limiting step from the release of THF with the wild type DHFR to the catalytic step with the mutant enzyme.

VI. SUMMARY AND CONCLUSIONS

Through the application of a wide variety of chemical and biochemical techniques, dihydrofolate reductase from *E. coli* has come to be one of the best understood enzymes with respect to its structure and catalytic function. Both the pterin and pyridine nucleotide substrates have specific binding subsites within the active site of the enzyme so that they can add in a random manner. The subsite for the pterin substrate contains only a single ionizing amino acid residue (Asp-27) which is not involved with the binding of substrate. Rather, in its acid form, it is the source of a proton for protonation of the N-5 nitrogen of DHF which precedes and facilitates hydride transfer from NADPH for the formation of THF. The carboxyl group of Asp-27 in the free enzyme, as well as in binary and ternary complexes, has an elevated pK_a value of about 6.5 This is because of its hydrophobic environment. The apparent pK_a value of the group can be raised to over 8.0. In the V/K profile this is due to the kinetic effects of the substrates. In the V profile the elevated value arises because it is not until higher pH values are reached that the catalytic rate falls to become comparable with and then less than the pH-independent rate of product release. A mutant form of enzyme, which cannot function as a proton donor, is still able to catalyze hydride transfer under conditions where DHF is chemically protonated. At neutral pH the chemistry of the reaction is much faster than product release and it is not until above pH 8.0 that the chemical step(s) becomes rate limiting. The dissociation of THF from the enzyme is much slower than NADP and can be increased through the formation of an enzyme-NADPH-THF complex. This is not the usual type of dead-end complex which simply dissociates to yield the components involved in its formation. Rather it is a complex that provides an alternative pathway for the release of THF. By contrast, the enzyme-NADP-DHF complex is a simple dead-end complex. The DHFR reaction at neutral pH is one in which sticky substrates add randomly to the enzyme, catalysis is not rate limiting and there is a preferred order of product release.

It is especially interestng that, while the function of Asp-27 is in catalysis and not in substrate binding, it is involved in the binding of inhibitors which are 2,4-diaminopteridines or 2,4-diaminopyrimidines. This is a consequence of the difference in the geometry of binding for DHF and these inhibitors. The interaction involves ion pair formation between protonated inhibitor and the ionized carboxyl group of Asp-27. It was considered that such an interaction in a hydrophobic environment was the prime reason for the tight-binding of MTX. But the results of studies with mutant enzyme forms do not support this proposal. It follows from the aforementioned findings that the variation with pH of the K_i value for a substrate analogue does not necessarily indicate that the ionizing group is also involved with substrate binding.

There are still several aspects of the DHFR reaction that require study. These include determinations of the rates of protonation and hydride transfer as well as the elucidation of the function of the various conserved amino acid residues within the subsites for NADPH and DHF. In addition, it is important to know why folate is reduced so slowly by DHFR from *E. coli* relative to the rate of reduction of DHF. The reasons for the faster rate of folate reduction by DHFR from chicken liver and the mechanism of its conversion to DHF also require elucidation. But perhaps the most interesting studies for the future will be undertaken with the tetrameric enzyme that is coded for by the R-plasmid of *E. coli*. The enzyme is a reasonably efficient catalyst, is poorly inhibited by MTX and TMP and its active site structure differs considerably from that of the chromosomal enzyme from *E. coli*. It is apparent that a different pathway for protonation of, and hydride transfer to, DHF must apply. Interest must also center on human DHFR which is available in relatively large amounts. The techniques and protocols that have been established should encourage more detailed investigations by those who are interested in DHFR, but who in the past, have been somewhat overawed by the kinetic approach to the understanding of enzyme action and inhibition.

REFERENCES

1. **Blakley, R. L.,** Dihydrofolate reductase, in *Folates and Pterins* Vol. 1, Blakley, R. L. and Benkovic, S. J., Eds., Wiley-Interscience, New York, 1984, chap. 5.
2. **Baccanari, D., Phillips, A., Smith, S., Sinski, D., and Burchall, J.,** Purification and properties of *Escherchia coli* dihydrofolate reductase, *Biochemistry,* 14, 5267, 1975.
3. **Stone, S. R. and Morrison, J. F.,** Mechanism of inhibition of dihydrofolate reductases from bacterial and vertebrate sources by various classes of folate analogues, *Biochem. Biophys. Acta,* 869, 275, 1986.
4. **Roth, B. and Cheng, C. C.,** Recent progress in the medicinal chemistry of 2,4-diaminopyrimidines, *Prog. Med. Chem.,* 19, 269, 1982.
5. **McCormack, J. J.,** Dihydrofolate reductase inhibitors as potential drugs, *Med. Res. Rev.,* 1, 303, 1981.
6. **Bolin, J. T., Filman, D. J., Matthews, D. A., Hamlin, R. C., and Kraut, J.,** Crystal structures of *Escherichia coli* and *Lactobacillus casei* dihydrofolate reductase refined at 1.7Å resolution. I. General features and binding of methotrexate, *J. Biol. Chem.,* 257, 13650, 1982.
7. **Filman, D. J., Bolin, J. T., Matthews, D. A., and Kraut, J.,** Crystal structures of *Escherichia coli* and *Lactobacillus casei* dihydrofolate reductase refined at 1.7Å resolution. II. Environment of bound NADPH and implications for catalysis, *J. Biol. Chem.,* 257, 13663, 1982.
8. **Volz, K. W., Matthews, D. A., Alden, R. A., Freer, S. T., Hansch, C., Kaufman, B. T., and Kraut, J.,** Crystal structure of avian dihydrofolate reductase containing phenyltriazine and NADPH, *J. Biol. Chem.,* 257, 2528, 1982.
9. **Matthews, D. A., Smith, S. L., Baccanari, D. P., Burchall, J. J., Oatley, S. J., and Kraut, J.,** Crystal structure of a novel trimethoprim-resistant dihydrofolate reductase specified in *Escherichia coli* by R-plasmid R67, *Biochemistry,* 25, 4194, 1986.

10. **Stuber, D., Bujard, H., Hochuli, E., Kocher, H. P., Kompis, I., Talmadge, K., Weibel, E. K., Winkler, F. K., and Then, R. L.,** Enzymatic characterization of recombinant human dihydrofolate reductase produced in *E. coli,* in *Chemistry and Biology of Pteridines,* Cooper, B. A. and Whitehead, V. M., Eds., Walter de Gruyter, Berlin, 1986, 839.
11. **Stone, S. R. and Morrison, J. F.,** Kinetic mechanism of the reaction catalyzed by dihydrofolate reductase from *Escherichia coli, Biochemistry,* 21, 3757, 1982.
12. **Morrison, J. F. and Stone, S. R.,** Effect of pH on reactions catalyzed by dihydrofolate reductase from chicken liver, in *Chemistry and Biology of Pteridines,* Cooper, B. A. and Whitehead, B. M., Eds., Walter de Gruyter, Berlin, 1986, 831.
13. **Stone, S. R., Mark, A., and Morrison, J. F.,** Interaction of analogues of nicotinamide adenine dinucleotide phosphate with dihydrofolate reductase from *Escherichia coli, Biochemistry,* 23, 4340, 1984.
14. **Baccanari, D. P., Daluge, S., and King, R. W.,** Inhibition of dihydrofolate reductase: effect of reduced nicotinamide adenine dinucleotide phosphate on the selectivity and affinity of diaminobenzylpyrimidines, *Biochemistry,* 21, 5068, 1982.
15. **Champness, J. N., Stammers, D. K., and Beddell, C. R.** Crystallographic investigation of the cooperative interaction between trimethoprim, reduced cofactor and dihydrofolate reductase, *FEBS Lett.,* 199, 61, 1986.
16. **Blakley, R. L. and Appleman, J. R.,** Recent advances in the study of dihydrofolate reductase, in *Chemistry and Biology of Pteridines,* Cooper, B. A. and Whitehead, B. M., Eds., Walter de Gruyter, Berlin, 1986, 769.
17. **Hyde, E. I., Birdsall, B., Roberts, G. C. K., Feeney, J., and Burgen, A. S. V.,** Proton nuclear magnetic resonance saturation transfer studies of coenzyme binding to *Lactobacillus casei* dihydrofolate reductase, *Biochemistry,* 19, 3738, 1980.
18. **Hyde, E. I., Birdsall, B., Roberts, G. C. K., Feeney, J., and Burgen, A. S. V.,** Phosphorus-31 nuclear magnetic resonance studies of the binding of oxidized coenzymes to *Lactobacillus casei* dihydrofolate reductase, *Biochemistry,* 19, 3746, 1980.
19. **Fontecilla-Camps, J. C., Bugg, C. E., Temple, C., Rose, J. D., Montgomery, J. A., and Kisliuk, R. L.,** Absolute configuration of biological tetrahydrofolates. A crystallographic determination, *J. Am. Chem. Soc.,* 101, 6114, 1979.
20. **Charlton, P. A., Young, D. W., Birdsall, B., Feeney, J., and Roberts, G. C. K.,** Stereochemistry of reduction of folic acid using dihydrofolate reductase, *J. Chem. Soc. Chem. Comm.,* 922, 1979.
21. **Mathews, C. K. and Huennekens, F. M.,** Further studies on dihydrofolate reductase. *J. Biol. Chem.,* 238, 3436, 1963.
22. **Nixon, P. F. and Blakley, R. L.,** Dihydrofolate reductase of *Streptococcus faecium, J. Biol. Chem.,* 243, 4722, 1968.
23. **Fierke, C. A., Johnson, K. A., and Benkovic, S. J.,** Construction and evaluation of the kinetic scheme associated with dihydrofolate reductase, *Biochemistry,* 26, 4085, 1987.
24. **Stone, S. R., Montgomery, J. A., and Morrison, J. F.,** Inhibition of dihydrofolate reductase from bacterial and vertebrate sources by folate, aminopterin, methotrexate and their 5-deaza analogues, *Biochem. Pharmacol.,* 33, 175, 1984.
25. **Morrison, J. F.,** The slow-binding and slow, tight-binding inhibition of enzyme-catalyzed reactions, *Trends Biochem. Sci.,* 7, 102, 1982.
26. **Cha, S.,** Tight-binding inhibitors. I. Kinetic behavior, *Biochem. Pharmacol.,* 24, 2177, 1975.
27. **Blakley, R. L. and Cocco, L.,** Role of isomerization of initial complexes in the binding of inhibitors to dihydrofolate reductase, *Biochemistry,* 24, 4772, 1985.
28. **Matthews, D. A., Bolin, J. T., Burridge, J. M., Filman, D. J., Volz, K. W., Kaufman, B. T., Beddell, C. R., Champness, J. N., Stammers, D. K., and Kraut, J.,** Refined crystal structures of *Escherichia coli* and chicken liver dihydrofolate reductase containing bound trimethoprim, *J. Biol. Chem.,* 260, 381, 1985.
29. **Matthews, D. A., Bolin, J. T., Burridge, J. M., Filman, D. J., Volz, K. W., and Kraut, J.,** Dihydrofolate reductase. The stereochemistry of inhibitor selectivity, *J. Biol. Chem.,* 260, 392, 1985.
30. **Bartlett, P. A. and Marlowe, C. K.,** Possible role for water dissociation in the slow binding of phosphorus-containing transition-state-analogue inhibitors of thermolysin, *Biochemistry,* 26, 8553, 1987.
31. **Baccanari, D. P., Stone, D., and Kuyper, L.,** Effect of a single amino acid substitution on *Escherichia coli* dihydrofolate reductase catalysis and ligand binding, *J. Biol. Chem.,* 256, 1738, 1981.
32. **Stone, S. R. and Morrison, J. F.,** Dihydrofolate reductase from *Escherichia coli:* the kinetic mechanism with NADPH and reduced acetylpyridine-adenine-dinucleotide phosphate as substrates, *Biochemistry,* 27, 5493, 1988.
33. **Morrison, J. F. and Stone, S. R.,** Mechanism of the reaction catalyzed by dihydrofolate reductase from *Escherichia coli*: pH and deuterium isotope effects with NADPH as the variable substrate, *Biochemistry,* 27, 5499, 1988.

34. **Cleland, W. W.,** Use of isotope effects to elucidate enzyme mechanisms, *Crit. Rev. Biochem.,* 13, 385, 1982.
35. **Cleland, W. W.,** Enzyme kinetics as a tool for determination of enzyme mechanisms in *Investigations of Rates and Mechanisms of Reactions,* Vol. 6, Bernasconi, C. F., Ed., John Wiley & Sons, New York, 1986, 791.
36. **Stone, S. R. and Morrison, J. F.,** Catalytic mechanism of dihydrofolate reductase as determined by pH studies, *Biochemistry,* 23, 2753, 1984.
37. **Stone, S. R. and Morrison, J. F.,** The pH-dependence of the binding of dihydrofolate and substrate analogues to dihydrofolate reductase from *Escherichia coli, Biochem. Biophys. Acta,* 745, 247, 1983.
38. **Cayley, P. J., Dunn, S. M., and King, R. W.,** Kinetics of substrate, coenzyme, and inhibitor binding to *Escherichia coli* dihydrofolate reductase, *Biochemistry,* 20, 874, 1981.
39. **Smith, D. R., Rood, J. I., Bird, P. I., Sneddon, M. K., Calvo, J. M., and Morrison, J. F.,** Amplification and modification of dihydrofolate reductase in *Escherichia coli, J. Biol. Chem.,* 257, 9043, 1982.
40. **Penner, M. H. and Frieden, C.,** Substrate-induced hysteresis in the activity of *Escherichia coli* dihydrofolate reductase, *J. Biol. Chem.,* 260, 5366, 1985.
41. **Chen, J.-T., Mayer, R. J., Fierke, C. A., and Benkovic, S. J.,** Site-specific mutagenesis of dihydrofolate reductase from *Escherichia coli, J. Cell. Biochem.,* 29, 73, 1985.
42. **Barshop, B. A., Wrenn, R. F., and Frieden, C.,** Analysis of numerical methods for computer simulation of kinetic processes: development of KINSIM — a flexible, portable system, *Anal. Biochem.,* 130, 134, 1983.
43. **Howell, E. E., Villafranca, J. E., Warren, M. S., Oatley, S. J., and Kraut, J.,** Functional role of aspartic acid-27 in dihydrofolate reductase revealed by mutagenesis, *Science,* 231, 1123, 1986.
44. **Lund, H.,** Electrochemistry of pteridines, in *Chemistry and Biology of Pteridines,* Vol. 5, Pfleiderer, W., Ed., Walter de Gruyter, Berlin, 1975, 645.
45. **Huennekens, F. M. and Scrimgeour, K. G.,** Tetrahydrofolic acid — the coenzyme of one-carbon metabolism, in *Pteridine Chemistry,* Pfleiderer, W. and Taylor, E. C., Eds., Pergamon Press, Elmsford, NY, 1964, 355.
46. **Gready, J. E.,** Theoretical studies on the activation of the pterin cofactor in the catalytic mechanism of dihydrofolate reductase, *Biochemistry,* 24, 4761, 1985.
47. **Baker, D. J., Beddell, C. R., Champness, J. N., Goodford, P. J., Norrington, F. E. A., Smith, D. R., and Stammers, D. K.,** The binding of trimethoprim to bacterial dihydrofolate reductase, *FEBS Lett.,* 126, 49, 1981.
48. **Cocco, L., Groff, J. P., Temple, C., Montgomery, J. A., London, R. E., Matwiyoff, N. A., and Blakley, R. L.,** Carbon-13 nuclear magnetic resonance study of protonation of methotrexate and aminopterin bound to dihydrofolate reductase, *Biochemistry,* 20, 3972, 1981.
49. **Cocco, L., Roth, B., Temple, C., Montgomery, J. A., London, R. E., and Blakley, R. L.,** Protonated state of methotrexate, trimethoprim and pyrimethamine bound to dihydrofolate reductase, *Arch. Biochem. Biophys.,* 226, 567, 1983.
50. **Gready, J. E.,** Dihydrofolate reductase: binding of substrates, inhibitors and catalytic mechanism, *Adv. Pharm. Chemother.,* 17, 37, 1980.
51. **Poe, M.,** Acidic dissociation constants of folic acid, dihydrofolic acid and methotrexate, *J. Biol. Chem.,* 252, 3724, 1977.
52. **Andrews, P. R., Sadek, M., Spark, M. J., and Winkler, D. A.,** Conformational energy calculations and electrostatic potentials of dihydrofolate reductase ligands: relevance to mode of binding and species specificity, *J. Med. Chem.,* 29, 698, 1986.
53. **London, R. E., Howell, E. E., Warren, M. S., Kraut, J., and Blakley, R. L.,** Nuclear magnetic resonance study of the state of protonation of inhibitors bound to mutant dihydrofolate reductase lacking the active-site carboxyl, *Biochemistry,* 25, 7229, 1986.
54. **Mayer, R. J., Chen, J.-T., Taira, K., Fierke, C. A., and Benkovic, S. J.,** Importance of a hydrophobic residue in binding and catalysis by dihydrofolate reductase, *Proc. Natl. Acad. Sci. U.S.A.,* 83, 7718, 1986.

Chapter 9

2-OXOACID DEHYDROGENASES

M. Hamada and H. Takenaka

TABLE OF CONTENTS

I. INTRODUCTION

Enzyme systems that catalyze a lipoic acid-mediated oxidative decarboxylation of 2-oxoacids have been isolated from microbial and eukaryotic cells as functional units with molecular weights in the millions.[1-4] Three types of complexes have been obtained, one specific for pyruvate (pyruvate dehydrogenase complex, PDC), a second specific for 2-oxoglutarate (2-oxoglutarate dehydrogenase complex, OGDC), and a third specific for the branched-chain 2-oxoacids, 2-oxoisovalerate, 2-oxoisocaproate and 2-oxo-4-methylvalerate (branched-chain 2-oxoacid dehydrogenase complex, BCOADC). Each complex is composed of multiple copies of three major components: a substrate-specific dehydrogenase (El); a dihydrolipoamide acyltransferase (E2) that is specific for each type of complex; and dihydrolipoamide dehydrogenase (E3), a FAD-containing protein that is a common component of the three types of multienzyme complexes. The complexes are organized about a core, consisting of the oligomeric E2, to which multiple copies of E1 and E3 are noncovalently bound.

The three component enzymes, acting in sequence, catalyze the reactions shown in Figure 1.[2,8] E1 catalyzes both the decarboxylation of the 2-oxoacid (reaction 1) and the subsequent reductive acylation of the lipoyl moiety (reaction 2), which is covalently bound to E2. E2 catalyzes the transacylation step (reaction 3), and E3 catalyzes the reoxidation of the dihydrolipoyl moiety with NAD^+ as the ultimate electron acceptor (Reactions 4 and 5). The PDCs from mammalian and avian tissues and *Neurospora crassa,* and the mammalian BCOADC also contain small amounts of two specific regulatory enzymes, pyruvate dehydrogenase kinase and phosphatase and branched-chain 2-oxoacid dehydrogenase kinase and phosphatase, respectively. The specific kinase and phosphatase modulate the activity of E1 by phosphorylation and dephosphorylation.[9-13] There are no reports that the PDC or the BCOADC in prokaryotic cells or the OGDC in eukaryotic or prokaryotic cells undergo phosphorylation/dephosphorylation.

Possible advantages of multienzyme complexes over individual activities in solution have been proposed.[14]

1. *Catalytic enhancement:* the reduction of diffusion time of an intermediate from one enzyme to the next.
2. *Substrate channeling:* the control over which biosynthetic route an intermediate should follow, by directing it to a specified enzyme rather than allowing competition from other enzymes in solution.
3. *Sequestration of reactive intermediates:* the protection of chemically unstable intermediates from aqueous solution.
4. *Servicing:* the rapid intramolecular acetyl transfer reactions that are observed in the 2-oxoacid dehydrogenase complexes.

This review discusses some aspects of the structural organization, reaction mechanisms, and regulatory properties of the 2-oxoacid dehydrogenase complexes.

II. SUBUNIT COMPOSITION AND STRUCTURE

A. *ESCHERICHIA COLI* PYRUVATE DEHYDROGENASE COMPLEX

The *E. coli* (Crookes strain) PDC (M_r = 4.6 million) has been separated into its three component enzymes, E1, E2 and E3, and the complex has been reconstituted from the isolated enzymes.[15,16] The subunit composition of the complex is shown in Table 1. The complex consists of 12 E1 dimers (M_r = 192,000), 24 E2 monomers ($M_r \simeq$ 1.6 million), and 6 E3 dimers (M_r = 112,000).[18,23] Each E2 chain contains at least two and possibly

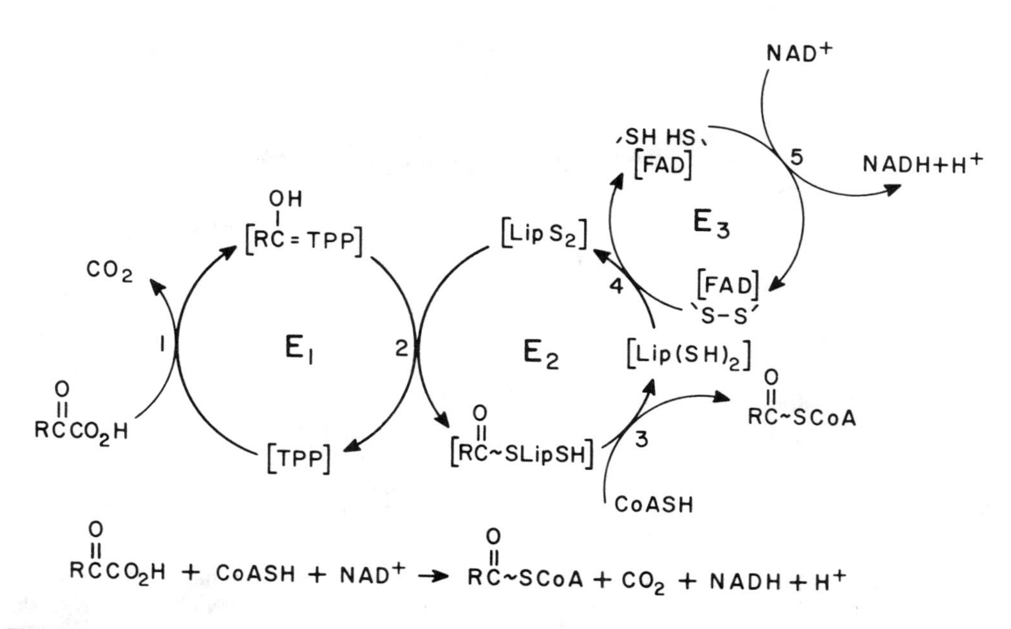

FIGURE 1. Reaction sequence in α-keto acid oxidation. The abbreviations used are TTP, thiamine pyrophosphate; $LipS_2$ and $Lip(LH)_2$, lipoyl moiety and its reduced form; CoASH, coenzyme A: FAD, flavin adenine dinucleotide; NAD^+ and NADH, nicotinamide adenine dinucleotide and its reduced form; E_1, pyruvate dehydrogenase; E_2, dihydrolipoamide acyltransferase; E_3, dihydrolipoamide dehydrogenase. (From Reed, L. J. and Yeaman, S. J., *The Enzymes*, Boyer, P. D. and Krebs, E. Eds., Academic Press, New York, in press.)

TABLE 1
Subunit Composition Of *E. Coli* Pyruvate and 2-Oxoglutarate Dehydrogenase Complexes

Enzyme	Mr	Subunits No.	Subunits Mr	Subunits per molecule of complex
E. coli PDC native complex	4,600,000			
E1	192,000	2	96,000	24
E2	1,548,000	24	64,500	24
E3	112,000	2	56,000	12
E. coli OGDC native complex	2,500,000			
E1	190,000	2	95,000	12
E2	1,000,000	24	42,000	24
E3	112,000	2	56,000	12

Note: The abbreviations used are PDC, pyruvate dehydrogenase complex; OGDC, 2-oxoglutarate dehydrogenase complex.

three molecules of covalently bound lipoic acid. Electron micrographs of the *E. coli* PDC show a polyhedral structure with a diameter about 300 Å. The acetyltransferase (E2) has the appearance of a cube, and its design appears to be based on octahedral 432 symmetry. The pyruvate dehydrogenase (E1) dimers are apparently located at the 12 twofold positions (i.e., on the edges) of the acetyltransferase cube, and the dihydrolipoamide dehydrogenase (E3) dimers are located at the 6 four-fold positions.[2,8]

Mass measurements of proteins by scanning transmission electron microscopy (STEM) are based on the electron scattering properties of unstained, lyophilized samples at low electron dosages. STEM mass measurements carried out on the PDC from *E. coli* show that

TABLE 2
Subunit Composition Of Bovine Heart Pyruvate Dehydrogenase Complex[8,30]

Enzyme	Mr	Subunits No.	Subunits Mr	Subunits per molecule of complex
Native complex (heart)	8,500,000			
E1	154,000	4		
E1α		2	41,000	60
E1β		2	36,000	60
E2	3,100,000	60	52,000	60
E3	110,000	2	55,000	12
Kinase	~100,000	1	48,000	
		1	45,000	
Phosphatase	~150,000	1	97,000	
		1	50,000	

the purified complex is homogeneous with respect to particle size and mass, that the M_r is $(5.28 \pm 0.40) \times 10^6$, and that the M_r values for the three component enzymes are $(2.06 \pm 0.26) \times 10^5$ for dimeric E1, $(1.15 \pm 0.17) \times 10^5$ for dimeric E3, and $(2.20 \pm 0.17) \times 10^6$ for E2, the 24-subunit core enzyme.[46] These M_r measurements support the subunit composition 24:24:12 for E1/E2/E3 that has been proposed on the basis of other data.[2,23,47,48]

B. MAMMALIAN PYRUVATE DEHYDROGENASE COMPLEX

The PDCs isolated from bovine kidney and heart mitochondria have molecular weights of about 7,000,000 and 8,500,000, respectively. The component enzymes of the two complexes are very similar, if not identical.[30] E1 has a M_r of about 154,000 and possesses the subunit composition $\alpha_2\beta_2$ (Table 2). The M_r of the two subunits are about 41,000 and 36,000, respectively. The core enzyme (E2) has a M_r of about 3,100,000 and consists of 60 apparently identical polypeptide chains of M_r about 52,000. Each E2 chain contains one covalently bound lipoyl moiety. The isolated E3 is a homodimer of M_r about 110,000 and contains two molecules of FAD.

The bovine kidney PDC contains about 20 E1 tetramers ($\alpha_2\beta_2$) and about 6 E3 dimers, whereas the heart complex contains about 30 E1 tetramers and 6 E3 dimers. The kidney complex can bind about 10 additional E1 tetramers, but neither complex can bind additional E3 dimers. The dissociation constant (K_d) of the E1-E2 subcomplex is about 13 n*M*, and the K_d of the E2-E3 subcomplex is about 3 n*M*.[31] In the electron microscope, E2 has the appearance of a pentagonal dodecahedron, and its design appears to be based on icosahedral 532 symmetry.[8] The E1 tetramers appear to be located on the 30 edges and the E3 dimers in the 12 faces of E2.

The kinase is tightly bound to E2 and copurifies with the PDC. The amount of endogenous kinase in the bovine kidney and heart complexes is small, about three molecules per molecule of kidney complex and less in the heart complex. The phosphatase also binds to E2, and this attachment requires the presence of Ca^{2+} ions.[32] There appear to be about five molecules of phosphatase per molecule of complex in bovine kidney mitochondria.

Recently, an additional component polypeptide of mammalian PDC, protein X, has been reported. This subunit has an apparent M_r of 50,000 and is capable of undergoing acetylation when the complex is incubated with pyruvate or with acetyl-CoA and NADH. This indicates the presence of a lipoic acid residue in protein X. It was generally thought for many years that protein X was a proteolytic fragment of E2, which copurified with the E2 core during resolution of the complex. However, immunological and biochemical studies have provided strong evidence that X is a distinct polypeptide.[38,39] Although the physiological role of protein X has yet to be established, a possible involvement in anchoring the kinase to E2 has been suggested.[40]

C. 2-OXOGLUTARATE DEHYDROGENASE COMPLEXES

The *E. coli* OGDC has been resolved into three enzymes, 2-oxoglutarate dehydrogenase (E1), dihydrolipoamide succinyltransferase (E2), and dihydrolipoamide dehydrogenase (E3).[41] The dihydrolipoamide dehydrogenase component of the OGDC and the flavoprotein isolated from the *E. coli* PDC are identical.[42,43,61] The subunit composition of the *E. coli* OGDC is summarized in Table 1. It consists of 12 E1 chains (six dimers), 24 succinyltransferase chains, and 12 flavoprotein chains (six dimers).[41,44] OGDC from pig heart[74] resembles the *E. coli* OGDC in subunit composition and appearance in the electron microscope.

The appearance of *E. coli* succinyltransferase in the electron microscope[8] is strikingly similar to that of the *E. coli* acetyltransferase (E2). The *E. coli* succinyltransferase "inner core" has been crystallized.[45] Electron micrographs and their optical diffraction patterns in combination with X-ray diffraction data were used to analyze the packing of succinyltransferase inner cores in the crystal.

It should be noted that the E2 components have a rather large cavity in their structures.[8] The physiological significance, if any, of this cavity has yet to be established. Another interesting feature of the structure of E2, revealed by limited proteolysis and electron microscopy, is that the E2 polypeptide consists of two different domains: a compact domain and a flexible, extended domain (Figure 2).[23,26] The compact domain contains the acyltransferase active site, and the assemblage of these domains constitutes the "inner core" of E2, conferring the cube-like or pentagonal dodecahedron-like appearance of E2 in the electron microscopic image. The extended domain, which is readily released from the inner core by limited proteolysis, contains the covalently bound lipoyl moiety or moieties (lipoyl domain). Protein nuclear magnetic resonance spectroscopy has provided evidence that the lipoyl domain is attached to the inner core by a highly mobile segment of the polypeptide chain.[27,28] This unique architectural feature is thought to facilitate interaction of the lipoyl moiety with successive active sites on the complex.

D. PRIMARY STRUCTURES OF COMPONENT ENZYMES OF THE *E. COLI* PYRUVATE AND 2-OXOGLUTARATE DEHYDROGENASE COMPLEXES

The primary structures of the E1, E2, and E3 components of the *E. coli* PDC and OGDC have been deduced from the nucleotide sequences of the corresponding structural genes.[66]

The PDC is encoded by three genes, *ace E* (Elp), *ace F* (E2p) and *lpd* (E3) comprising the *ace EF-lpd* operon (Figure 3),[50] which is transcribed from the *ace* promoter during PDC synthesis, and additionally from the *lpd* promoter to provide E3 components for synthesis of the analogous OGDC.[57,58]

The nucleotide sequence of the complete *ace EF-lpd* operon has been determined and the primary structure of E2p translated from the *ace F* structural gene.[59-61] This revealed the existence of three remarkably homologous segments of approximately 100 amino acid residues that are tandemly repeated in the N-terminal half of the E2p subunit chain.

Each repeat contains a potential lipoylation site in an invariant 18-residue sequence that includes the single 13-residue lipoyl-lysine containing peptide defined previously.[20,24] The repeating segments of E2p and the *ace F* gene are designated lip1, lip2 and lip3 and *lip1, lip2* and *lip3,* respectively (Figure 4).[50]

The three segments, lip1 to lip3, of E2p become at least partly acetylated in the complex in the presence of the substrate pyruvate and can be isolated as three distinct functional entities (domains) after limited proteolysis of the complex with *Staphylococcus aureus* V8 proteinase (Figure 4).[28] The repeating units contain lengthy C-terminal regions of polypeptide

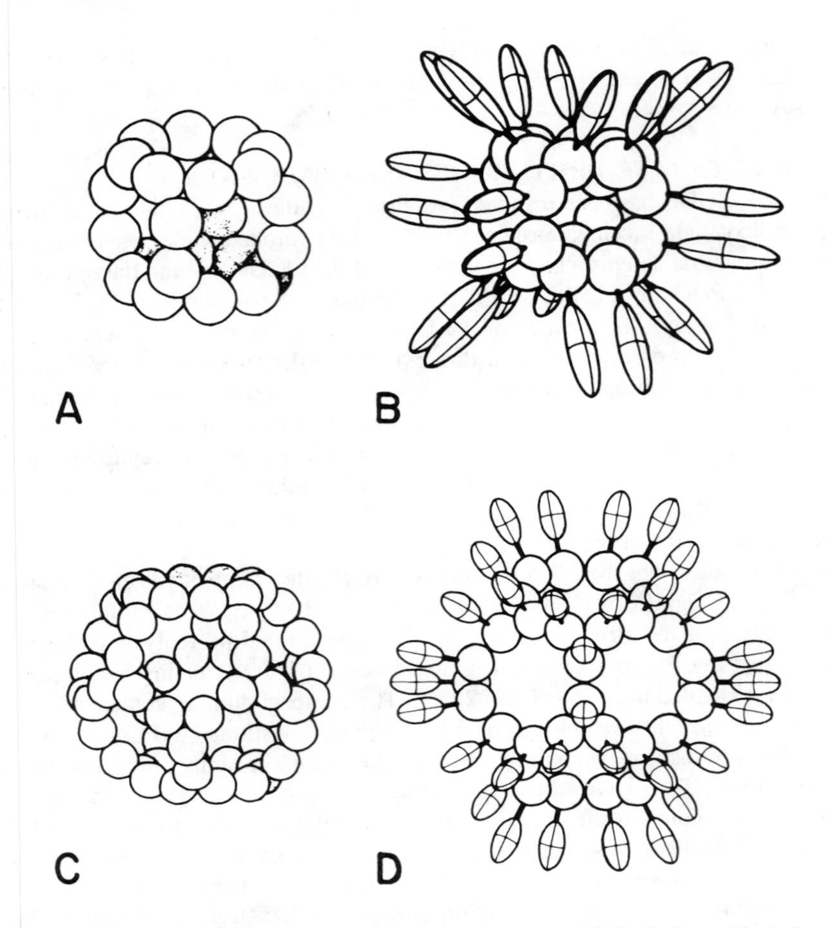

FIGURE 2. Interpretive models of the quaternary structure of dihydrolipoamide acyltransferases. (A) Model of those acyltransferases consisting of 24 subunits arranged in groups of 3 about the 8 vertices of a cube. (B) Model of the 24-subunit acyltransferases illustrating the proposed domain and subunit by one sphere and attached ellipsoid. The spheres represent the assemblage of compact domain (inner core), and the ellipsoids represent the extended lipoyl domains. (C) Model of those acyltransferases consisting of 60 subunits arranged in groups of 3 about the 20 vertices of a pentagonal dodecahedron. (D) Model of the 60-subunit acyltransferases illustrating the proposed domain and subunit structure. The figure is viewed down a twofold axis of symmetry. (From Reed, L. J. and Yeaman, S. J., *The Enzymes,* Boyer, P. D. and Krebs, E., Eds., Academic Press, New York, in press.)

chain (20 to 30 amino acid residues) that are unusually rich in residues of alanine, proline and charged amino acids. It is likely that these are the conformationally mobile regions predicted from ^{1}H NMR spectroscopy, which are removed from the complex by limited tryptic cleavage at Lys-316.[27,62] It is also clear from the amino acid sequence[60,63] that the C-terminal half of the E2p chain must constitute the large domain that aggregates to form the inner core of the complex and provides both the acetyltransferase active site and the binding sites for Elp and E3 subunits.[23,24] This segment of the E2p chain and the *ace F* (gene) is designated cat (*cat*). There is just one lipoyl domain in the E2p chain of the *E. coli* OGDC,[64,65] the amino acid sequence of which is homologous to those of the three lipoyl domains in the E2p chain.[63,66] The E2p chains of the PDC's of mammalian mitochondria and *Bacillus* species each contain only one lipoyl group[64,67,68] which, in concert with evidence from ^{1}H NMR spectroscopy, implies the existence of only one lipoyl domain per E2 chain in the complexes.[69]

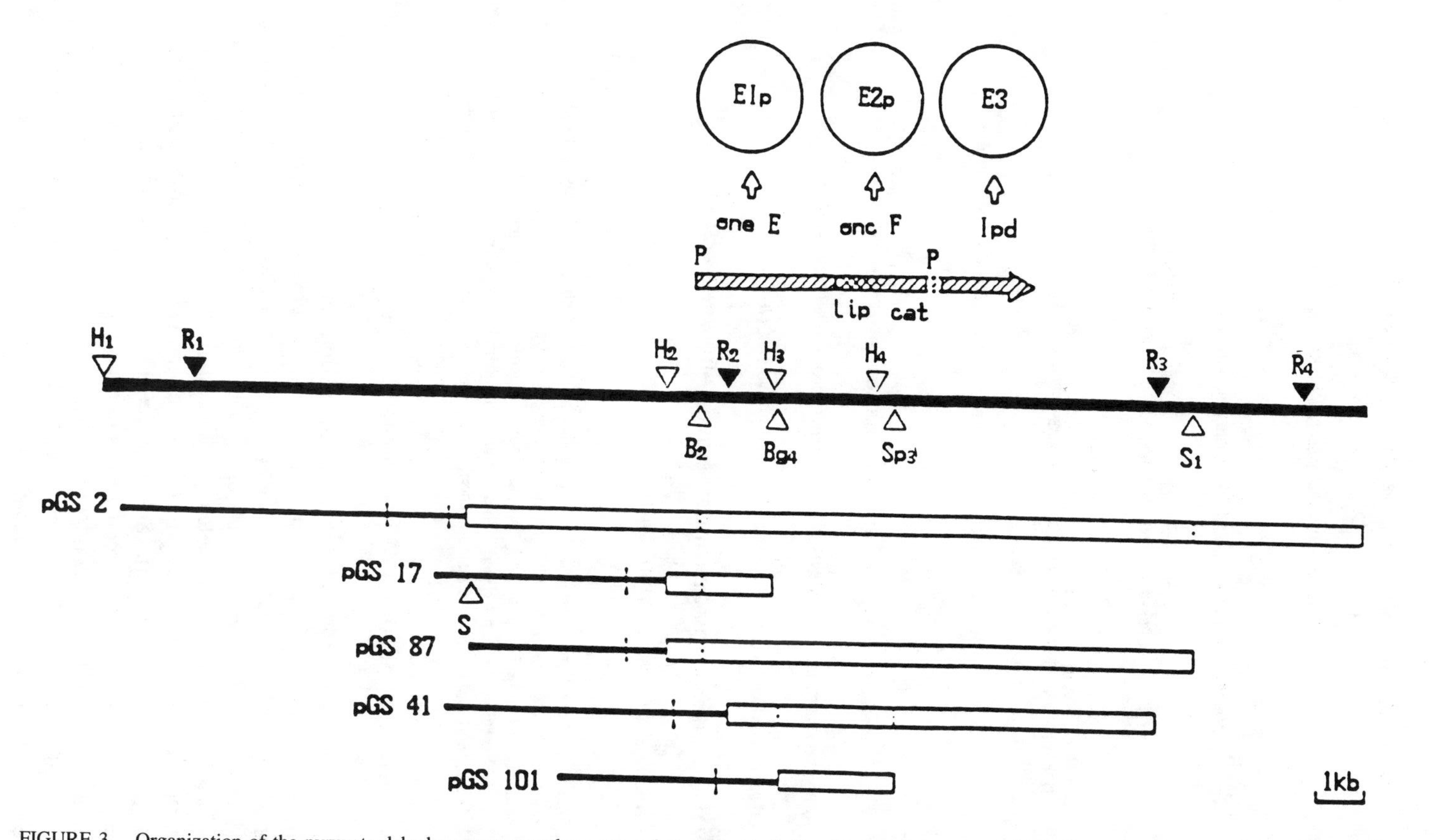

FIGURE 3. Organization of the pyruvate dehydrogenase complex genes of *E. coli* and of plasmids containing relevant segments of bacterial DNA. The genes are transcribed from left to right from one or both promoters (P), and the regions encoding the lipoyl and catalytic domains of the E_2P components are indicated. The line- (thin lines) and their polarities are defined by the vertical lines that denote the PstI sites in the vectors. Relevant restriction sites are numbered according to Guest et al.: B, BamHI; Bg, BglII; H. HindIII; K, KpnI; R. EcoRI; S. SalI; Sp, SphI. (From Guest, J. R., Lewis, H. M., Graham, L. D., Packman, L. C., and Perham, R. N., *J. Mol. Biol.*, 185, 743, 1985. With permission.)

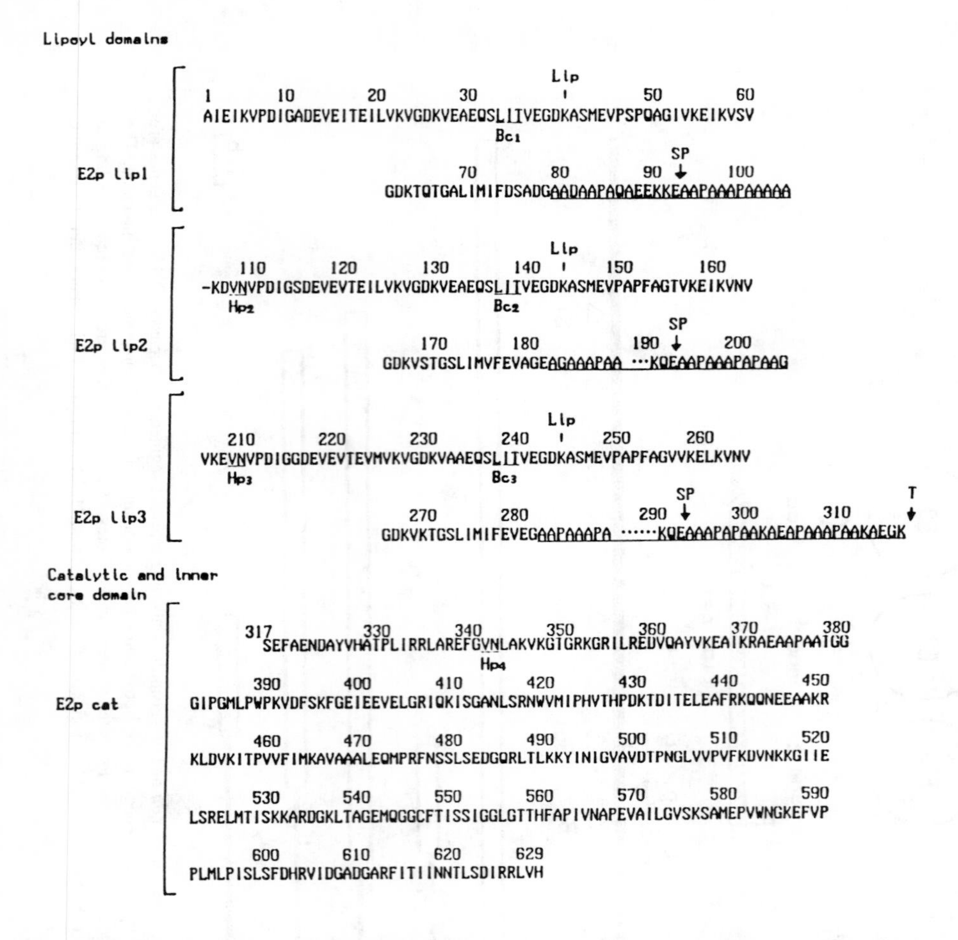

FIGURE 4. Primary structure (single-letter amino acid code) of the E2p chain of the pyruvate dehydrogenase complex of *E. coli* drawn to highlight the repeating lipoyl segments of domains and the inner core and catalytic domain. The covalently bound lipoyl cofactors are indicated (Lip) and the alanine and proline-rich regions are underlined. The residues encoded by the targets for restriction endonucleases BelI (Be) and HpaI (Hp) in the aceF structural gene are shown as are the sites for limited proteolysis by trypsin (T) and *S. aureus* V_8 proteinase (SP), which release the lipoyl domains from the inner core (catalytic domain). (From Guest, J. R., Lewis, H. M., Graham, L. D., Packman, L.C., and Perham, R. N., *J. Mol. Biol.*, 185, 743, 1985. With permission.)

It is puzzling that there should be more than one lipoyl domain in the E2p chain of the PDC from *E. coli*. Because the amino acid sequence homologies between the lipoyl domains derive from comparable homologies at the DNA level, several restriction sites are conserved at equivalent positions in the three *lip* segments of the *ace F* gene. These sites can be used to reconstruct the *ace F* gene, selectively deleting one or more of the lipoyl domains from the E2p chain it encodes in order to investigate whether individual lipoyl domains play different parts in the enzymic mechanism. Selective deletions were made *in vitro* in the dihydrolipoamide acetyltransferase gene (*ace F*) so as to excise one or two of the repeating sequences. This was facilitated by the high degree of homology in these sequences, which allowed the creation of hybrid lipoyl domains that closely resemble the originals. PDCs incorporating these genetically reconstructed E2p components were purified and their structures were confirmed.[50]

III. REACTION MECHANISM

It is evident from Figure 1 that the lipoyl moiety undergoes a cycle of transformations, i.e., reductive acylation, acyl transfer, and electron transfer. These transformations involve the interaction of the lipoyl moiety, which is covalently bound to E2, with 2-hydroxy-alkylthiamin pyrophosphate, which is bound to E1, and with FAD, which is bound to E3. The lipoyl moiety must also interact with CoA at a site on E2. Highly favorable alignment of the three enzymes and, by inference, of their coenzymes or prosthetic groups must be assumed in order to account for the occurrence of the overall reaction. A possible molecular basis of these interactions was suggested by the finding that the lipoyl moiety is bound in amide linkage to the ϵ-amino group of a lysyl residue in E2.[20] This attachment provides a flexible arm of about 14 Å for the reactive dithiolane ring, conceivably permitting the lipoyl moiety to rotate among the catalytic centers of the three different enzymes that comprise the complex — i.e., a "swinging arm" active-site coupling mechanism.[16] The results of kinetic analyses of the mammalian PDC[21] and OGDC,[72] revealing a three-site ping-pong mechanism, are consistent with this proposal. Experiments in which a spin label[7,52] or fluorescence probes[106,107] were attached to the lipoyl moieties in PDC and OGDC indicated considerable rotational mobility but did not distinguish between local rotational mobility of lipoyl moieties and actual rotation of lipoyl moieties between catalytic sites.[3] Fluorescence energy transfer measurements[109] indicated that the distances between catalytic sites in OGDC are consistent with a mechanism involving rotation of a single lipoyllysyl moiety between catalytic sites of E1, E2, and E3 during a normal catalytic cycle. However, the distances between catalytic sites in PDC were reported to be significantly larger (at least 40 Å) than the distance (~28 Å) that could be spanned by the rotation of a single lipoyllysyl moiety.[106] Measurements of time-resolved fluorescence polarization of fluorescence probes attached to lipoyl moieties in OGDC[110] were inferred to be consistent with rotation of lipoyl moieties between catalytic sites in the multienzyme complex. The sum of the association and dissociation rate constants for the rotation between sites is about $3 \times 10^6\ s^{-1}$.

The finding that the lipoyl moieties in PDC and OGDC are located on a large, protruding segment (lipoyl domain) of the E2 subunit[23,111] led to the suggestion that movement of lipoyl domains and not simply rotation of lipoyllysyl moieties provides the means to span the physical gaps between catalytic sites on PDC and OGDC.[27,56,112] On the basis of limited proteolysis and proton NMR experiments, Packman et al.[113] proposed that a substantial part of the lipoyl domains in PDC and OGDC has the characteristics of a random coil.

A computer modeling system was developed to analyze experimental data for inactivation of the *E. coli* OGDC and PDC accompanying enzymatic release of lipoyl moieties by lipoamidase and by trypsin.[56] The model studies indicate that the activities of OGDC and PDC are influenced by redundancies and random processes that is designated a multiple random coupling mechanism.[29,29a] In both complexes more than one lipoyl moiety services each E1 subunit, and an extensive lipoyl-lipoyl interaction network for exchange of electrons and possibly acyl groups must also be present.

The oxidative decarboxylation of pyruvate is thought to proceed[70] as described below. The positive charge on the nitrogen of the thiamin pyrophosphate

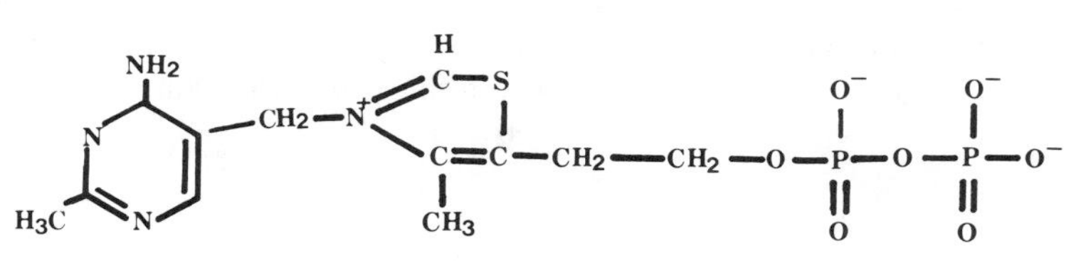

(Scheme I)

Scheme I promotes the ionization of the C-2 carbon by electrostatic stabilization. The ionized carbon is a potent nucleophile (Equation 1).

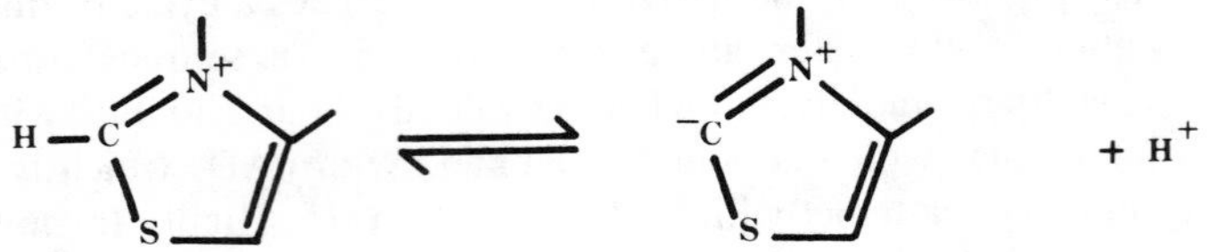

H—C N+ S ⇌ −C N+ S + H^+

(Equation 1)

The nitrogen atom can also stabilize by delocalizing a negative charge on the adduct of thiamin with many compounds, as, for example, in hydroxyethylthiamin pyrophosphate, a form in which much of the coenzyme is found *in vivo* (Equation 2).

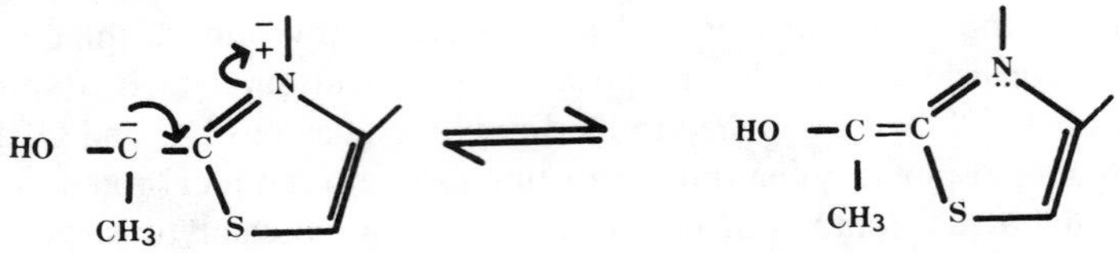

HO —C̄ —C N+ S CH_3 ⇌ HO —C =C N S CH_3

(Equation 2)

The combination of these reactions allows the decarboxylation of pyruvate by the route shown in Equation 3. Other carbon-carbon bonds adjacent to a carbonyl group may be cleaved in the same manner.

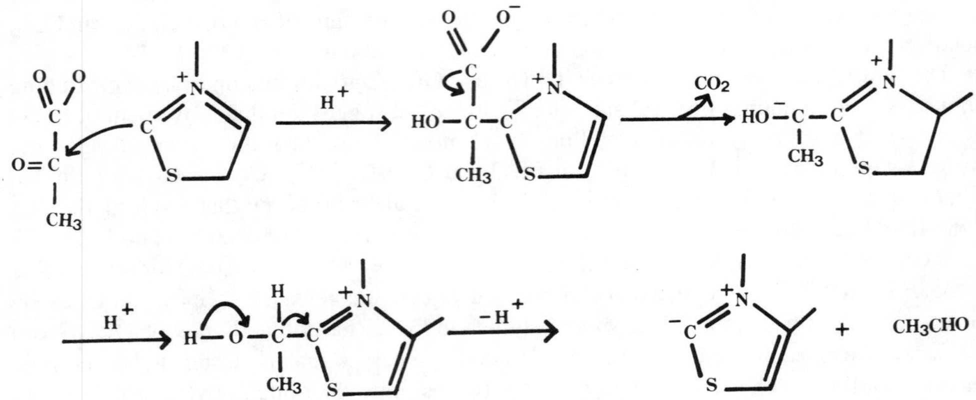

O O C O=C CH_3 + C +N S →(H^+) O O− C HO—C—C +N CH_3 S →(CO_2) HO—C̄—C +N CH_3 S

→(H^+) H H—O—C—C +N CH_3 S →($-H^+$) −C +N S + CH_3CHO

(Equation 3)

The hydroxyethylthiamin pyrophosphates are potent nucleophiles and may add to carbonyl compounds to form carbon-carbon bonds.

Hydroxyethylthiamin pyrophosphate is also nucleophilic toward lipoic acid. A hemithioacetal is formed and this decomposes to give a thioester (Equation 4):

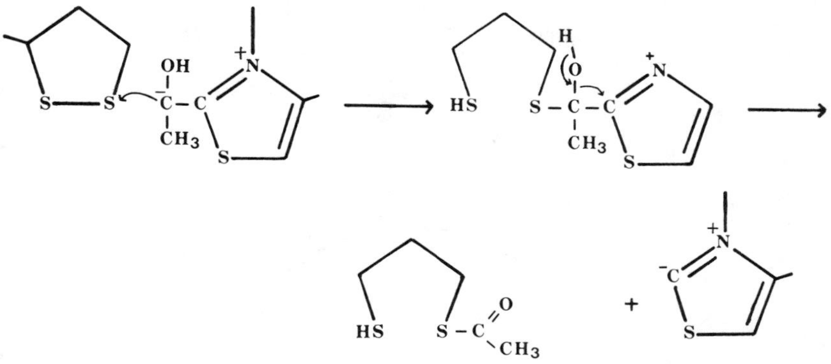

(Equation 4)

IV. REGULATORY PROPERTIES

2-Oxoacid dehydrogenase complexes occupy key positions in intermediary metabolism (Figure 5).[33] Deficiencies in the activities of the complexes lead to pathological states, including various forms of metabolic acidosis such as ''Maple Syrup Urine Disease'' in the case of BCOADC.[34] Furthermore, the activities of the complexes are under acute and long-term hormonal regulation. The OGDC is an enzyme of the tricarboxylic acid (TCA)-cycle, and the PDC catalyzes the committed reaction in the utilization of carbohydrate, with the resultant acetyl-CoA being either oxidized in the TCA-cycle or utilized as a precursor for various biosynthetic pathways. BCOADC catalyzes an irreversible step in the oxidation of the essential branched-chain amino acids and, as it also catalyzes the oxidative decarboxylation of 2-oxobutyrate[35] and 4-methylthio-2-oxobutyrate,[36] it may play a further role in the catabolism of methionine and threonine. All three complexes are located in mitochondria, within the inner membrane-matrix compartment.

The PDC is well designed for fine regulation of its activity. Interconversion of the active and inactive (phosphorylated) forms of pyruvate dehydrogenase (E1) is a dynamic process that leads rapidly to the establishment of steady states, in which the fraction of phosphorylated enzyme can be varied progressively over a wide range by changing the concentration or molar ratios of effectors that regulate activities of the kinase and phosphatase.[80,81] The pyruvate dehydrogenase complex, like other interconvertible enzyme systems, functions uniquely as a metabolic integration system. By means of multisite interactions with allosteric effectors, the kinase and phosphatase can sense simultaneous fluctuations in the intracellular concentrations of several metabolites and adjust the specific activity of pyruvate dehydrogenase. Figure 6 summarizes the control of pyruvate dehydrogenase kinase and pyruvate dehydrogenase phosphatase activities by effectors.[22]

Pyruvate dehydrogenase kinase has been purified about 2700-fold to apparent homogeneity from extracts of bovine kidney mitochondria.[82] The kinase has the subunit composition $\alpha\beta$. The two subunits have molecular weights of about 48,000 (α) and 45,000 (β). Kinase activity resides in the catalytic α-subunit; the function of β-subunit remains to be established. An attractive possibility is that it functions as a regulatory subunit. The turnover number (k_{cat}) of pyruvate dehydrogenase kinase is about 32 min^{-1}. The kinase appears to be specific for pyruvate dehydrogenase. It exhibits little activity, if any, toward rabbit skeletal muscle phosphorylase *b*, glycogen synthase *a*, histones, or casein. The kinase phosphorylates three serine residues on the α-subunit (M_r 41,000) of pyruvate dehydrogenase.[83,84,103] The amino acid sequence around the three phosphorylation sites is shown in Figure 7. Phosphorylation proceeds markedly faster at site 1 than at the other two sites, and phosphorylation at site 1 correlates closely with inactivation of pyruvate dehydrogenase.[83]

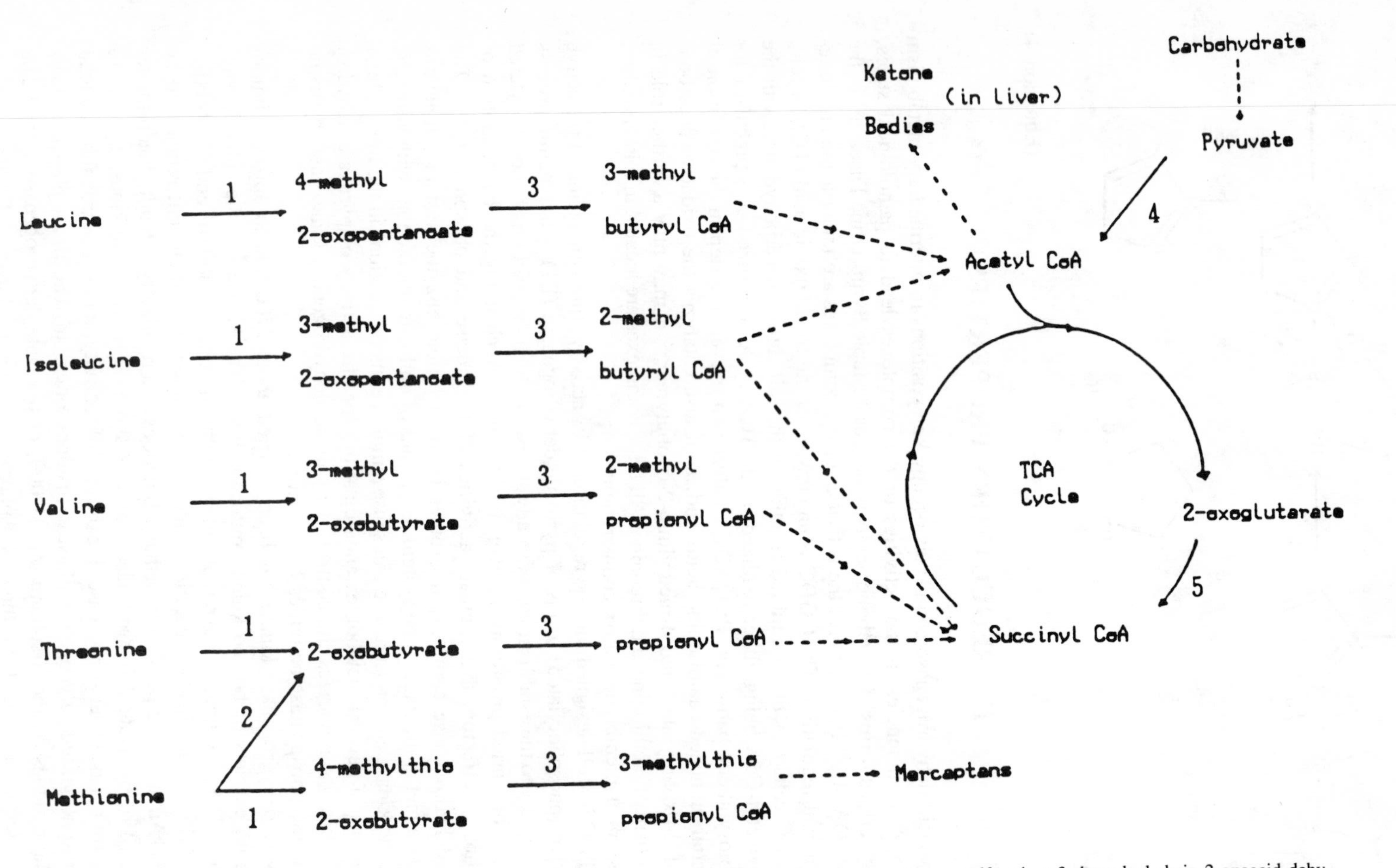

FIGURE 5. Outline of the metabolic role of the 2-oxoacid dehydrogenase complexes. 1, Transamination; 2, transsulfuration; 3, branched-chain 2-oxoacid dehydrogenase; 4, pyruvate dehydrogenase; 5, 2-oxoglutarate dehydrogenase. The broken lines indicate that several reaction steps are involved. (From Yeaman, S. J., *Trends Biochem. Sci.*, 11, 293, 1986. With permission.)

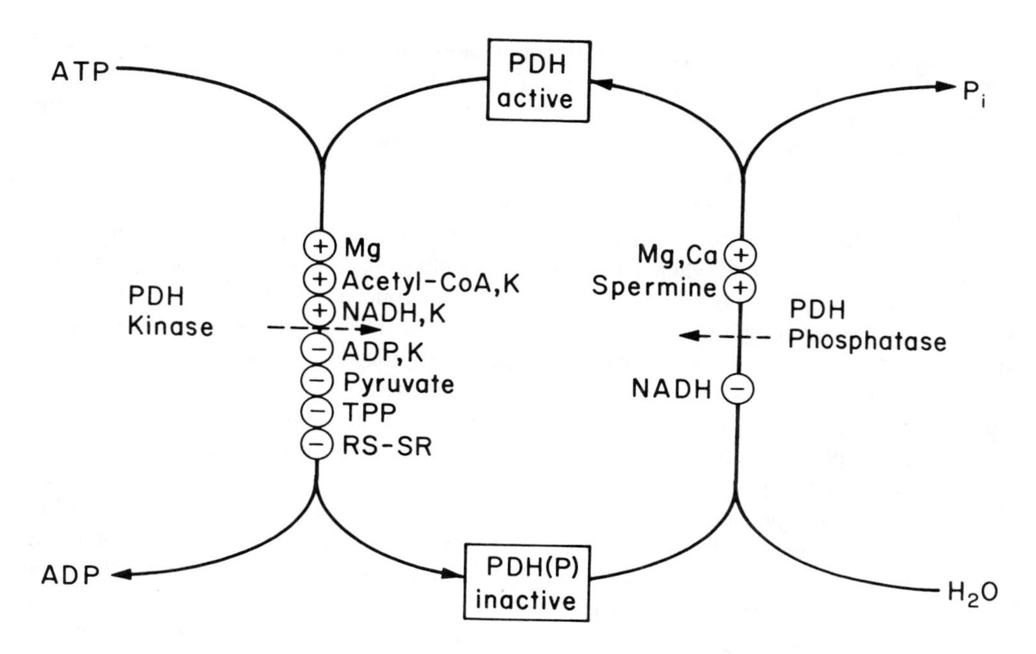

FIGURE 6. Covalent modification of pyruvate dehydrogenase and its control by effectors. (From Reed, L. J. and Yeaman, S. J., *The Enzymes,* Boyer, P. D. and Krebs, E., Eds., Academic Press, New York, in press.)

Tyr–His–Gly–His–Ser(P)–Met–Ser–Asp–Pro–Gly–Val–Ser(P)–Tyr–Arg (Site 1 = first Ser(P); Site 2 = second Ser(P))

Tyr–Gly–Met–Gly–Thr–Ser(P)–Val–Glu–Arg (Site 3 = Ser(P))

FIGURE 7. Phosphorylation sites on pyruvate dehydrogenase. (From Reed, L. J. and Yeaman, S. J., *The Enzymes,* Boyer, P. D. and Krebs, E. Eds., Academic Press, New York, in press.)

Pyruvate dehydrogenase kinase requires Mg^{2+} (apparent K_m, 0.02 m*M*) or Mn^{2+}.[10] Kinase activity is stimulated by acetyl-CoA and by NADH, products of pyruvate oxidation, provided K^+ or NH_4^+ ions are present, and kinase activity is inhibited by ADP and by pyruvate.[80,81] ADP is competitive with ATP, and this inhibition apparently requires the presence of monovalent cation. The coenzyme, thiamin pyrophosphate, inhibits kinase activity, apparently as a result of binding at the catalytic site of pyruvate dehydrogenase and thereby altering the conformation about phosphorylation site 1 (Figure 7) so that the serine hydroxyl group is less accessible to the kinase. Kinase activity is markedly inhibited by *N*-ethylmaleimide and by certain disulfides.[85] Inhibition by disulfides appears to be highly specific and is reversed by thiols. 5,5′-Dithiobis(2-nitrobenzoate) shows significant inhibition at 1 μ*M*. It appears that pyruvate dehydrogenase kinase contains a thiol group (or groups) that is involved in maintaining a conformation of the enzyme that facilitates phosphorylation of its protein substrate. Modulation of kinase activity by thiol-disulfide exchange may be an important physiological mechanism.

Pyruvate dehydrogenase phosphatase has been isolated in a homogeneous state from bovine heart and kidney mitochondria.[86,87] The phosphatase has an M_r of about 150,000 and consists of two different subunits of M_r = ~97,000 and ~50,000. The catalytic activity resides in the smaller subunit, which is sensitive to proteolysis. The k_{cat} of pyruvate dehydrogenase phosphatase with phosphorylated pyruvate dehydrogenase is about 300 min^{-1}.

The phosphatase is inactive toward *p*-nitrophenyl phosphate. It exhibits slight activity toward ^{32}P-labeled phosphorylase *a* from rabbit skeletal muscle[86] and ^{32}P-labeled branched-chain 2-oxoacid dehydrogenase complex,[88] i.e., about 10% and 0.5 to 1%, respectively, of the activity observed with ^{32}P-labeled PDC. Pyruvate dehydrogenase phosphatase activity is not inhibited by protein phosphatase inhibitor 1 or inhibitor 2, and the activity is not affected by addition of highly purified calmodulin from porcine brain.

Pyruvate dehydrogenase phosphatase has an absolute requirement for Mg^{2+} ions (apparent K_m, 2m*M*) or Mn^{2+}, when acting on both its physiological substrate (phosphorylated E1) and phosphopeptide substrates.[10,89] At physiological concentrations of Mg^{2+} (<1.0 m*M*), pyruvate dehydrogenase phosphatase activity is stimulated markedly by the polyamine spermine (0.5 m*M*), apparently by decreasing the apparent K_m for Mg^{2+}.[98,105] At saturating Mg^{2+} concentration, phosphatase activity toward its protein substrate is stimulated about 20-fold by micromolar concentrations of Ca^{2+} ions, provided the acetyltransferase (E2) is present.[32] However, phosphatase activity toward phosphopeptide substrates is not affected by Ca^{2+}, whether or not E2 is present.[89] These observations indicate that Ca^{2+} is not required for catalysis by the phosphatase. In the presence of Ca^{2+}, the phosphatase binds to E2, and its apparent K_m for phosphorylated E1 is decreased about 20-fold to 2.9 μ*M*.[32] $^{45}Ca^{2+}$-binding studies[86] have shown that the uncomplexed phosphatase binds one Ca^{2+} per molecule of the 150,000 M_r enzyme with a dissociation constant (K_d) of about 8 μ*M*. When both the phosphatase and E2 are present, two equivalent and independent Ca^{2+}-binding sites are detected with a K_d value of about 5 μ*M*. In the presence of 0.2 *M* KC1, which produces virtually complete inhibition of phosphatase activity, the enzyme binds only one Ca^{2+} per molecule even in the presence of E2. These results are interpreted to indicate that pyruvate dehydrogenase phosphatase possesses an "intrinsic" Ca^{2+}-binding site and that a second Ca^{2+}-binding site is produced when both the phosphatase and E2 are present. Ca^{2+} may act as a bridging ligand between phosphatase and E2.

Since pyruvate dehydrogenase phosphatase and the PDC are located in the mitochondrial matrix, changes in free Ca^{2+} concentrations in the matrix could play an important role in regulation of phosphatase activity and hence PDC activity. Denton and McCormack[90] and Hansford[91] estimated mitochondrial matrix free Ca^{2+} concentrations by measuring the activity in mitochondrial extracts of three Ca^{2+}-sensitive enzymes as a function of the Ca^{2+} concentration calculated from Ca^{2+}-EGTA buffers. Half-maximum activities were obtained at a calculated free Ca^{2+} concentration of about 1 μ*M* for pyruvate dehydrogenase phosphatase, 2-oxoglutarate dehydrogenase, and NAD^+-linked isocitrate dehydrogenase.

Both long-term and acute hormonal control of mammalian PDC have been studied extensively. The two most striking hormonal effects are the acute activation of the complex by adrenalin in heart and liver and activation by insulin in adipose tissue. It appears that these two effects are mediated by different mechanisms. Activation of the PDC by adrenalin is apparently mediated *via* activation of the phosphatase, resulting from an increase in the intramitochondrial concentration of free Ca^{2+} ions.[96] The rise in intramitochondrial free Ca^{2+} concentration is in turn thought to arise from the known increases in cytoplasmic Ca^{2+} concentration brought about by this hormone. This rise in intramitochondrial Ca^{2+} also leads to the allosteric activation of OGDC.[94,96] In contrast, the effects of insulin on PDC are apparently independent of changes in intramitochondrial Ca^{2+} concentration,[97] but the exact mechanism remains obscure. Insulin may act within mitochondria *via* elevation in the concentration of polyamines such as spermine, which increases the activity of pyruvate dehydrogenase phosphatase at sub-optimal concentrations of Mg^{2+} ions.[98] The mechanism of the acute and long-term hormonal control of PDC is an outstanding question of considerable importance.[33]

ACKNOWLEDGMENT

We deeply thank Dr. Lester Reed, University of Texas at Austin, for his encouragement and suggestions. We also thank Ms. Kaoru Wakamatsu for typing this manuscript.

REFERENCES

1. **Reed, L. J. and Cox, D. J.,** Multienzyme complexes, in *The Enzyme,* Vol. 1, 3rd ed., Boyer, P. D., Ed., Academic Press, New York, 1970, 213.
2. **Reed, L. J.,** Multienzyme complexes, *Acc. Chem. Res.,* 7, 40, 1974.
3. **Hammes, G. G.,** Processing of intermediates in multienzyme complexes, *Biochem. Soc. Symp.,* 46, 73, 1981.
4. **Roberts, G. C. K., Duckworth, H. W., Packman, L. C., and Perham, R. N.,** Mobility and active-site coupling in 2-oxo acid dehydrogenase complexes, *Ciba Foundation Symp.,* 93, 47, 1983.
5. **Collins, J. H. and Reed, L. J.,** Acyl group and electron pair relay system: a network of interacting lipoyl moities in the pyruvate and α-ketoglutarate dehydrogenase complexes from *Escherichia coli, Proc. Natl. Acad. Sci. U.S.A.,* 74, 4223, 1977.
6. **Danson, M. J., Fersht, A. R., and Perham, R. N.,** Rapid intramolecular coupling of active sites in the pyruvate dehdyrogenase complex of *Escherichia coil*: mechanism for rate enhancement in a multimeric structure, *Proc. Natl. Acad. Sci. U.S.A.,* 75, 5386, 1978.
7. **Ambrose, M. C. and Perham, R. N.,** Spin-label study of the mobility of enzyme bound lipoic acid in the pyruvate dehydrogenase multienzyme complexes of *Escherichia coli, Biochem. J.,* 155, 429, 1976.
8. **Oliver, R. M. and Reed, L. J.,** Multienzyme complexes, in *Electron Microscopy of Proteins,* Vol. 2, Harris, J. R., Ed., Academic Press, London, 1982, 1.
9. **Linn, T. C., Pettit, F. H., and Reed, L. J.,** α-Keto acid dehydrogenase complexes. X. Regulation of the activity of the pyruvate dehydrogenase complex from beef kidney mitochondria by phosphorylation and dephosphorylation, *Proc. Natl. Acad. Sci. U.S.A.,* 62, 234, 1969.
10. **Hucho, F., Randall, D. D., Roche, T.E., Burgett, M. W., Pelley, J. W., and Reed, L. J.,** α-Keto acid dehydrogenase complexes. XVII. Kinetic and regulatory properties of pyruvate dehydrogenase kinase and pyruvate dehydrogenase phosphatase from bovine kidney and heart, *Arch. Biochem. Biophys.,* 151, 328, 1972.
11. **Denton, R. M., Randle, R. J., Bridges, B. J., Cooper, R. H., Kerby, A. L., Pask, H. T., Severson, D. L., Stansbie, D., and Whitehouse, S.,** Regulation of the mammalian pyruvate dehydrogenase, *Mol. Cell. Biochem.,* 9, 27, 1975.
12. **Wieland, O. H.,** The mammalian pyruvate dehydrogenase complex: structure and regulation, *Rev. Physiol. Biochem. Pharmacol.,* 96, 123, 1983.
13. **Randle, P. J., Fatania, H. R., and Lau, K. S.,** Regulation of the mitochondrial branched-chain 2-oxoacid dehydrogenase complex of animal tissues by reversible phosphorylation, in *Molecular Aspects of Cell Regulation,* Vol. 3, Cohen, P., Ed., Elsevier, Amsterdam, 1984, chap. 1.
14. **Fersht, A.,** Three dimensional structure of enzymes, III. Reasons for multiple activities and multienzyme complexes in *Enzyme Structure and Mechanism,* Vol. 41, 2nd ed., W. H. Freeman, New York, 1985, chap. 1.
15. **Koike, M., Reed, L. J., and Carroll, W. R.,** α-Keto acid dehydrogenation complexes. I. Purification and properties of pyruvate and α-ketoglutarate dehydrogenation complexes of *Escherichia coli, J. Biol. Chem.,* 235, 1924, 1960.
16. **Koike, M., Reed, L. J., and Carrol, W. R.,** α-Keto acid dehydrogenation complexes. IV. Resolution and reconstitution of the *Escherichia coli* pyruvate dehydrogenase complex, *J. Biol. Chem.,* 238, 30, 1963.
17. **Willms, C. R., Oliver, R. M., Henney, H. R. Jr., Mukherjee, B. B., and Reed, L. J.,** α-Keto acid dehydrogenase complexes. VI. Dissociation and reconstitution of the dihydrolipoyl transacetylase of *Escherichia Coli, J. Biol. Chem.,* 242, 889, 1967.
18. **Eley, M. H., Namihira, G., Hamilton, L., Munk, P., and Reed, L. J.,** α-Keto acid dehydrogenase complexes. XVIII. Subunit composition of the *Escherichia coil* pyruvate dehydrogenase complex *Arch. Biochem. Biophys.,* 152, 655, 1972.
19. **Guest, J. R. and Creaghun, I. T.,** Gene-protein relationships of the α-Keto acid dehydrogenase complexes of *Escherichia coli* K12: isolation and characterization of lipoamide dehydrogenase mutants, *J. Gen. Microbiol.,* 75, 197, 1973.

20. **Nawa, H., Brady, W. T., Koike, M., and Reed, L. J.,** Studies on the nature of protein-bound lipoic acid, *J. Am. Chem. Soc.*, 82, 896, 1960.
21. **Tsai, C. S., Burgett, M. W., and Reed, L. J.,** α-Keto acid dehydrogenase complexes. XX. A kinetic study of the pyruvate Dehydrogenase complex from bovine kidney, *J. Biol. Chem.*, 248, 8348, 1973.
22. **Reed, L. J. and Yeaman, S. J.,** in *The Enzymes,* Boyer, P. D. and Krebs, E., Eds., Academic Press, New York, in press.
23. **Bleile, D. M., Munk, P., Oliver, R. M., and Reed, L. J.,** Subunit structure of dihydrolipoyl transacetylase component of pyruvate dehydrogenase complex from *Escherichia coli, Proc. Natl. Acad. Sci. U.S.A.*, 76, 4385, 1979.
24. **Hale, G. and Perham, R. N.,** Primary structure of the swinging arms of the pyruvate dehydrogenase complex of *Escherichia coli, FEBS Lett.*, 105, 263, 1979.
25. **Kresze, G. B. and Ronft, H.,** Bovine kidney pyruvate dehydrogenase complex limited proteolysis and molecular structure of the lipoate acetyltransferase component, *Eur. J. Biochem.*, 112, 589, 1980.
26. **Bleile, D. M., Hackert, M. L., Pettit, F. H., and Reed, L. J.,** Subunit structure of dihydrolipoyl transacetylase component of pyruvate dehydrogenase complex from bovine heart, *J. Biol. Chem.*, 256, 514, 1981.
27. **Perham, R. N., Duckworth, H. W., and Roberts, G. C. K.,** Mobility of polypeptide chain in the pyruvate dehydrogenase complex revealed by proton NMR. *Nature,* 292, 474, 1981.
28. **Packman, L. C., Perham, R. N., and Roberts, G. C. K.,** Domain structure and ^{1}H-n.m.r. spectroscopy of the pyruvate dehydrogenase complex of *Bacillus steatothermophilus, Biochem. J.*, 217, 219, 1984.
29. **Hackert, M. L., Oliver, R. N., and Reed, L. J.,** A computer model analysis of the active-site coupling mechanism in the pyruvate dehydrogenase multienzyme complex of *Escherichia coli, Proc. Natl. Acad. Sci. U.S.A.*, 80, 2907, 1983.

29a. **Hackert, M. L., Oliver, R. N., and Reed, L. J.,** Evidence for a multiple random coupling mechanism in the α-ketoglutarate dehydrogenase multienzyme complex of *Escherichia coli*: a computer model analysis, *Proc. Natl. Acad. Sci. U.S.A.*, 80, 2226, 1983.

30. **Barrera, C. R., Namihira, G., Hamilton, L., Munk, P. Eley, M. H., Linn, T. C., and Reed, L. J.,** α-Keto acid dehydrogenase complexes. XVI. Studies on the subunit structure of the pyruvate dehydrogenase complexes from bovine kidney and heart, *Arch. Bioch. Biophys.*, 148, 343, 1972.
31. **Wu, T.-L. and Reed, L. J.,** Subunit binding in the pyruvate dehydrogenase complex from bovine kidney and heart, *Biochemistry,* 23, 221, 1984.
32. **Pettit, F. H., Roche, T. E., and Reed, L. J.,** Function of calcium ions in pyruvate dehydrogenase phosphatase activity, *Biochem. Biophys. Res. Commun.*, 49, 563, 1972.
33. **Yeaman, S. J.,** The mammalian 2-oxoacid dehydrogenases: a complex family, *Trends Biochem. Sci.*, 11, 293, 1986.
34. **Block, K. P., Heywood, B. W., Buse, M. G., and Harper, A. E.,** Activation of rat liver branched-chain 2-oxo acid dehydrogenase *in vivo* by glucagon and adrenaline, *Biochem. J.*, 232, 593, 1985.
35. **Pettit, F. H., Yeaman, S. J., and Reed, L. J.,** Purification and characterization of branched chain α-keto acid dehydrogenase complex of bovine kidney, *Proc. Natl. Acad. Sci. U.S.A.*, 75, 4881, 1978.
36. **Jones, S. M. A. and Yeaman, S. J.,** Oxidative decarboxylation of 4-methylthio-2-oxobutyrate by branched-chain 2-oxo acid dehydrogenase complex, *Biochem. J.*, 237, 621, 1986.
37. **Yang, H., Hainfield, J. F., Wall, J. S., and Frey, P. A.,** Quaternary structure of pyruvate dehydrogenase complex from *Escherichia coli, J. Biol. Chem.*, 260, 16049, 1985.
38. **De Marcucci, O., and Lindsay, J. G.,** Component X: an immunologically distent polypeptide associated with mammalian pyruvate dehydrogenase multi-enzyme complex, *Eur. J. Biochem.*, 149, 641, 1985.
39. **Jilka, J. M., Rahmatullah, M., Kazema, M., and Roche, T. E.,** Properties of a newly characterized protein of the bovine kidney pyruvate dehydrogenase complex, *J. Biol. Chem.*, 261, 1858, 1986.
40. **Rahmatullah, M., Jilka, J. M., Radke, G. A., and Roche, T. E.,** Properties of the pyruvate dehydrogenase kinase bound to and separated from the dihydrolipoyl transacetylase-protein. X. Subcomplex and evidence for binding of the kinase to protein, *J. Biol. Chem.*, 261, 6515, 1986.
41. **Pettit, F. H., Hamilton, L., Munk, P., Namihira, G., Eley, M. H., Willms, C. R., and Reed, L. J.,** α-Keto acid dehydrogenase complexes. XIX. Subunit structure of the *Escherichia coli* α-ketoglutarate dehydrogenase complex, *J. Biol. Chem.*, 248, 5282, 1973.
42. **Pettit, F. H. and Reed, L. J.,** α-Keto acid dehydrogenase complexes: VIII. Comparison of dihydrolipoyl dehydrogenases from pyruvate and α-ketoglutarate dehydrogenase complexes of *Escherichia coli, Proc. Natl. Acad. Sci. U.S.A.*, 58, 1126, 1967.
43. **Brown, J. P. and Perham, R. N.,** An amino acid sequence in the active complex of *Escherichia coli, FEBS Lett.*, 26, 221, 1972.
44. **Koike, M., Hamada, M., Koike, K., Hiraoka, T., and Nakualua, Y.,** Purification and function of α-ketoacid dehydrogenases in the mammalian multienzyme complexes, in *Thiamin,* Gubler, C., Fujiwara, M., and Dreyfus, P., Eds., Wiley-Interscience, New York, 1975, 5.

45. **De Rosier, D. J., Oliver, R. M., and Reed, L. J.,** Crystallization and preliminary structural analysis of dihydrolipoyl transsuccinylase, the core of the 2-oxoglutarate dehydrogenase complex, *Proc. Natl. Acad. Sci. U.S.A.*, 68, 1135, 1971.
46. **Cajacob, C. A., Frey, P. A., Hainfeld, J. F., Wall, J. S., and Yang, H.,** *Escherichia coli* pyruvate dehydrogenase complex: particle masses of the complex and component enzymes measured by scanning transmission electron microscopy, *Biochemistry*, 24, 2425, 1985.
47. **Reed, L. J., Pettit, F. H., Eley, M. H., Hamilton, L., Collins, J. H., and Oliver, R. M.,** Reconstitution of the *Escherichia coli* pyruvate dehydrogenase complex, *Proc. Natl. Acad. Sci. U.S.A.*, 72, 3068, 1975.
48. **Anglides, K. J., Akiyama, S. K., and Hammes, G. G.,** Subunit stoichiometry and molecular weight of the pyruvate dehydrogenase multienzyme complex from *Escherichia coli*, *Proc. Natl. Acad. Sci. U.S.A.*, 76, 3279, 1979.
49. **Hunter, A. and Lindsay, J. G.,** Immunological and biosynthetic studies on the mammalian 2-oxoglutarate dehydrogenase multienzyme complex, *Eur. J. Biochem.*, 155, 103, 1986.
50. **Guest, J. R., Lewis, H. M., Graham, L. D., Packman, L. C., and Perham, R. N.,** Genetic reconstruction and functional analysis of the repeating lipoyl domains in the pyruvate dehydrogenase multienzyme complex of *Escherichia coli*, *J. Mol. Biol.*, 185, 743, 1985.
51. **Damson, M. J., Hale, G., Johnson, P., Perham, R. N., Smith, J., and Spragg, P.,** Molecular weight and symmetry of the pyruvate dehydrogenase multienzyme complex of *Escherichia coli*, *J. Mol. Biol.*, 129, 603, 1979.
52. **Grande, H. J., Van Telgen, H. J., and Veeger, C.,** Symmetry and asymmetry of the pyruvate dehydrogenase complexes from *Azotobactor vinelandii* and *Escherichia coli* as reflected by fluorescence and spin-label studies, *Eur. J. Biochem.*, 71, 509, 1976.
53. **Bates, D. L., Danson, M. J., Hale, G., Hooper, E. A., and Perham, R. N.,** Self-assembly and catalytic activity of the pyruvate dehydrogenase multienzyme complex of *Escherichia coli*, *Nature*, 268, 313, 1977.
54. **Packman, L. C., Stanley, C. J., and Perham, R. N.,** Temperature-dependence of intramolecular coupling of active sites in pyruvate dehydrogenase multienzyme complexes, *Biochem. J.*, 213, 331, 1983.
55. **Stepp, L. R., Bleile, D. M., McRorie, D. K., Pettit, F. H., and Reed, L. J.,** Use of trypsin and lipoamidase to study the role of lipoic acid moieties in the pyruvate and α-ketoglutarate dehydrogenase complexes of *Escherichia coli*, *Biochemistry*, 20, 4555, 1981.
56. **Guest, J. R.,** Aspects of the molecular biology of lipoamide dehydrogenase, *Adv. Neurol.*, 21, 219, 1978.
57. **Spencer, M. E. and Guest, J. R.,** Transcription analysis of the suc*AB*, ace*EF* and lpd genes of *Escherichia coli*, *Mol. Gen. Genet.*, 200, 145, 1985.
58. **Stephens, P. E., Darlison, M. G., Lewis, H. M., and Guest, J. R.,** The pyruvate dehydrogenase complex of *Escherichia coli* K12: nucleotide sequence encoding the dihydrolipoamide acetyltransferase component, *Eur. J. Biochem.*, 133, 481, 1983.
59. **Stephens, P. E., Darlison, M. G., Lewis, H. M., and Guest, J. R.,** The pyruvate dehydrogenase complex of *Escherichia coli* K12: nucleotide sequence encoding the pyruvate dehydrogenase component, *Eur. J. Biochem.*, 133, 155, 1983.
60. **Stephens, P. E., Lewis, H. M., Darlison, M. G., and Guest, J. R.,** Nucleotide sequence of the lipoamide dehydrogenase gene of *Escherichia coli* K12, *Eur. J. Biochem.*, 135, 519, 1983.
61. **Packman, L. C., Hale, G., and Perham, R. N.,** Repeating functional domains in the pyruvate dehydrogenase multienzyme complex of *Escherichia coli*, *EMBO J.*, 3, 1315, 1984.
62. **Spencer, M. E., Darlison, M. G., Stephens, P. E., Duckenfield, I. K., and Guest, J. R.,** Nucleotide sequence of the suc*B* gene encoding the dihydrolipoamide succinyltransferase of *Escherichia coli* K12 and homology with the corresponding acetyltransferase, *Eur. J. Biochem.*, 141, 361, 1984.
63. **White, R. H., Bleile, D. M., and Reed, L. J.,** Lipolc acid content of dihydrolipoyl transacylases determined by isotope dilution analysis, *Biochem. Biophys. Res. Commun.*, 94, 78, 1980.
64. **Perham, R. N. and Roberts, G. C. K.,** Limited proteolysis and proton n.m.r. spectroscopy of the 2-oxoglutarate dehydrogenase multienzyme complex of *Escherichia coli*, *Biochem. J.*, 199, 733, 1981.
65. **Guest, J. R., Darlison, M. G., Spencer, M. E., and Stephens, P. E.,** Cloning and sequence analysis of the pyruvate and 2-oxoglutarate dehydrogenase complex genes of *Escherichia coli*, *Biochem. Soc. Trans.*, 12, 220, 1984.
66. **Bleile, D. M., Hackert, M. L., Pettit, F. H., and Reed, L. J.,** Subunit structure of dihydrolipoyl transacetylase component of pyruvate dehydrogenase complex from bovine heart, *J. Biol. Chem.*, 256, 514, 1981.
67. **Stanley, C. J., Packman, L. C., Danson, M. J., Henderson, C. E., and Perham, R. N.,** Intramolecular coupling of active sites in the pyruvate dehydrogenase multienzyme complexes from bacterial and mammalian sources, *Biochemistry*, 195, 715, 1981.
68. **Fersht, A.,** Chemical catalysis, in *Enzyme Structure and Mechanism*, 2nd ed., W. H. Freeman, New York. 1985, 75.
69. **Krampitz, L. O.,** Pyruvate oxidase, in *Thiamin Diphosphate and Its Catalytic Function*, Marcel Dekker, New York. 32, 1970, chap. 4.

70. **Hamada, M., Koike, K., Nakaula, Y., Hiraoka, T., Koike, M., and Hashimoto, T.,** A kinetic study of the α-keto acid dehydrogenase complexes from pig heart mitochondria, *J. Biochem.*, 77, 1047, 1975.
71. **Hirashima, M., Hayakawa, T., and Koike, M.,** Mammalian α-keto acid dehydrogenase complexes. II. An improved procedure for the preparation of 2-oxoglutarate dehydrogenase complex from pig heart muscle, *J. Biol. Chem.*, 242, 902, 1967.
72. **Tanaka, N., Koike, K., Hamada, M., Otsuka, K.-I., Suematsu, T., and Koike, M.,** Mammalian α-Keto acid dehydrogenase complexes. VII. Resolution and reconstitution of the pig heart 2-oxoglutarate dehydrogenase complex, *J. Biol. Chem.*, 247, 4043, 1972.
73. **Tanaka, N., Koike, K., Otsuka, K.-I., Hamada, M., Ogasahara, K., and Koike, M.,** Mammalian α-keto acid dehydrogenase complexes. VIII. Properties and subunit composition of the pig heart lipoate succinyltransferase, *J. Biol. Chem.*, 249, 191, 1974.
74. **Koike, K., Hamada, M., Tanaka, N., Otsuka, K. -I., Ogasahara, K., and Koike, M.,** Properties and subunit composition of the pig heart 2-oxoglutarate dehydrogenase, *J. Biol. Chem.*, 249, 3836, 1974.
75. **Cleland, W. W.,** Derivation of rate equations for multisite ping-pong mechanisms with ping-pong reactions at one or more sites, *J. Biol. Chem.*, 248, 8353, 1973.
76. **Tsai, C. S., Burgett, M. W., and Reed, L. J.,** α-Keto acid dehydrogenase complexes. XX. A kinetic study of the pyruvate dehydrogenase complex from bovine kidney, *J. Biol. Chem.*, 248, 8348, 1973.
77. **Reed, L. J., Damuni, Z., and Merryfield, M. L.,** Regulation of mammalian pyruvate and branched-chain α-keto acid dehydrogenase complexes by phosphorylation-dephosphorylation, *Curr. Topics Cell Reg.*, 27, 41, 1985.
78. **Roche, T. E., and Reed, L. J.,** Monovalent cation requirement for ADP inhibition of pyruvate dehydrogenase kinase, *Biochem. Biophys. Res. Commun.*, 59, 1341, 1974.
79. **Pettit, F. H., Pelley, J. W., and Reed, L. J.,** Regulation of pyruvate dehydrogenase kinase and phosphatase by acetyl-CoA/CoA and NADH/NAD ratio, *Biochem. Biophys. Res. Commun.*, 65, 575, 1975.
80. **Stepp, L. R., Pettit, F. H., Yeaman, S. J., and Reed, L. J.,** Purification and properties of pyruvate dehydrogenase kinase from bovine kidney, *J. Biol. Chem.*, 258, 9454, 1983.
81. **Yeaman, S. J., Hutcheson, E. T., Roche, T. E., Pettit, F. H., Brown, J. R., Reed, L. J., Watson, D. C., and Dixon. G. H.,** Sites of phosphorylation on pyruvate dehydrogenase from bovine kidney and heart, *Biochemistry*, 17, 2364, 1978.
82. **Sugden, P. H., Kerbey, A. L., Randle, P. J., Waller, C. A., and Reid, K. B. M.,** Amino acid sequences around the sites of phosphorylation in the pig heart pyruvate dehydrogenase complex, *Biochem. J.*, 181, 419, 1979.
83. **Pettit, F. H., Humphreys, J., and Reed, L. J.,** Regulation of pyruvate dehydrogenase kinase activity by protein thiol-disulfide exchange, *Proc. Natl. Acad. Sci. U.S.A.*, 79, 3945, 1982.
84. **Teague, W. M., Pettit, F. H., Wu, T.-L., Silberman, S. R., and Reed, L. J.,** Purification and properties of pyruvate dehydrogenase phosphatase from bovine heart and kidney, *Biochemistry*, 21, 5585, 1982.
85. **Pratt, M. L., Maher, J. F., and Roche, T. E.,** Purification of bovine kidney and heart pyruvate dehydrogenase phosphatase on Sepharose derivatized with the pyruvate dehydrogenase complex, *Eur. J. Biochem.*, 125, 349, 1982.
86. **Damuni, Z., Merryfield, M. L. Humphreys, J. S., and Reed, L. J.,** Purification of bovine kidney and heart pyruvate dehydrogenase phosphatase on Sepharose derivatized with the pyruvate dehydrogenase complex, *Proc. Natl. Acad. Sci. U.S.A.*, 81, 4335, 1984.
87. **Davis, P. F., Pettit, F. H., and Reed, L. J.,** Peptides derived from pyruvate dehydrogenase as substrates for pyruvate dehydrogenase kinase and phosphatase, *Biochem. Biophys. Res. Commun.*, 75, 541, 1977.
88. **Denton, R. M. and McCormack, J. G.,** On the role of the calucium transport cycle in heart and other mammalian mitochondria, *FEBS Lett.*, 119, 1, 1980.
89. **Hansford, R. G.,** Effect of micromolar concentrations of free Ca^{2+} ions on pyruvate dehydrogenase interconversion in intact rat heart mitochondria, *Biochem. J.*, 194, 721, 1981.
90. **Coll, K. E., Joseph, S. K., Corkey, B. E., and Williamson, J. R.,** Determination of the matrix free Ca^{2+} concentration and kinetics of Ca^{2+} efflux in liver and heart mitochondria, *J. Biol. Chem.*, 257, 8696, 1982.
91. **McCormack, J. G., and Denton, R. M.,** Role of Ca^{2+} ions in the regulation of intramitochondrial metabolism in rat heart, *Biochem. J.* 218, 235, 1984.
92. **Denton, R. M. and McCormack, J. G.,** The effects of calcium ions and adenine nucleotides on the activity of pig heart 2-oxoglutrate dehydrogenase complex, *Biochem. J.*, 180, 533, 1979.
93. **Dumuni, Z., Humphreys, J. S., and Reed, L. J.,** A potent, heat-stable protein inhibitor of [branched-chain α-keto acid dehydrogenase]-phosphatase from bovine kidney mitochondria, *Proc. Natl. Acad. Sci. U.S.A.*, 83, 285, 1986.
94. **Denton, R. M. and McCormack, J. G.,** Ca^{2+} transport by mammalian mitochondria and its role in hormone action, *Am. J. Physiol.*, 249, E543, 1985.

95. **Marshall, S. E., McCormack, J. G., and Denton, R. M.,** Role of Ca^{2+} ions in the regulation of intramitochondrial metabolism in rat epididymal adipose tissue: evidence against a role for Ca^{2+} in the activation of pyruvate dehydrogenase by insulin, *Biochem. J.,* 218, 249, 1984.
96. **Dumuni, Z., Humphreys, J. S., and Reed, L. J.,** Stimulation of pyruvate dehydrogenase phosphatase activity by polyamines, *Biochem. Biophys. Res. Commun.,* 124, 95, 1984.
97. **Wagenmakers, A. J. M., Schepens, J. T. G., Veldhuizen, J. A. M., and Veerkamps, J. H.,** The activity state of the branched-chain 2-oxo acid dehydrogenase complex in rat tissues, *Biochem. J.,* 220, 273, 1984.
98. **Patston, P. A., Espinal, J., Shaw, J. M., and Rendle, P. J.,** Rat tissue concentration of branched-chain 2-oxo acid dehydrogenase complex: re-evaluation by immunoassay and bioassay, *Biochem. J.,* 235, 429, 1986.
99. **Kerby, A. L., Richardson, L. J., and Randle, P. J.,** The roles of intrinsic kinase and of kinase/activation protein in the enhanced phosphorylation of pyruvate dehydrogenase complex in starvation, *FEBS Lett.,* 176, 115, 1984.
100. **Mullinax, T. R., Stepp, L. R., Brown, J. R., and Reed, L. J.,** Synthetic peptide substrate for mammalian pyruvate dehydrogenase kinase and pyruvate dehydrogenase phosphatase, *Arch. Biochem. Biophys.,* 243, 655, 1985.
101. **Uhlinger, D. J., Yang, C. -Y., and Reed, L. J.,** Phosphorylation-dephosphorylation of pyruvate dehydrogenase from Baker's yeast, *Biochemistry,* 25, 5673, 1986.
102. **Thomas, A. P., Diggle, T. A., and Denton, R. M.,** Sensitivity of pyruvate dehydrogenase phosphate phosphatase to magnesium ions: similar effects of spermine and insulin, *Biochem. J.,* 238, 83, 1986.
103. **Shepherd, G. and Hammes, G. G.,** Flourescence energy transfer measurements in the pyruvate dehydrogenase multienzyme complex from *Escherichia coli* with chemically modified lipoic acid, *Biochemistry,* 16, 5234, 1977.
104. **Angelides, K. J. and Hammes, G. G.,** Fluorescence studies of the pyruvate dehydrogenase multienzyme complex from *Escherichia coli, Biochemistry,* 18, 1223, 1979.
105. **Angelides, K. J. and Hammes, G. G.,** Structural and mechanistic studies of the α-ketoglutarate dehydrogenase multienzyme complex from *Eschericha coli, Biochemistry,* 18, 5531, 1979.
106. **Waskiewicz, D. E. and Hammes, G. G.,** Fluorescence polarization study of the α-ketoglutarate dehydrogenase complex from *Escherichia coli, Biochemistry,* 21, 6489, 1982.
107. **Hale, G. and Perham, R. N.,** Limited proteolysis of the pyruvate dehydrogenase multienzyme complex of *Escherichia coli, Eur. J. Biochem.,* 94, 119, 1979.
108. **Fuller, C. C., Reed, L. J., Oliver, R. M., and Hackert, M. L.,** Crystallization of a dihydrolipoyl transacetylase-dihydrolipoyl dehydrogenase subcomplex and its implications regarding the subunit structure of the pyruvate dehydrogenase complex from *Escherichia coli, Biochem. Biophys. Res. Commun.,* 90, 431, 1979.
109. **Packman, L. C., Perham, R. N., and Roberts, G. C. K.** Cross-linking and ^{1}H n.m.r. spectroscopy of the pyruvate dehydrogenase complex of *Escherichia coli, Biochem. J.,* 205, 389, 1982.

Section V
Flavoprotein Catalysis

An assumed state becomes real when its properties are measured. G. N. Lewis and M. Kasha, *J. Am. Chem. Soc.*, 66, 2101, 1944.

Because several of the systems studied by Michaelis et al. resemble systems of which the oxidants are quinones Michaelis called the radicals "semiquinones." W. Mansfield Clark, in *Oxidation-Reduction Potentials of Organic Systems*, p. 185, The Williams & Wilkins Co., Baltimore (1960).

Chapter 10

OLD YELLOW ENZYME

Lawrence M. Schopfer and Vincent Massey

TABLE OF CONTENTS

I. INTRODUCTION

Old Yellow Enzyme (NADPH oxidoreductase, EC 1.6.99.1) occupies an important place in the early history of enzymology. In 1932, when Warburg and Christian first isolated this yellow enzyme from brewer bottom yeast[1], relatively little was known about enzymes, either structurally or mechanistically. In fact, a major issue at that time concerned the chemical nature of enzymes; whether enzymes were proteins, or not. Protein was typically associated with purified enzyme activities, but it was not clearly established whether the enzyme activity was associated with the protein. Old Yellow Enzyme (OYE) provided a critical test for this question.

In 1934, Theorell purified OYE to electrophoretic homogeneity.[2] The purified material was composed of protein and a yellow cofactor in a one-to-one stoichiometry. Separation of the cofactor from the protein caused complete loss of enzymatic activity. Recombination of the two components resulted in complete return of activity.[3] This was a clear, direct demonstration that protein was an essential structural element of an enzyme.

During this same period, studies on the structure of the yellow cofactor culminated in the demonstration that it was the phosphoric acid ester of riboflavin (vitamin B_2).[4,5] The essential growth factor, riboflavin, was thus shown to be a part of a cofactor which was essential for the expression of an enzymatic activity. A biochemical role for this vitamin was thereby established.

A more extensive description of the early work on OYE can be found in the Nobel lecture (1955) of Hugo Theorell[6] and the thorough review (1963) by Åkeson, Ehrenberg, and Theorell.[7] The seminal nature of Theorell's early work is best described in his Nobel lecture.[6]

> "The significance of these investigations on the yellow enzyme may be summarized as follows.
>
> 1. The reversible splitting of the yellow enzyme to apo-enzyme + coenzyme in the simple molecular relation 1:1 proved that we had here to do with a pure enzyme; the experiments would have been incomprehensible if the enzyme itself had been only an impurity.
> 2. This enzyme was thus demonstrably a protein. In the sequel all the enzymes which have been isolated have proved to be proteins.
> 3. The first coenzyme, FMN, was isolated and found to be a vitamin phosphoric acid ester. This has since proved to be something occurring widely in nature: the vitamins nicotinic acid amide, thiamine and pyridoxine form in an analogous way nucleotide-like coenzymes, which like the nucleic acids themselves combine reversibly with proteins."

The early work on OYE was exciting and important to enzymology in general. Develepements in OYE since 1956 have been equally significant for the more specialized area of flavin enzymology.

Throughout the past 30 years, long wavelength (500 to 800 nm) absorbing species have been described for an ever expanding number of flavoenzymes.[8] These bands have been variously ascribed to semiquinones,[9] paired-radical complexes,[10] and charge-transfer complexes.[8,11] OYE is the only case in which it has been possible to rigorously explore the physical basis for these absorbance bands. A wide variety of phenols bind to OYE, resulting in long wavelength absorbance bands.[12] All evidence points toward charge-transfer as the explanation for these new absorbance bands. From the OYE example, has grown a greater acceptance for the involvement of charge-transfer interactions in flavin enzymology.

The topography of the flavin binding site is another area in which OYE is making important contributions. Flavin enzymes catalyze a wide variety of reactions. Since the catalytically important portion of the flavin, the isoalloxazine ring, is the same in all flavin enzymes, the catalytic specificity of a given enzyme must derive from the protein groups around the flavin. This concept is widely held, but the details which differentiate one flavin enzyme from another are still not clear. Traditionally, the details of protein 3-dimensional structure have been the province of X-ray crystallography. However, preparation of X-ray quality crystals of proteins and the solution of the structure is an arduous task. Information on most flavin enzymes is unavailable. As an alternative, an extensive series of modified flavins has been developed.[13,14] These modified flavins serve as reporter groups, providing information on (1) the solvent accessibility of the flavin, (2) the location of reactive protein residues near the flavin, and (3) the polarity of the flavin environment. OYE, being readily available, has served as a test system in the developement of these probes. Consequently a complete set of data is available for OYE. The results of the modified flavin studies, in conjunction with ^{13}C and ^{15}N NMR studies on OYE, containing flavins enriched with these isotopes,[15-18] offers a picture of the flavin binding site the like of which was heretofore obtainable only by X-ray crystallography. Thus OYE is setting an example in the use of a new and potent methodology for flavoproteins.

The ready availability of OYE has also made it a popular candidate for studies using physical techniques, e.g., circular dichroism,[19,20] nuclear magnetic resonance,[15-18] electron paramagnetic resonance,[21] electron-nuclear double resonance,[22,23] and photochemically induced dynamic nuclear polarization.[24] As a result, a wealth of physical data exists on OYE, the value of which has yet to be fully developed.

II. PHYSICAL PROPERTIES

Old Yellow Enzyme is a dimer of identical subunits. The dimer molecular weight was found to be 102,000 to 106,000 by analytical ultracentrifugation[7] and 101,000 by Sephadex G-150 chromatography.[25] Subunit molecular weight was calculated to be 49,900 from the amino acid composition analysis,[26] and 49,000 from SDS acrylamide gel electrophoresis, under reducing conditions.[12] Comparison of SDS gels run under both reducing and non-reducing conditions showed that the subunits were not held together by a disulfide bond.[27]

There is one FMN cofactor bound per subunit, with a dissociation constant of about 10^{-10} *M* (0.1 *M* potassium phosphate buffer, pH 7.0, 25°).[28] The visible absorbance spectrum is characteric of an oxidized flavin, with maxima at 462 nm (ϵ = 10,600 M^{-1} cm^{-1}), 380 nm (ϵ = 10,100 M^{-1} cm^{-1}) and 278 nm (ϵ = 105,000 M^{-1} cm^{-1}),[12,26,29] Figure 1. Reduction of OYE by dithionite proceeds via the red, anionic semiquinone,[30,31] with maxima at 380 nm (ϵ = 13,700 M^{-1} cm^{-1}) and 481 nm (ϵ = 6,700 M^{-1} cm^{-1}).[30] Though formed in about 90% yield during the course of a reductive dithionite titration, the semiquinone will slowly disproportionate into oxidized and fully reduced OYE. Approach to a stable, thermodynamic equilibrium can take days unless the exchange of electrons is facilitated by a redox mediator such as methyl viologen. The maximal amount of semiquinone observed at equilibrium was only 15 to 20%.[32] The anionic semiquinone was also found by electron-nuclear double resonance spectroscopy (ENDOR).[22,23] Like the semiquinone, fully reduced enzyme is also anionic.[16,17] It has a single maximum in the near-UV at 332 nm (ϵ = 5,600 M^{-1} cm^{-1}).[30]

Binding FMN to apo-OYE results in complete quenching of the intrinsic flavin fluorescence.[33] A minor blue fluorescence (excitation 380 nm, emission 444 nm) develops during storage, over a period of months, even when the enzyme is frozen at −20° C.[19] The intensity of this fluorescence is markedly increased by brief exposure to white light.[19] This clearly reflects degradation of the enzyme. The oxidation product of tryptophan: α,β-dehydrotryptophan, has been shown to have similar fluorescence characteristics (excitation 340 nm, emission 420 nm).[34]

The circular dichroic (CD) spectrum showed a negative peak of ellipticity at 370 nm ($\Delta\epsilon \approx$ 2 to 3 M^{-1} cm^{-1}) with no peak in the 460 nm region,[19,20] Figure 2. A complete lack of ellipticity for one of the absorbance bands is unusual for flavoproteins, but not unprecedented. For example, the 460 nm absorbance band of FAD (free in solution) has no CD

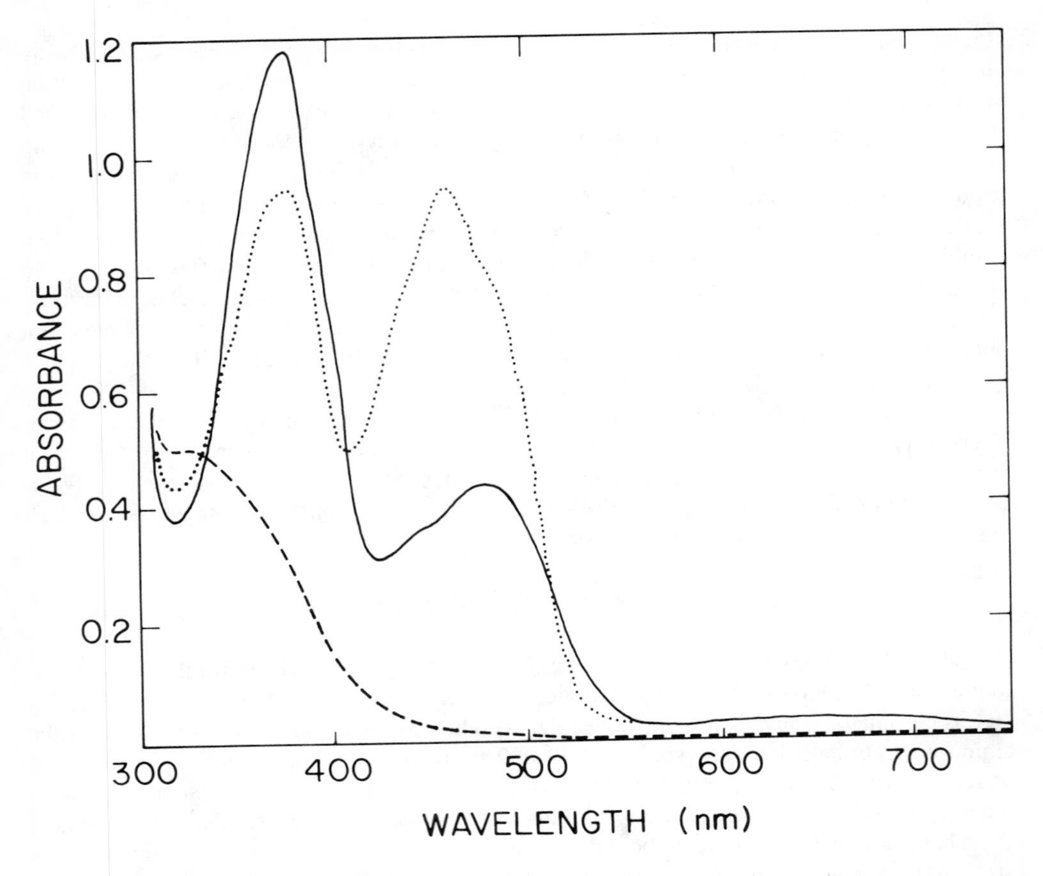

FIGURE 1. Absorbance spectra of Old Yellow Enzyme: oxidized (dotted line), red, anionic semiquinone (solid line), and fully reduced, anionic form (dashed line); in 0.1 *M* pyrophosphate, pH 8.5, 25°C.[30]

transition.[35] Lipoyl dehydrogenase[36] and L-lactate oxidase[27] have barely detectable ellipticity in this region. Binding ligands to native OYE or reconstituting apo-OYE with modified flavins (e.g., 7-bromo-FMN) caused a 460 nm CD transition to appear, typically as a positive peak ($\Delta\epsilon \approx 1$ to $3\ M^{-1}\ cm^{-1}$).[19]

The purified enzyme is devoid of carbohydrate[37] and of the metals [17,19,38] Ca, Mg, Zn, Fe, Mn, and Cu. It contains only one phosphate, that attached to the FMN.[37] The N-terminal sequence analysis gave a single sequence: H_2N-Pro-Phe-Val-Lys-Asp-Phe.[39]

The PKa for the N_3-proton in the oxidized enzyme is 10.5 or higher.[28]

Recent redox potential measurements indicate that the 2-electron midpoint potential is −235 mV (pH 7.0, 25° C).[32] The potential for the first electron is −-245 mV and for the second is −215 mV.[32] These values are considerably more negative than the −123 mV value reported earlier by Vestling.[40] However, when Vestling's experiments were done the ease of proteolysis of the enzyme had not been recognized, and in view of his report of nontheoretical behavior in his titrations beyond 50% reduction it must be assumed that his higher value is due to proteolysis.

OYE is readily reduced by NADPH,[7] dithionite,[30] or EDTA/light/deazaflavin.[41] Full reduction required 2-electrons per flavin.[30] NADPH is generally accepted as the physiological reductant for OYE. Reduction by NADPH does not produce semiquinone.[30,31] It is interesting to note that the uncommon α-anomer of NADPH is a slightly better reductant for OYE than is the β-anomer.[42] This is opposite to the finding with most other pyridine nucleotide dependent enzymes.[43] A more thorough discussion of the reduction kinetics is given below.

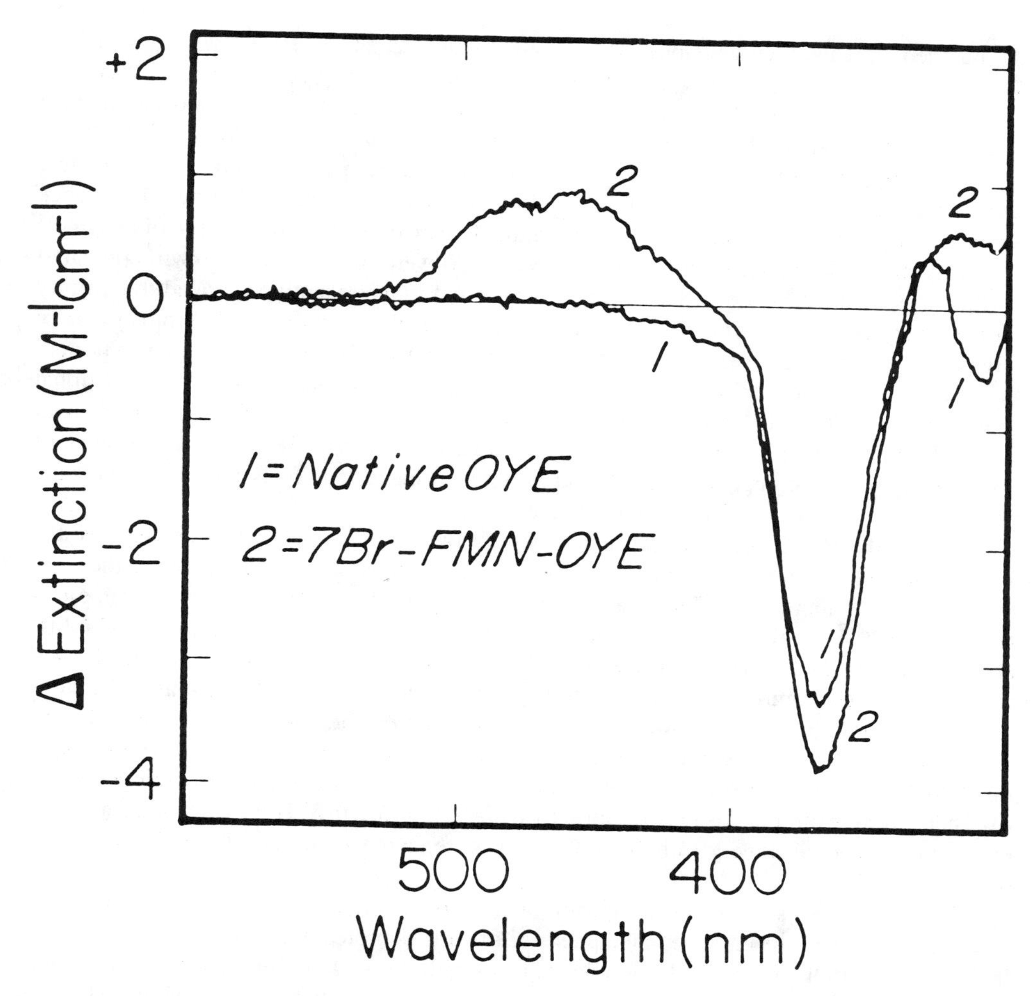

FIGURE 2. Circular dichroism spectra of oxidized, native Old Yellow Enzyme (trace 1) and oxidized 7-bromo-Old Yellow Enzyme (trace 2); in 0.1 *M* potassium phosphate, pH 7.0, 3° C.[19]

Reoxidation can be accomplished by a variety of oxidants including methylene blue,[7] Fe^{+3},[38] molecular oxygen,[7] quinones,[25,42,44] cytochrome C[7] and ferricyanide.[7] Oxygen is typically used as the acceptor for assay purposes. Reduction of oxygen by OYE results in the formation of both hydrogen peroxide[7] and superoxide,[45] in an approximate ratio of 4:1. A discussion of the possible physiological acceptor is given in the section on the "physiological electron acceptor".

III. PURIFICATION

Most workers in the area isolate OYE from brewer's bottom yeast. However, an enzyme with very similar properties has been found in the bacterium *Gluconobacter suboxydans* (IFO 12528; Institute for Fermentation, Osaka).[46]

The purification of OYE from brewer's bottom yeast has been periodically improved since 1934. The enzyme was crystallized in 1958.[29] The most recent purification procedure used affinity chromatography on N-(4-hydroxybenzoyl)aminohexyl Agarose to convienently prepare 100 to 150 mg of enzyme at a time.[12] Purification in the absence of the protease inhibitor phenylmethylsufonylfluoride (PMSF) resulted in limited proteolysis. Proteolysis of the purified enzyme even developed during storage at $-20°$ C, if PMSF was removed. A single cut was made which gave fragments of 35,000 and 13,000 mol wt, on SDS gel electrophoresis.[12] The clipped enzyme retained NADPH oxidase activity and exhibited a single band in acrylamide gel electrophoresis, under nondenaturing conditions.[12,26]

OYE prepared by the affinity chromatography procedure appeared to be homogeneous, based on gel electrophoresis (denaturing[12] and nondenaturing,[12,26]) and N-terminal sequence analysis.[39] Ligand binding studies were also characteristic of a single population of enzyme[12] All experiments described in this review (since 1976) were performed on enzyme prepared by this or equivalent methods. However, Miura et al.[15] have recently separated the affinity purified enzyme into five fractions by anion-exchange high pressure liquid chromatography (HPLC) on a Synchropak AX300 column. We have repeated this separation using fast protein liquid chromatography (FPLC) on a Mono-Q column, (quarternary ammonium functional group). It should be noted that the "green" form of OYE eluted from DEAE as four distinct bands.[47] At present, it is not possible to say whether this behavior is due to the presence of different ligands in complex with OYE as originally suggested, or is connected with the newly described heterogeneity of the enzyme.

Each of the HPLC fractions appears to be a distinct species of OYE. They all have the same molecular weight and visible absorption characteristics, described above for the mixture.[15] However there are differences in physical properties: (1) There are measurable differences ($\pm 20\%$) in the NADPH oxidase activity.[15] (2) Miura et al. found that ^{13}C-NMR of heterogenous OYE, reconstituted with FMN enriched with ^{13}C at the C_{4a}-position, showed two bands[15] (see Figure 6 for the flavin ring numbering scheme). The isolated fractions showed only one band, the position of which depended on the fraction examined. Beinert et al. found only one ^{13}C NMR band for the mixture enriched at C_{4a}.[17] Both groups found that NMR of the OYE mixture enriched at positions C_2, C_4, and C_{1a}, had no heterogeneity.[15] (3) There were several bands in isoelectric focusing (PI = 5.2 to 5.3).[16] (4) Detailed analysis of the rapid reaction kinetics of the unfractionated enzyme indicated at least two parallel reactions.[27]

The physical basis for this heterogeneity is unknown at this time. Proteolysis is unlikely since purification of the enzyme in the presence of various protease inhibitors gave no change in the HPLC elution profile.[15] The absence of carbohydrate on OYE[37] eliminates this common source of heterogeneity. Possible sources of heterogeneity still untested include: (1) five separate gene products, or (2) modification of some amino acid residue, eg. aspartate to asparagine or glutamate to glutamine, during growth, purification or storage. Further work will be required to identify the basis for the observed heterogeneity.

IV. LIGAND BINDING

Oxidized Old Yellow Enzyme has been found to form complexes with a wide variety of structurally diverse ligands (see Table 1): simple anions (both inorganic and organic); aromatic hydrocarbons (especially phenolate anions); and pyridine nucleotide derivatives. All three classes of ligand compete with one another for OYE,[42] suggesting that they occupy overlapping sites on the enzyme.

Ligand binding caused easily measured perturbations in the flavin absorbance spectrum. Two types of spectral perturbation were seen. The first was a shift (up to 20 nm) in the positions of the two major visible absorbance peaks (462 and 380 nm), usually accompanied by a 2 to 20% reduction in the peak height and a decrease in the sideband resolution. Shifts to either the red or the blue have been found, and each peak appears to move independently. The second was characterized by the appearance of two new absorbance bands: a broad, absorbance band in the 500 to 800 nm region ($\epsilon \approx 2000$ to $4000\ M^{-1}\ cm^{-1}$); and a band of variable extinction in the 300 to 425 nm region. The absorbance bands of the flavin were also perturbed, being shifted up to 20 nm and reduced in amplitude by as much as 40%

TABLE 1
Ligand Binding

Ligand	Binds (Y/N)	Kd (μM)	Long Wavelength (maximum)	pH	Temp (°C)	Ref.
Phenol	Y	30	610	7.0	4	12,19,27
4-F-phenol	Y		660			12
	Y	28	645	7.0	4	19,27
4-Cl-phenol	Y	0.46	645	9.45		12
	Y	≈1.0	645	7.0	4	19,27
4-Br-phenol	Y		no			12
4-NO_2-phenol	Y	0.50	545	7.0		12
	Y		550	7.0	4	19
4-CH_3-phenol	Y		660	7.0		12
	Y	23	650	7.0	4	19,27
4-CN-phenol	Y	<20	575	7.0	4	12,19,27
4-NH_2-phenol	Y	370	790	7.0	4	19,27
4-OCH_3-phenol	Y	70	700	7.0	4	19,27
4-Ethyl-phenol	Y		657	7.0	4	12,19
4-OH-phenol	Y	21	730	7.0	4	19,27
4-CH_2OH-phenol	Y		620			12,47
Penta-F-phenol	Y	0.15	560	7.0		12
Thiophenol	Y		620			12
Di-Cl-phenol indophenol	Y		700			12
Benzaldehyde	N		No			12,47
2-OH-benzaldehyde	Y	36	610	7.0		12
3-OH-benzaldehyde	Y		600			12
4-OH-benzaldehyde	Y	0.1	575	8.6		12,47
2,4-diOH-benzaldehyde	Y	0.4	550	8.5		12,47
2,5-diOH-benzaldehyde	Y	15	700	7.0		12
3,4-diOH-benzaldehyde	Y		600			12,47
2,3,4-triOH-benzaldehyde	Y		610			12
2,4,6-triOH-benzaldehyde	Y		No			12
4-OCH_3-benzaldehyde	Y	234	No	6.1		12
3-OCH_3,4-OH-benzaldehyde	Y		615			12,47
3,5-diOCH_3-benzaldehyde	N		No			12
2-OH-benzoate	Y	1000	No	8.5		12
4-OH-benzoate	Y	70	600	7.0		12,27,47
4-OH-benzoate methyl ester	Y	1.4	590	7.0		12
2-OH-acetophenone	Y		600			12,47
3-OH-acetophenone	Y		617			12,47
4-OH-acetophenone	Y		580			12,47
3-OCH_3,4-OH-acetophenone	Y		635			12,47
4-OH-cinnamate	Y		700			12,47
2-OH-cinnamate lactone	Y		No			12,47
2,4-diOH-cinnamate lactone	Y		No			12,47
2,7-diOH-cinnamate lactone	Y		615			12,47
4-OH-quinoline	N		No			26
8-OH-quinoline	N		No			26
3,4-diOH-quinoline	Y		710			26,12
Quinoline 2-carboxylate	N		No			26
4-OH-quinoline 2-carboxylate	Y		No	6.65	10	26
8-OH-quinoline 2-carboxylate	Y		560			26
4,8-diOH-quinoline 2-carboxylate	Y		660	7.3	10	26
8-OCH_3-quinoline 2-carboxylate	N		No			26
Picolinate	Y		No			12
Nicotinate	Y	240	No	7.0		12
4-OH-tryptamine-creatine	Y		No			26
Kojic acid	Y		620	7.0	4	27

TABLE 1
Ligand Binding (continued)

Ligand	Binds (Y/N)	Kd (μ*M*)	Long Wavelength (maximum)	pH	Temp (°C)	Ref.
4-OH-N-n-butylbenzamide	Y	0.2	620	8.0		12
NADPH	Y	90	No	8.5	25	10
(c-THN)TPN[a]	Y	≈0.44	No	7.0	4	42
(6-HTN)TPN[a]	Y	≈.0.2	No	7.0	4	42
NADH	Y	170	No	7.0	25	70
AAD[a]	Y	280	No	7.0	4	27
GDP[a]	Y	850	No	7.0	4	27
GTP[a]	Y	90	No	7.0	4	27
Acetate	Y	≈3000	No	7.0	4	42
Chloride	Y	≈8000	No	7.0	4	42
Azide	Y	280	No	7.0	4	42

[a] (c-THN)TPN = α-0²′-6B-cyclo-1,4,5,6-tetrahydro-NADP, (6-HTN)TPN = 6-OH-1,4,5,6-tetrahydro-NADP, AAD = 3-NH_2-adenine dinucleotide, GDP = guanosine diphosphate, and GTP = guanosine triphosphate.

A. SIMPLE ANIONS

Titration experiments[42] have shown that the simple monovalent anions: chloride, azide, and acetate bind to OYE. Cations did not appear to bind. Dissociation constants for this type of ligand were in the millimolar range. Of those studied, azide bound most tightly, with a dissociation constant of 0.28 m*M*.[42] These ligands caused the first type of spectral perturbation described above, a simple shifting of the existing absorbance maxima, Figure 3.

A large selection of anions was employed in a study by Rutter and Rolander[38] on the inhibitory effect of ions on the activity of OYE. All of the halogens, as well as nitrate, carbonate, and formate were effective inhibitors. The polyvalent anions phosphate, sulfate, citrate and EDTA were also effective in inhibiting the assay. However, Massey and co-workers found that the presence of 75 m*M* phosphate caused no spectral changes in OYE, nor was the apparent dissociation constant for azide altered.[27,42] This argues that phosphate and probably the other polyvalent anions do not bind to OYE. The activity study utilized a coupled assay (glucose-6-phosphate dehydrogenase). It is possible that some of the observations were biased by anion effects on the other components of the assay.

The type of spectral perturbation seen when monovalent anions bind to OYE are commonly found when ligands bind to flavoproteins. The physical basis for this effect is unclear. Similar observations have been made with free flavins in solution, when the hydrogen bonding capacity of the medium is altered.[48,49] However, the patterns of spectral change established with the free flavins are not strictly followed by OYE. One is left with the simple suggestion that the binding of these ligands appears to result in changes in the hydrogen bonding between the flavin and the protein/ligand. However effects due to changes in hydrophobicity/hydrophilicity or other factors may also be involved.

B. AROMATIC COMPOUNDS

Aromatic compounds compose the second class of ligand which binds to oxidized OYE. Of the compounds in this class, phenols are the most interesting in that on binding to OYE, they manifest new absorbance bands, in the 500 to 800 nm region and in the 300- to 425-nm region,[12,19,26,47] Figure 4. A naturally occuring complex of this type, involving p-hydroxybenzaldehyde, was responsible for the green form of OYE isolated by Matthews and co-workers.[26,47] Thiophenol and some condensed ring compounds, containing hydroxyl moie-

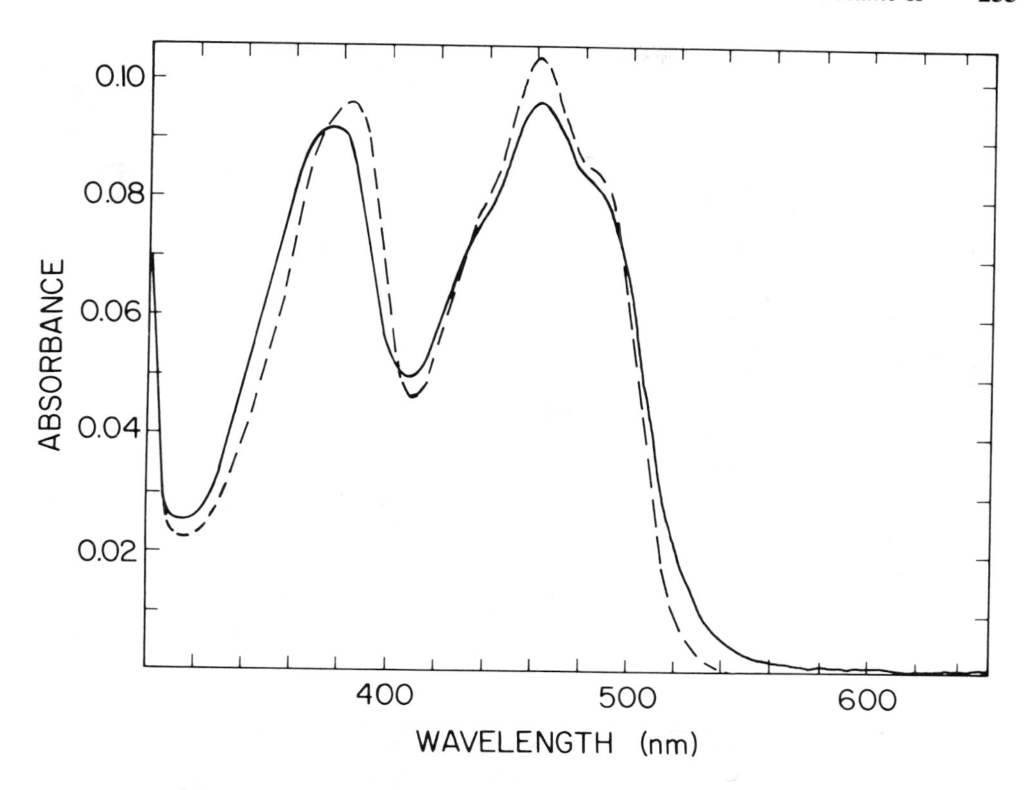

FIGURE 3. Absorbance spectra of oxidized Old Yellow Enzyme plus and minus potassium azide; oxidized OYE (dashed line), plus 8 mM azide (solid line); in 0.1 *M* potassium phosphate, pH 7.0, 3° C.

ties, also caused the appearance of long wavelength absorbance bands on binding to OYE.[12] Other aromatic compounds of similar structure (generally lacking the hydroxyl moiety) bound to OYE without causing a long wavelength band. These compounds perturbed the flavin spectrum in much the same manner as the simple, monovalent anions; however the size of the perturbation was typically larger.[12,26,47] Nonaromatic compounds gave no indication of binding,[12] nor did pteridine derivatives.[26] The common feature in all compounds of this class which caused long wavelength absorbance was the presence of an ionizable substituent,[12] suggesting that it is the phenolate (or comparable ionized substituent) which is essential for the appearance of this band. Five lines of evidence support this concept.

First, the pH dependence of binding, as illustrated by p-chloro-phenol, was consistent with the phenolate being in the complex which exhibited long wavelength absorbance.[28]

Second, binding pentafluorophenol to 8-hydroxy-FMN OYE resulted in the appearance of a long wavelength band and disappearance of the fluorescence of the 8-hydroxy-FMN anion.[50] Raising the pH at a fixed phenol concentration caused a loss of long wavelength absorbance and a return of the anion fluorescence. The dependence of the apparent pKa for 8-hydroxy-FMN OYE on pentafluorophenol concentration fits a model where only the phenolate anion in complex with the neutral flavin gave the long wavelength species.[50]

Third, the new absorbance band in the 300- to 425-nm region appears to be a red-shifted phenolate anion.[28] For every phenol examined, the position of this new band was approximately 22 nm to the red of the free phenolate absorbance maximum and the extinction of the new band was approximately the same as that of the free phenolate.[19] p-Nitro-phenol provided a particularly striking example of this phenomenon.[19] The red shift can be viewed as the result of complex formation.

Fourth, Beinert and coworkers using NMR of ^{13}C-labeled p-nitro-phenol in complex

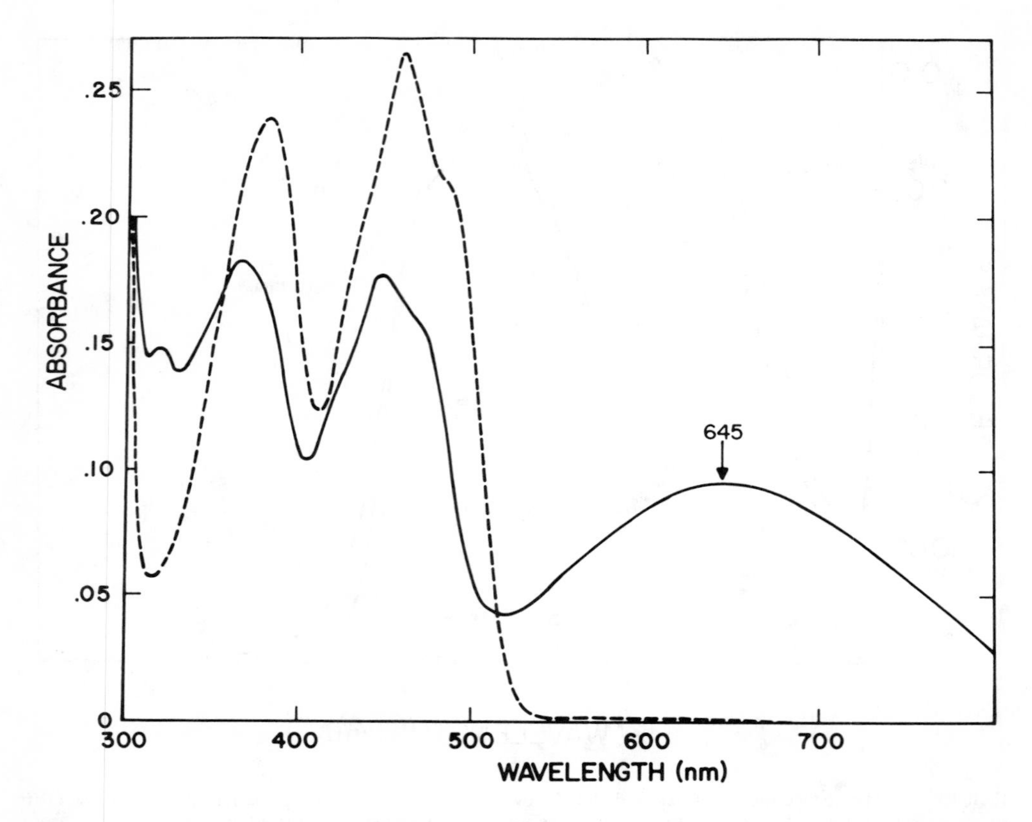

FIGURE 4. Absorbance spectra of oxidized Old Yellow Enzyme plus and minus p-chloro-phenol: oxidized OYE (dashed line), plus 70 μM p-chloro-phenol (solid line); in 0.1 M potassium phosphate, pH 6.5, 25°C.[28]

with OYE, found that the chemical shift of the C_1 resonance of the p-nitro-phenol was the same as that of the free phenolate anion.[17]

Fifth, there was a linear free energy relationship between the peak position of the long wavelength band (transition energy) and the Hammett σ_p value for a series of p-substituted phenols.[19,28] The positive slope of this correlation suggested that the phenol was the donor in a charge-transfer complex, with the flavin being the acceptor. The phenolate, being a more potent donor than the phenol, is the form most likely involved.

Matthews and Massey originally suggested that the long wavelength complexes between phenolate and oxidized OYE represented charge-transfer interactions, wherein the phenolate was the donor and the flavin the acceptor.[26] Though this position has been challenged,[20,51] the evidence, from a variety of physical studies, supports the original assignment.

The first strong evidence in support of the charge-transfer ascription came from the free energy correlations of Abramovitz and Massey,[12,28] described above. Such a correlation is characteristic of charge-transfer transitions.[52] Additional support came from experiments in which OYE, containing flavins of varying redox potential, was complexed to selected p-substituted phenols. A linear correlation was obtained between the position of the long wavelength band and the measured 1-electron redox potential of the OYE-bound flavin. Flavins which were more readily reduced (more positive potentials) generated long wavelength bands with peaks at longer wavelengths (lower energy) than flavins which were more difficult to reduce (more negative potentials).[32]

Additional support came from spontaneous resonance Raman studies.[53,54] The resonance Raman technique selects and amplifies the vibrational frequencies of the chromophore which absorbs at the frequency used for excitation of the Raman scattering. The OYE-phenolate long wavelength absorbance bands appear at wavelengths where neither the flavin nor the phenolate absorb. If this band is chosen as the excitation wavelength for the Raman scattering, vibrations from the flavin and phenolate will appear in the Raman spectrum only if they contribute to the chromophore, i.e., if they are partners in a charge-transfer interaction. The Raman spectrum did in fact contain vibrational bands for both flavin and phenol. The flavin band positions corresponded to those of the flavin in the uncomplexed enzyme.[55] Flavin vibrations associated with the N_5-C_{4a} locus were particularly intensified, suggesting that this region of the flavin is directly involved in the complex. The phenolate band positions varied, depending on the particular phenol used for the complex. Only in-plane ring vibrations of the phenol were resonance enhanced. This suggests that the Π-electrons of the benzene portion of the phenol interact with the flavin. Taken together, the Raman results describe a charge-transfer interaction in which the phenol and the pyrimidine portion of the flavin (isoalloxazine ring system) are stacked in a parallel fashion, such that there is appreciable overlap of their Π-orbitals.

Finally, ^{13}C and ^{15}N NMR studies of Beinert et al.[16,17] showed that complexing phenolates to OYE resulted in an increase in Π-electron density at positions N_5, N_{10}, and C_{4a} of the flavin. This was interpreted as a transfer of charge from the phenolate to the flavin via a Π-electronic interaction in this region.

C. PYRIDINE NUCLEOTIDES

Pyridine nucleotide derivatives constitute the third class of ligands for OYE. Since NADPH is a substrate for OYE, it is implicit that a complex is formed. However, the first direct observation of a complex between OYE and a pyridine nucleotide, was made by Haas.[56] He observed a change in the spectrum of oxidized OYE as a result of aerobic turnover with NADPH. The change consisted of a shift in the 462 nm peak to 473 nm and the broadening of this band so that absorbance tailed out to 600 nm. The 380 nm peak shifted slightly to the red. This was initially interpreted as the formation of a stable free radical. However, examination by EPR found only traces of radical and the compound was reinterpreted as a complex between NADP and the enzyme-bound FMN.[21] We subsequently observed that the change was dependent on the particular sample of NADPH used, suggesting that it was the result of a trace contaminant.[42] This was confirmed upon finding that the acid hydrolysis products of NADPH 6-hydroxy-1,4,5,6-tetrahydro-NADP ([6-OH-THN]TPN) and α-$O^{2'}$-6B-cyclo-1,4,5,6-tetrahydro-NADP([c-THN]TPN) bound tightly to OYE, causing the same type of spectral shift seen by Haas,[42] Figure 5.

The recognition that [c-THN]TPN, which has the α-configuration, binds tightly to the enzyme led to the testing of α-NADPH as a possible ligand, and the surprising discovery that it is an even better reductant than β-NADPH.[42]

Oxidized pyridine nucleotides do not bind to oxidized OYE,[42] though they do bind to the reduced enzyme.

V. STRUCTURE OF THE FLAVIN BINDING SITE

In recent years, many chemically modified flavins have been developed as probes for the structure of the flavin binding site in flavoproteins (see References 13 and 14 for reviews). Flavins with modifications at six positions around the the isoalloxazine ring (C_2, N_3, C_4, N_5, C_6, and C_8) are typically used. Modifications at positions N_1,C_7 and C_9 are also available. These probes provide information on (1) the accessibility of a particular position to solvent, (2) the presence of reactive amino acid side chains in the immediate vicinity of the position, (3) the electrostatic environment or hydrophobicity of the binding site in general, and (4) the steric constraints of the binding site, ie. the presence of protein bulk near the flavin. These probes complement techniques such as NMR and X-ray crystallography. Using the

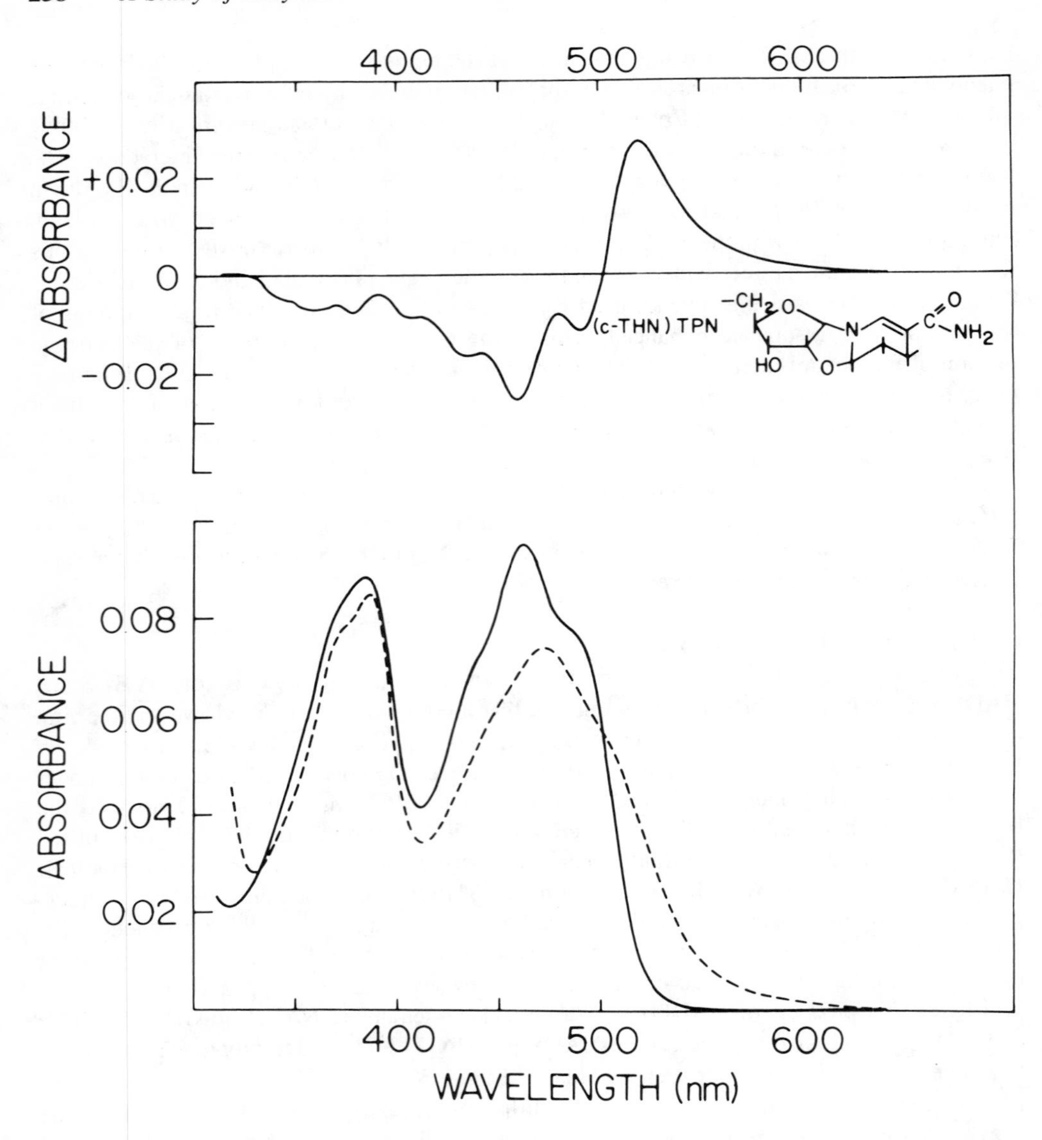

FIGURE 5. Absorbance spectra of oxidized Old Yellow Enzyme plus and minus α-0$^{2'}$-6B-cyclo-1,4,5,6-tetrahydro-NADP ([c-THN]TPN): oxidized OYE (solid line, lower panel), plus saturating [c-THN]TPN (dashed line, lower panel), difference spectrum (top panel); in 0.02 *M* potassium phosphate, pH 7.0, 4° C.[42]

first two techniques (X-ray quality crystals have not yet been obtained for OYE), the topography of the OYE flavin binding site has been extensively explored, see Figure 6.

A. POSITION C_8

8-Halogenated-flavins are susceptible to nucleophilic displacement, especially by thiols. 8-Chloro-FMN reconstituted OYE reacted more readily with solvent borne thiophenol ($k = 3.6 \times 10^4\ M^{-1}\ min^{-1}$) than did free 8-Chloro-FMN ($k = 3.2 \times 10^3\ M^{-1}min^{-1}$), clearly establishing the accessibility of the 8-position to solvent.[57] The same result was obtained from the reaction of iodoacetamide with 8-mercapto-FMN OYE ($k = 79\ M^{-1}\ min^{-1}$), versus free 8-mercapto FMN ($k = 47\ M^{-1}\ min^{-1}$).[57] Saturation of 8-chloro-FMN OYE with pentafluorophenol resulted in a complete loss of thiophenol reactivity. Azide and α-O$^{2'}$-6B-cyclo-1,4,5,6-tetrahydro-NADP were also effective inhibitors.[27] That all three compounds

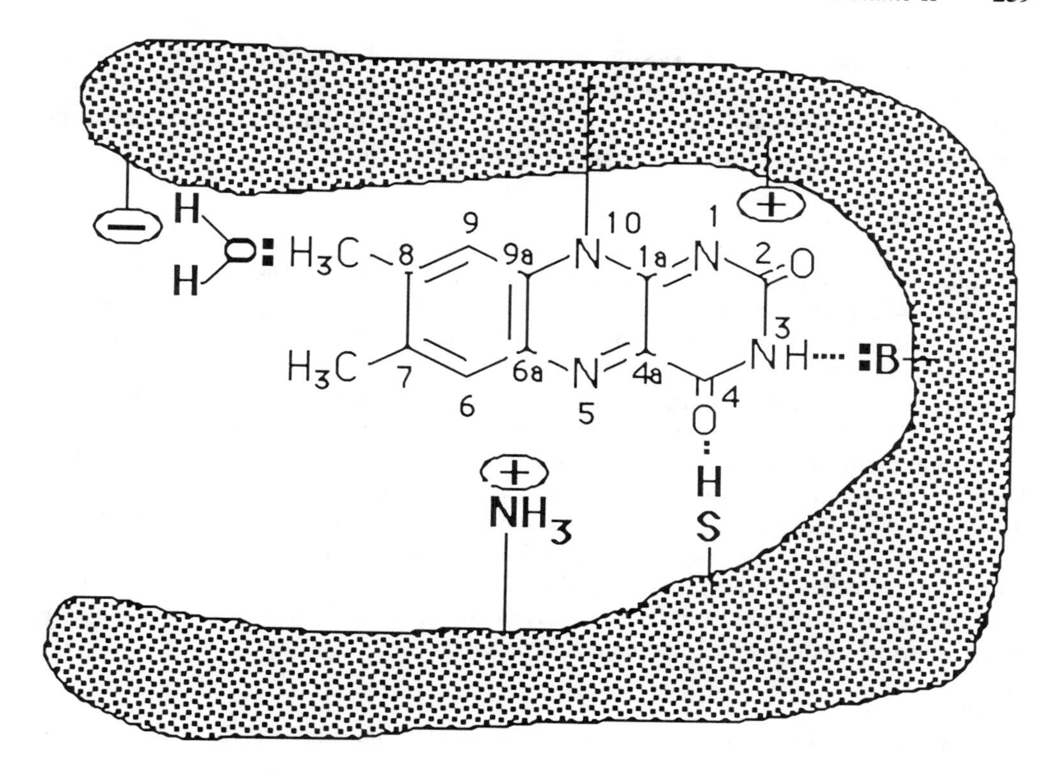

FIGURE 6. Schematic diagram of the flavin binding site of Old Yellow Enzyme.

(including the small azide anion) inhibited the reaction, argues against direct steric interference at the 8-position and suggests that OYE may undergo a conformational change on binding ligands.

Additional probes for protein residues, especially nucleophiles, in the vicinity of the 8-position include (1) the photo-reactive 8-azido-FMN, and (2) the highly reactive 8-fluoro-FMN. 8-Azido-FMN OYE decomposed to 8-amino-FMN OYE upon irradiation.[58] Irradiation caused the same reaction for free 8-azido-FMN in solution. Irradiation of azido compounds causes formation of nitrenes, which are extremely reactive.[59] That irradiation of 8-azido-FMN OYE did not result in covalent attachment of the flavin to the protein indicates that the nitrene reacted with water, confirming the ready availability of the 8-position to solvent.

8-Fluoro-FMN is very sensitive to nucleophilic displacement of the fluorine. However, 8-fluoro-FMN OYE did not undergo attack by protein nucleophiles. Rather it was hydrolyzed to 8-hydroxy-FMN OYE in a slow reaction.[27] Since the free 8-fluoro-FMN was stable in solution under the same conditions, this would indicate that the 8-fluoro-FMN OYE hydrolysis was catalyzed by the protein. A water molecule bound near the 8-position might effect such hydrolysis.

8-Mercapto-FMN OYE stabilized the blue, benzoquinoid resonance form (absorbance maximum at 598 nm) of 8-mercapto-FMN. This resonance form has the sulfur double-bonded to the 8-position with a formal negative charge in the N_1-C_2 locus of the isoalloxazine ring.[60] Localization of the negative charge at N_1-C_2 was taken as evidence for the existence of a positive charge on the protein in this region. The spectrum of the anionic form of 8-hydroxy-FMN OYE (absorbance maximum at 488 nm) supports this hypothesis.[50] A positive charge on the protein near N_1-C_2 would be expected to lower the pKa of 8-hydroxy-FMN bound to OYE, as it does for 8-hydroxy-FMN L-lactate oxidase.[50] However, 8-hydroxy-FMN OYE had a pKa = 6.2,[50] which was higher than that of free 8-hydroxy-FMN (pKa

= 4.8).[61] This unexpected observation can be rationalized if there is a negatively charged protein residue near the 8-position. Such a residue could also serve to hydrogen-bond a water molecule in position to hydrolyze 8-fluoro-FMN OYE, but this residue would be sufficiently removed from the flavin that it could not react directly.

B. POSITION C_2

The presence of a positive charge near C_2 or a hydrogen-bond to the C_2-oxygen was indicated by ^{13}C NMR studies.[17] The results with the modified flavins (at both the 6- and 8-positions) support the positive charge interpretation. A positive charge in this region also provides a rational explanation for the observed formation of stable anionic semiquinone and anionic reduced forms of native OYE.[16,17]

This position is protected from solvent. There was no reaction between 2-thio-FMN OYE and solvent borne methylmethanethiolsulfonate (MMTS).[16] The protein also seems to be close to the flavin in this area, since 2-thioacetamido-FMN did not bind to apo-OYE.[27] However, the flavin generally experiences a hydrophilic environment. This was shown by the spectrum of FMN-2-S-oxide OYE (absorbance maxima at 564 and 354 nm).[63] The spectrum of riboflavin-2-S-oxide has a pronounced dependence on solvent polarity. The absorbance maxima shift from 560 and 356 nm in water to 630 and 380 nm in benzene.[63] Binding pentafluorophenol to FMN-2-S-oxide OYE caused a slight decrease in the hydrophilic character of the flavin environment.

C. POSITION C_6

Accessibility of the 6-position to solvent was amply illustrated by the reactions of 6-mercapto-FMN OYE with MMTS and N-ethylmaleimide.[64] and 6-thiocyanato-FMN OYE with dithiothreitol.[64]

Confirmation of a positive charge in the N_1-C_2 region comes from 6-mercapto-FMN OYE[64] (absorbance maxima at 700, 456, and 350 nm) and 6-hydroxy-FMN OYE[27] (absorbance maxima at 640, 438, and 340 nm) both of which exhibit the characteristic benzoquinoid absorbance spectra.[64,65] Furthermore, the pKa of the 6-hydroxy-FMN OYE is shifted to 5.7 (relative to the 7.1 for the free flavin[65]) as expected for resonance stabilization of the negative charge.[27] Contrary to these observations, it was found that 6-sulfenato-FMN OYE (6-SOH-FMN) did not exist in its benzoquinoid form.[64] This could however be rationalized if there were a positively charged protein group near the 6-position, which could oppose the formation of a positive charge on the sulfur. Formation of a zwitterion with positive charge on the sulfur and negative charge at N_1-C_2 would be necessary for the appearance of the benzoquinoid form of the flavin 6-sulfenate.

Evidence in support of a positively charged protein residue near C_6 came from the photochemistry of 6-azido-FMN OYE. Irradiation of 6-azido-FMN OYE resulted in the formation of a stable covalent linkage between the flavin and the protein.[66] The stability of the bond suggests that the protein residue might be an amine.

D. POSITION N_5

Studies using ^{15}N NMR indicated the presence of a positive charge near the 5-position.[16] An alternative interpretation involving a hydrogen-bond to N_5 is unlikely in view of the low pKa = -3.8 of N_5 in oxidized flavin.[67] A single positively charged residue in this vicinity might suffice to explain the observations at C_6 as well.

A protein positive charge at the N_1-C_2, locus predicts that sulfite should form an adduct at N_5, as is the case with other flavoenzymes which have a positive charge at N_1-C_2.[13] This however did not occur.[68] The possibility that the lack of reaction was due to the inaccessibility of the 5-position to solvent was ruled out by the formation of an air stable N_5 photo-adduct (absorbance maximum at 312 nm) upon irradiating native OYE in the presence of trimethylacetate.[27] An alternative explanation is that the sulfite interacts with the positively charged group near N_5, thereby decreasing the nucleophilicity of the sulfite to a point where it can no longer attack the N_5 position of OYE.

E. POSITION C_4

4-Thio-FMN OYE reacted readily with hydrogen peroxide, indicating solvent accessibility for this position.[69] The area around the 4-position is sufficiently open to allow weak binding of 4-dimethylamino-FMN to apo-OYE, but not so open as to allow binding of 4-thioacetamido-FMN.[19] Binding pentafluorophenol to 4-thio-FMN OYE markedly decreased the hydrogen peroxide reaction rate.[27] In view of the similar protective effect for reactions at the 8–position, it appears likely that a significant conformational change occurs on binding ligands to OYE.

4-Thio-FMN OYE underwent spontaneous conversion to native OYE over a period of weeks.[69] Such a reaction is indicative of a protein thiol residue near the 4-position, but the rate was much slower than comparable reactions with 4-thio-FAD glucose oxidase and 4-thio-FMN L-lactate oxidase. It is possible that the thiolate is located farther from the 4-position in OYE, or the pKa of the protein thiol in OYE is higher than in the other two enzymes. Either case would result in less effective thiol catalysis. Evidence for a hydrogen bond to the C_4 oxygen has been found in the ^{13}C NMR work on native OYE.[17] A protonated thiol could be the donor for such an interaction.

F. POSITION N_3

The ^{15}N NMR studies[16] found the N_3-position to be hydrogen-bonded in native OYE, suggesting close contact between the flavin and the protein at this position. This is supported by finding a 4-order of magnitude decrease in the binding affinity of apo-OYE for flavin when the N_3-proton was replaced by CH_3, and a total loss of affinity when it was replaced by a CH_2COOH group.[28]

G. SUMMARY

Taken together, the foregoing results describe a flavin binding pocket that is accessible to solvent along one entire edge of the isoalloxazine ring system, from position-8 to position-4. The remainder of the isoalloxazine ring system is protected. The general flavin environment is hydrophilic. A negatively charged residue is located near the 8-position. Positively charged protein residues exist near the N_1-C_2 and C_6-N_5 loci. In the latter case, the positive charge is likely an amine. Twin positive charges in the flavin active site would provide ample attraction for the negatively charged ligands (phenolates and small anions) which bind to OYE. Hydrogen-bonds connect the protein to the flavin at the C_4-oxygen and the N_3-proton, the former possibly involving a protein thiol. This description constitutes a minimal picture of the flavin-protein interactions.

VI. KINETICS

Recent developments in the purification of OYE pose significant problems for the interpretation of all kinetic studies to date. Abramovitz and Massey[12] demonstrated that proteolysis could generate two populations of active enzyme. The proteolyzed enzyme exhibited a lower affinity for ligands and at best only one-half the steady state turnover of the intact enzyme.[12] All studies prior to 1976 were performed using enzyme which was purified without protease inhibitor protection.

More recent work has used nonproteolyzed enzyme; however Miura et al.[15] have very recently separated the apparently homogeneous OYE preparation of Abramovitz and Massey into five subfractions. The steady state turnover rates for the isolated subfractions varied from 21.5 to 32.9 min^{-1}, in a fixed concentration assay involving NADPH and O_2.[15] Detailed analysis of the rapid reaction kinetics (both the oxidative, NADP and O_2, and reductive, NADPH) indicated that at least two parallel reactions were present when the OYE mixture was used.[27,42] These factors make a precise, quantitative description of the reaction kinetics unwarranted at this time. However, the various kinetic studies in the literature are qualitatively similar, suggesting that the overriding qualitative nature of the kinetics is not obscured by the structural heterogeneities in the various enzyme preparations used. The following discussion will therefore concentrate on the qualitative aspects of the kinetics. References to the quantitative kinetic parameters must be considered approximate.

A. STEADY STATE

Under all conditions studied, linear parallel Linweaver-Burk plots (reciprocal apparent rate versus reciprocal substrate concentration were found. Conditions included (1) β-NADPH and O_2 (0.1 *M* potassium phosphate buffer pH 7.0, 3°),[42] (2) α-NADPH and O_2 (0.1 *M* potassium phosphate buffer pH 7.0, 3°),[42] (3) NADH and O_2 (0.1 *M* potassium phosphate buffer pH 7.0, 25°),[70] (4) NADH and ferricyanide (0.1 *M* potassium phosphate buffer pH 7.0, 25°),[70] (5) NADPH and ferricyanide (0.1 *M* pyrophosphate buffer pH 8.5, 25°),[10] and (6) *N*-propyldihydronicotinamide and O_2 (0.02 *M* Tris-HCI buffer pH 8.0, containing 0.2 *M* KCI, 25°).[71] Reactions were monitored either (1) spectrophotometrically at 340 nm, to follow oxidation of pyridine nucleotide or at 420 to follow oxidation of ferricyanide, or (2) polarographically (oxygen electrode) to follow reduction of oxygen, see Table 2 for representative turnover numbers). Parallel line patterns indicate the presence of an irreversible step, between the addition of the reductant and the addition of the oxidant, in the progress of the reaction. Since reduced OYE can be oxidized by sufficiently high concentrations of NADP[42] or NAD,[70] the irreversible step is most likely release of oxidized pyridine nucleotide. This argues that the mechanism is pong-pong Bi-Bi (Cleland's nomenclature).[72] Substrate Km values are approximately: 10 μ*M* for NADPH (α or β),[42] 200μ*M* for NADH,[70] 200 μ*M* for ferricyanide,[70] and 1 m*M* for O_2.[42,70]

Turnover experiments using [4-^{2}H] and [4-^{3}H]-NADPH[73] demonstrated that OYE abstracts the 4-R hydrogen from NADPH, and that the reductive step is at least partially rate limiting.

B. RAPID REACTION

Stopped-flow kinetic studies on the reduction of OYE have consistently found the transient appearance of a long wavelength intermediate.[10,30,31,42,70,71] In two cases, formation of this intermediate was preceded by another transient.[42,70]

The first intermediate appeared as a dead-time species in the stopped-flow spectrophotometer (intermediate 1 in Scheme 1). It caused a small decrease in absorbance in the 460 nm band and a shift of this maximum toward the red,[42] Figure 7. Such behavior is consistent with a simple binding perturbation, similar to those found when simple monovalent anions or nonphenolic, aromatic ligands bind to OYE.

The second intermediate was characterized by a long wavelength absorbance band extending to 700 nm and a red shift in the 462 nm absorbance peak, Figure 7. The presence of substantial absorbance at 470 nm was clear indication that the flavin remained oxidized. When NADPH was used as reductant, the decay of this intermediate was biphasic.[42]

The reverse reaction, oxidation of reduced OYE by NADP[42] (or NAD[70]) occurred readily. The rates showed a saturating dependence on the NADP concentration.[42] The final, stable, reoxidized enzyme had significant long wavelength absorbance, strongly resembling EFl_{ox}·NAD(P)H,[42,70] see Scheme 1. The sequence of events during reduction is represented by Scheme 1:

TABLE 2
Kinetic Parameters

Reductant	Kd_1 (μM)	Kd_2 (μM)	k_1 (min^{-1})	k_{-1} (min^{-1})	k_2 (min^{-1})	k_{-2} (min^{-1})	TN[a] (min^{-1})	Ref
β-NADPH[b]	Too low to measure	350	1250	Very small	18 71[c]	230 600[c]	30	42
β-NADPD[b] [4R-^{2}H]	Too low to measure	—	1040	—	1.4 7.4[c]	—	—	42
α-NADPH[b]	—	—	1550	—	17 99[c]	—	50	42
β-NADPH[d]	90	—	15000	—	80	—	24[e]	10
NADH[f]	170	—	4800	1800	42	300	42	70
N-propyl[g] dihydro nicotinamide	Too high to measure	—	Too fast to measure		720	—	—	71
N-propyl[g] [4,4′-(^{1}H,^{2}H)] dihydronicotinamide	—	—	—	—	280	—	—	71

[a] TN, steady-state turnover number, min^{-1}.
[b] 0.1 *M* potassium phosphate buffer, pH 7.0, 3°C.
[c] A biphasic change was observed for this step. Both values are given.
[d] 0.1 *M* pyrophospate buffer, pH 8.5, 25°C.
[e] Apparent TN in air saturated buffer.
[f] 0.1 *M* potassium phosphate buffer, pH 7.0, 25°C.
[g] 0.02 *M* Tris/CI buffer, pH 8.0, containing 0.2 *M* KCI, 25°C.

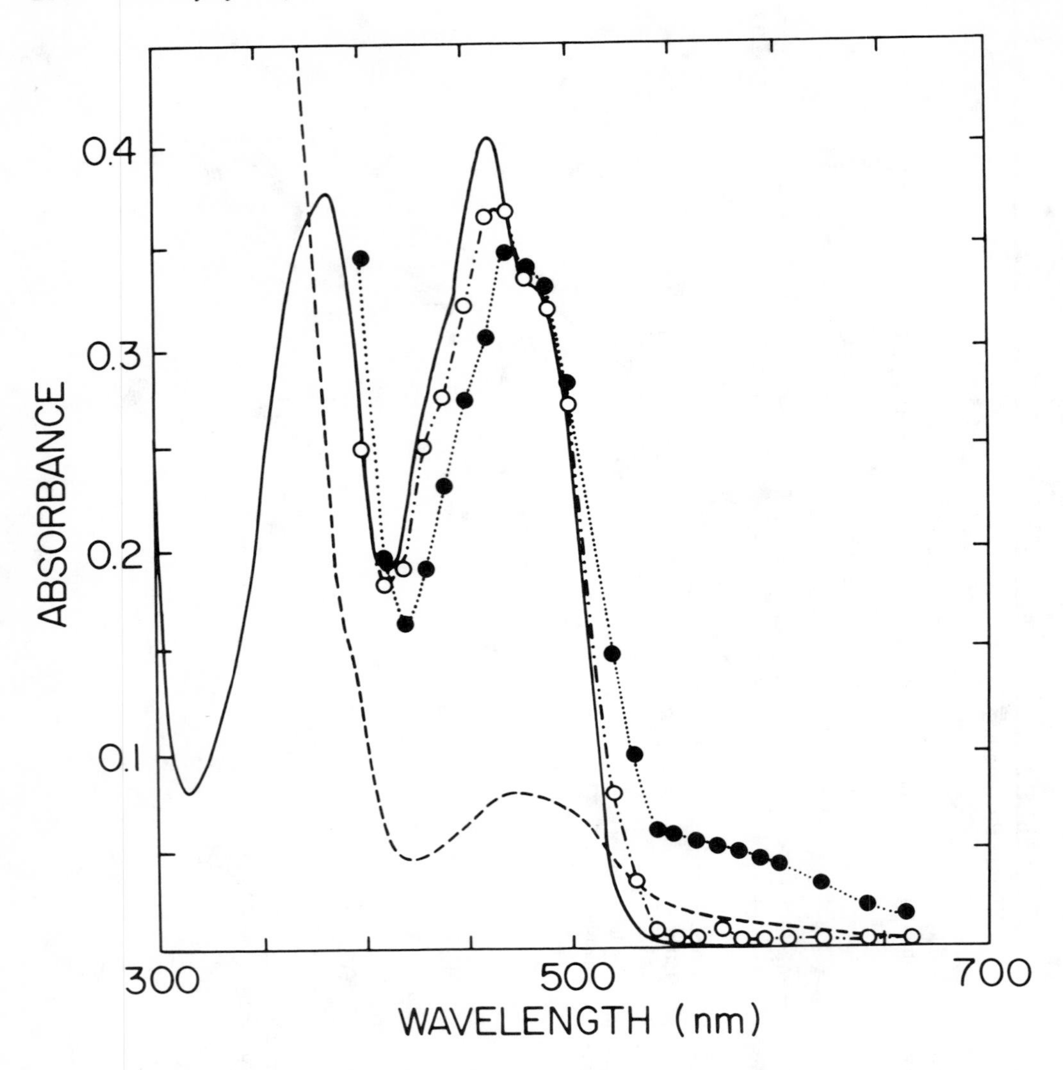

FIGURE 7. Absorbance spectra of oxidized Old Yellow Enzyme and the transient intermediates which appear during anaerobic reduction with β-NADPH: oxidized OYE (solid line), intermediate 1 (open circles), EFI_{ox}·NADPH (closed circles), and the final, stable reaction mixture (dashed line); in 0.02 *M* potassium phosphate, pH 7.0, 4°C.[42]

$$EFI_{ox} + NAD(P)H \overset{Kd_1}{\leftrightarrow} I \underset{k_{-1}}{\overset{k_1}{\rightleftarrows}} EFI_{ox} \cdot NAD(P)H \underset{k_{-2}}{\overset{k_2}{\rightleftarrows}} EFI_{red} \cdot NAD(P) \overset{Kd_2}{\leftrightarrow} EFI_{red} + NAD(P)$$

Steps labeled Kd_1 and Kd_2 are taken to be rapid equilibrium binding events. Steps labelled k_1, k_{-1}, k_2, and k_{-2} are measurable. Electron/proton transfer occurs at step k_2/k_{-2}. Kinetic parameters for Scheme 1, with six different reductants, are shown in Table 2. Comparing turnover number to the micro-rate constants indicates that the rate limiting step in turnover is k_2. Deuterated pyridine nucleotides caused a primary isotope effect in the k_2 rate,[42,71] indicating (1) that proton/electron transfer occurs at this step, and (2) that the long wavelength complex is composed of oxidized OYE and reduced pyridine nucleotide. This latter result is consistent with the visible absorbance spectrum of the long wavelength complexes. In view of these results and the established tendency for NAD(P)H and oxidized flavoproteins to form charge-transfer complexes,[8,10] it has been suggested that the long wavelength intermediate is a charge-transfer complex, wherein reduced pyridine nucleotide is the donor and oxidized OYE is the acceptor.[42,70,71] EPR studies support this concept, in that no free radical signal was found for the long wavelength complex.[70]

When followed in the stopped-flow spectrophotometer, oxidation of reduced OYE by NADP (α or β), molecular oxygen, or ferricyanide was in each case multiphasic. The NADP reaction was biphasic, with both phases showing a saturation dependence on NADP concentration.[42] Molecular oxygen gave a reaction with at least four phases.[27] Superoxide dismutase (SOD) included in the reaction eliminated two phases. The two major (SOD independent) phases had second order dependencies on oxygen concentration. Oxidation by ferricyanide showed two major phases, neither of which was dependent on ferricyanide concentration.[27]

The substrate dependence in the oxygen and NADP reactions argues strongly for a mixed population of OYE in the reaction. This is consistent with the NADPH reduction results in which the decay of the charge-transfer intermediate was biphasic, both phases showing a primary deuterium isotope effect.

VII. PHYSIOLOGICAL ELECTRON ACCEPTOR

Old Yellow Enzyme originally came to notice by virtue of the fact that it could be bleached by a system containing glucose-6-phosphate, NADP and a crude protein fraction from yeast (Zwischenferment).[1] The active component in the yeast extract was later identified as glucose-6-phosphate dehydrogenase (G6PD). The bleaching effect was due to NADPH generated by G6PD at the expense of glucose-6-phosphate. Since oxygen provided an adequate electron acceptor, OYE could be assayed, purified, and studied without knowledge of its normal reaction. With time, NADPH has come to be accepted as the physiological reductant. Molecular oxygen however, due to its slow rate of reaction (relative to other oxidases) is commonly viewed as an opportunistic oxidant, much like ferricyanide or methylene blue. Considerable effort has been expended in the search for the physiological oxidant and quinones are the only class of compounds which so far offer reasonable qualifications. Several simple quinones have been found which will turnover OYE: menadione, [25] coenzyme Q_1,[44] duroquinone (tetramethyl-p-benzoquinone),[27] and p-benzoquinone.[27]

Duroquinone is especially interesting.[27] It oxidized reduced OYE (0.1 *M* potassium phosphate buffer, pH 7.0, 3°) in a biphasic manner, consistent with the behavior of other oxidants. The rates of both phases were independent of duroquinone concentration down to 50 μ*M*, suggesting complex formation prior to oxidation with a dissociation constant in the micromolar range. Observed rates (1140 min^{-1} and 300 min^{-1})[27] were 10-fold faster than the reduction rates, using NADPH (saturating); and 100-fold faster than the oxidation rates using molecular oxygen (air saturation). The steady state turnover number (β-NADPH and duroquinone, anaerobically)[27] was 51 min^{-1}, consistent with reduction being rate limiting.

In complementary experiments, it was found that when p-benzoquinone was added to an aerobic assay mixture containing excess NADPH, oxygen consumption was stopped until all of the quinone was reduced, whereupon oxygen consumption resumed,[74] Figure 8. In every respect, quinones appear to outperform oxygen as an oxidant. In addition, a quinone as the physiological electron acceptor offers some explanation for the binding of phenolates to OYE. The phenolate becomes the reduced product, hydroquinone, which would be expected to have some affinity for the oxidized enzyme. That hydroquinones bind reasonably tightly to OYE (e.g., p-hydroxy-phenol with a dissociation constant of about 50 μ*M*) provides a feedback, product-inhibition control mechanism on the activity of OYE. In view of the twin positively charged residues in the flavin binding site, a p-quinone would be a particularly attractive candidate as the physiological oxidant.

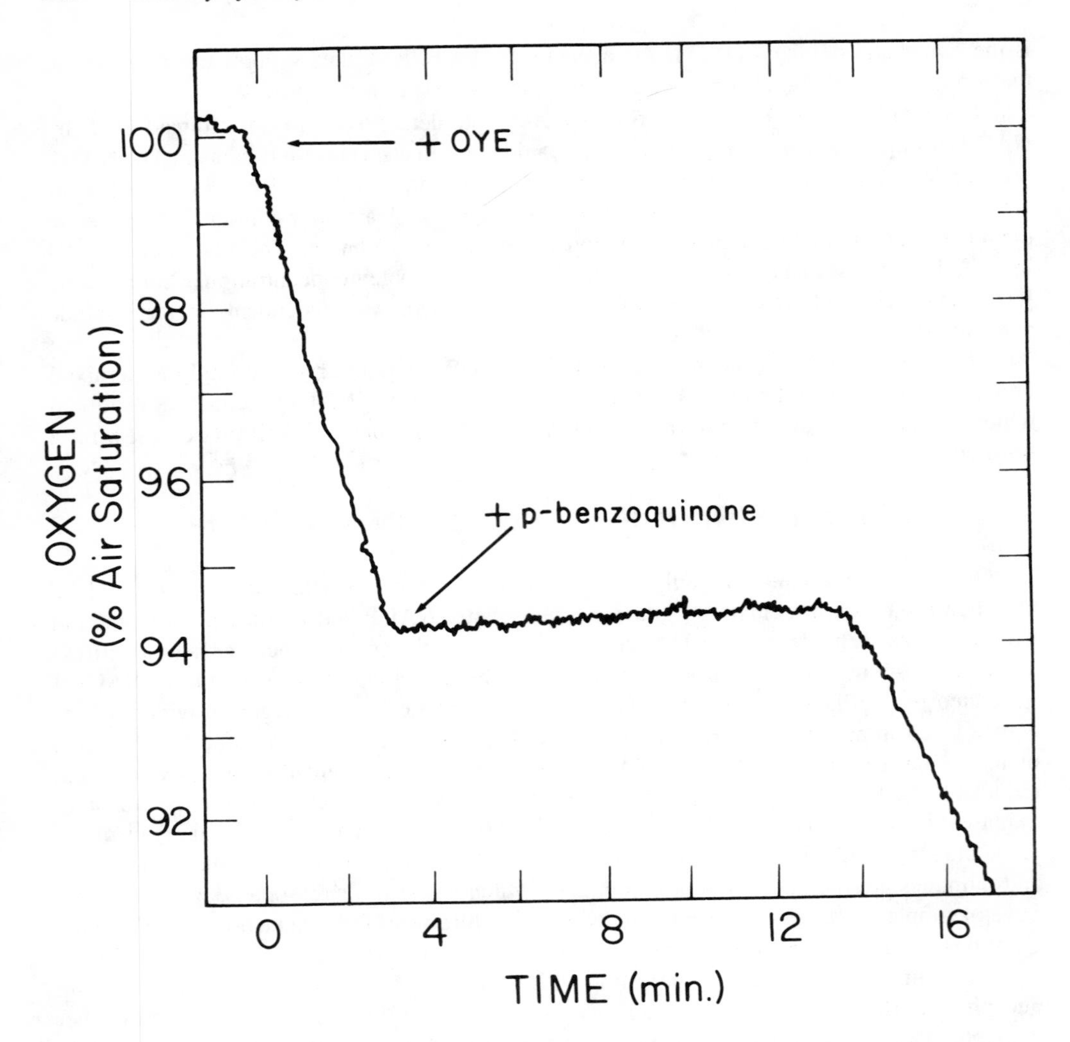

FIGURE 8. An oxygen electrode trace of an Old Yellow Enzyme assay (NADPH vs. molecular oxygen), illustrating the effect of *p*-benzoquinone on the time course of oxygen consumption. Conditions: the reaction chamber contained 200 μ*M* NADPH in 2 ml of air saturated (254 μ*M* oxygen) potassium phosphate buffer, 0.02 *M*, pH 7.0, 25° C. At time zero, 10 μl of OYE were injected, giving a final concentration of 0.2 μ*M* OYE. After 3 min, 30 μl of *p*-benzoquinone was injected, giving a final concentration of 150 μ*M* *p*-benzoquinone.

REFERENCES

1. **Warburg, O. and Christian, W.,** Ein Zweites Sauerstoffübertragendes Ferment und Sein Absorptionspektrum, *Naturwissenschaften,* 20, 688, 1932.
2. **Theorell, H.,** Reinigung des Gelben Atmungsfermentes mittels Elektrophorese, *Naturwissenschaften,* 22, 289, 1934.
3. **Theorell, H.,** Reindarstellung (Kristallisation) des Gelben Atmungsfermentes und die Reversible Spaltung Desselben, *Biochem Z.,* 272, 155, 1934.
4. **Kuhn, R., Rudy, H., and Weygand, F.,** Synthese der Lactoflavin-5′-phosphorsäure, *Ber,* 69, 1543, 1936.
5. **Theorell, H.,** Reindarstellung der Wirkungsgruppe des Gelben Ferments, *Biochem Z.,* 275, 344, 1935.
6. **Theorell, H.,** The nature and mode of action of oxidation enzymes, *Nobel Lecture,* 480, 1955.

7. **Åkeson, Å., Ehrenberg, A., and Theorell, H.,** Old yellow enzyme, in *The Enzymes,* Vol. 7, 2nd ed., Boyer, P. D., Lardy, H., and Myrbäck, K., Eds., Academic Press, New York, 1963, 477.
8. **Massey, V. and Ghisla, S.,** Role of charge-transfer interactions in flavoprotein catalysis, *Ann. N.Y. Acad. Sci.,* 227, 446, 1974.
9. **Beinert, H.,** Evidence for an intermediate in the oxidation-reduction of flavoproteins, *J. Biol. Chem.,* 225, 465, 1957.
10. **Massey, V., Matthews, R. G., Foust, G. P., Howell, L. G., Williams, C. H. Jr., Zanetti, G., and Ronchi, S.,** A new intermediate in TPNH-linked flavoproteins in *Pyridine Nucleotide Dependent Dehydrogenases,* Sund, H., Ed., Springer-Verlag, New York, 1970, 393.
11. **Kosower, E. M.,** The role of charge-transfer complexes in flavin chemistry and biochemistry, in *Flavins and Flavoproteins,* Slater, E. C., Ed., Elservier Publ. Co., New York, 1965, 1.
12. **Abramovitz, A. S. and Massey, V.,** Purification of intact old yellow enzyme using an affinity matrix for the sole chromatographic step, *J. Biol. Chem.,* 251, 5321, 1976.
13. **Massey, V. and Hemmerich, P.,** Active site probes of flavoproteins, *Biochem. Soc. Trans.,* 8, 246, 1980.
14. **Ghisla, S. and Massey, V.,** New flavins for old: artificial flavins as active site probes of flavoproteins, *Biochem. J.,* 239, 1, 1986.
15. **Miura, R., Yamano, T., and Miyake, Y.,** The heterogeneity of brewer's yeast old yellow enzyme, *J. Biochem.,* 99, 901, 1986.
16. **Beinert, W.-D., Rüterjans, H., and Müller, F.,** Nuclear magnetic resonance studies of the old yellow enzyme. I. ^{15}N NMR of the enzyme recombined with ^{15}N-labeled flavin mononucleotides, *Eur. J. Biochem.,* 152, 573, 1985.
17. **Beinert, W.-D., Rüterjans, H., Müller, F., and Bacher, A.,** Nuclear magnetic resonance studies of the old yellow enzyme. II. ^{13}C NMR of the enzyme recombined with ^{13}C-labeled flavin mononucleotides, *Eur. J. Biochem.,* 152, 581, 1985.
18. **Miura, R., Yamano, T., and Miyake, Y.,** ^{31}P- and ^{13}C-NMR studies on the flavin-protein and flavin-ligand interactions in brewer's yeast old yellow enzyme, *J. Biochem.,* 99, 907, 1986.
19. **Massey, V., Schopfer, L. M., and Dunham, W. R.,** On the enigma of old yellow enzyme's spectral properties, in *Flavins and Flavoproteins,* Bray, R. C., Engel, P. C., and Mayhew, S. G., Eds., Walter de Gruyter, New York, 1984, 191.
20. **Eweg, J. K., Müller, F. and vanBerkel, W. J. H.,** On the enigma of old yellow enzyme's spectral properties, *Eur. J. Biochem.,* 129, 303, 1982.
21. **Ehrenberg, A. and Ludwig, G. D.,** Free radical formation in reaction between old yellow enzyme and reduced triphosphopyridine nucleotide, *Science,* 127, 1177, 1958.
22. **Eriksson, L. E. G., and Ehrenberg, A.,** On the Powder ESR and ENDOR spectra of flavoprotein radicals, *Biochem. Biophys. Acta,* 293, 57, 1973.
23. **Kurreck, H., Bock, M., Bretz, N., Elsner, M., Kraus, H., Lubitz, W., Müller, F., Geissler, J., and Kroneck, P. M. H.,** Fluid solution and solid-state electron nuclear double resonance studies of flavin model compounds and flavoenzymes, *J. Am. Chem. Soc.,* 106, 737, 1984.
24. **van Schagen, C. G., Müller, F., and Kaptein, R.,** Photochemically induced dynamic nuclear polarization study on flavin adenine dinucleotide and flavoproteins, *Biochemistry,* 21, 402, 1982.
25. **Matthews, R. G.,** Free and Complexed Forms of Old Yellow Enzyme: Their Physical and Catalytic Properties, Ph.D. thesis, University of Michigan, Ann Arbor, 1969.
26. **Matthews, R. G. and Massey, V.,** Isolation of old yellow enzyme in free and complexed forms, *J. Biol. Chem.,* 244, 1779, 1969.
27. **Schopfer, L. M. and Massey, V.,** unpublished data, 1983-7.
28. **Abramovitz, A. S. and Massey, V.,** Interaction of phenols with old yellow enzyme: physical evidence for charge transfer complexes, *J. Biol. Chem.,* 251, 5327, 1976.
29. **Theorell, H. and Åkeson, Å.,** Molecular weight and FMN content of crystalline "old yellow enzyme", *Arch. Biochem. Biophys.,* 65, 439, 1956.
30. **Matthews, R. G. and Massey, V.,** Free and complexed forms of old yellow enzyme, in *Flavins and Flavoproteins,* Kamin, H., Ed., Univeristy Press, Baltimore, 1971, 329.
31. **Nakamura, T., Yoshimura, J., and Ogura, Y.,** Action mechanism of the old yellow enzyme, *J. Biochem.,* 57, 554, 1965.
32. **Stewart, R. C. and Massey, V.,** Potentiometric studies of native and flavin-substituted old yellow enzyme, *J. Biol. Chem.,* 260, 13639, 1985.
33. **Theorell, H. and Nygaard, A. P.,** Kinetics and equilibria in flavoprotein systems. I. A fluorescence recorder and its application to a study of the dissociation of the old yellow enzyme and its resynthesis from riboflavin phosphate and protein, *Acta Chem. Scand.,* 8, 877, 1954.
34. **Takai, K., Sasai, Y., Morimoto, H., Yamazaki, H., Yoshii, H., and Inoue, S.,** Enzymatic dehydrogenation of tryptophan residues of human globins in tryptophan side chain oxidase II, *J. Biol. Chem.,* 259, 4452, 1984.

35. **Miles, D. W., and Urry, D. W.,** Reciprocal relations and proximity of bases in flavin-adenine dinucleotide, *Biochemistry,* 7, 2791, 1968.
36. **Brady, A. H., and Beychok, S.,** Optical activity and conformation studies of pig heart lipoamide dehydrogenase, *J. Biol. Chem.,* 244, 4634, 1969.
37. **Perini, F., Schopfer, L. M., and Massey, V.,** unpublished data, 1986.
38. **Rutter, W. J., and Rolander, B.,** The effect of ions on the catalytic activity of enzymes: the old yellow enzyme, *Acta Chem. Scand.* 11, 1663, 1957.
39. **Lockridge, O., Schopfer, L. M., and Massey, V.,** unpublished data, 1985.
40. **Vestling, C. S.,** Kinetics and equilibria in flavoprotein systems IV. The standard potential of the old yellow enzyme of yeast, *Acta Chem. Scand.,* 9, 1600, 1955.
41. **Massey, V., and Hemmerich, P.,** Photoreduction of flavoproteins and other biological compounds catalyzed by deazaflavins, *Biochemistry,* 17, 9, 1978.
42. **Massey, V., and Schopfer, L. M.,** Reactivity of old yellow enzyme with α-NADPH and other pyridine nucleotide derivatives, *J. Biol. Chem.,* 261, 1215, 1986.
43. **Oppenheimer, N. J.,** Chemistry and solution conformation of the pyridine coenzymes, in *The Pyridine Nucleotide Coenzyme,* Everse, J., Anderson, B., and You, K.-S., Eds., Academic Press, New York, 1982, 51.
44. **Abramovitz, A. S.,** Interactins of Old Yellow Enzyme with Aromatic Compounds, Ph.D. thesis, University of Michigan, Ann Arbor, 1974.
45. **Massey, V., Strickland, S., Mayhew, S. G., Howell, L. G., Engel, P. C., Matthews, R. G., Schuman, M., and Sullivan, P. A.,** The production of superoxide anion radicals in the reaction of reduced flavins and flavoproteins with molecular oxygen, *Biochem. Biophys. Res. Commun.,* 36, 891, 1969.
46. **Adachi, O., Natsushita, K., Shinagawa, E.,and Ameyana, M.,** Occurrence of old yellow enzyme in *Gluconobacter suboxydans,* and the cyclic regeneration of NADP, *J. Biochem.,* 86, 699, 1979.
47. **Matthews, R. G., Massey, V., and Sweeley, C. C.,** Identification of p-Hydroxybenzaldehyde as the ligand in the green form of old yellow enzyme, *J. Biol. Chem.,* 250, 9294, 1975.
48. **Harbury, H. A., LaNoue, K. F., Loach, P. A., and Amick, R. M.,** Molecular interactions of isoalloxazine derivatives. II, *Proc. Natl. Acad. Sci. U.S.A.,* 45, 1708, 1959.
49. **Kotaki, A., Naoi, M., and Yagi, K.,** Effect of proton donors on the absorbtion spectrum of flavin compounds in apolar media, *J. Biochem.,* 68, 287, 1970.
50. **Ghisla, S., Massey, V., and Mayhew, S. G.,** Studies on the active centers of flavoproteins: binding of 8-hydroxy-FAD and 8-hydroxy-FMN to apoproteins, in *Flavins and Flavoproteins,* Singer, T. P., Ed., Elsevier, Amsterdam, 1976, 334.
51. **Eweg, J. K., Müller, F., vanBerkel, W. J. H., and Hesper, B.,** On the enigma of old yellow enzyme's spectral properties, in *Flavins and Flavoproteins,* Bray, R. C., Engel, P. C., and Mayhew, S. G., Eds., Walter de Gruyter, New York, 1984, 183.
52. **Foster, R.,** *Organic Charge Transfer Complexes,* Academic Press, New York, 1969, 60.
53. **Kitagawa, T., Nishina, Y., Shiga, K., Watari, H., Matsumura, Y., and Yamano, T.,** Resonance raman evidence for charge-transfer interactions of phenols with the flavin mononucleotide of old yellow enzyme, *J. Am. Chem. Soc.,* 101, 3376, 1979.
54. **Nishina, Y., Kitagawa, T., Shiga, K., Watari, H., and Yamano, T.,** Resonance raman study of flavoenzyme-inhibitor charge-transfer interactions, old yellow enzyme-phenol complexes, *J. Biochem.,* 87, 831, 1980.
55. **Bienstock, R. J., Schopfer, L. M., and Morris, M. D.,** Characterization of the flavin-protein interaction in L-lactate oxidase and old yellow enzyme by resonance inverse raman spectroscopy, *J. Raman Spectrosc.,* in press.
56. **Haas, E.,** Wirkungsweise des Proteins des Gelben Ferments, *Biochem. Z.,* 290, 291, 1937.
57. **Schopfer, L. M., Massey, V., and Claiborne, A.,** Active site probes of flavoproteins. Determination of the solvent accessibility of the flavin position 8 for a series of flavoproteins, *J. Biol. Chem.,* 256, 7329, 1981.
58. **Fitzpatrick, P. F., Ghisla, S., and Massey, V.,** 8-Azidoflavins as photoaffinity labels for flavoproteins, *J. Biol. Chem.,* 260, 8483, 1985.
59. **Bagley, H., and Knowles, J. R.,** Photoaffinity labeling, *Meth. Enzymol.,* 46, 69, 1977.
60. **Massey, V., Ghisla, S., and Moore, E. G.,** 8-Mercaptoflavins as active site probes of flavoenzymes, *J. Biol. Chem.,* 254, 9640, 1979.
61. **Ghisla, S., and Mayhew, S. G.,** Identification and properties of 8-hydroxyflavin-adenine dinucleotide in electron-transferring flavoprotein from *Peptostreptococcus elsdenii, Eur. J. Biochem.,* 63, 373, 1976.
62. **Claiborne, A., Massey, V., Fitzpatrick, P. F., and Schopfer, L. M.,** 2-Thioflavins as active site probes of flavoproteins, *J. Biol. Chem.,* 257, 174, 1982.
63. **Biemann, M., Claiborne, A., Ghisla, S., Massey, V., and Hemmerich, P.,** Oxidation of 2-thioflavins by peroxides. Formation of flavin 2-S-oxides, *J. Biol. Chem.,* 258, 5440, 1983.

64. **Massey, V., Ghisla, S., and Yagi, K.,** 6-Thiocyanatoflavins and 6-Mercaptoflavins as active-site probes of flavoproteins, *Biochemistry,* 25, 8103, 1986.
65. **Mayhew, S. G., Whitfield, C. D., Ghisla, S., and Schuman-Jorns, M.,** Identification and properties of new flavins in electron-transferring flavoprotein from *Peptostreptococcus elsdenii* and pig-liver glycolate oxidase, *Eur. J. Biochem.,* 44, 579, 1974.
66. **Massey, V., Ghisla, S., and Yagi, K.,** 6-Azido- and 6-aminoflavins as active-site probes of flavin enzymes, *Biochemistry,* 25, 8095, 1986.
67. **Eberlein, G. and Bruice, T. C.,** The chemistry of a 1,5-diblocked flavin. 1. Interconversion of the reduced, radical, and oxidized forms of 1,10-ethano-5-ethyllumiflavin, *J. Am. Chem. Soc.,* 105, 6679, 1983.
68. **Massey, V., Müller, F., Feldberg, R., Schuman, M., Sullivan, P. A., Howell, L. G., Mayhew, S. G., Matthews, R. G., and Foust, G. P.,** The reactivity of flavoproteins with sulfite. Possible relevance to the problem of oxygen reactivity, *J. Biol. Chem.,* 244, 3999, 1969.
69. **Massey, V., Claiborne, A., Biemann, M., and Ghisla, S.,** 4-Thioflavins as active site probes of flavoproteins. General properties, *J. Biol. Chem.,* 259, 9667, 1984.
70. **Honma, T., and Ogura, Y.,** Kinetic studies of the old yellow enzyme 1. The reaction mechanism of the enzyme and reduced nicotinamide adenine dinucleotide, *Biochim. Biophys. Acta,* 484, 9, 1977.
71. **Porter, D. J. T., and Bright, H. J.,** Oxidation of dihydronicotinamides by flavin in enzyme and model reactions. Old yellow enzyme and lumiflavin, *J. Biol. Chem.,* 255, 7362, 1980.
72. **Cleland, W. W.,** The kinetics of enzyme-catalyzed reactions with two or more substrates or products. I. Nomenclature and rate equations, *Biochim. Biophys. Acta,* 67, 104, 1963.
73. **Spencer, R. W.,** Flavin Analogs and Mechanisms of Flavoenzyme Catalysis, Ph.D. thesis, Massachusetts Institute of Technology, Boston, 1978.
74. **Haney, S., Schopfer, L. M., and Massey, V.,** unpublished data, 1985.

Chapter 11

D-AMINO ACID OXIDASE

Kunio Yagi

TABLE OF CONTENTS

I. INTRODUCTION

D-Amino acid oxidase [D-amino acid: oxygen oxidoreductase (deaminating) EC 1.4.3.3] attacks neutral or basic D-amino acid to yield the corresponding keto acid and ammonia. However, its catalytic action is restricted to the oxidation of the substrate D-amino acid to the corresponding imino acid and the formation of keto acid and ammonia is due to non-enzymatic hydrolysis. Although this action is common in the case of all these D-amino acids, the reaction mechanism regarding the intermediate complex is different between neutral and basic amino acids. In the case of neutral D amino acids, there appears an intermediate complex, the so-called purple intermediate, which was crystallized by us.[1] This intermediate is not detectable in the case of basic D-amino acids. These mechanisms are described in detail in this chapter.

As for the catalytic reaction of this enzyme, it is to be especially noted that the substrate specificity of the enzyme is low, though acidic D-amino acids are not oxidized by the enzyme. In addition to the above D-amino acids and glycine, structurally different substances such as L-proline and D-hydroxy acids are also oxidized by this enzyme. Such a low substrate specificity could be due to the flexibility of the apoenzyme of this oxidase. Different modes of action of the enzyme with different substrates afford valuable information on the reaction mechanism of flavin enzymes.

II. REACTION CATALYZED BY THE ENZYME

From reasonable chemical considerations, the initial oxidation product of amino acids in flavin enzyme catalysis has been considered to be an unstable α-imino acid which is hydrolyzed nonenzymatically to the corresponding keto acid and ammonia. In support of this hypothesis, the occurrence of the α,β-unsaturated intermediate (enamine) has been denied in the D-amino acid oxidase reaction.[2] We also presented supporting evidence by demonstrating the fact that H-D exchange did not occur at the β-position.[2] Thus, the reaction process is written as follows:

$$H_3\overset{+}{N}-\underset{COO^-}{\overset{R}{C}}-H + O_2 \rightarrow HN{=}\underset{COO^-}{\overset{R}{C}} + H_2O_2 + H^+ \quad (1)$$

$$HN{=}\underset{COO^-}{\overset{R}{C}} + H_2O + H^+ \rightarrow NH_4^+ + O{=}\underset{COO^-}{\overset{R}{C}} \quad (2)$$

To confirm this reaction process, we observed the changes in pH and oxygen consumption during the reaction of D-amino acid oxidase and D-leucine.[3] As shown in Figure 1, the time course of the pH decrease corresponds to that of the oxygen consumption. This indicates the validity of formula 1. Then, the pH increases up to the initial level, which indicates the validity of formula 2. Thus, the rate of hydrolysis of imino acid to the corresponding keto acid and ammonia is considered to be significantly slower than that of the oxidation of amino acid.

To confirm further the reaction process, we conducted an experiment using D-proline as substrate.[3] In the oxidation of D-proline, the substrate is a zwitterion; and the product,

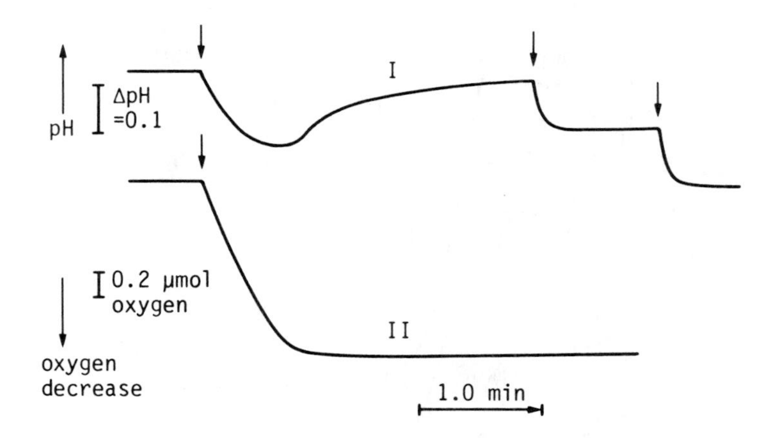

FIGURE 1. Simultaneous recording of pH and oxygen consumption during the reaction of D-amino acid oxidase and D-leucine. The reaction mixture containing 1 m*M* D-leucine, 10 μ*M* FAD, and 1.7 m*M* sodium pyrophosphate in 5 ml was adjusted to pH 8.3. The reaction was started by the addition of the enzyme (8.65 μ*M* with respect to the bound FAD) at 22°C. I, Change in pH; II, oxygen consumption. Left hand arrows show the addition of the enzyme and right hand arrows show that of each 0.188 μmol of HC1.

Δ^1-pyrroline 2-carboxylate, is practically unprotonated on its imino nitrogen atom (pK, approximately 6.0) over the range of pH at which the reaction takes place as follows:

$$\text{(pyrrolidine ring, }\overset{+}{N}H_2\text{)}-COO^- + O_2 \longrightarrow \text{(pyrroline ring, } N{=}C\text{)}-COO^- + H_2O_2 + H^+ \qquad (3)$$

Therefore, the neutralization of protons should not occur. In fact, this was found to be the case for the oxidation of D-proline by D-amino acid oxidase, as shown in Figure 2. From these data, we can safely conclude that D-amino acid oxidase catalyzes the oxidation of the substrate amino acid to produce its imino acid, and that the hydrolysis of the latter is a nonenzymatic process.

III. REACTION MECHANISM FOR NEUTRAL AMINO ACIDS

To approach the reaction mechanism of the enzyme, we intended to reveal the reaction intermediates. For this purpose, we tried to isolate the intermediate by crystallization. Since the reaction of this enzyme should be reversible under anaerobic conditions, we tried to accumulate the intermediate by reducing the substrate in the presence of the products.[1] Actually, the enzyme solution was mixed anaerobically with D-alanine, pyruvate, and ammonia at pH 7.0 to 7.2, brought to 5% saturation with respect to ammonium sulfate, and stood at 5°C in the dark; as a result, purple crystals were obtained.[1] The crystals were found to consist of equimolar amounts of the enzyme and substrate,[1] and to display strong charge-transfer interaction between the substrate and the coenzyme moieties.[4] Further, the crystals were spectroscopically identified with the rapidly appearing purple intermediate complex.[5] Thus, the crystals obtained marked the first crystallization of an enzyme-substrate complex.

Since the reflectance spectrum of the crystallized intermediate complex is obviously different from the spectra of the oxidized and reduced enzyme, as shown in Figure 3, the electronic state of the flavin in the complex was considered to be in between those of oxidized and fully reduced enzyme. The occurrence of the oxidized enzyme-substrate complex prior

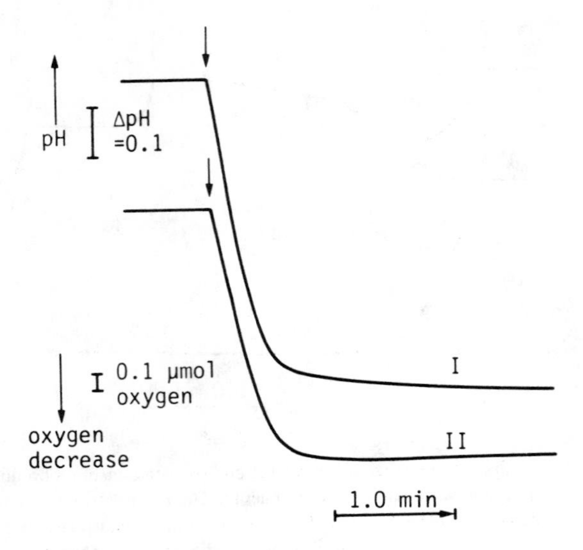

FIGURE 2. Simultaneous recording of pH and oxygen consumption during the reaction of D-amino acid oxidase and D-proline. The reaction mixture and procedure were the same as those specified in Figure 1 except that 2 m*M* DL-proline was used in place of D-leucine and final concentration of the enzyme was 2.63 μ*M* with respect to the bound FAD. I, Change in pH; II, oxygen consumption.

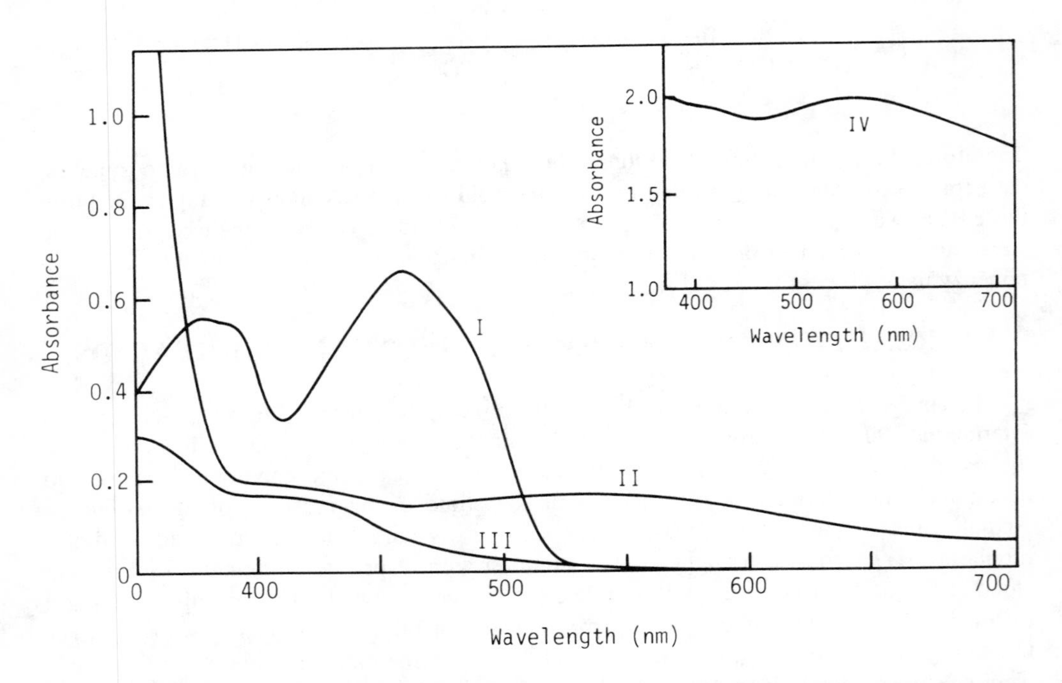

FIGURE 3. The absorption spectra of the oxidized enzyme, the purple intermediate complex, and the fully reduced enzyme of D-amino acid Oxidase. I, oxidized enzyme (58.6 μ*M* with respect to the bound FAD); II, the purple complex (I was mixed with 50 m*M* D-alanine in the presence of 50 m*M* ammonium sulfate and 0.1 *M* lithium pyruvate); III, the fully reduced enzyme (I was mixed with 33 m*M* D-alanine); IV, reflectance spectrum of the crystals of the purple complex.

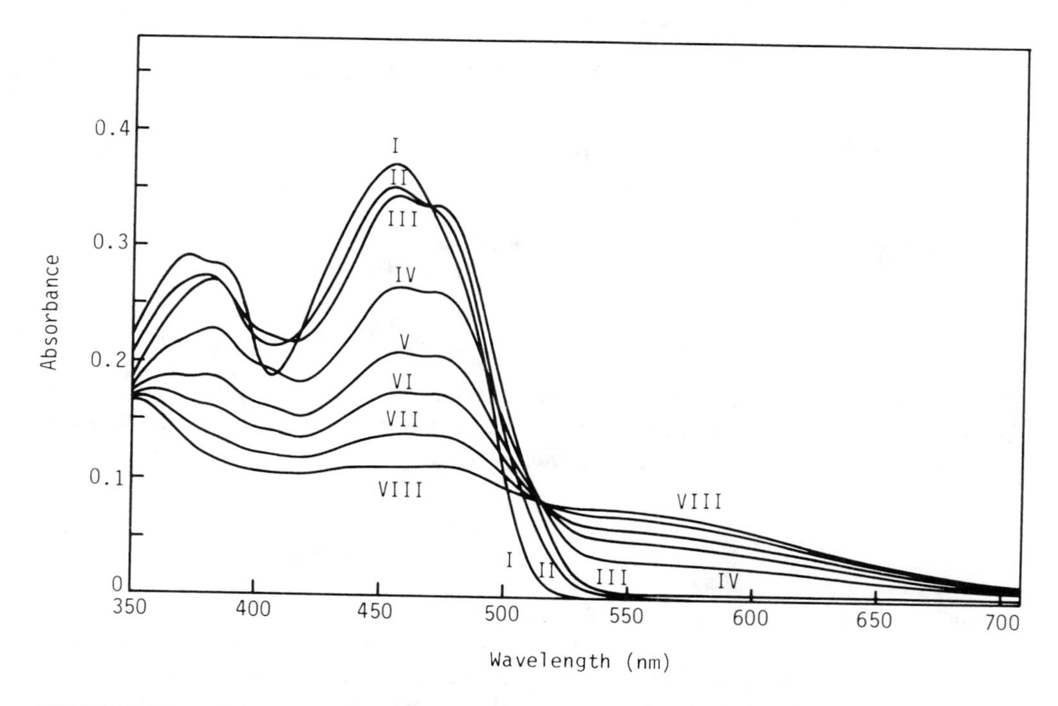

FIGURE 4. Spectral change upon complex formation between D-amino acid oxidase and D-α-aminobutyric acid at low temperature. D-Amino acid oxidase (31 μ*M* with respect to the bound FAD) and D-α-aminobutyric acid (10 m*M*) were mixed anaerobically in 100 m*M* sodium phosphate buffer (pH 6.5) containing 35% glycerol at − 10°C. I, The enzyme; II, immediately after the addition of the substrate; III, 20 s after mixing; VIII, 3 min after mixing.

to the purple intermediate was shown kinetically,[6,7] and was further confirmed spectrophotometrically by a low-temperature experiment,[8] as shown in Figure 4. Under anaerobic conditions, we also observed that the enzyme is reduced slowly to the fully reduced form through the purple intermediate. Accordingly, the reaction sequence of this enzyme under anaerobic conditions can be written as follows:

$$E_{ox} + S \rightleftharpoons E_{ox} \cdot S \rightleftharpoons E' \cdot S' \rightleftharpoons E_{red} \cdot P \rightleftharpoons E_{red} + P \quad (4)$$

where $E' \cdot S'$ represents the purple intermediate.

To understand the catalytic reaction of this enzyme under aerobic conditions, the intermediate which actually reacts with molecular oxygen should be determined. As to this problem, the purple intermediate is considered to be the very intermediate that reacts with molecular oxygen,[9] since the conversion of the purple intermediate to the fully reduced enzyme is so slow, as mentioned above, that the oxidation of the fully reduced enzyme with molecular oxygen cannot explain the large number of molecular activity of this enzyme. The direct reaction of the purple intermediate with molecular oxygen was demonstrated by Massey and Gibson[6] by observing a ''spike'' in the stopped-flow trace at 550 nm of the aerobic reaction of this enzyme with the substrate. The ''spike'' was interpreted to indicate the rapid formation of the purple intermediate followed by its oxidation with molecular oxygen. Thus the aerobic reaction of the enzyme with D-neutral amino acid is expressed by the following sequence of reactions:

$$E_{ox} + S \underset{k_{-1}}{\overset{k_{+1}}{\rightleftharpoons}} E_{ox} \cdot S \underset{k_{-2}}{\overset{k_{+2}}{\rightleftharpoons}} E' \cdot S' \quad (5)$$

$$E' \cdot S' + O_2 \xrightarrow{k_{+3}} E_{ox} \cdot P + H_2O_2 \tag{6}$$

$$E_{ox} \cdot P \underset{k_{-4}}{\overset{k_{+4}}{\rightleftharpoons}} E_{ox} + P \tag{7}$$

According to these formulae, the initial velocity is

$$\frac{e_0}{v} = \frac{1}{k_{+2}} + \frac{1}{k_{+4}} + \frac{k_{-1} + k_{+2}}{k_{+1} \cdot k_{+2}} \cdot \frac{1}{S} + \frac{k_{+2} + k_{-2}}{k_{+2} \cdot k_{+3}} \cdot \frac{1}{(O_2)} + \frac{k_{-1} + k_{-2}}{k_{+1} \cdot k_{+2} \cdot k_{+3} \cdot S \cdot (O_2)} \tag{8}$$

Since the last term can be practically neglected, and since the fact that the purple intermediate accumulates at a definite time after the anaerobic mixing of the enzyme with the substrate indicates $k_{+2} >> k_{-2}$, formula 8 can be reduced to

$$\frac{e_0}{v} = \frac{1}{k_{+2}} + \frac{1}{k_{+4}} + \frac{k_{-1} + k_{+2}}{k_{+1} \cdot k_{+2}} \cdot \frac{1}{S} + \frac{1}{k_{+3} \cdot (O_2)} \tag{9}$$

Therefore, the overall rate of the oxidation of the substrate by this enzyme should be dependent on the concentrations of both the substrate and oxygen. Experimental data fit the above formula, indicating that the purple intermediate reacts directly with oxygen.

All these data show that the purple intermediate is the key intermediate in the reaction of this enzyme with neutral D-amino acids, and the real entity of the complex is the object of elucidation. As mentioned before, a strong charge-transfer interaction occurs between the substrate and flavin moieties of the intermediate complex. This conclusion was deduced from the facts that the complex reveals a broad absorption band at around 550 nm (see Figure 3), which is assignable to charge-transfer interaction, and the complex converts slowly but definitely to the semiquinoid enzyme.[4] The latter phenomenon was interpreted to mean that the complex dissociates into semiquinoid enzyme and substrate radical. This can be expressed as

$$D + S \rightleftharpoons (D,A) \rightleftharpoons (D^+\text{-}A^-) \rightleftharpoons D^+ + A^- \tag{10}$$

where (D,A) indicates an outer complex (weak charge-transfer complex) and $(D^+\text{-}A^-)$, an inner complex (strong charge-transfer complex), according to the definition of Mulliken.[10] The purple complex corresponds to $(D^+\text{-}A^-)$.

To approach further the entity of the purple intermediate as well as the mechanism for its formation, we conducted several experiments. Observation of intermediate complexes by the so-called "slow reaction method"[11] (Figure 4) indicated that only one complex $(E_{ox} \cdot S)$ exists prior to the appearance of the purple complex, i.e., $E_{ox} \cdot S$ directly converts to the purple complex in the reaction sequence.[8]

For determination of the site in the substrate for the charge-transfer interaction with the flavin chromophore, the data on the formation of a charge-transfer complex of this enzyme with *o*-aminobenzoate or Δ^1-piperidine 2-carboxylate are quite suggestive. It is known that the mixing of *o*-aminobenzoate with this enzyme yields a green complex having a broad absorption band in the longer wavelength region, which is regarded as a charge-transfer band.[12,13] Δ^1-Piperidine 2-carboxylate gives a similar result.[12] These complexes are assigned as an outer complex (weak charge-transfer complex). However, mixing of benzoic acid with this enzyme yields a complex without the appearance of such a charge-transfer band. This indicates that a nitrogen lone pair is essential for the formation of the charge-transfer complex. Since the carboxyl group and the nitrogen atom of *o*-aminobenzoate or Δ^1-piperidine 2-

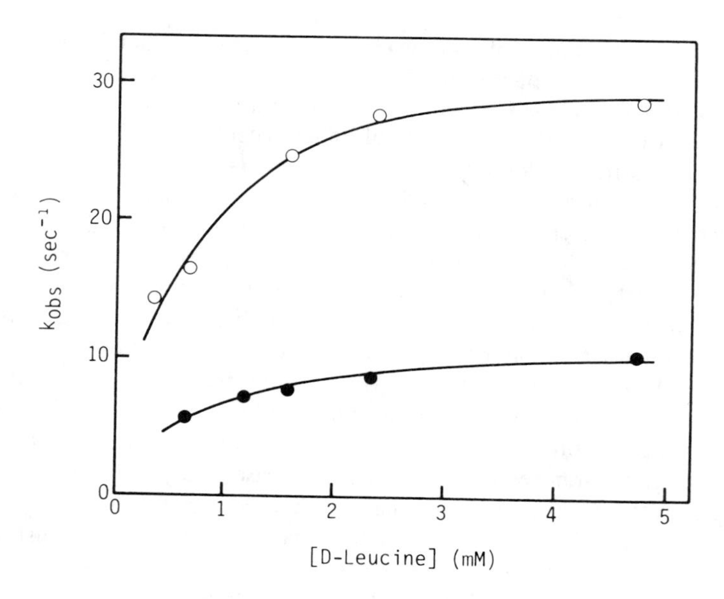

FIGURE 5. Kinetic isotope effect on the formation of the purple intermediate of D-amino acid oxidase with D-leucine. The formation of the purple intermediate at various concentrations of D-leucine was followed by measuring the absorbance change at 550 nm with a stopped-flow spectrophotometer at pH 8.3 and 20°C. D-Amino acid oxidase concentration was 14 μM with respect to the bound FAD. Open circles, DL-[α-^{1}H] leucine; closed circles, DL-[α-^{2}H] leucine.

carboxylate are located in sterically similar positions to those of the substrate, e.g., D-alanine, which also has a nitrogen lone pair, a similar interaction is expected to occur in the enzyme-substrate interaction. The interacting site of the flavin moiety of the enzyme is considered to be N(5) of the isoalloxazine nucleus of flavin, since N(5) has the largest coefficient in the lowest unoccupied molecular orbital.[14,15]

Accepting the fact that the nitrogen lone pair of the substrate or the above-mentioned inhibitors is the site for the charge-transfer interaction with N(5) of flavin, a question was naturally raised as to why the substrate forms an inner complex while the inhibitor forms an outer complex. To answer this question, it should be considered that the cleavage of the α-C-H bond of the substrate should occur during the oxidation of the substrate catalyzed by this enzyme, while such a cleavage does not occur with the inhibitor. Therefore, we considered it to be worthwhile to determine whether the α-C-H bond cleavage occurs before or after the initiation of the electronic interaction between the substrate nitrogen atom and the isoalloxazine nucleus of the coenzyme. For this purpose, we examined the kinetic isotope effect using an α-deuterated substrate, since it was known that H-D exchange of the substrate occurs at the α-carbon, but not at the β-carbon.[2] The rate of formation of the purple intermediate with DL-[α-^{2}H]leucine was compared with that of DL-[α-^{1}H]leucine. The pseudo-first-order rate constant (k_{obs}) of formation of the purple intermediate was plotted vs. the concentration of the substrate.[16,17] As shown in Figure 5, the rate of formation of the purple intermediate levelled off over the range of the substrate concentrations investigated, suggesting that the rate is controlled by the process of $E_{ox} \cdot S \rightarrow$ purple intermediate.

We found that the substitution of the α-hydrogen of leucine for deuterium reduced the rate of formation of the purple intermediate to approximately one-third of the original in the substrate concentration range where the rate levelled off. Such a kinetic isotope effect was not observed in the process of transformation of the purple intermediate to the fully reduced enzyme.[17] Therefore, we can conclude that the cleavage of α-C-H bond occurs before or at the same time as the achievement of a strong charge-transfer interaction between the substrate and the isoalloxazine nucleus of the coenzyme.

It is conceivable that the α-C-H bond cleavage results in an increased charge-transfer interaction. Apart from the simple example of donor-acceptor complex, in which only one electron is transferred from a donor to an acceptor, a two-electron transfer from a donor to an acceptor in usual oxidoreducation reactions of organic compounds occurs concomitantly with rearrangement of the chemical structure of the reactants. Although direct evidence is lacking as to whether the α-C-H bond cleavage occurs via proton, hydrogen, or hydride ion transfer, the base-catalyzed abstraction of an α-proton of the substrate may be predicted as suggested by Neims et al.[18] If this is the case, base-catalyzed proton abstraction preceding the oxidation facilitates the donor-acceptor interaction, which lowers the free energy of activation for the flow of two electrons. It was also found recently by a theoretical approach that attraction of electron(s) from the substrate nitrogen by N(5) of the isoalloxazine nucleus facilitates proton abstraction from the α-C-H bond by the base in the protein.[19] These data indicate that the charge-transfer interaction and the base-catalyzed proton abstraction are working cooperatively. Considering also the fact that these events are involved in a one-step reaction of $E_{ox} \cdot S \rightarrow E' \cdot S'$, we can conclude that the strong charge-transfer interaction between the nitrogen lone pair of the enzyme-bound substrate and the coenzyme occurs in concert with the cleavage of the α-C-H bond of the substrate.

Thus, the mechanism of formation of the purple intermediate and its entity can be schematically represented by Scheme I.

For the formation of the purple intermediate, the apoenzyme plays an important role. Besides its essentially important role of affording proximity to the coenzyme flavin and the substrate to allow their interaction, the apoenzyme abstracts the α-proton of the substrate as mentioned above. In addition to these roles, the apoenzyme also constructs a hydrophobic environment surrounding the flavin,[20] which stabilizes the purple intermediate, a strong charge-transfer complex, and affords the complex a sufficient lifetime to react with molecular oxygen.

The apoenzyme mainly contributes to the equilibrium between monomer and dimer of this enzyme. In the apoenzyme solution, the monomer is stable.[21] Upon complex formation with the coenzyme to form the holoenzyme, the equilibrium is shifted toward the dimer.[22,23] This is more so when the holoenzyme combines with the substrate or an inhibitor such as benzoate.[5,24,25] Both the monomer and the dimer of this enzyme are catalytically active. However, the reaction rates in the sequence of the reaction with neutral D-amino acid are different between the dimer and the monomer. The value of k_{+2} in formula 5 is larger in the dimer than in the monomer, while that of k_{+4} in formula 7 is larger in the monomer than in the dimer. Therefore, the appearance of the purple intermediate is faster in the dimer than in the monomer,[5] while the overall reaction rate is larger in the monomer than in the dimer.[26,27]

IV. REACTION MECHANISM FOR BASIC AMINO ACIDS

Differing from neutral D-amino acids, basic D-amino acids are oxidized by this enzyme without the appearance of any observable purple intermediate.[7,28] The stopped-flow traces of the reaction of the enzyme with D-arginine under anaerobic conditions revealed no change in the absorbance at 550 nm. However, the absorbance at 455 nm decreased, indicating the full reduction of the enzyme. These features are also seen in the reduction with D-ornithine or D-lysine.

A plot of the reciprocal of the rate of reduction of the enzyme against the reciprocal of the substrate concentration gave a straight line that passes through the origin, as shown in Figure 6. Accordingly, the concentration of any intermediate, if it exists, is always small

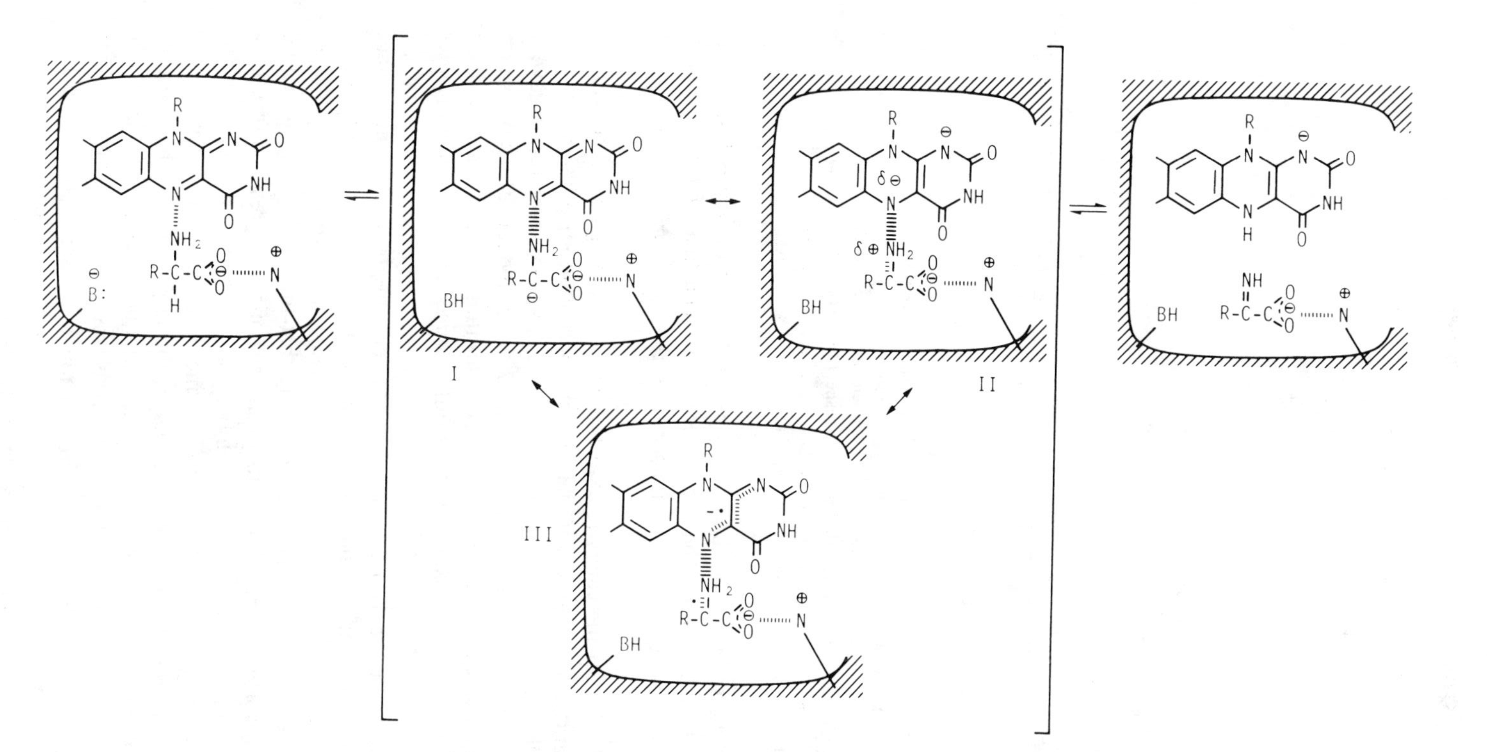

SCHEME I. A possible mechanism for the formation of the purple intermediate and its structure. The purple intermediate is schematically expressed as three resonance forms (I, II, and III). Electrons are localized mainly on amino acid (I), on FAD (II), or evenly on both amino acid and FAD (III).

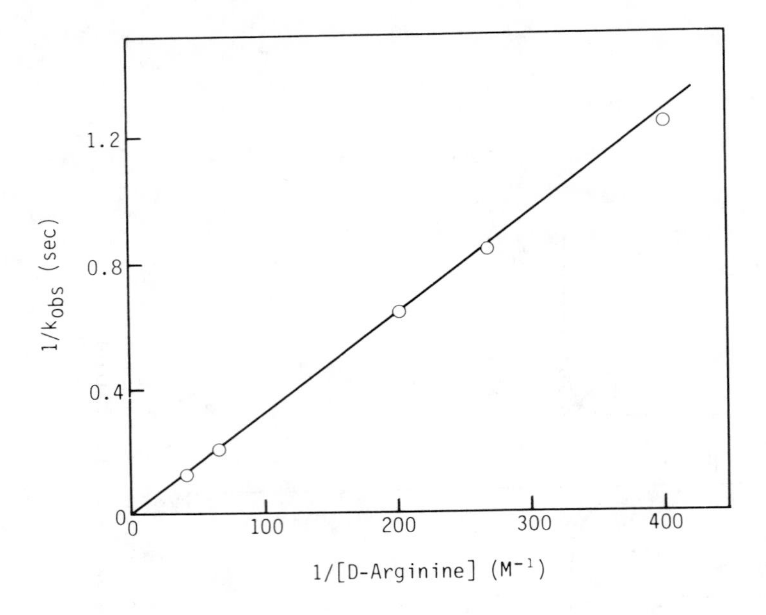

FIGURE 6. Double reciprocal plot for the full reduction of D-amino acid oxidase with D-arginine under anaerobic conditions. Transmittance changes at 455 nm were followed by stopped-flow spectrophotometer at pH 8.3 and 20°C. The final enzyme concentration was 46.8 μM with respect to the bound FAD. The reciprocal of the pseudo first-order rate constant ($1/k_{obs}$) was plotted against the reciprocal of the substrate concentration.

during the time course of the reaction and the reduction of the flavin moiety is faster than the formation of any kind of complex. The second-order rate constant of the anaerobic reduction of the enzyme with D-arginine was estimated to be $3.1 \times 10^2 M^{-1} s^{-1}$ at pH 8.3 and 20°C.

The Lineweaver-Burk plot for the catalytic oxidation of D-arginine gave the maximum rate of 4.3 s^{-1} in air-saturated solution (the initial oxygen concentration, 0.284 m*M*). The plot of the reciprocal of the rate against the reciprocal of the oxygen concentration gave a straight line passing through the origin, as shown in Figure 7. Since the rate of the oxidation step in air-saturated solution was found to be 5.5 s^{-1}, the rate-limiting step for the catalytic oxidation of D-arginine seems to be the oxidation of the fully reduced enzyme with molecular oxygen.

The reason why basic amino acids are attacked by this enzyme in a different manner should be ascribed to their structure, i.e., possession of a positive charge at the ω-position. In this connection, it should be noted that amino acids having negative charge at the ω-position cannot be attacked by this enzyme.[29] Considering that the alkyl group of neutral amino acids interacts with some hydrophobic locus of the enzyme, the presence of such a charged group at the ω-position may lead to repulsion between substrate and enzyme at this binding site, resulting in a possible instability of the enzyme-basic amino acid complex. In addition, the electrostatic nature of the charged group at the ω-position of the substrate should be considered. It is possible that the local electric field due to the positive charge at the ω-position of the substrate facilitates the flow of electrons from the substrate to the flavin moiety, whereas the local electric field due to negative charge at the same position opposes the flow of electrons.

Therefore, when the above-mentioned two kinds of effects provoked by the presence of a positive charge at the ω-position of the substrate are taken into account, we may suppose that in the case of basic amino acids, the enzyme-substrate complex, if it exists, is always

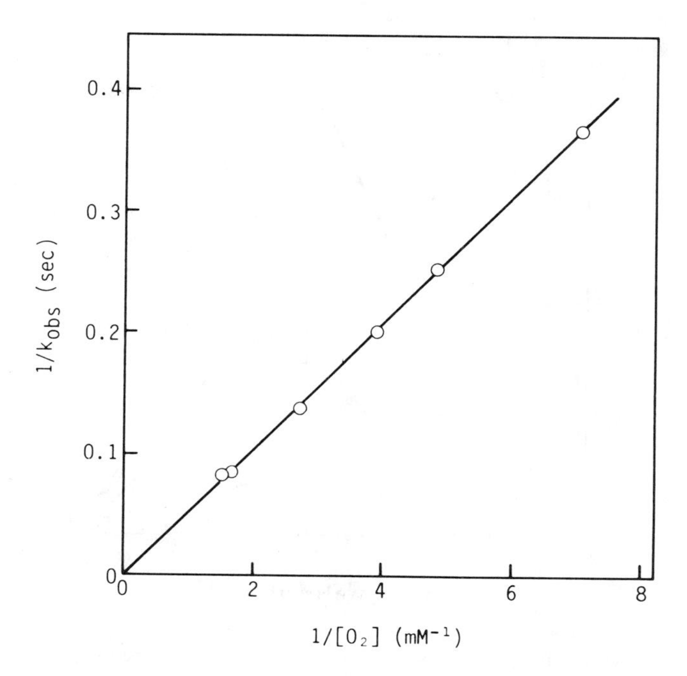

FIGURE 7. Double reciprocal plot for the oxidation of the fully reduced D-amino acid oxidase with molecular oxygen. The fully reduced enzyme was prepared by reducing the oxidized enzyme with slightly excess amount of D-arginine. Reaction traces were obtained by monitoring the transmittance changes at 455 nm, at pH 8.3 and 20°C. The reciprocal of the pseudo first-order rate constant obtained ($1/k_{obs}$) was plotted against the reciprocal of the oxygen concentration.

labile, but the reduction of the enzyme which would be influenced by the electric field is fast. On the other hand, with neutral amino acids, the enzyme-substrate complex (the purple intermediate) is relatively stable, but the reduction of the enzyme is slow presumably due to the absence of such an extra electrostatic effect. In fact, the overall rate of full reduction of the enzyme with neutral amino acid under anaerobic conditions is far slower than that with basic amino acid.

It might be mentioned that the enzymatic oxidation of D-lysine is interrupted immediately by Δ^1-piperidine 2 carboxylate, a strong inhibitor, formed from the reaction product through spontaneous cyclization.[30]

V. REACTION MECHANISM FOR OTHER SUBSTRATES

Besides neutral and basic amino acids, glycine is also oxidized by this enzyme. As to the reaction mechanism regarding the purple intermediate, the enzymatic oxidation of glycine is in between neutral and basic amino acids; in the case of glycine, the purple intermediate appears with a very short lifetime.[31]

One of the peculiar substrates for D-amino acid oxidase is L-proline. L-Proline is definitely oxidized by this enzyme.[32] Under anaerobic conditions, we observed a spectral change in the enzyme upon reaction with L-proline. As shown in Figure 8, the absorption spectrum of the oxidized enzyme (curve I) was changed (curve II) in the presence of L-proline. From a comparison of the two curves, a slight red shift of the absorption peak at the longer wavelengths and the disappearance of fine structure of the absorption band at the shorter wavelengths are found. On standing of the reaction mixture at room temperature (25°C),

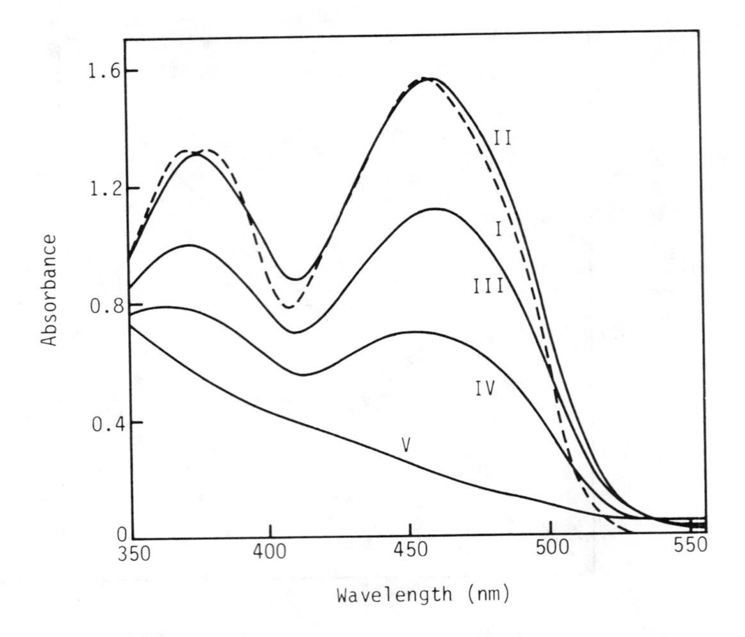

FIGURE 8. Reduction of D-amino acid oxidase with L-proline under anaerobic conditions. D-Amino acid oxidase (I, 0.14 m*M* with respect to the bound FAD) was mixed with 0.33 *M* L-proline at pH 8.3 and 25°C under anaerobic conditions. The absorption spectra were recorded 3 s (II), 45 s (III), 105 s (IV), and 10 min (V) after mixing.

the spectrum changed gradually into that of the fully reduced form of the enzyme, as shown by curve V, through the spectra shown by curves III and IV. It is clear that the purple intermediate does not appear as in the case of basic D-amino acids.

The rate of the reduction of the enzyme plotted against time gave a straight line, indicating that the reaction is of the first order. Plot of the reciprocal of the rate against the reciprocal of the L-proline concentration also gave a straight line.[32] From the intercept of the line, the rate constant of the decomposition of the enzyme-substrate complex was determined to be 0.80 min^{-1}, and the dissociation constant of the enzyme-substrate complex was estimated to be 82 m*M*.

Another peculiar substrate is D-hydroxy acids.[33] Using D-lactate, we can observe the absorption spectrum of the oxidized enzyme-substrate complex ($E_{ox} \cdot S$) even under aerobic conditions. The spectrum is similar to that observed with D-α-aminobutyric acid under anaerobic conditions at low temperature. Under anaerobic conditions, the spectrum of $E_{ox} \cdot S$ changes slowly to that of the fully reduced form without the appearance of the purple intermediate. In this case, the rate of formation of $E_{ox} \cdot S$ is obviously far faster than the reduction of the enzyme. A similar mechanism is also the case for D-mandelate.

REFERENCES

1. **Yagi, K. and Ozawa, T.,** Mechanism of enzyme action. I. Crystallization of Michaelis complex of D-amino acid oxidase, *Biochim. Biophys. Acta,* 81, 29, 1964.
2. **Yagi, K., Ohishi, N., and Nishikimi, M.,** Nuclear magnetic resonance study on D-amino acid oxidase reaction, *Biochim. Biophys. Acta,* 206, 181, 1970.
3. **Yagi, K., Nishikimi, M., Ohishi, N., and Takai, A.** Release of α-imino acid as primary product in D-amino acid oxidase reaction, *Biochim. Biophys. Acta,* 212, 243, 1970.

4. **Yagi, K., Okamura, K., Naoi, M., Sugiura, N., and Kotaki, A.,** Mechanism of enzyme action. II. Characterization of the purple intermediate in the anaerobic reaction of D-amino acid oxidase with D-alanine, *Biochim. Biophys. Acta,* 146, 77, 1967
5. **Yagi, K., Nishikimi, M., and Ohishi, N.,** Identity of the rapidly appearing purple intermediate of D-amino acid oxidase, *J. Biochem.,* 72, 1369, 1972.
6. **Massey, V. and Gibson, Q. H.,** Role of semiquinones in flavoprotein catalysis, *Fed. Proc.,* 23, 18, 1964.
7. **Yagi, K., Nishikimi, M., Ohishi, N., and Takai, A.,** Kinetic behavior of D-amino acid oxidase in its reaction with basic amino acids, *J. Biochem.,* 67, 153, 1970.
8. **Yagi, K., Lange, R., and Douzou, P.,** Spectroscopic demonstration of an initial state of the complex of D-amino acid oxidase and its substrate D-α-aminobutyric acid, *Biochem. Biophys. Res. Commun.,* 97, 370, 1980.
9. **Nakamura, T., Nakamura, S., and Ogura, Y.,** Semiquinone and enzyme kinetics of D-amino acid oxidase, *J. Biochem.,* 54, 512, 1963.
10. **Mulliken, R. S.,** Molecular compounds and their spectra. III. The interaction of electron donors and acceptors, *J. Phys. Chem.,* 56, 801, 1952.
11. **Yagi, K.,** Mechanism of enzyme action - An approach through the study of slow reactions, in *Advances in Enzymology,* Nord, F. F., Ed., Wiley-Interscience, New York, 1965, XXVII, 1.
12. **Massey, V. and Ganther, H.,** On the interpretation of the absorption spectra of flavoproteins with special reference to D-amino acid oxidase, *Biochemistry,* 4, 1161, 1965.
13. **Yagi, K., Naoi, M., Nishikimi, M., and Kotaki, A.,** Binding aspects of D-amino acid oxidase with various carboxylic acids, *J. Biochem.,* 68, 293, 1970.
14. **Song, P.-S., Choi, J. D., Fugate, R. D., and Yagi, K.,** Nature of the N-5 reactivity in enzymatic and photochemical oxidations by flavins, in *Flavins and Flavoproteins,* Singer, T. P., Ed., Elsevier, Amsterdam, 1976, 381.
15. **Nishimoto, K., Watanabe, Y., and Yagi, K.,** Hydrogen bonding of flavoprotein. I. Effect of hydrogen bonding on electronic spectra of flavoprotein, *Biochim. Biophys. Acta,* 526, 34, 1978.
16. **Yagi, K.,Nishikimi, M., Ohishi, N., and Takai, A.,** Kinetic isotope effect on the reaction of D-amino-acid oxidase. *FEBS Lett.,* 6, 22, 1970.
17. **Yagi, K., Nishikimi, M., Takai, A., and Ohishi, N.,** Mechanism of enzyme action. VI. Kinetic isotope effect on D-amino acid oxidase reaction, *Biochim. Biophys. Acta,* 321, 64, 1973.
18. **Neims, A. H., DeLuca, D. C., and Hellerman, L.,** Studies on crystalline D-amino acid oxidase. III. Substrate specificity and σ-ρ relationship, *Biochemistry,* 5, 203, 1966.
19. **Yagi, K., Kikuchi, T., Kobayashi, S., Ohishi, N., Tanaka, H., and Nishimoto, K.,** unpublished data.
20. **Naoi, M., Kotaki, A., and Yagi, K.,** Structure and function of D-amino acid oxidase. VII. Interaction of hydrophobic probes with D-amino acid oxidase, *J. Biochem.,* 74, 1097, 1973.
21. **Yagi, K., Ozawa, T., and Ohishi, N.,** Monomer-dimer equilibrium of D-amino acid oxidase apoenzyme, *J. Biochem.,* 64, 567, 1968.
22. **Yagi, K. and Ohishi, N.,** Structure and function of D-amino acid oxidase. IV. Electrophoretic and ultra-centrifugal approach to the monomer-dimer equilibrium, *J. Biochem.,* 71, 993, 1972.
23. **Tanaka, F. and Yagi, K.,** Cooperative binding of coenzyme in D-amino acid oxidase, *Biochemistry,* 18, 1531, 1972.
24. **Nishikimi, M., Osamura, M., and Yagi, K.,** Kinetic studies on the interaction of D-amino acid oxidase with benzoate and *o*-aminobenzoate, *J. Biochem.,* 70, 457, 1971.
25. **Yagi, K., Tanaka, F., Nakashima, N., and Yoshihara, K.,** Picosecond laser fluorometry of FAD of D-amino acid oxidase-benzoate complex, *J. Biol. Chem.,* 258, 3799, 1983.
26. **Shiga, K. and Shiga, T.,** The functional difference between the monomer and the dimer of D-amino acid oxidase, *Arch. Biochem. Biophys.,* 145, 701, 1971.
27. **Yagi, K., Sugiura, N., Ōhama, H., and Ohishi, N.,** Structure and function of D-amino acid oxidase. VI. Relation between the quaternary structure and the catalytic activity, *J. Biochem.,* 73, 909, 1973.
28. **Yagi, K., Nishikimi, M., Takai, A., and Ohishi, N.,** Mechanism of enzyme action. VII. Kinetic analysis of the reaction of D-amino acid oxidase with D-arginine, *Biochim. Biophys. Acta,* 341, 256, 1974.
29. **Dixon, M. and Kleppe, K.,** D-Amino acid oxidase. II. Specificity, competitive inhibition and reaction sequence, *Biochim. Biophys. Acta,* 96, 368, 1965.
30. **Yagi, K., Okamura, K., Naoi, M., Takai, A., and Kotaki, A.,** Reaction of D-amino acid oxidase with D-lysine, *J. Biochem.,* 66, 581, 1969.
31. **Yagi, K. and Ohishi, N.,** Lifetime of the purple intermediate of D-amino acid oxidase under anaerobic conditions, *J. Biochem.,* 84, 1653, 1978.
32. **Yagi, K. and Nishikimi, M.,** Kinetic study on the oxidation of L-proline by D-amino-acid oxidase, *J. Biochem.,* 64, 371, 1968.
33. **Yagi, K., Ozawa, T., and Naoi, M.,** Mechanism of enzyme action. V. Demonstration of an initial step of enzyme-substrate complex using D-amino acid oxidase and D-lactate or D-mandelate, *Biochim. Biophys. Acta,* 185, 31, 1969.

Chapter 12

CYTOCHROME *c*, QUINONE, AND CYTOCHROME P-450 REDUCTASES

Mark S. Johnson and Stephen A. Kuby

TABLE OF CONTENTS

I. INTRODUCTION

Included in this chapter, are summaries of four flavin reductases which share the ability to use NADPH as a substrate reductant. Although yeast[1] and mammalian[2] mitochondria appear capable of direct NADPH oxidation, the enzymes reviewed here are non-mitochondrial (quinone reductase may, however, be found mitochondrially in small amounts).[3,4] Properties of two cytochrome *c* reductases from a novel top-fermenting ale yeast [5-8] are reviewed in Section II. One of these [7,8] can utilize either pyridine nucleotide for reducing equivalents, has relatively low cytochrome *c* reductase activity and bears much resemblence to Old Yellow Enzyme (see Chapter 10). The other reductase from this yeast exhibits an exceptionally high specific activity with cytochrome *c* and is specific for NADPH.[5,6] Section III is a brief review of some kinetic and mechanistic studies on quinone reductase, the enzyme responsible for *in vivo* reduction of vitamin K_1,[9-11] and which has recently found notoriety from its perceived role in the inactivation of toxic and carcinogenic quinones.[12-15] The unique features of this reductase are its ability to function with either pyridine nucleotide and its requisite 2-electron transfer to acceptors. In Section IV is a summary of the mammalian microsomal cytochrome P-450 reductase, which transfers electrons from NADPH to cytochrome P-450 for the metabolism of various endogenous and xenobiotic compounds.[16,17] Of the four reductases reviewed, it has been the most actively studied and therefore has been a subject of several recent reviews.[18,19,20] The presence of two active flavin domains, one containing FMN and the other FAD, distinguishes this reductase from the others. Its microsomal origin and membrane association have made isolation and *in vitro* studies more difficult, but the understanding of ancillary problems such as the nature of microsomal lipid microheterogeneity and microsomal protein-protein interactions have been enhanced because of it.

II. PYRIDINE NUCLEOTIDE DEPENDENT CYTOCHROME *c* REDUCTASES FROM A TOP FERMENTING YEAST

One major NADH- and two major NADH(NADPH)-dependent nitrotetrazolium blue (NTB) reductases were present in a novel top ale yeast (*Saccharomyces cerevisiae*, Narragansett strain) which, due to growth conditions, possessed cytochromes P-450 and b_5 but lacked cytochromes *c* and a/a_3.[7] NTB activity stains on 7% nondenaturing polyacrylamide gels revealed three NADPH-dependent blue formazan bands at R_f = 0.47, 0.606, and 0.698, and two NADH-dependent bands at R_f = 0.47 and 0.606. Although the same three bands were evident for NADPH-dependent 2,6,-dichloroindophenol (DCIP) staining, only one band at R_f = 0.606 appeared with NADH as the reductant. One of these bands (R_f = 0.698) proved to be the NADPH-cytochrome *c* reductase earlier isolated by Tryon et al.,[5] a second (R_f = 0.47) was subsequently purified by Johnson and Kuby[7,8] and named NADH(NADPH)-cytochrome *c* reductase, and the third (R_f = 0.606), possibly Old Yellow Enzyme (see Chapter 10), remains uncharacterized.

A. NADH(NADPH)-CYTOCHROME *c* REDUCTASE

1. Properties[7]

Overall purification and yield of this soluble cytochrome *c* reductase was 67-fold and 2.3% respectively, although an estimate of 238-fold purification with an 8% yield was made when allowing for contributions made by the other two NTB reducing enzymes, both of which were capable of reducing cytochrome *c*. Separation from the other reductases was facilitated by the rather poor adsorption exhibited by this enzyme to DEAE cellulose at pH 8.5. Although the pI was only 5.25 (at 0.05 ionic strength), there was little change in total charge from pH 6 to 8.5 as evidenced by electrophoretic mobilities. The molecular weight was estimated to be 34,000 by SDS electrophoresis and 35,000 by sedimentation equilibrium in 4.0 *M* guanidinium chloride, while a molecular size of 70,000 Da was calculated under nondenaturing conditions by gel filtration. The flavin cofactor was FMN as determined by thin layer chromatography of the boiled supernatant and by fluorescence measurements of this fraction after acidification. As isolated, the reductase contained 0.63 moles FMN/subunit; association of the flavin to apoenzyme resulted in 96% fluorescence quenching.

2. Absorbance Spectra[7]

The absorbance spectrum for the purified reductase was typical of a flavoprotein,[30] with absorbance maxima at 278, 383, and 464 nm, and with molar extinction coefficients of ϵ_{464nm} = 9.88 mM^{-1} cm^{-1}, ϵ_{383nm} = 9.18 mM^{-1} cm^{-1}, and ϵ_{278nm} = 64.6 mM^{-1} cm^{-1}. The 464 nm absorbance was bleached by reduction with dithionite, or by anaerobic reduction with NADH or NADPH. Anaerobic titration with either pyridine nucleotide revealed a two electron reduction with no noticeable semiquinone intermediate. However, neither ESR measurements nor titrations in the presence of oxidized pyridine nucleotide (which might stabilize a semiquinone form)[21] were performed.

3. Dissociation Constant of FMN[7]

A kinetic method[5] was utilized for estimation of the flavin dissociation constant. Implicit in the determination, was the assumption that the reductase activity was directly proportional to the fraction of FMN bound, or fraction holoenzyme. If one FMN is bound per active subunit the following equation is applicable:

$$\mathrm{FMN/a} = (\mathrm{K'}_D/1 - \mathrm{a}) + \mathrm{E_t}$$

where a = the fraction activation, E_t = the total enzyme concentration, and K'_D = the dissociation constant of FMN. A plot of FMN/a vs. 1/(1-a) therefore yields a slope of K'_D and an intercept of E_t. A plot representing the activation of apoenzyme with FMN is shown in Figure 1A, in which a K'_D of 15 nM was determined, independent of pyridine nucleotide. A reciprocal plot for FMN is shown in Figure 1B, in which an estimate for the $K_{m,app}$ of FMN was 16 n*M*. Residual activity in the absence of exogenous FMN was less than 5% of V_{max}, indicating that nearly all of the flavin had been removed from the holoenzyme by the acidic treatment employed. Riboflavin at 9 μ*M* activated the NADH- and NADPH-dependent cyctochrome *c* reduction to a level of 39 and 31% of the holoenzyme respectively; FAD at similar concentrations showed at 64% enhancement with NADH but only a 33% activation with NADPH.

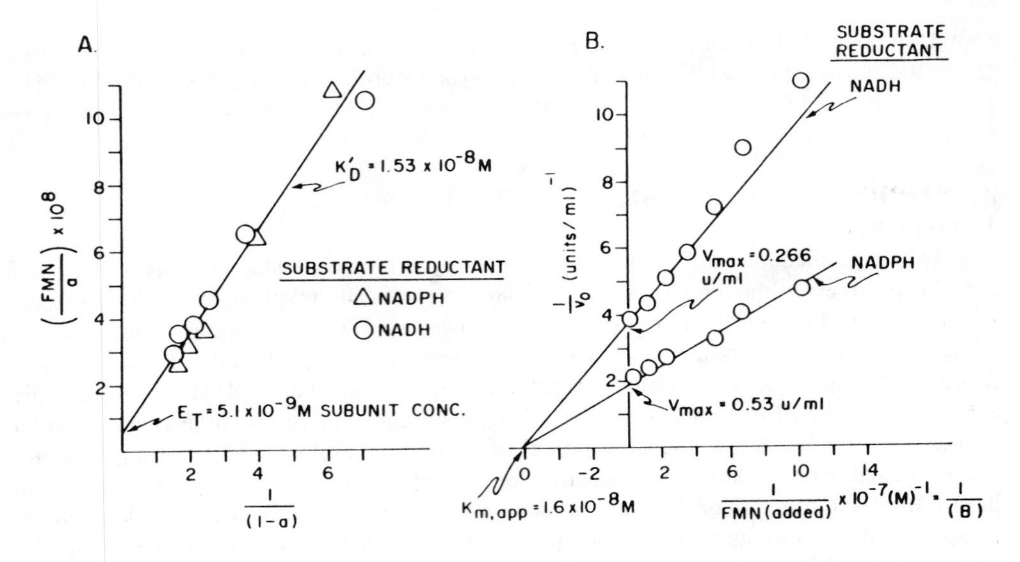

FIGURE 1. Kinetic determination of the dissociation constant (K'_D) and Michaelis constant (K_m) for FMN. (A) Apoenzyme was assayed for both NADH and NADPH-dependent cytochrome *c* reductase activities in the presence of various concentrations of FMN. K′D is determined from the following relationship: $^{FMN}/_a = {}^{K'}{}_D/_{(1-a)} + E_t$, where a = fraction activation and E_t = total enzyme concentration. NADH- and NADPH-dependent activities are plotted as open circles (○-○) and open triangles (△-△), respectively. (B) Reciprocal plots of NADH- and NADPH-dependent cytochrome *c* reduction are shown (as labeled on the figure) for the apoenzyme with exogenous addition of FMN. The Michaelis constant is designated as $K_{m,app}$ since its estimation is based on a single reciprocal plot. (From Reference 7.)

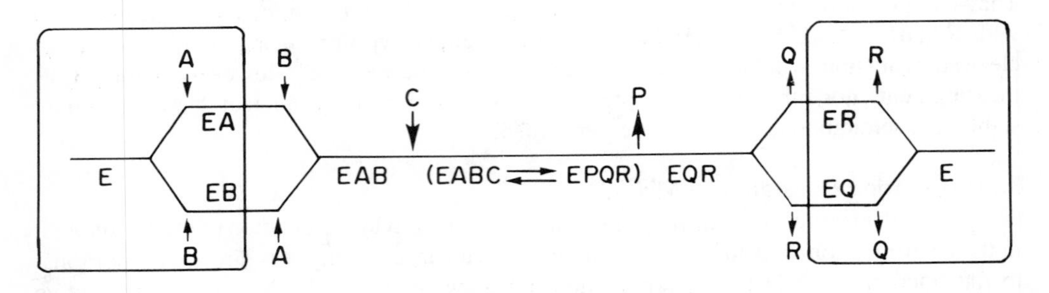

Scheme I. Ordered three-substrate system with a rapid equilibrium random in ligands A and B and in R and Q.

4. Amino Acid Composition[7]

The amino acid composition showed some similarity to Old Yellow Enzyme[22] but little homology with NADPH-cytochrome *c* reductase from the same top ale yeast. Although each subunit contained one cysteine, the thiol did not react with 4,4′-dithiodipyridine at pH 3.05. Activities of both holo- and apoenzyme were unaffected by the presence of 1mM 5,5′-dithiobis-(2-nitrobenzoic acid) (DTNB) for 16 h.

5. Steady-State Kinetics[8]

Steady-state kinetic parameters are shown in Table 1. Lineweaver-Burke plots showed qualitative differences between NADH- and NADPH-dependent cytochrome *c* reduction, with parallel lines for NADPH and converging lines for NADH as the substrate reductant. Both sets of data were consistent with a three-substrate (NAD(P)H, FMN, cytochrome *c*) hybrid random-ordered mechanism (see Scheme I), where the first two substrates are added in a random fashion, followed by the ordered addition of the third substrate. A three-substrate

TABLE 1
Values for the Steady-State Kinetic Parameters of NADH(NADPH)-Cytochrome *c* Reductase in the Presence of Menadione at pH 7.6 and 30°C [0.05 *M* Phosphate (K^+)]

Kinetic parameter	Value
$K_{m,NADPH}$	5.7 (±0.3) × 10^{-8} *M*
$K_{i,NADPH}$	4 (±0.4) × 10^{-9} *M* (7 (±2) × 10^{-7} M)[a]
$K_{m,cyt}$ c^{3+} (with (NADPH)	2.0 (±0.03 × 10^{-6} *M*
V_{max/E_t} (with NADPH and cyt c^{3+})	3.8 (±0.1) units/mg[b]
V_{max/E_t}	$1.3 \times 10^2 \frac{\text{moles}}{\text{min mole catalytic subunit}}$
$K_{m,NADH}$	5.9 (±0.05) × 10^{-6} *M*
$K_{i,NADH}$	6.0 (±0.2) × 10^{-6} *M*
$K_{m,cyt}$ c^{3+} (with NADH)	1.9 (±0.5) × 10^{-6} *M*
V_{max/E_t} (with NADH and cyt c^{3+})	2.2 (±0.4) units/mg[b]
V_{max/E_t}	$74 \frac{\text{moles}}{\text{min mole catalytic subunit}}$
$K_{i,NADP^+}$	9 × 10^{-4} *M* (competitive with NADPH)
K_{i,NAD^+}	1.5 × 10^{-2} *M* (competitive with NADH)
$K'_{D,FMN}$	1.5 × 10^{-8} *M*
$K_{m,FMN}$	1.6 × 10^{-8} *M*
$K_{m,app,menadione}$	1.3 × 10^{-7} *M* (with NADPH and cyt c^{3+})
	6.7 × 10^{-8} *M* (with NADH and cyt c^{3+})
	8 × 10^{-5} *M* (with catalase; without menadione)
$K_{m,oxygen}$	3 × 10^{-5} *M* (with menadione)
	3 × 10^{-6} *M* (with menadione and catalase)
$V_{max,oxygen^c}$ (extrapolated to 30°)	54 mu/mg (NADPH)
	544 mu/mg (NADPH + menadione)
	762 mu/mg (NADPH + menadione + catalase)
	56 mu/mg (NADH)
	302 mu/mg (NADH) + menadione)
	543 mu/mg (NADH + menadione + catalase)
$V_{max,oxygen/E_t}{}^d$	1.8 min^{-1} (NADPH)
	18.5 min^{-1} (NADPH + menadione)
	25.9 min^{-1} (NADPH) + menadione + catalase)
$V_{max,oxygen/E_t}{}^d$	1.9 min^{-1} (NADH)
	10.3 min^{-1} (NADPH + menadione)
	18.5 min^{-1} (NADH + menadione + catalase)

[a] Value in parentheses for $K_{i,NADPH}$ is the value calculated for a fully ordered three-substrate mechanism.
[b] Expressed in $\frac{\mu mol}{\text{min mg}}$
[c] Expressed in $\frac{\text{nmol}}{\text{min mg}}$
[d] Expressed in moles of O_2 reduced per minute per mole catalytic subunit.
[e] From Reference 8.

ordered mechanism could not be ruled out for the NADPH data. A change in this mechanism would affect only the estimated value for the binary dissociation constant, $K_{i,NADPH}$, as shown in parentheses in Table 1. If the term $K_{i,a}$ $K_{m,c/}$ (A) (C) reaches relatively low values under conditions where B is saturating, the equation for a hybrid random-ordered mechanism reduces to that of a so-called "ping-pong" Bi-Bi mechanism.[31] Thus, if $K_{i,NADPH}$ is significantly lower than the NADPH concentration present during the kinetic assays, a "collapse"

TABLE 2
Relative Rates of Reduction for Various Electron Acceptors (NADH-Cytochrome *c* Reduction = 100)

Electron acceptor	Electron donors			
	NADPH	+ Menadione	NADH	+ Menadione
Cytochrome *c*	77 (0.47 U/mg)	616 (3.8 U/mg)	100 (0.614 U/mg)	353(2.17 U/mg)
Ferricyanide	267 (1.64 U/mg)	No change	164 (1 U/mg)	no change
DCIP	177 (1.1 U/mg)	No change	4.7(0.09 U/mg)	no change
O_2	8.8(0.054 U/mg)	88.6(0.54 U/mg)	9.1(0.056 U/mg)	49(0.3 U/mg)
O_2 + catalase		124 (0.76 U/mg)		88(0.54 U/mg)
NAD-S	—	—	—	—
NADP-S	—	—	—	—
Anaerobic activities				
Cytochrome b_5	17.1(0.1 U/mg)	240 (1.47 U/mg)	11.7(0.072 U/mg)	186(1.1 U/mg)
Cytochrome *c*	81.6(0.5 U/mg)	677 (4.17 U/mg)	118 (0.714 U/mg)	504(3.1 U/mg)
DCIP	177 (1.1 U/mg)	no change	32.7(0.2 U/mg)	No change
Thionicotinamide analogs				
	NADP-SH	+ Menadione	NAD-SH	+ Menadione
Cytochrome *c*	153 (0.94 U/mg)	398 (2.5 U/mg)	36.7(0.12 U/mg)	165(1.02 U/mg)

[a] From Reference 8.

from converging lines to parallel lines will occur. Only at low concentrations of both NADPH and cytochrome *c* will convergence become evident, and indeed, this appeared to be the case for the reductase.[8]

Michaelis constants for cytochrome *c* were independent of the pyridine nucleotide, but K_m values for NADH and NADPH differed considerably, i.e., 5.9 μM vs. 57 nM, respectively. Further differences for the pyridine nucleotides were evident with other electron acceptors. For example, NADH rather than NADPH became the preferred electron donor in menadione mediated cytochrome *c* reduction (see Table 2).[8] Most strikingly, NADPH-dependent DCIP reduction was 37 times the NADH-dependent DCIP reduction rate. This indicates pyridine nucleotide control, and further argues that a ternary complex is present for one or both of the substrate reductants. The oxidized pyridine nucleotides were poor substrate inhibitiors of the reductase, with a $K_{i,NADP^+}$ of 0.9 mM and a K_{iNAD^+} of 15 mM.

6. Reactions with Oxygen[7,8]

The reductase reduced oxygen in the presence of either nucleotide and showed a 1:1 stoichiometry with NAD(P)H oxidation, a specific activity of 55 nmol 0_2 consumed min^{-1} mg^{-1} protein, and a $K_{m,oxygen}$ of 80 μM. Hydrogen peroxide was likely the end-product as the inclusion of catalase liberated O_2 and lowered respiration by one-half. However, whether hydrogen peroxide was formed directly or through the dismutation of the superoxide radical is not known (see Section II.A.8). Menadione facilitated oxygen consumption (fourfold for NADH- and tenfold for NADPH-dependent reduction) and lowered the $K_{m,oxygen}$ to 28 to 30 μM. The inclusion of catalase further decreased the $K_{m,oxygen}$ to 3 μM and paradoxically enhanced respiration. In the absence of catalase menadione was capable of reducing hydrogen peroxide, likely through hydroxyl radical,[23] but only after low concentrations of oxygen were reached. The proposed interactions of the reductase with oxygen that are consistent with these data are shown in Scheme II.

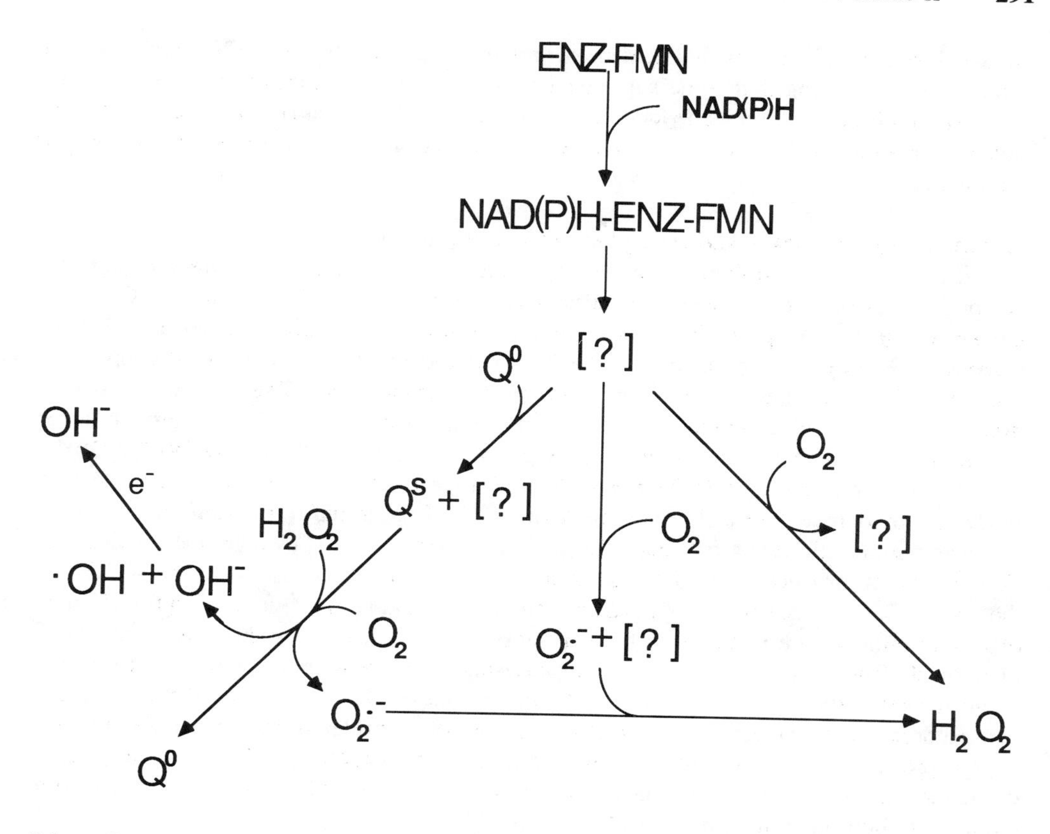

Scheme II. Proposed mechanism for the reduction of oxygen. After reduction by NAD(P)H, the enzyme reacts with oxygen to form either hydrogen peroxide and/or the superoxide radical, which then dismutates to hydrogen peroxide and oxygen. Oxidized menadione (Q^o), when present, is reduced to the semiquinone (Q^s), which reacts either with oxygen to form the superoxide radical, or (at low oxygen tension) with hydrogen peroxide to form hydroxide anion and hydroxyl radical. The hydroxyl radical can be further reduced, perhaps by the reductase, menadione or superoxide radical, to form the hydroxide anion. Unknown enzyme intermediates and/or complexes are indicated by a question mark [?].

7. Arrhenius Plots[8]

Arrhenius plots revealed considerable differences in the thermodynamic "activation" parameters. At 30°C, activation energies were 11.8 Kcal/mol for the NADPH-dependent and 6.0 cal/mol for the NADH-dependent reduction of cytochrome *c*. Free energies of activation were similar, thus indicating more favorable changes for the NADPH-dependent entropy of activation ($S_{act,\mathrm{NADPH}} = -24.6$ e.u. and $S_{act,\mathrm{NADH}} = -43.3$ e.u.). The slope of the NADPH-dependent reaction was nonlinear, which may reflect an inhibition by phosphate in the reaction mixture. The addition of menadione affected only the entropy of activation (i.e., the slope remained constant).

8. Electron acceptors[7,8]

The reactivity of the reductase with various electron acceptors is shown in Table 2. Low oxygen tension facilitated electron transfer, particularly with NADH as the electron donor. Cytochrome b_5 was reduced only under anaerobic conditions. Interestingly, oxygen apparently competed with DCIP for electrons only when NADH was the substrate reductant, as the rate of dye reduction increased sevenfold under low oxygen tension with NADH, but remained constant under the NADPH reducing system. No transhydrogenase activity was observed with thionicotinamide analogs, but these analogs were quite capable, when reduced by a glucose-6-phosphate dehydrogenase system, of acting as substrate reductants. Superoxide dismutase (20 units) inhibited cytochrome *c* reduction by >50% for both pyridine nucleotides indicating that some superoxide anion was formed. However, superoxide was not required for cytochrome *c* reduction as evidenced by the increase in the rate of reduction under anaerobic conditions where xanthine oxidase dependent cytochrome *c* reduction was eliminated.

9. Function of NADH(NADPH)-Cytochrome *c* Reductase

At present, the physiological function of the reductase is unknown. Cytochrome *c* was not present, cytochrome b_5 was reduced slowly, and vitamin K_1 and Coenzyme Q_{10} were ineffective mediators, possibly because of their insolubility.[9] Ainsworth et al.[24,25] have isolated a yeast cyanide-insensitive NAD(P)H oxidoreductase which contained copper and accounted for 70% of the respiration in the near anaerobic state. They suggested that its function was to balance the reduced/oxidized pyridine nucleotide ratio during growth under conditions of limited mitochondrial respiration. A similar role for the NADH(NADPH)-cytochrome *c* reductase could be postulated, as it respires at one third the level of the NAD(P)H oxidoreductase and represents about 50% of the initial respiration.

Many similarities exist between Old Yellow Enzyme and this cytochrome *c* reductase: The spectral absorbances,[21] the FMN prosthetic group,[21] the isoelectric point (pI = 5.22),[26] the loss of fluorescence on FMN binding,[27] the amino acid composition,[22] and the absence of reactive thiols[28] are common to both species. Various differences, however, have been mentioned.[7] These include dissimilar molecular weights, FMN affinities, rates of cytochrome *c* and oxygen reduction and pyridine nucleotide affinities. The possibility of the reductase being a proteolytic product[29] of Old Yellow enzyme was considered, but appeared unlikely.[7] Unique properties such as low affinity for DEAE-cellulose at pH 8.5 and R_f = 0.47 migration on nondenaturing gels were present immediately on extraction of the soluble portion from the acetone powder in the presence of 0.1 m*M* PMSF.

B. NADPH-CYTOCHROME *c* REDUCTASE

1. Properties[5]

Greater than 95% of the NADPH-cytochrome *c* reductase activity in this yeast was attributable to an NADPH specific cytochrome *c* reductase earlier isolated by Tryon et al. (R_f = 0.698 on 7% nondenaturing polyacrylamide gels[7]).[5,6] This reductase was extracted by autolysis from an acetone powder and purified 247-fold to a final specific activity of 152 units per milligram protein. The homogeneous protein, which had a pI of 5.2, gave by sedimentation equilibrium an estimated molecular weight of 68,000 under native conditions and 34,000 under denaturing conditions. As isolated, one FAD was present per 34,000 molecular weight subunit with no discernable FMN, although FMN was quite capable of reactivating the apoenzyme. The flavin dissociation constants estimated by the kinetic method discussed above were 4.7×10^{-8} *M* and 4.4×10^{-8} *M* for FAD and FMN, respectively.

2. Absorbance Spectra[5]

Absorbance peaks typical for a flavoprotein[30] were present at 275 nm (inflection at 277), 385 nm and 455 nm, with an extinction coefficient at 275 nm of $\epsilon_{275nm} = 8.5 \times 10^4$ M^{-1} cm^{-1}. On isolation, the reductase was in the oxidized form, and the ratio of 275 nm/455 nm varied from 8.4 to 9.2, depending on the loss of FAD during purification. Although sodium dithionite could fully reduce the enzyme, only halfway reduction with NADPH was possible under aerobic conditions. An air-stable intermediate with a small 630 nm peak, possibly a type A or blue semiquinone,[30] appeared on reoxidation.

3. Sulfhydryl Reactivity[5]

Titration with 4,4′-dithiodipyridine at pH 3.2 (30°C) yielded one reactive -SH/35,000 molecular weight. Inactivation by 2 to 4 m*M* DTNB (pH 7.9 at 32°C) was prevented by the addition of either NADPH or FAD. Thermal inactivation was also protracted with FAD or NADPH. From these data, Tryon et al.[5] concluded that FAD shielded a single -SH group and that NADPH interacted at or near this thiol.

4. Steady-State Kinetics[6]

Primary inverse plots with NADPH and cytochrome *c* showed parallel lines except at higher concentrations of fixed substrate, where an inversion occurred.[6] Although consistent with a ping pong Bi Bi mechanism,[31] a hybrid random-ordered mechanism as discussed above (Section II.A.5.) was also possible. A V_{max} value for cytochrome *c* reduction of 470 U/mg protein was determined, indicating a turnover number of 33,000 min^{-1}. Michaelis constants for NADPH and cytochrome *c* were 45 and 3.6 μM respectively. A $K_{m,NADPH}$ of 56 μM, a $K_{m,DCIP}$ of 145 μM and a V_{max} value of U/mg was estimated from inverse plots for the reduction of DCIP.

From the nonparallel slopes at higher concentrations of fixed substrate, apparent inhibitor constants of $K_{i,NADPH}$ = 1.1 m*M* and $K_{i,cytc}3+$ = 0.21 m*M* were calculated. Inhibition by NADPH was noncompetitive with respect to cytochrome *c* and vice versa. Product inhibition studies showed $NADP^+$ to be competitive with NADPH ($K_{i,NADP}{}^+$ = 33 μM) and noncompetitive with cytochrome *c* ($K_{i,NADP}{}^+$ = 290 μM).

The reduction of DCIP followed a different pH profile to that observed for cytochrome *c*. At pH 5.6, cytochrome *c* reduction was only 7% of that found at pH 7.6, whereas DCIP reduction increased by more than 50%. Moreover, the addition of catalase increased the rate of DCIP reduction by another 120% but had an inhibitory effect on cytochrome *c*.

The reduction of oxygen was relatively slow, with a V_{max} of 0.3 U/mg and a $K_{m,oxygen}$ of 150 μM.[6] The product of the reaction was hydrogen peroxide, which acted as a noncompetitive inhibitor (K_i = 0.5 m*M*) with respect to NADPH and a competitive inhibitor (K_i = 0.47 m*M*) with respect to cytochrome *c*. Superoxide anion did not appear to be released as a product as evidenced by the lack of inhibitory activity towards cytochrome *c* and DCIP reduction by superoxide dismutase, and by negative results from the Hodgson and Fridovich[32] luminescence test.

III. NAD(P)H:QUINONE REDUCTASE

A. INTRODUCTION

NAD(P)H:quinone reductase (EC 1.6.99.2), also known as DT diaphorase[4,33,34] and vitamin K reductase [3,35,36,37] is a flavin enzyme that supports a two electron quinone reduction by NADH or NADPH. The enzyme is found in a variety of animals,[3,4,36,37] tissues,[4,12,38] plants,[39] and microbial systems.[3] One physiological function of the mammalian reductase appears to be the reduction of vitamin K_1,[9-11] a necessary step in vitamin K dependent carboxylation,[40] which among other things is directly involved in the synthesis of four plasma clotting factors.[41,42] Another more general function, may be a role in protecting cells from the toxic and/or carcinogenic effects of quinones, which are products of oxidative aromatic hydrocarbon metabolism.[12] A wide variety of toxic polycyclic hydrocarbons and aromatic amines are known to induce the enzyme,[12] which might offer protection against reactive quinone products by affording a two-electron reduction to a stable hydroquinone.[13,14] For example, quinone reductase has been shown to inhibit the cytotoxic effects of menadione by reduction to the slowly oxidized hydroquinone and thereby preventing the formation of the toxic superoxide radical.[13,14]

B. PROPERTIES

Mammalian quinone reductase is a dimeric flavoprotein exhibiting a molecular weight (in rat and rabbit) of 54,000[43] and containing 2,[43,44] not 1 as earlier believed,[11,36,37,45,46] nondissociable[46] FAD molecules per dimer of identical subunit. Data from the rat liver enzyme, which has recently been cloned and sequenced, yielded a calculated subunit molecular weight of 28,564.[47] Spitsberg and Coscia have reported molecular weights for 17 plant reductases ranging from 38,000 to 53,000.[39]

Approximately 95% of the mammalian reductase is present in the soluble fraction, with small amounts in the microsomes (2 to 3%) and mitochondria (2 to 3%).[3,4] Although each subunit appears stable in SDS and 6M guanidine HCL, Wallin[44] reported a dissociation into two polypeptides with molecular weights of 9,000 and 18,000 after boiling the enzyme in the presence of SDS. Because they were not disulfide linked, Wallin suggested that they might interact by strong hydrophobic bonding.

Multiple isoelectric points have recently been reported for both rat and rabbit liver enzymes.[43] Purified rabbit liver reductase separated into 3 bands at pH 7.6, 7.2, and 6.5, whereas rat liver reductase contained a dominant band at pH 5.8 and several minor bands from pH 5.3 to 5.6. In contrast, earlier studies by Märki and Martius [36,37] indicated that the ox reductase was cathodic at pH 8. Although multiple pI values for the rat and rabbit liver enzyme may have represented deamidation or ampholyte complexation,[43] the authors allowed for the possibility of isoforms. Höjeberg et al.[48] have reported at least 3 antigenic forms of the rat reductase with different reactivities towards DCIP and menadione, and with one of the forms preferentially induced by 3-methylcholanthrene. Antigenicity may furthermore be species specific as immunochemical studies[44] showed no cross-reactivity of rat antibody towards the rabbit enzyme.

Characteristically, mammalian reductase is strongly inhibited by dicoumarol ($K_i = 10^{-9}$ to 10^{-8} *M*),[4] whereas with the exception of pea seed, a variety of plant reductases were insensitive to 0.5 m*M* dicoumarol.[39] Sulfhydryl inhibitors also inactivate the mammalian reductase,[34,43,44] but again the plant enzymes appear resistant.[3,39] Chemical modification of selected amino acid groups suggests that lysyl residues may be required for a functional rat liver reductase.[44]

C. ABSORBANCE AND CIRCULAR DICHROISM SPECTRA

Purified rat liver reductase showed three major absorption peaks[43] at 268, 366, and 446 nm, with an estimated extinction coefficient at 446 nm of $\epsilon_{446nm} = 32{,}000\ M^{-1}\ cm^{-1}$. Rabbit liver reductase absorption was similar,[43] with peaks at 268, 375, and 450 nm and $\epsilon_{450nm} = 21{,}000\ M^{-1}\ cm^{-1}$.

Circular dichroism spectra of the rat liver reductase exhibited three negative bands around 268, 304, and 377 nm and two positive bands near 283 and 460 nm.[43] The 325- to 500-nm region was similar to Shetna flavoprotein and Clostridial flavodoxin,[49] suggesting that all three proteins share similar flavin environments. The spectrum resembled that of cytochrome c_{552} and flavodoxin,[49,50] where a tyrosine rather than a histidine covalently binds the flavin.

Reducing the enzyme shifted the spectrum, with peaks now at 260, 322, 375, and a shoulder at 280 nm.[43] In particular, the shift of the 283-nm peak to 260 nm appeared suggestive of a change either in conformation or in oxidation state. Lastly, the spectrum revealed low intensity in the ultraviolet region which is characteristic of nonperiodic structure.

D. KINETIC STUDIES

1. Steady State

At low concentrations of substrates, inverse plots for either pyridine nucleotide and electron acceptor are parallel,[45,51] indicative of a Bi Bi ping-pong ordered mechanism.[31] Activity can vary with activators such as Bovine Serum Albumin, detergents, neutral Phospholipids,[4] and may be different for two strains of the same species.[44,52] For example, purified reductase from Sprague-Dawley and Wistar strains had V_{max} values (DCIP reduction) of 4673 μKat/mg and 2967 μKat/mg, respectively.[44] Turnover values of 44,000 min^{-1} were reported for ox and rat reductases[36,37,45] for NADH oxidation with menadione (+ cytochrome *c*) as the acceptor at 25°C. Wallin and Little[43] recently measured a DCIP reduction turnover number of 120 s^{-1} (7,200 min^{-1}) for the rabbit liver reductase compared with 310 s^{-1} (18,600 min^{-1}) for the rat liver enzyme. The kinetics of vitamin K_1 reduction has been analyzed by Fasco and Principe,[10] who performed the task of hydroquinone product quantitation by high performance liquid chromatography and reported a V_{max} of approximately 1.5% the rate of DCIP reduction. Calculated values for the pyridine nucleotide Michaelis constants have ranged from 12 to 500 μM,[10,36,37,43,51] depending on the species and study, while K_m values for the electron acceptors menadione, vitamin K_1 and DCIP have been reported as 0.696 to 4 μM,[3,51,23,24,45] 64.1 μM,[10] and 23.3 μM,[10] respectively.

2. Substrate and Product Inhibition

At higher substrate concentrations, inverse plots are no longer parallel and exhibit increasing slopes, indicative of substrate inhibition.[45,51] Menadione inhibition appears to be competitive with both NADH (K_i = 11 μM)[51] and NADPH (K_i = 5.4 μM),[45] whereas NADH is competitive with menadione (K_i = 2.3 mM),[51] but with a relatively high inhibition constant. DCIP has also been shown to inhibit competitively with respect to NADH.[51] The oxidized pyridine nucleotide products are weak inhibitors. Hall et al.[51] reported only slight inhibition of DCIP reduction with NAD^+, and in a similar study Hosoda et al.[45] found no competitive inhibition of NADPH with 1mM $NADP^+$. Hosoda et al.[45] did find, however, an apparent competition with menadione for reduced enzyme at this concentration. For the other products of the reaction, Hall et al.[51] reported that reduced DCIP was competitive with NADH; Hosoda et al.[45] reported some inhibition by reduced menadione.

3. Inhibitors

Dicoumarol and warfarin are competitive inhibitors with respect to the pyridine nucleotides and exhibit K_i values in the range of 10^{-8} to 10^{-9} M[3,4,34,51] and 10^{-5} M, respectively. Vitamin K_1 reduction, however, appeared unusually resistant to these anticoagulants.[10] Although dicoumarol was competitive with NADH and uncompetitive with respect to vitamin K_1, it was only 2.5% as inhibitory as with DCIP reduction. Surprisingly warfarin at 520 μM showed no inhibition of vitamin K_1 hydroquinone formation, suggesting that this anticoagulant may function elsewhere *in vivo*.

The class of inhibitor anticoagulant represented by 1,3-indanedione is also a competitive inhibitor with respect to NAD(P)H but binds to a site different from dicoumarol.[46,53] Rase et al.[46] observed nonlinear Dixon plots in the presence of dicoumarol and 2-pivaloyl-1,3-indandione that were described by second order terms of inhibitor concentrations. They concluded that either two alternative paths for EI or two or more mutually interacting binding sites were present. Scatchard and Hill plots were consistent with at least two binding sites with strong negative cooperativity.

4. Activators

Early studies by Ernster[34] showed that the V_{max} for rat liver reductase was increased by three- to tenfold in the presence of Bovine Serum Albumin, polyvinylpyrrolidone, and nonionic detergents. Pyridine nucleotide Michaelis constants decreased from 130 μM (180 μM) for NADH(NADPH) to 44 μM (88 μM) after BSA was included in the assay mixture.[34] In addition, BSA caused a rise in the inhibition constant for dicoumarol from 10^{-8} to 10^{-7} M. Hosoda et al.[45] reported that BSA inhibited the binding of both dicoumarol and menadione to the reductase. They explained these effects by the ability of BSA to bind dicoumarol (0.528 mole/mole BSA) and menadione (0.009 moles/mole BSA), both of which act as competitive inhibitors towards NAD(P)H. At low noninhibitory levels of menadione, BSA could not stimulate activity.[45]

The nonionic detergent, Tween 20, eliminated substrate inhibition by NADH but not by menadione.[51] Kinetic parameters showed a small increase in V_{max}, a decrease in $K_{m,NADH}$ from 111 to 85 μM and an increase in K_m,menadione from 2.3 to 7.3 μM in the presence of the detergent.[51] Inhibition constants increased for dicoumarol (from 2.65 n*M* to 16.7 n*M*), menadione (from 11μM to 22μM) and NADH (from 2.3 m*M* to infinity, i.e., no inhibition was detected), indicating to Hall et al.[51] that interactions between the donor and acceptor binding sites involved hydrophilic and hydrophobic regions of the enzyme molecule.

Although nonionic detergents and neutral phospholipids were generally stimulatory, Hollander et al.[53] found that they could enhance or inhibit the reaction rate and/or the substrate affinity depending on the electron acceptor and the relative concentrations of substrates or inhibitors. Furthermore, nonionic detergents and neutral lipids often affected a kinetic parameter in opposite ways.

5. Mechanism

Hosoda et al.[45] reported that enzyme reduction by NADPH followed second order kinetics with a rate constant of $k_{+1} = 3.13 \times 10^8 \ M^{-1} \ s^{-1}$. In the presence of dicoumarol, the reduction was inhibited and followed first order kinetics. Electron transfer to menadione (k_{+3}) exceeded the dead time of the stopped-flow apparatus, but preincubation of the oxidized enzyme with menadione prevented this rapid transfer even though reduction of the enzyme was not impaired. Reoxidation of this reduced complex was slow, which suggested that the positioning of menadione relative to the reduced flavin was faulty if binding occurred before the enzyme reduction step. From this observation, the authors proposed that a conformational change occurring after enzyme reduction was necessary for proper menadione:$FADH_2$ interaction. If menadione were to bind the enzyme before the reduction by NADPH, the conformational change could not rapidly occur. After making certain assumptions, Hosoda et al.[45] calculated a rate constant for the reduction of menadione by reduced enzyme of $k_{+3} = 1.66 \times 10^9 \ M^{-1} \ s^{-1}$ and a rate constant for the release of reduced menadione by the enzyme of $k_{+4} = 1146 \ s^{-1}$. They concluded that the enzyme followed a ping-pong mechanism, with the rate limiting step being the release of reduced menadione.

Utilizing a sepharose-menadione affinity column for enzyme purification, Wallin et al.[11] reported that NADH could elute the adsorbed enzyme, presumably after reduction of the quinone. Because this ternary complex appeared functional, they concluded that the reaction must occur by a randon-sequential mechanism.

Irregular, non-hyperbolic plots, which showed intermediate plateaus in the V vs. NADH plots at constant DCIP concentrations, and "troughs" or inverted peaks in the V vs. DCIP plots at constant levels of NADH, were shown by Hollander et al.[54] to fit more than one model. The irregularities were reproducible, but a reversible disappearance (or attenuation) occurred at either high or low temperatures. Potassium chloride (0.25 *M*) converted the V vs. DCIP plot ("trough" type) to one of an intermediate plateau, and the V vs. NADH plot (intermediate plateau type) to one with a "trough". The data fit most closely to a mathematical expression which was the sum of two rate equations, one of substrate inhibition and the other an enzyme with two active sites. Two other expressions which correlated well were one which described an enzyme with four different catalytic sites and another in which two enzymes, each with two catalytic sites, formed a dimeric complex. The authors favored the latter model, and included an additional scheme (consistent with the temperature data) whereby the dimer was in equilibrium between heat labile and cold labile forms, the interconversion of which would be provided by a conformational change.

6. Electron Acceptors

Relative activity of the reductase towards a variety of electron acceptors has been reported.[3,4,34] DCIP, benzo- and napthoquinones were good acceptors, methylene blue and ferricyanide were moderately good (approximately 50% of DCIP activity) and cytochrome *c* (without the electron mediator menadione) and vitamin K_1 were virtually nonreducible (about 0.01% of the DCIP activity).[4,34] Martius et al.[9] later showed that vitamin K_1 could be reduced if incorporated into liposomes. Ernster[34] reported that the most active quinones were those which lacked a "3" position side chain, with a decrease in activity as the chain length increased. Whereas methylation in the "2" position had little effect on activity, hydroxylation of the "2" or "3" position inactivated the acceptor. Menadione and other 1,4-napthoquinones were the best mediators for electron transfer to the normally inactive acceptor cytochrome *c* and coenzyme Q_{10}. Neither cytochrome b_5 nor nitrotetrazolium compounds were readily reduced.[4]

E. HOMOLOGY WITH OTHER PROTEINS

Bayney et al.[47] cloned and sequenced the rat liver reductase and compared the predicted amino acid sequence to that of NADPH-cytochrome *c* reductase, NADH-cytochrome b_5 reductase, ferridoxin-NADP$^+$ reductase, glutathione reductase, p-Hydroxybenzoate hydroxylase, Desulfovibrio vulgaris flavodoxin, Clostridium MP flavoprotein, pig kidney D-amino acid oxidase and Escherichia coli fumarate reductase. They found no significant amino acid sequence homology, and concluded that the quinone reductase evolved from an ancestral gene distinct from the others.

IV. NADPH-CYTOCHROME P-450 REDUCTASE

A. INTRODUCTION

NADPH-cytochrome P-450 reductase (EC 1.6.2.4), also known as NADPH-cytochrome *c* reductase,[55-58] and NADPH-ferrihemoprotein reductase,[59] is a microsomal flavoprotein which transfers reducing equivalents from NADPH to cytochrome P-450 (P-450) for metabolism of both endogenous and xenobiotic compounds.[16,17] Hemeoxygenase[60] (see Chapter 13), fatty acid desaturase,[61] and elongation systems[62] may also utilize electrons from the reductase.

Isolation of high activity flavin-containing NADPH-cytochrome *c* reductase from topale yeast was described by Haas et al.[63] in 1940 and from a hog liver acetone powder by Horecker[64] 10 years later. Purification and characterization of the pig liver enzyme was later reported in 1962 by Williams and Kamin[55] and by Phillips and Langdon,[56] who determined that its origin was microsomal not mitochondrial.[64] The ability of the endoplasmic reticulum to convert aromatics to polar products[65] and to incorporate molecular oxygen into substrates[66,67] had been demonstrated some years earlier. However, whereas the enzyme could reduce mitochondrial cytochrome *c*, it had lost the ability to reduce the putatative microsomal acceptor, cytochrome P-450. This was due to the proteolytic treatment initially employed to solubilize the protein during isolation and which cleaved a membrane bound N-terminal segment necessary for proper interaction with the integral membrane P-450 protein.[17,68,69] By avoiding proteolytic digestion, Lu and Coon[17] were with detergents able to solubilize, purify and retain P-450 reductase activity. Immunochemical studies [70-72] provided conclusive evidence that NADPH-cytochrome *c* reductase and P-450 reductase were equivalent. Rapid purification and high yield have been made possible by 2′,5′ADP-Sepharose[73,74] and NADP-Sepharose[75] affinity chromatography.

The reductase has been isolated from a variety of tissues and sources,[20] and is different from the mitochrondrial and bacterial types, which require an iron-sulfur protein intermediary to transfer electrons to P-450.[18] Although 2 reducing equivalents are ultimately required for P-450 activity, the heme-iron center can only accept one electron, and therefore transfer must be sequential.

B. GENERAL PROPERTIES

Cytochrome P-450 reductase contains 1 molecule each of FAD and FMN per polypeptide chain of 78,000 molecular weight.[69,73,76,77,78] The FMN flavin has a dissociation constant of 1.3×10^{-8} *M* and can be preferentially dissociated by dialysis against 2 *M* potassium bromide,[79] whereas the FAD, still intact under these conditions, can be dissociated at low pH (pH = 2.0).[80] Treatment with Trypsin cleaves an N-terminal domain (4,800[81] and 6,400[82] molecular weight for rabbit and rat reductase, respectively) responsible for attachment of the reductase to the microsomal membrane, and necessary for interaction with membrane bound cytochrome P-450.[68,69,81,82] The pig reductase is unusually sensitive to proteolytic cleavage during purification,[80] yielding 2 fragments of 20,000 and 60,000 molecular weight.[80,83,84] This may be due to a substitution of the amino acid arginine for glycine at residue number 504 in the porcine reductase versus the rabbit and rat reductases.[84]

Phospholipids are necessary to facilitate P-450:P-450 reductase interactions for maximal activity,[85-88] and their presence increases P-450 substrate binding and the rates of first and second electron transfer from the reductase.[89,90] In addition, they alter the conformation of the reductase as evidenced by circular dichroism spectra.[91-93] Nonionic detergents can to some extent replace this phospholipid requirement,[94,95] and full activation without phospholipid or detergent is possible if both proteins are preincubated at high concentrations of each enzyme.[96]

In the presence of phospholipid, purified reductase and P-450 form a 1:1 complex with a molecular weight of ~800,000[90] and a K_d in the range of 100 nM.[90,97] The catalytically active state in all lipid systems appears to be a binary complex between the reductase and P-450,[97] whereas the ratio of P-450 to the reductase in the microsomal membrane is approximately 20:1.[98]

Although several isozymes of cytochrome P-450 with overlapping substrate specificity have been isolated,[16,99,100] multiple forms of P-450 reductase do not occur.[93,101]

C. ABSORBANCE SPECTRA

Detergent solubilized rabbit reductase has absorption maxima in the oxidized state at 277, 382, and 456 nm, with extinction coefficients of 159 mM^{-1} cm^{-1}, 19.5 mM^{-1} cm^{-1}, and 21.4 mM^{-1} cm^{-1}, respectively,[93] and an A_{277nm}/A_{456nm} ratio of 7.4. The spectrum of rat reductase is similar, with an A_{277nm}/A_{456nm} ratio of 8.7.[102]

Titration with NADPH or dithionite yields an air stable semiquinone after introduction of 1 electron with a broad absorption at 585 nm (ϵ_{585nm} = 2.4 mM^{-1} cm^{-1})[102] and a shoulder at 630 nm. Further titration produces spectral changes that are too complex to detail here.[102-104] However, it is interesting to note that NADPH accumulates after the third electron is transferred due to the low midpoint potential of the fourth flavin couple.[102] Total reduction by the pyridine nucleotide can be accomplished by addition of an NADPH regenerating system or inclusion of NADase to hydrolyze $NADP^+$.[102]

D. OXIDATION-REDUCTION PROPERTIES OF THE FLAVIN MOIETIES

Although early spectrophotmetric and kinetic data suggested that the reductase cycled between fully oxidized and reduced forms,[57,58] subsequent studies indicated that cycling is between the 1-electron and 3-electron forms.[78,102,104] Iyanagi and Mason[76] were first to report that FAD and FMN were present in equimolar amounts, and that an air stable 1-electron semiquinone was formed. Potentiometric titrations yielded four flavin-couple midpoint potentials of −0.110 V, −0.270 V, −0.290 V, and −0.365 V,[105] but flavin assignments were not possible until Vermilion and Coon[79] developed a technique to selectively remove FMN, leading to the identification of FMN as the high potential flavin and FAD as the low potential flavin. NADPH reduction was possible with the FMN depleted reductase, and transfer of electrons from FAD to reintroduced FMN was thermodynamically favorable. This, combined with an estimated redox potential for FAD near that of the pyridine nucleotide couple and the failure of the FMN-depleted enzyme to reduce P-450LM_2, cytochrome *c*, DCIP and menadione, suggested that electron entry was through FAD and exit was through FMN.[79] The presence of ferricyanide reductase and transhydrogenase activity in the FMN depleted reductase, however, left open the possibility that FAD might also participate in 1-electron reductions.

E. FLAVIN DOMAINS

Flavin-protein interactions in various redox states with the detergent solubilized rabbit liver reductase have been studied by Raman spectroscopy.[106] Spectra were identical for the oxidized holoenzyme and FMN-depleted enzyme, supporting circular dichroism spectral data indicating that the flavins are in a similar environment.[93] The Raman spectra for the oxidized forms satisfied 3 of 4 criteria for hydrogen bonding.[106] Reduction to the semiquinone forms yielded different spectra for FADH· and FMNH·, consistent with the FADH· (semiquinone) hydrogen bonding to the protein through the N_1 atom, and the FMNH· (semiquinone) hydrogen bonding through the N_5-H proton of the isoalloxazine ring. No differences in carbonyl interactions were evident. (See Sections I and K for additional discussion on flavin-reductase interactions.)

F. KINETICS

1. Inverse Plots and Kinetic Parameters

Parallel,[57,107] converging,[75] and intersecting[108] double reciprocal plots, indicative of ping-pong, sequential, and ternary complex mechanisms, respectively, have been reported with the substrates, NADPH and cytochrome *c*. Dignam and Strobel[75] suggested that the low ionic strength of the assay conditions used by Masters et al.[57,58] may have prevented visualization of the converging lines. Kobayashi and Rikans,[108] however, concluded that the mechanism must be different for Guinea pig reductase, which showed intersecting lines in the third quadrant, consistent with formation of a ternary complex. Variations in ionic strength did not affect this mechanism.[108]

Maximum turnover numbers range from 1,200 to 5,400 min^{-1}[1,57,61,73,102,107-110] for the mammalian enzyme and approached 11,500 min^{-1} for the bakers yeast enzyme[111] with cytochrome *c* as the acceptor. Cytochrome P-450 reduction is highly variable depending on the isozyme, substrate and lipid system employed during the assay (e.g., compare References 112, 113, and 114). Michaelis constants for cytochrome *c* from 2.6 to 18 μM,[55-57,75,102,107,108,115,190] for NADPH from 1.7 to 26.7 μM[57,108,110,190] and for P-450 of ~90 nM[75] have been reported. The K_m value for NADH is relatively high (K_m = 1.1 mM),[110] whereas the inhibition constant for the oxidized pyridine nucleotide product, $NADP^+$, is low (K_m = 6.1 μM).[110]

Cytochrome b_5 can act as an acceptor, with a reported reductase turnover[61] of 1,500 min^{-1}, 5,100 min^{-1}, and 7,000 min^{-1} in 0.1% Renex, microsomes and phospholipid vesicles, respectively. Other electron acceptors that can be reduced by the reductase include ferricyanide, DCIP, menadione and NTB.[57,102,107,108,110]

2. Early Reduction of the P-450 Reductase

Stopped flow spectophotometric studies by Oprian and Coon[103] have revealed 3 phases for the reduction of cytochrome P-450 reductase. The first phase is a monophasic 2-electron reduction by NADPH. The association must be rapid and the binding constant low because the rate constant did not change at low concentrations (1:1 stoichiometry) of NADPH. Kamin et al.[116] had earlier observed fluorescence quenching of NADPH when mixed with protease solubilized reductase within the dead time of a stopped flow spectrophotometer. Spectral data[103] for the first phase paralleled the results of Masters et al.[57] for the protease solubilized enzyme. The formation of semiquinone species (an increase in 585 nm absorbance) was concomitant with the formation of a 2-electron reduced state (decrease in 502-nm absorbance), indicating that a rapid intramolecular electron transfer occurred, and likely that the reduction of FAD by NADPH was the rate limiting step for this phase.

At high concentrations of NADPH (9.5-fold excess), two monophasic events were evident. The first was a 2-electron reduction with a rate constant of 28 s^{-1}, whereas the second phase (5.4 s^{-1}) represented enzyme interaction with a second molecule of NADPH to form a metastable equilibrium among the various 2 electron states, the fully reduced state and NADPH. A third phase, which appeared after a longer interval, was proposed to be electron transfer between different protein molecules until the attainment of thermodynamic equilibrium.[103]

Utilizing a mathematical formulation developed by Olson et al.[117] for xanthine oxidase electron accepting centers, Oprian and Coon[103] reported a correlation between a "predicted" electronic distribution between the two flavins at various NADPH concentrations and that from actual spectral data.

3. Multiphasic Reduction of P-450

Gigon et al.[118] first reported the biphasic nature of microsomal P-450 reduction. Subsequently, confirmation of the kinetics, which occur by two first order reactions, has been forthcoming in both microsomal and reconstituted systems.[113,119-128] Gigon et al.[118] could not attribute the phases to differences in relative reduction rates of P-450 and P-448 isozymes, but did note both an effect of contaminating oxygen on the second phase, and a difference in the rate of P-450 reduction with the substrate bound versus the unbound form. Gillette[129] proposed that the biphasic nature might reflect a variable reductase/P-450 ratio in different microsomes and/or a rate limitation by protein diffusion in the viscous phospholipid environment.

More current explanations have focused on topologically separated P-450 populations,[119,128] reductase-specific attributes such as redox states,[126] and spin equilibrium between low- and high-spin P-450 forms.[114,121,122]

Peterson et al.[119] studied the early biphasic reduction of rat liver P-450 in microsomes at temperatures ranging from 4 to 37°C. A break in the Arrhenius plot attributable to a membrane phase transition occurred at 18°C for the slow but not the fast reduction phase. The fraction reduced by the fast phase increased from 0.4 at 5°C to 0.7 at 25°C, with no additional change at higher temperatures. From these and other data, they proposed a "cluster" model, whereby 8 to 12 P-450 molecules surround one reductase molecule for the rapid reduction phase (calculated from the fraction reduced in the fast phase and an estimated 20/1 ratio of P-450/reductase[98]), and an energy dependent replacement by a satellite P-450 occurs during the slow reduction phase. Peterson and Prough[18] more recently noted that this "cluster" is likely segregation of the P-450 into two kinetically separable populations caused by microheterogeneity of the microsomal lipid environment.

In a reconstituted system, Kominami et al.[128] attributed the fast phase of bovine adrenal microsomal P-450_{C21} reduction to electron transfer within the reductase-P450 complex, and the slow phase to electron transfer by a random collision between the two proteins.

Higher concentrations of NADPH increased the relative fraction of fast phase for P-$450LM_4$ reduction in a reconstituted system,[126] indicating that biphasic reduction might be related to properties inherent to the reductase, such as redox state, and not P-450. However, the 3- and 4-electron reduced states were kinetically indistinguishable as donors for P-$450LM_4$.[126] Three phases ($k_1 = 1\ s^{-1}$, $k_2 = 0.2\ s^{-1}$, $k_3 = {\sim}10^{-2}\ s^{-1}$) with preincubation and two phases (k_1 and k_2) without preincubation of P-$450LM_4$ and reductase were observed. Because kinetics for studies without preincubation were not second order, Oprian et al.[126] proposed the occurrence of a rapid bimolecular reaction followed by a slow intramolecular transformation prior to electron transfer.

Backes and Reker-Backes[112] have recently studied the effect of the reductase redox states on the initial rate of reduction for cytochrome P-$450LM_2$ reconstituted lipid vesicles. Using the technique of Oprian and Coon[103] for predicting the individual redox states at various NADPH concentration, the authors calculated rate constants for the various possible electron donating redox states, assuming electron transfer required the $FMNH_2$ form. Because early reduction kinetics was studied, and the formation of one- and three-electron containing species involves a slow multiphasic process where electronic exchange occurs intermolecularly, it was not unexpected that the best theoretical vs. experimental match occurred by assuming that only the FAD·$FMNH_2$ and $FADH_2$·$FMNH_2$ redox states were responsible for P-450 reduction. The rate constants in the absence of the substrate benzphetamine were 0.74 s^{-1} and 6.5 s^{-1} for the 2- and 4-electron containing species, respectively, and 2 s^{-1} and 17.5 s^{-1} for the 2- and 4-electron species in the presence of benzphetamine. The results indicated that under these conditions, the 4-electron containing species was the major reductant of cytochrome P-450. However, because the 3-electron species is known to reduce P-450,[126] the authors concluded that its possible importance *in vivo*, particularly with slowly reducing isoforms, could not be overlooked. Interestingly, the second phase of the reduction of P-450 reductase, as determined by Oprian and Coon,[103] was slower than the reduction of substrate-bound P-450LM_2 (5.4 s^{-1} vs. 17 s^{-1} at 10:1 NADPH/reductase), indicating that P-450 may affect the rate of electron transfer from NADPH to the reductase.[112]

A third explanation for multiphasic reduction brings into play the equilibrium between low spin (S = 1/2) and high spin (S = 5/2) P-450 forms, which are in part controlled by the substrate,[130,131] P-450 reductase, cytochrome b_5, and acidic phospholipids.[88,89] The ratio of these spin states can control P-450 function. Type I substrate binding, for example, shifts the equilibrium towards the high spin form,[132-138] increasing both the affinity for the reductase and the rate of P-450 reduction.[90,139] Similarly, a shift towards high spin by the formation of the cytochrome P-450 reductase:P-450 or cytochrome b_5:P-450 complex can increase the binding rate and affinity for the P-450 substrate.[90] The association of benzphetamine and P-450LM_2, for example, lowers the K_d for P-450 reductase (and cytochrome b_5) and vice versa.[90,140] Because high spin P-450 is preferentially reduced,[139] the specificity for the P-450 isoform in microsomes may in part be dictated by the substrate.[89] Whether the multiphasic nature of P-450 reduction can be explained by these spin states has been debated.[18,113,121-123,126,128,141] Factors which increase the level of the high spin form may also increase that fraction reduced in the fast phase. Although workers did not see a correspondence between the spin state ratios and relative absorbance changes for each phase in P-450LM_4[126] and adrenal microsomal P-450_{C21},[128] others reported a strong relationship between these two in P-450 PB-B and RLM_5.[113,122,140] On the basis of their own work and temperature jump studiesby Tsong and Yang,[142] Backes et al.[121,122] proposed that the rate limiting step for low spin P-450 reduction was its conversion to the high spin form. Based on kinetic measurements, Rein et al.[137] and Blanck[125] concluded that this was not a rate limiting step. Moreover, Raman laser temperature jump studies by Ziegler et al.[141] indicated that the equilibrium for the low spin to high spin conversion is in the order of 10^{-6} s, much faster than P-450 reduction. Lastly, spin state temperature dependence did not correlate with that for the fast and slow phase.[18]

Whatever the cause of the multiphasic nature, one must explain how in microsomes 50 to 70% of the reduction occurs in the fast phase.[19] If a preformed complex is necessary, one would expect only 5% of the reduction to occur in the fast phase, as the ratio of P-450 to reductase is 20/1.[98] In reconstituted systems, Blanck et al.[143,144] reported that the fast phase was responsible for ony 15% of the reduction under physiological conditions.

G. RIGID VS. NONRIGID P-450:P-450 REDUCTASE COMPLEXATION

Although the NADH-cytochrome b_5 reductase and cytochrome b_5 appear to diffuse freely in microsomes,[145-146] there remains a question as to whether NADPH-cytochrome P-450 reductase and cytochrome P-450 associate in a stable complex[90,119,128,147-150] or by random collisions through lateral diffusion in the membrane.[97,120,151-157]

Yang[151,158] first proposed a nonrigid, freely mobile organization of P-450 and the reductase in microsomes. On the basis of temperature dependence studies, Duppel and Ullrich[153] concluded that lateral mobility was necessary for function. Hydroxylation kinetics[120,156] and competitive antibody studies[159] further supported a nonrigid association. Because all microsomal P-450 could ultimately be reduced, even when much of the reductase was inhibited by treatment with trypsin or mersalyl, Yang et al.,[151,155] concluded that every P-450 molecule must be accessible to each reductase molecule. Furthermore, exogenous P-450 could restore function after denaturation of endogenous P-450.[152,155] Dilution of vesicles containing the reductase and P-450 decreased catalytic activity, suggesting that a transient and not a stable complex was formed.[97]

Franklin and Estabrook[147] proposed that P-450 and the reductase were associated in a rigid or "cluster" arrangement, which was later supported by the effects of temperature on the reduction phases.[119] EPR investigations by Schwarz et al.[160,161] further supported the possibility of cluster-like structures. Evidence for a stable complex has been demonstrated by gel filtration,[90,128] and by absorption anisotropy decay studies.[150,162] Crosslinking the reductase with antibodies prevented the rotational diffusion of cytochrome P-450 when the proteins were reconstituted at a 1:1 ratio in lipid vesicles.[150]

Sedimentation studies by Wagner et al.[163,164] suggest that a complex is not necessary for catalytic activity. Cytochrome P-450LM$_2$ and P-450LM$_4$ dependent hydroxylase activities were highest at detergent concentrations where no stable P-450:P-450 reductase complex was evident.[163,164] At higher detergent concentrations where P-450 aggregates (6-7 P-450 molecules) disrupted into dimers and monomers, hydroxylase activity stopped, even though P-450 reductase could still reduce P-450.[164] Thus, P-450 aggregates, which appear to occur in groups of 6 (at least in reconstituted membranes),[165] may be necessary for activity.

H. INTERACTION OF CYTOCHROME b_5 WITH THE P-450 SYSTEM

Microsomal cytochrome b_5 can act as an *in vivo* electron acceptor for both NADH-dependent cytochrome b_5 reductase and the NADPH-dependent P-450 reductase. In microsomes, the first order rate constant for the reduction of cytochrome b_5 by P-450 reductase is 5,100 min^{-1},[61] more than adequate to supply all of the electrons the microsomal cytochrome b_5 dependent stearyl CoA desaturase system needs at maximum turnover.[61] In addition, cytochrome b_5 can reduce P-450,[166] and, depending on the P-450 isozyme and substrate, electrons from NADH can be shuttled through cytochrome b_5 to P-450. Cohen and Estabrook[167-169] showed that NADH could enhance certain monooxygenase reactions. Both the rate of the reaction and the stoichiometry of product formation were enhanced,[167-169] and antibodies to cytochrome b_5 inhibited this low K_m NADH-dependent synergism.[101,170-173] Imai and Sato[174] reported that only 20 and 40% of the total NADPH electrons that were consumed were utilized for N-N-dimethylaniline and benzphetamine metabolism (measured by formaldehyde production) in the absence of cytochrome b_5. Although the addition of cytochrome b_5 did not change the net rate of formaldehyde production, the efficiency of electron utilization rose to 100%. The addition of NADH increased both the rate of pyridine nucleotide oxidation and formaldehyde production. Whereas the first electron was thought to be derived from NADPH exclusively, the second could be donated from either pyridine nucleotide.[174]

Whether there is no requirement,[175,176] a partial requirement[175,177,178] or an absolute requirement[176,179-181] for cytochrome b_5 in a P-450 dependent monooxygenase system appears to be contingent on the substrate and P-450 isozyme.[176,182] Waxman and Walsh,[183] for example, isolated a phenobarbital inducible P-450 isozyme, PB-1, which increased four- to sevenfold in monooxygenase activity in the presence of cytochrome b_5. Two other isozymes, PB-4 and PB-5, showed less than 2-fold stimulation. Canova-Davis and Waskel[181] found cytochrome b_5 essential in O-demethylation of the anaesthetic methoxyflurane, while Vastis et al.[179] reported a requirement for cytochrome b_5 in P-450LM$_2$-dependent prostaglandin metabolism and a facilitation by cytochrome b_5 in P-450LM$_4$-dependent prostaglandin hydroxylation. The greatest change in cytochrome b_5-dependent rate increases, at least in reconstituted systems,[184] has been observed at lower P-450 reductase/P-450 ratios.

Spectral studies show direct association of cytochrome b_5 and P-450. From absorbance changes in the 390- and 434-nm peaks, Bonfils[166] estimated a K_d of 2.3 μM for P-450:cytochrome b_5, which decreased to K_d = 0.5μM in the presence of benzphetamine.

Conversely, cytochrome b_5 changed the dissociation constant for P-450:benzphetamine from 200 to 50 μM. Binding was not competitive and therefore was at two separate sites.[166] Magnetic Circular Dichroism studies by Bosterling and Trudell[87] were consistent with either a ternary cytochrome P-450 reductase: P-450:cytochrome b_5 complex, or a cytochrome b_5 dependent shift in association equilibria between three possible dimeric and monomeric forms. Either a direct electron transfer from cytochrome b_5 to P-450[156, 166, 174] or a nonelectron transfering, but cytochrome b_5-dependent improved coupling between cytochrome P-450 reductase and P-450,[156, 174, 184, 185] or both,[182] may ultimately prove to be responsible for the cytochrome b_5 dependent rate increase.

I. SPECIFIC AMINO ACID FUNCTIONS

A cysteine requirement for reductase activity was suggested by early work with sulfhydryl modifying reagents by Phillips and Langdon,[56] Williams and Kamin,[55] and Masters et al.[57] Inactivation by mersalyl,[147] 5,5′dithiobis (2-nitrobenzoate),[186] *p*-chloromercuribenzoate (PCMB),[187] and other sulfhydryl reagents[80,83,188,189] is inhibited to various degrees by NADP(H) or 2′-AMP[80,110,147,186,187,190] and cytochrome *c*.[189] *In vivo* loss of liver mixed function oxidase activity by Thallium may be caused by the inhibition of -SH groups in the reductase.[191]

Lumper et al.[192] concluded that inactivation via modification of an essential cysteine in the pork liver reductase was due to inhibition of the attachment of NADP(H) by both steric hindrince and electrostatic interactions of the introduced groups with the NADPH binding site. The $K_{m,NADPH}$, but not the V_{max}, was affected by the modification, the effect of which was contigent on the size and charge of the modifying group. Modification of 2 sulfhydryl groups with PCMB was reported by Nisimoto and Shibata[187,188] to yield a paradoxical activation of rabbit liver reductase activity. Approximately 80% of the activity remained after 2 more -SH groups were modified, but catalytic function ceased after modification of the fifth.

An easily accessible sulfhydryl group may influence FMN binding (required for reduction of DCIP and cytochrome *c*), since more accessible groups appear in the FMN depleted enzyme.[188] Ferricyanide reductase activity (in the FMN depleted apoenzyme), which does not require the FMN cofactor, correlated with a critical sulfhydryl that was reactive only with excess PCMB.[188] This treatment (excess PCMB) caused a loss of FAD fluorescence quenching by the reductase when the fifth and sixth -SH group bound to PCMB, and caused a change in FAD redox properties that were attributed to a steric distortion between the flavin and the protein. Circular dichroism spectra indicated that FAD-protein interactions were decreased by PCMB.[188] Furthermore, the sedimentation coefficient changed from 4.28 s in the untreated protein to 3.65 s with excess PCMB, indicating a change in quaternary structure.[188] Unpublished results by Haniu et al.[84] suggest that Cys-471 is uniquely protected in the presence of FAD.

The location of the critical residue in pork liver reductase has recently been identified as Cys-565 (Cys-566 and Cys-567 in rat and rabbit, respectively).[80,83,193,194] Its function is not known, but it may be involved in a charge transfer intermediate[195] and/or in the intramolecular conversion of $FADH_2$·FMNH to FADH·$FMNH_2$.[187]

Tryptophan may interact with the FMN flavin,[188] as FMN depletion increased fluorescence intensity and shifted the emission maximum from 332nm to 336nm, indicative of a more hydrophilic environment. Ozone oxidation of indole rings increased the emission yield and released bound FMN.[188] Nisimoto and Shibata[188] further reported that the loss of two tryptophans inactivate cytochrome *c* reductase activity, perhaps due to a direct loss in interaction with the cytochrome *c* or to a conformational change in the protein.

Other studies by Nisimoto et al.[196] utilizing photochemically induced dynamic polarization indicated that a tyrosine residue is shielded in the presence of FMN and exposed in the absence of FMN. They concluded that either FMN shields the residue directly, or a conformational change in the protein occurs on the binding of FMN. Porter and Kasper[197,198] have more recently suggested that Tyr-140 and Tyr-178 of the rat liver enzyme might interact with the FMN moiety.

Treatment with 2,4,6 trinitrobenzene sulfonate has caused complete loss of cytochrome *c* reductase activity and partial loss of DCIP and ferricyanide reductase activity.[190] However, because NADPH protected against this inactivation, lysine may be required to stabilize the charge on the 2′-phosphate moiety of the cofactor. A total of three lysine residues were modified, two of which were accessible in the presence of NADPH.

Arginine may also be necessary for reductase activity.[20,199] Modification studies suggest that it may act at the active site to bind the 2′-phosphate moiety of NADPH. Inano and Tamaoki[190] reported that either aspartate or glutamate may be necessary for the binding of the reductase to cytochrome *c*.

J. IONIC INTERACTIONS OF P-450 REDUCTASE WITH CYTOCHROMES P-450 AND b_5

Ionic interactions are important for the electron transfer complexes between cytochrome P-450 reductase, cytochrome P-450, and cytochrome b_5.[140,200-202] Charge interactions appear to be: 1, cytochrome b_5 "−":"+" cytochrome P-450; 2, cytochrome P-450 reductase "−":"+" cytochrome P-450; 3, cytochrome b_5 "−":"+" cytochrome P-450 reductase; and 4, cytochrome P-450 reductase "−":"+" cytochrome *c*. Amino groups on cytochrome P-450 apparently complex with functionally conserved carboxyl groups on both cytochrome b_5 and cytochrome P-450 reductase. These P-450 reductase carboxyl groups likely similarly complex with conserved lysyl residues on cytochrome *c*.[203-205] Another complexation site, however, appears to interact with cytochrome b_5, as P-450 reductase cationic groups interact with carboxyl groups on this cytochrome.[202] Although methylamidation of 9 P-450 reductase carboxyl groups inhibited both the reduction of cytochrome *c* and the fast phase reduction of cytochrome P-450, the reduction of ferricyanide was not affected, indicating that electron transfer from NADPH may still proceed. Furthermore, the effects on the interaction with cytochrome P-450 are specifically related to functionality, as the V_{max} and not the K_m was perturbed.[200]

K. SEQUENCE HOMOLOGY

Recently, the reductases from rat,[197,198] rabbit,[194] and pig liver[84] have been sequenced. Rat and rabbit sequences were deduced from the nucleotide sequences of cDNA clones whereas that for the pig reductase was determined from sequence analysis of proteolytic fragments. Predicted molecular weights for the rat and rabbit reductases were 76,962[197] and 76,583[194] respectively, similar to that of 76,600[84] for the pig reductase calculated from the amino acid sequence (without FAD or FMN). The cloned sequences included an N-terminal methionine not present in the native pig protein, which had instead an acetylated glycine at its N-termius. Black and Coon[81] have shown that the mature rabbit protein is also acetylated at the N-terminus.

Functionally conserved regions of the reductase were inferred by analogy to proteins for which X-ray structural data is known. Porter and Kasper[198] assigned possible FMN phosphate binding functions to the hydroxyls of Ser-86, Thr-88 and Thr-90 and the main chain hydrogens Gln-87 and Ala-91 by homology to the FMN-containing Desulfovibrio vulgaris flavodoxin. They further speculated that Tyr-140 and Tyr-178 might, respectively, be the interior and exterior FMN-shielding residues, and that Glu-142, Asp-140 and Asp-147 could function in charge pairing with P-450. Residues 267-678 showed close similarity to the FAD containing ferredoxin-NADP$^+$ reductase, with the exception of a 120 amino acid spanning region of the P-450 reductase. Both enzymes were compared to glutathione reductase, the 3-dimensional structure of which is well characterized.[198] Three domains were proposed: an FAD-PP$_i$ binding segment between residues 292 and 326, an NADP-PP$_i$ binding region from residues 483 to 519 and an NADP-Ribose binding segment between residues

524 and 553.[198] Sequence identity between rat and rabbit[194] and rat and pig[84] was 91 and 90%, respectively. Three conserved glycyl residues at Gly-533, Gly-535 and Gly-537 in the rabbit (subtract 1 and 2, respectively, for the residue number in rat and pig) were apparent which might function in binding the AMP moiety of FAD or NADPH.[194] Approximately thirty residues after these glycyls was the earlier mentioned critical cysteine (Cys-565, pig; Cys-566, rat; Cys-567, rabbit) that was protected by NADPH from S-carboxymethylation.[80,83] Interestingly, a cysteine involved in NADH-binding in cytochrome b_5 reductase follows a glycine-rich region by 20 residues.[194] Sequence similarities between the rat reductase and both bacterial flavodoxins and ferredoxin-$NADP^+$ reductase led Porter and Kasper to propose that P-450 reductase arose through an ancestral fusion of these two genes.

V. SUMMARY

Some properties and mechanistic features of four flavoprotein reductases have been briefly summarized. All of the enzymes share the ability to utilize electrons from NADPH, whereas two of them can function equally well with NADH.

Two yeast cytochrome *c* reductases show marked differences in their pyridine nucleotide requirement, specific activities and flavin cofactor. One of the enzymes exhibits many features characteristic of Old Yellow Enzyme, and may function physiologically to balance NADH/NADPH ratios during conditions of catabolite repression. The other has an exceptionally high turnover with cytochrome *c* and may vary well be a cytochrome P-450 reductase with unusual flavin and molecular size characteristics.

Quinone reductase is unique from the others in that it can function only as a two electron mediator, the physiological importance of which may be to propitiate the toxic effects of quinones. Kinetic studies on this reductase indicate that complex interactions of a rare type between the protein and its substrates are occurring.[54]

Because cytochrome P-450 reductase possesses two flavin domains, a membrane anchor region, and sites for interaction with NADPH, cytochromes *c*, b_5, and P-450, it has become a rich reservoir for physicochemical studies. The presence of the reductase in microsomes has added to the complexity, but the many interactions occurring therein have inspired interesting and creative mechanistic proposals.

REFERENCES

1. **Djavadi, H. S., Moradi, M., and Djavadi-Ohaniance, L.,** Direct oxidation of NADPH by submitochondrial particles from *Saccharomyces cerevisiae, Eur. J. Biochem.,* 107, 501, 1980.
2. **Rydström, J., Montelius, J., Bäckström, D., and Ernster, L.,** The mechanism of oxidation of reduced nicotinamide dinucleotide phosphate by submitochondrial particles from beef heart, *Biochim. Biophys. Acta,* 501, 370, 1978.
3. **Martius, C.,** Quinone reductases, in *The Enzymes,* Vol. 7, 2nd ed., Boyer, P. D., Lardy, H., Myrbäck, K., Eds., Academic Press, New York, 1963, chap. 2.
4. **Ernster, L.,** DT diaphorase, in *Methods in Enzymology,* Vol. 10, Estabrook, R. W., and Pullman, M. E., Eds., Academic Press, New York, 1967, 309.
5. **Tryon, E., Cress, M. C., Hamada, M., and Kuby, S. A.,** Studies on NADPH-cytochrome c reductase. I. Isolation and several properties of crystalline enzyme from ale yeast, *Arch. Biochem. Biophys.,* 197, 104, 1979.
6. **Tryon, E., and Kuby, S. A.,** Studies on NADPH-cytochrome c reductase. II. Steady-state kinetic properties of the crystalline enzyme from ale yeast, *Enzyme,* 31, 197, 1984.
7. **Johnson, M. S., and Kuby, S. A.,** Studies on NADH(NADPH)-cytochrome c reductase (FMN-containing) from yeast. Isolation and physicochemical properties of the enzyme from top-fermenting ale yeast, *J. Biol. Chem.,* 260, 12341, 1985.
8. **Johnson, M. S., and Kuby, S. A.,** Studies on NADH(NADPH)-cytochrome *c* reductase (FMN-containing) from yeast: Steady state kinetic properties of the flavoenzyme from top-fermenting ale yeast, *Arch. Biochem. Biophys.,* 245, 271, 1986.

9. **Martius, C., Ganser, R., Viviani, A.,** The enzymatic reduction of K-vitamins incorporated in the membrane of liposomes, *FEBS Lett.*, 59, 13, 1975.
10. **Fasco, M. J., and Principe, L. M.,** Vitamin K_1 hydroquinone formation catalyzed by DT-diaphorase, *Biochem. Biophys. Res. Commun.*, 104, 187, 1982.
11. **Wallin, R., Gebhardt, O., Prydz, H.,** NAD(P)H dehydrogenase and its role in the vitamin K (2-methyl-3-phytyl-1,4-naphthaquinone)-dependent carboxylation reaction, *Biochem. J.*, 169, 95, 1978.
12. **Benson, A. M., Hunkler, M. J., and Talaley, P.,** Increase of NAD(P)H:quinone reductase by dietary antioxidants: Possible role in protection against carcinogenesis and toxicity, *Proc. Natl. Acad. Sci. U.S.A.*, 77, 5210, 1980.
13. **Thor, H., Smith, M. T., Hartzell, P., Bellomo, G., Jewell, S. A., and Orrenius, S.,** The metabolism of menadione (2-methyl-1,4-naphthoquinone) by isolated hepatocytes, *J. Biol. Chem.*, 257, 12419, 1982.
14. **Lind, C., Hochstein, P., and Ernster, L.,** DT-diaphorase as a quinone reductase: a cellular control device against semiquinone and superoxide radical formation, *Arch. Biochem. Biophys.*, 216, 178, 1982.
15. **Talalay, P., Balzinger, R. P., Benson, A. M., Bvedring, E. and Cha Y.-N.,** Biochemical studies on the mechanism by which dietary antioxidants suppress mutagenic activity, *Adv. Enzyme Regul.* 17, 23, 1979.
16. **White, R. E., and Coon,M. J.,** Oxygen activation by cytochrome P-450, *Ann. Rev. Biochem.*, 49, 315, 1980.
17. **Lu, A. Y. H., and Coon, M. J.,** Role of hemoprotein P-450 in fatty acid ω-hydroxylation in a soluble enxyme system from liver microsomes, *J. Biol. Chem.*, 243, 1331, 1968.
18. **Peterson, J. A., and Prough, R. A.,** Cytochrome P-450 reductase and cytochrome b_5 in cytochrome P-450 catalysis, in *Cytochrome P-450: Structure, Mechanism, and Biochemistry,* Ortiz de Montellano, P. R., Ed., Plenum Press, New York, 1986. chap. 4.
19. **Blanck, J., Smettan, G., and Greschner, S.,** The cytochrome P-450 reaction mechanism — kinetic aspects, in *Cytochrome P-450,* Ruckpaul, K., and Rein, H., Eds., Akademie-Verlag, Berlin, 1984, chap. 3.
20. **Jänig, G.-R., and Pfeil, D.,** Structure-function relationships of the essential components of the liver microsomal monooxygenase system, in *Cytochrome P-450,* Ruckpaul, K., and Rein H., Eds., Akademie-Verlag, Berlin, 1984, chap. 2.
21. **Nakamura, T., Yoshimura, J., and Ogura, Y.,** Action mechanism of the Old Yellow Enzyme, *J. Biochem.*, 57, 554, 1965.
22. **Matthews, R. G., and Massey, V.,** Isolation of Old Yellow Enzyme in free and complexed forms, *J. Biol. Chem.*, 244, 1779, 1969.
23. **Youngman, R. J., and Elstner, E. F.,** Generation of active-oxygen species by simple enzymatic redox systems, in *Handbook of Methods for Oxygen Radical Research,* Greenwald, R. A., Ed., CRC Press, Boca Raton, FL, 1985, 105.
24. **Ainsworth, P. J., Ball, A. J. S., and Tustanoff, E. R.,** Cyanide-resistant respiration in yeast. I. Isolation of a cyanide-insensitive NAD(P)H oxidoreductase, *Arch. Biochem. Biophys.*, 202, 172, 1980.
25. **Ainsworth, P. J., Ball, A. J. S., and Tustanoff, E. R.,** Cyanide-resistant respiration of yeast. II. Characterization of a cyanide-insensitive NAD(P)H oxidoreductase, *Arch. Biochem. Biophys.*, 202, 187, 1980.
26. **Åkeson, Å., Ehrenberg, A., and Theorell, H.,** Old Yellow Enzyme, in *The Enzymes,* Vol. 7, 2nd ed., Boyer, P. D., Lardy, H., and Myrbäck, K., Eds., Academic Press, New York, 1963, chap. 19.
27. **Theorell, H., and Nygaard, A. P.,** Kinetics and equilibria in flavoprotein systems. I. A fluorescence recorder and its application to a study of the dissociation of the Old Yellow Enzyme and its resynthesis from riboflavin phosphate and protein, *Acta Chem. Scand.*, 8, 877, 1954.
28. **Nygaard, A. P. and Theorell, H.,** On the chemical nature of the FMN-binding groups in the Old Yellow Enzyme, *Acta Chem. Scand.*, 8, 1489, 1954.
29. **Abramovitz, A. S. and Massey, V.,** Purification of intact old Yellow Enzyme using an affinity matrix for the sole chromatographic step, *J. Biol. Chem.*, 251, 5321, 1976.
30. **Palmer, G. and Massey, V.,** Mechanisms of flavoprotein catalysis, in *Biological Oxidations,* Singer, T. P., Ed., Wiley-Interscience, New York, 1968, 263.
31. **Cleland, W. W.,** The kinetics of enzyme-catalyzed reactions with two or more substrates or products. I. Nomenclature and rate equations, *Biochim. biophys. Acta,* 67, 104, 1963.
32. **Hodgson, E. K. and Fridovich, I.,** The mechanism of the activity-dependent luminescence of xanthine oxidase, *Arch. Biochem. Biophys.*, 172, 202, 1976.
33. **Ernster, L. and Navazio, F.,** Soluble diaphorase in animal tissues, *Acta Chem. Scand.*, 12, 595, 1958.
34. **Ernster, L., Danielson, L., and Ljunggren, M.,** DT diaphorase. I. Purification from the soluble fraction of rat-liver cytoplasm, and properties, *Biochim. Biophys. Acta,* 58, 171, 1962.
35. **Martius, C. and Strufe, R.,** Pyllochinonreduktase-vorlaufige mitteilung, *Biochem. Z.*, 326, 24, 1954.
36. **Märki, F. and Martius, C.,** Vitamin K reductase. Preparation and properties, *Biochem. Z.*, 333, 111, 1960.

37. **Märki, F. and Martius, C.,** Vitamin K reductases from cattle and rat liver, *Biochem. Z.*, 334, 293, 1961.
38. **Guiditta, A. and Strecker, H. J.,** Purification and some properties of a brain diaphorase, *Biochem. Biophys. Res. Commun.*, 2, 159, 1960.
39. **Spitsberg, V. L. and Coscia, C. J.,** Quinone reductases of higher plants, *Eur. J. Biochem.*, 127, 67, 1982.
40. **Wallin, R. and Hutson, S.,** Vitamin K-dependent carboxylation: evidence that at least two microsomal dehydrogenases reduce vitamin K_1 to support carboxylation, *J. Biol. Chem.*, 257, 1583, 1982.
41. **Suttie, J. W. and Jackson, C.,** Prothrombin structure, activation, and biosynthesis, *Physiol. Rev.*, 57, 1, 1977.
42. **Wallin, R. and Suttie, J. W.,** Vitamin K-dependent carboxylation and vitamin K epoxidation, *Biochem. J.*, 194, 983, 1981.
43. **Wallin, R. and Little, C.,** NAD(P)H dehydrogenase from rabbit and rat liver: Purification and some properties, *Int. J. Biochem.*, 16, 1099, 1984.
44. **Wallin, R.,** Some properties of NAD(P)H dehydrogenase from rat liver, *Biochem. J.*, 181, 127, 1979.
45. **Hosoda, S., Nakamura, W., and Hayashi, K.,** Properties and reaction mechanism of DT diaphorase from rat liver, *J. Biol. Chem.*, 249, 6416, 1974.
46. **Rase, B., Bartfai, T., and Ernster, L.,** Purification of DT-diaphorase by affinity chromatography, *Arch. Biochem. Biophys.*, 172, 380, 1976.
47. **Bayney, R. M., Rodkey, J. A., Bennett, C. D., Lu, A. Y. J., and Pickett, C. B.,** Rat liver NAD(P)H:Quinone reductase. Nucleotide sequence analysis of a quinone reductase cDNA clone and prediction of the amino acid sequence of the corresponding protein, *J. Biol. Chem.*, 262, 572, 1987.
48. **Höjeberg, B., Blomberg, K., Stenberg, S., and Lind, C.,** Biospecific absorption of hepatic DT-diaphorase on immobilized dicoumarol, *Arch. Biochem. Biophys.*, 207, 205, 1981.
49. **Edmonson, D. E. and Tollin, G.,** Circular dichroism studies of the flavin chromophore and of the relation between redox properties and flavin environment in oxidases and dehydrogenases, *Biochemistry*, 10, 113, 1971.
50. **Kenney, W. C., Edmonson, D. E., and Singer, T. P.,** The covalently-bound flavin of chromatium cytochrome c_{552}, *Eur. J. Biochem.*, 48, 449, 1974.
51. **Hall, J. M., Lind, C., Golvano, M. P., Rase, B., and Ernster, L.,** DT diaphorase - Reaction mechanism and metabolite function, in Structure and *Function of Oxidation-Reduction Enzymes*, Åkeson, Å., and Ehrenberg, A., Eds., Pergamon Press, Elmsford, NY, 1972, 433.
52. **Lind, C., Rase, B., Ernster, L., Townsend, M. G., and Martin, A. D.,** Strain differences in liver DT diaphorase activities, *FEBS Lett.*, 37, 147, 1973.
53. **Hollander, P. M. and Ernster, L.,** Studies on the reaction mechanism of DT diaphorase, *Arch. Biochem. Biophys.*, 169, 560, 1975.
54. **Hollander, P. M., Bartfai, T., and Gatt, S.,** Studies on the reaction mechanism of DT diaphorase, *Arch. Biochem. Biophys.*, 169, 568, 1975.
55. **Williams, C. H., Jr. and Kamin, H.,** Microsomal triphosphpyridine nucleotide cytochrome *c* reductase of liver, *J. Biol. Chem.*, 237, 587, 1962.
56. **Phillips, A. H. and Langdon, R. G.,** Hepatic triphosphopyridine nucleotide-cytochrome *c* reductase: isolation, characterization, and kinetic studies, *J. Biol. Chem.*, 237, 2652, 1962.
57. **Masters, B. S. S., Kamin, H., Gibson, Q. H., and Williams, C. H., Jr.,** Studies on the mechanism of microsomal triphosphopyridine nucleotide-cytochrome *c* reductase, *J. Biol. Chem.*, 240, 921, 1965.
58. **Masters, B. S. S., Bilimoria, M. H., Kamin, H., and Gibson, Q. H.,** The mechanism of 1- and 2-electron transfers catalyzed by reduced triphosphopyridine nucleotide cytochrome *c* reductase, *J. Biol. Chem.*, 240, 4081, 1965.
59. *Enzyme Nomenclature: Recommendations (1984) of the Nomenclature Committee of the International Union of Biochemistry*, preparation by Edwin C. Webb, Academic Press, New York, 1984.
60. **Schacter, B. A., Nelson, E. B., Marver, H. S., and Masters, B. S. S.,** Immunochemical evidence for an association of heme oxygenase with the microsomal electron transport system, *J. Biol. Chem.*, 247, 3601, 1972.
61. **Enoch, H. G. and Strittmatter, P.,** Cytochrome b_5 reduction by NADPH-cytochrome P-450 reductase, *J. Biol. Chem.*, 254, 8976, 1979.
62. **Ilan, Z., Ilan, R., and Cinti, D. L.,** Evidence for a new physiological role of hepatic NADPH:ferricytochrome (P-450) oxidoreductase, *J. Biol. Chem.*, 256, 10066, 1981.
63. **Haas, E., Horecker, B. L., and Hogness, T. R.,** The enzymatic reduction of cytochrome *c*. Cytochrome *c* reductase, *J. Biol. Chem.*, 136, 747, 1940.
64. **Horecker, B. L.,** Triphosphopyridine nucleotide-cytochrome *c* reductase in liver, *J. Biol. Chem.*, 183, 593, 1950.
65. **Brodie, B. B., Axelrod, J., Cooper, J. R., Gaudette, L., LaDu, B. N., Mitoma, C., and Udenfriend, S.,** Detoxication of drugs and other foreign compunds by liver microsomes, *Science*, 121, 603, 1955.

66. **Hayaishi, O., Katagiri, M., and Rothberg, S.,** Mechanism of the pyrocatechase reaction, *J. Am. Chem. Soc.,* 77, 5450, 1955.
67. **Mason, H. S., Fowlks, W. L., and Peterson, E.,** Oxygen transfer and electron transport by the phenolase complex, *J. Am. Chem. Soc.,* 77, 2914, 1955.
68. **Gum, J. R. and Strobel, H. W.,** Purified NADPH cytochrome P-450 reductase: interaction with hepatic microsomes and phospholipid vesicles, *J. Biol. Chem.,* 254, 4177, 1979.
69. **Black, S. D., French, J. S., Williams, C. H., Jr., and Coon, M. J.,** Role of a hydrophobic polypeptide in the N-terminal region of NADPH-cytochrome P-450 reductase in complex formation with P-450_{LM}, *Biochem. Biophys. Res. Commun.,* 91, 1528, 1979.
70. **Raftell, M. and Orrenius, S.,** Preparation of antisera against cytochrome b_5 and NADPH-cytochrome *c* reductase from rat liver microsomes, *Biochim. Biophys. Acta,* 233, 358, 1971.
71. **Masters, B. S. S., Baron, J., Taylor, W. E., Isaacson, E. L., and LoSpalluto, J.,** Immunochemical studies on electron transport chains involving cytochrome P-450. I. Effects of antibodies to pig liver microsomal reduced triphosphopyridine nucleotide-cytochrome *c* reductase and the non-heme iron protein from bovine adrenocortical mitochrondria, *J. Biol. Chem.,* 246, 4143, 1971.
72. **Oshino, N. and Omura, T.,** Immunochemical evidence for the participation of cytochrome b_5 in microsomal stearyl-CoA desaturation reaction, *Arch. Biochem. Biophys.,* 157, 395, 1973.
73. **Yasukochi, Y. and Masters, B. S. S.,** Some properties of a detergent-solubilized NADPH-cytochrome *c* (cytochrome P-450) reductase purified by biospecific affinity chromatography, *J. Biol. Chem.,* 251, 5337, 1976.
74. **Ardies, C. M., Lasker, J. M., Bloswick, B. P., and Lieber, C. S.,** Purificaiton of NADPH: cytochrome *c* (cytochrome P-450) reductase from hamster liver by detergent extraction and affinity chromatography, *Anal. Biochem.,* 162, 39, 1987.
75. **Dignam, J. D. and Strobel, H. W.,** NADPH-cytochrome P-450 reductase from rat liver: purification by affinity chromatography and characterization, *Biochemistry,* 16, 1116, 1977.
76. **Iyanagi, T. and Mason, H. W.,** Some properties of hepatic reduced nicotinamide adenine dinucleotide phosphate-cytochrome *c* reductase, *Biochemistry,* 12, 2297, 1973.
77. **Vermilion, J. L. and Coon, M. J.,** Highly purified detergent-solubilized NADPH-cytochrome P-450 reductase from phenobarbital-induced rat liver microsomes, *Biochem. Biophys. Res. Commun.,* 60, 1315, 1974.
78. **Iyanagi, T., Anan, F. K., Imai, Y., and Mason, H. S.,** Studies on the microsomal mixed function oxidase system: Redox properties of detergent-solubilized NADPH-cytochrome P-450 reductase, *Biochemistry,* 17, 2224, 1978.
79. **Vermilion, J. L. and Coon, M. J.,** Identification of the high and low potential flavins of liver microsomal NADPH-cytochrome P-450 reductase, *J. Biol. Chem.,* 253, 8812, 1978.
80. **Haniu, M., Iyanagi, T., Legesse, K., and Shively, J. E.,** Structural analysis of NADPH-cytochrome P-450 reductase from porcine hepatic mitochrondria: sequences of proteolytic fragments, cysteine-containing peptides, and a NADPH-protected cysteine peptide, *J. Biol. Chem.,* 259, 13703, 1984.
81. **Black, S. D. and Coon, M. J.,** Structural features of liver microsomal NADPH-cytochrome P-450 reductase, *J. Biol. Chem.,* 257, 5929, 1982.
82. **Gum, J. R. and Strobel, H. W.,** Isolation of the membrane binding peptide of NADPH-cytochrome P-450 reductase, *J. Biol. Chem.,* 256, 7478, 1981.
83. **Vogel, F., Kaiser, C., Witt, I., and Lumper, L.,** NADPH-cytochrome P-450 reductase (pig liver). Studies on the sequence of the cyanogen bromide peptides from the catalytic domain and on the reactivity of the thiol groups, *Biol. Chem. Hoppe-Seyler,* 366, 577, 1985.
84. **Haniu, M., Iyanagi, T. Miller, P., Lee, T. D., and Shively, J. E.,** Complete amino acid sequence of NADPH-cytochrome P-450 reductase from porcine hepatic microsomes, *Biochemistry,* 25, 7906, 1986.
85. **Lu, A. Y. H., Junk, K. W., Coon, M. J.,** Resolution of the cytochrome P-450-containing ω-hydroxylation system into three components, *J. Biol. Chem.,* 244, 3714, 1969.
86. **Strobel, H. W., Lu, A. Y. H., Heidema, J., and Coon, M. J.,** Phosphatidylcholine requirement in the enzymatic reduction of hemoprotein P-450 and in fatty acid, hydrocarbon, and drug hydroxylation, *J. Biol. Chem.,* 245, 4851, 1970.
87. **Bösterling, B. and Trudell, J. R.,** Association of cytochrome b_5 and cytochrome P-450 reductase with cytochrome P-450 in the membrane of reconstituted vesicles, *J. Biol. Chem.,* 257, 4783, 1982.
88. **Ingelman-Sundberg, M. J., Haaparanta, T., and Rydström, J.,** Membrane charge as effector of cytochrome P-450LM_2 catalyzed reactions in reconstituted liposomes, *Biochemistry,* 20, 4100, 1981.
89. **Ingelman-Sundberg, M.,** Cytochrome P-450 organization and membrane interaction, in *Cytochrome P-450: Structure, Mechanism and Biochemistry,* Ortiz de Montellano, P. R., Ed., Plenum Press, New York, 1986, chap. 5.
90. **French, J. S., Guengerich, F. P., and Coon, M. J.,** Interactions of cytochrome P-450, NADPH-cytochrome P-450 reductase, phospholipid, and substrate in reconstituted liver microsomal enzyme system, *J. Biol. Chem.,* 255, 4112, 1980.

91. **Magdalou, J., Thirion, C., Ballard, M., and Siest, G.,** Conformational studies of NADPH cytochrome P-450 reductase by circular dichroism: interaction with phospholipids, *Int. J. Biochem.,* 1103, 1985.
92. **Knapp, J. A., Digman, J. D., and Strobel, H. W.,** NADPH-cytochrome P-450 reductase. Circular dichroism and physical studies, *J. Biol. Chem.,* 252, 437, 1977.
93. **French, J. S. and Coon, M. J.,** Properties of NADPH-cytochrome P-450 reductase purification from rabbit liver microsomes, *Arch. Biochem. Biophys.,* 195, 565, 1979.
94. **Lu, A. Y. H., Levin, W., and Kuntzman, R.,** Reconstituted liver microsomal enzyme that hydroxylates drugs, other foreign compounds and endogenous sutstrates. VII. Stimulation of benzphetamine N-demethylation by lipid and detergent, *Biochem. Biophys. Res. Commun.,* 60, 266, 1974.
95. **Dean, W. L. and Gray, R. D.,** Relationship between state of aggregation and catalytic activity for cytochrome P-450LM_2 and NADPH-cytochrome P-450 reductase, *J. Biol. Chem.,* 257, 14679, 1982.
96. **Müller-Enoch, D., Churchill, P., Fleischer, S., and Guengerich, F. P.,** Interaction of liver microsomal cytochrome P-450 and NADPH-cytochrome P-450 and NADPH-cytochrome P-450 reductase in the presence and absence of lipid, *J. Biol. Chem.,* 259, 8174, 1984.
97. **Miwa, G. T. and Lu, A. H.,** The association of cytochrome P-450 and NADPH-cytochrome P-450 reductase in phospholipid membranes, *Arch. Biochem. Biophys.,* 234, 161, 1984.
98. **Estabrook, R. W., Franklin, M. R., Cohen, B., Shigamatzu, A., and Hildebrandt, A. G.,** Influence of hepatic microsomal mixed function oxidation reactions on cellular metabolic control, *Metabolism,* 20, 187, 1971.
99. **Black, S. and Coon, M. J.,** Comparative structures of P-450 cytochromes, in *Cytochrome P-450: Structure, Mechanism and Biochemistry,* Ortiz de Montellano, P. R., Ed., Plenum Press, New York, 1986, chap. 6.
100. **Welton, A. F. and Aust, S. D.,** Multiplicity of cytochrome P-450 hemoproteins in rat liver microsomes, *Biochem. Biophys. Res. Commun.,* 56, 898, 1974.
101. **Prough, R. A. and Burke, M. D.,** The role of NADPH-cytochrome P-450 reductase in microsomal hydroxylation reactions, *Arch. Biochem. Biophys.,* 170, 160, 1975.
102. **Vermilion, J. L. and Coon, M. J.,** Purified liver microsomal NADPH-cytochrome P-450 reductase: spectral characterization of oxidation-reduction states, *J. Biol. Chem.,* 253, 2694, 1978.
103. **Oprian, D. D. and Coon, M. J.,** Oxidation-reduction states of FMN and FAD in NADPH-cytochrome P-450 reductase during reduction by NADPH, *J. Biol. Chem.,* 257, 8935, 1982.
104. **Yasukochi, Y., Peterson, J. A., Masters, B. S. S.,** NADPH-cytochrome *c* (P-450) reductase: Spectrophotometric and stopped flow kinetic studies on the formation of reduced flavoprotein intermediates, *J. Biol. Chem.,* 254, 7097, 1979.
105. **Iyanagi, T. and Mason, H. S.,** Some properties of hepatic reduced nicotinamide adenine dinucleotide phosphate-cytochrome *c* reductase, *Biochemistry,* 12, 2297, 1974.
106. **Sugiyama, T., Nisimoto, Y., Mason, H. S., Loehr, T. M.,** Flavins of NADPH-cytochrome P-450 reductase: Evidence for structural alteration of flavins in their one-electron-reduced semiquinone states from resonance Raman spectroscopy, *Biochemistry,* 24, 3012, 1985.
107. **Mayer, R. T. and Durrant, J. L.,** Preparation of homogeneous NADPH cytochrome *c* (P-450) reductase from house flies using affinity chromatography techniques, *J. Biol. Chem.,* 254, 756, 1979.
108. **Kobayashi, S. and Rikans, L. E.,** Kinetic properties of guinea pig liver microsomal NADPH-cytochrome P-450 reductase, *Comp. Biochem. Physiol.,* 77B, 313, 1984.
109. **Strobel, H. W. and Dignam, J. D.,** Purification and properties of NADPH-cytochrome P-450 reductase, in *Methods in Enzymology, Vol.* 52, Part C, Fleischer, S., and Packer, L., Eds., Academic Press, New York, 1978, 89.
110. **Hiwatashi, A. and Ichikawa, Y.,** Physicochemical properties of reduced nicotinamide adenine dinucleotide phosphate-cytochrome P-450 reductase from bovine adrenocortical microsomes, *Biochim. Biophys. Acta,* 44, 1979.
111. **Aoyama, Y., Yoshida, Y., Kubota, S., Kumaoka, H., and Furumichi, A.,** NADPH-cytochrome P-450 reductase of yeast microsomes, *Arch. Biochem. Biophys.,* 185, 362, 1978.
112. **Backes, W. L., and Reker-Backes, C. E.,** The effect of NADPH concentration on the reduction of cytochrome P-450 LM_2, *J. Biol. Chem.,* 263, 247, 1988.
113. **Backes, W. L., Tamburini, P. P., Jansson, I., Gibson, G. G., Sligar, S. G., and Schenkman, J. B.,** Kinetics of cytochrome P-450 reduction: evidence for faster reduction of the high-spin ferric state, *Biochemistry,* 24, 5130, 1985.
114. **Tamburini, P. P., Gibson, G. G., Backes, W. L., Sligar, S. G., and Schenkman, J. B.,** Reduction kinetics of purified rat liver cytochrome P-450. Evidence for a sequential reaction mechanism dependent on the hemoprotein spin state, *Biochemistry,* 23, 4526, 1984.
115. **Crankshaw, D. L., Hetnarski, K., and Wilkinson, C. F.,** Purification and characterization of NADPH-cytochrome *c* reductase from the midgut of the southern armyworm *(Spodoptera eridania), Biochem. J.,* 181, 593, 1979.
116. **Kamin, H., Masters, B. S. S., and Gibson, Q. H.,** NADPH-cytochrome *c* oxidoreductase in *Flavins and Flavoproteins,* Slater, E. C., Ed., Elsevier, New York, 1966, 306.

117. **Olson, J. S., Ballou, D. P., Palmer, G., and Massey, V.**, The mechanism of action of xanthine oxidase, *J. Biol. Chem.*, 249, 4363, 1974.
118. **Gigon, P. L., Gram, T. E., and Gillette, J. R.**, Studies on the rate of reduction of hepatic microsomal cytochrome P-450 by reduced nicotinamide adenine dinucleotide phosphate: Effect of drug substrate, *Mol. Pharmacol.*, 5, 109, 1969.
119. **Peterson, J. A., Ebel, R. E., O'Keeffe, D. H., Matsubara, T., and Estabrook, R. W.**, Temperature dependence of cytochrome P-450 reduction, *J. Biol. Chem.*, 251, 4010, 1976.
120. **Taniguchi, H., Imai, Y., Iyanagi, T., and Sato, R.**, Interaction between NADPH-cytochrome P-450 reductase and cytochrome P-450 in the membrane of phosphatidylcholine vesicles, *Biochim. Biophys. Acta*, 550, 341, 1979.
121. **Backes, W. L., Sligar, S. G., and Schenkman, J. B.**, Cytochrome P-450 reduction exhibits burst kinetics, *Biochem. Biophys. Res. Commun.*, 97, 860, 1980.
122. **Backes, W. L., Sligar, S. G., and Schenkman, J. B.**, Kinetics of hepatic cytochrome P-450 reduction: Correlation with spin state of the ferric heme, *Biochemistry*, 21, 1324, 1982.
123. **Ruf, H. H.**, Reduction kinetics of microsomal P-450: A re-examination, in *Biochemistry, Biophysics and Regulation of Cytochrome P-450*, Gustafson, J. A., Carlstedt-Duke, J., Mode, A., and Rafter, J., Eds., Elsevier/North-Holland, New York, 1984, 355.
124. **Blanck, J., Rein, H., Sommer, M., Ristau, O., Smetten, G., and Ruckpaul, K.**, Correlations between spin equilibrium shift, reduction rate, and N-demethylation activity in liver microsomal cytochrome P-450 and a series of benzphetamine analogues as substrates, *Biochem. Pharmacol.*, 32, 1683, 1983.
125. **Blanck, J., Smetten, G., Ristau, O., Ingelman-Sundberg, M., and Ruckpaul, K.**, Mechanism of rate control of the NADPH-dependent reduction of cytochrome P-450 by lipids on reconstituted phospholipid vesicles, *Eur. J. Biochem.*, 144, 509, 1984.
126. **Oprian, D. D., Vatsis, K. P., and Coon, M. J.**, Kinetics of reduction of cytochrome P-450LM$_4$ in a reconstituted liver microsomal enzyme system, *J. Biol. Chem.*, 254, 8895, 1979.
127. **Kominami, S. and Takemori, S.**, Effect of spin state on reduction of cytochrome P-450 (P-450$_{c21}$) from bovine adrenocortical microsomes, *Biochim. Biophys. Acta*, 709, 147, 1982.
128. **Kominami, S., Hara, H., Ogishima, T., and Takemori, S.**, Interaction between cytochrome P-450 (P450$_{c21}$) and NADPH-cytochrome P-450 reductase from adrenocortical microsomes in a reconstituted system, *J. Biol. Chem.*, 259, 2991, 1984.
129. **Gillette, J. R.**, Effects of various inducers on electron transport system associated with drug metabolism by liver microsomes, *Metabolism*, 20, 215, 1971.
130. **Schenkman, J. B., Remmer, H., and Estabrook, R. W.**, Spectral studies of drug interaction with hepatic microsomal cytochrome, *Mol. Pharmacol.*, 3, 113, 1967.
131. **Jansson, I., Gibson, G. G., Sligar, S. G., Cinti, D. L., and Schenkman, J. B.**, Influence of substrates of hepatic mixed function oxidases on spin equilibrium of cytochrome P-450, in *Microsomes, Drug Oxidations and Chemical Carcinogenesis*, Vol. 1, Coon, M. J., Conney, A. H., Estabrook, R. W., Gelboin, H. V., Gillette, J. R. and O'Brien, P. J., Eds., Academic Press, New York, 1980, 139.
132. **Hildebrandt, A., Remmer, H., and Estabrook, R. W.**, Cytochrome P-450 of liver microsomes — one pigment or many, *Biochem. Biophys. Res. Commun.*, 30, 607, 1968.
133. **Jefcoate, C. R. E., Gaylor, J. L., and Calabrese, R. L.**, Ligand interactions with cytochrome P-450. I. Binding of primary amines, *Biochemistry*, 8, 3455, 1969.
134. **Waterman, M. R., Ullrich, V., and Estabrook, R. W.**, Effect of substrate on the spin state of cytochrome P-450 in hepatic microsomes, *Arch. Biochem. Biophys.*, 155, 355, 1973.
135. **Sligar, S. G.**, Coupling of spin, substrate and redox equilibria in cytochrome P-450, *Biochemistry*, 15, 5399, 1976.
136. **Lange, R., Bonfils, C., and Debey, P.**, The low-spin - high-spin transition of camphor-bound cytochrome P-450. Effects of medium and temperature on equilibrium data, *Eur. J. Biochem.*, 79, 623, 1977.
137. **Rein, H., Ristau, O., Friedrich, J., Jänig, G.-R., and Ruckpaul, K.**, Evidence for the existence of a high-spin-low spin equilibrium in liver microsomal cytochrome P-450, *FEBS Lett.*, 75, 19, 1977.
138. **Werringloer, J., Kawano, S., and Estabrook, R. W.**, Spin-state transitions of liver microsomal P-450., *Acta Biol. Med. Germ.*, 38, 163, 1979.
139. **Matsubara, T., Baron, J., Peterson, L. L., and Peterson, J. A.**, NADPH-cytochrome P-450 reductase, *Arch. Biochem. Biophys.*, 172, 463, 1976.
140. **Tamburini, P. P., White, R. E., and Schenkman, J. B.**, Chemical characterization of protein-protein interactions between cytochrome P-450 and cytochrome b_5, J. Biol. Chem., 260, 4007, 1985.
141. **Ziegler, M., Blanck, J., and Ruckpaul, K.**, Spin equilibrium relaxation kinetics of cytochrome P-450 LM$_2$, *FEBS Lett.*, 150, 219, 1984.
142. **Tsong, T.-Y. and Yang, C. S.**, Rapid conformational changes of cytochrome P-450: effect of dimyristoyl lecitihin, *Proc. Natl. Acad. Sci. U.S.A.*, 75, 5955, 1978.
143. **Blanck, J., Behlke, J., Jänig, G.-R., Pfeil, D., Ruckpaul, K.**, Kinetics of elementary steps in the cytochrome P-450 reaction sequence. III. NADPH reduction of cytochrome P-450LM at different integration levels, *Acta Biol. Med. Germ.*, 38, 11, 1979.

144. **Blanck, J., Rohde, K., and Ruckpaul, K.,** Kinetics of elementary steps in the cytochrome P-450 reaction sequence. IV. Mechanism of the NADPH reduction reaction of cytochrome P-450LM, *Acta Biol. Med. Germ,* 38, 23, 1979.
145. **Rogers, M. J. and Strittmatter, P.,** Evidence for random distribution and translational movement of cytochrome b_5 in endoplasmic reticulum, *J. Biol. Chem.,* 249, 895, 1974.
146. **Rogers, M. J. and Strittmatter, P.,** The binding of reduced nicotinamide adenine dinucleotide-cytochrome b_5 reductase to hepatic microsomes, *J. Biol. Chem.,* 249, 5565, 1974.
147. **Franklin, M. R., Estabrook, R. W.,** On the inhibitory action of mersalyl on microsomal drug oxidation: a rigid organization of the electron transport chain, *Arch. Biochem. Biophys.,* 143, 318, 1971.
148. **Stier, A. and Sackmann, E.,** Spin labels as enzyme substrates, *Biochim. Biophys. Acta,* 311, 400, 1973.
149. **Stier, A.,** Lipid structure and drug metabolizing enzymes, *Biochem. Pharmacol.,* 25, 109, 1976.
150. **Gut, J., Richter, C., Cherry, R. J., Winterhalter, K. H., and Kawato, S.,** Rotation of cytochrome P-450: complex formation of cytochrome P-450 with NADPH-cytochrome P-450 reductase in liposomes demonstrated by combining protein rotation with antibody-induced cross-linking, *J. Biol. Chem.,* 258, 8588, 1983.
151. **Yang, C. S.,** The association between cytochrome P-450 and NADPH-cytochrome P-450 and NADPH-cytochrome P-450 reductase in microsomal membranes, *FEBS Lett.,* 54, 61, 1975.
152. **Yang, C. S.,** Interactions between solubilized cytochrome P-450 and hepatic microsomes. Characterizations of the binding and the enhanced catalytic activities, *J. Biol. Chem.,* 252, 293, 1977.
153. **Duppel, W. and Ullrich, V.,** Membrane effects on drug monooxygenation activity in hepatic microsomes, *Biochim. Biophys. Acta,* 426, 399, 1976.
154. **Yang, C. S., Strickhart, F. S., and Kicha, L. P.,** The effect of temperature on monooxygenase reactions in the microsomal membrane, *Biochim. Biophys. Acta,* 465, 362, 1977.
155. **Yang, C. S., Strickhart, F. S., and Kicha, L. P.,** Interaction between NADPH-cytochrome P-450 reductase and hepatic microsomes, *Biochim. Biophys. Acta,* 1978.
156. **Ingelmann-Sundberg, M., and Johansson, I.,** Cytochrome b_5 as electron donor to rabbit liver cytochrome P-450LM$_2$ in reconstitued phospholipid systems, *Biochem. Biophys. Res. Commun.,* 97, 582, 1980.
157. **Archakov, A. I., Borodin, E. A., Davydov, D. R., Karyakin, A. I., and Borovyagin, V. L.,** Random distribution of NADPH-specific flavoprotein and cytochrome P-450 in liver microsomes, *Biochem. Biophys. Res. Commun.,* 109, 832, 1982.
158. **Yang, C. S. and Strickhart, F. S.,** Interactions between solublized cytochrome P-450 and hepatic microsomes, *J. Biol. Chem.,* 250, 7968, 1975.
159. **Omura, T., Noshiro, M. and Harada, N.,** Distribution of electron transfer components on the surface of microsomal membrane, in *Microsomes, Drug Oxidations, and Chemical Carcinogenesis,* Vol. 1, Coon, M. J., Conney, A. H., Estabrook, R. W., Gelboin, H. V., Gillette, J. R., and O'Brien, P. J., Eds., Academic Press, New York, 1980, 445.
160. **Schwarz, D., Pirrwitz, J., Ruckpaul, K.,** Rotational diffusion of cytochrome P-450 in the microsomal membrane — evidence for a clusterlike organization from saturation transfer electron paramagnetic resonance spectroscopy, *Arch. Biochem. Biophys.,* 216, 322, 1982.
161. **Schwarz, D., Pirrwitz, J., Coon, M. J., and Ruckpaul, K.,** Mobility and clusterlike organization of liposomal cytochrome P-450 LM2 — saturation transfer electron-paramagnetic studies, *Acta Biol. Med. Germ.* 41, 425, 1982.
162. **Gut, J., Richter, C., Cherry, R. J., Winterhalter, K. H., and Kawato, S.,** Rotation of cytochrome P-450. II. Specific interactions of cytochrome P-450 with NADPH-cytochrome P-450 reductase in phospholipid vesicles, *J. Biol. Chem.,* 257, 7030, 1982.
163. **Wagner, S., Dean, W. L., and Gray, R. D.,** Effect of a zwitterionic detergent on the state of aggregation and catalytic activity of cytochrome P-450LM$_2$ and NADPH-cytochrome P-450 reductase, *J. Biol. Chem.,* 259, 2390, 1984.
164. **Wagner, S. L., Dean, W. L., and Gray, R. D.,** Zwitterionic detergent mediated interaction of purified cytochrome P-450LM$_4$ from 5,6-benzoflavine-treated rabbits with NADPH-cytochrome P-450 reductase, *Biochemistry,* 26, 2343, 1987.
165. **Greinert, R., Finch, S. A. E., and Stier, A.,** Cytochrome P-450 rotamers control mixed-function oxygenation in reconstituted membranes. Rotational diffusion studied by delayed fluorescence depolarization, *Xenobiotica,* 12, 717, 1982.
166. **Bonfils, C., Balny, C., and Maurel, P.,** Direct evidence for electron transfer from ferrous cytochrome b_5 to the oxyferrous intermediate of liver microsomal cytochrome P-450LM$_2$, *J. Biol. Chem.,* 256, 9457, 1981.
167. **Cohen, B. S. and Estabrook, R. W.,** Microsomal electron transport reactions. I. Interaction of reduced triphosphopyridine nucleotide during the oxidative demethylation of aminopyrine and cytochrome b_5 reduction, *Arch. Biochem. Biophys.,* 143, 37, 1971.
168. **Cohen, B. S., and Estabrook, R. W.,** Microsomal electron transport reactions. II. The use of reduced triphosphopyridine nucleotide and/or reduced diphosphopyridine nucleotide for the oxidative N-demethylation of aminopyrine and other drug substrates, *Arch. Biochem. Biophys.,* 143, 46, 1971.

169. **Cohen, B. S. and Estabrook, R. W.,** Microsomal electron transport reactions. III. Cooperative interactions between reduced diphosphopyridine nucleotide and reduced triphosphopyridine nucleotide linked reactions, *Arch. Biochem. biophys.*, 143, 54, 1971.
170. **Sasame, H. A., Thorgeirsson, S. S., Mitchell, J. R., and Gillette, J. R.,** The possible involvement of cytochrome b_5 in the oxidation of lauric acid by microsomes from kidney cortex and liver of rats, *Life Sci.*, 14, 35, 1974.
171. **Mannering, G. J., Kuwahara, S., and Omura, T.,** Immunochemical evidence for the participation of cytochrome b_5 in the NADPH synergism of NADPH-dependent mono-oxidase system of hepatic microsomes, *Biochem. Biophys. Res. Commun.*, 57, 476, 1974.
172. **Noshiro, M. and Omura, T.,** Immunochemical study on the electron pathway from NADH to cytochrome P-450 of liver microsomes, *J. Biochem.*, 83, 61, 1978.
173. **Oshino, N.,** Cytochrome b_5 and its physiological significance, in *Hepatic Cytochrome P-450 Monooxygenase System,* Schenkman, J. B. and Kupfer, D., Eds., Pergamon Press, Elmsford, NY, 1982, chap. 16.
174. **Imai, Y. and Sato, R.,** The roles of cytochrome b_5 in a reconstituted N-demethylase system containing cytochrome P-450, *Biochem. Biophys. Res. Commun.*, 75, 420, 1977.
175. **Lu, A. Y. H., Levin, W., Selander, H., and Jerina, D. M.,** Liver microsomal electron transport systems. III. Involvement of cytochrome b_5 in the NADPH-supported cytochrome P-450 dependent hydroxylation of chlorobenzene, *Biochem. Biophys. Res. Commun.*, 61, 1348, 1974.
176. **Kuwahara, S. and Omura, T.,** Different requirement for cytochrome b_5 in NADPH-supported O-deethylation of p-nitrophenetole catalyzed by two types of microsomal cytochrome P-450, *Biochem. Biophys. Res. Commun.*, 96, 1562, 1980.
177. **Imai, Y.,** The roles of cytochrome b_5 in reconstituted monooxygenase systems containing various forms of hepatic microsomal cytochrome P-450, *J. Biochem.*, 89, 351, 1981.
178. **Chiang, J. Y. L.,** Interaction of purified microsomal cytochrome P-450 with cytochrome b_5, *Arch. Biochem. Biophys.*, 211, 662, 1981.
179. **Vatsis, K. P., Theoharides, A. D., Kupfer, D., and Coon, M. J.,** Hydroxylation of prostaglandins by inducible isozymes of rabbit liver microsomal cytochrome P-450, *J. Biol. Chem.*, 257, 11221, 1982.
180. **Sugiyama, T., Miki, N., and Yamano, T.,** NADH- and NADPH-dependent reconstituted p-nitroanisole O-demethylation system containing cytochrome P-450 with high affinity for cytochrome b_5, *J. Biochem.*, 87, 1457, 1980.
181. **Canova-Davis, E. and Waskell, L.,** The identification of the heat-stable microsomal protein required for methoxyflurane metabolism as cytochrome b_5, *J. Biol. Chem.*, 259, 2541, 1984.
182. **Morgan, E. T. and Coon, M. J.,** Effects of cytochrome b_5 on cytochrome P-450-catalyzed reactions: studies with manganese-substituted cytochrome $b)_5$ *Drug Metab. Dispos.* 2, 358, 1984.
183. **Waxman, D. J. and Walsh, C.,** Cytochrome P-450 isozyme 1 from phenobarbital induced rat liver: Purification, characterization and interactions with metyrapone and cytochrome b_5, *Biochemistry,* 22, 4846, 1983.
184. **Bösterling, B., Trudell, J. R., Trevor, A. J., and Bendix, M.,** Lipid-protein interactions as determinants of activation or inhibition by cytochrome b_5 of cytochrome P-450-mediated oxidations, *J. Biol. Chem.*, 257, 4375, 1982.
185. **Hlavica, P.,** On the function of cytochrome b_5 in the cytochrome P-450 dependent oxygenase system, *Arch. Biochem. Biophys.*, 228, 600, 1984.
186. **Lazar, T., Ehrig, H., and Lumper, C.,** The functional role of thiol groups in protease-solubilized NADPH-cytochrome *c* reductase from pork-liver microsomes, *Eur. J. Biochem.*, 76, 365, 1977.
187. **Nisimoto, Y. and Shibata, Y.,** Location of functinal -SH groups in NADPH-cytochrome P-450 reductase from rabbit liver microsomes, *Biochim. Biophys. Acta,* 662, 291, 1981.
188. **Nisimoto, Y. and Shibata, Y.,** Studies on FAD- and FMN-binding domains in NADPH-cytochrome P-450 reductase from rabbit liver microsomes, *J. Biol. Chem.*, 257, 12532, 1982.
189. **Lee, J. J. and Kaminsky, L. S.,** Fluorescence probing of the function-specific cysteines of rat microsomal NADPH-cytochrome P-450 reductase, *Biochem. Biophys. Res. Commun.*, 134, 393, 1986.
190. **Inano, H. and Tamaoki, B.-I.,** Purification of NADPH-cytochrome P-450 reductase from microsomal fraction of rat testis, and its chemical modification by tetranitromethane, *J. Steroid Biochem.*, 25, 21, 1986.
191. **Woods, J. S., Fowler, B. A., and Eaton, D. L.,** Studies on the mechanism of thallium-mediated inhibition of hepatic mixed function oxidase activity. Correlation with inhibition of NADPH-cytochrome *c* (P-450) reductase, *Biochem. Pharmacol.*, 33, 571, 1984.
192. **Lumper, L., Busch, F., Dzelic', S. Henning, J. and Lazar, T.,** Studies on the cosubstrate site of protease solubilized NADPH-cytochrome P-450 reductase, *Int. J. Pept. Protein Res.*, 16, 83, 1980.
193. **Vogel, F. and Lumper, L.,** Fluorescence labelling of NADPH-cytochrome P-450 reductase with the monobromomethyl derivative of Syn-9,10-dioxabimane, *Biochem. J.*, 215, 159, 1983.
194. **Katagiri, M., Murakami, H., Yabusaki, Y., Sugiyama, T., Okamoto, Y., and Ohkawa, H.,** Molecular cloning and sequence analysis of full-length cDNA for rabbit liver NADPH-cytochrome P-450 reductase mRNA, *J. Biochem. (Tokyo),* 100, 945, 1986.

195. **Kamin, H. and Lambeth, J. D.,** Role of flavins and iron-sulfur proteins in the reduction of cytochrome P-450, in *Flavins and Flavoproteins,* Massey, V., Williams, C. H., Jr., Eds., Elsevier/North-Holland, New York, 1981, 655.
196. **Nisimoto, Y., Hayashi, F., Akutsu, H., Kyogoku, Y., and Shibata, Y.,** Photochemically induced dynamic nuclear polarization study on microsomal NADPH-cytochrome P-450 reductase, *J. Biol. Chem.,* 259, 2480, 1984.
197. **Porter, T. D., and Kasper, C. B.,** Coding nucleotide sequence of rat NADPH-cytochrome P-450 oxidoreductase cDNA and identification of flavin-binding domains, *Proc. Natl. Acad. Sci. U.S.A.,* 82, 973, 1985.
198. **Porter, T. D. and Kasper, C. B.,** NADPH-cytochrome P-450 oxidoreductase: Flavin mononucleotide and flavin adenine nucleotide domains evolved from different flavoproteins, *Biochemistry,* 25, 1682, 1986.
199. **Ebel, R. E.,** Inactivation of NADPH-cytochrome P-450(*c*) reductase by 2,3- butanedione and phenylglyoxal. Evidence for an essential arginine, in *Microsomes, Drug Oxidations and Chemical Carcinogenesis,* Vol. 1, Coon, M. J., Conney, A. H., Estabrook, R. W., Gelboin, H. V., Gillette, J. R., O'Brien, P. J., Academic Press, New York, 1980, 315.
200. **Tamburini, P. P. and Schenkman, J. B.,** Differences in the mechanism of functional interaction between NADPH-cytochrome P-450 reductase and its redox partners, *Mol. Pharmacol.,* 30, 178, 1986.
201. **Tamburini, P. P. and Schenkman, J. B.,** Mechanism of interaction between cytochromes P-450LM$_5$ and b_5: evidence for an electrostatic mechanism involving cytochrome b_5 heme propionate groups, *Arch. Biochem. Biophys.,* 245, 512, 1986.
202. **Tamburini, P. P., MacFarquhar, S., and Schenkman, J. B.,** Evidence of binary complex formations between cytochrome P-450, cytochrome b_5, and NADPH-cytochrome P-450 reductase of hepatic microsomes, *Biochem. Biophys. Res. Commun.,* 134, 519, 1986.
203. **Ng, S., Smith, M. B., Smith, H. T., and Millet, F.,** Effect of modification of individual cytochrome *c* lysines on the reaction with cytochrome b_5, *Biochemistry,* 16, 4975, 1977.
204. **Rieder, R. and Bosshard, H. R.,** Comparison of the binding sites on cytochrome *c* for cytochrome *c* oxidase, cytochrome bc_1 and cytochrome c_1. Differential acetylation of lysyl residues in free and complexed cytochrome *c, J. Biol. Chem.,* 255, 4732, 1980.
205. **Bechtold, R. and Bossard, H. R.,** Structure of an electron transfer complex. II. Chemical modification of carboxyl groups of cytochrome *c* peroxidase in presence and absence of cytochrome *c, J. Biol. Chem.,* 260, 5191, 1985.

Section VI
Activated Oxygen Reactions

I hope I do not err if I assume as many kinds of air as experiment reveals to me. Carl Wilhelm Scheele (1742—1786), the Swedish chemist noted for a number of discoveries which included the gases, ammonia, chlorine and oxygen. (See Alembic Club Reprint No. 8).

Chapter 13

DIOXYGENASES AND MONOOXYGENASES

Shozo Yamamoto and Yuzuru Ishimura

TABLE OF CONTENTS

I. INTRODUCTION

Oxygenases are the enzymes which incorporate one or both atoms of molecular oxygen (O_2) directly into organic substrates. The enzymatic oxygenation is classified into two categories, monooxygenation and dioxygenation, depending on the number of oxygen atom to be incorporated. Dioxygenases and monooxygenases are listed in the categories of oxidoreductases "acting on single donors with incorporation of molecular oxygen (1.13)" and "acting on paired donors with incorporation of molecular oxygen (1.14)", respectively, in the "Enzyme Nomenclature 1984" recommended by the Nomenclature Committee of the International Union of Biochemistry. Oxygenase reactions are encountered in the metabolic pathways of organic compounds including aromatic xenobiotics, sugars, amino acids, fatty acids and steroids.

Oxygenases are distinguishable from oxidases which also consume molecular oxygen as a substrate. The oxidase is a dehydrogenase (Equation 1) with molecular oxygen as a hydrogen acceptor. The molecular oxygen is reduced to H_2O_2 or H_2O merely as a hydrogen acceptor (Equations 2 and 2′), and oxygen atoms derived from the molecular oxygen do not appear in the product. In contrast, dioxygenase inserts both atoms of molecular oxygen into the substrate(s). They are classified further into two subgroups, i.e., single and two substrate reactions, depending on the number of substrates which receive oxygen atoms (Equations 3 and 3′). Most dioxygenases do not require exogenous hydrogen donors such as NAD(P)H and ascorbate. One of a few exceptions in this respect is the anthranilate dioxygenase from a *Pseudomaonas* species.

$$\text{Dehydrogenase} \quad SH_2 + X = S + XH_2 \tag{1}$$

$$\text{Oxidase} \quad SH_2 + O_2 = S + H_2O_2 \tag{2}$$

$$2SH_2 + O_2 = 2S + 2H_2O \tag{2'}$$

$$\text{Dioxygenase} \quad S + O_2 = SO_2 \tag{3}$$

$$S + S' + O_2 = SO + S'O \tag{3'}$$

$$\text{Monooxygenase} \quad S + O_2 + XH_2 = SO + X + H_2O \tag{4}$$

$$SH_2 + O_2 = SO + H_2O \tag{4'}$$

Monooxygenase incorporates only one atom of molecular oxygen, and the other oxygen atom is reduced to water at the expense of an equimolar amount of hydrogen donor. Some monooxygenases require exogenous hydrogen donors such as NADH, NADPH and ascorbic acid (Equation 4), while other monooxygenases utilize hydrogen atoms of the substrate itself (Equation 4′). Monooxygenases are also referred to as mixed function oxidase, indicating that the enzyme also has a nature of oxidase.

Most oxygenases require iron, copper, heme, flavin or pteridine as an essential coenzyme. Molecular oxygen is usually in a triplet ground state, while most of organic substrates which accept oxygen are largely in a singlet state. For this reason, either molecular oxygen or organic substrate or both are assumed to be activated prior to the incorporation of oxygen atoms. Thus, a conceptual term "oxygen activation" often appears in the discussion of oxygenase mechanisms, and the role of a coenzyme, particularly of heme and nonheme iron and flavin has been extensively analyzed and discussed in relation to the oxygen activation mechanisms.

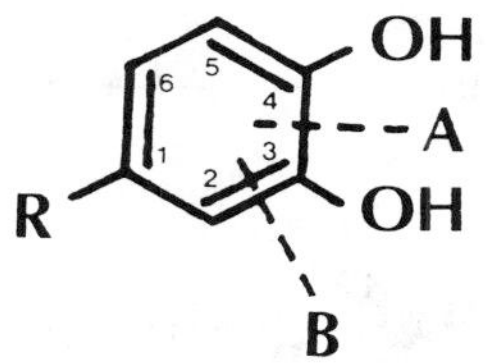

FIGURE 1. Mode of ring fission in catechol dioxygenase reactions.

The following are references for the oxygenase in general: general monographs[1,2] proceedings for symposia,[3,4] and review articles.[5-8]

II. CATECHOL DIOXYGENASES

Catechol dioxygenases are historically most famous oxygenases, the discovery of which by Hayaishi et al.[9] together with that of monooxygenase by Mason et al.[10] in 1955 had opened the field of oxygenase research. By the insertion of both atoms of molecular oxygen, catechol dioxygenase cleaves the aromatic ring of catechol and its derivatives to give the corresponding muconic acid derivative or its semialdehyde depending on the site of dioxygen insertion (A and B in Figure 1). The enzyme which cleaves the aromatic ring between the two carbon atoms bearing hydroxyl groups are called as an intradiol type, whereas those cleave the ring at the adjacent position to the two carbon atoms are called as an extradiol type dioxygenase.[7,11-13] Physiological role of these dioxygenases are to cleave the aromatic ring which is otherwise resistent to enzymatic and nonenzymatic degradations.

Several catechol dioxygenases were obtained in crystalline forms or in homogeneous states, and their molecular properties have been extensively studied.[7,11-13] Interestingly, the differences in intradiol and extradiol types cover not only the mode of the ring fission but also the state of their essential cofactor, iron in the enzyme. Intradiol type enzyme contains exclusively trivalent iron (Fe^{3+}), whereas the extradiol type contains only divalent iron (Fe^{2+}). For this reason, the former type of enzymes show a deep red color, while the latter enzymes are colorless. Most of these catechol dioxygenases so far studied are of bacterial origin and are inducible by their substrates or related substances.

A. INTRADIOL-TYPE CATECHOL DIOXYGENASE

Pyrocatechase (catechol 1,2-dioxygenase) and protocatechuate 3,4-dioxygenase, which are typical examples of the intradiol type dioxygenase, catalyze the reactions shown below.

Protochatechuate 3,4-dioxygenase has been obtained in a crystalline form from cells of *Psuedomonas aeruginosa*.[14] The enzyme contains 8-g atoms of ferric iron (Fe^{3+}) per mole of enzyme based on the molecular weight of 700,000, which consists of 32 subunits with an $(\alpha_2\beta_2)_8$ structure.[15] Molecular weights of the α and β subunits are 22,500 and 25,000, respectively. Protocatechuate 3,4-dioxygenases with similar properties have also been purified from other bacteria and were used to study the reaction mechanism[16,17] On the other hand, a homogenous preparation of pyrocatechase, another intradiol type of catechol dioxygenase, purified from *Pseudomonas arvilla* strain C-1, contains a single ferric iron per mole of enzyme based on the molecular weight of 63,000.[15] It has also an $\alpha\beta$ structure, in which molecular weights of the α and β subunits are 30,000 and 32,000, respectively.[18] Thus, intradiol type of catechol dioxygenases so far studied has the same fundamental protein structure composed of α and β type subunits, although the number of subunits and iron content are quite variable.

The structure of the iron center in these intradiol catechol dioxygenases has been extensively studied by the aid of various spectroscopic techniques.[16,17,19] The results have shown that the iron in the resting enzymes is in a high spin ferric state and coordinates to

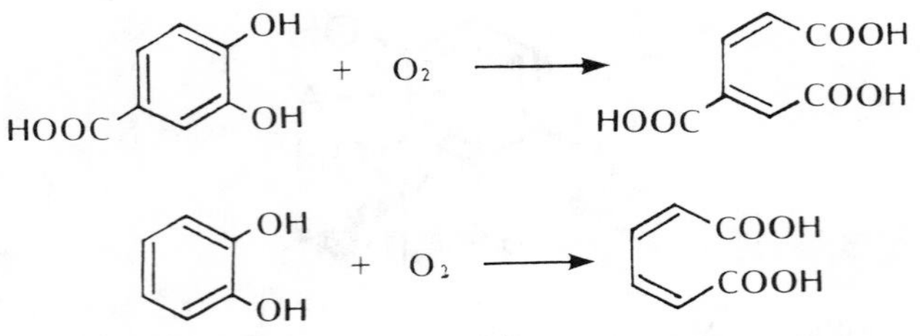

FIGURE 2. Protocatechuate 3,4-dioxygenase and pyrocatechase reactions.

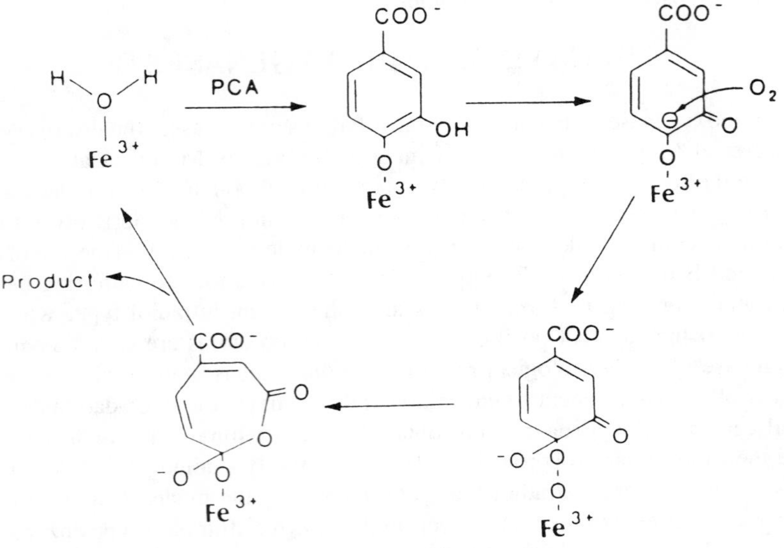

FIGURE 3. A proposed reaction mechanism for the intradiol catechol dioxygenase reaction.

two tyrosinates, two histidines and a water. These enzymes show optical and EPR spectra similar to those of rubredoxin but can be distinguished from the latter by their sensitivities to the changes in electronic environments caused by the bindings of substrates, inhibitors and other ligands to the iron. In fact, the deep red brown color of the native enzymes turns to grayish blue together with a decrease in the EPR signal of $g = 4.3$ upon binding of the substrate. Also known is that the substrate binds directly to the iron and displaces the water, while the tyrosinate and histidine ligations persist in both substrate and inhibitor complexes. The substrate ligated to the ferric iron, however, does not reduce it, and thus the iron in this type of catechol dioxygenase appears to remain in a ferric state throughout the catalytic cycle. Based on these results the following catalytic reaction mechanism has been proposed (Figure 3).[16]

In the above scheme, the substrate loses both of catecholate protons upon coordination to the iron and becomes susceptible to the reaction with oxygen to yield a peroxide intermediate, which then decomposes into the product. Thus, the proposed mechanism involves the activation of substrate rather than the oxygen activation. For more details, readers are referred to other articles such as references.[15-19]

B. EXTRADIOL-TYPE CATECHOL DIOXYGENASES

Metapyrocatechase (catechol 2,3-dioxygenase) and protocatechuate 4,5-dioxygenase are typical examples of extradiol dioxygenases, which catalyze the following reactions (Figure 4).

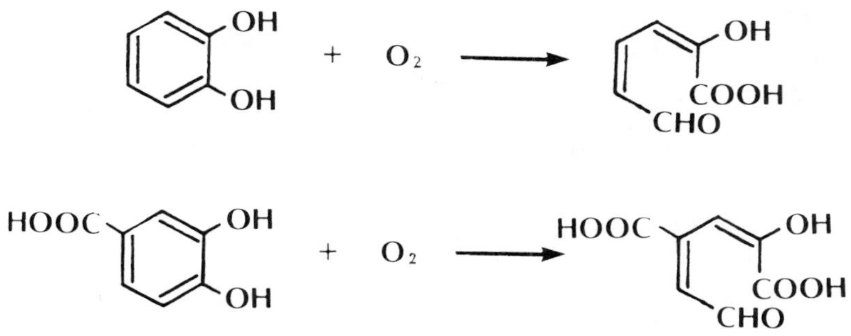

FIGURE 4. Metapyrocatechase and protocatechuate 4,5-dioxygenase reactions.

Metapyrocatechase from *Pseudomonas arvilla* is the first oxygenase which was obtained in a crystalline form.[20] The enzyme has a molecular weight of 140,000 with 4 identical subunits of 35,000 and contains 4 gram atoms of ferrous iron.[21] The primary structure of the enzyme has been reported.[8] Protocatechuate 4,5-dioxygenase from *Pseudomonas testosteroni*,[22] which also contains ferrous iron, has a molecular weight of 140,000 with the subunit strucure of $(\alpha\beta)_2$.[23] In contrast to intradiol-type enzymes, these extradiol-type oxygenases are colorless and exhibit no useful EPR signal to characterize the iron catalytic center. Only Mössbauer spectroscopy has been successfully used to identify the state of iron in native enzymes as a ferrous high spin form.[24] Recently, however, studies on NO complexes of extradiol catechol dioxygenase by Lipscomb and his coworkers[25,26] have indicated that at least 2 sites in iron ligation can be available for exogenous ligands and oxygen binding may also occur at the same site(s). It has been also suggested that one of the iron ligand in the resting enzyme is water which is displaced upon binding of the substrate.[26] Figure 5 is the reaction mechanism for the extradiol-type of catechol dioxygenase proposed by the same authors.[25,26] In support of this proposal, kinetic experiments have indicated that the reaction proceeded in an ordered Bi-Uni mechanism which in this respect is the same to that of intradiol type of dioxygenase.[27]

Similar mechanism has been expected for other extradiol-type catechol dioxygenases including 3,4-dihydroxyphenylacetate (homo-protocatechuate) 2,3-dioxygenases.[28] It is interesting to note, however, that a 3,4-dihydroxyphenylacetate 2,3-dioxygenase from *Bacillus brevis* contains manganese ion in place of iron in ordinary catechol dioxygenases.[29]

III. DIOXYGENASES-CONTAINING HEME

Dioxygenases which contain heme as the essential cofactor are relatively few. In fact only two groups of heme-containing dioxygenases have been found in literature; the tryptophan 2,3-dioxygenase family which includes L-tryptophan 2,3-dioxygenase, indoleamine 2,3-dioxygenase and possibly pyrrooxygenase, and the lipooxygenase family which includes prostaglandin cyclooxygenase (see next section) and *Fusarium* lipooxygenases.[30] We discuss here only the former group of heme-containing dioxygenases, because the role of heme in their reactions has been unambiguously established.

A. L-TRYPTOPHAN 2,3-DIOXYGENASE

The enzyme is also referred to as tryptophan pyrrolase and in systemic name as L-tryptophan: oxygen 2,3-oxidoreductase (decycling). As indicated in these names, it cleaves the pyrrole ring of L-tryptophan by the insertion of 2 atoms of oxygen from molecular oxygen (O_2) giving L-formylkynurenine as the reaction product[31] (Figure 6).

The enzyme has been purified from cells of *Pseudomonas acidovorans* (ATCC 11299b),[32] and from livers of rats and mice[33-35] which received simultaneous administrations of L-

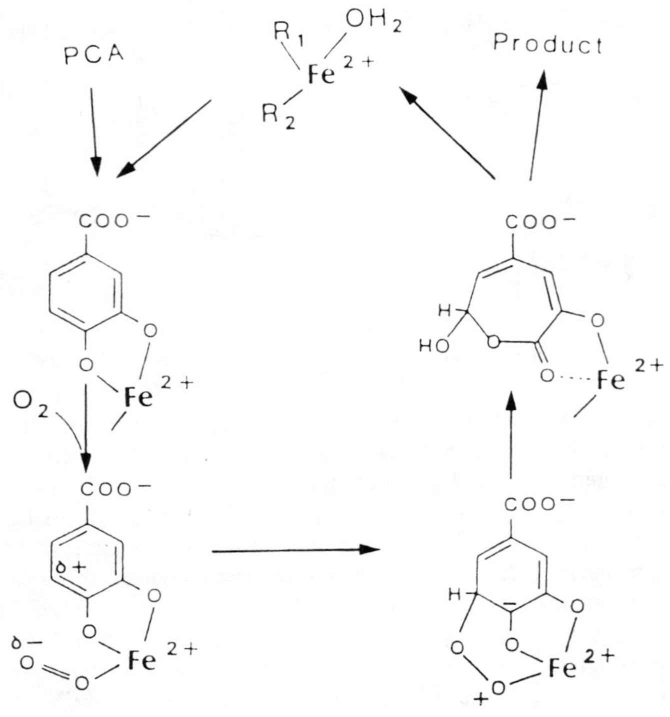

FIGURE 5. A proposed reaction mechanism for the extradiol catechol dioxygenase reaction.

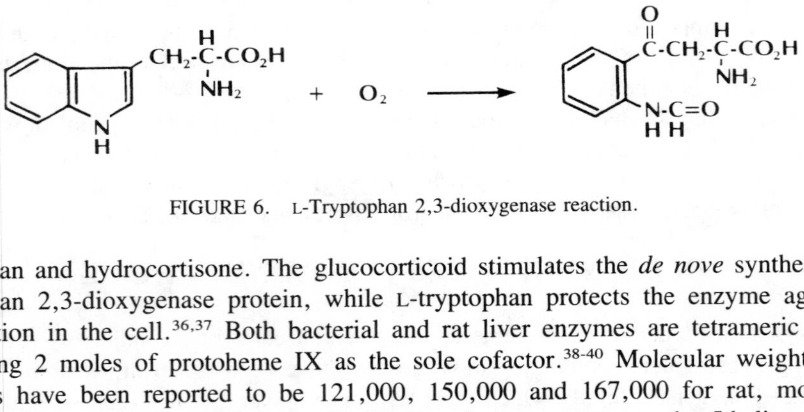

FIGURE 6. L-Tryptophan 2,3-dioxygenase reaction.

tryptophan and hydrocortisone. The glucocorticoid stimulates the *de nove* synthesis of L-tryptophan 2,3-dioxygenase protein, while L-tryptophan protects the enzyme against its degradation in the cell.[36,37] Both bacterial and rat liver enzymes are tetrameric proteins containing 2 moles of protoheme IX as the sole cofactor.[38-40] Molecular weights of the enzymes have been reported to be 121,000, 150,000 and 167,000 for rat, mouse and *Pseudomonas* enzymes, respectively. In the *Pseudomonas* enzyme, the 5th ligand of the heme has been identified as a histidinyl nitrogen.[41] At a neutral pH in the absence of L-tryptophan, the ferric form of the enzymes exhibits a high spin state ($S = 2/5$), which is convertible to a low spin state ($S = 1/2$) upon binding of L-tryptophan or upon raising the pH to above 8.[42] Optical spectrum of the *Pseudomonas* enzyme is shown in Figure 7.

The catalytically active form of the enzymes has been shown to be the ferrous form (Fe^{2+}), while the ferric form (Fe^{3+}) of the enzyme is incapable of catalyzing the reaction.[43] No copper is involved in the purified enzymes.[38-41] The ferrous form combines first with L-tryptophan and then with molecular oxygen to form a ternary complex of the enzyme, tryptophan and oxygen.[43] In the ternary complex, either L-tryptophan or oxygen or both are activated to yield the reaction product, L-formylkynurenine.[43] Figure 8 illustrates a reaction mechanism proposed by Hayaishi in 1965.[44]

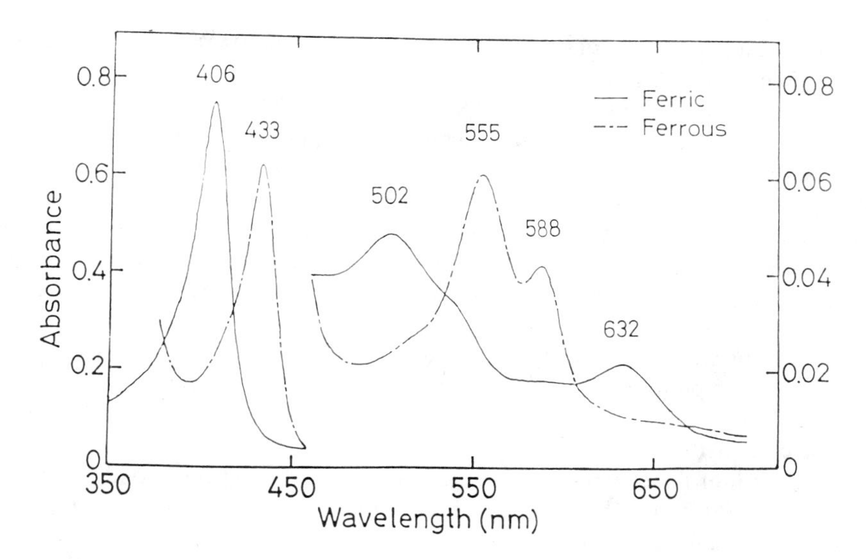

FIGURE 7. Absorption spectra of ferric and ferrous forms of *Pseudomonas* L-tryptophan 2,3-dioxygenase in the absence of L-tryptophan at pH 7.3. The numerals denote the positions of absorption maxima in nanometers.

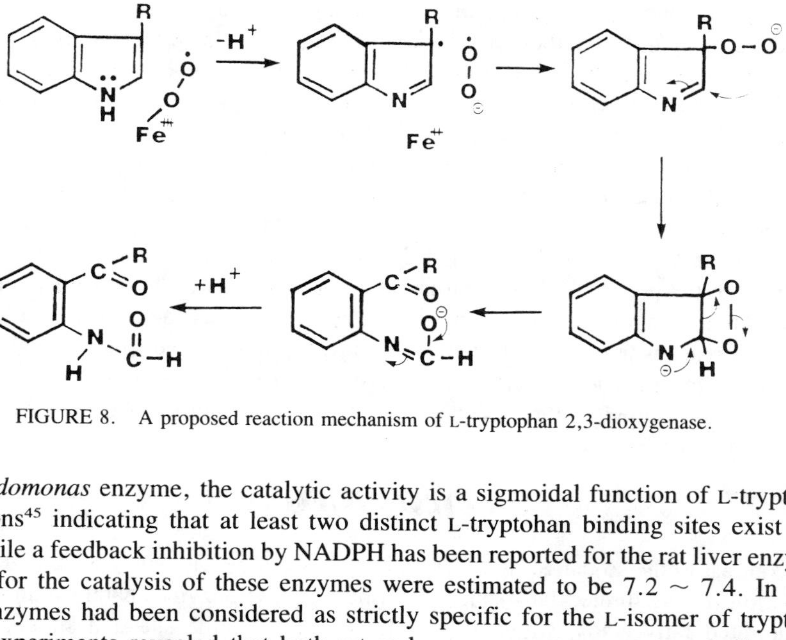

FIGURE 8. A proposed reaction mechanism of L-tryptophan 2,3-dioxygenase.

In *Pseudomonas* enzyme, the catalytic activity is a sigmoidal function of L-tryptophan concentrations[45] indicating that at least two distinct L-tryptohan binding sites exist in the enzyme, while a feedback inhibition by NADPH has been reported for the rat liver enzyme.[46] pH optima for the catalysis of these enzymes were estimated to be 7.2 ~ 7.4. In earlier days, the enzymes had been considered as strictly specific for the L-isomer of tryptophan but recent experiments revealed that both rat and mouse enzymes can metabolize the D-isomer at a significant rate.[34] Amino acid compositions of the rat liver[33] and *Pseudomanas*[32] enzymes have been reported. In addition, presence of a unique enzyme which specifically acts on L-tryptophan in poly- and oligopeptides has been reported, although the details for its property and physiological significance is not known.[47]

B. INDOLEAMINE 2,3-DIOXYGENASE

The enzyme catalyzes the reaction identical to that of L-tryptophan 2,3-dioxygenase but has a broader substrate specificicty. It metabolizes various indoleamine derivatives including not only L-and D-tryptophan but also 5-hydroxytryptophan, tryptamine, and serotonin, and gives the corresponding formylkynurenine derivative as the product.[48-50] For this reason, the enzyme is called as indoleamine 2,3-dioxygenase. The enzyme is also unique in that it utilizes both molecular oxygen (O_2) and superoxide anion (O_2^-) as the oxygenating agent. Its protein structure is also different from that of L-tryptophan 2,3-dioxygenase.[51]

In contrast to L-tryptophan 2,3-dioxygenase which is only found in liver, indoleamine 2,3-dioxygenase is distributed in almost all kinds of animal tissues other than liver. The tissue most abundant in this enzyme is probably the ileocecal region of small intestine of domestic rabbit.[51] Of special interest is that the enzyme in lung is inducible by the administration of bacterial endotoxin or interferon, and by the infection of influenza virus.[51-54]

The most highly purified preparation of the enzyme has been obtained form the ileocecal region of rabbit intestine to a homogeneous state.[51] The enzyme is a monomeric protein containing a mole of protoheme IX, of which molecular weight is approximately 42,000. The fifth ligand of the heme is a histidine.[50] At least three isozymes exist, and all of them contain some carbohydrates.[50]

The ferric form of the enzyme is in a high spin state, which is converted to a low spin state upon binding of L-tryptophan.[55,56] When used at high concentrations, L-tryptophan also exhibits a substrate inhibition of the reaction.[55] With an appropriate concentration of L-tryptophan, however, the enzyme forms a ternary complex of oxygen, tryptophan and the ferrous enzyme, i.e., the oxygenated intermediate, which degrades into the reaction products, formylkynurenine and the ferrous enzyme. It should be noted however that, unlike that of L-tryptophan 2,3-dioxygenase, the oxygenated form of this enzyme is formed both in the presence and absence of L-tryptophan or its analogues, and can be formed by the binding of molecular oxygen (O_2) to the ferrous (Fe^{2+}) enzyme and also by the binding of superoxide anion (O_2^-) to the ferric (Fe^{3+}) enzyme.[57-59] In the other words, both molecular oxygen and superoxide anion can serve as the substrate. Spectroscopic properties of the oxygenated intermediate has recently been examined in detail both in the presence and absence of L-tryptophan at subzero temperatures.[60] Incorporation of both atoms of oxygen in superoxide anion (O_2^- to the reaction product has been confirmed with the use of ^{18}O-labeled superoxide anion.[61] Such a unique role of superoxide anion as the oxygenating agent has been confirmed also in living cells.[62] Also known is that the enzyme can catalyze the monooxygenase type of reactions in the presence of an appropriate amount of H_2O_2.[63] In benzphetamine demethylation and aniline hydroxylation reactions, the monooxygenase activity of indoleamine 2,3-dioxygenase reaches to an almost comparable level to that of liver microsomal cytochrome P-450, and a mechanism similar to that of hydroperoxide-dependent monooxygnenation reaction of cytochrome P-450 has been postulated for the reaction.[63] Amino acid composition of the enzyme has been described.[50]

IV. LIPOXYGENASES

In contrast to the so-called lipid peroxidation, which is a nonenzymatic and nonspecific oxygenation of unsaturated fatty acids, the lipoxygenase enzymes catalyze an oxygenation of unsaturated fatty acids, which is specific in terms of the site and stereochemistry of oxygenation. Lipoxygenase, earlier referred to as lipoxidase, was previously believed to be found only in the plant kingdom.[64] However, it is now well known that lipoxygenases are distributed in many animal tissues and responsible for the biosyntheses of a variety of bioactive compounds such as prostaglandins, thromboxane and leukotrienes. As shown in Figure 9, lipoxygenase recognizes a methylene-interrupted double bonds (1-*cis*-4-*cis*-pentadiene) of unsaturated fatty acids including arachidonic, linoleic, and linolenic acids. One molecule of oxygen is attached to one of the carbon atoms constituting the double bond, and a hydroperoxy unsaturated fatty acid is produced. Concomitantly one of the two hydrogen atoms of the methylene carbon is lost. A double bond of *cis*-form is isomerized to *trans*-

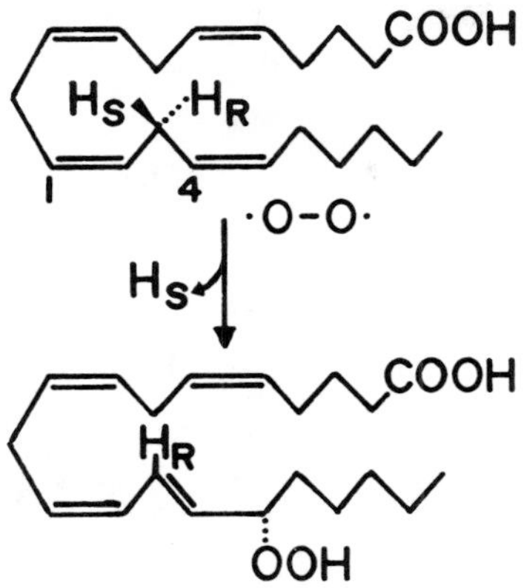

FIGURE 9. Lipoxygenase reaction.

form, giving a conjugated diene. The site of oxygenation is different from enzyme to enzyme, and each lipoxygenase is referred to with a numerical prefix which indicates the carbon number counted starting from the carboxyl carbon of arachidonic acid, for example, 5-lipoxygenase. References, 65, 65a, and 65b, are review articles on lipoxygenases.

A. ARACHIDONATE 5-LIPOXYGENASE

A pathway starting with the reaction of 5-lipoxygenase is concerned with the biosynthesis of leukotrienes.[66] They are a group of bioactive oxyeicosanoids with a conjugated triene, and types A-F have so far been isolated (Figure 10). The leukotrienes are considered as chemical mediators for anaphylaxis and inflammation.

5-Lipoxygenase incorporates a molecular oxygen into the C-5 of arachidonic acid, and produces 5-hydroperoxy-6,8,11,14-eicosatetraenoic acid. The hydroperoxy group is of *S*-configuration. The 5-lipoxygenase is found in a larger amount in leukocytes than other animal tissues, and has been purified to an apparent homogeneity from the cytosol of leukocytes of various animals.[67-69] A similar enzyme has been purified from potato.[70] Calcium ion at millimolar concentrations is required for the enzyme activity, and the addition of ATP in the order of mM stimulates the calcium-dependent 5-oxygenation. The mechanism of the activation by calcium and ATP is still unclarified. 8,11,14-Eicosatrienoic, 5,8,11-eicosatrienoic, and 5,8,11,14,17-eicosapentaenoic acids are also oxygenated by 5-lipoxygenase at C-8, C-5, and C-5, respectively. Linoleic and linolenic acids are almost inactive as substrate.

As a unique property of lipoxygenases in general, the enzyme reaction does not proceed in a linear time course and slows down as soon as the reaction is initiated even if the enzyme is still saturated with substrate. The enzyme is sometimes nicknamed as a suicide enzyme. The mechanism of the self-catalyzed inactivation of lipoxygenase has not been clarified. Furthermore, some lipoxygenase preparations require a hydroperoxide for its activity. In the absence of the hydroperoxide the enzyme particularly in a small amount shows a lag phase in its reaction.

The hydroperoxy acid produced by 5-lipoxygenase is the substrate for the next step of leukotriene biosynthesis. The bond between the two oxygen atoms of the 5-hydroperoxy group is cleaved, and the oxygen atom remaining in the fatty acid skeleton forms a 5,6-epoxide. Comcomitantly, the 6-*trans* and 8-*cis* double bonds are isomerized to 7-*trans* and 9-*trans* form, respectively, producing a conjugated triene (7-*trans*, 9-*trans*, 11-*cis*). The product with the 5,6-epoxide and the conjugated triene is referred to as leukotriene A_4. In the purified enzymes of potato and leukocytes the leukotriene synthase activity is tightly coupled with the 5-lipoxygenase activity as judged by several experimental observations,

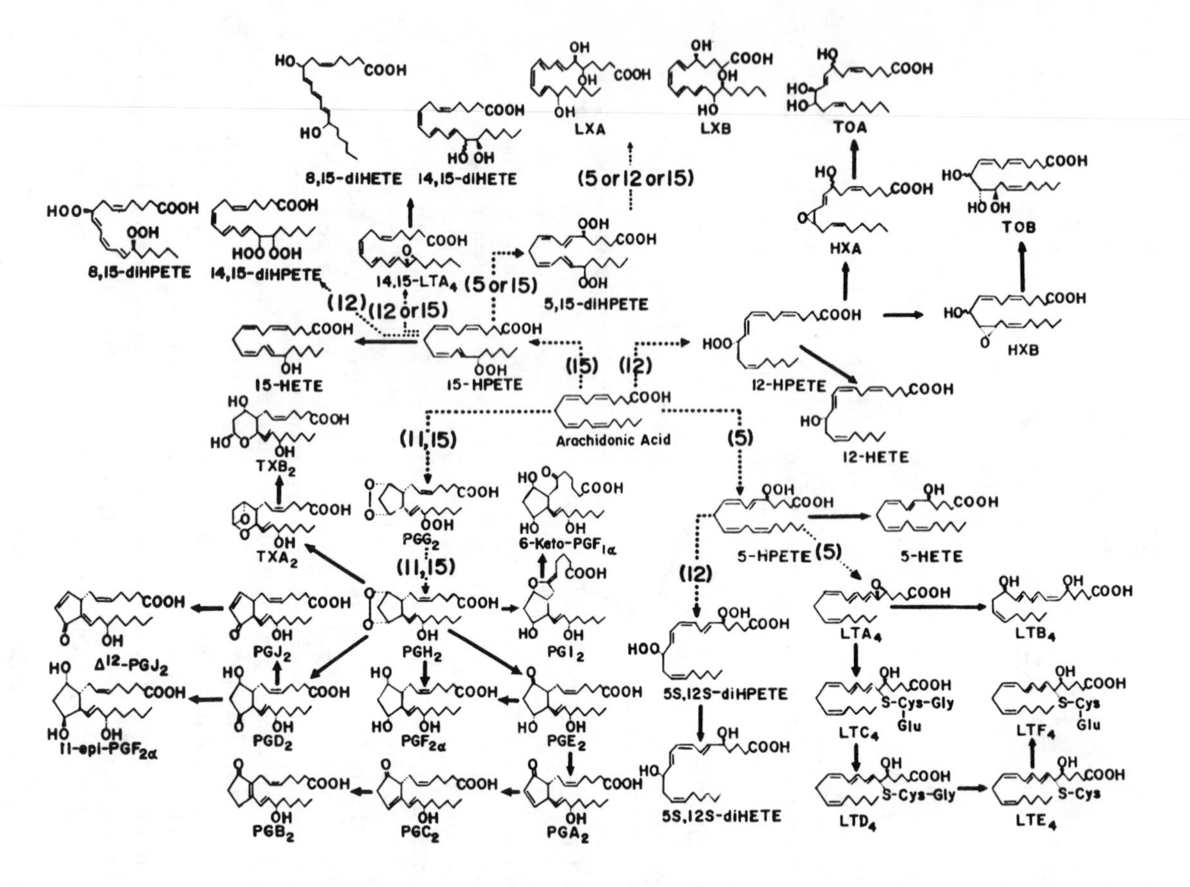

FIGURE 10. The arachidonate cascade, oxygenations of arachidonic acid by lipoxygenases, and further transformations of oxygenated products. PG, prostaglandin; TX, thromboxane; LT, leukotriene; LX, lipoxin; HX, hepoxilin; TO, trioxilin; HETE, hydroxyeicosatetraenoic acid; HPETE, hydroperoxyeicosatetraenoic acid; diHPETE, dihydroperoxy-eicosatetraenoic acid.

for example, copurification of the two enzyme activities, parallel inactivation of both activities by heat treatment and by enzyme inhibitors, and requirement of both calcium and ATP. Thus the leukotriene A synthase is now considered to be an integral part of 5-lipoxygenase.

The leukotriene A synthase activity of 5-lipoxygenase is also involved in the synthesis of lipoxins,[71] which are another group of recently discovered oxyeicosanoids.[72] Lipoxin has a characteristic structure with three hydroxyl groups and a conjugated tetraene (Figure 10). Lipoxins A and B and their isomers have so far been isolated by incubation of human leukocytes with 15*S*-hydroperoxy-5,8,11,13-eicosatetraenoic acid, which is produced from arachidonic acid by 15-lipoxygenase. The 15-hydroperoxy acid can be a substrate for 5-lipoxygenase to produce 5*S*,15*S*-dihydroperoxy-6,8,11,13-eicosa-tetraenoic acid, which in turn is a substrate for leukotriene A synthase and converted to 5,6-epoxy-15-hydroperoxy-7,9,11,13-eicosatetraenoic acid (15-hydroperoxy-leukotriene A_4). The latter compound is unstable and hydrolyzed to 5,6,15-trihydroxy-7,9,11,13-eicosatetraenoic acid (lipoxin A) and its isomers.

In view of the pathological roles of leukotrienes, a variety of selective inhibitors of 5-lipoxygenase have been developed, for example, a benzoquinone derivative (AA-861), caffeic acid derivatives (TMK-series), a flavonoid (cirsiliol), 5,6- and 11,12-dehydroarachidonic acids, and 5,6-methano-leukotriene A_4.

B. ARACHIDONATE 12-LIPOXYGENASE

The enzyme oxygenates the C-12 of arachidonic acid and produces 12*S*-hydroperoxy-5,8,10,14-eicosatetraenoic acid (Figure 10). 12-Lipoxygenase is found in platelets, leukocytes and many other tissues of various animals. The physiological roles of 12-lipoxygenase have not yet been so well established as 5-lipoxygenase.

An enzyme of the cytosol of porcine leukocytes was purified to an apparent homogeneity, and its reactivity was extensively studied.[73] Arachidonic acid as substrate gives an abnormal Michaelis-Menten curve, and an apparent substrate inhibition is observed at higher concentrations of arachidonic acid. The enzyme, especially in a small amount, shows a lag phase for 1 to 3 min before the reaction starts. The lag phase is abolished by the addition of a low concentration of 12-hydroperoxy acid.

The substrate specificity of porcine leukocyte 12-lipoxygenase is unique as compared with 5-lipoxygenase and cyclooxygenase. The enzyme is active with octadecaenoic acid (linoleic, α- and γ-linolenic acids) as well as eicosaenoic acids. Oxygen molecule is incorporated into C-13 of linoleic and α-linolenic acids and C-10 of γ-linolenic acid. Hydroperoxy products of other lipoxygenases are also substrates of 12-lipoxygenase of porcine leukocytes. 5-Hydroperoxy-6,8,11,14-eicosatetraenoic acid is oxygenated at its C-12 and transformed to 5*S*,12*S*-dihydroperoxy-6,8,10,14 eicosatetraenoic acid (Figure 10). When the enzyme is allowed to react with 15*S*- hydroperoxy-5,8,11,13-eicosatetraenoic acid, the product profile is complicated.[73] As shown in Figure 10, part of the 15-hydroperoxy acid is oxygenated at C-8 or C-14, giving 8*S*,15*S*- and 14*R*,15*S*-dihydroperoxy acids, respectively. Another major product is 14,15-epoxy-5,8,10,12-eicosatetraenoic acid, which is an unstable leukotriene A-type compound and non-enzymatically hydrolyzed to a mixture of 8,15- and 14,15-dihydroxy acids of various stereoconfigurations (Figure 10). Thus, 12-lipoxygenase like 5-lipoxygenase has the leukotriene A synthase activity in addition of the oxygenase activity. These activities of 12-lipoxygenase are also involved in the syntheses of lipoxins.[71] By the oxygenase activity 5,15-dihydroperoxy acid is oxygenated at C-14, giving a 14*R*-hydroperoxy derivative of lipoxin B. On the other hand, part of the 5,15-dihydroperoxy acid is transformed to a 5*S*-hydroperoxy derivative of 14,15-leukotriene A4, which is non-enzymatically hydrolyzed to various isomers of lipoxins A and B.

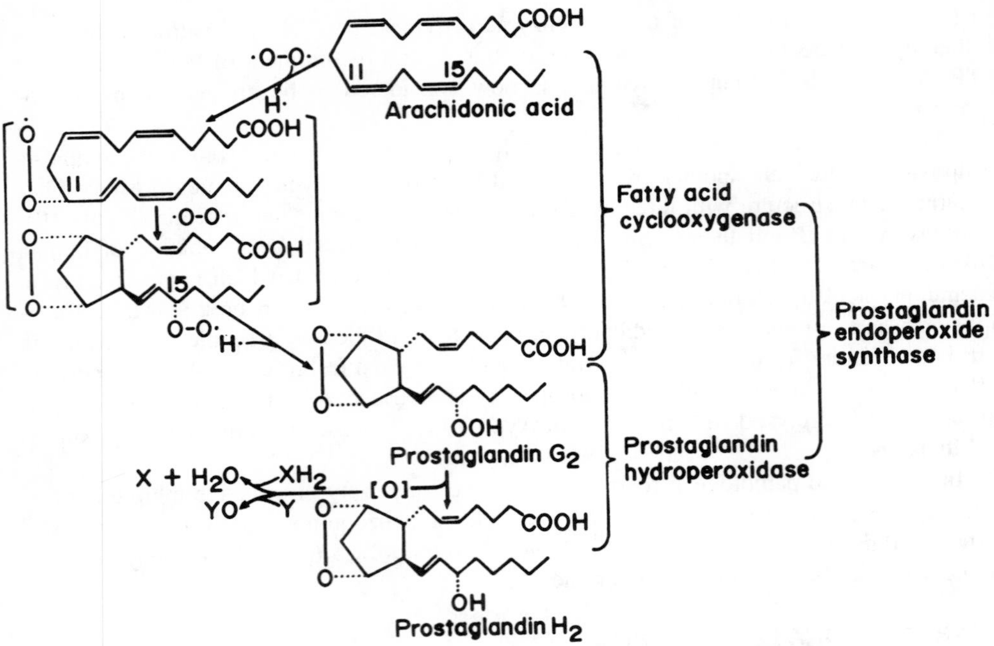

FIGURE 11. Reactions of fatty acid cyclooxygenase and prostaglandin hydroperoxidase.

C. ARACHIDONATE 15-LIPOXYGENASE

A classical enzyme purified from soybean catalyzes oxygenation of octadeca- and eicosapolyenoic acids. Arachidonic acid is oxygenated at its C-15.[74] The most extensively studied 15-lipoxygenase of mammalian origin is an enzyme purified from rabbit reticulocytes.[75] The enzyme contains 1.76 mol of iron per mol of enzyme. Polyunsaturated fatty acids esterified to phospholipids are also substrates of the enzyme. Thus, the enzyme is considered to play a role in destruction of mitochondria during the maturation of reticulocytes to erythrocytes. The enzyme oxygenates the C-15 and partly the C-12 of arachidonic acid. The 15*S*-hydroperoxy product is subjected to the 5*S*- and 8*S*-oxygenase activities and the 14,15-leukotriene A synthase activity of the same enzyme (Figure 10). 15-Lipoxygenase of soybean also shows the 5*S*- and 8*S*-oxygenase activities with 15-hydroperoxy acid as substrate.[76]

D. FATTY ACID CYCLOOXYGENASE

This enzyme is an atypical lipoxygenase incorporating two molecules of oxygen into arachidonic acid, and produces prostaglandin G_2 with 9,11-endoperoxide and 15-hydroperoxide. According to an earlier hypothetical mechanism,[77] an 11-peroxy intermediate is produced by 11-oxygenase reaction, followed by 15-oxygenase reaction and cyclization of carbon chain to produce prostaglandin G_2 (Figure 11).

The enzyme is found in almost all mammalian tissues and a homogenous enzyme preparation is available from the microsomes of bovine and ovine vesicular gland,[78] which is the richest enzyme source. The purified enzyme requires micromolar concentration of heme for its activity. Interestingly, the heme can be replaced by manganese protoporphyrin. A suicide-type enzyme inactivation is also observed in the cyclooxygenase reaction. As a characteristic nature of the enzyme, prostaglandin G_2 is further metabolized by the same enzyme. The bond between the two oxygen atoms of the 15-hydroperoxy group is cleaved, and prostaglandin H_2 with 15-hydroxy group is produced. As shown in Figure 11, one of the hydroperoxy oxygens is either reduced by a certain hydrogen donor (XH_2) or trapped by a certain oxygen acceptor (Y). A number of peroxidase substrates serve as a hydrogen donor, for example, guaiacol, benzidine, monophenol, and epinephrine. These hydrogen donors can be replaced by tryptophan and other indole compounds, although a stoichiometric consumption of the latter compounds is not clearly demonstrated. Thus, the reaction mechanism of the conversion of prostaglandin G_2 to H_2 has not yet been fully understood. However, the reaction is apparently a peroxidase reaction, and the enzyme responsible for this reaction is referred to as prostaglandin hydroperoxidase. It should be noted that the hydroperoxidase activity also requires heme. Therefore, arachidonic acid is transformed to prostaglandin H_2 (not G_2) if the enzyme is supplied with heme and tryptophan. It is of interest that manganese protophophyrin is inactive in the conversion of prostaglandin G_2 to H_2. Several lines of enzymological evidence (the enzyme suicide, optimal pH of enzyme activities, profile of heat denaturation of enzyme and isoelectric point) indicate that both the cyclooxygenase and hydroperoxidase activities are attributed to the same single enzyme protein, which is referred to as prostaglandin endoperoxide synthase or prostaglandin H synthase.

Another important aspect of the enzyme is its selective inhibition by nonsteroidal antiinflammatory drugs such as aspirin, indomethacin, flurbiprofen and ketoprofen. It is now well known that the cyclooxygenase reaction is inhibited by these drugs.[79] The cyclooxygenase protein is believed to be acetylated with aspirin, while indomethacin is not covalently bound to the enzyme. The prostaglandin hydroperoxidase activity of the enzyme is not affected by these drugs.

E. STEREOSPECIFIC HYDROGEN ABSTRACTION IN LIPOXYGENASE REACTIONS

In all the lipoxygenase reactions a hydrogen atom is lost from a methylene group which is close to the oxygenation site. Out of the two methylene hydrogens, one (either *pro-R* or *pro-S*) is removed stereospecifically. The lipoxygenase reaction is believed to start with the stereospecific hydrogen removal. Presumably a fatty acid radical is produced. A *cis*-double bond close to the oxygenation site migrates, giving a *trans*-double bond, which forms a conjugated diene together with another *cis*-double bond. A carbon atom with an unpaired electron may be attacked by a molecular oxygen, producing a hydroperoxy fatty acid.

A stereospecifc hydrogen abstraction was first demonstrated with soybean lipoxygenase.[74] The enzyme was incubated with 8,11,14-eicosatrienoic acid in which either *pro-R* or *pro-S* hydrogen at C-13 was labeled with ^3H, and it was found that *pro-S* ^3H was lost while *pro-R* ^3H was retained (Figure 12). Similarly, *pro-S* hydrogen at C-7 is liberated in the 5-lipoxygenase reaction, and *pro-S* hydrogen at C-10 is lost in the 12-lipoxygenase reaction. The cyclooxygenase reaction also shows a stereoselective removal of *pro-S* hydrogen from C-13 concomitant with oxygenation at C-11 and C-15 and cyclization of carbon skeleton. It should be noted that in all these lipoxygenase reactions an antarafacial relationship is observed between the hydrogen abstraction and the oxygen incorporation[80] (Figure 13). Namely, when the hydrogen atom leaves to this side, the oxygen molecule enters from that side.

As shown in Figure 12, in the leukotriene A synthase reactions by 5- and 12-lipoxygenases in which oxygenation is not accompanied, the stereoselective abstraction of hydrogen atom is also observed. *pro-R* Hydrogen at C-10 is lost in the 5,6-leukotriene A synthase reaction. In contrast, *pro-S* hydrogen is liberated from the same carbon in the 14.15-leukotriene A synthase reaction.

V. FLAVOPROTEIN OXYGENASES

Flavin coenzyme is well known as a prosthetic group of various dehydrogenase and oxidase enzymes. In addition, a number of monooxygenases have been identified to be flavoproteins (Table 1). These flavoprotein monooxygenases can be classified into two groups in terms of the origin of the electron donor, i.e., external and internal monooxygenases.

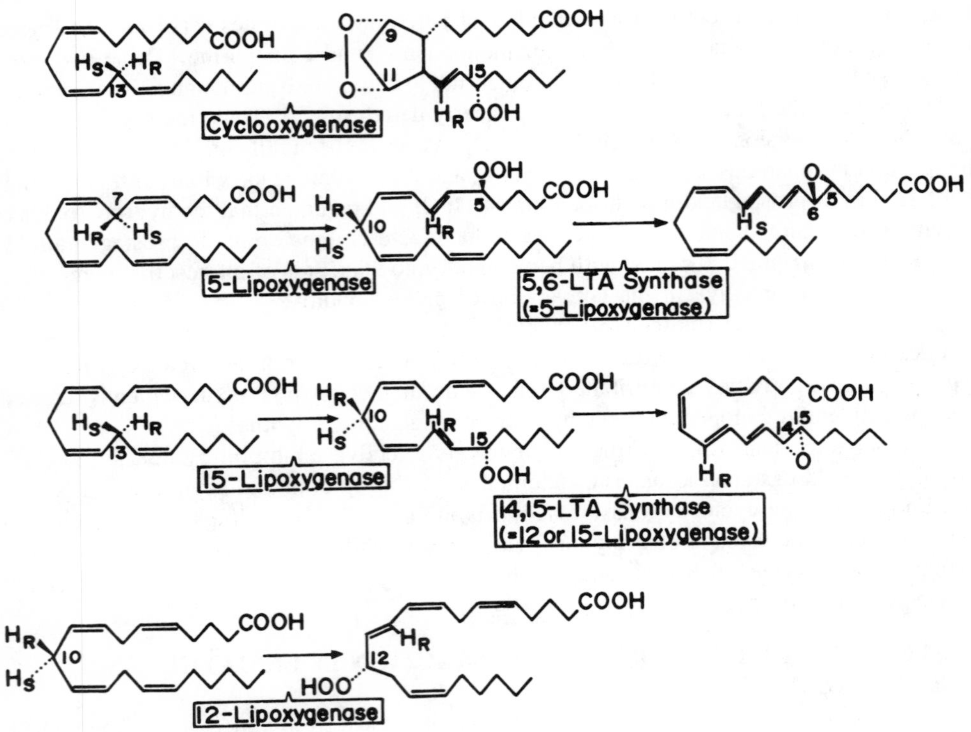

FIGURE 12. Stereospecific hydrogen abstraction in lipoxygenase and leukotriene A synthase reactions.

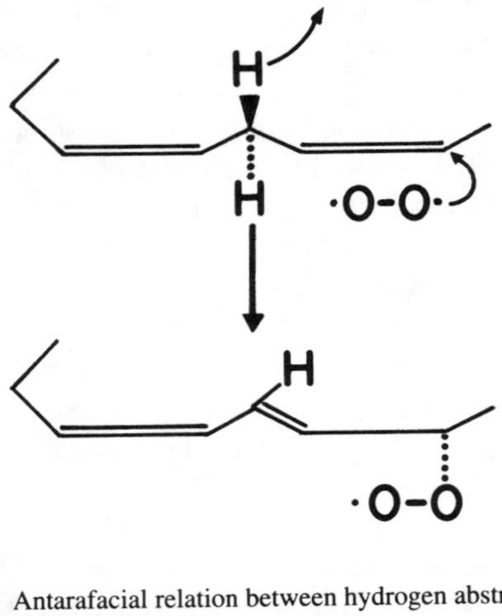

FIGURE 13. Antarafacial relation between hydrogen abstraction and oxygen incorporation in lipoxygenase reactions.

The lower part of Table 1 includes external monooxygenases. For example, p-hydroxybenzoate hydroxylase incorporates one atom of molecular oxygen into the substrate to produce 3,4-dihydroxybenzoate. The other atom of oxygen is reduced to water consuming an equimolar amount of NADPH as an exogenous hydrogen donor. Salicylate hydroxylase is also an external monooxygenase. The hydroxylation of salicylate is accompanied by decarboxylation, and catechol is produced.

In contrast, internal monooxygenases require no exogenous reductant. Hydrogen atoms

TABLE 1
Flavoprotein monooxygenases

Enzyme	Reaction	Coenzyme	Hydrogen donor	
Lactate monooxygenase	$CH_3CH(OH)COOH + O_2 \rightarrow CH_3COOH + CO_2 + H_2O$	FMN	Lactate	81,82
Lysine monooxygenase	$H_2N(CH_2)_4CH(NH_2)COOH + O_2 \rightarrow H_2N(CH_2)_4CONH_2 + CO_2 + H_2O$	FAD	Lysine	83,84
Arginine monooxygenase	$H_2N(HN{=})C\text{-}NH\text{-}(CH_2)_3CH(NH_2)COOH + O_2 \rightarrow H_2N(HN{=})C\text{-}NH\text{-}(CH_2)_3CONH_2 + CO_2 + H_2O$	FAD	Arginine	85
Salicylate hydroxylase	$C_6H_4(COOH)(OH) + O_2 + NADH + H^+ \rightarrow C_6H_4(OH)_2 + CO_2 + H_2O + NAD^+$	FAD	NADH	86,87
p-Hydroxybenzoate hydroxylase	$C_6H_4(COOH)(OH) + O_2 + NADPH + H^+ \rightarrow C_6H_3(COOH)(OH)_2 + H_2O + NADP^+$	FAD	NADPH	88,89
Melilotate hydroxylase	$C_6H_4(CH_2CH_2COOH)(OH) + O_2 + NADH + H^+ \rightarrow C_6H_3(CH_2CH_2COOH)(OH)_2 + H_2O + NAD^+$	FAD	NADH	90,91
Imidazoleacetate hydroxylase	$HC{=}C\text{-}CH_2COOH$ (HN, N, CH ring) $+ O_2 + NADH + H^+ \rightarrow HO\text{-}C{=}C\text{-}CH_2COOH$ (HN, N, CH ring) $+ H_2O + NAD^+$	FAD	NADH	92
Phenol hydroxylase	$C_6H_5OH + O_2 + NADPH + H^+ \rightarrow C_6H_4(OH)_2 + H_2O + NADP^+$	FAD	NADPH	93

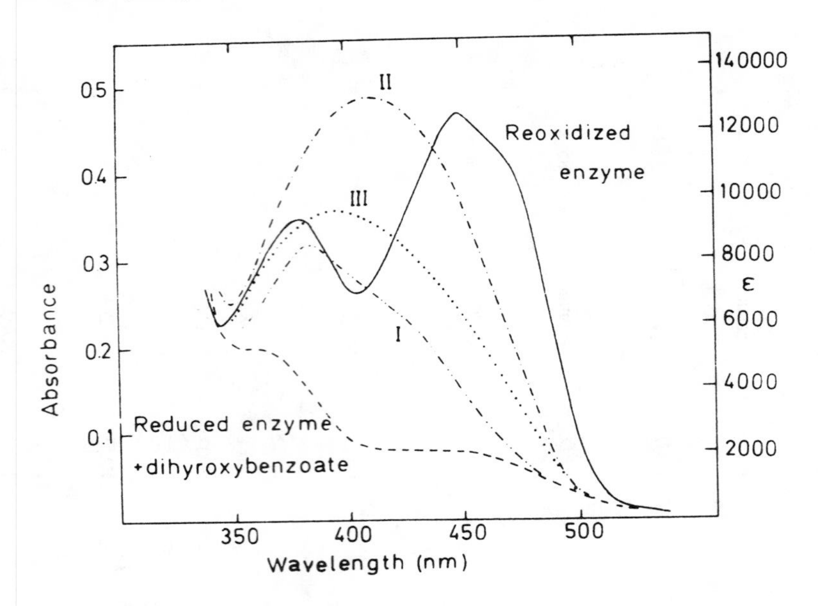

FIGURE 14. Intermediate spectral species in the reaction of molecular oxygen with the reduced p-hydroxybenzoate hydroxylase complexed with 2,4-dihydroxbenzoate. (Citation from Reference 106 with permission from Professor Vincent Massey.)

to reduce one atom of oxygen are derived from the substrate itself. As shown in the upper part of Table 1, lactic acid is monooxygenated to acetic acid concomitant with dehydrogenation and decarboxylation. Similarly, lysine is monooxygenated to δ-amino-*n*-valeramide. The other oxygen atom is reduced to water by the use of hydrogen atoms of lysine concomitant with decarboxylation.

These monooxygenases are flavoproteins containing FAD or FMN. These flavin coenzymes are considered to be involved not only in the dehydrogenation but also in the oxygenation. This view is supported by an analytical finding that neither organic nor inorganic iron was detected in an equimolar quantity in several purified enzymes including salicylate hydroxylase, imidazoleacetate hydroxylase and lysine monooxygenase.[94]

Since the flavin coenzyme has a characteristic absorption spectrum (absorption maxima around 375 and 450 nm), spectrophotometric studies on the enzyme-bound flavin were extensively carried out with several flavoprotein monooxygenases, and brought about several important findings to elucidate the reaction mechanism of the enzymes. In external monooxygenases like salicylate hydroxylase,[95] the addition of its aromatic substrate caused a spectral change in the oxidazed form of the enzyme-bound flavin. A marked shoulder appeared around 480 nm. the spectral change was attributed to the formation of an enzyme-substrate complex. It was also demonstrated that the enzyme-bound $FADH_2$ (produced with either NAD(P)H or dithionite) was a direct electron donor to react with molecular oxygen.[96] Furthermore, when the reaction of the reduced enzyme-substrate complex with molecular oxygen was studied by a rapid reaction technique, three characteristic spectral species of p-hydroxybenzoate hydroxylase were detected as shown in Figure 14.[97] The intermediate I with an absorption maximum at 385 nm was attributed to a covalent adduct of reduced flavin and O_2 (Figure 15). Similar spectral intermediates were also reported for mellilotate and phenol hydroxylases.

On the bases of these spectrophotometric observations together with detailed kinetical studies, the following catalytic pathway may be deduced for external monooxygenases.

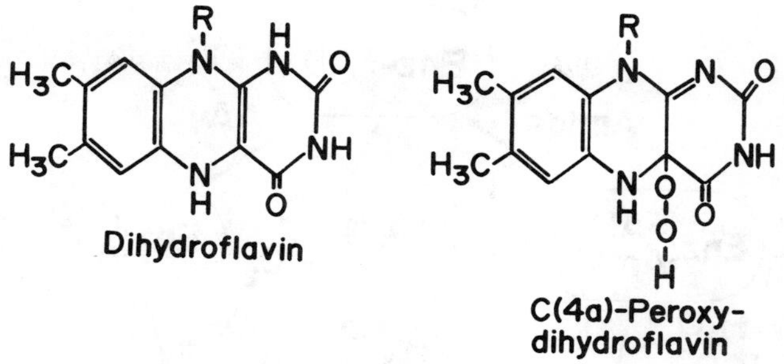

Dihydroflavin

C(4a)-Peroxy-dihydroflavin

FIGURE 15. Dihydroflavin and C(4a)-peroxydihydroflavin.

$$\text{Enzyme-FAD} + \text{substrate} \rightarrow \text{Enzyme}\begin{matrix}\diagup \text{FAD} \\ \diagdown \text{substrate}\end{matrix}$$

$$\text{Enzyme}\begin{matrix}\diagup \text{FAD} \\ \diagdown \text{substrate}\end{matrix} + \text{NADH} + \text{H}^+ \rightarrow \text{Enzyme}\begin{matrix}\diagup \text{FADH}_2 \\ \diagdown \text{substrate}\end{matrix} + \text{NAD}^+$$

$$\text{Enzyme}\begin{matrix}\diagup \text{FADH}_2 \\ \diagdown \text{substrate}\end{matrix} + \text{O}_2 \rightarrow \text{Enzyme-FAD} + \text{oxygenated product} + \text{H}_2\text{O} \qquad \text{(Str 1)}$$

A tight coupling of the reductant dehydrogenation and the substrate oxygenation is a characteristic nature of monooxygenase in general. For example, the NADH oxidation of salicylate hydroxylase is negligible in the absence of salicylate, and markedly stimulated by the addition of salicylate.[98]

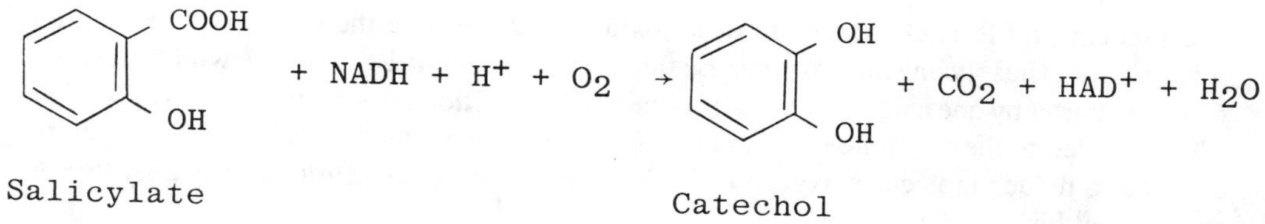

$$\text{Salicylate (COOH, OH)} + \text{NADH} + \text{H}^+ + \text{O}_2 \rightarrow \text{Catechol (OH, OH)} + \text{CO}_2 + \text{HAD}^+ + \text{H}_2\text{O}$$

Salicylate Catechol

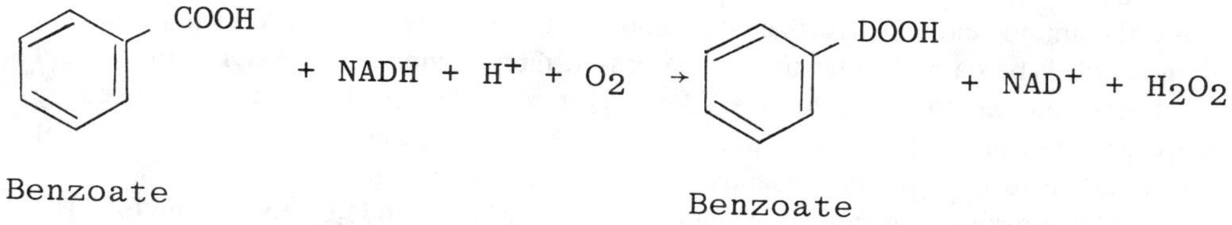

$$\text{Benzoate (COOH)} + \text{NADH} + \text{H}^+ + \text{O}_2 \rightarrow \text{Benzoate (DOOH)} + \text{NAD}^+ + \text{H}_2\text{O}_2$$

Benzoate Benzoate

However, the enzyme shows an NADH oxidase activity producing H_2O_2 (not H_2O) when benzoate (instead of salicylate) is added as an nonmetabolizable substrate analogue.[99] The observation can be interpreted as an uncoupling of the NADH dehydrogenation and the substrate oxygenation.

The uncoupling phenomenon is also observed with internal monooxygenases. Lysine monooxygenase converting lysine (C_6-carboxylic acid with two amino groups at α- and ϵ-positions) to α-amino-n-valeramide (C_5-acid amide with δ-amino group), also reacts with ornithine (C_5-carboxylic acid with two amino groups at α- and δ-positions) at about 10% the rate of lysine oxygenation. However, the reaction product is not an acid amide, but an α-keto acid. H_2O_2 (not H_2O) and NH_3 are produced without decarboxylation of ornithine.[100]

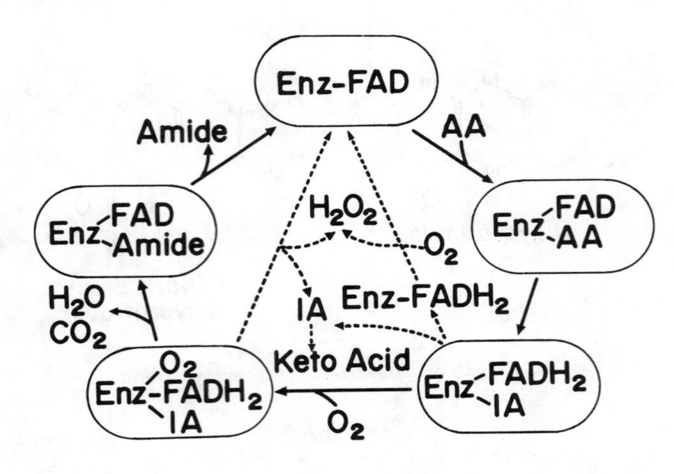

FIGURE 16. A possible reaction mechanism of lysine monooxygenase.

$$\underset{\text{Lysine}}{H_2N(CH_2)_4\underset{\displaystyle NH_2}{\underset{|}{C}}HCOOH} + O_2 \rightarrow \underset{\delta\text{-Amino-n-valeramide}}{H_2N(CH_2)_4\underset{\displaystyle NH_2}{\underset{|}{C}}{=}O} + CO_2 + H_2O$$

$$\underset{\text{Ornithine}}{H_2N(CH_2)_3\underset{\displaystyle NH_2}{\underset{|}{C}}HCOOH} + O_2 \rightarrow \underset{\alpha\text{-Keto-}\delta\text{-amino-n-valerate}}{H_2N(CH_2)_3\underset{\displaystyle O}{\underset{\|}{C}}COOH} + NH_3 + H_2O_2$$

The product profile is characteristic of an oxidase reaction like the catalysis by an amino acid oxidase. Thus, lysine monooxygenase functions as an amino acid oxidase with a substrate which is shorter by one carbon atom than lysine. In this reaction the substrate dehydrogenation is not coupled to the substrate oxygenation. The hydrogen atoms derived from the substrate are used to reduce molecular oxygen to H_2O_2. The molecular oxygen is merely a hydrogen acceptor in this reaction.

As illustrated in Figure 16, when the amino acid substrate is bound to lysine monooxygenase, the amino acid is dehydrogenated reducing the enzyme-bound FAD and producing an imino acid. When a molecular oxygen comes in, three components ($FADH_2$, imino acid and O_2) react one another in a ternary complex, and then an acid amide is produced by monooxygenation of the imino acid intermediate. However, as in the case of ornithine, if the substrate melecule misfits the active site of the enzyme, the imino acid is merely hydrolyzed to an α-keto acid and the oxygen molecule is reduced to H_2O_2 consuming $FADH_2$.

Representative references for individual enzymes are listed in Table 1. References 101 to 103 are general review articles for flavoprotein monooxoygenases.

VI. PTERIN-LINKED MONOOXYGENASES

Three monooxygenases to hydroxylate phenylalanine, tyrosine and tryptophan, respectively, are involved in the biosyntheses of catecholamines and indoleamines. The three physiologically important enzymes require tetrahydrobiopterin as an electron donor. Sterochemically the active form of the cofactor is (6R)-L-erythro-5,6,7,8-tetrahydrobiopterin(BH_4), which is biosynthesized from guanosine 5′-triphosphate (Figure 17).

In contrast to a tight binding of the flavin coenzyme to the flavoprotein monooxygenases,

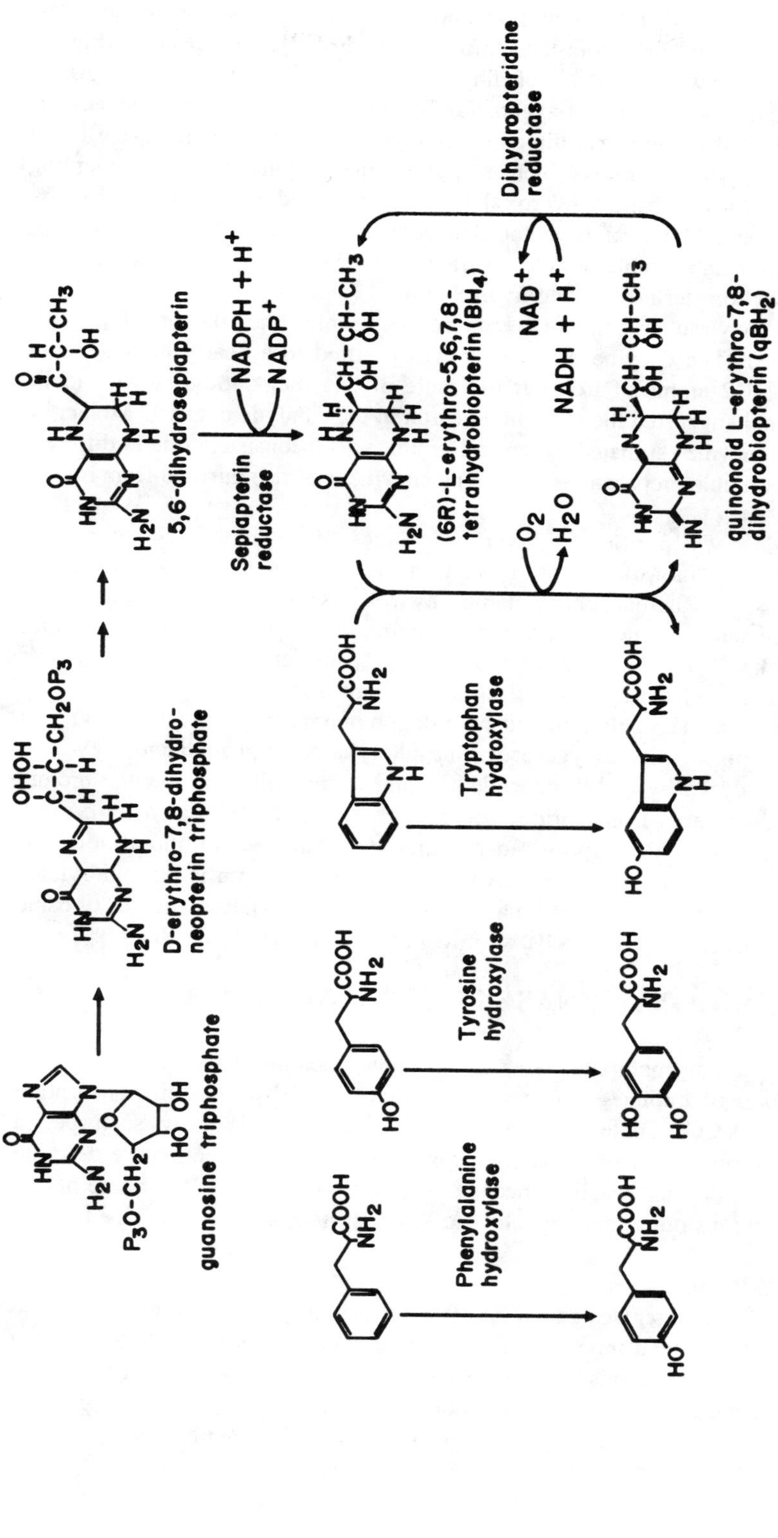

FIGURE 17. Hydroxylases requiring tetrahydrobiopterin.

the pteridin cofactor is bound weakly to the enzyme proteins, and considered as a substrate rather than a prosthetic group. As shown in figure 17, one atom of molecular oxygen is incorporated into the product, and the other atom is reduced to water at the expense of BH_4. The dehydrogenation of BH_4 is coupled with the monooxygenation of aromatic amino acids, and the BH_4 is transformed to quinonoid-L-erythro-7,8-dihydrobiopterin (qBH_2), which is then reduced back to BH_4 by the catalysis of dihydropterin reductase consuming NADH. Thus, the hydrogen required for the monooxygenation is supplied indirectly from the reduced pyridine nucleotide. 6-Methyl- and 6,7,-dimethyl-tetrahydropteridines can replace BH_4, a naturally occurring coenzyme, as a convenient reagent in the enzyme assays. Uncoupling of BH_4 dehydrogenation and substrate hydroxylation was reported for phenylalanine hydroxylase.[104] The rat liver enzyme catalyzes the BH_4 dehydrogenation in the presence of p-tyrosine, which is a nonmetabolizable pseudosubstrate. References 105 and 106 are general (but not recent) reveiws for pterin-requiring hydroxylases.

Phenylalanine hydroxylase was the first enzyme with which the role of BH_4[107] and qBH_4[108] was established. The enzyme of rat liver was purified to a nearly homogeneous preparation containing 1-2 atoms of iron per molecule of enzyme.[109] Studies with metal chelators and EPR signal suggested the reduction of iron in the catalytic cycle.[109] Molecular weight of the rat liver enzyme estimated by several groups was inconsistent due to different subunit structures of available preparations.[110-112] The enzyme was also purified from human liver.[113]

Tyrosine hydroxylase was purified from various sources.[114-117] The bovine adrenal enzyme contains 0.50 to 0.75 mol of iron per mol of enzyme.[114] Sequencing of cDNAs for rat tyrosine hydroxylase and human phenylalanine hydroxylase showed a high degree of homology in the amino acid sequence.[118] Tryptophan hydroxylase was purified from several sources.[119-122] Phenylalanine is also an active substrate at a considerable reaction rate.[119-122] The enzyme activity is stimulated by the addition of divalent iron.[119,121,122]

As a mechanism of enzyme regulation, phosphorylation of enzyme protein is now known to activate the pterin-requiring monooxygenases. Phosphorylation of phenylalanine hydroxylase of rat liver is catalyzed by cAMP-dependent[123] and calmodulin-dependent[124] protein kinases. Tyrosine hydroxylases from various sources are also phosphorylated by cAMP-dependent,[116,125,126], calcium/phospholipid-dependent,[127] and calcium/calmodulin-dependent[128,129] protein kinases. Tryptophan hydroxylase from rat brain is phosphorylated by calcium/calmodulin dependent protein kinase.[129] Four phosphorylation sites in tyrosine hydroxylase of rat pheochromocytoma cells were identified to be serine residues.[130]

VII. MONOOXYGENASES CONTAINING HEME

Most of heme-containing monoxygenases so far known belong to the family of cytochrome P-450, a group of thiolate-heme proteins characterized by the Soret absorption maximum of their ferrous CO complexes around 450 nm.[131-135] Figure 18 shows such spectra of a cytochrome P-450 obtained from *Pseudomonas putida*.[134] In this section we describe briefly the general properties and reaction mechnism(s) of cytochrome P-450. Some heme-containing monooxygenases other than cytochrome P-450 system are also described.

A. CYTOCHROME P-450

Cytochrome P-450 (hereafter designated as P450) distributes widely in aerobic living organisms and act as the terminal monooxygenase in a multienzyme system (P450 system) responsible for the oxidative metabolism of a variety of lipophilic substances such as steroids, bile acids, fatty acids, hydrocarbons, and xenobiotics.[133-135] In mammalian cells, P450s are always associated with the membrane fractions of the cell, particularly with microsomal and mitochondrial membranes.[133] The P450s are especially rich in liver, kidney, lung, and

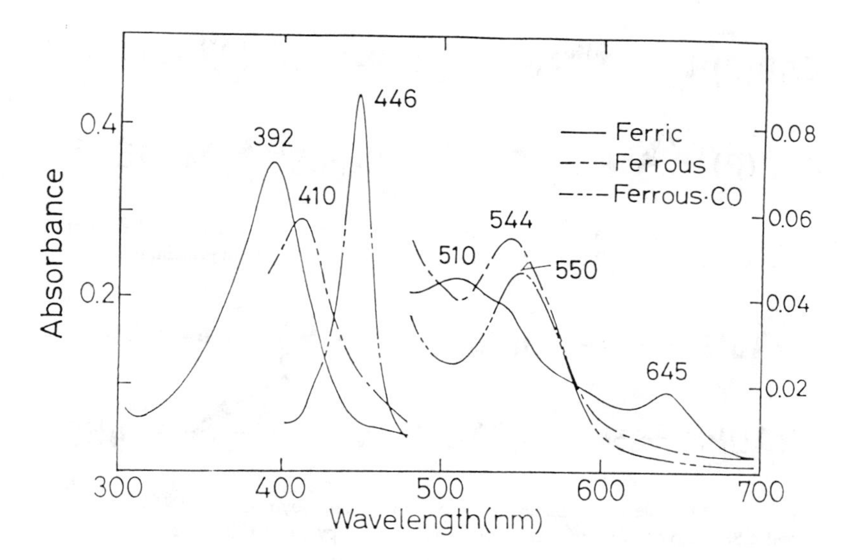

FIGURE 18. Absorption spectra of cytochrome P-450 from *P. putida* in their ferric, ferrous and ferrous CO forms in the presence of D-camphor.

steroidogenic organs including adrenal cortex, ovary and placenta. They are also detectable in other types of cells albeit smaller in quantity.

A typical reaction catalyzed by a P450-dependent monooxygenase system can be described as follows.

$$RH + O_2 + NAD(P)H + H^+ \rightarrow ROH + H_2O + NAD(P)^+ \quad (5)$$

where RH denotes a substrate to be monooxygenated.[133] This type of reactions are often called as hydroxylase reactions because of the formation of a hydroxyl group in the reaction product. Other types of monooxygenase reactions catalyzed by P450 systems include epoxidation, S- and N-oxide formation, dealkylation, desulfuration, deamination, dehalogenation, and peroxidation. Despite such seemingly diverse reactions, they are all monooxygenase reactions where one atom of molecular oxygen (O_2) is incorporated into the substrate, and the other atom is reduced to water. On the other hand, some P450 systems also catalyzes the reduction of epoxides, azo groups, nitro groups and certain N-oxides. In addition, certain P450s such as $P450_{LM2}$ can catalyze the oxidase reaction[135] as depicted below.

$$NADPH + O_2 + H^+ \rightarrow NADP^+ + H_2O_2 \quad (6)$$

In any reaction mentioned above, one or two additional proteins which convey electrons from a reduced pyridine nucleotide to P450 are necessary. The electron transfering protein in the microsomal P450 system is the NADPH-P450 reductase which contains both FAD and FMN,[136,137] while those in mitochondrion are NADPH-adrenodoxin reductase (an FAD-containing flavoprotein) and adrenodoxin (an iron-sulfur protein).[138] The latter is also called as adrenal ferredoxin and transfers electrons from the flavoprotein to mitochondrial P450s (Figure 19). The electron donor for these mammalian P450 systems is NADPH. Similar electron transfering components are also neccesary for nonmammalian P450 systems, which again require NADPH as the electron donor. The only known exception in this regard is the P450cam system[133,134] which requires NADH rather than NADPH.

In addition to these electron transfer systems, some microsomal P450s such as $P450_{BI}$ from rabbit liver require additional electron transfering components consisted of NADH-

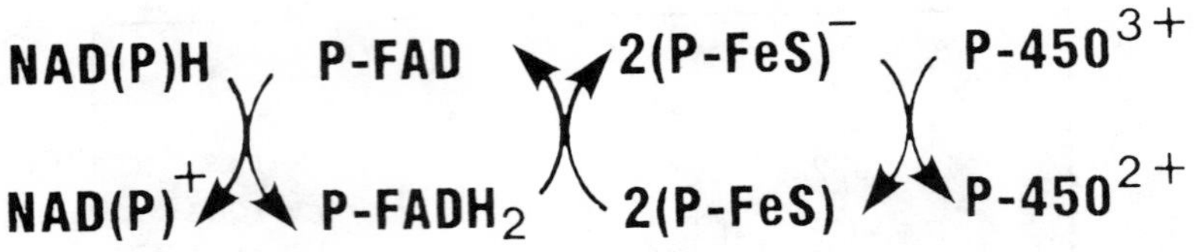

NAD(P)H P-FAD 2(P-FeS)⁻ P-450^{3+}

NAD(P)$^{+}$ P-FADH$_2$ 2(P-FeS) P-450^{2+}

FIGURE 19. Electron transfer chain for the cytochrome P-450 system in adrenocortical mitochondria. P-FAD and P-FeS denote a flavoprotein and an iron-sulfur protein, respectively.

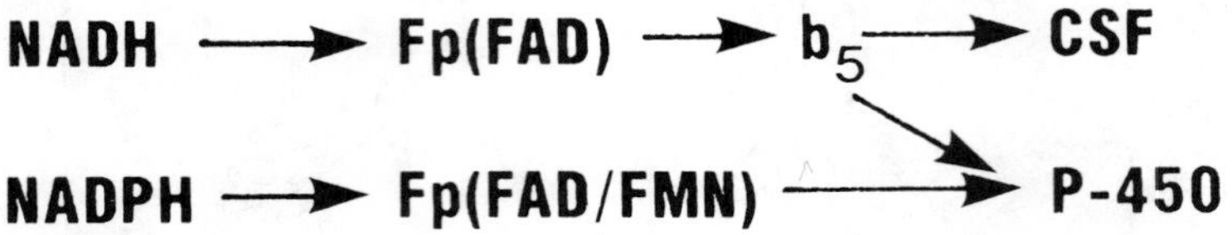

NADH → Fp(FAD) → b_5 → CSF

NADPH → Fp(FAD/FMN) → P-450

FIGURE 20. Electron transfer system for microsomal cytochrome P-450. Fp, b_5, and CSF denote a flavoprotein, cytochrome b_5 and cynide-sensitive factor, respectively.

cytochrome b_5 reductase and cytochrome b_5.[139] Available evidence suggests that, in the latter case, the first electron which reduces ferric P450 to a ferrous state comes from NADPH and the second electron to reduce the oxygenated intermediate (see below) is derived from NADH *via* cytochrome b_5.

All the P450s contain iron protoporphyrin IX in their active center for dioxygen activation. As has been demonstrated by X-ray studies on P450cam,[140,141] the heme is bound to the protein by a coordination bond through an axial thiolate of cysteine residue in addition to the hydrophobic and Coulombic interactions between the porphyrin and apoproteins. The presence of a thiolate anion at the 5th ligation position is rather unique in hemoproteins; except for P450 and chloroperoxidase, the 5th ligand of the heme-iron is always an imidazole nitrogen in various hemoproteins including hemoglobin, myoglobin, peroxidases and heme-containing dioxygenases such as L-tryptophan 2,3-dioxygenases and indoleamine 2,3-dioxygenase. On the other hand, the sixth coordination site of the heme-iron in P450, which in its resting ferric state is occupied by a labile ligand such as water, serves as the site of dioxygen coordination when reduced to a ferrous state. The site also allows coordination of carbon monoxide (CO), nitric oxide and various nitrogenous bases including metyrapone in the ferrous state, and also of many nitrogenous bases such as imidazole, pyridine and their derivatives in the ferric state.[133-135] Unlike other hemoproteins, the affinity of ferric P450 for cyanide is weak, and hence P450-catalyzed reactions are not usually inhibited by a low concentration of cyanide. As mentioned previously, the CO complex of P450 gives rise to a characteristic Soret band (λmax around 450 nm), a feature which has created the nomenclature cytochrome P-450.[131-135]

Extensive studies on the mechanism(s) of P450-catalyzed reactions have been made by using various P450 systems as well as by using model systems. A partial list of recent reviews and original articles on the reaction mechanisms include References 135 and 140 to 146 for native P450-catalyzed reactions and 147 to 152 for the model reactions. Figure 21 illustrates the reaction cycle of cytochrome P450-catalyzed reaction, which is initially deduced from studies on P450cam[134,142,153,154] but is probably applicable to those catalyzed by other P450-systems as well.

In the above scheme, Fe^{3+}, Fe^{2+}, RH, ROH denote the ferric and ferrous forms of P450, substrate and hydroxylated substrate, respectively. As shown, the reaction begins with the binding of a substrate to the ferric resting state of P450 *(step 1)*, which causes a considerable rise in the redox potential of the heme-iron. This facilitates the one electron

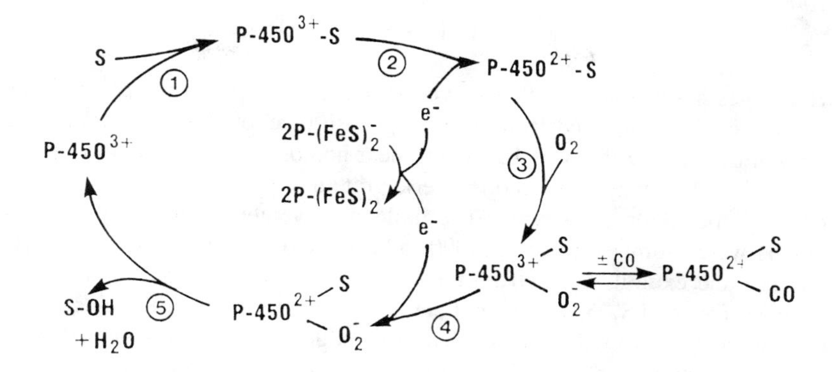

FIGURE 21. A proposed reaction cycle for the hydroxylation of steroids in adrenocortical mitochondria. S and FeS denote substrate and iron-sulfur, respectively.

reduction of P450 to give a ferrous substrate-bound form *(step 2)*, which then reacts with molecular oxygen to form a ternary complex of oxygen and substrate with the ferrous form of P450, of which structure may be written as $Fe^{3+}O_2^-$ (RH) *(step 3)*. Finally, the transfer of one more electron to the dioxygen complex *(step 4)* is mandatory to yield the reaction products, ROH and water, and ferric P450 *(step 5)*.

The reaction may proceed through an intermediacy of the ferric peroxo complex such as $Fe^{3+}O_2^{2-}$ (RH) *(step 5)*, which is equivalent in the oxidation state to that of compound I of peroxidase. The formulation as a $Fe^{3+}O_2^{2-}$ or $Fe^{2+}O_2^-$ is however a subject of debates, but the former formulation was supported by the experiments using liver microsomes or microsomal P450; hydroxylation of aromatic hydrocarbon proceeded by the additon of various peroxides instead of oxygen to the ferric form of P450.[155,156] The results have been confirmed by many workers,[157-159] and suggested the following intermediate.

$$P450^{3+}(RH) + R'OOH \rightarrow P450^{3+}\text{–O–O–}R^{7}(RH) + H^{+} \qquad (7)$$

It should be noted, however, that the organoperoxo intermediate as $P450^{3+}$-O_2^{2-} (RH) or $P450^{3+}$-O-O-R^7 (RH) has never been detected nor isolated as a discernible entity in enzymic reactions. Then two possible pathways may be envisaged for the decomposition of this still hypothetical peroxo intermediate; one is heterolytic cleavage of the O-O bond, which may be called as an oxenoid pathway, and the other is a homolytic cleavage of the bond, a quasi-Fenton pathway.[135] In fact, this step is the least understood process in the mechanism of P450 catalyzed-reaction and hence has been the subject of discussion.[160-162]

On the other hand, recent model experiments using synthetic iron-porphyrin complexes indicated that an oxo-iron complex in two oxidation level above the Fe^{3+} porphyrin system, i.e., [Fe^{4+}=O(porph)$^+$] could efficiently mimic the P450 reaction such as epoxidation of olefin.[149-151] These complexes are equivalent in oxidation state to the compound I of peroxidases having remarkably high oxidizing power. Such an oxo-iron complex produced by a homolytic cleavage of O-O bond acts in fact as the direct oxygenating species during the P450-catalyzed reaction remained to be elucidated.

For other aspects of cytochrome P-450 and related subjects including genetic aspects, readers are refered to recent review articles such as References 163 and 164.

B. PEROXYGENASE

The enzyme catalyzes the transfer of one atom of oxygen from a peroxide to certain organic substrates such as indole as illustrated below, and hence can be regarded as an atypical type of monooxygenase.[165,166]

$$SH + RO^*O^*OH \rightarrow S^*OH + R^*OH \quad (8)$$

where SH denotes a substrate which receives an oxygen atom from a peroxide molecule (ROOH) which could be derived from a lipoxygenase reaction. The asterisks are to trace the oxygen atoms. The enzyme was purified to near homogeniety from microsomes of peas *(Pisum sativum).*[165] The enzyme is an oligomeric protein of $\alpha_3\beta_1$ structure containing a mole of protoheme IX per mole of enzyme. The molecular weights of the oligomeric form α- and β-subunits were estimated to be 230,000, 62,500, and 41,000, respectively. The ferric form of the enzyme exhibits a spectrum typical to a high spin protohemoprotein with the absorption maxima at 408, 504, and 636 nm, and the ferrous form binds with CO.[166] Hydroperoxides such as H_2O_2 and linoleic hydroperoxide can serve as oxygen donors, while indole, phenol and coumarine are active as oxygen acceptor. The ferric enzyme is reducible by H_2O_2.[165,166]

C. HEME OXYGENASE

The overall reaction catalyzed by this enzyme is the conversion of protoheme IX to biliverdine, in which a single heme serves as both substrate and cofactor.[167,168] The purified enzyme is free from the heme, but combines with the protoheme stoichiometrically to form a heme oxygenase-heme complex.[169,170] The heme in the complex is converted autocatalytically to hydroxyheme which then decompose into biliverdine through an intermediate compound designated as 688 nm substance. Beside the enzyme protein and protoheme, the reaction requires the presence of molecular oxygen, NADPH and NADPH-P450 reductase. Thus the enzyme can be regarded as a kind of monooxygenase containing the heme. For details readers are refered to elsewhere in this book or original articles.[167-174]

Molecular weights of the spleen and liver enzymes were reported to be both 32,000.[169,170] Protoheme binds to the enzyme stoichiometrically to form an enzyme substrate complex, which exhibits a spectrum of a high spin ferric protohemoprotein with absorption maxima at 405, 497, and 631 nm.[170] The enzyme in the ferric state is then reduced to a ferrous state by the action of P450 reductase and reacts with molecular oxygen to form an oxyferroheme complex of the enzyme.[171] Finally the oxyheme complex decomposes via the so called 688-nm substance to the reaction product, biliverdine.[172,173] Carbon monoxide (CO) combines with the heme iron and inhibits the reaction, in which the inhibition is reversed by light.[174] The action spectrum for the light-reversible CO inhibition has been reported. The 688-nm substance appeared during the conversion of hydroxyheme to biliverdine has been shown to be an obligatory intermediate of the reaction.[171-174]

REFERENCES

1. **Hayaishi, O.,** Ed., *Oxygenases,* Academic Press, New York, 1962.
2. **Hayaishi, O.,** Ed., *Molecular Mechanisms of Oxygen Activation,* Academic Press, New York, 1974.
3. **Bloch, K. and Hayaishi, O.,** Eds., *Biological and Chemical Aspects of Oxygenases,* Maruzen Company, Tokyo, 1966.
4. **Nozaki, M., Yamamoto, S., Ishimura, Y., Coon, M. J., Ernster, L., and Estabrook, R. W.,** Eds., *Oxygenases and Oxygen Metabolism,* Academic Press, New York, 1982.
5. **Mason, H. S.,** Comparative biochemistry of the phenolase complex, *Adv. Enzymol.,* 16, 105, 1955.
6. **Hayaishi, O.,** Enzymic hydroxylation, *Annu. Rev. Biochem.,* 38, 21, 1969.
7. **Hayaishi, O. and Nozaki, M.,** *Science,* 164, 389, 1969.
8. **Nozaki, M.,** Oxygenases and dioxygenases, *Topics Curr. Chem.,* 78, 145, 1979.
9. **Hayaishi, O., Katagiri, M., and Rothberg, S.,** Mechanism of the pyrocatechase reaction, *J. Am. Chem. Soc.,* 77, 5450, 1955.
10. **Mason, H. S., Fowlks, W. L., and Peterson, E.,** Oxygen transfer and electron transport by the phenolase complex, *J. Am. Chem. Soc.,* 77, 2914, 1955.

11. **Nozaki, M., and Ishimura, Y., Oxygenases,** in *Microbial Iron Metabolism,* Nielands, J. B., Ed., Academic Press, New York, 1974, 417.
12. **Nozaki, M.,** Oxygenases and dioxygenases, *Topic Curr. Chem.,* 78, 145, 1979.
13. **Que, L., Jr.,** The catechol dioxygenase, *Adv. Inorg. Biochem.,* 5, 167, 1983.
14. **Fujisawa, H. and Hayaishi, O.,** Protocatechuate 3,4-dioxygenase, *J. Biol. Chem.,* 243, 2673, 1968.
15. **Nozaki, M., Iwaki, M., Nakai, C., Saeki, Y., Horiike, K., Kagamiyama, H., Nakazawa, T., Ebina, Y., Inoue, S., and Nakazawa, A.,** Primary structures of intradiol and extradiol dioxygenase, in *Oxygenases and Oxygen Metabolism,* Nozaki, M., Yamamoto, S., Ishimura, Y., Coon, M. J., Ernster, L. and Estabrook, R. W., Eds., Academic Press, New York, 1982, 15.
16. **Whittaker, J. D. and Lipscomb, J. D.,** Transition state analogs for protocatechuate 3,4-dioxygenase. Spectroscopic and kinetic studies of the binding reactions of ketonized substrate analogues. *J. Biol. Chem.,* 259, 4476, 1984.
17. **Whittaker, J. D. and Lipscomb, J. D.,** ^{17}O-Water and cyanide ligation by the active site iron of protocatechuate 3,4-dioxygenase. Evidence for displaceable ligands in the native enzyme and in complexes with inhibitors or transition state analogs. *J. Biol. Chem.,* 259, 4487, 1984.
18. **Nakai, C., Kagamiyama, H., Saeki, Y., and Nozaki, M.,** Nonidentical subunits of pyrocatechase from *pseudomonas arvilla* C-1, *Arch. Biochem. Biophys.,* 195, 12, 1979.
19. **Que, L., Jr., Kolanczyk, R. C., and White, L. S.,** Functional models for catechol 1,2-dioxygnenase. Structure, reactivity, and mechanism. *J. Am. Chem. Soc.,* in press.
20. **Nozaki, M., Kagamiyama, H., and Hayaishi, O.** Metapyrocathecase. I. Purification, crystallization and some properties. *Biochem. Z.,* 338, 582, 1963.
21. **Nakai, C., Hori, K., Kagamiyama, H., Nakazawa, T., and Nozaki, M.,** Purification, subunit structure, and partial amino acid sequence of metapyrocatechase, *J. Biol. Chem.,* 258, 2916, 1983.
22. **Lipscomb, J. D., Whittaker, J. D., and Arciero, D. M.,** Comparative studies of intra and extradiol protocatechuate dioxygenases, *Oxygenases and Oxygen Metabolism,* Nozaki, M., Yamamoto, S., Ishimura, Y., Coon, M. J., Ernster, L. and Estabrook, R. W., Eds., Academic Press, New York, 1982, 27.
23. **Zabinski, R., Munck, E., Champion, P. M., and Wood, J. M.,** Kinetic and Mössbauer studies on the mechanism of protocatechuic acid 4,5-dioxygenase, *Biochemistry,* 11, 3212, 1972.
24. **Tatsuno, Y., Saeki, Y., Nozaki, M., Otsuka, S., and Maeda, Y.,** Mössbauer spectra of metapyrocatechase, *FEBS Letter,* 112, 83, 1980.
25. **Arciero, D. M., Orville, A. M., and Lipscomb, J. D.,** [^{17}O] Water and nitric oxide bindings by protocatechuate 4,5-dioxygenase and catechol 2,3-dioxygenase. Evidence for binding of exogenous ligands to the active site Fe^{2+} of extradiol dioxygenase, *J. Biol. Chem.,* 260, 14035, 1985.
26. **Arciero, D. M. and Lipscomb, J. D.,** Binding of ^{17}O-labeled substrate and inhibitors to protocatechuate 4,5-dioxygenase-nitrosyl complex. Evidence for direct substrate binding to the active site Fe^{2+} of extradiol oxygenases. *J. Biol. Chem.,* 261, 2170, 1986.
27. **Hori, K., Hashimoto, T., and Nozaki, M.,** Kinetic studies on the reaction mechanism of dioxygenases *J. Biochem. (Tokyo),* 74, 375, 1973.
28. **Kita, H., Miyake, Y., Kamimoto, M., Senoh, S., and Yamano, T.,** Studies on oxygenases. I. Comparative studies on 3,4-dihydroxyphenylacetate-2,3-oxygenase and pyrocatechase by electron spin resonance spectroscopy, *J. Biochem. (Tokyo),* 66, 45, 1969.
29. **Que, L., Jr., Widom, J., and Crawford, R. L.,** 3,4-Dihydroxy-phenylacetate 2,3-dioxygenase. A manganese dioxygenase from *Bacillus brevis, J. Biol. Chem.,* 256, 10941, 1981.
30. **Beppu, T., Shoun, H., Sudo, Y., and Seto, Y.,** Heme-containing lipooxygenase from a fungus *Fusarium oxysporum,* in *Oxygenases and Oxygen Metabolism,* Nozaki, M., Yamamoto, S., Ishimura, Y., Coon, M. J., Ernster, L., and Estabrook, R. W., Eds., Academic Press, New York, 1982, 581.
31. **Hayaishi, O., Rothberg, S., Mehler, A. H., and Saito, Y.,** Studies on oxygenases; enzymatic formation of kynurenine from tryptophan, *J. Biol. Chem.,* 229, 889, 1957.
32. **Poillon, W. N., Maeno, H., Koike, K., and Feigelson, P.,** Tryptophan oxygenase of *Pseudomonas acidovorans;* purification, composition, and subunit structure, *J. Biol. Chem.,* 244, 3447, 1969.
33. **Schutz, G. and Feigelson, P.,** Purification and properties of rat liver tryptophan oxygenase, *J. Biol. Chem.,* 247, 5327, 1972.
34. **Watanabe, Y., Fujiwara, M., Yoshida, R., and Hayaishi, O.,** Stereospecificity of hepatic L-tryptophan 2,3-dioxygenase, *Biochem. J.,* 189, 393, 1980.
35. **Feigelson, P. and Brady, F. O.,** Heme-containing dioxygenases, in *Molecular Mechanisms of Oxygen Activation,* Hayaishi, O., Ed., Academic Press, New York, 1974, 87.
36. **Schimke, R. T., Sweeny, E. W., and Berlin, C. M.,** An analysis of the kinetics of rat liver tryptophan pyrrolase induction, *Biochem. Biophys. Res. Commun.,* 15, 214, 1964.
37. **Schimke, R. T., Sweeny, E. W., and Berlin, C. M.,** Studies of the stability *in vivo* and *in vitro* of rat liver tryptophan pyrrolase, *J. Biol. Chem.,* 240, 4609, 1965.
38. **Ishimura, Y. and Hayaishi, O.,** Noninvolvement of copper in the L-tryptophan 2,3-dioxygenase reaction, *J. Biol. Chem.,* 248, 8610, 1973.

39. **Makino, R. and Ishimura, Y.,** Negligible amount of copper in hepatic L-tryptophan 2,3-dioxygenase, *J. Biol. Chem.*, 251, 7722, 1976.
40. **Ishimura, Y., Makino, R., Ueno, R., Sakaguchi, K., Brady, F. O., Feigelson, P., Aisen, P., and Hayaishi, O.,** Copper is not essential for the catalytic activity of L-tryptophan 2,3-dioxygenase, *J. Biol. Chem.*, 255, 3835, 1980.
41. **Henry, Y., Ishimura, Y., and Peisach, J.,** Binding of nitric oxide to reduced L-tryptophan-2,3-dioxygenase as studied by electron paramagnetic resonance, *J. Biol. Chem.*, 251, 1578, 1976.
42. **Makino, R., Sakaguchi, K., Iizuka, T., and Ishimura, Y.,** Acid-alkaline transition and thermal spin equilibrium of the heme in ferric L-tryptophan 2,3-dioxygenases, *J. Biol. Chem.*, 255, 11883, 1980.
43. **Ishimura, Y., Nozaki, M., Hayaishi, O., Tamura, M., Nakamura, T., and Yamazaki, I.,** The oxygenated form of L-tryptophan 2,3-dioxygenase as reaction intermediate, *J. Biol. Chem.*, 245, 3593, 1970.
44. **Hayaishi, O.,** Oxygenases, *Proceedings of the Plenary Session of the Sixth International Congress of Biochemistry, New York, IUB Vol. 33*, 31, 1964.
45. **Feigelson, P. and Maeno, H.,** Studies on enzyme-substrate interactions in the regulation of tryptophan oxygenase activity, *Biochem. Biophys. Res. Commun.*, 28, 289, 1967.
46. **Cho-Chung, Y. S. and Pitot, H. C.,** Feedback control of rat liver tryptophan pyrrolase; I. End product inhibition of tryptophan pyrolase activity, *J. Biol. Chem.*, 242, 1192, 1967.
47. **Frydman, R. B., Tomaro, M. L., and Frydman, B.,** Pyrrolooxygenases: isolation, properties, and products formed, *Biochim. Biophys. Acta*, 284, 63, 1972.
48. **Yamamoto, S. and Hayaishi, O.,** Tryptophan pyrrolase of rabbit intestine; D- and L-tryptophan-cleaving enzyme or enzymes, *J. Biol. Chem.*, 242, 5260, 1967.
49. **Hirata, F. and Hayaishi, O.,** New degradative routes of 5-hydroxytryptophan and serotonin by intestinal tryptophan 2,3-dioxygenase, *Biochem. Biophys. Res. Commun.*, 47, 1112, 1972.
50. **Shimizu, T., Nomiyama, S., Hirata, F., and Hayaishi, O.,** Indoleamine 2,3-dioxygenase; purification and some properties, *J. Biol. Chem.*, 253, 4700, 1978.
51. **Hayaishi, O.,** Newer aspects of tryptophan metabolism, in *Development in Biochemistry*, Vol. 16, Hayaishi, O., Ishimura, Y., and Kido, R., Eds., Elsevier, Amsterdam, 1980, 15.
52. **Yoshida, R. and Hayaishi, O.,** Induction of pulmonary indoleamine 2,3-dioxygenase by intraperitoneal injection of bacterial lipopolysaccharide, *Proc. Natl. Acad. Sci. U.S.A.*, *75*, 3998, 1978.
53. **Yoshida, R., Urade, Y., Tokuda, M., and Hayaishi, O.,** Induction of indoleamine 2,3-dioxygenase in mouse lung during virus infection, *Proc. Natl. Acad. Sci. U.S.A.*, 76, 4084, 1979.
54. **Yoshida, R., Imanishi, J., Oku, T., Kishida, T., and Hayaishi, O.,** Induction of pulmonary indoleamine 2,3-dioxygenase by interferon, *Proc. Natl. Acad. Sci. U.S.A.*, 78, 129, 1981.
55. **Ohnishi, T., Hirata, F., and Hayaishi, O.,** Indoleamine 2,3-dioxygenase; potassium superoxide as substrate, *J. Biol. Chem.*, 252, 4643, 1967.
56. **Sono, M., Taniguchi, T., and Hayaishi, O.,** Indoleamine 2,3-dioxygenase; equilibrium studies of the tryptophan binding to the ferric, ferrous, and CO-bound enzymes, *J. Biol. Chem.*, 255, 1339, 1980.
57. **Hirata, F. and Hayaishi, O.,** Possible participation of superoxide anion in the intestinal tryptophan 2,3-dioxygenase reaction, *J. Biol. Chem.*, 246, 7825, 1971.
58. **Hirata, F., Ohnishi, T., and Hayaishi, O.,** Indoleamine 2,3-dioxygenase; characterization and properties of enzyme·O_2^- complex, *J. Biol. Chem.*, 252, 4637, 1977.
59. **Taniguchi, T., Sono, M., Hirata, F., Hayaishi, O., Tamura, M., Hayashi, K., Iizuka, T., and Ishimura, Y.,** Indoleamine 2,3-dioxygenase; kinetic studies on the binding of superoxide anion and molecular oxygen to enzyme, *J. Biol. Chem.*, 254, 3288, 1979.
60. **Sono, M.,** Spectroscopic and equilibrium properties of the indoleaminee 2,3-dioxygenase-tryptophan-O_2 ternary complex and of analogous enzyme derivatives; tryptophan binding to ferrous enzyme adducts with dioxygen, nitric oxide, and carbon monoxide, *Biochemistry*, 25, 6089, 1986.
61. **Hayaishi, O., Hirata, F., Ohnishi, T., Henry, J.-P., Rosenthal, I., and Katoh, A.,** Indoleamine 2,3-dioxygenase; incorporation of $^{18}O_2^-$ and $^{18}O_2$ into the reaction products, *J. Biol. Chem.*, 252, 3548, 1977.
62. **Taniguchi, T., Hirata, F., and Hayaishi, O.,** Intracellular utilization of superoxide anion by indoleamine 2,3-dioxygenase of rabbit enterocytes, *J. Biol. Chem.*, 252, 2774, 1977.
63. **Takikawa, O., Yoshida, R., and Hayaishi, O.,** Monooxygenase activities of dioxygenases; benzphetamine demethylation and aniline hydroxylation reactions catalyzed by indoleamine 2,3-dioxygenase, *J. Biol. Chem.*, 258, 6808, 1983.
64. **Tappel, A. L.,** Lipoxidase, in *The Enzymes*, Vol. 8, 2nd ed., Boyer, P. D., Lardy, H., and Myrbäck, K., Eds., Academic Press, New York, 1963, 275.
65. **Kühn, H., Schewe, T., and Rapoport, S. M.,** The stereochemistry of the reactions of lipoxygenases and their metabolites. Proposed nomenclature of lipoxygenases and related enzymes, in *Advances in Enzymology*, Vol. 58, Meister, A., Ed., John Wiley & Sons, New York, 1986, 273.

65a. **Shimizu, T.,** Enzymes functional in the syntheses of leukotrienes and related compounds, *Int. J. Biochem.*, 20, 661, 1988.

65b. **Yamamoto, S.,** Mammalian lipoxygenases; molecular and catalytic properties, *Prostaglandin Leukotrienes and Essential Fatty Acids Rev.*, 35, 219, 1989.
66. **Samuelsson, B.,** Leukotrienes: mediators of immediate hypersensitivity reactions and inflammation, *Science*, 220, 568, 1983.
67. **Rouzer, C. A., Matsumoto, T., and Samuelsson, B.,** Single protein from human leukocytes possesses 5-lipoxygenase and leukotriene A_4 synthase activities, *Proc. Natl. Acad. Sci. U.S.A.*, 83, 857, 1986.
68. **Ueda, N., Kaneko, S., Yoshimoto, T., and Yamamoto, S.,** Purification of arachidonate 5-lipoxygenase from porcine leukocytes and its reactivity with hydroperoxyeicosatetraenoic acids, *J. Biol. Chem.*, 261, 7982, 1986.
69. **Shimizu, T., Izumi, T., Seyama, Y., Tadokoro, K., Rådmark, O., and Samuelsson, B.,** Characterization of leukotriene A_4 synthase from murine mast cells: evidence for its identity to arachidonate 5-lipoxygenase, *Proc. Natl. Acad. Sci. U.S.A.*, 83, 4175, 1986.
70. **Shimizu, T., Radmark, O., and Samuelsson, B.,** Enzyme with dual lipoxygenase activities catalyzes leukotriene A_4 synthesis from arachidonic acid, *Proc. Natl. Acad. Sci. U.S.A.*, 81, 689, 1984.
71. **Yamamoto, S., Ueda, N., Yokoyama, C., Fitzsimmons, B. J., Rokach, J., Oates, J. A., and Brash, A. R.,** Lipoxin syntheses by arachidonate 12- and 5-lipoxygenases purified from porcine leukocytes, in Lipoxins; Biosynthesis, Chemistry, and Biological Activities, Wong, P. Y.-K., and Serhan, C. N., Eds, Plenum Press, New York, 1988, 15.
72. **Serhan, C.N., Hamberg, M., and Samuelsson, B.,** Lipoxins: novel series of biologically active compounds formed from arachidonic acid in human leukocytes, *Proc. Natl. Acad. Sci. U.S.A.*, 81, 5335, 1984.
73. **Yokoyama, C., Shinjo, F., Yoshimoto, T., Yamamoto, S., Oates, J. A., and Brash, A. R.,** Arachidonate 12-lipoxygenase purified from porcine leukocytes by immunoaffinity chromatography and its reactivity with hydroperoxy-eicosatetraenoic acids, *J. Biol. Chem.*, 261, 16714, 1986.
74. **Hamberg, M. and Samuelsson, B.,** On the specificity of the oxygenation of unsaturated fatty acids catalyzed by soybean lipoxidase, *J.Biol. Chem.*, 242, 5329, 1967.
75. **Schewe, T., Rapoport, S. M., and Kühn, H.,** Enzymology and physiology of reticulocyte lipoxygenase: comparison with other lipoxygenases, in *Advances in Enzymology*, Vol. 58, Meister, A., Ed., John Wiley & Sons, New York, 1986, 191.
76. **Van Os, C.P.A., Rijke-Schilder, G. P. M., Van Halbeek, H., Verhagen, J. and Vliegenthart, J. F. G.,** Double dioxygenation of arachidonic acid by soybean lipoxygenase. I. Kinetics and regio-stereospecificities of the reaction steps, *Biochim. Biophys. Acta*, 663, 177, 1981.
77. **Hamberg, M., and Samuelsson, B.,** On the mechanism of the biosynthesis of prostaglandins E_1 and $F_{1\alpha}$, *J. Biol. Chem.*, 242, 5336, 1967.
78. **Yamamoto, S.,** Enzymes in the arachidonic acid cascade, in *Prostaglandins and Related Substances*, Pace-Asciak, C. and Granström, E., Eds., Elsevier, Amsterdam, 1983, 171.
79. **Lands, W. E. M. and Hanel, A. M.,** Inhibitors and activators of prostaglandin biosynthesis, in *Prostaglandins and Related Substances*, Pace-Asciak, C. and Granström, E., Eds., Elsevier, Amsterdam, 1983, 203.
80. **Hamberg, M. and Hamberg, G.,** On the mechanism of the oxygenation of arachidonic acid by human platelet lipoxygenase, *Biochem. Biophys. Res. Commun.*, 95, 1090, 1980.
81. **Takemori, S., Nakazawa, K., Nakai, Y., Suzuki, K., and Katagiri, M.,** A lactate oxygenase from Mycobacterium phlei, *J. Biol. Chem.*, 243, 313, 1968.
82. **Sullivan, P. A.,** Crystallizataion and properties of L-lactate oxidase from Mycobacterium smegmatis, *Biochem. J.*, 110, 363, 1968.
83. **Takeda, H., Yamamoto, S., Kojima, Y., and Hayaishi, O.,** Studies on monooxygenases, I. General properties of crystalline L-lysine monooxygenase, *J. Biol. Chem.*, 244, 2935, 1969.
84. **Flashner, M. I. S., and Massey, V.,** Purification and properties of L-lysine monooxygenase form Pseudomonas fluorescens, *J. Biol. Chem.*, 249, 2579, 1974.
85. **Olomucki, A., Pho, D. B., Lebar, R., Delcambe, L., and Thoai, N. V.,** Arginine oxygenase decarboxylante. V. Purification et nature flavinique, *Biochim. Biophys. Acta*, 151, 353, 1968.
86. **Yamamoto, S., Katagiri, M., Maeno, H., and Hayaishi, O.,** Salicylate hydroxylase, a monooxygenase requiring flavin adenine dinucleotide, I. Purification and general properties, *J. Biol. Chem.*, 240, 3408, 1965.
87. **Katagiri, M., Maeno, H., Yamamoto, S., and Hayaishi, O.,** Salicylate hydroxylase, a monooxygenase requiring flavin adenine dinucleotide. II. The mechanism of salicylate hydroxylation to catechol, *J. Biol. Chem.*, 240, 3414, 1965.
88. **Hosokawa, K. and Stanier, R. Y.,** Crystallization and properties of p-hydroxybenzoate hydroxylase from *Pseudomonas putida*, *J. Biol. Chem.*, 241, 2453, 1966.
89. **Howell, L. G., Spector, T., and Massey, V.,** Purification and properties of p-hydroxybenzoate hydroxylase from *Pseudomonas fluorescens*, *J. Biol. Chem.*, 247, 4340, 1972.
90. **Levy, C. C.,** Melilotate hydroxylase, *J. Biol. Chem.*, 242, 747, 1967.
91. **Strickland, S. and Massey, V.,** The purification and properties of the flavoprotein melilotate hydroxylase, *J. Biol. Chem.*, 248, 2944, 1973.

92. **Maki, Y., Yamamoto, S., Nozaki, M., and Hayaishi, O.**, Studies on monooxygenases, II. Crystalization and some properties of imidazole acetate monooxygenase, *J. Biol. Chem.*, 244, 2942, 1969.
93. **Neujahr, H. Y. and Gall, A.**, Phenol hydroxylase from yeast, *Eur. J. Biochem.*, 35, 386, 1973.
94. **Yamamoto, S., Takeda, H., Maki, Y., and Hayaishi, O.**, Studies on monooxygenases. III. Examinations of metal participation in flavoprotein monooxygenases of pseudomonads, *J. Biol. Chem.*, 244, 2951, 1969.
95. **Takemori, S., Yasuda, H., Mihara, K., Suzuki, K., and Katagiri, M.**, Mechanism of the salicylate hydroxylase reaction. II. The enzyme-substrate complex, *Biochim. Biophys. Acta*, 191, 58, 1969.
96. **Takemori, S., Yasuda, H., Mihara, K., Suzuki, K., and Katagiri, M.**, Mechanism of the salicylate hydroxylase reaction. III. Characterization and reactivity of chemically or photochemically reduced enzyme-flavin, *Biochim. Biophys. Acta*, 191, 69, 1969.
97. **Spector, T. and Massey, V.**, p-Hydroxybenzoate hydroxylase from Pseudomonas fluorescens, *J. Biol. Chem.*, 247, 7123, 1972.
98. **Takemori, S., Nakamura, M., Suzuki, K., Katagiri, M., and Nakamura, T.**, Mechanism of the salicylate hydroxylase reaction, V. Kinetic analyses, *Biochim. Biophys. Acta*, 284, 382, 1972.
99. **White-Stevens, R. H., and Kamin, H.**, Studies of a flavoprotein, salicylate hydroxylase, I. Preparation, properties and the uncoupling of oxygen reduction from hydroxylation, *J. Biol. Chem.*, 247, 2358, 1972.
100. **Nakazawa, T., Hori, K., and Hayaishi, O.**, Studies on monooxygenases, V. Manifestation of amino acid oxidase activity by L-lysine monooxygenase, *J. Biol. Chem.*, 247, 3439, 1972.
101. **Flashner, M. S. and Massey, V.**, Flavoprotein oxygenases, in *Molecular Mechanism in Oxygen Activation*, Hayaishi, O., Ed., Academic Press, New York, 1974, 245.
102. **Massey, V., and Hemmerich, P.**, Flavin and pteridine monooxygenases, in *The Enzymes*, 3rd ed., Volume 12, Boyer, P. D., Ed., Academic Press, New York, 1975, 191.
103. **Yamamoto, S., Ohnishi, T., Maruyama, K., Yamauchi, T., and Hayaishi, O.**, *Oxygenation and Oxidation Catalyzed by Lysine Monooxygenase, a Flavoprotein*, Singer, T. P., Ed., Elsevier, Amsterdam, 1976, 169.
104. **Fisher, D. B. and Kaufman, S.**, Tetrahydropterin oxidation without hydroxylation catalyzed by rat liver phenylalanine hydroxylase, *J. Biol. Chem.*, 248, 4300, 1973.
105. **Kaufman, S. and Fisher, D. B.**, Pterin-requiring aromatic amino acid hydroxylases, in *Molecular Mechanism in Oxygen Activation*, Hayaishi, O., Ed., Academic Press, New York and London, 1974, 285.
106. **Massey, V. and Hemmerich, P.**, Flavin and pteridine monooxygenases, in *The Enzymes*, 3rd ed., Vol. 12, Boyer, P. D., Ed., Academic Press, New York, 1975, 191.
107. **Kaufman, S.**, The structure of the phenylalanine-hydroxylation cofactor, *Biochemistry*, 50, 1085, 1963.
108. **Kaufman, S.**, Studies on the structure of the primary oxidation product formed from tetrahydropteridines during phenylalanine hydroxylation, *J. Biol. Chem.*, 239, 332, 1964.
109. **Fisher, D. B., Kirkwood, R., and Kaufman, S.**, Rat liver phenylalanine hydroxylase, an iron enzyme, *J. Biol. Chem.*, 247, 5161, 1972.
110. **Kaufman, S. and Fisher, D. B.**, Purification and some physical properties of phenylalanine hydroxylase from rat liver, *J. Biol. Chem.*, 245, 4745, 1970.
111. **Gullam, S. S., Woo, S. L. C., and Woolf, L. I.**, The isolation and properties of phenylalanine hydroxylase from rat liver, *Biochem. J.*, 139, 731, 1974.
112. **Nakata, H. and Fujisawa, H.**, Purification and characterization of phenylalanine 4-monooxygenase from rat liver, *Biochim. Biophys. Acta*, 614, 313, 1980.
113. **Woo, S. L. C., Gillam, S. S., and Woolf, L. I.**, The isolation and properties of phenylalanine hydroxylase from human liver, *Biochem. J.*, 139, 741, 1974.
114. **Hoeldtke, R. and Kaufman, S.**, Bovine adrenal tyrosine hydroxylase, *J. Biol. Chem.*, 252, 3160, 1977.
115. **Okuno, S. and Fujisawa, H.**, Purificaiton and some properties of tyrosine 3-monooxygenase from rat adrenal, *Eur. J. Biochem.*, 122, 49, 1982.
116. **Richtand, N. M., Inagami, T., Misono, K., and Kuczenski, R.**, Purification and characterization of rat striatal tyrosine hydroxylase, *J. Biol. Chem.*, 260, 8465, 1985.
117. **Markey, K. A., Kondo, S., Shenkman, L., and Goldstein, M.**, Purification and characterization of tyrosine hydroxylase from a clonal pheochromocytoma cell line, *Mol. Pharmacol.*, 17, 79, 1980.
118. **Ledley, F. D., DiLella, A. G., Kwok, S. C. M., and Woo, S. L. C.**, Homology between phenylalanine and tyrosine hydroxylases reveals common structural and functional domains, *Biochemistry*, 24, 3389, 1985.
119. **Nukiwa, T., Tohyama, C., Okita, C., Kataoka, T., and Ichiyama, A.**, Purification and some properties of bovine pineal tryptophan 5-monooxygenase, *Biochem. Biophys. Res. Commun.*, 60, 1029, 1974.
120. **Tong, J. H. and Kaufman, S.**, Tryptophan hydroxylase, *J. Biol. Chem.*, 250, 4152, 1975.
121. **Nakata, H. and Fujisawa, H.**, Purification and properties of tryptophan 5-monooxygenase from rat brain-stem, *Eur. J. Biochem.*, 122, 41, 1982.
122. **Nakata, H. and Fujisawa, H.**, Tryptophan 5-monooxygenase from mouse mastocytoma P815, *Eur. J. Biochem.*, 124, 595, 1982.
123. **Abita, J.-P., Milstien, S., Chang, N., and Kaufman, S.**, *In vitro* activation of rat liver phenylalanine hydroxylase by phosphorylation, *J. Biol. Chem.*, 251, 5310, 1976.

124. **Døskeland, A. P., Schworer, C. M., Døskeland, S. O., Chrisman, T. D., Soderling, T. R., Corbin, J. D., and Flatmark, T.,** Some aspects of the phosphorylation of phenylalanine 4-monooxygenase by a calcium-dependent and calmodulin-dependent protein kinase, *Eur. J. Biochem.,* 145, 31, 1984.
125. **Lazar, M. A., Lockfeld, A. J., Truscott, R. J. W., and Barchas, J. D.,** Tyrosine hydroxylase from bovine striatum: catalytic properties of the phosphorylated and nonphosphorylated forms of the purified enzyme, *J. Neurochem.,* 39, 409, 1982.
126. **Okuno, S., and Fujisawa, H.,** A new mechanism for regulation of tyrosine 3-monooxygenase by end product and cyclic AMP-dependent protein kinase, *J. Biol. Chem.,* 260, 2633, 1985.
127. **Albert, K. A., Helmer-Matyjek, E., Nairn, A. C., Muller, T. H., Haycock, J. W., Greene, L. A., Goldstein, M., and Greengard, P.,** Calcium/phospholipid-dependent protein kinase (protein kinase C) phosphorylates and activates tyrosine hydroxylase, *Proc. Natl. Acad. Sci. U.S.A.,* 81, 7713, 1984.
128. **Yamauchi, T. and Fujisawa, H.,** Tyrosine 3-monooxygenase is phosphorylated by Ca^{2+}-, calmodulin-dependent protein kinase, followed by activation by activator protein, *Biochem. Biophys. Res. Commun.,* 100, 807, 1981.
129. **Yamauchi, T., Nakata, H., and Fujisawa, H.,** A new activator protein that activates tryptophan 5-monooxygenase and tyrosine 3-monooxygenase in the presence of Ca^{2+}-, calmodulin-dependent protein kinase, *J. Biol. Chem.,* 256, 5404, 1981.
130. **Campbell, D. G., Hardie, D. G., and Vulliet, P. R.,** Identification of four phosphorylation sites in the N-terminal region of tyrosine hydroxylase, *J. Biol. Chem.,* 261, 10489, 1986.
131. **Omura, T. and Sato, R.,** A new cytochrome in liver microsomes, *J. Biol. Chem.,* 237, PC 1375, 1962.
132. **Omura, T. and Sato, R.,** The carbon monoxide-binding pigment of liver microsomes, *J. Biol. Chem.,* 239, 2370, 1964.
133. **Sato, R.,** Cytochrome P-450-containing oxygenase systems, in *Cytochrome P-450,* Sato, R., and Omura, T. Eds., Kodansha-Academic Press, Tokyo and New York, 1978, 137.
134. **Gunsalus, I. C., Meeks, J. R., Lipscomb, J.D., Debrunner, P. and Münck, E.,** Bacterial monoxygenases; the P450 cytochrome system, in *Molecular Mechanisms of Oxygen Activation,* Hayaishi, O., Ed., Academic Press, New York, 1974, 561.
135. **White, R. E. and Coon, M. J.,** Oxygen activation by cytochrome P-450, *Ann. Rev. Biochem.,* 49, 315 and references therein cited, 1980.
136. **Iyanagi, T. and Mason, H. S.,** Some properties of hepatic reduced nicotinamide adenine dinucleotide phosphate-cytochrome *c* reductase, *Biochemistry,* 12, 2297, 1973.
137. **Iyanagi, T., Makino, N., and Mason, H. S.,** Redox properties of the reduced nicotinamide adenine dinucleotide-cytochrome P-450 and reduced nicotinamide adenine dinucleotide-cytochrome b_5 reductases, *Biochemsitry,* 13, 1701, 1974.
138. **Kimura, T. and Suziki, K.,** Components of the electron transport system in adrenal steroid hydroxylase, *J. Biol. Chem.,* 242, 485, 1972.
139. **Sugiyama, T., Miki, N., and Yanano, T.,** NADH- and NADPH-dependent reconstituted *p*-nitroanisole *o*-demethylation system containing cytochrome P-450 with high affinity for cytochrome b_5, *J. Biochem. (Tokyo),* 87, 1457, 1978.
140. **Poulos, T. L., Finzel, B. C., Gunsalus, I. C., Wagner, G. C., and Kraut, J.,** The 2.6-A crystal structure of *Pseudomonas putida* cytochrome P-450, *J. Biol. Chem.,* 260, 16122, 1985.
141. **Poulos, T. L., Finzel, B. C., and Howard, A. J.,** Crystal structure of substrate-free *Pseudomonas putida* cytochrome P-450, *Biochemistry,* 25, 5314, 1986.
142. **Ishimura, Y., Ullrich, V., and Peterson, J. A.,** Oxygenated cytochrome P-450 and its possible role in enzymic hydroxylation, *Biochem. Biophys. Res. Commun.,* 42, 140, 1971.
143. **Coon, M. J., and White, R. E.,** in *Metal Ion Activation of Dioxygen,* Spiro, T. G., Ed., John Wiley & Sons, New York, 1980, 73.
144. **Makino, R., Iizuka, T., Sakaguchi, K., and Ishimura, Y.,** Effects of heme substitution on the activity of heme-containing oxygenases, in *Oxygenases and Oxygen Metabolism,* Nozaki, M., Yamamoto, S., Ishimura, Y., Coon, M. J., Ernster, L., and Estabrook, R. W., Eds., Academic Press, New York, 1982, 467.
145. **McMurry, T. J., and Groves, J. T.,** Metalloporphyrin models for cytochrome P-450, in *Cytochrome P-450: Structure, Mechanism, and Biochemistry,* Ortiz de Montellano, P. R., Ed., Plenum Press, New York, 1986, 1.
146. **Heimbrook, D. C., Murray, R. I., Egeberg, K. D., and Sligar, S. G.,** Demethylation of N,N-dimethylaniline and p-cyano-N,N-dimethylaniline and their N-oxides by cytochromes $P450_{LM2}$ and $P450_{CAM}$, *J. Am. Chem. Soc.,* 106, 1514, 1984.
147. **Stearns, R. A. and Ortiz de Montellano, P. R.,** Cytochrome P-450 catalyzed oxidation of quadricyclane; evidence for a radical cation intermediate, *J. Am. Chem. Soc.,* 107, 4081, 1985.
148. **Groves, J. T. and Suburamanian, D. V.,** Hydroxylation by cytochrome P-450 and metalloporphyrin models; evidence for allylic rearrangment, *J. Am. Chem. Soc.,* 106, 2177, 1984.

149. **Collman, J. P., Brauman, J. I., Meunier, B., Hayashi, T., Kodadek, T., and Raybuck, S. A.,** Epoxidation of olefins by cytochrome P-450 model compounds; kinetics and stereochemistry of oxygen atom transfer and origin of shape selectivity, *J. Am. Chem. Soc.*, 107, 2000, 1985.
150. **Collman, J. P., Kodadek, T., Raybuck, S. A., Brauman, J. I., and Papazian, S. A.,** Mechanism of oxygen atom transfer from high valent iron porphyrins to olefins; implications to the biological epoxidation of olefins by cytochrome P-450, *J. Am. Chem. Soc.*, 107, 4343, 1985.
151. **Groves, J. T. and Watanabe, Y.,** On the mechanism of olefin epoxidation by oxo-iron porphyrins; direct observation of an intermediate, *J.Am. Chem. Soc.*, 108, 507, 1986.
152. **Dicken, C. M., Woon, T. C., and Bruice, T. C.,** Kinetics and mechanisms of oxygen transfer in the reaction of p-cyano-N,N-dimethylaniline N-oxide with metalloporphyrin salts. III. Catalysis by (meso-Tetrakis(2,6-dichlorophenyl)porphinato)iron (III) chloride, *J. Am. Chem. Soc.*, 108, 1636, 1986.
153. **Tyson, C., Lipscomb, J. D., and Gunsalus, I. C.,** The roles of putidaredoxin and $P450_{cam}$ in methylene hydroxylation, *J. Biol. Chem.*, 247, 5777, 1972.
154. **Peterson, J. A., Ishimura, Y., and Griffin, B. W.,** *Pseudomonas putida* cytochrome P-450; characterization of an oxygenated form of the hemoprotein, *Arch. Biochem. Biophys.*, 149, 197, 1972.
155. **Rahimtula, A. D. and O'Brien, P. J.,** Hydroperoxide catalyzed liver microsomal aromatic hydroxylation reactions involving cytochrome P-450, *Biochem. Biophys. Res. Commun.*, 60, 440, 1974.
156. **Rahimtula, A. D. and O'Brien, P. J.,** The role of cytochrome P-450 in the hydroperoxide-catalyzed oxidation of alcohols by rat liver microsomes, *Europ. J. Biochem.*, 77, 201, 1978.
157. **Nordblom, G. D., White, R. E., and Coon, M. J.,** Studies on hydroperoxide-dependent substrate hydroxylation by purified liver microsomal cytochrome P-450, *Arch. Biochem. Biophys.*, 175, 524, 1976.
158. **Lichtenberger, F., Nastainczyk, W., and Ullrich, V.,** Cytochrome P450 as an oxene transferase, *Biochem. Biophys. Res. Commun.*, 70, 939, 1976.
159. **Berg, A., Ingelman-Sundberg, M., and Gustafsson, J. A.,** Purification and characterization of cytochrome $P\text{-}450_{meg}$, *J. Biol. Chem.* 254, 5264, 1979.
160. **Sligar, S. G., Kennedy, K. A., and Pearson, D. C.,** Chemical mechanisms for cytochrome P-450 hydroxylation; evidence for acylation of heme-bound dioxygen, *Proc. Natl. Acad. Sci. U.S.A.*, 77, 1240, 1980.
161. **White, R. E., Sligar, S. G., and Coon, M. J.,** Evidence for a homolytic mechanism of peroxide oxygen; oxygen bond cleavage during substrate hydroxylation by cytochrome P-450, *J. Biol. Chem.*, 255, 11108, 1980.
162. **Blake, R. C. and Coon, M. J.,** On the mechanism of action of cytochrome P-450, *J. Biol. Chem.*, 256, 5755, 1981.
163. **Ortiz de Montellano, P. R.,** Ed., *Cytochrome P-450; Structure, Mechanism, and Biochemistry,* Plenum Press, New York, 1986.
164. **Whitelock, J. P., Jr.,** The regulation of cytochrome P-450 gene expression, *Annu. Rev. Pharmacol. Toxicol.*, 26, 333, 1986.
165. **Ishimaru, A. and Yamazaki, I.,** The carbon monoxide-binding hemoprotein reducible by hydrogen peroxide in microsomal fractions of pea seeds, *J. Biol. Chem.*, 252, 199, 1977.
166. **Ishimaru, A.,** Purification and characterization of solubilized peroxygenase from microsomes of pea seeds, *J. Biol. Chem.*, 254, 8427, 1979.
167. **Yoshida, T. and Kikuchi, G.,** Features of the reaction of heme degradation catalyzed by the reconstituted microsomal heme oxygenase system, *J. Biol. Chem.*, 253, 4230, 1978.
168. **Yoshida, T., Noguchi, M., Kikuchi, G., and Sano, S.,** Degradation of mesoheme and hydroxymesoheme catalyzed by the heme oxygenase system: involvement of hydroxyheme in the sequence of heme catabolism, *J. Biochem.*, 90, 125, 1981.
169. **Yoshida, T. and Kikuchi, G.,** Purification and properties of heme oxygenase from pig spleen microsomes, *J. Biol Chem.*, 253, 4224, 1978.
170. **Yoshida, T. and Kikuchi, G.,** Purification and properties of heme oxygenase from rat liver microsomes, *J. Biol. Chem.*, 254, 4487, 1979.
171. **Yoshida, T., Noguchi, M., and Kikuchi, G.,** Oxygenated form of heme-heme oxygenase complex and requirement for second electron to initiate heme degradation from the oxygenated complex, *J. Biol. Chem.*, 255, 4418, 1980.
172. **Noguchi, M., Yoshida, T., and Kikuchi, G.,** Degradation of mesoheme and hydroxymesoheme catalyzed by the heme oxygenase system; involvement of hydroxyheme in the sequence of heme catabolism, *J. Biochem. (Tokyo),* 90, 1961, 1981.
173. **Yoshida, T., Noguchi, M., and Kikuchi, G.,** A new intermediate of heme degradation catalyzed by the heme oxygenase system, *J. Biochem. (Tokyo),* 88, 557, 1980.
174. **Yoshida, T., Noguchi, M., and Kikuchi, G.,** Inability of the NADH-cytochrome b_5 reductase system to initiate heme degradation yielding biliverdin IXα from the oxygenated form of heme-heme oxygenase complex, *FEBS Lett.*, 115, 278, 1980.

Section VII
ATP-ases, Nucleotide Binding Carriers and Nucleotidyl Transferases

. . . *as the outcome of intensive study of the intermediate reactions in fermentation and the relation between muscular action and metabolism, it later become evident that the primary phosphate ester bond of hexose changes metabolically into a new type of energy-rich phosphate bond.* (K. Lohmann and O. Meyerhof, Biochem. Z., 273, 60, 1934). Fritz Lipmann, Metabolic Generation and Utilization of Phosphate Bond Energy, p. 100, in *Advances in Enzymology and Related Subject* (F. F. Nord and C. H. Werkman, editors). Vol. I, Interscience Publishers, Inc., New York, 1941.

Chapter 14

Ca^{2+}, Mg^{2+}-ATPase (MICROSOMAL)

Hitoshi Takenaka, Michihiro Sumida, and Minoru Hamada

TABLE OF CONTENTS

I. INTRODUCTION

Calcium ion (Ca^{2+}) is now recognized to play a major role in regulation of various biological processes, that include muscular contraction, protein phosphorylation, and cellular death.[1-3] The Ca^{2+} concentration in the cytosol of muscle cells is generally kept as low as 10^{-8} *M* at the resting state.[4-6] This is made possible by the actions of Ca^{2+}-specific transporting devices embedded in the phospholipid bilayer of the surface membrane (plasma membrane, PM), the intracellular membrane (endoplasmic reticulum, ER, or sarcoplasmic reticulum, SR), and mitochondria. The ER often acts as the intracellular Ca^{2+} pool, which releases Ca^{2+} to raise the cytosolic Ca^{2+} level, to the order of 10^{-5} *M* in muscle cells, responding to various stimuli. The PM and ER differ each other in terms of their localization and the chemical nature,[7] although both systems function similarly to isolate the cytosolic milieu from extracellular or intrareticular spaces. Among the devices, to translocate Ca^{2+}, the adenosine triphosphatase (ATPase), which utilizes the energy of ATP hydrolysis to exclude Ca^{2+} from the cytosol, is the most well-characterized. As the Ca^{2+} pumping ATPase depends upon the Ca^{2+} concentration and requires magnesium ions (Mg^{2+}) as the cofactor, it is often called Ca^{2+},Mg^{2+}-ATPase, although we will use Ca^{2+}-ATPase for simplification in this chapter.

The studies on structure and function of the Ca^{2+}-ATPase have been carried out using mainly the microsomal fractions of skeletal muscle, skeletal sarcoplasmic reticulum (SkSR). The research has recently been extended to cardiac sarcoplasmic reticulum (CSR), and ERs of smooth muscles as well as non-muscle cells. We will start this review by summarizing the properties of SkSR Ca^{2+}-ATPase. We will then briefly compare the characteristics of Ca^{2+}-ATPases in other membrane systems in conjunction with regulatory mechanisms, which are becoming a major concern in the field.

There are many excellent reviews which have been published in the past decade, and about various aspects of the Ca^{2+}-ATPase.[2,6-15] Readers are suggested to refer to those reviews for a further understanding of the enzyme.

II. SKELETAL SARCOPLASMIC RETICULUM Ca^{2+}-ATPase

A body of studies have demonstrated that the organization of sarcoplasmic reticulum (SR) is in a good agreement with the fluid mosaic model. Isolated SR vesicles show the right side out orientation, in which Ca^{2+}-ATPase molecules project the cytosolic face toward the outer surface of the vesicles. Moreover, electron microscopic observations and fluorescence energy transfer experiments revealed that the Ca^{2+}-ATPase molecules form ordered arrays in the plane of the membrane.[16,17]

Lipids make up approximately 80 mol% of SkSR, the majority of which are phospholipids:[18,19] phosphatidylcholine (~70 mol%), phosphatidylinositol (~16 mol%), phosphatidylserine (9 mol%), sphingomyelin (~4 mol%), and cardiolipin (~0.3 %). SR contains several major proteins;[20-23] the Ca^{2+}-ATPase (Mr = 100,000 to 113,000), calsequestrin (Mr = 44,000 or 63,000, depending on the electrophoretic conditions), high-affinity Ca^{2+}-binding protein (M_{55}, Mr = 55,000), intrinsic glycoprotein (Mr = 53,000 and 160,000), and proteolipid (Mr = 12,000). Centrifugation with a discontinuous concentration (density) gradient of sucrose has been successfully applied for separation of the crude SkSR into three major fractions.[24,25] Vesicles of the lowest density (LSR) is SkSR derived from the longitudinal SR, while those showing the highest density (HSR) from the terminal cisternae. LSR mainly uptake Ca^{2+} from the cytosol, while HSR is considered to participate in Ca^{2+} release into the cytosol to initiate muscular contraction. Properties of the intermediate fraction of SkSR have not been defined yet. The amount of Ca^{2+}-ATPase in LSR is as high as 90% of the total SkSR protein.

A. STRUCTURE OF SkSR Ca^{2+}-ATPase

Ca^{2+}-ATPase molecule was found to be a complete transmembranous protein, the shape of which was predicted by Herbette et al.[26] using X-ray and neutron diffraction analyses. The molecular weight of Ca^{2+}-ATPase is estimated to be 100,000 to 113,000 Da, which is made up by a single polypeptide. The recent achievement of MacLennan et al.,[27] who deduced the amino acid sequence from its complementary DNA sequence, revealed that Ca^{2+}-ATPase has a mass of 109,763 Da with 997 amino acids. According to them, the Ca^{2+}-ATPase molecule has three cytoplasmic domains joined to a set of ten transmembrane helices. They postulated that ATP bound to one cytoplasmic domain would phosphorylate an aspartic residue in an adjoining cytoplasmic domain. Luminescence energy transfer measurement by Scott[28] predicted that the two Ca^{2+} binding sites were close to each other, while the Ca^{2+} and ATP sites were separated by 3.5 to 4.7 nm. Highsmith and Murphy[29] recently reported that the distance between ATP and Ca^{2+} sites would be greater than 2.6 nm. Gangola and Shamoo[30] suggested that a certain portion in the Ca^{2+}-ATPase would have ionophoric properties.

B. MOLECULAR ARCHITECTURE OF Ca^{2+} PUMP

1. ATPase-ATPase Interaction

In order to define a mechanism which is responsible for Ca transport, many investigators have made an effort to identify whether or not Ca^{2+}-ATPase can perform both ATP hydrolysis and Ca^{2+} translocation. It is now agreed that the monomeric ATPase obtained by disruption of membrane with detergent has full activity in ATP hydrolysis.[31-33] A series of cross-linking experiments have strongly suggested that Ca^{2+}-ATPase molecules are closely packed in the membrane.[34-36] Kinetic analyses of Ca^{2+} transport reaction has also postulated the occurrence of ATPase-ATPase interactions by which two types of EPs with high and low Ca^{2+} affinity are produced in separate halves of an enzyme dimer.[37,38] Analysis by fluorescence energy transfer between Ca^{2+}-ATPase molecules indicated that the distance between two enzymes would be less than 5 nm.[39]

In contrast, Møller et al. showed that monomeric enzyme by itself is capable of participating in energy transduction including Ca^{2+}-dependent ATP hydrolysis, Ca^{2+} binding, and ATP binding at high affinity sites, although modulation by Low Mg^{2+}-ATP concentration would be due to protein-protein interaction.[40-42] However, Yantorno et al.[43] later confirmed the necessity of oligomeric interactions in the SkSR membrane to transport Ca^{2+}. Anatomical approaches to look at the molecular shape of the Ca^{2+} pump in the SkSR membrane were also attempted. Hymel et al.[44] using a radiation inactivation technique concluded that the minimal molecular weight for Ca^{2+} pumping would be 210,000—250,000 Da, which corresponds to a dimer of Ca^{2+}-ATPase. Napolitano et al.[45] analyzed optical diffraction profiles of electron micrographs of SkSR. They found that one particle in the SkSR membrane occupies about 12,000Å^2, which indicated that the particle is a dimer of the ATPase.

2. Conformational Change Associated with Ca^{2+} Transport Reaction

Conformational changes of Ca^{2+}-ATPase during Ca^{2+} transport have been analyzed mainly by physical techniques, tryptic digestions, and chemical modifications of specific amino acid residues.[8,10,46-48] The shape of a Ca^{2+} pump unit undergoes a significant change during the reaction, as found by Blasie et al.[49] They analyzed X-ray and neutron diffraction profiles after starting a synchronized reaction by flash photolysis of the caged ATP. However, the method sums up all the protein motions occurring in the plane of the membrane. From nitrophenylation of lysine residues, Yamamoto and Tonomura postulated that Ca^{2+}-ATPase would undergo a rotational movement.[8,48] In contrast, Froehlich and Taylor[37,50] proposed a model in which two ATPase monomers interact with each other in a flip-flop motion. Saturation transfer of ESR by Thomas and Hidalgo[51] and phosphorescence transfer by Restall et al.[52] have shown that Ca^{2+}-ATPase exhibits rotation, which is probably a rotation about the axis normal to the plane of membrane.

Yamada and Ikemoto[53] and Coan and Inesi[54] reported that binding of Ca^{2+} to the high affinity site induces a conformational change of the enzyme. Coan et al.[55] felt that a large conformational change of Ca^{2+}-ATPase took place on binding Ca^{2+} and ATP, but that there was no significant difference in the conformations of the two phosphorylated enzyme species, E_1P and E_2P.

It is of great interest that vanadate causes an ordered array of Ca^{2+}-ATPase in the plane of the SkSR with a stabilization of the E_2 enzyme form. Dux and Martonosi[6,56] found that treatment of SkSR with vanadate, Na_3VO_4, caused Ca^{2+}-ATPase molecules to form crystalline arrays in the membrane, which was inhibited by Ca^{2+} or ATP, but not by ADP nor by nonhydrolyzable ATP analogs. Pick[57] later reported vanadate binds to the low affinity ATP binding site in E_2. Jona and Martonosi[58] later added a fact that the crystallization is promoted by an inside positive membrane potential but inhibited by a negative potential. Dux et al.[59] recently suggested that inside positive membrane potentials or lanthanide ions increase the E_1 conformation or the monomeric fraction, while the inside negative potential or Ca^{2+} favors E_2 or the dimer conformation. Hasselbach et al.[60] indicated that vanadate binding induces a transition of the external (high affinity) to internal (low affinity) Ca^{2+} site, and it causes asymmetrical arrangement of the electron dense protein particles to be symmetrically distributed.

C. FUNCTIONS OF Ca^{2+}-ATPase

1. Forward Reaction of Ca^{2+}-ATPase

The Ca^{2+}-ATPase utilizes the energy liberated from ATP hydrolysis for transporting Ca^{2+} into the intravesicular lumen. The simplified reaction scheme is presented below. Suffixes for Ca^{2+} stand for the number and those for E represent the possible conformer.

$2Ca^{2+}$(out) ATP

Ca_2E_1 ADP

E_1 Ca_2E_1ATP ⟷ $[Ca_2E_1ATP]$ ⟷ Ca_2E_1P

E_1ATP

ATP $2Ca^{2+}$(out)

E_2 ⟷ E_2Pi ⟷ $E_2{\cdot}P$ ⟷ E_2P ⟷ $[Ca_2E_2P]$

Pi $2Ca^{2+}$(in) (Str. 1)

The forward reaction of the Ca pump begins when substrates, two Ca^{2+} and one ATP, bind to the Ca^{2+}-ATPase at the outside of the vesicles to form Ca_2E_1ATP complex. After hydrolyzing ATP and liberating ADP into the extravesicular space, the enzyme forms a phosphorylated intermediate (EP), Ca_2E_1P. The phosphoryl group attacks a β-carboxyl group of an aspartyl residue to form an acylphosphate linkage as shown by Degani and Boyer.[61] The amount of EPs can be detected as the acid-stable phosphorylated protein after quenching the ATPase reaction by an addition of acid, such as trichloroacetic acid. An addition of ADP readily decomposes Ca_2E_1P with the synthesis of ATP, so that this intermediate has a high energy phosphoryl bond. The next reaction step includes the conversion of Ca_2E_1P

to a transient intermediate, $[Ca_2E_2P]$, in which the Ca^{2+}-binding sites face the intravesicular lumen with low affinity. Ca_2E_2P liberates Ca^{2+} into the intravesicular space when H^+ replaces the site for Ca^{2+}. However, the site and amount of Mg^{2+} binding to the ATPase molecule are still controversial. Neither Ca_2E_2P nor E_2P decomposes when only ADP is added to the Ca^{2+}-filled SkSR vesicles, while a simultaneous addition of ADP with EGTA, a Ca^{2+} chelating agent, propels the reaction backward to form ATP. These intermediates are considered to have a low energy phosphate bond. Regarding the sensitivity against ADP, Ca_2E_1P and E_2P are also called ADP-sensitive and ADP-insensitive EP, respectively. The conversion step is thus regarded as the very step at which Ca^{2+} is translocated across the membrane. After releasing Pi to the extravesicular space, the enzyme forms E_2, a "refractory" state of enzyme that must be transformed to E_1 before the cycle starts again. There is general agreement that the rate-determining step in the Ca^{2+} pump reaction takes place at one of the steps of EP degradation and/or E_2 to E_1 transformation. In the following, each elementary step will be further discussed.

a. Binding of Ca^{2+} and ATP

The beginning of the Ca^{2+} uptake reaction is the step at which Ca^{2+}-ATPase binds to its substrates. It is of importance to determine which binds first to the enzyme, Ca^{2+} or ATP. Tonomura et al.,[8,9] based on their series of kinetic analyses of the Ca^{2+}-ATPase reaction, proposed a model in which Ca^{2+} and ATP bind to the enzyme in a random rapid equilibrium manner. However, an hypothesis in which the Ca^{2+} binding precedes the ATP binding has recently been proposed.[62] In addition to acceleration of the ATP binding by Ca^{2+}, Ca^{2+} binding is accelerated several times by Mg^{2+}-ATP.[63]

Ikemoto,[64] using an equilibrium dialysis method, found three classes of Ca^{2+} binding sites in the purified ATPase of SkSR: one with a high (α), one with an intermediate (β), and three with low (γ) affinity for Ca^{2+} at 0°C. Raising temperature to 22°C shifted β to α without affecting the γ sites. Temperature dependent profiles of the Ca^{2+}-ATPase activity have been shown by many investigators.[8-14]

The mode of ATP binding has also been studied using ATP analogs.[65-73] From a structural viewpoint of ATP, the hydroxyl group at the 3′-position of ribose ring is essential for ATP hydrolysis and for further activation of ATPase reaction by ATP as well.[67] Both of the two stereoisomers (R and S types) of adenosine 5′-O-(2-thiotriphosphate) ($ATP_\beta S$) supported both of Ca^{2+} uptake and nucleotide hydrolytic activities, while $ATP_\beta S$-$ADP_\beta S$ and ATP-$ADP_\beta S$ exchange were slowed, but EP formation was potentiated.[68] Yasuoka et al.[69] found that adenosine 5′-O-(3-thiotriphosphate) formed phosphoenzyme without its hydrolysis at 0°C. Affinity labelling of Ca^{2+}-ATPase with photoreactive ATP analogs demonstrated that Ca^{2+}-ATPase possesses more than one ATP binding site due to different affinities for ATP.[70] 2′,3′-O-(2,4,6-Trinitrophenyl) ATP, TNP-ATP, has often been used as a reporter group for the detection of conformational changes of the ATPase. The amount of TNP-ATP binding revealed that each Ca^{2+}-ATPase molecule has one nucleotide site.[71] Watanabe and Inesi[71] pointed out that the TNP-ATP binding site is composed of two distinct sites with different affinities. Bishop et al.[72] and Dupont et al.[73] later supported a possibility that TNP-ATP binds to a noncatalytic site, to which ATP binding accelerates ATP hydrolysis.

Fluorescein 5′-isothiocyanate (FITC) has also been used for analyses of the ATP binding sites. Pick and Bassilian[74] succeeded in inhibiting completely the Ca^{2+}-ATPase activity by labelling the two Ca^{2+}-ATPase polypeptides with one FITC; but FITC did not inhibit Ca^{2+} uptake supported by acetylphosphate. Pick[75] later suggested that FITC is directed towards only one ATP binding site in two enzyme conformations, the phosphorylation site in the E_1 form and the 'regulatory' site in the E_2 form. Mitchinson et al.[76] isolated the FITC labelled peptide made of 164 residues from the phosphorylation site. They also found that complete inhibition of the Ca^{2+}-ATPase required 4.5 nmol/mg of SR but 7.4 nmol/mg of purified ATPase, possibly due to aggregation of Ca^{2+}-ATPase to form a dimer in the membrane.

Froud and Lee[77] recently analyzed the effect of Ca^{2+} on the fluorescence derived from FITC and concluded that the Ca^{2+}-ATPase exists as either E_1 or E_2, with the equilibrium between E_1 and E_2 shifted toward E_1 as the pH was raised.

b. EP Formation and E_1P-E_2P Conversion

Since the formation and decomposition of the EPs are the key reactions of Ca^{2+} transport, extensive studies have been conducted from various points of view. Sumida and Tonomura[78] obtained evidence that the EP formation and its decomposition were directly coupled to Ca^{2+} transport. Shigekawa et al.[79-81] found that the EP formed from ATP in a low KCl medium did not decompose when ADP was added (ADP-insensitive EP, E_2P), while most of EP formed in the presence of high KCl concentration was ADP-sensitive EP, E_1P. They also found that conversion from E_1P to E_2P was accelerated by Mg^{2+} but inhibited by Ca^{2+}, whereas acceleration by Ca^{2+} of the E_2P to E_1P conversion was not affected by Mg^{2+} significantly. These facts indicated that E_1P and E_2P occur sequentially in the Ca^{2+}-ATPase reaction. Nakamura and Tonomura[82] and Takisawa and Makinose[83] showed that the two Ca^{2+} bound to E_1P were found to be occluded and released when E_1P was converted to E_2P. Occlusion of Ca^{2+} in a solubilized monomeric Ca^{2+}-ATPase was recently suggested by Klemens et al.[84] who analyzed the ESR spectra derived from gadlinium ions occupying two Ca^{2+} sites. Nakamura et al.[85] recently reported that the Ca_2E_1ATP complex, after hydrolysis of ATP, proceeded via two alternate pathways to form Ca_2E_2P:

$$\text{i)}\quad Ca_2EATP \rightarrow Ca_2E_1P\ ADP \rightarrow Ca_2E_2P\ ADP \underset{}{\overset{ADP}{\rightleftharpoons}} Ca_2E_2P$$

$$\text{ii)}\quad Ca_2EATP \rightarrow Ca_2E_1P\ ADP \overset{ADP}{\rightleftharpoons} Ca_2E_1P \rightarrow Ca_2E_2P \qquad \text{(Str. 2)}$$

They indicated that the upper pathway is slow and effective when the intravesicular Ca^{2+} is sufficiently high to increase the amount of EP, while the lower path is rapid and is prevalent at low intravesicular Ca^{2+} concentration.

Dynamic equilibria between E_1P and E_2P have been suggested to correlate with rapid Ca^{2+} exchange across the membrane. Takakuwa and Kanazawa[86,87] indicated an occurrence of Ca^{2+} exchange mediated by E_1P without energy consumption. Takenaka et al.[88] later found that rapid Ca^{2+} exchange was accompanied with the increase in the E_1P/E_2P ratio. Chiesi and Wen[89] suggested that rapid shuttling between two EPs allows Ca^{2+} release from the inside of SR. Gerdes and Møller[90] indicated that the outward movement of Ca^{2+} could not be carried out by the Ca^{2+}-ATPase.

c. EP Decomposition and E_1 to E_2 Conversion

Ca_2E_2P has a low affinity for Ca^{2+} and results in the liberation of Ca^{2+} into the intravesicular space. Liberating Ca^{2+}, the Ca^{2+}-ATPase forms EPi, in which no covalent bond between the enzyme and Pi is formed. The presteady state reaction of Ca^{2+}-ATPase of SkSR was analyzed by Froehlich and Taylor,[37,50] who found three phases in the time course of Pi liberation, a lag phase, a rapid burst phase, and the steady state phase. They postulated an occurrence of a phosphate-bound intermediate, which would be labile in an acidic condition. Sumida et al.[91] later confirmed this model by a computer simulation. Takisawa and Tonomura[9,92] later examined the transient kinetics further to conclude that the complicated timecourses of Pi liberation and EP formation could be explained if the acid-labile E_2·P or E_2Pi had been formed through at least two acid-stable EP intermediates.

After liberation of Pi, the enzyme forms the last intermediate, E_2. In contrast to E_1, E_2 has a low affinity for Ca^{2+} while it is the starting enzyme for resynthesis of ATP by a backward reaction.

d. Activation of Ca^{2+}-ATPase by ATP

Tonomura et al.[8,9] found that a double-reciprocal plot of the ATPase activity against the ATP concentration diverged downwards from a straight line at the higher ATP concentration range. Dupont[93] later observed two classes of ATP binding sites in the absence of Ca^{2+}. Transient state kinetic analyses by Froehlich and Taylor[37,50] revealed that the apparent presteady state rate of EP formation and Pi liberation was accelerated by an increase in ATP concentration, and predicted the occurrence of a second ATP binding site for activation. Taylor and Hattan[94] and Verjovski-Almeida and Inesi[95] later added evidence of a substrate activation. Nakamura and Tonomura[96] analyzed further the ATP binding in the presence and absence of Ca^{2+} and found that the addition of Ca^{2+} accelerated ATP hydrolysis to form EP, while no significant decrease was associated with the amount of ATP on the regulatory site. Furthermore, ATP binding to the regulatory site shifts the equilibrium on the catalytic site from E + ATP toward EP + ADP. Cable et al.,[97] however, recently hypothesized that stimulation of monomeric ATPase activity is caused by the binding of ATP to the catalytic site but not to the regulatory site after ADP dissociation but before Pi dissociation.

e. Roles of Mg^{2+}

Mg^{2+} shows two distinct roles in the Ca^{2+}-ATPase reaction: (1) Mg^{2+} forms an Mg^{2+}-ATP complex which is the substrate for the Ca^{2+}-ATPase reaction and (2) Mg^{2+} accelerates EP decomposition to activate the reaction cycle. Yamada and Ikemoto[98] suggested from kinetic studies that Mg^{2+} permits the transconformation of E_1P to E_2P. Guillain et al.[99] later reported that Mg^{2+} competed with Ca^{2+} for the high affinity Ca^{2+} binding site to make the E_2 form stable, from their titration analysis of intrinsic fluorescence changes of Ca^{2+}-ATPase. Chiesi and Inesi,[100] however, reported that the activation of Ca^{2+}-ATPase by Mg^{2+} was due to Mg^{2+} occupancy of an allosteric site, other than Ca^{2+} binding site, on the outer surface of SR vesicles, and not due to acceleration of an antiport mechanism. Nakamura[101] reported that Ca^{2+} bound to EP was released by Mg^{2+} before decomposition of EP. Shigekawa et al.[102] later indicated that the Mg^{2+} which activated EP decomposition was the Mg^{2+} derived from the Mg^{2+}-ATP complex. Daiho et al.[103] and Salalma and Scarpa[104] provided evidence that Mg^{2+} was not countertransported with Ca^{2+} during the reaction.

2. Effect of pH on Ca^{2+}-ATPase Reaction

Ca^{2+} transport reaction is significantly affected by pH,[105-111] KCl concentration,[79-81,106,107,112] osmotic pressure,[113] temperature,[111,114] and dimethyl sulfoxide.[112,115,116] These are attributed to the electrogenic properties of the reaction,[117-119] and the reaction depends upon the water activity as well. Verjovski-Almeida and de Meis[105] indicated that the Ca^{2+} binding is influenced by the H^+ concentration in both the outside and the inside of the vesicles. Inesi and Hill,[108] in a theoretical treatment, summarized the cooperative binding of Ca^{2+} and H^+, and assumed that the Ca^{2+} pump units would have two interacting domains permitting competitive binding of two cations. Yamaguchi and Kanazawa[109] recently demonstrated that increased H^+ concentration induced H^+ binding from the inside with coincident acceleration of Ca^{2+} dissociation from the low Ca^{2+} affinity site in E_2. This was later confirmed by Fassold and Hasselbach[110] who found that Ca^{2+} transport was facilitated by low intravesicular pH, while Ca^{2+} exchange across the membrane and ATP-ADP exchange were both rapid at an intermediate pH. This indicates that pH affects a reaction step which occurs later in the cycle.

3. Alternative Pathway for ATP Hydrolysis

We have described the Ca^{2+} uptake reaction according to the scheme shown above. However, there are indications that ATP and its analogs can be hydrolyzed via alternate pathways. Nakamura and Tonomura[120] found that p-nitrophenylphosphate could be hydrolyzed via two routes, one of which formed EP, while the other formed an E(p-nitrophenol) complex. Nakamura[121] recently showed that EP decomposition at 0°C proceeded via two routes with different rate constants. His results indicated that mM concentration of Ca^{2+} favored the slower route of EP decomposition. Van Winkle et al.[122] using GTP as the substrate showed that in the absence of Ca^{2+}, GTP hydrolysis was accompanied without EP formation.

SR ATPase activity is usually associated with ATPase activity without the Ca^{2+} dependency, which is called the Ca^{2+}-independent or basal ATPase activity. The basal ATPase has not been well characterized yet, although Inesi et al.[123] implied that the basal and Ca^{2+}-dependent ATPases would be in an equilibrium which relates to the physical state of the enzyme molecules in its membrane environment. It is interesting that that $ATP_{\beta}S$ strongly attenuated the basal ATPase activity.[68]

4. The Backward Reaction of Ca^{2+}-ATPase

All the steps in the Ca^{2+}-ATPase are reversible.[6-13,124-136] It is evidenced by the findings of rapid ATP-Pi exchange, where medium Pi is incorporated into the γ-phosphoryl group of ATP, and ATP-ADP exchange.

Yamada and Tonomura[125] and Makinose[124,126] showed that ATP resynthesis utilized the chemiosmotic energy due to the concentration gradient of Ca^{2+} across the membrane, and the efflux of two Ca^{2+} were found to be coupled with formation of one ATP. Yamada et al.,[127] furthermore, demonstrated that phosphorylation took place on a 100,000-Da protein, which showed the same characteristics as EP formed in the forward direction, indicating that the occurrence of the backward reaction mediated by Ca^{2+}-ATPase. deMeis's group has also accumulated evidence that Ca^{2+}-ATPase catalyses the backward reaction to resynthesize ATP from ADP and Pi.[128-132] Masuda and deMeis[129] showed that ATP formation was activated in the presence of a high Pi concentration at a low pH, but it was inhibited strongly by Ca^{2+}. This fact was explained by a mechanism in which phosphorylation by Pi of the enzyme having a low affinity for Ca^{2+} (E_2) was followed by transfer of a phosphate to ADP in the presence of Ca^{2+}. Kanazawa and Boyer[134] found that the SkSR completely devoid of Ca^{2+} catalyzed ^{18}O exchange between Pi and water, which was activated by Mg^{2+} while inhibited by external Ca^{2+} at a concentration similar to that for activation of EP formation from E and ATP. They concluded from these findings that Mg^{2+} activation occurred at the outer surface of the membrane. Mg^{2+} was later found to facilitate the E_2P formation. Kanazawa[135] later found that solubilized Ca^{2+}-ATPase also resynthesized ATP from EP and ADP. The EP formation in this system required a high concentration of Mg^{2+} and was significantly enhanced by raising the temperature.

Fernandez-Belda and Inesi[136] recently suggested that high-affinity binding and low-affinity binding of Ca^{2+} would occur in different locations in the protein, indicating that Ca^{2+} translocates from one to another as a result of EP formation. After transfer of a phosphoryl group in EP to ADP as described before, Ca_2E_1ATP is formed.

Shigekawa and Kanazawa recently found that ATP formed from ADP and EP could dissociate slowly.[137] This indicated that an additional intermediate, in which ATP is occluded, was formed. Jencks et al.[138,139] also found an occurrence of ATP occlusion and later predicted that a different Ca_2E_1ATP intermediate should be formed in the forward and backward reactions.

III. CARDIAC SARCOPLASMIC RETICULUM Ca^{2+}-ATPase

Cardiac sarcoplasmic reticulum (CSR) plays a role similar to SkSR in muscular contraction-relaxation.[140] The lipid composition of CSR is not the same as that of SkSR,[141] and CSR contains more associated proteins than the SkSR. DeFoor et al.[142] pointed out that the Ca^{2+}-ATPases of SkSR and CSR are immunologically dissimilar. Jones and Cala[143] separated the crude CSR into two fractions, which resembled the LSR and HSR in SkSR. Radiation inactivation analysis of CSR by Chamberlain et al.[144] indicated that a Ca^{2+} pump unit was composed of two Ca^{2+}-ATPase, which suggested that the Ca^{2+} pump in CSR is a dimeric structure as in the SkSR.

Shigekawa et al.[145] compared the Ca^{2+} pump activities of CSR and SkSR. They found that (1) the Ca^{2+}-ATPase of CSR formed EP with similar characteristics as that of the SkSR, (2) the CSR had a lower affinity for Ca^{2+} than SkSR, while their K_M values for Mg^{2+}-ATP were the same, and (3) no difference in the turnover number (ATPase activity/EP amount) was seen between them.

The most significant difference between SkSR and CSR is that CSR contains a specific protein named phospholamban.[146-150] Phosphorylation of phospholamban, which activates the Ca^{2+}-ATPase and Ca^{2+} transport reaction, is likely to be involved in regulation of cardiac muscle contractility.

Tada et al.[146] found that a cyclic adenosine 3′,5′-monophosphate (cAMP)-dependent protein kinase phosphorylated a single microsomal component of approximately 22,000 Da, which was named phospholamban and resulted in activation of Ca^{2+} transport of CSR. It did not phosphorylate the Ca^{2+}-ATPase at 100,000 Da. Louis et al.[151] later showed that phospholamban could be separated into its subunits by heating in sodium dodecylsulfate (SDS), while reassociation took place after cooling. This indicated that phospholamban protomer would have high hydrophobic properties. Imagawa et al.[152] later found that phospholamban phosphorylation occurred in five discrete steps, indicating that phospholamban should be composed of five subunits.

Hicks et al.[153] reported that phospholamban phosphorylation by cAMP-dependent protein kinase increased the Ca^{2+} sensitivity of Ca^{2+}-ATPase. Tada et al.[154] found that phosphorylation significantly increased the rate constant of EP decomposition (ATPase activity/EP amount). Sumida et al.[91,155] compared the kinetic properties of the Ca^{2+}-ATPase reaction of CSR containing a non-phosphorylated phospholamban, with that of the SkSR. With a quench-flow technique, they found that the presteady state rate of EP formation was the same for both SkSR and CSR. However, the rates of Ca^{2+} dissociation, phosphorylation of Ca^{2+}-free CSR, and dephosphorylation from EP were found to be slower than those of SkSR. Tada et al.,[156] also using a quench-flow technique, showed that phosphorylation accelerated Ca^{2+} binding to the enzyme, which resulted in an increase in the EP amount in the presence of ATP at high concentration, while the accelerated EP decomposition apparently lowered the EP amount at low ATP. Phospholamban was later found to be phosphorylated by a Ca^{2+}-calmodulin-dependent protein kinase as well. Chiesi et al.[157] and Imagawa et al.[152] showed that the sites for phosphorylation by cAMP-dependent and calmodulin-dependent protein kinases were distinct. Tada et al.[158] later showed that the effects of phosphorylation of both sites were additive on the EP formation, ATPase activity, the affinity for Ca^{2+}, and Ca^{2+} transport activity.

Simmerman et al.[159] recently identified a 36 amino acid sequence near the NH_2-terminus, which has a serine and a threonine residue and which are the sites for the phosphorylation by cAMP-dependent and calmodulin-dependent protein kinases, respectively. More recently, Tada's group[160,161] determined the complete amino acid sequence deduced from the complementary DNA sequence. Their results showed that the phospholamban monomer was 6,080 Da polypeptide. A majority of its residues appeared hydrophobic, and could hold the molecule in a lipid bilayer. The remaining portion (30 residues) has a hydrophilic nature, with a proline at the 21st position. Phosphorylatable serine and threonine residues were located at the 16th and 17th position, respectively. They hypothesized that phospholamban resides in the membrane as a pentamer.

Inui et al.[162] and Kirchberger et al.[163] recently reported that phosphorylation of phospholamban released the inhibitory action by the cytosolic part of unphosphorylated phospholamban on the Ca^{2+}-ATPase reaction.

Physiological significance of phospholamban phosphorylation has been presented. Lindemann et al.[164] found that administration of isopreterenol, which is a positive inotropic agent effective as the β-agonist, increased the amount of phosphorylated phospholamban, suggesting that cAMP-dependent protein kinase activated myocardiac functions. However, Lindemann and Watanabe[165] reported that an increased extracellular Ca^{2+} concentration induced increased contractility but not phosphorylation of phospholamban, which would have been expected through activation of calmodulin-dependent protein kinase.

IV. Ca^{2+}-ATPase AND Ca^{2+} REGULATION OF SMOOTH MUSCLE CELLS

A. Ca^{2+}-ATPase OF ARTERIAL SMOOTH MUSCLE

Like skeletal and cardiac muscles, the contraction-relaxation of arterial smooth muscle is also very likely regulated by the cytosolic Ca^{2+} concentration. Electron microscopic observations revealed that arterial smooth muscle contains intracellular membrane systems like SR in skeletal and cardiac muscles.[166] Microsomes isolated from aortic muscles possesses an ATP-dependent Ca^{2+} uptake activity, indicating that the endoplasmic reticulum participates in the regulation of the cytosolic Ca^{2+} concentration in the Ca^{2+} pool.[167,168] This hypothesis has been supported by a fact that the arterial muscle tissue is contracted after the addition of adrenaline, even after external Ca^{2+} had been removed by addition of EGTA.[169] Furthermore, leaky cells, which was skinned by a treatment with surfactant such as saponin, were capable of reducing the medium Ca^{2+}, which had been increased after addition of inositol trisphosphate.[170] However, the low purity and poor yield of microsomal preparations from the smooth muscle tissues have inhibited the research, in detail, on the mechanism of Ca^{2+} uptake reaction.

We recently characterized the properties of a Ca^{2+}-ATPase in aortic microsomes which forms an acid-stable phosphoenzyme like SkSR and CSR.[171-173] The results are shown in Table 1: (1) Ca^{2+}-ATPase activity depends upon the same concentration range of Ca^{2+} and ATP, (2) phosphorylation takes place in a 100,000 to 110,000-Da protein, (3) the phosphoprotein is of the acylphosphate type, as it decomposes in an alkaline condition and in the presence of hydroxylamine, (4) a time course of phosphoprotein formation is as rapid as that of EP of SkSR (see Reference 91), (5) the temperature dependence of phosphoprotein decomposition, which is the rate-limiting step, shows a break at 16°C, and the activation energies calculated are 25.6 and 14.6 kcal/mol below and above 16°C, respectively (see Reference 145), (6) phosphoprotein formation is strongly inhibited by Cd^{2+} and Hg^{2+} but not by NaN_3 or ouabain. These properties strongly indicate that smooth muscle ER also forms EP, although the amount of EP formed is markedly lower ($\sim$50 pmol/mg) than that of SkSR ($>$1 nmol/mg). These are probably due to the impure nature of the microsomal samples and/or abundance in associated proteins in the vesicles. Furukawa et al.[174] attempted to purify the Ca^{2+}-ATPase from the plasma membrane (PM) of bovine aortic smooth muscle by using a calmodulin affinity column chromatography. The eluted fraction contained the calmodulin-sensitive Ca^{2+}-ATPase, while our samples have shown no sensitivity to calmodulin. The size of their ATPase was 135,000 Da, which was characteristic of the erythrocyte PM.[175]

B. Ca^{2+}-ATPase IN OTHER SMOOTH MUSCLES

Research on Ca^{2+}-ATPase in microsomes have been extended to bovine uterine[176] and porcine gastric[177] muscles. These Ca^{2+}-ATPases were also dependent on Ca^{2+} concentration and phosphoproteins had been isolated as well. However, detailed analyses on Ca^{2+} transport of these microsomes and the reaction mechanism concerning formation and decomposition of phosphoproteins have just begun. Sumida et al. (unpublished observation) recently found that microsomal fractions prepared from chicken gizzard formed EP at the same level (about

TABLE 1
Characteristics of Phosphorylated Intermediates of the Ca^{2+}-ATPase from Various Tissues

Tissues	MW ($\times 10^3$ Da)	K_{Ca}[a] (μM)	NH_2OH sensitivity	Calmodulin activation	Location[b]	Other properties	Ref.
Muscle							
Rabbit skeletal	105	1.3	+	−	SR		8, 9
Canine cardiac	100—110	4.7	+	−	SR	d	8
Bovine aorta	100—110	0.4	+	−	ER		171—173
Bovine aorta	135	N.D.[b]	+	+	PM	e	174
Bovine uterus	100—110	1.7	+	N.D.	ER ?		176
Porcine stomach	100	N.D.	N.D.	−binding	ER		177
	135	N.D.	N.D.	+binding	PM		177
Nonmuscle animal							
Human erythrocyte	138	0.5	+	+[c]	PM		175
Porcine thyroid	105	1	+	N.D.	ER		181
Bovine aderenocortex	100—110	1	+	−	ER		182
Rat parotid	110	0.65	+	N.D.	ER		183
Rat pancreatic	100	0.025	+	N.D.	ER		184
Rat liver	118	0.35	+	N.D.	ER		185, 186
	110	21	N.D.	N.D.	PM	f	189
Bovine lens	105	N.D.	+	N.D.	ER		187
	105	N.D.	+	N.D.	PM		187
Murine erythroleukemic cells	138	0.13	N.D.	+	PM		187
Rat adipocyte	110	0.14	N.D.	+[c]	PM	g	190
Plants and yeast							
Oat root	105	N.D.	+	N.D.	PM		191
Corn root	N.D.	N.D.	N.D.	N.D.	PM		192
Yeast (*S. cerevisiae*)	100	N.D.	+	N.D.	PM		193

Note: K_{Ca} stands for the Ca^{2+} concentration required for half-maximal formation of phosphorylated intermediate and calmodulin sensitivity is also in terms of phosphorylated intermediate formation, unless otherwise specified in the table.

[a] Ca^{2+} concentration required for half-maximal activation of Ca^{2+}-ATPase.

TABLE 1 (continued)
Characteristics of Phosphorylated Intermediates of the Ca^{2+}-ATPase from Various Tissues

[b] Abbreviations used are: SR, sarcoplasmic reticulum; ER, endoplasmic reticulum; and PM, plasma membrane (erythrocyte type). N.D.: Not determined.
[c] Stimulation by calmodulin was observed on the Ca^{2+}-ATPase activity.
[d] Phospholamban phosphorylation activated the Ca^{2+} uptake reaction.
[e] The Ca^{2+}-ATPase was purified with a calmodulin affinity column.
[f] Glucagon inhibited the decomposition of phosphorylated intermediate.
[g] Insulin stimulated the Ca^{2+}-ATPase activity.

40 pmol/mg) as bovine aortic ER. It is thus likely that every kind of muscle contains the intracellular membrane system which regulates the cytosolic Ca^{2+} concentration in a similar manner.

V. Ca^{2+}-ATPase IN NONMUSCULAR ANIMAL TISSUES

The cytosolic Ca^{2+} concentrations in nonmuscular tissues are also kept below 10^{-6} *M* by Ca^{2+}-ATPase embedded in ER and PM.[178,179] Change in the intracellular Ca^{2+} concentration is considered to influence the metabolic conditions such as energy supplying the glycolytic pathways.[180] Phosphorylation reactions like EP formation by Ca^{2+}-ATPase of SkSR have been detected in various tissues such as the thyroid, adrenocortex, liver, and adipose tissues as shown in the Table 1. Although these phosphoproteins have similar characteristics to the EP of SkSR, i.e., which form an acid-stable and hydroxylamine-sensitive acylphosphate, it may be adequate to classify them into two categories: (1) the SkSR type and (2) the erythrocyte PM type (see References 8 and 175). The common properties of EP of the SkSR type are (1) a 100,000- to 110,000-Da protein which is responsible for phosphorylation, (2) phosphorylation is not sensitive to calmodulin, and (3) phosphorylation takes place in the ER fractions. On the other hand, erythrocyte PM type is characterized by (1) a 130,000 to 140,000 Da protein responsible for phosphorylation, (2) calmodulin activates the phosphorylation, (3) it is associated with the PM fractions, and (4) the phosphorylation is highly sensitive to inhibition by vanadate. Both of them require 0.5 to 1.0 μM of Ca^{2+} for half-maximal activation (K_{Ca}) of phosphoprotein formation, which suggests that both types of enzyme regulate the cytosolic Ca^{2+} concentration.

Detailed kinetic analyses on these phosphorylated enzymes have not as yet been done, though Nakamura et al.[181] recently determined the K_{Ca} value for phosphoenzyme formation in microsomes isolated from porcine thyroid to be about 1 μM (see Table 1). According to our studies on the formation and decomposition of EP by bovine adrenocortical microsomes (BAC), the EP formation is as rapid as SkSR, while the decomposition is ten times slower than SkSR.[182] In spite of the discrepancy between two types of EPs, BAC EP can be classified as the SkSR type, because it is formed on the 100,000 Da protein and is not as sensitive as the erythrocyte PM type against vanadate inhibition. EP formation of the SkSR type has been found in microsomes isolated from rat parotid gland,[183] pancreas,[184] and liver.[185,186]

Microsomes of the erythrocyte PM type have been isolated with a higher purity than those of the SkSR type,[175] that has been made possible by facts that they are activated either by calmodulin, addition of acidic phospholipids, or partial tryptic digestion. Microsomes of the erythrocyte PM type have been isolated from bovine lenses,[187] murine erythroleukemic cells,[188] and smooth muscle of bovine aorta[174] and porcine stomach.[177]

Some reports described microsomes isolated from rat liver[189] and rat adipose tissues[190] which were characteristic of the mixed type of SkSR and erythrocyte PM, since phosphorylation took place on a 100,000 to 110,000 Da protein in the PM fraction. However, it should be pointed out that most of microsome preparations are identified only from the activity measurements of marker enzymes specific for either ER or PM, and possible cross-contamination of either type of microsomes in their preparations have not been excluded.

VI. Ca^{2+}-ATPase OF PLANT AND YEAST

Ca^{2+}-ATPase activity and associated 'PM' phosphoproteins have been found in plant tissues and yeast cells as well. Phosphorylation has been detected in 100,000 to 110,000-Da proteins in oat root,[191] corn root,[192] and yeast.[193] The concentration of ATP required for half-maximal activation has been estimated to be as high as 0.2 to 0.3 m*M*. Detailed analyses on the mechanism of Ca^{2+}-ATPase are still expected to provide information concerning the physiological importance of such biological systems.

VII. CONCLUDING REMARKS

This review has aimed at summarizing briefly the information concerning the mechanisms of Ca^{2+} transport reaction, which is mediated by a Ca^{2+}-ATPase in various membrane systems. The following major topics were emphasized: (1) the structure of SkSR Ca^{2+}-ATPase and the molecular architecture of the Ca^{2+} pump, (2) the mechanism of Ca^{2+}-ATPase reaction, (3) the functional difference of two possible enzyme conformers, (4) Ca^{2+} transport in cardiac SR, (5) the regulation of cardiac Ca^{2+}-ATPase by phospholamban, and (6) Ca^{2+}-ATPase in other membrane systems. Based on a large body of accumulated knowledge, SR Ca^{2+}-ATPase transports Ca^{2+} across the membrane with use of the hydrolytic energy of ATP; and this energy serves to induce a conformational change of the enzyme(s), which is the very reaction required for the chemiosmotic energy coupling.

Although we did not describe the thermodynamic concepts on Ca^{2+}-ATPase reaction, it is a very important area for the understanding of the membrane transport system as a chemiosmotic energy transducing system. We would suggest readers to refer to publications on the thermodynamic consideration on Ca^{2+}-ATPase reactions by Tanford[194,195] and Pickart and Jencks,[196] in addition to reviews cited in this article. We did not discuss on the effects of lipids on the Ca^{2+}-ATPase in this article, either. An excellent review by Sandermann[197] may provide important information on this subject.

In the past decade, the interest in the Ca^{2+} regulation by Ca^{2+}-ATPase has spread over various membrane systems other than SkSR, such as cardiac SR and smooth muscle ER. Analyses have concerned mostly the pathophysiological functions of ERs. Discovery of phospholamban in CSR is one of historical achievement, as its function clearly demonstrates the relation between the external signals and intracellular responses. Further understanding on the effects of phospholamban phosphorylation on cardiac inotropy is expected. In contrast to SkSR, ER from smooth muscle tissue is contaminated with other membrane systems, that makes investigation extremely difficult. It is now urged that a method be developed by which one can prepare relatively pure ER from various tissues.

Authors are grateful to Dr. S. A. Kuby of University of Utah for his kindness in reviewing and criticizing the manuscript. We also thank Dr. Y. Nakamura of Miyazaki Medical College for his helpful discussion and valuable suggestions for the preparation of this manuscript.

REFERENCES

1. **Rasmussen, H.,** The calcium messenger system, *N. Engl. J. Med.*, 314, 1094 and 1164, 1986.
2. **Tonomura, Y.,** *Energy-transducing ATPase — Structure and kinetics,* Cambridge University Press, London, 1986.
3. **Ganote, C. E. and Nayler, W. G.,** Contracture and calcium paradox, *J. Mol. Cell. Cardiol.*, 17, 733, 1985.
4. **Endo, M.,** Calcium release from the sarcoplasmic reticulum, *Physiol. Rev.*, 57, 71, 1977.
5. **Fabiato, A.,** Calcium-induced release of calcium from the sarcoplasmic reticulum, *J. Gen. Physiol.*, 85, 189, 1985.
6. **Martonosi, A.,** Mechanism of Ca^{2+} release from sarcoplasmic reticulum of skeletal muscle, *Physiol. Rev.*, 64, 1240, 1984.
7. **Fauvel, J., Chap, H., Roques, V., Levy-Toledano, S., and Douste-Blazy, L.,** Biochemical characterization of plasma membranes and intracellular membranes isolated from human platelets using Percoll gradients, *Biochim. Biophys. Acta,* 856, 155, 1986.
8. **Tada, M., Yamamoto, T., and Tonomura, Y.,** Molecular mechanism of active calcium transport by sarcoplasmic reticulum, *Physiol. Rev.*, 58, 1, 1978.
9. **Yamamoto, T., Takisawa, H., and Tonomura, Y.,** Reaction mechanisms for ATP hydrolysis and synthesis in the sarcoplasmic reticulum, *Curr. Topics Bioenerg.*, 9, 179, 1979.

10. **Ikemoto, N.,** Structure and function of the calcium pump protein of sarcoplasmic reticulum, *Annu. Rev. Physiol.,* 44, 297, 1982.
11. **Inesi, G.,** The sarcoplasmic reticulum of skeletal and cardiac muscle, in *Cell and Muscle Motility,* Dowben, R. M. and Shay, J. W., Eds., Plenum Press, New York, 1981, 63.
12. **Hasselbach, W.,** Calcium-activated ATPase of the sarcoplasmic reticulum membrane, in *Membrane Transport,* Bonting, S. L. and De Pont, J. J. H. H. M., Eds., Elsevier/North-Holland, New York, 1981, 183.
13. **Haynes, D. H.,** Mechanism of Ca^{2+} transport by Ca^{2+}-Mg^{2+}-ATPase pump: analysis of major states and pathways, *Am. J. Physiol.,* 244, G3, 1983.
14. **Berman, M. C.,** Energy coupling and uncoupling of active calcium transport by sarcoplasmic reticulum membranes, *Biochim. Biophys. Acta,* 694, 95, 1982.
15. **Stekhoven, F. S. and Bonting, S. L.,** Transport adenosine triphosphatase: properties and functions, *Physiol. Rev.,* 61, 1, 1981.
16. **Franzini-Armstrong, C. and Ferguson, D. G.,** Density and disposition of Ca^{2+}-ATPase in sarcoplasmic reticulum membrane as determined by shadowing techniques, *Biophys. J.,* 48, 607, 1985.
17. **Ferguson, D. G., Franzini-Armstrong, C., Castellani, L., Hardwicke, P. M. D., and Kenny, L. J.,** Ordered arrays of Ca^{2+}-ATPase on the cytoplasmic surface of isolated sarcoplasmic reticulum, *Biophys. J.,* 48, 597, 1985.
18. **MacLennan, D. H., Seeman, P., Iles, G. H., and Yip, C. C.,** Membrane formation by the adenosine triphosphatase of sarcoplasmic reticulum, *J. Biol. Chem.,* 246, 2702, 1971.
19. **Owens, K. I. R., Rugh, C., and Weglicki, W. B.,** Lipid composition of purified fragmented sarcoplasmic reticulum of the rabbit, *Biochim. Biophys. Acta,* 288, 479, 1972.
20. **MacLennan, D. H. and Holland, P. C.,** Calcium transport in sarcoplasmic reticulum, *Annu. Rev. Biophys. Bioengineer.,* 4, 377, 1975.
21. **Michalak, M., Campbell, K. P., and MacLennan, D. H.,** Localization of the high affinity calcium binding protein and an intrinsic glycoprotein in sarcoplasmic reticulum membranes, *J. Biol. Chem.,* 255, 1317, 1980.
22. **Campbell, K. P. and MacLennan, D. H.,** Purification and characterization of the 53,000-dalton glycoprotein from the sarcoplasmic reticulum, *J. Biol. Chem.,* 256, 4626, 1981.
23. **Maurer, A., Tanaka, M., Ozawa, T., and Fleischer, S.,** Purification and crystallization of the calcium binding protein of sarcoplasmic reticulum from skeletal muscle, *Proc. Natl. Acad. Sci. U.S.A.,* 82, 4036, 1985.
24. **Meissner, G.,** Density gradient fractionation of sarcoplasmic reticulum vesicles from rabbit skeletal muscle, in *Methodological Surveys in Biochemistry,* Vol. 6, Roid, E. Ed., Ellis Horwood, Chichester, 1977, 17.
25. **Watras, J. and Katz, A. M.,** Calcium release from two fractions of sarcoplasmic reticulum from rabbit skeletal muscle, *Biochim. Biophys. Acta,* 769, 429, 1984.
26. **Herbette, L., DeFoor, P., Fleischer, S., Pascolini, D., Scarpa, A., and Blasie, J. K.,** The separate profile structures of the functional calcium pump protein and the phospholipid bilayer within isolated sarcoplasmic reticulum membranes determined by X-ray and neutron diffraction, *Biochim. Biophys. Acta,* 817, 103, 1985.
27. **MacLennan, D. H., Brandl, C. J., Korczak, B., and Green, N. M.,** Amino-acid sequence of a $Ca^{2+} + Mg^{2+}$-dependent ATPase from rabbit muscle sarcoplasmic reticulum, deduced from its complementary DNA sequence, *Nature,* 316, 696, 1985.
28. **Scott, T. L.,** Distance between the functional sites of the ($Ca^{2+} + Mg^{2+}$)-ATPase of sarcoplasmic reticulum, *J. Biol. Chem.,* 260, 14421, 1985.
29. **Highsmith, S. and Murphy, A. J.,** Nd^{3+} and Co^{2+} binding to sarcoplasmic reticulum CaATPase, *J. Biol. Chem.,* 259, 14651, 1984.
30. **Gangola, P. and Shamoo, A. E.,** Synthesis and characterization of a peptide segment of ($Ca^{2+} + Mg^{2+}$)-ATPase, *J. Biol. Chem.,* 261, 8601, 1986.
31. **Yamamoto, T. and Tonomura, Y.,** Ca^{2+}/Mg^{2+}-dependent ATPase in sarcoplasmic reticulum, in *Membrane and Transport,* Vol. 1, Martonosi, A., Ed., Plenum Press, New York, 1982, 573.
32. **Møller, J. V., Lind, K. E., and Andersen, J. P.,** Enzyme kinetics and substrate stabilization of detergent-solubilized and membranous ($Ca^{2+} + Mg^{2+}$)-activated ATPase from sarcoplasmic reticulum, *J. Biol. Chem.,* 255, 1912, 1980.
33. **Lüdi, H. and Hasselbach, W.,** Preparation of a highly concentrated, completely monomeric, active sarcoplasmic reticulum Ca^{2+}-ATPase, *Biochim. Biophys. Acta,* 821, 137, 1985.
34. **Louis, C. F., Saunders, M. L., and Holroyd, J. A.,** The cross-linking of rabbit skeletal muscle sarcoplasmic reticulum protein, *Biochim. Biophys. Acta,* 493, 78, 1977.
35. **Baskin, R. J. and Hanna, S.,** Cross-linking of the ($Ca^{2+} + Mg^{2+}$)-ATPase protein, *Biochim. Biophys. Acta,* 576, 61, 1979.
36. **Bailin, G.,** Crosslinking of sarcoplasmic reticulum ATPase protein with 1,5-difluoro 2,4-dinitrobenzene, *Biochim. Biophys. Acta,* 624, 511, 1980.

37. **Froehlich, J. P. and Taylor, E. W.,** Transient state kinetic effects of calcium ion on sarcoplasmic reticulum adenosine triphosphatase, *J. Biol. Chem.*, 251, 2307, 1976.
38. **Ikemoto, N. and Nelson, R. W.,** Oligomeric regulation of the later reaction steps of the sarcoplasmic reticulum calcium ATPase, *J. Biol. Chem.*, 259, 11790, 1984.
39. **Highsmith, S. and Cohen, J. A.,** Spatial organization of CaATPase molecules in sarcoplasmic reticulum vesicles, *Biochemistry*, 26, 154, 1987.
40. **Jørgensen, K. E., Lind, K. E., Røigaard-Petersen, H., and Møller, J. V.,** The functional unit of calcium-plus-magnesium-ion-dependent adenosine triphosphatase from sarcoplasmic reticulum, *Biochem. J.*, 169, 489, 1978.
41. **Andersen, J. P., Lassen, K., and Møller, J. V.,** Change in Ca^{2+} affinity related to conformational transitions in the phosphorylated state of soluble monomeric Ca^{2+}-ATPase from sarcoplasmic reticulum, *J. Biol. Chem.*, 260, 371, 1985.
42. **Andersen, J. P., Jørgensen, P. L., and Møller, J. V.,** Direct demonstration of structural changes in soluble, monomeric Ca^{2+}-ATPase associated with Ca^{2+} release during the transport cycle, *Proc. Natl. Acad. Sci. U.S.A.*, 82, 4573, 1985.
43. **Yantorno, R. E., Yamamoto, T., and Tonomura, Y.,** Energy transfer between fluorescent dyes attached to Ca^{2+},Mg^{2+}-ATPase in the sarcoplasmic reticulum, *J. Biochem.*, 94, 1137, 1983.
44. **Hymel, L., Maurer, A., Berenski, C., Jung, C. Y., and Fleischer, S.,** Target size of calcium pump protein from skeletal muscle sarcoplasmic reticulum, *J. Biol. Chem.*, 259, 4890, 1984.
45. **Napolitano, C., Cooke, P., Segelman, K., and Herbette, L.,** Organization of calcium pump protein dimers in the isolated sarcoplasmic reticulum membrane, *Biophys. J.*, 42, 119, 1983.
46. **Fagan, M. H. and Dewey, T. G.,** Resonance energy transfer study of membrane-bound aggregates of the sarcoplasmic reticulum calcium ATPase, *J. Biol. Chem.*, 261, 3654, 1986.
47. **Imamura, Y., Saito, K., and Kawakita, M.,** Conformational change of Ca^{2+},Mg^{2+}-adenosine triphosphatase of sarcoplasmic reticulum upon binding of Ca^{2+} and adenyl-5′-yl-imidodiphosphate as detected by trypsin sensitivity analysis, *J. Biochem.*, 95, 1305, 1984.
48. **Yamamoto, T. and Tonomura, Y.,** Chemical modification of the Ca^{2+}-dependent ATPase of sarcoplasmic reticulum from skeletal muscle, *J. Biochem.*, 82, 653, 1977.
49. **Blasie, J. K., Herbette, L., Pierce, D., Pascolini, D., Scarpa, A., and Fleischer, S.,** Static and time-resolved structural studies of the Ca^{2+}-ATPase of isolated sarcoplasmic reticulum, *Annu. N.Y. Acad. Sci.*, 402, 478, 1982.
50. **Froehlich, J. P. and Taylor, E. W.,** Transient state kinetic studies of sarcoplasmic reticulum adenosine triphosphatase, *J. Biol. Chem.*, 250, 2013, 1975.
51. **Thomas, D. D. and Hidalgo, C.,** Rotational motion of the sarcoplasmic reticulum Ca^{2+}-ATPase, *Proc. Natl. Acad. Sci. U.S.A.*, 75, 5488, 1978.
52. **Restall, C. J., Dale, R. E., Murray, E. K., Gilbert, C. W., and Chapman, D.,** Rotational diffusion of calcium-dependent adenosine-5′-triphosphatase in sarcoplasmic reticulum: a detailed study, *Biochemistry*, 23, 6765, 1984.
53. **Yamada, S. and Ikemoto, N.,** Distinction of thiols involved in the specific reaction steps of the Ca^{2+}-ATPase of the sarcoplasmic reticulum, *J. Biol. Chem.*, 253, 6801, 1978.
54. **Coan, C. and Inesi, G.,** Ca^{2+}-dependent effect of ATP on spin-labeled sarcoplasmic reticulum, *J. Biol. Chem.*, 252, 3044, 1977.
55. **Coan, C., Verjovsky-Almeida, S., and Inesi, G.,** Ca^{2+} regulation of conformational states in the transport cycle of spin-labeled sarcoplasmic reticulum ATPase, *J. Biol. Chem.*, 254, 2968, 1979.
56. **Dux, L. and Martonosi, A.,** Two-dimensional arrays of proteins in sarcoplasmic reticulum and purified Ca^{2+}-ATPase vesicles treated with vanadate, *J. Biol. Chem.*, 258, 2599, 1983.
57. **Pick, U.,** The interaction of vanadate ions with the Ca-ATPase from sarcoplasmic reticulum, *J. Biol. Chem.*, 257, 6111, 1982.
58. **Jona, I. and Martonosi, A.,** The effects of membrane potential and lanthanides on the conformation of the Ca^{2+}-transport ATPase in sarcoplasmic reticulum, *Biochem. J.*, 234, 363, 1986.
59. **Dux, L., Taylor, K. A., Ting-Beall, H. P., and Martonosi, A.,** Crystallization of the Ca^{2+}-ATPase of sarcoplasmic reticulum by calcium and lanthanide ions, *J. Biol. Chem.*, 260, 11730, 1985.
60. **Hasselbach, W., Medda, P., Migala, A., and Agostini, B.,** A conformational transition of the sarcoplasmic reticulum calcium transport ATPase induced by vanadata, *Z. Naturforsch.*, 38c, 1015, 1983.
61. **Degani, C. and Boyer, P. D.,** A borohydride reduction method for characterization of the acyl phosphate linkage in proteins and its application to sarcoplasmic reticulum adenosine triphosphatase, *J. Biol. Chem.*, 248, 8222, 1973.
62. **Ogawa, Y. and Harafuji, H.,** Transient rate of the formation of phosphorylated intermediate in the sarcoplasmic reticulum vesicles isolated from bullfrog skeletal muscle: its dependence on the sequence of ligand additions, presented at Int. Symp. Bioenergetics, Nagoya, May 1 to 4, 1986, 30.
63. **Stahl, N. and Jencks, W. P.,** Adenosine 5′-triphosphate at the active site accelerates binding of calcium to calcium adenosinetriphosphatase, *Biochemistry*, 23, 5389, 1984.

64. **Ikemoto, N.,** Transport and inhibitory Ca^{2+} binding sites on the ATPase enzyme isolated from the sarcoplasmic reticulum, *J. Biol. Chem.*, 250, 7219, 1975.
65. **Makinose, M. and The, R.,** Calcium-Akkumulation und Nucleosidtriphosphat-Spaltung durch die Vesikel des sarkoplasmatischen Reticulum, *Biochem. Z.*, 343, 383, 1965.
66. **Schoner, W., Serpersu, E. H., Pauls, H., Patzelt-Wenczler, R., Kreickmann, H., and Rempeters, G.,** Comparative studies on the ATP-binding sites in Ca^{2+}-ATPase and $(Na^+ + K^+)$-ATPase by the use of ATP-analogues, *Z. Naturforsch.*, 37c, 692, 1982.
67. **Anderson, K. W. and Murphy, A. J.,** Alterations in the structure of the ribose moiety of ATP reduced its effectiveness as a substrate for the sarcoplasmic reticulum ATPase, *J. Biol. Chem.*, 258, 14276, 1983.
68. **Pintado, E., Scarpa, A., and Cohn, M.,** Calcium transport and ATPase activities of sarcoplasmic reticulum with adenosine 5′-O-(2-thiotriphosphate) diastereomers as substrates, *J. Biol. Chem.*, 257, 11346, 1982.
69. **Yasuoka, K., Kawakita, M., and Kaziro, Y.,** Interaction of adenosine-5′-*O*-(3-thiotriphosphate) with Ca^{2+},Mg^{2+}-adenosine triphosphatase of sarcoplasmic reticulum, *J. Biochem.*, 91, 1629, 1982.
70. **Calvalho-Alves, P. C., Oliveira, C. R., and Verjovski-Almeida, S.,** Stoichiometric photolabeling of two distinct low and high affinity nucleotide sites in sarcoplasmic reticulum ATPase, *J. Biol. Chem.*, 260, 4282, 1985.
71. **Watanabe, T. and Inesi, G.,** The use of 2′,3′-O-(2,4,6-trinitrophenyl) adenosine 5′-triphosphate for studies of nucleotide interaction with sarcoplasmic reticulum vesicles, *J. Biol. Chem.*, 257, 11510, 1982.
72. **Bishop, J. E., Johnson, J. D., and Berman, M. C.,** Transient kinetic analysis of turnover-dependent fluorescence of 2′,3′-O-(2,4,6-trinitrophenyl)-ATP bound to Ca^{2+}-ATPase of sarcoplasmic reticulum, *J. Biol. Chem.*, 259, 15163, 1984.
73. **Dupont, Y., Pougeois, R., Ronjat, M., and Verjovsky-Almeida, S.,** Two distinct classes of nucleotide binding sites in sarcoplasmic reticulum Ca-ATPase revealed by 2′,3′-O-(2,4,6-trinitrocyclohexadienylidene)-ATP, *J. Biol. Chem.*, 260, 7241, 1985.
74. **Pick, U. and Bassilian, S.,** Modification of the ATP binding site of the Ca^{2+}-ATPase from sarcoplasmic reticulum by fluorescein isothiocyanate, *FEBS Letters*, 123, 127, 1981.
75. **Pick, U.,** Interaction of fluorescein isothiocyanate with nucleotide-binding sites of the Ca-ATPase from sarcoplasmic reticulum, *Eur. J. Biochem.*, 121, 187, 1981.
76. **Mitchinson, C., Wilderspin, A. F., Trinnaman, B. J., and Green, N. M.,** Identification of a labelled peptide after stoichiometric reaction of fluorescein isothiocyanate with the Ca^{2+}-dependent adenosine triphosphatase of sarcoplasmic reticulum, *FEBS Lett.*, 146, 87, 1982.
77. **Froud, R. J. and Lee, A. G.,** Conformational transitions in the $Ca^{2+} + Mg^{2+}$-activated ATPase and the binding of Ca^{2+} ions, *Biochem. J.*, 237, 197, 1986.
78. **Sumida, M. and Tonomura, Y.,** Reaction mechanism of the Ca^{2+}-dependent ATPase of sarcoplasmic reticulum from skeletal muscle. X. Direct evidence for Ca^{2+} translocation coupled with formation of a phosphorylated intermediate, *J. Biochem.*, 75, 283, 1974.
79. **Shigekawa, M., Dougherty, J. P., and Katz, A. M.,** Reaction mechanism of Ca^{2+}-dependent ATP hydrolysis by skeletal muscle sarcoplasmic reticulum in the absence of added alkali metal salts, *J. Biol. Chem.*, 253, 1442, 1978.
80. **Shigekawa, M. and Dougherty, J. P.,** Reaction mechanism of Ca^{2+}-dependent ATP hydrolysis by skeletal muscle sarcoplasmic reticulum in the absence of added alkali metal salts, *J. Biol. Chem.*, 253, 1451 and 1458, 1978.
81. **Shigekawa, M. and Akowitz, A. A.,** On the mechanism of Ca^{2+}-dependent adenosine triphosphatase of sarcoplasmic reticulum, *J. Biol. Chem.*, 254, 4726, 1979.
82. **Nakamura, Y. and Tonomura, Y.,** Changes in affinity for calcium ions with the formation of two kinds of phosphoenzyme in the Ca^{2+},Mg^{2+}-dependent ATPase of sarcoplasmic reticulum, *J. Biochem.*, 91, 449, 1982.
83. **Takisawa, H. and Makinose, M.,** Occlusion in the ADP-sensitive phosphoenzyme of the adenosine triphosphatase of sarcoplasmic reticulum, *J. Biol. Chem.*, 258, 2986, 1983.
84. **Klemens, M. R., Andersen, J. P., and Grisham, C. M.,** Occluded calcium sites in soluble sarcoplasmic reticulum Ca^{2+}-ATPase, *J. Biol. Chem.*, 261, 1495, 1986.
85. **Nakamura, Y., Kurzmack, M., and Inesi, G.,** Kinetic effects of calcium and ADP on the phosphorylated intermediate of sarcoplasmic reticulum ATPase, *J. Biol. Chem.*, 261, 3090, 1986.
86. **Takakuwa, Y. and Kanazawa, T.,** Reaction mechanism of (Ca^{2+},Mg^{2+})-ATPase of sarcoplasmic reticulum vesicles. I. Phosphoenzyme with bound Ca^{2+} which is exposed to the external medium, *J. Biol. Chem.*, 256, 2691, 1981.
87. **Takakuwa, Y. and Kanazawa, T.,** Reaction mechanism of (Ca^{2+},Mg^{2+})-ATPase of sarcoplasmic reticulum vesicles. II. (ATP,ADP)-dependent Ca^{2+}-Ca^{2+} exchange across the membrane, *J. Biol. Chem.*, 256, 2696, 1981.
88. **Takenaka, H., Adler, P. A., and Katz, A. M.,** Calcium fluxes across the membrane of sarcoplasmic reticulum vesicles, *J. Biol. Chem.*, 257, 12649, 1982.

89. **Chiesi, M. and Wen, Y. S.,** A phosphorylated conformational state of the (Ca^{2+}-Mg^{2+})-ATPase of fast skeletal muscle sarcoplasmic reticulum can mediate rapid Ca^{2+} release, *J. Biol. Chem.*, 258, 6078, 1983.
90. **Gerdes, U. and Moller, J. V.,** The Ca^{2+} permeability of sarcoplasmic reticulum vesicles. II. Ca^{2+} efflux in the energized state of the calcium pump, *Biochim. Biophys. Acta,* 734, 191, 1983.
91. **Sumida, M., Wang, T., Mandel, F., Froehlich, J. P., and Schwartz, A.,** Transient kinetics of Ca^{2+} transport of sarcoplasmic reticulum, *J. Biol. Chem.*, 253, 8772, 1978.
92. **Takisawa, H. and Tonomura, Y.,** Factor affecting the transient phase of the Ca^{2+},Mg^{2+}-dependent ATPase reaction of sarcoplasmic reticulum from skeletal muscle, *J. Biochem.*, 83, 1275, 1978.
93. **Dupont, Y.,** Kinetics and regulation of sarcoplasmic reticulum ATPase, *Eur. J. Biochem.*, 72, 185, 1977.
94. **Taylor, J. S. and Hattan, D.,** Biphasic kinetics of ATP hydrolysis by calcium-dependent ATPase of the sarcoplasmic reticulum of skeletal muscle, *J. Biol. Chem.*, 254, 4402, 1979.
95. **Verjovsky-Almeida, S. and Inesi, G.,** Fast-kinetic evidence for an activating effect of ATP on the Ca^{2+} transport of sarcoplasmic reticulum ATPase, *J. Biol. Chem.*, 254, 18, 1979.
96. **Nakamura, Y. and Tonomura, Y.,** The binding of ATP to the catalytic and the regulatory site of Ca^{2+},Mg^{2+}-dependent ATPase of the sarcoplasmic reticulum, *J. Bioenerg. Biomemb.*, 14, 307, 1982.
97. **Cable, M. B., Feher, J. J., and Briggs, F. N.,** Mechanism of allosteric regulation of the Ca,Mg-ATPase of sarcoplasmic reticulum: studies with 5′-adenylyl methylenediphosphate, *Biochemistry,* 24, 5612, 1985.
98. **Yamada, S. and Ikemoto, N.,** Reaction mechanism of calcium-ATPase of sarcoplasmic reticulum, *J. Biol. Chem.*, 255, 3108, 1980.
99. **Guillain, F., Gingold, M. P., and Champeil, P.,** Direct fluorescence measurements of Mg^{2+} binding to sarcoplasmic reticulum ATPase, *J. Biol. Chem.*, 257, 7366, 1982.
100. **Chiesi, M. and Inesi, G.,** Mg^{2+} and Mn^{2+} modulation of Ca^{2+} transport and ATPase activity in sarcoplasmic reticulum vesicles, *Arch. Biochem. Biophys.*, 208, 586, 1981.
101. **Nakamura, J.,** The ADP- and Mg^{2+}-reactive calcium complex of the phosphoenzyme in skeletal sarcoplasmic reticulum Ca^{2+}-ATPase, *Biochim. Biophys. Acta,* 723, 182, 1983.
102. **Shigekawa, M., Wakabayashi, S., and Nakamura, H.,** Effect of divalent cation bound to the ATPase of sarcoplasmic reticulum, *J. Biol. Chem.*, 258, 14157, 1983.
103. **Daiho, T., Takisawa, H., and Yamamoto, T.,** Inhibition of hydrolysis of phosphorylated Ca^{2+},Mg^{2+}-ATPase of the sarcoplasmic reticulum by Ca^{2+} inside and outside the vesicles, *J. Biochem.*, 97, 643, 1985.
104. **Salama, G. and Scarpa, A.,** Magnesium permeability of sarcoplasmic reticulum, *J. Biol. Chem.*, 260, 11697, 1985.
105. **Verjovsky-Almeida, S. and de Meis, L.,** pH-Induced changes in the reactions controlled by the low- and high-affinity Ca^{2+}-binding sites in sarcoplasmic reticulum, *Biochemistry,* 16, 329, 1977.
106. **Haynes, D. H.,** Relationship between H^+, anion, and monovalent cation movements and Ca^{2+} transport in sarcoplasmic reticulum: further proof of a cation exchange mechanism for the Ca^{2+}-Mg^{2+}-ATPase pump, *Arch. Biochem. Biophys.*, 215, 444, 1982.
107. **Meissner, G.,** Calcium transport and monovalent cation and proton fluxes in sarcoplasmic reticulum vesicles, *J. Biol. Chem.*, 256, 636, 1981.
108. **Inesi, G. and Hill, T. L.,** Calcium and proton dependence of sarcoplasmic reticulum ATPase, *Biophys. J.*, 44, 271, 1983.
109. **Yamaguchi, M. and Kanazawa, T.,** Protonation of the sarcoplasmic reticulum Ca-ATPase during ATP hydrolysis, *J. Biol. Chem.*, 259, 9526, 1984.
110. **Fassold, E. and Hasselbach, W.,** The dependence on internal pH of Ca^{2+}-fluxes across sarcoplasmic reticulum vesicular membranes, *Eur. J. Biochem.*, 154, 7, 1986.
111. **Meltzer, S. and Berman, M. C.,** Effects of pH, temperature, and calcium concentration on the stoichiometry of the calcium pump of sarcoplasmic reticulum, *J. Biol. Chem.*, 259, 4244, 1984.
112. **Wakabayashi, S., Ogurusu, T., and Shigekawa, M.,** Factor influencing calcium release from the ADP-sensitive phosphoenzyme intermediate of the sarcoplasmic reticulum ATPase, *J. Biol. Chem.*, 261, 9762, 1986.
113. **Beeler, T.,** Osmotic changes of sarcoplasmic reticulum vesicles during Ca^{2+} uptake, *J. Memb. Biol.*, 76, 165, 1983.
114. **Anzai, K., Kirino, Y., and Shimizu, H.,** Temperature-induced change in the Ca^{2+}-dependent ATPase activity and in the state of the ATPase protein of sarcoplasmic reticulum membrane, *J. Biochem.*, 84, 815, 1978.
115. **Coan, C.,** Sensitivity of spin-labeled sarcoplasmic reticulum to the phosphorylation state of the catalytic site in aqueous media and dimethyl sulfoxide, *Biochemistry,* 22, 5826, 1983.
116. **Watanabe, T., Lewis, D., Nakamoto, R., Kurzmack, M., Fronticell, C., and Inesi, G.,** Modulation of calcium binding in sarcoplasmic reticulum adenosinetriphosphatase, *Biochemistry,* 20, 6617, 1981.
117. **Wiggins, P. M.,** The effect of the Ca^{2+}-ATPase of sarcoplasmic reticulum upon activities of Na^+, K^+, and H_3O^+ ions, *J. Biol. Chem.*, 255, 11365, 1980.
118. **Beeler, T., Russell, J. T., and Martonosi, A.,** Optical probe responses on sarcoplasmic reticulum: oxacarbocyanines as probe of membrane potential, *Eur. J. Biochem.*, 95, 579, 1979.

119. **Hasselbach, W. and Oetliker, H.,** Energetics and electrogenicity of the sarcoplasmic reticulum calcium pump, *Annu. Rev. Physiol.*, 45, 325, 1983.
120. **Nakamura, Y. and Tonomura, Y.,** Reaction mechanism of *p*-nitrophenylphosphatase of sarcoplasmic reticulum, *J. Biochem.*, 83, 571, 1978.
121. **Nakamura, Y.,** Two alternate kinetic routes for the decomposition of the phosphorylated intermediate of sarcoplasmic reticulum Ca^{2+}-ATPase, *J. Biol. Chem.*, 259, 8183, 1984.
122. **Van Winkle, W. B., Tate, C. A., Bick, R. J., and Entman, M. L.,** Nucleotide triphosphate utilization by cardiac and skeletal muscle sarcoplasmic reticulum, *J. Biol. Chem.*, 256, 2268, 1981.
123. **Inesi, G., Cohen, J. A., and Coan, C. R.,** Two functional states of sarcoplasmic reticulum ATPase, *Biochemistry*, 15, 5293, 1976.
124. **Makinose, M. and Boll, W.,** The role of magnesium on the sarcoplasmic calcium pump, in *Cation Flux across Biomembrane*, Mukohata, Y. and Packer, L. Eds., Academic Press, New York, 1979, 89.
125. **Yamada, S. and Tonomura, Y.,** Phosphorylation of the Ca^{2+}-Mg^{2+}-dependent ATPase of the sarcoplasmic reticulum coupled with cation translocation, *J. Biochem.*, 71, 1101, 1972.
126. **Makinose, M.,** Phosphoprotein formation during osmo-chemical energy conversion in the membrane of the sarcoplasmic reticulum, *FEBS Lett.*, 25, 113, 1972.
127. **Yamada, S., Sumida, M., and Tonomura, Y.,** Reaction mechanism of the Ca^{2+}-dependent ATPase of sarcoplasmic reticulum from skeletal muscle, III. Molecular mechanism of the conversion of osmotic energy to chemical energy in the sarcoplasmic reticulum, *J. Biochem.*, 72, 1537, 1972.
128. **deMeis, L. and Vianna, A. L.,** Energy interconversion by the Ca^{++}-dependent ATPase of the sarcoplasmic reticulum, *Annu. Rev. Biochem.*, 48, 275, 1979.
129. **Masuda, H. and deMeis, L.,** Phosphorylation of the sarcoplasmic reticulum membrane by orthophosphate. Inhibition by calcium ions, *Biochemistry*, 12, 4581, 1973.
130. **deMeis, L., and Calvalho, M. G. C.,** Role of the Ca^{2+} concentration gradient in the adenosine 5′-triphosphate-inorganic phosphate exchange catalyzed by sarcoplasmic reticulum, *Biochemistry*, 13, 5032, 1974.
131. **deMeis, L. and Tume, R. K.,** A new mechanism by which an H^{+} concentration gradient drives the synthesis of adenosine triphosphate, pH jump, and adenosine triphosphate synthesis by the Ca^{2+}-dependent adenosine triphosphate of sarcoplasmic reticulum, *Biochemistry*, 16, 4455, 1977.
132. **deMeis, L. and Inesi, G.,** ATP synthesis by sarcoplasmic reticulum ATPase following Ca^{2+}, pH, temperature, and water activity jumps, *J. Biol. Chem.*, 257, 1289, 1982.
133. **Lacapere, J.-J., Gingold, M. P., Champeil, P., and Guillain, F.,** Sarcoplasmic reticulum ATPase phosphorylation from inorganic phosphate in the absence of calcium gradient, *J. Biol. Chem.*, 256, 2302, 1981.
134. **Kanazawa, T. and Boyer, P. D.,** Occurrence and characteristics of a rapid exchange of phosphate oxygens catalyzed by sarcoplasmic reticulum vesicles, *J. Biol. Chem.*, 248, 3163, 1973.
135. **Kanazawa, T.,** Phosphorylation of solubilized sarcoplasmic reticulum by orthophosphate and its thermodynamic characteristics. The dominant role of entropy in the phosphorylation, *J. Biol. Chem.*, 250, 113, 1975.
136. **Fernandez-Belda, F. and Inesi, G.,** Transmembrane gradient and ligand-induced mechanisms of adenosine 5′-triphosphate synthesis by sarcoplasmic reticulum adenosinetriphosphatase, *Biochemistry*, 25, 8083, 1986.
137. **Shigekawa, M. and Kanazawa, T.,** Phosphoenzyme formation from ATP in the ATPase of sarcoplasmic reticulum, *J. Biol. Chem.*, 257, 7657, 1982.
138. **Pickart, C. M. and Jencks, W. P.,** Slow dissociation of ATP from the calcium ATPase, *J. Biol. Chem.*, 257, 5319, 1982.
139. **Petithory, J. R. and Jencks, W. P.,** Phosphorylation of the calcium adenosinetriphosphatase of sarcoplasmic reticulum: rate-limiting conformational change followed by rapid phosphoryl transfer, *Biochemistry*, 25, 4493, 1986.
140. **Katz, A. M.,** *Physiology of the Heart*, Raven Press, New York, 1977.
141. **Weglicki, W. B., Stam, A. C. and Sonnenblick, E. H.,** Structural lipid content of organelles of the well-oxygenated canine myocardium, *J. Mol. Cell. Cardiol.*, 1, 131, 1970.
142. **DeFoor, P. H., Levitsky, D., Biryukova, T., and Fleischer, S.,** Immunological dissimilarity of the calcium pump protein of skeletal and cardiac muscle sarcoplasmic reticulum, *Arch. Biochem. Biophys.*, 200, 196, 1980.
143. **Jones, L. R. and Cala, S. E.,** Biochemical evidence for functional heterogeneity of cardiac sarcoplasmic reticulum vesicles, *J. Biol. Chem.*, 256, 11809, 1981.
144. **Chamberlain, B. K., Berenski, C. J., Jung, C. Y., and Fleischer, S.,** Determination of the oligomeric structure of the Ca^{2+} pump protein in canine cardiac sarcoplasmic reticulum membranes using radiation inactivation analysis, *J. Biol. Chem.*, 258, 11997, 1983.
145. **Shigekawa, M., Finegan, J. M., and Katz, A. M.,** Calcium transport ATPase of canine cardiac sarcoplasmic reticulum, *J. Biol. Chem.*, 251, 6894, 1976.

146. **Tada, M. and Katz, A. M.,** Phosphorylation of the sarcoplasmic reticulum and sarcolemma, *Annu. Rev. Physiol.*, 44, 401, 1982.
147. **Tada, M., Yamada, M., Kadoma, M., Inui, M., and Ohmori, F.,** Calcium transport by cardiac sarcoplasmic reticulum and phosphorylation of phospholamban, *Mol. Cell. Biochem.*, 46, 73, 1982.
148. **Tada, M., Kirchberger, M. A., and Katz, A. M.,** Phosphorylation of a 22,000-dalton component of the cardiac sarcoplasmic reticulum by adenosine 3′:5′-monophosphate-dependent protein kinase, *J. Biol. Chem.*, 250, 2640, 1975.
149. **Schwartz, A., Entman, M. L., Kaniike, K., Lane, L. K., Van Winkle, W. B., and Bornet, E. P.,** The rate of calcium uptake into sarcoplasmic reticulum of cardiac muscle and skeletal muscle. Effects of cyclic AMP-dependent protein kinase and phosphorylase b kinase, *Biochim. Biophys. Acta*, 426, 57, 1976.
150. **Mandel, F., Kranias, E. G., and Schwartz, A.,** The effect of cAMP-dependent protein kinase phosphorylation on the external Ca^{2+} binding sites of cardiac sarcoplasmic reticulum, *J. Bioenerg. Biomemb.*, 15, 179, 1983.
151. **Louis, C. F., Maffitt, M., and Jarvis, B.,** Factors that modify the molecular size of phospholamban, the 23,000-dalton cardiac sarcoplasmic reticulum phosphoprotein, *J. Biol. Chem.*, 257, 15182, 1982.
152. **Imagawa, T., Watanabe, T., and Nakamura, T.,** Subunit structure and multiple phosphorylation site of phospholamban, *J. Biochem.*, 99, 41, 1986.
153. **Hicks, M. J., Shigekawa, M., and Katz, A. M.,** Mechanism by which cyclic adenosine 3′:5′-monophosphate-dependent protein kinase stimulates calcium transport in cardiac sarcoplasmic reticulum, *Circ. Res.*, 44, 384, 1979.
154. **Tada, M., Ohmori, F., Yamada, M., and Abe, H.,** Mechanism of the stimulation of Ca^{2+}-dependent ATPase of cardiac sarcoplasmic reticulum by adenosine 3′:5′-monophosphate-dependent protein kinase, *J. Biol. Chem.*, 254, 319, 1979.
155. **Sumida, M., Wang, T., Schwartz, A., Younkin, C., and Froehlich, J. P.,** The Ca^{2+}-ATPase partial reactions in cardiac and skeletal sarcoplasmic reticulum, *J. Biol. Chem.*, 255, 1497, 1980.
156. **Tada, M., Yamada, M., Ohmori, F., Kuzuya, T., Inui, M., and Abe, H.,** Transient state kinetic studies of Ca^{2+}-dependent ATPase and calcium transport by cardiac sarcoplasmic reticulum, *J. Biol. Chem.*, 255, 1985, 1980.
157. **Chiesi, M., Gasser, J., and Carafoli, E.,** Phospholamban of cardiac sarcoplasmic reticulum consists of two functionally distinct proteolipids, *FEBS Letters*, 160, 61, 1983.
158. **Tada, M., Inui, M., Yamada, M., Kadoma, M., Kuzuya, T., Abe, H., and Kakiuchi, S.,** Effects of phospholamban phosphorylation catalyzed by adenosine 3′:5′-monophosphate- and calmodulin-dependent protein kinases on calcium transport ATPase of cardiac sarcoplasmic reticulum, *J. Mol. Cell. Cardiol.*, 15, 335, 1983.
159. **Simmerman, H. K. B., Collins, J. H., Theibert, J. L., Wegener, A. D., and Jones, L. R.,** Sequence analysis of phospholamban, *J. Biol. Chem.*, 261, 13333, 1986.
160. **Fujii, J., Ueno, A., Kitano, K., Tanaka, S., Kadoma, M., and Tada, M.,** Complete complementary DNA-derived amino acid sequence of canine cardiac phospholamban, *J. Clin. Invest.*, 79, 301, 1987.
161. **Kadoma, M., Fujii, J., Kimura, Y., Kijima, Y., and Tada, M.,** Molecular structure of phospholamban deduced from chemical and cDNA analyses, *J. Mol. Cell. Cardiol.*, 19 (suppl. 1), S34, 1987.
162. **Inui, M., Chamberlain, B. K., Saito, A., and Fleischer, S.,** The nature of the modulation of Ca^{2+} transport as studied by reconstitution of cardiac sarcoplasmic reticulum, *J. Biol. Chem.*, 261, 1794, 1986.
163. **Kirchberger, M. A., Borchman, D., and Kasinathan, C.,** Proteolytic activation of the canine cardiac sarcoplasmic reticulum calcium pump, *Biochemistry*, 5, 5484, 1986.
164. **Lindemann, J. P., Jones, L. R., Hathaway, D. R., Henry, B. G., and Watanabe, A. M.,** β-Adrenergic stimulation of phospholamban phosphorylation and Ca^{2+}-ATPase activity in guinea pig ventricles, *J. Biol. Chem.*, 258, 464, 1983.
165. **Lindemann, J. P. and Watanabe, A. M.,** Phosphorylation of phospholamban in intact myocardium, *J. Biol. Chem.*, 260, 4516, 1985.
166. **Gabella, G.,** Structure of smooth muscles, in *Smooth Muscle: An Assessment of Current Knowledge*, Bulbring, Brading, Jones, and Tomita, Eds., Edward Arnold, London, 1980, 1.
167. **Daniel, E. E., Grover, A. K., and Kwan, C. Y.,** Calcium, in *Biochemistry of Smooth Muscle*, Vol. 3, Stephans, N. L., Ed., CRC Press, Inc., FL, 1983, 58.
168. **Carsten, M. E. and Miller, J. D.,** Purification and characterization of microsomal fractions from smooth muscle, in *Excitation-Contraction Coupling in Smooth Muscle*, Casteels, R., Godfraind, T., and Rüegg, J. C., Eds., Elsevier/North-Holland Biomedical Press, New York, 1977, 155.
169. **Karaki, H., Kubota, H., and Urakawa, N.,** Mobilization of stored calcium for phasic construction induced by norepinephrine in rabbit aorta, *Eur. J. Pharmacol.*, 56, 237, 1979.
170. **Suematsu, E., Hirata, M., Hashimoto, T., and Kuriyama, H.,** Inositol 1,4,5-trisphosphate releases Ca^{2+} from intracellular store sites in skinned single cells of porcine coronary artery, *Biochem. Biophys. Res. Commun.*, 120, 481, 1984.

171. **Sumida, M., Okuda, H., and Hamada, M.,** Ca^{2+},Mg^{2+}-ATPase of microsomal membranes from bovine aortic smooth muscle. Identification and characterization of an acid-stable phosphorylated intermediate of the Ca^{2+},Mg^{2+}-ATPase, *J. Biochem.*, 96, 1365, 1984.
172. **Sumida, M., Hamada, M., Takenaka, H., Hirata, Y., Nishigauchi, K., and Okuda, H.,** Ca^{2+},Mg^{2+}-ATPase of microsomal membranes from bovine aortic smooth muscle: Effects of Sr^{2+} and Cd^{2+} on Ca^{2+} uptake and formation of the phosphorylated intermediate of the Ca^{2+},Mg^{2+}-ATPase, *J. Biochem.*, 100, 765, M1986.
173. **Sumida, M., Okuda, H., Hamada, M., Takenaka, H., Watras, J. M., Sarmiento, J. G., and Froehlich, J. P.,** Pre-steady state kinetics of E~P formation and decomposition by Ca^{2+},Mg^{2+}-ATPase in bovine aorta microsomes, in *Structure and Function of Sarcoplasmic Reticulum,* Tonomura, Y. and Fleischer, S., Eds., Academic Press, New York, 1984, 279.
174. **Furukawa, K. and Nakamura, H.,** Characterization of the (Ca^{2+}-Mg^{2+})-ATPase purified by calmodulin-affinity chromatography from bovine aortic smooth muscle, *J. Biochem.*, 96, 1343, 1984.
175. **Carafoli, E. and Zurini, M.,** The Ca^{2+}-pumping ATPase of plasma membranes — purification, reconstitution, and properties, *Biochim. Biophys. Acta,* 683, 279, 1982.
176. **Carsten, M. E. and Miller, J. D.,** Properties of a phosphorylated intermediates of the Ca,Mg-activated ATPase of microsomal vesicles from uterine smooth muscle, *Biochim. Biophys. Acta,* 683, 279, 1984.
177. **Wuytack, F., Raeymaekers, L., Verbist, J., de Smedt, H., and Casteels, R.,** Evidence for the presence in smooth muscle of two types of Ca^{2+}-transport ATPase, *Biochem. J.*, 224, 445, 1984.
178. **Schatzmann, H. J.,** The plasma membrane calcium pump of erythrocytes and other animal cells, in *Membrane Transport of Calcium,* Carafoli, E., Ed., Academic Press, London, 1982, 41.
179. **Rasmussen, H. and Barrett, P. Q.,** Calcium messenger system: an integrated view, *Physiol. Rev.*, 64, 938, 1984.
180. **Williamson, J. R., Cooper, R. H., and Hoek, J. B.,** Role of calcium in the hormonal regulation of liver metabolism, *Biochim. Biophys. Acta,* 639, 234, 1981.
181. **Nakamura, Y., Miyamoto, T., Koono, M., and Ohtaki, S.,** Active calcium transport by porcine thyroid microsomes, *Endocrinology,* 119, 2058, 1986.
182. **Sumida, M., Hamada, M., and Okuda, H.,** Ca^{2+}-uptake by bovine adrenocortex microsomes, presented at Int. Symp. Bioenerg, Nagoya, May 1 to 4, 1986, 13.
183. **Bonis, D., Chambaut-Guertin, A. M., and Rossignol, B.,** ATP-dependent calcium transport in rat parotid microsomes. I. Localization, properties, Ca^{2+}-ATPase activity and phosphoenzyme formation, *Biol. Cell,* 55, 55, 1985.
184. **Imamura, K. and Schultz, I.,** Phosphorylated intermediate of (Ca^{2+} + Mg^{2+})-stimulated Mg^{2+}-dependent transport ATPase in endoplasmic reticulum from rat pancreatic acinar cells, *J. Biol. Chem.*, 260, 11339, 1985.
185. **Heilman, C., Spamer, C., and Gerok, W.,** The phosphoprotein intermediate of a Ca^{2+} transport ATPase in rat liver endoplasmic reticulum, *Biochem. Biophys. Res. Commun.*, 114, 584, 1983.
186. **Heilman, C., Spamer, C., and Gerok, W.,** The calcium pump in rat liver endoplasmic reticulum, *J. Biol. Chem.*, 259, 11139, 1984.
187. **Chiesa, R., Sredy, J., and Spector, A.,** Phosphorylated intermediate of two Ca^{2+}-ATPase in membrane preparations from lens epithelial cells, *Current Eye Res.*, 4, 897, 1985.
188. **Debetto, P. and Cantley, L.,** Characterization of a Ca^{2+}-stimulated Mg^{2+}-dependent adenosine triphosphatase in friend murine erythroleukemia cell plasma membranes, *J. Biol. Chem.*, 259, 13824, 1984.
189. **Lotersztajn, S., Espand, R. M., Mallat, A., and Pecker, F.,** Inhibition by glucagon of the calcium pump in liver plasma membranes, *J. Biol. Chem.*, 259, 8195, 1984.
190. **McDonald, J. M., Chan, K. M., Goewert, R. R., Mooney, R. A.,and Pershadsingh, H. A.,** The (Ca^{2+} + Mg^{2+})-ATPase of adipocyte plasma membrane: Regulation by calmodulin and insulin, *Ann. N.Y. Acad. Sci.*, 402, 381, 1982.
191. **Vara, F. and Serrano, R.,** Phosphorylated intermediate of the ATPase of plasma membranes, *J. Biol. Chem.*, 258, 5334, 1983.
192. **Briskin, D. P. and Leonard, R. T.,** Partial characterization of a phosphorylated intermediate associated with the plasma membrane ATPase of corn roots, *Proc. Natl. Acad. Sci. U.S.A.*, 79, 6922, 1982.
193. **Malpartida, F. and Serrano, R.,** Phosphorylated intermediate of the ATPase from the plasma membrane of yeast, *Eur. J. Biochem.*, 116, 413, 1981.
194. **Tanford, C.,** The sarcoplasmic reticulum calcium pump, *FEBS Lett.*, 166, 1, 1984.
195. **Tanford, C.,** Mechanism of active transport: Free energy dissipation and free energy transduction, *Proc. Natl. Acad. Sci. U.S.A.*, 79, 6527, 1982.
196. **Pickart, C. M. and Jencks, W. P.,** Energetics of the calcium-transporting ATPase, *J. Biol. Chem.*, 259, 1629, 1984.
197. **Sandermann, H., Jr.,** Regulation of membrane enzymes by lipids, *Biochim. Biophys. Acta,* 515, 209, 1978.

Chapter 15

MECHANISTIC AND ENERGETIC ASPECTS OF CARRIER CATALYSIS — EXEMPLIFIED WITH MITOCHONDRIAL TRANSLOCATORS*

M. Klingenberg

TABLE OF CONTENTS

* This manuscript was submitted in final version in September 1987.

I. INTRODUCTION

The specific solute transport is a central function of biomembranes for facilitating the translocation of solutes such as metabolites and ions. To this purpose biomembranes are equipped with specific catalysts, carriers, or translocators which catalyze the transport of these solutes. These are proteins which span the phospholipid bilayer of the membranes. It is useful to differentiate right in the beginning between carriers, which catalyze transport in well-defined catalytic cycles and which transport only one or few molecules, and channels and pores, which at certain time intervals permit the variable flux of large amounts of ions. Carriers and translocators thus resemble enzymes in that they interact with one substrate molecule during a well-defined catalytic cycle. However, the catalytic mechanism of carriers is less well understood than that of enzymes. In the following we shall concentrate on various aspects of transport catalysis by carriers rather than by ion channels and pores.

In order to simplify the treatment as much as possible and to focus on the actual translocation events, we are going to center on solute carriers which do not have a built-in energy transducing machinery such as the ATP-driven ion pumps. Here the energy transduction is superimposed on the actual translocation events. Simple solute carriers can catalyze transport in a way which has formerly been called "facilitated diffusion." The transport may be driven unidirectionally by superimposed membrane potential, often in conjunction with accompanying ions. This is not obligatory and one can separately treat the mere catalysis of these transport systems by disregarding these "secondary active" transports.

The description and analysis of the transport catalysis by biomembrane carriers is a multilayered approach. There is the structural understanding of the molecular events during the catalytic process, and there is the kinetic and thermodynamic description of the energetic conditions and events during the catalysis. The two approaches are interlinked so that the structural analysis should allow an ultimate understanding of the molecular forces which are involved and which control the kinetic and thermodynamic events.

In the following treatment we use both approaches, being aware of the great limitations still prevalent in our knowledge of carrier catalysis. We shall try to elucidate the problems involved and to raise the questions which eventually will have to be experimentally answered. In this undertaking we shall use as examples mitochondrial solute carriers, in particular those used for translocating ATP and ADP, the ADP/ATP carrier (AAC) and the uncoupling protein (UCP) which transports H^+.

II. ANALOGY AND DIFFERENCE BETWEEN CARRIER AND ENZYME CATALYSIS

Most useful for understanding carrier mechanisms is the comparison to enzymes. Thus the problems and peculiarities of carrier analysis can be more clearly elucidated. We shall emphasize this approach throughout the following treatise.

For introductory purposes we may compare the simple catalytic reaction cycle between enzyme and carrier (Figure 1). The first common step for both systems is the binding of the substrate. At this point the binding is not further segregated, although it is, in fact, a complex sequence of events, including transitions of the conformation of the substrate and of the protein. The second step for enzymatic catalysis is the chemical transformation of the substrate which is composed of substeps, including the activation of the substrate and conformation changes of the enzyme. This then is followed by the dissociation of the product and the release of the enzyme. The free enzyme is now ready to enter a new catalytic cycle.

The obvious difference between carrier and enzyme is the catalytic reaction step which in the carrier does not imply a chemical reaction, but a translocation event. One might also describe the enzymatic reaction as essentially scalar and the carrier-catalyzed reaction as

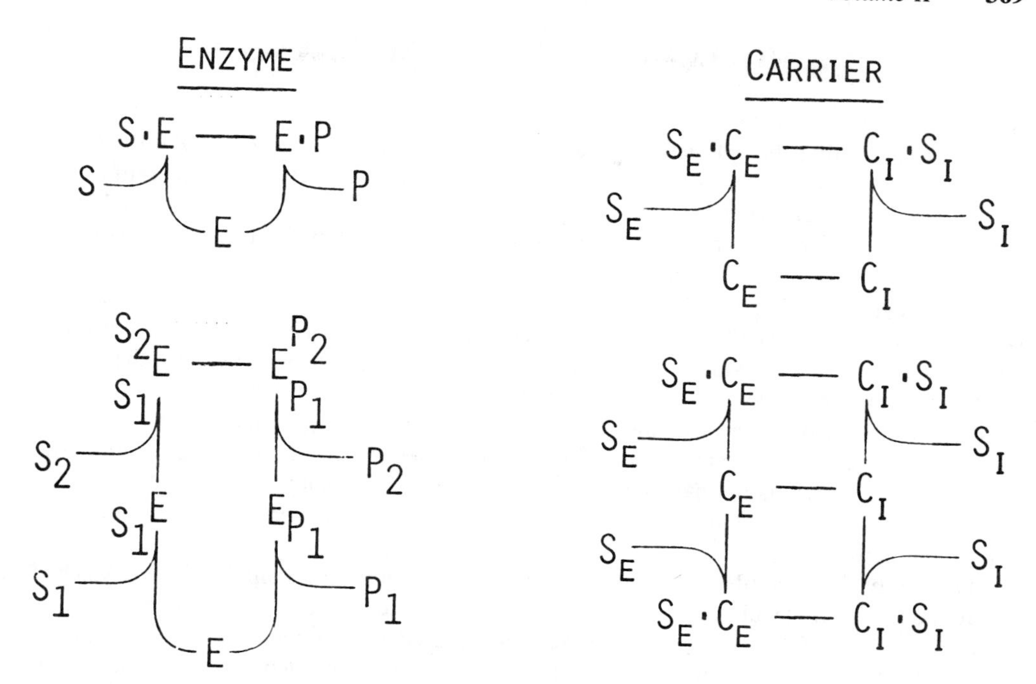

FIGURE 1. Comparison of the catalytic cycles of enzyme and biomembrane carriers. In enzymes after completion of the chemical reaction, the free catalyst is regenerated into the original state, whereas in carriers after release of the substrate on the other side of the membrane the carrier is in a different state. Two cases of carrier catalysis are shown, unidirectional transport and exchange type of transport.

vectorial. The carrier catalyzed translocation step involves intermediate stages which we shall discuss below. The principal difference of the substrate role has major consequences for the protein catalyst. After release of the substrate, the carrier has only completed half of the catalytic cycle, because it is now in a state ''opposite'' to the original one. We may describe these as ''external'' and ''internal'' states. Before going into further detail, it suffices to say that these differences reflect conformation changes of the carrier protein, in particular around the binding site.

There are two principal possibilities for the return of the carrier into the original state. $C_i \rightarrow C_e$ or $S_iC_i \rightarrow C_eS_i$. The carrier may return unloaded, or with a ligand, taken up from the inside, which may be different from the one transported in the forward step. In the latter case, there is a counter exchange between exogenous and endogenous substrates, whereas in the unidirectional transport a net uptake can be achieved. Whether the return is facultative or obligatory with a substrate load depends on the size of the kinetic barrier for the return of the unloaded carrier. In the asymmetrical unidirectional transport the unloaded translocation rate is in general lower than the substrate translocation. In the obligatory counter exchange carrier system, the unloaded return step may be forbiddingly low.

For a comparison to the enzymatic cycle, the return reaction can be described as a relaxation from the i- to the e-state. In enzymes, after the product release, the binding center would relax from the product, like conformation into the ground state. However, this process should be very fast as compared to carriers. Moreover, the conformation change involved in the enzyme is only minor as compared to the major difference between these states in the carrier.

III. MOLECULAR CARRIER MECHANISM

A. GENERAL

Before elucidating the energetics of carrier catalysis in detail, we would like to specify the molecular mechanism of carrier translocation. There were a variety of proposals of carrier action during the premolecular era of transport studies.[1] They were derived from the kinetic interpretation of transport reactions which were scrutinized often in a quite sophisticated manner. However, most of these concepts are nowadays superseded by the advance of molecular evidence. The first breakthrough and evidence for a molecular mechanism was obtained with the ADP/ATP carrier of mitochondria.[2,3] This was based on the fortunate occurrence of membrane-side specific inhibitor ligands by which one could fix the carrier in the two opposite states, C_i and C_e, and thus directly demonstrate the existence of these states. Unusual at that time was the assumption of only one single binding center instead of more; for example, two binding sites located along the translocation path. Nowadays the single binding center model has gained general acceptance for solute carriers. A particular feature of the single site model is that it allows the use of a meaningful analogy between carriers and enzymes.

Originally, and complementary to enzyme cycles, rotatory mechanisms were envisaged where the carrier molecule turns around with its binding center such that it can face the in- or outside (Figure 2A). In this case the carrier required only a single binding site in the unidirectional case. The counter-exchange could be accomplished by the rotation of two oppositely located binding centers. The rotatory mechanism has been dismissed for at least two reasons. First, both the membrane and also the membrane proteins are asymmetrically constructed and therefore only one location within the membrane is thermodynamically favored.[4] Second, a transfer of the hydrophilic surface regions of the carrier protein through the lipid bilayer is energetically difficult.[5] A stationary membrane protein which provides a translocation path for the substrate now forms the basis of all carrier models.

There is another basic difference of the substrate-protein relation between the rotatory and the stationary model. Just as in enzymes, the substrate-carrier geometry is relatively static in the rotary model when the substrate is delivered through the membrane. In a stationary carrier it is essentially the substrate which moves vis-a-vis the protein.[6] In this case, therefore, more than one binding site along the translocation path seems logical. However, evidence first obtained for the AAC and then also for other carriers supports only one binding center. Paradoxically this evidence was obtained from binding studies with chemically widely different ligands which at first sight would, in fact, suggest two or more different sites. However, these tools actually demonstrated the variability of the single binding center which is the essence of the translocation process, as will be discussed below.

The postulated conformational changes of the binding center, required for coping with the various ligands, can be rationalized as a result of the vectorial substrate translocation. As a consequence of the asymmetrical topological relation between substrate and protein, the binding interface has to be different when the binding center opens to the in or to the outside (Figure 2A and B).[7] During this translocation a major portion of the residues interacting with the substrate are shared in the two states. Some additional residues, however, participate in the binding either in the external or internal state. This scheme is distinctly different from the proposal of two distinct binding centers.

Essential for the rearrangement of the binding center are what we call the ''gates'' at the inner and outer aspects of the center. These have to open and close alternatively during the reorientation process. Certain mobile residues either obstruct or open the access to the binding center. These gating movements — essential for the carrier process — are part of the total coordinated conformation change in the protein (see below Figure 4).

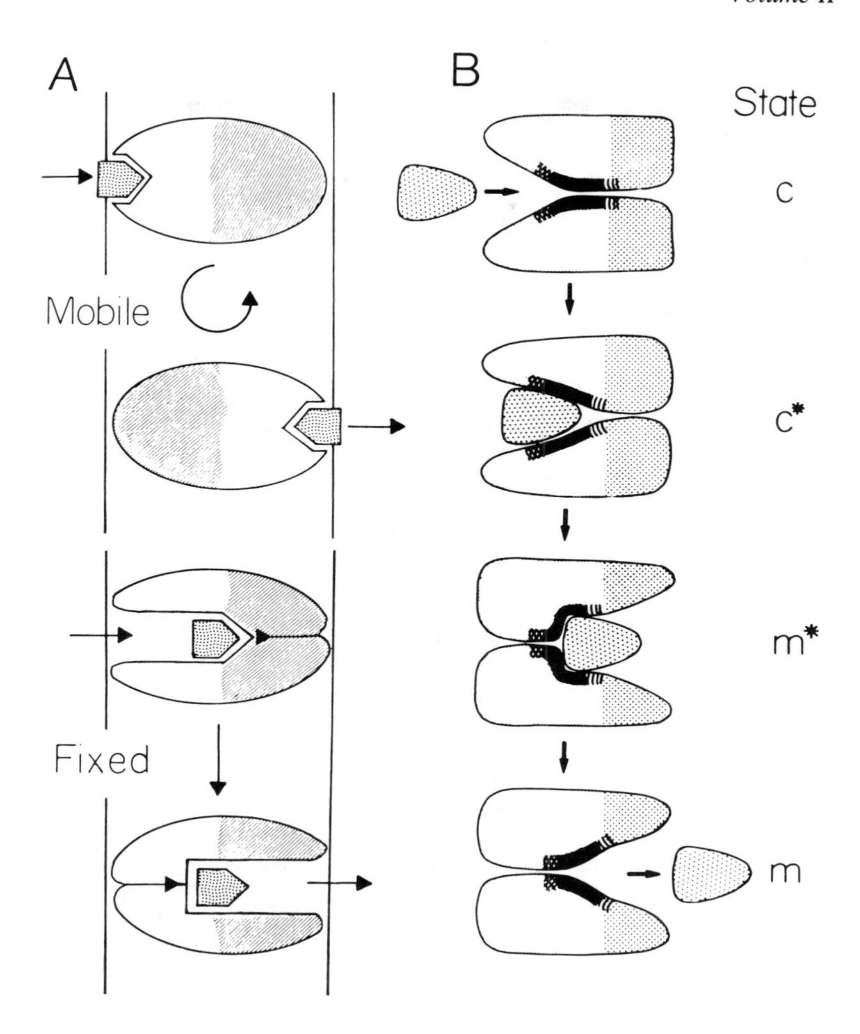

FIGURE 2. (A) Comparison of the binding center configuration in a rotational and in a fixed type of carrier. The rotational carrier binding site remains unchanged whereas in the fixed type of carrier, substrate interaction between the internal and external state is different.[2] (B) The single binding center gated pore model. On changing from the external (cytosol "c") to the external (matrix "m") state there remains a central binding domain common for both states, and there are different domains in the two states.[6]

B. EVIDENCE FOR THE "SINGLE BINDING CENTER GATED PORE" (SBCGP) MECHANISM PROVIDED BY THE AAC

A brief description of the experimental evidence leading to the SBCGP concept as elaborated for the AAC will be given (Figure 3). In a first step, by scrutinizing the nucleotide binding sites at the AAC in mitochondria, a switching of binding sites to the inside was demonstrated.[8,9] Elaborate binding studies based on the differentiation with atractylate had revealed two families of binding sites with high and low affinity.[10] The 30% high affinity sites were interpreted and shown to reflect binding sites which captured nucleotides from the outside and brought them inside. They mimick a high affinity because the nucleotides are mixed with the internal pool. The portion of the site directed outside (about 65%), being in direct equilibrium with the added nucleotides, shows the true, lower affinity. The switching of the binding sites in equilibrium with the external nucleotides thus results in the apparent dual affinities:

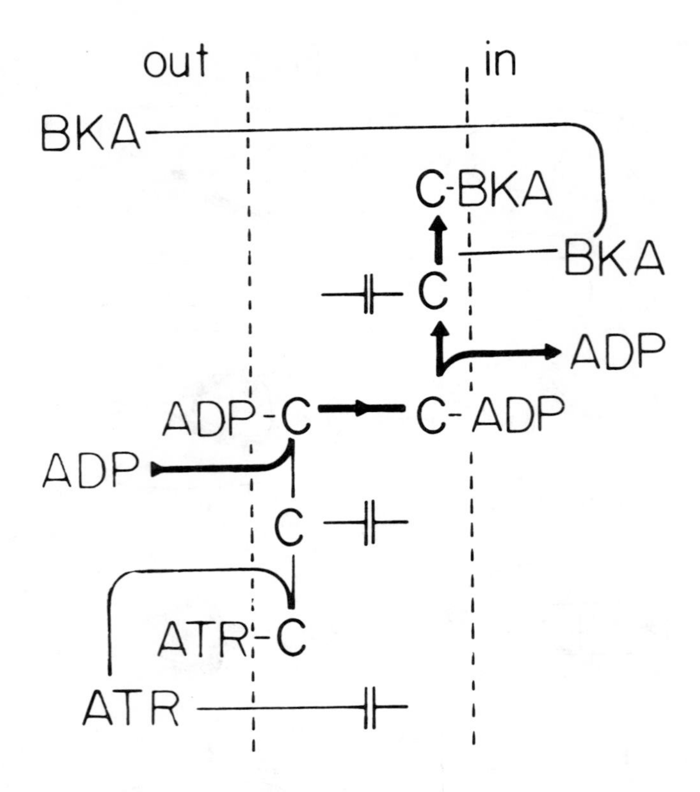

FIGURE 3. Demonstration on a molecular level of the reorientation of the single binding center of the ADP/ATP carrier. Atractylate (ATR) traps the binding center only at the outside whereas bongkrekate (BKA) traps the binding center only from the inside. BKA is a permeant inhibitor whereas ATR is impermeant. ADP or ATP are required for the translocation step.[2] For more details see text.

$$N_e + C_e \rightleftarrows (NC)_e \rightleftarrows (CN)_i \; (CN)_i \rightleftarrows N_i \text{ (endogenous nucleotide pool)} + C_i$$

The inhibitor used for discriminating the nucleotide binding to the carrier, atractylate,[8] was found to bind only to the outside form C_e. The advent of an inhibitor which binds only from the inside to the AAC, bongkrekate (BKA) then provided a unique tool for demonstrating the dynamics of the translocation process for the first time at the molecular level.[9] With bongkrekate it was possible to increase in mitochondria the share of normally about 30% "high affinity" nucleotide sites to nearly 100%.[8] Sites carrying nucleotides were pulled inside because BKA captured these sites due to its high affinity and displaced the nucleotides which were trapped inside.

$$N_e + C_e \rightarrow (NC)_e \rightarrow (NC)_i \rightarrow N_i + C_i + \text{BKA} \rightarrow (C_i\text{BKA}) \qquad \text{(Str 1)}$$

Binding studies with radioactively labeled bongkrekate, atractylate and substrates showed that all these ligands can replace each other and that in no instance two ligands can bind simultaneously to the carrier.[11,12] The large differences in the chemical structure of bongkrekate and atractylate were rationalized in reflecting the asymmetry of the binding center in the inner and outer state which for the special case of the AAC were called the m-(matrix) and c-(cytosolic) states.[2]

Our analysis was strikingly unconventional at that time and it is not surprising that other authors invoked different explanations using separate sites for inhibitors and substrate and allosteric interactions between these sites.[13-15]

C. THE SUBSTATES OF THE CARRIER DURING CATALYSIS

So far essentially two types of carrier conformations are differentiated where the binding center is either directed outside or inside. Each of these two states can be expected to have a number of substates according to the flexibility of the binding center. Evidently the carrier is in different states when occupied by different ligands. Thus there may be at least three different substates of the carrier: free, loaded with substrate, and with inhibitor. According to their similarities, these three states should be aligned according to the sequence CI-C-CS which also reflects the degree of flexibility of the binding center. The carrier inhibitor complex CI is fixed and immobile, the free carrier C has a more or less impaired mobility and the carrier substrate complex CS is active in translocation and highly flexible. In principle, the same type of conformational substate distribution should hold for the internal or external state of the carrier. Speaking in terms of the molecular structure, all the external substates have in common that the internal gate is closed and the external gate open; vice versa for the group of the internal substates.

The variety of ligands reflects the flexibility of the binding center which is the key element for the translocation catalysis. The flexibility is exploited by the ligand fitting to the binding center through the ligand protein interaction. Starting with the unloaded binding center, the ligand binding drives the conformation in opposite directions when the transported substrate or the inhibitor ligand is bound. Thus the conformation of the binding center in the ground state is neither substrate- nor inhibitor-like. For reasons which will become more evident after the energetic treatment of the conformation changes, the empty binding state seems to be more inhibitor- than substrate-like.[16] Thus in an exchange-type of carrier the unloaded side is largely inhibited from switching to the other side and therefore similar to the still more restricted state with the inhibitor bound. Substrate binding, on the other hand, induces a major change leading to the internal-external state.

D. THE GATING

The most important event in the carrier conformation change is the transition of the binding center between the external and internal state. In the SBCGP model the most obvious changes are the opening and closing of the gates on both sides of the binding center (Figure 4). The gates limit the access of the substrates to the binding center. But they also have to prevent that the diffusion path, which is laid out in the structure of the carrier, forms at any moment a continuous channel even for the diffusion of small ions. This prerequisite in the construction of carriers should prevent the electrical breakdown of the membrane potential, for example, by a leakage of small ions. The total diffusion path of the substrate across the carrier should also encompass two diffusional paths or funnels outside of both surfaces to the gates. How the diffusion in this path is controlled by surface potentials, etc., will not be discussed in this article. However, it should be mentioned that this funnel will influence the local concentration and even orientation of the substrate molecules before they hit the binding center.

The closure and opening of the gates requires a coordination within the carrier structure. Nothing is known about the molecular nature of the gates. Movement of amino acid residues along with positional rearrangement of α-structures might be invoked. In the AAC, ion bridge forming groups such as lysine have been suggested to be part of the gating process. Charged residues have the advantage of providing a better obstruction of the channel against the movement of cations or anions. The size of the gate permitting passage of the substrate is an important parameter to reckon with in any model.

In an exchange type of carrier the opening and closing of the two gates requires substrate binding. The substrate protein interaction facilitates this change as will be analyzed from

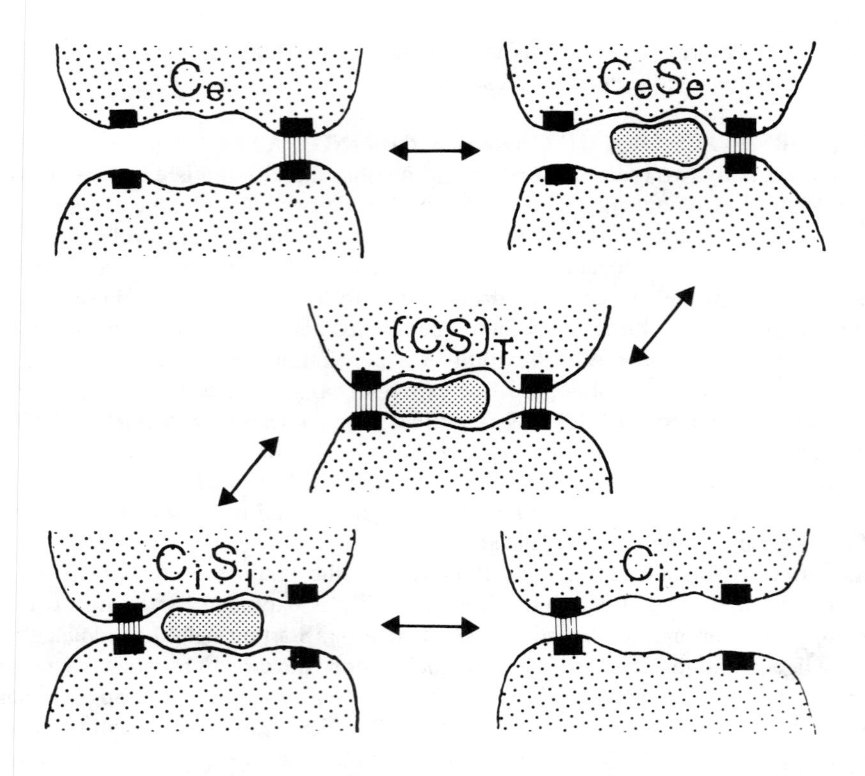

FIGURE 4. The gating and the transition state of the carrier with the single binding center.

the viewpoint of the energetics below. It may suffice to state here that the energization of the gating is closely linked to the induction of a substrate-like conformation of the binding center.

At some stage during the transition period the substrate may have an intermediate location, with both gates completely or partially closed. This would correspond to the transition state in enzymes where the substrate configuration is intermediate to that of the original substrate and product. No clear evidence for this intermediate C_T-state has been obtained because it would be difficult to be differentiated from the $(CS)_i$- and $(CS)_e$-complexes due to the rapid fluctuation between these states. Criteria which differentiate between the e- and i-conformation (see below) will not be distinctly applicable when the carrier is associated with an actually transported substrate.

One of the most fascinating aspects is the delicate energetic balance which is maintained during the large conformational changes between the i- and e-state of the carrier. It is often not appreciated that basically the carrier transport is a purely catalytic one, without dependence on external energy. This elementary part of the catalytic process is to be understood before a superimposed driving force for an uphill transport is incorporated in the analysis of carrier action. It should also be understood that on binding of the substrate the carrier is mobilized and can oscillate between the two states. This is the more remarkable since the conformation change involved in the transition is often a major one and much larger than necessary in enzyme catalysis. However, it seems that the carrier protein machinery is constructed in such a way that it permits the maintenance of a nearly equal free energy profile during this transition. This requires intramolecular binding forces to be coordinated in such a way that they permit the tightening and opening at the opposite end of the binding center. We shall return to this aspect after a treatment of the catalytic energy balance.

IV. SUBSTRATE PROTEIN INTERACTION

A. GENERAL

In our task of trying to understand the catalytic function of the carrier, i.e., the acceleration of transport through a membrane, it is useful to compare enzymes and carriers. The most important feature common to enzymes and carriers is the substrate protein interaction which is the source for the catalytic energy, i.e., the breakdown of the activation barrier. The highly ordered and specific macromolecular structure of proteins is a prerequisite for this intricate task. However, the results of the substrate protein interaction are quite different: In enzymes the substrate is chemically converted, while in carriers it is spatially translocated. The fact that such divergent functions can be provided from the common source of substrate protein interaction is not surprising in view of the enormous variability provided in protein structure.

B. IN ENZYMES

A large amount of data is available about substrate protein interaction in enzymes. Beginning with studies of the enzyme kinetics it is particularly the investigation of the substrate binding and its dependence on various parameters, as pH, temperature, specificity, as well as the influence of amino acid reagents, which yields information about the substrate interaction. In some cases the exact location of substrates within the enzyme structure in atomic resolution has been determined. This allows elucidation of the various types of bonds of the substrate with the residues within the binding center. Already kinetic and binding studies have indicated that the binding center is not rigid and may respond to the substrate with configuration changes at the binding center. X-ray and NMR-studies have in the meantime shown that proteins have an internal flexibility which could be exploited by the ligand protein interaction. It will be stressed below that this flexibility must have been developed to the extreme for the translocation function of carriers.

In enzymes the catalytic acceleration has been explained by a variety of factors and contributions which may vary according to the conditions and the type of enzymatic reaction. Two major factors are discussed.[18,19] First, and most important, the preparation of the substrate for the transition state. This requires the distortion and stretching of bonds in the transition state in preparation for the following reaction. The energy required for this contribution is thought to be provided by the intrinsic binding energy. Second, it is advantageous if the active site in the enzyme is a priori in a configuration which is more similar to the transition state of the substrate rather than to the original substrate. In this way the intrinsic binding energy of the substrate protein interaction is channelled more into the rate acceleration since not much additional energy is required for the change of the active site configuration. The substrate distorted for the transition state will be well fitted into the binding center. The resulting high intrinsic binding energy between the distorted substrate and the enzyme is largely consumed in the prior distortion of the substrate and results in a loose overall binding. The preformation in the enzyme of a transition state like binding site attracts the transformation of the substrate into the transition state and therefore provides a stronger rate acceleration. How close the conformation of the active site of the enzyme will be a priori to the transition state of the substrate, may depend on the type of the reaction. Because of the energetic and catalytic advantages, many enzymes seem to provide active center configuration which would preempt the substrate configuration in the transition state.

C. IN CARRIER VS. ENZYMES

The demands on carrier catalysis are quite different from those on enzymes. Therefore the pathway and mechanism of substrate protein interaction have characteristics quite distinct from those seen in enzymes. Most important, the substrate is not chemically changed as in enzymes. Instead the prime task of the catalytic process is the selective translocation of substrate from one space into the other. Different from the situation in enzymes, in the majority of cases an external force may ''distort'' the purely catalytic effect and enforce an uphill transport and if the transport does not require an external energy source it would correspond to what has been formerly called ''facilitated diffusion''. For the sake of clarity in the treatment of the catalytic activation process and also to clarify the distinction to enzymes, we like to segregate these contributions by postulating that the external energy force is superimposed on the basic catalytic process. However, it seems probable that many carriers are constructed such that they integrate the antenna for the external forces also in the active site. The fact that often carriers are able to facilitate transport without external forces justifies a separate treatment of the non-energy-dependent transport. It permits us to stress the role of substrate protein interaction in carriers vis-a-vis to that in enzymes.

The specificity of the carrier catalyzed transport is an essential and integral aspect of great biological importance. The selectivity may even be higher than in enzymes for mechanistic reasons which will be elucidated below. Selectivity of transport also means that the carrier must be maintained in a tight state during all stages of the transport cycle. In other words, the passage way must always be closed to freely moving solutes.

D. PROTEIN-SUBSTRATE INTERACTION AND THE SINGLE SITE GATED PORE MODEL

The following discussion of the carrier activation process will be confined entirely to the ''single binding center gated pore'' model, as elucidated above. As already mentioned, this greatly facilitates the analysis since only one binding center is involved, which also allows an easier comparison with the understanding of enzyme catalysis. What are these specific requirements imposed by the carrier catalysis on the energetics of the substrate protein interaction? The substrate, after being inserted in the binding center, will induce a sequence of reaction steps required for the translocation process. In order to understand the nature of these events, let us first define the requirements in the ''single binding center gated pore'' mechanism. As a result of the binding the substrate induces the rearrangement within the binding center which produces the closure of the entrance and the opening of the exit gate. There would be an intermediate or transition state where both gates are in the same state, either closed or half open.[4,5] In this symmetrical state of the binding center the forces responsible for the rearrangement must be well balanced. Beginning with the first encounter of the substrate with the protein, the activation pathway for carriers must be quite different from that for enzymes. In fact we must postulate that already the unliganded binding center in the carrier has a configuration vis-a-vis the substrate quite different from that in enzymes.

From these considerations it is clear that in carrier catalysis the main demand on the substrate protein interaction is the conformation change of the protein rather than of the substrate. Paradoxical as it may at first seem, this can more easily be explained with the counter exchange system than with a unidirectional transport. The reason is that in a counter-exchange system the activation of a translocation step by substrate binding is more stringent than in a unidirectional system. In the latter case, the binding site can also translocate in the unliganded state, but still the unliganded translocation branch is mostly much slower and more restricted than the substrate-linked branch. We may conclude that in the counter exchange system the unloaded carrier has a very low probability of existing in the transition state. As a consequence the two unliganded states of the carrier, C_e and C_i are well segregated by a high activation barrier.

E. SUBSTRATE AND BINDING CENTER CONFIGURATIONS

In enzymes, as already elucidated above, the active site will have little differences once it is ready to bind a substrate or release a product. There might be small differences of the binding center when it emerges from the dissociation either of P or of S. However, they are probably minor and not segregated by an important activation barrier. As a result, in the time average the enzyme active site should have a configuration compromising between the substrate and product likeness. In fact, the transition state configuration should represent a medium state between the substrate and product type. Energetic arguments for the optimalization of the enzyme catalysis had arrived at the conclusion that it is more favorable if the active site has a configuration which resembles the transition state of the substrate.[18,19] Thus less energy would be consumed when driving the active site into the transition state. Furthermore, in the transition state the interaction between the enzyme and the distorted substrate is at a maximum. As a result of this energy gradient the substrate is pulled into the transition state conformation and the activation barrier becomes maximally diminished.

In transport catalysis by carriers the events and therefore the energetic requirements for catalysis are quite different from those in enzymes. As emphasized above, the common feature is the substrate protein interaction which we can consider in both cases as the source of what we call "catalytic energy." However, in carrier catalysis this catalytic energy is not to be invested in the activation of the substrate for chemical bond change. Accordingly, bonds within the substrate do not have to be distorted or strained in a transition state in preparation for the chemical reaction. The absence of a substrate transition state in carrier catalysis therefore also requires that the substrate binding center has configuration requirements very different from those in enzymes. The flexibility of the binding center, which is inherent in the single site gated pore mechanism elucidated above, is the central issue in the carrier mechanism, rather than the straining and configuration change of the substrate. Therefore the intrinsic substrate protein interaction energy is primarily invested into the strong conformation change of the protein or, more specifically, of the binding center, rather than of the substrates. On the basis of these fundamental differences the configuration of the binding center vis-a-vis the substrate exists in states uncommon to those in enzymes. We shall discuss in the following firstly the binding center configuration and then the energetic requirements involved.

In contrast to the enzymes the carriers should exhibit a drastic configuration change of the binding center which is clearly defined by the gating process. As a result there are two ground states of the binding center without the substrate, where either the inner or outer gate is open (Figure 4). Evidently the binding center in these states is distinctly different from a central transition state, in contrast to the case in enzymes. There are other arguments that the carrier binding center is different from the transition state, in particular for exchange type of carriers, but also when an external energy source is required to accelerate the carrier. The unloaded binding center will switch through the transition state into the opposite gating configuration only with a low probability, obviously because it is quite distant from the transition state. Closely related is the energetic argument that without the substrate the transition state is at a comparatively higher energy level than the ground state.

The problem then amounts to the question, what is the configuration of the carrier binding center as compared to the substrate in the ground state and in the transition state? We shall discuss these questions first on the basis of principle considerations before presenting the experimental evidence. First we may ask the question; if the ground state is so much different from the transition state, to what extent is it substrate-like? The configuration of the carrier binding center with substrate-likeness would be the usual picture which is generally used in the literature. However, the substrate-likeness of the ground state would be disadvantageous to the translocation catalysis since, on binding the substrate (Figure 5), the carrier would remain in that state as a tight ligand-protein complex and would not reach the transition state.

The carrier alone has no energetic incentive to go from the ground to the transition state, because of the large energy required for the conformation change. The change into the transition state would be greatly facilitated by substrate binding if the transition state rather than the ground state is substrate-like. In this case, the binding center in the ground state

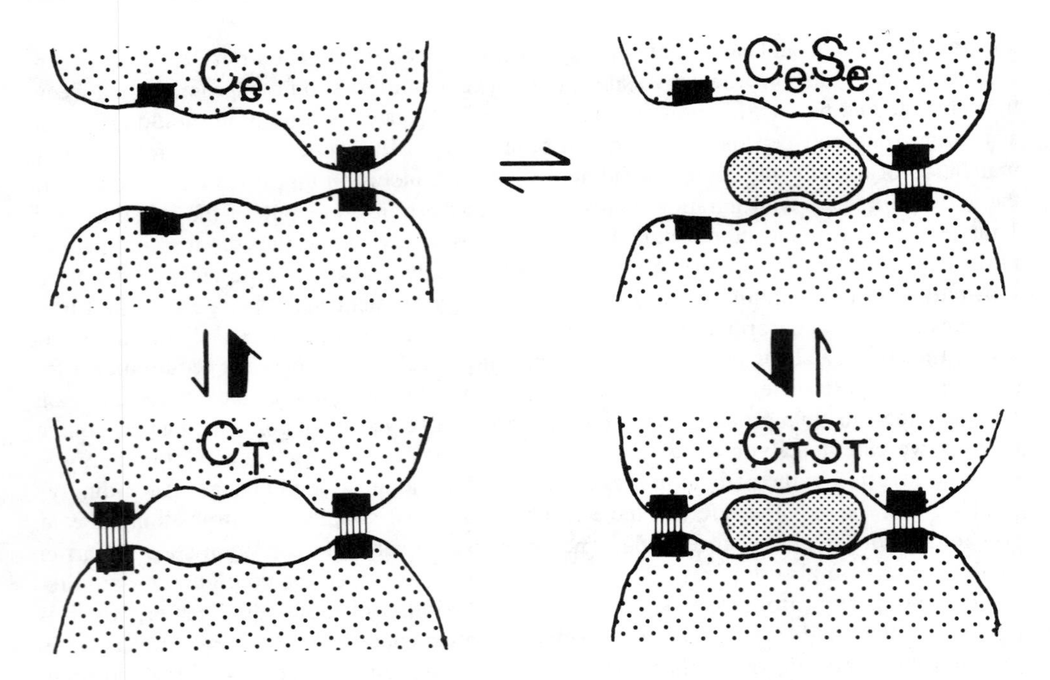

FIGURE 5. The induced transition fit concept of carrier activation. In the ground state the carrier is not substrate like. However, in the transition state, the substrate fits well to the binding center. The binding center without substrate prefers the substrate-unlike configuration, whereas with substrate the transition state configuration may have a good residence time.

will meet the incoming substrate with an imperfect fit and this initial carrier substrate complex will easily move into the transition state. From here it will then smoothly go into the opposite ground state. Only by this arrangement of the binding center configuration the whole translocation process seems to be feasible.

V. ENERGETICS OF CARRIER CATALYSIS

A. GENERAL

With this mechanistic picture of the binding center configuration during the translocation events we have now a basis for understanding the energetics in the carrier catalysis. It is clear that the substrate protein binding energy is the source for the catalytic energy which lowers the activation barrier. This is illustrated in the activation energy profile in Figure 6. The initial binding of the substrate in the ground state of the carrier is comparatively loose because of the imperfect interactions of the substrate with the binding center. The energy level of the ground state complexes is not much lower than that of the free carrier, depending also on the concentration and other environmental factors. Independent of those influences, the intrinsic binding energy will be limited in the ground state because of the poor fit. In the transition state the binding center configuration will provide a good fit and therefore a high intrinsic binding energy. This large energy source is used to facilitate the important configuration changes in the protein from the ground into the transition state.

As illustrated in Figure 6, without a substrate the energy level of the transition is relatively high, resulting in a poor probability of translocation. The strong substrate protein interaction in the transition state largely removes this activation barrier. The energy required for the configuration change of the active binding center is furnished by the high intrinsic binding energy in the transition state. Conversely the tight intrinsic binding in the transition state is relieved by the energy required for the conformation change. The high intrinsic binding

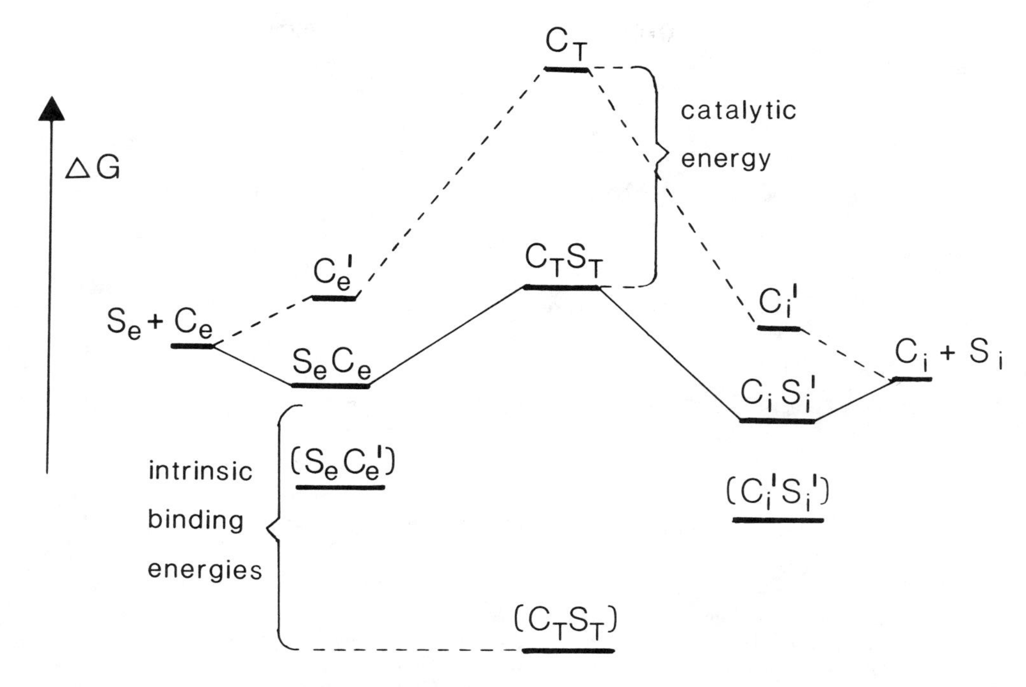

FIGURE 6. Energy profile for the carrier catalyzed solute translocation. The same values as detailed in Figure 7 are given in this energy profile for the translocation cycle between the external and the internal ground states (C_e, C_i). The diagram also illustrates that the "catalytic energy", i.e., the decrease of the energy level of the transition state by the substrate binding, is provided by the intrinsic binding energy.

energy would only be expressed if the carrier were fixed into the transition state prior to the substrate binding. Any tight binding state would be detrimental to the catalytic process since it would withdraw a large proportion of the carrier molecules into that state and therefore slow down the overall reaction. A relatively smooth energy profile with small differences of the intermediate energy levels is important for catalysis.

A mutual compensation of the intrinsic binding energies and the conformation energies of the carrier binding center can be more clearly illustrated in an energy level diagram which segregates the contributions from the substrate and carrier (Figure 7). The energy values given in this diagram are arbitrary, but should reflect the relative magnitudes. Saturating substrate concentrations are employed in order to simplify the energy profile. Limiting substrate concentrations would elevate the energy level of the substrate complexes. The energy difference between the binding partners and the initial substrate carrier complex is broken down into the contributions required first to bring the substrate and the carrier into the binding states, $S_e \rightarrow S_e'$ and $C_e \rightarrow C_e'$, and second to form the complex from those preformed states, $C_e' + S_e' \rightarrow S_e' C_e'$. For the substrate the energy investment ΔG_S is mainly of entropic nature by restricting the translational vibrational and rotational movement. However, because of the poor fit to the binding center at this stage, the substrate S_e' still contains some freedom and therefore the energy requirement is relatively small. The binding center conformation $C_e \rightarrow C_e'$ is considered to undergo only minor changes at this level requiring also little energy ΔG_C. The intrinsic binding energy ΔG^i_{SC} is defined by energy released on binding to the carrier when both are predisposed in their binding conformations S_e' and C_e'. Although significant, it is comparatively small and may just exceed the sum of the energy input for the change of the two binding components ($-\Delta G^i_{SC} \geqslant \Delta G_S + \Delta G_C$), bringing the energy level $\Delta G_{SC} = \Delta G_S + \Delta G_C + \Delta G^i_{SC}$ on the initial complex slightly below the energy level of the unliganded components.

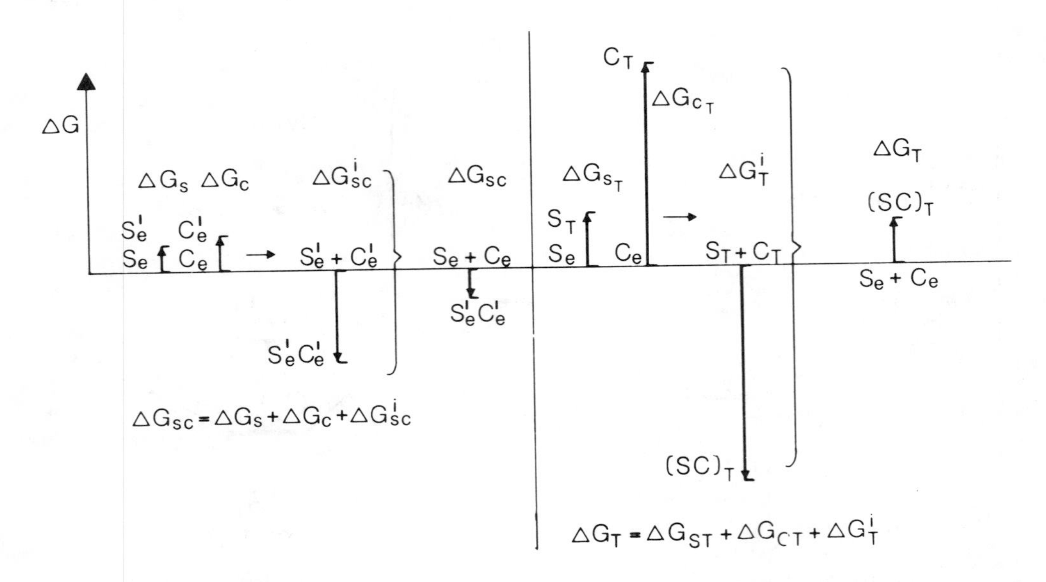

FIGURE 7. Internal energy transfer requirements for carrier catalysis. The breakdown of the energy contributions for the conformation changes of substrate and carrier and for the intrinsic binding energy of the preformed substrate and carrier. Firstly, the breakdown for the binding energy of the carrier-substrate complex in the ground state is given. The total binding energy ΔG_{SC} is the sum of the conformational energies $\Delta G_S + \Delta G_C$ and the intrinsic binding energy $\Delta G'_{SC}$. Secondly, the energy contribution for the function of the transition complex is given. The total binding energy ΔG_T is the sum of the conformation energy changes of the transition state $\Delta G_{ST} + G_{CT}$ and the intrinsic binding energy ΔG_T.

The energy change for the subsequent formation of the transition states is more important. Now, due to the exact fit of the binding center, the substrate loses nearly all internal freedom and assumes a precise conformation which requires a significant entropic energy input. This amount ΔG_T is still considerably smaller than in enzymes where a greater amount of energy is to be invested into the substrate in the transition state because, in addition to the entropic contributions, energy is required for the substrate distortion, bond stretching, etc. The transformation of the carrier binding center into the transition state requires a major energy input ΔG_{CT} because of the large conformation changes between the initial binding complex and the transition state. In the transition state the full intrinsic binding energy ΔG^i_T is released in the imaginary case when both the substrate and the carrier are already in the transition state conformation. This intrinsic binding energy should be high and thus be able to largely compensate the energy consumption of the substrate and carrier binding center deformations $(-\Delta G^i_T \approx \Delta G_{CT} + \Delta G_{ST})$.

B. COUNTER EXCHANGE VS. UNIDIRECTIONAL TRANSPORT

One consequence of this treatment is an understanding of the relation between the size of the activation barriers resulting from the substrate-protein interaction and the discrimination between unidirectional and counter exchange transport. High activation barriers could be matched by high intrinsic binding energies and high activation barriers can be expected in a binding center constructed for relatively large molecules, because of larger conformation changes involved. Correspondingly, in general, also the interaction energy will increase with the size of the substrate. Conversely, the smaller the activation energy barrier for the substrate-free carriers, the higher the probability for translocational transition without substrate. The higher activation barrier would largely prevent the free carrier transition and therefore enforce a counter exchange type of transport. In this case, unidirectional transport would be facilitated, although the substrate-free branch of the transport cycle would still be a greatly limiting step. We shall demonstrate below some examples for the validity of this concept.

C. INDUCED TRANSITION FIT

The "ligand-induced transition fit", as we henceforth will call the theory of carrier catalysis, allows another comparative look at enzymes. Since in both cases we are dealing with protein substrate interaction, it is not surprising that we find a fragmentary part of this mechanism in enzymes, termed "induced fit".[20] Originally it was introduced as a more solid counter-concept to the diffuse allostery theory of regulatory enzymes. Now it is more specifically understood to have advantages for the substrate-induced regulation in certain enzymes. It has also been pointed out, however, that induced fit is detrimental to the catalytic capacity of enzymes.[18] Part of the substrate-protein interaction energy is consumed in the protein conformation change and therefore subtracted from the energy required for the substrate deformation. In the catalytically ideal case the enzyme is fully predisposed in the transition state.

With an understanding of the energetics of the "ligand-induced transition fit" let us revisit the problem of the nature of the ground state of the carrier. The configuration of the binding center in the ground state will have some substrate likeness in order to recognize the substrate. An important requirement for the ground state configuration is that it should be so disposed as to permit a substrate elicited induction of the transition state. Probably a quite precise configuration of the binding center in the ground state is required to fulfill this task, i.e., the degrees of freedom are quite limited. The gating process is an integral part of the induced transition fit. The binding should induce the gating concomitant with the conformation transition. Both types of ground states, the external and the internal, have to be ready to be induced to the transition state.

D. INHIBITORS OF TRANSLOCATION

The structure of the substrate with respect to the ground state configuration must be such that it does not obstruct the transformation into the transition state. If the ligand has a structure which allows binding, but obstructs transition, inhibition of the transport would result.[16] Such a ligand would be an effective inhibitor if it binds tightly enough. Thus good inhibitors should precisely fit the ground state and not the transition state. In fact, we can postulate that any ligand which binds tightly to the ground state of the binding center will be more or less inhibitory. Since tight binding is possible only when the ground state configuration is well matched by ligands, we may conclude that inhibitors are constructed such as to mimic the ground state configuration, in contrast to substrates.[21]

The energy relationships are illustrated in Figure 8. On forming the inhibitor carrier complex, due to the good fit the binding center configuration will undergo usually only a small change, requiring small energy input. The high intrinsic binding energy between the inhibitor and the carrier is only little reduced by the energy requirements for the change of the inhibitor and carrier structures. As a result, the total binding energy of the inhibitor with the carrier will be high. Thus the ligand carrier complex sits in an energy trap from which the passage into the transition state becomes difficult. Conversely, as argued above, formation of the substrate carrier complex does not tolerate an energy trap in order to permit the easy passage to the transition state.

Our original question as to the configuration of the ground state can now be answered by stating that this configuration may be mirrored in inhibitory ligands. Actually the inhibitory state may be slightly different from the ground state, since ligands may induce a small but definite conformation change of the carrier site for a better match with their own structure. A most important consequence is that there should be two types of structurally different inhibitors which reflect the widely different two ground states of the carrier, the external

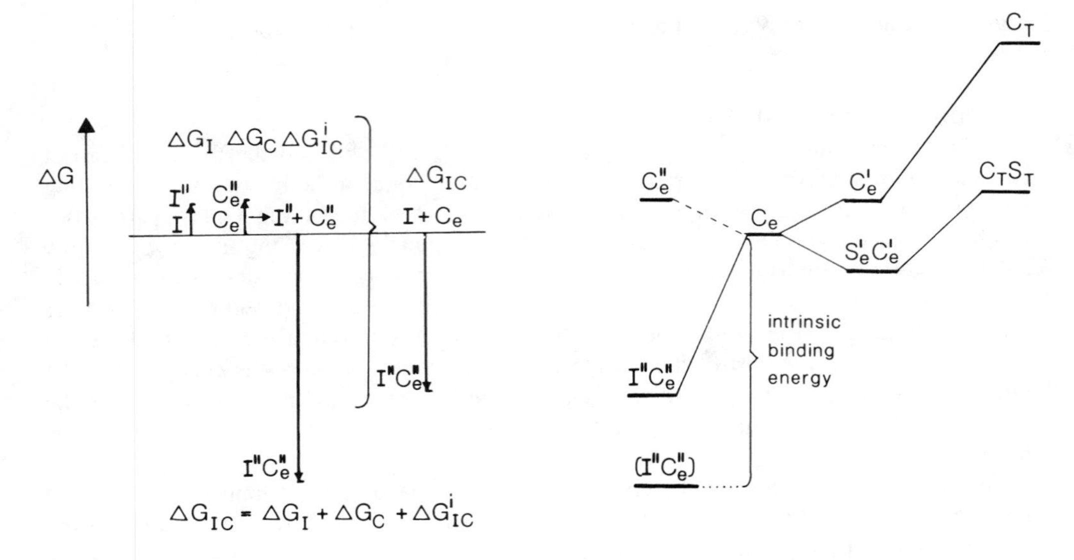

FIGURE 8. The application of internal energy transfer analysis to the binding of substrate site specific inhibitors of carriers. The inhibitor binds to the carrier in a conformation C_e'' close to the ground state C_e (small ΔG_C). The intrinsic binding energy ΔG_{IC} is high and largely expressed in the overall binding (ΔG_{IC}). As a result, binding of inhibitors is strong, in contrast to weak binding of substrate. The "abortive" ground state fixed by the inhibitor (C'') has a conformation most different from the transition state (C_T).

and the internal one. Whereas the substrate can bind both to the external and internal states by virtue of the induced transition fit, the good match of the inhibitors to the ground states requires that they cannot be exchanged between the two states. This conclusion can also be drawn from the consideration that the binding center has different configurations when it meets the substrates from opposite sides, as elucidated above (Figure 2B).

The structural differences of the internal and external inhibitors may, however, go beyond the differences of the binding center only by their use of additional proximal areas for the interaction with the protein. As illustrated in Figure 9, this range may stretch into the gating or beyond the gating residues. It would have the additional advantage of obstructing the gating and thus the transformation into the transition state. On this basis, we may expect to find two types of inhibitors, those which depend on tight binding only, and those which depend on the obstruction of gating. The most powerful inhibitor would incorporate both types of effects, precise matching of the ground state and the obstruction of the transformation into the transition state.

If there is any change of the ground state of the binding center on binding an inhibitor, it should be away from the transition state configuration. This would even widen the structural differences between inhibitors for the internal or external state. As a consequence they may have only little resemblance with the substrates and there should be still less structural similarity between the internal and external type inhibitors. In a diagram the relation of the conformation change energies and the intrinsic binding energy are shown (Figure 8). The energy required for the conformation change on binding of both components should be small. The large intrinsic binding energy is therefore reduced only to a small extent and the result is a high overall binding energy. In the adjacent energy profile diagram, the energy levels and the reaction position of the inhibitor carrier complex is compared to the transition state. With the inhibitor the binding center is in an energy trap, in contrast to the carrier substrate complex in the transition state.

In carriers, the way inhibitors interact with the binding center is in carriers exactly opposite to that in enzymes. In both cases effective inhibitors will be those which match

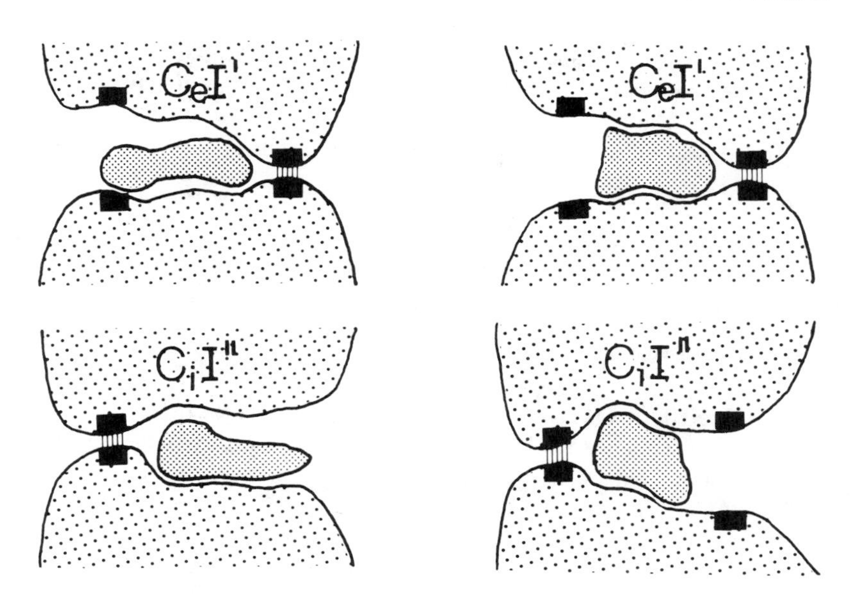

FIGURE 9. Illustration of the side specificity of inhibitors and the possible existence of gate obstructing inhibitors apart from ground state fitting inhibitors.

the ground state in order to bring the protein inhibitor complex into an energy trap. However, in enzymes, since the ground state resembles the transition state, the structure of the inhibitor should mimic the configuration of the binding center in the transition state. These well-described transition state analogues immobilize the enzyme in a state which is positioned centrally in the catalytic cycle, in contrast to the inhibitors of carriers. In both enzyme and carrier, these inhibitors reach the same purpose, the substrate is effectively prevented from binding, because the inhibitor brings the protein into a deep energy trap. The situation may also be described in another manner. Due to the different effects of the substrate protein interaction on the configuration of the two binding partners, the ground state configurations in enzymes and carriers are in a contrary position in the catalytic cycle and consequently inhibitors utilizing those ground states fix the carriers into these contrary conformations.

VI. THE LIGAND-INDUCED FIT TRANSITION EXPERIMENTALLY EXEMPLIFIED IN THE ADP/ATP CARRIER AND IN THE UNCOUPLING PROTEIN

A. GENERAL

The ''single binding center gated pore'' mechanism was based on the experimental evidence produced by the scrutiny of ligand binding to the ADP/ATP carrier (AAC). The major conclusion, as discussed above, was that the carrier exists in two states, with the binding center inside and outside, and that the transition between these two states occurs only on binding of ADP or ATP. The energetics of the ligand induced transition fit elucidated above provide a further tool to analyse the extensive experimental data available on the ligand-protein interaction of the AAC. The unique suitability of the AAC for this type of analysis is based on several particular qualities. First, isolated protein is available in large amounts and therefore permits a wide range of experimentation in ligand protein interaction. Second, the substrate for the carrier is relatively large, very well defined as a nucleotide and exerts a highly specific interaction. A vast arsenal of nucleotide analogues is available and can be exploited. Thirdly, there is a unique availability of natural compounds type of highly specific and tightly binding inhibitors which allow the AAC to be fixed either in the ''abortive'' cytosolic (c-) or matrix (m-) state.

TABLE 1
Contrasts of Nucleotide Protein Interaction in Two Homologous Carriers

	ADP/ATP carrier	Uncoupling protein
Nucleotide	Substrate	Inhibitor
Function	Activation of own transport	Of H^+ transport
Affinity	Low ($<10^5$ M^{-1})	High (10^6 to 10^9 M^{-1})
Specificity	Exclusive ADP/ATP	G/A-DP/TP
Influence on carrier structure	Destabilizing	Stabilizing
Trypsin degradation	Facilitated	Inhibited
Sidedness of nucleotide binding	Both sides	Only external

B. BINDING OF NUCLEOTIDES AND NUCLEOTIDE ANALOGUES

An important consequence of the ligand-induced transition fit is the apparently loose binding of the substrate. This consequence is strikingly demonstrated for the ADP or ATP binding to the AAC.[10] The affinity is relatively small, i.e., for the mitochondrial bound AAC it is about 10^5 M^{-1}, and for the isolated purified carrier in solution or in the reconstituted membrane it is only about 2 to 5 times 10^4 M (unpublished data). The intrinsic affinity should be much higher, in the range of 10^9 to 10^{12} M. The low affinity observed is highly suggestive of an internal energy compensation due to the ligand induced transition fit. An intriguing hint for the intrinsic high affinity is the high specificity of the AAC for its substrates. Only ADP and ATP are transported, whereas minor modifications in the base and phosphate portions are not tolerated.[2] Thus the specificity of the AAC is higher than that of most nucleotide utilizing enzymes. The high specificity betrays an intimate interaction of the protein with the ATP or ADP, tracing all sections of the molecule, which should add up to a high intrinsic binding energy.

A strong support for these conclusions comes from the nucleotide protein interaction observed in the uncoupling protein (UCP) of brown adipose tissue mitochondria. The comparability to the AAC is based on two reasons. First, both belong to the rare species which interact with the free nucleotides and not with the Mg^{++} nucleotide complexes,[23,24] and secondly, AAC and UCP are structurally related.[25] The primary structures of the two proteins have some homology and, more strikingly, the peculiar internal triplicate construction of the AAC is also found in the UCP. The different features of the nucleotide-protein interaction in the two proteins are compared in Table 1.[26] In UCP the nucleotide is an inhibitor of the H^+ transport activity of this protein. The apparent affinity of the nucleotide binding is higher than 10^6 M, but strongly depends on the H^+ concentration. Extrapolating to H^+ saturation at the binding center, one arrives at an intrinsic affinity of 10^9 M.[24] Although it has a more than 10^4 higher affinity than the AAC, the specificity of UCP towards nucleotides is lower, since also guanosine nucleotides are accepted with equal strength. This apparently paradoxical finding can be rationalized by maintaining that the intrinsic affinity in the AAC is higher than that in the UCP, but that in the UCP the binding of the nucleotides to the protein is not compensated by a ligand induced transition fit, since it is not substrate but inhibitor. In UCP the intrinsic binding energy is largely expressed in the observed external binding affinity. In other words, the nucleotide binding to the UCP reveals a high intrinsic binding affinity of a Mg^{++}-free nucleotide NTP^{4-} or NDP^{3-} type interaction which is probably even higher to the similar AAC.

Further support for the ligand-induced fit interaction in the AAC comes from the conformational instability induced in the AAC. This is demonstrated by the sensitivity against tryptic digestion which is strongly increased by ADP and ATP. In UCP, to the contrary, the nucleotide stabilizes the conformation in line with its inhibitory effect.

Certain types of nucleotide derivatives seem still to bind to the AAC but are not active in translocation. In fact, these derivatives can even be inhibitory as a result of relatively high affinity as compared to the ADP and ATP. Because of the technical difficulties of binding measurements to the AAC, the data in this field are rather vague. For example, it is suggested that 8-Br-ATP is bound, but not transported.[27] The only precise evidence comes from the fluorescent 3′-*O*-dimethylamino naphthoyl (''DAN'')-substituted ADP or ATP above mentioned which bind primarily to the m-side with a relatively high affinity of about 10^6 *M* and with a much smaller affinity to the c-side.[28] Also this analogue is not transported. The high fluorescence of the ''DAN''-ATP in the m-state was interpreted to indicate a hydrophobic interaction of the DAN-group with the protein. This would support the ground state configuration concept that the DAN-substituted ATP binds with a better fit to the binding center than the ATP. Moreover, the asymmetry of DAN-ATP binding illustrates the difference of the ground state configuration in the m- vs. the c-state. The 3′O-naphthoyl-substitution at ATP or ADP results in analogues which seem to bind to the AAC both in the c- and m-state with less affinity than the DAN-derivatives.[29] The dimethylamino-group is obviously important for the strong interaction with the m-state binding center.

C. BINDING OF INHIBITORS

Although the nucleotide derivatives may be quite revealing about the binding center because of their homology to the actual substrate, they seem to be only crude ground state ligands as compared to the obviously much more refined inhibitors generated by nature against the ADP/ATP carrier, i.e., the compounds of the atractylate and of the bongkrekate groups.

At first sight the structure of these compounds seems to be widely different from that of the nucleotides (Figure 10). Therefore, our first interpretation was that they affect the substrate binding from a separate site, in particular, when the ''cooperative'' effect of BKA on the ATP/ADP binding was found.[30] Then it was realized that the data on the ADP-BKA-interaction reveal for the first time at a molecular level the movement of the binding center between the out- and inside.[9] This implied that the inhibitors and nucleotides bind to the same center at the AAC. It is understandable that this early assumption was contested on several types of evidence.[14,15,32] However, the majority of experimental evidence provided from our laboratory has consistently supported the early assumption that these ligands attach to the substrate binding center.

One common feature in the chemical structure of the nucleotides and the inhibitors is the occurrence of a cluster of three or four negative charges (see Figure 10). It is assumed that these negative charges attach to the same positively charged residues in the binding center, whether they use ATP^{4-} or ADP^{3-} and $carboxyatractylate^{4-}$ or $bongkrekate^{3-}$. The strong structural differences between these two groups of inhibitors could be explained by the discovery that the atractylate group only binds from the c-side to the carrier and the bongkrekate from the m-side.[8,33] As explained above, the binding center should interact quite differently with the nucleotide in the c- or m-state. These large differences in the ''ground states'' are then exploited by the different structures of the two types of inhibitor groups, resulting in what we early called the ''abortive'' ground state.[11] On the basis of the idea mentioned above that the ground states are more inhibitor than substrate-like, these two types of inhibitors should match the binding center in the ground state much more completely than the substrates ADP or ATP, resulting in a high affinity.

The concept has been forwarded by us[16] that the binding center in the carrier undergoes a certain well defined trajectory of configurations from the extreme external to the extreme internal state. In these conformation changes the involved residues change the relative coordinates within the whole carrier backbone. The carrier is specifically designed for this precisely defined flexibility which should fulfill requirements of the ligand-induced transition fit and the concomitant gating process. The fact that the structures of atractylate and bongkrekate must be precisely maintained for them to be good ligands to the ADP/ATP carrier, reflects the precision of the ground state configurations, paired with their flexibility within

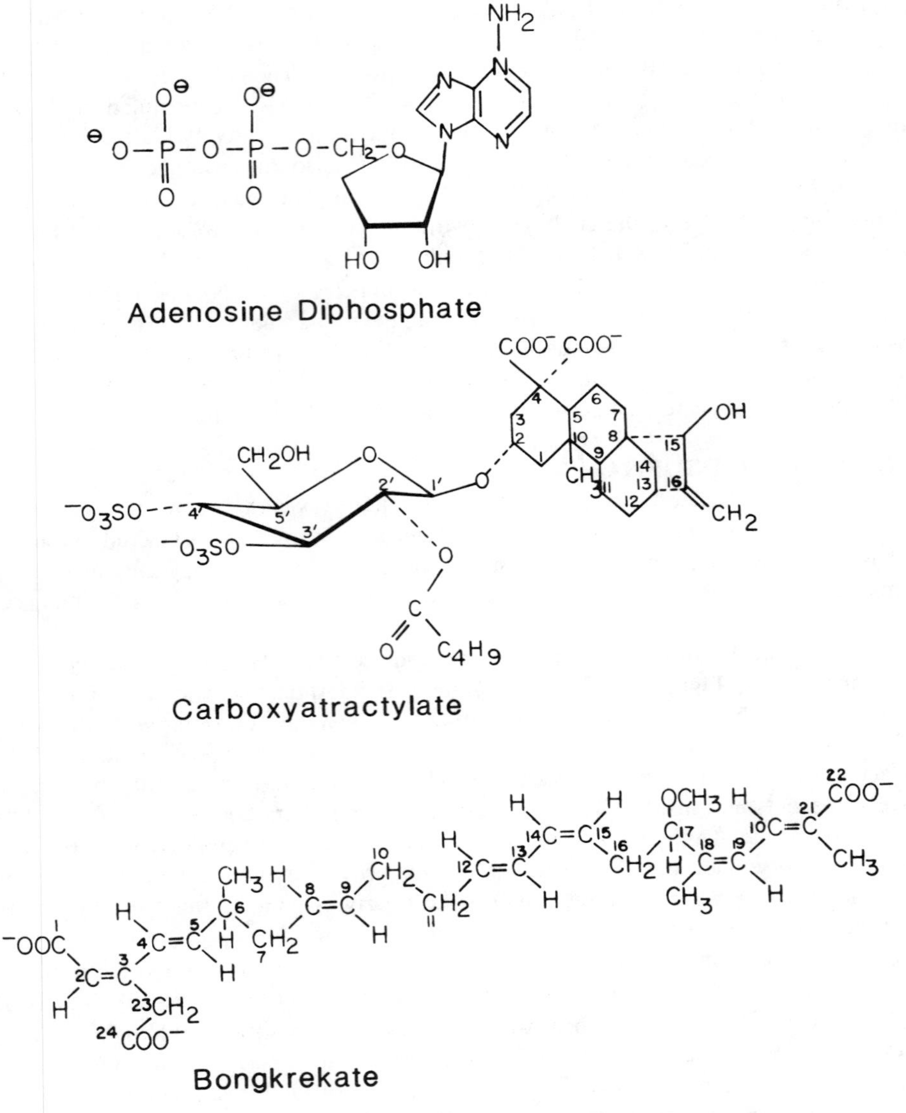

Adenosine Diphosphate

Carboxyatractylate

Bongkrekate

FIGURE 10. Comparison of chemical structures of ligands to the ADP/ATP carrier. The only common denominator is the existence of three or four anionic charges.

the translocation process. The nearly absolute membrane-sided specificity of these inhibitors — in contrast to the nucleotides — is a fundamental consequence of the parallel drastic configuration changes.

D. APPLICATION OF THE LIGAND INDUCED FIT TRANSITION THEORY TO THE EXCHANGE TYPE VS. UNIDIRECTIONAL TRANSPORT PROBLEM

The comparison between the ADP/ATP carrier and the UCP provides a unique possibility to exemplify the energetic consequences for the problem of exchange type versus unidirectional transport, as predicted by the ligand-induced transition fit theory. In an exchange type of transport the interaction energy of the substrate with the carrier should be much larger than in a unidirectional type of transport. Whereas the ADP/ATP carrier transports about the largest solute known in an exchange, UCP translocates unidirectionally about the smallest solute, H^+.[34,35] This comparison is the more significant since the two proteins seem to be constructed according to a similar building block principle.[25]

With a large substrate such as ADP and ATP the binding center should undergo a particularly large conformation change. The ground states should be separated therefore by high activation barriers which nearly completely segregate the internal and external states. The high intrinsic binding energy provided by nucleotide protein interaction will largely abolish the activation barrier and thus enable the substrate catalyzed transition between the c- and m-state. As a result, the AAC is an exchange type of carrier system which does not permit the transition of the unoccupied binding center between the c- and m-state (see Figure 2). In the UCP the substrate binding consists of protonation probably of only one group at the binding center. The interaction energy is essentially equivalent to only one ionic bond and therefore low. Concomitantly, only a small activation barrier should separate the two ground states. Therefore also the unprotonized state can quite readily translocate through the membrane, resulting in a unidirectional H^+ or opposite OH^- transport.

Another well known example which illustrates the activation of the trans-location by substrate binding, is the glucose transport through plasma membranes in animal cells.[36] The physiological function of this carrier is the uptake of glucose into the cell which operates generally not without a large gradient and therefore is not driven by superimposed energy force. In this best studied case of "facilitated diffusion", the translocation of the unloaded carrier is much slower than with glucose. Thus the exchange between glucose on both sides of the membranes is more than ten times faster than the net transport.[37]

The situation is more complex in the well studied case of H^+ galactoside cotransport by the "lac permease" of *E. coli*. Again we are only considering transport without a superimposed membrane potential. In this case the exchange rate does not show a major increase over the net transport rate.[38] However we may consider that also in the net transport mode there exists a kind of exchange system. The carrier can be seen to have two substrates, galactoside and H^+. The translocation of the carrier is blocked when it is loaded with only one of the substrates, H^+ (CH^+) or galactoside (CG), but it is active when loaded with both (CH^+G) or when unloaded (C). In the H^+ cotransport system the protonization may be considered to be rather like an unloading, as it neutralizes a negatively charged group in the binding center ($C^- + H^+ \rightarrow CH$). Obviously, on deprotonization ($CH \rightarrow C^- + H^+$) and formation of this negatively charged species the transition state barrier is strongly lowered. On the combined binding of substrate and deprotonation the activation processes seem to cancel each other, resulting in a blocked carrier (C^-G). By these energetic provisions the physiologically important obligatory cotransport function has evolved.

VII. CONCLUSIONS

Solute transport through biomembranes must have evolved in such a manner that the various requirements for this fundamental process are fulfilled. The more important ones are the selectivity for specific metabolites, the preservation of electrical insulation of the membrane, the catalytic optimalization of the translocation process, the responsiveness to external energy forces and regulatory capabilities. The single binding center in a transmembrane protein seems to comply with these tasks. The gating on both sides of the binding center maintains a continuous electrical insulation and provides the internal energy transfer for the translocational catalysis. The single binding center also permits a meaningful comparison of the carrier catalysis with enzyme catalysis, in particular with respect to the substrate-protein interaction.

Whereas in enzymes the binding center changes only slightly during a catalytic cycle, the single binding center in carriers is distinguished by a high flexibility. Thus on completion of the reaction, the enzyme is delivered in the original form, whereas the carrier is in a conformational state different from the original state at substrate uptake. The two states are segregated by a high activation barrier which is suppressed by the catalytic energy provided by the substrate-carrier interaction. In both catalysts the configuration of the unoccupied binding center is not substrate-like, however, in opposite manner for carrier and enzyme. In carriers the ground state is widely different from the transition state, whereas in enzymes, the catalytic center has optimally a configuration like the substrate in the transition state. Around the binding center of the carrier in the ground state one gate is open and the other closed, and in the transition state both gates are closed or half-open.

The substrate protein interaction provided by these binding center configurations provides the internal energy transfer which forms the basis for the translocational catalysis. The substrate-unlike configuration in the ground state results in weak substrate-protein interaction. In the transition state the binding center assumes substrate likeness and produces the high intrinsic binding energy which is the source for the energy requirement of the strong conformation changes in the binding center. The catalytic advantage of a high intrinsic binding energy in the transition state as opposed to the situation in the ground state is the near equalization of the energy levels of the substrate-carrier complex in the three states, i.e., the relatively high level of the carrier substrate complexes in the two ground states and the relatively low level in the transition state. This ''induced transition fit'' of carrier catalysis is reminiscent of the ''induced fit'' in enzymes where it is, however, catalytically unproductive, but of regulatory importance.

Among several important consequences of the ''induced transition fit'' concept is the effect of inhibitors on carriers. The substrate-unlike configuration of the ground state is utilized by those ground state fitting inhibitors. As a result of this good fit, intrinsic energy is largely expressed in tight binding which generates what has been called the ''abortive'' ground state. The inhibitors are therefore extremely side-specific, i.e., chemically quite different types of inhibitors bind specifically either only to the external or to the internal state of the binding center. In contrast, in enzymes there are transition state analog inhibitors. Thus, although inhibitors bind to the same site as a substrate, they may actually have a quite different structure. In fact, their structure should reflect quite faithfully the ground state configuration of the binding center. Nature has obviously made use of this possibility by producing strong and highly specific inhibitors against carriers as xenobiotic defense substances.

ACKNOWLEDGMENTS

This work was supported by grant K1 134/25 from the Deutsche Forschungsgemeinschaft.

REFERENCES

1. **Stein, W. D.,** *The Movement of Molecules Across Membranes,* Academic Press, New York, 1967.
2. **Klingenberg, M.,** The ADP-ATP carrier in mitochondrial membranes, in *The Enzymes of Biological Membranes: Membrane Transport,* Vol. 3, Martonosi, A. N., Ed., Plenum Press, Elmford, NY, 1976, 383.
3. **Klingenberg, M.,** The ADP/ATP carrier in mitochondrial membranes, in *The Enzymes of Biological Membranes,* Vol. 4, Martonosi, A. N., Ed., Plenum Press, Elmsford, NY, 1985, 511.
4. **Klingenberg, M.,** Membrane protein oligomeric structure and transport function, *Nature,* 290, 449, 1981.
5. **Singer, S. J.,** The molecular organization of membranes, *Annu. Rev. Biochem.*, 43, 805, 1974.
6. **Klingenberg, M., Riccio, P., Aquila, H., Buchanan, B. B., and Grebe, K.,** Mechanism of carrier transport and the ADP,ATP carrier, in *Proc. Int. Symp. The Structural Basis of Membrane Function,* Tehran, Hatefi, Y., and Djavadi-Ohaniance, L., Eds., Academic Press, New York, 1976, 293.

7. **Klingenberg, M., Aquila, H., Krämer, R., Babel, W., and Feckl, J.,** The ADP/ATP translocation and its catalyst, in FEBS Symp. No. 42, *Biochemistry of Membrane Transport,* Zürich, Semenza, G. and Carafoli, E., Eds., Springer Verlag, Berlin, 1977, 567.
8. **Erdelt, H., Weidemann, M. J., Buchholz, M., and Klingenberg, M.,** Some principle effects of bongkrekic acid on the binding of adenine nucleotides to mitochondrial membranes, *Eur. J. Biochem.,* 30, 107, 1972.
9. **Klingenberg, M. and Buchholz, M.,** On the mechanism of bongrekate effect on the mitochondrial adenine-nucleotide carrier as studied through the binding of ADP, *Eur. J. Biochem.,* 38, 346, 1973.
10. **Klingenberg, M.,** Metabolite transport in mitochondria: an example for intracellular membrane function, in *Essays in Biochemistry,* Vol. 6, Campbell, P. N., and Dickens, F., Eds., Academic Press, New York/London, 1970, 119.
11. **Klingenberg, M., Grebe, K., and Scherer, B.,** The binding of atractylate to mitochondria, *Eur. J. Biochem.,* 52, 351, 1975.
12. **Klingenberg, M., Appel, M., Babel, W., and Aquila, H.,** The binding of bongkrekate to mitochondria, *Eur. J. Biochem.,* 131, 647, 1983.
13. **Lauquin, G. J. M. and Vignais, P. V.,** Interaction of (^{3}H) bongkrekic acid with the mitochondrial adenine nucleotide translocator, *Biochemistry,* 15, 2316, 1976.
14. **Block, M. R., Lauquin, G. J. M., and Vignais, P. V. C.,** Chemical modifications of atractyloside and bongkrekic acid binding sites of the mitochondrial adenine nucleotide carrier. Are there distinct binding sites?, *Biochemistry,* 20, 2692, 1981.
15. **Vignais, P. V., Block, M. R., Boulay, F., Brandolin, G., and Lauquin, G. J. M.,** Molecular aspects of structure-function relationships in mitochondrial adenine nucleotide carrier, in *Structure and Properties of Cell Membrane,* Vol. 2, Bengha, G., Ed., CRC Press, Boca Raton, FL, 1985, 139.
16. **Klingenberg, M.,** Catalytic energy and carrier-catalyzed solute transport in biomembranes, in *Achievement and Perspectives of Mitochondrial Research,* Vol. 1, Quagliariello, E., Slater, E. C., Palmieri, F., Saccone, C., and Kroon, A. M., Eds., Elsevier, Amsterdam, 1985, 303.
17. **Bogner, W., Aquila, H., and Klingenberg, M.,** The transmembrane arrangement of the ADP/ATP carrier as elucidated by the lysine reagent pyridoxal 5-phosphate, *Eur. J. Biochem.,* 161, 611, 1986.
18. **Jencks, W. P.,** Binding energy, specificity and enzymatic catalysis: the circe effect, in *Advances in Enzymology,* Vol. 43, Meister, A., Ed., John Wiley & Sons, New York, 1975, 219.
19. **Fersht, A.,** *Enzyme Structure and Mechanism,* W. H. Freeman, San Francisco, 1977.
20. **Koshland, D. E., Jr. and Neet, K. E.,** The catalytic and regulatory properties of enzymes, *Annu. Rev. Biochem.,* 37, 359, 1967.
21. **Klingenberg, M.,** The mechanism of the mitochondrial ADP,ATP carrier as studied by the kinetics of ligand binding, in *Dynamics of Energy-Transducing Membranes,* Ernster, L., Estabrook, R. W., and Slater, E. C., Eds., Elsevier, Amsterdam, 1974, 511.
22. **Lienhard, G. E.,** Transition state analogue inhibitors, *Science,* 188, 149, 1973.
23. **Pfaff, E. and Klingenberg, M.,** Adenine nucleotide translocation of mitochondria. I. Specificity and control, *Eur. J. Biochem.,* 6, 66, 1968.
24. **Klingenberg, M.,** Nucleotide binding to uncoupling protein. Mechanism of control by protonation, *Biochemistry,* 27, 781, 1988.
25. **Aquila, H., Link, T. A., and Klingenberg, M.,** The uncoupling protein from brown fat mitochondria is related to the mitochondrial ADP/ATP carrier. Analysis of sequence homologies and of folding of the protein in the membrane, *EMBO J.,* 4, 2369, 1985.
26. **Klingenberg, M.,** Principles of carrier catalysis elucidated by comparing two similar membrane translocators from mitochondria, the ADP/ATP carrier and the uncoupling protein, *Ann. N.Y. Acad. Sci.,* 456, 279, 1985.
27. **Schlimme, E., Boos, K. S., and deGroot, E. J.,** Adenosine di- and triphosphate transport in mitochondria. Role of the amidine region for substrate binding and transport, *Biochemistry,* 19, 5569, 1980.
28. **Klingenberg, M., Mayer, I., and Dahms, A. S.,** *Biochemistry,* 23, 2442, 1984.
29. **Block, M. R., Lauquin, G. J. M., and Vignais, P. V.,** Interaction of 3′-O-(1-naphthoyl)adenosine 5′-diphosphate, a fluorescent adenosine 5′-diphosphate analogue, with the adenosine 5′-diphosphate/adenosine 5′-triphosphate carrier protein in the mitochondrial membrane, *Biochemistry,* 21, 5451, 1982.
30. **Weidemann, M. J., Erdelt, H., and Klingenberg, M.,** Effect of bongkrekic acid on the adenine nucleotide carrier in mitochondria: Tightening of adenine nucleotide binding and differentiation between inner and outer sites, *Biochem. Biophys. Res. Commun.,* 39, 363, 1970.
31. **Vignais, P. V.,** Molecular and physiological aspects of adenine nucleotide transport in mitochondria, *Biochim. Biophys. Acta,* 456, 1, 1976.
32. **Block, M. R., Lauquin, G. J. M., and Vignais, P. V.,** Use of 3′-O-naphthoyladenosine 5′-diphosphate to probe distinct conformational states of membrane-bound adenosine 5′-diphosphate/adenosine 5′-triphosphate carrier, *Biochemistry,* 22, 2202, 1983.

33. **Scherer, B. and Klingenberg, M.,** Demonstration of the relationship between the adenine nucleotide carrier and the structural changes of mitochondria as induced by adenosine 5′-diphosphate, *Biochemistry,* 13, 161, 1974.
34. **Nicholls, D. G.,** Brown adipose tissue mitochondria, *Biochim. Biophys. Acta,* 549, 1, 1979.
35. **Klingenberg, M. and Winkler, E.,** The reconstituted isolated uncoupling protein is a membrane potential driven H^+ translocator, *EMBO J,* 4, 3087, 1985.
36. **Stein, W. D.,** *Transport and Diffusion Across Cell Membranes,* Academic Press, Orlando, FL, 1986.
37. **Lieb, W. R.,** A kinetic approach to transport studies, in *Red Cell Membrane: A Methodological Approach,* Ellory, J. C. and Young, J. D., Eds., Academic Press, London, 1982, 135.
38. **Wright, J. K., Dornmair, K., Mitaku, S., Möröy, T., Neuhaus, J. M., Seckler, R., Vogel, H., Weigel, U., Jähnig, F., and Overath, P.,** Lactose: H^+ carrier of *Escherichia coli*: kinetic mechanism, purification, and structure, *Ann. N.Y. Acad. Sci.,* 456, 326, 1985.

Chapter 16

TERMINAL DEOXYNUCLEOTIDYL TRANSFERASE

Mukund J. Modak and Virendra N. Pandey

TABLE OF CONTENTS

I. INTRODUCTION

Terminal deoxynucleotidyl transferase (TdT) is a unique enzyme that catalyzes DNA synthesis without template direction using one or all four deoxynucleoside triphosphates (dNTPs) as the substrate. The reaction is characteristic of classical nucleotidyl transferase and is strictly dependent on the presence of a single-stranded DNA initiator with a free 3′-OH terminus.

TdT was discovered about three decades ago during the search for DNA polymerase in eukaryotic organisms.[1,2] The choice of the thymus gland as a source of DNA polymerase was both fortuitous and fortunate since this organ is now known to be the only major site of this enzyme.[3-5] On the other hand, a similar type of enzyme activity has not been found in prokaryotes. A number of early investigations were directed towards defining the reaction requirements and ionic conditions necessary for the synthesis of a specific DNA product, with the hope of finding some biological utility for such an enzyme activity.[6-8] The view that TdT is merely a proteolytic breakdown product of some normal DNA polymerase persisted for some time until a lack of its cross-reactivity with anti-DNA polymerase antibody was established.[9] Nevertheless, extensive characterization of the reaction catalyzed by TdT has made this enzyme an important tool in molecular biology for identifying DNA molecules by adding homopolymer tails to their 3′-OH terminus.[10-12] Similarities between the DNA polymerase reaction and TdT catalysis have made the latter an ideal model for probing the nature of the interaction of the enzyme with both dNTP substrate and the initiator molecule, with the ultimate goal of understanding the mechanism of the nucleotidyl transfer reaction.

II. ENZYMOLOGICAL PROPERTIES

A typical TdT reaction may be represented as

$$\text{(ssDNA)–OH} + \text{dNTPs} \xrightarrow{Me^{2+}} \text{(ssDNA)}_n\text{–OH} + \text{PPi} \qquad \text{(Str 1)}$$

where a single-stranded DNA of any sequence or length containing a free 3′-OH terminus[(ssDNA)-OH] is an initiator, dNTP may be any of the four dNTPs and Me^{2+} is a divalent cation such as Mg^{2+}, Mn^{2+}, Co^{2+}, or Zn^{2+}. The product of the reaction is ssDNA, which is now extended by one or more nucleotide residues (n) and an equal number of PPi. The addition of nucleotide residues to the 3′-OH terminus of initiator DNA proceeds in a 5′-3′ direction.

The basic information regarding the properties of and requirements for the TdT reaction, the kinetic constants for various substrates and initiator DNAs, and the influence of the mono and divalent cationic environment on the rates of polymerization has been gathered using highly purified TdT from calf thymus.[13] Subsequent studies with homogeneous preparations of different molecular weight forms of calf TdT(14,15) as well as TdT obtained from human leukemic cells[16] have confirmed these properties. An extensive account of these properties as well as the cell biology and molecular biology of TdT have been the subject of many reviews.[5,12,13] Only the salient features of the catalytic properties of TdT will be narrated here. TdT has an absolute requirement for dNTP substrate; generally purine dNTPs are the most preferred substrates.[13,14] However, when Co^{2+} is substituted for Mg^{2+}/Mn^{2+}, pyrimidine dNTP substrates are preferred instead.[12,13] Substitutions in the base moiety of the nucleotide are well tolerated in the polymerase reaction,[13,17,18] but modification of the sugar moiety renders it ineffective as a substrate.[19,20]

A single-stranded DNA (but not RNA) consisting of at least 3 bases with a free 3′-OH group can be used as an initiator.[6,13] However, a linear increase in the rate of polymerization has been noted with an increase in the chain length of initiator DNA.[6,13,21] Substitutions at the 5′ end of the initiator do not affect the polymerase reaction[13]. Under standard conditions of reaction, TdT prefers purine over pyrimidine nucleotide containing initiators;[6,12] double stranded DNA is poorly used as an initiator.[22] However, the latter can also be effectively used when Co^{2+} is used as a divalent cation.[10,11] Under catalytic conditions, TdT recognizes and binds 3′-OH ends of either oligomeric or polymeric initiator, whereas under noncatalytic conditions it binds to ssDNA in a cooperative manner with a site size of approximately 10-11 bases. The cooperative mode of binding of TdT to DNA is independent of the 3′-OH concentration but is dependent on and proportional to the chain length of ssDNA.[15]

TdT has also been shown to catalyze both pyrophosphate exchange and pyrophosphorolytic reactions.[23] These reactions require the presence of a divalent cation (either Mg^{2+} or Co^{2+} but not Mn^{2+}), and a single or double stranded oligomeric or polymeric DNA or RNA. In addition, the PPi exchange reaction requires dNTP or rNTP. Rigorous kinetic analysis of the TdT reaction and inhibitor studies suggest that this enzyme has two distinct sites; one for dNTP and the other for an initiator DNA.[6,12,15] The catalytic reactions conform to rapid equilibrium and random mechanism models, suggesting that dNTP and initiator binding functions are independent and are not influenced by each other.[6,15] TdT has been shown to be a zinc metalloenzyme and intrinsic zinc is postulated to coordinate the binding of the 3′-OH group of initiator DNA.[24] The enzyme has also been shown to require the presence of reduced sulfhydryl groups, for SH-alkylating agents have a strong inhibitory effect on TdT catalysis. The presence of lysine residues in the active center has likewise been indicated by the sensitivity of TdT to pyridoxal phosphate, a substrate binding site directed inhibitor in a number of DNA polymerases.[25,26] TdT catalyzes DNA synthesis in a distributive manner.[12] Therefore, synthesis of the desired number of nucleotide residues can be controlled by using the appropriate ratio of initiator to substrate dNTPs.

A. SUBSTRATE BINDING SITE DIRECTED INHIBITORS OF TdT

Additional information is available on the process of substrate binding in TdT, mainly through the study of specific inhibitors such as ATP,[20,27] Ap4A,[28,29] and Ap5A.[29,30] One of the unusual properties of TdT, in contrast to other DNA polymerases, is its ability to bind ribonucleotides with high affinity in place of dNTPs.[20] Since ribonucleotide monomers are not polymerized, enzyme inhibition ensues. Thus, ATP inhibits DNA synthesis catalyzed by TdT but not that catalyzed by replicative DNA polymerases. This observation has been effectively used in the detection of TdT in cell extracts which also contain other DNA polymerases.[27] Detailed mechanistic studies of TdT inhibition by ATP have shown that ATP is a competitive inhibitor of TdT with respect to substrate dNTPs but not to initiator DNA. Furthermore, the addition of ATP to an ongoing TdT reaction results in an instant inhibition of catalysis. Since under certain reaction conditions, limited addition of ATP to the DNA primer terminus can be effected, it has been proposed that the inhibitory action of ATP is due to incorporation of one or two rATP monomers onto the 3′OH terminus of the initiator DNA, which results in termination of the reaction. However, this interpretation has been ruled out by the fact that an addition of excess initiator molecules to an inhibited reaction does not enable the enzyme to catalyze new DNA synthesis (Reference 20 and Figure 1). These studies are indicative of the formation of a dead end complex consisting of TdT-ATP and initiator DNA. Similarly, other rNTPs are also inhibitors of TdT catalysis and their inhibition kinetics are identical to ATP.[20] Many rNTP analogues with a substitution in the base moiety, such as azido ATP and its photolysed product[31] and azido UTP (Haley, B. unpublished results), or with a substitution in the sugar moiety, such as the arabinofuranosyl derivative of ATP or CTP,[19] have also been shown to compete for the substrate binding site in TdT. Enzyme inhibition by these derivatives also appears to result from the formation of a dead end complex with initiator bound TdT.

Another series of interesting analogues of ATP are diadenosine phosphates, linked in 5′-5′ fashion with varying phosphate length. These analogues are also competitive inhibitors

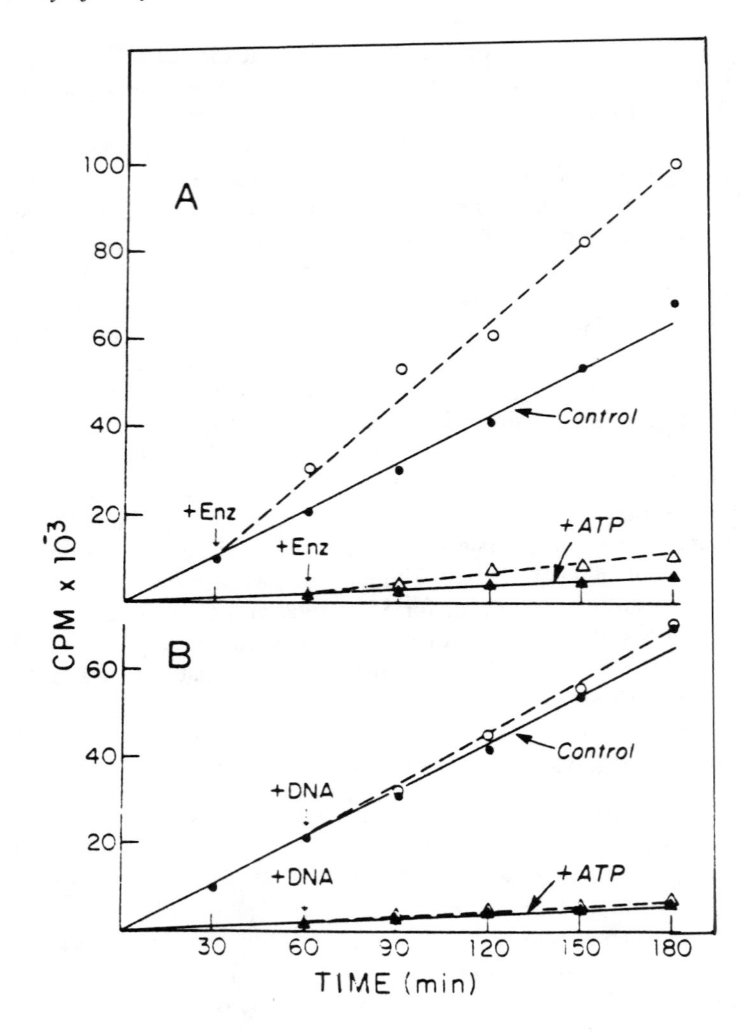

FIGURE 1. Effect of addition of excess enzyme (panel A) or initiator (panel B) to an ongoing reaction in the presence and absence of ATP. The arrows in Panel A indicate the time of addition of enzyme or initiator in a volume of 5 μl. In panel B excess DNA was added at 60 min. (From Modak, *Biochemistry*, 17, 3116, 1978. With permission.)

of TdT with respect to dNTP substrates.[28-30] The inhibitory effect of these analogues, like that of ATP, is selective and restricted to TdT among DNA polymerases; hence, replicative DNA polymerases such as *E. coli* DNA polymerase I, eukaryotic DNA polymerase α and β and viral reverse transcriptase are insensitive to these analogues.[28-30] Among the various diadenosine phosphate derivatives, the inhibitory action of Ap5A is rather unique in that it competes with both dNTPs and initiator DNA for binding to TdT (Figures 2 and 3).

Kinetic analysis of Ap5A inhibition further indicates that an initiator saturated or substrate saturated enzyme can still be inhibited by Ap5A with similar inhibition constants. However, in the absence of both initiator and dNTP substrates, a sixfold decrease in the dissociation constant of the Ap5A-TdT complex has been noted.[30] Therefore, Ap5A may be considered a most potent inhibitor of TdT catalysis. An estimate of the binding stoichiometry of Ap5A to TdT has been obtained with the use of an oxidation product of Ap5A (oAp5A). An oxidation product, presumably a tetra-aldehyde derivative of the hydroxyls at positions 2 and 3 of the sugar moieties, probably reacts with an active site lysine residue in TdT, via

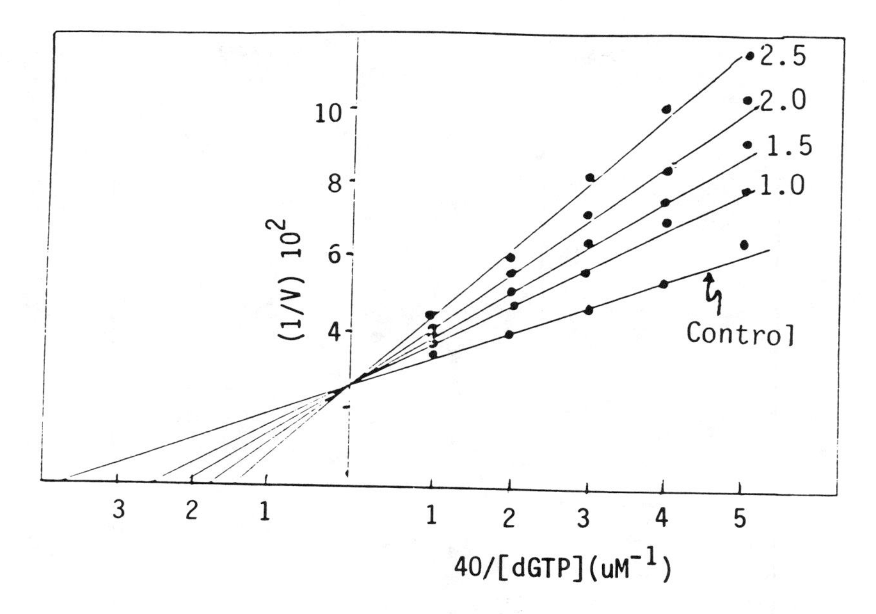

FIGURE 2. Competitive inhibition of the TdT catalyzed reaction by Ap5A with varying concentrations of dGTP as the substrate and a fixed concentration of activated DNA as the initiator. The concentration of Ap5A is shown on the line of each double reciprocal plot. (From Pandy, V. N. and Modak, M. J., *Biochemistry,* 26, 2033, 1987. With permission.)

Schiff base formation, which can be reduced and quantitated with borohydride. The incorporation of oAp5A into TdT in this manner results in irreversible inactivation of the enzyme. The degree of inactivation has been shown to be proportional to the amount of oAp5A incorporation into TdT (Table 1). Complete inactivation of TdT results when one mole of oAp5A is incorporated per mol of TdT. This observation suggests that a single molecule of Ap5A spans the distance between the substrate dNTP and initiator binding domain of TdT.[30] Since the length of the Ap5A molecule is close to 25 Å, when fully stretched, the distance between the substrate and primer binding sites in the three dimensional structure of TdT is probably close to this number.

III. AFFINITY LABELING OF THE NUCLEOTIDE BINDING SITE

The inhibition studies described above have shown that ATP and its analogues, as well as dNTP substrates, share the same binding domain in TdT. However, these conclusions are based on interpretation of kinetic results which have their own limitations. In order to identify the actual domain involved in the binding of dNTP substrate or its inhibitory analogue such as ATP, isolation and sequencing of this region is essential. An important step towards this goal has been the development of an affinity labeling approach which utilizes photoaffinity analogues of nucleotides as labeling agents.[31] The availability of highly reactive photoaffinity nucleotide analogues such as 8-azido ATP made it possible to demonstrate the specific labeling of the substrate binding site in TdT.[28] The technological problems associated with analysis of nucleotide linked protein prevented further identification of the analogue binding site. Simultaneously, protocols for covalently linking unsubstituted dNTP substrates or ATP to TdT were also developed.[29] The use of unsubstituted NTPs as an affinity label is more attractive since it is expected to reveal the true binding site in the enzyme, provided that the labeling reaction exhibits characteristics identical to the binding of substrate to enzyme (i.e., E-S formation). Such a protocol was successfully developed based on the observation that TdT forms a covalent cross-link with substrate dNTP or ATP when exposed

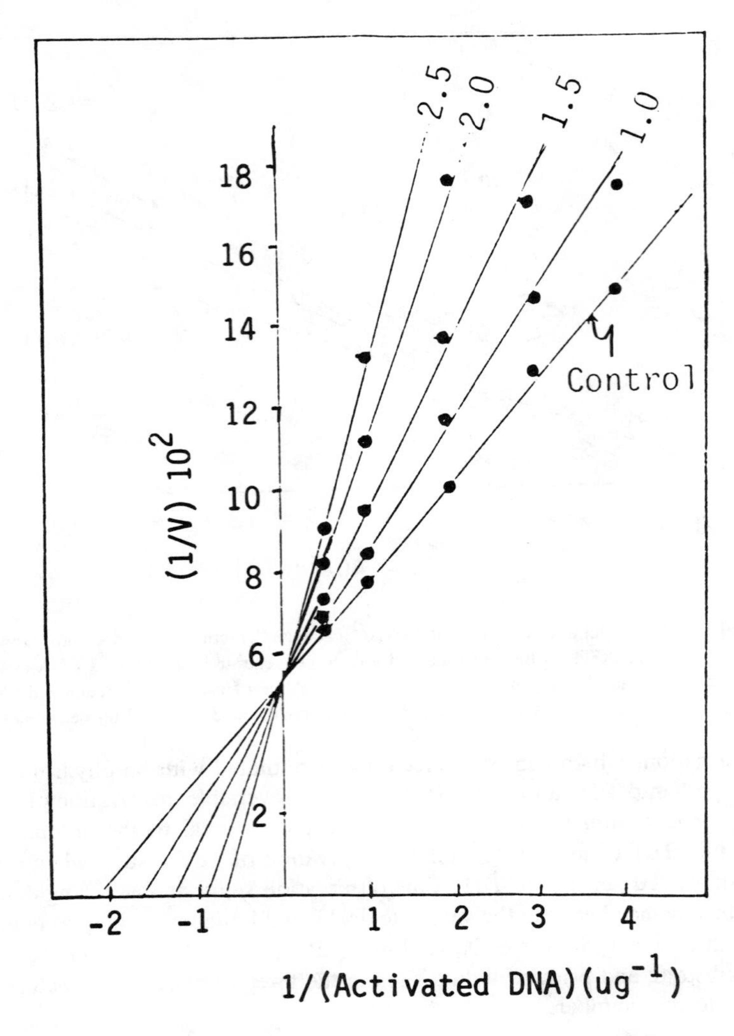

FIGURE 3. Competitive inhibition of the TdT catalyzed reaction by Ap5A with varying concentrations of activated DNA as the initiator and a fixed concentration of dGTP as the substrate. (From Pandy, V. N. and Modak, M. J., *Biochemistry,* 26, 2033, 1987. With permission.)

to ultraviolet (UV) energy.[32] The specificity of the cross-linking reaction is indicated by the observation that the crosslinking is dependent on the native conformation of the enzyme and the presence of dNTP in its metal chelate form.[31-33] The specificity of labeling of dTTP and ATP[33,34] is further supported by the demonstration that photoaffinity labeling of TdT with these nucleotides is inhibited competitively in the presence of other dNTPs, ATP and its analogues, Ap4A and Ap5A (Figure 4). Furthermore, the concentration of these nucleotides required to reduce the extent of labeling to approximately 50% is similar to the kinetic constant determined by steady state analysis of the TdT reaction.[14,29,33] Both ribo- and deoxyribonucleosides, as well as their monophosphate derivatives, which are not inhibitors of TdT catalysis, do not affect the rates or extent of labeling of either dTTP or ATP to TdT.[33,34] These results clearly demonstrate the specificity of the affinity labeling procedure and have allowed identification of domains involved in the cross-linking of dTTP and ATP.

TABLE 1
Stoichiometry of o-Ap5A Binding to TdT

Enzyme incubated with	% Enzyme bound to o-Ap5A (or inactivated enzyme)	% Enzyme free or unbound to o-Ap5A	Mol of o-Ap5A bound per mole of enzyme
None	0	100	0
^{3}H o-Ap5A			
4 μ*M*	79	21	0.8
2 μ*M*	68	32	0.7
1 μ*M*	59	41	0.58
0.33 μ*M*	51	49	0.51
0.25 μ*M*	38	62	0.42
0.2 μ*M*	37	63	0.37

Note: 4.5 μg of TdT (0.1 nmol) in a total volume of 100 μl containing 50 m*M* HEPES-KOH buffer pH 7.8, 0.5 m*M* Mn^{2+} and the indicated concentration of ^{3}H-o-Ap5A (80,000 cpm/nmol) were incubated for 4 h on ice. A freshly prepared solution of borohydride in 10 m*M* NaOH was added to a final concentration of 500 μ*M* and individual tubes were further incubated for an hour on ice; 10 μl aliquots from individual tubes were removed for enzyme activity and the remainder was precipitated directly with 10% TCA. The enzyme activity data was used to determine the percentage of free (active) and o-Ap5A-bound enzyme (inactive species). Trichloroacetic acid insoluble radioactivity present in enzyme-o-Ap5A complex was used to determine the mol fraction of Ap5A bound to enzyme. (From Pandley, V. N. and Modak, M. J., *Biochemistry,* 26, 2033, 1987. With permission.)

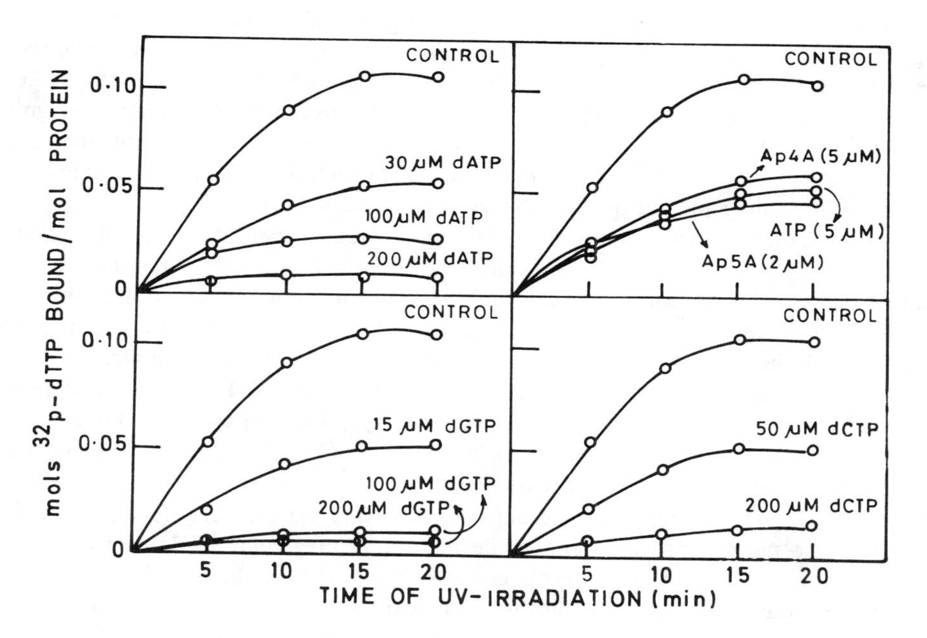

FIGURE 4. Inhibition of crosslinking of α [^{32}P]-dTTP to TdT by other dNTPs and by substrate analogues. The crosslinking reaction contained 90 μ*M* of α [^{32}P]-dTTP (10 μCi/μmol) in its metal chelate form, 30 μ*M* of TdT, 50 m*M* Tris HCL pH 7.8, 0.5 m*M* dithiothreitol and indicated concentrations of dNTP, ATP or its analogues. UV-mediated affinity labeling and assay of covalent crosslinks were carried out according to standard protocol. (From Pandey, V. N. and Modak, M. J., *J. Biol. Chem.,* 263, 3744, 1988. With permission.)

IV. UNITY OF THE dNTP AND ATP BINDING DOMAIN IN TdT

The unity of the dNTP and ATP binding domain in TdT has been demonstrated through the peptide mapping and isolation and sequencing of dNTP-linked peptide. A 29 amino acid long domain corresponding to residues 221 to 249 in the primary amino acid sequence of calf TdT[35] has been identified as a region that contains the site for nucleotide binding in this enzyme.[33,34] By amino acid composition and sequence analysis, two cysteine residues at positions 227 and 234 in calf TdT have been shown to provide the actual contact points for the cross-linking of dTTP upon UV irradiation of the enzyme substrate complex.[33,34] The investigation to identify the site of ATP binding, using the same approach, has clearly shown that ATP binding in calf TdT occurs in the same general domain and that the same two cysteine residues are involved in the binding of ATP to TdT. The two Cys residues detected as being responsible for the binding of dNTP and ATP reside only seven amino acids apart in an environment that lacks aromatic amino acids but is enriched in glutamic and isoleucine residues (Figure 6).

V. MODEL FOR NUCLEOTIDE BINDING IN TdT

A schematic representation of nucleotide binding in a 29 amino acid long domain involving Cys 227 and Cys 234 is depicted in Figure 5. In this model, dTTP or ATP is oriented and stabilized in the substrate binding domain such that C(6) and C(2) oxogroups of the pyrimidine moiety of dTTP or the N(1) and N(3) atoms of the purine ring of ATP are in close proximity to Cys 227 and 234 (33). Hydrogen bond formation between pyrimidine oxogroups and cysteine sulfhydryl or purine N(1) or N(3) atoms and sulfhydryl groups appears likely. The high sensitivity of the enzyme-nucleotide interaction to N-ethylmaleimide, an -SH alkylating reagent, supports the above conclusion that Cys sulfhydryl groups are involved in the nucleotide binding function of TdT. An examination of amino acid sequences around the two cysteine residues in calf as well as in mouse and human TdT provides an intriguing picture of the substrate binding domain of TdT in these 3 species. Figure 6 shows the primary amino acid sequence of this region translated from the nucleotide sequence of genes coding for TdT from calf,[35] mouse,[35] and human.[36] Given that TdT from all three sources exhibits nearly identical catalytic properties, conservation of the amino acid sequence of the active site domain is expected. Indeed, all the residues from 221 to 249 are conserved with the exception of 4 residues, including residue 234. The number assignment for amino acid residues refers to calf TdT; mouse and human TdT have different residue numbers for the same region. In calf TdT, #234 is Cys, while in mouse and human TdT, this residue corresponds to Ser and Gly, respectively. If Cys 234 in calf TdT is indeed an H bond donor, then a similar function may have to be satisfied by serine or glycine in mouse and human TdTs. The serine hydroxyl group can effectively share its proton with a pyrimidine oxo group or a purine nitrogen atom, while the ability of glycine to do so has not been documented. The fact that Cys 234 is not conserved in mouse and human TdTs raises doubt regarding the true involvement of residue 234 in the substrate binding function of TdT. However, the presence of a nucleophylic residue in the form of Lys 233, as a preceding amino acid, together with Cys 227 may provide appropriate contact points for nucleotide binding in TdT from all 3 sources. The ϵ amino group of lysine residues is an excellent proton donor and has been shown to play an important role in a number of nucleotide binding enzymes including DNA polymerases.[25,26] It may be further speculated that an additional binding, particularly to the sugar phosphate moiety of substrate nucleotides, may be provided by other residues in this domain. Clarification of these aspects will have to await site-specific mutagenesis of the desired residue in the gene and an examination of the properties of the resulting mutant enzyme. Resolution of the crystal structure of TdT will aid in the construction

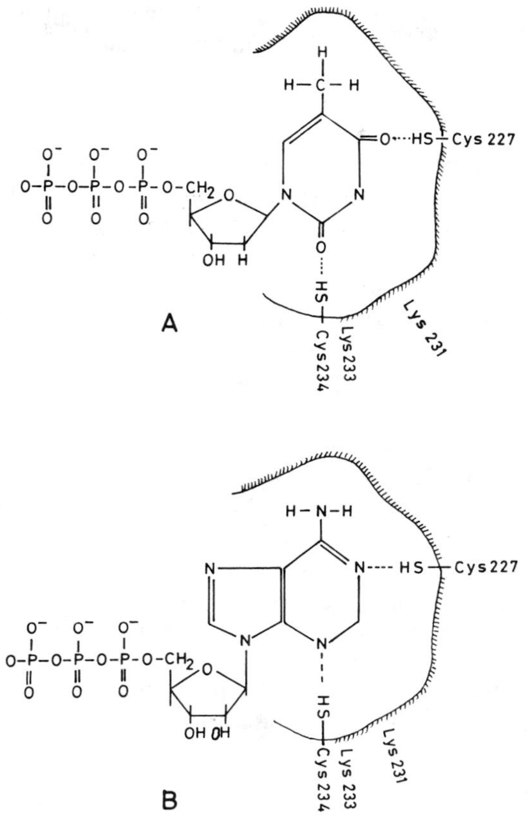

FIGURE 5. Schematic representation of the nucleotide binding domain of TdT. A and B represent binding of dTTP and ATP respectively, within the 29 residue domain of TdT. Cys 227 and Cys 234 are shown in close proximity of the C(6) and C(2) oxogroups of dTTP and to the N(1) and N(3) atoms of ATP. The broken lines represent possible hydrogen bonding between the Cys residues and the pyrimidine or purine moiety of the nucleotide substrate or its analogues. (From Pandey, V. N. and Modak, M. J., *J. Biol. Chem.*, 263, 3744, 1988. With permission.)

```
                221                     227                       234
Bovine TdT...  |Asp-Thr-Glu-Gly-Ile-Pro-Cys-Leu-Gly|-Asp-|Lys-Val-Lys|-Cys
Mouse  TdT...  |Asp-Thr-Glu-Gly-Ile-Pro-Cys-Leu-Gly|-Asp-|Lys-Val-Lys|-Ser
Human  TdT...  |Asp-Thr-Glu-Gly-Ile-Pro-Cys-Leu-Gly|-Ser-|Lys-Val-Lys|-Gly

                235             241                                249
               |Ile-Ile-Glu|-Glu-|Ile-Ile-Glu-Asp-Gly-Glu-Ser-Ser-Glu|-Val-|Lys|
               |Ile-Ile-Glu|-Gly-|Ile-Ile-Glu-Asp-Gly-Glu-Ser-Ser-Glu|-Ala-|Lys|
               |Ile-Ile-Glu|-Glu-|Ile-Ile-Glu-Asp-Gly-Glu-Ser-Ser-Glu|-Val-|Lys|
```

FIGURE 6. Amino acid sequence of the 29 residue long substrate binding domain of bovine, mouse and human TdTs. The amino acid sequences in the box are common among the three TdTs.

of models of the catalytic domain which should give some indication of the importance and contribution of individual amino acid residues in this domain.

VI. FUTURE STUDIES

Two distinct trends in TdT research are beginning to emerge; one relates to the issue of biological function, while the other addresses the molecular enzymology. On the biological front, recent advances in the understanding of immunoglobulin gene recombination, as predicted by nucleotide sequence rearrangements, are beginning to provide some clue to the biological function of TdT. The specific nucleotide sequences found in the joining region of immunoglobulin as well as T cell receptor genes appear to be synthesized by TdT-like enzyme activity.[37,38] Since gene recombination process entails involvement of a number of enzymes and DNA binding proteins, TdT may be expected to be one of the many components which work in concert. Detailed investigations of the enzymatic machinery involved in this process will clarify the site of TdT action as well as factors which control its expression. On the enzymological front, the possibility of obtaining large quantities of enzyme protein via gene cloning and monoclonal antibody column technology will stimulate structure-function studies of TdT, particularly to define the structural domains which participate in various stages of catalysis. Thus, domains responsible for primer and substrate binding together with identification of amino acid residues involved in the binding of these components may be clarified. Additionally, identification of residues involved in the overall polymerization reaction, such as stabilization of enzyme initiator complex and movement of enzyme along the initiator, PPi exchange and pyrophosphorolysis reactions, will provide an insight into the dynamics of the polymerase reaction by model building. In this regard, attempts to crystallize TdT and resolve its 3-dimensional structure may be expected. An excellent complement to these studies will be afforded by higher power NMR spectroscopy of enzyme-substrate and enzyme-initiator complexes. In summary, the prospects for clarification of the mechanisms of enzymatic DNA synthesis using TdT as a model enzyme appear quite bright.

REFERENCES

1. **Bollum, F. J.,** Oligodeoxynucleotide primers for calf thymus polymerase, *J. Biol. Chem.*, 235, PC18, 1960.
2. **Krakow, J. S., Coutsogeorgopoulos, C., and Canellakis, E. S.,** Incorporation of deoxyribonucleotides into terminal positions of DNA, *Biochem. Biophys. Res. Commun.*, 5, 477, 1961.
3. **Chang, L. M. S.,** Development of terminal deoxynucleotidyl transferase in embryonic calf thymus gland, *Biochem. Biophys. Res. Commun.*, 44, 124, 1971.
4. **Gregoire, K. E., Goldschneider, I., Barton, R. W., and Bollum, F. J.,** Ontogeny of terminal deoxynucleotidyl transferase positive cells in lymphohemopoietic tissues of rat and mouse, *J. Immunol.*, 123, 1347, 1979.
5. **Bollum, F. J.,** Terminal deoxynucleotidyl transferase as a hematopoietic cell marker, *Blood*, 54, 1203, 1979.
6. **Kato, K., Goncalves, J. M., Houts, G. E., and Bollum, F. J.,** Deoxynucleotide polymerizing enzymes of calf thymus glands. II. Properties of the terminal deoxynucleotidyl transferase, *J. Biol. Chem.*, 242, 2780, 1967.
7. **Hansbury, E., Kerr, V. N., Mitchel, V. E., Ratliff, R. L., Smith, D. A., Williams, D. I., and Hayes, F. N.,** Synthesis of polydeoxynucleotides using chemically modified subunits, *Biochim. Biophys. Acta*, 199, 322, 1970.
8. **Bollum, F. J.,** Terminal deoxynucleotidyl transferase: biological studies, in *Advances in Enzymology*, Vol. 47, Meister, A., Ed., John Wiley & Sons, New York, 1978, 347.

9. **Chang, L. M. S. and Bollum, F. J.,** Antigenic relationship in mammalian DNA polymerase, *Science,* 175, 1116, 1972.
10. **Roychoudhury, R., Jay, E., and Wu, R.,** Terminal labeling and addition of homopolymer tracts to duplex DNA fragments by terminal deoxynucleotidyl transferase, *Nucleic Acid Res.,* 3, 363, 1976.
11. **Roychoudhury, R. and Wu, R.,** Terminal transferase catalyzed addition of nucleotides to the 3′ terminal of DNA, *Methods Enzymol.,* 65, 42, 1980.
12. **Chang, L. M. S. and Bollum, F. J.,** Molecular Biology of terminal transferase, *Crit. Rev. Biochem.,* 21, 27, 1987.
13. **Bollum, F. J.,** Terminal deoxynucleotidyl transferase, in *The Enzymes,* Vol. 10, Boyer, P. D., Ed., Academic Press, New York, 1974, 145.
14. **Pandey, V. N. and Modak, M. J.,** Purification of high molecular mass species of calf thymus terminal deoxynucleotidyl transferase, *Prep. Biochem.,* 17, 359, 1987.
15. **Robbins, D. J., Barkley, M. D., and Coleman, M. S.,** Interaction of terminal transferase with single-stranded DNA, *J. Biol. Chem.,* 260, 9494, 1987.
16. **Deibel, M. R., Jr., Coleman, M. S., Acree, K., and Hutton, J. J.,** Biochemical and immunological properties of human terminal deoxynucleotidyl transferase purified from blasts of acute lymphoblastic and chronic myelogenous leukemia, *J. Clin. Invest.,* 67, 725, 1981.
17. **Deibel, M. R., Jr., Liu, C. G., and Barkley, M. D.,** Fluorimetric assay for terminal deoxynucleotidyl transferase activity, *Anal. Biochem.,* 144, 336, 1985.
18. **Zmudzka, B., Bollum, F. J., and Shugar, D.,** Polydeoxyribouridylic acid and its complexes with polyribo- and deoxyriboadenylic acids, *J. Mol. Biol.,* 46, 169, 1969.
19. **Dicioccio, R. N. and Srivastava, B. I. S.,** Kinetics of inhibition of deoxynucleotide polymerizing enzyme activities from normal and leukemic human cells by 9-β-D-arabinofuranosyl adenine 5′-triphosphate and 1-β-D-arabinofuranosyl cytosine 5′-triphosphate, *Eur. J. Biochem.,* 79, 411, 1977.
20. **Modak, M. J.,** Biochemistry of terminal deoxynucleotidyl transferase: Mechanism of adenosine 5′-triphosphate inhibition of TdT, *Biochemistry,* 17, 3116, 1978.
21. **Deibel, M. R. and Coleman, M. S.,** Biological properties of purified human terminal deoxynucleotidyl transferase, *J. Biol. Chem.,* 255, 4206, 1980.
22. **Michelson, A. M. and Orkin, S. H.,** Characterization of the homopolymer tailing reaction catalyzed by terminal deoxynucleotidyl transferase- implication for the cloning of cDNA, *J. Biol. Chem.,* 257, 14773, (1982).
23. **Srivastava, A. and Modak, M. J.,** Biochemistry of terminal deoxynucleotidyl transferase: Identification, characterization and active site involvement in the catalysis of associated pyrophosphate exchange and pyrophosphorolytic activity, *Biochemistry,* 19, 3270, 1980.
24. **Chang, L. M. S. and Bollum, F. J.,** Deoxynucleotide polymerizing enzymes of calf thymus gland IV. Inhibition of terminal deoxynucleotidyl transferase by metal ligands, *Proc. Natl. Acad. Sci. U.S.A.,* 65, 1041, 1970.
25. **Modak, M. J. and Dumaswala, U. J.,** Divalent cation-dependent pyridoxal 5-′ phosphate inhibition of Rauscher Leukemia Virus DNA polymerase, *Biochim. Biophys. Acta,* 654, 227, 1981.
26. **Basu, A. and Modak, M. J.,** Identification and amino acid sequence of deoxynucleotide triphosphate binding site in *E. coli* DNA polymerase I., *Biochemistry,* 26, 1704, 1987.
27. **Bhalla, R. B., Schwartz, M. K., and Modak, M. J.,** Selective inhibition of terminal deoxynucleotidyl transferase (TdT) by adenosine ribonucleoside triphosphate and its application in the detection of TdT in human leukemia, *Biochem. Biophys. Res. Commun.,* 76, 1056, 1977.
28. **Ono, K., Iwata, Y., Nakamura, H., and Matsukage, A.,** Selective inhibition of terminal deoxynucleotidyl transferase by diadenosine 5,′ 5″-P,[1],[4] tetraphosphate, *Biochem. Biophys. Res. Commun.,* 95, 34, 1980.
29. **Pandey, V. N., Amrute, S. B., Satav, J. G., and Modak, M. J.,** Inhibition of terminal deoxynucleotidyl transferase by adenine dinucleotides: unique inhibitory action by Ap5A, *FEBS Lett.,* 213, 204, 1987.
30. **Pandey, V. N. and Modak, M. J.,** Biochemistry of terminal deoxynucleotidyl transferase (TdT): characterization and mechanism of inhibition of TdT by p1,p5-bis-(5′-adenosyl)pentaphosphate, *Biochemistry,* 26, 2033, 1987.
31. **Abraham, K. I., Haley, B., and Modak, M. J.,** Biochemistry of terminal deoxynucleotidyl transferase: Affinity labeling of substrate binding site with azido-ATP, *Biochemistry,* 22, 4197, 1983.
32. **Modak, M. J. and Gillerman-Cox, E.,** Biochemistry of terminal deoxynucleotidyl transferase: conditions for the ultraviolet light mediated cross-linking of substrate to TdT, *J. Biol. Chem.,* 257, 15105, 1982.
33. **Pandey, V. N. and Modak, M. J.,** Biochemistry of terminal deoxynucleotidyl transferase: affinity labeling and identification of deoxynucleotide triphosphate binding domain of TdT, *J. Biol. Chem.,* 263, 3744, 1988.
34. **Pandey, V. N. and Modak, M. J.,** Identification and unity of ribo- and deoxyribonucleoside triphosphate binding site in terminal deoxynucleotidyl transferase, *Biochemistry,* (communicated), 1988.

35. **Koiwai, O., Yokota, T., Kageyama, T., Hirose, T., Yoshida, S., and Arai, K.,** Isolation and characterization of bovine and mouse terminal deoxynucleotidyl transferase cDNA expressible in mammalian cells, *Nucleic Acid Res.,* 14, 5777, 1986.
36. **Peterson, R. C., Cheung, L. C., Mattaliano, R. J., White, S., Chang, L. M. S., and Bollum, F. J.,** Expression of human terminal deoxynucleotidyl transferase in *E. coli, J. Biol. Chem.,* 260, 10495, 1985.
37. **Desiderio, S. V., Yancopoulos, G. D., Paskind, M., Thomas, E., Boss, M. A., Landau, N., Alt, F. W., and Baltimore, D.,** Insertion of N-region into heavy-chain genes is correlated with expression of terminal deoxynucleotidyl transferase in B cells, *Nature,* 311, 752C, 1984.
38. **Alt, F. W., Blackwell, T. K., and Yancopoulos, G. D.,** Development of the primary antibody repertoire, *Science,* 228, 1079, 1987.

Section VIII
Phosphotransferases — Structure and Mechanisms

It seems likely that the responses to variation in energy charge of several ATP-regenerating sequences and of many ATP-utilizing sequences, as modulated in each case by other metabolite modifiers, are of central importance in the maintenance of metabolic homeostasis, Daniel E. Atkinson, Adenine Nucleotides as Stoichiometric Coupling Agents in Metabolism and as Regulatory Modifiers: The Adenylate Energy Charge, Chapter 1, p. 20, in *Metabolic Regulation*, Vol. V, (Henry J. Vogel, editor), in *Metabolic Pathways*, Third Edition (David M. Greenberg, Editor-in-Chief), Academic Press, New York, 1971.

Chapter 17

ADENYLATE KINASE

Minoru Hamada, Hitoshi Takenaka, Michihiro Sumida, and Stephen A. Kuby

TABLE OF CONTENTS

I. INTRODUCTION

The isolation of rabbit muscle myokinase (ATP-AMP transphosphorylase, adenylate kinase) in crystalline form was first described by Noda and Kuby,[1,2] who also determined several of its physicochemical properties.[3] Adenylate kinase is an ubiquitous enzyme,[57,58] and it has since been prepared in either crystalline or homogeneous form and characterized from a variety of sources, which include rabbit,[1-3,17] porcine,[4] carp,[5] bovine liver mitochondria,[7,8] rat muscle,[9] rat liver,[9,10,11] porcine liver,[12] rat hepatomas,[13] rat brain,[14] human erythrocytes,[15] bakers' yeast,[16] calf muscle and liver,[17] human liver mitochondria,[18] porcine heart,[19] human muscle[6,20] and liver,[20] normal human serum and an aberrant form in human dystrophic serum,[21] human FSH (Fascio-scapulo-humeral type) dystrophic muscle and liver,[20a] bovine heart mitochondria,[22] and *Escherichia coli*.[40]

Adenylate kinase has been reported to exist in two major sets of multiple or polymorphic forms in populations of the human erythrocyte which were presumably genetically determined.[24,25] These two major genetic variants of human erythrocyte adenylate kinase, which had been designated AK_1 and AK_2,[24] were separated and partially purified;[26,27] a species which was designated "AK_a" was isolated,[15] and one species which was designated AK_1 was sequenced.[28] In addition to the genetically determined variants, tissue-determined variations of adenylate kinase in man were described early[29] and these studies were extended by Russell et al.[30]

At least in mammals, there may be two major forms of isozymes of adenylate kinase: the cytoplasmic type, present largely in the skeletal muscle, and the mitochondrial type as illustrated by the liver. Although there is a very high degree of homology between the muscle-types in mammals, there is apparently a sufficient number of differences between the muscle-type and liver-type to permit their distinction by immunological means.[17,30]

Recently, an aberrant adenylate kinase isozyme was detected[55] and isolated from the serum of patients with Duchenne Muscular Dystrophy,[21] which resembled the isolated human liver mitochondrial adenylate kinase,[18,31] in its kinetics, stability, and electrophoretic properties,[32] but by immunological tests it appeared structurally to be a muscle type.[31,32] It would be of interest from a both genetic and a pathogenetic stand-point to understand precisely which covalent alterations may have taken place in this human variant adenylate kinase isozyme and whether or not it represents a fetal species.

The amino acid sequences have been deduced for the porcine muscle adenylate kinase,[33] for one species of human muscle adenylate kinase designated AK_1,[28] for calf and rabbit muscle adenylate kinase,[51] for chicken muscle adenylate kinase,[43] for adult bovine heart mitochondria designated AK_2,[22,39] and from *E. coli*.[40,41] The X-ray crystal structures of form A and B of the porcine adenylate kinase[34-36] and of the yeast cytosolic adenylate kinase[104] and of the *E. coli* enzyme[115] were elucidated, and recently the human, porcine, and rabbit muscle adenylate kinases were studied by ^1H nuclear magnetic resonance spectroscopy.[37,38,67,116]

Recently, the amino acid analyses were presented[20] for the crystalline normal human liver, calf liver, and rabbit liver adenylate kinases, and compared with the normal human muscle, calf muscle, and rabbit muscle myokinases. The liver-types as a group and the muscle-types as a group show a great deal of homology, but some distinct differences are evident between the liver and muscle enzyme groups, especially in the number of residues of His, Pro, and half-cystine and in the presence of Trp in the liver enzymes.[20] Porcine AK_1 and adult bovine heart AK_2 display some degree of homology especially near the amino terminus.[22,39] Recently, the *adk* gene encoding adenylate kinase in *E. coli* was cloned in pBR322, and the primary structure of the *E. coli* adenylate kinase was deduced from the nucleotide sequence of the *adk* gene.[41] Gilles et al.[42] then showed that a thermosensitive adenylate kinase was due to the substitution of a serine residue for a proline residue at position-87; they also reported a circular dichroism study of both species.[79] In the case of the chick, a cDNA clone for muscle adenylate kinase was isolated from a cDNA library of chick skeletal muscle poly $(A)^{+}$RNA, and the DNA sequence was determined by Kishi et al.,[43] who thereby deduced its primary structure.

The MgATP-binding site of rabbit muscle adenylate kinase was located by combining the NMR data[67] with the X-ray diffraction data.[34-36] It is near three protein segments, five to seven amino acids in length, which are homologous in sequence to segments found in other nucleotide-binding phosphotransferases, such as myosin and F_1-ATPase, ras p21 and transducin GTPases, and cAMP-dependent and src protein kinases, suggesting that equivalent mechanistic roles of these segments exist in all of these proteins.[44] Recently, the AMP-binding site was located also by NMR studies,[86] and a final deduced structure for the ternary complex was presented. A detailed analysis of the steady-state kinetic reaction mechanism was presented earlier by Hamada et al.[45] and together with substrate binding data on the muscle-type adenylate kinases, and derived peptide fragments,[46,90] some mechanistic ideas surfaced, especially in pinpointing the two sites for binding of MgATP/MgADP and of AMP/ADP.

For an excellent review on the adenylate kinases prior to 1973, please refer to Reference 58.

II. KINETICS AND REACTION MECHANISM

A. STEADY-STATE KINETIC PROPERTIES OF ADENYLATE KINASE

The kinetic properties of the adenylate kinases from several sources have been the subject of a number of investigations (e.g., see the summary in Reference 58), with the first definitive attempt by Noda[80] on the rabbit muscle enzyme. However, because in the early investigations metal complexation by the nucleotide substrates were not quantitatively considered, conflicting interpretations of the steady-state data were often encountered (e.g., see References 57 and 87). By a careful study of the isotope exchange rates at equilibrium, Rhoads and Lowenstein[56] were able to exclude an ordered mechanism, as well as a mechanism involving enzyme covalent intermediates for the rabbit muscle enzyme. Moreover, they added support for the hypothesis[57] that there were two binding sites for adenylate kinase, one for the magnesium nucleotide complexes and one for the uncomplexed nucleotides as shown in Equation 1:

$$MgATP^{2-} + AMP^{2-} \rightleftharpoons MgADP^{-} + ADP^{3-} \qquad (1)$$

Similarly, Su and Russell[81] concluded that although the yeast adenylate kinase followed a random mechanism, a rate-limiting step existed at the interconversion of the ternary complexes, whereas, Rhoads and Lowenstein[56] felt that phosphate transfer was not at the site of the rate-limiting reaction. This conclusion of Rhoads and Lowenstein[66] seemed to be confirmed by ^{31}P-NMR studies on the porcine muscle adenylate kinase.[82] Markland and Wadkins[8] had tentatively interpreted their isotope exchange studies of the bovine mitochondrial adenylate kinase as consistent with an ordered mechanism. In a systematic kinetic study on two genetic variants of the human erythrocyte adenylate kinase, with a careful examination of the effects of the metal complexes of the substrates on the reaction velocity, i.e., of $MgATP^{2-}$ and of $MgADP^{-}$, Brownson and Spencer[27] concluded that a rapid equilibrium random mechanism was followed by these enzymes, with a rate-limiting step at the interconversion of the ternary complexes, but for a case without independent binding of the substrates.

Since the two crystalline isoenzymes from calf muscle and calf liver had been obtained,[17] the opportunity presented itself to compare them kinetically together with the rabbit muscle enzyme[45,47] whose structure-equilibrium substrate binding relationships were being reexamined.[46,47]

$$E + MA \underset{K_1}{\rightleftharpoons} E\cdot MA;\quad E\cdot MA + B \underset{K_3}{\rightleftharpoons} E\cdot MA\cdot B;\quad E + B \underset{K_2}{\rightleftharpoons} E\cdot B;\quad E\cdot B + MA \underset{K_4}{\rightleftharpoons} E\cdot MA\cdot B$$

$$E\cdot MA\cdot B \underset{V^r_{max}/E_t}{\overset{V^f_{max}/E_t}{\rightleftharpoons}} E\cdot MC\cdot C$$

$$E\cdot MC\cdot C \underset{K_9}{\rightleftharpoons} E\cdot MC + C;\quad E\cdot MC \underset{K_7}{\rightleftharpoons} MC + E;\quad E\cdot MC\cdot C \underset{K_8}{\rightleftharpoons} E\cdot C + MC;\quad E\cdot C \underset{K_6}{\rightleftharpoons} E + C$$

SCHEME I

$MA = MgATP^{2-}$; $B = AMP^{2-}$; $MC = MgADP^{1-}$; $C = ADP^{3-}$

$A_0 = \sum(A + MA + \cdots)$

$C_0 = \sum(C + MC + \cdots)$

$M_0 = \sum(M + MA + \cdots + MC + \cdots)$

SET: $(MC) = (C) = 0$

$$v^f_0 = \frac{V^f_{max}}{1 + \frac{K_4}{(MA)} + \frac{K_3}{B} + \frac{K_1K_3}{(MA)(B)}}$$

$$K_1 \cdot K_3 = K_2 \cdot K_4$$

$(MA) = (B) = 0$

$$v^r_0 = \frac{V^r_{max}}{1 + \frac{K_8}{(MC)} + \frac{K_9}{(C)} + \frac{K_7K_9}{(MC)(C)}}$$

$$K_7 \cdot K_9 = K_6 \cdot K_8$$

SCHEME 1. Postulated kinetic mechanism of ATP-AMP transphosphorylase (myokinase) from rabbit muscle.[45]

The steady-state kinetics, at pH 7.4, 25°C, and essentially fixed ($\Gamma/2$) of 0.16 to 0.18, of the rabbit muscle, calf muscle, and calf liver adenylate kinases all seem to be adequately expressed by a random quasi-equilibrium type of mechanism with a rate-limiting step largely at the interconversion of the ternary complexes (Scheme 1). It should be remarked at this point, that the data presented in Reference 45, by themselves, will not exclude the possibility of a random mechanism in which product release is rate limiting. The treatment of the data was described in detail in Reference 45. In the case of the rabbit muscle enzyme, five substrate pairs for the forward reaction ($MgATP^{2-}/AMP^{2-}$; $MgdATP^{2-}/dAMP^{2-}$; $MgATP^{2-}/dAMP^{2-}$; $MgdATP^{2-}/AMP^{2-}$; $Mg\epsilon ATP^{2-}/AMP^{2-}$) and two substrate pairs for the reverse reaction ($MgADP^-/ADP^{3-}$; $MgdADP^-/dADP^{3-}$) were studied under conditions where a careful and systematic control was exercised over the metal complex nucleotide species. For the forward reactions between $MgATP^{2-}$ + AMP^{2-}, and $MgdATP^{2-}$ + $dAMP^{2-}$, the estimated values for $\overline{K}_{s,1}$ and $K_{s,1}$ (i.e., intrinsic dissociation constants of the substrate from the binary and ternary complexes) differed, pointing to substrate(s) induced conformational changes in the ternary complexes (Table 1). That similar derived values either for $\overline{K}_{AMP}{}^{2-}$ and for $K_{AMP}{}^{2-}$ could be estimated for each of the above three sets of substrate pairs in the forward reaction supports the conclusion that a common rate-limiting step exists and is likely in the interconversion of the ternary complexes. ϵAMP^{2-} was found to be a competitive inhibitor of AMP^{2-} (at fixed $MgATP^{2-}$) and $MgdADP^-$ to be a non-competitive inhibitor of ADP^{3-} (at fixed $MgADP^-$), in support of the conclusion of Rhoads and Lowenstein[56] that a separate site exists for the magnesium complexes of the nucleotide substrates ($MgATP^{2-}$ and $MgADP^-$) and for the uncomplexed nucleotide substrates (ADP^{3-} and AMP^{2-}). ϵAMP^{2-}, although apparently not a substrate, binds fairly tightly to the enzyme, and in a competitive fashion with respect to AMP^{2-} (with $K_i \approx 10^{-4}$ *M*), pointing

TABLE 1
Comparison of Values for Derived Kinetic Parameters at pH 7.4, 25°C, and 0.16 (r/2) for the Forward and Reverse Reactions, Catalyzed by ATP-AMP Transphosphorylase from Rabbit Muscle

$K_1 = (\bar{K}_{MgATP^{2-}})$[a] $2.0\ (\pm 0.4) \times 10^{-5}$ (M)	$(\bar{K}_{MgdATP^{2-}})$ $2.1_5\ (\pm 0.9_8) \times 10^{-4}$	$(\bar{K}_{MgATP^{2-}})$ $1.6\ (\pm 0.3) \times 10^{-4}$	$(\bar{K}_{MgdATP^{2-}})$ $2.2\ (\pm 1.5) \times 10^{-4}$	$(\bar{K}_{Mg\epsilon ATP^{2-}})$ $9.8\ (\pm 0.8) \times 10^{-5}$
$K_2 = (\bar{K}_{AMP^{2-}})$ $3.5\ (\pm 1.1) \times 10^{-5}$	$(\bar{K}_{JAMP^{2-}})$ $1.4_4\ (\pm 0.0_4) \times 10^{-4}$	$(\bar{K}_{JAMP^{2-}})$ $6.5\ (\pm 2.5) \times 10^{-5}$	$(\bar{K}_{AMP^{2-}})$ $6.2\ (\pm 4.8) \times 10^{-5}$	$(\bar{K}_{AMP^{2-}})$ $8.0\ (\pm 0.8) \times 10^{-5}$ $K_1\ (\bar{K}_{\epsilon AMP^{2-}}$ $2.29\ (\pm 0.10) \times 10^{-4}$
$K_3 = (K_{AMP^{2-}})$ $9.4\ (\pm 1.0) \times 10^{-5}$	$(K_{JAMP^{2-}})$ $3.6_3\ (\pm 1.5_9) \times 10^{-5}$	$(K_{JAMP^{2-}})$ $5.3\ (\pm 0.7) \times 10^{-5}$	$(K_{AMP^{2-}})$ $3.0\ (\pm 1.3) \times 10^{-5}$	$(K_{AMP^{2-}})$ $8.1\ (\pm 1.6) \times 10^{-5}$
$K_4 = (K_{MgATP^{2-}})$[a] $5.4\ (\pm 1.1) \times 10^{-5}$	$(K_{JAMP^{2-}})$ $4.8_7\ (\pm 0.0_4) \times 10^{-5}$	$(K_{MgATP^{2-}})$ $1.4\ (\pm 0.4) \times 10^{-4}$	$(K_{MgdATP^{2-}})$ $1.1\ (\pm 0.5) \times 10^{-4}$	$(K_{Mg\epsilon ATP^{2-}})$ $9.9\ (\pm 1.7) \times 10^{-5}$
$K_6 = (\bar{K}_{ADP^{3-}})$ $4.0\ (\pm 0.9) \times 10^{-5}$	$(\bar{K}_{dADP^{3-}})$ $1.3_3\ (\pm 0.9_3) \times 10^{-4}$			
$K_7 = (\bar{K}_{MgADP^{-}})$ $1.3_4\ (\pm 0.52) \times 10^{-4}$	$(\bar{K}_{MgdADP^{-}})$ $3.0_7\ (\pm 0.4_4) \times 10^{-5}$ $K_1 = 2.2_7\ (\pm 0.02) \times 10^{-5}$			
$K_8 = (K_{MgADP^{-}})$ $5.1\ (\pm 0.9) \times 10^{-5}$	$(K_{MgdADP^{-}})$ $0.9_9\ (\pm 0.5_9) \times 10^{-5}$			
$K_9 = (K_{ADP^{3-}})$ $1.6\ (\pm 0.7) \times 10^{-5}$	$(K_{dADP^{3-}})$ $3.4_8\ (\pm 0.9_0) \times 10^{-5}$			
$V_{max}^{forward}/E_t$ $5.5\ (\pm 0.5) \times 10^2$ μmol $min^{-1}mg^{-1}$ $1.1_8 \times 10^4$ mol·min^{-1} (mol protein)$^{-1}$ $1.9_6 \times 10^2\ sec^{-1}$	$0.17\ (\pm 0.01) \times 10^2$	$0.63\ (\pm 0.06) \times 10^2$	$4.3\ (\pm 0.1) \times 10^2$	$3.7\ (\pm 0.1) \times 10^2$
$V_{max}^{reverse}/E_t$ $1.3\ (\pm 0.1) \times 10^3$ μmol $min^{-1}mg^{-1}$ $2.7_8 \times 10^4$ mol·min^{-1} (mol protein)$^{-1}$ $4.6_4 \times 10^2\ sec^{-1}$	$0.18\ (\pm 0.04) \times 10^2$			

TABLE 1 (continued)
Comparison of Values for Derived Kinetic Parameters at pH 7.4, 25°C, and 0.16 (r/2) for the Forward and Reverse Reactions, Catalyzed by ATP-AMP Transphosphorylase from Rabbit Muscle

[a] $\bar{K}_5$ denotes an intrinsic dissociation constant of the particular substrate from the binary complex. K_s denotes a dissociation constant of the particular substrate from the ternary complex.

From Hamada, M. and Kuby, S. A., *Arch. Biochem. Biophys.*, 190, 772, 1978. With permission.

to an individual binding site for uncomplexed AMP^{2-} and illustrating the high degree of substrate specificity shown in myokinase for the uncomplexed nucleotide, i.e., for the 6-NH_2 group of the adenine moiety. Less stringent, however, are the requirements for the magnesium complexed nucleotide, where $Mg\epsilon ATP^{2-}$ is a fairly good substrate (last column of Table 1). In Table 3 of Reference 21 some of the phosphate donor substrate specificities are summarized of the adenylate kinases derived from human tissues. With AMP as the acceptor, ATP, GTP, UTP, ITP, and CTP to varying degrees acted as phosphate donors for the normal muscle and liver enzymes; in fact GTP and UTP were excellent phosphate donors for the normal liver enzyme in contrast to their relatively poor donor ability for the muscle enzyme. In Reference 18, where the liver mitochondrial enzyme is again compared with the normal muscle adenylate kinase, with AMP, catalytic activity was observed in the presence of adenosine-5′-(3-thio) triphosphate; whereas, the substrate pair ATP and adenosine-5′-thiophosphate proved to be catalytically active, $MgATP^{2-}\gamma S$ and $AMP^{2-}\alpha S$ did not. Thus, there appear to be relatively few phosphate acceptors besides AMP (and possibly dAMP, and 8,5′-cycloAMP[62]) which appear to be relatively effective substrates for the mammalian enzymes. However, in the case of the *E. coli* enzyme,[40] it appears to have a very much broader substrate specificity for the nucleoside monophosphates; since not only 2′dAMP, but also 3′dAMP, araAMP, TuMP (tubercidine monophosphate) were able to replace 5′AMP as phosphate acceptors. In fact, it appears as if the *E. coli* enzyme may be a nonspecific nucleoside monophosphokinase, rather than a specific adenylate kinase.

An enzymatic and ^{31}P nuclear magnetic resonance study of the adenylate kinase-catalyzed stereospecific phosphorylation of adenosine 5′-phosphorothioate (Ado-5′[thio-P]) was conducted by Sheu and Frey,[66] who found that the apparent Km for the rabbit muscle adenylate kinase catalyzed phosphorylation of Ado-5′[thio-P] is 31 m*M* compared to a an apparent Km for AMP of 0.11 m*M*, at pH 8.0 and 27°C. The apparent V_{max} found for AMP was 39 times that for Ado-5′[thio-P]. The adenylate kinase-catalyzed phosphorylation in the presence of pyruvate kinase and phosphoenolpyruvate produced adenosine 5′-O-(1-thiotriphosphate), whereas, in the presence of arginine phosphokinase and phosphoarginine, the product was largely adenosine 5′-O-(thiodiphosphate). The adenylate kinase-catalyzed phosphorylation of Ado-5′-[thio-P] appeared to be stereospecific, with the diastereoisomer A of adenosine 5′-O-(1-thiodiphosphate) as the product, and this reaction likely involved enzymatic binding and rotational immobilization of the thiophosphoryl group.

In the case of the calf muscle and liver isoenzymes,[45] and the human muscle[21] and liver mitochondrial isoenzymes,[18] similar kinetic reaction mechanisms appeared to be followed, in the steady state. Some distinguishing kinetic features of the calf liver adenylate kinase over the calf muscle isoenzymes were found to be

(i) V^r_{max} (liver) $\ll$ V^r_{max} (muscle); but V^f_{max} (liver) $>$ V^f_{max} (muscle)
(ii) K_{ADP}^{3-} (liver), $\bar{K}_{ADP}^{3-}$ (liver $<$ K_{ADP}^{3-} (muscle),$\bar{K}_{ADP}^{3-}$ (muscle)
(iii) inhibition of the liver enzyme by phosphoenolpyruvate; and
(iv) curiously, a relatively weak inhibition by p^1, p^5 - di-(adenosine-5′) pentaphosphate (AP_5A) of the liver enzyme compared to the muscle enzyme.[17]

The huge differences in V_{max} between the calf isoenzymes, with V^r_{max} being smaller for the liver isoenzyme, may be of physiological importance, if the mitochondrial adenylate kinase plays any regulatory role in oxidative phosphorylation.

The kinetically inferred conformational changes mentioned above[45] are also consistent with the two X-ray deduced crystal structures,[34-36] i.e., for crystal forms A and B of the porcine muscle adenylate kinase. That Ap_5A is such a poor inhibitor for the liver isoenzyme, relatively speaking, compared to the muscle enzyme, might imply that either the atomic distances, in which two binding sites much approach one another, differ in the liver isoenzymes, compared to the muscle isoenzyme; or that possibly the liver isoenzyme may exist

TABLE 2
Kinetically Derived Values for the Thermodynamic Equilibrium Constant at 25°C, and 0.16—0.18 (Γ/2)

Haldane expression	Calf muscle adenylate kinase	Calf liver adenylate kinase	Rabbit muscle adenylate kinase
$K_{eq} = \frac{V^f_{max}}{V^r_{max}} \frac{K_7K_9}{K_1K_3} =$	2.03×10^{-1}	1.98×10^{-1}	4.88×10^{-1}
$K_{eq} = \frac{V^f_{max}}{V^r_{max}} \frac{K_6K_8}{K_2K_4} =$	2.74×10^{-1}	2.07×10^{-1}	4.48×10^{-1}
$K_{eq} = \frac{V^f_{max}}{V^r_{max}} \frac{K_7K_9}{K_2K_4} =$	2.66×10^{-1}	1.96×10^{-1}	4.78×10^{-1}
$K_{eq} = \frac{V^f_{max}}{V^r_{max}} \frac{K_6K_8}{K_1K_3} =$	1.80×10^{-1}	2.09×10^{-1}	4.57×10^{-1}
Average K_{eq} =	2.31×10^{-1}	2.03×10^{-1}	4.68×10^{-1}

Overall average $K_{eq} = 3.01\ (\pm 1.46)^a \times 10^{-1}$; where $K_{eq} = \frac{(MgADP^-)\ (ADP^{3-})}{(MgATP^{2-})\ (AMP^{2-})}$. (5)

[a] Standard deviation $= \sigma = [\Sigma\ (X - \bar{X})^2/(n - 1)]^{1/2}$.

From Hamada, M. and Kuby, S. A., *Arch. Biochem. Biophys.*, 190, 772, 1978. With permission.

in a less flexible structure to allow for the necessary binding and required juxtaposition of the two binding sites. $V^r_{max} \ll V^f_{max}$ in the liver isoenzyme seems to indicate that the liver enzyme is already partially fixed in some preferred conformation so as to facilitate phosphoryl transfer to AMP^{2-}; this may be its important role in the mitochondrion.

These same features were displayed by the human liver, muscle, erythrocyte, and serum enzymes.[18,21,55] A thermodynamic value for $K_{eq} = \frac{(MgADP^-)(ADP^{3-})}{(MgATP^{2-})(AMP^{2-})} = 0.27(\pm 0.06)$ was also arrived at in Reference 45 for 25°C (see References 56 and 82) which compared favorably with the average value derived from its kinetic estimation via four Haldane relations for three adenylate kinase sets of data (i.e., from rabbit muscle, calf muscle, and liver adenylate kinase see Table 2). Kinetically derived values for K_{eq} obtained from the human isoenzyme-catalyzed reactions[18,21] were also in agreement with the thermodynamic value, attesting to the self-consistent nature of the kinetic analyses and lending further credence to the proposed reaction of Equation 1.

Vasavada et al.[65] analyzed the ^{31}P NMR spectra of enzyme-bound reactants and products of adenylate kinase, with the use of a density matrix theory of chemical exchange and computer simulations. For the rates of exchange with the experimental spectra for porcine adenylate kinase at pH 7.0 and 4°C, the following characteristic rates were determined: interconversion rates, 375 ($\pm$ 30)s^{-1} for ATP formation and 600 ($\pm$ 59)s^{-1} for ADP formation; interchange rates of donor and acceptor ADP's, 100 ($\pm$ 30)s^{-1} in the presence of optimal Mg^{2+} concentration, 1500 ($\pm$ 100)s^{-1} in the absence of Mg^{2+}. These data implied that the interchange rate represented the lower limit for the dissociation rate of ADP (or of MgADP from the acceptor site, when Mg^{2+} was present) from the enzyme complexes. The significance of these interchange rates and their values relative to the interconversion rates were clarified with special reference to the role of Mg^{2+} ion in differentiating between the two nucleotide binding sites on the enzyme molecule and provide additional evidence for two separate nucleotide binding sites.

Recently, Ray et al.[110] presented ^{31}P spin-relaxation measurements on enzyme-bound substrate complexes of porcine muscle adenylate kinase with Mn(II) and Co(II), in an effort to estimate cation-^{31}P distances in different enzyme complexes with ATP, GTP, GDP, and

AMP. They emphasized the necessity for a correct assessment of the contribution of the exchange rate to the observed relaxation rate, and concluded that a value of 5.0 Å appears appropriate for the Mn(II)-P(AMP) distance in the complexes E·MnGDP·AMP and E·CoGDP·AMP. The authors[110] noted that their results may only appear to be in agreement with the result of 5.9 ± 0.3 Å obtained by Fry et al.[86] for the inhibitor complex E·CrAMPPCP·AMP of rabbit muscle adenylate kinase (see below).

B. EQUILIBRIUM SUBSTRATE BINDING PROPERTIES OF THE ADENYLATE KINASE AND DERIVED PEPTIDE FRAGMENTS

The X-ray structure for the porcine muscle enzyme was elucidated by Schultz et al.[34] at a 3-Å resolution; the same enzyme was sequenced by Heil et al.[33] Unfortunately, attempts by Schulz et al. to bind substrate or substrate analogs to their crystals [later defined as crystal form A[35]] apparently failed, and therefore, X-ray crystallographic evidence for the substrate binding sites were not convincing. Largely, by comparison with the results of $A_{P5}A$ bound to crystals A, Pai et al.[36] indirectly assigned the sites for binding of AMP and ATP to crystal form B, by X-ray diffraction analysis. However, no bound AMP could be detected at their assigned "AMP-site," although $MnATP^{2-}$ could be made to bind to it; no ATP could be made to bind to their assigned "ATP-binding site" (also, see Reference 116).

In fact, Rao et al.,[82] largely by ^{31}P-NMR studies seemed to exclude $MgATP^{2-}$ from the putative "AMP-binding site" of the porcine enzyme, casting doubt therefore on the identity of the AMP-binding site occupied by $MnATP^{2-}$ which had been assigned by X-ray diffraction analyses.[36] It is of interest in this regard to note that $A_{P5}A$ was found to act as a competitive inhibitor with respect to either substrate of the forward reaction (i.e., $MgATP^{2-}$ + AMP^{2-}) catalyzed by both calf and rabbit muscle myokinase, but acted as a noncompetitive inhibitor of either substrate of the reverse reaction (i.e., $MgADP^{3-}$ + ADP^{3-}),[17] a finding which might tend to make any unambiguous assignment of $A_{P5}A$ to the myokinase substrate binding sites difficult.

An early attempt was made[83] to shed light on the interactions between the rabbit muscle adenylate kinase and the many and diverse metal complex and ionic species of the adenine nucleotides that may exist in solution.[88]

Measurements were made by a sedimentation gradient procedure (at pH 7.9, 3°C, and at an (Γ/2) of about 0.2) of the equilibrium binding to rabbit muscle adenylate kinase of ADP^{3-}, $MgATP^{2-}$, ATP^{4-}, AMP^{2-}, and Mg^{2+}; where $MgADP^{-}$ could not be directly measured for obvious reasons. The extrapolated value of $n_{ADP}{}^{3}$ (maximal number of moles of ADP^{3}-bound per mole enzyme) was only about 1.6 (at a 90% confidence level) and, accordingly, it was felt[83] that it would "also be possible to fit the data to the equation for two distinct sets of binding sites (with n = 1 for each set);" but because of the limited number of experimental points in the critical range at relatively high values for $\bar{r}_A{}'$ (average number of moles of A bound per mole of total protein) approaching the x-intercept, a more complicated nonlinear plot to distinguish the two binding constants did not seem justified at the time. Also, under the conditions of measurement[83] an extrapolated value of $n \approx 1.8$, as an average, was estimated for both ATP^{4-} and $MgATP^{2-}$.

By nuclear magnetic relaxation and electron paramagnetic resonance techniques, Price et al.[89] had deduced that there is only one binding site for MgATP or ATP per mole of porcine muscle adenylate kinase. On the other hand, by ^{31}P-NMR Rao et al.[82] deduced that both ATP and ADP can bind at both nucleotide binding sites of porcine myokinase, and one site (the so-called AMP site) binds only the uncomplexed nucleotides and the other site binds to ADP and ATP with or without the presence of magnesium. A reinvestigation of the substrate binding properties of adenylate kinase seemed worthwhile at this point, with the recent availability of suitable analogs of the adenine nucleotides[84] and the detailed kinetics study[45] on their substrate or inhibitor properties of these analogs (see Section II.A).

Further, during the course of determination of the amino acid sequences of rabbit and

TABLE 3
Equilibrium Binding of 1,N^6-Etheno Analogs of Adenine Nucleotides to Rabbit Muscle Myokinase

	εAMP	εADP		εATP
		$n_1 + n_2 = 2.00\ (\pm 0.05)$		
Without Mg^{2+}	$n = 1.12\ (\pm 0.08)$[a]	$n_1 = 0.61\ (\pm 0.13)$	$n_2 = 1.3\ (\pm 0.30)$	$n = 0.81\ (\pm 0.09)$
	$K'_a = 7.5_2\ (\pm 0.77) \times 10^3\ M^{-1}$	$K'_{a1} = 1.7_1\ (\pm 0.36) \times 10^4\ M^{-1}$	$K'_{a2} = 1.3\ (\pm 0.4) \times 10^3\ M^{-1}$	$K'_a = 3.3_6\ (\pm 0.42) \times 10^4\ M^{-1}$
	$K'_d = 1.3_3\ (\pm 0.14) \times 10^{-4}\ M$	$K'_{d1} = 5.8_7\ (\pm 1.24) \times 10^{-5}\ M$	$K'_{d2} = 7.8\ (\pm 2.4) \times 10^{-4}\ M^{-1}$	$K'_d = 2.9_8\ (\pm 0.37) \times 10^{-5}\ M$
With Mg^{2+}	$K'_a \leq 6 \times 10^2\ M^{-1}$	$n = 0.93\ (\pm 0.40)$		$n = 0.97\ (\pm 0.04)$
		$K'_a = 1.87\ (\pm 0.47) \times 10^4\ M^{-1}$		$K'_a = 3.8_0\ (\pm 0.20) \times 10^4\ M^{-1}$
		$K'_d = 5.3_3\ (\pm 1.33) \times 10^{-5}\ M$		$K'_d = 2.6_3\ (\pm 0.14) \times 10^{-5}\ M$

[a] n = maximal number of moles bound per mole protein; K'_a = intrinsic association constant; K'_d = intrinsic dissociation constant.

From Hamada, M., Palmieri, R. H., Russell, G. A., and Kuby, S. A., *Arch. Biochem. Biophys.*, 195, 155, 1979. With permission.

calf muscle myokinases[51] (Scheme II), certain relatively large peptide fragments became available for ligand binding studies; coupled with the synthesis of several ATP-binding peptides (see below, Table 6[90]) which contained His-36 (see Scheme II, below), which had been implicated McDonald et al.[85] in the case of the porcine enzyme as interacting with ATP, a clearer picture of the two distinct substrate binding sites but one catalytic site for the enzyme has now emerged.[46,47,90]

Both a fluorescence-quenching technique and a UV-difference spectral method were used to study the binding of 1,N^6-etheno analogs of the adenine nucleotides (εATP, εADP, εAMP)[84] to crystalline rabbit and calf muscle ATP-AMP transphosphorylase in the presence and absence of Mg^{2+}, at 0.16 (Γ/2), 25°C, and pH 7.4.[46] In addition, the binding of the ε-analogs of the adenine nucleotides was studied to two S-[^{14}C] carboxymethylated peptide fragments of the rabbit muscle enzyme ($Res_{1\text{-}44}$ = MT-I; $Res_{172\text{-}194}$ = MT-XII), as well as to several synthetic peptides (Table 6, see also Reference 90) including one, $I_{1\text{-}45}$, which corresponds to residues 1-45 of the rabbit muscle enzyme, and was used extensively in the NMR studies (e.g., Reference 67) described below. In the case of the rabbit and calf enzymes $Mg\epsilon ATP^{2-}$, ϵATP^{4-}, $Mg\epsilon ADP^{-}$, ϵAMP^{2-} are bound stoichiometrically ($n \approx 1$), MgAMP is insignificantly bound, and $n \approx 2$ for ADP^{3-} (n = maximal number of moles bound per mole of protein) (e.g., Table 3). In the case of the S-carboxymethylated peptide fragment: MT-I binds stoichiometrically to $Mg\epsilon ATP^{2-}$, ϵATP^{4-}, $Mg\epsilon ADP^{-}$, and ϵADP^{3-} with values of $n \approx 1$; but MT-I does not bind to ϵAMP^{2-} significantly (Table 4). MT-XII binds stoichiometrically to uncomplexed AMP^{2-} or to uncomplexed ADP^{3-} (both with $n \approx 1$); the binding of $Mg\epsilon ADP^{-}$, $Mg\epsilon ATP^{2-}$, and MgεAMP to MT-XII are comparatively insignificant.

Other peptide fragments in the molecule, i.e., fragment MT-IV (Res 77-96) or MT-VI (Res 106-126) did not bind significantly to any of the etheno analogs, nor did insulin, nor, e.g., did bovine serum albumin (Table 4). The binding of the etheno analogs was also studied to an equimolar mixture of peptides MT-I + MT-XII, which qualitatively duplicated the binding pattern of the entire native molecule, and except for ϵATP^{4-} or $Mg\epsilon ATP^{2-}$ (which are bound more tightly to the entire native molecule), even quantitatively (Table 5); the charges depicted for the etheno analogs are those residing in the phosphoryl groups only.

TABLE 4
Equilibrium Binding of 1,N^6-Etheno Analogs of Adenine Nucleotides to *S*-[^{14}C] Carboxymethylated Peptides Derived from Rabbit Muscle Myokinase (i.e., *S*-Carboxymethyl-Cys^{25} and *S*-Carboxymethyl-Cys^{187})

		εAMP	εADP	εATP
MT-1 (residues 1 to 44)	Without Mg^{2+}	$K'_a \leq 2 \times 10^2$ M^{-1}	n = 0.90 (±0.12)[a] $K'_a = 3.3_4$ (±0.33) × 10^4 M^{-1} $K'_d = 2.9_9$ (±0.30) × 10^{-5} M	n = 0.98 (±0.05) $K'_a = 3.6_9$ (±0.30) × 10^3 M^{-1} $K'_d = 2.7_1$ (±0.22) × 10^{-4} M
	With Mg^{2+}	—	n = 1.10 (±0.10) $K'_a = 3.6_2$ (±0.37) × 10^4 M^{-1} $K'_d = 2.7_6$ (±0.28) × 10^{-5} M	n = 0.94 (±0.07) $K'_a = 5.1_2$ (±0.48) × 10^3 M^{-1} $K'_d = 1.9_5$ (±0.18) × 10^4 M
MT-XII (residues 172 to 194)	Without Mg^{2+}	n = 1.00 (±0.09) $K'_a = 4.3_4$ (±0.73) × 10^3 M^{-1} $K'_d = 2.3_0$ (±0.39) × 10^{-4} M	n = 1.02 (±0.12) $K'_a = 2.1_7$ (±0.35) × 10^3 M^{-1} $K'_d = 4.6_1$ (±0.74) × 10^{-4} M	$K'_a \leq 2 \times 10^2$ M^{-1}
	With Mg^{2+}	$K'_a \leq 3 \times 10^2$ M^{-1}	$K'_a \leq 2 \times 10^2$ M^{-1}	—
MT-IV + MT-VI (residues 77 to 96 + 106 to 126)	Without Mg^{2+}	$K'_a \leq 3 \times 10^2$ M^{-1}	$K'_a \leq 3 \times 10^2$ M^{-1}	$K'_a \leq 2 \times 10^2$ M^{-1}
Insulin B-chain	With or without Mg^{2+}	$K'_a \leq 2 \times 10^2$ M^{-1}	$K'_a \leq 2 \times 10^2$ M^{-1}	$K'_a \leq 2 \times 10^2$ M^{-1}
Bovine serum albumin	With or without Mg^{2+}	$K'_a \leq 3 \times 10^2$ M^{-1}	$K'_a \leq 3 \times 10^2$ M^{-1}	$K'_a \leq 3 \times 10^2$ M^{-1}

[a] n = maximal number of moles bound per mole of protein or peptide.

From Hamada, M., Palmieri, R. H., Russell, G. A., and Kuby, S. A., *Arch. Biochem. Biophys.*, 195, 155, 1979. With permission.

These binding data support the idea that there are two separate sites for the binding of either (1) the *complexed nucleotide* substrate ($MgATP^{4-}$ or $MgADP^-$) residing in part in the sequence of MT-I (Res. 1-44) or (2) the *uncomplexed nucleotide* substrate (AMP^{2-} or ADP^{3-}) incorporating the sequence of MT-XII (Res. 172-194) or of CB-VI (Res. 126-194; Table 6) of the rabbit muscle enzyme.

Figure 1 presents in schematic fashion one set of postulates as to the binding interactions between MT-I and MgϵATP^{2-} (left-hand side) and between MT-XII uncomplexed ϵADP^{3-} (right-hand side). This proposal was made early, before the topography was mapped out in the remarkable series of NMR experiments of Fry et al.[44,67,86] — vide infra. However, many of the binding interactions appear to have been verified. The proposed binding interactions are depicted by dotted lines. In the case of MT-I and MgϵATP^{2-} the position shown for Mg^{2+} is arbitrarily located between the terminal two phosphate groups, but it could as well lie between the α- and β-phosphates; also, the role of water molecules in the coordination sphere of Mg^{2+} is not depicted. The specific ligand binding at MT-I or MT-XII is envisioned as largely through specific sites for H-bonding.

The binding properties of these unusual peptide fragments (MT-I and MT-XII or CB-VI) derived from the head and from the tail of the molecule, respectively, now have confirmed the hypothesis of Rhoads and Lowenstein[56] deduced kinetically, that two separate and distinct binding sites exist (1) for the magnesium complexes of the nucleotide substrates, $MgAMP^{2-}$ and $MgADP^-$, and (2) for the uncomplexed nucleotide substrates, AMP^{2-} and ADP,[45,46] but with one overall catalytic site.

TABLE 5
A Comparison of the Equilibrium Binding of 1,N^6-Etheno Analogs of Adenine Nucleotides to an Equimolar Mixture of *S*-[^{14}C]Carboxymethylated Peptides Derived from Rabbit Muscle Myokinase (i.e., *S*-Carboxymethyl-Cys^{25} and *S*-Carboxy-thyl-Cys^{187}) with Native Rabbit Muscle Myokinase

		εAMP	εADP	εATP
An equimolar muxture of MT-I (residues 1-44) and MT-XII (residues 172-194)	Without Mg^{2+}	$n = 0.99\ (\pm 0.04)$ $K'_a = 6.8_0\ (\pm 0.60) \times 10^3\ M^{-1}$ $K'_d = 1.4_7\ (\pm 0.13) \times 10^{-4}\ M$	$n_1 + n_2 = 1.9_8\ (\pm 0.23)$ $n_1 = 0.60\ (\pm 0.19)$; $n^2 = 1.40\ (\pm 0.25)$ $K'_{a_1} = 2.6_0\ (\pm 1.09) \times 10^4\ M^{-1}$; $K'_{a_2} = 2.0\ (\pm 0.4) \times 10^3\ M^{-1}$ $K'_{d_1} = 3.8_5\ (\pm 1.61) \times 10^{-5}\ M$; $K'_{d_2} = 5.0\ (\pm 1) \times 10^{-4}\ M$	$n = 0.99\ (\pm 0.08)$ $K'_a = 3.1_8\ (\pm 0.53) \times 10^3\ M^{-1}$ $K'_d = 3.1_4\ (\pm 0.52) \times 10^{-4}\ M$
	With Mg^{2+}	$K'_a \leqslant 4 \times 10^2\ M^{-1}$	$n = 1.0_0\ (\pm 0.10)$ $K'_a = 1.1_1\ (\pm 0.12) \times 10^4\ M^{-1}$ $K'_d = 9.0_1\ (\pm 0.97) \times 10^{-5}\ M$	$n = 1.0_0\ (\pm 0.08)$ $K'_a = 2.3_1\ (\pm 0.30) \times 10^3\ M^{-1}$ $K'_d = 4.3_3\ (\pm 0.56) \times 10^{-4}\ M$
Native rabbit muscle myokinase	Without Mg^{2+}	$n = 1.1_2\ (\pm 0.08)$ $K'_a = 7.5_2\ (\pm 0.77) \times 10^3\ M^{-1}$ $K'_d = 1.3_3\ (\pm 0.14) \times 10^{-4}\ M$	$n_1 + n_2 = 2.0_0\ (\pm 0.05)$ $n_1 = 0.61\ (\pm 0.13)$; $n_2 = 1.3_9\ (\pm 0.30)$ $K'_{a_1} = 1.7_1\ (\pm 0.36) \times 10^4\ M^{-1}$; $K'_{a_2} = 1.3\ (\pm 0.4) \times 10^3\ M^{-1}$ $K'_d = 5.8_7\ (\pm 1.24) \times 10^{-5}\ M$; $K'_d = 7.8\ (\pm 2.4) \times 10^{-4}\ M$	$n = 0.81\ (\pm 0.09)$ $K'_a = 3.3_6\ (\pm 0.42) \times 10^4\ M^{-1}$ $K'_d = 2.9_8\ (\pm 0.37) \times 10^{-5}\ M$
	With Mg^{2+}	$K'_a \leqslant 6 \times 10^2\ M^{-1}$	$n = 0.93\ (\pm 0.40)$ $K'_a = 1.8_7\ (\pm 0.47) \times 10^4\ M^{-1}$ $K'_d = 5.3_3\ (\pm 1.33) \times 10^{-5}\ M$	$n = 0.97\ (\pm 0.04)$ $K'_a = 3.8_0\ (\pm 0.20) \times 10^4\ M^{-1}$ $K'_d = 2.6_3\ (\pm 0.14) \times 10^{-5}\ M$

From Hamada, M., Palmieri, R. H., Russell, G. A., and Kuby, S. A., *Arch. Biochem. Biophys.*, 195, 155, 1979. With permission.

TABLE 6
Binding of Etheno Analogs of Adenine Nucleotides by Synthetic Peptides

Peptide	MgεATP			εATP			MgεADP			εADP		
	n	Ka′	Kd′	n	Ka′	Kd′	n	Ka′	Kd′	n	Ka′	Kd′
$I_{31\text{-}45}$	1.0	2.2×10^4	4.5×10^{-5}	0.98	2.8×10^3	3.6×10^{-4}	0.99	2.9×10^4	3.4×10^{-5}	1.0	2.6×10^3	3.8×10^{-4}
$I_{20\text{-}45}$	1.0	2.3×10^4	4.3×10^{-5}	1.0	2.3×10^4	4.3×10^{-5}	0.99	1.5×10^4	6.7×10^{-5}	n.m.	n.m.	n.m.
$I_{1\text{-}45}$	1.0	0.75×10^4	1.3×10^{-4}	1.0	0.64×10^4	1.6×10^{-4}	0.99	1.1×10^4	9.1×10^{-5}	1.0	1.3×10^4	7.4×10^{-5}
$MT\text{-}I_{(1\text{-}44)}$	0.94	0.51×10^4	2.0×10^{-4}	0.98	0.37×10^4	2.7×10^{-4}	1.1	3.6×10^4	2.8×10^{-5}	0.90	3.3×10^4	3.0×10^{-5}
Native rabbit muscle myokinase	0.97	3.8×10^4	2.6×10^{-5}	0.81	3.3×10^4	3.0×10^{-5}	0.93	1.9×10^4	5.3×10^{-5}	(see Table 3)		

Peptide	εAMP			εADP		
	n	Ka′	Kd′	n	Ka′	Kd′
$F'_{178\text{-}192}$	—	$\leqq 10^2$	$\geqq 10^{-2}$	—	$\leqq 10^{-2}$	$\geqq 10^{-2}$
$H'_{178\text{-}194}$	0.95	5.4×10^3	1.9×10^{-4}	n.m.[a]	n.m.	n.m.
$H'_{172\text{-}194}$	1.0	3.0×10^3	3.3×10^{-4}	n.m.	n.m.	n.m.
$MT\text{-}XII_{(172\text{-}194)}$	1.0	4.3×10^3	2.3×10^{-4}	1.0	2.2×10^3	4.5×10^{-4}
$CV\ VI_{(126\text{-}194)}$	1.0	7.1×10^3	1.4×10^{-4}	—	—	—
Native rabbit muscle myokinase	1.1	7.5×10^3	1.3×10^{-4}	see Table 3		

[a] n.m. — not measured.

From Kuby, S. A. et. al., *J. Protein Chem.*, 8, 549, 1989. With permission.

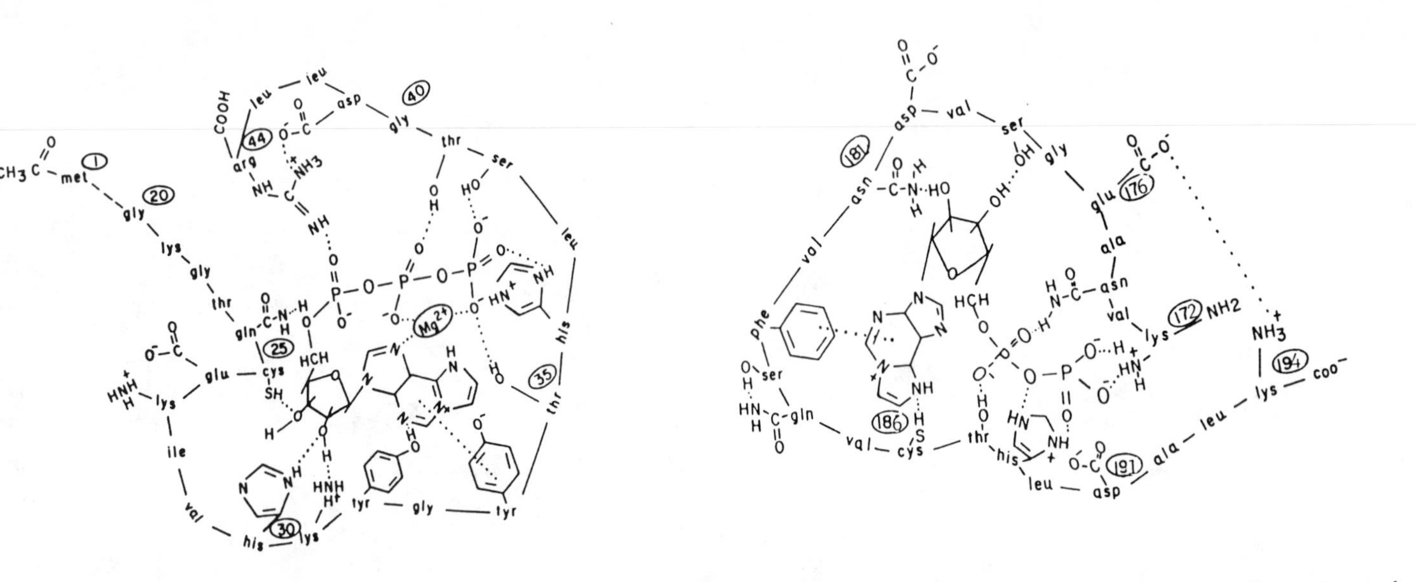

FIGURE 1. Left-hand side: binding of Peptide MT-I (residues 1-44 of rabbit muscle myokinase) to MgϵATP^{2-} as visualized by a schematic diagram (Proposed binding interactions are depicted by dotted lines. Role of water molecules in coordination sphere of Mg^{2+} not depicted). Right-hand side: binding of Peptide MT-XII (Residues 172-194 of rabbit muscle myokinase) to ϵADP^{3-} as visualized by a schematic diagram (proposed binding interactions are depicted by dotted lines). (From Hamada, M., Palmieri, R. H., Russell, G. A., and Kuby, S. A., *Arch. Biochem. Biophys.*, 195, 155, 1979. With permission.)

Therefore, the catalyzed reaction for adenylate kinase may actually be written as given in Equation 1; the idea that catalysis by the adenylate kinases requires an interaction between the two binding sites (see References 17, 45, 46, and 90) is consistent with the X-ray deduced structure[34-36] as shown dramatically in Steitz's computer-generated drawing[92] of the space-filling model of crystal form "B" of the porcine adenylate kinase. It is of course necessary to postulate a conformational change that may bring both binding sites into juxtaposition (Hamada et al.[46]) in the case of the muscle-type adenylate kinases. Such a conformational change has been implicated by NMR studies as a result of the binding of $A_{P5}A$ to human muscle adenylate kinase.[37]

Following the conformational change, phosphoryl group transfer should then be facilitated to allow catalysis of the overall reaction, i.e., Equation 1.[45] This will be described in much finer detail in Sections III.C and D as a result of elegant studies of Fry et al.[44,67,86]

Recently, a series of peptides have been synthesized[90] around the ATP-binding site; these have included peptides with an increasing length of the fragment up to 45 residues. Preliminary estimates indicate that peptide $I_{1\text{-}45}$ compares favorably with the fragment MT-I (1-44), and possesses a free -SH group at Cys-25 in contrast to the carboxymethyl group blocking the thiol in MT-I (see Table 6). It is noteworthy that $I_{1\text{-}45}$ possesses an ordered globular tertiary structure, as first noted by Fry et al.[67] in its proton spectrum, since denaturation of the peptide with acetone-d_6 narrowed the individual resonance widths. This was confirmed by FTIR and CD measurements and 2-D NMR[91] and will be discussed in Section III.

In addition, several peptides have been synthesized around the AMP-binding site. The peptide corresponding to residues 178-192 ($F'_{178\text{-}192}$), which lacks Lys 172 or Lys 194, was found to lack the ability to bind either ϵ-AMP^{2-} or ϵ-ADP^{3-}. However, the synthetic peptide corresponding to residues 172-194 ($H'_{172\text{-}194}$) did possess the binding properties of MT-XII (Table 3). Thus, either lysine 172 or 194 are apparently required for binding of AMP^{2-}. Moreover, fragment CB-VI (residues 126-194) binds ϵAMP (or AMP) with an affinity comparable to the intact native enzyme (Table 6 and Reference 90). It is evident that a chemical method is at hand for outlining the requirements for binding of either site, and that a powerful tool is now available for the systematic mapping of those residues or sequences of residues involved in the binding of the substrates by rabbit muscle myokinase.

From the largest normalized value of $1/T_{2P}$ (transverse relaxation rate) of the synthetic peptide $I_{1\text{-}45}$[90] (Table 6), i.e., that of the aromatic peak #1, Fry et al.[67] estimated the rate constant for Cr^{3+}ATP dissociating from the peptide, i.e., $k_{off} \geq 544 \pm 300\ s^{-1}$, at pH = 5.6. This lower limit value of k_{off} is comparable to the k_{cat} (or $V^{reverse}_{max_{Et}}$, see Table 1) of the enzyme in the direction of MgATP formation, i.e., 464 s^{-1} (Reference 45), at pH 7.4. From the limiting value of k_{off} of Cr^{3+}ATP and its measured dissociation constant (K_D = 35 ± 20 μM) from titration curves of $1/T_{1P}$ (the longitudinal relaxation rate) of selected resonances of the peptide vs. added Cr^{3+}ATP a lower limiting value of k_{on}, the rate constant for Cr^{3+}ATP binding to the peptide could be estimated to be $k_{on} \geq 1.6 \times 10^7\ M^{-1}\ s^{-1}$, which is well within the diffusion limit.[67]

III. RELATIONSHIP OF THE STRUCTURE OF ADENYLATE KINASE TO THE TOPOGRAPHY OF ITS MgATP AND AMP BINDING SITES

A. PRIMARY STRUCTURE OF THE ADENYLATE KINASES AND SOLUTION STRUCTURE OF THE MgATP-BINDING SYNTHETIC PEPTIDE FRAGMENT

The total amino acid sequences of rabbit and calf muscle adenylate kinases were presented,[51] each with a single polypeptide chain of 194 amino acid residues ($M_r \approx$21,400),

1 2 3 4 5 6 7 8 9 10 11 12 13 14 15 16 17 18 19 20 21 22 23 24
RABBIT: AC-MET-GLU-GLU-LYS-LEU-LYS-LYS-ALA-LYS-ILE-ILE-PHE-VAL-VAL-GLY-GLY-PRO-GLY-SER-GLY-LYS-GLY-THR-GLN-
MT-I
CALF: -ALA-
HUMAN: -THR-
PORCINE: -SER-

25 26 27 28 29 30 31 32 33 34 35 36 37 38 39 40 41 42 43 44 45 46 47 48 49
RABBIT: CYS-GLU-LYS-ILE-VAL-HIS-LYS-TYR-GLY-TYR-THR-HIS-LEU-SER-THR-GLY-ASP-LEU-LEU-ARG-ALA-GLU-VAL-SER-SER-
CALF: -GLN- -ALA-
HUMAN: -GLN- -SER-
PORCINE: -GLN- -ALA-

50 51 52 53 54 55 56 57 58 59 60 61 62 63 64 65 66 67 68 69 70 71 72 73 74
RABBIT: GLY-SER-ALA-ARG-GLY-LYS-LYS-LEU-SER-GLU-ILE-MET-GLU-LYS-GLY-GLN-LEU-VAL-PRO-LEU-GLU-THR-VAL-LEU-ASP-
CALF: -MET-
HUMAN: -LYS-
PORCINE: -MET-

75 76 77 78 79 80 81 82 83 84 85 86 87 88 89 90 91 92 93 94 95 96 97 98 99
RABBIT: MET-LEU-ARG-ASP-ALA-MET-VAL-ALA-LYS-ALA-ASP-THR-SER-LYS-GLY-PHE-LEU-ILE-ASP-GLY-TYR-PRO-ARG-GLN-VAL-
CALF: -VAL-ASN- -GLN-
HUMAN: -VAL-ASN- -GLU-
PORCINE: -VAL-ASP- -GLU-

100 101 102 103 104 105 106 107 108 109 110 111 112 113 114 115 116 117 118 119 120 121 122 123 124
RABBIT: GLN-GLN-GLY-GLU-GLU-PHE-GLU-ARG-ARG-ILE-ALA-GLN-PRO-THR-LEU-LEU-LEU-TYR-VAL-ASP-ALA-GLY-PRO-GLU-THR-
CALF: -ARG-ILE-ALA-
HUMAN: -ARG-ILE-GLY-
PORCINE: -LYS-ILE-GLY-

125 126 127 128 129 130 131 132 133 134 135 136 137 138 139 140 141 142 143 144 145 146 147 148 149
RABBIT: MET-GLN-LYS-ARG-LEU-LEU-LYS-ARG-GLY-GLU-THR-SER-GLY-ARG-VAL-ASP-ASP-ASN-GLU-GLU-THR-ILE-LYS-LYS-ARG-
CALF: GLN-LYS
HUMAN: THR-ARG
PORCINE: THR-LYS

150 151 152 153 154 155 156 157 158 159 160 161 162 163 164 165 166 167 168 169 170 171 172 173 174 175
RABBIT: LEU-GLU-THR-TYR-TYR-LYS-ALA-THR-GLU-PRO-VAL-ILE-ALA-PHE-TYR-GLU-LYS-ARG-GLY-ILE-VAL-ARG-LYS-VAL-ASN-ALA-
CALF:
HUMAN:
PORCINE:

176 177 178 179 180 181 182 183 184 185 186 187 188 189 190 191 192 193 194
RABBIT: GLU-GLY-SER-VAL-ASP-ASN-VAL-PHE-SER-GLN-VAL-CYS-THR-HIS-LEU-ASP-ALA-LEU-LYS.
MT-XII
CALF: -ASN-
HUMAN: -GLU-
PORCINE: -ASP- -THR-

SCHEME 2. Amino acid sequence of rabbit muscle ATP-AMP transphosphorylase (myokinase) compared with calf muscle myokinase, porcine muscle adenylate kinase (Reference 33), and human muscle adenylate kinase (Reference 28).

together with a comparison of the four muscle-type adenylate kinases whose covalent structure were determined at that time, i.e., rabbit, calf, porcine[33] and human;[28] and the comparison demonstrated an extraordinary degree of homology (Scheme 2). Almost 93% of the structure of muscle-type adenylate kinases designated AK_1[24,28,33] has been conserved in these four mammalian species.

The sequence analysis of the 238 amino acid residue polypeptide chain of adenylate kinase isoenzyme (designated AK_2[50]) from bovine heart mitochondria was also completed.[49] Two of the four cysteine residues of AK_2, Cys-41 and Cys-233, were assigned as free thiols, which could be carboxymethylated in the intact protein without loss of enzymatic activity, whereas chemical and model-building studies suggested that Cys-43/Cys-93 forms a disulfide in native AK_2 (a similar case of two thiol groups and one disulfide group has also been reported for the rat liver enzyme).[61] The relative molecular mass of AK_2, as deduced from the sequence, is 26,104. However, according to other methods, including titration of thiol groups, sedimentation equilibrium ultracentrifugation and gel filtration, Mr values in the range of 26,000 to 31,500 were obtained, with each value dependent on the method of determination. This Mr value of 26K is in good agreement with that obtained for the calf liver enzyme,[17] see below.

Bovine heart AK_2 contains 44 residues more than the homologous isozyme AK_1 (cytosolic, see Scheme II); since all (except one) single insertions and deletions cancel, the higher Mr of AK_2 is due to 9 residues preceding the N-terminus of AK_1, a stretch of 30 residues in the middle and 6 residues at the end.[49] AK_2 and AK_1 appear similar in their active-site geometry according to Reference 49; by contrast, however, AK_2 does not possess any of the three antigenic sites of AK_1, which is consistent with the lack of immunological cross-reactivity between AK_1 and AK_2 (see also Reference 17). From Scheme II, the 194 residue (M_r ≈21,400) adenylate kinases from rabbit and calf muscle,[51] and from porcine[33] and human muscle,[28] all contain only two cysteines(Cys-25,Cys-187), which may be readily titrated under acidic conditions with a 4,4′dithiodipyridine, 6 to 7 Tyr residues, but no Trp. On the other hand, although the calf liver adenylate kinase has a measured M_r of 25,600 ± 200 (Reference 17) and is identical with the estimated measured M_r for AK_{2b},[108] there appear to be a few significant differences in its reported amino acid composition from that of AK_2 from heart mitochondria,[49] i.e., in the two Cys, five Tyr, and 1 Trp residue per mole.[20] It is not known how closely these two bovine adenylate kinase species are related, but Kishi et al.[108] recently reported the isolation of two types of cDNA for bovine mitochondrial adenylate kinase, which code for the 241-residue AK_{2A} (see Reference 49), and 234-residue AK_{2B} with cys-234 replaced by ser. One may note that of the other two liver-type adenylate kinases whose amino acid compositions were reported,[20] the human and rabbit liver enzymes, these two enzymes also contained 1 Trp and 5 Tyr, but 4 half-cysteines which are not readily titrated by 5,5′-dithiobis (nitrobenzoic acid) under neutral conditions, but may be titrated with 4,4′-dithiodipyridine under acidic conditions.[20]

GTP:AMP phosphotransferase from mammalian mitochondria has been designated as a third isoenzyme, AK_3,[54] and the beef heart mitochondrial enzyme has been sequenced, except for a segment of 33 residues in the middle of the polypeptide chain.[54] It shows, however, only about 33% homology with AK_1, with all of the residues aligned.

Mention should be made of sequences of peptide fragments MT-I and MT-XII, derived from the head and from the tail of the rabbit muscle adenylate kinase molecule;[51] in Scheme II, they are underlined. Both fragments, interestingly, contain an identical tripeptide sequence, i.e., Thr^{35}-His^{36}-Leu^{37} and Thr^{188}-His^{189}-$Leu.^{190}$ His-36 and His-189 have been the subject of extensive NMR investigations in the porcine muscle adenylate kinase[38,48] and in human muscle adenylate kinase[37] and His-36 was implicated early as interacting with ATP.[48] The substrate-binding properties of these unusual peptide fragments, MT-I and MT-XII, were described above,[46] as well as that of CB-VI (Table 6). The longer peptide CB-VI (res. 126-194) may well satisfy the rigid substrate conformational requirements deduced by Fry et al.[86] for the "AMP site," since it binds either AMP or εAMP[90] with a K_d value (see Table 6) in good agreement with the K_i for εAMP estimated kinetically for the intact enzyme.[45]

A cDNA clone for chicken muscle adenylate kinase was recently isolated from a cDNA library of chicken skeletal muscle poly $(A)^+$RNA, and the DNA sequence was determined.[43] The cDNA insert had 854 nucleotides, which consisted of the 5′-untranslated sequence of 57 nucleotides, the sequence of 582 nucleotides coding for 194 amino acid, the 3′-untranslated sequence of 163 nucleotides, and the poly$(A)^+$ tail of 52 nucleotides. The amino acid sequence predicted from the nucleotide sequence was highly homologous with the reported sequences of human, calf, porcine, and rabbit muscle adenylate kinases.[51] RNA blot analysis of poly$(A)^+$ RNA from various chicken tissues revealed a single species of mRNA of approximately 850 nucleotides and its tissue-specific distribution. The induction of muscle adenylate kinase mRNA synthesis during chick embryogenesis was studied by the blot analysis, and southern blot analysis indicated a single gene for the muscle enzyme in the chicken genome.[43]

The isolation and characterization of thermosensitive mutants of *E. coli* exhibiting a complex phenotype (defective synthesis of phospholipids, RNA, and ATP at nonpermissive temperatures) allowed localization of the mutation to the *adk* gene.[76,77] Eventually, this brought the study to the molecular cloning of the *adk* gene and to the deduction of its primary structure from the nucleotide sequence.[41] The *adk* gene encoding adenylate kinase in *E. coli* was cloned in pBR322.[40] Adenylate kinase represented about 4% of the total protein in extracts of cells containing the pBR322: *adk* plasmid. The primary structure of *E. coli* adenylate kinase (of 214 amino acid residues) as deduced from the *adk* gene nucleotide sequence[41] was confirmed by the sequencing of tryptic peptides.[40] Of all the adenylate kinases the *E. coli* enzyme displays the lowest degree of homology to the muscle types (Scheme II) with only about 30% homology, which is largely at the N-terminus; there is also a replacement of His-36 by a glutamine. A much broader substrate specificity for the AMP site is also evident in the bacterial enzyme,[40] and one wonders if the *E. coli* enzyme is not a general nucleoside monophosphokinase with little resemblance to the mammalian ATP-AMP transphosphorylases.

Curiously, there is only a single unreactive cysteine in the *E. coli* adenylate kinase which may react only in 2 to 4 *M* urea with 5′-dithiobis-2-nitrobenzoic acid and does not appear to be required for enzymatic activity.[40] In fact, the protein may be cyanylated and cleaved at this cysteine leaving two peptide residues (1-76 and 77-214). Interestingly, a mixture of these purified peptides tended to reassociate with a restoration apparently of the correct conformation of its nucleotide binding sites and about 13% of its specific activity.[40]

An identification was made of an amino acid substitution (Pro-87 to Ser) to be responsible for thermosensitivity of adenylate kinase of *E. coli* CR341 T28,[40,42] with a significant change in the secondary structure, i.e., reduction in the α-helix content (39%) of mutant protein as compared to wild-type enzyme (50%). Such an altered conformation of thermosensitive adenylate kinase was manifested by an increase in susceptibility to proteolysis by trypsin. Ap_5A and ATP, which may induce important conformational changes in the adenylate kinases, protected the mutant enzyme against inactivation by trypsin. They concluded that there was ''loosening'' of the steric structure of the *E. coli* enzyme by the Pro to Ser substitution which might be largely compensated when an enzyme-ATP or enzyme-$A_{P5}A$ complex is produced.

Monnot et al.[79] investigated the secondary structures of *E. coli* adenylate kinase by circular dichroism (CD) spectroscopy in the far- and near-ultraviolet. A study was made of the wild-type protein, the S-carboxymethylated (Cys-77) enzyme, and the thermosensitive enzyme which differs from the wild type in that a serine is substituted for a proline at position 87. The secondary structure composition of the wild-type bacterial adenylate kinase (50% α-helix and 15% β-sheet) was close to that derived from X-ray analysis of porcine enzyme.[34]

Carboxymethylation of the wild-type enzyme did not greatly affect the CD-spectrum. The secondary structure of the thermosensitive adenylate kinase was observed to be significantly different from that of the wild-type protein with reduction in α-helix content to 39%. Changes in ellipticities at 222 nm as a function of temperature indicated that the melting temperature for thermosensitive adenylate kinase was 38°C, and that for the wild-type protein was 54°C.

Isolated chemically cleaved peptides, $C_1(Res_1 \sim Res_{76})$ and C_2 $(Res_{77} \sim Res_{214})$, had a large proportion of unordered structures. When mixed, C_1 and C_2 peptides reassociated into structures resembling the native, uncleaved adenylate kinase molecule. The recovery of ordered structures, as indicated by CD-spectroscopic pattern, also paralleled the recovery of catalytic activity.

1. Solution Structure of the 45-Residue ATP-Binding Peptide of Adenylate Kinase

From the X-ray structures of crystalline porcine muscle adenylate kinase[34-36] one would deduce that synthetic peptide I_{1-45} (see Table 6 and Reference 90) would exist as 47% α-helix, 22% β-structure (strand and turns), and 31% coil or other structures. The solution structure of synthetic peptide I_{1-45} which corresponds to residues 1-45 in the rabbit muscle adenylate kinase, and which largely constitutes the MgATP binding site[67] (see Section III.C), was studied by three independent spectroscopic methods.[91,107] Globularity of the peptide was shown by its broad NMR resonances which narrow on denaturation,[67] and by its ability to bind MgATP with a similar affinity (Table 6) and conformation[67] as does the intact enzyme. COSY and NOESY 2D NMR methods at 250 and 500 MHz revealed proximities among NH, Cα, and Cβ protons indicative of 24% α-helix and 38% β-structure and 38% other structures. Correlation of regions of secondary structure with the primary sequence by 2D NMR indicates that there are at least two α-helical regions (Res 4 to 7 and Res 23 to 29) and three stretches of β-strands (Res 8 to 15, 30 to 32, and 35 to 40[91,107]). The broad amide I band in the deconvoluted FTIR spectrum could be fitted as the sum of five peaks due to specific secondary structures,[93] and indicative of ≤10% α-helix, 35% β-structure, and about 55% turns and other structures. The CD spectrum, from 185 to 250 nm, interpreted with a three parameter basis set, yielded ≤5% α-helix, 35% β-structure, and 50% other structures. The inability of the high-frequency FTIR and CD methods to detect helices in the amount *found* by NMR may be the result of their short lengths as well as static and dynamic disorder in the peptide. It is evident, however, that the solution structure of synthetic peptide I_{1-45} resembles that of residues 1-45 in the crystalline structure.[91,107] Upon binding of MgATP, a number of conformational changes in the backbone of the peptide are detected by NMR, with smaller alterations in the overall secondary structures detected by FTIR and CD. Overall, the detailed assignments of resonances in the peptide spectrum and intermolecular NOEs between protons of bound MgATP and those of the peptide[107] were consistent with the previously proposed binding site for MgATP on adenylate kinase.[67]

B. X-RAY DIFFRACTION ANALYSES OF CRYSTALLINE ADENYLATE KINASE

Porcine muscle adenylate kinase was crystallized in three crystal forms, denoted A, B, and C,[63] which are interconvertible, e.g., by pH changes.[35] For crystal form A, the most stable crystal form,[64] X-ray data at 3-Å resolution together with its amino acid sequence yielded the enzyme structure of one molecular conformation in atomic detail.[33,34] A stereo drawing of the polypeptide chain is given in Reference 34. This view exposes the deep cleft which divides the molecule into two domains; the domain on the right side consists of three helical rods which enclose a small hydrophobic core. The domain on the left side contains a parallel pleated sheet formed by five strands, each about five residues long which has a right-handed twist like that found in many other proteins. This sheet is the "architectural center of this domain." There are two hydrophobic cores, one on each side of the sheet which are formed between the sheet and covering helices. About 55% of all the residues are in a right-handed helical configuration and 13% are in the sheet; therefore, almost two thirds of all the residues are involved in secondary structures. All of the charged side-chains are on the surface, except for Asp-93, which is situated deep in the cleft. An unusual piece of chain containing five glycines and one proline that exists between residues 15 and 22 (this will be dealt with in Section III.C) is situated in the cleft forming a crooked loop as shown in Figure 2. Sachsenheimer and Schulz[35] also reported the X-ray diffraction analysis of crystal form B at a 4.7-Å resolution and presented a comparison with the A form. During the transition from crystal form A and B, the packing arrangement of the molecules changes slightly and the molecule undergoes an appreciable conformational change. By displacing a chain segment of seven residues and two adjacent α-helices, a hydrophobic pocket is

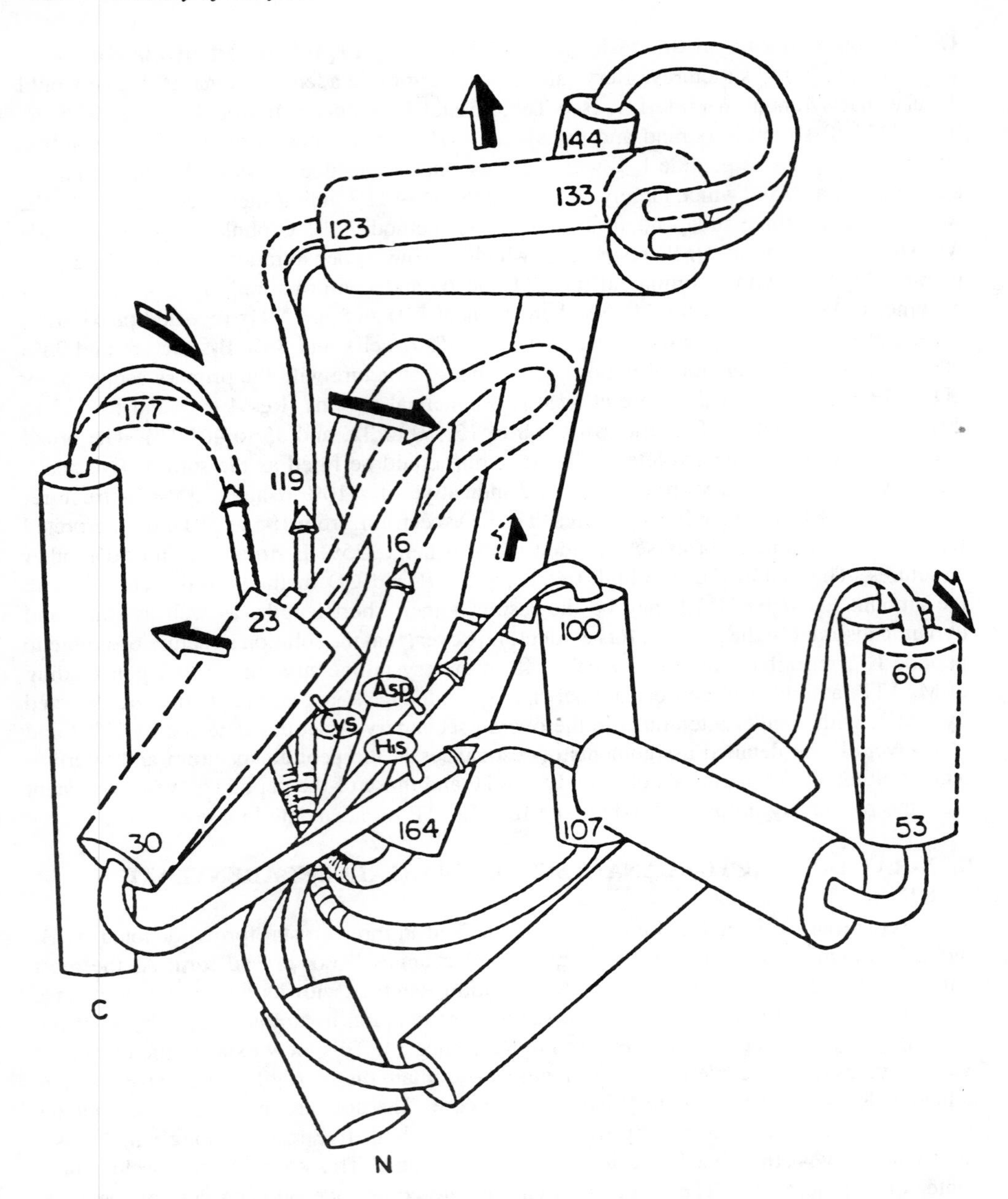

FIGURE 2. Sketch of adenylate kinase in the A (solid lines) and B (broken lines) conformations which are present in crystal forms A and B, respectively. Helices are represented by cylinders. The five strands forming the twisted parallel pleated sheet in the center of the molecule are marked by arrow-heads. Several residue numbers as well as the side-chain triplet, Cys-25, His-36, Asp-93, are given. Arrows show in which direction and to what extent structural elements move during the transition from the A to the B conformation. The reverse transition is equally possible. The largest movement of about 6 Å occurs at loop 16-22. The shift of helix 144-164 is less than 1 Å and not conclusive. For clarity movements near residues 15, 119, 174 leading to an extension of the pleated sheet have been omitted. (From Sachsenheimer, W. and Schulz, G. E., *J. Mol. Biol.*, 114, 23, 1977. With permission.)

opened deep in the cleft near the center of the molecule (Figure 2). Concomitantly, the β-pleated sheet is enlarged by about four hydrogen bonds in the B form.

The putative binding positions of ATP and AMP in porcine muscle adenylate kinase were deduced early by X-ray diffraction analyses[36] (Figure 3). In the case of crystal form A, no ATP-binding site was observed in any experiment where the crystals were "soaked"

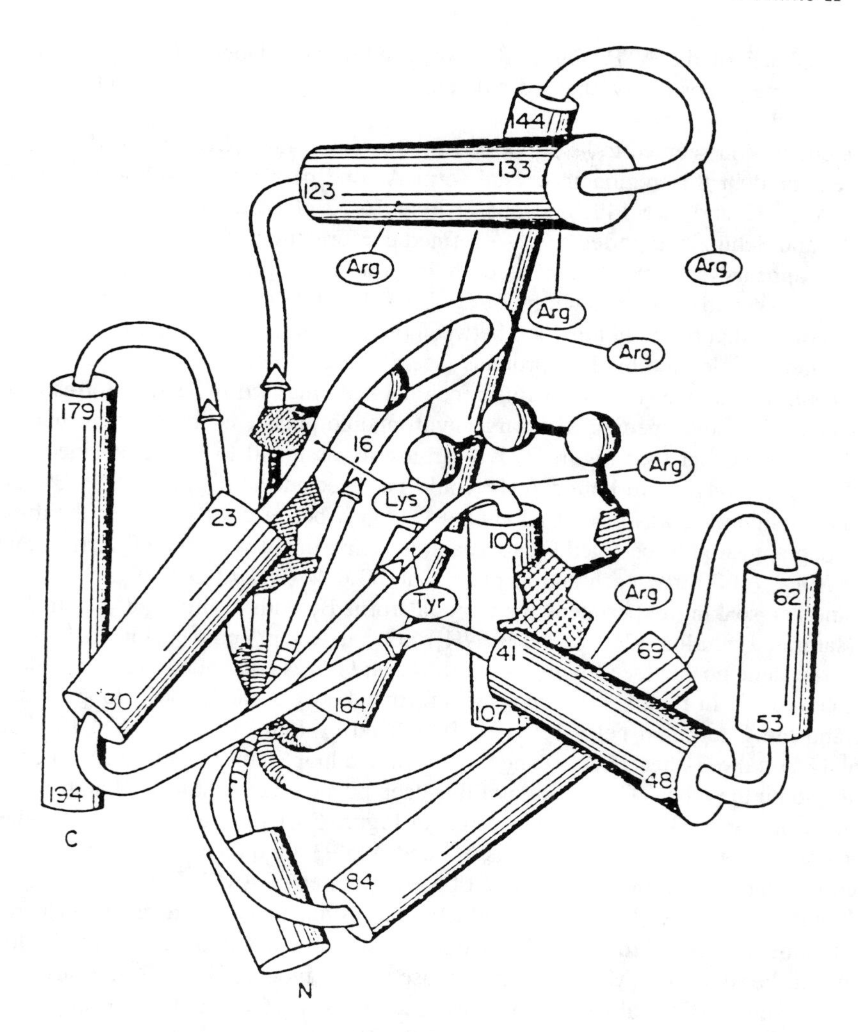

FIGURE 3. Sketch of adenylate kinase. Helices are represented by cylinders and the five strands of the pleated sheet by arrows. Substrate positions as derived from difference Fourier maps at 6-Å resolution are inserted. The side-chains of Lys-21, Arginines -44, -97, -128, -132, -138, and -149, and Tyr-95 are shown. (From Pai, E. F., Sachsenheimer, W., Schirmer, R. H., and Schulz, G. E., *J. Mol. Biol.*, 114, 37, 1977. With permission.)

with solutions of ATP in 3 *M* $(NH_4)_2SO_4$ (pH 6.2 to 6.9); however, ''soaking'' experiments with $A_{P5}A$ or with salicylate seemed to locate one ''adenosine'' pocket which was assigned to the ''ATP-site.'' In Figure 3, taken from Reference 36, this assigned pocket was located between helices 69-84 and 100-107 and was lined by Val-67, Leu-69, Val-72, Leu-73, and Leu-76 on one side and Ile-92, Tyr-95,* Arg-97, and Gln-101 on the other. Crystals of A form soaked in a 10% solution of decavanadate, at pH 7.7 (and 3 *M* $(NH_4)_2SO_4$), appeared to reveal a binding site in the deep cleft of the enzyme which was assigned to the region

* By site-directed mutagenesis, Tyr-95 was replaced by Phe in the human muscle-type adenylate kinase, expressed in *E. coli*[109] the mutant adenylate kinase possessed a k_{cat} $^{(V_{max})}/_{(E_t)}$ which was slightly greater than its wild-type counterpart, making it unlikely that Tyr-95 interacts with the adenosine moiety, as proposed in Reference 36. Arg-97, however, when replaced by an Ala still retained about 4% of its activity, i.e., expressed as $^{V_{max}}/_{E_t}$, whereas, its $K_{m,app}$ for ATP actually decreased to about one half the value for the wild type, suggesting that Arg-97 is not required for MgATP binding, although it is important (but not essential), for the enzymatic activity. To investigate further the requirements of all of the arginines in the so-called ''catalytic cleft'' (Figure 3) of adenylate kinase, studies are underway by site-directed mutagenesis.[114]

of the phosphates of the ATP (Figure 3). This binding site, labeled the "catalytic cleft," is lined with many positively charged side-chains, i.e., Lys-21, and Arg-44, -97, -128, -132, -138, and -149.

This conclusion seemed to be supported by the observation that three sulfate ions from the solvent bind in this region in crystal form A: one between Lys-21 and Arg-97, one between Arg-132 and Arg-149, and one at loop 16-22 close to Arg-132. More recently, Dreusicke and Schulz[105] reported a 2.1-Å refined conformation of the glycine-rich loop (15-22), which appears to form a "giant anion hole" for one of the sulfate ions in crystal form A, with its two negative charges within 3 Å of Lys-21 and Arg-132. They also drew attention to the homology that exists in this loop between the other adenylate kinases, the p^{21} proteins and some mononucleotide-binding proteins (see also Reference 106).

As mentioned above, the major difference between the two molecular forms of crystal form A and B[35] is the opening of a large hydrophobic pocket in the B conformation (see Figure 2). This pocket (see Figure 3) was arbitrarily assigned to the adenosine moiety of the AMP-site although "no bound AMP could be detected there" as Pai et al.[36] carefully noted. The assignment was made by Pai et al.[36] largely because the pocket in the difference Fourier map appeared to be filled by a "soaking" experiment with crystal form B and with ATP + Mn^{2+} (at 30 m*M* each and at pH 6.1 and 3.0 *M* $(NH_4)_2SO_4$).

In another soaking experiment with crystal form B, with 1-anilino-8-naphthalene-sulfonate (saturated solution, pH 5.8, 3.0 *M* $(NH_4)_2SO_4$), a reagent known to bind to hydrophobic pockets, the same pocket (see Figure 3) was filled and it was concluded to be "an adenosine-binding centre".[36] In Figure 3, this pocket is formed between the β-sheet, loop 16-22, helix 23-30, and the C-terminal helix 179-194. It is lined by Leu-91, Ile-11, Val-13, Leu-116, and Val-118 on the β-sheet, Ser-19 and Gly-20 on the loop, Gln-24 and Ile-28 on one helix, and Val-186 on the other. The location of this deep pocket in crystal form B and the putative assignment of the AMP site is shown better in Figure 2 of Reference 36 itself, where the positions of Ser-19, Lys-21, Cys-25, His-36, and Asp-93 are indicated, residues which were assumed to interact with the adenine and ribose structures of AMP. It was remarked by Pai et al.[36] that this pocket (Figure 3) would be accessible to substrate analogues in the B conformation, whereas in the A conformation, its entrance would be closed by the loop 16-22,[35] and further that the AMP conformation used in the model (Figure 3) is nearly identical to the 8,5′-cyclo AMP analogue, which had been shown interestingly by Hampton et al.[62] to be a better substrate than AMP itself. Finally, their model and the relationship of their assigned positions for the ATP and AMP binding sites (Figure 3) consequently suggested to them that the B conformation was related to the structure of the free enzyme before a substrate-induced conformational change, whereas the A conformation appeared to be related to a second stable conformation, after such a change. Thus, in the "two-state induced-fit" model of Koshland[94] and Jencks,[95] adenylate kinase would undergo a substrate-induced conformational change, with the substrate (e.g., AMP) triggering the change.[36] We shall see, *vide infra*, that, whereas the assignments of the ATP and AMP binding sites may have been reversed by Pai et al.,[36] as discussed by Hamada et al.[46] and by Smith and Mildvan,[38] certainly many of the ideas and experimental data of Schulz et al. and of Pai et al., in their pioneering attempt at describing the active ternary complex of adenylate kinase, could be incorporated in the novel assignments by NMR (see below) of the MgATP binding site and of the AMP binding site within the ternary complex of the enzyme.

C. NMR STUDIES AND THE TOPOGRAPHY OF THE MgATP-BINDING SITE OF ADENYLATE KINASE

As mentioned above, a reassessment of the nucleotide binding sites of adenylate kinase proved necessary by the remarkable finding of Hamada et al.,[46] i.e., that two fragments derived from a tryptic digest of maleylated S-carboxymethylated rabbit muscle adenylate kinase, and demaleylated, selectively bound magnesium 1,N^6-ethenoadenosine 5′-triphosphate (MgϵATP) and ϵAMP, with affinities comparable to that of the native enzyme. The ATP-binding fragment, consisting of residues 1-44, and the AMP-binding fragment, corresponding to residues 172-194, were located in regions of the molecule that differed from the binding sites proposed on the basis of the X-ray data as described above. These fragments offered an attractive opportunity for the NMR studies on adenylate kinase by individually examining what appeared to be the isolated substrate binding sites. [We now know[86] that a longer peptide, e.g., CB-VI (res. 126-194) may satisfy the stringent substrate conformational requirements deduced for the "AMP site."] A careful study of the MgATP binding site of rabbit muscle adenylate kinase, and of a synthetic 45-residue peptide fragment (I_{1-45}, see Reference 90) was carried out by Fry et al.[67] Proton NMR was used to study the interaction of β,γ-bidentate Cr^{3+}ATP and of MgATP with the intact enzyme, and with the synthetic globular peptide I_{1-45}, which was shown to bind MgϵATP[90] (see Table 6), and also to bind Cr^{3+}ATP competitively with MgATP [K'_D (Cr^{3+}ATP) = 35 μM], and comparable to that of the native enzyme [K'_I(Cr^{3+}ATP) = 12 μM[23]]. Time-dependent nuclear Overhauser effects (NOEs) were used to measure interproton distances on the enzyme- and on the peptide-bound MgATP. The correlation time was measured directly for peptide-bound MgATP by studying the frequency dependence of the NOEs at 250 and 500 MHz. The H2′ to H1′ distance obtained (3.07 Å) was within the range established by X-ray and model building studies of nucleotides (2.9 ± 0.2 Å). Interproton distances yielded the conformations of the enzyme- and peptide-bound MgATP (see Figure 4) with indistinguishable *anti*-glycosyl torsional angles (X = 63 ± 12°) and 3′-endo/O1′-endo ribose puckers (δ = 96 ± 12°). The enzyme- and the peptide-bound MgATP molecules exhibited different C4′-C5′ torsional angles (γ) of 170 and 50°, respectively.

Ten intermolecular NOEs from protons of the enzyme and four such NOEs from protons of the peptide to protons of bound MgATP were detected, which indicated proximity of the adenine ribose moiety to the same residues on both the enzyme and the peptide. Paramagnetic effects of β,γ-bidentate Cr^{3+}ATP on the longitudinal relaxation rates of protons of the peptide provided a set of distances to the side chains of five residues, which allowed the location of the bound Cr^{3+} atom to be uniquely defined. Distances from enzyme-bound Cr^{3+}ATP to the side chains of three residues of the protein agreed with those measured for the peptide. The mutual consistency of interproton and Cr^{3+} to protein distances obtained in the metal-ATP complexes of both the enzyme and the peptide suggested that the conformation of the peptide was very similar to that of residues 1-45 of the enzyme. If this were assumed to be the case and with the use of molecular models and a computer graphics system MgATP could be fit into the X-ray structure of adenylate kinase in a unique manner such that all of the distances determined by NMR were accommodated. In Figure 5,[67] computer drawings are presented showing the MgATP-binding sites of intact adenylate kinase and of the synthetic peptide I_{1-45}. The adenine ribose moiety is bound in a hydrophobic pocket consisting of residues Ile-28, Val-29, His-36, Leu-37, and Leu-91, while the Mg^{2+}-triphosphate portion binds near Lys-21 and Lys-27 and Gln-24. In this complex, the γ-phosphoryl group of MgATP is directed toward residues 172-194 (see Figure 8) which as MT-XII (or $H_{172-194}$, Table 6) had previously been shown to bind ϵAMP by Hamada et al.[46]

The MgATP molecule appears highly extended and cuts across the helix, bend, and a strand of β-sheet formed by residues 23-37 (Plate 1, Figures 6 and 7).* Adenine H2 is near residues His-36 and Leu-37, and the α-, β-, and γ-phosphates are near the α-helix 23-30, at the level of residue Thr-23. The ribose ring is between the side chains of Gln-24 and Val-13, with its H1′-H4′ face directed away from Gln-24. The ϵ-NH_3^+ group of Lys-21 and the side chain of Gln-24 can be positioned very close to the α-phosphate, and Lys-27 can also closely approach the β- and γ-phosphates. Positioning the MgATP in this manner results in the orientation of its terminal phosphoryl group toward the strand of β-structure,

* Plate 1 appears after page 430.

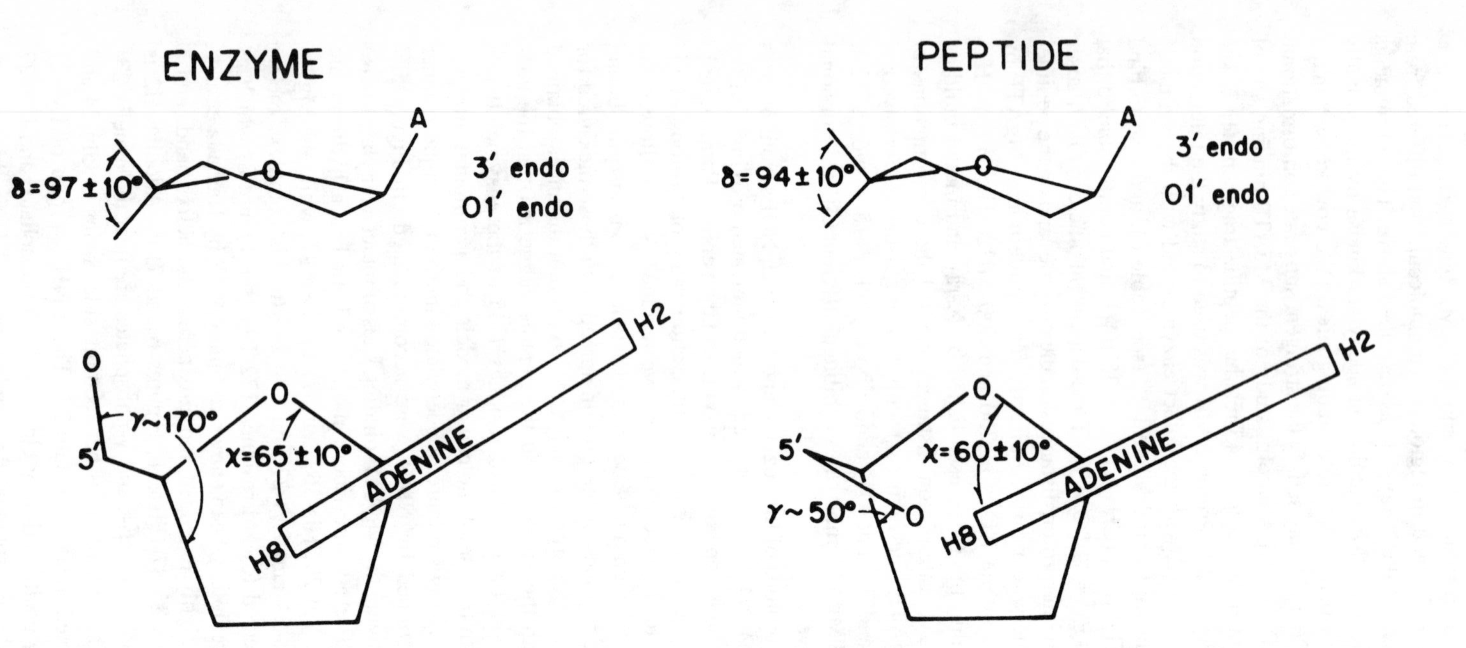

FIGURE 4. Conformation of bound MgATP on rabbit muscle adenylate kinase and on the MgATP-binding peptide. Conformations were determined by model building using the distances obtained from intramolecular NOEs. (From Fry, D. C., Kuby, S. A., and Mildvan, A. S., *Biochemistry,* 24, 4680, 1985. With permission.)

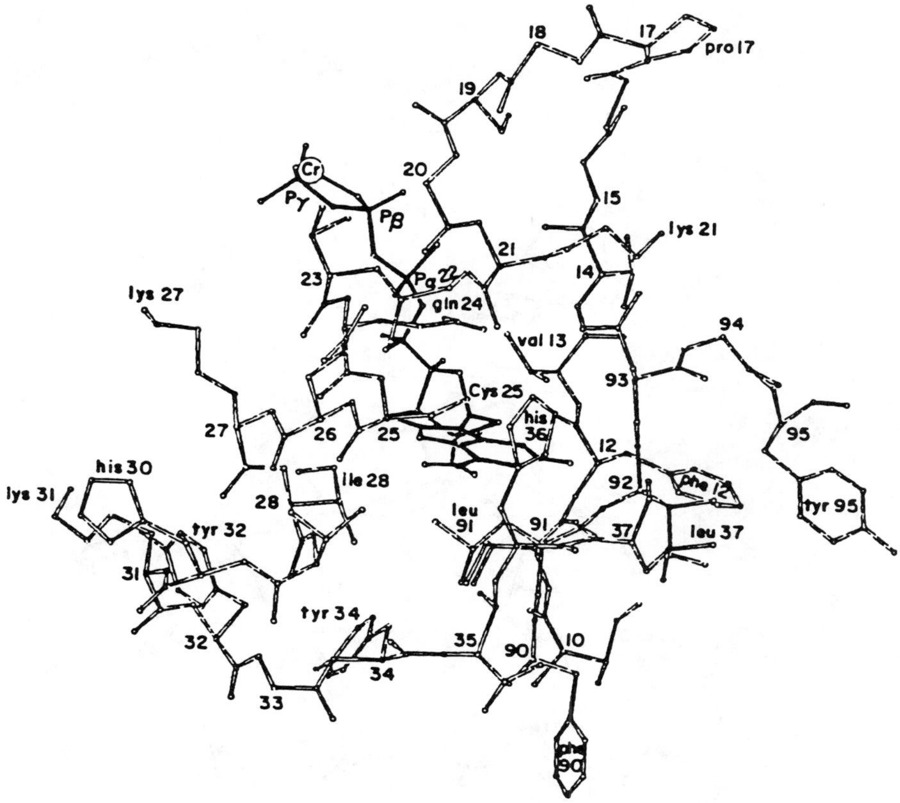

A

FIGURE 5. ORTEP drawings showing the MgATP-binding sites of adenylate kinase (A) and the 1-45-residue peptide (B). Enzyme and peptide structures were drawn by using the X-ray coordinates of Sachsenheimer and Schulz.[35] Residues 10-37 and 90-95 are shown in (A); residues 10-37 are shown in (B). (From Fry, D. C., Kuby, S. A., and Mildvan, A. S., *Biochemistry,* 24, 4680, 1985. With permission.)

β-turn, and α-helix 179-194 formed by the terminal 23 residues of adenylate kinase (and see above, MT-XII, or $H_{172\text{-}194}$, or CB-VI (126-194) Table 6, bind εAMP with reasonable affinities[46]). Thus, the model building also pinpoints the location of the other substrate, AMP, and this will be described in detail below. The NMR assigned position of MgATP (Figures 5 to 7 and Plate 1) may be compared with that deduced by Pai et al.[36] (see Figure 3). The "deep pocket" in crystal form B which had been assigned to the AMP site would now seem, in part, to contain those residues which may be assigned to the MgATP site. In addition, the MgATP-binding site, as determined by NMR means is also consistent with the results of previously reported chemical modification studies (e.g., see References 96, 98, and 103), and kinetic studies on the relatively broad specifity for the ATP substrate site (e.g., see References 45 and 72 and see above Table 1), since the hydrophobic pocket surrounding the adenine and ribose moieties of the bound MgATP seems to be sufficiently spacious to accommodate a variety of ring structures. This may account for the relatively lower selectivity exhibited by this substrate site,[58] which includes the relatively low K_m and high V_{max} observed with MgϵATP^{2-} (Table 1, Reference 45).

In this model, the most sterically confined portion of the enzyme-bound MgATP would appear to be at the edge of the adenine ring near H2, since this hydrogen is located near

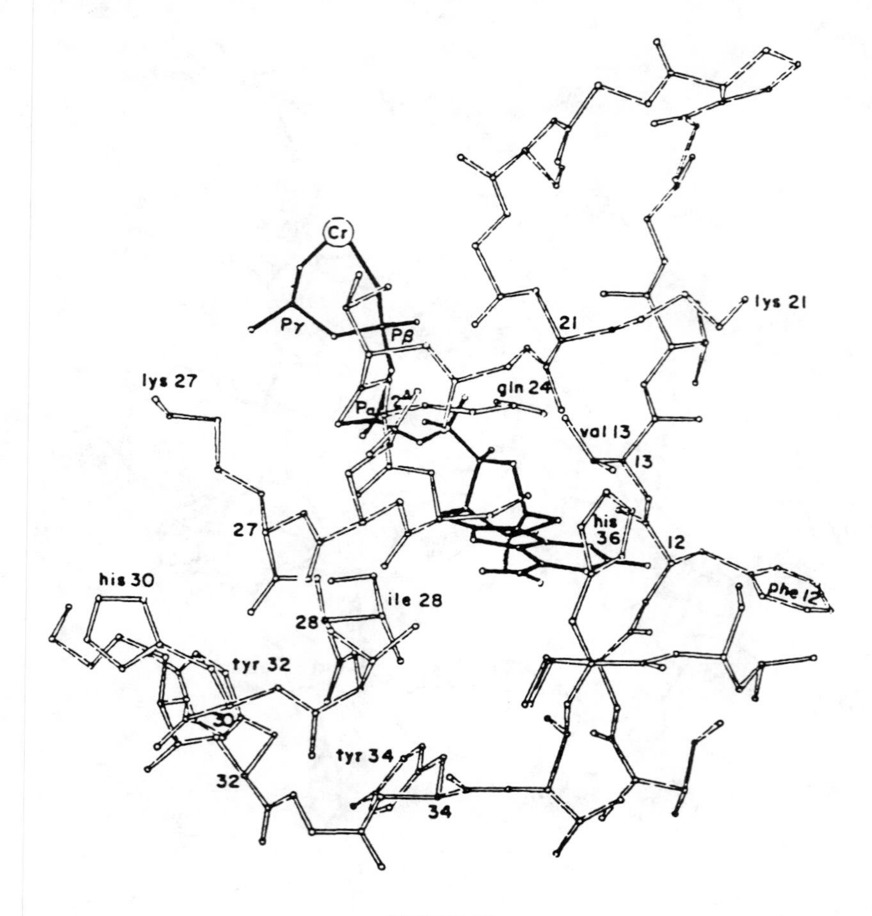

FIGURE 5B

the peptide carbonyl groups of Leu-37 and Ile-92.[67] The bulky C2 amino group of GTP and C2 carbonyl groups of CTP, UTP, and ITP may explain the order of magnitude higher Km values of these substrates.[72] Also, in this model, the region with least opportunity for specific enzyme-substrate contact is the ribose, a point which may explain the relatively low K_m and high $V_{max}^{forward}$ measured for $MgdATP^2$, with the latter value comparable to that found for $MgATP^{2-}$ (Table 1 and References 45 and 72). Finally, the proposed structure is supported by the structural and sequence homologies present in many nucleotide binding proteins (see Scheme 3). It was shown by Walker et al.[68] that such ATPases as F_1, myosin, and recA protein contain a homologous sequence of eight residues in length which is associated with the ATP-binding site. Fry et al.[44,67] extended the list to several other proteins which share this sequence (Scheme 3). Adenylate kinase also contains this sequence (Reference 51; Schemes 2 and 3), and exhibits a subsequent five-residue sequence that shows a slight homology to the above proteins and a strong homology to the ATP-binding regions of cAMP- and cGMP-dependent protein kinases[69,70] and to regions of other proteins. This list includes rabbit muscle creatine kinase,[71] whose reactive cysteinyl residue (Cys-282) would correspond to Cys-25 of rabbit muscle adenylate kinase; also, an identical triplet of His-Leu-Ser- appears in both adenylate kinase (residue 36-38) and in creatine kinase (residues 300-302). In Figure 7, the homologous regions as shown by the stippling are all found to be at or near the ATP-binding sites.

As reported by Fry et al.[44] there are, in addition, certain mechanistic implications which may be drawn from these homologies of the MgATP-binding site of adenylate kinase and

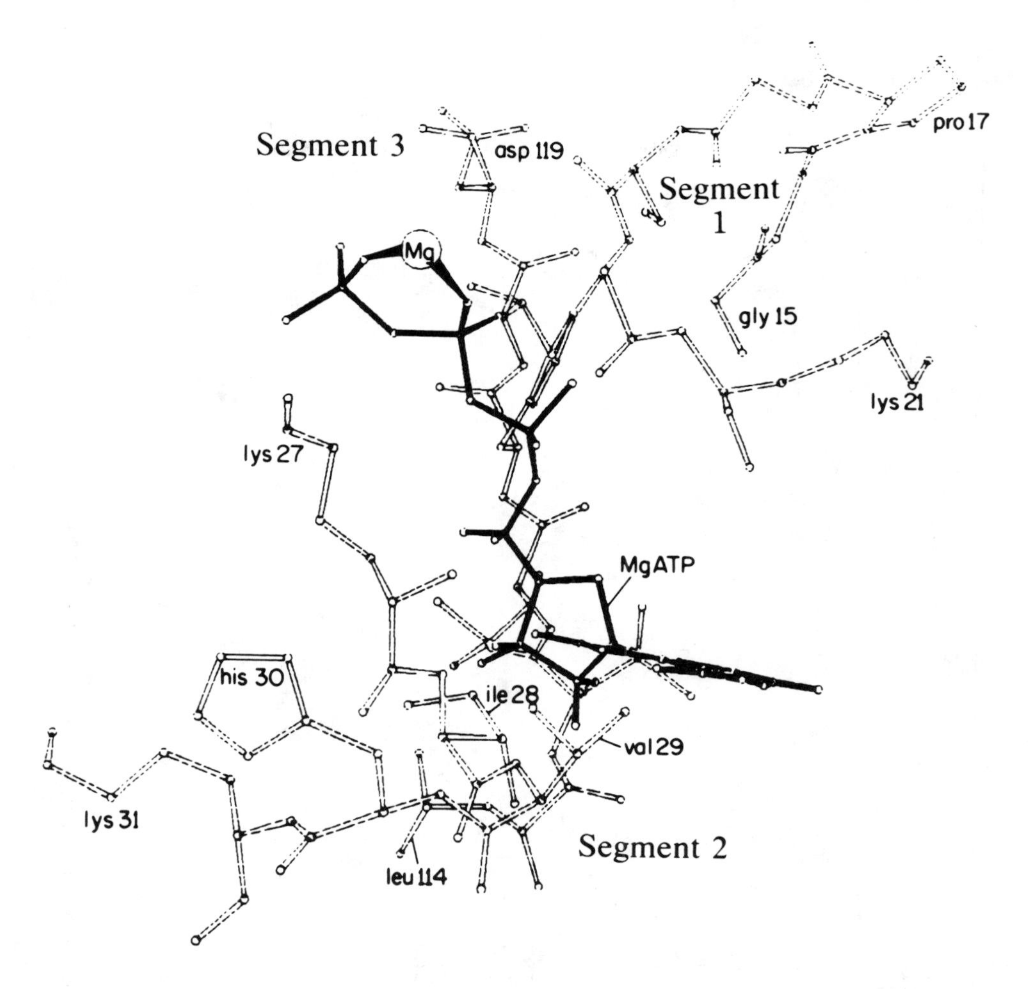

FIGURE 6. ORTEP (computer graphic program) representation showing MgATP and the three homologous segments of adenylate kinase. (From Fry, D. C., Kuby, S. A., and Mildvan, A. S., *Proc. Natl. Acad. Sci. U.S.A.*, 83, 907, 1986. With permission.)

several nucleotide-binding proteins. Thus, the MgATP-binding site of adenylate kinase, located by a combination of NMR and X-ray diffraction procedures, as described above, is near three protein segments (Figures 6 to 7, Plate 1, and Scheme 3), five to seven amino acids in length, that are homologous in sequence to segments found in the other nucleotide-binding phosphotransferases, such as myosin and F_1-ATPase, ras p21 and transducing GTPases, and cAMP-dependent and src protein kinases, suggesting that there are equivalent mechanistic roles of these segments in all of these proteins.

Segment 1 is a glycine-rich flexible loop that, in the case of adenylate kinase, may control access to the ATP-binding site by changing its conformation. Segment 2 is an α-helix with two hydrophobic residues which may interact with the adenine-ribose moiety of ATP, and a lysyl residue which may bind to the β- and γ-phosphates of ATP. Segment 3 is a hydrophobic strand of parallel β-pleated sheet, ending in a an aspartyl residue, which flanks the triphosphate binding site. Included in Scheme 3 are the homologies noted by Walker and co-workers[68,68a] largely in segments 1 and 2; some of the proteins listed show only one or two of the three homologous segments. Segment 1 (residues 15-21 in adenylate kinase: Gly-Gly-Pro Gly-Ser-Gly-Lys) is shared by several ATPases and GTPases as well as by DNA A protein, Epstein-Barr virus protein, thymidine kinase, cAMP-dependent protein

Protein	Sequence 15 20 25 30	Sequence 110 115 120
Adenylate kinase	G-G-P -G-S-G-K-G-T-Q-C-E-K -I-V-H-K	G-Q-P-T-L-L-L-Y-V-D-A-G
F_1-ATPase		
α(*E. coli*)	G-D-R -Q- -G-K-T- -A-I-	G- -A-L-I-I-Y-D-D-
β(*E. coli*)	G-G-A -G- -G-K-T- -L-I-	G- -V-L-L-F-V-D-
β(bovine)	G-G-A -G- -G-K-T- -L-I-	G-Q- -V-L-L-F-I-D-
Myosin		
nematode	-G-G -G-G-G-K- -V-	
	G- -G- -G-K-T- -K -V-I-	
rabbit	G- -G- -G-K-T- -R -V-I-	
Thymidine kinase	G- -G- -G-K-T-T- -L-V-	
RecA protein	G- -S-G-K-T-T- -V-I-	
Transducin α	G- -G- -S-G-K- -T- -K-	
G_o protein α	G- -G- -S-G-K- -T- -K-	
ras p21	G- -G -G- -G-K- -L-I	G- -L-L-D-I-L-D-T-A
DnaA protein	G-G- -G- -G-K-T- -V-	
Epstein–Barr virus protein	G-G- -G-K-G- -A-	
Glycogen phosphorylase	G- -G -G- -G-R- -C-	
Phospholipase A_2	G- -G -G- -G-R-	
Protein kinase		
cAMP-dependent	G- -G- -G-R- -/10/-K -I-L- -K	
cGMP-dependent	-K -I-L- -K	
src protein	G- -G- -G- -/10/-K -T-L-K-	
Nitrogenase (Fe protein)	G- -G -G- -G-K- -T- -/11/-K -I-L-	
Transcarboxylase biotin subunit	G-G- -G- -G-K- -/10/-K -I-L- -K	G-Q-T-V-L-V-L-E -Bct-
ATP/ADP translocase		G- -V-L-V-L-Y-D
Phosphofructokinase		G- -L-V-V-I- -D-

(Taken from Ref. 72)

SCHEME 3. Sequence homologies among the ATP-binding region of adenylate kinase and segments of other proteins.

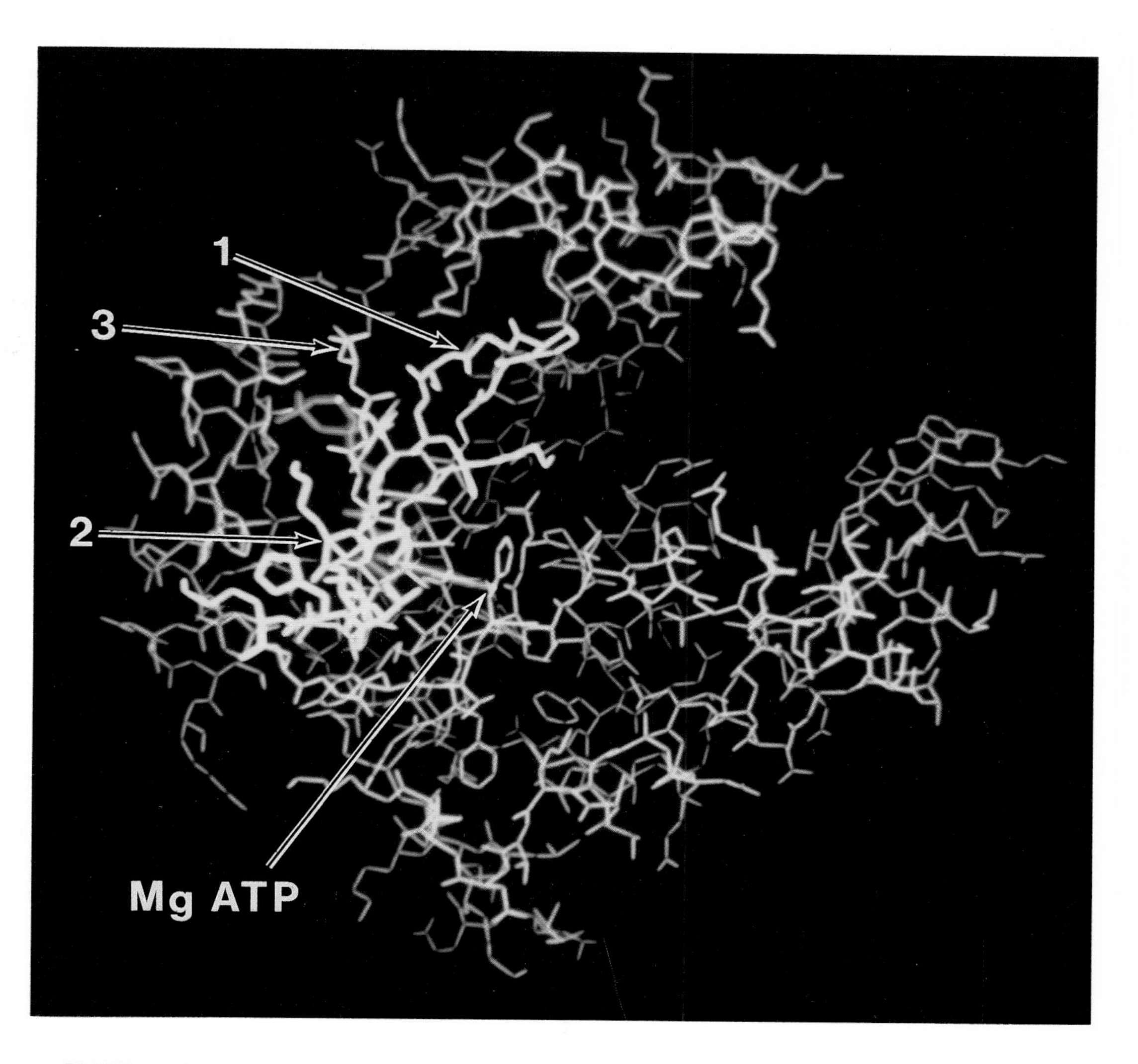

PLATE 1. Computer graphics representation of rabbit muscle adenylate kinase showing the location of bound metal-ATP. The three segments of the enzyme exhibiting sequence homology to other proteins (Scheme III) are shown in pink, and the ATP molecule is shown in red. The X-ray coordinates of confromation A of porcine adenylate kinase[34] were used, with substitution of a histidine residue for glutamine at position 30. Metal-ATP was fit into the enzyme structure using a set of distances obtained by NMR.[67] Segments 1, 2, and 3 and MgATP are identified. (From Fry, D. C., Kuby, S. A., and Mildvan, A. S., *Proc. Natl. Acad. Sci. U.S.A.*, 83, 907, 1986. With permission.)

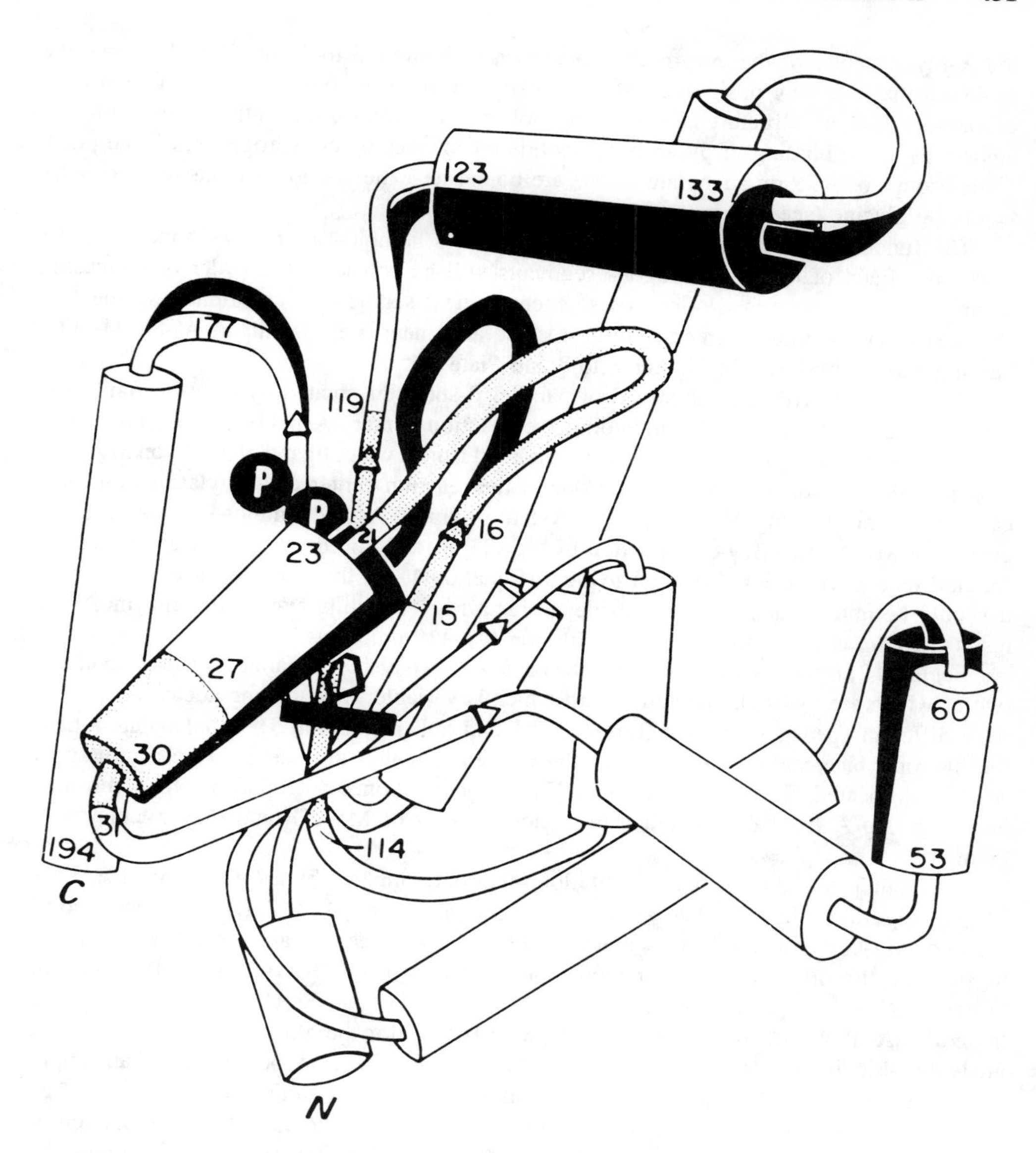

FIGURE 7. The metal-ATP binding site of adenylate kinase is shown with respect to the two crystal forms of the enzyme: A (solid) and B (open). The binding of Mn^{2+}ATP changes the conformation from B to A.[35,36] The three segments of homology (residues 15-21, 27-31, and 114-119) are depicted by stippling in conformation B. The drawings of the enzyme are based on the X-ray structures.[34,36] The position of metal-ATP was determined by NMR.[67] (From Fry, D. C., Kuby, S. A., and Mildvan, A. S., *Proc. Natl. Acad. Sci. U.S.A.*, 83, 907, 1986. With permission.)

kinase, phospholipase A2, glycogen phosphorylase, nitrogenase, and the biotin-containing subunit of transcarboxylase. Slightly further along in the sequence is segment 2 (residues 27-31: Lys-Ile-Val-His-Lys) which bears a weak homology to corresponding portions of many of these proteins but a strong homology to regions of the protein kinases and transcarboxylase. Regions of near homology to segments 1 and 2 are also found in phosphoglycerate kinase[99,100] although it is not listed. Segment 3 (residues 114-119:Leu-Leu-Leu-Tyr-Val-Asp) has counterparts in the sequences of F_1-ATPase, *ras* p21, ATP/ADP translocase, phosphofructokinase, and transcarboxylase, and to creatine kinase,[71] also not listed.

The MgATP binding site of adenylate kinase, as determined by NMP[67] and fitted into the X-ray structure of the enzyme,[34-36] is shown in Figures 5 to 7 and Plate 1. It may be observed that this new position of MgATP places it in close contact with the five strands of parallel β-sheet structure and with one of the interconnecting helices, more closely analogous to the binding of pyridine nucleotide coenzymes by dehydrogenases;[73] the phosphate groups of ATP on adenylate kinase are bound near the amino terminus of an α-helix begun by glycine (see Reference 74).

The functional roles in adenylate kinase of the ''homologous regions'' may now be discussed. Each of these homologous segments will be considered in order of increasing complexity of its presumed role: first segment 3, next segment 2, and finally segment 1. One notes that the three segments are all located at or near (i.e., within 11 Å) the MgATP binding site, as is shown in Figures 6 to 7 and Plate 1.[44]

Segment 3, a hydrophobic strand of parallel β-sheet terminated by an Asp, flanks the triphosphate chain of MgATP, including the reaction center. Asp-119 at the end of this segment (see Figure 6) may accept a hydrogen bond from a water ligand of Mg^{2+} on MgATP, a reasonable role for a carboxylate residue at the reaction center of adenylate kinase. Or, by directly coordinating Mg^{2+}, the carboxylate group of Asp-119 might facilitate the migration of Mg^{2+} from β,γ-coordination in MgATP to α,β-coordination in MgADP, as the forward reaction proceeds from left to right (Equation 1), in the transition state. However, it should be noted that there is no evidence for a Mg^{2+}-binding site on the enzyme;[83] the results of Kim et al.[117] now make any role for Asp-119 unlikely.''

Segment 2, is mainly an α-helix which consists of two lysines separated by three residues, two of which are hydrophobic and a third which is variable, even among adenylate kinases from different species (e.g., see Reference 51 and Schemes 2 and 3). In adenylate kinase the hydrophobic residues form part of the pocket in which the adenine-ribose moiety of MgATP is located. The first lysine, Lys-27, can be positioned such that its NH_3^+ nitrogen is 5.7 ± 2.0 Å from the β- and γ-phosphorus atoms of MgATP and may interact with them.

Segment 1 is a glycine-rich flexible loop that is terminated by a cationic residue, Lys-21 (see Figure 7). In adenylate kinase, Lys-21 is near (i.e., 4 ± 2 Å) the α-phosphoryl group of MgATP and could easily interact with it. Pro-17 near the apex of the loop appears to stabilize this loop by occupying position 2 of a type IV β-turn, as is often seen in proteins.[75] With the exception of Lys-21, the loop makes no direct interactions with the bound MgATP. Segment 1, therefore, although believed to be a sequence diagnostic of a nucleotide-binding site (e.g., see Reference 68), appears to have a function other than simply binding a nucleotide. The function of this loop may involve its ability to undergo a change in conformation (possibly as a result of substrate binding) as evidenced by a comparison of the two crystal forms of porcine adenylate kinase (A and B) (Reference 35 and Figure 2). A transition from B to A, corresponding to the binding of Mn^{2+}ATP, involves structural changes in four regions, with the largest displacement (6 Å) occurring at the glycine-rich loop (Figures 2 and 7). From a study of adenylate kinase, and the information available in the structure and mechanisms of the other homologous proteins, three possible roles are suggested for a conformational change at segment 1.[44,67] These are (1) control of accessibility to the substrate binding sites, (2) modification of binding site affinities, and (3) relocation of catalytic groups toward the reaction center of the bound substrate. Control of access to the MgATP-binding site by the segment 1 loop is suggested by its location, for it is situated in the middle of a cleft that leads from the bound MgATP molecule to the surface of the enzyme (Figures 2 and 7). The change in conformation of this loop observed in the two crystal forms of the enzyme shows that it is capable of alternately blocking and opening the cleft (Figure 2). Lys-21 may interact with the α-phosphoryl group of bound MgATP. A different orientation of this lysyl residue could favor MgADP at this site, since the negative charge distribution among the phosphoryl groups is different for the two substrates. Such a change in the orientation of Lys-21 is one plausible way in which a conformational change affecting segment 1 could alter binding site affinity. The third possible role of a conformational change at segment 1 is to bring catalytic groups into the reaction center. If by a conformational change this loop were to approach the triphosphate chain of bound MgATP, the phosphoryl transfer location could be facilitated by the hydroxyl group of Ser-19 and the amide NH protons of the back-bone functioning as H-bond donors as proposed for phosphoglycerate kinase.[99] It is of interest that in the X-ray studies of adenylate kinase[35] movement of the segment 1 loop brings it into close proximity with another section of the polypeptide chain near residues 120-123 (see Figures 2 and 7). If the homology between *ras* p21 and adenylate kinase at segment 3 is extended, residues 120-123 of adenylate kinase (-Ala-Gly-Pro-Glu-) corresponds to residues 59-62 of p21 (-Ala-Gly-Gln-Glu) that include the mutagenic sites. The presence of certain side chains in this portion of p21 may result in a hindered movement of the loop, inhibiting the hydrolysis of GTP, and possibly providing an explanation for the transforming variants of the *ras* protein.

Recently, studies by site-directed mutagenesis have been made to exchange amino acid residues in the glycine-rich loop (residues 15-22 of the rabbit muscle enzyme, i.e., segment 1 of Reference 44). Tagaya et al.[111] replaced in chicken adenylate kinase Pro-17 by a Gly or Val. From their results they suggested that Pro-17 was important for binding of the substrates, but not for "catalytic efficiency," and that it did not directly interact with the substrates. Earlier they had shown by labeling Lys-21 with adenosine diphosphopyridoxal[112] that the ε-amino group of Lys-21 is located in the ATP-binding site of the enzyme.

Similarly, Reinstein et al.[113] replaced Lys-13 in the *E. coli* enzyme, which corresponds to Lys-21 in the rabbit muscle enzyme, by a Gln, yielding a nearly inactive enzyme, and thus pointed to the catalytic requirement for this lysine.

Also, Pro-9 (equivalent to Pro-17 in the rabbit muscle) and Gly-10 (equivalent to Gly-18 in the rabbit muscle) were replaced by a Leu and a Val, respectively. Both replacements resulted in increased Kms for their substrates (ATP and AMP), but their V_{max}s were similar to that of the wild type. They regarded the loop as important not only for the binding of the substrates, but "as an element in structural rearrangements that are necessary for the function of the enzyme."

D. NMR STUDIES ON THE AMP BINDING SITE OF ADENYLATE KINASE AND A RATIONALE FOR ITS CATALYTIC MECHANISM

The metal-ATP binding site determined by NMR means[44,67] (e.g., see Figures 5 to 7 and Plate 1) proved to be substantially different from that based on X-ray studies of the binding site of salicylate.[36] This MgATP-binding site more closely resembled the site which had been designated as the putative AMP-binding site (Figure 3), and was based on X-ray studies of the crystals (form B) and after MnATP was allowed to diffuse and into the crystals and with the assumption MnATP had been converted to AMP during this process.[36] Thus, it was necessary to reevaluate the location of the uncomplexed AMP-binding site. This study was made by NMR means in two ways,[86] which may be summarized as follows: first, its orientation was determined with respect to enzyme-bound Cr^{3+}AMPPCP, by using the paramagnetic probe-T_1 method to measure the distances from Cr^{3+} to the phosphorus and protons of AMP (Figure 8). These intersubstrate distances also permitted an estimate of the lower limit reaction coordinate distance from the entering phosphate oxygen of AMP to the γ-phosphorus of metal-ATP, and, in turn, provided valuable mechanistic information. Second, nuclear Overhauser experiments were used to refine the conformation of bound AMP (Figure 9) and to obtain a set of proximate distances between the protons of AMP and those of the side chains of the enzyme. With the use of the X-ray structure of the enzyme by Sachsenheimer and Schulz[35] and the previously determined binding site for Cr^{3+}ATP,[67] AMP could be docked into a unique binding-site which satisfied the constraints provided by the NMR data (Figures 10 and 11).

The location on the rabbit muscle enzyme and the conformation of bound-AMP was

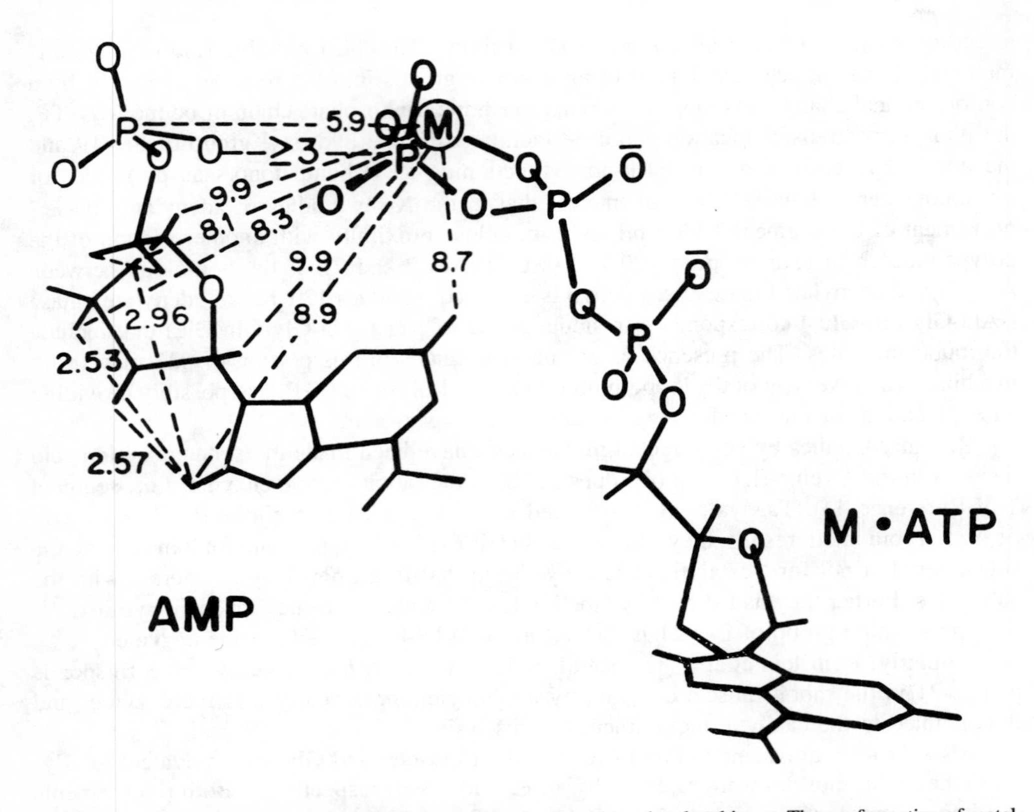

FIGURE 8. Conformations and arrangement of substrates bound to adenylate kinase. The conformation of metal-ATP[67] and of AMP[86] are shown with intersubstrate and interproton distances given in Angstrom units. Also shown is the lower limit reaction coordinate distance (≥3 Å) between the entering oxygen of AMP and the γ-phosphorus of metal ATP determined by model building. (From Fry, D. C., Kuby, S. A., and Mildvan, A. S., *Biochemistry*, 26, 1645, 1987. With permission.)

determined by measuring the distances with the paramagnetic probe-T_1 method from Cr^{3+}AMPPCP (a linear competitive inhibitor with respect to MgATP) to *six* protons and to the phosphorus atom of AMP bound to adenylate kinase (see Figure 8). Time-dependent NOEs were used to measure *five* interproton distances on the enzyme-bound AMP (Figure 8). These distances permitted an assignment of the most plausible conformation of bound AMP, in addition to its position with respect to metal-ATP. Thus, this conformation of enzyme-bound AMP (Figure 9) appeared to possess a high anti-glycosyl torsional angle (X = 110 ± 10°), a 3′-endo/2′-exo ribose pucker (δ = 105 ± 10°), and gauche-trans orientations about the C4′-C5′bond (γ = 180 ± 10°) and about the C5′-05′ bond (β = 170 ± 20°). The distance from Cr^{3+} to the phosphorus atom of AMP was estimated to be 5.9 ± 0.3 Å, which yields a reaction coordinate distance of ≥3 Å (Figure 8), consistent with an associative S_N2 mechanism for phosphoryl transfer (see below).[95,102] This conformation of bound-AMP at the active site of adenylate kinase appears to be a somewhat unusual combination of a high anti-glycosyl torsional angle with a 3′-endo 2′-exo ribose pucker;[101] therefore, this conformation could be considered distorted or strained, perhaps as a result of the relatively strict substrate requirements of the AMP-site.

Ten intermolecular NOEs from the enzyme protons to those of AMP were detected, indicative of the proximity of at least 3 hydrophobic amino acid residues to bound AMP. These constraints, the unusual conformation of bound AMP, and the intersubstrate distances were utilized to position AMP into the X-ray structure of adenylate kinase which led to the unique location of the AMP-site (Figures 10 and 11). This location of the AMP-site places

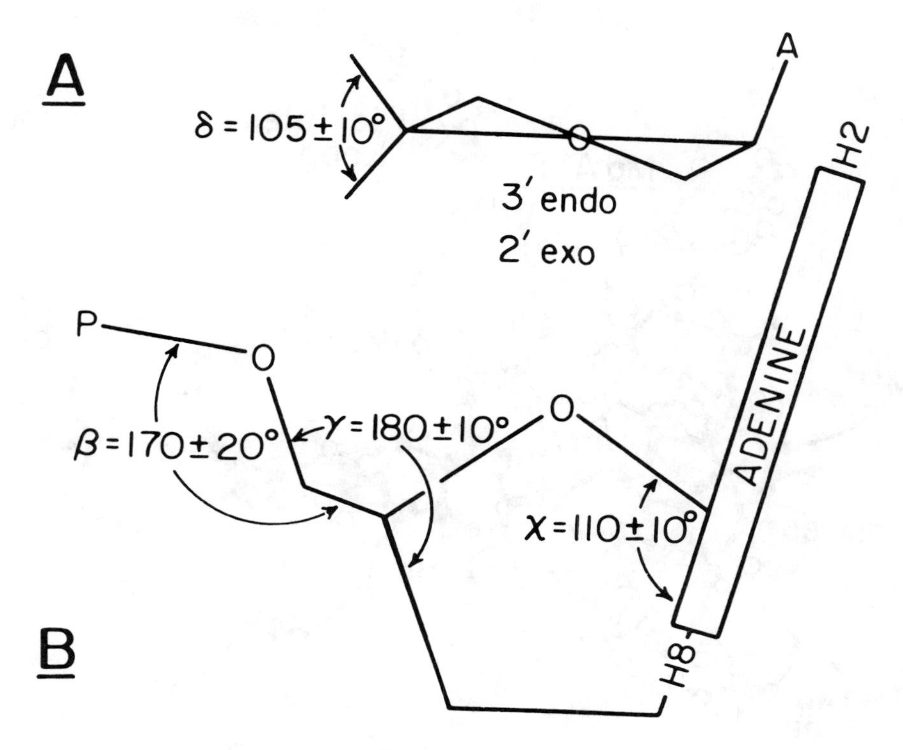

FIGURE 9. Detailed conformation of adenylate kinase-bound AMP showing the four conformational angles. (From Fry, D. C., Kuby, S. A., and Mildvan, A. S., *Biochemistry,* 26, 1645, 1987. With permission.)

it within ≤4 Å of Leu-116, Arg-171, Val-173, Val-182, and Leu-190; all of these residues are invariant in the muscle-type rabbit, calf, human, and porcine adenylate kinases (Reference 51 and Scheme 2) and in chicken muscle adenylate kinase.[43] One notes that the last 23 residues of adenylate kinase, including the helix 179-194 and the terminal bend and β structure are included within the binding domain of AMP (compare peptide fragment MT-XII, $H'_{172\text{-}194}$ or CB-VI; see Scheme 2, which was found to bind εAMP (Reference 46 and Table 6)).

NOE studies have suggested at least two conformations for peptide $H'_{172\text{-}194}$-bound AMP, probably due to the incompleteness of the AMP binding site. Further studies with larger peptides from this region (e.g., CB-VI (126-194), see Table 6) should better approximate the AMP binding site of the enzyme.

If one now studies the substrate-binding sites in the complete enzyme, i.e., within the postulated ternary complex (Figures 10 and 11), in the final formation of the ternary complex via the rapid equilibrium random mechanism of Hamada and Kuby[45] (Scheme 1), it is possible that AMP may enter the active site from the left, in the view of Figure 11; MgATP would then be required to enter from the right.[86] However, a conformational change would also be required to eliminate the barrier to MgATP binding imposed possibly by the flexible loop 15-22, the helix 23-30, and the bend in the vicinity of 114-119 (see Figure 7), as well as the correct positioning of AMP between helix 179-194 and helix 22-30 and near the bend 171-177 (see also Figure 2). Such a conformational change was deduced kinetically by Hamada and Kuby[45] *vida supra*, by the X-ray analysis of Pai et al.[36] and by Sachsenheimer and Schulz[35] on the binding of MnATP (see above), and by the ^{1}H NMR studies of Kalbitzer et al.[37] on the binding of Ap_5A.

Finally, these results provide information as to the reaction mechanism of the adenylate

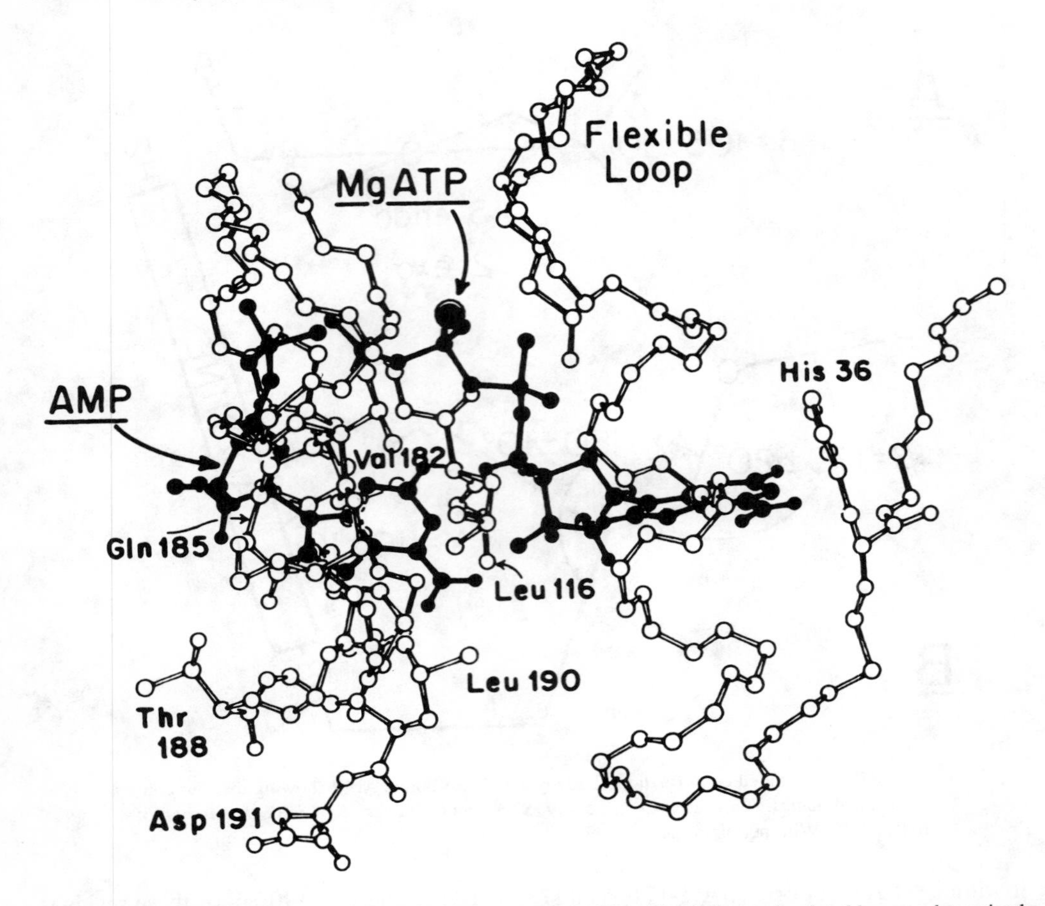

FIGURE 10. ORTEP drawing showing the binding sites for AMP and MgATP on adenylate kinase as determined by "NMR-docking." The MgATP binding site was taken from Fry et al.[67] The AMP site was located with an Evans and Sutherland PS 300 computer graphics system, based on the conformation of the enzyme-bound AMP, the intersubstrate distances (see Figures 8 and 9), and the intermolecular nuclear Overhauser effects from protons of the enzyme to those of AMP. (From Fry, D. C., Kuby, S. A., and Mildvan, A. S., *Biochemistry,* 26, 1645, 1987. With permission.)

kinase-catalyzed reaction. Similar to carbon chemistry, nucleophilic displacements on phosphorous have been grouped into two classes of mechanisms, the dissociative or S_N1 mechanism (Equation 2), and the associative mechanism, which, in its extreme form, is S_N2 (Equation 3).[95,102]

X — P(=O)(O⁻)(O⁻) → X: + PO₃⁻ (P, O, O⁻, O) —:Y→ Y — P(=O)(O⁻)(O⁻) (2)

X — P(=O)(O⁻)(O⁻) + :Y → X — P(O⁻)(O⁻)(O) — Y → X :+ P(O⁻)(O⁻)(=O) — Y (3)

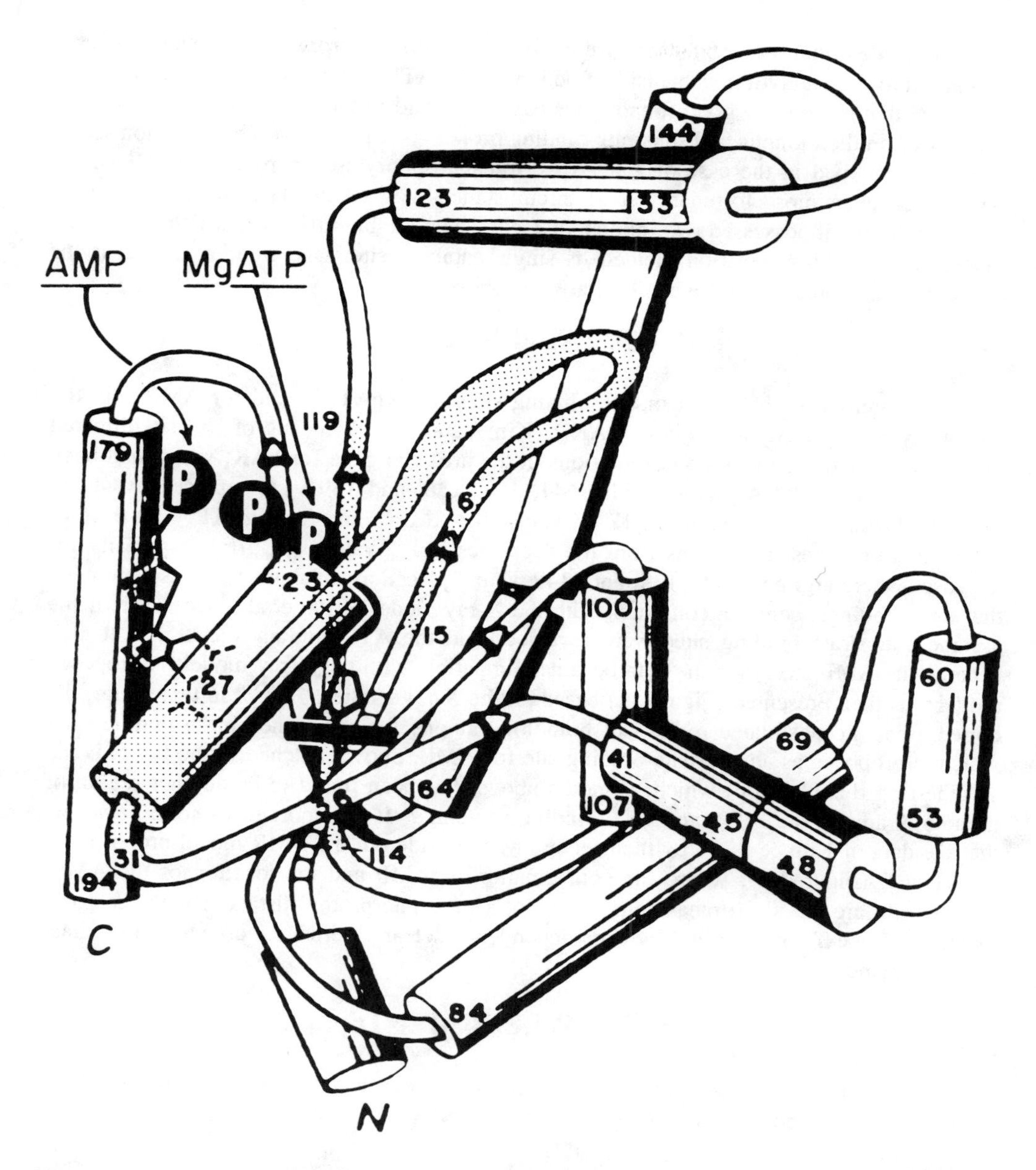

FIGURE 11. Simplified representation of the crystal structure of the entire adenylate kinase molecule showing the location of bound AMP and bound MgATP in the ternary enzyme complex, as determined by "NMR-docking." The stippling shows regions of amino acid sequence homology with other nucleotide binding enzymes.[44] (From Fry, D. C., Kuby, S. A., and Mildvan, A. S., *Biochemistry,* 26, 1645, 1987. With permission.)

One criterion for distinguishing between these mechanisms is the intersubstrate distance on the enzyme. For an associative mechanism, a molecular contact distance of $r \leq 3.3$ Å between the entering atom Y and the phosphoryl phosphorus might be expected, whereas a dissociative mechanism would require a greater distance, i.e., $r \geq 4.9$ Å, to allow space for the intermediate monomeric metaphosphate to exist between the leaving group X and the entering group Y. Thus, a reaction coordinate distance of $\geq 3 \pm 1$ Å between the entering oxygen of AMP and the γ-phosphorus of bound metal ATP, derived from the positioning of AMP in the ternary complex as shown in Figure 11,[86] (refer also to the discussion in Reference 78) lends support to the idea of an associative mechanism for adenylate kinase.

It is evident that a combination of the X-ray and NMR approaches may yield valuable insights into the catalytic mechanism of action of the ATP-transphosphorylases, provided, however, that reasonably good foundations have been laid in their most fundamental properties, i.e., in their kinetic and substrate-binding properties. Such a favorable situation seems to have developed in the case of ATP-AMP transphosphorylase or myokinase. However, perhaps an even more fortunate set of circumstances had taken place in the myokinase molecule in that it possessed two, almost discrete, substrate-binding sites, which after being placed in proper juxtaposition yielded its single catalytic site. One wonders if that is the one common property of all the ATP-transphosphorylases.

IV. APPENDIX

In an attempt to shed light on the binding sites for $MgATP^{2-}$ and for AMP^{2-} in the human cytosolic adenylate kinase (hAK_1), Kim et al.,[117] by means of the site-directed mutagenesis method, made point mutations in the artificial gene for hAK_1 which had been synthesized earlier.[109] Arginine residues 44, 132, 138, and 149 were replaced by alanyl residues. From the increases in the $K_{m,app}$ values for AMP^{2-} and for $MgATP^{2-}$ for each of the mutant enzymes, as well as from the decreases in k_{cat} values (V_{max}/E_t), especially for the cases where arginine-132, -138, and -149 were replaced by an alanine, it was concluded that their findings were not consistent with the X-ray model of Pai et al.[36] However, if the assigned substrate binding sites were reversed, and the ATP-binding site of Pai et al.[36] became the AMP-site, then the data became compatible with the X-ray model. A proposed model was then presented (Figure 12) based on the X-ray studies.[35,118,119] In this figure, the dotted areas are postulated regions of homology among the adenylate kinase families and which is proposed as an alternate binding site for AMP. Shown hatched in the figure is the glycine-rich flexible loop, which, as noted above, has been found to be another region of homology among the ATP- and GTP-binding proteins.[44] It is important to note, however, that the data of Kim et al.,[117] by themselves, will not rule out the NMR model proposed by Fry et al.[86] (Figure 11), where the conformation of AMP and the location of the AMP-binding site are based on measurements of intra- and interproton distances with the paramagnetic probe-T_1 method and by time-dependent nuclear Overhauser effects in the intact native enzyme.

ACKNOWLEDGMENTS

We thank Drs. Danchin and Saint-Girons for their preprints. Research in Stephen Kuby's laboratory was supported, in part, by a grant from NIH, DK 07824 (formerly, AM07824).

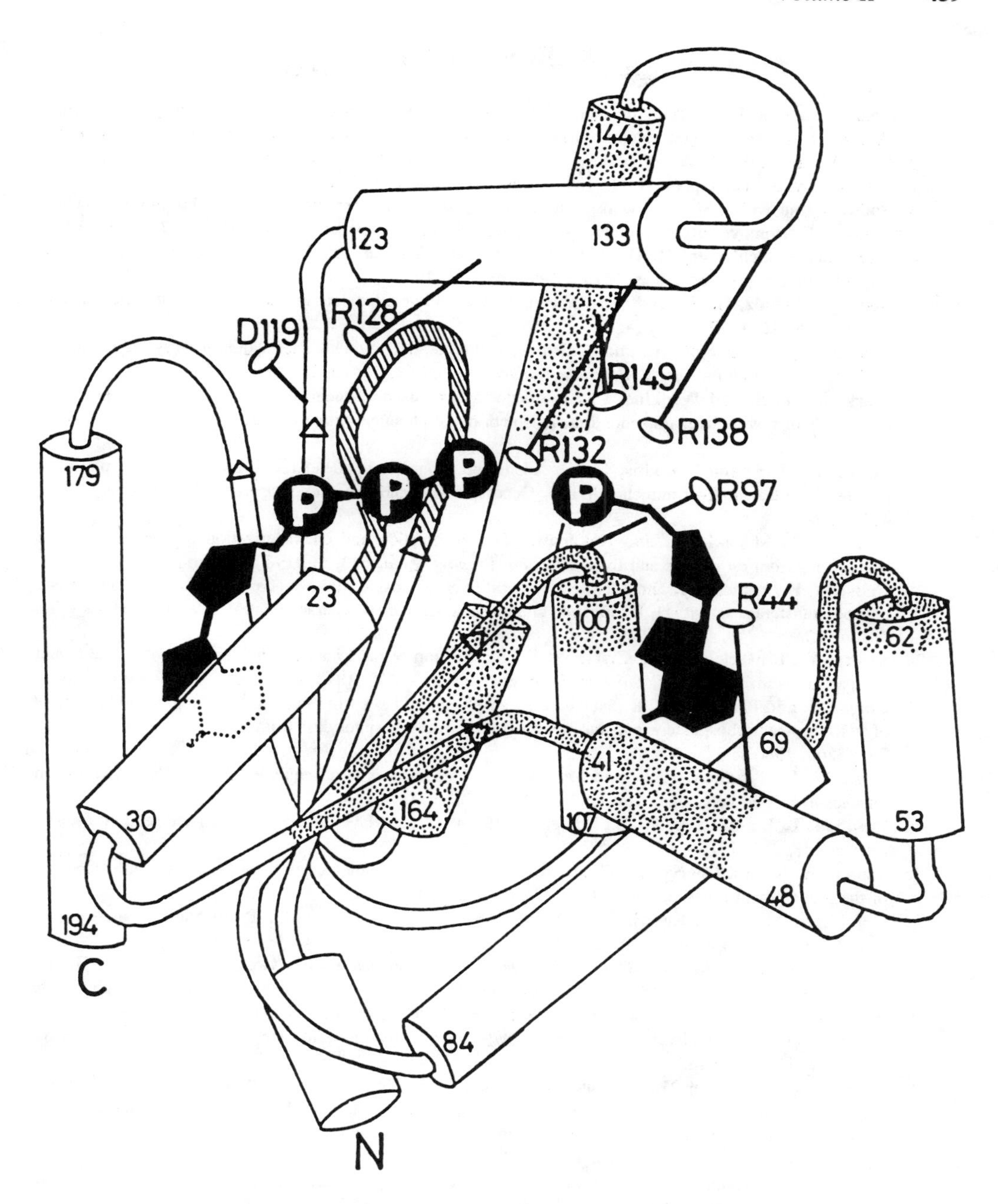

FIGURE 12. A proposed model of the binding sites for AMP and MgATP in adenylate kinase, constructed by interchanging the substrate binding sites of Pai et al.[36] (see Figure 3) plus some modifications. The model is presented in a schematic form of crystal form A (the substrate-bound form of AK_1) of porcine adenylate kinase, taken in modified form from the X-ray studies of Sachsenheimer and Schulz.[35] The side chains of Arg-44, -97, -128, -132, -138, and -149 are depicted. The direction of the arginine residues, replaced by alanyl residues in the mutant enzymes, is based on the refined X-ray structure of porcine AK_1 at 2.1-Å resolution.[118,119] The dotted areas indicate the postulated regions of a high degree of homology among the AK families and which may constitute the substrate-binding site for AMP; the glycine-rich flexible loop, which is also a region of a high degree of homology, is shown as hatched in the figure (see Figure 7 and Plate 1, Scheme 3, and References 44, 67, and 68). (From Kim, H. J., Nishikawa, S., Tokutomi, Y., Takenaka, H., Hamada, M., Kuby, S. A., and Uesugi, S., *Biochemistry,* 29, 1107, 1990. With permission.)

REFERENCES

1. **Noda, L. and Kuby, S. A.,** Adenosine triphosphate-adenosine monophosphate transphosphorylase (myokinase). I. Isolation of the crystalline enzyme from rabbit skeletal muscle, *J. Biol. Chem.,* 226, 541, 1957.
2. **Noda, L. and Kuby, S. A.,** Myokinase (ATP-AMP transphosphorylase), in *Methods in Enzymology,* Vol. 6, Colowick, S. and Kaplan, N. O., Eds., Academic Press, New York, 1963, 223.
3. **Noda, L. and Kuby, S. A.,** Adenosine triphosphate-adenosine monophosphate transphosphorylases (Myokinase). II. Homogeneity measurements and physiochemical properties, *J. Biol. Chem.,* 226, 551, 1957.
4. **Schirmer, I., Schirmer, R. H., Schulz, G. E., and Thuma, E.,** Purification, characterization and crystallization of pork myokinase, *FEBS Lett.,* 10, 333, 1970.
5. **Noda, L., Schulz, G. E., and von Zabern, I.,** Crystalline adenylate kinase from carp muscle, *Eur. J. Biochem.,* 51, 229, 1975.
6. **Thuma, E., Schirmer, R. H., and Schirmer, I.,** Preparation and characterization of a crystalline human ATP: AMP phosphotransferase, *Biochim. Biophys. Acta,* 268, 81, 1972.
7. **Markland, F. S. and Wadkins, C. L.,** Adenosine triphosphate-adenosine 5′-monophosphate phosphotransferase of bovine liver mitochondria. I. Isolation and chemical properties, *J. Biol. Chem.,* 241, 4124, 1966.
8. **Markland, F. S. and Wadkins, C. L.,** Adenosine triphosphate-adenosine 5′-monophosphate phosphotransferase of bovine liver mitochondria. II. General kinetic and structural properties, *J. Biol. Chem.,* 241, 4136, 1966.
9. **Tamura, T., Shiraki, H., and Nakagawa, H.,** Purification and characterization of adenylate kinase isoenzymes from rat muscle and liver, *Biochim. Biophys. Acta,* 612, 56, 1980.
10. **Criss, W. E., Sapico, V., and Litwack, G.,** Rat liver adenosine triphosphate: adenosine monophosphate phosphotransferase activity. I. Purification and physical and kinetic characterization of adenylate kinase III, *J. Biol. Chem.,* 245, 6346, 1970.
11. **Sapico, V., Litwack, G., and Criss, W. E.,** Purification of rat liver adenylate kinase isoenzyme II and comparison with isozyme III, *Biochim. Biophys. Acta,* 258, 436, 1972.
12. **Chiga, M. and Plaut, G. W. E.,** Nucleotide transphosphorylases from liver. I. Purification and properties of an adenosine triphosphate-adenosine monophosphate transphosphorylase from swine liver, *J. Biol. Chem.,* 235, 3260, 1960.
13. **Criss, W. E., Pradhan, T. J., and Morris, H. P.,** Physical protein regulation of adenylate kinase from muscle, liver, and hepatoma, *Cancer Res.,* 34, 3062, 1974.
14. **Pradhan, T. J. and Criss, W. E.,** Three major forms of adenylate kinase from adult and fetal rat tissues, *Enzyme,* 21, 337, 1976.
15. **Tsuboi, K. K. and Chervenka, C. H.,** Adenylate kinase of human erythrocyte. Isolation and properties of the predominant inherited form, *J. Biol. Chem.,* 250, 132, 1975.
16. **Chiu, C. S., Su, S., and Russell, P. J.,** Adenylate kinase from baker's yeast. I. Purification and intracellular location, *Biochim. Biophys. Acta,* 132, 361, 1967.
17. **Kuby, S. A., Hamada, M., Gerber, D., Tsai, W.-C., Jacobs, H. K., Cress, M. C., Chua, G. K., Fleming, G., Wu, L. H., Fischer, A. H., Frischat, A., and Maland, L.,** Studies on adenosine triphosphate transphosphorylases. Isolation and several properties of the crystalline calf transphosphorylases (adenylate kinase) from muscle and liver and some observations on the rabbit muscle adenylate kinase, *Arch. Biochem. Biophys.,* 187, 34, 1978.
18. **Hamada, M., Sumida, M., Okuda, H., Watanabe, T., Nojima, M., and Kuby, S. A.,** Adenosine triphosphate adenosine-5′-monophosphate phosphotransferase from normal human liver mitochondria. Isolation, chemical properties, immunochemical comparison with duchenne dystrophic serum aberrant adenylate kinase, *J. Biol. Chem.,* 257, 13120, 1982.
19. **Itakura, T., Watanabe, K., Shiokawa, H., and Kubo, S.,** Purification and characterization of acidic adenylate kinase in porcine heart, *Eur. J. Biochem.,* 82, 431, 1978.
20. **Kuby, S. A., Fleming, G., Frischat, A., Cress, M. C., and Hamada, M.,** Studies on adenosine triphosphate transphosphorylases. Human isoenzymes of adenylate kinase: isolation and physicochemical comparison of the crystalline human ATP-AMP transphosphorylases from muscle and liver, *J. Biol. Chem.,* 258, 1901, 1983.

20a. **Kuby, S. A. and Fleming, G.,** unpublished observations, 1986.

21. **Hamada, M., Sumida, M., Kurokawa, Y., Sunayashiki-Kusuzaki, K., Okuda, H., Watanabe, T., and Kuby, S. A.,** Studies on the adenylate kinase isoenzymes from the serum and erythrocyte of normal and duchenne dystrophic patients. Isolation, physicochemical properties and several comparisons with the duchenne dystrophic aberrant enzyme, *J. Biol. Chem.,* 260, 11595, 1985.
22. **Frank, R., Trosin, M., Tomasselli, A. G., Noda, L., Siegel, R. L., and Schirmer, R. H.,** Mitochondrial adenylate kinase (AK_2) from bovine heart. The complete primary structure, *Eur. J. Biochem.,* 154, 205, 1986.

23. **Dunaway-Mariano, D. and Cleland, W. W.**, Investigations of substrate specificity and reaction mechanism of several kinases using chromium (III) adenosine 5′-triphosphate and chromium (III) adenosine 5′-diphosphate, *Biochemistry,* 19, 1506, 1980.
24. **Fildes, R. A. and Harris, H.**, Genetically determined variation of adenylate kinase in man, *Nature,* 209, 261, 1966.
25. **Giblet, E. R.**, in *Genetic Markers in Human Blood,* F. A. Davis, Philadelphia, 1969, 512.
26. **Brownson, C. and Spencer, N.**, Partial purification and properties of the two common inherited forms of human erythrocyte adenylate kinase, *Biochem. J.,* 130, 797, 1972.
27. **Brownson, C. and Spencer, N.**, Kinetic studies on the two common inherited forms of human erythrocyte adenylate kinase, *Biochem. J.,* 130, 805, 1972.
28. **Von Zabern, I., Wittmann-Liebold, B., Untucht-Grau, R., Schirmer, R. H., and Pai, E. F.**, Primary and tertiary structure of the principal human adenylate kinase, *Eur. J. Biochem.,* 68, 281, 1976.
29. **Klethi, J. and Mandel, P.**, Tissue determined variations of adenylate kinase, *Nature,* 218, 467, 1968.
30. **Russell, P. J., Horenstein, J. M., Goins, L., Jones, S., and Laver, M.**, Adenylate kinase in human tissues. I. Organ specificity of adenylate kinase isoenzymes, *J. Biol. Chem.,* 249, 1874, 1974.
31. **Hamada, M., Takenaka, H., Fukumoto, K., Fukamachi, S., Yamaguchi, T., Sumida, M., Shiosaka, T., Kurokawa, Y., Okuda, H., and Kuby, S. A.**, Structure and function of adenylate kinase isoenzymes in normal humans and muscular dystrophy patients, in *Isoenzymes,* Vol. 16, Rattazzi, M. C., Scandalios, J. G., and Whitt, G. S., Eds., Alan R. Liss, New York, 1987, 81.
32. **Kuby, S. A., Fleming, G., Frischat, A., Cress, M. C., Hamada, M., Sumida, M., Okuda, H., Watanabe, T., and Nojima, M.**, Human isoenzymes of adenylate kinase: isolation and physicochemical comparison of the normal human enzymes with the aberrant adenylate kinase from duchenne dystrophic serum, *Fed. Proc., Fed. Am. Soc. Exp. Biol.,* 41, 903, 1982.
33. **Heil, S., Müller, G., Noda, L., Pinder, T., Schirmer, R. H., Schirmer, I., and Von Zabern, I.**, The amino acid sequence of porcine adenylate kinase from skeletal muscle, *Eur. J. Biochem.,* 43, 131, 1974.
34. **Schulz, G. H., Elzinga, M., Marx, F., and Schirmer, R. H.**, Three-dimensional structure of adenylate kinase, *Nature,* 250, 120, 1974.
35. **Sachsenheimer, W. and Schulz, G. E.**, Two conformations of crystalline adenylate kinase, *J. Mol. Biol.,* 114, 23, 1977.
36. **Pai, E. F., Sachsenheimer, W., Schirmer, R. H., and Schulz, G. E.**, Substrate positions and induced-fit in crystalline adenylate kinase, *J. Mol. Biol.,* 114, 37, 1977.
37. **Kalbitzer, H. R., Marquetant, R., Rösch, P., Schirmer, R. H.**, Structural isomerization of human-muscle adenylate kinase as studied by ^{1}H nuclear magnetic resonance, *Eur. J. Biochem.,* 126, 531, 1982.
38. **Smith, G. M. and Mildvan, A. S.**, Nuclear magnetic resonance studies of the nucleotide binding sites of porcine adenylate kinase, *Biochemistry,* 21, 6119, 1982.
39. **Frank, R., Trosin, M., Tomasselli, A. G., Schulz, G. E., and Schirmer, R. H.**, Mitochondrial adenylate kinase (AK_2) from bovine heart. Homology with the cytosolic isoenzyme in the catalytic region, *Eur. J. Biochem.,* 141, 629, 1984.
40. **Saint-Girons, I., Gilles, A.-M., Margarita, D., Michelson, S., Monnot, M., Fermandjian, S., Danchin, A., and Barzu, O.**, Structural and catalytic characteristics of *E. coli* adenylate kinase, *J. Biol. Chem.,* 262, 622, 1987.
41. **Brune, M., Schumann, R., and Wittinghofer, F.**, Cloning and sequencing of the adenylate kinase gene (adk) of *Escherichia coli, Nucl. Acids. Res.,* 13, 7139, 1985.
42. **Gilles, A.-M., Saint-Girons, I., Monnot, M., Fermandjian, S., Michelson, S., and Barzu, O.**, Substitution of a serine residue for proline-87 reduces catalytic activity and increases susceptibility to proteolysis of *Escherichia coli* adenylate kinase, *Proc. Natl. Acad. Sci. U.S.A.,* 83, 5798, 1986.
43. **Kishi, F., Maruyama, M., Tanizawa, Y., and Nakazawa, A.**, Isolation and characterization of cDNA for chicken muscle adenylate kinase, *J. Biol. Chem.,* 261, 2942, 1986.
44. **Fry, D. C., Kuby, S. A., and Mildvan, A. S.**, NMR studies of the MgATP binding site of adenylate kinase and of a 45 residue peptide fragment of the enzyme, *Proc. Natl. Acad. Sci. U.S.A.,* 83, 907, 1986.
45. **Hamada, M. and Kuby, S. A.**, Studies on adenosine triphosphate transphosphorylases. XIII. Kinetic properties of the crystalline rabbit muscle ATP-AMP transphosphorylase (adenylate kinase) and a comparison with the crystalline calf muscle and liver adenylate kinases, *Arch. Biochem. Biophys.,* 190, 772, 1978.
46. **Hamada, M., Palmieri, R. H., Russell, G. A., and Kuby, S. A.**, Studies on adenosine triphosphate transphosphorylases. XIV. Equilibrium binding properties of the crystalline rabbit and calf muscle ATP-AMP transphosphorylase (adenylate kinase) and derived peptide fragments, *Arch. Biochem. Biophys.,* 195, 155, 1979.
47. **Kuby, S. A., Hamada, M., Palmieri, R. H., Tsai, W.-C., Jacobs, H. K., Maland, L., Wu, L., and Fischer, A.**, ATP-AMP transphosphorylase (myokinase) from rabbit and calf muscle-structure and equilibrium binding, *Fed. Proc. Fed. Am. Soc. Exp. Biol.,* 35 (Abstr.), 1629, 1976.
48. **McDonald, G. C., Cohn, M., and Noda, L.**, Proton magnetic resonance spectra of porcine muscle adenylate kinase and substrate complexes, *J. Biol. Chem.,* 241, 4136, 1975.

49. **Frank, R., Trosin, M., Tomasselli, A. G., Noda, L., Krauth-Siegel, R. L., and Schirmer, R. H.,** Mitochondrial adenylate kinase (AK_2) from bovine heart, the complete primary structure, *Eur. J. Biochem.*, 154, 205—211, 1986.
50. **Frank, R., Trosin, M., Tomasselli, A. G., Schulz, G. E., and Schirmer, R. H.,** Mitochondrial adenylate kinase (AK_2) from bovine heart. Homology with the cytosolic enzyme in the catalytic region, *Eur. J. Biochem.*, 141, 629, 1984.
51. **Kuby, S. A., Palmieri, R. H., Frischat, A., Fischer, A. H., Wu, L. H., Maland, L., and Manship, M.,** Studies on adenosine triphosphate transphosphorylases. XVI. Amino acid sequence of rabbit muscle ATP-AMP transphosphorylase (adenylate kinase), *Biochemistry,* 23, 2393, 1984.
52. **Mahowald, T. A., Noltmann, E. A., and Kuby, S. A.,** Studies on adenosine triphosphate transphosphorylases. I. Amino acid composition of adenosine triphosphate-adenosine 5′-phosphate transphosphorylase (myokinase), *J. Biol. Chem.*, 239, 1138, 1962.
53. **Olson, O. E. and Kuby, S. A.,** Studies on adenosine triphosphate transphosphorylases. V. Carboxyl-terminal sequences of adenosine triphosphate-creatine transphosphorylase and of adenosine 5′-phosphate transphosphorylase (myokinase), *J. Biol. Chem.*, 239, 460, 1953.
54. **Wieland, B., Tomasselli, A. G., Noda, L., Frank, R., and Schulz, G. E.,** The amino acid sequence of GTP:AMP phosphotransferase from beef-heart mitochondria. Extensive homology with cytosolic adenylate kinase, *Eur. J. Biochem.*, 143, 331, 1984.
55. **Hamada, M., Okuda, H., Oka, K., Watanabe, T., Ueda, K., Nojima, M., and Kuby, S. A., Manship, M., Tyler, F. H., and Ziter, F. A.,** An aberrant adenylate kinase isoenzyme from the serum of patients with duchenne muscular dystrophy, *Biochem. Biophys. Acta,* 660, 227, 1981.
56. **Rhodes, D. G. and Lowenstein, J. M.,** Initial velocity and equilibrium kinetics of myokinase, *J. Biol. Chem.*, 243, 3963, 1968.
57. **Noda, L.,** Nucleotide triphosphate-nucleoside monophosphokinases, in *Enzymes,* Vol. 6, 2nd ed., Boyer, P., Lardy, H., and Myrbäck, K., Eds., Academic Press, New York, 1962, 139.
58. **Noda, L.,** Adenylate kinase in *Enzymes,* Vol. 8, 3rd ed., Boyer, P. D., Ed., 1973, 279.
59. **Krohne-Ehrich, G., Schirmer, R. H., and Untucht-Grau, R.,** Glutathione reductase from human erythrocytes. Isolation of the enzyme and sequence analysis of the redox-active peptide, *Eur. J. Biochem.*, 80, 65, 1977.
60. **Mannervik, B., Axelsson, K. A., Sundewall, A. C., Holmgren, A.,** Relative contributions of thioltransferase- and thioredoxin-dependent systems in reduction of low-molecular-mass and protein disulfides, *Biochem. J.*, 213, 519, 1983.
61. **Watanabe, K., Kinoshita, T., Kawai, N., Tashiro, N., Mori, T., Kubo, S., and Yamamoto, S.,** Adenylate kinase from rat liver — molecular properties and structural comparison with yeast enzyme, *Jpn. J. Vet. Sci.*, 47, 63, 1985.
62. **Hampton, A., Harper, P. J., and Sasaki, T.,** Evidence for the conformation of enzyme-bound adenosine 5′-phosphate. Substrate and inhibitor properties of 8,5′-cycloadenosine 5′-phosphate with adenylate kinase, adenylate aminohydrolase, adenylosuccinate lyase, and 5′-nucleotidase, *Biochemistry,* 11, 4965, 1972.
63. **Schulz, G. E., Biedermann, K., Kabsch, W., and Schirmer, R. H.,** Low resolution structure of adenylate kinase, *J. Mol. Biol.*, 80, 857, 1973.
64. **Schirmer, I., Schirmer, R. H., Schulz, G. E., and Tuma, E.,** Purification, characterization and crystallization of pork myokinase, *FEBS Lett.*, 10, 333, 1970.
65. **Vasavada, K. V., Kaplan, J. I., and Nageswara Rao, B. D.,** Analysis of ^{31}P NMR spectra of enzyme-bound reactants and products of adenylate kinase using density matrix theory of chemical exchange, *Biochemistry,* 23, 961, 1984.
66. **Rex Sheu, K.-F. and Frey, P. A.,** Enzymatic ^{31}P nuclear magnetic resonance study of adenylate kinase-catalyzed stereospecific phosphorylation of adenosine 5′-phosphorothioate, *J. Biol. Chem.*, 252, 4445, 1977.
67. **Fry, D. C., Kuby, S. A., and Mildvan, A. S.,** NMR studies of the MgATP binding site of adenylate kinase and of a 45 residue fragment of the enzyme, *Biochemistry,* 24, 4680, 1985.
68. **Walker, J. E., Saraste, M., Runswick, M. J., and Gay, N. J.,** Distantly related sequences in the α and β-subunits of ATP synthase, myosin, kinases and other ATP-requiring enzymes and a common nucleotide binding fold, *EMBO J.*, 1, 945, 1982.

68a. **Gay, N. J. and Walker, J. E.,** Homology between human bladder carcinoma oncogene product and mitochondrial ATP-synthase, *Nature,* 301, 262, 1983.

69. **Nelson, N. C., Zoeller, M. J., and Taylor, S. S.,** Mapping of the ATP binding site of catalytic subunit from cAMP dependent protein kinase, *Fed. Proc., Fed. Am. Soc. Exp. Biol.*, 40 (Abstr.), 1609, 1981.
70. **Hashimoto, E., Takio, K., and Krebs, E. G.,** Amino acid sequence at the ATP-binding site of cGMP-dependent protein kinase, *J. Biol. Chem.*, 257, 727, 1982.
71. **Putney, S., Herlihy, W., Royal, N., Pang, H., Aposhian, H. V., Pickering, L., Belagaje, R., Biemann, K., Page, D., Kuby, S. A., and Schimmel, P.,** Rabbit muscle creatine phosphokinase: cDNA cloning, primary structure, and detection of human homologues, *J. Biol. Chem.*, 259, 14317, 1984.

72. **O'Sullivan, W. J. and Noda, L.,** Magnetic resonance and kinetic studies related to the manganese activation of the adenylate kinase reaction, *J. Biol. Chem.*, 243, 1424, 1968.
73. **Rossman, M. G., Liljas, A., Branden, C. I., and Banaszak, L. J.,** Evolutionary and structural relationships among dehydrogenases, *The Enzymes*, Vol. 11, 3rd ed., Boyer, P. D., Ed., Academic Press, New York, 1975, 61.
74. **Wierenga, R. K., DeMaeyer, M. C. H., and Hol, W. G. J.,** Interaction of pyrophosphate moieties with α-helices in dinucleotide binding proteins, *Biochemistry*, 24, 1346, 1985.
75. **Richardson, J. S.,** The anatomy and taxonomy of protein structure, *Adv. Protein. Chem.*, 34, 167, 1981.
76. **Cousin, D. and Suttin, G.,** Mutants Thermosensibles D'Escherichia Coli K12 III.-Une Mutation Létale D'E. Coli Affectant L' Activité De L' Adénylate-Kinase, *Ann. Inst. Pasteur*, 117, 612, 1969.
77. **Esmon, B. D., Kensil, C. R., Cheng, C.-C. H., and Glaser, M.,** Genetic Analysis of *Escherichia coli* Mutants Defective in Adenylate Kinase and sn-Glycerol 3-Phosphate Acyltransferase, *J. Bacteriol.*, 141, 405, 1980.
78. **Mildvan, A. S. and Fry, D. C.,** NMR Studies of the Mechanism of Enzyme Action, *Advances in Enzymology*, and Related Areas of Molecular Biology, Vol. 59, Meister, Ed., John Wiley & Sons, 1987, 241.
79. **Monnot, M., Gilles, A.-M., Saint-Girons, I., Michelson, S., Bârzu, O., and Fermandjian, S.,** Circular dichroism investigation of *Escherichia coli* adenylate kinase, *J. Biol. Chem.*, 262, 2502, 1987.
80. **Noda, L.,** Adenosine triphosphate-adenosine monophosphate transphosphorylase. III. Kinetic studies, *J. Biol. Chem.*, 232, 237, 1958.
81. **Su, S. and Russell, P. J.,** Adenylate kinase from baker's yeast. III. Equilibrium exchange and mechanism, *J. Biol. Chem.*, 243, 3826, 1968.
82. **Rao, B. D. N., Cohn, M., and Noda, L.,** Differentiation of nucleotide binding sites and role of metal ion in the adenylate kinase reaction by ^{31}P NMR. Equilibria, interconversion rates, and NMR parameters of bound substrates, *J. Biol. Chem.*, 253, 1149, 1978.
83. **Kuby, S. A., Mahowald, T. A., and Noltmann, E. A.,** Studies on adenosine triphosphate transphosphorylases. IV. Enzyme-substrate interactions, *Biochemistry*, 1, 748, 1962.
84. **Secrist, J. A., III, Barrio, J. R., Leonard, N. J., and Weber, G.,** Fluorescent modification of adenosine-containing coenzymes. Biological activities and spectroscopic properties, *Biochemistry*, 11, 3499, 1972.
85. **McDonald, G. C., Cohn, M., and Noda, L.,** Proton magnetic resonance spectra of porcine muscle adenylate kinase and substrate complexes, *J. Biol. Chem.*, 250, 6947, 1975.
86. **Fry, D. C., Kuby, S. A., and Mildvan, A. S.,** NMR studies of the AMP-binding site and mechanism of adenylate kinase, *Biochemistry*, 26, 1645, 1987.
87. **Callaghan, O. H. and Weber, G.,** Kinetic studies on rabbit myokinase, *Biochem. J.*, 73, 473, 1959.
88. **Kuby, S. A. and Noltmann, E. A.,** ATP-creatine transphosphorylase, in *The Enzymes*, Vol. 6, 2nd ed., Boyer, P. D., Lardy, H., and Myrbäck, K., Eds., Academic Press, New York, 1962, 515.
89. **Price, N. C., Reed, G. H., and Cohn, M.,** Magnetic resonance studies of substrate and inhibitor binding to porcine muscle adenylate kinase, *Biochemistry*, 12, 3322, 1973.
90. **Kuby, S. A., Hamada, M., Johnson, M. S., Russell, G. A., Manship, M., Palmieri, R. H., Bredt, D. S., and Mildvan, A. S.,** Studies on adenosine triphosphate transphosphorylases. XVIII. Synthesis and preparation of peptides and peptide fragments of rabbit muscle ATP-AMP transphosphorylase (adenylate kinase) and their nucleotide-binding properties, *J. Protein Chem.*, 8, 549, 1989.
91. **Fry, D. C., Byler, D. M., Susi, H., Brown, E. M., Kuby, S. A., and Mildvan, A. S.,** Solution structure of the 45-residue ATP-binding peptide of adenylate kinase as determined by 2-D NMR, FTIR, and CD spectroscopy, *Fed. Proc. (ASBC/DBC,ACS) Abstr.*, 45, 1517, 1986.
92. **Anderson, C. M., Zucker, F. H., and Steitz, T. A.,** Space-filling models of kinase clefts and conformation changes. Comparison of the surface-structures of kinase enzymes implicates closing in their mechanism, *Science*, 204, 375, 1979.
93. **Susi, H. and Byler, D. M.,** Protein structure by fourier transform infrared spectroscopy: second derivative structure, *Biochem. Biophys. Res. Commun.*, 115, 391, 1983.
94. **Koshland, D. E., Jr.,** Application of a theory of enzyme specificity to protein synthesis, *Proc. Natl. Acad. Sci. U.S.A.*, 44, 98, 1958.
95. **Jencks, W. P.,** Binding energy, specificity, and enzymic catalysis: the circe effect, in *Advances in Enzymology*, Vol. 43, John Wiley & Sons, New York, 1975, 219.
96. **Schirmer, R. H., Schirmer, I., and Noda, L.,** Studies of histidine residues of rabbit muscle myokinase, *Biochim. Biophys. Acta*, 207, 165, 1970.
97. **Price, N. C.,** The sulfhydryl groups of porcine muscle adenylate kinase, *Fed. Proc., Fed. Am. Soc. Exp. Biol.*, 31 (Abstr.), 501, 1972.
98. **Crivellone, M. D., Hermodson, M., and Axelrod, B.,** Inactivation of muscle adenylate kinase by site-specific destruction of tyrosine 95 using potassium ferrate, *J. Biol. Chem.*, 260, 2657, 1985.

99. **Watson, H. C., Walker, N. P. C., Shaw, P. J., Bryant, T. N., Wendell, P. L., Fothergill, L. A., Perkins, R. E., Conroy, S. C., Dobson, M. J., Tuite, M. F., Kingsman, A. J., and Kingsman, S. M.,** Sequence and structure of yeast phosphoglycerate kinase, *EMBO J.*, 1, 1635, 1982.
100. **Banks, R. D., Blake, C. C. F., Evans, P. R., Haser, R., Rice, D. W., Hardy, G. W., Merrett, M., and Phillips, A. W.,** Structure and activity of phosphoglycerate kinase: a possible hinge-binding enzyme, *Nature*, 279, 773, 1979.
101. **de Leeuw, H. P. M., Haasnoot, A. C., and Altana, C.,** Empirical correlations between conformational parameters in β-D-furanoside fragments derived from a statistical survey of crystal structures of nucleic acid constituents. Full description of nucleoside molecular geometries in terms of four parameters, *Isr. J. Chem.*, 20, 108, 1980.
102. **Mildvan, A. S.,** The role of metals in the mechanisms of enzyme-catalyzed nucleophilic substitutions at each of the three phosphorus atoms of ATP, in *NMR and Biochemistry*, Opella, S. J. and Lu, P., Eds., Marcel Dekker, New York, 1979, 301.
103. **Tagaya, M., Yagami, T., and Fukui, T.,** Affinity labeling of adenylate kinase with adenosine diphosphopyridoxal. Presence of Lys^{21} in the ATP-binding site, *J. Biol. Chem.*, 262, 8257, 1987.
104. **Egner, U., Tomasseli, A. G., and Schulz, G. E.,** Structure of the complex of yeast adenylate kinase with the inhibitor P^1, P^5-di(adenosine-5′) pentaphosphate at 2.6 Å resolution, *J. Mol. Biol.*, 195, 649, 1987.
105. **Dreusicke, D. and Schulz, G. E.,** The glycine-rich loop of adenylate kinase forms a giant anion hole, *FEBS Lett.*, 208, 301, 1986.
106. **Schulz, G. E., Schiltz, E., Tomasselli, A. G., Frank, R., Brune, M., Wittinghofer, A., and Schirmer, R. H.,** Structural relationships in the adenylate kinase family, *Eur. J. Biochem.*, 161, 127, 1986.
107. **Fry, D. C., Byler, D. M., Susi, H., Brown, E. M., Kuby, S. A., and Mildvan, A. S.,** Solution structure of the 45-residue MgATP-binding peptide of adenylate kinase as examined by 2-D NMR, FTIR, and CD spectroscopy, *Biochemistry*, 27, 3588, 1988.
108. **Kishi, F., Tanizawa, Y., and Nakazawa, A.,** Isolation and characterization of two types of cDNA for mitochondrial adenylate kinase and their expression in *Escherichia coli*, *J. Biol. Chem.*, 262, 11785, 1987.
109. **Kim, H. J., Nishikawa, S., Tanaka, T., Uesugi, S., Takenaka, H., Hamada, M., and Kuby, S. A.,** Synthetic genes for human muscle-type adenylate kinase in *Escherichia coli*, *Protein Eng.*, 2, 379, 1989.
110. **Ray, B. D., Rösch, P. and Nageswara Rao, B. D.,** ^{31}P NMR Studies of the structure of cation-nucleotide complexes bound to porcine muscle adenylate kinase, *Biochemistry*, 27, 8669, 1988.
111. **Tagaya, M., Yagami, T., Noumi, T., Futai, M., Kishi, F., Nakazawa, A., and Fukui, T.,** Site-directed mutagenesis of pro-17 located in the glycine-rich region of adenylate kinase, *J. Biol. Chem.*, 264, 990, 1989.
112. **Tagaya, M., Yagami, T., and Fukui, T.,** Affinity labeling of adenylate kinase with adenosine diphosphopyridoxal. Presence of Lys^{21} in the ATP-binding site, *J. Biol. Chem.*, 262, 8257, 1987.
113. **Reinstein, J., Brune, M., and Wittinghofer, A.,** Mutations in the nucleotide binding loop of adenylate kinase of *Escherichia coli*, *Biochemistry*, 27, 4712, 1988.
114. **Kim, H. J., Tokutomi, Y., Nishikawa, S., Takenaka, H., Hamada, M., Kuby, S. A., and Uesugi, S.,** *In vitro* mutagenesis of arginine residues at the catalytic cleft of human adenylate kinase, 6th Int. Conference on Isozymes, Toyama, Japan, May 29 to June 1, 1989.
115. **Müeller, C. W. and Schulz, G. E.,** Structure of the complex of adenylate kinase from *Escherichia coli* with the inhibitor, P^1 P^5-bis(5′-adenosyl) pentaphosphate, *J. Mol. Biol.*, 202, 909, 1988.
116. **Rösch, P., Klaus, W., Auer, M., and Goody, R. S.,** Nucleotide and AP_5A complexes of porcine adenylate kinase: a ^1H and ^{19}F NMR study, *Biochemistry*, 28, 4318, 1989.
117. **Kim, H. J., Nishikawa, S., Tokutomi, Y., Takenaka, H., Hamada, M., Kuby, S. A., and Uesugi, S.,** *In vitro* mutagenesis studies at the arginine residues of adenylate kinase. A revised binding site for AMP in the X-ray deduced model, *Biochemistry*, 29, 1107, 1990.
118. **Dreusicke, D., Karplus, P. A., and Schulz, G. E.,** The fine structure of porcine cytosolic adenylate kinase at 2.1 Å resolution, *J. Mol. Biol.*, 199, 359, 1988.
119. **Dreusicke, D. and Schulz, G. E.,** The switch between two conformations of adenylate kinase, *J. Mol. Biol.*, 203, 1021, 1988.

Chapter 18

PHOSPHOFRUCTOKINASE AND FRUCTOSE 6-PHOSPHATE, 2-KINASE: FRUCTOSE 2,6-BISPHOSPHATASE

Kosaku Uyeda

TABLE OF CONTENTS

I. INTRODUCTION

Phosphofructokinase, "PFK" (ATP:D-fructose 6-phosphate 1-phosphotransferase, EC 2.6.1.11), catalyzes the phosphorylation of fructose 6-phosphate (F6P) in the presence of ATP to form fructose 1,6-bisphosphate (FBP) and ADP (Equation 1).

$$\text{Fructose } 6\text{–}P^{2-} + MgATP^{2-} \rightleftharpoons \text{Fructose } 1,6\text{–}P_2^{4-} + MgADP^{1-} + H^{+} \quad (1)$$

The enzyme activity is affected by a multitude of metabolites, and it plays a key regulatory role in glycolysis in nearly all cells. Since the structures and other properties of PFKs from several sources have been reviewed previously[1-4] this article will focus only on the reaction mechanism.

This review also includes the structure and kinetic properties of an interesting bifunctional enzyme, fructose-6-phosphate,2-kinase:fructose-2,6-bisphosphatase which synthesizes and degrades a newly discovered activator of phosphofructokinase, fructose-2,6-bisphosphate.

II. PHOSPHOFRUCTOKINASE, REACTION MECHANISM

Study of the reaction mechanism of PFK is complicated by the fact that the activity is affected by the substrates, products, and other effectors. To overcome these difficulties, investigators have used high pH, high concentrations of Mg, and noninhibitory substrates with varying degrees of success.

A. RABBIT MUSCLE PFK

Among various PFKs, rabbit muscle PFK is the most extensively studied, consequently this enzyme receives the most attention.

1. Initial Velocity Studies

The initial velocity studies with PFKs from yeast,[5] calf lens,[6] human skeletal muscle and erythrocytes,[7] and rabbit muscle[8] have shown parallel double reciprocal velocity plots suggesting that the reaction mechanism may be a ping-pong pathway in which the first product is released from the enzyme before the second substrate binds. Lowry and Passoneau,[9] however, had reported intersecting velocity patterns with partially purified sheep brain enzyme. These conflicting results were due partly to differences in the experimental conditions and assay methods employed by these investigators.

Uyeda[10] reinvestigated the initial rate of rabbit muscle PFK using fructose 1-P (F1P) as a substrate and found that the reciprocal plots were converging. Similar intersecting lines were obtained by Hanson et al.[11] who used a high concentration of Mg^{2+}. They demonstrated that the reaction in the reverse direction also yields a family of converging lines. These results clearly rule out the ping-pong mechanism for the PFK reaction and support a sequential mechanism. The parallel lines were observed because K_{ia} (5 μM) in the rate equation (Equation 2) is considerably smaller than K_a (20 μM) and K_b (21 μM) (Table 1).[11]

$$v = \frac{VAB}{K_{ia}K_b + K_{aB} + K_bA + AB} \quad (2)$$

2. Product Inhibition Studies

To address the question of whether the PFK reaction is a random or ordered mechanism, product inhibition studies were carried out (Table 2). Uyeda[10] found that ADP vs. ATP is competitive, but all others including ADP vs. F1P, FBP vs. F1P, and FBP vs. ATP are noncompetitive. The same inhibition patterns were obtained by Hanson et al.[11] using F6P

TABLE 1
Kinetic Constants of PFKs (μM)

Source	K_{F6P}	K_{ATP}	K_{ia}	Ref.
Rabbit Muscle				
Forward	70	27	2	12
	45	53	1	12
	21	20	5.1	11
	44	61		13
	K_{FBP}	K_{ADP}	K_{ia}	Ref.
Reverse	282	29	800	12
	420	20	69	11
Rat liver	K_{F6P}	K_{ATP}	K_{ia}	Ref.
Forward	77	111	12	15
	K_{FBP}	K_{ADP}	K_{ia}	Ref.
Reverse	427	140	1600	15
E. coli	K_{F6P}	K_{ADP}	K_{ia}	Ref.
Pfk2	32	20	90	16
Pfk2*	23	38	40	16

TABLE 2
Production Inhibition Patterns of Rabbit Muscle PFK

		Inhibition pattern	
Substrate	**Inhibitor**	**Ref. 13 and 14**	**Ref. 10 and 11**
F6P (F1P)	ADP	NC	NC
F6P (F1P)	FBP	C	NC
ATP	ADP	C	C
ATP	FBP	NC	NC

as the substrate. These results appear to favor an ordered mechanism. Kee and Griffin,[12] however, who studied the forward reaction by following H^+ production with a pH stat, found that the product inhibition patterns are the same as above except that FBP is a competitive inhibitor of F6P; therefore, they suggested a rapid equilibrium random mechanism with dead-end complexes including E·F6P·MgADP and E·MgATP·FBP. Bar-Tana and Cleland[13] also found that nucleotides are mutually competitive, as are sugar-phosphates, but nucleotides are noncompetitive with sugar phosphates and vice versa. The only difference between the random and the ordered mechanisms is the inhibition by FBP with respect to F6P. Part of the reason why FBP inhibition is difficult to determine is that this sugar-phosphate appears to have multiple binding sites on the enzyme and is a potent activator of PFK. More definitive evidence in favor of the random mechanism for this reaction was provided by the use of substrate analogs. Hanson et al.[11] found that D-arabinose 5-P, which is not a substrate, is competitive with F6P and FBP but noncompetitive with ADP and ATP. If it were an ordered mechanism, with ATP being the first substrate to bind, then arabinose 5-P would be an uncompetitive inhibitor of ATP rather than a noncompetitive as they found. Furthermore, they showed that GDP is a noncompetitive inhibitor of FBP, again consistent with a random mechanism.

Bar-Tana and Cleland[14] carried out more extensive investigations of the reaction mechanism using various substrate analogs. They observed that 1-deoxyfructose 6-P inhibits competitively both F6P in the forward reaction and FBP in the reverse reaction, and it is noncompetitive with adenine nucleotides. CrATP, which is a dead-end inhibitor, is a competitive inhibitor of MgATP or MgADP and a noncompetitive inhibitor of sugar-phosphates. Although these results support the random mechanism for the PFK reaction, these authors conclude that the assumption for rapid equilibrium reaction as originally suggested by Kee and Griffin[12] is not valid for the forward reaction. Quantitative analysis of the above inhibition and comparison with the initial velocity data indicate that the dissociation constants of F6P and ATP are only 20 to 30% of the Michaelis constants, suggesting that the dissociation of these substrates is slower than the V_{max}. However, since the rate of the reverse reaction is only 1/10 that of the forward reaction, it is in rapid equilibrium in the reverse direction. Furthermore, they believe that since the release of the substrates is slower than V_{max}, the reciprocal plots are slightly concave upward. In the region of the substrate concentrations where the kinetics are usually determined, it will appear parallel. Thus, there is overwhelming evidence now that the reaction mechanism for rabbit muscle PFK is random addition of the substrates and release of the products (Scheme I).

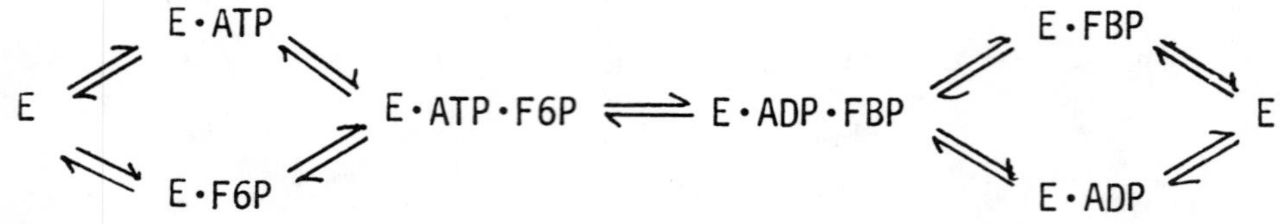

$$
\begin{array}{ccccccccc}
 & & E\cdot ATP & & & & E\cdot FBP & & \\
 & \rightleftharpoons & & \rightleftharpoons & & \rightleftharpoons & & \rightleftharpoons & \\
E & & & & E\cdot ATP\cdot F6P \rightleftharpoons E\cdot ADP\cdot FBP & & & & E \\
 & \rightleftharpoons & & \rightleftharpoons & & \rightleftharpoons & & \rightleftharpoons & \\
 & & E\cdot F6P & & & & E\cdot ADP & &
\end{array}
$$

Scheme I

3. Isotope Exchange Studies

In 1970, Uyeda[8] showed that rabbit muscle PFK catalyzes the isotope exchange reaction between ATP-ADP in the absence of F6P, and that this exchange activity is dependent on Mg^{2+}. He also observed an exchange reaction between F6P and FBP in the absence of adenine nucleotide. Hanson et al.[11] and Hume and Tipton[18] also demonstrated these isotope exchange reactions. These results tend to support a ping-pong mechanism and formation of a phosphoryl enzyme intermediate. However, the rates of these isotope exchange reactions are only 1% of the rate of the overall forward reaction indicating that it is not a part of the overall reaction pathway and that phosphoryl enzyme, if formed, is not an important reaction intermediate.

4. ATP and FBP Phosphohydrolase Activities

Rabbit muscle PFK catalyzes the hydrolysis of MgATP and FBP at rates 0.06 and 0.4% of the forward and the reverse reactions, respectively.[8] These hydrolytic activities are not due to any contaminating enzyme since both PFK and the hydrolases copurify, undergo the same degree of inactivation by various reagents, and have the same immunoreactivity.[8] The ATPase activity has been observed with other kinases such as hexokinase, and Zewe et al.[19] have suggested that this results from the action of H_2O occupying the active site of the enzyme where the 6-hydroxyl of glucose would normally bind to the enzyme. A similar explanation may hold for the ATPase activity of PFK in which the binding region for the 1-OH of F6P is occupied by an H_2O molecule.

B. RAT LIVER PFK

In contrast to rabbit muscle PFK, a different mechanism has been suggested for rat liver PFK by Brand and Soling.[15] They found that both sugar phosphates are competitive in both the forward and the reverse directions and that ADP vs. F6P and FBP vs. ATP are noncompetitive as is observed with rabbit muscle enzyme. ADP, however, is a noncompetitive

TABLE 3
Product Inhibition of *E. coli* PFKs[17]

Substrate	Inhibitor	pfk2	pfk2*
F6P	ADP	NC	NC
F6P	FDP	C	NC
ATP	ADP	NC	C
ATP	FDP	NC	NC

inhibitor of ATP at subsaturating and saturating concentrations of F6P, which is in contrast to the results observed with rabbit muscle PFK. Thus, the authors suggest that the mechanism is an ordered addition of substrates with F6P being the first substrate to bind to the enzyme followed by ATP; ADP is the first product to be released from the enzyme followed by FBP (Scheme II). Two additional lines of evidence seem to support this ordered mechanism: (1) FBP inhibition vs. ATP inhibition is released by F6P, and (2) the noncompetitive inhibition of ADP by F6P becomes uncompetitive with saturating concentration of ATP.

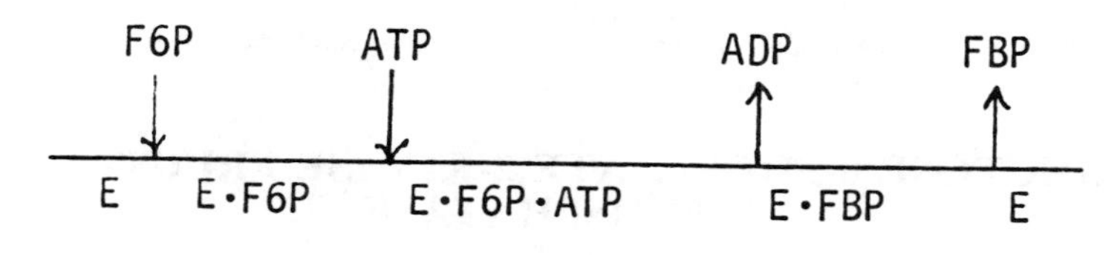

Scheme II

C. *E. COLI* PFK

More recently Campos et al.[16] investigated the kinetic mechanism of two isozymic forms of *E. coli* PFK termed pfk2 and pfk2*. Pfk2 is a minor isozymic form of PFK in the bacterium, and pfk1 is the major isozyme. Although pfk2* is usually described as a variant enzyme that is present only in certain *E. coli* mutants, it has been shown recently that its amino acid sequence is completely different from that of pfk2.[17] According to their results of initial velocity studies, the reaction catalyzed by both isozymes is sequential as are all other PFKs. Interestingly, however, the product inhibition patterns of pfk2 and pfk2* are different. For pfk2, FBP is a competitive inhibitor of F6P and a noncompetitive inhibitor of ATP (Table 3) while ADP is a noncompetitive inhibitor of F6P and ATP. They show also that adenylylimido-P forms a dead-end complex and that it is an uncompetitive inhibitor of F6P.

In contrast to pfk2, the product inhibition patterns for pfk2* show that ADP is a competitive inhibitor of ATP and a noncompetitive inhibitor vs. F6P. FBP is noncompetitive vs. both substrates (Table 3). Furthermore, sorbitol 6-P, a dead-end inhibitor, is competitive with F6P but uncompetitive with ATP. Thus, they conclude from these data that pfk2 and pfk2* show an ordered reaction mechanism, but the order of the addition of the substrates and release of the products is different. For pfk2, F6P is the first substrate added, and FBP is the last product to be released, same as rat liver PFK (Scheme II). On the other hand, for pfk2*, ATP is the first substrate to bind, and ADP is the last product to dissociate from the enzyme (Scheme III). These differences may not be so surprising because the amino acid sequences of these PFK isozymes have recently been shown to be completely different.[17]

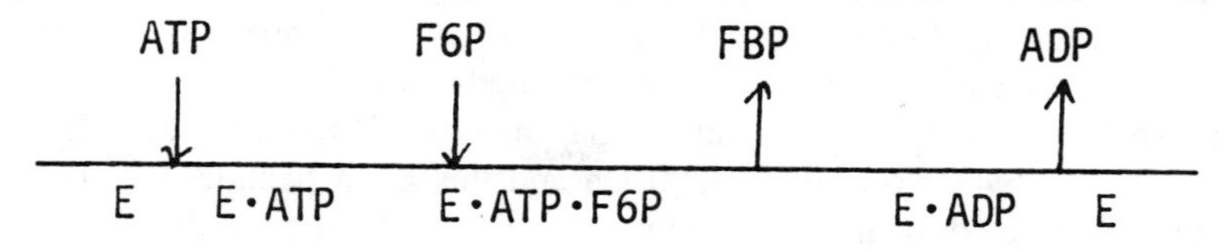

Scheme III

TABLE 4
Molecular Weight of Fru-6-P,2-kinase:Fru-2,6-bisphosphatase

Source	Mr Enzyme	Mr Subunit	Ref.
Rat liver	85,000 —90,000		27
	100,000	55,000	31
	101,000[a]	55,000[a]	32
	109,000	55,000	33
	107,000[a]		33
	112,500	58,000	34
Bovine liver	102,000	49,000	35
Chicken liver	110,000	54,000	36
Bovine heart	97,000	52,000	34
		58,000	37
Pigeon muscle	110,000	53,000	36

[a] Sedimentation equilibrium + sedimentation equilibrium in 6 *M* guanidine. All others are by gel filtration.

III. FRUCTOSE 6-PHOSPHATE,2-KINASE:FRUCTOSE 2,6-BISPHOSPHATASE

A unique sugar phosphate, $F2,6P_2$, was discovered several years ago.[20-22] It was shown to be the most potent activator of PFK known and appears to play an important role in regulating PFK activity in liver, in various other mammalian tissues[23] as well as in plants[24] and in yeast.[25] Synthesis of $F2,6P_2$ is catalyzed by an enzyme variously called ''Fructose-6-P,2-kinase'',[26] ''6-phosphofructo-2-kinase'',[27] or ''PFK-2''[28] (Equation 3).

$$F6P^{2-} + MgATP^{2-} \rightleftharpoons F2,6P_2^{4-} + MgADP^{1-} + H^+ \qquad (3)$$

The degradation of $F2,6P_2$ is catalyzed by F2,6-bisphosphatase (Equation 4), and interestingly, both the kinase and bisphase activities reside in the same protein, i.e., a bifunctional enzyme.[28-30]

$$F26P_2^{4-} + H_2O \rightarrow F6P^{2-} + Pi^{2-} \qquad (4)$$

A. STRUCTURE

F6P,2-kinase:F2,6-bisphosphatase has been purified to varying degrees of purity from a variety of sources. The molecular weight of the enzymes from these mammalian sources is 100 to 110K (Table 4). The variation in these values is partly due to the methods of determination and differences in the enzyme preparations. The differences in molecular weight as determined by sedimentation equilibrium is due to the difference in assumed $\bar{v}$, since 1% difference in the $\bar{v}$ affects the Mr by 3%. The subunit molecular weights are 50 to 58K indicating that these enzymes are dimers. The subunits of rat liver enzyme are very similar, if not identical. F6P,2-kinase is particularly susceptible to proteolytic digestion which probably accounts for some of the variation in the reported molecular weights. For example, bovine heart[34] and pigeon muscle[36] enzymes have been reported to be smaller than that of rat liver enzyme and could not be phosphorylated by cAMP-dependent protein kinase. However, Kitamura and Uyeda[37] have shown that the bovine enzyme is actually larger (Mr = 58K) than the rat liver enzyme by 3000 Da and this extra peptide does contain a site that can be phosphorylated by the protein kinase. These results suggest that the heart enzyme

TABLE 5
Amino Acid Composition of Rat Liver F6P,2-kinase:F2,6-bisphosphatase

Amino acid	Residues/55,000 Da Ref. 32	Residues/55,000 Da Ref. 38
Aspartic acid	45.6 ± 1.1	38.5
Threonine	23.8 ± 0.9	21.1
Serine	26.1 ± 2.6	66.9
Glutamic acid	75.0 ± 1.6	82.5
Proline	20.0 ± 0.1	13.8
Glycine	28.8 ± 2.9	81.6
Alanine	31.1 ± 2.4	38.5
Valine	31.3 ± 1.3	23.8
Methionine	4.8 ± 0.1	5.5
Isoleucine	28.8 ± 2.0	14.7
Leucine	44.8 ± 2.1	22.0
Tyrosine	25.2 ± 0.7	13.8
Phenylalanine	14.0 ± 0.5	9.2
Tryptophan	ND	9.2
Lysine	27.0 ± 1.1	33.9
Histidine	14.0 ± 1.5	14.7
Arginine	35.9 ± 0.5	24.8
Half-cystine	3.7 ± 0.1	8.3

is susceptible to protease digestion and a peptide containing the phosphorylation site is readily cleaved.

The amino acid composition of rat liver enzyme has been reported by two groups and is significantly different (Table 5). Similarly, the isoelectric point of the enzyme has been reported as 4.7[32] and 6.6.[29,39]

B. REVERSIBILITY AND THE EQUILIBRIUM CONSTANT

As demonstrated, the reaction catalyzed by Fru-6-P,2-kinase is reversible, and the rate of the reverse reaction is approximately one half that of the forward reaction as demonstrated by Kitajima et al.[40] The reverse reaction is slowed due to inhibition by F6P which accumulates during the course of the reaction, and in order to obtain a linear rate for the reaction in that direction, a F6P depleting system is required. The equilibrium constant may be estimated from these forward and reverse rates by the Haldane relationship[41]

$$\frac{V_f}{V_r} = \frac{K' \cdot K_{ADP} \cdot K_{F26P_2}}{K_{ATP}K_{F6P}}$$

where V_f is the forward rate, V_r is the reverse rate, K' is the apparent equilibrium constant, and other Ks are the Michaelis constants. The apparent equilibrium constant is estimated as 16 at pH 7.4, and the standard free energy change for the reaction at 30°C is -2 kcal. The free energy of hydrolysis of 2-phosphate of F2,6P_2, therefore, is calculated as -5.6 kcal mol^{-1}.

C. INITIAL RATE STUDIES

The initial rate studies carried out by Kitajima et al.[40] demonstrate that the extrapolated lines intersect to the left of the 1/v axis. Similar converging lines were obtained in the reverse direction which indicates that the F6P,2-kinase reaction is sequential and rules out participation of any kinetically important, covalent intermediate. The pertinent kinetic constants are summarized in Table 6.

TABLE 6
Kinetic Constants of Rat Liver F6P,2-kinase:F2,6-bisphosphatase[40]

	Km (μM)			
Conditions	**ATP**	**F6P**	**ADP**	**F2,6-P$_2$**
F6P,2-kinase				
Forward	150	16		
Reverse			62	
Exchange	120		66	8
F2,6-bisphosphatase				1

	Ki			
F6P,2-kinase				
Inhibitor	ADP	ADP	F2,6-P$_2$	F2,6-P$_2$
Varied substrate	ATP	F6-P	ATP	F6P
Ki (slope)	290 μM	1.2 mM	55 μM	22 μM
Ki (intercept)		1.2 mM	38 μM	47 μM
F2,6-bisphosphatase				
Inhibitor		F6P		
Ki (slope)		1.2 μM		
Ki (intercept)		1.5 μM		

D. PRODUCT INHIBITION STUDIES

In systematic product-inhibition studies, Kitajima et al.[40] found that ADP is a competitive inhibitor of ATP and a noncompetitive inhibitor of F6P. F2,6P$_2$ is a noncompetitive inhibitor versus ATP and versus F6P. These product inhibition patterns may be interpreted as a sequential order of addition of substrates in which ATP binds to the enzyme first, followed with F6P. However, the product inhibition patterns alone would not be sufficient to decide whether the mechanism is random or ordered. The mechanism for F6P,2-kinase reaction may indeed be the random addition of the substrates as we have seen with the PFK reaction. Additional experiments such as the use of analogs of the substrates or the products or dead-end inhibitors are required to distinguish between these two mechanisms.

E. ISOTOPE EXCHANGE ACTIVITY

Pilkis et al.[43] reported F6P,2-kinase catalyzes ATP-ADP exchange as well as F6P-F2,6P$_2$ exchange reactions in the absence of the other substrates, and that the rates of these exchange activities are at least 50% of the overall kinase activity. In fact, the F6P-F2,6P$_2$ exchange rate was comparable to that of the kinase reaction, and ADP stimulated this exchange rate three-fold.

In contrast to these results, however, Kitajima et al.[40] demonstrated that the rate of ATP-ADP exchange activity is only 3% of the forward kinase reaction. They could not observe any F6P-F2,6P$_2$ exchange activity. This low adenine nucleotide exchange activity is an integral part of the kinase reaction for a number of reasons. The exchange activity remains at 3% even after several affinity chromatography of the purified enzyme.[40] Chemical modification of the enzyme with pyridoxal-P followed with borohydride reduction results in inactivation of all three activities (the forward, the reverse, and the adenine nucleotide exchange) to the same degree. In addition, ligands protect all three activities against this inactivation to the same extent, and limited proteolysis with trypsin produces the same extent of inactivation of all three enzyme activities. Thus, the above evidence strongly indicates that the slow adenine nucleotide exchange is an integral part of F6P,2-kinase activity. Like many other kinases, as discussed above, it is not surprising that this enzyme also catalyzes this slow exchange reaction.

The high exchange activity of adenine nucleotides and fructose-P esters observed[43] could be due to a contaminating enzyme. Since the specific activity of F6P,2-kinase is 1/1000 of that of usual enzymes, a trace contamination with other enzymes could contribute significantly to the exchange activity. Furthermore, a later report by the Pilkis group[42] indicates that, with a new preparation of the enzyme, the adenine nucleotide exchange rate is 15% rather than 50% reported earlier which is still 5-fold higher than the value reported by Kitajima et al.[40] El-Maghrabi et al.[44] have shown that thermolysin-digested enzyme loses the kinase and the fructose-P ester exchange activities but retains the ATP-ADP exchange. Moreover, the sulfhydryl reagents inactivate the kinase without affecting the exchange activity. They conclude that there is little correlation between the kinase activity and the ATP-ADP exchange or sugar-P exchange, which is in direct contrast to the conclusion reached by Kitajima et al.[40] El-Maghrabi et al.[44] interpret their results by suggesting that these modifications affect different catalytic sites. However, such an interpretation implies that the ATP-ADP (or fructose-P) binding site for the exchange activity is distinct from that for the kinase activity. This is unlikely since the Km values for the forward and the reverse reactions and the exchange activities are essentially the same (Table 6) as shown by Kitajima et al.[40] Moreover, the ligand binding studies demonstrate that the enzyme contains only one ATP binding site/protomer.[45] Since the enzyme prepared by El-Maghrabi et al.[42] still shows exchange activity 5-fold higher than that of Kitajima et al.,[40] it is possible that this exchange activity may be due to a contaminating enzyme. Clearly there are considerable discrepancies in the results reported by these two groups, but this may be due in large part to the purity of the enzyme preparations. Nevertheless, Pilkis and his co-workers now agree with the conclusion reached earlier by Kitajima et al.[40] that there is no kinetically important phosphorylenzyme intermediate.

F. ATPase ACTIVITY

F6P,2-kinase:F2,6Pase shows ATPase activity like many of the known kinases. The activity has been estimated as less than 10% of the overall reaction.[40] Pilkis et al.[43] initially reported that the ATPase activity was as high as 100% of the 2-kinase reaction. Part of this activity, however, was apparently due to the presence of small amounts of F6P in the enzyme preparation. This F6P serves as a substrate for the kinase reaction, and its product, F2,6P_2, is hydrolyzed by F26Pase resulting in a net overall hydrolysis of ATP (Equations 3 and 4). When a F6P depletion system consisting of phosphoglucose isomerase and glucose 6-P dehydrogenase was used, the ATPase activity was significantly reduced.

It is unlikely that the ATPase activity occurs as a result of phosphorylenzyme intermediate formation during the kinase reaction since the evidence against such a mechanism is overwhelming, as discussed above. The possible explanation for the intrinsic ATPase activity in F6P,2-kinase is that it results from the action of H_2O residing at the site 2-OH of F6P normally occupies.

G. F2,6-BISPHOSPHATASE

Unlike the F1,6Pase activity of PFK, the F2,6Pase activity of this bifunctional enzyme is about 83% of F6P,2-kinase at V_{max}.[30]

Product inhibition patterns indicate that F6P is a noncompetitive inhibitor of F2,6P_2,[40] while phosphate is an activator. These results are consistent with either a ping-pong mechanism or an ordered sequential mechanism. Evidence against the ping-pong mechanism and the formation of phosphorylenzyme is provided by Kitajima et al.[40] who showed that when a stoichiometric amount of the enzyme was incubated with F2,6P_2 and the F6P formation was determined with a coupled enzyme system consisting of P-glucose isomerase and glucose 6-P dehydrogenase, no rapid release of F6P was detected. These negative results seem to

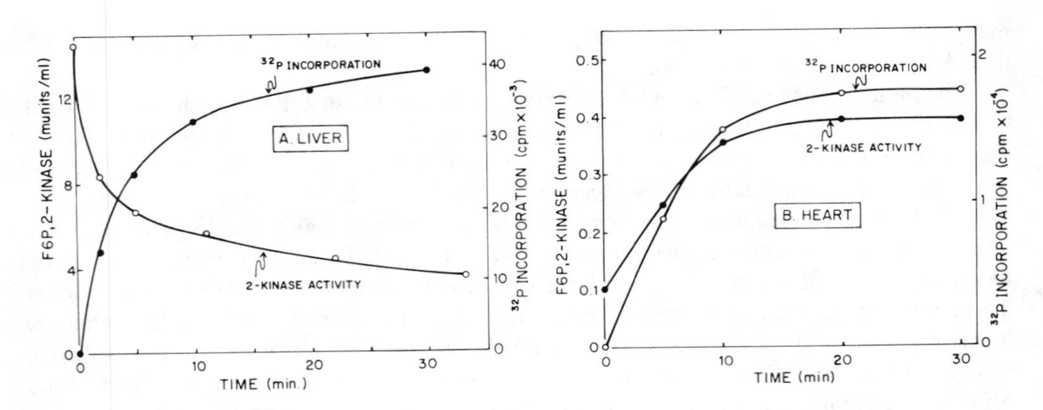

FIGURE 1. The effect of phosphorylation on the activities of liver[30] (A) and muscle[51] (B) F6P2-kinase by cAMP dependent protein kinase.

weaken the formation of a phosphoryl enzyme intermediate, but they do not rule out its existence because the dissociation of F6P from the enzyme may be slow even though it may precede the phosphate release. In support of a ping-pong mechanism, Stewart et al.[46] showed that when the enzyme is incubated with F2,6P$_2$, phosphorylation of the enzyme occurs, and the rates of the phosphorylation and dephosphorylation of the enzyme are sufficiently rapid to be a reaction intermediate.[46] Thus, the reaction catalyzed by F2,6-Pase appears to involve phosphorylenzyme intermediate, and the release of F6P may precede that of Pi.

H. POSSIBLE ROLES OF F2,6P$_2$ IN THE REGULATION OF GLYCOLYSIS IN LIVER AND HEART

Administration of glucagon or epinephrine to liver leads to a decrease in F2,6P$_2$ as a result of phosphorylation of F6P,2-kinase:F2,6-Pase by cAMP-dependent protein kinase (reviewed in References 47 to 49). The most interesting aspect of this phosphorylation is that the regulatory properties of F6P,2-kinase and F2,6-Pase are altered in opposite directions, i.e., the kinase activity decreases at the physiological concentration of F6P (Figure 1A) because the Km for F6P shifts to a higher Km value while F2,6-Pase activity increases because the substrate saturation curve for F2,6P$_2$ changes from negative cooperativity to normal Michaelis-Menton kinetics. The net result is a decrease in F2,6P$_2$ concentration, which inhibits PFK and ultimately decreases the glycolytic rate in liver so that more glucose is released into circulation.

In contrast to liver, epinephrine treatment of perfused heart leads to an increase in F2,6P$_2$ due to activation of F6P,2-kinase in this tissue.[50] Further investigation into the mechanism of this activation reveals that the isolated heart enzyme also can be phosphorylated *in vitro* by cAMP-dependent protein kinase, which leads to activation (Figure 1B) by lowering the Km for its substrates, F6P and ATP.[51] F2,6-Pase activity, however, is not affected by the phosphorylation. Thus, these results are in direct contrast to the effect of the phosphorylation on the liver enzymes. These opposite effects on two isozymes following phosphorylation by the same protein kinase represent a unique regulatory mechanism. In all other cases, the differences in regulatory properties may differ quantitatively but are always in the same direction, i.e., inhibition or activation. Obviously, the different isozymes of F6P,2-kinase: F2,6-Pase have different allosteric regulatory properties in their respective tissues so that changes in F2,6P$_2$ concentration are consistent with the specialized functions of each tissue. Epinephrine stimulates both glycogenolysis and glycolysis, and the activation of the glycolytic pathway involves activation of PFK, in part, by increased F2,6P$_2$ via activation of F6P,2-kinase.

REFERENCES

1. **Bloxham, D. P. and Lardy, H. A.,** Phosphofructokinase, in *The Enzymes,* Vol. 8, Academic Press, New York, 1973, 239.
2. **Hofmann, E.,** The significance of phosphofructokinase to the regulation of carbohydrate metabolism, in *Rev. Physiol. Biochem. Pharmacol.,* Vol. 75, Springer-Verlag, Berlin, 1976.
3. **Uyeda, K.,** Phosphofructokinase, in *Advances in Enzymology and Related Areas of Molecular Biology,* Vol. 48, Meister, A., Ed., John Wiley & Sons, New York, 1979, 193.
4. **Goldhammer, A. R. and Paradies, H. H.,** Phosphofructokinase: structure and function, in *Current Topics in Cellular Regulation,* Vol. 15, Horecker, B. L. and Strodtman, E. R., Eds., Academic Press, New York, 1979, 109.
5. **Vinuela, E., Salas, M. L., and Sols, A.,** End-product inhibition of yeast phosphofructokinase by ATP, *Biochem. Biophys. Res. Commun.,* 12, 140, 1963.
6. **Lou, M. F. and Kinoshita, J. H.,** Control of lens glycolysis, *Biophys. Acta,* 141, 547, 1967.
7. **Layzer, R. B., Rowland, L. P., and Bank, W. J.,** Physical and kinetic properties of human phosphofructokinase from skeletal muscle and erythrocytes, *J. Biol. Chem.,* 244, 3823, 1969.
8. **Uyeda, K.,** Studies on the reaction mechanism of skeletal muscle phosphofructokinase, *J. Biol. Chem.,* 245, 2268, 1970.
9. **Lowry, O. H. and Passonneau, J. V.,** Kinetic evidence for multiple binding sites on phosphofructokinase, *J. Biol. Chem.,* 241, 2268, 1966.
10. **Uyeda, K.,** Studies on the Fructose 1-phosphate kinase activity of rabbit muscle phosphofructokinase, *J. Biol. Chem.,* 247, 1692, 1972.
11. **Hanson, R. L., Rudolph, F. B., and Lardy, H. A.,** Rabbit muscle phosphofructokinase, the kinetic mechanism of action and the equilibrium constant, *J. Biol. Chem.,* 248, 7852, 1973.
12. **Kee, A. and Griffin, C. C.,** Kinetic studies of rabbit muscle phosphofructokinase, *Arch. Biochem. Biophys.,* 149, 361, 1972.
13. **Bar-Tana, J. and Cleland, W. W.,** Rabbit muscle phosphofructokinase. I. Anomeric specificity, initial velocity kinetics, *J. Biol. Chem.,* 249, 1263, 1974.
14. **Bar-Tana, J. and Cleland, W. W.,** Rabbit muscle phosphofructokinase II. Product and dead-end inhibition, *J. Biol. Chem.,* 248, 1271, 1974.
15. **Brand, I. A. and Soling, H. D.,** Rat liver phosphofructokinase purification and characterization of its reaction mechanism, *J. Biol. Chem.,* 249, 7824, 1974.
16. **Campos, G., Guixe, V., and Babul, J.,** Kinetic mechanism of phosphofructokinase-2 from *Escherichia coli,* a mutant enzyme with a different mechanism, *J. Biol. Chem.,* 259, 6147, 1984.
17. **Hellinga, H. W. and Evans, P. R.,** Nucleotide sequence and high-level expression of the major *Escherichia coli* phosphofructokinase, *Eur. J. Biochem.,* 149, 363, 1985.
18. **Hume, E. C. and Tipton, K. F.,** The isotope-exchange reactions of ox heart phosphofructokinase, *Biochem. J.,* 122, 181, 1971.
19. **Zewe, V., Fromm, H. J., and Fabiano, R.,** The effect of manganous ion on the kinetics and mechanism of the yeast hexokinase reaction, *J. Biol. Chem.,* 239, 1625, 1964.
20. **Furuya, E. and Uyeda, K.,** An activation factor of liver phosphofructokinase, *Proc. Natl. Acad. Sci. U.S.A.,* 77, 5861, 1980.
21. **Van Schaftingen, E., Hue, L., and Hers, H. G.,** Control of the fructose-6-phosphate/fructose-1,6-bisphosphate cycle in isolated hepatocytes by glucose and glucagon: role of a low molecular weight stimulator of phosphofructokinase, *Biochem. J.,* 192, 887, 1980.
22. **Claus, T. H., Schlumpf, J., Pilkis, J., Johnson, R. A., and Pilkis, S. J.,** Evidence for a new activator of rat liver phosphofructokinase, *Biochem. Biophys. Res. Commun.,* 98, 359, 1981.
23. **Kuwajima, M. and Uyeda, K.,** The tissue distribution of Fructose-2,6-P_2 and Fructose-6-P,2-kinase in rats and the effect of starvation, diabetes, and lypoglycemia on hepatic fructose-2,6-P_2 and fructose-P,2-kinase, *Biochem. Biophys. Res. Commun.,* 104, 84, 1982.
24. **Sabularse, D. C. and Anderson, R. L.,** D-Fructose 2,6-bisphosphate: a naturally occurring activator for inorganic pyrophosphate:D-fructose-6-phosphotransferase in plants, *Biochem. Biophys. Res. Commun.,* 103, 848, 1981.
25. **Lederer, B., Vissers, S., Van Schaftingen, E., and Hers, H. G.,** Fructose-2,6-bisphosphate in yeast, *Biochem. Biophys. Res. Commun.,* 103, 1281, 1981.
26. **Furuya, E. and Uyeda, K.,** A novel enzyme catalyzes the synthesis of activation factor from ATP and D-Fructose-6-P, *J. Biol. Chem.,* 256, 7109, 1981.
27. **El-Maghrabi, M. R., Claus, T. H., Pilkis, J., and Pilkis, S. J.,** Partial purification of a rat liver enzyme that catalyzes the formation of fructose-2,6-bisphosphate, *Biochem. Biophys. Res. Commun.,* 101, 1071, 1981.
28. **Van Schaftingen, E., Davies, D. R., and Hers, H. G.,** Fructose 2,6-bisphosphatase from rat liver, *Eur. J. Biochem.,* 124, 143, 1982.

29. **El-Maghrabi, M. R., Claus, T. H., Pilkis, J., Fox, E., and Pilkis, S. J.,** Regulation of rat liver fructose 2,6-bisphosphatase, *J. Biol. Chem.*, 257, 7603, 1982.
30. **Sakakibara, R., Kitajima, S., and Uyeda, K.,** Differences in kinetic properties of phospho and dephospho forms of fructose-6-phosphate,2-kinase and fructose-2,6-bisphosphatase, *J. Biol. Chem.*, 259, 41, 1984.
31. **Furuya, E., Yokoyama, M., and Uyeda, K.,** Regulation of fructose-6-phosphate, 2-kinase by phosphorylation and dephosphorylation: possible mechanism for coordinated control of glycolysis and glycogenolysis, *Proc. Natl. Acad. Sci. U.S.A.*, 79, 325, 1982.
32. **Sakakibara, R., Tanaka, T., Uyeda, K., Richards, E. G., Thomas, H., Kangawa, K., and Matsuo, H.,** Studies of the structure of fructose 6-phosphate 2-kinase:fructose 2,6-bisphosphatase, *Biochemistry*, 24, 6818, 1985.
33. **El-Maghrabi, M. R., Coneia, J. J., Heil, P. J., Pate, T. M., Cobb, C. E., and Pilkis, S. J.,** Tissue distribution, immunoreactivity, and physical properties of 6-phosphofructo 2-kinase/fructose-2,6-bisphosphatase, *Proc. Natl. Acad. Sci. U.S.A.*, 83, 5005, 1986.
34. **Rider, M. H., Foret, D., and Hue, L.,** Comparison of purified heart and rat liver 6-phosphofructo-2,kinase: Evidence for distinct enzymes, *Biochem. J.*, 231, 193, 1985.
35. **Kountz, P. D., El-Maghrabi, M. R., and Pilkis, S. J.,** Isolation and characterization of 6-phosphofructokinase/fructose-2,6-bisphosphatase, *Arch. Biochem. Biophys.*, 238, 531, 1985.
36. **Van Schaftingen, E. and Hers, H. G.,** Purification and properties of phosphofructokinase-2/fructose-2,6-bisphosphatase from chicken liver and from pigeon muscle, *Eur. J. Biochem.*, 159, 359, 1986.
37. **Kitamura, K. and Uyeda, K.,** The mechanism of activation of heart fructose-6-phosphate,2-kinase:fructose-2,6-bisphosphatase, *J. Biol. Chem.*, 262, 679, 1987.
38. **El-Maghrabi, M. R., Fox, E., Pilkis, J., and Pilkis, S. J.,** Cyclic-AMP dependent phosphorylation of rat liver 6-phosphofructo-2-kinase/fructose 2,6-bisphosphatase, *Biochem. Biophys. Res. Commun.*, 106, 794, 1982.
39. **Garrison, J. C. and Wagner, J. D.,** Glucagon and Ca^{2+}-linked hormones, angiotensin II, norepinephrine, and vasopressin stimulate the phosphorylation of distinct substrates in intact hepatocytes, *J. Biol. Chem.*, 257, 1313, 1982.
40. **Kitajima, S., Sakakibara, R., and Uyeda, K.,** Kinetic studies of fructose-6-phosphate,2-kinase and fructose-2,6-bisphosphatase, *J. Biol. Chem.*, 259, 6896, 1984.
41. **Haldane, J. B. S.,** *Enzymes,* MIT Press, Cambridge, MA, 1930.
42. **El-Maghrabi, M. R., Pate, T. M., and Pilkis, S. J.,** Characterization of the exchange reactions of rat liver 6-phosphofructo 2-kinase/fructose 2,6-bisphosphatase, *Biochem. Biophys. Res. Commun.*, 123, 749, 1984.
43. **Pilkis, S. J., Chrisman, T., Burgess, B., McGrow, M., Colosia, A., Pilkis, J., Claus, T. H., and El-Maghrabi, M. R.,** Rat hepatic 6-phosphofructo-2-kinase/fructose-2,6-bisphosphatase: a unique bifunctional enzyme, *Adv. Enzyme Regul.*, 21, 147, 1983.
44. **El-Maghrabi, M. R., Pate, T. M., Murray, K. J., and Pilkis, S. J.,** Differential effects of proteolysis and protein modification on the activities of 6-phosphofructo-2-kinase/fructose-2,6-bisphosphatase, *J. Biol. Chem.*, 259, 13096, 1984.
45. **Thomas, H., Taniyama, M., Kitajima, S., and Uyeda, K.,** Binding of ligands to Fructose-6-phosphate,2-kinase:Fructose-2,6-bisphosphatase, *Fed. Proc.*, 45, 1013, 1986.
46. **Stewart, H. B., El-Maghrabi, M. R., and Pilkis, S. J.,** Evidence for a phosphorylenzyme intermediate in the reaction pathway of rat hepatic Fructose-2,6-bisphosphatase, *J. Biol. Chem.*, 260, 12935, 1985.
47. **Uyeda, K., Furuya, E., Richards, C. S., and Yokoyama, M.,** Fructose 2,6-P_2, chemistry and biological function, *Mol. Cell. Biochem.*, 48, 97, 1982.
48. **Pilkis, S. J., Christman, T., Burgress, B., McGrane, M. M., Colosia, M. M., Pilkis, J., Claus, T. H., and El-Maghrabi, M. R.,** Rat hepatic 6-phosphofructo-2-kinase/fructose 2,6-bisphosphatase: a unique bifunctional enzyme, *Adv. Enzyme Regul.*, 21, 147, 1983.
49. **Van Schaftingen, E.,** Fructose-2,6-bisphosphate, in *Advances in Enzymology and Related Areas of Molecular Biology,* Vol. 59, Meister, A., Ed., 1987, 315.
50. **Narabayashi, H., Lawson, J. W. R., and Uyeda, K.,** Regulation of phosphofructokinase in perfused rat heart — requirement for fructose 2,6-bisphosphate and a covalent modification, *J. Biol. Chem.*, 260, 9750, 1985.
51. **Kitamura, K. and Uyeda, K.,** The mechanism of activation of heart fructose 6-phosphate,2-kinase:fructose 2,6-bisphosphatase, *J. Biol. Chem.*, 262, 679, 1987.

Section IX
Emden-Meyerhof Reactions — Mechanism of Action

Therefore our pride in the progress achieved in the last decades must be tempered by confession of ignorance regarding many crucial points. There are still many problems for this generation of research workers to solve. O. Meyerhof in a discussion of Intermedicate Carbohydrate Metabolism, given at *A Symposium on Respiratory Enzymes*, p. 14, The University of Wisconsin Press, Madison, 1942.

Chapter 19

TRIOSE- AND HEXOSE-PHOSPHATE ISOMERASES*

K. Ümit Yüksel and Robert W. Gracy

TABLE OF CONTENTS

* This review is dedicated to the memory of Professor Ernst A. Noltmann (1931—1986) who devoted most of his scientific career to studies on the aldose-ketose isomerases.

I. INTRODUCTION

A. BACKGROUND

Around the turn of the century the nonenzymatic aldose-ketose interconversion of sugars was recognized to occur in alkaline solution and was termed the "Lobry de Bruyn-Alberda van Ekenstein transformation" (for a thorough review see Reference 1). The enzyme-catalyzed aldose-ketose isomerizations have been studied over the past 4 decades. The aldol-ketol isomerases (EC 5.3.1.X) catalyze essential reactions of carbohydrate metabolism, and their metabolites participate in diverse cellular activities such as the formation of glycoproteins and the activation of proteases.[2] Due to their relative simplicity (single substrate-single product), their exceedingly high catalytic efficiency (triosephosphate isomerase has been termed "the perfect catalyst"[3,4]), and their absolute stereospecificity, the isomerases became favorite subjects of enzymologists. Several excellent reviews, e.g., by Topper,[5] Noltmann,[6] and Rose,[7-9] have been presented previously. We will not attempt to duplicate their efforts, but focus primarily on the information gained since their reviews, and how this new information impacts on the structural-functional understanding of the mechanisms by which these reactions occur. Furthermore, we will restrict this review to triosephosphate isomerase (TPI, EC 5.3.1.1), glucosephosphate isomerase (GPI, EC 5.3.1.9), and mannosephosphate isomerase (MPI, EC 5.3.1.8) which catalyze the reactions shown in Figure 1.

Although they catalyze the same types of reactions, these enzymes differ from each other substantially. TPI, composed of the shortest peptide chain of the three isomerases, is the best characterized. Its amino acid sequence has been established from several species,[10-20] and its X-ray crystal structure has been determined at high resolution.[13,21,22] The active site,[23] the origin of isozymes,[24] and a detailed kinetic and thermodynamic profile of the catalytic mechanism are all well established. GPI, like TPI, is a homodimer, requiring no coenzymes or cofactors, showing no cooperativity; however, it is a larger protein than TPI and not as well characterized. MPI, a monomeric zinc metalloenzyme,[25] is the least characterized of the three isomerases. Table 1 compares some of the basic properties of the three aldose-ketose isomerases.

B. GENERAL MECHANISM OF ISOMERIZATIONS

The isomerization reactions obey a base catalysis mechanism and are believed to pass through a common enediol(ate) intermediate (Figure 2). The mechanism of catalysis of TPI was studied by Rieder and Rose[40,41] and Bloom and Topper[42] using tritium exchange experiments. It was proposed that the reaction proceeds through an enediol rather than by a concerted hydride shift.[8] Isomerization of dihydroxyacetone phosphate (DHAP) by TPI in ^3H_2O revealed that tritium was stereospecifically introduced into the trioses.[41] The same study also indicated the improbability of a hydride transfer mechanism, and outlined an enediol mechanism. The mechanism proposed in both of these studies[41,42] involved the loss of the proton from the substrate to the medium (exchange). The basis of this exchange reaction has been the subject of some controversy. For example, Noltmann[6] pointed out that the dissociation of the proton had to be faster than a diffusion controlled reaction rate and that the proposed reaction scheme[8] had to be altered. He suggested that " . . . the reaction may instead proceed as a competition for the base proton between the solvent and enediolate ion bound to the enzyme active site . . . ". Studies on the conformation of the transition state led to proposals that (1) the reaction proceeds through a *cis*-enediol intermediate[42-44] and (2) the *cis*-enediol intermediate might be the common form for all aldose-ketose isomerase reactions.[7]

Support for the enolization reaction for GPI also derived from the classic experiments of Topper[42,45] and Rose,[40,43] who studied the incorporation of radiolabel from water into the products of the isomerization reactions. In the case of GPI, when the reaction was conducted

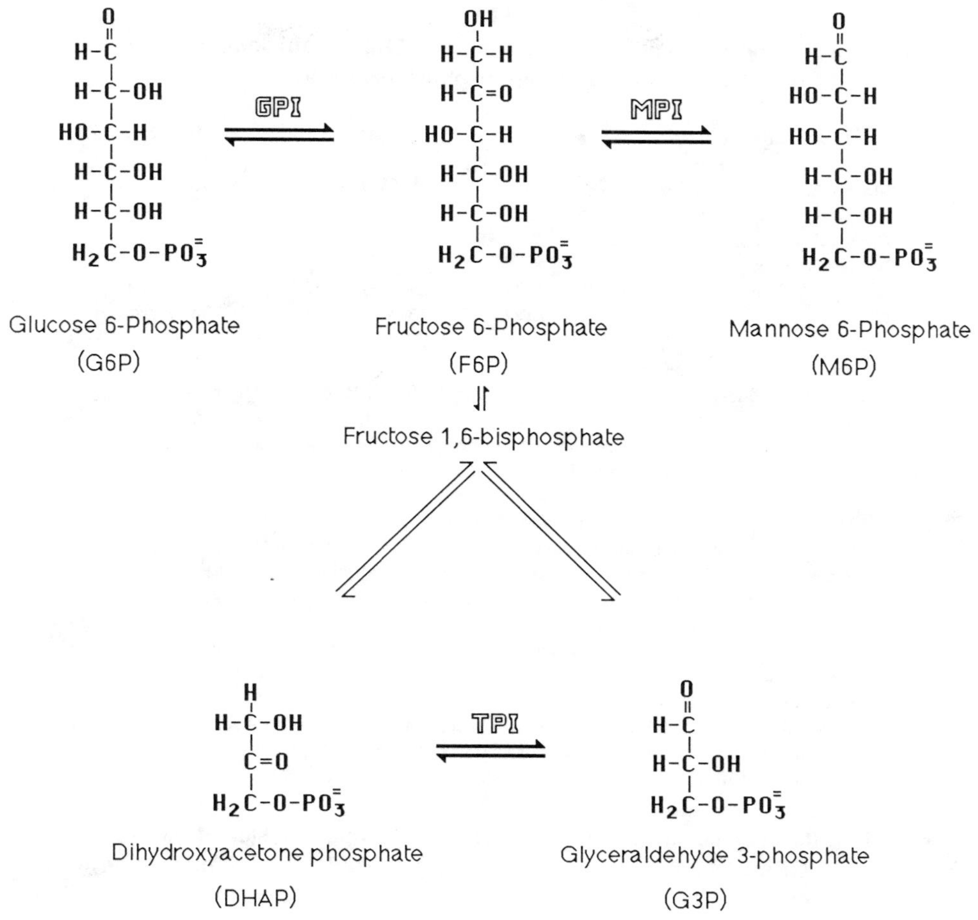

GPI MPI

Glucose 6-Phosphate (G6P)

Fructose 6-Phosphate (F6P)

Mannose 6-Phosphate (M6P)

Fructose 1,6-bisphosphate

TPI

Dihydroxyacetone phosphate (DHAP)

Glyceraldehyde 3-phosphate (G3P)

FIGURE 1. Isomerization reactions catalyzed by triosephosphate isomerase, glucosephosphate isomerase, and mannosephosphate isomerase. The hexosephosphates are shown in their open chain configurations.

in the direction fructose 6-phosphate → glucose 6-phosphate in ^3H_2O, tritium was found both in the product and in the substrate. This was explained by the nucleophilic involvement of the enzyme in proton abstraction and the conjugate acid being able to exchange protons with the medium before the proton is transferred onto the adjacent carbon atom[43] (Figure 2). The route of isomerization was further clarified by experiments with the rabbit and spinach enzymes.[43,46] The ratio (T/E) of the direct transfer (T) of a proton from C-1 to C-2 of the substrate vs. the exchange (E) of the abstracted proton into the solvent was determined to be temperature dependent.** The inverse relationship between T/E and temperature was explained as a higher activation energy for the proton exchange with the medium than for H-transfer between the adjacent carbon atoms of the substrate.[41] These observations were further interpreted to mean that the isomerization reaction could proceed by either route. The stereospecific transfer of the proton between C-1 and C-2 further implied that the same nucleophile was responsible for proton abstraction from either position.[44]

Isotope exchange experiments with MPI indicated the stereospecificity of this enzyme,[45] but suggested that MPI mobilized a different set of protons than GPI. Hence, MPI uses the 1S/2S hydrogens instead of the 1R/2R hydrogens used by GPI and, to be able to go through

** T/E value for enzymes from different species varies by about a factor of 2.[43]

TABLE 1
Properties of Triosephosphate Isomerase, Glucosephosphate Isomerase, and Mannosephosphate Isomerase

	TPI	GPI (31)	MPI (25,36,37)
Molecular weight	2 × 26,750 (10)	2 × 63,000	45,000
Amino terminus	A (10)	blocked[a]	Q,I[b]
Carboxyl terminus	Q(10)	Q	L,R[c]
Isoelectric point	5.8[d]	9.25	4.5—5.0
$K_{m(aldol)}$	2.5×10^{-4} *M* (27)	1.25×10^{-4} *M* (33)	1.35×10^{-3} *M*
$K_{m(ketol)}$	3.7×10^{-4} *M* (27)	7.13×10^{-5} *M* (33)	Not determined
$K_{eq\ (ketol/aldol)}$	22[e]	0.27[f]	1.03
pH optimum	7.8 (30)	8.1—8.7 (33)	7.0
Active site residues	E, K, H, S	E, K, H, R W (35)	H, Zn^{++}

Note: The numbers in parentheses are the references for the data. Data are shown for human TPI and GPI and for yeast MPI except where noted.

[a] The N-terminus of the rabbit enzyme was reported to be N-acetyl alanine (32).
[b] Determined from nucleotide sequencing of *E. coli* (38) and *Pseudomonas aerginosa* (39) genes, respectively. It should be noted that there are two possible initation sites in *P. aerginosa* (39). If the second initiation site is considered, the N-terminus becomes glutamine as in *E. coli*.
[c] Determined from nucleotide sequencing of *E. coli* (38) and *Pseudomonas aerginosa* (39) genes, respectively.
[d] pI of the most basic isozyme (26).
[e] This is the originally published value (28). When corrected for the nonhydrated species being the only substrates, a value of 367 is found (29).
[f] Data shown for the rabbit enzyme at 25°C, pH 8.5, I = 0.12 (34).

a *cis*-enediol, the enzyme attacks the substrate from the opposite side of the sugar molecule from TPI and GPI. Deuterium exchange studies exhibited 3 to 7% direct transfer of the proton between C-1 and C-2 positions of the substrate.[47] The low level exchange, in comparison to GPI, was thought to be the result of slight differences in transition state geometry since MPI is a zinc metalloenzyme and GPI is not.

II. TRIOSEPHOSPHATE ISOMERASE

A. KINETIC STUDIES

In the previous section we have summarized the studies proposing a mechanism for the enzymatic interconversion of dihydroxyacetone phosphate (DHAP) and glyceraldehyde 3-phosphate (G3P). Although this reaction can take place spontaneously (i.e., nonenzymatically via the Lobry de Bruyn-Alberda van Ekenstein transformation), the enzyme-catalyzed reaction is 10^9 times faster.[3]

Rose and co-workers[28] proposed the following reaction scheme:

$$\text{DHAP} + \text{E} \leftrightharpoons \text{E} \cdot \text{DHAP} \leftrightharpoons \text{E} \cdot \text{X} \leftrightharpoons \text{E} \cdot \text{G3P} \leftrightharpoons \text{E} + \text{G3P}$$

Although it is not certain if the intermediate EX is an enediol or enediolate, the mechanism was predicted as shown in Figure 2.

Knowles and co-workers[48-54] studied the energy barrier of each step of the TPI catalyzed isomerization, based on the reaction scheme shown above. In this mechanism, only the enediol(ate) intermediate can exchange protons with the surroundings. Thus, the incorporation of tritium into and exchange on the substrate-product pair was studied. The overall reaction in the presence of isotopes was written as

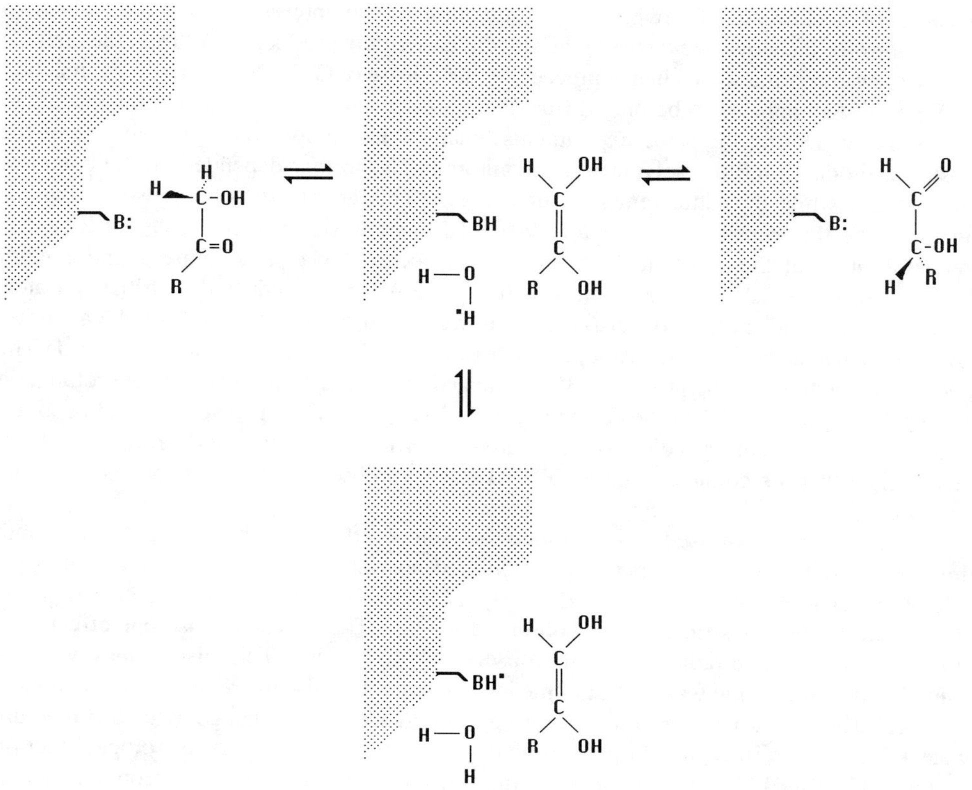

FIGURE 2. General mechanism of the aldose-ketose isomerases. The mechanism is based on the proposal of Rieder and Rose.[41] In the horizontal direction, the stereospecific transfer of proton from C-1 position of the substrate to C-2 position of the product is shown. The vertical direction depicts the exchange of the proton abstracted by the enzyme, with solvent.

$$E + S \underset{k_{-1}}{\overset{k_1}{\rightleftharpoons}} E \cdot S \underset{k_{-2}}{\overset{k_2}{\rightleftharpoons}} E \cdot X \underset{k_{-3}}{\overset{k_3}{\rightleftharpoons}} E \cdot P \underset{k_{-4}}{\overset{k_4}{\rightleftharpoons}} E + P$$

$$k_5 \downarrow\uparrow k_{-5}$$

$$E + S' \underset{-1'}{\overset{k_{1'}}{\rightleftharpoons}} E \cdot S' \underset{k_{-2'}}{\overset{k_{2'}}{\rightleftharpoons}} E \cdot X' \underset{k_{-3'}}{\overset{k_{3'}}{\rightleftharpoons}} E \cdot P' \underset{k_{-4'}}{\overset{k_{4'}}{\rightleftharpoons}} E + P'$$

where E = enzyme, S = substrate, P = product, ES = enzyme substrate complex, EX = enzyme intermediate complex, EP = enzyme product complex, k_x and k_{-x} = the rates in the forward and reverse directions. The symbol (′) indicates the involvement of isotopes.

In the reaction DHAP → G3P, the low level incorporation of label from R-[1-^{3}H]-DHAP into the product led to the conclusions that (1) the proton-abstracting enzymic base was essentially in complete equilibrium with the solvent and (2) additional (direct) proton transfer between C-1 and C-2, without exchange, did occur to a limited extent.[49] The partition of intermediate (EX) into product or back to the substrate was determined by running the same reaction with unlabeled DHAP in tritiated water.[50] It was found that after exchanging tritium the intermediate returned to the substrate three times more frequently as it formed the product, G3P. There was only a 1.3-fold discrimination against tritium, suggesting the presence of a slow step following the protonation of the intermediate at C-2. The two possibilities for the slow step were (1) the slow release of the product (G3P) from the enzyme or (2) a slow conformational change preceeding the release of G3P. Since the reverse reaction (G3P → DHAP) appears to be only diffusion controlled, the second explanation appears to be less likely. During the same experiments, changes in the specific radioactivities of the substrate, product, and the solvent were monitored. The former depended on the extent of the reaction, while the latter remained unchanged. This appeared to be consistent with a model where the transition energy for ES → EX was lower than for EX → EP, and a preequilibration of the substrate with the solvent occurred via the enzyme bound enediol (EX) intermediate.[50] When the reaction G3P → DHAP was conducted in tritiated water, the label was found in the product (DHAP) as well as in the substrate (G3P).[51] This time, however, considerable primary isotope effect (ninefold discrimination) was observed. The specific radioactivity of the product (13%) indicated the kinetic importance of the protonation of the intermediate at C-1 in the formation of DHAP from G3P. The specific radioactivity of the substrate remaining indicated three times faster formation of DHAP from the enediol intermediate than its collapse back to G3P. These two observations are in accord with the previous report.[50]

Deuterium was also used to investigate the isotope effects on the TPI reaction in both directions.[52] Initial velocity experiments exhibited no isotope effect for the reaction *D*-[2-^{2}H]G3P → DHAP. For the reaction *R*-[1-^{2}H]DHAP → G3P, no isotope effect was found on K_m, but a 2.9-fold difference was observed for k_{cat}. The absence of isotope effect with deuterium appears contradictory to that observed with tritium. This discrepancy was explained as the label being washed out of the intermediate, because the reaction was conducted in water. The enediol intermediate undergoes rapid isotope exchange with the medium (water), and is also in rapid equilibrium with the substrate. The observed isotope effect on k_{cat} for R-[1-^{2}H]DHAP was explained by the difference in the proton/deuteron abstraction rates. The extent of the transfer of label from *R*-[1-^{2}H]DHAP to G3P was found to be similar to that of R-[1-^{3}H]DHAP → G3P.[53] This has been taken to indicate that the exchange process did not involve "a rate limiting transition state in which the isotope is in flight". Three possible modes of proton exchange between the solvent (water) and the catalytic base were considered: (1) ionization-solvent exchange-reprotonation, (2) proton exchange followed by solvent reorganization within the ion pair, and (3) rapid exchange of isotope on the enediol into a limited pool of water molecules sequestered from solution.[53] Experimental findings appeared to support the first two pathways while the last one was perceived to be less likely. The data obtained in the studies above[48-53] were analyzed according to the model proposed earlier,[48] and the Gibbs free-energy profile for the TPI catalyzed reaction was constructed[54] (Figure 3).

Recently *ab initio* calculations were used to determine the energy profile of the enzyme catalyzed reaction:[56] These calculations indicate that during the nonenzymatic isomerization of DHAP → G3P a high energy barrier must be overcome. The enzyme was proposed to function by lowering the energy barrier via the stabilization of the enediolate intermediate with Glu-165 (protonated) with respect to DHAP and the Glu-165 anion. It was suggested that His-95 and Lys-13/Glu-97 ion pair was important in this stabilization. It was further noted that an enediol intermediate would not be inconsistent with the proposed model, and proton abstraction from DHAP and protonation of the enediolate would not have to take place in a concerted reaction. Speculative calculations were carried out on the possible effects of a His-95 → Gln-95 mutation.[56] It was suggested that this enzyme, in theory, could stabilize the enediolate better than the wild-type enzyme. However, it would probably be inactive since Gln-95 appeared to form a H-bond with Glu-165, thus impairing the proton abstracting ability of Glu-165.[56] Similar calculations were performed on the enzyme bearing a Glu-165 → Asp-165 mutation.[57] These calculations indicated that the Glu-165 → Asp-

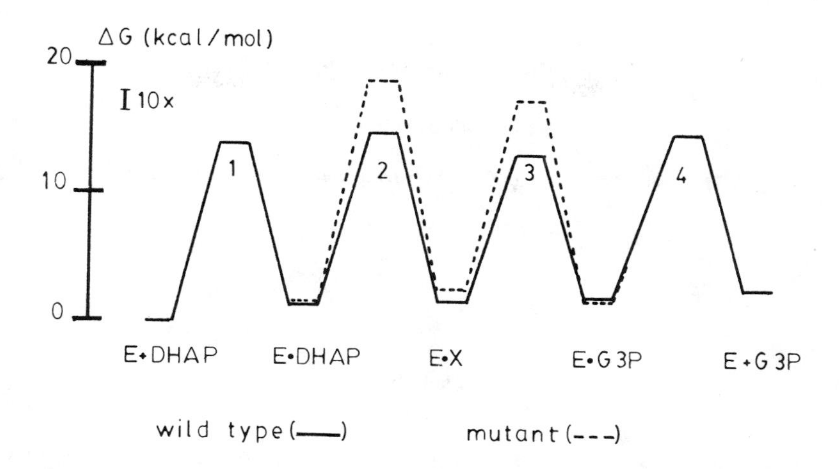

FIGURE 3. Energetics of the triosephosphate isomerase reaction. The energy barriers of the wild type enzyme (—) and the enzyme bearing a Glu-165 → Asp-165 mutation (- -) are shown.[55]

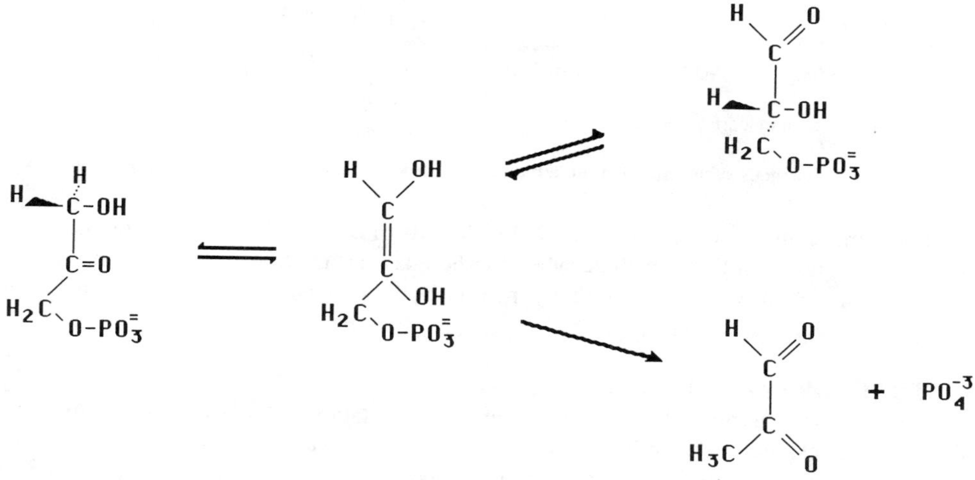

FIGURE 4. Phosphatase activity of TPI. The elimination reaction ascribed to TPI yielding methylglyoxal and inorganic phosphate from the enzyme bound enediol is shown.

165 mutation would create longer distances between the side chain carbonyl of Glu-165 and the pro-*R*-hydrogen of DHAP, hence increasing the energy barrier by about 4 kcal and decreasing the enzyme activity about 1000-fold.[57] This is indeed the value observed by Raines and Knowles.[58] Kinetic studies on the mutated enzyme[55] were carried out in a similar fashion as was done for the wild-type enzyme (*vide supra*). It was found that the mutant enzyme retained its stereospecificity and, in the reaction $R[1\text{-}^3H]$DHAP → G3P, there was essentially no change in the proton exchange rates of the intermediate and the medium.

As in the case of fructose 1,6-bisphosphate aldolase,[59] TPI possesses a phosphatase activity.[60,61] This phosphatase activity yields methylglyoxal and inorganic phosphate (Figure 4). The reaction does not appear to be due to a contaminating protein, as it was eliminated by specific inactivation of TPI, and was not an uncatalyzed breakdown of G3P.[60] Alkaline lability of DHAP (i.e., generation of methylglyoxal and P_i via enediol formation and subsequent 1,4-elimination) is known (e.g., see Reference 62). Studies on aldolase[59,63] led to

TABLE 2
Interactions at the Active Site of TPI

Enzyme-Substrate Interactions

TPI residue #	Atom	Substrate atom	Distance (nm)
Lys-13	ε-NH_2	carbonyl oxygen	0.29
		O-2	0.33
		O-3	
His-95	ε-N	O-1	0.32
		O-2	0.35

Protein Chain Interactions

TPI residue	Atom	TPI residue	Atom	Type of interaction
Lys-13 (A)[a]	ε-NH_2	Glu-97 (B)	ε-O-1	
Glu-97 (A)	ε-O-2	Phe-74 (B)	amide nitrogen	v.d. Waal's
		Thr-75 (B)	amide nitrogen	
	ε-O-2	His-95 (A)	imidazole	
His-95 (A)	δ-N-2	Glu-97 (A)	main chain NH	H-bond
		Ser-96 (A)	δ-O	
Ser-96 (A)	γ-O	Glu-165 (A)	ε-O-1	

Note: Data compiled from Reference 79.

[a] The homologous subunits of TPI are arbitrarily named (A) and (B) for identification purposes.

a proposed mechanism analogous to the alkaline lability of DHAP: β-elimination of phosphate from the enzyme-bound intermediate whereby the enzyme functions as a base. In the case of TPI, this reaction was not a simple hydrolysis of the phosphate group as methylglyoxal and P_i were formed at the same rate.[64]

B. STRUCTURE OF THE ACTIVE CENTER

Structural studies on TPI using carboxyamidomethylation[65] suggested the involvement of a histidine residue in the active center. Later, a lysine residue was also proposed to aid the substrate binding through interactions with the phosphate group of the substrate.[66] Specific active-site reagents, e.g., haloacetol phosphates[67] and phosphoglycolate,[68,69] were synthesized and proved extremely useful[23,67,70-73] in elucidating the active center of TPI. These active site reagents primarily modified a single glutamic acid residue.[23,72] Although some cysteines[74] and tyrosines[73] were labeled, these were shown to be due to side reactions, and that a glutamate was the primary active-site nucleophile. Support for a glutamate at the active site came from the studies of Rose and co-workers,[75-77] who utilized the 1-2-epoxypropanol 3-phosphate (glycidol phosphate) as active-site label. Finally, the amino acid sequence around the modified glutamate of the rabbit enzyme was elucidated by Hartman[23] and identified to be Glu-165. Homology of the active site residue of the human TPI was subsequently reported by Hartman and Gracy.[78]

High resolution X-ray crystallographic studies along with site-specific mutagenesis have further delineated the active site of TPI. Early studies on TPI crystals suggested that the active site was at the subunit interface and in a protected pocket. Subsequent studies on TPI crystals diffused with DHAP revealed interactions between several residues of the enzyme and the substrate.[79] Besides confirming the active site glutamate (Glu-165), participation of Lys-13 and His-95 were shown (Table 2). These observations are in agreement with the earlier kinetic predictions.[65,66] It should also be noted that Glu-165 and His-95 are contributed by one subunit while Lys-13 is from the other subunit. The active site of TPI appears to undergo conformational changes upon substrate binding. The loop composed of residues 169-176 (specifically Thr-172) appears to move by about 0.8 nm in the chicken enzyme.[79] Similarly, in the yeast enzyme, residues 168-177 move, with the main chain of Thr-172 being displaced by 1.04 nm.[79] This conformational change has been interpreted to be the result of the interaction between the enzyme and the phosphate group of the substrate. Recent chemical modification studies with ferrate ion showed loss of activity concurrent with the destruction of Trp-168, Tyr-191, and His-248, and limited destruction of Tyr-148.[80] Of these, only Trp-168 is believed to be important in catalysis. Although no specific role has been assigned to the highly conserved Trp-168 and to the highly conserved amino acids surrounding it, they appear to be essential in the catalytic action of TPI.

Our understanding of the catalytic mechanism of TPI has benefited from the advances in molecular biology. The chicken TPI gene was isolated,[81] cloned into an *E. coli* strain-deficient of TPI and observed to complement this deficiency.[82] A mutation (Glu-165 $\rightarrow$ Asp-165) was introduced to the same deficient strain.[83] This mutation decreased $k_{cat(G3P)}$ to approximately 0.07% of the wild-type enzyme and the $K_{m(G3P)}$ was increased by 3.6-fold. Values for $k_{cat(DHAP)}$ and $K_{m(DHAP)}$ were 0.4% and 2-fold that of the wild type enzyme, respectively.[83] The study concluded that mutation of Glu-165 to Asp-165 affected the free energy of the transition state(s) of the catalytic reaction.[83] Subsequent kinetic and active site modification studies of the mutant enzyme revealed that (1) the mechanism of proton abstraction was not changed, i.e., Asp-165 as well as Glu-165 abstracts the proton directly, without a water molecule acting as a relay[58] and (2) the stabilities of the transition states were dramatically reduced.[55] Further site-specific mutagenesis studies have modified oth-eressential residues and probed their effects.[84,85] An initial report[85] indicated that mutations of His-95 and Lys-13 to Gln-95 and Gln-13 produced enzymes with k_{cat} values of 1 and 0% of the wild-type enzyme, respectively. These findings are consistent with the importance of His-95 and Lys-13 in polarizing the carboxyl oxygens of the substrate and stabilizing the enediol(ate) intermediate.[79]

C. POSTSYNTHETIC MODIFICATIONS OF TPI

Multiple electrophoretic and chromatographic forms of TPI from many species have been reported by several laboratories. For example, the crystalline enzyme from human erythrocytes exhibits several catalytically active species which can be resolved by electrophoresis or isoelectric focusing.[86] A comparison of the electrophoretic properties of TPI from various species revealed that the multiple electrophoretic forms are present in all human tissues, and that most vertebrates also exhibit a similar multiplicity.[87] The first clue to the molecular basis for the multiple forms came with the observation that the more acidic isozymes accumulate in old erythrocytes, suggesting that the basis was of a postsynthetic nature.[88] Later, these more acidic isozymes were shown to accumulate in skin fibroblasts from old individuals,[89] fibroblasts aged in culture,[89] fibroblasts from patients with premature aging diseases,[90] and in the older cells in the lens of the eye.[91] Determination of the complete primary structure of the human TPI[10] and comparative structural studies on the "young" and "old" isozymes resulted in the elucidation of the molecular basis as the specific deamidations of Asn-15 and Asn-71 (24). Asn-15 and Asn-71 of each subunit are juxtaposed and located in the subunit-subunit interface. The deamidation and consequent introduction of four negative charges into the subunit-subunit interface destabilizes the dimer. It is believed that this spontaneous, specific deamidation process may be the first step in the normal catabolism of the enzyme.[24] The accumulation of the deamidated forms in old cells has been reviewed elsewhere.[91,92] The accumulation of the deamidated forms in aging cells is believed to be due to a defect in the ability of old cells to "recognize" and/or proteolytically degrade such postsynthetically modified proteins.

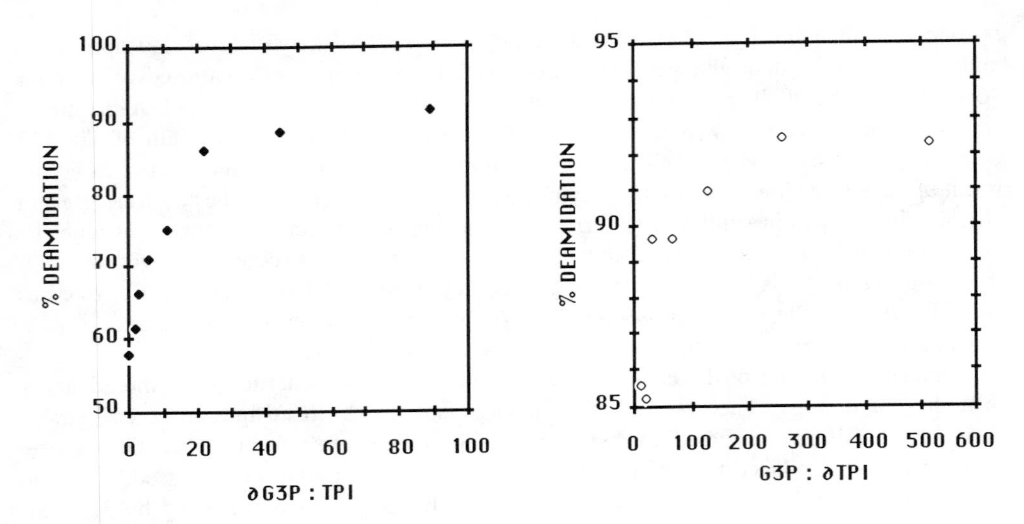

FIGURE 5. Deamidation of TPI in the presence of the substrate. Left: constant TPI concentration (4.68 μ*M*) with varying G3P concentrations (0 to 419 μ*M*). Right: constant G3P concentration (300 μ*M*) with varying TPI concentrations (0.6 to 37.4 μ*M*). For details see Reference 93.

Recently we have found that substrate enhances the rate of deamidation of TPI.[93] When TPI was incubated in the presence of the substrate(s), markedly increased rates of deamidation were observed, and deamidation exhibited a saturation behavior with respect to the molar ratio of the substrate to the enzyme (Figure 5). Only compounds that are substrates enhanced deamidation (i.e., inhibitors which do not turn over were ineffective). It is suggested that catalytic turnover is required for enhanced deamidation, and that the conformational changes in the active site during substrate binding[79,84] are instrumental in this process. Accelerated deamidation in the presence of substrate may also explain the rapid deamidation of TPI observed *in vivo*.

D. THE TPI GENE

The gene(s) for human TPI is located on chromosome 12. The functional transcription unit and processed pseudogenes have been isolated and characterized from human[11,94] and chimpanzee.[95] The functional gene is 3.5 kb with characteristic RNA polymerase II promoter elements: a TATA box at -26 to -21 and a CAAT box (CCAAT) at -73 to -70. In addition, the 5′ flanking region also contains a GC-rich sequence consistent with the Sp1 binding site. Seven exons exist and they are highly homologous in the human and chimpanzee genes. Although the introns were not sequenced in the case of the human, they have been completed in the chimpanzee.[95] Site-specific mutagenesis studies on TPI have provided further information on the mechanism of catalysis and other structural-functional features of the enzyme (*vide supra*).

III. GLUCOSE PHOSPHATE ISOMERASE

A. KINETIC STUDIES

From a mechanistic point of view it is obviously essential to know the structures of the actual substrates. GPI and MPI exhibit absolute selectivity for their substrates and absolute stereospecificity with regard to the structure at C-1 and C-2 of the substrates. That is to say GPI converts fructose 6-phosphate only to glucose 6-phosphate, while MPI catalyzes the reaction of fructose 6-phosphate only to mannose 6-phosphate. Moreover, the isomerases require the phosphorylated sugars and exhibit essentially no reaction with the nonphos-

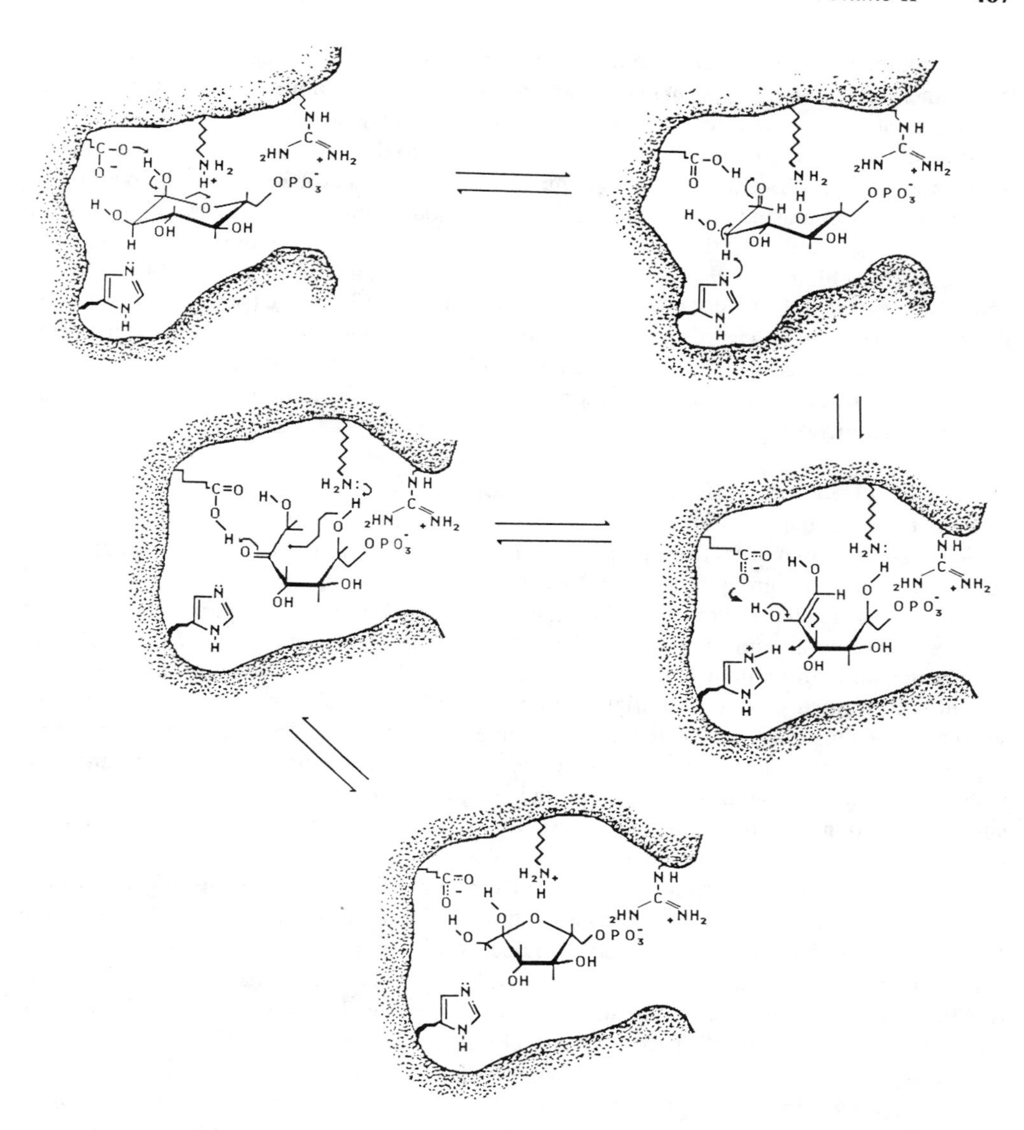

FIGURE 6. Mechanism of the GPI reaction. Stepwise catalysis of the GPI reaction according to Dyson and Noltmann[35] is shown. Arg and His are added as the phosphate stabilizing group, and as the general acid base, respectively.

phorylated analogues. In solution, hexoses exist both as the cyclic α and β anomers and the acyclic forms.[96] Although the ratios of these forms are subject to discussion,[97-100] the amount of the acyclic form appears to be, at most, 2.5% for F6P.[101] Since acyclic analogues of the substrates exhibit higher affinity for the enzyme than the largely cyclic substrates,[102-104] the open chain forms were thought to be the actual substrates.[102] However, it is clear that the cyclic forms must be able to react with the enzyme because of the anomerase activity of GPI observed with α-glucose 6-phosphate.[102,105,106] The ring opening step required in the conversion of the α-form into the β-anomer may explain how the cyclic forms can also be substrates. Since only the β-G6P was found to undergo isomerization with GPI, the reaction mechanism proposed by Dyson and Noltmann[34] contained an obligatory ring opening/anomerization step as an integral part of the isomerase reaction (Figure 6).

The anomerase reaction of GPI has been studied independently.[105,106] K_m and K_i values of the anomerase reaction were found to be significantly lower than the values observed for the isomerase reaction.[106] When the reactions were carried out in deuterated water, the magnitude of the isotope effect was small for either enzymatic activity.[105] The isotope effect associated with the anomerase reaction was comparable to the isotope effect observed during the anomerization of D-glucose either spontaneously or by mutarotase. Thus, the possibility of a common mechanism was pointed out. Chemical modification experiments with pyridoxal 5′-phosphate allowed selective inhibition of the two activities: (1) anomerase activity was preferentially inhibited if the reaction was carried out under incandescent light, (2) the converse was true if the reaction was carried out in dark, and (3) both activities were protected if 6-phosphogluconate was present.[106] On the basis of this and other chemical modification experiments it was concluded that, in accord with the previous proposal,[34] a lysine group was involved in the ring opening/anomerase activity, and a histidine was important for the isomerase activity.[106]

B. STRUCTURE OF THE ACTIVE CENTER

The nature of the specific amino acids of GPI involved in catalysis were first considered by Rose and O'Connell.[43] They proposed a nucleophile with no exchangeable hydrogens, such as a carboxylate group or a nonprotonated nitrogen of an imidazole ring. They reasoned that if a protonated base, like the ϵ-NH_2 group of lysine, abstracted the hydrogen, one would observe the same amount of label on the enzyme as in the C-2 position of the substrate. The participation of an unprotonated base was supported by specific activity calculations of the initially formed product.[43] Initial chemical modification experiments suggested the involvement of a histidine residue in the active center.[107-109] pH titration data[110] and temperature dependence of the pH effects[34] supported the possible involvement of histidine, and also suggested lysine as an additional likely participant in the catalytic center. Based on the active site modification and the substrate specificity data, Dyson and Noltman[34] proposed the reaction mechanism shown in Figure 6.

Attempts to covalently modify the catalytic center with the goal of identifying the specific active-site nucleophile spanned 2 decades. Studies with substrate analogs[111] and active-site labeling with pyridoxal 5′-phosphate[112] provided further evidence for the involvement of a carboxyl group (Glu) and a lysine, respectively. *N*-bromoacetylethanolamine phosphate (BAEP) was finally found to be the most effective site-specific agent for covalently labeling the active center of GPI.[101] This analogue closely resembles the open chain competitive inhibitors of GPI, such as ribulose 5-phosphate and 5-phosphoarabinonate (Figure 7). *N*-Bromoacetylethanolamine phosphate was found to rapidly inactivate GPI and to exhibit all the criteria of a specific covalent affinity label. This reagent has the additional advantage that, after modification of an active-site nucleophile, acid hydrolysis results in conversion of the nucleophile into the stable carboxymethylated derivative. Following reaction of the enzyme with this substrate analogue and tryptic digestion, a peptide containing the active site-labeled residue was obtained and sequenced.[110] The label was found on the histidine located on the third residue from the amino terminus of this peptide:

Val–Leu–His*–Ala–Glu–Asn–Val–Asp–(Gly, Thr, Ser)–Glu–Ile–

(Thr, Gly, His, Lys, Glx)–Tyr–Phe–(Ser, Glx, Gly, Gly)–Arg

Interestingly, there is a similarity between this sequence and the active site of TPI (*vide infra*). These structural studies suggest that a histidine serves as the nucleophile in the catalytic center as proposed by the kinetic studies of Noltmann and others. It is not known if the glutamate located two residues away from the active-site histidine represents the glutamate originally proposed by Rose and O'Connel.[114]

A covalently and stoichiometrically modifiable lysine in the active center of GPI was

```
O
‖
H-C
H-C-OH
HO-C-H
H-C-OH
H-C-OH
H2-C-O-PO3=
```

glucose 6-phosphate (G6P)

```
OH
H-C-H
C=O
HO-C-H
H-C-OH
H-C-OH
H2-C-O-PO3=
```

fructose 6-phosphate (F6P)

```
H-C
   >O
H-C
HO-C-H
H-C-OH
H-C-OH
H2-C-O-PO3=
```

1,2-epoxymannitol 6-phosphate

```
O
‖
C-OH
H-C-OH
HO-C-H
H-C-OH
H-C-OH
H2-C-O-PO3=
```

6-phosphogluconate

```
H
H-C-Cl
C=O
HO-C-H
H-C-OH
H-C-OH
H2-C-O-PO3=
```

1-chloro-2-oxohexanol phosphate

```
H
H-C-Br
C=O
N-H
H-C-H
H2-C-O-PO3=
```

N-bromoacetylethanol-amine phosphate

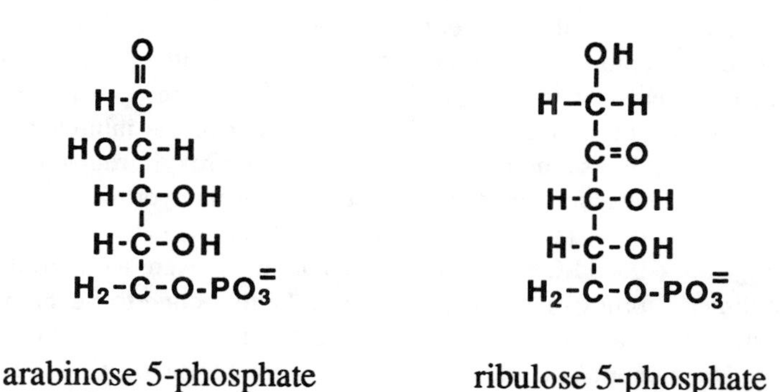

FIGURE 7. Structures and K_m (K_i) values for the substrates (analogues) of GPI. The brackets denote the species from which the enzyme was obtained. $K_{mG6P\ [HUMAN]} = 1.25 \times 10^{-4}\ M$,[33] $K_{mF6P\ [HUMAN]} = 7.13 \times 10^{-5}\ M$,[33] $K_{i\ 1,2\text{-epoxymannitol 6-phosphate [YEAST]}} = 2.7 \times 10^{-4}\ M$,[114] $K_{i6\text{-phosphogluconate [HUMAN]}} = 6.31 \times 10^{-5}\ M$,[33] $K_{i\ 1\text{-chloro-2-oxohexanol phosphate [RABBIT]}} = 1.43 \times 10^{-2}\ M$,[113] $K_{i\ N\text{-bromoacetylethanolamine phosphate [HUMAN]}} = 5.6 \times 10^{-5}\ M$,[110] $K_{i\ 5\text{-phosphoarabionate [YEAST]}} = 1 \text{ to } 2 \times 10^{-6}\ M$,[102] and $K_{i\ \text{erythrose 4-phosphate [YEAST]}} = 5 \times 10^{-5}\ M$.[102]

also found. This essential lysine was modified with pyridoxal 5′-phosphate and subsequent reduction of the Schiff base intermediate with borohydride.[112] The relative location of these two covalently modified residues in the active center of GPI posed several problems. In view of the large number of lysine, arginine, and methionine residues in GPI, conventional approaches to peptide fragmentation were difficult. However, Lu and Gracy[115] cleaved the

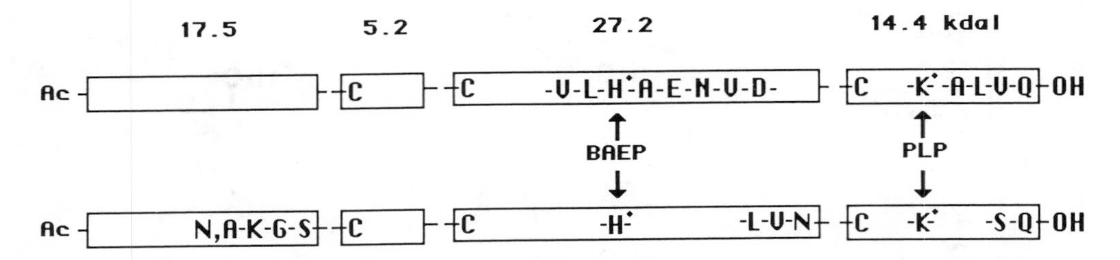

FIGURE 8. Summary of the active site and primary structure data for GPI. Top: human, bottom: swine enzyme. The active site residues histidine and lysine are marked by asterisks. Letters designate the amino acids in their standard one letter codes except, BAEP: *N*-bromoacetylethanolamine phosphate, PLP: pyridoxal 5′-phosphate, Ac (at the amino terminus): acetyl group, OH (at the carboxyl terminus): hydroxyl. The numbers at the very top are the sizes of the peptides in kilodaltons. Cysteine cleavage data and carboxyl terminal data are from Reference 115; other amino acid sequence data are from Reference 110.

human enzyme at the amino terminal sides of the three cysteines using 2-nitro-5-thio-cyanobenzoic acid. These studies revealed the relative locations of the active site histidine, labeled with BAEP, and the active site lysine, labeled with PLP (Figure 8).[115]

A specific arginine and a tryptophan also appear to be essential for the catalytic activity of the enzyme.[35] The critical arginine is stoichiometrically modified with either 2,3-butadione or cyclohexandione, and the essential tryptophan is stoichiometrically modified by *N*-bromosuccinimide. These modifications do not change the quartenary structure and have little effect on the secondary or the tertiary structure. The precise locations of these residues are currently not known.

C. CRYSTALLOGRAPHIC AND MATRIX-BOUND STUDIES

In spite of the lack of complete amino acid sequence data, some structural features of the active site have been deduced from X-ray crystallographic studies. Initial studies were carried out on the swine enzyme at 0.6-nm resolution[116] and later refined at 0.35 nm.[117,118] The data showed that the subunits were related to each other by a dyad axis[116] and that each had two β-structure domains, one containing four, the other six parallel strands.[117] Crystal structure studies in the presence of the competitive inhibitor arabinonate 5′-phosphate helped to delineate the active site region. The binding site appears to be in a protected pocket and formed from elements of each of the two subunits.[117] Furthermore, it was observed that portions of the dimeric enzyme moved together upon binding of the inhibitor, and it was proposed an amino acid on one subunit readily reacts with hydroxyl groups at C-1 and C-2 of the substrate, while a residue from the other subunit forms hydrogen bonds with the hydroxyls on C-3 and C-4, thus stabilizing the *cis*-enediol intermediate.[117]

Studies utilizing matrix-bound GPI also indicated that dimers were required for catalytic activity.[119,120] The dimeric form of the enzyme was covalently bound to Sepharose 4B and determined to be catalytically active. Dissociation of the subunits rendered the matrix-bound monomers inactive. The monomers could be reactivated upon redimerization. About half of the activity was regained if matrix-bound monomers were hybridized with covalently inactivated soluble monomers indicating that dimers are required for catalytic activity and that the two catalytic centers function independently.[119,120]

More recently the X-ray structure of the swine enzyme was resolved at 0.26 nm.[121] Data from crystallographic studies[116-118,121] confirm that the active site is at the subunit interface and at the domain interfaces.[121] High-resolution X-ray data along with some amino acid sequence data allowed the construction of a tentative active site for the pig enzyme.[121]

D. POSTSYNTHETIC MODIFICATIONS OF GPI

Multiple electrophoretic forms of GPI have been observed in a variety of species. These have been attributed to genetic heterogeneity, sulfhydryl oxidation, proteolysis, aggregation, dissociation, and isolation artifacts (for review see Reference 31). Recent studies on the multiple forms of GPI in bovine tissues have revealed that they are due to the postsynthetic deamidation of specific residues, strikingly resembling the deamidated forms of TPI.[122] Of particular interest is a parallel with the accumulation of the deamidated forms of TPI in aging cells. In the case of GPI, young mitotically active cells (e.g., intestinal mucosa and epithelial layer cells of the eye lens) contain only the most basic (nondeamidated) isozyme while the older cells contain increasingly higher levels of the deamidated isozymes. It is intriguing that the deamidations of GPI also cause a decrease in the stability of the dimer, similar to that of TPI. Preliminary results also suggest that these specific labile residues in GPI are Asn-Gly sequences similar or identical to the deamidation sites of TPI. It will be of particular interest to determine if the sites of deamidation in GPI and TPI bear primary sequence homologies, and if they are also located at the subunit-subunit interfaces.

E. THE GPI GENE

In human, the GPI gene is located on the long arm of chromosome number 19.[123] The structure of this gene has not been elucidated at the present time, but such studies are underway. Unlike TPI, the human GPI genome appears to be quite tolerant of mutations and many variants exist in the population. Most of the mutations are not of clinical significance, but many render the enzyme less stable.[124,125] One frequently occurring mutant form found in most populations exhibits increased stability.[33] While most of these variants are due to point mutations, a catalytically active form which is the result of a deletion has been characterized.[126] Clinically important variants with significantly lowered GPI activity exist, and GPI deficiency disease is the third most common of the erythroenzymopathies (after glucose 6-phosphate dehydrogenase deficiency and pyruvate kinase deficiency).

IV. MANNOSEPHOSPHATE ISOMERASE

A. KINETIC STUDIES

Although GPI had been identified as early as 1933 by Lohmann,[127] it was not until 1950 that the existence of a separate isomerase catalyzing the interconversion of fructose 6-phosphate to mannose 6-phosphate was suggested by Slein.[128] The existence of two distinct hexosephosphate isomerases was finally proven in 1958 when Noltmann and Bruns[129] obtained a preparation of MPI from brewers yeast which was free of GPI activity. MPI was first isolated to homogeneity 10 years later.[36] Subsequent studies characterized the yeast enzyme as a zinc metalloenzyme[25] and proposed a mechanism for catalysis.[37]

MPI exhibits strict specificity for the interconversion of fructose 6-phosphate and mannose 6-phosphate and utilizes only the β-anomers of mannose 6-phosphate[130] and fructose 6-phosphate.[131] Unlike GPI, it does not appear to possess an anomerase activity. On the other hand, GPI appears to anomerize mannose 6-phosphate[130] although this sugar phosphate is not a substrate for GPI (*vide supra*). MPI, like the other aldose-ketose isomerases, obeys Michaelis-Menten saturation kinetics. The yeast enzyme exhibits a K_m for mannose 6-phosphate of 1.35 m*M*, and it is significant that even the binding affinity of the epimer, glucose 6-phosphate, is an order of magnitude less (K_{iG6P} = 13.6 m*M*).[37] Most of the 5 and 6 carbon sugar phosphates and their alcohol derivatives are competitive inhibitors of MPI. Similar to GPI, the acyclic forms appear to exhibit higher affinities (e.g., mannitol phosphate exhibits a K_i of 0.8 m*M*[37]).

B. THE STRUCTURE OF THE ACTIVE CENTER

Yeast MPI was shown by atomic absorption spectroscopy to contain 1 g atom of zinc per mole of enzyme. Substrates protect against inactivation of the enzyme by chelating agents, and the zinc is believed to be directly involved in catalysis. The apoenzyme is totally without catalytic activity, but readdition of zinc or other metals such as cobalt reactivate the enzyme. Such transition metals have also been shown to catalyze the nonenzymatic aldose-ketose isomerizations.[37,132] Based on a series of studies of the pH dependence of the kinetic parameters, pK values of 6.6 and 7.8 for the free enzyme and 6.4 and 8.1 for the enzyme-substrate complex were determined for two ionizable groups believed to be involved in substrate binding and catalysis.[37] The pK values of 6.6 and 6.4 were assigned to the imidazole of a histidine residue. Structural analyses on MPI to locate the site of this residue have not yet been conducted, nor is it known which amino acids contribute the ligands to the active site zinc. The active site residue with the higher pK (7.8 to 8.1) is unknown. Since mannose and other nonphosphorylated analogs are not inhibitors of the enzyme, it appears that there is an essential requirement for the phosphate for binding to the enzyme, and thus a positively charged moiety in the catalytic center has been proposed.[37] Based on the above studies the mechanism shown in Figure 9 was proposed. On the basis of tritium exchange studies Simon and Medina[47] speculated that SH groups may be involved in complex formation between zinc and the enzyme.

C. STRUCTURAL STUDIES

The most striking features which distinguish MPI from GPI and TPI are that MPI is a monomeric protein while the other isomerases are isologous dimers, and the fact that MPI is a metalloenzyme. Detailed physical studies on yeast MPI showed a molecular weight of 45,000 Da which is distinctly different from the subunit molecular weights of either GPI or TPI (see Table 1). While it is unfortunate that detailed structural analyses of the MPI protein have not been conducted, nucleotide sequence analysis of two MPI genes have been recently reported. First, Miles and Guest[38] have deduced the sequence of 390 amino acid residues comprising the MPI monomer from *E. coli* with a predicted M_r of 42,716. This value is in good agreement with the M_r of the *E. coli* protein of 42,000 previously reported and with the yeast enzyme mentioned above. Second, the nucleotide sequence of the MPI gene from *Pseudomonas aeruginosa* was reported.[39] The gene was defined by the start and stop codons and an open reading frame, corresponding to a predicted polypeptide product of Mr = 52,860.[39] These authors failed to demonstrate any significant homology at the nucleotide level between the *P. aeruginosa* MPI gene and the *E. coli* MPI gene. In contrast to the nucleotide sequence homology studies, we observed that the predicted secondary structures appear to exhibit several similarities (Figure 10). Both enzymes are predicted to be largely helical in nature (50 and 45% for *E. coli* and *P. aeruginosa*, respectively) and some of these helices are rather long, i.e., *E. coli* residues 131-153, 167-206, and *P. aeruginosa*, residues 89-116, 198-224, 313-347. The rest of the other helices are on the average 7 amino acids long (range 2-15), while the β-sheets are considerably shorter (average length 4 residues, range 2-8).

V. SIMILARITIES AND DIFFERENCES IN THE ISOMERASES: CONVERGENCE OR DIVERGENCE?

TPI and GPI exhibit extremely high catalytic turnover numbers. In the case of TPI, the enzymatic reaction is 10^9 times faster than the nonenzymatic reaction.[62,135,136] The rate of the enzymatic reaction appears to be controlled only by the diffusion of the product away from the enzyme[103,136] and thus TPI has been termed "the perfect catalyst".[3,20] Kinetic studies as well as *ab initio* calculations suggest that, on the basis of the reaction being diffusion controlled, there should be no evolutionary pressure on the enzyme to "improve" its active site.[3,56] This is corroborated when the primary structures of TPI are compared (Figure 11): in all the species investigated so far, from *E. coli* to the human, the active site residues Lys-13, His-95, Gln-97, and Glu-165 are totally conserved. In addition to these critical residues and the residues surrounding them, the mobile loop of amino acids[166-178]

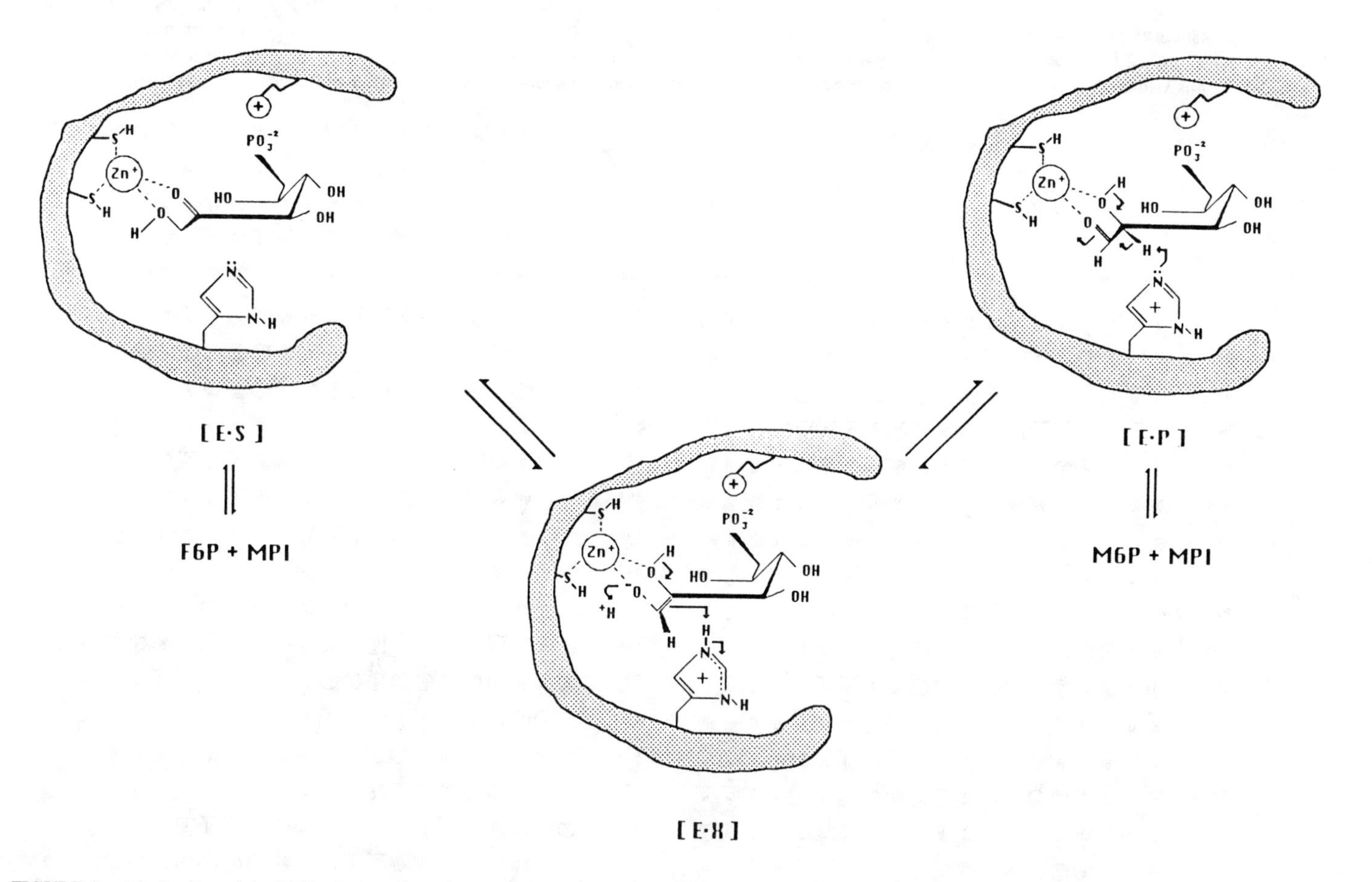

FIGURE 9. Mechanism of the MPI reaction. Stepwise catalysis of the isomerization reaction is according to Gracy and Noltmann.[37] The -SH groups believed to bind zinc are according to Reference 47.

SECONDARY STRUCTURE OF MPI P. aeruginosa

```
M I P V I L S G G S G L A L W P L S R K    20
Q Y P K Q F L A L T G D D T L F Q Q T L    40
K R L A F D G M Q A P L L V C N K A H R    60
F I V Q E Q L E A Q N L A S Q A I L L E    80
P F G R N T A P A V A I A A M K L V A E   100
G A D E L L I L P A D H V L E D Q A A F   120
Q Q A L A L A T N A A E K G E M V L F G   140
I P A S R P E T G Y G Y I A A T A D A Q   160
L P E G V S R V Q S F V E K P D E A R A   180
A E F V A A G G Y Y W N S G M F L F A A   200
S R Y L E E L K K H D A D I Y D T C L L   220
A L E R S Q H D G D L V N I D A A T F E   240
C C P D N S I D Y A V M E K T S A A C V   260
V P L S A G W N D V G S W S S I W D V H   280
A K D A N G N V T K G D V L V H D S H N   300
C L V H G H G K L V S V I G L E D I V V   320
V E T K D A M M I A H K D R V Q D V K H   340
V V K D L D A Q G R S E T Q N H C E L Y   360
R P W G S Y D S V D M G G R F Q V K H I   380
T V K P G A R L S L Q M H H H R A E H W   400
I V V S G T A Q V T C D D K T F L L T E   420
N Q S T Y I P I A S V H A L A N P G K I   440
P L E I I E V Q S G S Y L G E D D I E R   460
L E D V Y G R T A E P A L Q V V A A S R   480
```

SECONDARY STRUCTURE OF MPI E. Coli

```
M Q K L I N S V Q N Y A W G S K T A L T    20
E L Y G M E N P S S Q P M A E L W M G A    40
H P K S S S R V Q N A A G D I V S L R D    60
V I E S D K S T L L G E A V A K R F G E    80
L P F L F K V L C A A Q P L S I Q V H P   100
N K H N S E I G F A K E N A A G I P M D   120
A A E R N Y K D P N H K P E L V F A L T   140
P F L A M N A F R E F S E I V S L L Q P   160
V A G A H P A I A H F L Q Q P D A E R L   180
S E L F A S L L N M Q G E E K S R A L A   200
I L K S A L D S Q Q G E P W Q T I R L I   220
S E F Y P E D S G L F S P L L L N V V K   240
L N P G E A M F L F A E T P H A Y L Q G   260
V A L E V M A N S D N V L A A G L T P K   280
Y I D I P E L V A N V K F E A K P A N Q   300
L L T Q P V K Q G A E L D F P I P V D D   320
F A F S L H D L S D K E T T I S Q Q S A   340
A I L F C V E G D A T L W K G S Q Q L Q   360
L K P G E S A F I A A N E S P V T V K G   380
H G R L A R V Y N K L
```

FIGURE 10. Secondary structure predictions for mannosephosphate isomerase from *E. coli* and *Pseudomonas aeruginosa*. The primary structure data were taken from References 39 and 38, respectively. Secondary structures were predicted according to Garnier et al.,[133] hydropathies predicted according to Kyte and Doolittle,[134] with prediction windows of n = 17 and 11, respectively. Legend: circle: α-helix, triangle: β-sheet, square: β-turn, underlined: random coil, shaded/dark: inside, light: outside.

```
H= HUMAN  R=RABBIT   C=CHIKEN L=CEOLACANTH   Y=YEAST E=E.COLI B=B.STEAROTHERMOPHILUS T=T. BRUCEI
M=MAIZE   N=ASPERGILLUS NIDULANS

*         5         10        5         20        5         30        5         40        5         50
H A P S R K F F U G G N W K M N G R K Q N L G E L I S T L Q G A K U P A D T E U U C I G P T A - Y L D F
R A P S R K F F U G G N W K M N G R K K N L G E L I T T L N A A K U P A D T E U U C A P P T A - Y I D F
C A P - R K F F U G G N W K M N G K R K S L G E L I H T L D G A K L S A D T E U U C G A P S I - Y L D F
L A P - R K F F U G G N W K M N G D K K S L G E L I Q T L N A A K U P F T G E I U C A P P E A - Y L D F
Y A - - R T F F U G G N F K L N G S K Q S I K E ! U E R L N T A S I P E N U E U U I C P P A T - Y L D Y
E - - - R H P L U M G N W K L N G S R H M U H E L U S N L R K E L A G U A G C A U A I A P P E M Y I D M
B - - M R K P I I A G N W K M H K T L A E A U Q F U E D U K G H U P P A D E U D S U U A P F L F R L U Q
T - S K P Q P I A A A N W K C N G S Q Q S L S E L I D L F N S T S I N H D U Q C U U A S T F U H - L A M
M - - G R K F F U G G N W K C N G T T D Q U E K I U K T L N E G Q U P P S D U U E U U U S P P - Y U F L
N - P - R K F F U G G N F K M N G N A E S T T S I I K N L N S A N L D K S U E U U U S P P A L - Y L L Q

*         5         60        5         70        5         80        5         90        5         100
H - - - A R Q K L D Q K I A A G A Q N C Y K U T N G A F T G E I S P G M I K D C G A T W U U L G H S E R R H
R - - - A R Q K L D P K I A U A A Q N C Y K U T N G A F T G E I S P G M I K D C G A T W U U L G H S E R R H
C - - - A R Q K L D A K I G U A A Q N C Y K U P K G A F T G E I S P A M I K D I G A A W U I L G H S E R R H
L - - - A R L K U D P K F G U A A Q N C Y K U S K G A F T G E I S P A M I K D C G U T W U I L G H S E R R H
Y - - S U S L U K K P Q U T U G A Q N A Y L K A S G A F T G E N S U D Q I K D U G A K W U I L G H S E R R S
E - A K R R E A E G S H I M L G A Q N U N L N L S G A F T G E T S A A M L K D I G A Q Y I I I G H S E R R T
B - A A D G T D L - Q K I - - G A Q T M H F A B Z G A Y T G E U S P U M L K D L G U T Y U I L G H S E R R H
T - T K E R L S H - P K F U I A A Q N A I A K - S G A F T G E U S L P I L K D F G U N W I U L G H S E R R A
M P U U K S Q L R Q E F H - U A A Q N C W U K K G G A F T G E U S A E M L U N L G U P W U I L G H S E - - -
N - - - A R E U A N K E I G U A A Q N U F D K P N G A F T G E I S U Q Q L R E A N I D W T I L G H S E R R U

*         5         110       5         120       5         130       5         140       5         150
H U F G E S D E L I G Q K U A H A L N E G L G U I A C I G E K L D E R E A G I T E K U U F E Q T K U I
R U F G E S D E L I G Q K U A H A L S E G L G U I A C I G E K L D E R E A G I T E K U U F E Q T K U I
C U F G E S D E L I G Q K U A H A L A E G L G U I A C I G E K L D E R E A G I T E K U U F Q E T K A I
L U F G E S D E L I G Q K U S H A L S E G L G U U A C I G E K L D E R E A G I T E G U U F E U T E U I
Y Y F H E D D K F I A D K T K F A L G Q G U G U I L C I G E T L E E K K A G K T L D U U E R Q L N A U
E Y H K E S D E L I A K K F A U L K E Q G L T P U L C I G E T E A E N E A G K T E E U C A R Q I D A U
B M F A Z T B Z T U B K K U L A A F T R G L I P I I C C G E S L E E R Q A G E T D A U U S Q U E K A L
T Y Y G E T N E I U A D K U A A A U A S G F M U I A C I G E T L Q E R E S G R T A U U U L T Q I A A I
M - - G E S N E F U G D K U A Y A L S Q G L K U I A C U G E T L E Q R E A G S T M D U U A A Q T K A I
N I L K E T D E F I A R K T K A A I E G G L Q U I F C I G E T L E E R E A N K T I D U U T R Q L N A A
```

FIGURE 11. Comparison of the primary structures of TPI. The amino acids are aligned to achieve maximum homology; spaces were added where needed. The amino acids are numbered in reference to the rabbit TPI.[12] Sequence data are taken from the following references: human,[10] rabbit,[12] chicken,[13] corrected according to Reference 18, yeast,[18] *Bacillus stearothermophilus*,[14] *E. coli*,[20] *Trypnosoma brucei*,[19] maize,[15] and *Aspergillus nidulans*.[16] Human sequence is corrected at position 19 according to our observations and of Reference 11, and at position 166 according to References 11 and 20.

```
*         5       160             5       170         5       180         5       190         5       200
H A D D V K - - D W S K V V L A Y E P V W A I G T G K T A T P Q Q A Q E V H E K L R G W L N S N V S D A V
R A D N V K - - D W S K V V L A Y E P V W A I G T G K T A T P Q Q A Q E V H E K L R G W L K S N V S D A V
C A D N V K - - D W S K V V L A Y E P V W A I G T G K T A T P Q Q A Q E V H E K L R G W L K T H V S D A V
L A D D V K - - D W S K V V L A Y E P V W A I G T G K T A S P Q Q S Q E L H G K L R K W L K E N V S E T V
Y L E E V K - - D W T N V V V A Y E P V W A I G T G L A A T P E D A Q D I H A S I R K F L A S K L G D K A
E L K T Q G A A A F E G A V I A Y E P V W A I G T G K S A T P A Q A Q A V H K F I R D H I A K V D A N I A
B A G L T P - - Q E V K I I L A Y E P L W A I G T G K S S T P Z B A B S V C G H I R S V V S R L F G P E A
T A K K L K K A D W A K V V I A Y E P V W A I G T G K V A T P Q Q A Q E A H A L I R S W V S S K I G A D V
M A E K I K     D W S N V V V A Y E P V W A I G T G K V A T P A Q A Q E V H A S L R D W L K T N A S P E V
N A K E L S K E Q W A K V V I A Y E P V W A I G T G K V A T T E Q A Q E V H S A I R K W L K D A I S A E A

*         5       210         5       220         5       230         5       240                   5       250
H A Q Q T R I I Y G G S V T G A T C K E L A S Q P D V D G F L V G G A S L K P E F V D L I - - - - - N A K Q
R A Q S T R I I Y G G S V T G A T C K E L A S Q P D V D G F L V G G A S L K P E F V D I I - - - - - N A K Q
C A V Q S R I I Y G G S V T G G N C K E L A S Q H D V D G F L V G G A S L K P E F V D I I - - - - - N A K H
L A D S V R I I Y G G S V T G A T C K E L A S E P D V D G F L V G G A S L K P E F V E Y K - - - - - D V R Q
Y A S E L R I L Y G G S A N G S N A V T F K D K A D V D G F L V G G A S L K P E F V D I I - - - - - N S R N
E E Q V - I I Q Y G G S V N A S N A A E L F A Q P D I D G A L V G G A S L K A D A F A V I V K A A E A A K Q
B A E A I R I Q Y G G S V K P D N I R D F L A Z Z Z I B G A L V G G A S L E P A S F L Q L V Q A G - - - R H E
T R G E L R I L Y G G S V N G K N A R T L Y Q Q R D V N G F L V G G A S L K P E F V D I I - - - - - K A T Q
M A E S T R I I Y G G S V T A A N C K E L A A Q P D V D G F L V G G A S L K P E F I D I I - N A A T V K S A
N A E N T R I I Y G G S V S E K N C K D L A K E A D I D G F L V G G A S L K P A F V D I - V N A R L
```

FIGURE 11 (continued)

Ancesteral isomerase	Val-Leu-His-Ala-Tyr-Glu- X -Val
GPI	Val-Leu-His-Ala- Glu-Asn-Val
TPI	Val-Leu- Ala-Tyr-Glu-Pro-Val

FIGURE 12. Comparison of the active site peptides from TPI and GPI. (From Gibson, D. R., Gracy, R. W., and Hartman, F. C., *J. Biol. Chem.*, 255, 9369, 1980. With permission.)

and the regions of residues 205-212 and 228-238 are also highly conserved. (The side chain hydroxyl group of Ser-211 and the amide hydrogens of Gly-232 and Gly-233 appear to interact with sulfate ion, a structural analog of the phosphate group[79]). Thus, the active site of TPI appears to have evolved a very long time ago and has changed little over the eons. As predicted, mutations induced in the active-site region result in enzymes with negligible catalytic activities (*vide supra*). These findings are in good agreement with the thesis that the mammalian species exhibit a very low frequency of allelic variants. It is thus proposed that essentially any mutation effecting the ''perfectly evolved'' enzyme would be selected against or result in null mutations. The human population may contain a large number of null alleles.[137]

How do GPI and MPI compare with TPI in terms of changes during evolution? Unfortunately, primary sequence data on these two isomerases are incomplete. Preliminary data suggest that they may tolerate a higher number of mutations than TPI. On the other hand, amino acid sequence data of the GPI active site peptide showed striking resemblance to the active-site peptide from TPI (Figure 12).[110] Thus, the idea of a common ancesteral gene has been discussed.[138] We have also utilized computer-assisted programs to search the MPI sequences from *E. coli* and *P. aeruginosa* for a conserved active-site sequence similar to those of TPI and GPI. We did not find a region homologous to the conserved active-sites of TPI and GPI.

One other point of similarity between TPI and GPI appears to be the postsynthetic modification of these isomerases. TPI has been known for some time to exhibit several catalytically active forms. Some of these were determined to be of postsynthetic origin, specifically, due to the deamidations of Asn-15 and Asn-71.[24] These forms were shown to accumulate in cells from old donors[89] or with premature aging diseases like the Gilford-Hutchinson progeria syndrome or the Werner's syndrome.[90] Deamidation renders TPI more labile to heat or chemical denaturation;[27] thus, it has been viewed as one of the first steps in the normal catabolism of the enzyme.[24] Recently, GPI has been observed to exhibit multiple forms which are very much like the pseudoisozymes of TPI.[122] Indeed, these multiple forms of GPI also appear to result from the specific deamidations of asparagines located in Asn-Gly sequences.[139] The deamidated forms of GPI also appear to accumulate in cells from old tissues.[91]

Similarities or differences in the primary structure are considered as evidence for recent evolutions, while secondary and tertiary structure changes are believed to reflect more distant evolutionary events.[140] In this sense, a number of glycolytic enzymes share a large number of common structural features;[141] TPI consists of a barrel of eight parallel β-sheets with surrounding helices.[142] Similarly, GPI contains two domains with six and four helices, respectively.[117] Interestingly, similar domains were observed in other enzymes not related to glycolysis, and this structural feature cannot be used conclusively. In reviewing the evolution of the glycolytic pathway from a protein structure point of view, Fothergill-Gilmore concluded that TPI and GPI were not evolutionally related.[143] On the other hand, in light of the similarities of the active site peptides, tertiary structures, composition of the active sites from both of the two homologous subunits, in the transition states, and the post synthetic modifications, one is clearly drawn to the structural and functional similarities of TPI and GPI. At this time, however, it is not possible to determine if these similarities are due to divergence or convergence. MPI, being not as well characterized, appears to be evolutionally less related to either GPI or TPI.

NOTES ADDED IN PROOF

1. Site directed mutagenesis of the yeast enzyme (Glu-165 → Asp-165, a very sluggish mutant in terms of catalysis) and viscogenic agents were utilized to provide further proof that the TPI catalyzed reaction is diffusion controlled and that the reaction is limited by the rate at which glyceraldehyde 3-phosphate is bound or released from the active site (Blacklow, S. C., Raines, R. T., Lim, W. A., Zamore, P. D., and Knowles, J. R., Triosephosphate catalysis is diffusion controlled, *Biochemistry*, 27, 1158, 1988).

2. Tracer pertubation methods have shown that TPI exists in two unliganded forms with a rate constant of interconversion of $10^6\ s^{-1}$. While the molecular nature of the two forms were not determined, protein shuffling between Lys-13 and His-95, two electrophiles believed to facilitate the enolization step, have been postulated (Raines, R. T., and Knowles, J. R., Enzyme relaxation in the reaction catalyzed by triosephosphate: detection and kinetic characterization of two unliganded forms, *Biochemistry*, 26, 7014, 1987.)

3. To analyze the role of the electrophile His-95, the yeast enzyme was first expressed in a TPI deficient strain of *E. coli*, and characterized (Nickbarg, E. B. and Knowles, J. R., Triosephosphate isomerase: energetics of the reaction catalyzed by the yeast enzyme expressed in *Escherlchia coli*, *Biochemistry*, 27, 5939, 1988). The kinetic parameters of the yeast enyzme were found to be very similar to the chicken enzyme (*vide supra*). Next, a His-95 → Gln-95 substitution was introduced and the mutant enzyme characterized (Nickbarg, E. B. and Knowles, J. R., Triosephosphate isomerase: removal of a putatively electrophilic histidine residue results in a subtle change in catalytic mechanism, *Biochemistry*, 27, 5948, 1988). The specific catalytic activity of the His-95 → Gln-95 mutant was decreased about 400-fold, and it did not catalyze the exchange of protons between the solvent and the remaining substrate and the solvent. The conclusions drawn from this study were that His-95 was necessary to stabilize the reaction intermediate, while it could not be demonstrated whether the His-95 functioned as a proton donor or as an hydrogen bond donor in stabilizing the intermediate. Also, a new mechanism was proposed for the behavior of the His-95 → Gln-95 mutant.

4. Finally, four consecutive residues (Ile-170-Gly-Thr-Gly-173), two of which were previously implicated in binding of the phosphate group were removed from the putative loop of the triosephosphate isomerase to examine their role in catalysis (Pompliano, D. L., Peyman, A., and Knowles, J. R., Stabilization of a reaction intermediate as a catalytic device: definition of the functional role of the flexible loop in triosephosphate isomerase, *Biochemistry*, 29, 3186, 1990). The properties of this loop have been discussed in the text. The mutant enzyme had a lower specific catalytic activity by a factor of 10^5-fold, and essentially catalyzed the elimination of the phosphate group from the enediol(ate) intermediate. These data confirm the importance of the mobile loop of TPI in sequestering the intermediate in a favorable conformation and excluding the solvent out of the active site and thereby minimizing futile β-elimination of the phosphate group which results in the extremely reactive and toxic metabolite methylglyoxal.

5. The amino acid sequence for GPI from *E. coli* (Froman, B. E., Tait, R. C., and Gottlieb, L. D., Isolation and characterization of the phosphoglucose isomerase gene from *Escherichia coli, Mol. Gen. Genet.*, 217, 126, 1989), two strains of yeast (Wésoloviski-Louvel, M., Goffrini, P., and Ferrero, I., The *RAG2* gene of the yeast *Kluyveromyces lactis* codes for a putative phosphoglucose isomerase, *Nucleic Acids Res.*, 16, 8714, 1988; Green, J. B. A., Wright, A. P. H., Cheung, W. Y., Lancashire W. E., and Hartley, B. S., The structure and regulation of phosphoglucose isomerase in *Saccharamyces cerevisiae, Mol. Gen. Genet.*, 215, 100, 1988) and pig (Chaput, M., Claes, V., Portetelle, D., Cluds, I., Cravador, A., Burny, A., Gras, H., Tartar, A., The neurotrophic factor neuroleukin is 90% homologous with phosphohexose isomerase, *Nature,* 332, 454, 1988) have been deduced from nucleic acid sequencing. While homology to the monomeric neurotrophic factor neurokeukin (Chaput et al., 1988; Faik, P., Walker, J. I. H., Redmill, A. A. M., Morgan, M. J., Mouse glucose-6-phosphate isomerase and neuroleukin have identical 3′ sequences, *Nature,* 332, 445, 1988) and to the surface glycoprotein gp-120 of the human immunodeficiency virus have been postulated, the catalytic mechanism of GPI has not been been further elucidated. We have isolated a new form of GPI from human placenta which substantially differs from the previously characterized form in its specific catalytic activity, amino acid composition and molecular weight (Sun, A. Q., Yüksel, K. U., Jacobson, T. M., and Gracy, R. W., Human glucose 6-phosphate isomerase isoforms containing two different size subunits: isolation, characterization and comparison with neuroleukin, submitted for publication).

ACKNOWLEDGEMENTS

Some of the work reported here was sponsored by research grants from the National Institutes of Health (AM14638, AG01274), the R. A. Welch Foundation (B-502), and the Texas Advanced Technology and Research Program (Applied Enzymology).

REFERENCES

1. **Speck, J. C.,** The Lobry de Bruyn-Alberda van Ekenstein Transformation, *Adv. Carbohydrate Chem.*, 13, 63, 1958.
2. **von Figura, K. and Hasilik, A.,** Lysosomal enzymes and their receptors, *Annu. Rev. Biochem.*, 55, 167, 1986.
3. **Albery, W. J. and Knowles, J. R.,** Evolution of enzyme function and development of catalytic efficiency, *Biochemistry,* 15, 5631, 1976.
4. **Albery, W. J. and Knowles, J. R.,** Efficiency and evolution of enzyme catalysis, *Angew. Chem. Intl. Engl. Ed.*, 16, 285, 1977.
5. **Topper, Y. J.,** Aldose-ketose transformations, in *The Enzymes,* Vol. 5, 2nd ed., Boyer, P. D., Ed., Academic Press, New York, 1961, 429.
6. **Noltman, E. A.,** Aldose ketose isomerases, in *The Enzymes,* Vol. 6, 3rd ed., Boyer, P. D., Ed., Academic Press, New York, 1972, 271.
7. **Rose, I. A.,** Mechanism of aldose-ketose isomerase reactions, in *Advances in Enzymology,* Vol. 43, Meister, A., Ed., John Wiley & Sons, New York, 1975, 491.
8. **Rose, I. A.,** Enzymology of proton abstraction and transfer reactions, in *The Enzymes,* Vol. 2, 3rd ed., Boyer, P. D., Ed., Academic Press, New York, 1970, 281.
9. **Rose, I. A.,** Mechanism of enzyme action, *Annu. Rev. Biochem.*, 35, 23, 1966.
10. **Lu, H. S., Yuan, P. M., and Gracy, R. W.,** Primary structure of human triosephosphate isomerase, *J. Biol. Chem.*, 259, 11958, 1984.
11. **Maquat, L. E., Chilkote, R., and Ryan, P. M.,** Human triosephosphate isomerase cDNA and protein structure, *J. Biol. Chem.*, 260, 3748, 1985.

12. **Corran, P. H. and Waley, S. G.,** The amino acid sequence of rabbit muscle triose phosphate isomerase, *Biochem. J.*, 145, 335, 1975.
13. **Banner, D. V., Bloomer, A. C., Petsko, G. A., Phillips, D. C., Pogson, I. A., Corran, P. H., Furth, A. J., Millman, D. J., Offord, R. E., Priddle, J. D., and Waley, S. G.,** Structure of chicken muscle triose phosphate isomerase determined crystallographically at 2.5 Å resolution using amino acid sequence data, *Nature*, 255, 609, 1975.
14. **Artavanis-Tsakonas, S., and Harris, J. I.,** Primary structure of triosephosphate isomerase from *Bacillus stearothermophilus, Eur. J. Biochem.*, 108, 599, 1980.
15. **Marchionni, M. and Gilbert, W.,** The triosephosphate isomerase gene from maize: introns antedate the plant-animal divergence, *Cell*, 46, 133, 1986.
16. **Mc Knight, G. L., O'Hara, P. J., and Parker, M. L.,** Nucleotide sequence of triosephosphate isomerase gene from *Aspergillus nidulans*: implications for a differential loss of introns, *Cell*, 46, 143, 1986.
17. **Kolb, E., Harris, J. I., and Bridgen, J.,** Triose phosphate isomerase from the *Ceolacanth, Biochem. J.*, 137, 195, 1974.
18. **Alber, T. and Kawaski, G.,** Nucleotide sequence of triose phosphate isomerase of *Saccharomyces cerevisiae, J. Mol. App. Genet.*, 1, 419, 1982.
19. **Swinkels, B. W., Gibson, W. C., Osinga, K. A., Kramer, R., Veeneman, G. H., van Boom, J. H., and Borst, P.,** Characterization of the gene for the microbody (glycosomal) triosephosphate isomerase of *Trypanosoma brucei, EMBO J.*, 5, 1291, 1986.
20. **Pichersky, E., Gottlieb, L. D., and Hess, J. F.,** Nucleotide sequence of the triose phosphate isomerase gene of *Escherichia coli, Mol. Gen. Genet.*, 195, 314, 1984.
21. **Alber, T., Hartman, F. C., Johnson, R. M., Petsko, G. A., and Tsernoglou, D.,** Crystallization of yeast triose phosphate isomerase from polyethylene glycol, *J. Biol. Chem.*, 256, 1356, 1981.
22. **Banner, D. V., Bloomer, A. C., Petsko, G. A., Phillips, D. C., and Wilson, I. A.,** Atomic coordinates for triose phosphate isomerase from chicken muscle, *Biochem. Biophys. Res. Commun.*, 72, 146, 1976.
23. **Hartman, F. C.,** Haloacetol phosphates: characterization of the active site of the rabbit muscle triose phosphate isomerase, *Biochemistry*, 10, 146, 1971.
24. **Yuan, P. M., Talent, J. M., and Gracy, R. W.,** Molecular basis for the accumulation of acidic isozymes of triosephosphate isomerase on aging, *Mech. Ageing Dev.*, 17, 151, 1981.
25. **Gracy, R. W. and Noltmann, E. A.,** Studies on phosphomannose isomerase. II. Characterization as a zinc metalloenzyme, *J. Biol. Chem.*, 243, 4109, 1968.
26. **Yuan, P. M., Dewan, R. N., Thompson, R. E., and Gracy, R. W.,** Isolation and characterization of triosephosphate isozymes from human placenta, *Arch. Biochem. Biophys.*, 198, 42, 1979.
27. **Sawyer, T. H. and Gracy, R. W.,** Ligand binding and denaturation titration of free and matrix-bound triosephosphate isomerases, *Arch. Biochem. Biophys.*, 169, 51, 1975.
28. **Veech, R. L., Raijman, L., Dalziel, K., and Krebs, H. A.,** Disequilibrium in the triosephosphate isomerase system in rat liver, *Biochem. J.*, 115, 837, 1969.
29. **Reynolds, S. J., Yates, D. W., and Pogson, C. I.,** Dihydroxyacetone phosphate. Its structure and reactivity with α-glycerophosphate dehydrogenase, aldolase, triose phosphate isomerase and some possible metabolic implications, *Biochem. J.*, 122, 285, 1971.
30. **Gracy, R. W.,** Nature of multiple forms of glucosephosphate and triosephosphate isomerases, in *Isozymes: Molecular Structure*, Vol. 1, Markert, C. L., Ed., Academic Press, New York, 1975, 471.
31. **Gracy, R. W.,** Glucosephosphate and triosephosphate isomerases: significance of isozyme structural differences in evolution, physiology, and aging, in *Isozymes: Current Topics in Biological and Medical Research*, Vol. 6, Rattazzi, M. C., Scandalios, J. G., and Witt, G. S., Eds., Alan R. Liss, New York, 1982, 169.
32. **James, G. T. and Noltmann, E. A.,** Chemical studies on the subunit structure of rabbit muscle phosphoglucose isomerase, *J. Biol. Chem.*, 248, 730, 1973.
33. **Tilley, B. E., Gracy, R. W., and Welch, S. G.,** A point mutation increasing the stability of human phosphoglucose isomerase, *J. Biol. Chem.*, 249, 4571, 1974.
34. **Dyson, J. E. D. and Noltman, E. A.,** The effect of pH and temperature on the kinetic parameters of phosphoglucose isomerase, *J. Biol. Chem.*, 243, 1401, 1968.
35. **Lu, H. S., Talent, J. M., and Gracy, R. W.,** Chemical modification of critical catalytic residues of lysine, arginine, and tryptophan in human glucose phosphate isomerase, *J. Biol. Chem.*, 256, 785, 1981.
36. **Gracy, R. W. and Noltmann, E. A.,** Studies on phosphomannose isomerase. I. Isolation, homogeneity measurements and determination of some physical properties, *J. Biol. Chem.*, 243, 3161, 1968.
37. **Gracy, R. W. and Noltmann, E. A.,** Studies on phosphomannose isomerase. III. A mechanism for catalysis and for the role of zinc in the enzymatic and the nonenzymatic isomerization, *J. Biol. Chem.*, 243, 5410, 1968.
38. **Miles, J. S. and Guest, J. R.,** Nucleotide sequence and transcription start point of the phosphomannose isomerase gene (ManA) of *E. coli, Gene*, 32, 41, 1984.

39. **Darzins, A., Frantz, B., Vanags, R. I., and Chakrabarty, A. M.,** Nucleotide sequence analysis of the phosphomannose isomerase gene (pmi) of *Pseudomonas aeruginosa* and comparison with the corresponding *Escherichia coli* gene manA, *Gene,* 42, 293, 1986.
40. **Rieder, S. V. and Rose, I. A.,** Mechanism of action of muscle triose phosphate isomerase, *Fed. Proc.,* 15, 337, 1956.
41. **Rieder, S. V. and Rose, I. A.,** Mechanism of the triose phosphate isomerase reaction, *J. Biol. Chem.,* 234, 1007, 1959.
42. **Bloom, B. and Topper, Y. J.,** Mechanism of action of aldose and phosphotriose isomerase, *Science,* 124, 982, 1956.
43. **Rose, I. A. and O'Connel, E. L.,** Intramolecular hydrogen transfer in the phosphoglucose isomerase reaction, *J. Biol. Chem.,* 236, 3086, 1961.
44. **Rose, I. A. and O'Connel, E. L.,** Stereospecificity of the sugar-phosphate isomerase reactions; a uniformity, *Biochim. Biophys. Acta,* 42, 159, 1960.
45. **Topper, Y. J.,** On the mechanism of action of phosphoglucose isomerase and phosphomannose isomerase, *J. Biol. Chem.,* 225, 419, 1957.
46. **Fedtke, C.,** Intramolecular hydrogen transfer in isomersation reactions of sugar phosphates in the Calvin cycle, *Progr. Photosyn. Res.,* 3, 1597, 1969.
47. **Simon, H. and Medina, R.,** Untersuchung der Mannose-6-Phosphate-Isomerase-Reaktion mit Mannose-2-T, *Z. Naturforschung,* 21b, 496, 1966.
48. **Albery, W. J. and Knowles, J. R.,** Deuterium and tritium exchange in enzyme kinetics, *Biochemistry,* 15, 5588, 1976.
49. **Herlihy, J. M., Maister, S. G., Albery, W. J., and Knowles, J. R.,** Energetics of triosephosphate isomerase: the fate of the 1-(R)-^{3}H label of tritiated dihydroxyacetone phosphate in the isomerase reaction, *Biochemistry,* 15, 5601, 1976.
50. **Maister, S. G., Pett, C. P., Albery, W. J., and Knowles, J. R.,** Energetics of triosephosphate isomerase: the appearance of solvent tritium in substrate dihydroxyacetone phosphate and in the product, *Biochemistry,* 15, 5607, 1976.
51. **Fletcher, S. J., Herlihy, J. M., Albery, W. J., and Knowles, J. R.,** Energetics of triosephosphate isomerase: the appearance of solvent tritium in substrate glyceraldehyde 3-phosphate and in the product, *Biochemistry,* 15, 5612, 1976.
52. **Leadlay, P. F., Albery, W. J., and Knowles, J. R.,** Energetics of triosephosphate isomerase: deuterium isotope effects in the enzyme catalyzed reaction, *Biochemistry,* 15, 5617, 1976.
53. **Fisher, L. M., Albery, W. J., and Knowles, J. R.,** Energetics of triosephosphate isomerase: the nature of proton transfer between the catalytic base and solvent water, *Biochemistry,* 15, 5621, 1976.
54. **Albery, W. J. and Knowles, J. R.,** Energetics of triosephosphate isomerase: free energy profile for the reaction catalyzed by triosephosphate isomerase, *Biochemistry,* 15, 5627, 1976.
55. **Raines, R. T., Sutton, E. L., Straus, D. R., Gilbert, W., and Knowles, J. R.,** Reaction energetics of a mutant triosephosphate isomerase in which the active-site glutamate has been changed to aspartate, *Biochemistry,* 25, 7142, 1986.
56. **Alagona, G., Desmeules, P., Ghio, C., and Kollman, P. A.,** Quantum mechanical and molecular studies on a model for the dihydroxyacetone phosphate-glyceraldehyde phosphate isomerization catalyzed by triosephosphate isomerase (TIM), *J. Am. Chem. Soc.,* 106, 3623, 1984.
57. **Alagona, G., Ghio, C., and Kollman, P. A.,** A simple model for the effect of Glu165→Asp165 mutation on the rate of catalysis in triose phosphate isomerase (TIM), *J. Mol. Biol.,* 191, 23, 1986.
58. **Raines, R. T. and Knowles, J. R.,** The mechanistic pathway of a mutant triosephosphate isomerase, *Ann. N.Y. Acad. Sci.,* 471, 266, 1986.
59. **Grazi, E. and Trombetta, G.,** A new intermediate of the aldolase reaction, the pyruvaldehyde-aldolase-orthophosphate complex, *Biochem. J.,* 175, 361, 1978.
60. **Webb, M. R., Standring, D. N., and Knowles, J. R.,** Phosphorus-31 nuclear magnetic resonance of dihydroxyacetone phosphate in the presence of triosephosphate isomerase. The question of nonproductive binding of substrate hydrate, *Biochemistry,* 16, 2738, 1977.
61. **Campbell, I. D., Jones, R. B., Kiener, P. A., and Waley, S. G.,** Enzyme-substrate and enzyme-inhibitor complexes of triose phosphate isomerase studied by ^{31}P nuclear magnetic resonance, *Biochem. J.,* 179, 607, 1979.
62. **Richard, J. P.,** Acid-base catalysis of the elimination and isomerization reactions of triose phosphates, *J. Am. Chem. Soc.,* 106, 4926, 1984.
63. **Iyengar, R. and Rose, I. A.,** Liberation of the triosephosphate isomerase reaction intermediate and its trapping by isomerase, yeast aldolase, and methylglyoxal synthase, *Biochemistry,* 20, 1229, 1981.
64. **Iyengar, R. and Rose, I. A.,** Concentration of activated intermediates of the fructose 1,6-bisphosphate aldolase and triosephosphate isomerase reactions, *Biochemistry,* 20, 1223, 1981.
65. **Burton, P. M. and Waley, S. G.,** The active centre of triosephosphate isomerase, *Biochem. J.,* 100, 702, 1966.

66. **Burton, P. M. and Waley, S. G.,** The reaction of triosephosphate isomerase and aldolase with limited amounts of diazonium salts, *Biochem. J.*, 104, 3P, 1967.
67. **Hartman, F. C.,** A potential active site reagent for aldolase, triose phosphate isomerase and glycerol 1-phosphate dehydrogenase, *Fed. Proc.*, 27, 654, 1968.
68. **Wolfenden, R.,** Transition state analogues for enzyme catalysis, *Nature*, 233, 704, 1969.
69. **Wolfenden, R.,** Binding of substrate and transition state analogs to TPI, *Biochemistry*, 9, 3404, 1970.
70. **Hartman, F. C.,** Haloacetolphosphates. Potential active site reagents for aldolase, triose phosphate isomerase, and glycerophosphate dehydrogenase. II. Inactivation of aldolase, *Biochemistry*, 9, 1783, 1970.
71. **Hartman, F. C.,** Irreversible inactivation of triose phosphate isomerase by 1-hydroxy-3-iodo-2-propane phosphate, *Biochem. Biophys. Res. Commun.*, 33, 888, 1968.
72. **Hartman, F. C.,** Isolation and characterization of an active-site peptide from triose phosphate isomerase, *J. Am. Chem. Soc.*, 92, 2170, 1970.
73. **Coulson, A. F. W., Knowles, J. R., Priddle, J. D., and Offord, R. E.,** Uniquely labelled active site sequence in chicken muscle triose phosphate isomerase, *Nature*, 227, 180, 1970.
74. **Hartman, F. C.,** Haloacetolphosphates. Potential active site reagents for aldolase, triose phosphate isomerase, and glycerophosphate dehydrogenase. I. Preparation and properties, *Biochemistry*, 9, 1776, 1970.
75. **Rose, I. A. and O'Connell, E. L.,** Inactivation and labeling of triose phosphate isomerase and enolase by glycidol phosphate, *J. Biol. Chem.*, 244, 6548, 1969.
76. **Waley, S. G., Miller, J. C., Rose, I. A., and O'Connell, E. L.,** Identification of site in triose phosphate isomerase labelled by glycidol phosphate, *Nature*, 227, 181, 1970.
77. **Miller, J. C. and Waley, S. G.,** The active centre of rabbit muscle triose phosphate isomerase, *Biochem. J.*, 123, 163, 1971.
78. **Hartman, F. C. and Gracy, R. W.,** An active-site peptide from human triose phosphate isomerase, *Biochem. Biophys. Res. Commun.*, 52, 388, 1973.
79. **Alber, T., Banner, D. V., Bloomer, A. C., Petsko, G. A., Phillips, R. S. D., Rivers, P. S., and Wilson, I. A.,** On the three dimensional structure and catalytic mechanism of triose phosphate isomerase, *Phil. Trans. R. Soc. London*, 293B, 159, 1981.
80. **Steczko, J., Hermodson, M., Axelrod, B., and Dziember-Kentzer, E.,** Identification of the target amino acids in the site-specific inactivation of triose phosphate isomerase by ferrate ion, *J. Biol. Chem.*, 258, 13148, 1983.
81. **Straus, D. and Gilbert, W.,** Genetic engineering in the precambrian: structure of the chicken triosephosphate isomerase gene, *Mol. Cell. Biol.*, 5, 3497, 1985.
82. **Straus, D. and Gilbert, W.,** Chicken triosephosphate isomerase complements an *Escherichia coli* deficiency, *Proc. Natl. Acad. Sci. U.S.A.*, 82, 2014, 1985.
83. **Straus, D., Raines, R., Kawashima, E., Knowles, J. R., and Gilbert, W.,** Active site of triosephosphate isomerase: *in vitro* mutagenesis and characterization of an altered enzyme, *Proc. Natl. Acad. Sci. U.S.A.*, 82, 2272, 1985.
84. **Petsko, G. A., Davenport, R. C., Frankel, D., and RaiBhandary, U. L.,** Probing the catalytic mechanism of yeast triose phosphate isomerase by site-specific mutagenesis, *Biochem. Soc. Trans.*, 12, 229, 1984.
85. **Davenport, R. C. and Petsko, G. A.,** Site-directed mutagenesis of yeast triosephosphate isomerase, *Biochemistry*, 24, 3373, 1985.
86. **Rozacky, E. E., Sawyer, T. H., Barton, R. A., and Gracy, R. W.,** Studies on human triosephosphate isomerase. I. Isolation and properties of the enzyme from erythrocytes, *Arch. Biochem. Biophys.*, 146, 312, 1971.
87. **Snapka, R. M., Sawyer, T. H., Barton, R. A., and Gracy, R. W.,** Comparison of the electrophoretic properties of triosephosphate isomerases of various tissues and species, *Comp. Biochem. Physiol., B*, 49, 733, 1974.
88. **Gracy, R. W.,** Epigenetic formation of isozymes: the effect of aging, in *Isozymes: Current Topics in Biological and Medical Research*, Vol. 7, Rattazzi, M. C., Scandalios, J. G., and Whitt, G. S., Eds., Alan R. Liss, New York, 1983, 187.
89. **Tollefsbol, T. O. and Gracy, R. W.,** Premature aging diseases: cellular and molecular changes, *BioScience*, 33, 634, 1983.
90. **Tollefsbol, T. O., Zaun, R. M., and Gracy, R. W.,** Increased lability of triosephosphate isomerase in progeria and Werner's syndrome fibroblasts, *Mech. Ageing Dev.*, 20, 93, 1982.
91. **Gracy, R. W., Yüksel, K. Ü., Chapman, M. L., Cini, J. K., Jahani, M., Lu, H. S., Oray, B., and Talent, J. M.,** Impaired protein degradation may account for the accumulation of "abnormal" protein in aging cells, in *Modification of Proteins During Aging*, Adelman, R. C. and Dekker, E. E., Eds., Alan R. Liss, New York, 1985, 1.
92. **Gracy, R. W., Chapman, M. L., Cini, J. K., Jahani, M., Tollefsbol, T. O., and Yüksel, K. Ü.,** Accumulation of abnormal proteins in progeria and aging fibroblasts, in *Molecular Biology of Aging*, Woodhead, A. D., Blackett, A. D., and Hollaender, A., Eds., Plenum Press, New York, 1985, 427.
93. **Yüksel, K. Ü. and Gracy, R. W.,** *In vitro* deamidation of human triosephosphate isomerase, *Arch. Biochem. Biophys.*, 228, 452, 1986.

94. **Brown, J. R., Daar, I. O., Krug, J. R., and Maquat, L. E.,** Characterization of the functional gene and several processed pseudogenes in the human triosephosphate isomerase gene family, *Mol. Cell. Biol.*, 5, 1694, 1985.
95. **Craig, L., Pirtle, I., Gracy, R. W., and Pirtle, R. M.,** Structure of the transcription unit and two processed pseudogenes of chimpanzee triosephosphate isomerase, submitted for publication.
96. **Angyal, S. J.,** The composition and conformation of sugars in solution, *Angew. Chem. Intl. Ed. Eng.*, 8, 157, 1969.
97. **Los, J. M., Simpson, L. B., and Wiesner, K.,** The kinetics of mutarotation of D-glucose with consideration for intermediate free-aldehyde form, *J. Am. Chem. Soc.*, 78, 1564, 1956.
98. **Cantor, S. M. and Peniston, Q. P.,** The reduction of aldoses at the dropping mercury electrode: estimation of the aldehydro structure in aqueous solutions, *J. Am. Chem. Soc.*, 62, 2113, 1940.
99. **Gray, G. R. and Barber, R.,** Studies on the substrates of *D*-fructose 1,6-diphosphate aldolase in solution, *Biochemistry*, 9, 2459, 1970.
100. **Swenson, C. A. and Barber, R.,** Proportion of keto and aldelhydo forms in solutions of sugars and sugar phosphates, *Biochemistry*, 10, 3151, 1971.
101. **Hartman, F. C., Suh, B., Welch, M. H., and Barker, R.,** Inactivation of class I fructose diphosphate aldolase by the substrate analog *N*-bromoacetylethanolamine phosphate, *J. Biol. Chem.*, 248, 8233, 1973.
102. **Salas, M., Vinuela, E., and Sols, A.,** Spontaneous and enzymatically catalyzed anomerization of glucose 6-phosphate and anomeric specificity of related enzymes, *J. Biol. Chem.*, 240, 561, 1965.
103. **Parr, C. N. and Whittaker, M.,** Some observations on phosphotriose isomerase, *Biochem. J.*, 74, 34P, 1960.
104. **Sols, A.,** Carbohydrate metabolism, *Annu. Rev. Biochem.*, 30, 213, 1961.
105. **Schray, K. J. and Howell, E. E.,** Anomerization of glucose 6-phosphate: catalysis of phosphoglucose isomerase, *Arch. Biochem. Biophys.*, 192, 241, 1979.
106. **Howell, E. E. and Schray, K. J.,** Comparative inactivation and inhibition of the anomerase and isomerase activities of phosphoglucose isomerase, *Mol. Cell. Biochem.*, 37, 101, 1981.
107. **Chatterjee, G. C. and Noltmann, E. A.,** Dye-sensitized photooxidation as a tool for the elucidation of critical amino acid residues in phosphoglucose isomerase, *Eur. J. Biochem.*, 2, 9, 1967.
108. **Chatterjee, G. C. and Noltmann, E. A.,** Reaction of sulfhydryl groups in phosphoglucose isomerase with organic mercurials, *J. Biol. Chem.*, 242, 3440, 1967.
109. **Schnackerz, K. D. and Noltmann, E. A.,** Carboxyamidomethylation and inactivation of rabbit muscle phosphoglucose isomerase, *J. Biol. Chem.*, 245, 6417, 1970.
110. **Gibson, D. R., Gracy, R. W., and Hartman, F. C.,** Affinity labeling and characterization of the active site histidine of glucosephosphate isomerase, *J. Biol. Chem.*, 255, 9369, 1980.
111. **Rose, I. A. and O'Connel, E. L.,** Active site reagent for P-glucose isomerase, *Fed. Proc.*, 30, 1158, 1971.
112. **Schnackerz, K. D. and Noltmann, E. A.,** Pyridoxal 5′-phosphate as a site-specific protein reagent for a catalytically critical lysine residue in rabbit muscle phosphoglucose isomerase, *Biochemistry*, 10, 4837, 1971.
113. **Schnackerz, K. D., Chirgwin, J. M., and Noltmann, E. A.,** Synthesis of 1-chloro-2-oxohexanol 6-phosphate, a covalent active site reagent for phosphoglucose isomerase, *Biochemistry*, 20, 1756, 1981.
114. **O'Connell, E. L. and Rose, I. A.,** Affinity labeling of phosphoglucose isomerase by 1,2-anhydrohexitol 6-phosphates, *J. Biol. Chem.*, 248, 2225, 1973.
115. **Lu, H. S. and Gracy, R. W.,** Specific cleavage of glucosephosphate isomerases at cysteinyl residues, using 2-nitro-5-thiocyanobenzoic acid: analyses of peptides eluted from polyacrylamide gels, and localization of active site histidyl and lysyl residues, *Arch. Biochem. Biophys.*, 212, 347, 1981.
116. **Muirhead, H. and Shaw, P. J.,** Three-dimensional structure of pig muscle phosphoglucose isomerase at 6 Å resolution, *J. Mol. Biol.*, 89, 195, 1974.
117. **Shaw, P. J. and Muirhead, H.,** The active site of glucose phosphate isomerase, *FEBS Lett.*, 65, 50, 1976.
118. **Muirhead, H. and Shaw, P. J.,** Crystallographic structure analysis of glucose 6-phosphate isomerase at 3.5 Å resolution, *J. Mol. Biol.*, 109, 475, 1977.
119. **Bruch, P., Schnackerz, K. D., and Gracy, R. W.,** Matrix bound phosphoglucose isomerase, *Eur. J. Biochem.*, 68, 153, 1976.
120. **Gracy, R. W. and Schnackerz, K. D.,** Subunit complementation of glucosephosphate isomerases with matrix-bound monomers, *Tex. J. Sci. Spec. Publ.*, 3, 27, 1977.
121. **Achari, A., Marshall, S. E., Muirhead, H., Palmieri, R. H., and Noltmann, E. A.,** Glucose 6-phosphate isomerase, *Phil. Trans. R. Soc. London, B*, 293, 145, 1981.
122. **Cini, J. K. and Gracy, R. W.,** Molecular basis of the isozymes of bovine glucose 6-phosphate isomerase, *Arch. Biochem. Biophys.*, 249, 500, 1986.

123. **Lusis, A. J., Heinzmann, C., Sparkes, R. S., Scott, J., Knott, T. J., Geller, R., Sparkes, M. C., and Mohandas, T.,** Regional mapping of human chromosome 19: organization of genes for plasma lipid transport (APOC1, -C2, and -E and LDLR) and the genes C3, PEPD, and GPI, *Proc. Natl. Acad. Sci. U.S.A.*, 83, 3929, 1986.
124. **Gibson, D. R., Talent, J. M., Gracy, R. W., Schnackerz, K. D., and Ishimoto, G.,** A preparative method for the isolation of genetic variant forms of glucosephosphate isomerase and a study of five variants G., *Clin. Chim. Acta.*, 89, 355, 1978.
125. **Purdy, K. L., Jones, S., Tai, H. H., and Gracy, R. W.,** A radioimmunoassay for human glucosephosphate isomerase: measurement of immunoreactive enzyme from genetic variants, *Arch. Biochem. Biophys.*, 200, 485, 1980.
126. **Yuan, P. M., Zaun, M. R., Kester, M. V., Snider, C. E., Johnson, M., and Gracy, R. W.,** A deletion mutation in glucosephosphate isomerase (GPI Denton), *Clin. Chim. Acta*, 92, 481, 1979.
127. **Lohmann, K.,** Über Phosphorilierung und Dephosphorilierung. Bildung der natürlichen Hexosemono-phosphorsäure aus ihren Komponenten, *Biochem. Z.*, 262, 137, 1933.
128. **Slein, M. W.,** Phosphomannose isomerase, *J. Biol. Chem.*, 186, 753, 1950.
129. **Noltmann, E. A. and Bruns, F. H.,** Phosphomannose-isomerase II. Anreicherung des Enzyms aus Hefe und Abtrennung von Phosphoglucose-isomerase durch Säulenchromatographie an Hydroxylapatit, *Biochem. Z.*, 330, 514, 1958.
130. **Rose, I. A., O'Connell, E. L., and Schray, K. J.,** Mannose 6-phosphate: anomeric form used by phosphomannose isomerase and its 1-epimerization by phosphoglucose isomerase, *J. Biol. Chem.*, 248, 2232, 1973.
131. **Schray, K. J., Waud, J. M., and Howell, E. E.,** Phosphomannose isomerase. Isomerization of the predicted β-*D*-fructose 6-phosphate, *Arch. Biochem. Biophys.*, 189, 106, 1978.
132. **Tilley, B. E., Porter, D. W., and Gracy, R. W.,** Metal ion catalysis of aldose-ketose isomerizations in acidic solutions, *Carbohydrate Res.*, 27, 289, 1973.
133. **Garnier, J.,** Analysis of the accuracy and implications of simple methods for predicting the secondary structure of globular proteins, *J. Mol. Biol.*, 120, 97, 1978.
134. **Kyte, J. and Doolittle, R. F.,** A simple method for displaying the hydropathic character of a protein, *J. Mol. Biol.*, 157, 105, 1982.
135. **Putman, S. J., Coulson, A. F. W., Farley, I. R. T., Riddleston, B., and Knowles, J. R.,** Specificity and kinetics of triose phosphate isomerase from chicken muscle, *Biochem. J.*, 129, 301, 1972.
136. **Hall, A. and Knowles, J. R.,** The uncatalyzed rates of enolization of dihydroxyacetone phosphate and of glyceraldehyde 3-phosphate in neutral aqueous solution. The quantitative assessment of the effectiveness of an enzyme catalyst, *Biochemistry*, 14, 4348, 1975.
137. **Mohrenweiser, H. W. and Neel, J. V.,** Frequency of thermostability variants: estimation of the total "rare" variant frequency in human populations, *Proc. Natl. Acad. Sci. U.S.A.*, 78, 5729, 1981.
138. **Gibson, D. R., Gracy, R. W., and Hartman, F. C.,** Affinity labeling and characterization of the active site histidine of glucosephosphate isomerase — sequence homology with triosephosphate isomerase, *J. Biol. Chem.*, 255, 9369, 1980.
139. **Cini, J. K., Cook, P. F., and Gracy, R. W.,** Molecular basis for the isozymes of bovine glucose 6-phosphate isomerase, *Arch. Biochem. Biophys.*, 263, 96, 1988.
140. **Rossman, M. G.,** Evolution of glycolytic enzymes, *Phil. Trans. R. Soc. London, B*, 293, 191, 1981.
141. **Sternberg, M. J. E., Cohen, F. E., Taylor, W. R., and Feldman, R. J.,** Analysis and prediction of structural motifs in the glycolytic enzymes, *Phil. Trans. R. Soc. London, B*, 293, 177, 1981.
142. **Phillips, D. C., Sternberg, M. J. E., Thornton, J. M., and Wilson, I. A.,** An analysis of the structure of triose phosphate isomerase and its comparison to lactate dehydrogenase, *J. Mol. Biol.*, 119, 923, 1978.
143. **Fothergill-Gilmore, L. A.,** The evolution of the glycolytic pathway, *Trends Biol. Sci.*, 11, 47, 1986.

Chapter 20

MECHANISTIC STUDIES OF MUSCLE ALDOLASE

D. J. Hupe

TABLE OF CONTENTS

I. INTRODUCTION

Aldolases are one of the oldest groups of enzymes to be studied mechanistically. They are responsible for the carbon-carbon bond cleavage reactions within glycolysis resulting in the conversion of six carbon sugars to three carbon units. Although there are two broad classes of these enzymes, the type I which catalyze the reaction through the intermediacy of an enzyme-substrate imine adduct, and the type II which use metal ion activation, it is the first of these which has received most mechanistic attention and which is the subject of this review. There are few cases in the chemistry-enzymology literature where such a broad array of physical organic chemistry has been so neatly interwoven with a large number of studies on the enzyme to give a sense of quantitative understanding of how the enzymatic rate enhancement occurs.

This paper represents an update of previous reviews on the more general subject of imine formation in enzymatic catalysis.[1-3]

II. TYPE 1 IMINE FORMING ALDOLASES

A. GENERAL MECHANISM

Rabbit muscle aldolase (E.C. 4.1.2.13) is the archtypical type 1 (imine forming) aldolase and constitutes 3% of the soluble protein from this source.[4] It is responsible for catalyzing the cleavage or synthesis of the carbon-carbon bond which joins carbons 3 and 4 of fructose 1,6-diphosphate (F-1,6P_2) to form dihydroxyacetone phosphate (DHAP), and glyceraldehye-3-phosphate, (G-3P). The mechanism involves (in the synthesis direction) the binding of DHAP as an iminium ion in the first step (1 in Figure 1). An α-proton abstraction then occurs stereospecifically (2), and after this G-3P binds (3). The aldol condensation reaction then occurs to form the new C-C bond (4), resulting in the iminium ion of F-1,6-P_2, which can then hydrolyze (5). The evidence for this is presented below.

B. PHYSICAL-ORGANIC CHEMISTRY OF IMINIUM ION FORMATION AND CARBONYL GROUP ACTIVATION

In order to understand the reasons for the selection of this complicated pathway in which covalent intermediates are formed between substrate and enzyme, a brief description of the physical-organic chemistry of these systems is required. An equilibrium is established rapidly in aqueous solution upon addition of a primary amine and a ketone, which is dependent upon pH and involves the formation of the imine and its protonated form, the iminium ion.[1] As shown in Equation 1, the equilibria for these

$$\begin{array}{ccccc} RNH_2 & C{=}O & \overset{K_{im}}{\rightleftarrows} & RN{=}C & \\ K_a^{RNH_3+}\ \uparrow\downarrow & & & \uparrow\downarrow\ K_a^{imh+} & \\ RNH_3^+ & C{=}O & \overset{K_{imh+}}{\leftrightarrows} & R\overset{H}{\underset{+}{N}}{=}C & \end{array} \qquad (1)$$

processes are related, and in general equilibrium constants for formation of imine from the carbonyl and amine (K_{im}) are larger than equilibrium constants for formation of iminium ion from the carbonyl and the ammonium ion (K_{imh+}). These constants are related since the ratio of K_{imh+}/K_{im} is equal to the ratio $K_a^{RNH}3^+/K_a^{imh+}$, which is about 10^{-3}. Therefore, at neutral pH values the concentration of iminium ion is 1/1000th that of the imine. Typically for a ketone and a primary aliphatic amine, $pK_a^{RNH}3^+ = 10$, $pK_a^{IMH+} = 7$, $K_{im} = 10^{-1}$, and $K_{imh+} = 10^{-4}$. Shown in Figure 2 is a plot of the concentrations of the different species

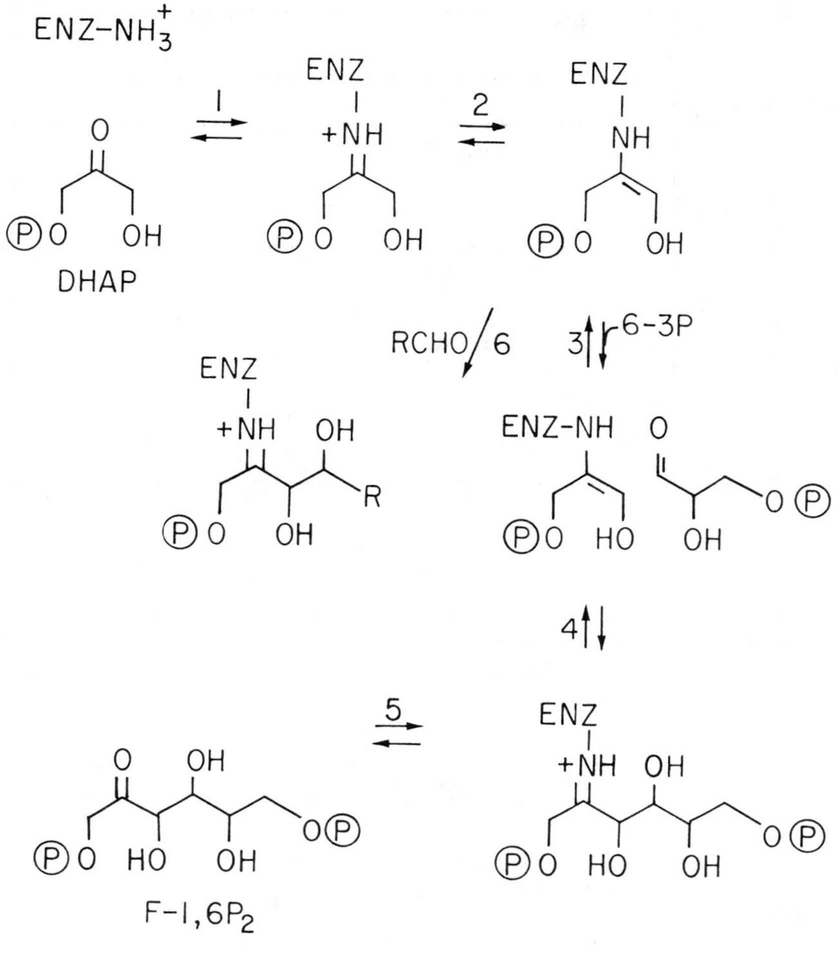

FIGURE 1. A general mechanism for rabbit muscle aldolase demonstrating that DHAP binds first (1) and must be deprotonated (2) prior to productive binding of G-3P (3), followed by the aldol condensation step (4), and hydrolysis of product F-1,6 P_2. In the cleavage direction carboxypeptidase treated aldolase can not proton transfer effectively (2), but can get as far as eneamine, and can still effectively transfer the DHAP unit to a cosubstrate aldehyde (6), so that the modified enzyme behaves like a transaldolase.

at equilibrium calculated using these values at arbitrarily chosen concentrations of 0.1 *M* ketone and 0.01 *M* amine. Only a small fraction of either amine or ketone is tied up as imine or iminium ion at any pH value. The concentration of iminium ion is linked to the concentration of ammonium ion and its concentration dependence does not show an apparent pK_a equal to its real value, but rather depends upon the pK_a of the parent amine. In practical terms, if a rapid equilibrium formation of iminium ion occurs prior to other rate-determining steps in an enzymatic or nonenzymatic reaction, the rate equation will reflect the pK_a of the parent primary amine rather than the pK_a of the iminium ion which is formed. These considerations help in understanding the reasons for the abnormally low pK_a values found for an imine-forming lysine amino group in acetoacetate decarboxylase, for example.[3]

It is interesting to compare the equilibrium constants for the formation of iminium ion from simple aliphatic carbonyl compounds and amines with the same reaction catalyzed by an enzyme. Values of K_m for substrate DHAP in the micromolar range have been measured, so that for the enzyme substrate iminium ion $K_{imh+} = 10^6$. This contrasts dramatically with

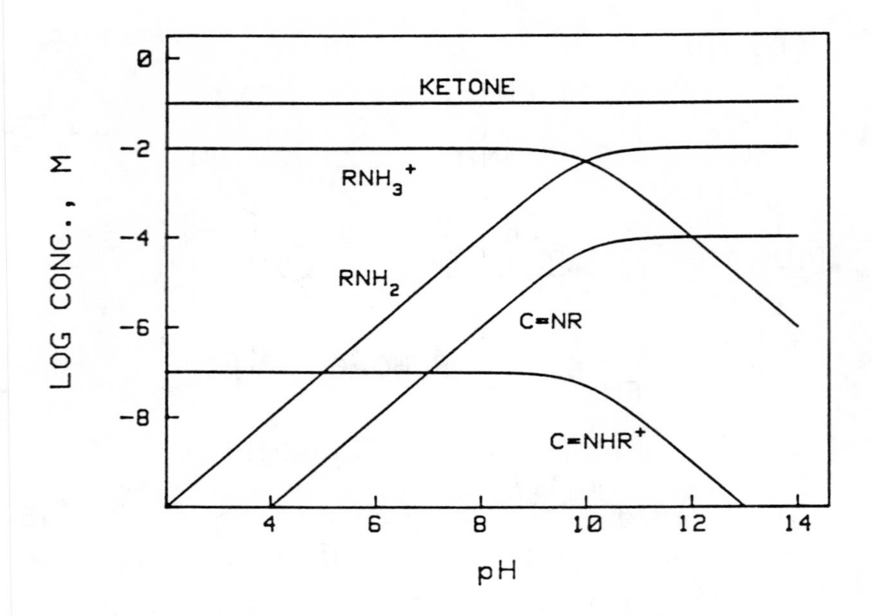

FIGURE 2. The pH dependence for the equilibrium that exists between a typical ketone, primary amine, imine, and iminium ion in aqueous solution under the conditions described in the text.

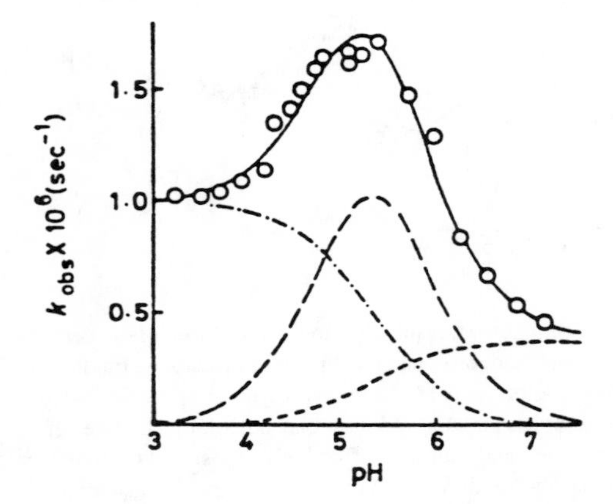

FIGURE 3. A pH rate profile for an α-proton abstraction reaction in aqueous solution catalyzed by a primary amine with a pK_a value of 5.34. There are substantial terms in the rate expression proportional to free amine, protonated amine, and free times protonated amine.

the K_{imh+} value of 10^{-4} described above. It is apparent that a considerable fraction of the binding energy comes from the usual noncovalent associations between substrate and active site rather than because of the stability of the covalent imine bond being formed.

Once an iminium ion has been formed, it converts the carbonyl compound into a more active entity, particularly with respect to reactions, such as deprotonations, decarboxylations, or de-aldolization reactions which leave a formal negative charge on the α-carbon. These types of reactions are catalyzed by primary amines in aqueous solution and by enzymes relying on iminium ion formation.[2] Typically a deprotonation reaction will show complex behavior such as that shown in Figure 3, where the rate expression due to amine was k_{obs}

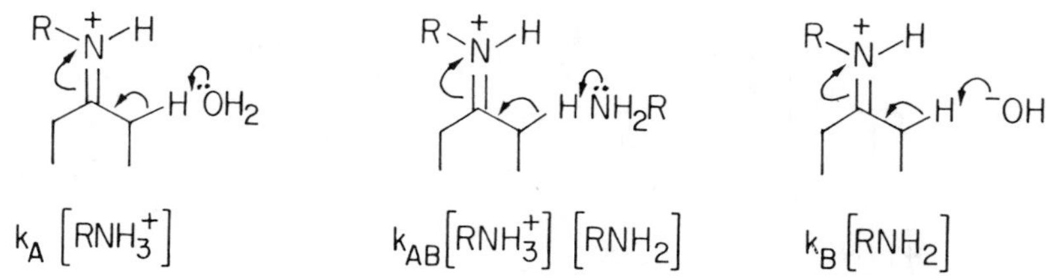

$R{-}\overset{+}{N}{-}H$ ... $H\ddot{O}H_2$ $R{-}\overset{+}{N}{-}H$ ... $H\ddot{N}H_2R$ $R{-}\overset{+}{N}{-}H$ ... $H\ ^{-}OH$

$k_A\,[RNH_3^+]$ $k_{AB}[RNH_3^+]\,[RNH_2]$ $k_B\,[RNH_2]$

FIGURE 4. The ascribed mechanism for each of the kinetic terms demonstrated in the pH rate profile in Figure 3.

$= k_a[RNH_3^+] + k_{ab}[RNH_3^+]\,[RNH_2] + k_b[RNH_2]$. Each of these terms was absent for tertiary amines which cannot form iminium ions.[6,7] As shown in Figure 4 the terms are each due to the more reactive iminium ion reacting with the bases water, free amine, or hydroxide ion. The contribution to the reactivity of these systems from both the base and the iminium ion forming amine has been analyzed. It has been concluded that the highest rate of reaction will be obtained when the pK_a of the amine equals the pK_a of the proton abstracting base, and these equal the pH of the solution at which the reaction occurs.[2] Interestingly, there is considerable data to suggest that, through evolutionary selection, the enzymes that operate *via* an iminium ion have altered the properties of the catalytic groups involved to match these constraints.

C. CHEMISTRY OF ACTIVE SITE RESIDUES

Because aldolase is essential to glycolysis, readily available and mechanistically interesting, it has been the subject of a large number of enzymatic studies. These include determinations of active site residues, order of substrate addition, stereochemistry, chemical mechanism, mechanism based inhibitor design, and X-ray structure studies. An attempt will be made to integrate some of this vast amount of information into a coherent picture of the active site mechanics of the enzyme.

Following the study by Westheimer and Cohen[8] of the primary and secondary amine-catalyzed dealdolization of diacetone alcohol which was proposed to occur through an essential iminium ion intermediate, Speck and Forist[9] proposed that the same class of reaction may occur in aldolase. The first characterization of a stable covalent entity formed between substrate and enzyme was done by Horecker et al.,[10] on transaldolase. This enzyme, which forms a stable complex with dihydroxy acetone formed from the cleavage of fructose-6-P, was found to be susceptible to reduction by sodium borohydride. It was demonstrated that this adduct was formed irreversibly in the presence of substrate and borohydride and that irreversible enzymatic inactivation occurred at the same rate as incorporation of $[^{32}P]$-DHAP. Grazi et al.[11] were responsible for demonstrating that a similar adduct was formed between DHAP and rabbit muscle aldolase in the presence of borohydride, as shown in Figure 5, and by using $[^{14}C]$-DHAP, they isolated the radioactive derivative of lysine that was expected, N^6-β-glyceryllysine.[12,13] The conclusion was that both transaldolase and muscle aldolase formed an imine linkage between the ε-amino group of an active site lysine and the carbonyl group of the substrate. Further evidence for the formation of this linkage, and its presence on the mechanistic pathway, arose from the inhibition of enzyme in the presence of substrate with cyanide, because of an adduct formed by addition of cyanide to the iminium ion.[14] Also, experiments demonstrating ^{18}O exchange during the course of substrate turnover provided further supporting evidence.[15] These experiments described above are the archtypical experimental means of demonstrating an imine intermediate in an enzymatic reaction. DiIasio et al.[16] have also shown that the reduction reaction with sodium borohydride is, in the case of liver aldolase, a stereospecific process resulting in the addition of hydride to the si face of the iminium ion.

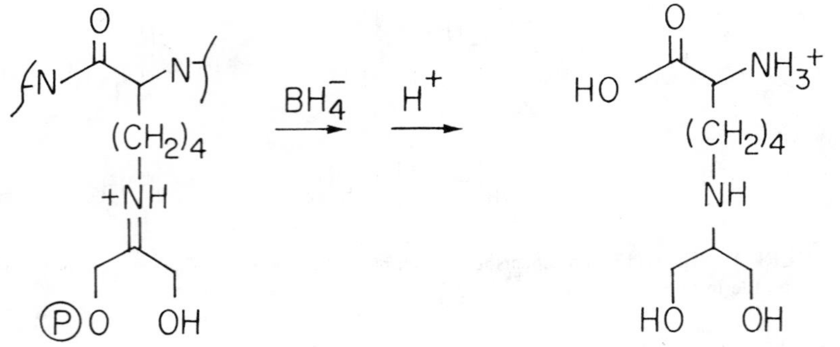

FIGURE 5. The sodium borohydride reduction of the enzyme lysine adduct of dihydroxyacetone phosphate gives, upon hydrolysis, N^6-ε-glyceryl lysine as the product.

The primary sequence of aldolase has been determined, and the enzyme has a molecular weight of 160,000 with 4 equivalent subunits.[17-18] Some aldolases are found as a tetramer of α and β subunits which differ only by the conversion of Asn 360 to an aspartate residue. The thiol groups present in aldolase have been studied extensively. Four of the total of eight cysteine thiols are particularly reactive toward thiol reagents. These are Cys-72, 239, 289, and 338, whereas the four other cysteines, Cys-134, 149, 177, and 201 are buried in the native protein and are modified only after denaturation. This behavior is exemplified by the modification of aldolase with substituted sulfenyl-sulfoxides.[19] In this case the rapid modification occurred to Cys-72 and 338 in the absence of substrate but did not in the presence of substrate. Cys-239 and 289 were modified in the presence or absence of substrate. In a similar manner, phenanthroline and copper oxidized Cys-72 and 338 to form an inactive disulfide, and this modification was prevented by the presence of substrate.[20] Also, bromoacetate attacks an active-site thiol with substrate protection, whereas other thiol groups are not protected by substrate. Other reagents that have been used to quantify and separate thiol groups into active-site or nonactive-site categories are chlorodinitrobenzene, p-mercuribenzoate, N-ethyl-maleimide, and carboxyethyldisulfide.[21] The last of these is interesting because it is reversed by mercaptoethanol through a thiol disulfide exchange reaction. Thus, one can make an inactive derivative of aldolase which is a mixed disulfide of the active site thiol and β-mercaptopropionic acid and reactivate this protein by treatment with a low molecular weight thiol.

The α-haloacetol-phosphates were created as potential active-site-directed inhibitors of enzymes using DHAP since alkylative trapping of the base required for proton abstraction is likely.[22] Whereas this did occur with triosephosphate isomerase, the corresponding trapping of the proton abstracting base in aldolase did not occur.[23] α-Iodoacetol phosphate was responsible for inhibition of the enzyme by oxidizing thiol groups.[24] α-Chloroacetol phosphate was found to react with an active-site thiol group only at high pH values.[25] Interestingly, in this study as in others, it has not been possible to prove that modification of this thiol group alone results in complete abolition of activity. Thus, the thiol group may have an auxiliary function in catalysis, rather than some fundamental catalytic role. The inactivation of rabbit muscle aldolase by 2-keto-3-butenyl phosphate discussed below has been proposed to occur by reaction with both the essential lysine amino group and the active site thiol group[5,26,27] and is prevented by formation of mixed disulfide as described above. The thiol data suggests that there are eight per subunit, with four buried, and four exposed, and with two of these protectable by substrate.

Compound 1 in Figure 6 was synthesized with the thought that it might bind as the imine, form eneamine and eliminate p-nitrophenol to unmask an electrophilic eneiminium ion able to react with an active site nucleophile such as Cys-338.[5,26,27] Although 1 did bind,

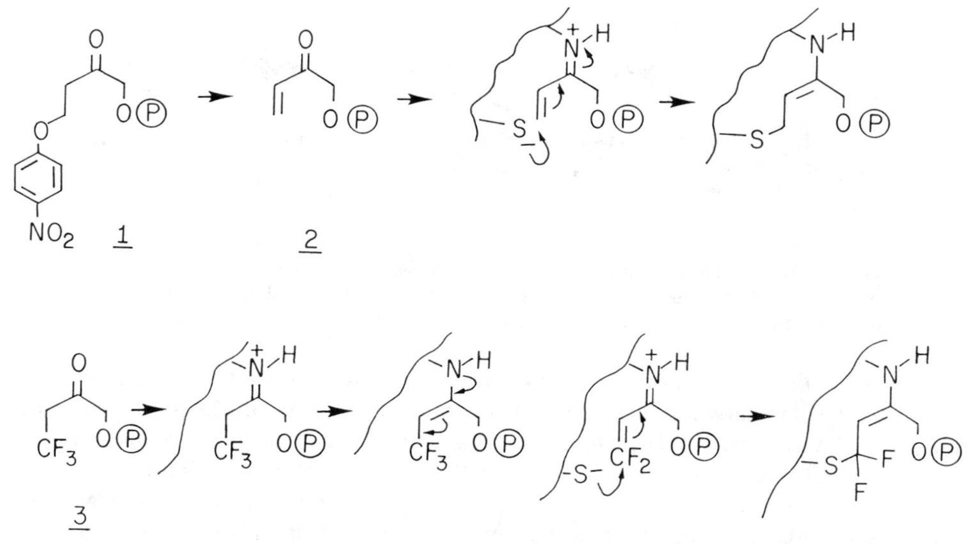

FIGURE 6. Nonenzymic elimination of p-nitrophenol from 1 resulted in the production of 2, which was a potent inhibitor of aldolase. The enzyme catalyzed the elimination of HF from the suicide inactivator 3 to give a similar adduct.

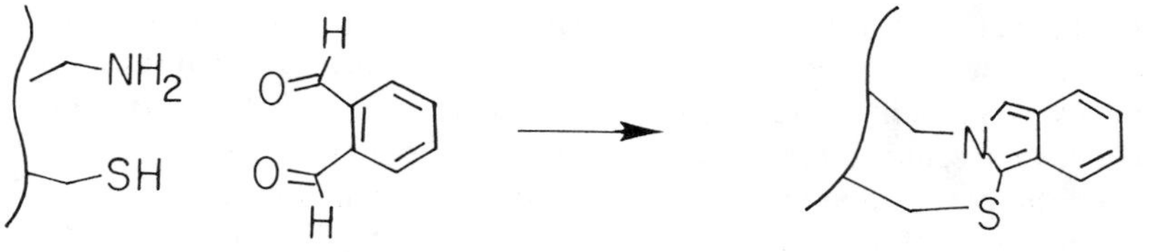

FIGURE 7. o-Phthalaldehyde caused the cross linking of an active site lysine and cysteine.

proton abstraction could not occur for steric reasons. However, the elimination product, 2, obtained nonenzymatically, was a potent time-dependent inhibitor which would not affect enzyme protected by substrate or which had the thiol protected as mixed disulfide, implying the formation of the lysine-cysteine-inhibitor adduct shown. Recently a French group has demonstrated that 2-oxo-4,4,4-trifluorobutyl phosphate, 3, does bind and eliminate and thereby acts as a suicide substrate for the enzyme.[28] Apparently the smaller leaving group allows the necessary conformation for deprotonation which was not possible for 1. The inhibitor had a dissociation constant of 1.4 m*M*, and a rate constant of 0.023 s^{-1} for inhibition and a low partition ratio since 1 to 2 mol of F^- were eliminated for each mole of enzyme inhibited. Substrate protection against inhibition occurred and the loss of one reactive thiol group was demonstrated by DTNB titration. Palczewski has demonstrated that o-phthalaldehyde also cross-links the active site Lys-229 with Cys-338 as an isoindole resulting from iminium ion formation followed by thiol addition,[29] as shown in Figure 7.

In addition to the critical Lys-229 and proximal Cys-338, aldolase contains several other active-site residues which have been shown by trapping experiments to behave as if they are near the active site, such as Lys-146 which reacts with N-bromoacetylethanol amine phosphate.[30] The ATP analog 5′[p-(fluorosulphonyl)-benzoyl]-1,N^6-ethenoadenosine results in inactivation with substrate protection, of Tyr-363 and lysine 107.[31] Other chemical modifications include inhibition of a lysine by pyridoxal-6-phosphate which is probably involved in the binding of the 6-phosphate since F-1,6-P_2 protects better than does DHAP.[32] It has

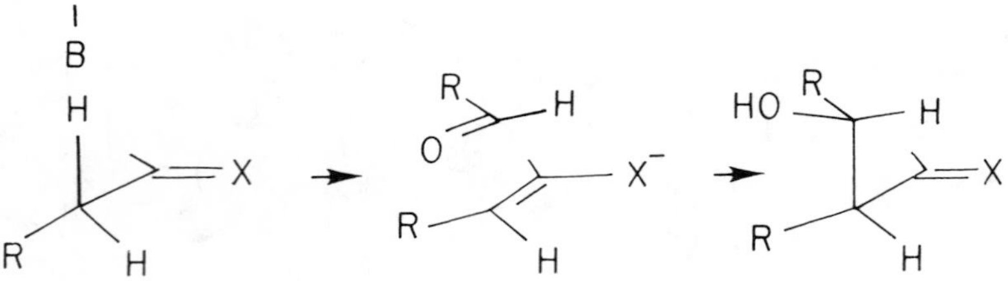

FIGURE 8. All aldolases catalyze the reaction by the same stereochemically cryptic pathway involving a retention mechanism in which the proton removal and aldehyde condensation occur from the same face of the intermediate.

been proposed that the lowering of the pK_a of the Lys 146 may be due to the proximity of an arginine residue known to be at the active site, and which would be positively charged throughout the pH range.[32] The presence of this arginine side chain was demonstrated by inhibition of aldolase with phenylglyoxal in the presence of borate, with protection by substrate or phosphate.[33] The presence of His-361 was demonstrated by its alkylation with N-bromoacetylethanolamine phosphate at low pH values.[34] Throughout all of these studies no apparent candidate for the proton abstracting base was evident.

The modification of either histidine or tyrosine residues produces similar changes in enzyme activity. Photooxygenation of about half of the 40 histidine residues in aldolase, sensitized by Rose Bengal, results in reduced activity.[35] The modified enzyme can still form the imine with substrate, and in fact has little change in the rate of cleavage of F-1,6-P_2. The most dramatic change in the enzyme is in the rate of proton transfer to the eneamine. As shown in Figure 1, the change occurs in step 2 in which proton transfer occurs, rather than in the cleavage step. Interestingly, the rate of cleavage of F-1,6-P_2 is stimulated by the addition of other acceptor aldehydes (step 6), because the eneamine is trapped by the aldehyde even though the protonation reaction is prevented by the modification. Removal of the C-terminal tyrosine, Tyr-363, by carboxypeptidase treatment results in very similar behavior.[36] Acetylation of the tyrosine by acetylimidazole results in a similar modification of enzymatic activity.[37] The net effect of these changes is to convert aldolase into an enzyme having transaldolase activity.[38,39] Histidine and tyrosine residues are therefore apparently involved in the proton transfer, even though trapping studies suggest that they are not themselves the proton abstracting bases.

D. STEREOCHEMISTRY OF THE REACTION

Hanson and Rose have pointed out that all aldolases, regardless of mechanism, operate with a retention mechanism.[40] As shown in Figure 8 a proton is removed from the face of an incipient enolate anion or eneamine and then the base doing the proton abstraction must swing out of the way so that the site may be occupied by the aldehyde co-substrate in a subsequent step, followed by C-C bond formation. In muscle aldolase, for example, it is known that the pro-S proton is removed and replaced with a carbon on the same face of the eneamine in order to create the appropriate stereochemistry for fructose. That all aldolases operate via this same mechanism suggests that this stereochemically cryptic pathway is an advantageous one that has been selected through evolutionary pressure because it serves a useful function.

Clearly an inversion mechanism can be envisioned which would lead to the same product from the same substrates and yet natural selection has avoided that pathway in all cases. This contrasts dramatically with Claissen-type condensations, which invariably operate via an inversion mechanism. One rationale proposed for this behavior is that this might allow proton recycling, so that the same proton abstracted from C-3 could be used to protonate the oxygen on C-4 during the condensation reaction. In enzymes with rapid turnover, the loss or acquisition of a proton from solvent may be slow. In this case, however, the relatively slow turnover of aldolase would seem to preclude this explanation. We have proposed another explanation for this generally observed behavior.[5] If an enzyme is required to cleave a carbon-carbon bond in the same position in the active site where the subsequent protonation reaction must occur, the retention mechanism automatically affords protection against the deprotonation of ketones other than the intended substrate. This complex mechanism, requiring the movement of the proton-abstracting base, therefore provides insurance against the indiscriminate enolization, dehydration, or tautomerization of other ketonic metabolites. The evolutionary pressure, therefore, for selection of this pathway would have come from a need to avoid unwanted reactions, rather than because of a drive to produce a more rapid rate of the desired reaction.

E. KINETIC CONSEQUENCES OF MODIFICATION

Rose et al.[38] have demonstrated, by isotope exchange experiments, that aldolase operates with an ordered mechanism, as outlined in Figure 1, in which the formation of an iminium ion with DHAP is followed by abstraction at C-3 of the pro-S proton. Only after this step may G-3-P bind productively. A condensation step then can occur to produce the iminium ion of the product fructose. In the reverse reaction the cleavage step of the fructose creates the eneamine of DHAP, but the eneamine may not be protonated unless the G-3-P has left the surface of the enzyme. Any acceptable proposed mechanism for aldolase must include this order of addition of substrates.

Grazi et al.[42] have found that dihydroxyacetone sulfate (DHAS) is capable of forming an iminium ion at the active site which reacts with borohydride or with cyanide, but which is incapable of undergoing proton abstraction. We have shown that the rate of irreversible inhibition of aldolase by an enone-type inhibitor bound as the iminium ion is independent of whether the charged group is phosphate or sulfate.[5] Therefore, the change from phosphate to sulfate does not disrupt the electron sink activity of the iminium ion. It has been demonstrated in model system studies that α-ketophosphates undergo an intramolecular proton abstraction in which the phosphate removes a proton from the α-carbon through a seven-membered transition state.[5] Because of this data, and because of the fact that active-site-directed alkylating agents were unable to trap the proton abstracting base as they had in other enzymes, we proposed that the substrate phosphate is responsible for the proton abstraction reaction rather than a base on the enzyme. This would account also for the fact that DHAS does not undergo proton exchange since the lower pK_a oxygens on sulfate would not catalyze proton abstraction even after formation of the iminium ion.

F. A PROPOSED MECHANISM

In Figure 9 is shown a proposed mechanism for the action of aldolase which incorporates all of the data described above. Although features of this mechanism are speculative, it does provide a working model which incorporates the stereochemical, chemical, kinetic, and other features elucidated by the many experiments described above.

In this figure the protein cleft is approximated by the box drawn over the atoms which is open to solvent water on the top and far sides. In step A the conformation drawn is consistent with the fact that only the pro-S proton on C-3 is exchanged with solvent.[42] The pro-S C-H bond has been drawn so that it may overlap with the π orbital of the iminium ion during the proton abstraction. The face of the imininium ion drawn so as to be exposed to solvent has been shown by Dilasio et al.[16] to be the side attacked by borohydride, and therefore also by water. The E isomer of the iminium ion has been drawn because this allows hydrogen bonding between the C-3 hydroxyl and the iminium ion. This would explain the fact that the iminium ion generated by the condensation of hydroxy acetone phosphate (bearing no OH on C-3) with the lysine group is less stable than that formed with DHAP.[43,44]

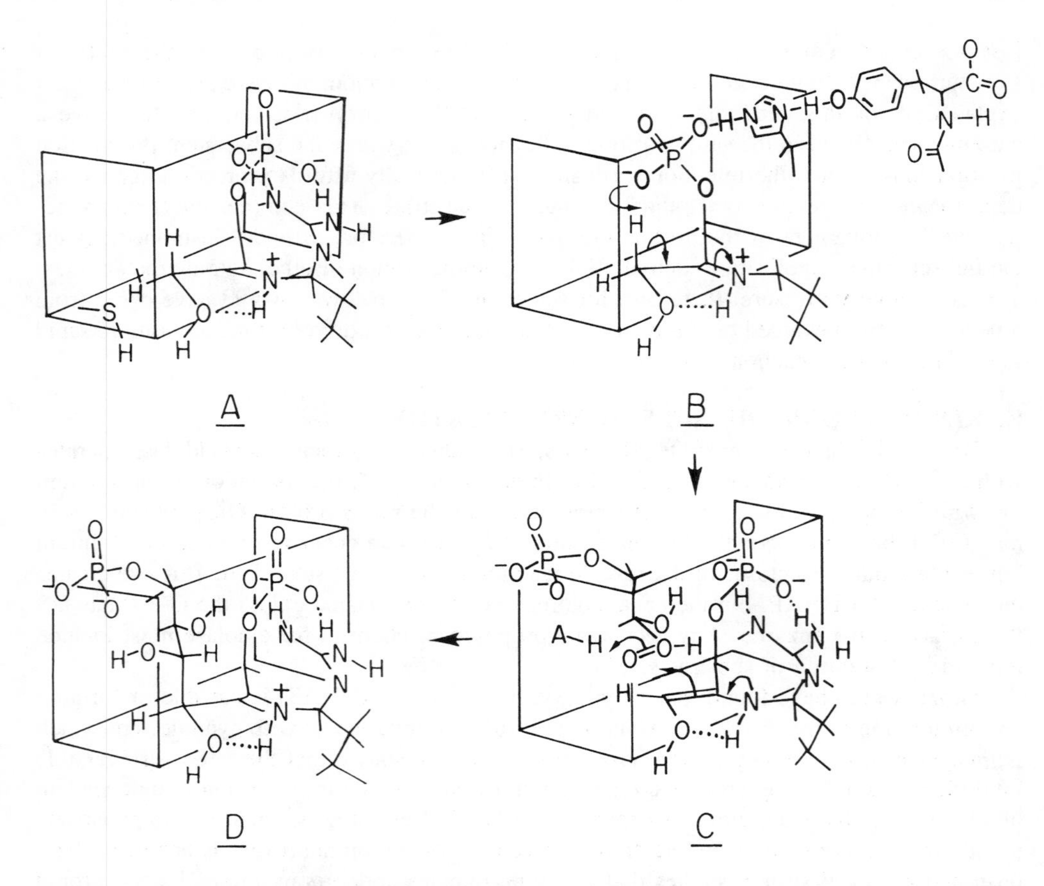

FIGURE 9. A proposed mechanism for aldolase which incorporates the chemical, kinetic, stereochemical, inhibitor, and structural features described in the text.

The C-1 phosphate has been drawn in a position hydrogen bonded with an arginine residue. The iminium ion is sufficiently exposed to solvent so that borohydride, cyanide, or tetranitromethane[45] reactions with the bound substrate are possible. A thiol group has been drawn in a position where the enone phosphate-type inhibitors shown in Figure 6 or ortho phthalaldehyde could react with it and the lysine.

In step B the phosphate group has been rotated to a position where proton abstraction may occur in an intramolecular fashion. The pK_a of the phosphate group (approximately 7) is ideal for proton abstraction as discussed earlier, and the rotation of the phosphate around the C-1 C-0 bond provides the two positions absolutely required for the base doing the proton abstraction since a retention mechanism prevails.

The positioning of the phosphate during the proton abstraction is maintained by the C-terminal residues His-361 and Tyr-363, in a manner such that disruption of either of these residues would produce a lesion in the proton transfer process, although not in the cleavage or transaldolase type of reactions. In the cleavage direction, with modified enzyme, the eneamine would be protonated only very slowly and reversal of the cleavage reaction could occur in the presence of other aldehydes, giving the transaldolase activity.

In step C, the acceptor aldehyde binds after the phosphate has rotated back to its original position, thus insuring the retention mechanism. The aldehyde carbonyl must be protonated by a stereospecific catalyst rather than by water since tagatose 1,6-bisphosphate (which is identical to F-1,$6P_2$, except for an inverted configuration at C-4) is not cleaved.[46]

Following step D, after formation of the iminium ion of F-1,$6P_2$, water would attack

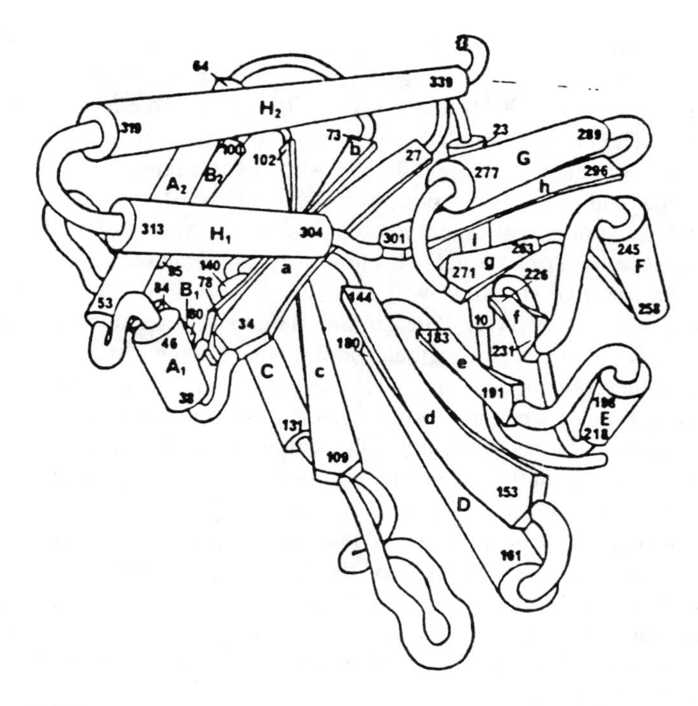

FIGURE 10. A schematic diagram of the X-ray structure for rabbit muscle aldolase, in which β-sheet structure regions are represented with arrow-like bars, and α-helical segments are tubes. (From Sygusch, J., Beaudry, D., and Allaire, M., *Proc. Natl. Acad. Sci. U.S.A.*, 84, 7846, 1987. With permission.)

the re face to form the carbinolamine which can then eliminate the lysine amine to generate the keto form of F-1,6P_2. This is, however, a necessarily oversimplified view of the hydrolysis step since it has been demonstrated that it is the β-D-fructose anomer that is the substrate for the enzyme in the cleavage direction.[47] The corresponding α-anomer does not act as a substrate. In the hydrolysis direction this would require an intramolecular nucleophilic attack by the C-5 hydroxyl to form the β-anomer prior to departure of the F-1,6-P_2. It is clear from the geometrical arrangement described in this mechanism that this catalyst would be extremely efficient for enolizing DHAP, but would be ineffective at enolizing other substrates which had large groups in the position occupied by C-4, C-5, and C-6. It is also comforting that at least one self-consistent mechanism which incorporates all of these disparate data can be proposed.

G. X-RAY STRUCTURE

Recently, Sygusch et al. have published a study[48] of the molecular architecture of rabbit muscle aldolase based on a 2.7 Å resolution X-ray diffraction map generated using phase information from a single isomorphous $Pt(CN)_4^{2-}$ derivative. Shown in Figure 10 is a schematic diagram of the polypeptide from a single subunit, wherein the conformation corresponds to a singly wound β-barrel, with the residues comprising the active site located in the center of the β-barrel. Compared to previous β-barrel structures whose amino acid sequence is known, aldolase is unusual because the interior of the β-barrel has, in addition to the expected hydrophobic residues, a number of potentially charged residues. Some of these are the same residues identified in the experiments described above, which were found in chemical modification studies.

The residues Asp-33, Lys-107, Lys-146, Glu-187, and Lys-229 are all found in the interior cavity of the β-barrel, and the charged moieties are approximately colinear with alternating acidic and basic groups, suggesting internal neutralization of charges within the hydrophobic interior. The three lysine residues 107, 146, and 229 are located in a line spaced 4A apart. The critical Lys-229 which forms the imine with substrate projects into the center of the β-barrel from the middle of β strand f in Figure 10.

The C-terminal end of the polypeptide contains the His-361 and Tyr-363 residues which were implicated in the proton abstractions step, and which were proposed in Figure 6 to be involved in the positioning of the DHAP phosphate for intramolecular proton abstraction. This region is an arm-like structure with a flexible elbow joint comprising residues 346-354, which folds over α-helix H_2. The terminal amino acid residues appear to cover the active site. Presumably, this conformational mobility plays a role in the ability of this segment to position the proton abstracting base when necessary, and then to be swung out of the way in the subsequent step when the cosubstrate aldehyde must bind and react. This mobility also might contribute to the susceptability of the region to proteolysis. Because of the apparent disorder caused by the flexibility of this region, unambiguous assignment of the residues in this region could not be made and they are absent from Figure 10. Although all of the active-site amino acid residues have been conserved throughout human, rabbit, maize, trypanosome, and drosophila type 1 aldolases, only 4 of the 22 residues in the COOH terminal region have been conserved. This suggests that, although the role of those groups participating in the condensation and cleavage reactions may be constant, the nature of the means by which the C-terminal region influences the proton transfer step may vary considerably depending upon the type of aldolase.

REFERENCES

1. **Hupe, D. J.,** Enzyme reactions involving imine formation, in *The Chemistry Enzyme Action,* Page, M., Ed., Elsevier, Amsterdam, 1984, chap. 8.
2. **Spencer, T. A.,** Amine catalysis via iminium ion formation, in *Bio Organic Chemistry,* van Tamelen, E. E., Ed., Academic Press, New York, 1977, chap. 13.
3. **Hupe, D. J.,** Imine formation in enzymatic reactions, in *Enzyme Mechanisms,* Page, M. I. and Williams, A., Eds., Royal Society of Chemistry, London, 1987, chap. 17.
4. **Horecker, B., Tsolas, O., and Lai, C.,** The aldolases, in *The Enzymes,* Vol. 7, 3rd ed., Boyer, P., Ed., Academic Press, New York, 1972, chap. 6.
5. **Periana, R. A., Motiu-DeGrood, R., Chiang, Y., and Hupe, D.,** Does substrate rather than protein provide the catalyst for α-proton abstraction in aldolase?, *J. Am. Chem. Soc.,* 102, 3923, 1980.
6. **Hupe, D. J., Kendall, M., and Spencer, T. A.,** Amine catalysis of β-ketol dehydration. II. Catalysis via iminium ion formation. General analysis of nucleophilic amine catalysis, *J. Am. Chem. Soc.,* 95, 2271, 1972.
7. **Hupe, D. J., Kendall, M., and Spencer, T. A.,** Amine catalysis of elimination from a β-acetoxy ketone. A study of catalysis via iminium ion formation, *J. Am. Chem. Soc.,* 94, 1254, 1972.
8. **Westheimer, F. and Cohen, H.,** Amine catalysis of the dealdolization of diacetone alcohol, *J. Am. Chem. Soc.,* 60, 90, 1938.
9. **Speck, J. and Forist, A.,** Kinetics of the amino acid catalyzed dealdolization of diacetone alcohol, *J. Am. Chem. Soc.,* 79, 4659, 1957.
10. **Horecker, B., Pontremoli, S., Ricci, C., and Cheng, T.,** The nature of the transaldolase dihydroxyacetone complex, *Proc. Natl. Acad. Sci. U.S.A.,* 47, 1949, 1961.
11. **Grazi, E., Cheng, T., and Horecker, B.,** The formation of a stable aldolase dihydroxyacetone phosphate complex, *Biochem. Biophys. Res. Commun.,* 7, 250, 1962.
12. **Grazi, E., Rowley, P., Cheng, T., Tchola, O. and Horecker, B.,** Mechanism of action of aldolases. III. Schiff base formation with lysine, *Biochem. Biophys. Res. Commun.,* 9, 38, 1962.
13. **Speck, J., Rowley, P., and Horecker, B.,** Identity of synthetic N^6-β-glyceryl-lysine and the ^{14}C- labelled amino acid obtained on sodium borohydride reduction and hydrolysis of a complex from ^{14}C-fructose-6-phosphate transaldolase interaction, *J. Am. Chem. Soc.,* 85, 1012, 1963.

14. **Cash, D. and Wilson, I.,** The cyanide adduct of the aldolase dihydroxyacetone phosphate imine, *J. Biol. Chem.,* 241, 4290, 1966.
15. **Model, P., Ponticarvo, L., and Rittenberg, D.,** Catalysis of an oxygen-exchange reaction of fructose 1,6 diphosphate and fructose 1-phosphate with water by rabbit muscle aldolase, *Biochemistry,* 7, 1339, 1978.
16. **DiIasio, A., Trombetta, G., and Grazi, E.,** Fructose 1,6 bisphosphate aldolase from liver: the absolute configuration of the intermediate carbinolamine, *FEBS Lett.,* 73, 244, 1977.
17. **Lai, C. Y., Nakai, N., Chang, D.,** Amino acid sequence of rabbit muscle aldolase and the structure of the active center, *Science,* 137, 1204, 1974.
18. **Benfield, P. A., Forcina, B. G., Gibbons, I., and Perham, R. N.,** Extended amino acid sequences around the active-site lysine residue of class-I fructose 1,6-bisphosphate aldolases from rabbit muscle, sturgeon muscle, trout muscle and ox liver, *Biochem. J.,* 183, 429, 1979.
19. **Steinman, H., and Richards, F.,** Participation of cysteinyl residues in the structure and function of muscle aldolase. Characterization of mixed disulfide derivatives, *Biochemistry,* 9, 4360, 1970.
20. **Kobashi, K. and Horecker, B.,** Reversible inactivation of rabbit muscle aldolase by o-phenanthroline, *Arch. Biochem. Biophys.,* 121, 178, 1967.
21. **Kowal, J., Cremona, T., and Horecker, B.,** The mechanism of action of aldolases, *J. Biol. Chem.,* 240, 2485, 1965.
22. **Hartman, F.,** Haloacetol phosphates. Potential active-site reagents for aldolase, triose phosphate isomerase, and glycerophosphate dehydrogenase. I. Preparation and properties, *Biochemistry,* 9, 1776, 1970.
23. **Hartman, F.,** Haloacetol phosphate characterization of the active site of rabbit muscle triose phosphate isomerase, *Biochemistry,* 10, 146, 1971.
24. **Hartman, F.,** Haloacetol phosphates. Potential active-site reagents for aldolase, triose phosphate isomerase, and glycerophosphate dehydrogenase. II. Inactivation of aldolase, *Biochemistry,* 9, 1783, 1970.
25. **Paterson, M., Norton, I., and Hartman, F.,** Haloacetol phosphates. Selective alkylation of sulfhydryl groups of rabbit muscle aldolase by chloroacetol phosphate, *Biochemistry,* 11, 2070, 1972.
26. **Wilde, J., Hunt, W., and Hupe, D.,** Rates and equilibria for the inactivation of muscle aldolase by an active site directed Michael reaction, *J. Am. Chem. Soc.,* 101, 2182, 1979.
27. **Wilde, J., Hunt, W., and Hupe, D.,** An inhibitor for aldolase, *J. Am. Chem. Soc.,* 101, 2182.
28. **Magnien, A., LeClef, B., and Biellmann,** Suicide inactivation of fructose-1,6-bisphosphate aldolase, *Biochemistry,* 23, 6858, 1984.
29. **Palczewski, P., Hargrave, P., and Kochman, M.,** o-Phthalaldehyde, a fluorescence probe of aldolase active site, *Eur. J. Biochem.,* 137, 429, 1983.
30. **Hartman, F. and Brown, J.,** Affinity labeling of a previously undetected essential lysyl residue in class I fructose bisphosphate aldolase, *J. Biol. Chem.,* 251, 3057, 1970.
31. **Palczewski, K., Hargrave, E. J., Folta, E., and Kockman, M.,** Affinity labelling of rabbit muscle fructose 1,6-bisphosphate aldolase with 5′[p-(fluorosulfonyl)benzoyl]-1-N^6-ethenoadenosine, *Eur. J. Biochem.,* 146, 309, 1985.
32. **Shapiro, S., Esner, M., Pugh, E., and Horecker, B.,** Effect of pyridoxal phosphate on rabbit muscle aldolase, *Arch. Biochem. Biophys.,* 128, 554, 1968.
33. **Lobb, R., Stokes, A., Hill, H., and Riordan, J.,** Arginine as the C-1 phosphate binding site in rabbit muscle aldolase, *FEBS Lett.,* 54, 70, 1975.
34. **Hartman, F., Suh, B., Welch, M., and Barker, R.,** Inactivation of class I fructose diphosphate aldolases by the substrate analog N-bromoacetylethanolamine phosphate, *J. Biol. Chem.,* 248, 8233, 1973.
35. **Hoffee, P., Lai, C., Pugh, E., and Horecker, B.,** The function of histidine residues in rabbit muscle aldolase, *Proc. Natl. Acad. Sci. U.S.A.,* 57, 107, 1967.
36. **Drescher, E., Boyer, P., and Kowalsky, A.,** The catalytic activity of carboxypeptidase-degraded aldolase, *J. Biol. Chem.,* 235, 604, 1960.
37. **Pugh, E. and Horecker, B.,** The function of tyrosine residues in rabbit muscle aldolase, *Arch. Biochem. Biophys.,* 122, 196, 1967.
38. **Rose, I., O'Connell, E., and Mehler, A.,** The mechanism of the aldolase reaction, *J. Biol. Chem.,* 240, 1758, 1965.
39. **Spolter, P., Adelman, R., and Weinhouse, S.,** Distinctive properties of native and carboxypeptidase-treated aldolases of rabbit muscle and liver, *J. Biol. Chem.,* 240, 1327, 1965.
40. **Hanson, K. and Rose, I.,** Interpretations of enzyme reaction stereospecificity, *Acc. Chem. Res.,* 8, 1, 1975.
41. **Grazi, E., Sivieri-Pecorari, C., Gagliano, R., and Trombetta, G.,** Complexes of Fructose diphosphate aldolase with dihydroxyacetone phosphate and dihydroxyacetone sulfate, *Biochemistry,* 12, 2583, 1973.
42. **Reider, S. and Rose, I.,** The mechanism of the triose phosphate isomerase reaction, *J. Biol. Chem.,* 234, 1007, 1959.
43. **Rose, I. and O'Connell,** Studies on the interaction of aldolase with substrate analogs, *Eur. J. Biol. Chem.,* 244, 126, 1969.

44. **Pratt, R.,** Rabbit muscle aldolase catalyzed proton exchange of hydroxyacetone phosphate with solvent, *Biochemistry,* 16, 3988, 1977.
45. **Healy, M. and Cristen, P.,** Reaction of carbanionic aldolase substrate intermediate with tetranitromethane. Identification of products hydroxypyruvaldehyde phosphate and D-5-ketofructose 1,6-diphosphate, *J. Am. Chem. Soc.,* 94, 7911, 1972.
46. **Tung, T., Ling, K., Byrne, W., and Lardy, H.,** Substrate specificity of muscle aldolase, *Biochim. Biophys. Acta,* 14, 488, 1954.
47. **Schray, K. J., Fishbein, R., Bullard, W. P., and Benkovic, S. J.,** The anomeric form of D-Fructose 1,6-bisphosphate used as substrate in the muscle and yeast aldolase reactions, *J. Biol. Chem.,* 250, 4883, 1975.
48. **Sygusch, J., Beaudry, D., and Allaire, M.,** Molecular architecture of rabbit skeletal muscle aldolase at 2.7Å Resolution, *Proc. Natl. Acad. Sci. U.S.A.,* 84, 7846, 1987.

Section X
Metalloenzymes

The fundamental assumption of the trace substance-enzyme thesis is that there is no rational explanation available of how traces of some substance can exert profound biological activity except in terms of enzymic phenomena. D. E. Green, Enzymes and Trace Substances, p. 196, in *Advances in Enzymology and Related Subjects*, (F. F. Nord and C. H. Werkman, editors), Vol. I, Interscience Publishers, Inc., New York, 1941.

Chapter 21

MECHANISM AND STRUCTURE OF SUPEROXIDE DISMUTASES

John A. Tainer, Victoria A. Roberts, Cindy L. Fisher, Robert A. Hallewell, and Elizabeth D. Getzoff

TABLE OF CONTENTS

I. INTRODUCTION

Superoxide dismutases (SODs) are ubiquitous enzymes in oxygen-tolerant organisms. A family of SOD enzymes having Cu and Zn (Cu,Zn SOD) or either Mn (Mn SOD) or Fe (Fe SOD) at the active site protects the organism against the toxic effects of the superoxide radical ($O_2^{\cdot -}$) by catalyzing its dismutation to molecular oxygen and hydrogen peroxide; this reaction occurs through the alternate reduction and oxidation of the active site metal ion (Cu, Mn, or Fe), giving the overall reaction: $O_2^{\cdot -} + O_2^{\cdot -} + 2H^+ \rightarrow H_2O_2 + O_2$.[1]

There are already several excellent reviews covering aspects of the biochemistry and biological activity of SODs.[1-3] This review presents new structural and theoretical information in the context of previous work with a focus on Cu,Zn SOD, for which there is more biochemical and structural data. (Although X-ray diffraction data to atomic resolution is available for Fe and Mn SODs, complete crystallographic analysis has been hampered by limited sequence information on the enzymes from the species being studied.)

A full understanding of enzymatic activity requires a detailed knowledge of the enzyme's structural basis at a level that approaches atomic resolution. Due to its biochemical properties and activity, SOD has been widely studied using a broad spectrum of techniques, including X-ray crystallography. The three-dimensional X-ray structures of Cu,Zn SOD,[4-6] Fe SOD,[7] and Mn SOD[8] have contributed directly to understanding both the activity and biochemical properties of SODs, and have provided specific models for the synthesis and integration of the studies on these enzymes. Studies using structural information from crystallography in combination with computational and biochemical results have brought an increased understanding of the mechanism and activity of SODs.

An example of the issues that can be approached with structural information is enzymatic catalysis. The catalytic rate of Cu,Zn SOD is very rapid (2×10^{-9} M^{-1} s^{-1}),[9] suggesting the evolution of an optimal active site for the recognition and chemical catalysis of $O_2^{\cdot -}$. It is now well established that substrates bind with high affinity to enzyme active sites when their respective structures are complementary. Empirical evidence for this comes from the very tight binding of transition state analogs, inhibitor molecules that mimic the transition state of a chemical reaction but cannot themselves undergo the complete reaction. In addition, minor chemical modifications to the substrate structures often have large effects on the catalytic rate constants. The modification of a single active site amino acid residue of the enzyme can also greatly reduce the activity, as is the case for site-directed mutants of SOD. Understanding the detailed molecular events that lead to enzymatic catalysis requires an accurate definition of the atomic positions in the active site, interpreted in the light of known biochemistry.

Other issues concerning SOD structure and mechanism that can now be addressed using the knowledge of the precise geometry of residues in the active site of Cu,Zn SOD include the relevance of specific hydrogen-bonding patterns to the proposed mechanism, the size and likelihood of movements of atoms and residues during the two catalytic steps, the source and geometry of the protons available for reduction of superoxide to hydrogen peroxide, the structural and catalytic importance of the Zn, and the implications of the active site structure for inner-sphere (direct binding to the catalytic Cu ion) vs. outer-sphere (not involving direct binding) catalysis. Although none of these questions has been answered definitively for Cu,Zn SOD by current experimental results, a surprising amount of information relating to these and other questions has come from the combination of biochemical, structural, and computational results.

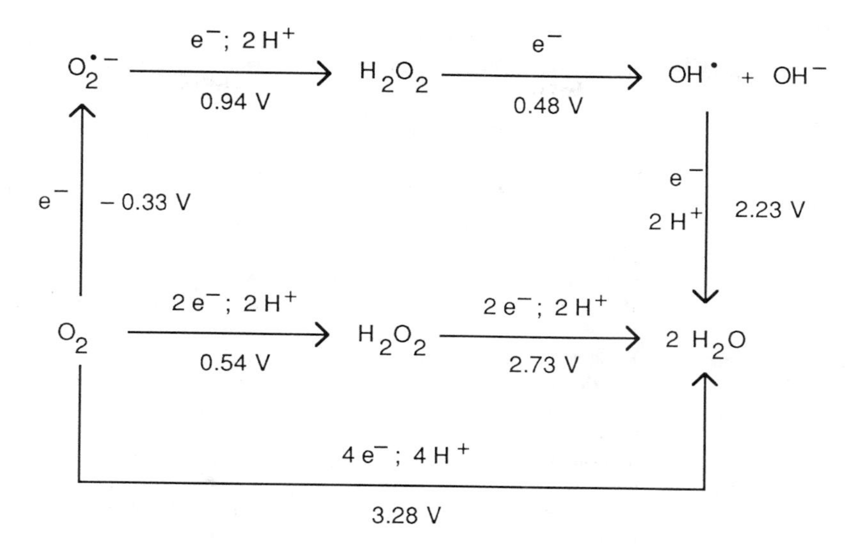

FIGURE 1. The reduction potentials, nE_o' for dioxygen species at pH 7.0 (so the half-reaction $H^+ + e^- \rightarrow {}^1/_2\, H_2$ has a reduction potential of −0.42 V). Values for each half-reaction were taken from Reference 13 or 223 and multiplied by the number of electrons involved.

II. BIOCHEMICAL AND BIOLOGICAL ACTIVITY

A. GENERATION AND TOXICITY OF SUPEROXIDE

In this review, we will consider the state of knowledge on structure and mechanism of SOD enzymes. Before considering the enzymes in detail, it is useful to briefly review properties of dioxygen that are responsible for the existence of superoxide and hence SODs *in vivo*.

Ground state dioxygen is in the triplet state (two unpaired electrons with like spin), so most two-electron reductions are spin-forbidden. This spin restriction hinders the divalent reduction of O_2 and is responsible for the relatively low activity of free oxygen,[10] thereby allowing the coexistence of free oxygen and organic molecules.[1] Molecular oxygen, however, is quite reactive in the presence of metals, suggesting that *d* orbital participation may allow reactions to occur that are otherwise spin- forbidden. The spin-allowed univalent reduction of O_2 during respiration involves reactive intermediates, including superoxide, that require biological defenses. Figure 1 diagrams the various O_2 reduction potentials and thus the relationship between the various dioxygen species.

The superoxide radical $O_2^{\cdot -}$ is the conjugate base of the perhydroxy radical $HO_2^{\cdot}$, a weak acid having a pK_a of 4.8.[11] Superoxide has the potential of acting as a reductant with an E_o' for $O_2/O_2^{\cdot -}$ of about −0.33 V and acting as an oxidant with an E_o' for $H_2O_2/O_2^{\cdot -}$ of 0.94 V.[12,13] Superoxide anion is a good nucleophile, a good reductant in all media, and a poor oxidant in aprotic media. Thus, $O_2^{\cdot -}$ is fairly stable in aprotic solvents.[14]

The presence of endogenous SODs prevents the measurement of $O_2^{\cdot -}$ produced *in vivo*. In *Streptococcus faecalis* with antibody-suppressed Cu,Zn SOD activity, about 17% of the O_2 utilized produces $O_2^{\cdot -}$.[15] Although it is difficult to identify the most important biological sources of $O_2^{\cdot -}$ for a given cell type, superoxide is clearly generated in many biological reactions involving molecular oxygen. Superoxide anion production has been directly detected by ESR from turnover of xanthine oxidase,[16] reoxidation of reduced flavins and ferrodoxins,[17,18] and illumination of chloroplasts.[19] Biological oxidations that result in significant superoxide production are presented in Table 1. In addition to superoxide production by these oxidative reactions, some enzymes may actually use $O_2^{\cdot -}$ as a cofactor. For example,

TABLE 1
Some Biological Oxidations Known to Produce Superoxide

Type	System	Method of detection[a]
Cells	Polymorphonuclear leukocytes	CYT c[188]
Cell organelles	Chloroplasts	Spin trapping[19]
	SMP (*Arum maculatum*)	Adrenochrome[189]
	SMP and Plant mitochondria	Adrenochrome[190]
	SMP and Mammalian mitochondria	Adrenochrome[191]
	Microsomes	Adrenochrome,[192] NBT reduction[193]
	Nuclei	Adrenochrome[194]
Enzymes and proteins	Xanthine oxidase	ESR,[16] CYT c[72,195]
	Peroxidases	ESR,[196] Adrenochrome[197]
	Diamine oxidase	CYT c[198]
	Flavoproteins	CYT c[199]
	Iron-sulfur proteins	ESR,[196] Adrenochrome[200]
	P450	Reaction inhibition,[201] CYT c[202]
	Hemoglobin	CYT c,[203] Adrenochrome[204]
	Rubredoxin	Adrenochrome[205]
Small Molecules	Flavins	ESR,[18] Adrenochrome[206]
	Quinones	PR,[207] Adrenochrome,[206] CYT c[208]
	Bipyridilium herbicides	PR,[209] CYT c and Adrenochrome[210]
	Dialuric acid	CYT c[211]
	Thiols	Adrenochrome[212]

Note: A similar and more comprehensive table has been compiled by Fee.[213] Here only the most complete reference is cited.

[a] Methods of detection have been abbreviated as follows: CYT c, reduction of cytochrome c; Adrenochrome, formation of adrenochrome; NBT reduction, reduction of nitroblue tetrazolium; Reaction inhibition, inhibition of $O_2^{\cdot -}$ requiring reaction; and PR, pulse radiolysis. Other abbreviations are SMP, submitochondrial particle; P450, cytochrome P450.

the enzyme 2,3 dioxygenase, a regulator of neurotransmitter and hallucinogen concentration, may be active only in the presence of $O_2^{\cdot -}$. An enzyme-$O_2^{\cdot -}$ complex has been detected spectroscopically.[20]

The discovery of the protective role of Cu,Zn SOD against oxygen toxicity suggested that superoxide is an important mediator of oxygen toxicity *in vivo*.[21] Aside from the reactivity of $O_2^{\cdot -}$ itself, superoxide may be a precursor for the more reactive hydroxyl radical, $OH^{\cdot}$.[1] *In vivo,* the formation of the hydroxyl radical may depend upon the Haber-Weiss reaction, which describes the interaction of $O_2^{\cdot -}$ with endogenous iron compounds:

$$O_2^{\cdot -} + Fe(III) \rightarrow O_2 + Fe(II) \quad (1)$$

$$Fe(II) + H_2O_2 \rightarrow Fe(III) + OH^- + OH\cdot \quad (2)$$

$$O_2^{\cdot -} + H_2O_2 \rightarrow O_2 + OH^- + OH\cdot \quad (3)$$

In this reaction, the iron facilitates electron transfer from $O_2^{\cdot -}$ to H_2O_2.[1] While the actual process of the *in vivo* generation of more potent oxidants from $O_2^{\cdot -}$ is as yet poorly understood, a wealth of data indicates the deleterious biological effects of $O_2^{\cdot -}$ and the resultant protection by SOD.[2,22] Damage from $O_2^{\cdot -}$ includes the oxidation of DNA, proteins fatty acids, and other molecular species.

B. BIOLOGICAL DISTRIBUTION AND ROLE OF SUPEROXIDE DISMUTASE ENZYMES

A family of SOD enzymes having either Cu and Zn, or Mn, or Fe at the active site show the same catalytic activity for superoxide *in vitro,* and biological studies outlined below suggest that these enzymes share similar roles *in vivo.*[23] The Mn and Fe SODs, which occur in bacteria and mitochondria, share high sequence and structural homology,[7,24] but have a completely different primary and tertiary structure[25] from the Cu,Zn SODs,[4,26] which primarily occur in the eucaryote cytosol. Recently, a secreted Cu,Zn enzyme, which is glycosylated and contains additional amino acids at both the amino and carboxyl termini, was characterized from extracellular fluids[27,28] of vertebrates. Thus, the respective protein structures reflect a need to be sequestered in different cellular environments, while their active sites share features relevant to their common catalytic activity.

Initial surveys suggested that respiring organisms possessed SOD enzymes and obligate anaerobes did not.[21] Later it was discovered that some anaerobes do have SODs.[29-32] The initial hypothesis that SOD is essential to protect against the toxic effects of dioxygen is consistent with these later results. Obligate anaerobes will not grow in the presence of dioxygen, but they are often subject to at least transient exposure to oxygen. The oxygen tolerance and SOD content of these species are correlated.[1,33]

SOD activity can be induced by exposure to dioxygen,[1] the precursor of $O_2^{\cdot-}$. Intracellular levels of SOD are increased after exposure to oxygen in facultative species such as *S. faecalis*[34,35] and *Saccharomyces cerevisiae,*[36] indicating that higher levels of SOD were reflected as higher resistance to oxygen toxicity. The induction of SOD and resulting increased protection against oxygen toxicity has been verified for numerous organisms, cells, and tissues, includng *Photobacterium leiognathi,*[37,38] *Euglena gracilis,*[39] macrophages,[40-42] leukocytes,[42] and rat lung.[43,44]

$O_2^{\cdot-}$ producing agents also increase the levels of SOD enzymes. The herbicide methyl viologen can easily be reduced to a cation radical that is reoxidized in the presence of O_2 with a resulting formation of $O_2^{\cdot-}$.[31] In *Escherichia coli,* methyl viologen has no effect in the absence of O_2, but is a powerful inducer of Mn SOD in the presence of O_2.[45] Similarly, light is known to generate $O_2^{\cdot-}$ in chloroplasts,[19] and light induces SOD synthesis in Euglena.[39]

For most organisms, there are no endogenous nonprotein catalysts with SOD activity. The high concentration of Mn (possibly in a complex with phosphates) that results in substantial SOD activity in *Lactobacillus plantarum* is unusual.[46] Usually, a few proteins in crude cell extracts have SOD activity, all of which belong to one of the two general classes (either Cu,Zn or Fe/Mn) of SODs.[1] Dialysis of such crude extracts does not decrease their activity, indicating that low molecular weight metal complexes with SOD activity are not usually found *in vivo.*

Table 2 shows that the level of $O^{\cdot}{}_2^-$ dismutation activity found for SODs is unique to these enzymes. Other metalloenzymes (e.g., cytochrome oxidase, Fe-porphyrin, and hemoglobin) and most copper-histidine compounds exhibit little or no $O^{\cdot}{}_2^-$ dismutation activity. Aqueous Cu(II), which has a tremendous advantage over the protein in terms of the ratio of catalytic to noncatalytic surface area, is about four times as efficient as the SOD enzymes, but does not exist *in vivo.*

Compelling evidence for similar biological roles for different types of SODs comes from studies showing the balance between levels of Cu,Zn SOD activity and Mn SOD activity. Copper deficiency in the fungus *Dactylium dendroides* results in decreased levels of Cu,Zn SOD activity, compensated for by a corresponding increase in the levels of Mn SOD activity.[47,48] Consequently, the total level of SOD activity is preserved. Similarly, manganese-deficient chickens had decreased Mn SOD activity, but elevated Cu,Zn SOD activity in the liver.[49] This relationship between the Cu,Zn and Mn SODs, such that a decrease in

TABLE 2
Relative Activities of Catalysts of Superoxide Dismutase

Catalyst	Relative efficiency
Cu,Zn SOD	1[72]
Fe SOD	1[214]
Mn SOD	1[215]
$Cu(II)_{aqueous}$	4[216]
Cu-amino acid complexes	0.001—0.1[217,218]
Cytochrome oxidase	0.01—0.03[219]
Fe-porphyrin	0.001[220]
Hemoglobin	0.001[221]
Mn-EDTA	0[213]

From Fee, J. A., *Oxidases and Related Oxidation-Reduction Systems,* King, T. E., Mason, H. S., and Morrison, M., Eds., Pergamon Press, Oxford, 1979. With permission.

one type of enzyme results in a compensatory increase in the other, argues directly for the importance of the shared *in vivo* enzymatic activity of these two proteins.

C. DATA FROM ENZYME-DEFICIENT MUTANTS AND CLONED SOD GENES

The cloning and characterization of the genes or cDNA for *E. coli* Fe SODs,[50] *E. coli,*[51] yeast,[52] and mouse[53] Mn SODs, and *P. leiognathi,*[54] maize,[55] and human[56-58] Cu,Zn SODs have greatly enhanced the understanding of the *in vivo* functions of the enzymes by allowing the increase or decrease of SOD levels.

Studies of organisms producing little or no SOD have shown that SOD is required for normal aerobic growth. A mutant *E. coli* strain, *sodAsodB*, has been constructed that is deficient in both Mn and Fe SODs and therefore lacks any endogenous SOD activity.[59] Growth of this strain in air shows increased sensitivity to paraquat (which causes the production of superoxide in the presence of oxygen[60] and hydrogen peroxide (which is in equilibrium with superoxide). This sensitivity is reduced to normal levels by the reintroduction of plasmids that encode for expression of Mn or Fe SODs[59] or for the structurally distinct human Cu,Zn SOD.[61] Thus, SODs are essential for protection of the *E. coli* cell against superoxide-mediated stress. In addition, protection can still be provided by an SOD (i.e., the human Cu,Zn SOD) not normally present in the organism and of a very different structure. The *sodAsodB* mutant *E. coli* are unable to grow aerobically on minimal media, possibly because one or more of the amino acid biosynthetic enzymes may be adversely affected by superoxide. Thus, the choice of media can be used as a powerful tool to select rapidly for mutants of *sodAsodB* that have been engineered to reexpress SOD.[61]

Mutants lacking Mn SOD have been constructed using the cloned yeast gene,[52] while yeast strains that lack an active Cu,Zn SOD have been obtained using conventional genetic methods.[62] These yeast mutants exhibit many of the properties observed in the *E. coli* mutants discussed above, including hypersensitivity to paraquat and an inability to grow aerobically on minimal media.

Overproduction of SOD has been investigated in several organisms. Excess Fe SOD in the *E. coli sodAsodB* strain increases $O_2^{\cdot -}$ sensitivity, probably mediated by SOD-catalyzed H_2O_2 production from superoxide.[63] In agreement with these findings, 50-fold overproduction of human Cu,Zn SOD in yeast[64] results in cells that are slightly more sensitive to paraquat (R. Hallewell, unpublished results). Resistance to copper and zinc is increased in this strain, but the level of resistance is low compared to that conferred by metal-resistance proteins such as copper chelatin. Therefore, Cu,Zn SOD does not have an important role in metal homeostasis as prevously suggested; rather, its major role lies in the protection of the organism from superoxide.

At present, studies of mammalian cells, using either mutants or the cloned genes, have been restricted to overproduction of Cu,Zn SOD. A three- to fivefold overexpression of the cloned human Cu,Zn SOD gene or cDNA in fibroblast or CHO cells made them more resistant to paraquat.[60] but increased levels of paraquat-dependent lipid peroxidation.[65] Lipid peroxidation resulting from accumulation of SOD-derived hydrogen peroxide has been implicated as a contributing factor in Down's syndrome. This disease has been associated with an increased level of Cu,Zn SOD, which is encoded by chromosome 21.[66] Down's syndrome patients have either an extra copy of this chromosome, or a trisomy of the particular band of chromosome 21 that is responsible for encoding SOD[67] (among other proteins).[57] Studies of overproduction of Cu,Zn SOD in both mouse L cells and in NS20Y neuroblastoma cells showed a concomitant increase in the selenium-containing glutathione peroxidase, which is responsible for protection against hydrogen peroxide in mammalian cells.[68] Thus, the regulation of both SOD and peroxidase activities are important in preventing oxygen toxicity.

D. PHARMACOLOGICAL AND CLINICAL ASPECTS

Although their complete biological role is not understood, SODs have diverse biochemical and clinical uses. SOD inhibits the peroxidation of fatty acids in inflammation and hemostasis,[69,70] and is anti-inflammatory in animal models.[71-73] Due to the capacity of $O_2^{\cdot -}$ to form the hydroxy radical ($OH^{\cdot}$) via the Fe-catalyzed Haber-Weiss reaction (Equations 1 to 3), $OH^{\cdot}$ may be a powerful potentiator of inflammation.[74,75] *In vivo* studies using SOD and catalase, which was made long lasting by complexation to polyethylene glycol and other polymers, showed that SOD decreases leukocyte infiltration into the inflamed site. Thus the $OH^{\cdot}$ generated via the iron-catalyzed Haber-Weiss reaction from $O_2^{\cdot -}$ and H_2O_2 is not participating in the inflammatory response.[76]

SOD protects against cell damage. Current evidence suggests that free radical activity, a common characteristic of cell damage, acts to promote a systemic response to the injury: oxidation products of the type resulting from interaction of substances with $O_2^{\cdot -}$ have a significant role in chemotaxis, phagocytosis, adherence, and anaphylaxis. For example, nonenzymatic oxidation of arachidonic acid by treatment of silica-bound arachidonate with UV radiation resulted in chemotactic activity.[77] This cytotaxin was shown to result from the action of the lipoxygenase enzyme of platelets on arachidonic acid.[78] Since this discovery several studies have shown the chemotactic activity of lipoxygenase products.[79,80] Treatment of serum with a superoxide-generating system produces an active chemotactic agent that is lipid-soluble,[81] and the interaction of arachidonic acid with superoxide generates a potent chemotactic agent.[82] Another example is the significant tissue damage resulting from a period of tissue ischemia, followed by reperfusion. This damage may actually occur during the reperfusion phase, rather than during the period of ischemia, and may be caused by oxygen-derived free radicals.[83] Recently, antioxidant enzymes including SOD have been shown to protect against the damage that occurs when the oxygen flow is reestablished to a given site.[84-92]

The ability to analyze the structure and mechanism of SODs (described below) together with genetic tools to construct site-specific mutants promises to better define their pharmacological activities.

III. THREE-DIMENSIONAL STRUCTURES OF SUPEROXIDE DISMUTASES

The three-dimensional crystal structures of enzymes at high resolution provide a wealth of information about the active sites and the mechanism of reaction. The structure of bovine

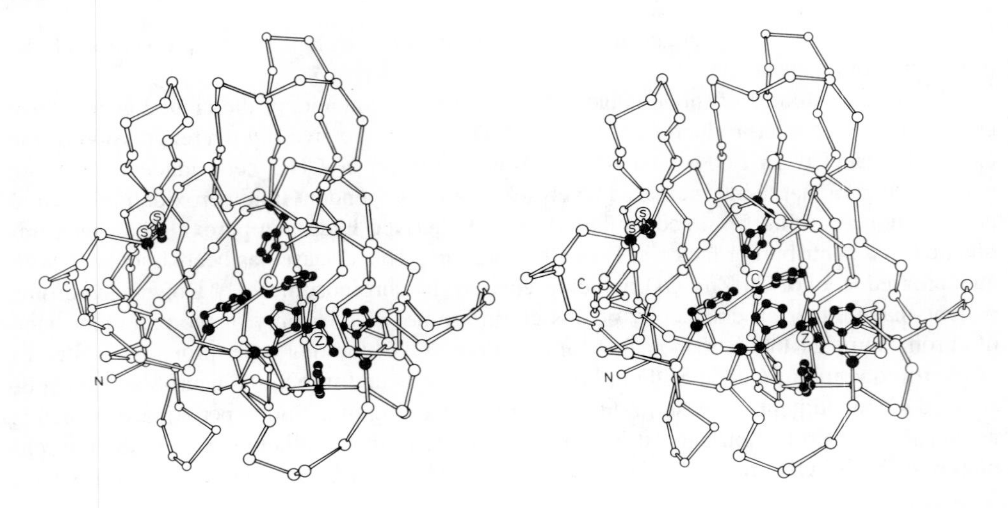

FIGURE 2. Stereo representation of the bovine Cu,Zn SOD subunit showing the active site side chains. This stereo drawing shows the C_α backbone of the SOD subunit viewed down the axial direction of the Cu from the solvent. Solid spheres represent the C_α and side chain atoms of the Cu ligand residues (His 44, 46, 61, and 118), the Zn ligand residues (Asp 81 and His 61, 69, and 78), and the nonsulfur atoms in the disulfide bond (Cys 55 and 144). The sulfur atoms of the disulfide bond (upper left) are shown with open spheres labeled S. The side chain of His 61 bridges the Cu (circled C) and the Zn (circled Z). Residue 1 (labeled N) is in the lower left with residue 151 (labeled C) above it. (From Tainer, J. A., Getzoff, E. D., Beem, K. M., Richardson, J. C., and Richardson, D. C., *J. Molec. Biol.*, 160, 181, 1982. With permission from Academic Press, Ltd.)

Cu,Zn SOD with Cu(II) has been solved (R-factor of 0.14, which suggests a positional error of less than 0.2 Å for atoms in the active site around the metal ions).[4–6] The four subunits in the asymmetric unit of the crystal were solved independently to provide an additional check of the accuracy of the structure. RMS comparisons of these subunits also suggest an error of less than 0.2 Å. The area around the active site is particularly well resolved with low temperature factors (suggesting little local mobility) and clear definition in the electron density maps. Thus, the metal ligands and residues that make up the active site channel are close-packed and stable without significant dynamic or static (all four subunits show the same conformation) structural variations. The *E. coli* (3.1-Å resolution),[7] and *Pseudomonas ovalis* (2.9-Å resolution)[93] Fe SOD and *Thermus thermophilus* Mn SOD (2.4-Å resolution)[8] crystal structures have been determined but not fully solved, because the entire sequences are not shown for the Fe SODs.[94] The *T. therophilus* Mn SOD sequence has only recently been determined.[95] It is clear, however, that the overall foldings of the Fe and Mn enzymes are similar to one another, but are very different from that of the Cu,Zn enzyme.

A. MAJOR STRUCTURAL FEATURES

1. Cu,Zn SOD

Cu,Zn SOD is a dimer made up of two identical single domain subunits of the antiparallel β category.[4] The two identical subunits of the dimer are related by a noncrystallographic twofold axis that orients the two active sites facing away from each other with the Cu atoms 33.8 Å apart.[6] As illustrated by the C_α backbone shown in Figure 2, each 16,000-Da subunit (151 amino acid residues) is composed primarily of eight antiparallel β strands, which form a flattened cylinder, plus three external loops. The SOD β barrel structure consists of four very regular, extended β strands on one side (the back in Figure 2) and four β strands in more twisted or irregular conformations on the other side (the front in Figure 2). The two sides of the β barrel might also be interpreted as two β sheets forming a sandwich. The loops contain tight turns and are primarily as well defined and ordered as the β strands. The active site, which contains one Cu and one Zn ion, is made of two large loops that connect

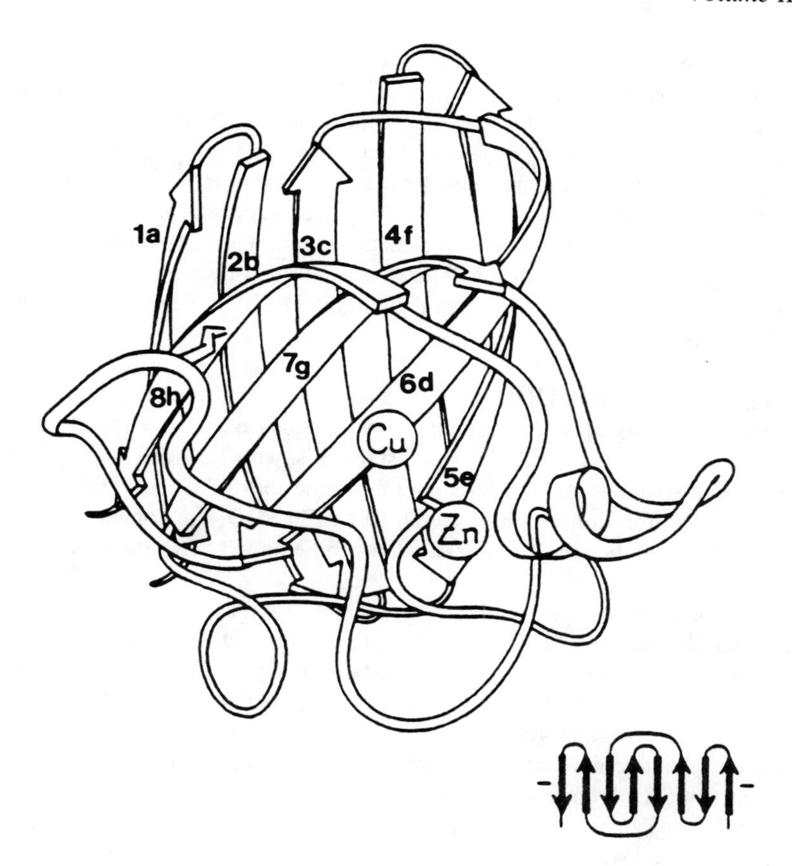

FIGURE 3. Schematic backbone drawing of the SOD subunit (viewed from the same direction as Figure 2). The β strands are shown as arrows and the disulfide as a zigzag. Two long loops extend forward from pairs of β strands to form the upper and lower sides of the active site channel. The Cu and Zn lie at the bottom of the channel, with the Cu accessible to solvent from the viewing direction. β Strands are identified by a number beginning at the N-terminus and proceeding clockwise around the top of the β barrel, and also by a letter assigned in order of amino acid sequence. The small diagram (lower right) illustrates the Greek key toplogy of the β barrel, spread open and viewed from the outside. (From Tainer, J. A., Getzoff, E. D., Beem, K. M., Richardson, J. C., and Richardson, D. C., *J. Molec. Biol.*, 160, 181, 1982. With permission from Academic Press, Ltd.)

the β strands. The single disulfide bridge forms a left-handed spiral conformation and ties one loop to the β barrel. Within the major category of antiparallel β domains, the overall topology of the SOD subunit classifies the enzyme as a Greek-key β barrel, a very common folding motif also found in immunoglobulins, plant lectins, and the capsid subunits of spherical viruses.[96]

The behavior and stability of a folding domain depends upon the interaction of intra-subunit structural elements or subdomains. Figure 3 shows a schematic of the subdomains forming the overall structural fold of SOD. Our analysis of the subunit structure[4,97] is based on a total of 6 subdomains: β strands 1 to 4 (the regular side of the β sandwich); β strands 5 to 8 (the irregular side of the β sandwich); Greek key loop 4,7; Zn ligand region of loop 6,5; disulfide region of loop 6,5; and active site lid loop 7,8.

2. Fe/Mn SOD

The overall structure of the Fe and Mn enzymes is very different from that determined for the Cu,Zn enzyme. The Fe/Mn SODs are either dimeric or tetrameric and composed of subunits with molecular weights of about 20,000 (180 to 200 amino acids) and one Mn or one Fe each.[94] X-ray structures of Fe SOD[7,93] show a fold with over 50% helicity. The structure of Mn SOD from *T. thermophilus*[25] indicates a helical fold closely analogous to that of Fe SOD, with equivalent metal and ligand positions. Monomers of both Fe and Mn SODs are composed of two domains with six major α helices and three strands of antiparallel β sheet. The N-terminal domain secondary structure is formed primarily by two long antiparallel helices, each of which contains a metal ligand. The C-terminal domain is a three-layered, mixed α/β structure with the central layer formed by a β sheet. The end of this β sheet contributes the remaining two metal ligands. The active center lies at the interface between these two domains.

B. METAL SITE GEOMETRIES

The geometry of ligands around metal ions depends on the coordination of the metal and the type of ligand involved. Of interest in this work are the four-, five-, and six-coordinate systems found around the copper and zinc ions of Cu,Zn SOD, the manganese ion in Mn SOD and the iron ion in Fe SOD. In Cu,Zn SOD, the copper alternates between Cu(I) and Cu(II) while the zinc remains divalent. In Fe or Mn SODs, the metal ion alternates from divalent to trivalent.

There are three basic four-coordinate geometries found in metal complexes. The most prevalent is tetrahedral, especially around nontransition metals. Square planar, the second most common, is found in many transition metal complexes. With some metals (including copper), interconversion between these two geometries is rapid, as the difference in stability is small. Finally, the irregular geometry is found in some compounds with one extra electron pair.[98] Five-coordinate complexes can take on two basic geometries — trigonal bipyramidal and square pyramidal. A large number of five-coordinate compounds have structures intermediate to these two forms, as they are easily deformable and interconvertible. This gives five-coordinate complexes an important stereochemical nonrigidity.[98] The most common coordination number in metallostructures, including metal-peptide complexes, is six. Almost all six-coordinate compounds have an octahedral geometry, which may be distorted in several ways. The most common distortion is tetragonal, also termed ''Jahn-Teller distortion'', an elongation or contraction of the two axial ligands.[98]

An excellent review on metal-peptide complexes by Freeman[99] discusses in detail geometries found in a number of transition metal-peptide crystal structures. Some of the salient points of this review relevant to SOD metal geometry follow.

In typical inorganic and organometallic complexes, the ligands around the metal conform fairly well with the rigid coordination geometries described above. In contrast, in both biological systems and metal-peptide complexes, the metal geometry is often distorted from the ideal because of restrictions in the ligand positions arising from the steric preferences of the metal orbitals, the accommodation of strains in the ligands directly bonded to the metal, and the bonding network of the ligands with the rest of the surrounding structure. Protein-metal-substrate complexes must be sufficiently stable to hold the substrate in place while it undergoes catalysis, but be sufficiently flexible enough to allow the perturbations required to release the substrate. Copper(II) and zinc(II) complexes especially have several ligand configurations that facilitate interconversion between geometries, coordination numbers, and stereochemistries while stablizing the intermediate structures.[99]

Typically, copper(I) ions, such as that in reduced Cu,Zn SOD, form two-, three-, or four-coordinate complexes. Four-coordinate cuprous compounds usually form tetrahedra, ideal or distorted.[98] Regardless of the number of ligands, commonly four, five, or six, Cu(II)-peptide complexes form a coordination square, often distorted, of four donors bound to the metal. Vacancies may exist at one or both of the axial positions, and the ligands that do bind axially show more variation in distances than those in the coordination square. In five-

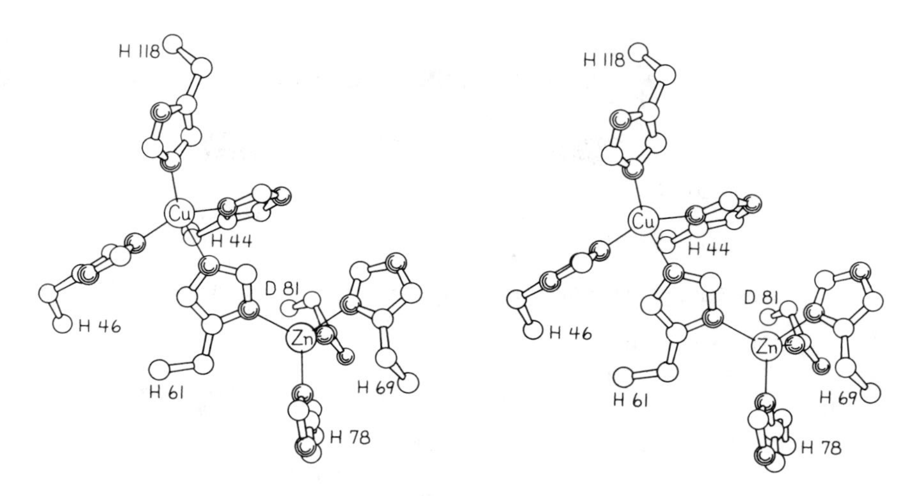

FIGURE 4. Copper and zinc ligand geometry in bovine Cu,Zn SOD. This stereo drawing shows the geometry of the active site metals and their ligands as viewed from the solvent. Nitrogen and oxygen atoms are shaded. Amino acid ligands are labeled with one-letter code; H for histidine, D for aspartic acid. (From Tainer, J. A., Getzoff, E. D., Beem, K. M., Richardson, J. C., and Richardson, D. C., *J. Molec. Biol.*, 160, 181, 1982. With permission from Academic Press, Ltd.)

coordinate cupric systems, square-pyramidal geometries are preferred, although trigonal-bipyramidal complexes are observed; both are typically nonideal. Almost all known small molecule, five-coordinate copper(II)-histidine and copper(II)-imidazole compounds form square pyramids. The distance of the apical ligand from the copper atom in the square pyramid is characteristically longer than those from the coordination square, and depends on the ligand field of the four atoms. Six-coordinate octahedra are almost always distorted in copper(II)-peptide systems, usually tetragonally.[99] Zinc(II)-peptide complexes are commonly found in the four- and six-coordination states, although five-coordinate systems have been found. Tetrahedral coordination is preferred in four-coordinate zinc(II), especially when the ligands are easily polarized. Divalent zinc can also be found in distorted trigonal bipyramids, square pyramids, and octahedra.[99]

Manganese in SOD alternates between the Mn(II) and Mn(III) states upon reduction and oxidation. Manganese(II) compounds normally exist as octahedra, although tetrahedral and square-planar complexes are not uncommon. Five-, seven-, and eight-coordinate complexes are also found. Octahedra are also typical for manganese(III) compounds, but square-pyramidal and trigonal-bipyramidal complexes are observed as well.[98] The important iron oxidation states in SOD are Fe(II) and Fe(III). Divalent iron is most commonly of octahedral geometry, although both square-pryamidal and trigonal-bipyramidal complexes and tetrahedral complexes are found as well. Iron(III) compounds are also usually octahedral, although trigonal-bipyramidal, square-planar, and tetrahedral geometries are also seen.[98]

1. Cu,Zn SOD

The geometry of the active site metals of bovine Cu,Zn SOD and their ligands in the oxidized enzyme, as viewed from the solvent, is shown in Figure 4.[4] The active site copper ion, which is alternately oxidized and reduced during the enzymatic reaction, is ligated by the imidazole nitrogens of four histidine residues: ND1 (the proximal N) of His 44, and NE2 (the distal N) of His 46, 61, and 118. The ligands show a tetrahedral distortion from a square plane. The bonds to His 61 and 118 are about 10° in front of the mean ligand plane in the solvent direction, while the bonds to His 44 and 46 lie about 25° behind the plane in the direction away from the solvent. This uneven distortion, combined with the His ring orientations, makes the axial position of the Cu much more open on the solvent side than

TABLE 3
Active Site ESR Parameters

Species	Measure	Value
$Cu(II)_2Zn(II)_2$[102]	$g_{\parallel}$	2.259 ± 0.002[a]
	$A_{\parallel}$	134 ± 1 G[b]
$Cu(II)_2E_2$ (pH 6.4)[136]	$g_{\parallel}$	2.265 ± 0.001
	$A_{\parallel}$	139 ± 1 G
$Cu(II)_2E_2$ (pH 3.6)[136]	$g_{\parallel}$	2.260
	$A_{\parallel}$	149 ± 1 G
$Ag(I)_2Cu(II)_2$ (−100°C)[104]	g_m	2.125
	$g_{\parallel}$	2.312
	$A_{\parallel}$	97 G

[a] Normal for 4 N donor ligands.[103]
[b] Low value may be due to the distortion away from square planar and toward tetrahedral geometry.

on the protein side, allowing the possibility of ligation of a solvent molecule with the metal. A solvent peak in the fragment difference map (F_o-F_c), representing electron density not accounted for by the protein, is located about 2.5 Å from the Cu(II), suggesting that a water molecule is coordinated to the axial position. Difference maps also show other peaks in the active site channel near the Cu ion consistent with a network of hydrogen-bonded water molecules. The existence of at least one water bound to the Cu has also been indicated by NMR data.[100]

The Zn ion is ligated by three histidines and one oxygen from the side chain of Asp 81.[4] His 61, 69, and 78 all ligate the Zn(II) with ND1 (the proximal N). The Zn ligand geometry is tetrahedral with a strong distortion toward a trigonal pyramid with Asp 81 at the apex. The three histidine rings are located slightly in front of the Zn in the direction of the active channel, with Asp 81 behind and completely buried.

Following refitting to difference maps, the Cu-N and Zn-N bond lengths of the ligating histidines averaged 2.1 Å for all four subunits, while the Zn to Asp 81 OD1 bond length averaged just under 2 Å.[6] These values are within the normal range for Cu-N bonds in organometallic compounds.[101] The longer bond to the bound water (2.5 Å) would be expected for an axial ligand.[99]

Electron spin resonance (ESR) studies of both native and metal-substituted SODs are a powerful aid to understanding the geometry around metal binding sites. Zinc-substituted derivatives of Cu,Zn SOD are often used for both ESR and NMR, since the zinc can be substituted without affecting the activity of the enzyme. To designate these metal-site substitutions, the metal in the copper site is conventionally listed first and the zinc site second. Empty metal sites are indicated with an "E." Thus, the Cu,Zn SOD dimer derivative E_2Co_2SOD has no metal ions in the two copper sites and cobalt ions in the two zinc sites. ESR parameters for native and derivative SODs are collected in Table 3. The ESR spectrum of $Cu(II)_2E_2SOD$ shows an increase in $A_{\parallel}$ compared to $Cu(II)_2M(II)_2SOD$ (M = Zn, Cd, or Hg); this suggests that the Cu-ligand geometry is more tetragonal in the Zn-free enzyme.[102] At low pH the ESR spectrum of $Cu(II)_2E_2SOD$ has the appearance of a nitrogen-ligated tetrahedral Cu(II).[103] The spectrum of $Ag(II)_2Cu(II)_2SOD$ is consistent with either a flattened tetrahedral geometry or a five-coordinate geometry for the Cu(II) in the Zn site.[104]

The ESR spectrum of the native enzyme clearly lacks full axial symmetry,[102] indicating that the Cu(II) site does not have completely regular square planar geometry. Peisach and Blumberg[103] have grouped Cu proteins based upon the relationship between $g_{\parallel}$ and $A_{\parallel}$. Due to its relatively low $A_{\parallel}$, native Cu,Zn SOD falls below the region where all nitrogen-ligated and mixed oxygen- and nitrogen-ligated model compounds overlap. The decrease in $A_{\parallel}$ is

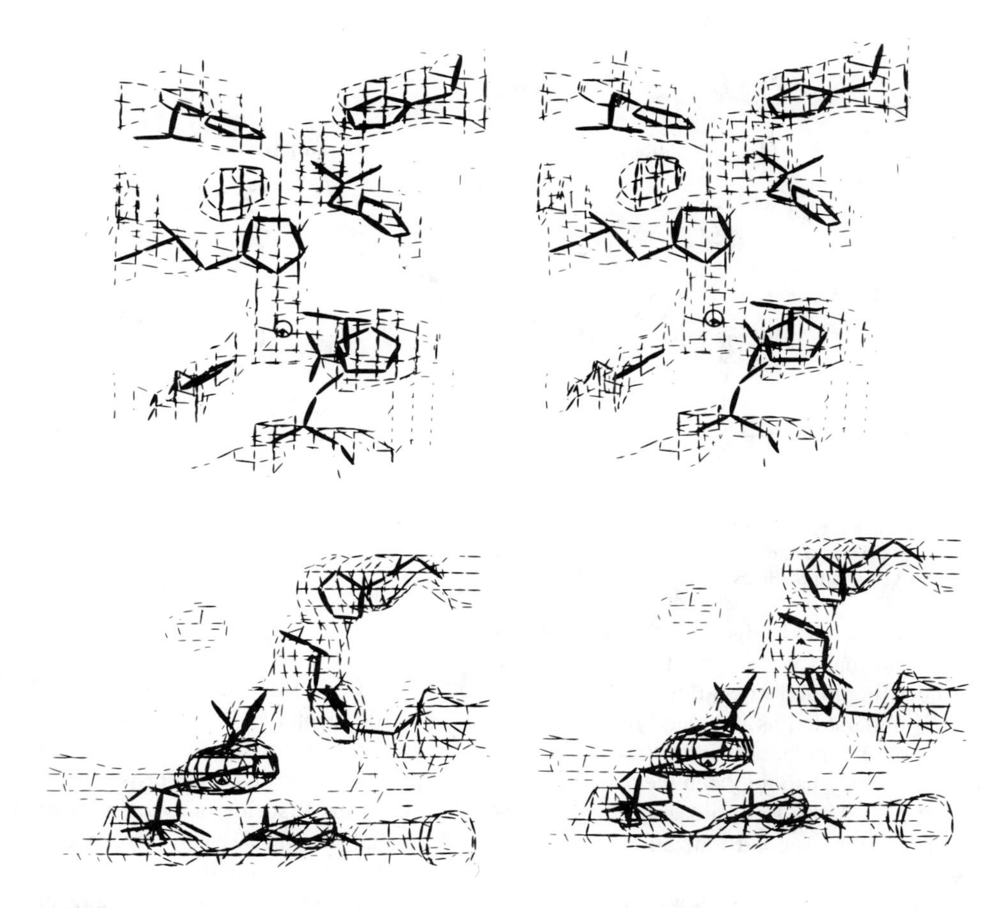

FIGURE 5. Electron density map of copper, zinc, and active site ligands in bovine Cu,Zn SOD at 2Å resolution. These two stereo views show the metals and their ligands after refinement with fragment difference electron density maps contoured at about 0.7 $eÅ^{-3}$, which encloses the top 7% of the electron density. The upper stereo pair looks into the active site from the solvent. A solvent peak appears in front and slightly to the left of the copper (density at top center). The imidazole ring of His 61 (central pentagon) bridges from the copper to the zinc (circled at lower center). Asp 81 is located behind the zinc in this view. The lower stereo pair shows the view along the ring plane of His 61 (center), which is shown on edge and positioned in the continuous density that bridges from the copper above to the zinc below.

a feature observed in Cu(II) complexes with a tetrahedral distortion of the ligand field away from the normal tetragonal arrangement.[105]

A unique aspect of Cu,Zn SOD is the bridge formed by ligand His 61 between the copper and zinc ions, with the imidazole ring approximately planar to both metals (Figure 5A and B). The metal ligand geometry and the histidine ring orientations are clearly seen in the high-resolution X-ray structure. ESR and NMR studies indicate that the metal ion pairs within each subunit of Cu,Zn SOD dimer derivatives ($Cu(II)_2Co(II)_2$SOD and $Cu(II)_2Cu(II)_2$SOD) are antiferromagnetically coupled, which suggests the existence of a ligand bridge between the metals occupying the Cu and Zn sites.[106-108] The $Cu(II)_2Cu(II)_2$SOD ESR spectrum is dominated by the magnetic coupling between the two Cu ions and strongly resembles spectra of small, imidazolate-bridged Cu compounds.[109,110] In apparent conflict with both crystallographic and ESR models, perturbed angular correlation (PAC) of gamma rays in Cd derivatives of yeast Cu,Zn SOD have been interpreted to exclude a bridging imidazolate.[111] There are several problems with this interpretation. First, the negative charge on His 61 assumed in the analysis may be substantially dissipated by the two positively

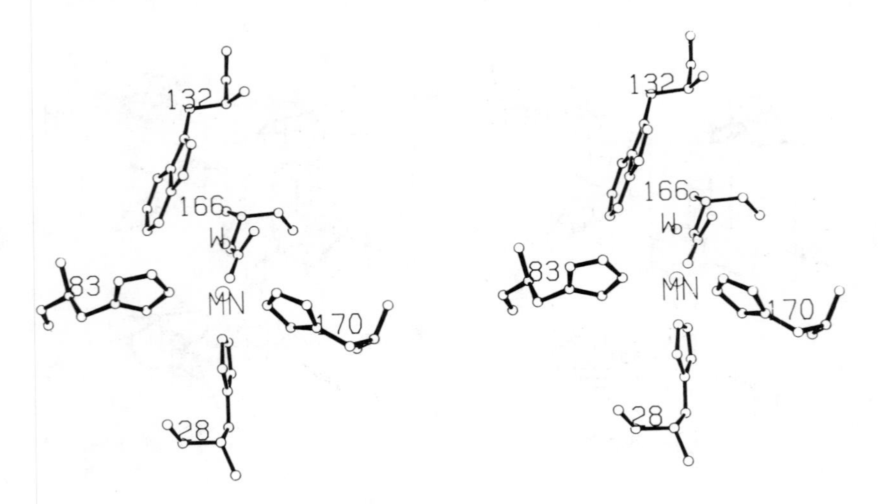

FIGURE 6. Manganese ligand geometry in Mn SOD. The stereo view shows the geometry of the active site metal and its ligands. W indicates water ligand. This figure kindly provided by William Stallings.

charged metal ions bridged by the imidazole ring. Second, Cd may bind in a site distinct from the normal Cu and Zn sites; seven discrete heavy metal binding sites in Cu,Zn SOD have been identified crystallographically.[112,113] Indeed, Cd SOD derivatives are currently not well-understood as indicated by conflicting ^{113}Cd NMR results.[114,115]

Although imidazole rings that bridge between two metals are found in small molecules,[116-119] they had not been identified previously in proteins. Normally, the proximal ND1 and distal NE2 of a His imidazole ring have pK_as of 6.0 for the first ring nitrogen ionization and 14.4 for the second ionization. The binding of a metal ion to ND1 can lower the pK_a of NE2 by about two log units; this second ionization is also greatly facilitated by changing the solvent from water to 50% ethanol or dioxane. Consequently, soluble small molecule compounds with bridging histidine rings occur only at relatively high pH or in nonaqueous solvents.[119] They can, however, be further stabilized by the surrounding structure, as in the macrocycle synthesized by Coughlin et al.[120] In the oxidized Cu,Zn SOD structure, His 61 bridges between the Cu and Zn metal atoms at the crystallization pH of 7.5. Spectrophotometric data of the oxidized enzyme of the oxidized enzyme suggest that this bridge remains unbroken over the pH range from 5 to 9.5 in which Cu,Zn SOD activity level is constant.[9] Hodgson and Fridovich[121] have postulated that this bridging His imidazolate ring may function as a proton carrier to facilitate protonation of the $O_2^{\cdot -}$ that interacts with the reduced enzyme.

2. Fe/Mn SOD

Despite large differences in overall structure compared with the Cu,Zn enzyme, the active sites of the Fe and Mn enzymes appear to resemble the Cu,Zn enzyme. In the oxidized Mn SOD, the ligands have a distorted trigonal pyramidal geometry around the Mn as shown in Figure 6. His 28 and a water molecule are the apical ligands with His 83, His 170, and Asp 166 making up the rest of the pentacoordinate complex.[8] Although the Fe SOD sequence information is incomplete, rigid-body transformations of the Mn SOD model onto the Fe SOD electron density maps shows that the Fe SOD ligand geometry closely matches the pentacoordinate geometry seen for Mn SOD.[122-124] Using a partial sequence for Fe SOD from *P. ovalis,* it has been determined that three of the ligands are two histidines corresponding to His 28 and His 170 in the Mn SOD structure and one is an aspartate corresponding to Asp 166.[123] The structure is not sufficiently resolved to see a water ligand, but NMR

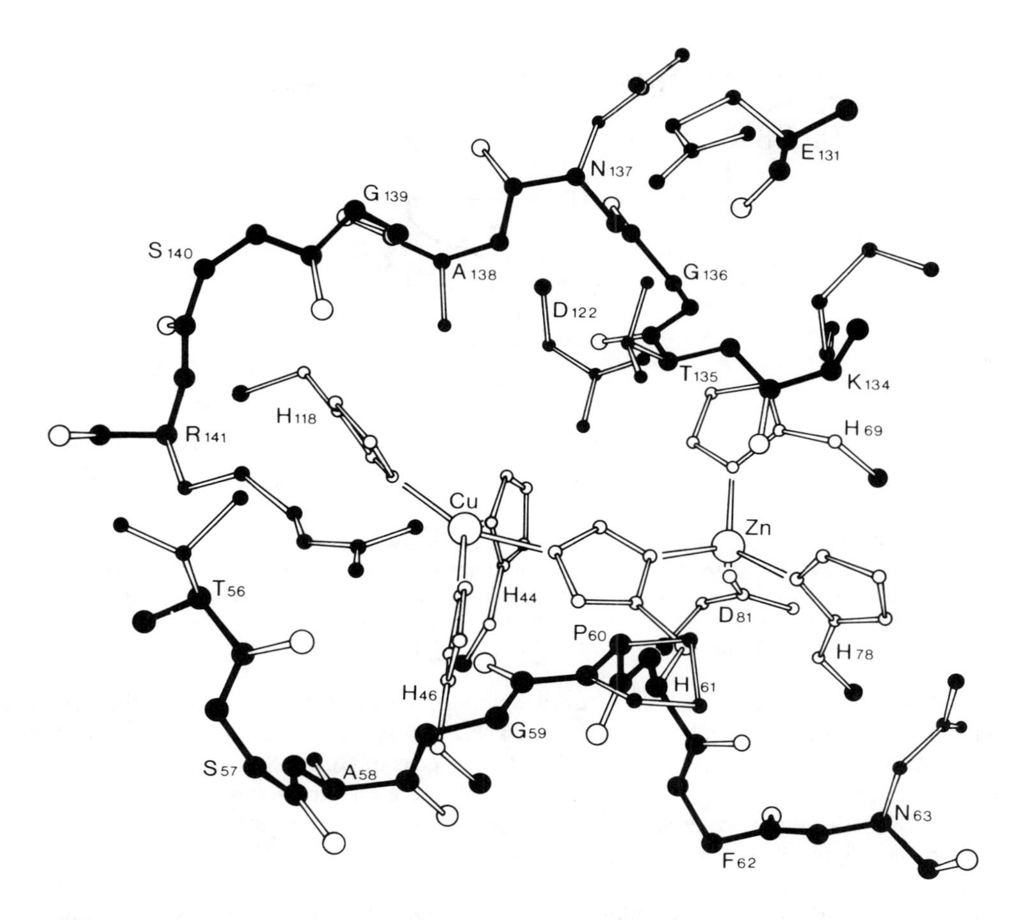

FIGURE 7. Critical active site channel residues of bovine Cu,Zn SOD as viewed from the solvent. The main chain is shown by solid black bonds and atoms, the ligand side chains by open bonds and atoms, and the other side chains by solid atoms and open bonds. Main-chain atoms only of residues 57, 62, and 140 have also been included in the drawing for continuity. (From Tainer, J. A., Getzoff, E. D., Richardson, J. S., and Richardson, D. C., *Nature,* 306, 284, 1983. With permission from Macmillan Magazines, Ltd.)

indicates a water ligand to the Fe does exist.[125] The identity of the fifth ligand has not been determined. The geometries of the Fe and Mn enzymes thus show similarities to the Cu,Zn enzyme in at least three respects (C. L. Fisher and J. A. Tainer, manuscript in preparation): (1) four protein side chains act as metal ligands, (2) a single water molecule makes a fifth metal ligand, and (3) the overall geometry places the water molecule and a protein side chain in trans positions (His 44 could be described as being trans to the axial water ligand in Cu,Zn SOD). Thus, the general arrangement of the ligands in all three enzymes could be described as a tetrahedral distortion from a square plane of protein ligands with an axial water ligand.

C. ACTIVE SITE ENVIRONMENTS

1. Cu,Zn SOD

The active site area of Cu,Zn SOD is made up of highly conserved, closely packed residues.[6] The active site Cu(II) and Zn(II) lie 6.3 Å apart at the bottom of a long channel between two large loop regions on the external surface of β barrel strands 5e, 6d, and 7g (Figure 3). Twenty-one residues contribute to the active site channel structure (Figure 7). Residues Glu 131, Lys 134, Thr 135, Gly 136, Asn 137, Ala 138, Gly 139, and Arg 141 form one rim of the channel and residues Thr 56, Ala 58, Gly 59, Pro 60, and Asn 63 form

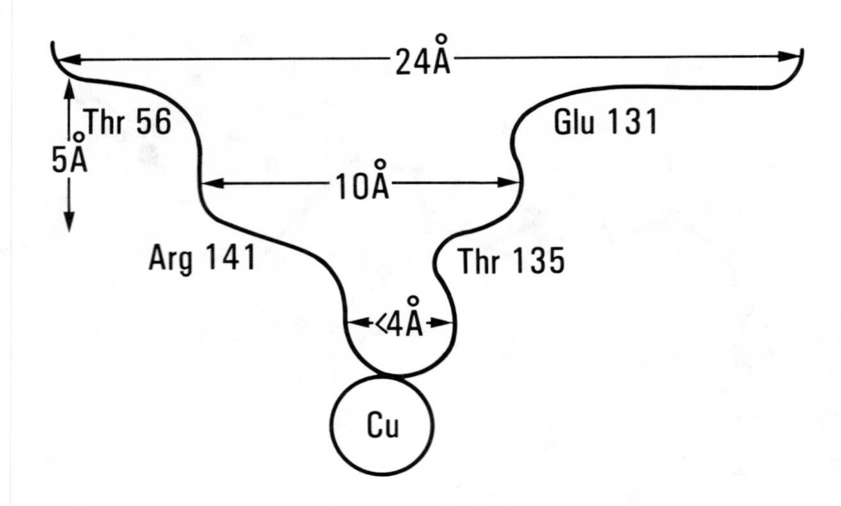

FIGURE 8. Schematic view of the shape and dimensions of the active site channel of bovine Cu,Zn Sod in cross-section. The amino acids forming the channel wall are indicated. Lys 134, located forward of the plane of the paper and adjacent to Glu 131, also contributes structurally to channel architecture. (From Getzoff, E. D. and Tainer, J. A., *Ion Channel Reconstitution,* Miller, C., Ed., Plenum Press, New York, 1986, 57. With permission.)

the other rim. The metals and their seven ligating residues form the floor of the channel with Asp 122 buried beneath. Compared with the rest of the molecule, the residues of the channel are highly conserved in the published sequences of Cu,Zn SODs,[126] suggesting that the topography of the channel's surface is critical to the enzyme's function.[6] This channel is shaped like a funnel, wide at the outside, and narrowing down as it leads inward to the Cu ion (see Figure 8). At the bottom of the channel is a site complementary to superoxide that would place the superoxide molecule in the proper geometry to interact directly with the copper. Thus, the high resolution structure of bovine Cu,Zn SOD presents a picture of an enzyme that has developed to fit its substrate optimally in the active site.

The orientations of both the side chains and the main chains of the Cu- and Zn-ligating residues are stabilized by a extensive network of hydrogen bonds.[6] Most of the atoms of the seven metal-ligating residues that could potentially form hydrogen bonds do so with other protein atoms. The two exceptions are the main chain nitrogen atom of His 61, which appears to hydrogen bond to a water molecule in the active site channel, and the main chain carbonyl oxygen atom of His 69, which appears to hydrogen-bond to a water molecule external to the channel.

The Cu and Zn differ greatly in the number of structural elements involved in the covalent and hydrogen-bonding patterns of their ligands.[6] The Zn ligands are all located in a continuous stretch of chain that forms the second half of a loop between β strands 6d and 5e (Figures 2 and 3). Residues that hydrogen-bond to the Zn ligands are also located primarily in the Zn-binding region of this loop. Except for His 61, the Cu ligands are located in the β barrel (His 44 and 46 from strand 6d and His 118 from strand 7g), joining the two large loop regions that help form the active site channel.[4] The larger number of different structural elements involved in the Cu network vs. the Zn network may account for the larger effect of Cu site occupancy on the thermal stability of the enzyme.[127] Comparative analysis of the known metalloprotein structures[128] suggests that the involvement of multiple structural elements and the wide separation of the Cu ligands in the amino acid sequence (up to 74 residues) are expected for a catalytic metal site as opposed to a structural or storage site.

The buried carboxyl of Asp 122, as with the bridging ligand His 61, is in a position to function as a link between the Cu oxidation state and Zn ligand geometry.[6] This carboxyl group, which is completely inaccessible to solvent, forms hydrogen bonds with both Zn ligand His 69 and Cu ligand His 44 to make an indirect bridge between the Cu and Zn ions. In the electron density maps of all four subunits, these hydrogen bonds are well defined and consistently short, suggesting that they are charged hydrogen bonds, which are energetically more favorable than neutral ones.[129] In model compounds, the NMR chemical shift for the C2 proton of a histidine bound to a Cu ion (about 8.0 ppm) lies between that for a neutral histidine (7.7 ppm) and that for a protonated histidine (8.7 ppm).[130,131] The NMR chemical shifts for the C2 protons of His 44 and 69 (7.7 ppm) are similar to those of neutral histidines, and smaller than expected for metal-ligating histidines.[132] This is consistent with strong hydrogen-bonding interactions between the two histidines and Asp 122 that reduce the partial positive charge on the imidazole rings caused by their interaction with the Cu ion. The correlation of enzyme activity with this hydrogen-bonding network around the metal ligands would be facilitated by knowledge of the structure of the reduced Cu(I) enzyme.

In summary, the orientations of both the side chains and the main chains of the Cu- and Zn-ligating residues are stabilized by an interlocked network of hydrogen bonds. The positions of the ligand side chains are also stabilized by extensive nonbonded contacts. These hydrogen-bonding and tight packing interactions in the active channel suggest that, regardless of any possible strain implied by distortions of the metal-ligand geometry, large movements or rearrangements of the ligands during catalysis are unlikely. The two metal ions in each subunit interact directly through His 61 and indirectly through the hydrogen bond network involving Asp 122. The low temperature factors (suggesting little local mobility) and clear definition in the electron density maps of the active site channel residues are consistent with this area being quite stable and ordered without significant dynamic or static structural variations.

2. Fe/Mn SOD

Detailed information is available on the active site environment for the Mn SOD from *T. thermophilus*.[8] As in Cu,Zn SOD, the active site ligands are held in their orientations by a combination of hydrogen-bonding and packing interactions. In addition to the ligands, a group of aromatic side chains, including Tyr 36, Phe 86, Trp 132, and Trp 168, are positioned within 7 Å of the Mn ion and five additional aromatic side chains (His 32, His 33, Trp 87, Tyr 172, and Tyr 183) are only slightly farther away. Six of these aromatic side chains (Tyr 36, Phe 86, Trp 87, Trp 132, Trp 168, and Tyr 183) provide a hydrophobic lining around most of the metal-ligand center. Although it is premature to assign complete hydrogen-bonding interactions at this stage of the structure determination, ligand side chains appear to be stabilized by a hydrogen-bonding network to the backbone in a fashion analogous to that found in the Cu,Zn SOD structure.

The overall folding and active sites appear similar for both Fe and Mn SODs.[25,94] The Fe structure has been superimposed onto the more highly resolved Mn structure to reveal these similarities.[25] At the current resolution of these structures, there are no obvious differences that might explain why some of these enzymes are inactivated by replacement of Fe for Mn or vice versa.[94]

The interaction between metal ions in the Fe/Mn SOD dimers is less clear than in the Cu,Zn enzyme. Although the two metal ions in the dimer are about 18 Å apart, some of the metal ligands and their adjacent residues make close contacts across the dimer interface.[94] In addition, the active sites of Fe/Mn SOD are exposed to the dimer interface. The metal ions of the dimer could interact with each other through contacting ligands across the dimer interface. Conversely, they may be too far apart to interact, making the two active sites in the dimer independent of each other.

IV. THE REACTION MECHANISM

A. REACTION EQUATION FOR SUPEROXIDE DISMUTASE

The superoxide radical is unstable in an aqueous environment and spontaneously dismutes to O_2 and H_2O_2. Spontaneous dismutation of $O_2^{\cdot -}$ is pH dependent with the maximum rate occurring at pH 4.8, the pK_a of the conjugate acid of $O_2^{\cdot -}$. At pH 7.4, the rate of spontaneous dismutation is about $2 \times 10^5\ M^{-1}\ s^{-1}$.[71] Since uncatalyzed dismutation involves the direct reaction of two superoxide molecules, it is hindered by electrostatic repulsion between these anions. The *in vivo* dismutation rate without catalysis is also affected by the low steady-state concentration of superoxide (10^{-10} *M*).[71] SOD solves these problems by the alternate reduction and oxidation of the enzyme during successive encounters with superoxide, thereby catalyzing electron transfer between two $O_2^{\cdot -}$ anions without requiring their direct interaction. Comparison of the rate constant for the spontaneous dismutation[71] with that for the enzymatic reaction[133] indicates that, at physiological pH, the Cu,Zn SOD enzyme increases the rate over noncatalyzed $O_2^{\cdot -}$ breakdown about 10,000 fold.

The requirements for metal-ion-catalyzed dismutation in all of the SODs include (1) the presence, in both valence states, of an open coordination position on the metal ion occupied by a water molecule, (2) the ability of the metal in its protein environment to be oxidized and reduced by superoxide, and (3) the formation of metal-superoxide complexes within which electron transfer can occur.[134] The evidence considered in the following sections is consistent with these three requirements for both Cu,Zn and Fe/Mn SODs. Since less is known about the Fe/Mn SOD family in terms of structure and mechanism, we will discuss the Cu,Zn enzyme in detail, with a short discussion of the Fe/Mn enzymes.

The equation for the mechanism of Cu,Zn SOD catalysis has been determined by pulse radiolysis.[9,133,135] The absorbance at 680 nm due to the Cu(II) undergoes a cyclical bleaching by $O_2^{\cdot -}$ without a loss of activity. After initial reduction of the enzyme to Cu(I) by H_2O_2, an $O_2^{\cdot -}$ pulse could restore the absorbance at 680 nm, indicating the reoxidation of the Cu(I) to Cu(II). The enzyme thus undergoes a reduction and reoxidation of the active site Cu via a two-step catalysis:

$$\text{E–Cu(II)} + O_2^{\cdot -} \rightarrow \text{E–Cu(I)} + O_2 \tag{4}$$

$$\text{E–Cu(I)} + O_2^{\cdot -} + 2H^+ \rightarrow \text{E–Cu(II)} + H_2O_2 \tag{5}$$

$$2O_2^{\cdot -} + 2H^+ \rightarrow O_2 + H_2O_2 \tag{6}$$

In an aqueous environment, the breakdown of the relatively unstable superoxide anion is inorganically catalyzed by free Cu(II) and other transition state metals; this is prevented *in vivo* by the multitude of potential metal-binding substances.

The activity of the Cu,Zn SOD family is distinguished from that of the Mn and Fe enzymes by being insensitive to pH over the broad range: $5.0 < pH < 9.5$.[9] UV-visible and ESR spectra of Cu,Zn SOD similarly show little variation over this pH range.[136] The enzyme's conformational stability at low pH has been confirmed by comparisons of NMR spectra between pH 6.9 and pH 3.4.[137] Above pH 9, spectral changes suggest the deprotonation of a water ligated to the Cu(II).[138] The bovine enzyme is apparently irreversibly denatured above pH 12.2.[138] Below pH 4.5, spectral changes occur that reflect the loss of the Zn(II) ion.[136]

While the catalytic mechanism for both Cu,Zn SOD and the Fe/Mn SODs involves alternate reduction and reoxidation of the active site metal, details from kinetic studies suggest that the mechanism in the Fe/Mn SODs is different from that in the Cu,Zn enzymes.[139,140] The dismutation of superoxide by Cu,Zn SOD occurs by a diffusion-controlled process and shows no evidence of saturation.[140] The second-order rate constant for the reaction is therefore at least $2 \times 10^9 M^{-1} s^{-1}$.[133,140] Fe SOD from *E. coli* and Mn SOD from *T. thermophilus* do show saturation kinetics,[140] but the reasons for this are not understood.

B. STRUCTURAL, COMPUTATIONAL, AND BIOCHEMICAL IMPLICATIONS FOR THE REACTION MECHANISM OF CU,ZN DISMUTASES

1. Structural Anatomy of the Active Sites

One of the most powerful tools for the structural analysis of intermolecular recognition has been the use of a molecular surface algorithm.[141] The solvent-accessible molecular surface of the active site channel of Cu,Zn SOD has been determined by rolling a 1.4-Å radius probe over the van der Waals surface obtained from the crystallographic coordinates.[4,6] The Cu(II) ion exposes about 5 $Å^2$ of its surface area to the probe, but the Zn(II) ion is completely buried. The molecular surface of the active channel comproses approximately 610 $Å^2$ or about 10% of the total exposed surface area. The molecular surface reveals two pits forming specific binding sites in the floor of the active channel. The Cu pit is located directly above the Cu(II) and involves the Cu ion, His 61, His 118, Thr 135, and Arg 141; the water pit is adjacent to this position above and between the Cu ligands His 44 and 118 and is surrounded by Thr 135, Gly 136, Ala 138, and Gly 139. Both pits lie in the particularly narrow part of the channel between the guanidinium group of Arg 141 and the side chain of Thr 135 (see Figure 8).

Using interactive computer graphics, independently surfaced $O_2^{\cdot -}$ and CN^- ions were manually docked into each of the two pits of the active site channel molecular surface.[6] $O_2^{\cdot -}$ was positioned in the Cu pit with an excellent fit to the surface, such that one oxygen was bound to the Cu (about 2 Å away) and the other formed a hydrogen bond with Arg 141 NH1 (about 3 Å away). The fit of the $O_2^{\cdot -}$ into the adjacent water pit is poorer and makes less chemical sense in terms of oxygen-bonding geometry and Cu to $O_2^{\cdot -}$ distance (about 3.4 Å). Neglecting the somewhat lengthy distance from the water pit to the Cu, CN^- can be fit into either site with the required linear geometry.

Both of the pits in the active channel surface were found to contain water molecules (temperature factors < 10 and occupancies > 0.9) in the refined crystallographic structure.[6] Two water molecules appear to form a ghost of an $O_2^{\cdot -}$ positioned in the Cu pit; one water binds 2.8 Å from the Cu(II) and the other 3.3 Å away from a guanidinium N of Arg 141. These two waters are positioned to form a 120° angle relative to the Cu ion as expected for the geometry of a bound $O_2^{\cdot -}$. The water pit holds a bound water in the correct geometry to form a hydrogen bond to the Cu-bound oxygen of an $O_2^{\cdot -}$ in the Cu pit. Since this water is about 3.4 Å from the Cu, it does not appear to be an inner-sphere ligand in the crystal structure.

2. A Mechanism for Dismutation with Supporting Biochemical Data

Various researchers have proposed that Cu,Zn SOD acts through an inner-sphere mechanism,[121,142] while others have proposed an outer-sphere mechanism.[143,144] The extremely rapid catalysis of Cu,Zn SOD has so far prevented direct study of an enzyme-substrate complex; however, the detailed structure of the active site at atomic resolution is consistent with direct binding of the superoxide to the Cu(II).[6] The details of the active site geometry (including His 61 bridging the Cu(II) and Zn(II), the tightly bound water molecules, and the complex network involving Asp 122 described earlier) have been used to suggest a specific model for the enzyme mechanism shown in Figure 9.[6] This model (outlined below), based on an inner-sphere mechanism, explains the structural features, the chemical reaction

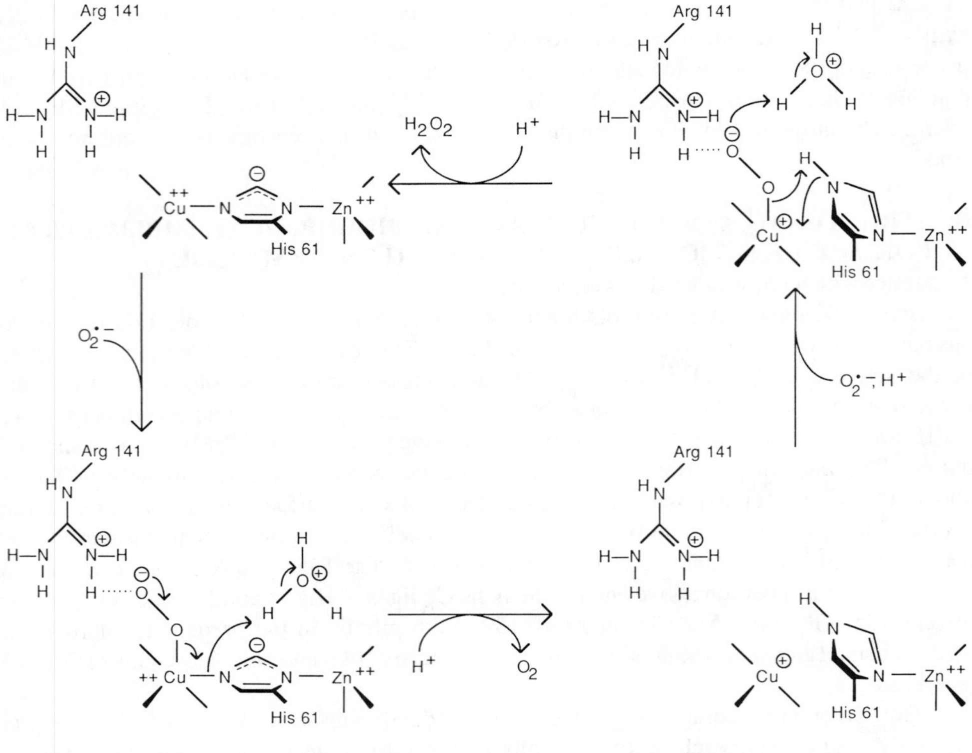

FIGURE 9. Schematic mechanism of the oxidation-reduction pathway in Cu,Zn SOD.

equation, and the known biochemistry of Cu,Zn SOD including the pH independence of the catalytic rate between pH 5 and pH 9.5.

As discussed above, the molecular surface of the active site channel shows a single, definite complementary binding position for superoxide; $O_2^{\cdot -}$ docked here binds the Cu(II) and the important Arg 141 with the correct geometry. If $O_2^{\cdot -}$ does bind directly to the Cu, as implied by the perfect fit of a superoxide into the pit above the Cu(II), what active site geometry would result during the different steps in the reaction equation? When the $O^{\cdot}{}_2^-$ binds, there is a virtual shift in the mean ligand plane so that $O_2^{\cdot -}$ binding is equatorial. One superoxide oxygen atom binds to the positively charged Cu(II) while the other oxygen atom is positioned to form a hydrogen bond with the positively charged guanidinium nitrogen on Arg 141. Bound $O_2^{\cdot -}$ reduces the Cu(II) to Cu(I) with the simultaneous breaking of the bond between His 61 and the Cu, and O_2 is released. The His 61 side chain is forced out of the plane of the two metals to allow a more tetrahedral geometry around the Zn(II). Once the Zn(II) has relaxed to this geometry, the hydrogen on His 61 NE2 is in a perfect position for donation to the second superoxide anion. The total movement of His 61 need not be large; a rotation about the C_β-C_γ bond is sufficient to move His 61 NE2 out of the plane of the metals about 0.6 Å. The movement of His 61 out of the plane also allows the Cu(I) to become more tetrahedral, the preferred geometry for a Cu(I) site. Further movements around the Cu(I) site might involve the buried Asp 122 carboxyl, which forms hydrogen bonds to the Cu ligand His 44 and the Zn ligand His 69. In addition, a shift in the position of His 44 toward a more tetrahedral environment around the Cu(I) might be transmitted through this hydrogen-bond bridge to reduce distortion of the tetrahedral Zn site also.

Binding of a second $O_2^{\cdot -}$ directly to the Cu(I) places one superoxide oxygen atom in position to form a short, charged hydrogen bond to the hydrogen atom on His 61 NE2, while the second oxygen atom of superoxide can still be positioned to form a hydrogen bond to the positively charged Arg 141 guanidinium group, as in the first step of catalysis. While His 61 ND1 is bound to the Zn(II) ion but not to Cu(I), the pK_a for the dissociation of the proton on NE2 is above 9,[119] so NE2 will always be protonated at physiological pH. Based upon the superoxide coordinates from docking $O_2^{\cdot -}$ into the best fit in the active site, the distance from rotated His 61 NE2 to the nearer superoxide oxygen atom is about 2.5 Å, which is consistent with a charged hydrogen bond involving partial proton transfer to the superoxide.[6,132,] The proton is transferred to the superoxide with spontaneous oxidation of the Cu and reformation of the bond between His 61 and the Cu, resulting in production of a hydrogen peroxide molecule.

Spectral and kinetic evidence support the breaking of the His 61-Cu bond during the enzymatic catalysis,[139] causing the uncoupling of the metal sites[106] and the uptake of the one proton per subunit[145] by the His 61 ring nitrogen not ligated to the Zn(II).[146] The removal of the His 61 imidazole nitrogen from the Cu coordination changes the Zn(II) X-ray absorption spectrum[147] and the visible spectrum of the Co in the Zn site of $Cu(I)_2$ $Co(II)_2$ derivatives,[137] indicating coupling of the Cu ionization state to the Zn-ligand geometry. The proton NMR spectrum of the $Cu(I)_2Co(II)_2$ derivative[108] shows that all three histidine residues ligated to the Zn ion each have one exchangeable proton, which indicates that the His 61 bond to the Cu(I) must be broken and His 61 must be protonated in the reduced enzyme. X-ray absorption spectroscopy suggests that the Cu site becomes more tetrahedral upon reduction.[147] The very fast rate of the reaction implies that no large rearrangements of the active site can occur during the reaction. All of the movements discussed above — the breaking of the histidine bond to the Cu ion and the corresponding movements of the ligands to more tetrahedral arrangements around the metals upon reduction — involve small movements, mainly rotations, of the histidine rings.

In Cu,Zn SOD, the presence of the Zn, the imidazolate bridge between the Cu(II) and Zn(II), and the tetrahedral distortion of the Cu(II) site increase the redox potential of the Cu(II) to the observed 0.42 V,[145,146] which is very high relative to that of aqueous Cu(II) (0.17 V) or Cu(II)-histidine complexes (about 0.01 V).[148] Unlike the native enzyme, the Zn-free enzyme is not reduced by ferrocyanide.[146] The Zn-free enzyme prepared from apo Cu,Zn SOD at pH 3.8 (the preferred method)[149] is only about 20% as active as the native enzyme.[104] The native enzyme's catalytic rate is independent of pH over the range 5 to 9.5,[9] indicating that the protons required by the reaction stoichiometry do not come from bulk water. In contrast, the activity of the Zn-free derivative is pH dependent with an apparent pK_a of 6.9,[12] which may reflect the pK_a of His 61.

This inner-sphere mechanism, proposed from implications from the X-ray structure of the Cu,Zn enzyme, suggests a catalytic role for the Zn,[6] in addition to the previously proposed structural role.[149] In the first step of the dismutation reaction, relaxation of the distortions in the Zn-ligand geometry present in the Cu(II) enzyme (mediated through the His 61 imidazolate bridge and the pair of hydrogen bonds to Asp 122) may stabilize the transition state. In the second step of the reaction, the roles of the Zn are to assure protonation of His 61 NE2 and to correctly position this proton for hydrogen bonding and subsequent transfer to the second incoming $O_2^{\cdot -}$, thus completing the catalytic cycle. The Zn would enhance the enzymatic activity without being essential, since His 61 NE2 will be at least partially protonated at pH 7 and the necessary protons are also available from the water molecules in the active channel.

Finally, evidence from chemical modification of Arg 141 supports its catalytic importance in the complementary fit of the $O_2^{\cdot -}$ into the active channel. At least a 90% decrease in activity results from modification of this single residue with phenylglyoxal in both bovine and yeast Cu,Zn SODs.[150,151] (See also discussion on site-directed mutants of Arg 141 below).

3. Quantum Chemical Calculations

A second mechanism involving both inner- and outer-sphere interactions has been proposed based on quantum chemical calculations.[152] To allow more than 20 atoms to be included, the active site model was calculated with the restricted open Hartree-Fock approach using the effective core potential approximation with a split valence basis set. The active site model included the copper ion, the complete bridging histidine ring (His 61), three ammonia groups representing the three remaining histidine ligands of the copper ion, and a proton representing the zinc ion. The calculation was done both with and without an ammonium ion representing Arg 141. Without the inclusion of the Arg 141 ammonium ion, the presence of $O_2^{\cdot -}$ adjacent to the copper ion results in a charge distribution where the copper ion is reduced to Cu(I) and the $O_2^{\cdot -}$ is oxidized to O_2, indicating that the electron transfer from the $O_2^{\cdot -}$ to copper is favorable. Addition of the ammonium ion representing Arg 141 affects the charge distribution by stabilizing the negative charge on the $O_2^{\cdot -}$, while maintaining the oxidized state of the copper. The authors propose that this Cu-superoxide complex then oxidizes another $O_2^{\cdot -}$ molecule to O_2 by an outer-sphere mechanism, resulting in reduction of the copper. The initial $O_2^{\cdot -}$ molecule is then reduced by Cu(I) and protonated by Arg 141, to release HO_2^-.

This is indeed an interesting possibility, that the superoxide-copper complex serves as the oxidizing species. Unfortunately, the calculations are not sufficiently extensive to prove the new mechanism or disprove the standard mechanism. In particular, conformational changes as suggested above for His 61 are not included in the calculation. The breaking of the His 61-copper bond upon reduction of the copper would clearly affect the charge distribution of the active site complex. In addition, the positive charge from Arg 141 that stabilizes the superoxide in the transition complex is probably greatly overestimated in the model. The ammonium ion representation does not allow delocalization of the charge over the guanidinium group or its hydrogen-bonded neighbors. So while this calculation suggests a possible new mechanism, proof for it remains to be shown.

Ab initio Hartree-Fock calculations on a smaller system where His 61 is modeled by an ammonia molecule and Arg 141 is modeled by an ammonium ion also indicated that the complex of superoxide hydrogen-bonding to Arg 141 is a stable intermediate.[153] Without detailed treatment of the Arg 141 side chain and the breaking of the His 61 bridge, it cannot be ascertained whether this intermediate is the actual oxidizing species for the next superoxide molecule or whether the intermediate exists only until the His 61 bridge is broken and the electron is transferred from superoxide to the copper ion.

4. Binding of Anion Inhibitors

Examination of a complex of SOD with superoxide has been impossible because the rate of reaction is so great. The binding of several anion inhibitors of Cu,Zn SOD has been examined in the hopes of learning more about the mechanism. For the Cu(II) enzyme, CN^-, N_3^-, SCN^-, OCN^-, and halides are reported to bind to the copper.[154-156] The narrow width of the active site channel (<4 Å) just above the Cu (Figure 8) is consistent with the accessibility of the Cu to Cl^- (diameter of 3.62 Å) and Br^- (diameter of 3.90 Å), but not I^- (diameter of 4.32 Å).[157] The competitive anion inhibitors (CN^-, N_3^-, and F^-) apparently bind by displacing a water molecule,[138] as was proposed for $O_2^{\cdot -}$ binding from the atomic structure of the Cu,Zn enzyme.[6]

Cyanide binds reversibly in a one-to-one ratio with the Cu(II) ion[135] to inhibit Cu,Zn SOD.[158] ESR studies of the $^{13}CN^-$-enzyme complex indicate that cyanide binds equatorially[159,154,160] with the carbon atom bound to the Cu(II).[160] The ESR spectrum of the enzyme with CN^- bound is also more axial, suggesting a less distorted Cu ion geometry than in the native enzyme alone.

These data have been interpreted to indicate that cyanide promotes a major reorganization of the Cu(II) ligands, including displacement of a histidine ligand.[149,161,162] Currently, electron spin echo studies suggest that, contrary to earlier proposals, the His imidazole bridge between

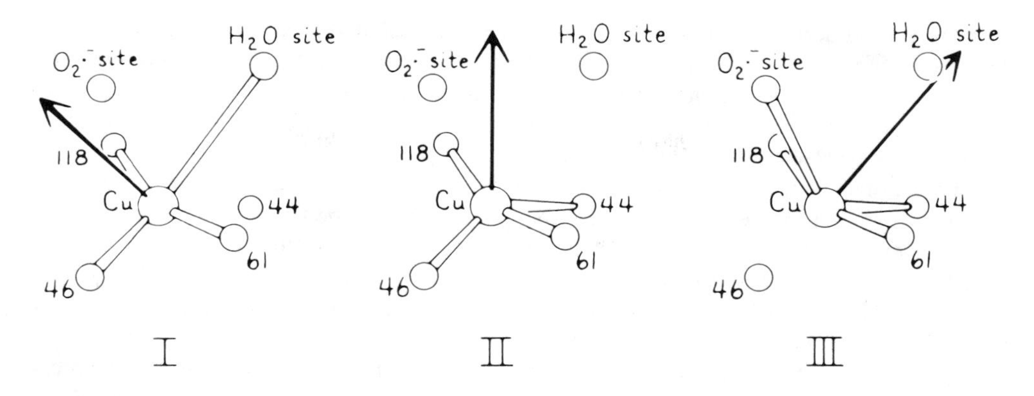

FIGURE 10. Anion binding sites and the interconversion of axial and equatorial directions of copper coordination. The normal to the least squares plane of the four histidine ligands around the Cu is shown as an arrow in II. The shift upon anion binding in the water site (I) and the $O_2^{\cdot -}$ site (III) reflects an interconversion of the equatorial plane. The RMS deviations from the least squares plane are 0.32 Å for I, 0.42 Å for II, and 0.34 Å for III. The shift in the normal vector direction is about 37° for II → I and 53° for II → III. Allowing for error, small realignments, and potential differences between the electron spin resonance parameters and the planes calculated from coordinates, docking of CN^- in the water site (I) is consistent with the single crystal ESR data,[164] and NMR data on azide binding.[167,168] (From Tainer, J. A., Getzoff, E. D., Richardson, J. S., and Richardson, D. C., *Nature*, 306, 284, 1983. With permission from Macmillan Magazines, Ltd.)

the Cu(II) and the Zn(II) remains intact on binding both CN^- and N_3^-.[149,161] Instead, either His 44 or 46 is thought to be displaced when CN^- binds. The ligand geometry and lack of solvent accessibility of these two histidines make it unlikely that there is large rearrangement of these tightly packed residues by CN^-.

Based upon the geometry and molecular surface obtained from the crystal structure, it is clear that replacement of the axial water by a stronger ligand could allow a virtual rearrangement of the mean equatorial plane, such that the solvent-accessible binding position becomes part of the new equatorial plane.[6] Therefore, a CN^- ion could bind in either the superoxide anion site or the water site (see Figure 10), and shift the equatorial plane without allowing electron transfer. As a result of this virtual rearrangement, the Cu site geometry would be more axial, as observed by ESR, because the new equatorial plane is less distorted. In this model His 46 or His 44 becomes axial, with only small movements required. Interconversion of axial and equatorial directions in aqueous Cu complexes is very rapid,[148] but studies of small molecule Cu complexes suggests that significant physical changes in Cu ligand geometry would probably pose a large reorganizational energy barrier resulting in rates of electron exchange considerably slower than that known for Cu,Zn SOD.[163] The axial-equatorial interconversion model for anion binding to Cu,Zn SOD explains the data on anion inhibitors without requiring the large physical rearrangements in the active site that others have proposed.[164,165]

Similar to CN^-, azide binds reversibly to the Cu(II),[104] but is a less effective inhibitor. For enzyme concentrations of about 10^{-9} *M* at pH 8.2, 50 μ*M* cyanide causes 50% inhibition of human Cu,Zn SOD,[135] whereas 32 m*M* azide is required.[166] The visible spectrum of azide SOD has a new band at 370 nm,[159] which is characteristic of an azide-Cu(II) complex.[146] As for cyanide, the ESR spectrum of azide SOD is axial.

NMR studies on $Cu(II)_2Co(II)_2SOD$[167,168] validate the axial-to-equatorial interconversion for anion binding. The replacement of zinc with cobalt gives a derivative with activity similar to the native enzyme, but with dramatically different NMR properties. Magnetic exchange between the cobalt and copper allows the proton NMR signals to be observed for the histidines ligated to either metal. Very large shifts of the His 44 resonances from the paramagnetic region to the diamagnetic region are seen as azide is bound.[167,168] The distance of His 44 from the Cu ion derived from the relaxation times[168] revealed very small movements of the His ligand, predominantly a rotation about the C_β-C_γ bond. Similar results were obtained for NCO^- and NCS^- anion binding.[167] Thus, the anion binding causes a change in the mean equatorial plane of ligands around the Cu ion (see Figure 10, I), but with little movement in this tightly packed region of the protein.

Interactions of Cu(I) SOD with anions have not been well characterized. Fee and Ward[169] find direct competition between Cl^- and CN^- for binding to the Cu(I) enzyme. Data from X-ray absorption also suggest that CN^- binds to the Cu(I).[170] Binding of anions to reduced yeast SOD results in perturbation of the chemical shift of three of the five His C-2 proton resonances observed in the NMR spectrum. The perturbation order determined was $Cl^- = Br^- \geqslant I^- \geqslant F^-$, with the halides rapidly exchanging between bulk solution and the protein binding site.[171] Valentine and Pantoliano[149] have suggested that these results do not necessarily show anion binding to Cu(I), since they could result from binding to a site adjacent to the copper.

From the above evidence, it appears that binding of anion inhibitors cannot tell us much about the mechanism for superoxide binding, since different histidines are involved in bond breaking with the Cu ion in the two processes. It appears, however, that the anions do bind directly to the Cu(II) ion thus indicating that model studies of direct superoxide binding are reasonable and supporting the inner-sphere mechanism. Whether the anions inhibit the SOD reaction by direct competition or by changing the redox potential of the Cu ion is unclear. The studies on anions also show that small movements of the ligands around metals can result in the breaking of histidine-Cu bonds and in changes in the mean ligand plane around the metal.

C. STRUCTURAL AND BIOCHEMICAL IMPLICATIONS FOR THE REACTION MECHANISM OF FE/MN DISMUTASES

Superoxide reacts much more slowly with Fe and Mn SODs than with Cu,Zn SOD, making it possible to measure the rate of reaction.[140] Fe SOD from *E. coli* showed saturation kinetics with k_{cat} being independent of pH from pH 7 to pH 10, while K_m is constant below pH 9.[140] These kinetic studies also showed uptake of one proton upon reduction of the metal, and a mechanism similar to the inner-sphere mechanism described above for Cu,Zn SOD has been suggested, with an ionizable ligand to the Fe(III) that dissociates from Fe(II) to take up a proton. An incoming molecule of superoxide would then take up that proton, oxidize the metal, and cause the reformation of the ligand metal bond, resulting in production of hydrogen peroxide. For Fe SOD, this second step is apparently irreversible, because the rapid addition of a high concentration of H_2O_2 (>1 *M*) has no effect on the activity.[122] The chemically plausible inner-sphere mechanism outlined above for the Cu,Zn enzyme appears consistent with the currently available data for the Fe/Mn SODs, with the Fe or Mn undergoing the cyclical reduction and oxidation shown in Equations 4 and 5 for the Cu,Zn enzyme. It has been suggested that the rate-determining step of the reaction may be either a proton transfer from water to the protein[8] or electron transfer from superoxide to Fe(III).[122]

Structural changes in the active site upon reduction of the metal have been studied by X-ray crystallography. Crystals of oxidized Fe SOD or Mn SOD were reduced with dithionite and a Fourier difference map between the oxidized and reduced forms was examined.[123] This map was essentially featureless around the active site, indicating no ligand movement upon reduction. The unrefined 2.4-Å structure, however, is probably not sufficiently well resolved to indicate small movements of the ligands, such as the His ring rotations suggested above for Cu,Zn SOD (His 61 during the reaction and His 44 upon anion binding).

The binding of anions to Fe and Mn SODs has been studied, but, as for Cu,Zn SOD, this may cause structural changes different from those caused by the binding of superoxide. Comparison of the crystal structure of Fe SOD, with azide bound, to the more highly resolved Mn SOD structure indicated that the anion does not bind to the water coordination site.[8]

Instead, the coordination number apparently increases from five to six with the inhibitor binding at an exposed face of the metal ion between His 83 and His 170 in a cavity formed by the van der Waals surfaces of His 32 and Tyr 36.[124] Thus, the similarity in overall metal site geometry noted between the Cu,Zn enzyme and the Fe/Mn (C. L. Fisher and J. A. Tainer, Manuscript in preparation) enzymes extends to the existence of two potential binding sites for the superoxide anion, with anion inhibitors being bound in a second site rather than replacing the water ligand to the active site metal. This closely matches the predictions made for the binding of inhibitors in a second site of Cu,Zn SOD described above.[6]

An additional complication in Fe SODs is the indication of another anion binding site far from the metal.[140] Optical and ESR spectroscopy of Fe SOD show no changes in the metal properties upon treatment with the inhibitors SCN^-, ClO_4^-, Cl^-, and SO_4^{-2}, indicating a binding site distant from the metal. Some inhibitors, such as azide, appear to bind to both sites. This is a phenomenon that has not been observed in Cu,Zn SODs.

The differences observed between SODs in activity and anion binding, even among the apparently similar Fe and Mn SODs, indicate that there are at least subtle differences in their reaction mechanisms. Further crystallographic studies and refinement of existing structures will be instrumental in revealing the detailed relationship between structure and mechanism for SODs.

V. ELECTROSTATIC RECOGNITION AND REACTION RATE

A. ELECTROSTATIC RECOGNITION BETWEEN ENZYME AND SUBSTRATE

The dismutation of superoxide ($O_2^{\cdot -}$) by Cu,Zn SOD is almost as fast as that by aqueous Cu ion. This enzymatic rate is surprising, since Cu,Zn SOD must overcome a large handicap to compete with free Cu; $O_2^{\cdot -}$ can react with aqueous Cu from any coordinate position, but must find the tiny fraction (5 $Å^2$) of the molecular surface of Cu,Zn SOD (6500 $Å^2$) that forms the free Cu coordination site.[5] The chances of productive collision are improved by the dimer formation; the two active sites are oriented so as to allow productive collisions from opposing directions, with about 1000 $Å^2$ of the molecular surface of each monomer that would lead to unproductive collision buried in the dimer contact.[172] The long-range forces responsible for the precollision guidance to the active sites must be primarily electrostatic.

Biochemical data on Cu,Zn SOD implicate electrostatic forces in the recognition of the substrate $O_2^{\cdot -}$. At physiological pH, Cu,Zn SOD has a net negative charge that varies with species; the pI value for bovine enzyme is 4.9, while other species exhibit pI values ranging from 4.6 to 6.8.[173] This net negative charge would be expected to create an electrostatic barrier for incoming $O_2^{\cdot -}$, which should be diminished with increasing ionic strength, thus accelerating the reaction. Instead, raising the ionic strength decreases the reaction rate,[174] implying that attractive electrostatic forces facilitate the reaction; local positive charges on the enzyme may actually attract $O_2^{\cdot -}$ to the active site.[175,176]

The effects of ionic strength on residual enzymatic activity (10 to 15%) following two different neutralizations of charged residues have also been examined.[174] Acylation of seven to eight lysine residues per subunit reverses the response to ionic strength; raising the ionic strength then increases the reaction. In contrast, modification of Arg 141 by phenylglyoxal leaves the ionic strength response for the enzyme's residual activity unchanged.

Calculations implicate both long- and short-range electrostatic forces in $O_2^{\cdot -}$ recognition in Cu,Zn SOD.[5,175-177] Using the atomic positions from the refined model, the role of electrostatic forces in enzyme-substrate recognition has been examined through the calculation and display of the electrostatic potential (Figure 11) and the potential gradient (Figure 12).[5] The electrostatic potential and field around the active site has been calculated from partial charges of the atoms in the dimer using either a constant or a linearly distance-

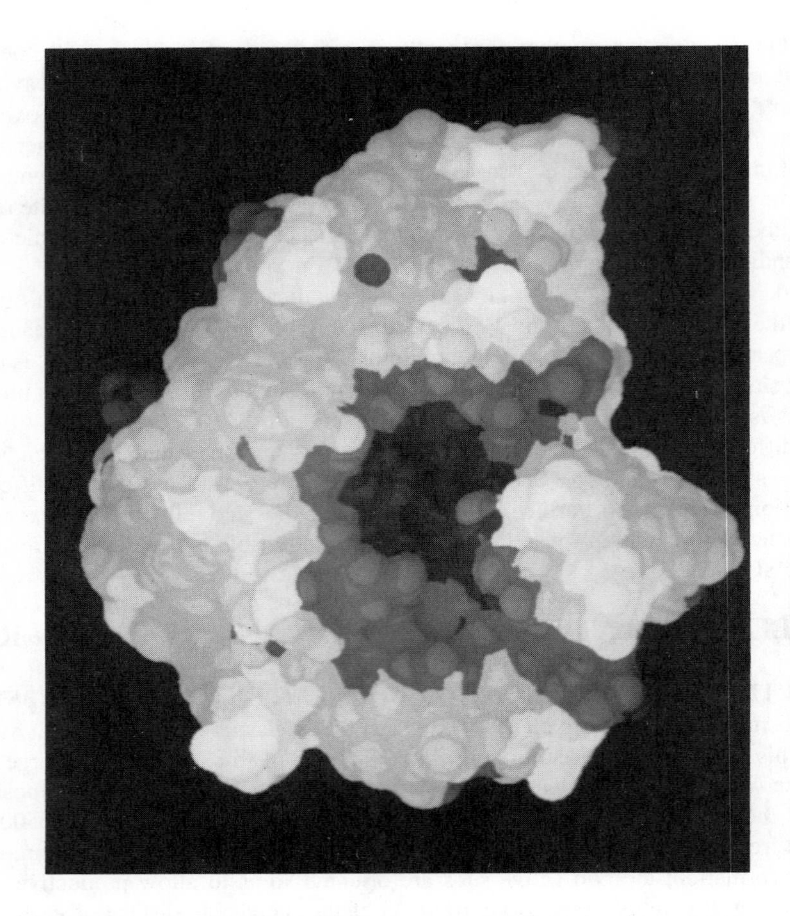

FIGURE 11. The electrostatically color-coded active site face of the bovine Cu,Zn SOD subunit. The electrostatic potential is divided into five categories coded by a radiating body brightness scale with the most negative potential being white to the most positive potential being black (white < -21 kcal mol^{-1}; light gray -21 to -7 kcal mol^{-1}; medium gray -7 to $+7$ kcal mol^{-1}; dark gray $+7$ to $+21$ kcal mol^{-1}; black > $+21$ kcal mol^{-1}). In SOD, this striking pattern of positive electrostatic potential is unique to the active site channel. The solid external molecular surface was coded by the electrostatic potentials seen by a water molecule from partial charges assigned to the atomic positions of the refined crystal structure of the bovine erythrocyte enzyme. (From Getzoff, E. D. and Tainer, J. A., *Ion Channel Reconstitution,* Miller, C., Ed., Plenum Press, New York, 1986, 57. With permission.)

dependent dielectric (Figure 12). This method allows the contribution of individual residues to the electrostatic potential to be analyzed (Table 4). These theoretical calculations indicate that, even though bovine erythrocyte Cu,Zn SOD is negatively charged at physiological pH, the electrostatic field around the active site is important for directing the $O_2^{\cdot -}$ to the active site.

The electrostatic potential surface of bovine Cu,Zn SOD shows no organized pattern over most of the surface, but reveals a positive region that extends over the long, deep channel above the Cu ion (Figure 11). The positive potential near the Cu is very high, forming a complementary binding site for the negatively charged $O_2^{\cdot -}$ substrate. In addition, several residues around the active site appear to guide the substrate into the active site.[5] The electrostatic potential within and beyond the active site is positive due to contributions from Lys 120, Lys 134, and Arg 141, as well as from the Zn and Cu ions. In other regions of

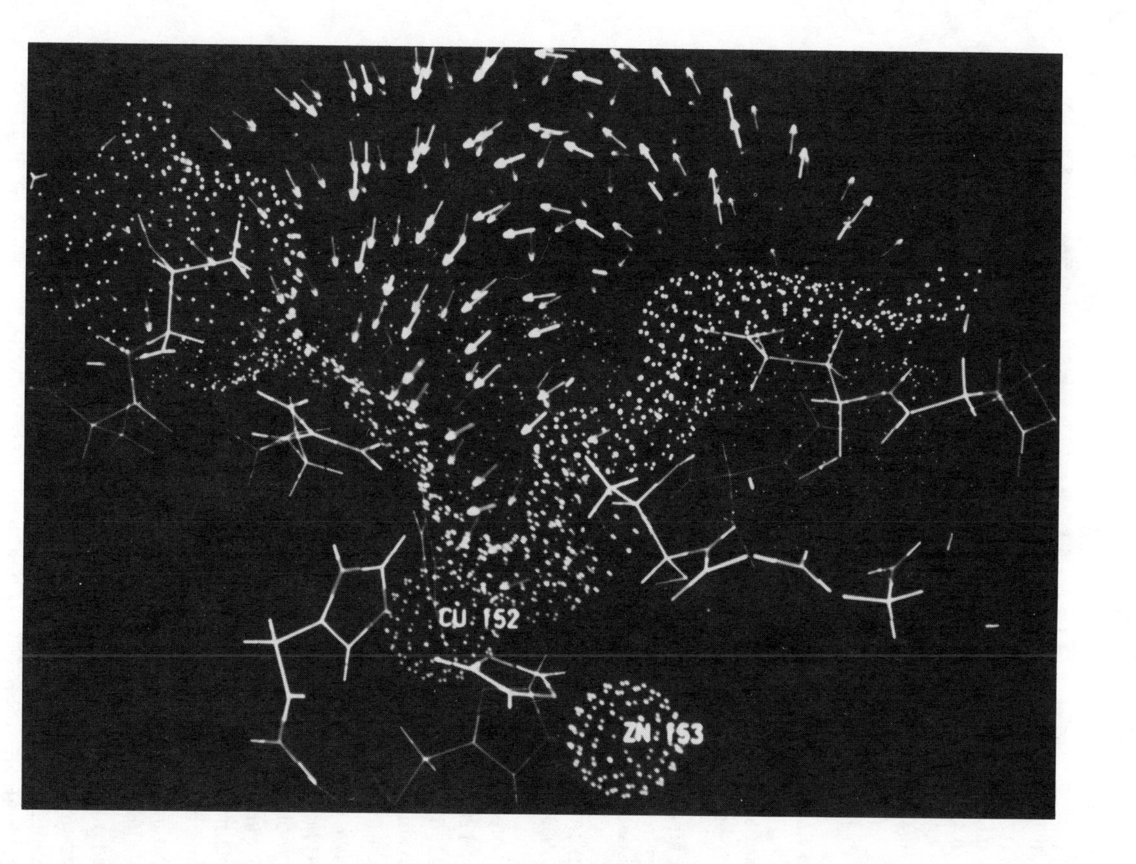

FIGURE 12. Computer graphics view of a slice through the deep active site channel of bovine Cu,Zn SOD with electrostatic field vectors. The active site channel is seen in cross section. The vectors were calculated at points outside the molecular boundary on concentric spheres of up to 14-Å radius centered on the axial water ligand of the copper ion. The skeletal model Cu ligands and active site Arg 141 (left center) can be seen along with the Cu (bottom center) in the most positive part of the active channel. The electrostatic field or electrostatic potential gradient (the magnitude and direction of maximum change in the potential) was calculated by evaluating the partial derivative of the potential with respect to each of the three coordinate axes.[5] By convention, electrostatic field vectors indicate the direction a positive charge would move in the field; we have reversed this convention here to show the directional force on the negatively charged superoxide substrate. (From Getzoff, E. D., Tainer, J. A., Weiner, P. K., Kollman, P. A., Richardson, J. S., and Richardson, D. C., *Nature*, 306, 287, 1983. With permission from Macmillan Magazines Ltd.)

TABLE 4
Effects of Individual Residues on the Electrostatic Field Direction for Cu,Zn SOD: Average Angular Change (Degrees) in Electrostatic Field Direction Produced in Each of the Eight Shells

Residue(s) neutralized[a]	0 Å	2 Å	4 Å	6 Å	8 Å	10 Å	12 Å	14 Å	All
Zn and His 61	1.3	4.1	1.9	2.1	3.1	3.9	4.3	5.8	4.4
Glu 119	0.6	1.7	3.1	5.5	7.1	8.8	8.6	8.7	8.1
Lys 120	0.7	2.0	5.3	7.5	18.0	22.6	26.3	24.5	22.4
Glu 130	0.2	0.5	1.0	2.0	3.9	6.1	9.7	19.0	10.7
Glu 131	0.6	3.0	10.1	23.1	30.7	37.4	37.9	36.9	34.7
Lys 134	0.4	1.2	2.7	5.6	11.9	20.3	31.0	37.3	26.7
Arg 141	7.4	16.0	31.9	29.3	21.7	23.9	24.5	20.7	23.2
Glu 119 and Lys 120	0.4	1.4	4.2	4.8	13.4	16.8	19.6	17.1	16.3
Glu 131 and Lys 134	0.3	2.1	7.9	17.5	25.0	34.2	43.4	44.2	37.3
D = 80[b]	4.8	6.4	13.6	9.7	12.8	14.3	167	18.2	15.8

Note: Shells, named by the radial distance from the position of the axial water bound to the Cu(II), with the number of vectors calculated in each successive shell being 1, 5, 10, 14, 32, 65, 105, and 107.

[a] Mathematically neutralized for calculating the electrostatic field.
[b] D = 80, changes resulting from the use of the bulk water constant dielectric of 80 instead of a dielectric model with a linear distance dependence.

From Getzoff, E. D., Tainer, J. A., Weiner, P. K., Kollman, P. A., Richardson, J. S., and Richardson, D. C., *Nature,* 306, 287, 1983. With permission from Macmillan Magazines, Ltd.

the protein, the potential is predominantly negative, reflecting the overall −4 charge on the dimer. Both Lys 120 and Lys 134 form salt bridges with negatively charged residues (Glu 119 and Glu 131, respectively). This may prevent nonproductive binding from occurring with these two lysine residues, which are at the top of the channel. Arg 141 mainly affects the field direction near the Cu ion, which is consistent with its local role in $O_2^{\cdot -}$ binding.

More complex electrostatic calculations have been done using a continuum model of the dielectric properties of the enzyme.[178] The Poisson-Boltzmann equation was solved numerically for molecules of arbitrary shape by approximation of this continuous function by distinct values at points on a grid (with 1.5-Å spacings) that includes and surrounds the protein. The protein was assigned a dielectric constant of 2, while the solvent had a dielectric constant of 80 and contained ions as represented by the Debye-Hückel model. Varying the Debye-Hückel parameter allowed the ionic strength to change. Calculation of the electrostatic potential in the absence of salt showed that the positive potential around the active site channel extended out to cover about 25 to 30% of the surface area of the protein. Increasing the salt concentration resulted in a decrease of this target area, which is consistent with experimental results of decreasing Cu,Zn SOD activity with increasing ionic strength.

Qualitatively, both approaches show that, except for the active site channel, the enzyme is surrounded by a negative potential. The positive field about the copper ion allows the superoxide to be attracted to a much larger area of the protein than just the available surface area of the copper alone. The continuum method with two dielectric constants, one for the protein and one for the solvent, predicts a target surface area of positive potential two to three times larger than that predicted by the first method.

B. BROWNIAN DYNAMICS AND FACILITATED DIFFUSION

Given the electrostatic potential around an enzyme, the diffusion process of the substrate can be modeled using Brownian dynamics to give an association rate for the enzyme-substrate complex. Cu,Zn SOD is an excellent system for such theoretical studies because the experimentally measured rate is very fast (about one-tenth the calculated diffusion-controlled rate), the electrostatic forces appear to be predominantly responsible for the rapid rate, and the enzyme does not undergo significant conformational change to interact with the substrate. Thus, the electrostatic potential of the enzyme can be modeled from the atomic positions observed in the crystal structure and the time scale of the reaction is accessible by current molecular modeling techniques.

The first such models involved the approach of superoxide, modeled as a sphere of radius 1.5 Å, to the enzyme, modeled as a sphere of 30-Å radius with a charge of −4. The active site was represented by two ten-degree patches on the sphere or two recessed patches.[179] Trajectories of the substrate were then calculated to allow an association rate to be obtained for the reaction. This model gave rates of reaction lower than those determined experimentally. A more detailed model included the shape of the enzyme by calculation of the electrostatic potential from point charges on the 76 ionized groups for points on a cubic grid about the protein.[180] The inner grid (from −50 to +50 Å centered about the protein) had grid spacings of 2 Å, while an outer grid (from −100 to + 100 Å) had grid spacings of 5 Å. The electrostatic force of Cu,Zn SOD, calculated for each grid point, was used to obtain the force on the substrate from the nearest grid point at each time step of the trajectory. Use of the lattice method allows more complex or more computationally expensive representations of the electrostatic potential of the enzyme without huge increases in computer time, because the full potential is only calculated once.

The inclusion of salt effects into the model led to a calculated association constant about five to ten times larger than the measured maximum rate for the enzyme.[181,182] The extremely rapid rate of the dismutation of superoxide by Cu,Zn SOD, its lack of saturation kinetics,[140] studies of the dependence of the rate on temperature,[183] viscosity,[135] and ionic strength[184] indicate that the rate is diffusion-controlled; that is, the association rate and reaction rate are the same. The Brownian dynamics results suggest that the rate constant for the reaction may be significantly slower than the association rate. It is probably premature to start looking for ways to make the enzyme more efficient based on this, however. One problem is that the choice of several parameters, such as the dielectric constants (in the solvent, near the protein, and in the protein), the size of the grid spacing, the reaction distance (the distance between the copper ion and the superoxide molecule that corresponds to a successful trajectory), the time step size for the trajectory, and the starting and exit radii appears to have a profound effect on the final rate constant.[185] In addition, internal fluctuations of the protein structure, especially in the active site channel, may facilitate the movement of the substrate, while motions of water molecules in the channel may cause the substrate to leave the channel. Thus, dynamic considerations of the protein and water molecules may also affect the association rate.

VI. PROBING THE ENZYME RATE AND MECHANISM WITH MUTANTS

A. REACTIVITY OF SITE-DIRECTED MUTANT ENZYMES

Biochemical, crystallographic, and computational evidence presented above for Cu,Zn SOD suggest functions for key residues in the enzymatic mechanism. Site-directed mutagenesis can be used to test our theories on the roles of specific residues in the SOD active site (Figure 13) and to further probe their function.[177] These mutagenesis experiments have been done on human Cu,Zn SOD, which is highly homologous to bovine SOD.

An important sequence-invariant residue in all known Cu,Zn SOD sequences, Arg 143 (Arg 141 in bovine SOD) forms part of the superoxide binding site and has been implicated in local electrostatic stabilization of the substrate as discussed above.[5,6] Chemical modification of bovine SOD Arg 141 with phenylglyoxal causes a 90% reduction in activity.[150]

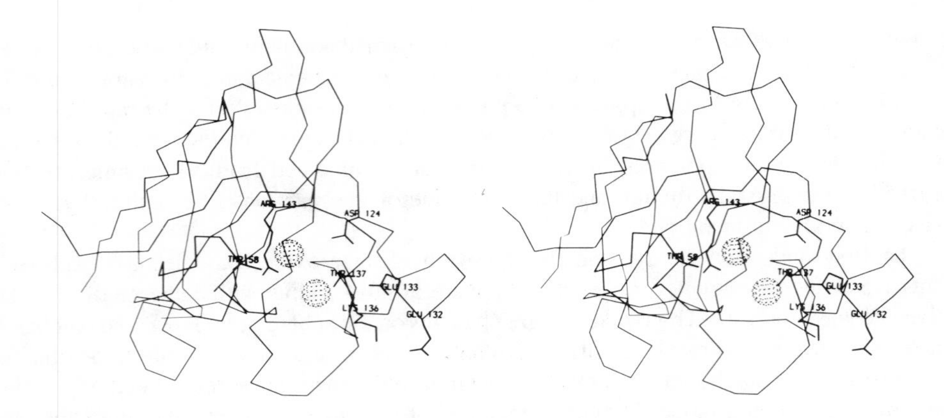

FIGURE 13. A stereo view of a model structure of the human Cu,Zn SOD showing the location of various site-directed mutants. The C_α backbone is shown with side chains and residue labels for the mutant sites. The structure for the human enzyme was derived from the bovine structure by implementing sequence changes, model-building two inserted residues, and performing energy minimization and molecular dynamics calculations. (J. A. Tainer, J. Sayre, E. D. Getzoff, unpublished results.)

TABLE 5
Activity of Site-Directed Mutants of Human Cu,Zn SOD

Type		Mutation(s)[a]			% Activity[b]
Active site	Arg 143	→Lys(*)			50%
		→Ile(*)			10%
		→Glu			≤1%
		→Asp			≤1%
		→Ala			30%
	Thr 137	→Ile			≤1%
		→Ala			10%
		→Arg			5%
		→Arg,	Arg 143	→Ala	
	Thr 58	→Ile			
	Asp 124	→Leu,	Asp 125	→Asn	≤1%
		→Asn,		→Asn	≤1%
Electrostatic recognition	Glu 133	→Gln			50%
		→Gln,	Glu 132	→Gln[c]	20%
		→Lys,		→Gln[c]	10%
	Glu 132	→Gln[c]			80%

[a] All mutations were made in the thermostable $Cys^6 \rightarrow$ Ala, $Cys^{111} \rightarrow$ Ser mutant unless marked with an asterisk (see Reference 222).
[b] % Activity was estimated from native activity gels.
[c] Adjacent residue Glu 132 was also mutated to test its role in replacing the electrostatic function of Glu 133.

Mutations of the human SOD Arg 143 → Lys, Ile, Glu, Asp, or Ala, all produce mutants with significantly reduced activity (see Table 5).[186] As expected for a residue important for electrostatic recognition, mutations to like charges (Lys^+) show lesser effects than mutations to uncharged (Ala, Ile) or oppositely charged residues (Glu^-, Asp^-). The Ile mutant of human SOD, although it has considerably less activity than the wild type, does display increased stability. Examination of the bovine structure shows that Arg 141 overlies a hydrophobic cavity between a loop region and the β barrel. The hydrophobic, branched isoleucine side chain may fill the cavity.

Other human SOD residues in the binding site that have been mutated to date include two threonines and an aspartate. Thr 137 (bovine 135), which forms the opposite side of the superoxide binding pocket from Arg 143 (see Figure 10), has been changed to Ile and Ala to investigate the importance of removing the polar group. It has also been mutated to Arg to investigate the effects of increasing the positive charge in the active site. All of these mutations have resulted in decreased activity. We are currently investigating a mutant with Thr 137 mutated to Arg and Arg 143 mutated to Ala to see if having an Arg on the other side of the pocket will result in an active enzyme. Thr 58 (bovine 56), which helps control access at the top of the active site channel, has been mutated to Ile. Asp 124 (bovine 122), which forms two important charged hydrogen bonds stabilizing the Cu ligands, has been changed to Leu and Asn (see Table 5; Figure 13). Although much of the analysis remains to be done, this type of experiment promises to directly probe the role of specific residues in the mechanism of superoxide dismutation.

The long-range electrostatic forces involved in recognition and orientation of the negatively charged superoxide substrate were examined by mutations of Lys 136 (bovine 134) and Glu 133 (bovine 131), whose side chains form a salt-bridge. Calculations show that neutralization of these two residues individually or together resulted in the largest changes in the electrostatic field of any of the charged residues near the active site.[5] Brownian dynamics simulations on the bovine enzyme suggest that mutation of either Glu 133 (bovine 131) or Glu 121 (bovine 119), the negative residue that forms a salt-bridge with Lys 120, to a lysine would double the rate constant.[181] (Changes in conformation were not considered in this calculation). Instead, mutation of Glu 133 in human SOD to either a Gln or Lys resulted in a significant decrease in activity (R. A. Hallewell, unpublished results). Therefore, Glu 133 may be important in preventing nonproductive binding of superoxide with Lys 136.

B. FUTURE GOALS AND PROSPECTS

Advances in techniques for molecular genetics, spectroscopy, X-ray crystallography, and computational analysis of macromolecules promise to allow the mechanisms for the enzyme-substrate interaction and catalysis in both Cu,Zn and Fe/Mn SODs to be determined in considerable detail. This determination will necessarily depend upon the combined results of many laboratories and research disciplines. The outline of the current state of knowledge presented in this review provides a road map of the well-defined areas and also directs our attention to questions that need future attention.

As revealed in the above sections, SOD is an attractive and in some ways unique enzyme for the application of these new techniques. The recent development of a genetic selection for SOD activity using *sodAsodB E. coli* will facilitate investigations by allowing the selection of active SOD mutants from inactive or partially active ones. The ability to use genetic selection in the *E. coli* SOD^- mutants will allow millions of possible mutations to be scanned relatively rapidly. The existence of high-resolution structures and the ability to produce cloned SOD mutants in quantities sufficient for X-ray studies[187] should allow a complete determination of the structural role of the critical residues in the active site. The combination of structural and biochemical analysis on mutant enzymes with computational methods now being developed will provide increasingly accurate information on the catalytic role of individual atoms, individual residues, and the enzyme's overall structure and surface topography.

REFERENCES

1. **Fridovich, I.,** Superoxide and superoxide dismutase, in *Advances in Inorganic Biochemistry,* Eichhorn, G. L. and Marzilli, L. G., Eds., Elsevier/North-Holland, New York, 1979, 67.
2. **Fridovich, I.,** Superoxide dismutases, in *Advances in Enzymology,* Vol. 58, Meister, A., Ed., John Wiley & Sons, New York, 1986, 61.

3. **Steinman, H. M.,** Superoxide dismutases: protein chemistry and structure-function relationships, in *Superoxide Dismutase,* Vol. 1, Oberley, L. W., Ed., CRC Press, Boca Raton, FL, 1982, chap. 2.
4. **Tainer, J. A., Getzoff, E. D., Beem, K. M., Richardson, J. S., and Richardson, D. C.,** Determination and analysis of the 2Å structure of copper, zinc superoxide dismutase, *J. Mol. Biol.,* 160, 181, 1982.
5. **Getzoff, E.D., Tainer, J. A., Weiner, P. K., Kollman, P. A., Richardson, J. S., and Richardson, D. C.,** Electrostatic recognition between superoxide and copper, zinc superoxide dismutase, *Nature,* 306, 287, 1983.
6. **Tainer, J. A., Getzoff, E. D., Richardson, J. S., and Richardson, D. C.,** Structure and mechanism of copper, zinc superoxide dismutase, *Nature,* 306, 284, 1983.
7. **Stallings, W. C., Powers, T. B., Pattridge, K. A., Fee, J. A., and Ludwig, M. L.,** Iron superoxide dismutase from *Escherichia coli* at 3.1-Å resolution: a structure unlike that of copper/zinc protein at both monomer and dimer levels, *Proc. Natl. Acad. Sci. U.S.A.,* 80, 3884, 1983.
8. **Stallings, W. C., Pattridge, K. A., Strong, R. K., and Ludwig, M. L.,** The structure of manganese superoxide dismutase from *Thermus thermophilus HB8* at 2.4-Å resolution, *J. Biol. Chem.,* 260, 16424, 1985.
9. **Klug, D., Rabani, J., and Fridovich, I.,** A direct demonstration of the catalytic action of superoxide dismutase through the use of pulse radiolysis, *J. Biol. Chem.,* 247, 4839, 1972.
10. **Taube, H.,** Mechanisms of oxidation with oxygen, *J. Gen. Physiol. Part 2,* 49, 29, 1965.
11. **Behar, D., Czapski, G., Rabani, J., Dorfman, L. M., and Schwarz, H. A.,** The acid dissociation constant and decay kinetics of the perhydroxyl radical, *J. Phys. Chem.,* 74, 3209, 1970.
12. **Burger, A. R.,** Physical and Chemical Studies on the Role of Zinc in Superoxide Dismutase, Ph.D. thesis, Columbia University, New York, 1979.
13. **Koppenol, W. H.,** Reactions involving singlet oxygen and the superoxide anion, *Nature,* 262, 420, 1976.
14. **Valentine, J. S. and Curtis, A. B.,** A convenient preparation of solutions of superoxide anion and the reaction of superoxide anion with a copper(II) complex, *J. Am. Chem. Soc.,* 97, 224, 1975.
15. **Britton, L., Malinowski, D. P., and Fridovich, I.,** Superoxide dismutase and oxygen metabolism in *Streptococcus faecalis* and comparisons with other organisms, *J. Bacteriol.,* 134, 229, 1978.
16. **Knowles, P. F., Gibson, J. F., Pick, F. M., and Bray, R. C.,** Electron-spin-resonance evidence for enzymic reduction of oxygen to a free radical, the superoxide ion, *Biochem. J.,* 111, 53, 1969.
17. **Bray, R. C., Pick, F. M., and Samuel, D.,** Oxygen-17 hyperfine splitting in the electron paramagnetic resonance spectrum of enzymically generated superoxide, *Eur. J. Biochem.,* 15, 352, 1970.
18. **Ballou, D., Palmer, G., and Massey, V.,** Direct demonstration of superoxide anion production during the oxidation of reduced flavin and of its catalytic decompositon by erythrocuprein, *Biochem. Biophys. Res. Commun.,* 36, 898, 1969.
19. **Harbour, J. R. and Bolton, J. R.,** Superoxide formation in spinach chloroplasts: electron spin resonance detection by spin trapping, *Biochem. Biophys. Res. Commun.,* 64, 803, 1975.
20. **Hirata, F. and Hayaishi, O.,** New degradative routes of 5-hydroxytryptophan and serotonin by intestinal tryptophan 2,3-dioxygenase, *Biochem. Biophys. Res. Commun.,* 47, 1112, 1972.
21. **McCord, J. M., Keele, B. B., Jr., and Fridovich, I.,** An enzyme-based theory of obligate anaerobiosis: the physiological function of superoxide dismutase, *Proc. Natl. Acad. Sci. U.S.A.,* 68, 1024, 1971.
22. **Rotilio, G.,** *Superoxide and Superoxide Dismutase in Chemistry, Biology and Medicine,* Elsevier Science, Amsterdam, 1986.
23. **Fridovich, I.,** The biology of superoxide and of superoxide dismutases-in brief, *Prog. Clin. Biol. Res.,* 51, 153, 1981.
24. **Steinman, H. M. and Hill, R. L.,** Sequence homologies among bacterial and mitochondrial superoxide dismutases, *Proc. Natl. Acad. Sci. U.S.A.,* 70, 3725, 1973.
25. **Stallings, W. C., Pattridge, K. A., Strong, R. K., and Ludwig, M. L.,** Manganese and iron superoxide dismutases are structural homologs, *J. Biol. Chem.,* 259, 10695, 1984.
26. **Steinman, H. M., Naik, V. R., Abernethy, J. L., and Hill, R. L.,** Bovine erythrocyte superoxide dismutase. Complete amino acid sequence, *J. Biol. Chem.,* 249, 7326, 1974.
27. **Hjalmarsson, K., Marklund, S. L., Engström, Å., and Edlund, T.,** Isolation and sequence of complementary DNA encoding human extracellular superoxide dismutase, *Proc. Natl. Acad. Sci. U.S.A.,* 84, 6340, 1987.
28. **Tibell, L., Hjalmarsson, K., Edlund, T., Skogman, G., Engström, Å., and Marklund, S. L.,** Expression of human extracellular superoxide dismutase in Chinese hamster ovary cells and characterization of the product, *Proc. Natl. Acad. Sci., U.S.A.,* 84, 6634, 1987.
29. **Hewitt, J. and Morris, J. G.,** Superoxide dismutase in some obligately anaerobic bacteria, *FEBS Lett.,* 50, 315, 1975.
30. **Morris, J. G.,** Fifth Stenhouse-Williams Memorial Lecture: oxygen and the obligate anaerobe, *J. Appl. Bacteriol.,* 40, 229, 1976.
31. **Gregory, E. M., Kowalski, J. B., and Holderman, L. V.,** Production and some properties of catalase and superoxide dismutase from the anaerobe *Bacteroides distasonis, J. Bacteriol.,* 129, 1298, 1977.

32. **Hatchikian, E. C. and Henry, Y. A.,** An iron-containing superoxide dismutase from the strict anaerobe *Desulfovibrio desulfuricans* (Norway 4), *Biochimie*, 59, 153, 1977.
33. **Tally, F. P., Goldin, B. R., Jacobus, N. V., and Gorbach, S. L.,** Superoxide dismutase in anaerobic bacteria of clinical significance, *Infect. Immun.*, 16, 20, 1977.
34. **Gregory, E. M., Yost, F. J., and Fridovich, I.,** Superoxide dismutases of *Escherichia coli*: intracellular localization and functions, *J. Bacteriol.*, 115, 987, 1973.
35. **Gregory, E. M. and Fridovich, I.,** Induction of superoxide dismutase by molecular oxygen, *J. Bacteriol.*, 114, 543, 1973.
36. **Gregory, E. M., Goscin, S. A., and Fridovich, I.,** Superoxide dismutase and oxygen toxicity in a eukaryote, *J. Bacteriol.*, 117, 456, 1974.
37. **Puget, K. and Michelson, A. M.,** Isolation of a new copper-containing superoxide dismutase, bacteriocuprein, *Biochem. Biophys. Res. Commun.*, 58, 830, 1974.
38. **Puget, K. and Michelson, A. M.,** Iron containing superoxide dismutases from luminous bacteria, *Biochimie*, 56, 1255, 1974.
39. **Asada, K., Kanematsu, S. Takahashi, M., and Kona, Y.,** Superoxide dismutases in photosynthetic organisms, in *Advances in Experimental Medicine and Biology: Iron and Copper Proteins*, Vol. 74, Yasunobu, K. T., Mower, H. I., and Hayaishi, O., Eds., Plenum Press, New York, 1976, 551.
40. **Stevens, J. B. and Autor, A. P.,** Oxygen-induced synthesis of superoxide dismutase and catalase in pulmonary macrophages of neonatal rats, *Lab. Invest.*, 37, 470, 1977.
41. **Simon, L. M., Liu, J., Theodore, J., and Robin, E. D.,** Effect of hyperoxia, hypoxia, and maturation on superoxide dismutase activity in isolated alveolar macrophages, *Am. Rev. Resp. Dis.*, 115, 279, 1977.
42. **Rister, M. and Baehner, R. L.,** Induction of superoxide dismutase (SOD) activity *in vivo* by oxygen (O_2) in polymorphonuclear leukocytes (PMNS) and alveolar macrophages (AM), *Blood*, 46, 1016, 1975.
43. **Stevens, J. B. and Autor, A. P.,** Induction of superoxide dismutase by oxygen in neonatal rat lung, *J. Biol. Chem.*, 252, 3509, 1977.
44. **Autor, A. P. and Stevens, J. B.,** Mechanism of oxygen detoxification in neonatal rat lung tissue, *Photochem. Photobiol.*, 28, 775, 1978.
45. **Hassan, H. M. and Fridovich, I.,** Regulation of the synthesis of superoxide dismutase in *Escherichia coli*. Induction by methyl viologen, *J. Biol. Chem.*, 252, 7667, 1977.
46. **Archibald, F. S. and Fridovich, I.,** Manganese and defenses against oxygen toxicity in *Lactobacillus plantarum*, *J. Bacteriol.*, 145, 442, 1981.
47. **Shatzman, A. R. and Kosman, D. J.,** The utilization of copper and its role in the biosynthesis of copper-containing proteins in the fungus, *Dactylium dendroides*, *Biochim. Biophys. Acta*, 544, 163, 1978.
48. **Shatzman, A. R. and Kosman, D. J.,** Biosynthesis and cellular distribution of the two superoxide dismutases of *Dactylium dendroides*, *J. Bacteriol.*, 137, 313, 1979.
49. **DeRosa, G., Keen, C. L., Leach, R. M., and Hurley, L. S.,** Regulation of superoxide dismutase activity by dietary manganese, *J. Nutrit.*, 110, 795, 1980.
50. **Sakamoto, H. and Touati, D.,** Cloning of the iron superoxide dismutase gene *(sodB)* in *Escherichia coli* K-12, *J. Bacteriol.*, 159, 418, 1984.
51. **Touati, D.,** Cloning and mapping of the manganese superoxide dismutase gene *(sodA)* of *Escherichia coli* K-12, *J. Bacteriol.*, 155, 1078, 1983.
52. **van Loon, A. P. G. M., Pesold-Hurt, B., and Schatz, G.,** A yeast mutant lacking mitochondrial manganese-superoxide dismutase is hypersensitive to oxygen, *Proc. Natl. Acad. Sci. U.S.A.*, 83, 3820, 1986.
53. **Hallewell, R. A., Mullenbach, G. T., Stempien, M. M., and Bell, G. I.,** Sequence of a cDNA coding for mouse manganese superoxide dismutase, *Nucl. Acids Res.*, 14, 9539, 1986.
54. **Steinman, H.,** Bacteriocuprein superoxide dismutase of *Photobacterium leiognathi*. Isolation and sequence of the gene and evidence for a precursor form, *J. Biol. Chem.*, 262, 1882, 1987.
55. **Cannon, R. E., White, J. A., and Scandalios, J. G.,** Cloning of cDNA for maize superoxide dismutase 2 (SOD2), *Proc. Natl. Acad. Sci. U.S.A.* 84, 179, 1987.
56. **Hallewell, R. A., Masiarz, F. R., Najarian, R. C., Puma, J. P., Quiroga, M. R., Ranolph, A., Sanchez-Pescador, R., Scandella, C. J., Smith, B., Steimer, K. S., and Mullenbach, G. T.,** Human Cu/Zn superoxide dismutase cDNA: isolation of clones synthesising high levels of active or inactive enzyme from an expression library, *Nucl. Acids Res.*, 13, 2017, 1985.
57. **Levanon, D., Lieman-Hurwitz, J., Dafni, N., Wigderson, M., Sherman, L., Bernstein, Y., Laver-Rudich, Z., Danciger, E., Stein, O., and Groner, Y.,** Architecture and anatomy of the chromosomal locus in human chromosome 21 encoding the Cu/Zn superoxide dismutase, *EMBO J.*, 4, 77, 1985.
58. **Hallewell, R. A., Puma, J. P., Mullenbach, G. T., and Najarian, R. C.,** Structure of the human Cu/Zn SOD gene, in *Superoxide and Superoxide Dismutase in Chemistry, Biology and Medicine*, Rotilio, G., Ed., Elsevier Science, Amsterdam, 1986, 249.
59. **Carlioz, A. and Touati, D.,** Isolation of superoxide dismutase mutants in *Escherichia coli:* is superoxide dismutase necessary for aerobic life?, *EMBO J.*, 5, 623, 1986.

60. **Krall, J., Bagley, A. C., Mullenbach, G.T., Hallewell, R. A., and Lynch, R. E.,** Superoxide mediates the toxicity of paraquat for cultured mammalian cells, *J. Biol. Chem.*, 263, 1910, 1988.
61. **Natvig, D. O., Imlay, K., Touati, D., and Hallewell, R. A.,** Human copper-zinc superoxide dismutase complements superoxide dismutase-deficient *E. coli* mutants, *J. Biol. Chem.*, 262, 14697, 1987.
62. **Biliński, T., Krawiec, Z., Litwńska, A., and Liczmanński, J.,** Is hydroxyl radical generated by the Fenton reaction *in vivo?*, *Biochem. Biophys. Res. Commun.*, 130, 533, 1985.
63. **Scott, M. D., Meshnick, S. R., and Eaton, J. W.,** Superoxide dismutase-rich bacteria. Paradoxical increase in oxidant toxicity, *J. Biol. Chem.*, 262, 3640, 1987.
64. **Hallewell, R. A., Mills, R., Tekamp-Olson, P., Blacher, R., Rosenberg, S., Ötting, F., Masiarz, F. R., and Scandella, C. J.,** Amino terminal acetylation of authentic human Cu,Zn superoxide dismutase produced in yeast, *Biotechnology*, 5, 363, 1987.
65. **Elroy-Stein, O., Bernstein, Y., and Groner, Y.,** Overproduction of human Cu/Zn-superoxide dismutase in transfected cells: extenuation of paraquat-mediated cytotoxicity and enhancement of lipid peroxidation, *EMBO J.*, 5, 615, 1986.
66. **Tan, Y. H., Tischfield, J., and Ruddle, F. H.,** The linkage of genes for the human interferon-induced lantiviral protein and indophenol oxidase-B traits to chromosome G-21, *J. Exp. Med.*, 137, 317, 1973.
67. **Sinet, P.-M., Couturier, J., Dutrillaux, B., Poissonnier, M., Raoul, O., Rethoré, M.-O., Allard, D., Lejeune, J., and Jerome, H.,** Trisomie 21 et superoxyde dismutase-I (IPO-A), *Exp. Cell Res.*, 97, 47, 1975.
68. **Ceballos, I., Delabar, J. M., Nicole, A., Lynch, R. E., Hallewell, R. A., Kamoun, P., and Sinet, P. M.,** Expression of transfected human CuZn superoxide dismutase gene in mouse L cells and NS20Y in neuroblastoma cells induces enhancement of glutathione peroxidase activity, *Biochim. Biophys. Acta*, 949, 58, 1988.
69. **Richter, C., Wendel, A., Weser, U., and Azzi, A.,** Inhibition by superoxide dismutase of linoleic acid peroxidation induced by lipoxidase, *FEBS Lett.*, 51, 300, 1975.
70. **Carlin, G. and Arfors, K.-E.,** Lipid peroxidation promoted by human polymorphonuclear leukocytes, in *Superoxide and Superoxide Dismutase in Chemistry, Biology and Medicine*, Rotilio, G., Ed., Elsevier Science, Amsterdam, 1986, 22.
71. **Fridovich, I.,** Superoxide dismutases, *Annu. Rev. Biochem.*, 44, 147, 1975.
72. **McCord, J. M. and Fridovich, I.,** Superoxide dismutase. An enzymatic function for erythrocuprein (hemocuprein), *J. Biol. Chem.*, 244, 6049, 1969.
73. **Oyanagui, Y.,** Participation of superoxide anions at the prostaglandin phase of carrageenan foot-oedema, *Biochem. Pharmacol.*, 25, 1465, 1976.
74. **McCord, J. M. and Day, E. D., Jr.,** Superoxide-dependent production of hydroxyl radical catalyzed by iron-EDTA complex, *FEBS Lett.*, 86, 139, 1978.
75. **Halliwell, B., Gutteridge, J. M. C., and Blake, D.,** Metal ions and oxygen radical reactions in human inflammatory joint disease, *Phil. Trans. R. Soc. London B*, 311, 659, 1985.
76. **McCord, J. M., Stokes, S. H., and Wong, K.,** Superoxide radical as a phagocyte-produced chemical mediator of inflammation, *Adv. Inflamm. Res.*, 1, 273, 1979.
77. **Turner, S. R., Campbell, J. A., and Lynn, W. S.,** Polymorphonuclear leukocyte chemotaxis toward oxidized lipid components of cell membranes, *J. Exp. Med.*, 141, 1437, 1975.
78. **Turner, S. R., Tainer, J. A., and Lynn, W. S.,** Biogenesis of chemotactic molecules by the arachidonate lipoxygenase system of platelets, *Nature*, 257, 680, 1975.
79. **Ford-Hutchinson, A. W., Bray, M. A., Dois, M. V., Shipley, M. E., and Smith, M. J. H.,** Leukotriene B, a potent chemokinetic and aggregating substance released from polymorphonuclear leukocytes, *Nature*, 286, 264, 1980.
80. **Goetzl, E. J., Hill, H. R., and Gorman, R. R.,** Unique aspects of the modulation of human neutrophil function by 12-L-hydroperoxy-5,8,10,14-eicosatetraenoic acid, *Prostaglandins*, 19, 71, 1980.
81. **Petrone, W. F., English, D. K., Wong, K., and McCord, J. M.,** Free radicals and inflammation: superoxide-dependent activation of a neutrophil chemotactic factor in plasma, *Proc. Natl. Acad. Sci. U.S.A.*, 77, 1159, 1980.
82. **Weissmann, G., Korchak, H. M., Perez, H. D., Smolen, J. E., Goldstein, I. M., and Hoffstein, S.T.,** Leukocytes as secretory organs of inflammation, *Adv. Inflamm. Res.*, 1, 95, 1979.
83. **Guarnieri, C., Flamigni, F., and Caldarera, C. M.,** Role of oxygen in the cellular damage induced by re-oxygenation of hypoxic heart, *J. Mol. Cell Cardiol.*, 12, 797, 1980.
84. **Zweir, J. L., Flaherty, J. T., and Weisfeldt, M. L.,** Direct measurement of free radical generation following reperfusion of ischemic myocardium, *Proc. Natl. Acad. Sci. U.S.A.*, 84, 1404, 1987.
85. **McCord, J. M.,** Oxygen-derived free radicals in postischemic tissue injury, *N. Engl. J. Med.*, 312, 159, 1985.
86. **Meerson, F. Z., Kagan, V. E., Kozlov, Y. P., Belkina, L. M., and Arkhipenko, Y. V.,** The role of lipid peroxidation in pathogenesis of ischemic damage and the antioxidant protection of the heart, *Basic Res. Cardiol.*, 77, 465, 1982.

87. **Hess, M. L. and Manson, N. H.,** Molecular oxygen: friend and foe. The role of the oxygen free radical system in the calcium paradox, the oxygen paradox, and ischemia/reperfusion injury, *J. Mol. Cell Cardiol.*, 16, 969, 1984.
88. **Myers, C. L., Weiss, S. J., Kirsh, M. M., and Shlafer, M.,** Involvement of hydrogen peroxide and hydroxyl radical in the 'oxygen paradox': reduction of creatine kinase release by catalase, allopurinal or deferoxamine, but not by superoxide dismutase, *J. Mol. Cell Cardiol.*, 17, 675, 1984.
89. **Ambrosio, G., Weisfeldt, M. L., Jacobus, W. E., and Flaherty, J. T.,** Evidence for a reversible oxygen radical-mediated component of reperfusion injury: reduction by recombinant human superoxide dismutase administered at the time of reflow, *Circulation*, 75, 282, 1987.
90. **Das, D. K., Engelman, R. M., Rousou, J. A., Breyer, R. H., Otani, H., and Lemeshow, S.,** Pathophysiology of superoxide radical as potential mediator of reperfusion injury in pig heart, *Basic Res.Cardiol.*, 81, 155, 1986.
91. **Otani, H., Engelman, R. M., Rousou, J. A., Breyer, R. H., Lemeshow, S., and Das, D. K.,** Cardiac performance during reperfusion improved by pretreatment with oxygen free-radical scavengers, *J. Thorac. Cardiovasc. Surg.*, 91, 290, 1986.
92. **Granger, D. N., Rutili, G., and McCord, J. M.,** Superoxide radicals in feline intestinal ischemia, *Gastroenterology*, 81, 22, 1981.
93. **Ringe, D., Petsko, G. A., Yamakura, F., Suzuki, K., and Ohmori, D.,** Structure of iron superoxide dismutase from *Pseudomonas ovalis* at 2.9-Å resolution, *Proc. Natl. Acad. Sci. U.S.A.*, 80, 3879, 1983.
94. **Ludwig, M. L., Pattridge, K. A., and Stallings, W. C.,** Manganese superoxide dismutase: structure and properties, in *Manganese in Metabolism and Enzyme Function*, Schramm, V. L. and Wedler, F. C., Eds., Academic Press, Orlando, 1986, 405.
95. **Sato, S., Nakada, Y., and Nakazawa-Tomizawa, K.,** Amino-acid sequence of a tetrameric, manganese superoxide dismutase from *Thermus thermophilus* HB8, *Biochim. Biophys. Acta*, 912, 178, 1987.
96. **Richardson, J. S.,** The anatomy and taxonomy of protein structure, in *Advances in Protein Chemistry*, Vol. 34, Anfinsen, C. B., Edsall, J. T., and Richards, F. M., Eds., Academic Press, New York, 1981, 167.
97. **Getzoff, E. D.,** The Refined 2Å Structure of Copper, Zinc, Superoxide Dismutase: Implications for Stability and Catalysis, Ph.D. thesis, Duke University, Durham, NC, 1982.
98. **Cotton, F. A. and Wilkinson, G.,** *Advanced Inorganic Chemistry*, 4th ed., John Wiley & Sons, New York, 1980.
99. **Freeman, H. C.,** Crystal structures of metal-peptide complexes, *Adv. Prot. Chem.*, 22, 257, 1967.
100. **Gaber, B. P., Brown, R. D., Koenig, S. H., and Fee, J. A.,** Nuclear magnetic relaxation dispersion in protein solutions. V. Bovine erythrocyte superoxide dismutase, *Biochim. Biophys. Acta*, 271, 1, 1972.
101. **Kennard, O., Watson, D. G., Allen, F. H., Isaacs, N. W., Motherwell, W. D. S., Petterson, R. C. and Town, W. G., Eds.,** *Molecular Structures and Dimensions*, Vol. A1, International Union of Crystallography, Utrecht, Netherlands, 1972.
102. **Beem, K. M., Rich, W. E., and Rajagopalan, K. V.,** Total reconstitution of copper-zinc superoxide dismutase, *J. Biol. Chem.*, 249, 7298, 1974.
103. **Peisach, J. and Blumberg, W. E.,** Structural implications derived from the analysis of electron paramagnetic resonance spectra of natural and artificial copper proteins, *Arch. Biochem. Biophys.*, 165, 691, 1974.
104. **Beem, K. M., Richardson, D. C., and Rajagopalan, K. V.,** Metal sites of copper-zinc superoxide dismutase, *Biochemistry*, 16, 1930, 1977.
105. **Vänngård, T.,** Copper proteins, in *Biological Applications of Electron Spin Resonances*, Swartz, H. M., Bolton, J. R., and Borg, D. C., Eds., Wiley-Interscience, New York, 1972, 411.
106. **Rotilio, G., Calabrese, L., Mondovi, B., and Blumberg, W. E.,** Electron paramagnetic resonance studies of cobalt-copper bovine superoxide dismutase, *J. Biol. Chem.*, 249, 3157, 1974.
107. **Fee, J. A. and Briggs, R. G.,** Studies on the reconstitution of bovine erythrocyte superoxide dismutase. V. Preparation and properties of derivatives in which both zinc and copper sites contain copper, *Biochim. Biophys. Acta*, 400, 439, 1975.
108. **Bertini, I., Luchinat, C. and Monnanni,R.,** Evidence of the breaking of the copper-imidazolate bridge in copper/cobalt-substituted superoxide dismutase upon reduction of the copper(II) centers, *J. Am. Chem. Soc.*, 107, 2178, 1985.
109. **O'Young, C.-L., Dewan, J. C., Lilienthal, H. R., and Lippard, S. J.,** Electron spin resonance, magnetic, and x-ray crystallographic studies of a binuclear, imidazolate bridged copper(II) complex, $[(TMDT)_2Cu_2(im)(ClO_4)_2](ClO_4)$, *J. Am. Chem. Soc.*, 100, 7291, 1978.
110. **O'Young, C.-L.,** Imidazolate Bridged Dicopper(II) Complexes and the Reactions of Superoxide Anion with Copper(II) Complexes, Ph.D. thesis, Columbia University, New York, 1980.
111. **Bauer, R., Demeter, I., Hasemann, V., and Johansen, J. T.,** Structural properties of the zinc site in Cu,Zn-superoxide dismutase; perturbed angular correlation of gamma ray spectroscopy on the Cu,^{111}Cd-superoxide dismutase derivative, *Biochem. Biophys. Res. Commun.*, 94, 1296, 1980.

112. **Thomas, K. A., Rubin, B. H., Bier, J. C., Richardson, J. S., and Richardson, D. C.,** The crystal structure of bovine Cu^{2+},Zn^{2+} superoxide dismutase at 5.5Å resolution, *J. Biol. Chem.*, 249, 5677, 1974.
113. **Thomas, K. A., Jr.,** The X-Ray Crystal Structure of Bovine Erythrocyte Superoxide Dismutase, Ph.D. thesis, Duke University, Durham, NC, 1974.
114. **Bailey, D. B., Ellis P. D., and Fee, J. A.,** Cadmium-113 nuclear magnetic resonance studies of cadmium-substituted derivatives of bovine superoxide dismutase, *Biochemistry,* 19, 591, 1980.
115. **Armitage, I. M., Schoot Uiterkamp, A. J. M., Chlebowski, J. F., and Coleman, J. E.,** Cd NMR as a probe of the active sites of metalloenzymes, *J. Magn. Reson.*, 29, 375, 1978.
116. **Kolks, G. and Lippard, S. J.,** Magnetic exchange interactions in imidazolate bridged copper(II) complexes, *J. Am. Chem. Soc.*, 99, 5804, 1977.
117. **Kolks, G., Frihart, C. R., Rabinowitz, H. N., and Lippard, S. J.,** Imidazolate-bridged complexes of copper(II), *J. Am. Chem. Soc.*, 98, 5720, 1976.
118. **Lundberg, B. K. S.,** Metal complexes with mixed ligands. III. The crystal structure of an imidazolato-bridged polynuclear copper(II)-imidazole chloride complex, $Cu(C_3H_3N_2)$-$(C_3H_4N_2)_2Cl$, *Acta Chem. Scand.*, 26, 3902, 1972.
119. **Sundberg, R. J. and Martin, R. B.,** Interactions of histidine and other imidazole derivatives with transition metal ions in chemical and biological systems, *Chem. Rev.*, 74, 471, 1974.
120. **Coughlin, P. K., Lippard, S. J., Martin, A. E., and Bulkowski, J. E.,** Enhanced stability of the imidazolate-bridged dicopper(II) ion in a binucleating macrocycle, *J. Am. Chem. Soc.*, 102, 7616, 1980.
121. **Hodgson, E. K. and Fridovich, I.,** The interaction of bovine erythrocyte superoxide dismutase with hydrogen peroxide: inactivation of the enzyme, *Biochemistry,* 14, 5294, 1975.
122. **Stallings, W. C., Bull, C., Pattridge, K. A., Powers, T. B., Fee, J. A., Ludwig, M. L., Ringe, D., and Petsko, G. A.,** The three-dimensional structure of iron superoxide dismutase: kinetic and structural comparisons with Cu/Zn and Mn dismutases, in *Oxygen Radicals in Chemistry and Biology,* Bors, W., Saran, M., and Tait, D., Eds., Walter de Gruyter, Berlin, 1984, 779.
123. **Stallings, W. C., Pattridge, K. A. and Ludwig, M. L.,** The active centers of *T. thermophilus* Mn superoxide dismutase and *E. coli* Fe superoxide dismutase: current status of the crystallography, in *Superoxide and Superoxide Dismutase in Chemistry, Biology and Medicine,* Rotilio, G., Ed., Elsevier Science, Amsterdam, 1986, 195.
124. **Stallings, W. C., Pattridge, K. A., Strong, R. K., Ludwig, M. L., Yamakura, F., Isobe, T., and Steinman, H. M.** Active site homology in iron and manganese superoxide dismutases, in *Patterson and Pattersons: Fifty Years of the Patterson Function,* Glusker, J. P., Patterson, B. K., and Rossi, M., Eds., Oxford University, Oxford, 1987, 505.
125. **Villafranca, J. J.,** EPR spectra of Fe(III)-superoxide dismutase with special reference to the electron spin relaxation time of Fe(III), *FEBS, Lett.*, 62, 230, 1976.
126. **Steffens, G., Frankus, E., Hering, K., Kim, S. M. A., Ötting, F., Schwertner, E., Puget, K., Michelson, A. M., and Flohe, L.,** Sequencing confirms normal divergent evolution of Cu/Zn SOD, in *Superoxide and Superoxide Dismutase in Chemistry, Biology and Medicine,* Rotilio, G., Ed., Elsevier Science, Amsterdam, 1986, 246.
127. **Forman, H. J. and Fridovich, I.,** On the stability of bovine superoxide dismutase. The effects of metals, *J. Biol. Chem.*, 248, 2645, 1973.
128. **Lipscomb, W. N.,** in *Advances in Inorganic Biochemistry: Methods for Determining Metal Ion Environments in Proteins. Structure and Function of Metalloproteins,* Darnall, D. W. and Wilkins, R. G., Eds., Elsevier/North-Holland, New York, 1980, 265.
129. **Leatherbarrow, R. J. and Fersht, A.R.,** Protein engineering, *Prot. Eng.*, 1, 7, 1986.
130. **Temussi, P. A. and Vitagliano, A.,** Model ligands for copper proteins. Proton magnetic resonance study of acetylhistamine and acetylhistidine complexes with copper(I), *J. Am. Chem. Soc.*, 97, 1572, 1975.
131. **Sugiura, Y.,** Proton-magnetic-resonance study of copper(I) complexes with peptides containing sulfhydryl and imidazole groups as possible model ligands for copper proteins, *Eur. J. Biochem.*, 78, 431, 1977.
132. **Stoesz, J. D., Malinowski, D. P., and Redfield, A. G.,** Nuclear magnetic resonance study of solvent exchange and nuclear Overhauser effect of the histidine protons of bovine superoxide dismutase, *Biochemistry,* 18, 4669, 1979.
133. **Klug-Roth, D., Fridovich, I., and Rabani, J.,** Pulse radiolytic investigations of superoxide catalyzed disproportionation. Mechanism for bovine superoxide dismutase, *J. Am. Chem. Soc,* 95, 2786, 1973.
134. **Fee, J. A., McClune, G. J., Lees, A. C., Zidovetski, R., and Pecht, I.,** The pH dependence of the spectral and anion binding properties of iron containing superoxide dismutase from *E. coli* B: an explanation for the azide inhibition of dismutase activity, *Isr. J. Chem.*, 21, 54, 1981.
135. **Rotilio, G., Bray, R. C., and Fielden, E. M.,** A pulse radiolysis study of superoxide dismutase, *Biochim. Biophys. Acta,* 268, 605, 1972.
136. **Pantoliano, M. W., McDonnell, P. J., and Valentine, J. S.,** Reversible loss of metal ions from the zinc binding site of copper-zinc superoxide dismutase. The low pH transition, *J. Am. Chem. Soc.*, 101, 6454, 1979.

137. **Fee, J. A. and Phillips, W. D.,** The behavior of holo- and apo-forms of bovine superoxide dismutase at low pH, *Biochim. Biophys. Acta,* 412, 26, 1975.
138. **Boden, N., Holmes, M. C., and Knowles, P. F.,** Properties of the cupric sites in bovine superoxide dismutase studied by nuclear-magnetic-relaxation measurements, *Biochem. J.,* 177, 303, 1979.
139. **McAdam, M. E., Fielden, E. M., Lavelle, F., Calabrese, L., Cocco, D., and Rotilio, G.,** The involvement of the bridging imidazolate in the catalytic mechanism of action of bovine superoxide dismutase, *Biochem. J.,* 167, 271, 1977.
140. **Bull, C. and Fee, J. A.,** Steady-state kinetic studies of superoxide dismutases: properties of the iron containing protein from *Escherichia coli, J. Am. Chem. Soc.,* 107, 3295, 1985.
141. **Connolly, M. L.,** Solvent-accessible surfaces of proteins and nucleic acids, *Science,* 221, 709, 1983.
142. **Rigo, A., Viglino, P., and Rotilio, G.,** Kinetic study of O_2^- dismutation by bovine superoxide dismutase. Evidence for saturation of the catalytic site by O_2^-, *Biochem. Biophys. Res. Commun.,* 63, 1013, 1975.
143. **Fee, J. A.** Structure-function relationships in superoxide dismutases, in *Superoxide and Superoxide Dismutases,* Michelson, A. M., McCord, J. M., and Fridovich, I., Eds., Academic Press, London, 1977, 173.
144. **Plonka, A. Metodiewa, D., and Gasyna, Z.,** ESR studies on oxidation state changes of copper in superoxide dismutase during reactions with water radiolysis products at cryogenic temperatures, *Biochim. Biophys. Acta,* 612, 299, 1980.
145. **Lawrence, G. D. and Sawyer, D. T.,** Potentiometric titrations and oxidation-reduction potentials of manganese and copper-zinc superoxide dismutases, *Biochemistry,* 18, 3045, 1979.
146. **Morpurgo, L., Giovagnoli, C., and Rotilio, G.,** Studies of the metal sites of copper proteins V. A model compound for the copper site of superoxide dismutase, *Biochim. Biophys. Acta,* 322, 204, 1973.
147. **Blumberg, W. E., Peisach, J., Eisenberger, P., and Fee, J. A.,** Superoxide dismutase, a study of the electronic properties of the copper and zinc by X-ray absorption spectroscopy, *Biochemistry,* 17, 1842, 1978.
148. **Margerum, D. W., Cayley, G. R., Weatherburn, D. C., and Pagenkoph, G. K.,** Kinetics and mechanisms of complex formation and ligand exchange, in *Coordination Chemistry,* Vol. 2, Martell, A. E., Ed., American Chemical Society, Washington, D.C., 1978, 1.
149. **Valentine, J. S. and Pantoliano, M. W.,** Protein-metal ion interactions in cuprozinc protein (superoxide dismutase), a major intracellular repository for copper and zinc in the eukaryotic cell, in *Copper Proteins: Metal Ions in Biology,* Vol. 3, Spiro, T. G., Ed., John Wiley & Sons, New York, 1981, 291.
150. **Malinowski, D. P. and Fridovich, I.,** Chemical modification of arginine at the active site of the bovine erythrocyte superoxide dismutase, *Biochemistry,* 18, 5909, 1979.
151. **Borders, C. L., Jr., and Johansen, J. T.,** Identification of Arg-143 as the essential arginyl residue in yeast Cu,Zn superoxide dismutase by use of a chromophoric arginine reagent, *Biochem. Biophys. Res. Commun.,* 96, 1071, 1980.
152. **Osman, R. and Basch, H.,** On the mechanism of action of superoxide dismutase: a theoretical study, *J. Am. Chem. Soc.,* 106, 5710, 1984.
153. **Rosi, M., Sgamellotti, A., Tarantelli, F., Bertini, I., and Luchinat, C.,** Ab initio calculations of the Cu^{2+}-O_2 interaction as a model for the mechanism of copper/zinc superoxide dismutase, *Inorg. Chem.,* 25, 1005, 1986.
154. **Rotilio, G., Finazzi Agro, A., Calabrese, L., Bossa, F., Guerrieri, P., and Mondovi, B.,** Studies of the metal sites of copper proteins. Ligands of copper in hemocuprein, *Biochemistry,* 10, 616, 1971.
155. **Fee, J. A. and Gaber, B. P.,** Anion binding to bovine erythrocyte superoxide dismutase. Evidence for multiple binding sites with qualitatively different properties, *J. Biol. Chem.,* 247, 60, 1972.
156. **Rigo, A., Viglino, P., Calabrese, L., Cocco, D., and Rotilio, G.,** The binding of copper ions to copper-free bovine superoxide dismutase, *Biochem. J.,* 161, 27, 1977.
157. **Rigo, A., Stevanato, R., Viglino, P., and Rotilio, G.,** Competitive inhibition of Cu, Zn superoxide dismutase by monovalent anions, *Biochem. Biophys. Res. Commun.,* 79, 776, 1977.
158. **Rotilio, G., Calabrese, L., Bossa, F., Barra, D., Finazzi Agro, A., and Mondovi, B.,** Properties of the apoprotein and role of copper and zinc in protein conformation and enzyme activity of bovine superoxide dismutase, *Biochemistry,* 11, 2182, 1972.
159. **Rotilio, G., Morpurgo, L., Giovagnoli, C., Calabrese, L., and Mondovi, B.,** Studies of the metal sites of copper proteins. Symmetry of copper in bovine superoxide dismutase and its functional significance, *Biochemistry,* 11, 2187, 1972.
160. **Haffner, P. H., and Coleman, J. E.,** Cu(II)-carbon bonding in cyanide complexes of copper enzymes. ^{13}C splitting of the Cu(II) electron spin resonance, *J. Biol. Chem.,* 248, 6626, 1973.
161. **Fee, J. A., Lieberman, R. A., Van Camp, H., Peisach, J., and Mims, W. B.,** The mechanism of bovine Zn/Cu superoxide dismutase (BSD): configurational changes at the Cu^{2+} upon anion binding and its relevance to catalytic dismutation, *Fed. Proc.,* 39, 1769, 1980.
162. **Lieberman, R. A.,** Electron Paramagnetic Resonance of Bovine Superoxide Dismutase, Ph.D. thesis, University of Michigan, Ann Arbor, MI, 1981.

163. **Augustin, M. A. and Yandell, J. K.,** Rates of electron-transfer reactions of some copper(II)-phenanthroline complexes with cytochrome *c*(II) and tris(phenanthroline) cobalt(II) ion, *Inorg. Chem.*, 18, 577, 1979.
164. **Lieberman, R. A., Sands, R. H., and Fee, J. A.,** A study of the electron paramagnetic resonance properties of single monoclinic crystals of bovine superoxide dismutase, *J. Biol. Chem.*, 257, 336, 1982.
165. **Van Camp, H. L., Sands, R. H., and Fee, J. A.,** An examination of the cyanide derivative of bovine superoxide dismutase with electron-nuclear double resonance, *Biochim. Biophys. Acta,* 704, 75, 1982.
166. **Misra, H. P. and Fridovich, I.,** Inhibition of superoxide dismutases by azide, *Arch. Biochem. Biophys.*, 189, 317, 1978.
167. **Bertini, I., Lanini, G., Luchinat, C., Messori, L., Monnanni, R., and Scozzafava, A.,** Investigation of Cu_2Co_2SOD and its anion derivatives. ^{1}H NMR and electronic spectra, *J. Am. Chem. Soc.*, 107, 4391, 1985.
168. **Banci, L., Bertini, I., Luchinat, C., and Scozzafava, A.,** Nuclear relaxation in the magnetic coupled system Cu_2Co_2SOD. Histidine-44 is detached upon anion binding, *J. Am. Chem. Soc.*, 109, 2328, 1987.
169. **Fee, J. A. and Ward, R. L.,** Evidence for a coordination position available to solute molecules on one of the metals at the active center of reduced bovine superoxide dismutase, *Biochem. Biophys. Res. Commun.*, 71, 427, 1976.
170. **Blumberg, W. E., Peisach, J., and Fee, J. A.,** An x-ray absorption edge study of copper and zinc in bovine superoxide dismutase, *Fed. Proc.*, 39, 1976, 1980.
171. **Cass, A. E. G., Hill, H. A. O., Hasemann, V., and Johansen, J. T.,** ^{1}H nuclear magnetic resonance spectrocopy of yeast copper-zinc superoxide dismutase. Structural homology with the bovine enzyme, *Carlsberg Res. Commun.*, 43, 439, 1978.
172. **Getzoff, E. D., Tainer, J. A., and Olson, A. J.,** Recognition and interactions controlling the assemblies of β barrel domains, *Biophys. J.*, 49, 191, 1986.
173. **Salin, M. L. and Wilson, W. W.,** Porcine superoxide dismutase, *Mol. Cell. Biochem.*, 36, 157, 1981.
174. **Cudd, A. and Fridovich, I.,** Electrostatic interactions in the reaction mechanism of bovine erythrocyte superoxide dismutase, *J. Biol. Chem.*, 257, 11443, 1982.
175. **Koppenol, W. H.,** The physiological role of the charge distribution on superoxide dismutase, in *Oxygen and Oxy-Radicals in Chemistry and Biology,* Rodgers, M. A. J., and Powers, E. L., Eds., Academic Press, New York, 1981, 671.
176. **Koppenol, W. H.,** On the reactivity of the superoxide anion and the biological function of superoxide dismutase, in *Oxidases and Related Redox Systems,* King, T. E., Mason, H. S., and Morrison, M., Eds., Pergamon Press, Oxford, 1982, 127.
177. **Getzoff, E. D. and Tainer, J. A.,** Superoxide dismutase as a model ion channel, in *Ion Channel Reconstitution,* Miller, C., Ed., Plenum Press, New York, 1986, 57.
178. **Klapper, I., Hagstrom, R., Fine, R., Sharp, K., and Honig, B.,** Focusing of electric fields in the active site of Cu-Zn superoxide dismutase: effects of ionic strength and amino-acid modification, *Prot. Struct. Funct. Gen.*, 1, 47, 1986.
179. **Allison, S. A., Ganti, G., and McCammon, J. A.,** Simulation of the diffusion-controlled reaction between superoxide and superoxide dismutase. I. Simple models, *Biopolymers,* 24, 1323, 1985.
180. **Ganti, G., McCammon, J. A., and Allison, S. A.,** Brownian dynamics of diffusion-controlled reactions: the lattice method, *J. Phys. Chem.*, 89, 3899, 1985.
181. **Sharp, K., Fine, R., and Honig, B.,** Computer simulations of the diffusion of a substrate to an active site of an enzyme, *Science,* 236, 1460, 1987.
182. **Allison, S. A., Bacquet, R. J., and McCammon, J. A.,** Simulation of the diffusion controlled reaction between superoxide and superoxide dismutase II. Detailed models, *Biopolymers,* 27, 251, 1988.
183. **Fielden, E. M., Roberts, P. B., Bray, R. C., Lowe, D. J., Mautner, G. N., Rotilio, G., and Calabrese, L.,** The mechanism of action of superoxide dismutase from pulse radiolysis and electron paramagnetic resonance, *Biochem. J.*, 139, 49, 1974.
184. **Argese, E., Viglino, P., Rotilio, G., Scarpa, M., and Rigo, A.,** Electrostatic control of the rate-determining step of the copper, zinc superoxide dismutase catalytic reaction, *Biochemistry,* 26, 3224, 1987.
185. **McCammon, J. A., Bacquet, R. J., Allison, S. A., and Northrup, S. H.,** Trajectory simulation studies of diffusion-controlled reactions, *Faraday Disc. Chem. Soc.*, 83, 213, 1987.
186. **Beyer, W. F., Jr., Fridovich, I., Mullenbach, G. T., and Hallewell, R.,** Examination of the role of arginine-143 in the human copper and zinc superoxide dismutase by site-specific mutagenesis, *J. Biol. Chem.*, 262, 11182, 1987.
187. **Parge, H. E., Getzoff, E. D., Scandella, C. S., Hallewell, R. A., and Tainer, J. A.,** Crystallographic characterization of recombinant human CuZn superoxide dismutase, *J. Biol. Chem.*, 261, 16215, 1986.
188. **Drath, D. B. and Karnovsky, M. L.,** Superoxide production by phagocytic leukocytes, *J. Exp. Med.*, 141, 257, 1975.
189. **Huq, S. and Palmer, J. M.,** Superoxide and hydrogen peroxide production in cyanide resistant *Arum maculatum* mitochondria, *Plant Sci. Lett.*, 11, 351, 1978.
190. **Rich, P. R. and Bonner, W. D., Jr.,** The sites of superoxide anion generation in higher plant mitochondria, *Arch. Biochem. Biophys.*, 188, 206, 1978.

191. **Nohl, H. and Hegner, D.,** Do mitochondria produce oxygen radicals *in vivo?, Eur. J. Biochem.,* 82, 563, 1978.
192. **Bartoli, G. M., Galeotti, T., Palombini, G., Parisi, G., and Azzi, A.,** Different contribution of rat liver microsomal pigments in the formation of superoxide anions and hydrogen peroxide during development, *Arch. Biochem. Biophys.,* 184, 276, 1977.
193. **Auclair, C., de Prost, D., and Hakim, J.,** Superoxide anion production by liver microsomes from phenobarbital treated rat, *Biochem. Pharm.,* 27, 355, 1978.
194. **Bartoli, G. M., Galeotti, T., and Azzi, A.,** Production of superoxide anions and hydrogen peroxide in Ehrlich ascites tumour cell nuclei, *Biochim. Biophys. Acta,* 497, 622, 1977.
195. **Fridovich, I. and Handler, P.,** Xanthine oxidase. V. Differential inhibition of the reduction of various electron acceptors, *J. Biol. Chem.,* 237, 916, 1962.
196. **Nilsson, R., Pick, F. M., and Bray, R. C.,** EPR studies on reduction of oxygen to superoxide by some biochemical systems, *Biochim. Biophys. Acta,* 192, 145, 1969.
197. **Rotilio, G., Falcioni, G., Fioretti, E., and Brunori, M.,** Decay of oxyperoxidase and oxygen radicals: a possible role for myeloperoxidase, *Biochem. J.,* 145, 405, 1975.
198. **Rotilio, G., Calabrese, L., Finnazzi Agrò, A., and Mondovi, B.,** Indirect evidence for the production of superoxide anion radicals by pig kidney diamine oxidase, *Biochim. Biophys. Acta,* 198, 618, 1970.
199. **Forman, H. J. and Kennedy, J. A.,** Role of superoxide radical in mitochondrial dehydrogenase reactions, *Biochem. Biophys. Res. Commun.,* 60, 1044, 1974.
200. **Misra, H. P. and Fridovich, I.,** The generation of superoxide radical during the autoxidation of ferredoxins, *J. Biol. Chem.,* 246, 6886, 1971.
201. **Strobel, H. W. and Coon, M. J.,** Effect of superoxide generation and dismutation on hydroxylation reactions catalyzed by liver microsomal cytochrome P-450, *J. Biol. Chem.,* 246, 7826, 1971.
202. **Kuthan, H., Tsuji, H., Graf, H., Ullrich, V., Werringloer, J., and Estabrook, R. W.,** Generation of superoxide anion as a source of hydrogen peroxide in a reconstituted monooxygenase system, *FEBS Lett.,* 91, 343, 1978.
203. **Demma, L. S. and Salhany, J. M.,** Direct generation of superoxide anions by flash photolysis of human oxyhemoglobin, *J. Biol. Chem.,* 252, 1226, 1977.
204. **Misra, H. P. and Fridovich, I.,** The generation of superoxide radical during the autoxidation of hemoglobin, *J. Biol. Chem.,* 247, 6960, 1972.
205. **May, S. W., Abbott, B. J., and Felix, A.,** On the role of superoxide in reactions catalyzed by rubredoxin of *Pseudomonas oleovorans, Biochem. Biophys. Res. Commun.,* 54, 1540, 1973.
206. **Misra, H. P. and Fridovich, I.,** The univalent reduction of oxygen by reduced flavins and quinones, *J. Biol. Chem.,* 247, 188, 1972.
207. **Patel, K. B. and Willson, R. L.,** Semiquinone free radicals and oxygen: pulse radiolysis study of one electron transfer equilibria, *J. Chem. Soc. Faraday Trans. I,* 69, 814, 1973.
208. **Heikkila, R. E. and Cohen, G.,** 6-Hydroxydopamine: evidence for superoxide radical as an oxidative intermediate, *Science,* 181, 456, 1973.
209. **Farrington, J. A., Ebert, M., Land, E. J., and Fletcher, K.,** Bipyridylium quaternary salts and related compounds. V. Pulse radiolysis studies of the reaction of paraquat radical with oxygen. Implications for the mode of action of bipyridyl herbicides, *Biochim. Biophys. Acta,* 314, 372, 1973.
210. **Stancliffe, T. C. and Pirie, A.,** The production of superoxide radicals in reactions of the herbicide diquat, *FEBS Lett.,* 17, 297, 1971.
211. **Fee, J. A., Bergamini, R., and Briggs, R. G.,** Observations on the mechanism of the oxygen/dialuric acid-induced hemolysis of vitamin E-deficient rat red blood cells and the protective roles of catalase and speroxide dismutase, *Arch. Biochem. Biophys.,* 169, 160, 1975.
212. **Misra, H. P.,** Generation of superoxide free radical during the autoxidation of thiols., *J. Biol. Chem.,* 249, 2151, 1974.
213. **Fee, J. A.,** in *Oxidases and Related Oxidation-Reduction Systems,* King, T. E., Mason, H. S., and Morrison, M., Eds., Pergamon Press, Oxford, 1979.
214. **Slykhouse, T. O. and Fee, J. A.,** Physical and chemical studies on bacterial superoxide dismutases. Purification and some anion binding properties of the iron-containing protein of *Escherichia coli, J. Biol. Chem.,* 251, 5472, 1976.
215. **Keele, B. B., Jr., McCord, J. M., and Fridovich, I.,** Superoxide dismutase from *Escherichia coli* B. A new manganese-containing enzyme, *J. Biol. Chem.,* 245, 6176, 1970.
216. **Rabani, J., Klug-Roth, D., and Lilie, J.,** Pulse radiolytic investigations of the catalyzed disproportionation of peroxy radicals. Aqueous cupric ions, *J. Phys. Chem.,* 77, 1169, 1973.
217. **Klug-Roth, D. and Rabani, J.,** Pulse radiolytic studies on reactions of aqueous superoxide radicals with copper(II) complexes, *J. Phys. Chem.,* 80, 588, 1976.
218. **Younes, M. and Weser, U.,** Inhibition of nitroblue tetrazolium reduction by cuprein (superoxide dismutase), $Cu(tyr)_2$ and $Cu(lys)_2$, *FEBS Lett.,* 61, 209, 1976.
219. **Markossian, K. A., Poghossian, A. A., Paitian, N. A., and Nalbadyan, R. M.,** Superoxide dismutase activity of cytochrome oxidase, *Biochem. Biophys. Res. Commun.,* 81, 1336, 1978.

220. **Pasternack, R. F. and Halliwell, B.,** Superoxide dismutase activities of an iron porphyrin and other iron complexes, *J. Am. Chem. Soc.,* 101, 1026, 1979.
221. **Winterbourn, C. C., Hawkins, R. E., Brian, M., and Carrell, R. W.,** The estimation of red cell superoxide dismutase activity, *J. Lab. Clin. Med.,* 85, 337, 1975.
222. **Hallewell, R. A., Imlay, K. C., Laria, I., Gallegos, C., Irvine, B., Fong, N. M., Cabelli, D. E., Bielski, B. H. J., Getzoff, E. D., Tainer, J. A., Olson, P., Cousens, L. S., and Mullenbach, G. T.,** Thermostabilization of recombinant CuZn superoxide dismutases by removal of free cysteines, *Proteins: Struct. Func. Gen.,* in press.
223. **Weast, R. C., Ed.,** *Handbook of Chemistry and Physics,* 60th ed., CRC Press, Boca Raton, FL, 1979.

Section XI
Hydrolases — Mechanisms

Enzyme chemistry has expanded rapidly in recent years and has now reached a stage when it is possible to think in more precise terms about the chemical groups and the physical forces which contribute to the binding of the substrate to the enzyme and its subsequent transformation. D. R. Davies and A. L. Green, in The Mechanism of Hydrolysis by Cholinestersase and Related Enzymes, p. 283, Vol. XX, in *Advances in Enzymology and Related Subjects of Biochemistry*, (F. F. Nord, editor), Interscience Publishers, Inc., New York, 1958.

Chapter 22

PHOSPHOLIPASES

Gilbert Arthur and Patrick C. Choy

TABLE OF CONTENTS

I. INTRODUCTION

Phospholipases are a group of enzymes that catalyse the hydrolysis of ester bonds in glycerophospholipids and therefore function as esterases. Phospholipases differ from other esterases in their requirement for a lipid-water interface to express full activity. These enzymes are ubiquitously distributed in both procaryotes and eucaryotes and within cells are found in virtually every subcellular fraction. The widespread distribution of phospholipases is not surprising in view of their involvement in the metabolism of glycerophospholipids, a major component of all biological membranes. A number of excellent reviews[1-8a] on phospholipases have appeared in recent years which contain a wealth of information on these enzymes. The aim of this chapter is not to duplicate these efforts but to give an overview of the mechanism of action of phospholipases. Since most of the studies on the mechanism of phospholipases have mainly utilized extracellular phospholipase A_2, this chapter will focus on this enzyme.

In the last 2 decades, considerable effort has been expended on elucidating the mechanism of phospholipase A_2 activity. There is as yet no general concensus on the mechanism of this enzyme. One reason for this is undoubtedly the numerous systems used by different laboratories to investigate the kinetics of the enzyme. The variety of systems used stems from difficulties in working with substrates that are insoluble in aqueous medium and an enzyme that preferentially hydrolyses substrates at the lipid-water interface. A brief description of the substrate seems to be an appropriate point to begin this chapter.

II. GLYCEROPHOSPHOLIPIDS

The substrates hydrolyzed by phospholipases are glycerophospholipids. Their structure is typified by that of ethanolamine-GPL shown in Figure 1. They consist of a glycerol backbone to which long chain hydrocarbon units are attached at the *sn*-1 and *sn*-2 positions. The linkage of the hydrocarbon at the *sn*-1 position can be an ester, alkyl, or alkenyl bond while the long chain hydrocarbon at the *sn*-2 position is attached via an ester bond. The *sn*-3 carbon is esterified with a phosphate group which is attached to a polar base. The nature of the polar base determines the glycerophospholipid class while the type of linkage at the *sn*-1 position gives rise to three subclasses: 1,2-diacyl-*sn*-glycero-3-phospholipid, 1-alkyl-2-acyl-*sn*-glycero-3-phospholipid, and 1-alkenyl-2-acyl-*sn*-3-glycerophospholipid. The latter two subclasses make up the ether glycerophospholipids. Monoacylglycerophospholipids are commonly known as lysophospholipids.

Glycerophospholipids are mainly insoluble in aqueous systems. At very low concentrations they exist as monomers but, above their critical micellar concentrations (CMC), they aggregate into a number of polymorphic phases.[9] The type of aggregate formed depends on the phospholipid class as well as the presence or absence of ions. In addition, phospholipids can also undergo thermotropic phase changes that reflect the fluidity of the fatty acyl chains.[10] Three phases have been identified. The order assumed with increasing temperature are a gel phase β, β′ and a liquid crystalline state (Figure 2). Transition temperatures from one state to another are ill defined in phospholipids of mixed acyl chains isolated from natural sources, but very well defined in phospholipid molecules with homogenous saturated acyl chains. A number of physical studies with phospholipids indicate they take up a preferred conformation regardless of class.[4,11-15] In general, the sn-1 chain is extended perpendicular to the bilayer at all segments of the molecule. The sn-2 chain starts off parallel to the membrane surface but is bent perpendicular to it after the C2 segment (Figure 3). The *sn*-2 ester bond is therefore more exposed in the interface than the *sn*-1 bond.

One fundamental property exhibited by all phospholipases is their preference for aggregated substrates. Unfortunately, a clear understanding of changes in phospholipid con-

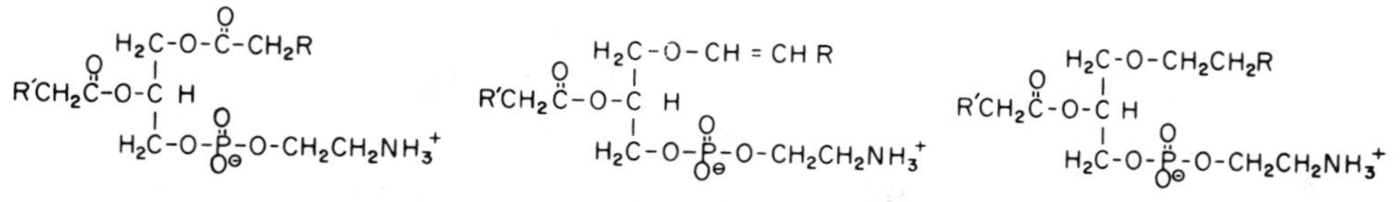

1,2-diacyl-GPE   1-alk-1'-enyl-2-acyl-GPE   1-alkyl-2-acyl-GPE

FIGURE 1. The structures of ethanolamine glycerophospholipids. GPE represents *sn*-glycero-3-phosphoethanolamine.

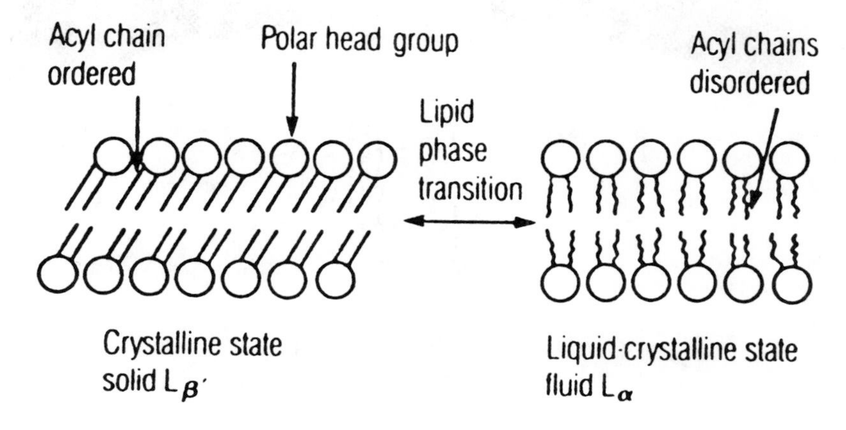

FIGURE 2. Lipid phase transition between crystalline state (gel phase) and liquid-crystalline state.

formation and packing between the different forms of phospholipid aggregates, as well as between different phospholipid classes and molecular species at the interface is lacking. The concept of the "quality of the interface" has been advanced by De Haas and colleagues[1-3,16,17] to account for these poorly understood parameters that influence phospholipase A_2 activity. Additional factors such as the surface pressure and surface charge may affect both the specificity and activity of the phospholipases.[2,3,5,10] In view of all these variables, it is not surprising that our comprehension of factors that dictate the specificity of the phospholipases is rather limited. Recently, Jain et al.[18] undertook studies with over 50 synthesized compounds in an attempt to understand the underlying factors responsible for the substrate specificity of pig pancreatic phospholipase A_2. The conclusions of the study, though tentative, suggested that increased hydrophobicity, due to either an increase in chain length or the hydrophobicity of the head group, resulted in a decrease in the rate of hydrolysis. The charge of the phospholipid also appears to play an important role in determining the substrate specificity of the enzyme. Thuren et al.[19] reported that porcine and bovine pancreatic phospholipase A_2 as well as porcine intestinal enzyme utilized acidic phospholipid liposomes (PM $>$ PS $>$ PG) better than neutral phospholipids (PC $>$ PE). In contrast, the enzyme from *Crotalus atrox* had a higher activity with PC than with the acidic phospholipids. The lowest rates of hydrolysis exhibited by all the enzymes examined was obtained with PE liposomes. This observation was attributed to the formation of nonbilayer structure by PE. When the phospholipids were presented to the pancreatic phospholipase A_2 as monolayers, the highest activities were again obtained with the acidic phospholipids, but PE was found to be a better substrate than PC. These results might be caused by the enhancement of the interaction of the enzyme to the interface. The interaction of the *Naja naja naja* phospholipase A_2 with both acyl chains of the phospholipid is required for efficient catalysis of substrates.[20]

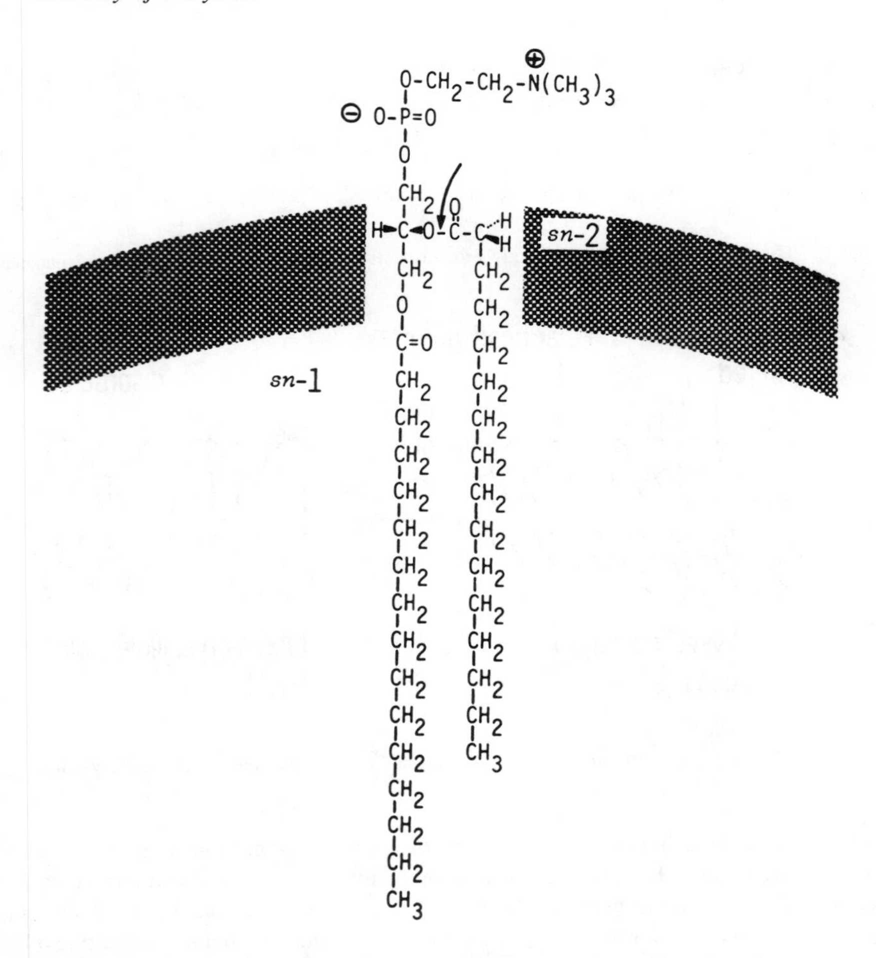

FIGURE 3. Schematic representation of a phospholipid molecule in lipid-water interphase. Note the difference in the carbonyl groups of the two fatty acyl chains. (From Dennis, E. A., *Enzymes,* 16, 308, 1983. With permission.)

III. TYPES OF PHOSPHOLIPASES

The site specificity of the phospholipases has given rise to their classification into phospholipases A_1, A_2, C, and D (Figure 4). Both the ether and diacyl-GPL can be potentially hydrolysed by phospholipases A_2, C, and D but only the diacyl GPL are susceptible to phospholipase A_1 hydrolysis. Enzymes that hydrolyze the acyl ester bond of monoacyl glycerophospholipids are generally called lysophospholipases. A lysophospholipase D that removes the base specifically from 1-alkyl-GPC has also been described.[21] The term phospholipase B describes phospholipases that hydrolyse both the *sn*-1 and *sn*-2 acyl esters. Such an enzyme has been purified from *Pennicilin notatum*[22] and can be perceived as possessing both phospholipase A (A_1 and A_2) and lysophospholipase activities.[5] It should be noted that there are some enzymes which are not classified as phospholipases, but have the ability to hydrolyze phospholipids. For example, sphingomyelinase[23] catalyzes the hydrolysis of an ester bond in sphingomyelin, which is a phospholipid that has a sphingosine rather than a glycerol base.

Phospholipases can be broadly divided into two groups: the extracellular phospholipases and the intracellular phospholipases. The extracellular enzymes are mainly of the A_2 type and include the enzymes found in snake and bee venoms, as well as the phospholipases

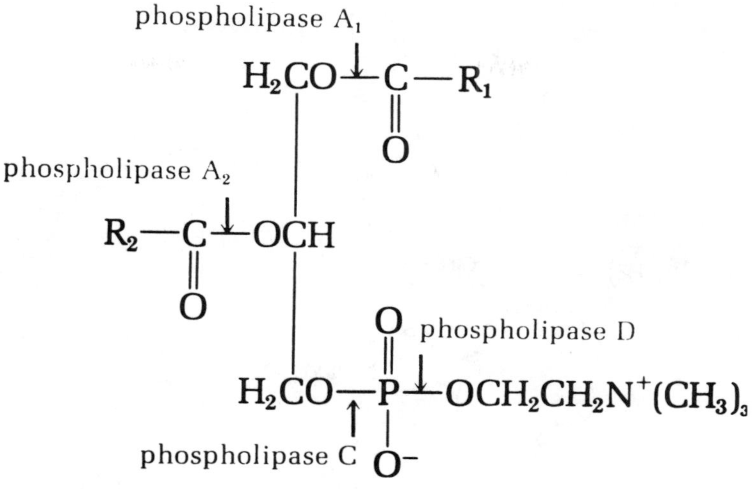

phospholipase A_1

$H_2CO \downarrow C - R_1$ (C=O)

phospholipase A_2

$R_2 - C \downarrow OCH$ (C=O)

O phospholipase D

$H_2CO - P \downarrow OCH_2CH_2N^+(CH_3)_3$

phospholipase C O^-

FIGURE 4. Site of action of phospholipases on phosphatidylcholine (PC).

secreted in digestive juices. Intracellular phospholipases encompass all the nonsecretory phospholipases and include all the types mentioned above (A_1, A_2, C, D, and lysophospholipase). They participate in a variety of cellular functions and play a major role in the turnover of the phospholipids.[8,8a,8b]

IV. EXTRACELLULAR PHOSPHOLIPASES

Extracellular phospholipases comprise the phospholipases secreted in the digestive juice of animals and those excreted by bacteria, as well as the phospholipases found in the venom of snakes and bees. They are mainly of the A_2 type. The phospholipases of the digestive juices obviously have a digestive function. There are reports that intestinal phospholipases may be involved in controlling the bacterial flora in the gut.[24,25] While the functions of the venom enzymes are not so obvious, they contribute to immobilizing the prey through hemolysis of the cells as a result of the lytic effects of lysophospholipids produced. Some venom enzymes have also been shown to be neurotoxic.[8b] A major difference between the pancreatic and the venom enzymes is that the former is synthesized as a proenzyme which is activated via proteolysis to remove a heptapeptide from the NH_2-terminal.

V. STRUCTURE OF PHOSPHOLIPASE A_2

Detailed analysis of the primary structures of a large number of extracellular phospholipase A_2 has been reviewed[3] and the evolutionary aspects of these enzymes have been discussed.[26] A large number of snake venom, as well as bee venom and pancreatic phospholipase A_2 have been purified and their primary structure determined. They are all stable, small molecular weight proteins that are dependent on calcium for activity. Comparative studies of the primary structure indicate a large degree of homology among the snake venom and pancreatic phospholipases A_2 suggesting that they evolved from a common ancestor.[2,3,26] The bee venom enzyme has little homology with the other enzymes and presumably evolved independently.[3] The primary structure of bovine pancreatic phospholipase A_2 is shown in Figure 5.

With very few exceptions, phospholipase A_2 contains 7 disulfide bridges which contribute to maintaining the conformation required for activity. Reduction of these disulfide bridges results in loss of enzyme activity which can only be restored after reoxidation in the presence

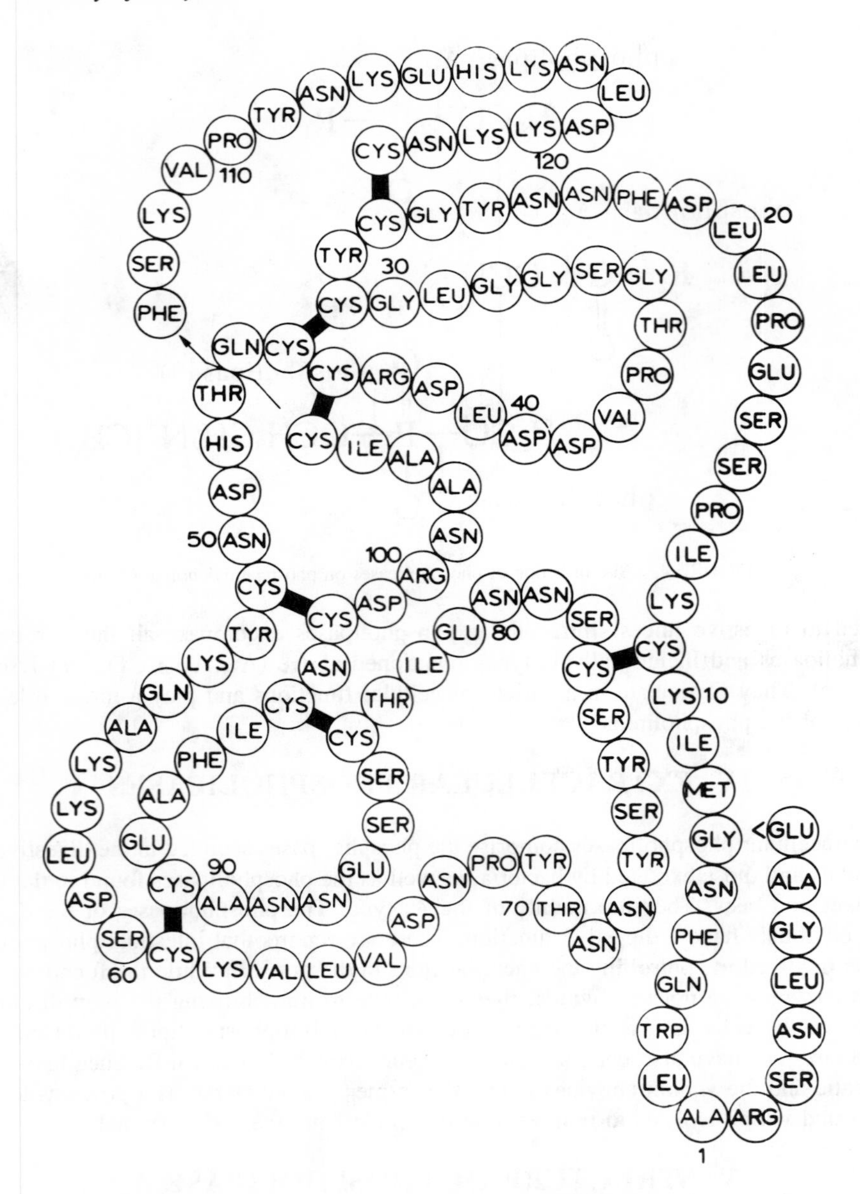

FIGURE 5. The amino acid sequence of bovine prophospholipase A_2 and the intramolecular disulfide bridges. (From Slotboom, A. J., Verheij, H. M., and de Haas, G. H., *Phospholipids*, Hawthorne, J. N. and Ansell, G. B., Eds., Elsevier/North-Holland, Amsterdam, 1982, 359. With permission.)

of thiol reagents.[27] In spite of the large homology between the phospholipase A_2 from different sources, there are some significant structural differences which form the basis of their classification into two major types.[28,29] Enzymes from mammalian pancreatic juices and venoms of snake families Elapidae and Hydrophidae have a half cystine at amino acid 11 and 77, and comprise the group I enzymes. The group II enzymes consist of those in the snake families Viperidae and Crotalidae. They lack the half cystines at positions 11 and 77, but have half cystines at positions 50 and at the C-terminus. In addition, the group II enzymes have extensions of about 6-8 residues beyond the corresponding C-terminus in group I enzymes. The group II enzymes also have a large deletion between amino acid 57 and 66.[28]

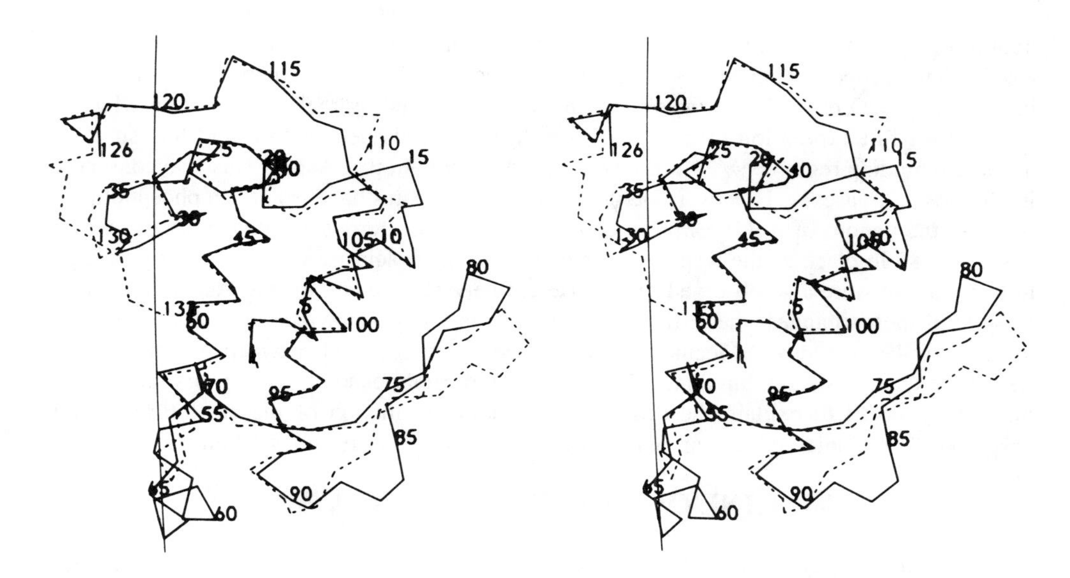

FIGURE 6. Comparison of the C-backbone of bovine and *C. atrox* phospholipase A_2.A core structure was constructed from six segments of phospholipase A_2 from bovine pancreas and *C. atrox* which had been identified as homologous between the two enzymes. The C atoms of the bovine enzyme (continuous line) were superimposed on both the right and left subunits of the *C. atrox* enzyme (dotted line). The vertical line represents the molecular dyad of the *C. atrox* enzyme. (From Renetseder, R., Brunei, S., Dijkstra, B. W., Drenth, J., and Sigler, P. B., *J. Biol. Chem.*, 260, 11627, 1985. With permission.)

Differences in the aggregated states of purified phospholipases A_2 have been reported. The pancreatic phospholipase A_2 are isolated as monomers and do not aggregate except in the presence of very strong anionic detergents.[30-32] On the other hand, enzymes from *Crotalidae* are dimers with very little tendency to dissociate into monomers.[33-37] The *Naja* enzymes are reported to undergo a concentration and lipid dependent aggregation.[38-41]

The tertiary structure of a number of groups I and II phospholipase A_2 have been established by X-ray crystallography.[5,29,33,41-48] All the evidence indicates that regardless of the group, the tertiary structures are remarkably similar. Renetseder et al.[33] recently compared phospholipase A_2 from bovine pancreas (Gp I and monomer) and *C. atrox,* (Gp II and dimer). The backbone conformation of the two enzymes was virtually superimposable (Figure 6). The differences in the primary structure between the two types of phospholipases, translated into significant changes in the conformation in three regions. The loss of the Cys 11-Cys 77 disulfide bond and its replacement with one between Cys 50 and Cys 133 resulted in the *Crotalus* enzyme (Gp II) assuming a more open orientation in the β-wings (residue 74-84) compared to the pancreatic enzyme (Gp-I). The deletions between residue 57-66 resulted in a missing loop in the venom enzyme and it was speculated that the extra loop in the mammalian enzyme prevents dimerization of the pancreatic phospholipase. Whether this is a general phenomenon will have to await the determination of the complete tertiary structure of venom enzymes from *Elapidae* and *Hydrophidae*. The extension at the C-terminal end of the *Crotalid* enzyme was obviously not present in the pancreatic phospholipase A_2. The functional significance of this extension is still obscure.

A number of functional regions have been identified in the tertiary structure of phospholipase A_2. The amino acids postulated to participate in the catalytic process are His 48, and Asp 99.[3,43,33] Together with Tyr 52 and 73, these amino acids form a hydrogen bonded network that in all likelihood participates in catalysis.[33] These residues are absolutely conserved in all phospholipases A_2.[44-46,48] The tertiary structures of this network in both group I and II phospholipase A_2 are virtually identical.[33] In addition, a variable region comprised mainly of hydrophobic amino acid residues that partially encompass the catalytic site has also been identified in both groups of phospholipases A_2.[3] These side chains may be involved in anchoring the enzyme to the interface (the interfacial binding region). Calcium, which is required for full expression of phospholipase A_2 activity, is believed to bind to Asp 49 in a calcium binding region.[3,43,44] Until recently, it was thought that Asp 49 was also conserved in all phospholipase A_2. However, Heinrickson and associates have isolated a phospholipase A_2 from the venom of *Agkistrodon piscivorus piscivorus,* a group II phospholipase, which has a Lys 49 in place of the Asp.[28,49] A number of amino acid residues previously thought to be invariant were also observed to be different between the Lys 49 and Asp 49 enzymes. In spite of these changes, the tertiary structure of the Lys 49 enzyme was still very similar to the Asp 49 enzymes.[29] Mechanistic differences were reported between the two groups with regards to calcium binding. Maragamore and Heinrickson[50] have proposed a very interesting model to explain the ordered and random formation of the catalytic complex based on the cationic and anionic nature of the side chains at residues 49 and 53.

VI. KINETICS OF PHOSPHOLIPASE A_2

One distinctive property shared by both venom and pancreatic phospholipase A_2 is the dramatic difference in their rate of hydrolysing micellar aggregates compared to monomeric forms[2,3,5,11] (Figure 7). This observation is pertinent in the *in vivo* milieu since phospholipids invariably exist there as aggregates.

An essential requirement for any kinetic model of phospholipase A_2 is its ability to explain the enhanced increase in reaction rates with aggregated substrate compared to monomers. Unfortunately kinetic studies with these enzymes yield complicated rate curves which are difficult to interpret and to which the normal Michaelis-Menten equations are not applicable. In addition, the reproducibility of the assay systems presents another problem. For example, Tinker et al.[51] reported difficulties in obtaining quantitatively reproducible reaction rates with different dispersions of DMPC and DPPC, and concluded from their studies that the extent of substrate dispersion profoundly affects the reaction. These observations are supported by the studies of Upreti and Jain,[52] who showed that the extent of sonication can affect the profile of the enzymatic hydrolysis. They were able to demonstrate that unilamellar vesicles, unannealed multilamellar vesicles, and annealed multilamellar vesicles are hydrolyzed at different rates which probably reflect the differences in molecular packing in the bilayers or defects within the bilayer. Therefore, a mixed population of vesicles will give different kinetic profiles from a pure population. Phospholipase A_2 activity is affected by compounds such as fatty acids, LPC, and anions. The mode of addition of LPC to PC vesicles also affects the kinetic profile.[53] These studies demonstrate the importance of using substrates of known composition and well-defined protocols to assay the enzyme in order to obtain relevant results. The various techniques used in assaying phospholipase A_2 have been reviewed by Slotboom et al.[3]

A fundamental question that remains unanswered is whether all phospholipases A_2 regardless of origin share the same reaction mechanism. One reason why it has been difficult to answer the above question is that the assay systems used to study the kinetics of a number of phospholipase A_2 are not uniform. For a variety of reasons, monolayers,[54-58] phospholipid monomers,[59,60] short chain micelles,[61,62] phospholipid bilayers,[51,52,63,64] mixed micelles of detergent, and phospholipids[40,65-69] have all been used. Not surprisingly this has led to difficulties in correlating the work from one laboratory to another. These different forms of substrates present interfaces that may be susceptible to different physical parameters. For example, monolayers are greatly influenced by surface pressures whereas bilayers are not. Since the phospholipase A_2 activity is intimately associated with the interface, one must therefore be aware of potential kinetic differences that could result from the use of different

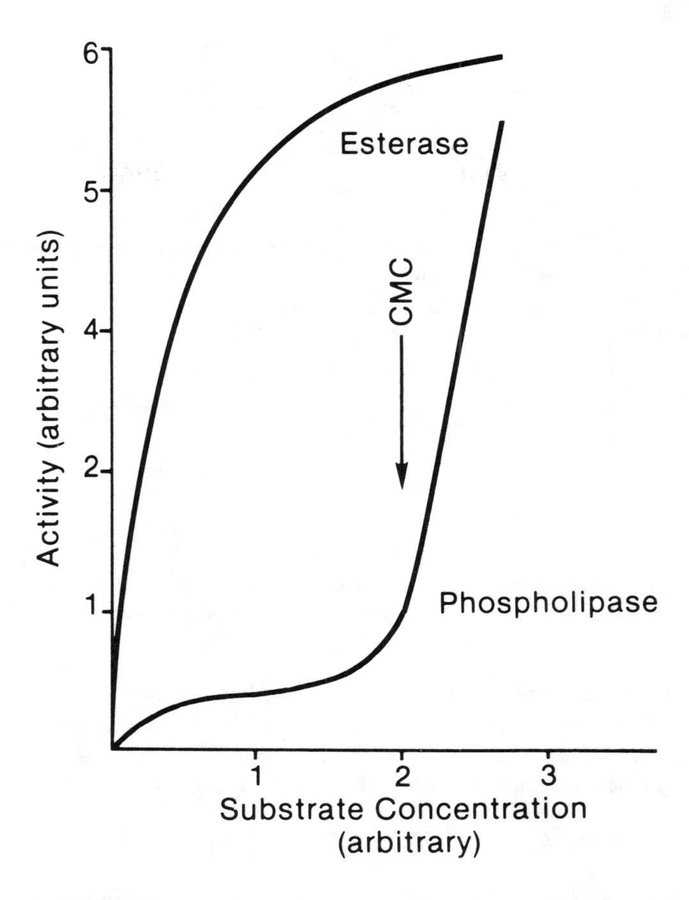

FIGURE 7. The effect of substrate concentration on phospholipase A_2 and nonspecific esterase activities. CMC represents the "critical micelles concentration". (From Waite, M., *Biochemistry of Lipids and Membranes*, Vance, D. E. and Vance, J. E., Eds., Benjamin/Cummings, Menlo Park, CA, 1985, 299. With permission.)

phospholipid dispersions. Recently, Jain et al.[70] have observed that pig pancreatic phospholipase A_2 hydrolyzes phosphatidylmethanol vesicles without complicated kinetics or complex reaction progress curves. This substrate has the potential of being extremely useful in the elucidation of the mechanism of phospholipase A_2 hydrolysis.

Four theories have been proposed to explain the activation of phospholipases A_2 at water-lipid interfaces.[2,3,5,71]

1. The enzyme theory — This theory assumes that upon interacting with lipid-water interface or activators, there is a conformational change in the enzyme molecule that is eventually translated into increased activity.
2. The substrate theory — This theory suggests changes in the phospholipid molecule in its transformation from a monomer to an aggregated form, makes it more susceptible to hydrolysis by the enzyme.
3. The product theory — In this theory, the rate limiting step of the reaction is the release of the products which is slow in water and is increased at the lipid-water interface.
4. The concentration effect — This hypothesis suggests that at the lipid-water interface, the enzyme in effect is exposed to a higher substrate concentration that brings the enzyme closer to saturation.

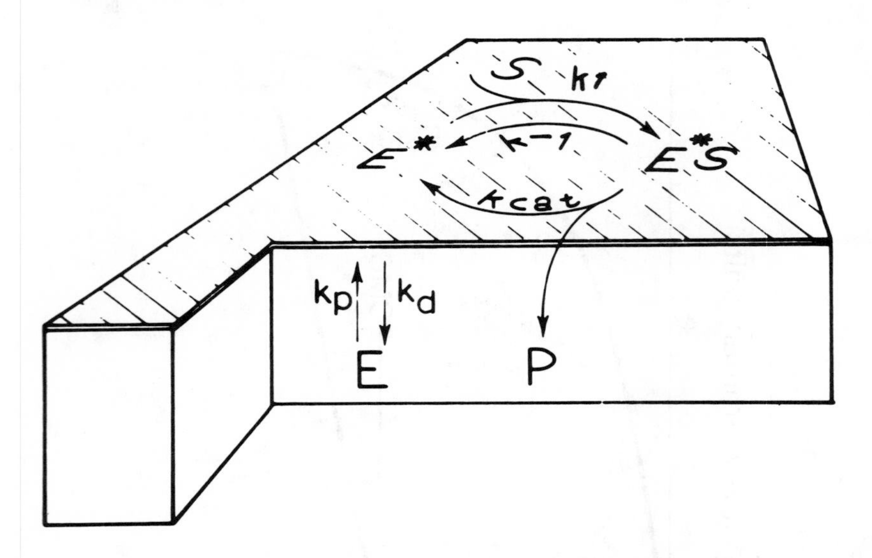

FIGURE 8. Proposed IRS model for the action of phospholipase A_2 at an interface. (From Verger, R., Mieras, M. C. E., and de Haas, G. H., *J. Biol. Chem.*, 248, 4023, 1973. With permission.)

Three kinetic models have been proposed to explain the activation of phospholipases that occurs at the lipid-water interphase. All the models are based largely on the enzyme theory, but other factors are also taken into consideration. The three models are the interface recognition site model, the dual phospholipid model, and the Tinker model.

VII. INTERFACE RECOGNITION SITE (IRS) MODEL

This model was proposed by de Haas and co-workers[2,3,72-74] for the pancreatic phospholipase A_2 more than 10 years ago and is based on the enzyme theory. The model was initially derived from work with monolayers but other phospholipid dispersions have been utilised in subsequent development of the model which is illustrated in Figure 8. The active phospholipase A_2 possesses a region termed the interface recognition site through which it binds reversibly to the interface. This binding results in a conformational change that results in optimal alignment or exposure of the active site residues to the substrate leading to enhanced hydrolysis. The model proposes the existence of two successive equilibria.

1. $E \leftrightarrow E^*$
2. $E^* + S \leftrightarrow E^*S$
3. $E^*S \rightarrow E + P$ (Str 1)

The first of the two equilibrium steps, the binding of the enzyme to the interface and subsequent conformational change, was postulated as the rate-limiting step. This was due mainly to the long lag times observed between addition of enzyme and active hydrolysis of the substrate. The nature of the interaction between the substrate and enzyme is still not clearly defined and may represent a number of interactions such as penetration, anchoring, or adsorbtion.[2,3] Recent evidence suggests that both hydrophobic and ionic interactions are involved.[75]

The model evolved from kinetic comparisons between the inactive zymogen and the active enzyme. The observation that both the zymogen and active enzyme hydrolyze phospholipid monomers with comparable activities suggested the existence of a functional catalytic site in both forms. The inability of the zymogen to hydrolyze the micelles led to the supposition that it could not recognize the organised lipid-water interface, and consequently lacked or could not expose the interaction sites (IRS) required for recognition and binding. Evidence for the topographic separation of the active site from the IRS comes from studies in which the active site of the enzyme was irreversibly modified without impairing the enzyme's ability to bind to lipid-water interface.[73] Nuclear magnetic relaxation[76,77] and X-ray crystallographic studies[43,44] also support such a distinction. A similar conclusion was drawn from studies with the enzyme from *C. atrox*.[78]

X-ray crystallographic studies of the active enzyme and the zymogen have shown that their structures are virtually identical and have revealed the existence of an active site in both forms.[47,79] Furthermore, a hydrophobic region in the molecule that may well serve as the IRS was identified. In the bovine pancreatic enzyme, this region was postulated to comprise the following residues: Leu 2, Trp 3, Asn 6, Glu 17, Leu 19, Leu 20, Asn 23, Asn 24, Leu 31, Lys 56, Val 65, Asn 67, Tyr 69, Thr 70, Asn 72, Lys 116, Asn 117, Asp 119, Lys 120, and Lys 121.[79] A number of these residues were observed to be more mobile in the zymogen (with its seven extra residues, Figure 5) than in the active enzyme.[47,79] The disordered IRS site in the proenzyme may lead to an impaired affinity of the enzyme for the interface, and, consequently, the inability of the zymogen to undergo the required conformational change that optimizes the active site.

Since its initial postulation, there has been a steady stream of evidence in support of the IRS model. Protein modification studies have revealed that a protonated α-amino group at the N-terminal is essential for the hydrolysis of aggregated phospholipids by phospholipase A_2. This requirement for the α-amino group is found in mammalian pancreatic enzymes,[79,80] as well as venom enzymes.[81] In order to delineate the role of the α-amino group, Dijkstra et al.[79] transaminated a number of mammalian pancreatic phospholipases A_2 and examined the catalytic and binding properties of these enzymes, along with their tertiary structure. The modified enzymes retained their ability to hydrolyze monomeric phospholipids but lost their catalytic property towards phospholipid aggregates thus behaving like the zymogen. The structural studies showed the active sites and the Ca^{++} binding region in the modified enzyme were identical to that of the native enzyme. However, a high degree of disorder in residues 1-3 and 63-67 were observed. Interestingly, these residues form part of the postulated IRS residues and were those that had been reported to show a high degree of disorder in the proenzyme.[47,79] Based on these studies the authors concluded that the function of the α-amino group is to immobilize part of the IRS into a conformation that facilitates binding to the interface. This is achieved by hydrogen bonding between the α-amino group of Ala and the carbonyl oxygen of Asn 71. The binding thus results in the conformational change required for hydrolysis of micelles. The involvement of the N-terminal region in the binding of the enzyme to the interface has also recently been implicated in N-terminal modification studies.[82] Transamination of venom enzymes also resulted in a loss of their ability to hydrolyse phospholipid aggregates but not monomers.[3,81]

Further evidence that a distorted IRS prevents the optimization of the active site of the zymogen was obtained in studies with negatively charged detergents and substrates. De Haas and co-workers[31,32,75,83,84] showed that both the proenzyme and the active enzyme bind strongly to negatively charged detergents with very high affinities to form high molecular weight complexes of protein and detergents. With s-n-alkanoylthiolglycol sulfate, a detergent which is also hydrolyzed by the enzyme, they were able to show that below 0.08 × cmc, both the zymogen and active enzyme bound the detergent and showed similar low activities. However, within a narrow range of detergent concentration 0.08 × cmc to 0.12 × cmc, a protein detergent complex comprising of about 6 protein and 40 detergent molecules was formed. This complex formation correlated with an increased hydrolysis of the substrate by the active enzyme. Although the proenzyme also formed similar high molecular weight complexes, there was no increased hyrolysis of the substrate. The authors postulate that the aggregation of the enzyme causes the formation of a pseudomicellar detergent core that results in the increased hydrolysis of the substrate. It should be noted that similar structures are also formed with the proenzyme, and hence, complex formation per se does not lead to increased hydrolysis.

Phospholipase A_2 interaction with the interface was investigated by examining the effect of negative charges at the lipid-water interface either on the substrate or as a nonhydrolyzable component of the interface.[75] Phosphatidylglycerol, which is negatively charged at pH 8, was synthesized from PC and the kinetics of phospholipase A_2 hydrolysis of both phospholipids compared. The effect of negatively charged detergents on the hydrolysis of both substrates was also determined. Higher rates of hydrolysis were obtained with PG than with PC in all cases. The presence of the detergents at the interface resulted in higher hydrolytic rates for both phospholipids. It was concluded that maximum rates of hydrolysis occur at a negatively charged surface. Kinetic analysis suggest that when the charge is on the substrate, the greater rate of reaction is due to both a better binding of the enzyme to the interface and an acceleration of a step in the reaction mechanism after the enzyme has bound to the interface. When the charge is on the detergent, only the improved binding of the enzyme to the interface contributes to the increased hydrolysis.[75]

The interaction of the pancreatic phospholipase A_2 with the interface has also been investigated using DMPMe as substrate[82] From this study, an anionic site on the enzyme is thought to be involved in the interaction with the interface. In contrast to the hydrolysis of phospholipids such as DMPC, DMPMe is hydrolyzed by phospholipase A_2 without any lag phase, even in the absence of additives. The reaction curves obtained were simple and could be described by first order rate equations.[70] The presence of anions in the aqueous phase caused a decrease in the initial rate of hydrolysis of the DMPMe vesicles, but promoted the extent of hydrolysis of the vesicles. These seemingly contradictory observations were explained by the effect of the anions on the residence time of the enzyme at the interface. The anions were postulated to bind to an anionic binding site on the enzyme and weaken the binding of the enzyme to the interface. This results in a decrease in the residence time of the enzyme and effectively reduces the rate of hydrolysis of the substrate. However, by promoting desorption, the anions in effect increase the concentration of the enzyme in the bulk aqueous phase, thus promoting intervesicular exchange and the amount of substrate hydrolyzed.[82,85] These observations led to the supposition that both electrostatic and hydrophobic interactions contribute to the formation of the interfacial complex required to generate the conformational change necessary for the observed increase in reaction rates by the pancreatic phospholipase A_2. There is currently no direct evidence for the postulated conformational change.

Recent work of Jain et al.[18,70,82,85] on the hydrolysis of DMPMe provides further support of the IRS model. The major difference between the scheme of these investigators and the IRS model was in the proposed rate-limiting step. In the original model, this step was postulated to be the interaction of the enzyme with the interface and the formation of the primed enzyme ($E \leftrightarrow E^*$); Jain et al.[70,85] suggest that the rate-limiting step is the formation of the E*S complex ($E^* + S \leftrightarrow E^*S$) because under the appropriate conditions, binding of the enzyme to the interface is not rate limiting.

VIII. DUAL PHOSPHOLIPID MODEL (DPM)

This kinetic model for phospholipase A_2 hydrolysis has been postulated by Dennis and co-workers.[5,86-89] In contrast to the work of de Haas, this theory has been developed mainly with the venom enzyme from *N. naja naja* with Triton X-100/phospholipid mixed micelles as substrate. The DPM model stems from the enzyme theory which assumes a change in the enzyme on interacting with the lipid-water interface. It also incorporates the concentration effect. It was shown that, as the concentration of detergent increased relative to that of the phospholipid, there was a decrease in enzyme activity even though the bulk concentration was held constant.[5,90,91] The observation was interpreted to indicate a dilution of the phospholipid molecules at the interface by the detergent. Therefore, the rate of hydrolysis of the cobra enzyme is dependent not only on the bulk substrate but also on the concentration at the interface.

Another observation which led to the DPM model was the putative half site reactivity of the cobra phospholipase A_2.[92] Although this supposition has been subsequently withdrawn,[93] other evidence[66,67,89,94,95] suggesting two distinct substrate sites in the enzyme has been used in support of this theory. The DPM model proposes that two phospholipid molecules are required for a single cycle of enzyme catalysis. The monomers must first bind Ca^{++} before they can bind the phospholipid.[5] Subsequently the binding of one phospholipid molecule results in sequestration of the enzyme to the interface followed by the binding of another substrate to the functional active site that results in hydrolysis. Within this framework of the requirement for two phospholipid molecules, several variations of the DPM model have been proposed.[5] The current view is that the monomeric enzyme aggregates to a dimer or higher aggregated form subsequent to the binding of the first phospholipid and prior to catalysis (Figure 9). The existence of an asymmetric aggregated enzyme will circumvent the difficulty of assigning two substrate sites to a small enzyme with a molecular weight of only 11 kDa.[40]

Support for the existence of two sites came from studies which demonstrated that the hydrolysis of PE and other phospholipids were activated by a host of compounds such as PC, SM, LPC, and oleic acid.[66,67,89,94,95] These studies also showed that substrate activators like PC could protect the active site from irreversible inactivation, while the nonsubstrate activators such as LPC and SM could not.[66] It was concluded that there were two sites in the molecule, an activator site (which required a lipid containing the phosphorylcholine group and a fatty acid) and a catalytic site. The ability of the enzyme to interact with single phospholipid molecules was implicated by the activation of phospholipase A_2 hydrolysis of PE by dibutyryl-PC.[5,94] Dibutyryl-PC was assumed to exist solely as a monomer which was not sequestered in the micelle. However, this assumption was questioned[2] and it was subsequently acknowledged[89,96] that dibutyryl-PC may well be incorporated into micelles albeit at very high concentrations.

Evidence for the existence of an active dimeric or aggregated enzyme includes the report by Roberts et al.[86] In the presence of cations (Ba^{++}) and phospholipid, *N. naja naja* phospholipase A_2 was cross linked with suberimidate to form both dimers and trimers. In another study,[71] the enzyme was immobilized on agarose gel and its characteristics compared with the soluble enzyme. The binding of the soluble to the immobilized enzyme was investigated. Immobilization of the enzyme did not affect the active site of the enzyme or its binding to micelles. Binding of monomeric enzyme to immobilized enzyme in the presence of dodecylphosphorylcholine, an activator of phospholipase A_2, was observed. Binding also occurred in the presence of Zwittergent 3-14 which indicates that the binding was probably nonspecific. In view of the nonspecificity of the aggregation, the conclusion that interaction of cobra venom phospholipase A_2 with activators leads to specific aggregation (dimerization) appears to be quite speculative. There was no evidence to indicate if these aggregates hydrolyzed micelles at higher rates than the immobilized enzyme. The aggregation presumably removed the one constraint that was postulated to prevent aggregation of the immobilized enzyme. Perhaps the best evidence for the formation of aggregates subsequent to the binding of one substrate comes from the report of Pluckthun and Dennis.[97] When n-dodecyl phosphorylcholine (an activator) was included in the elution buffer in gel filtration chromatography, a marked increase in the apparent molecular weight of the enzyme was identified.

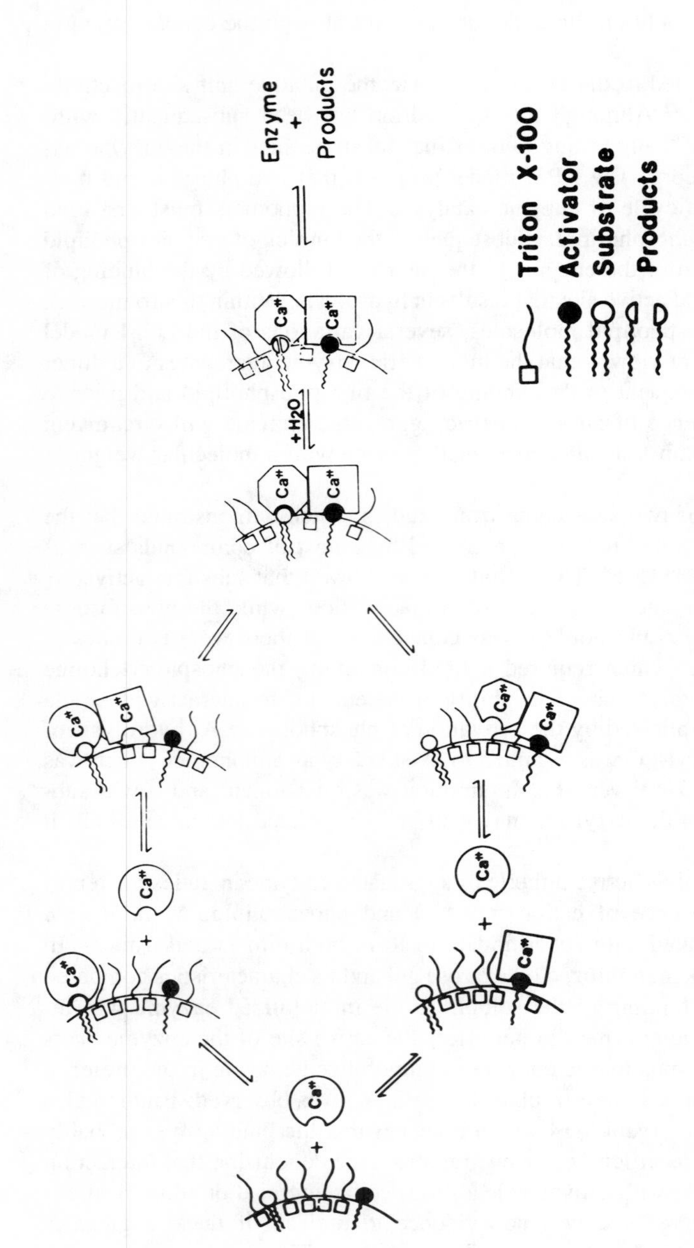

FIGURE 9. Proposed two-site dimer of the DPM model for *N. naja naja* phospholipase A_2. The activator monomer (square) or the inactive catalytic monomer (circle) can bind its phospholipid with subsequent binding of and interaction with the other monomer, thus activating the catalytic monomer (octagon). (From Dennis, E. A., *Enzymes*, 16, 308, 1983. With permission.)

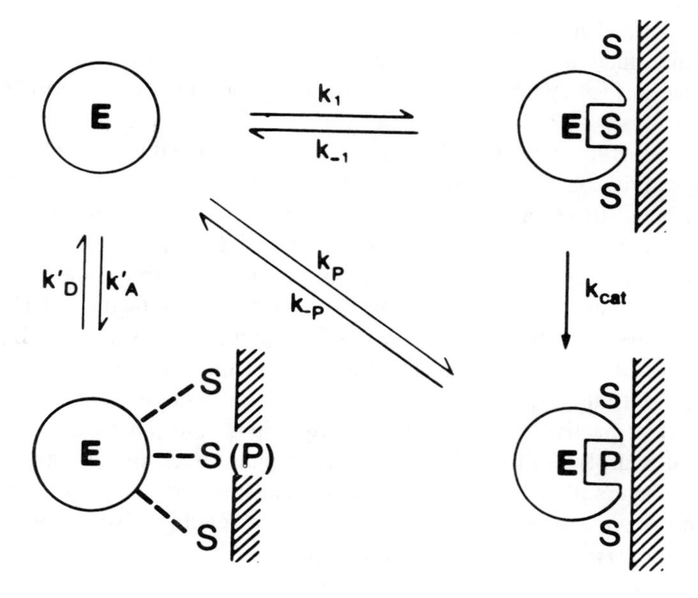

FIGURE 10. The Tinker model for the hydrolysis of lipid (S) at the surface of an insoluble phase catalysed by a water soluble enzyme (E). (P) represents the product of the reaction. (From Tinker, D. O., Low, R., and Lucassen, M., *Can. J. Biochem.*, 58, 898, 1980. With permission.)

Unfortunately, this aggregation effect was not produced by dibutyryl-PC, an activator of the enzyme.

IX. TINKER MODEL

This model was proposed by Tinker and co-workers[51,78,98] to explain the hydrolysis of saturated phosphatidylcholine (DMPC and DPPC bilayers) by *C. atrox* phospholipase A_2. The model is based largely on the enzyme theory; however, the products of the reaction are postulated to play a major role in the progress of the reaction. DMPC and DPPC bilayers undergo thermotropic phase changes with well-defined transition points, and the kinetics of hydrolysis of both phases were utilized in the development of this model. The progress of the reaction with the gel phase differed from those obtained with the liquid crystalline phase. The reaction profiles of the liquid crystalline phase hydrolysis were characterized by an initial burst of activity followed by a long lag period of very slow reaction and a subsequent dramatic increase in the rate of hydrolysis. The lag phase could be abolished by the addition of LPC or fatty acid (10 mol%). Hydrolysis of the gel phase followed a hyperbolic function which was dependent on the bulk lipid concentration. At the pretransition temperature ($\beta \rightarrow \beta'$), a large increase in the rate of hydrolysis was observed without a change in the affinity of the enzyme for the substrate (Km). The gel phase was also reported to be hydrolyzed at a higher rate compared to the liquid crystalline phase. When both the liquid crystalline and gel phases were present in bilayers, the liquid crystalline phase was preferentially hydrolyzed.[51] Based on gel filtration binding studies, Tinker et al.[78] reported that adsorption of the enzyme to the lipid surface was not dependent on the presence of catalytic activity nor on the binding of Ca^{++}.

The above observations led to the postulation of the model shown in Figure 10. In this model, the enzyme binds a lipid molecule to form an ES complex, then undergoes a conformational change to expose the hydrophobic residues that allow the complex to bind the interface, followed by catalysis. A reversible adsorbtion of the free enzyme at the interface was later incorporated into the model in order to account for an observed binding of the enzyme to the interface in the absence of catalysis.[78] After hydrolysis, the enzyme can either desorb from the surface (hopping) or it can diffuse along the surface to another substrate (scooting).

The high rates of hydrolysis of the gel phase was attributed to the hopping mechanism, while the low activity observed during the lag phase of hydrolysis of the liquid crystalline phase was attributed to scooting.[51] The rationale was that the higher viscosity of the lipid impedes the movement of the enzyme in the plane of the bilayer (scooting), making it a much slower process than desorption and binding (hopping) through the aqueous media. In addition, it was suggested that the tight packing of the lipids in the gel phase would not permit scooting. The initial burst of activity observed in the hydrolysis of the liquid crystalline phase was explained as being due to the hydrolysis of the ES complex as it settled on the surface.[51] After settling, a lag phase occurs which represents a stage where the adsorbed enzyme species is "inactive" and the rate-limiting step is desorption from the lipid interface. The preferential hydrolysis of the liquid crystalline phase over the gel phase when both are present in the bilayers also contributed to delineating the scooting and hopping pathways. In scooting, the enzyme can be considered as being irreversibly adsorbed to the lipid bilayer while this is obviously not the situation with hopping. Although the enzyme hydrolyzes the gel phase substrate at greater rates, it is this "irreversible" binding to the liquid crystalline phase that is responsible for the observed preferential hydrolysis.

The effect of the hydrolytic products (LPC and fatty acids) which stimulate the reaction was explained as being due to product-stimulated desorption.[78] When the enzyme cannot encounter substrates, it undergoes a reverse conformational change in which the hydrophobic areas are shielded, resulting in desorption of the enzyme from the interface back into the solution and a shift towards the hopping pathway with its higher rates of hydrolysis. Whereas fatty acids stimulate both the initial burst and the subsequent lag, LPC inhibits the initial burst but stimulates desorption. The composite effect of LPC and fatty acids was to slightly inhibit the initial burst and stimulate the subsequent slow reaction.[78] In summary, the basic features of the Tinker model are the formation of an ES complex prior to binding to the interface. The binding results in catalysis and the stimulation of desorption of the enzyme by the reaction products in turn causes higher rates of hydrolysis via the hopping mechanism.

X. COMPARISON OF THE THREE KINETIC MODELS

The three kinetic models of phospholipase A_2 activity have been obtained primarily with different combinations of substrates and enzyme. De Haas and associates have worked mainly with micelles and pancreatic phospholipase A_2, while Dennis' group utilize mixed micelles of Triton X-100 and phospholipids with the cobra enzyme and Tinker and co-workers have utilized bilayers of disaturated phospholipids and the *Crotalus* enzyme. The unifying feature of all three models is their primary reliance on the enzyme theory to explain the interfacial activation. We shall first compare the Tinker model with the interface recognition site model since both were derived from assays that did not include detergents. Subsequently, these two will be compared with the dual phospholipid model.

XI. IRS MODEL VS TINKER MODEL

One major difference between the IRS and the Tinker model is the form of the enzyme that binds at the interface. In the IRS model, the enzyme interacts with the interface, prior to forming an ES complex. In the Tinker model, the enzyme forms an ES complex prior to binding to the interface. The postulation that an ES complex is formed prior to binding stems essentially from the pre-steady-state burst of activity observed with liquid crystalline bilayers. Such an event was not predicted by the IRS model where a lag phase would be expected.[51] Indeed such a lag phase has been observed for the hydrolysis of pure annealed DMPC vesicles or bilayers by porcine pancreas phospholipase A_2.[53,99] This lag phase could be abolished by the addition of lysophospholipids.[53,99,100] Fluorescent enhancement studies showed that the enzyme was bound to the vesicles under these conditions.[53] When the diacylphospholipid was co-dispersed with the lysophospholipid there was no hydrolysis or binding of the enzyme to the substrate.[97,100] These observations support the concept that phospholipase A_2 does not readily bind the bilayers, but the introduction of defects or cracks may enhance the binding and consequently the hydrolysis of the substrate. This concept was advanced by Upreti and Jain[52] following their studies on the hydrolysis of bilayers by bee venom phospholipase A_2. The early burst of hydrolysis depends on factors that modulate the organization of the bilayer. The extent of hydrolysis correlated with increasing disorder in the bilayer induced by factors such as sonication, osmotic shock, or phase transition. The site of action of the enzyme in the pre-steady-state phase appears to be the defective sites in the bilayer structure; therefore, the formation of an ES complex may not be required to explain the initial burst of hydrolysis. That an ES complex is not required for binding to the interface is indeed conceded by Tinker et al.[78] In their evaluation of the Tinker model, Slotboom et al.[3] allude to the unusual affinity of phospholipase A_2 to all types of surfaces. Such affinity makes the requirement of an ordered ES complex for binding superflous.

In the IRS model, a conformational change that optimizes the active site when the enzyme binds to the interface is postulated as being responsible for the enhanced hydrolysis. Once bound, the enzyme is not expected to desorb and scooting is the expected mode of engaging new substrate. In the Tinker model, accelerated hydrolysis is due to a hopping of the enzyme from the interface to the bulk solution subsequent to each catalytic cycle and back to the interface to engage new substrate. Prolonged stay at the interface (scooting) yields lower activities. While there is no direct evidence for the postulated change in conformation of the enzyme (as in the IRS model), it has been demonstrated that interfacial binding stimulates the inhibition of phospholipase A_2 activity by p-bromophenacylbromide, a compound that reacts with His 48 at the active site of the enzyme.[73] This observation suggests binding of the enzyme to the interface provides a better access to the active site of the enzyme. Recent studies by Jain and colleagues[17,70,82,85] have provided compelling evidence that scooting rather than hopping is the primary mode by which the pancreatic phospholipase A_2 engages new substrate. These studies indicate that an enzyme bound to DMPMe vesicle can hydrolyse all the substrate molecules of the outer monolayer without desorption. After hydrolyzing the outer monolayer, the enzyme does not desorb to engage new substrate in the absence of additives that promote desorption.[70] Slotboom et al.[3] have also pointed out that the postulated slow diffusion of the enzyme within the plane of the bilayer does not take into account the high rate of diffusion of free phospholipid molecules within the plane of the bilayer.

In the Tinker model, the effective hydrolysis of phospholipids in the gel phase, along with the accelerated hydrolysis due to the presence of reaction products, are explained as being due to product induced desorption from the interface. The IRS model explains these as the results of product-facilitated adsorbtion due to defects introduced into the interface by the products. Direct binding studies indicate that the presence of LPC and fatty acids promotes binding.[53,99,100] Jain et al.[70] found little if any free enzyme in the bulk aqueous phase regardless of the presence of product-containing vesivles. With substrates such as DMPC, in the absence of any additives or cracks, the observed lag phase is probably due to the inability of the enzyme to bind the substrate. Indeed, it is quite interesting to note that in binding studies, Tinker et al.[78] also observed that a higher fraction of the enzyme bound to the DMPC bilayers in the presence of 10 mol% reaction products. This was attributed to an increase in surface area of lipid exposed due to the presence of these products. Hence, the available evidence does not seem to support the product enhanced desorption proposed by Tinker.

XII. DUAL PHOSPHOLIPID MODEL VS. OTHER MODELS

Difficulties in comparing the DPM with other models stem from the fact that most of the studies that have been used to develop the DPM model involved Triton X-100/phospholipid micelles while the others were developed mainly without detergents. Although Triton X-100 is not a competitive inhibitor of the phospholipid molecules,[5] one cannot rule out the possibility of its interference in the observed hydrolysis through its effect on the ''quality'' of the interface.[2,3]

The basic difference between the DPM and other models is the postulated binding of an activator to an enzyme monomer at the interface resulting in a conformational change that allows the binding of another monomer to form the active dimer or even higher aggregates. Presently, there is no unambiguous proof that aggregated *N. naja naja* enzyme rather than its monomers are the active species. There is, however, ample evidence to indicate that the monomers of the pancreatic phospholipase A_2 are the active species and there is no requirement for a dimer.[70,85] It appears that aggregation per se does not lead to activation as has been demonstrated by De Haas and co-workers.[31,32,75]

In one of the few comparative studies, Pluckthun and Dennis[97] compared the properties of a number of phospholipases A_2 by the Triton X-100/phospholipid mixed micelle system. The pancreatic enzyme was activated by fatty acids which is in agreement with previous reports, but the reactions of both the *N. naja naja* and *C. adamenteus* enzymes in the Triton X-100 system were inhibited by fatty acids. This is in contrast to the observation[78] that fatty acids stimulated the hydrolysis of DPPC by *C. atrox*. In addition, Pluckthun and Dennis[97] did not observe any effect of LPC on the lag period of any of the enzymes, contrary to those obtained by other groups.[53,78] While such discrepancies could be due to the mode of addition of LPC,[53] a direct effect of Triton X-100 on the reaction cannot be ruled out. It was suggested that fatty acids (and bile salts) may activate the pancreatic phospholipase A_2 and a parallel could be drawn with the activation of cobra enzyme by phosphorylcholine-containing compounds.[97] However, current evidence[31,32,75,83] suggests fatty acids and negatively charged detergents are more likely to affect the reaction by facilitating the binding of the enzyme to the interface rather than a direct binding to the enzyme. Hydrolysis of phospholipids by the pancreatic enzyme can proceed without a lag phase in the absence of additives.[18,53,70,78,82] Furthermore, aggregation of the pancreatic phospholipase A_2 zymogen does not lead to activation.[31,32,75] Therefore, the differences between the two enzymes appears to be more than one between the chemical nature of their activators. It is interesting that the results of Pluckthun and Dennis[97] obtained for the pancreatic enzyme using Triton X-100/phospholipid mixed micelles can be accomodated by the IRS model, whereas those obtained with the cobra enzyme cannot. Based on these observations, there is a distinct possibility that different phospholipases may not have the same kinetic mechanism.

XIII. CATALYTIC MECHANISM

The catalytic mechanism of phospholipase A_2 was largely elucidated by X-ray crystallography studies. The majority of the work has been focused on the bovine pancreatic enzyme and the subject has been recently reviewed.[3,5,10] Analysis of the structure at 2.4Å resolution indicates that the active site is in a depression at the molecular surface. The essential residue (His 48), which is directly involved in the catalysis, is located in a cleft in this region flanked by the absolutely conserved side-chains of Asp-49, Tyr-52 and Asp-99. The catalytic mechanism is illustrated in Figure 11. A water molecule about 3 Å away from the N-1 nitrogen

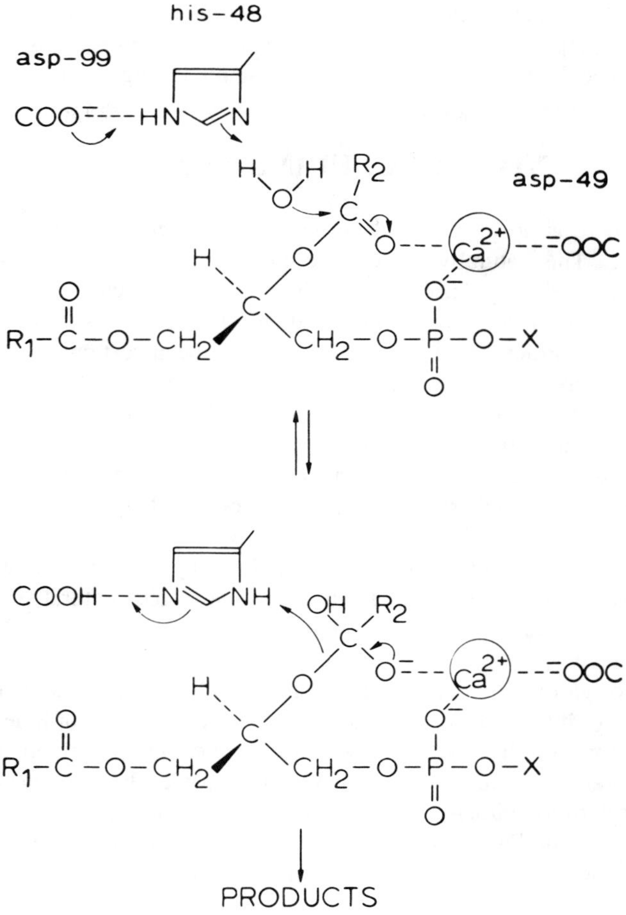

FIGURE 11. Proposed catalytic mechanism of phospholipase A_2. (From Slotboom, A. J., Verheij, H. M., and de Haas, G. H., *Phospholipids,* Hawthorne, A. J. and Ansell, G. B., Eds., Elsevier/North-Holland, Amsterdam, 1982, 359. With permission.)

of His-48 initiates the nucleophilic attact on the substrate carbonyl carbon atom. The imidazole ring of His-48 gains a proton from the water and the proton is subsequently donated to the alkoxy oxygen. The hydrolytic mechanism is analogous to the action of serine esterases.[3]

Since phospholipase A_2 exhibits higher activity towards aggregates, Dijkstra et al.[43] postulate that the enzyme has its active site directed towards the micellar surface with three distinct regions of interaction. One region comprises residues in a ring surrounding the entrance to the active site, and the other two regions are at the extremes of the molecule. A number of residues, including those at positions 3, 6, 19, 31, and 69, have been implicated in the interaction with micellar substrates. Although none of these residues are conserved, it appears that they are largely basic. This suggests that salt bridges may be formed with the phosphate groups in the micelle which would orient the enzyme with respect to the micelle surface. Studies with X-ray crystallography and NMR reveal that phospholipid molecules in micelles have a unique conformation.[42] The main features are parallel and extended fatty acid chains with a kink in the second chain at the C-2 position (see section on *glycerophospholipids*). The shape of the enzyme active site is such that a substrate molecule in this conformation can easily enter the active site and approach His-48 with its scissile bond. The hypothesis utilizes a second entropic advantage besides the orientation of the enzyme molecule in order to explain the substantially higher enzyme activity towards micelles than monomers.

XIV. CONCLUDING REMARKS

At present, there is no general concensus on the origin of the interfacial activation of phospholipases A_2. This naturally leads to the question as to which one of the three kinetic models, if any, can best explain the mechanism of phospholipases A_2. Before addressing this point, one needs to determine whether there should be a single mechanism for all phospholipase A_2 regardless of their origin. Most of the investigators in the field do not think the idea of different phospholipase A_2 having different mechanisms a tenable one.[2,3,5,97] This is due to the fact that, regardless of the source of the enzyme, they all catalyze identical reactions and are remarkably similar with respect to both their primary and tertiary structures. If all phospholipase A_2 act according to one single kinetic model, it is not unreasonable to expect a convergence of the experimental results obtained by various investigators. The results to date do not suggest such a convergence. While this lack of convergence may due to the variety of enzymes and the assay systems used, it may also signify that all phospholipase A_2 do not share a common kinetic mechanism, despite their structural similarities. In order to resolve this issue, more comparative studies with enzymes from different sources are needed. The protocol of Jain et al.[70] appears to be a useful one for such studies. In their assay system, no additives are required to promote the reaction, and the simple kinetic profiles obtained from the reaction should allow the determination of kinetic constants. The system has been successfully used to show that the pancreatic phospholipase A_2 is active as a monomer. The same system can be applied to studies with the Naja enzyme. Such studies should yield useful information as to whether the enzyme is active as monomers or aggregates — a central point of the DPM model. Without comparative studies, the validity of any of the proposed kinetic models outside the system used in its development cannot be readily discerned.

ACKNOWLEDGMENT

This work was supported by the Manitoba Heart Foundation. We thank L. Page for proofreading the manuscript.

ABBREVIATIONS

PA-phosphatidic acid; PC-phosphatidylcholine; PE-phosphatidylethanolamine; PG-phosphatidylglycerol; PI-phosphatidylinositol; PM-phosphatidylmethanol; PS-phosphatidylserine; SM-sphingomyelin; GPL-glycerophospholipid. LPC-lysophosphatidylcholine; DMPC-dimyristoylphosphatidylcholine; DPPC-dipalmitoylphosphatidylcholine; and DMPMe-dimyristoylphosphatidylmethanol.

REFERENCES

1. **Verger, R.,** Enzyme kinetics of lipolysis, *Methods Enzymol.,* 64, 340, 1980.
2. **Verheij, H. M., Slotboom, A. J., and de Haas, G. H.,** Structure and function of phospholipase A_2, *Rev. Physiol. Biochem. Pharmacol.,* 91, 91, 1981.

3. **Slotboom, A. J., Verheij, H. M., and de Haas, G. H.,** On the mechanism of phospholipase A_2, in *Phospholipids,* Hawthorne, J. N. and Ansell, G. B., Eds., Elsevier/North-Holland, Amsterdam, 1982, 359.
4. **Dennis, E. A., Darke, P. L., Deems, R. A., Kensil, C. R., and Pluckthun, A.,** Cobra venom phospholipase A_2: a review of its action toward lipid/water interfaces, *Mol. Cell Biochem.* 36, 37, 1981.
5. **Dennis, E. A.,** Phospholipases, *Enzymes,* 16, 308, 1983.
6. **Volwerk, J. J. and de Haas, G. H.,** in *Lipid Protein Interactions,* Vol. 1, Jost, P. C. and Griffith, O. H., Eds., John Wiley & Sons, New York, 1982, 69.
7. **Heller, M.,** Phospholipase D, *Adv. Lipid Res.,* 16, 267, 1978.
8. **Van den Bosch, H.,** Intracellular phospholipase A, *Biochim. Biophys. Acta,* 604, 191, 1980.

8a. **Van den Bosch, H.,** Phospholipases, in *Phospholipids,* Hawthorne, J. N. and Ansell, G. B., Eds., Elsevier/North-Holland, Amsterdam, 1982, 313.

8b. **Hostetler, K. Y.,** The role of phospholipases in human diseases, in *Phospholipids and Cellular Regulation,* Vol. 1, Kuo, J. F., Ed., CRC Press, Boca Raton, FL, 1985, 181.

9. **Cullis, P. R., Hope, M. J., de Kruijff, B., Verkleij, A. J., and Tilcock, C. P. S.,** Structural properties and functional roles of phospholipids in biological membranes, in *Phospholipids and Cellular Regulation,* Kuo, J. F., Ed., CRC Press, Boca Raton, FL, 1985, 1.
10. **Chapman, D.,** Phase transitions and fluidity characteristics of lipids and cell membranes, *Q. Rev. Biophys.,* 8, 185, 1975.
11. **Waite, M.,** Phospholipases, in *Biochemistry of Lipids and Membranes,* Vance, D. E. and Vance, J. E., Eds., Benjamin/Cummings, Menlo Park, CA, 1985, 299.
12. **Hitchcock, P. B., Mason, R., Thomas, K. M., and Shipley, G. G.,** Structural chemistry of 1,2 dilauroyl-DL-phosphatidylethanolamine: molecular conformation and intermolecular packing of phospholipids, *Proc. Natl. Acad. Sci. U.S.A.,* 71, 3036, 1974.
13. **Pearson, R. H. and Pasher, I.,** The molecular structure of lecithin dihydrate, *Nature,* 281, 499, 1979.
14. **Roberts, M. F., Bothner, A. A., and Dennis, E. A.,** Magnetic nonequivalence within the fatty acyl chains of phospholipids in membrane models: 1H nuclear magnetic resonance studies of the a methylene groups, *Biochemistry,* 17, 935, 1978.
15. **De Bony, J. and Dennis, E. A.,** Magnetic nonequivalence of the two fatty acyl chains in phospholipids of small unilammelar vesicles and mixed micelles, *Biochemistry,* 20, 5256, 1981.
16. **Zografi, G., Verger, R., and de Haas, G. H.,** Kinetic analysis of the hydrolysis of lecithin monolayers by phospholipase A, *Chem. Phys. Lipids,* 7, 185, 1971.
17. **Verger, R. and de Haas, G. H.,** Interfacial enzyme kinetics of lipolysis, *Annu. Rev. Biophys. Bioeng.,* 5, 77, 1976.
18. **Jain, M. H., Rogers, J., Maracek, J. F., Ramirez, F., and Eibl, H.,** Effect of the structure of phospholipid on the kinetics of intravesicle scooting of phospholipase A_2, *Biochim. Biophys. Acta,* 860, 462, 1986.
19. **Thuren, T., Virtanen, J. A., Verger, R., and Kinnunen, P. K. J.,** Hydrolysis of 1-palmitoyl-2-[6-(pyren-1-yl)]hexanoyl-sn-glycero-3-phospholipids by phospholipase APV2PV: effect of the polar head group, *Biochim. Biophys. Acta,* 917, 411, 1987.
20. **DeBose, C. D., Burns, Jr., R. A., Donovan, J. M., and Roberts, M. F.,** Methyl branching in short-chain lecithins: are both chains important for effective phospholipase A_2 activity?, *Biochemistry,* 24, 1298, 1985.
21. **Wykle, R. L., Kraemer, W. F., and Schremmer, J. M.,** Specificity of lysophospholipase D, *Biochim. Biophys. Acta,* 619, 58, 1980.
22. **Okumura, T., Kimura, S., and Saito, K.,** A novel purification procedure for *Penicillum notatum* phospholipase B and evidence for a modification of phospholipase B activity by the action of an endogenous protease, *Biochim. Biophys. Acta,* 617, 264, 1980.
23. **Barenholz, Y. and Gatt, S.,** Sphingomyelin metabolism, chemical synthesis, chemical and physical properties, in *Phospholipids,* Hawthorne, J. N. and Ansell, G. B., Eds., Elsevier/North-Holland, Amsterdam, 1982, 129.
24. **Mansbach, C. M., Jr., Pieroni, G., and Verger, R.,** Intestinal phospholipase, a novel enzyme, *J. Clin. Invest.,* 69, 368, 1982.
25. **Verger, R., Ferrato, F., Mansbach, C. M., Jr., and Pieroni, G.,** Novel intestinal phosphoilipase A_2: purification and some molecular characteristics, *Biochemistry,* 21, 6883, 1982.
26. **Dufton, M. J. and Hider, R. C.,** Classification of phospholipase A_2 according to sequences. Evolutionary and pharmacological implications, *Eur. J. Biochem.,* 137, 545, 1983.
27. **Van Scharrenburg, G. J. M., de Haas, G. H., and Slotboom, A. J.,** Regeneration of full enzymatic activity by reoxidation of reduced pancreatic phospholipase A_2, *Hoppes Seylers' Z. Physiol. Chem.,* 361, 571, 1980.
28. **Heinrikson, R. L., Krueger, E. T., and Keim, P. S.,** Amino acid sequence of phospholipase A_2 from the venom of crotalus adamenteus. A new classification of phospholipase A_2 based upon structural determinants, *J. Biol. Chem.,* 252, 4913, 1977.

29. **Maragamore, J. M. and Heinrikson, R. L.**, The Lysine-49 phospholipase A_2 from the venom of *Agkistrodon piscivorus piscivorus, J. Biol. Chem.*, 261, 4797, 1986.
30. **Soares de Araujo, P., Rossenen, M. Y., Kremer, J. M. A., Van Zoolen, E. J. J., and de Haas, G. H.**, Structure and thermodynamic properties of complexes between phospholipase A_2 and lipid micelles, *Biochemistry*, 18, 580, 1979.
31. **Hille, J. D. R., Egmond, M. R., Dijkman, R., van Oort, M. G., Jirgensons, B., and de Haas, G. H.**, Aggregation of porcine pancreatic phospholipase A_2 and its zymogen induced by submicellar concentrations of negatively charged detergents, *Biochemistry*, 22, 5347, 1983.
32. **Hille, J. D. R., Egmond, M. R., Dijkman, R., van Oort, M. G., Sauve, P., and de Haas, G. H.**, Unusual kinetic behaviour of porcine pancreatic (pro) phospholipase A_2 on negatively charged substrates at submicellar concentrations, *Biochemistry*, 22, 5353, 1983.
33. **Renetseder, R., Brunei, S.,Dijkstra, B. W., Drenth, J., and Sigler, P. B.**, A comparison of the crystal structures of phospholipase A_2 from bovine pancreas and *Crotalus atrox* venom, *J. Biol. Chem.*, 260, 11627, 1985.
34. **Brunie, S., Bolin, J., Gewirth, D., and Sigler, P. B.**, The refined crystal structure of dimeric phospholipase A_2 at 2.5 Å, *J. Biol. Chem.*, 260, 9742, 1985.
35. **Smith, C. M. and Wells, M. A.**, A further examination of the active form of *Crotalus adamenteus* phospholipase A_2, *Biochim. Biophys. Acta*, 663, 687, 1981.
36. **Wells, M. A. and Hanahan, D. J.**, Studies on phospholipase A. I. Isolation and characterisation of two enzymes from crotalus adamenteus venom, *Biochemistry*, 8, 414, 1971.
37. **Hachimora, Y., Wells, M. A., and Hanahan, D. J.**, Observations on phospholipase A2 of *Crotalus atrox* molecular weight and other properties, *Biochemistry*, 10, 4084, 1971.
38. **Joubert, F. J. and van der Walt, S. J.**, *Naja melanoleuca* (forest cobra) venom purification and some properties of phospholipase, *Biochim. Biophys. Acta*, 379, 317, 1975.
39. **Yang, C. C. and King, K.**, Chemical modification of the Histidine residue in basic phospholipase A_2 from the venom of *Naja nigcricollis, Biochim. Biophys. Acta*, 614, 373, 1980.
40. **Deems, R. A. and Dennis, E. A.**, Characterisation of the major form of cobra venom *(Naja naja naja)* phosphjolipase A_2 that has a molecular weight of 11,000, *J. Biol. Chem.*, 250, 9008, 1975.
41. **Adamich, M., Roberts, M. F., and Dennis, E. A.**, Phospholipid activation of cobra venom phospholipase A_2. II. Characterisation of the phospholipid enzyme interaction, *Biochemistry*, 18, 3308, 1979.
42. **Dijkstra, B. W., Drenth, J., Kalk, K. H., and Vandermaelen, P. J.**, Three dimensional structure and disulfide bond connections in bovine pancreatic phospholipase A_2, *J. Mol. Biol.*, 124, 53, 1978.
43. **Dijkistra, B. W., Drenth, J., and Kalk, K. H.**, Active site and catalytic mechanism of phospholipase A_2, *Nature*, 289, 604, 1981.
44. **Dijkstra, B. W., Kalk, K. H., Hol, W. G. J., and Drenth, J.**, Structure of bovine pancreatic phospholipase A_2 at 1.72Å resolution, *J. Mol. Biol.*, 147, 97, 1981.
45. **Keith, C., Feldman, D. S., Deganello, S., Glick, J., Ward, K. B., Jones, E. B., and Sigler, P. B.**, A crystal structure of a dimeric phospholipase A_2 from the venom of *Crotalus atrox, J. Biol. Chem.*, 256, 8602, 1981.
46. **Dijkstra, B.W., van Nes, G. J. H., Brendenburg, N. P., Hol, W. G. J., and Drenth, J.**, The structure of bovine pancreatic prophospholipase A_2 at 3.0Å resolution, *Acta Cryst. Sect. B*, 38, 793-799, 1982.
47. **Dijkstra, B. W., Renetseder, R., Kalk, K. H., Hol, W. G. H., and Drenth, J.**, Structure of porcine pancreatic phospholipase A_2 at 2.6Å resolution and comparison with the bovine phospholipase A_2, *J. Mol. Biol.*, 168, 163, 1983.
48. **Dijkstra, B. W., Weijer, W. J., and Wierenga, R. K.**, Polypeptide chains with similar amino acid sequences but a distinctly different conformation, *FEBS Lett.*, 164, 25, 1983.
49. **Maragamore, J. M., Merutka, G., Cho, W., Welches, W., Kezdy, F. J., and Heinrikson, R. L.**, A new class of phospholipase A_2 with a Lys in place of Asp 49, *J. Biol. Chem.*, 259, 13839, 1984.
50. **Maragamore, J. M. and Heinrikson, R. L.**, The role of lysyl residues of phospholipase A_2 in the formation of the catalytic complex, *Biochem. Biophys. Res. Commun.*, 131, 129, 1985.
51. **Tinker, D. O., Purdon, A. D., Wei, J., and Mason, E.**, Kinetics of hydrolysis of dispersions of saturated phosphatidylcholine by *Crotalus atrox* phospholipase A_2, *Can. J. Biochem.*, 56, 552, 1978.
52. **Upreti, G. C. and Jain, M. K.**, Action of phospholipase A_2 on unmodified phosphatidylcholine bilayers: organisational defects are preferred sites of action, *J. Membr. Biol.*, 55, 113, 1980.
53. **Jain, M. K. and de Haas, G. H.**, Activation of phospholipase A_2 by freshly added lysophospholipids, *Biochim. Biophys. Acta*, 736, 157, 1983.
54. **Verger, R., van Dam-Mieras, M. C. E., and de Haas, G. H.**, Action of phospholipase A_2 at interfaces, *J. Biol. Chem.*, 248, 4023, 1973.
55. **Verger, R., Rietsch, J., van Dam Mieras, M. C. E., and de Haas, G. H.**, Comparative studies of lipase and phospholipase A_2 acting on substrate monolayers, *J. Biol. Chem.*, 251, 3128, 1976.
56. **Pattus, F., Slotboom, A. J., and de Haas, G. H.**, Regulation of phospholipase A_2 activity by the lipid water interface: a monolayer approach, *Biochemistry*, 18, 2691, 1979.

57. **Pattus, F., Slotboom, A. J., and de Haas, G. H.,** Regulation of the interaction of pancreatic phospholipase A_2 with lipid-water interface by Ca^{2+} ions: a monolayer study, *Biochemistry,* 18, 2698, 1979.
58. **Pattus,F., Slotboom, A. J., and de Haas, G. H.,** Amino acid substitutions of the NH2-terminal Ala of porcine pancreatic phospholipase A_2: a monolayer study, *Biochemistry,* 18, 2703, 1979.
59. **Wells, M. A.,** A kinetic study of the phospholipase A_2 *(Crotalus adamenteus)* catalysed hydrolysis of 1,2, dibutyryl-sn-glycero-3-phosphorylcholine, *Biochemistry,* 11, 1030, 1972.
60. **Zhelkovskii, A. M., Dyakov, V. D., Ginodman, L. M., and Antonov, V. K.,** Active centre of phospholipase A_2 from the venom of the central Asian cobra. A catalytically active carboxyl group, *Bioorgh. Khim.,* 4, 1665, 1978.
61. **De Haas, G. H., Bonsen, P. P. M., and Pieterson, W. A., and van Deenen, L. L. M.,** Studies on phospholipase A and its zymogen from porcine pancreas. III. Action of the enzyme on short chain lecithins, *Biochim. Biophys. Acta,* 239, 252, 1971.
62. **Wells, M. A.,** The mechanism of interfacial activation of phospholipase A_2, *Biochemistry,* 13, 2248, 1974.
63. **Wilschut, J. C., Regts, J., Westenberg, H., and Scherphof, G.,** Hydrolysis of phosphatidylcholine liposomes by phospholipase A_2. Effects of local anaesthetic dibucaine, *Biochim. Biophys. Acta,* 433, 20, 1976.
64. **Wilschut, J. C., Regts, J., Westenberg, H., and Scherphof, G.,** Action of phospholipase A_2 on phosphatidylcholine bilayers, *Biochim. Biophys. Acta,* 508, 185, 1978.
65. **Roberts, M. F., Otaness, A.-B., Kensil, C. A., and Dennis, E. A.,** The specificity of phospholipase A_2 and phospholipase C in a mixed micellar system, *J. Biol. Chem.,* 253, 1252, 1978.
66. **Adamich, M., Roberts, M. F., and Dennis, E. A.,** Phospholipid activation of cobra venom phospholipase A_2. Characterisation of phospholipid-enzyme interaction, *Biochemistry,* 18, 3304, 1979.
67. **Roberts, M. F., Adamich, M., Robson, R. J., and Dennis, E. A.,** Phospholipid activation of cobra venom phospholipase A_2. I. Lipid-Lipid or Lipid-enzyme interaction, *Biochemistry,* 18, 3301, 1979.
68. **Adamich, M. and Dennis, E. A.,** Exploring the action and specificity of cobra venom phospholipase A_2 toward human erythrocytes, ghost membranes and lipid mixtures, *J. Biol. Chem.,* 253, 5121, 1978.
69. **Deems, R. A., Eaton, B. R., and Dennis, E. A.,** Kinetic analysis of phospholipase A_2 activity towards mixed micelles and its implications for the study of lipolytic enzymes, *J. Biol. Chem.,* 9013, 1975.
70. **Jain, M. K., Rogers, J., Jahagirdar, D. V., Maracek, J. F., and Ramirez, F.,** Kinetics of interfacial catalysis by phospholipase A_2 in intravesicle scooting mode and heterofusion of anionic and zwitterionic vesicles, *Biochim. Biophys. Acta,* 860, 435, 1986.
71. **Lombardo, D. and Dennis, E. A.,** Immobilised phospholipase A_2 from cobra venom; prevention of substrate interfacial and activator effects, *J. Biol. Chem.,* 260, 16114, 1985.
72. **Pieterson, W. A., Vidal, J. C., Volwerk, J. J., and de Haas, G. H.,** Zymogen catalysed hydrolysis of monomeric substrates and the presence of a recognition site for lipid-water interfaces in phospholipase A_2, *Biochemistry,* 13, 1455, 1974.
73. **Volwerk, J. J., Pieterson, W. A., and de Haas, G. H.,** Histidine at the active site of phospholipase A_2, *Biochemistry,* 13, 1446, 1974.
74. **Verger, R., Mieras, M. C. E., and de Haas, G. H.,** Action of phospholipase A at interfaces, *J. Biol. Chem.,* 248, 4023, 1973.
75. **Volwerk, J. J., Jost, P. C., de Haas, G. H., and Griffith, O. H.,** Activation of porcine pancreatic phospholipase A_2 by the presence of negative charges at the lipid-water interface, *Biochemistry,* 25, 1726, 1986.
76. **Hershberg, R. D., Reed, G. H., Slotboom, A. J., and de Haas, G. H.,** Nuclear magnetic resonance studies of the aggregation of dihexanoyllecithin and of diheptanoyllecithin in aqueous solution, *Biochim. Biophys. Acta,* 424, 73-81, 1976.
77. **Hershberg, R. D., Reed, G. H., Slotboom, A. J., and de Haas, G. H.,** Phospholipase A2 complexes with gadolinium(III) and interaction of the enzyme-metal ion complex with monomeric and micellar alkyl phosphorylcholines. Water proton magnetic relaxation studies, *Biochemistry,* 15, 2268, 1976.
78. **Tinker, D. O., Low, R., and Lucassen, M.,** Heterogenous catalysis by phospholipase A_2: mechanism of hydrolysis of gel phase phosphatidylcholine, *Can. J. Biochem.,* 58, 898, 1980.
79. **Dijkstra, B. W., Kalk, K. H., Drenth, J., de Haas, G. H., Egmond, M. R., and Slotboom, A. J.,** Role of N terminus in the interaction of pancreatic phospholipase A_2 with aggregated substrates. Properties and crystal structure of transaminated phospholipase A2, *Biochemistry,* 23, 2759, 1984.
80. **Van Scharenberg, G. J. M., Jansen, E. H. J. M., Egmond, M. R., de Haas, G. H., and Slotboom, A. J.,** Structural importance of the amino terminal residue of pancreatic phospholipase A_2, *Biochemistry,* 23, 6285, 1984.
81. **Verheij, H. M., Egmond, M. R., and de Haas, G. H.,** Chemical modification of the a amino group in snake venom phospholipase A_2. A comparison of the interaction of pancreatic and venom phospholipase with lipid-water interfaces, *Biochemistry,* 20, 94, 1981.
82. **Jain, M. H., Maliwal, B. P., de Haas, G. H., and Slotboom, A. J.,** Anchoring of phospholipase A_2: the effect of anions and deuterated water, the role of the N-terminus region, *Biochim. Biophys. Acta,* 860, 448, 1986.

83. **Van Oort, M. G., Dijkman, R., Hille, J. D. R., and de Haas, G. H.,** Kinetic behaviour of porcine pancreatic phospholipase A_2 on zwitterion and negatively charged double chain substrates, *Biochemistry,* 24, 7987, 1985.
84. **Van Oort, M. G., Dijkman, R., Hille, J. D. R., and de Haas G. H.,** Kinetic behaviour of porcine pancreatic phospholipase A_2 on zwitterion and negatively charged single chain substrates, *Biochemistry,* 24, 7993, 1985.
85. **Jain, M. K., de Haas, G. H., Maracek, J. F., and Ramirez, F.,** The affinity of phospholipase A_2 for the interface of the substrate analogs, *Biochim. Biophys. Acta,* 860, 475, 1986.
86. **Roberts, M. F., Deems, R. A., and Dennis, E. A.,** Dual role of interfacial phospholipid in phospholipase A_2 catalysis, *Proc. Natl. Acad. Sci. U.S.A.,* 74, 1950, 1977.
87. **Dennis, E. A., Darke, P. L., Deems, R. A., Kensil, C. R., and Pluckthun, A.,** Cobra venom phospholipase A_2: a review of its action towards lipid/water interfaces, *Mol. Cell Biochem.,* 36, 37, 1981.
88. **Hendrickson, H. S. and Dennis, E. A.,** Kinetic analysis of the dual phospholipid model for phospholipase A_2 action, *J. Biol. Chem.,* 259, 5734, 1984.
89. **Hendrickson, H. S. and Dennis, E. A.,** Analysis of the kinetics of phospholipid activation of cobra venom phospholipase A_2, *J. Biol. Chem.,* 259, 5740, 1984.
90. **Dennis, E. A.,** Kinetic dependence of phospholipase A_2 activity on the detergent Triton X-100, *J. Lipid Res.,* 14, 152, 1973.
91. **Dennis, E. A.,** Phospholipase A_2 activity towards phosphatidylcholine in mixed micelles: surface dilution kinetics and the effect of thermotropic phase transitions, *Arch. Bioch. Biophys.,* 158, 485, 1973.
92. **Roberts, M. F., Deems, R.A., Mincey, T. C., and Dennis, E. A.,** Chemical modification of the histidine residues in phospholipase A_2 *(Naja naja naja).* A case of half site reactivity, *J. Biol. Chem.,* 252, 2405, 1977.
93. **Darke, P. L., Jarvis, A. A., Deems, R. A., and Dennis, E. A.,** Further characterisation and N-terminal sequence of cobra venom phospholipase A_2, *Biochim. Biophys. Acta,* 626, 154, 1980.
94. **Pluckthun, A. and Dennis, E. A.,** Role of monomeric activators in cobra venom phospholipase A_2 action, *Biochemistry,* 21, 1750, 1982.
95. **Adamich, M. and Dennis, E. A.,** Specificity reversal in phospholipase A_2 hydrolysis of lipid mixtures, *Biochem. Biophys. Res. Commun.,* 80, 424, 1978.
96. **Pluckthun, A. and Dennis, E. A.,** 31P Nuclear magnetic resonance study on the incorporation of monomeric phospholipids into nonionic detergent micelles, *J. Phys. Chem.,* 85, 678, 1981.
97. **Pluckthun, A. and Dennis, E. A.,** Activation, aggregation and product inhibition of cobra venom phospholipase A_2 and comparison with other phospholipases, *J. Biol. Chem.,* 260, 11099, 1985.
98. **Tinker, D. O. and Wei, J.,** Heterogenous catalysis by phospholipase A_2: formulation of a kinetic description of surface effects, *Can. J. Biochem.,* 57, 97, 1979.
99. **Jain, M. K., Egmond, M. R., Verheij, H. M., Apitz-Castro, R., Dijkman, R., and de Haas, G. H.,** Interaction of phospholipase A_2 and phospholipid bilayers, *Biochim. Biophys. Acta,* 688, 341, 1982.
100. **Apitz-Castro, R., Jain, M. K., and de Haas, G. H.,** Origin of latency phosphatidylcholine vesicles, *Biochim. Biophys. Acta,* 688, 349, 1983.

Chapter 23

ESTERASES (INCLUDING CHOLINE ESTERASES)

Hiromichi Okuda

TABLE OF CONTENTS

I. INTRODUCTION

Unspecific carboxylesterases (EC 3.1.1.1), which hydrolyze a large number of carboxylic esters, are widely distributed in animals, plants and bacteria. In contrast to lipases, their action is generally restricted to short chain fatty acid esters, or water soluble substrates. In 1953, Aldridge[1] proposed a classification of esterases (A-esterases and B-esterases) based on their behaviors with organophosphorus compounds, such as diethyl p-nitrophenyl phosphate (= paraoxon; E600). A-Esterases (EC 3.1.1.2) are not inhibited by organophosphorus compounds, but hydrolyze them. They also split aromatic esters such as phenyl acetate and, therefore, have been designated as arylesterases. In contrast, B-esterases (EC 3.1.1.1) are inhibited stoichiometrically by organophosphates without hydrolyzing them. These enzymes are known as aliesterases or unspecific esterases, because of their wide range of substrates.

However, these classifications of esterases are not completely satisfactory and unambigous. For example, B-esterases, which are inhibited by organophosphates, include a large number of hydrolases such as choline esterase, acetylcholine esterase, chymotrypsin, trypsin, esterase, thrombin, and subtilisin in addition to unspecific esterases, or aliesterase. Furthermore, the unspecific esterase from human serum was recently found to hydrolyze choline esters and o-nitroacetanilide in addition to methylbutyrate, suggesting that the activities of unspecific esterase, choline esterase (EC 3.1.1.8), and acylamidase (EC 3.5.1.4) in human serum are due to the same enzyme molecule.[2-5] Therefore, esterases should not be classified by their inhibitors and substrates, but by other criteria. For example, it may suitable to clarify them on the basis of their physiological significances. In this connection, it must be noted that elucidation of the physiological significances of esterases should be the final goal of research on these enzymes.

II. TISSUE ESTERASES

A. RAT LIVER ESTERASES

Esterases are widely distributed in vertebrate tissues,[6-9] serum,[3] insects,[10,11] plants,[12] citrus fruits,[13,14] mycobacteria,[15] and fungi.[16] In mammals, the highest activities are found in the liver, kidney, duodenum, brain, and adipose tissue.[17-19] The esterases are found predominantly in the microsomal fraction of cell homogenates, where they are present in large amounts. Kunert and Heymann[20] reported that esterases constitute about 70% of the protein in pig liver microsomes.

Rat liver esterases of the serine hydrolase type are separable by preparative isoelectric focussing into five forms with pI values of 5.2, 5.6, 6.0, 6.2 and 6.4.[21] These esterases can be purified from microsomes of rat liver after their extraction by freeze thawing. The solubilized esterases are purified by successive ammonium sulfate fractionation, gel filtration on a Sephadex G-150 column, DEAE-Sephacel column chromatography, and isoelectric focusing followed by Sephacryl S-200 column chromatography. By these procedures, the five esterases have been completely separated from each other as can be demonstrated by analytical polyacrylamide gel electrophoresis and isoelectric focusing on flat gels stained for esterase activity.[22] The esterase of pI 5.2 was separated into two forms which were found to be very similar proteins.[7] The esterase of pI 5.6 consists of at least five active forms differing slightly in isoelectric points.[22] Only the enzyme of pI 6.0 was obtained as a homogeneus protein, although artifactual microheterogeneity was observed after extensive handling of the purified enzyme.[23] The esterases of pI 6.2 and 6.4 are proved to be variants of one protein. Antibody directed against the esterase of pI 6.2 precipitated the esterases of both pI 6.2 and 6.4 but not the esterases of pI 5.2, 5.6, and 6.0.[23] The esterases of pI 6.2 and 6.4 are both labeled with radioactive bis (4-nitrophenyl phosphate), which is an irreversible inhibitor directed to their active site.[65] After these labeled esterases have been split

TABLE 1
Molecular Properties of the Individual Hydrolases[23]

	Method	Hydrolase pI 5.2	5.6	6.0	6.2	6.4
M_r[a]	Gel chromatography	60,000	60,000	180,000	60,000	60,000
Subunit M^r	Dodecyl sulfate electrophoresis	61,000	61,000	58,000	61,000	61,000
Equivalent weight	Titration with paraoxon	ND	60,000	62,000	56,000	58,000
Total hexose (mol/mol)	Phenol/sulfuric acid	6	15	8	60	26
N-Terminal amino acid	Dansylation	ND	—[b]	Tyr	Asx	Asx
C-Terminus	Carboxypeptidase Y	ND	Leu(?)	-Ala-Val-Leu	-Glu-Val-Ser[c] (6.2 and 6.4)	

[a] Data cited from References 1, 6, and 23.
[b] Note detectable after dansylation and hydrolysis; ND, not determined.
[c] Ser could not be distinguished from Asn and Gln.

by proteases or by BrCN, two radioactive peptides can be separated by electrophoresis or thin-layer chromatography. This analysis indicates that these two esterases have similar peptide patterns. In addition, these esterases have identical N- and C-termini that differ from those of other esterases (Table 1).[23] Antibodies against the esterases of pI 5.2 precipitate these enzymes, but they do not inhibit their activities.[7] In addition, they do not react with the enzymes of pI 6.0 and pI 6.4, but partially recognize one of the pI 5.6 enzymes.[7] Anti-esterase of pI 5.6 can precipitate only this enzyme. Antiserum against the pI 6.0 esterase reacts with this enzyme and also reacts weakly with the esterases of pI 6.2 and 6.4.[23] The fingerprint patterns of active site-labeled peptides from the rat liver esterases of pI 5.2, 5.6, 6.0, and 6.2 are clearly different, suggesting that these enzymes have different primary structures. All these findings indicate that rat liver microsomes contain four distinct esterases (pI 5.2, 5.6, 6.0, and 6.2 esterases) with different amino acid sequences. These esterases have various subunits with similar molecular weights (58,000 to 61,000)[7,22,24,68] and have one active site per subunit, as can be shown by active-site titration with their inhibitors bis (4-nitrophenyl phosphate),[23] paraoxan,[66] or 4-nitrophenyl dimethylcarbamate.[67] Most of them exist as monomers except the pI 6.0 esterase, which is found to be a trimer. All esterases contain small amounts of hexoses as shown in Table 1.4-Nitrophenyl acetate has frequently been used as a standard substrate for esterases. The kinetic parameters of the five purified enzymes with this substrate are shown in Table 2. The esterases of pI 5.6 and 6.0 show different kinetics at low and high substrate concentrations of less than and more than 0.05 m*M*. This type of difference is known for esterases and has usually been explained by substrate activation.[25-27]

B. PHYSIOLOGICAL SIGNIFICANCE OF RAT LIVER ESTERASES

An examination of substrate specificities is essential for elucidating the physiological significance of esterases. Systemic studies have been made on the activities of these enzymes with various classes of natural lipoids, including palmitoyl-L-carnitine, monoglycerides, CoA esters, cholesterol oleate, lysophospholipids, and phospholipids (Table 3). The esterase of pI 6.0 is very active on 1-monoglycerides of medium chain length, whereas long chain monoglycerides are predominantly hydrolyzed by the esterases of pI 6.2 and 6.4. None of the esterases hydrolyzes dipalmitoyl glycerol or dioctanoyl glycerol appreciably. The hydrolysis of palmitoyl-CoA is probably the most significant feature of the esterases of pI 6.2

TABLE 2
Kinetics of Hydrolysis of 4-Nitrophenyl Acetate at pH 8.0/37°C[24]

pI of esterase	Range of substrate concentration (m*M*)	K_m (m*M*)	V (U/mg)	V_{1mM} (U/mg)
5.2	0.1—1.6	0.74	78	45
5.6	0.003—0.05[a]	0.020	10	25.9
	0.05—1.6	0.16	30	
6.0	0.003—0.05[a]	0.012	73	69[b]
	0.02—0.2[b]	0.087	161	
6.2	0.05—1.6	0.17	151	129
6.4	0.05—1.6	0.24	176	142

[a] Substrate activation above 0.05 m*M*.
[b] Substrate inhibition above 0.2 m*M*.

and 6.4. These esterases do not hydrolyze acetyl-CoA and hydrolyze oleoyl-CoA only very slowly. In fact oleoyl-CoA is a strong competitive inhibitor of these esterases (Ki ≈ 1 μ*M*), possibly because it is bound tightly but hydrolyzed only slowly. Acetyl-coA and CoA-SH have no inhibitory effect. The esterase of pI 5.6 has the highest activity on palmitoyl carnitine; the apparent Km value is 16 μ*M* and V is 154 m U/mg in the concentration range of 0.001 to 2 m*M*. Palmitoyl carnitine acts as a strong competitive inhibitor of the arylesterase activity of this esterase. Liver microsomes are reported to contain a cholesterol oleate-hydrolyzing enzyme[28,29] that is distinct from the cytoplasmic[30] and lysosomal[31] cholesterol esterases. Of the purified esterases, only that of pI 6.2 has some activity on cholesterol-oleate. In contrast to the cytoplasmic cholesterol oleate-hydrolyzing enzyme, the microsomal enzyme and the purified esterase of pI 6.2 are inhibited by bis (4-nitrophenyl) phosphate.[24] All the purified esterases show low fatty acid-releasing activity with lysophospholipids and with phospholipids. The esterase of pI 5.6 has the highest activities toward these substrates; its apparent Km value with 1-palmitoyl-Sn-glycero-3-phosphocholine as substrate is 50 μ*M*, when assayed with this substrate in the concentration range of 0.02 to 0.5 m*M* at 37°C and pH 7.4.[24]

Based on these and other experimental results, Mentlein et al.[23] suggest that the esterases of pI 6.4 and 6.2 represent long-chain acyl-CoA hydrolases or long-chain monoacylglycerol hydrolases, the esterase of pI 6.0 is a medium-chain monoacylglycerol hydrolase, the esterase of pI 5.6 is a palmitoyl carnitine hydrolase, and the esterase of pI 5.2 is rather specific for medium chain acyl carnitines and diglycerides. Ikeda et al.[32] also demonstrated that the purified esterase of pI 6.0 preferentially hydrolyzes medium-chain-length 1 monoacylglycerols, such as 1-monocaprylylglycerol.

In 1968, Okuda and Fujii[33] reported an experiment on the conversion of liver esterase to lipase in a rather crude preparation. For this, 20 g of fresh rat liver was homogenized in 150 ml of cold acetone and the mixture was centrifuged. The precipitate was washed successively with ethanol (50 ml) and ether (50 ml), and the resulting acetone, ethanol, and ether extracts were combined and concentrated *in vacuo,* and the residual syrupy material was used as the lipid fraction. The acetone precipitate was extracted with 80 ml of 5-m*M* phosphate buffer (pH 8.0) and the extract is subjected to acetone fractionation. The fraction precipitated at 40 to 55% acetone saturation was further extracted with 20 ml of 5 m*M* phosphate buffer (pH 8.0) and the extract was used as the liver esterase preparation. A sample of 8 ml of this esterase solution was mixed with the lipid fraction and sonicated at 10 KC for a total of 45 s, in three sonication periods of 15 s, with cooling for about 1 min after each sonication. The sonicated mixture is applied to a Sephadex G-200 column (3 × 60 cm). As a control, 8 ml of liver esterase was subjected to gel filtration on a Sephadex

TABLE 3
Hydrolysis of Natural Lipoids by Rat Liver Esterases at pH 7.4/37°C[24]

Substrate	Substrate concentration (m*M*)	Activity (mU/mg) with esterase (pI)				
		5.2	5.6	6.0	6.2	6.4
Palmitoyl-L-carnitine						
Radioactive assay	1		282	31		
Estimate of free fatty acid	0.5	14	149	5	7	7
1-Palmitoyl glycerol (rac)	2	100	200	200	6120	6200
1-Caproyl glycerol (rac)	2	800	6400	31300	3200	2900
Palmitoyl-CoA	0.040	<10	30	440	4500	7680
Oleoyl-CoA	0.040	<10	<10	<10	260	370
Oleoyl cholesterol	2.8	<0.2	<0.2	<0.2	1.04	<0.2
1-Palmitoyl-sn-glycero-3-phosphocholine	0.5	15	70	<1	19	11
1-Palmitoyl-sn-glycero-3-phosphoethanolamine	0.5		48	3	14	19
1-Stearoyl-sn-glycero-3-phospho-L-serine	0.5		22	<1	32	34
1-Palmitoyl-sn-glycero-3-phospho-rac-glycerol	0.5		36	3	9	17
1-Palmitoyl-sn-glycero-3-phosphate	0.5		29	2	9	7
1-Palmitoyl-2-oleoyl-sn-glycero-3-phosphocholine	0.5		15	<1	3	3
1,2-Dipalmitoyl-sn-glycero-3-phosphocholine	0.5		9	<1	3	3
1,2-Dioctanoyl-sn-glycero-3-phosphocholine	0.5		13	12	3	3
1,2-Dipalmitoyl-sn-glycero-3-phosphate	0.5		9	<1	5	8

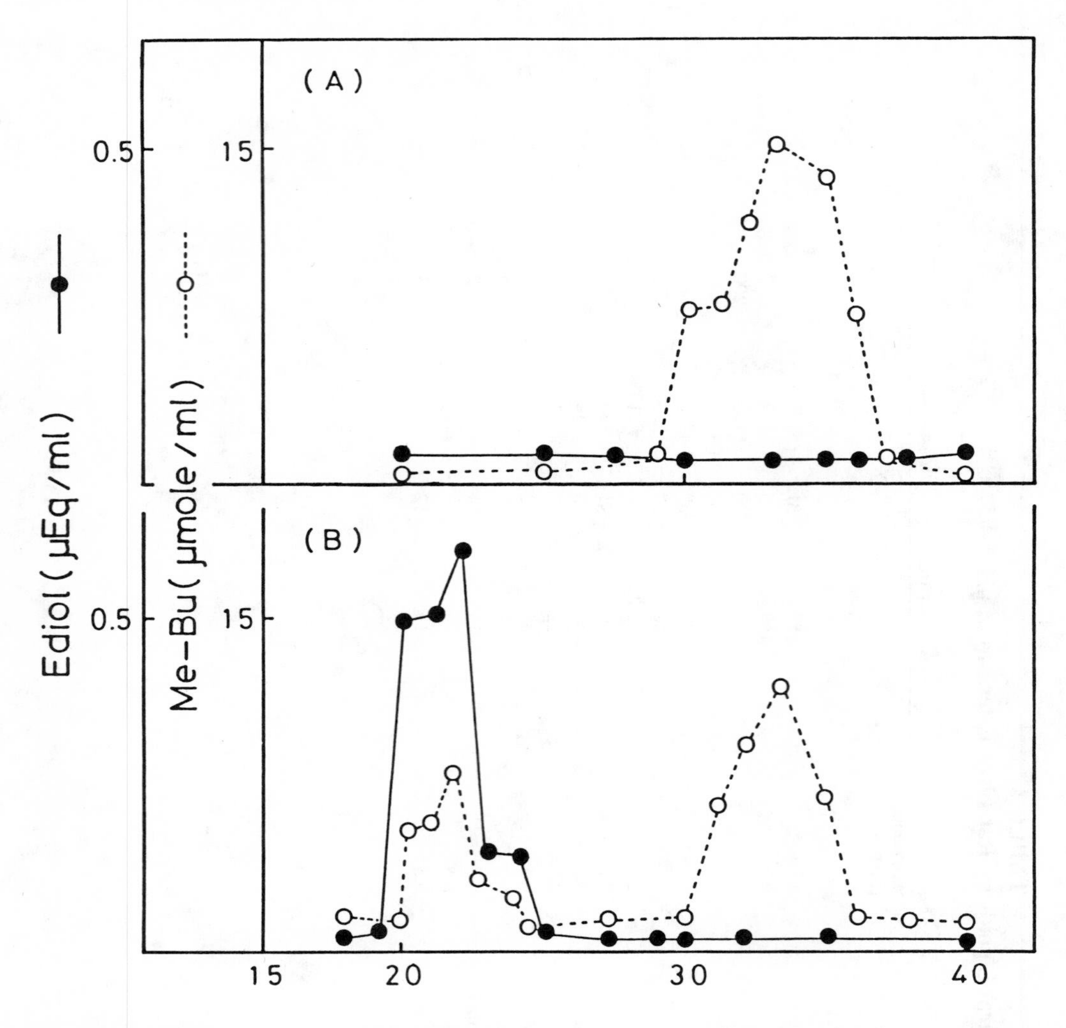

FIGURE 1. Conversion of liver esterase to lipase. (A) Gel filtration of liver esterase. (B) Gel filtration of a sonicated mixture of liver lipid and esterase

G-200 column, and the effluent was collected in 6-ml fractions. The gel filtration pattern showed no Ediol (coconut oil emulsion) hydrolyzing activity and a peak of methylbutyrate hydrolyzing activity in fraction No. 33 shown in Figure 1. On the other hand, on gel filtration, the sonicated mixture of the lipid and the esterase gave a peak of Ediol hydrolyzing activity in fraction No. 22 and two peaks of methyl butyrate hydrolyzing activity in fractions No. 22 and 33, respectively. Of the methyl butyrate activity, 25% was found in the former peak and the rest (75%) in the latter. These results suggest that 25% of the liver esterase was modified to hydrolyze Ediol, that is, it was converted to lipase by this procedure. Therefore, it seems likely that liver lipase is a complex of liver esterase and lipid. The activity of the esterase on insoluble substrates or the esters of long chain fatty acids may be increased by formation of a lipid-enzyme complex as shown in Figure 2.

Recently, evidence have been outlined that human serum cholinesterase can be converted to lipase after modification with acidic phospholipids.[62,63] These results also suggest that an investigation on modified forms of esterases in microsomes may be important for elucidating the physiological significance of these enzymes.

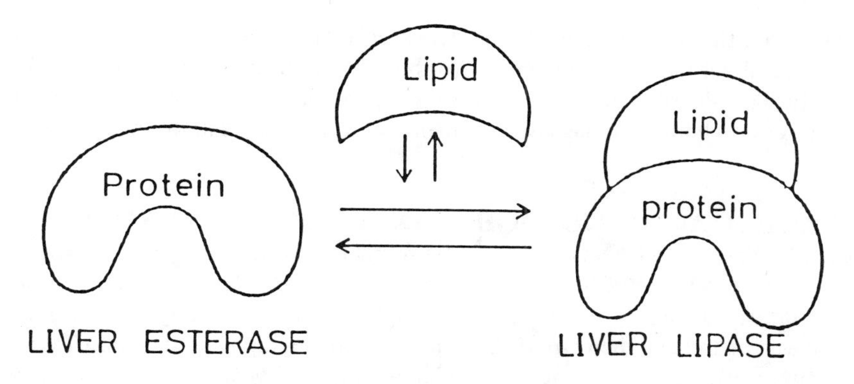

FIGURE 2. Scheme for the relationship between liver lipase and esterase.

C. OTHER TISSUE ESTERASE

As in the case of the esterase of pI 6.0 of rat liver microsomes, trimeric forms of esterases have been found in pig liver,[34-36,69-71] pig kidney,[36,72] ox liver,[36,73,74] human liver,[6,37] rabbit liver,[8,75] and adipose tissue.[19,76] The molecular weights of these esterases are about 180,000 and their subunit weights are about 60,000. In general, they are composed of three subunits of equal size, with one active site per subunit. For example, the molecular weight of the microsomal esterase from human liver was estimated to be 181,000 to 186,000 by gel filtration, disc electrophoresis, and analytical ultracentrifugation, and the molecular weight of its subunits was determined to be 61,500 by sodium dodecylsulfate polyacrylamide gel electrophoresis. The equivalent weight of the enzyme was estimated to be 62,500 from the stoichiometry of its reaction with diethyl-p-nitrophenyl-phosphate. After partial cross-linking of the subunits with dimethyl suberimidate, the preparation gave three bands with molecular weights of 60,000, 120,000, and 180,000 on sodium dodecylsulfate polyacrylamide gel electrophoresis. All these results indicated that the human liver esterase is a trimeric protein composed of three subunits of equal size with one active site per subunit.[37] The molecular weight of a trimeric esterase purified from rabbit liver was found to be 209,000. The esterase dissociated into an active monomeric subunit at high dilution or on treatment at pH 4.35 or with 1.0 *M* KCl. The molecular weight of the active dissociated subunit was 74,500 as determined by gel filtration.[8] Tsujita et al.[19] purified an esterase from rat epididymal adipose tissue to an electrophoretically homogeneus form. Its molecular weight was determined by gel filtration on Sephadex G-200 to be 195,000. The monomeric molecular weight of the enzyme is 65,000 and the enzyme associates to form a trimer. Its amino terminal residue is glycine. An antibody raised in rabbits against the highly purified enzyme strongly inhibited the esterase of rat adipose tissue, but did not inhibit the esterase of rat liver. This esterase catalyzes the hydrolysis of short chain triacylglycerols, such as tributyrin, and medium chain monoacylglycerols, such as monocaprin, but its physiological significance is unknown.

III. CHOLINESTERASES

A. ACETYLCHOLINESTERASE AND BUTYRYLCHOLINESTERASE

The enzymes known as cholinesterases have been of interest for more than 50 years, and seem to be increasingly relevant to neurobiology, medicine, and agriculture. Cholinesterases are thought to belong to two families, true cholinesterases and pseudocholinesterases, or preferably acetylcholinesterases (EC 3.1.1.7) and butyrylcholinesterase (EC 3.1.1.8). Both families are widely distributed throughout the body where they occur in a variety of molecular forms.[38,39] Acetylcholinesterase is known to be essential for cholinergic neurotransmission, and the physiological significance of its polymorphism is just beginning to be understood, but the role of acetylcholinesterase in noncholinergic sites is unknown. Even less is known about the physiological role of butyrylcholinesterase (BuChE). Both types of cholinesterase are distributed in the brain, muscle, and blood.[40] Human serum contains only BuChE activity, whereas human red cell membranes contain only acetylcholinesterase activity.

B. CATALYTIC ACTIVITIES OF HUMAN SERUM CHOLINESTERASE

Recently, unspecific esterase activity was found to be due to the same enzyme molecule as the enzyme known as BuChE. The cholinesterase with BuChE activity was purified from human serum by affinity column chromatographies with p-trimethylammoniumanilinium dichloride as a ligand and then DEAE-Sephadex chromatography as shown in Table 4.[3] This purified enzyme appeared homogeneous electrophoretically and its specific activity was approximately 3000-fold that in human serum. The ratio of cholinesterase activity with benzoylcholine as substrate to unspecific activity with tributyrin remained almost constant during the purification procedure, indicating that these two hydrolytic activities are due to a single enzyme molecule.[2,3] Human serum aryl acylamidase activity is also found to be associated with the cholinesterase.[4] During purification of unspecific esterase from human serum, the ratios of the specific activities of unspecific esterase, cholinesterase, and aryl acylamidase remained approximately constant and the percentage recoveries of the three enzyme activities were about the same. Moreover, the elution profiles of the three enzymes on column chromatographies, including affinity chromatography, were also identical.[5] Thus human serum unspecific esterase, cholinesterase, and aryl acylamidase activities are due to a single enzyme molecule. The active site of the cholinesterase is known to possess an esteratic and an anionic site.[41,42] The esteratic site probably possesses activity for hydrolysis of only neutral esters, but the hydrolysis of positively charged esters such as benzoylcholine may require both the anionic site and the esteratic one. The anionic site may neutralize the positive charge of the ester, and so accelerate hydrolysis by the esteratic site (Figure 3). p-Trimethyl ammonium anilinium dichloride, the affinity ligand described above, has a positive charge and no ester bond in its molecule. This affinity ligand inhibits the hydrolysis of benzoylcholine competitively, but inhibits that of tributyrin noncompetitively. These facts suggest that the positively charged inhibitor and benzoylcholine associate with the anionic site and that both benzoylcholine and tributyrin bind to the esteratic site, but that the inhibitor cannot bind to the esteratic site.[3] Tryptophan, lysine, and histidine residues of purified human serum cholinesterase were suggested to be included in a catalytic site for both cholinesterase and aryl acylamidase activities.[43]

C. MOLECULAR PROPERTIES OF HUMAN SERUM CHOLINESTERASE

The molecular size of cholinesterase in human serum is about 320,000 Da. The enzyme is composed of a tetramer of apparently identical subunits (MW 80,000).[44,45] The four subunits are held together by noncovalent, hydrophobic bonds, as well as by disulfide bonds. The interchain disulfide bonds contribute to the stability of the molecular structure, as judged from the results of heat-stability studies. The disulfide bonds are not essential for maintaining the tetrameric structure of the molecule or for its catalytic activity. The interchain disulfide bonds can be selectively reduced and alkylated without dissociation of the subunits.[45] Mild proteolytic treatment with trypsin cleaves off a very small peptide (less than 5000 daltons) from one end of the subunits and causes the tetramer to dissociate into monomers, dimers, and trimers.[46] This small peptide has two functions: it is responsible for the hydrophobic bonding between subunits, and it contains the interchain disulfide bond. About 24% of the molecular weight of the cholinesterase is due to carbohydrate.[47] Analysis of the pure protein indicates that there are 580 amino acids per subunit.[48] Human serum cholinesterase is known to be a serine enzyme that is inhibited by diisopropylfluorophosphate. The serine residue in

TABLE 4
Purification of Cholinesterase and Unspecific Esterase from Human Serum[3]

Purification step	Total protein (mg)	Cholinesterase				Unspecific esterase			
		Total activity (μmol/min)	Yield (%)	Specific activity (μmol/mg per min)	Fold	Total activity (μmol/min)	Yield (%)	Specific activity (μmol/mg per min)	Fold
Serum	23400	260	100	0.011	1	10.6	100	0.00045	1
First affinity chromatography	64	148	57.2	2.33	212	6.32	59.6	0.099	220
Second affinity chromatography	10.2	70	26.9	6.90	627	3.05	28.8	0.173	664
DEAE-Sephadex column chromatography	0.57	14.6	5.6	30.6	2782	0.67	6.3	1.42	3155

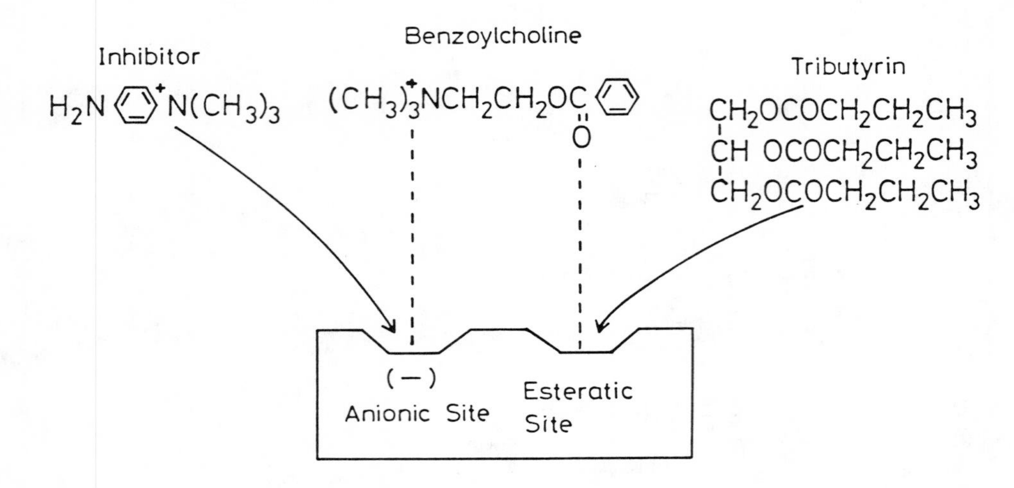

FIGURE 3. Diagram of interaction of esterase with benzoylcholine, tributyrin and inhibitor.

the active site is specifically alkylated with [^{3}H] diisopropylfluorophosphate. On analysis of the tryptic peptides of the cholinesterase after labeling, only one peptide fragment was found to contain a labeled serine residue. This finding supports the view that the active sites of the four subunits of the enzyme have the same structure.[49] Recently, the active site isolated from the tryptic peptides was suggested to have the following sequence: Ser Val Thr Leu Phe Gly Glu Ser* Ala Gly^{10} Ala Ala Ser Val Ser Leu His Leu Leu Ser^{20} Pro Gly Ser His Ser Leu Phe Thr Arg,^{29} where essential active-site serine residue is shown by an asterisk.[49] A comparison of the sequences of the active-sites of human serum cholinesterase and Torpedo acetylcholinesterase[50] shows identity of 17 residues and conservative replacement of 5 residues. This high degree of sequence homology in the active sites of a true acetylcholinesterase of Torpedo and the cholinesterase of human serum is remarkable.

```
                      *
Human ChE      S V T L F G E S A G A A S V S L H L L S P G S H
                 | |   | | | | | |   | | |     |   | | | | |     (Str 1)
Torpedo AChE   T V T I F G E S A G G A S V G M H I L S P G S R
```

The active site of human serum cholinesterase shows much less homology with those of some other serine proteases, such as trypsin, chymotrypsin, and thrombin:

Human ChE	SVTLFGES*A
	GAASVSLHLLSPGSHSLFTR
Bovine trypsin	KDSCQGDSGGPVVCS
Bovine chymotrypsin	VSSCMGDSGGPLVCK
Bovine thrombin	GDACEGDSGGPFVMK

(Str 2)

There are, at present, at least five recognized genes that participate in directing human serum cholinesterase biosynthesis.[51] There are the usual (E_1^u), atypical (E_1^a), fluoride resistant (E_1^f), silen (E_1^s), and $C_5(E_2)$ genes. The atypical, fluoride resistant, and silent genes are allelomorphic variants of the usual enzyme. The C_5 variant is nonallelic and is assigned to a different locus, the E_2 locus. All but the C_5 variant have been implicated in the exaggerated clinical response to the relaxant effect of succinyldicholine, the most serious complication being respiratory apnea.

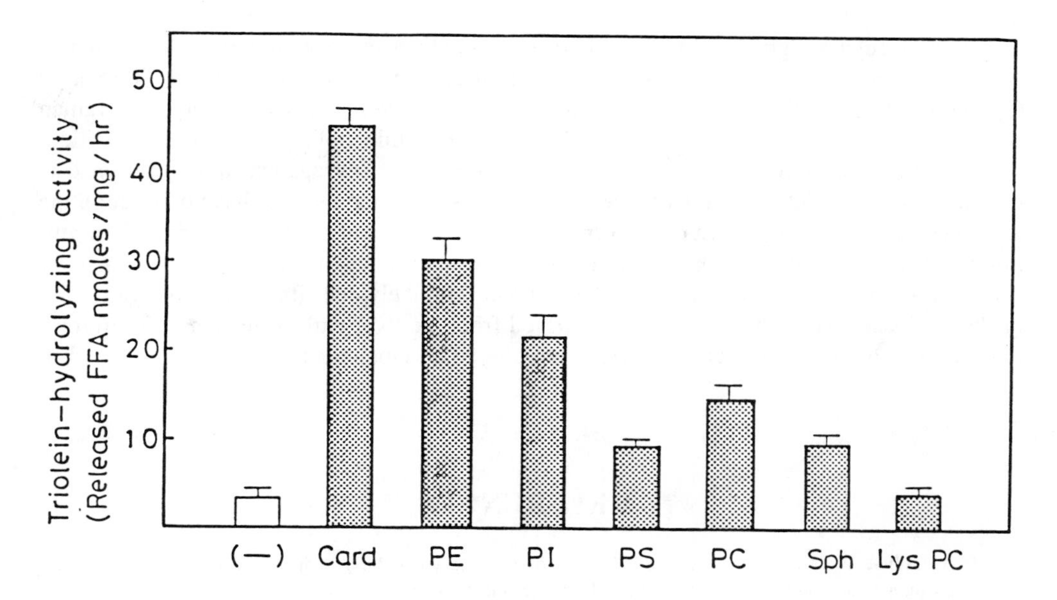

FIGURE 4. Effects of various phospholipids on triolein-hydrolyzing activity of human serum cholinesterase.[63] Card: cardiolipin, PE: Phosphatidyle-thanolamine, PI: Phosphatidylinositol, PS: phosphatidylserine, PC: phosphatidylcholine, Sph: sphingomyelin, LysPC: lysophosphatyidylcholine

D. PHYSIOLOGICAL SIGNIFICANCE OF HUMAN SERUM CHOLINESTERASE

Serum cholinesterase activity is routinely measured clinically as a liver-function test, although the physiological significance of the enzyme is uncertain. Serum cholinesterase decreases in various disorders of the liver,[52] in carcinoma,[53] and after administration of anticholinesterase drugs.[53] Slightly increased levels are often associated with diabetes mellitus, fatty liver, the nephrotic syndrome, obesity, hyperlipidemia, coronary artery disease, hyperthyroidism, and other disorders.[53-58,77-79] A number of functions of serum cholinesterase have been suggested. The enzyme has been proposed to be associated with fatty acid metabolism.[59] regulation of the plasma choline level,[60] and lipoprotein metabolism.[58,61] Cholinesterase activity is found to increase proportionally with increases in low density lipoprotein, cholesterol, and triglycerides both in the serum and in the low density lipoprotein fraction. The cholinesterase activity in the low-density lipoprotein fraction of serum is increased in patients with types IIa, IIb, and IV hyperlipoproteinemia, but the serum cholinesterase activity is increased only in patients with types IIb and IV.[58,80] These findings suggest that cholinesterase functions in lipid and lipoprotein metabolism. Recently, cholinesterase was shown to hydrolyze neutral esters of long-chain fatty acids such as triolein after addition of a hydrophobic substance present in human serum.[62] The hydrophobic substance that enhanced triolein hydrolysis was identified as acidic phospholipids such as cardiolipin, phosphatidylethanolamine, and phosphatidylinositol (Figure 4).[63] These findings suggest that the cholinesterase in a modified form can hydrolyze neutral esters of long chain fatty acids.

E. ORIGIN OF HUMAN SERUM CHOLINESTERASE

There is strong evidence that cholinesterase in human serum is produced by and released from the liver and that its serum level is proportional to the synthetic activity of liver cells, like that of serum albumin. It is noteworthy, however, that the cholinesterase level in human liver is very low, whereas the carboxylesterase activity in the liver microsomal fraction is high.[37] The carboxylesterases in human liver and serum differ in the following characters: (1) human liver carboxylesterase is composed of three subunits of equal size (60,000 Da) and its molecular weight is 180,000, whereas the cholinesterase with carboxylester hydrolyzing activity (MW 320,000) is composed of four subunit of identical size (80,000 Da.)[64] (2) The pI values of the liver and serum enzymes are 6.3 and 4.0, respectively. (3) Human serum carboxylesterase catalyzes the hydrolysis of butyrylthiocholine and o-nitroacetanilide, whereas the liver enzyme does not. (4) Anti-γ-globulin raised against human serum carboxylesterase completely inhibits the activity of the serum enzyme, but has no effect on the carboxylesterase activity of human liver. (5) Neostigmine and p-trimethylammonium anilinium dichloride strongly inhibit serum carboxylesterase, but do not inhibit human liver carboxylesterase activity. From these results, it seems unlikely that either the carboxylesterase or the cholinesterase in human serum is derived from the liver carboxylesterase. Therefore, the origin of human serum cholinesterase remains to be elucidated.

REFERENCES

1. **Aldridge, W. N.,** Serum esterases; 2 types of esterase (A and B) hydrolyzing p-nitrophenyl acetate, propionate and butyrate and method for their determination, *Biochem. J.*, 53, 110, 1953.
2. **Muraoka, T., and Okuda, H.,** Identification of acylcholine acyl-hydrolase with carboxylic ester-hydrolase in human serum, *J. Biochem.*, 82, 207, 1977.
3. **Tsujita, T., Nagai, K., and Okuda, H.,** Purification and properties of human serum esterase, *Biochim. Biophys Acta,* 570, 88, 1979.
4. **George, S. T., and Balasubramanian, A. S.,** The aryl acylamidases and their relationship to cholinesterases in human serum, erythrocyte and liver, *Eur. J. Biochem.*, 121, 177, 1981.
5. **Tsujita, T., and Okuda, H.,** Carboxylesterases in rat and human sera and their relationship to serum aryl acylamidases and cholinesterases, *Eur. J. Biochem.*, 133, 215, 1983.
6. **Tsujita, T., and Okuda, H.,** Human liver carboxylesterase. Properties and comparison with human serum carboxylesterase, *J. Biochem.*, 94, 793, 1983.
7. **Robbi, M., and Beaufay, H.,** Purification and characterization of various esterases from rat liver, *Eur. J. Biochem.*, 137, 293, 1983.
8. **Miller, S. K., Main, A. R., and Rush, R. S.,** Purification and physical properties of oligomeric and monomeric carboxylesterases from rabbit liver, *J. Biol. Chem.*, 255, 7161, 1980.
9. **Keough, D. T., de Jersey, J., and Zermer, B.,** The relationship between the carboxylesterase and monoacylglycerol lipase activities of chicken liver microsomes, *Biochim. Biophys. Acta,* 829, 164, 1985.
10. **Beckendorf, G. W. and Stephen, W. P.,** The effect of aging on the multiple molecular esterase forms taken from tissues of periplaneta americana, *Biochim. Biophys. Acta,* 201, 101, 1970.
11. **Stephen, W. P., and Cheldelin, I. H.,** Characterization of soluble esterases from the thoracic muscle of the american cockroach, periplaneta americana, *Biochim. Biophys. Acta,* 201, 109, 1970.
12. **Singer, T. P.,** On the mechanism of enzyme inhibition by sulfhydryl reagents, *J. Biol. Chem.*, 174, 11, 1948.
13. **Jansen, E. F., Jang, R., and MacDonald, L. R.,** Citrus acetylesterase, *Arch. Biochem. Biophys.*, 15, 415, 1947.
14. **Blakeley, R. L., de Jersey, J., Webb, E. C., and Zermer, B.,** On the homology of the active-site peptides of liver carboxylesterases, *Biochim. Biophys. Acta,* 139, 208, 1967.
15. **Myers, D. K., Tol, J. W., and de Jonge, M. H. T.,** Studies on ali-esterases. V. Substrate specificity of the esterases of some saprophytic mycobacteria, *Biochem. J.*, 65, 223, 1957.
16. **Byrde, R. J. W., and Fielding, A. H.,** Studies on the acetylesterase of sclerotinia laxa, *Biochem. J.*, 61, 337, 1955.
17. **Benöhr, H. C., Franz, W., and Krish, K.,** Carboxylesterasen der mikrosomenfraktion. Hydrolyse von tributyrin, procain und phenacetin durch einige organe und lebermikrosomen verschiedener tierarten, *Arch. Pharmacol. Exp. Pathol.*, 255, 163, 1966.
18. **Holmes, R. S., and Masters, C. J.,** A comparative study of the multiplicity of mammalian esterases, *Biochim. Biophys. Acta,* 151, 147, 1968.
19. **Tsujita, T., Okuda, H., and Yamasaki, N.,** Purification and some properties of carboxylesterase of rat adipose tissue, *Biochim. Biophys. Acta,* 715, 181, 1982.
20. **Kunert, M., and Heymann, E.,** The equivalent weight of pig liver carboxylesterase (EC 3.1.1.1) and the esterase content of microsomes, *FEBS Lett.*, 49, 292, 1975.

21. **Heymann, E. and Mentlein, R.,** Carboxylesterases-amidases, *Methods Enzymol.,* 77, 333, 1981.
22. **Mentlein, R., Heiland, S., and Heymann, E.,** Simultaneous purification and comparative characterization of six serine hydrolases from rat liver microsomes, *Arch. Biochem. Biophys.,* 200, 547, 1980.
23. **Mentlein, R., Schumann, M., and Heymann, E.,** Comparative chemical and immunological characterization of five lipolytic enzymes (carboxylesterases) from rat liver microsomes, *Arch. Biochem. Biophys.* 234, 612, 1984.
24. **Mentlein, R., Suttorp, M., and Heymann, E.,** Specificity of purified monoacylglycerol lipase, palmitoyl-CoA hydrolase, palmitoyl-carnitine hydrolase, and nonspecific carboxylesterase from rat liver microsomes, *Arch. Biochem. Biophys.,* 228, 230, 1984.
25. **Barker, D. L. and Jenks, W. P.,** Pig liver esterase. Physical properties. Some kinetic properties, *Biochemistry,* 8, 3879; 3890, 1969.
26. **Stoops, J. K., Horgan, D. J., Runnegar, M. C. T., de Jersey, J., Webb, E. C., and Zerner, B.,** Carboxylesterases (EC 3.1.1). Kinetic studies on carboxylesterases, *Biochemistry,* 8, 2026, 1969.
27. **Heymann, E., Junge, W., and Krisch, K.,** Reaktion mit phenylmethansulfonylfluorid und nachweis von isoenzymen, *Hoppe-Seyler's Z. Physiol. Chem.,* 353, 576, 1972.
28. **Deykin, D. and Goodman, D. S.,** The hydrolysis of long-chain fatty acid esters of cholesterol with rat liver enzymes, *J. Biol. Chem.,* 237, 3649, 1962.
29. **Nilsson, A.,** Hydrolysis of chyle cholesterol esters with cell-free preparations of rat liver, *Biochim. Biophys. Acta,* 450, 379, 1976.
30. **Taháčková, Z., Kríz, O., and Hradec, J.,** Purification and some properties of a cholesterol esterase from rat liver, *Biochim. Biophys. Acta,* 617, 439, 1980.
31. **Brown, W. J. and Sgoutas, D. S.,** Purification of rat liver lysosomal cholesteryl ester hydrolase, *Biochim. Biophys. Acta,* 617, 305, 1980.
32 **Ikeda, Y., Okamura, K., Arima, T., and Fujii, S.,** Purification and characterization of two types of esterase from rat liver microsomes, *Biochim. Biophys. Acta,* 487, 189, 1977.
33. **Okuda, H. and Fujii, S.,** Relationship between lipase and esterase, *J. Biochem. (Tokyo),* 64, 377, 1968.
34. **Adler, A. J and Kistiakowsky, G. B.,** Isolation and properties of pig liver esterase, *J. Biol. Chem.,* 236, 3240, 1961.
35. **Ocken, P. R. and Levy, M.,** The nature of the modifier site of pig liver esterase, *Biochim. Biophys. Acta,* 212, 450, 1970.
36. **Heymann, E., Junge, W., and Krisch, K.,** Subunit weight and N-terminal groups of liver and kidney carboxylesterases (EC 3.1.1.1), *FEBS Lett.,* 12, 189, 1971.
37. **Junge, W., Heymann, E., and Krisch, K.,** Human liver carboxylesterase purification and molecular properties, *Arch. Biochem. Biophys.,* 165, 749, 1974.
38. **Brimijoin, S.,** Molecular forms of acetylcholinesterase in brain, nerve and muscle: nature, localization and dynamics, *Prog. Neurobiol.,* 21, 291, 1983.
39. **Massoulié, J. and Bon, S.,** The molecular forms of cholinesterase and acetylcholinesterase in vertebrates, *Annu. Rev. Neurosci.,* 5, 57, 1982.
40. **Ord, M. G. and Thompson, R. H. S.,** Distribution of cholinesterase types in mammalian tissues, *Biochem. J.,* 46, 346, 1950.
41. **Bergman, F. and Wurzel, M.,** The structure of the active surface of serum cholinesterase, *Biochim. Biophys. Acta,* 13, 251, 1954.
42. **Bergman, F., Segal, R., Skimoni, A., and Wurzel, M.,** The pH-dependence of enzymic ester hydrolysis, *Biochem. J.,* 63, 684, 1956.
43. **Boopathy, R. and Balasubramanian, A. S.,** Chemical modification of the bifunctional human serum pseudocholinesterase. Effect on the pseudocholinesterase and aryl acylamidase activities, *Eur. J. Biochem.,* 151, 351, 1985.
44. **Lockridge, O. and La Du, B. N.,** Comparison of atypical and usual human serum cholinesterase, *J. Biol. Chem.,* 253, 361, 1978.
45. **Lockridge, O., Eckerson, H. W., and La Du, B. N.,** Interchain disulfide bonds and subunit organization in human serum cholinesterase, *J. Biol. Chem.,* 254, 8324, 1979.
46. **Lockridge, O. and La Du, B. N.,** Loss of the interchain disulfide peptide and dissociation of the tetramer following limited proteolysis of native human serum cholinesterase, *J. Biol. Chem.,* 257, 12012, 1982.
47. **Haupt, H., Heide, K., Zwisler, O., and Schruick, H. G.,** Isolierung und physikalische-chemische charaklerisierung der cholinesterase aus humanserum, *Blut,* 14, 65, 1966.
48. **Lockridge, O.,** *Cholinesterases: Fundamental and Applied Aspects, Walter de Gruyter, Berlin, 1984, 5.*
49. **La Du B. N. and Lockridge, O.,** Molecular biology of human serum cholinesterase, *Fed. Proc.,* 45, 2965, 1986.
50. **MacPhee-Quigley, K., Taylor, P., and Taylor, S.,** Primary structures of the catalytic subunits from two molecular forms of acetylcholinesterase A comparison of NH_2-terminal and active center sequences, *J. Biol. Chem.,* 260, 12185, 1985.

51. **LaMotta, R. V. and Woronick, C. L.,** Molecular heterogeneity of human serum cholinesterase, *Clin. Chem.,* 17, 135, 1971.
52. **Vorhaus, L. J., II, Scudamore, H. H., and Kark, R. M.,** Measurement of serum cholinesterase activity in study of diseases of liver and biliary system, *Gastroenterology,* 15, 304, 1950.
53. **Neitlich, H. W.,** Increased plasma cholinesterase activity and succinylcholine resistance: a genetic variant, *J. Clin. Invest.,* 45, 380, 1966.
54. **Thompson, R. H. and Trounce, J. R.,** Serum cholinesterase levels in diabetes mellitus, *Lancet,* 1, 656, 1956.
55. **Shibata, H., Toyama, K., and Watanabe, Y.,** Study on the hypercholinesterasemia. II. Value of hypercholinesterasemia in diagnosis of fatty liver or fatty change in the liver, *Acta Hepatol. Jpn.,* 17, 755, 1976.
56. **Lehtonen, A., Marniemi, J., Inberg, M., Maatela, J., Alanen, E., and Niittymäki, K.,** Levels of serum lipids, apolipoproteins A-I and B and pseudocholinesterase activity and their discriminative values in patients with coronary by-pass operation, *Atherosclerosis,* 59, 215, 1986.
57. **Kutty, K. M., Kean, K. T., Jain, R., and Huang, S.,** Plasma pseudocholinesterase: a potential marker for early detection of obesity, *Nutrition Res.,* 3, 211, 1983.
58. **Chu, M. I., Fontaine, P., Kutty, K. M., Murphy, D., and Redheendran, R.,** Cholinesterase in serum and low density lipoprotein of hyperlipidemic patients, *Clin.Chim. Acta,* 85, 55, 1978.
59. **Cucuianu, M., Popescu, T. A., Opincaru, A., and Haragus, S.,** Serum pseudocholinesterase and ceruloplasmin in various types of hyperlipoproteinemia, *Clin. Chim. Acta,* 59, 19, 1975.
60. **Funnel, H. S. and Oliver, W. T.,** Proposed physiological function for plasma cholinesterase, *Nature,* 208, 689, 1965.
61. **Kutty, K. M., Rowden, G., and Cox, A. R.,** Interrelationship between serum lipoprotein and cholinesterase, *Can. J. Biochem.,* 51, 883, 1973.
62. **Shirai, K., Ohsawa, J., Ishikawa, Y., Saito, Y., and Yoshida, S.,** Human plasma carboxyl esterase-catalyzed triolein hydrolysis. Existence of promoting factor in serum, *J. Biol. Chem.,* 260, 5225, 1985.
63. **Shirai, K., Ohsawa, J., Saito, Y., and Yoshida, S.,** Human plasma carboxyl esterase-catalyzed triolein hydrolysis. Existence of promoting factor in serum and its significance, *Shishitsu Seikagaku Kenkyu,* 27, 29, 1985.
64. **Muensch, H., Goedde, H. W., and Yoshida, H.,** Human-serum cholinesterase subunits and number of active sites of the major component, *Eur. J. Biochem.,* 70, 217, 1976.
65. **Heymann, E. and Krisch, K.,** Phosphorsäure-bis (p-nitrophenylester), ein neuer hammstoff mikrosomaler carboxylesterasen, *Hoppe-Seyler's Z. Physiol. Chem.,* 348, 609, 1967.
66. **Krisch, K.,** Reaction of a microsomal esterase from hog-liver with dielhyl rho-nitrophenyl phosphate, *Biochim. Biophys. Acta,* 122, 265, 1966.
67. **Dudman, N. P. B. and Zerner, B.,** Carboxylesterases from pig and ox liver, in *Methods in Enzymology,* Vol. 35, Lowenstein, J. M., Ed., Academic Press, New York, 1975, 190.
68. **Brandt, E., Heymann, E. and Mentlein, R.,** Selective inhibition of rat liver carboxylesterases by various organophosphorus diesters *in vivo* and *in vitro, Biochem. Pharmacol.,* 29, 1927, 1980.
69. **Junge, W. and Krisch, K.,** Current problems on the structure and classification of mammalian liver carboxylesterases, *Mol. Cell. Biochem.,* 1, 41, 1973.
70. **Junge, W. and Krisch, K.,** The carboxylesterases/amidases of mammalian liver and their possible significance, *Crit. Rev. Toxicol.,* 3, 371, 1975.
71. **Levine, L., Baer, A., and Jencks, W. P.,** Dissociation of pig liver carboxylesterase measured by quantitative micro-complement fixation, *Arch. Biochem. Biophys.,* 203, 236, 1980.
72. **Heymann, E., Krisch, K., and Pahlich, E.,** Zur struktur des aktiven zentrums von mikrosomalen carboxylesterasen aus schweineniere und-lleber, *Z. Physiol. Chem.,* 351, 931, 1970.
73. **Augusteyn, R. C., de Jersey, J., Webb, E. C., and Zerner, B.,** On the homology of the active-site peptides of liver carboxylesterases, *Biochim. Biophys. Acta,* 171, 128, 1969.
74. **Runnegar, M. T. C., Webb, E. C., and Zerner, B.,** Carboxylesterases (EC 3.1.1). Dissociation of ox liver carboxylesterase, *Biochemistry,* 8, 2018, 1969.
75. **Rush, R. S., Main, A. R., Miller, S. K., and Kilpatrick, B. F.,** Resolution and purification of two monomeric butyrylcholinesterases from rabbit liver, *J. Biol. Chem.,* 255, 7155, 1980.
76. **Taniguchi, A., Okuda, H., Oseko, F., Nagata, I., Kono, T., Tsujita, T., Kataoka, K., and Imura, H.,** Regional differences in carboxylesterase activity between human subcutaneous and omental adipose tissue, *Life Sci.,* 36, 1465, 1985.
77. **Antopol, W., Schifrim, A., and Tuchman, L.,** Decreased cholinesterase activity of serum in jaundice and in biliary disease, *Proc. Soc. Exp. Biol.,* 38, 363, 1938.
78. **Haragus, S., Cucuianu, M., and Missits, P.,** Serum cholinesterase and euglobulin lysis time in chronic cor pulmonale, *Br. Heart J.,* 31, 447, 1969.

79. **Goedde, H. W., Doenicke, A., and Altland, K.,** *Pseudocholinesterasen, Pharmakogenetik, Biochemic, Klinik,* Springer Verlag, Berlin, Heidelberg,
80. **Kutty, K. M., Chandra, R. K., and Chandra, S.,** Acetylcholinesterase in erythrocytes and lymphocytes: its contribution to cell membrane structure and function, *Experientia,* 32, 289, 1976.

Chapter 24

LIPASES

Hiromichi Okuda

TABLE OF CONTENTS

I. INTRODUCTION

In 1856, Claude Bernard first discovered a lipase (EC 3.1.1.3) in pancreatic juice. He recognized the lipase as an enzyme that hydrolyzed insoluble oil droplets and converted them to soluble products. In 1896, Hanriot[1] found a monobutyrin-hydrolyzing enzyme in serum and called it a lipase. However, Arthus et al.[2] and Deyon et al.[3] considered a lipase to be defined as an enzyme that hydrolyzed water-insoluble esters, not water-soluble esters, and that an enzyme which hydrolyzed water-soluble esters should be called "esterases". In 1958, Sarda and Desnuelle[4] proposed another classification to differentiate between liver esterase and pancreatic lipase based on differences in their activations by interfaces of substrates. This difference is illustrated in Figure 1. The esterase shows normal Michaelis-Menten activity dependency on triacetin concentration, whereas the lipase has scarcely any activity with the same substrate in the monomeric state, but its activity increases sharply when the triacetin concentration is increased above its solubility limit. Apparently, the esterase is active only on molecularly dispersed substrate molecules such as triacetin and methylbutyrate, whereas the lipase rapidly hydrolyzes substrates in their insoluble forms. Although the definition of Sarda and Desnuelle has been widely accepted for the last 30 years, it was not consistent with our findings with purified enzymes,[5-7] as shown in Figure 2. Human liver and serum esterases, and porcine pancreatic lipase, respectively, exhibit typical esterase and lipase patterns of substrate dependency as defined by Sarda and Desnuelle (1 and 4 in Figure 2). However, the following findings are not consistent with their definition: the activity of rat hepatic lipase does not increase above the saturation concentration of methylbutyrate, although this enzyme is known to hydrolyze triacylglycerols of lipoproteins.[40] The substrate dependency of rat hepatic lipase is essentially the same as that of human serum esterase (cholinesterase) (2 in Figure 2). Furthermore, the pattern of rat adipose tissue esterase is quite different from that of human liver esterase (1 and 3 in Figure 2); the rate of hydrolysis of methylbutyrate by rat adipose tissue esterase is maximal at a substrate concentration where the substrate is in a soluble state and decreases at higher substrate concentrations, whereas that of human liver esterase remains at the maximal level at higher substrate concentrations. These facts indicate that rat hepatic lipase and adipose tissue esterase cannot be classified by the definition of Sarda and Desnuelle.[4]

Therefore, it seems more reasonable to use the classical definition for lipases and esterases; namely, a lipase is an enzyme that acts on water-insoluble neutral esters and an esterase is one that acts on water-soluble neutral esters.

II. PANCREATIC LIPASE

Dietary triglycerides must be partially hydrolyzed before they can be absorbed from the intestinal tract and this hydrolysis is performed by digestive lipases that are produced in the pancreas of mammals. Pancreatic lipase was one of the first enzymes to be recognized; Claude Bernard described its action in 1856. Subsequently this enzyme has received much attention from physiologists. The first intensive effort to purify and characterize it was made by Willstätter and his group in the 1920s.[8,9] Thereafter, there have been extensive studies on its purification, molecular properties, and catalytic mechanism, and findings have been thoroughly reviewed.[10-18] Thus, this section gives mainly a brief review of the catalytic mechanism of pancreatic lipase.

Recently, Rover's group[19] determined the primary structure of porcine pancreatic lipase. Porcine pancreatic lipase is composed of a single chain of 449 amino acids. The calculated molecular weight of the protein is 49,859, to which that of the glycan moiety (about 2000)[20] should be added, resulting in a total molecular weight of about 52,000. As mentioned above, pancreatic lipase is fully active with substrates in an aggregated form. The enzyme shows

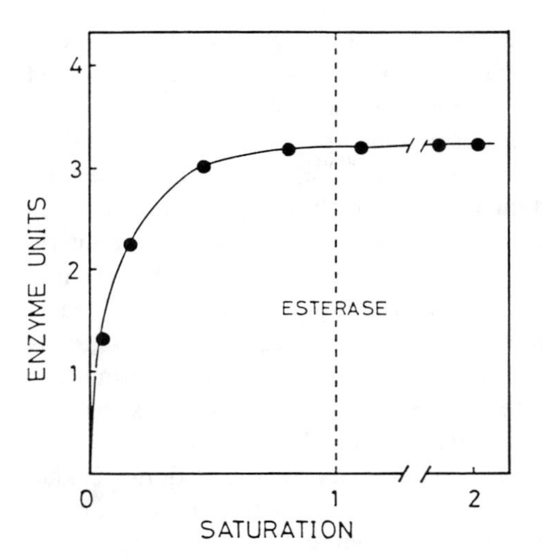

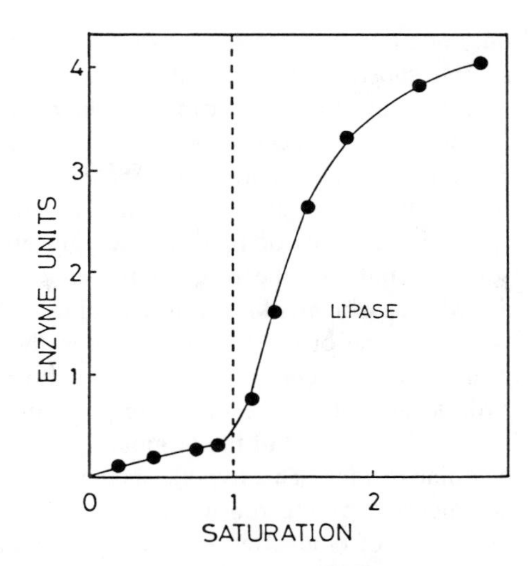

FIGURE 1. Hydrolysis of triacetin by horse liver esterase and porcine pancreatic lipase. Reaction rates are shown as functions of substrate concentration, which is expressed in multiples of saturation. (From Sarda, L. and Desnuelle, P., *Biochim. Biophys. Acta,* 30, 513, 1958. With permission.)

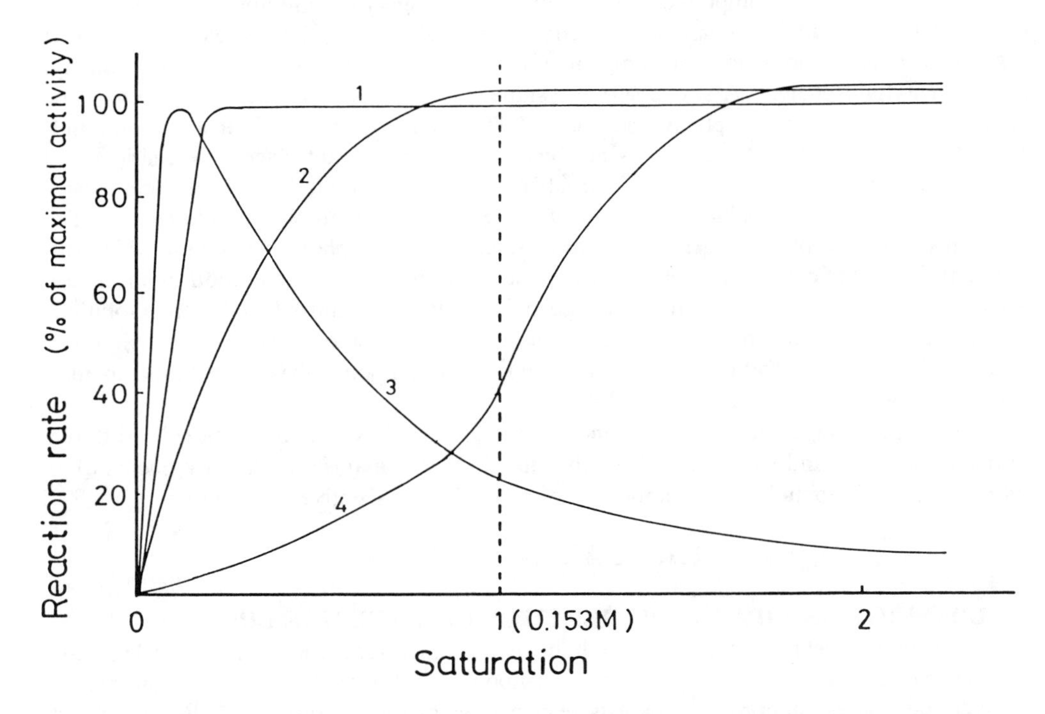

FIGURE 2. Hydrolysis of methyl-butyrate by various lipases and esterases. Reaction rates are plotted as functions of substrate concentration, which is expressed in multiples of saturation. 1: human liver,[5] 2: human serum esterase[5] and rat hepatic lipase,[6] 3: rat adipose tissue esterase,[7] and 4: porcine pancreatic lipase.

very low activity towards monomeric triacetin in solution[13,18,21] (0.3% of the lipase activity in the presence of micellar or emulsified substrate solution), and this activity is enhanced about 20-fold with the same substrate by addition of a low concentration of water-miscible organic solvents. Pancreatic lipase hydrolyzes monomeric solutions of p-nitrophenyl acetate with an organic solvent such as 4% acetonitrite.[22] Chapus et al.[22] concluded that there is only one esteratic site for substrates both in the monomeric and the aggregated form. With soluble substrates this esteratic site shows low efficiency, whereas, when lipase is adsorbed to a substrate emulsion or micelle interface by its binding site, the esteratic site acquires its full activity. A recent report[23] on limited proteolysis of porcine lipase showed that the Phe-335-Ala-336 bond is preferentially split by chymotrypsin, thus giving rise to two polypeptide fragments (1-335) and (336-449). The latter fragment does not show any activity towards specific substrates of lipase, triacylglycerols, either in the aggregate form or monomeric solution, but like the original lipase it hydrolyzes p-nitrophenyl acetate.[24] Ethoxyformic anhydride, which is known to combine with histidine residues of pancreatic lipase,[25] reacts with only one of the two histidines in the fragment. The activity of the fragment towards p-nitrophenyl acetate is inhibited by modification of the fragment with ethoxyformic anhydride as in the case of the original lipase. These facts lead to the hypothesis that the fragment (336-449) of the original lipase includes the esteratic site that contains an essential histidine residue. This hypothesis that the esteratic site contains an essential histidine residue is supported by the following facts. (1) The Kcat and Km values of the lipase-catalyzed hydrolysis of emulsified tributyrin change at pH values at which an ionizable group in the lipase with a pK of 5.8 involves.[26] (2) The first-order rate constant of enzyme inactivation by photooxidation is best correlated with the degree of modification of the essential histidine.[26]

Organophosphorus compounds, commonly used as specific inhibitors of serine hydrolases, do not abolish pancreatic lipase activity on substrates in the soluble form;[13] these organophosphorus compounds are only inhibitory as emulsions, in particular with diethyl p-nitrophenylphosphate.[27] A single serine residue, Ser 152, is modified by reaction of the enzyme with diethyl p-nitrophenylphosphate.[28] The modified enzyme is inactive on emulsified triglyceride substrates and does not bind to hydrophobic interfaces, i.e., siliconized glass beads, but retains activity on p-nitrophenyl acetate solutions.[25,27] These facts clearly implicate the Ser 152 residue in binding of the enzyme to hydrophobic interfaces.[25] The reaction of porcine pancreatic lipase with another organophosphorus compound, bis-p-nitrophenyl methylphosphonate, results in complete and irreversible inhibition of the lipase activity on tributyrin emulsion, but no change in the activity on p-nitrophenyl acetate solution. After modification with this organophosphorus compound, the enzyme does not bind to a hydrophobic interface (siliconized glass beads). This compound has been found to react selectively with Tyr 49 in the lipase.[27,29]

These facts suggest that porcine pancreatic lipase consists of two functionally different domains, esteratic and water-insoluble substrate binding domains, and that fragments (336-449) and (1-335) include the esteratic and binding sites, respectively.

III. PLASMA LIPASES

A. LIPOPROTEIN LIPASE AND HEPATIC TRIGLYCERIDE LIPASE

Intravenous injection of heparin into humans and other vertebrates results in release into the plasma of at least two lipolytic enzymes: lipoprotein lipase (EC 3.1.1.34) and hepatic triglyceride lipase. Lipoprotein lipase is responsible for the hydrolyses of di- and triacylglycerol constituents of very low density lipoproteins and chylomicrons.[30] It functions as an exoenzyme on the luminal surface of capillary endothelial cells in extrahepatic tissues.[31] In cell cultures, lipoprotein lipase has been shown to bind to endothelial cell heparan sulfate proteoglycans with an association constant of $0.7 \times 10^7 \, M^{-1}$.[32,33]

Hepatic triglyceride lipase catalyzes hydrolysis of triacylglycerols in lipoproteins *in vitro*.[40] However, the exact function of hepatic triglyceride lipase is still controversial. Intravenous injection of antibody against this enzyme into rats results in increases of cholesterol and phospholipids in high- and low-density lipoprotein fractions,[34,35] suggesting that hepatic triglyceride lipase may play a role in the clearance of phospholipid and cholesterol associated with these lipoproteins. Some investigators have also suggested a role of this enzyme in the metabolism of intermediate density lipoproteins.[36,37]

In post heparin plasma, of patients with cancers and of tumor-bearing animals, hepatic triglyceride lipase activity is decreased, whereas lipoprotein lipase activity is increased.[54,55] Recently, Cheng et al.[38] purified human lipoprotein lipase and hepatic lipase from the post-heparin plasma 250,000- and 100,000-fold and obtained homogeneous preparations in yields of 27 ± 15 and 19 ± 6%, respectively. The molecular weights of lipoprotein lipase and hepatic triglyceride lipase determined by polyacrylamide gel electrophoresis in the presence of sodium dodecyl sulfate and reducing agents were 60,500 ± 1,800 and 65,200 ± 400, respectively.[38] These lipase preparations were shown to be free of detectable antithrombin by measuring its activity and by probing Western blots of lipases with a monospecific antibody against antithrombin.[38] The rates of thermal inactivation of these enzymes at 40°C were significantly different.[38] Moreover, amino acid analysis showed significant differences in their percentage compositions of ten amino acids.[38,39] A monoclonal antibody to hepatic triglyceride lipase immunoprecipitated the parent enzyme but not lipoprotein lipase.[38] All these findings support the view that the two enzymes are different proteins.

B. MOLECULAR PROPERTIES OF LIPOPROTEIN LIPASE

Lipoprotein lipase is known to catalyze the hydrolyses of plasma tri- and diacylglycerols, phosphatidylcholines, and phosphatidylethanolamines transported in plasma chylomicrons and very low-density lipoproteins.[41-43] Lipoprotein lipase does not circulate in the blood but is immobilized on endothelial cell surfaces by its interaction with membrane-associated heparan sulfates. Intravenous administration of heparin releases the enzyme from cell surface binding sites.

The characteristic properties of the enzyme are (1) its inhibition by 1 *M* NaCl and protamine sulfate, (2) enhancement of its activity by apolipoprotein C-II (apo C-II), a protein constituent of triglyceride-rich lipoproteins and high-density lipoproteins (HDL), and (3) its pH optimum of 8 to 9 for triacylglycerol substrates.

A schematic representation of the lipoprotein lipase reaction on the surface of endothelial cell is shown in Figure 3.[43] In this model, the lipoprotein lipase molecule has four functional sites; (1) a heparan sulfate-binding site that anchors the enzyme to the endothelial cell surface, (2) a lipid-binding site that interacts with the surface of the lipoprotein interface, (3) on apo C-II binding site, and (4) a catalytic site.

Removal of heparan sulfate from the surface of cultured endothelial cells with a specific platelet endoglucuronidase[32] or heparinase[44] inhibits the binding of lipoprotein lipase to the cell surfaces. Competitive binding studies with the enzyme immobilized on Sepharose showed that heparin and apo C-II bind to different sites on the lipoprotein lipase molecule.[45] The immobilized enzyme retained enzyme activity for trioleylglycerol and was stimulated sevenfold by apo C-II. The amount of ^{125}I-labeled apo C-II bound to the immobilized enzyme was not significantly altered in the presence of heparin, and the amount of [^{125}I] heparin bound was not appreciably decreased by the addition of apo C-II. Furthermore, lipoprotein lipase bound to heparin-Sepharose retained its ability to bind apo C-II. Apo C-II is a single peptide consisting of 78 amino acid residues and contains no cysteine, cystine or histidine.[46] The minimum sequence of apo C-II required for activation of lipoprotein lipase and the role of functionally important regions of apo C-II with both native and synthetic fragments of apo C-II have been determined. The carboxyl terminal CNBr fragment containing residues 60-78 stimulated triglyceride hydrolysis fourfold, whereas the same concentration of intact apo C-II caused an average of ninefold activation.[47] A synthetic fragment 55-78 enhanced the hydrolysis 12-fold, which was comparable to the activation by intact apo C-II. In constrast, a synthetic peptide containing residues 66-78 did not activate the enzyme. Removel

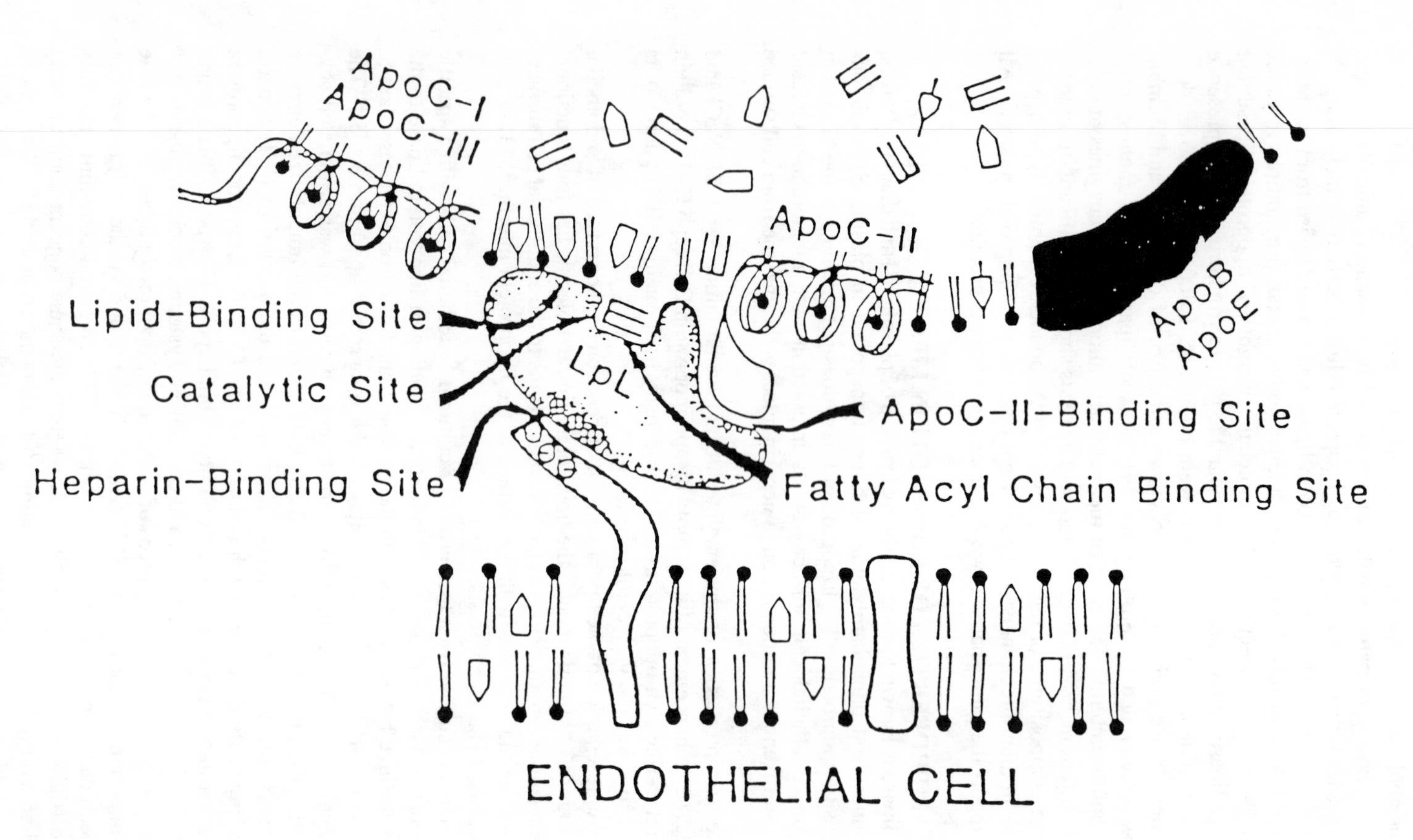

FIGURE 3. Schematic representation of the lipoprotein lipase reaction on the surface of endothelial cells. (From McLean, L. R., Demel, R. A., Socorro, L., Shinomiya, M., and Jackson, R. L., *Methods in Enzymology*, Vol. 129, Academic Press, New York, 1986, 738. With permission.)

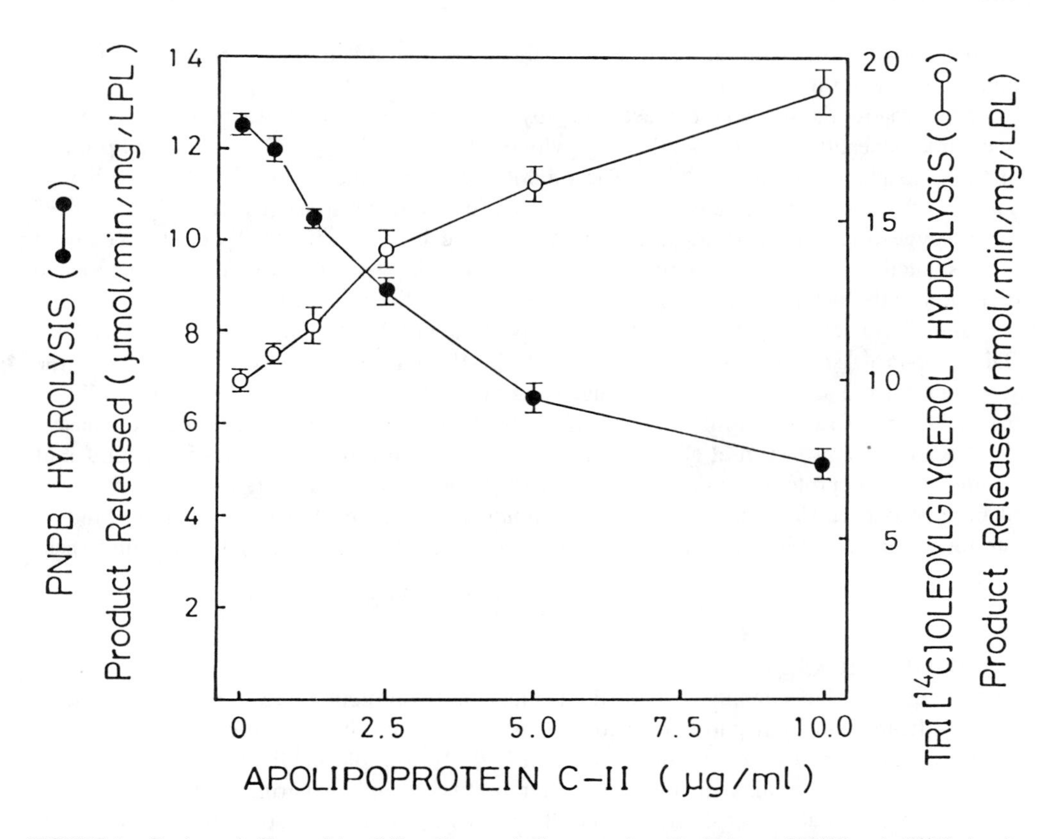

FIGURE 4. Reciprocal effects of ApoC-II on lipoprotein lipase-catalyzed hydrolyses of PNPB and tri[^{14}C] oleoylglycerol. (From Shirai, K., Jackson, R. L., and Quinn, D. M., *J. Biol. Chem.*, 257, 10200, 1982. With permission.)

of the three COOH-terminal residues, Gly-Glu-Glu, from fragment 60-78 decreased its ability to activate the enzyme.[47] From these results, Kinnunen et al.[47] suggested that residues 55-78 of apo C-II are the minimal sequence for maximal activation of LPL.

In addition to physiological lipids such as triacylglycerols and phospholipids, lipoprotein lipase can catalyze the hydrolyses of nonphysiological short chain fatty acyl ester substrates such as p-nitrophenyl acetate[48] and p-nitrophenyl butyrate (PNPB). Apo C-II is required for the maximal rate of hydrolysis of long chain fatty acid esters by the lipase. On the other hand, apo C-II does not enhance the hydrolysis of short chain fatty acyl esters such as tributyrin[49] by the enzyme. Furthermore, apo C-II depresses the hydrolysis of PNPB by lipoprotein lipase with concomitant stimulation of the hydrolysis of trioleoylglycerol, as shown in Figure 4.[56] This reciprocal effect of apo C-II on the hydrolyses of PNPB and trioleoylglycerol suggests that the same type of lipase-apo C-II interaction is responsible for apo C-II inhibition of PNPB hydrolysis.

These and other experimental results suggest that apo C-II causes a change in enzyme conformation that increases the affinity of the lipid-binding site for water-insoluble substrates such as long chain fatty acyl esters.[85,86] With water-soluble esters such as PNPB, the rate-determining factor is the organization of the active site itself, not the association between the lipid-binding site and the substrate. On the other hand, with water-insoluble substrates such as long chain fatty acyl esters, the lipase must bind to the substrate-water interface at its lipid-binding site, and its active site must become properly oriented with respect to the interface where the substrate molecules are. This process may be facilitated through interaction with apo C-II. In connection with this possibility, it is noteworthy that tryptic digestion of hepatic triglyceride lipase reduces the rate of hydrolysis of long chain fatty acid esters but not that of short chain fatty acid esters, suggesting that this treatment removes the lipid-binding site from the enzyme.[6]

Lipoprotein lipase can be inactivated by classical serine esterase inhibitors such as phenylmethanesulfonyl fluoride, diisopropylphosphorofluoridate (DFP), and diethyl p-nitrophenyl phosphate (paraoxon).[41,50,51] These inhibitors inhibit the hydrolysis of both a water-soluble substrate (p-nitrophenyl butyrate) and an insoluble substrate (trioleoylglycerol),[51,52] again suggesting that these substrates are hydrolyzed by the same catalytic site. Reddy et al.[52] isolated a catalytic site peptide after cleavage with cyanogen bromide and successive digestion with bacterial proteinase of [^{3}H] DFP labeled lipoprotein lipase. They found that the amino acid sequence of this catalytic site peptide was Ala-Ile-Gly-Ile-His-Trp-Gly-Gly-(DIP)-Ser-Pro-Asn-Gln-Lys-Asn-Gly-Ala-Val-Phe-Ile-Asn-(Ser, Lue)-Glu. Analysis of the sequence for the secondary structure suggested that the reactive serine of lipoprotein lipase is in a β-turn, a structure similar to those of the catalytic sites of most other serine proteinases.

Recently, Ben-Avram et al.[53] analyzed the amino acid sequences of tryptic peptides of bovine milk lipoprotein lipase. Most of these peptides showed close homology with those of porcine pancreatic lipase. Based on this homology, they postulated a potential binding site for water-insoluble substrates and a carbohydrate-binding domain for lipoprotein lipase.

IV. TISSUE LIPASES

A. HEART LIPASES

Rat heart tissue contains two distinct pools of lipoprotein lipase.[57,58] One is readily released from heart by heparin perfusion and is localized on the surface of endothelial cells of capillaries where it is directly involved in the hydrolysis of plasma lipoprotein triglyceride. The other, which remains in the tissue after perfusion with heparin, is thought to be a precursor or an intracellular storage pool of the former lipase and is responsible for hydrolysis of intracellular triglyceride. By rapid sequential perfusion of several rat hearts[59-63] with heparin a soluble, partially purified and relatively stable lipoprotein lipase preparation can be obtained.[64] This preparation was inhibited by antiserum against rat adipose tissue lipoprotein lipase and was identified as a single protein with an apparent Mr of 69,000 by immunoblotting.[65] An inverse relationship between increase in the intracellular lipoprotein lipase and decrease of the triglyceride content of the heart was observed in rats in conditions such as exercise.[66]fasting, and cold exposure.[63]

Interestingly, the cyclic AMP concentration of the heart was found to remain unchanged under these conditions. This observation rules out the possibility that cyclic AMP is involved in endogenous lipolysis. In addition to the intracellular lipoprotein lipase, cardiac tissue also contains triglyceride lipases with neutral (pH 7 to 7.5) and acid (pH 4.5 to 5) pH optima.[67-72]

Recently, Stam et al.[73] suggest that the intracellular lipoprotein lipase and the neutral lipase may not be the only determinants of endogenous lipolysis in rat heart and that lipolysis may also be catalyzed by a lysosomal, acid lipase.

B. HORMONE-SENSITIVE LIPOLYTIC ACTIVITY IN ADIPOSE TISSUE

The lipolytic process in adipose tissue plays a key role in energy metabolism in whole animals. The lipolytic process is thought to be catalyzed by hormone-sensitive lipase, the activity of which is regulated by a variety of hormones including epinephrine. Hormone-sensitive lipase has been known to be one of the lipolytic enzymes in fat calls since the early 1960s.[74-79] Unlike lipoprotein lipase, the enzyme has a neutral (pH 6.8) pH optimum, is inhibited by sodium fluoride, and is not activated by serum.[80,81] Fredrikson et al.[82] purified hormone-sensitive lipase from rat adipose tissue and found that the enzyme has a subunit

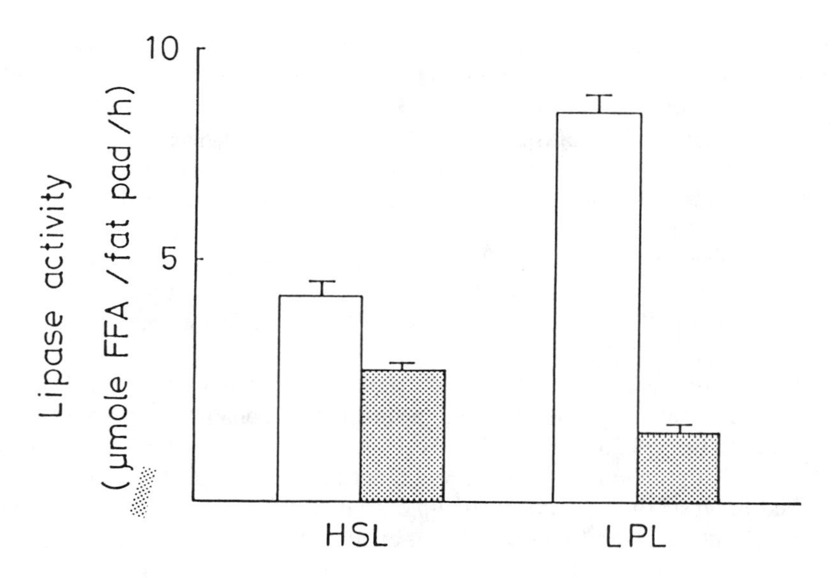

FIGURE 5. Effects of starvation on lipase activities in rat epididymal adipose tissue. HSL: Hormone-sensitive lipase. LPL: Lipoprotein lipase. □: Fed rats. ▒: Starved rats (48H)

of Mr 84,000. This subunit has two functionally distinct serine residues, basal and regulatory sites, which are phosphorylated by cyclic AMP-independent and -dependent protein kinases, respectively.[83] Cyclic AMP is now thought to play a key role in lipolytic hormone-mediated lipolysis in fat cells. Namely, lipolytic hormones such as epinephrine and adrenocorticotropic hormone stimulate adenyl cyclase in the fat cell membrane and increase the cyclic AMP level of the cells. This increased level of cyclic AMP level of the cells. This increased level of cyclic AMP stimulates protein kinase activity, which in turn activates hormone-sensitive lipase by phosphorylation of the regulatory site of the enzyme subunit, resulting in stimulation of triglyceride hydrolysis.[76-78,83]

Although the above cyclic AMP cascade system is widely accepted, some contradictory findings have been reported.[90,91,100-102] The expression of hormone-sensitive lipase activity per wet weight or per fat pad has resulted in serious confusion. For example, Rizack[74] expressed the lipase activity as n mole fatty acids released per 50 mg wet weight tissue per hour and proposed that the lipase activity in starving rats was significantly higher than that in fed rats. In contrast, if the lipase activity is expressed as micromole fatty aci‹'s released per epididymal fat pad per hour, its activity in starving rats is less than in fed rats, as shown in Figure 5.[84] Taking into account the metabolic state of whole bodies, expression of lipase activity per fat pad seems preferable. On this basis, the conclusion that hormone-sensitive lipase decreases during starvation may be complicated by the fact that fatty acid release from adipose tissue increases during starvation.[87] In this connection, it is of importance to note that the lipase activity assayed *in vitro* does not necessarily reflect that in adipose tissue or fat cells *in vivo*. For resolution of this problem, various attempts have been made to characterize the lipase reaction in fat cells.[90,91,100-102] Results have shown that the catalytic ability of the lipase localized in fat cells is influenced not only by its content, but also by other factors such as the availability of insoluble substrate (endogenous lipid droplets) for the enzyme. The latter conclusion is supported by the following observation.

The lipase activity is possibly enough to release fatty acids, even when it is extracted from fat cells without pretreatment with epinephrine and the activity is not changed even though the cells are incubated with sufficient hormone to stimulate their lipolysis.[88,89] In this connection, more detailed examinations have been carried out to elucidate the role of substrate on lipolytic activity, using endogenous lipid droplets or exogenous lipid micelles.[99]

Endogenous lipid droplets were prepared by subjecting fat cells to hypotonic shock and treatment with Triton X-100. The resulting endogenous lipid droplets consisted of triglyceride (98.2%), phospholipid (0.2%), cholesterol (0.3%), and protein(1.3%).[99] Lipolysis was not enhanced by addition of ATP plus cyclic AMP and protein kinase to the endogenous lipid droplets in the presence of adipose tissue lipase, but was stimulated by use of exogenous lipid micelles (Triolein emulsified with arabic gum) as substrate instead of droplets under the same experimental conditions.[99] In addition, lipolysis was clearly stimulated by addition of epinephrine to the endogenous lipid droplets in the presence of adipose tissue lipase, while it was not stimulated by addition of the hormone to a mixture of exogenous lipid micelles and the lipase.[99] These findings indicated that the substrate plays a crucial role in regulation of lipolytic activity.

That lipolysis of the endogenous lipid droplets was not stimulated by ATP plus cyclic AMP and protein kinase suggested that the mechanism of epinephrine-induced lipolysis of the droplets cannot be explained simply by a cyclic AMP cascade system. Additional evidence for this point is that protein kinase inhibitor did not inhibit epinephrine- or dibutyryl cyclic AMP-induced lipolysis of the endogenous lipid droplets.[100]

What then could account for enhancement of lipolysis of the endogenous lipid droplets by epinephrine? In 1966, Okuda et al.[90] first suggested the existence of another mechanism of epinephrine-induced lipolysis and proposed the hypothetical scheme shown in Figure 6.[91-99] The lipase concerned with lipolysis of the endogenous lipid droplets is mainly associated with the membranous structure, possibly the endoplasmic reticulum (ER) in fat cells.[93,103] Association between the lipase and triglyceride portion of the droplets might be affected by their surface phospholipids (PL), especially their hydrophilic groups. When epinephrine or dibutyryl cyclic AMP (DBcAMP) binds to hydrophilic groups, the hydrophobicity on the surface of the lipid droplets increases, resulting in association between the lipase in the membrane and the triglyceride portion of the lipid droplets, and eliciting hydrolysis of the triglyceride.[88,93,94,95,97,99,106] The above hypothesis is supported by the fact that pretreatment of the lipid droplets with phospholipase C inhibits epinephrine- and DBcAMP-induced lipolysis of the lipid droplets by adipose tissue lipase.[100] The existence of this other mechanism besides the cyclic AMP cascade system is also supported by the results of Mosinger,[104,105] Wise and Jungas,[101] and Oschry and Shapiro.[102]

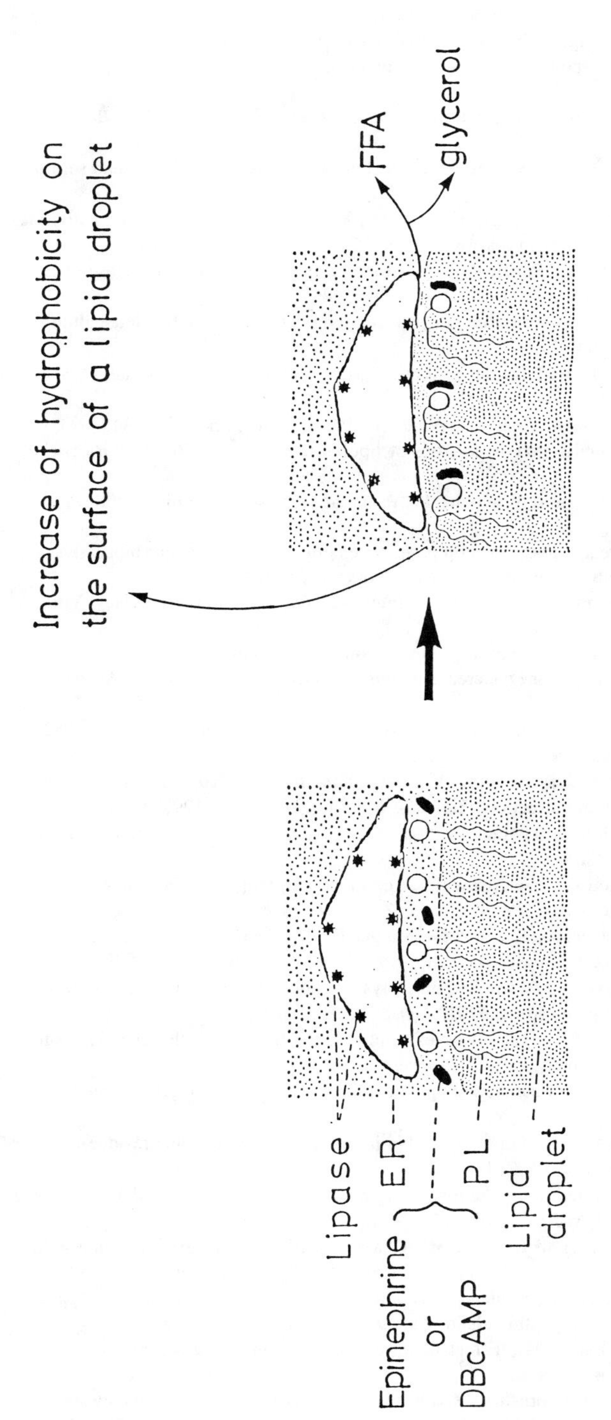

FIGURE 6. Mechanism of action of adrenaline in lipolysis of fat cells.

REFERENCES

1. **Hanriot, M.,** Sur un nouveau du sang, C. R. *Soc. Biol.*, 48, 925, 1896.
2. **Arthus, M.,** Sur la monobutyrinase du sang, *J. Physiol. Pathol. Gen.*, 4, 455, 1902.
3. **Deyon, M. M. and Morel, A.,** Action saponifiante du sérum sur les éthers, C. R. *Soc. Biol.*, 55, 682, 1903.
4. **Sarda, L. and Desnuelle, P.,** Action de la lipase pancréatique sur les esters en émulsion, *Biochim. Biophys. Acta*, 30, 513, 1958.
5. **Tsujita, T. and Okuda, H.,** Human liver carboxylesterase. Properties and comparison with human serum carboxylesterase, *J. Biochem.*, 94, 793, 1983.
6. **Tsujita, T., Nakagawa, A., Sirai, K., Saito, Y., and Okuda, H.,** Methyl butyrate-hydrolyzing activity of hepatic triglyceride lipase from rat post-heparin plasma, *J. Biol. Chem.*, 259, 11215, 1984.
7. **Tsujita, T., Okuda, H., and Yamasaki, N.,** Purification and some properties of carboxylesterase of rat adipose tissue, *Biochim. Biophys. Acta,* 715, 181, 1982.
8. **Willstätter, R., Waldschmitz-Leitz, E., and Memmen, F.,** Bestimmung der pankreatischen fettspaltung, *Hoppe-Seyler's Z. Physiol. Chem.*, 125, 93, 1923.
9. **Willstätter, R. and Memmen, F.,** Ũber die wirkung der pankreaslipase auf verschiedene substrate, *Hoppe-Seyler's Z. Physiol. Chem.*, 133, 229, 1928.
10. **Brockerhoff, H. and Jensen, R. G.,** *Lipolytic Enzymes,* Academic Press, New York, 1974, 34.
11. **Verger, R. and De Haas, G. H.,** Interfacial enzyme kinetics of lipolysis, *Annu. Rev. Biophys. Bioeng.*, 5, 77, 1976.
12. **Borgström, B., Erlanson-Albertsson, C., and Wieloch, T.,** Pancreatic colipase: chemistry and physiology, *J. Lipid Res.*, 20, 805, 1979.
13. **Sémériva, M. and Desnuelle, P.,** Pancreatic lipase and copipase. An example of heterogenous biocatalysis, in *Advances in Enzymology,* Vol. 48, John Wiley & Sons, New York, 1979, 319.
14. **Verger, R.,** Enzyme kinetics of lipolysis, in *Methods in Enzymology,* Vol. 64, Academic Press, New York, 1980, 340.
15. **Desnuelle, P. and Figarella, C.,** *Le Pancréas Exocrine,* Flammarion, Paris, 1980, 94.
16. **Brockman, H. L.,** Triglyceride lipase from porcine pancreas, in *Methods in Enzymology,* Vol. 71, Academic Press, New York, 1981, 619.
17. **Verger, R. and Pattus, F.,** Lipid-protein interactions in monolayers, *Chem. Phys. Lipid,* 30, 189, 1982.
18. **Verger, R.,** *Lipases,* Elsevier, Amsterdam, New York. Oxford, 1984, 121.
19. **De Caro, J., Boudouard, M., Bonicel, J., Guidoni, A., Desnuelle, P., and Rovery, M.,** Porcine pancreatic lipase. Completion of the primary struction, *Biochim. Biophys. Acta,* 671, 129, 1981.
20. **Plummer, T. H. and Sarda, L.,** Isolation and characterization of the glycopeptides of porcine pancreatic lipase L_A and L_B. *J. Biol. Chem.*, 248, 7865, 1973.
21. **Entressangles, B. and Desnuelle, P.,** Action of pancreatic lipase on monomeric tripropionin in the presence of water-miscible organic compounds, *Biochim. Biophys. Acta,* 341, 437, 1974.
22. **Chapus, C., Sémériva, M., Bovier-Lapierre, C., and Desnuelle, P.,** Mechanism of pancreatic lipase action. 1. Interfactial activation of pancreatic lipase, Biochemistry, 15, 4980, 1976.
23. **Bousset-Risso, M., Bonicel, J., and Rovery, M.,** Limited proteolysis of porcine pancreatic lipase. Lability of the Phe 335-Ala 336 bond towards chymotrypsin, *FEBS Lett.*, 182, 323, 1985.
24. **De Caro, J., Rouimi, P., and Rovery, M.,** Hydrolysis of p-nitrophenyl acetate by the peptide chain fragment (336-449) of porcine pancreatic lipase, *Eur. J. Biochem.*, 158, 601, 1986.
25. **Chapsu, C. and Sémériva, M.,** Mechanism of pancreatic lipase action. 2. Catalytic properties of modified lipases, *Biochemistry,* 15, 4988, 1976.
26. **Sémériva, M., Dufour, C., and Desnuelle, P.,** On the probable involvement of a histidine residue in the active site of pancreatic lipase, *Biochemistry,* 10, 2143, 1971.
27. **Maylié, M. F., Charles, M., and Desnuelle, P.,** Action of organophosphates and sulfonyl halides on porcine pancreatic lipase, *Biochim. Biophys. Acta,* 276, 162, 1972.
28. **Guidoni, A., Benkouka, F., De Caro, J., and Rovery, M.,** Characterization of the serine reacting with diethyl p-nitrophenyl phosphate in porcine pancreatic lipase, *Biochim. Biophys. Acta,* 660, 148, 1981.
29. **Bianchetta, J. D., Bidaud, J., Guidoni, A. A., Bonicel, J. J. and Rovery, M.,** Porcine pancreatic lipase. Sequence of the first 234 amino acids of the peptide chain, *Eur. J. Biochem.*, 97, 395, 1979.
30. **Kompiang, P., Bensadoun, A., and Yang, M.,** Effect of an anti-lipoprotein lipase serum on plasma triglyceride removal, *J. Lipid Res.*, 17, 498, 1976.
31. **Scow, R. O., Blanchette-Mackie, E. J., and Smith, L. C.,** Role of capillary endothelium in the clearance of chylomicrons. A model for lipid transport from blood by lateral diffusion in cell membranes, *Circ. Res.*, 39, 149, 1976.
32. **Cheng, C. F., Oosta, G. M., Bensadoun, A., and Rosenberg, R. D.,** Binding of lipoprotein lipase to endothelial cells in culture, *J. Biol. Chem.*, 256, 12893, 1981.

33. **Wang-Iverson, P., Jaffe, F. A., and Brown, W. V.,** *Atherosclerosis,* Vol. 5, Springer-Verlag, Berlin, 1980, 375.
34. **Kuusi, T., Kinnunen, P. K. J., and Nikkila, E. A.,** Hepatic endothelial lipase anterserum influences rat plasma low and high density lipoproteins *in vivo, FEBS Lett.,* 104, 384, 1979.
35. **Jansen, H., van Tol, A., and Hülsmann, W. C.,** On the metabolic function of heparin-releasable liver lipase, *Biochem. Biophys. Res. Commun.,* 92, 53, 1980.
36. **Murase, T., and Itakura, H.,** Accumulation of intermediate density lipoprotein in plasma after intravenous administration of hepatic triglyceride lipase antibody in rats, *Atherosclerosis,* 39, 293, 1981.
37. **Goldberg, I. J., Le, N. A., Paterniti, J. R., and Ginsberg, H. N.,** Lipoprotein metabolism during acute inhibition of hepatic triglyceride lipase in the cynomolgus monkey, *J. Clin. Invest.,* 70, 1184, 1982.
38. **Cheng, C. F., Bensadoun, A., Bersot, T., Hsu, J. S. T., and Melford, K. H.,** Purification and characterization of human lipoprotein lipase and hepatic triglyceride lipase, *J. Biol. Chem.,* 260, 10720, 1985.
39. **Östlung-Lindquist, A. M.,** Properties of salt-resistant lipase and lipoprotein lipase purified from human post-heparin plasma, *Biochem. J.,* 179, 555, 1979.
40. **Masuno, H. and Okuda, H.,** Hepatic triacylglycerol lipase in circulating blood of normal and tumor-bearing mice and its hydrolysis of very-low-density lipoportein and synthetic acylglycerols, *Biochim. Biophys. Acta,* 879, 339, 1986.
41. **Qinn, D., Shirai, K., and Jackson, R. L.,** Lipoprotein lipase: mechanism of action and role in lipoprotein metabolism, *Prog. Lipid Res.,* 22, 35, 1983.
42. **Smith, L. C. and Pownall, H. J.,** *Lipases,* Borgstrom, B. and Brockman, H. L., Eds., Elsevier, Amsterdam, 1984, 263.
43. **McLean, L. R., Demel, R. A., Socorro, L., Shinomiya, M., and Jackson, R. L.,** Mechanism of action of lipoprotein lipase, in *Methods in Enzymology,* Vol. 129, Academic Press, New York, 1986, 738.
44. **Shimada, K., Gill, P. J., Silbert, J. E., Douglas, W. H. J., and Fanburg, B. L.,** Involvement of cell surface heparan sulfate in the binding of lipoprotein lipase to cultured bovine endothelial cells, *J. Clin. Invest.,* 68, 995, 1981.
45. **Matsuoka, N., Shirai, K., and Jackson, R. L.,** Preparation and properties of immobilized lipoprotein lipase, *Biochim. Biophys. Acta,* 620, 308, 1980.
46. **Jackson, R. L., Baker, H. N., Gilliam, E. B., and Gotto, A. M., Jr.,** Primary structure of very low density apolipoprotein C-II of human plasma, *Proc. Natl. Acad. Sci. U.S.A.,* 74, 1942, 1977.
47. **Kinnunen, P. K. J., Jackson, R. L., Smith, L. C., Gotto, A. M., Jr., and Sparrow, J. T.,** Activation of lipoprotein lipase by native and synthetic fragments of human plasma apolipoprotein C-II, *Proc. Natl. Acad. Sci. U.S.A.,* 74, 4848, 1977.
48. **Egelrud, T. and Olivecrona, T.,** Purified bovine milk (lipoprotein) lipase: activity against lipid substrates in the absence of exogenous serum factors, *Biochim. Biophys. Acta,* 306, 115, 1973.
49. **Rapp, D. and Olivecrona, T.,** Kinetics of milk lipoprotein lipase. Studies with tributyrin, *Eur. J. Biochim.,* 91, 379, 1978.
50. **Chung, J. and Scanu, A. M.,** Isolation, molecular properties, and kinetic characterization of lipoprotein lipase from rat heart, *J. Biol. Chem.,* 252, 4202, 1977.
51. **Quinn, D. M.,** Diethyl-p-nitrophenyl phosphate: an active site titrant for lipoprotein lipase, *Biochim. Biophys. Acta,* 834, 267, 1985.
52. **Reddy, M. N., Maraganore, J. M., Meredith, S. C., Heinrikson, R. L., and Kézdy, F. J.,** Isolation of an active-site peptide of lipoprotein lipase from bovine milk and determination of its amino acid sequence, *J. Biol. Chem.,* 261, 9678, 1986.
53. **Ben-Avram, C. M., Ben-Zeev, O., Lee, T. D., Haaga, K., Shively, J. E., Goers, J., Pedersen, M. E., Reeve, J. R., Jr., and Schotz, M. C.,** Homology of lipoprotein lipase to pancreatic lipase, *Proc. Natl. Acad. Sci. U.S.A.,* 83, 4185, 1986.
54. **Masuno, H., Shiosaka, T., Itoh, Y., Onji, M., Ohta, Y., and Okuda, H.,** Hepatic triglyceride lipase and lipoprotein lipase activities in post-heparin plasma of patients with various cancers, *Jpn. J. Cancer Res. (Gann),* 76, 202, 1985.
55. **Masuno, H., Tsujita, T., Nakanishi, H., Yoshida, A., Fukunishi, R., and Okuda, H.,** Lipoprotein lipase-like activity in the liver of mice with sarcoma 180, *J. Lipid Res.,* 25, 419, 1984.
56. **Shirai, K., Jackson, R. L., and Quinn, D. M.,** Reciprocal effect of apolipoprotein C-II on the lipoprotein lipase-catalyzed hydrolysis of p-nitrophenyl butyrate and trioleolyglycerol, *J. Biol. Chem.,* 257, 10200, 1982.
57. **Robinson, D. S.,** Assimilation, distribution, and storage. C. Function of the plasma triglycerides in fatty acid transport, *Compr. Biochem.,* 18, 51, 1970.
58. **Borensztajn, J.,** *The Biochemistry of Atherosclerosis,* Scanu, A. M., Wissler, R. W., and Getz, G. S., Eds., Marcel Dekker, New York, 1979, 231.
59. **Oscai, L. B.,** Role of lipoprotein lipase in regulating endogenous triacylglycerols in rat heart, *Biochem. Biophys. Res. Commun.,* 91, 227, 1979.

60. **Palmer, W. K., Caruso, R. A., and Oscai, L. B.,** Possible role of lipoprotein lipase in the regulation of endogenous triacylglycerols in the rat heat, *Biochem. J.,* 198, 159, 1981.
61. **Palmer, W. K. and Kane, T. A.,** Hormonal activation of type-L hormone-sensitive lipase measured in defatted heart powders, *Biochem. J.,* 212, 379, 1983.
62. **Biliński, T., Krawiec, Z., Litwińska, A., and Liczmanński, J.,** Is hydroxyl radical generated by the Fenton reaction *in vivo?, Biochem. Biophys. Res. Commun.,* 130, 533, 1985.
63. **Miller, W. C. and Oscai, L. B.,** Relationship between type-L hormone-sensitive lipase and endogenous triacylglycerol in rat heart, *Am. J. Physiol.,* 247, R621, 1984.
64. **Reardon, C. A., Hay, R. V., Gordon, J. I., and Getz, G. S.,** Processing of rat liver apolipoprotein E primary translation product, *J. Lipid Res.,* 25, 348, 1984.
65. **Friedman, G., Chajek-Shaul, T., Etienne, J., Stein, O., and Stein, Y.,** Characterization of the lipoprotein lipase in the functional pool of rat heart by immunoblotting, *Biochim. Biophys. Acta,* 875, 397, 1986.
66. **Oscai, L. B., Caruso, R. A., and Wergeles, A. C.,** Lipoprotein lipase hydrolyzes endogenous triacylglycerols in muscle of exercised rats, *J. Appl. Physiol. Respir. Environ. Exercise. Physiol.,* 52, 1059, 1982.
67. **Severson, D. L.,** Characterization of triglyceride lipase activities in rat heart, *J. Mol. Cell. Cardiol.,* 11, 569, 1979.
68. **Hulsmann, W. C., Stam, H., and Breeman, W. A. P.,** Acid- and neutral lipases involved in endogenous lipolysis in small intestine and heart, *Biochem. Biophys. Res. Commun.,* 102, 440, 1981.
69. **Rosen, P., Budde, T., and Reinauer, H.,** Triglyceride lipase activity in the diabetic rat heart, *J. Mol. Cell Cardiol.,* 13, 539, 1981.
70. **Severson, D. L., Sloan, S. K., and Kryski, A., Jr.,** Acid and neutral triacylglycerol ester hydrolases in rat heart, *Biochem. Biophys. Res. Commun.,* 100, 247, 1981.
71. **Stam, H. and Hulsmann, W. C.,** Comparison of heparin-releasable lipase and tissue neutral lipase activity of rat heart, *Biochem. Int.,* 7, 187, 1983.
72. **Ramirez, I., Kryski, A., Jr., Ben-Zeev, O., Shotz, M. C., and Severson, D. L.,** Characterization of triacylglycerol hydrolase activities in isolated myocardial cells from rat heart, *Biochem. J.,* 232, 229, 1985.
73. **Stam, H., Broekhoven-Schokker, S., and Hülsmann, W. C.,** Studies on the involvement of lipolytic enzymes in endogenous lipolysis of the isolated rat heart, *Biochem. Biophys. Acta,* 875, 87, 1986.
74. **Rizack, M. A.,** An epinephrine-sensitive lipolytic activity in adipose tissue, *J. Biol. Chem.,* 236, 657, 1961.
75. **Vaughan, M., Berger, J. E., and Steinberg, D.,** Hormone-sensitive lipase and monoglyceride lipase activities in adipose tissue, *J. Biol. Chem.,* 239, 401, 1964.
76. **Butcher, R. W., Ho, R. J., Meng, H. C., and Sutherland, E. W.,** Adenosine 3′,5′-monophosphate in biological materials. II. The measurement of adenosine 3′,5′-monophosphate in tissues and the role of the cyclic nucleotide in the lipolytic response of fat to epinephrine, *J. Biol. Chem.,* 240, 4515, 1965.
77. **Khoo, J. C. and Steinberg, D.,** Reversible protein kinase activation of hormone-sensitive lipase from chicken adipose tissue, *J. Lipid Res.,* 15, 602, 1974.
78. **Khoo, J. C., Steinberg, D., Huang, J. J., and Vagelos, P. R.,** Triglyceride, diglyceride, monoglyceride and cholesterol ester hydrolases in chicken adipose tissue activated by adenosine 3′:5′ monophosphate-dependent protein kinase, *J. Biol. Chem.,* 251, 2882, 1976.
79. **Severson, D. L.,** Regulation of lipid metabolism in adipose tissue and heart, *Can. J. Physiol. Pharmacol.,* 57, 923, 1979.
80. **Hollenberg, C. H., Raben, M. S., and Astwood, E. B.,** The lipolytic response to corticotropin, *Endocrinology,* 68, 589, 1961.
81. **Björntorp, P. and Furman, R. H.,** Lipolytic activity in rat epididymal fat pads, *Am. J. Physiol.,* 203, 316, 1962.
82. **Fredrickson, G., Strålfors, P., Nilsson, N. Ö., and Belfrage, P.,** Hormone-sensitive lipase of rat adipose tissue. Purification and some properties, *J. Biol. Chem.,* 256, 6311, 1981.
83. **Belfrage, P., Fredrickson, G., Strålfors, P., and Tornquist, H.,** *Lipases,* Borgström, B. and Brockman, H. L., Eds., Elsevier, Amsterdam, New York, Oxford, 1984, 366.
84. **Okuda, H., Yanagi, I., Sek, F. J., and Fujii, S.,** Studies on the hormone-sensitive lipase of rat epididymal adipose tissue, *J. Biochem.,* 68, 199, 1970.
85. **Matsuoka, N., Shirai, K., Johnson, J. D., Kashyap, M. L., Srivastava, L. S., Yamamura, T., Yamamoto, Y., Saito, Y., Kumagai, A., and Jackson, R. L.,** Effects of apolipoprotein C-II (apoC-II) on the lipolysis of very low density lipoproteins from apoC-II deficient patients, *Metabolism,* 30, 818, 1981.
86. **Quinn, D. M., Shirai, K., Jackson, R. L., and Harmony, J. A. K.,** Lipoprotein lipase catalyzed hydrolysis of water-soluble p-nitrophenyl esters. Inhibition by apolipoprotein C-II, *Biochemistry,* 21, 6872, 1982.
87. **Fredrickson, D. S. and Gordon, R. S., Jr.,** Transport and fatty acids, *Physiol. Rev.,* 38, 585, 1958.

88. **Okuda, H.,** Machanism of actions of adrenaline and ACTH in fat mobilization, *Pharmacol. Biochem. Behav.* 3 (Suppl. 1), 149, 1975.
89. **Saito, Y., Matsuoka, N., Kumagai, A., Okuda, H., and Fujii, S.,** Studies on reduction of lipolysis in adipose tissue on freezing and thawing, *Endocrinol.,* Jpn., 25, 13, 1978.
90. **Okuda, H., Yanagi, I., and Fujii, S.,** The mechanism of in vitro stimulation of lipolysis by adrenaline, *J. Biochem.,* 59, 438, 1966.
91. **Yanagi, I., Okuda, H., and Fujii, S.,** Further studies on the mechanism of in vitro stimulation of lipolysis by adrenaline, *J. Biochem.,* 63, 249, 1983.
92. **Okuda, H., Sek, F. J., and Fujii, S.,** Studies on temperature sensitivity on adrenaline-induced lipolysis in rat adipose tissue, *J. Biochem.,* 69, 677, 1971.
93. **Okuda, H., Saito, Y., Matsuoka, N., and Fujii, S.,** Mechanism of adrenaline-induced lipolysis in adipose tissue, *J. Biochem.,* 75, 131, 1974.
94. **Saito, Y., Okuda, H., and Fujii, S.,** Studies on lipolysis on lipid micelles isolated from adipose tissue, *J. Biochem.,* 75, 1327, 1974.
95. **Saito, Y., Matsuoka, N., Okuda, H., and Fujii, S.,** Studies on hormone-sensitive lipolytic activity, *J. Biochem.,* 76, 1061, 1974.
96. **Saito, Y., Matsuoka, N., Okuda, H., and Fujii, S.,** Further studies on the mechanism of adrenaline-induced lipolysis in lipid micelles, *J. Biochem.,* 80, 929, 1976.
97. **Okuda, H., Saito, Y., Matsuoka, N., Takeda, E. and Kumagai, A.,** Role of phospholipid in adrenaline-induced lipolysis and cyclic AMP production, *J. Biochem.,* 83, 887, 1978.
98. **Miyoshi, Y., Uchida, K., Takeda, E., Nagai, K., and Okuda, H.,** The mechanism of the lipolytic action of theophylline in fat cells, *Pharmacol. Biochem. Behav.,* 14, 701, 1981.
99. **Okuda, H., Tsujita, T., Sumida, M., Takahashi, Y., Shimizu, D., and Fujii, S.,** Role of endogenous lipid droplets of fat cells in epinephrine-induced lipolysis, *J. Biochem.,* 93, 575, 1983.
100. **Okuda, H., Tsujita, T., and Kinutani, M.,** Studies on a protein kinase inhibitor-insensitive, phospholipid C-sensitive pathway of lipolysis in rat adipocytes, *Pharmacol. Res. Commun.,* 18, 877, 1986.
101. **Wise, L. S. and Jungas, R. L.,** Evidence for a dual mechanism of lipolysis activation by epinephrine in rat adipose tissue, *J. Biol. Chem.,* 253, 2624, 1978.
102. **Oschry, Y. and Shapiro, B.,** Lipolytic activity in adipocyte cell fractions, *Biochim. Biophys. Acta,* 618, 293, 1980.
103. **Matsuoka, N., Saito, Y., Okuda, H., and Fujii, S.,** Studies on triglyceride synthesis in lipid micelles from adipose tissue, *J. Biochem.,* 76, 359, 1974.
104. **Mosinger, B.,** Lipolytic action of EDTA and catecholamines in intact and homogenized adipose tissue, *Life Sci.,* 8, Part II, 137, 1969.
105. **Mosinger, B.,** Regulation of lipolysis in adipose tissue homogenate: activating effect of catecholamine, thyroxin, serotonin, EDTA, pyrophosphate and other factors in unsupplemented homogenate, *Arch. Int. Physiol. Biochem.,* 80, 79, 1972.
106. **Okuda, H., Takeda, E., Saito, Y., and Matsuoka, N.,** Studies on adrenaline-induced lipolysis in artificial lipid micelles, *J. Biochem.,* 81, 1633, 1977.

Index

INDEX

A

B

D

E

F

G

H

M

N

O

P

Q

R

U

V

W

X

Y

Z